ARISTOTELIAN METEOROLOGY IN SYRIAC

ARISTOTELES SEMITICO-LATINUS

founded by H.J. Drossaart Lulofs

is prepared under the supervision of the ROYAL NETHERLANDS ACADEMY OF ARTS AND SCIENCES as part of the CORPUS PHILOSOPHORUM MEDII AEVI project of the UNION ACADÉMIQUE INTERNATIONALE.

The Aristoteles Semitico-Latinus project envisages the publication of the Syriac, Arabic and Hebrew translations of Aristotle's works, of the Latin translations of those translations, and of the mediaeval paraphrases and commentaries made in the context of this translation tradition.

VOLUME 15

ARISTOTELIAN METEOROLOGY IN SYRIAC

Barhebraeus, Butyrum Sapientiae, *Books of Mineralogy and Meteorology*

BY

HIDEMI TAKAHASHI

BRILL
LEIDEN · BOSTON
2004

This book is printed on acid-free paper

Library of Congress Cataloging-in-Publication Data

The Library of Congress Cataloging-in-Publication Data is also available on
http://catalog.loc.gov

ISSN 0927-4103
ISBN 90 04 13031 4

PRINTED IN THE NETHERLANDS

Coniugi

CONTENTS

Preface xiii
Abbreviations xvii
Note on the Transcription xviii
Note on the Syriac Font xix

INTRODUCTION

1. *Butyrum sapientiae*: Overview of the Work 3
1.1. On the Author 3
1.2. Scope and Title of the Work 4
1.3. Date of Composition 9
1.4. Divisions of the Work 10
1.5. Models of *Butyrum sapientiae* 11

2. Manuscripts of *Butyrum sapientiae* 15
A. Manuscripts which Contain the Whole Work or whose Exact Contents are Unknown 15
B. Manuscripts Containing Whole or Part of the Section on Logic ... 19
C. Manuscripts of the Sections on Natural Sciences, Metaphysics and Practical Philosophy 25
D. Manuscripts Containing Excerpts or Fragments of the Section on Natural Sciences 28
E. Further Excerpts etc. 31
F. Manuscript Families 32

3. Books of Mineralogy and Meteorology 36
3.1. Overview of the Books 36
3.2. Sources of Books of Mineralogy and Meteorology 37
3.2.1. Aristotelian Meteorology in Syriac 37
3.2.2. Mineralogy-Meteorology in Earlier Works of Barhebraeus 42
3.2.3. Sources of Mineralogy-Meteorology in *Butyrum sapientiae* 48
3.3. Language and Style 60
3.4. Barhebraeus' Meteorology: an Assessment 65

TEXT AND TRANSLATION

Notes on the Text and Translation 73

Book of Mineralogy

Chapter I: On Mountains and Springs 78
Section i: On Formation of Stones 78
Section ii: On Formation of Mountains and their Uses 82
Section iii: On Springs 86

Chapter II: On Earthquakes 90
Section i: On the Opinions of the Ancients on Earthquakes 90
Section ii: On the True Cause of the Earthquake 92
Section: On Times when Movements Occur More 96

Chapter III: On Mineral Bodies 100
Section i: On the Categories of Minerals 100
Section ii: On the Fact that the Element of the Malleable Minerals is Mercury 104
Section iii: On Imitation of these Bodies by Means of Each Other 106

Chapter IV: On Matters concerning the Habitable World 110
Section i: On the Natural Position of the Earth 110
Section ii: On Climates of [Various] Regions. 114
Section iii: Refutation of the Avicennian Arguments 118
Section iv: On the Cause of Heat and Cold according to Time and Place 122

Chapter V: On Matters concerning the Sea 126
Section i: On the Position of the Seas 126
Section ii: On the Salinity of Seawater 128
Section iii: On the Migration of the Sea 132
Section iv: On Tartarus 134

Book of Meteorology

Chapter I: On Clouds and Things which Descend from them 138
Section i: On How Clouds are Formed 138
Section ii: On Rain 140
Section iii: On Dew, Frost, Snow, Hail, Fog and Mist 144

Chapter II: On Illusory Matters which are Seen on Clouds 150
Section i: Introduction to the Teaching on the Efficient Cause of Cloud-Related Illusions 150
Section ii: On Halo 156

Section iii: On Rainbow 158
Section iv: On Mock Suns and Lances 164

Chapter III: On Winds 168
Section i: On Generation of Winds 168
Section ii: Definition of the Directions of Winds 170
Section iii: On the Variation of Winds according to Places and Seasons 174
Section iv: On Cloud Wind, Whirlwind, Tornado and Eddy Wind 178

Chapter IV: On the Remaining Products of the Smoky Exhalations 182
Section i: On Thunder and Lightning 182
Section ii: On Thunderbolt and Firewind 186
Section iii: On Δοκίδες, Shooting Stars and Comets 188

Chapter V: On Great Events which Take Place in the World 192
Section i: On Deluges 192
Section ii: On the Fact that Plants and Animals are also Generated without Reproduction 194
Section iii: On the Fact that Skills, Languages and Writing too are Renewed 196

COMMENTARY

Preliminary Remarks 203

Book of Mineralogy 205

Chapter I 205
Section i 206
Section ii 222
Section iii 234

Chapter II 243
Section i 244
Section ii 254
Section iii 269

Chapter III 281
Section i 282
Section ii 300
Section iii 312

Chapter IV 321
Section i 321
Section ii 336
Section iii 351
Section iv 360

Chapter V 369
Section i 369
Section ii 385
Section iii 398
Section iv 411

Book of Meteorology 419

Chapter I 419
Section i 420
Section ii 425
Section iii 433

Chapter II 444
Section i 444
Section ii 459
Section iii 465
Section iv 481

Chapter III 487
Section i 487
Section ii 499
Section iii 504
Section iv 517

Chapter IV 527
Section i 527
Section ii 541
Section iii 547

Chapter V 561
Section i 564
Section ii 571
Section iii 579

A.M.D.G.

PREFACE

The place Gregory Barhebraeus (1225/6-1286) occupies as one of the greatest figures in Syriac literature is uncontested. The work of which a part is edited below, the *Butyrum sapientiae* (*Ḥêwat ḥekmtā*, *Cream of Wisdom*), is the one major work of that author which has yet to be published in full. The reasons for the neglect of this work are not difficult to imagine. One is its sheer length, which would have worked as a deterrence on any prospective editor. Another is its derivative character, whereby it was not expected that this work would offer much in the way of "original" thought and developments.

It has long been known that the *Butyrum sapientiae* is modelled on Ibn Sīnā's *Kitāb al-šifā*ʾ, but it is only in recent years that a closer examination began to be made of other sources used by Barhebraeus in composing this work. The most important discovery was that made by H.J. Drossaart Lulofs, who found that in composing the *Butyrum* Barhebraeus had made extensive use of the Syriac version of the *Compendium* of Aristotelian philosophy by Nicolaus Damascenus, a work that is almost completely lost in the original Greek and survives in Syriac too only in a fragmentary state. It was also demonstrated by M. Zonta in his "provisional" edition of the Book of Economics in the *Butyrum* that the last part the work was modelled not on the *Šifā*ʾ but on a Persian work by a near contemporary of Barhebraeus, the *Aḫlāq-i nāṣirī* of Naṣīr al-Dīn al-Ṭūsī.

The study made in the course of preparing the edition of the books on mineralogy and meteorology shows that the situation is even more complicated. Even in those parts of the work which appear at first sight to be based on the *Šifā*ʾ, a closer examination often reveals that the immediate source used is not the *Šifā*ʾ but the works of more recent authors such as Abū al-Barakāt al-Baghdādī and Fakhr al-Dīn al-Rāzī. At the same time, Barhebraeus did not neglect materials which had been made available in Syriac in earlier times. Besides Nicolaus' *Compendium*, he has used the Syriac version of the Pseudo-Aristotelian *De mundo* and traces of other Syriac and Greco-Syriac works are also to be found in scattered places, as for example, in the quotation from Theo of Alexandria's *Small Commentary* on Ptolemy's *Handy Tables*.

The work of which a part is edited here has its importance in the first place as a product of the so-called "Syriac Renaissance", in which the author, the foremost representative of that period of Renaissance, made a valiant attempt to create a new type of scientific writing in Syriac through the synthesis of materials gathered from the more recent Arabo-Persian literature with materials taken from older Syriac works. At the same time, the work will be of interest to different groups of people for different reasons, to the classicist, for example, as a basis for the reconstruction of the lost *Compendium* of Nicolaus, to the Islamic philosopher as a relatively rare instance of the reception of Ibn Sīnā in Syriac and to the Syriac lexicographer as a work containing numerous instances where the significance of the words as intended by the author may be ascertained through comparison with its Arabic sources.

A book such as this is never the work of a single person. Even if the name of one person appears on its front page and it is he alone who bears the responsibility for the numerous errors that remain, it is in reality the fruit of the generous assistance provided by countless people with whom that one person has had the privilege to be in contact.

What is published here constituted the main part of a doctoral dissertation submitted to the Johann Wolfgang Goethe-Universität Frankfurt am Main (Fachbereich 9 Sprach- und Kulturwissenschaften).[1] It is difficult now to distinguish clearly between those to whom I am indebted for the parts of my dissertation which are published here and the parts which are not. Acknowledgement is first due to Prof. Emer. Masaaki Kubo (Tokyo), whose suggestion it was that launched me on the course leading to the composition of this work. The greatest debt is owed to my supervisor, Prof. Hans Daiber (Frankfurt), who had the boldness (verging on folly) of accepting a classicist with little knowledge of the Semitic languages as a candidate for preparing an edition of a Syriac work based to a large extent on Arabic sources and who had to suffer the consequences of his decision in the following years, always with patience and kindness. Among my teachers at the Orientalisches

---

[1] The part of the introduction to the dissertation which dealt with the "life and works" of Barhebraeus and a part of the appendices dealing, again, with the works of Barhebraeus in general are due to be published separately under the title *Barhebraeus (Bar ʿEbroyo): a Bio-Bibliography.*

Seminar in Frankfurt, it is with a note of sadness that I remember the late Dr. Hussam Saghir, who never lacked the time to explain to me the intricacies (so they seemed at the time) of his beloved Arabic language and whose far too early departure to the next was a painfully felt loss for all those who knew him. I am indebted in all kinds of different ways to my other teachers, colleagues and friends at the same institute, Doctors Wim Raven, Peter Joosse, Eva-Maria Kluge, Armin Schopen, Isabel Stümpel, Mohsin Zakeri, Mr. Lotfi Toumi and Ms. Anna Akasoy.

It is with sincere gratitude that I remember my teachers and friends elsewhere, whose generous advice and help were indispensable in the preparation of the dissertation and whom I allow myself simply to cite by name here: H.E. Metropolitan Gregorios Yohanna Ibrahim (Aleppo), Abuna Hanna Aydin (Warburg), Fr. Antun Deli-Afo (Aleppo), Fr. Joseph Tobji (Aleppo), Fr. Baby Varghese (Kottayam), Fr. Emmanuel Youkhanna (Wiesbaden), Mr. Gabriel Afram (Skarholmen), Mr. Elias Assad (London), Prof. Françoise Briquel-Chatonnet (Paris), Dr. Sebastian Brock (Oxford), Mr. Georges Chami and the Fondation Georges et Mathilde Salem (Aleppo), Mr. Jean Fathi-Chelhod (Riyadh), Dr. Theresia Hainthaler (Frankfurt), Dr. Taro Hyuga (Tokyo), Dr. Thomas Joseph (Los Angeles), Dr. Andreas Juckel (Münster), Prof. Hideo Katayama (Tokyo), Mr. Yousef Kouriyhe (Berlin), Prof. Ian Kidd (St. Andrews), Prof. David King and the staff of the Institut für Geschichte der Naturwissenschaften (Frankfurt), Prof. Rokuro Makabe (Frankfurt), dayroyo Mattias Nayis (Maᶜarrat Saidnaya), Dr. Heleen Murre-van den Berg (Leiden), the Orient-Institut der DMG (Beirut), Dr. Karl Pinggéra (Marburg), Mr. Gabriel Rabo (Göttingen), Prof. Jamil Ragep (Oklahoma), Mr. Sami Salameh (Louaizé), Dr. Assad Sauma (Spånga), Prof. Fuat Sezgin and the staff of the Institut für Geschichte der Arabisch-Islamischen Wissenschaften (Frankfurt), Prof. Gotthard Strohmaier (Berlin), Mr. Mesut Tan (Göppingen), Prof. Anne Tihon (Louvain), Dr. Carsten Walbiner (Bonn), Dr. John Watt (Cardiff), Dr. Dorothea Weltecke (Göttingen), Prof. Witold Witakowski (Uppsala) and Dr. Helen Younansardaroud (Berlin). To the external examiners of my dissertation, Prof. Hubert Kaufhold (Munich) and Prof. Herman Teule (Nijmegen), I owe a good number of corrections and useful suggestions. Thanks are also due to Ms. Trudy Kamperveen of Brill Academic Publishers for her patience and kindness at the stage of bringing this work to print, to Mr. Takumi Ishiwata (Tokyo) for his

help in the preparation of the indices and to my students Emi Takenaka, Takami Uda and Keitaro Manzen for their help in the preparation of the final manuscript.

For the means required for keeping my body and soul together during much of the period devoted to the preparation of the present work I am indebted to the Japan Society for Promotion of Science.

Last but not least, I thank my ever-patient wife Minako and my children, Miho, Hideyuki and Mika, without whose loving support the sailor would never have reached harbour, to use a metaphor beloved of Syriac scribes.

Hidemi Takahashi
Feast of St. Ignatius of Antioch, 2003
Chuo University, Tokyo

ABBREVIATIONS

For the abbreviations used for names of ancient and medieval authors and their works, see the first part of the Bibliography.

1. Periodicals, Serial Publications and Reference Works

The abbreviations for journals and serial publications, other than those given below, are as in S.M. Schwertner, *Internationales Abkürzungsverzeichnis für Theologie und Grenzgebiete*, 2nd ed. (IATG^{2}), Berlin-New York, 1992.

ARAM	ARAM Periodical, Oxford, 1989-
ASL	Aristoteles Semitico-Latinus, Amsterdam, then Leiden etc.
Centaurus	Centaurus, Copenhagen, 1950-
EHAS	R. Rashed (ed.), *Encyclopedia of the History of Arabic Science*, London-New York 1996
Hugoye	Hugoye. Journal of Syriac Studies, www.syrcom.cua.edu/Hugoye, 1998-
JIASyr	Journal of the Iraq Academy, Syriac Corporation, Baghdad, 5.1979-
JSyrA	Journal of the Syriac Academy, Baghdad, 1.1975-4.1978
LSJ	H. G. Liddell, R. Scott, H.S. Jones et al., *A Greek-English Lexicon, ... With a Supplement*, Oxford 1968
MMMA	Mağallat Ma^{c}had al-maḫṭūṭāt al-^{c}arabīya, Cairo, 1955-
MMIA	Mağallat al-Mağma^{c} al-^{c}ilmī al-^{c}arabī, Damascus, 1.1921-35.1960
PS	R. Payne Smith, *Thesaurus syriacus*, Oxford 1879-1901
SPMSE	Sitzungsberichte der physicalisch-medicinischen Societät zu Erlangen, Erlangen, 1871-
Symp. Syr.	proceedings of Symposia Syriaca I-VII = OCA 197 (1974), 205 (1978), 221 (1983), 229 (1987), 236 (1990), 247 (1994), 256 (1998)
VIGAIW	Veröffentlichungen des Institutes für Geschichte der arabisch-islamischen Wissenschaften, Frankfurt
ZGAIW	Zeitschrift für Geschichte der arabisch-islamischen Wissenschaften, Frankfurt, 1984-

2. Others

A.Gr.	*anno Graecorum*
A.H.	*anno Hegirae*
arab.	*arabus*, Arabic
b. (in personal names)	*bar*, *ibn*, son of
corr.	corresponding to
ed.	edition of, edited by
gr.	*graecus*, Greek
h. (in dates)	*hiğrī*, *anno Hegirae*
h.š. (in dates)	*hiğrī šamsī*
heb.	*hebraicus*, Hebrew
K. (in work titles)	*ktā̱ḇā*, *kitāb*
L. (in work titles)	*liber*
lat.	*latinus*, Latin
ob.	*obiit*
ms(s).	manuscript(s)
syr.	*syriacus*, Syriac
sec.	*secundum*, according to
tr.	translation (of), translated by

Note on the Transcription

The system normally used for transcription of Arabic and other Oriental languages on the Continent has the advantage that each letter of the original language is represented by a single letter in transcription and so is less prone to ambiguities. I have found it difficult, on the other hand, to regard such letters as *ḫ* and *ġ* as a part of the English language. As a compromise, a distinction is made in what follows between translation and transliteration. Personal and place names are normally "translated" (e.g. Ikhwān al-Ṣafāʾ), while book titles etc. are "transliterated" (e.g. *Rasāʾil Iḫwān al-ṣafāʾ*). Persian is transcribed according to the same system as Arabic.

Transcription of Syriac is a minefield in which it is difficult even for the expert to be completely free of blame, let alone for someone like the present writer without any formal training in the language. An attempt has been made to follow the rules given in such classical grammars as those of Nöldeke. While it has seemed strange at times to be quoting the words of a 13th century West Syrian author in forms in

which he himself would never have pronounced them, it is hoped that that author, the foremost "ecumenist" in the history of the Syrian Churches, will look upon what is found here with indulgence (the same applies for the use of the Estrangelo script).

Note on the Syriac Font

The computer font used in the course of preparing the present work had a number of shortcomings, the most conspicuous of these being the way in which it constantly recognised the plural sign (*syāmē*) as word end. The reader is requested to look with indulgence on such instances as ܡܪ̈ܐ and ܪ̈ܘܚܐ.

INTRODUCTION

CHAPTER ONE

BUTYRUM SAPIENTIAE: OVERVIEW OF THE WORK

1.1. *On the Author*

The Syrian (Syriac) Orthodox prelate and polymath, Gregory Abū al-Faraj Bar ᶜEḇrāyā (Bar ᶜEbroyo), commonly known in the West as Barhebraeus, is the foremost representative of the so-called "Syriac Renaissance" of the 12th-13th centuries. He was born in 1225/6 A.D. in Melitene, a city then under the rule of the Rūm Seljūks, but with a large and prestigious community of Syrian Orthodox Christians. After periods of study in Antioch, Tripoli (both then still in the hands of the Franks) and, possibly, Damascus, he was raised to the episcopate at the age of twenty in 1246 and was appointed, successively, to the sees of Gubos and Laqabin in the vicinity of Melitene, before being translated ca. 1253 to Aleppo, where he was to witness the fall of the city to the Mongols in 1260. In 1264 he was raised to the office of the Maphrian of the East, the second highest office in the Syrian Orthodox Church with jurisdiction over those areas which had been under Persian rule in pre-Islamic times. As maphrian, his normal place of residence was Mosul and the nearby Monastery of Mar Mattai, but a significant part of his maphrianate was spent in Marāgha and Tabrīz, the new centres of power under the Mongol Īl-Khāns, where he befriended the leading Muslim scholars of the day. He died in Marāgha on 29/30th July 1286.

Barhebraeus has left us a large corpus of works whose number could be as high as forty and whose subjects range from biblical exegesis and dogmatic and moral-mystical theology to jurisprudence, philosophy, historiography, belles lettres, grammar, lexicography, the exact sciences and medicine. Most of these are in Syriac, although some are in Arabic. Typical of Barhebraeus is the manner in which he took an Arabic (occasionally Persian) work as his model and structured his own work around it. He then incorporated into this framework materials taken from both Arabic and Syriac sources, thus making a new synthesis out the older Syriac and the more recent Arabic materials. In his works on moral-mystical theology, for example, he stands under the influence of Ghazālī (1058-1111), while in his works on the exact

sciences he is influenced by Naṣīr al-Dīn al-Ṭūsī (1201-74), a near contemporary with whom he was most probably personally acquainted.

In his philosophical works Barhebraeus is strongly influenced by Ibn Sīnā (980-1038). His works in this field include, besides some short works on logic and psychology, three works of different lengths covering all areas of Aristotelian philosophy. The earliest of these is the medium-length *Tractatus tractatuum* (*Têḡrat_ têḡrāṯā*), probably composed in the first years of his maphrianate. This was followed, perhaps in the middle years of the maphrianate, by *Sermo sapientiae* (*Swāḏ sōp̄iya*), the shortest of the three, and, at the end of his life, by *Butyrum sapientiae* (*Ḥêwat_ ḥeḵmṯā*), his longest work on philosophy and the work with which we are concerned here.[1]

1.2. *Scope and Title of the Work*

In considering the scope and title of the work, we might begin by looking at what the author himself has to say on the matter in his proem to the work.

Ms. Florence, Laur. or. 69 [= F] 2v a 4-b 11:

ܠܟ ܐܠܗܐ ܚܘܒܐ ܕܡܫܩܦ ܠܢܘܗܪܐ ܡܫܒ̈ܚܢ ܒܡܫܒܚܬܟ. ܘܡܫܒܚܢܘܬܐ ܐܠܗܝܬܐ ܡܫܬܡܫܢ̈ܝܢ
ܕܗ ܩܕܝܫ ܡܢܐ ܠܘܬ ܗܢܐ ܦܘܩܐ ܒܡܫܒܚܬܟ. ܡܫܬܐܕܐ ܠܥܕܐ ܒܠܝܘܢܝܬ ܠܢܦܫܟ.
ܘܡܬܠܟ ܡܝܠܬܐ ܡܫܬܟܠܢܐ ܝܨܝܦܬ ܠܢܦܫܟ. ܠܟ ܚܝ ܡܝܬܝܢܐ ܕܡܒܘܥܐ
ܕܝܬܐ ܒܢܦܫ ܗܘ. ܘܢܦܫ ܡܝܬܝܢܐ ܕܚܝܬܐ ܒܡܫܡܫܢܐ ܒܢܦܫ ܗܘ. ܦܠܝܛܘܬܗ
ܐܫܘܬܗ ܕܝܠܦܬܟ. ܦܠܝܛܘܬܗ ܒܝܫܘܬܗ ܠܡܫܡܫܬܟ. ܡܢ ܗܘܢܐ ܕܗܘܠܐ ܕܠܐ
ܘܐܝܬܝܘܬܗ. ܘܟܠ ܕܡܬܩܢ ܗܘܢܐ ܕܠܐ ܘܐܬܝܘܬܗ. ܕܐܝܬ ܗܘ ܒܪܝܐ ܕܘܟܐ
ܘܡܬܩܢܐ. ܘܗܢ ܒܡܫܡܫܐ ܕܡܠܝܐ ܡܫܠܡܢܐ ܘܡܫܠܡܢܐ. ܘܕܠܟ ܠܟ ܡܫܒܚܐ
ܘܐܝܬܝܐ ܒܪܘܚ ܗܘܬܟ. ܘܬܫܒܘܚܬܐ ܡܠܟܘܬܐ ܘܗܐ ܕܐܠܗܘܬܐ ܐܝܬܘܬܟ ܀ ܘܗܘ
ܐܘܟܝܐ ܢܘܗܪܐ ܦܫܝܛܐ ܐܘ̈ܢܐ ܦܣܬܝܩܝܬܐ. ܘܡܫܬܡܫܐ ܐܝܬܝܩܝܬܐ.
ܕܐܦܢ ܠܝܬܘܬܐ ܡܠܠܬܐ ܘܐܝܬܝܐ ܦܠܝܛܘܬܐ ܘܐܬܝܬܘܬܐ ܘܐܦܝܛܘܬܐ
ܡܠܝܘܬܐ ܘܐܝܬܝܘܬܐ ܐܝܬܘܬܐ. ܘܡܐܝܬܝܐ ܒܪܘܚ ܡܢܗܘܢ ܚܝܐ ܠܢܦܫܬܐ
ܫܘܬܦܐ ܢܦܫ ܟܠܝܗܘܢ ܐܝܬܘܬܐ ܘܐܝܬܝܘܬܐ ܐܝܬܘܬܐ.

[O = Bodl. Hunt. 1; f = Laur. or. 37; B = Beirut, Université Saint Joseph, syr. 48; L = British Library, Or. 4079; V = Vat. syr. 603; R = Manchester, John Rylands, syr. 56; N = Mingana syr. 326 || 1 ܠܟ: ܠܗ N; om. BLVR || ܐܠܗܐ BLVR || 2 ܩܕܝܫ: add. ܩܕܝܫ BV (cf. Jes. 6.3) || ܡܫܬܐܕܐ: ܡܫܬܐܕܝܢܐ O; ܡܫܬܐܕܢܐ R || ܠܥܕܐ: om. L || 3 ܘܡܬܠܟ: ܘܡܢ supra lin. B; om. V || 4 ܦܠܝܛܘܬܗ FfL || 5 ܕܗܘܠܐ BLVR || ܕܠܐ: ܕܠܐ FO;

---

[1] For a more detailed account of Barhebraeus' "life and works", see Takahashi, *Bio-Biblio*.

ܕܐܠܐ f || 6 ܗܘܝܐ f || ܕܐܝܬ ܗܘ : ܕܐܝܬܘ BLV || 8 ܡܫܡܫܢܐ BV || ܡܬܠܝܢ BLVR || 9 ܚܒܝܪܐ ܘܢܘܗܪܐ BLVR || ܦܬܓܡܐ B || ܦܝܪܝܦܐܛܝܩܝܐ (hic, nec ad ܐܦܘܦܐܢܛܝܩܝܐ): add. gl. ܡܚܘܝܢܝܬܐ in marg. L^{2}R || 10 ܡܠܠܐ BLVR || ܦܝܠܠܘܣܘܦܘܬܐ OR; ܦܝܠܘܣܘܦܘܬܐ BLV || 12 ܐܪܝܬܝܬ: add. ܐܝܟ BV]

For thee, our mighty God - whose coals enkindle the fiery Seraphim [Jes. 6.6][2] and whose priests, the Seraphim, officiate flying towards one another in the Threefold Sanctuary[3] - the soul, a spark of thy fire, yearns. Towards thee reason, a flash of thy light, is led. For it thirsts for thee, with whom is the source of life [Ps. 36.10][4]; and it delights in thee, with whom is the precious one among the peoples. Bind it in the fetters of thy love; nourish it with the ***cream of thy wisdom***. From the abyss of matter draw and raise it; upon the height of the intelligences let it rise and stand. Thou who art truly the saviour and purifier, the one who lifts and raises from the depth of the nether regions. Thine are all glory and honour in the greatness of thy beauty; and of the hypostases, the word and the life of the necessity of thy being.

These [i.e. what follow], friends, are the Peripatetic sayings and declaratory words,[5] in which I have brought [together], fully and clearly, the techniques of logic and the opinions of theoretical and practical philosophy, and (in which) I have expounded, manifestly and plainly, how vigilant minds may acquire correct teachings through them.

The scope of the work is specified in the last part of this proem. It is intended to be a clear and systematic exposition of Peripatetic philosophy.[6] - In one of his earlier philosophical works, Barhebraeus

---

[2] The reference is to the burning coal of Isaiah 6.6, understood in the Syrian tradition as a type of the consecrated host (see, e.g., Renaudot, *LOC* II.63; Brightman [1896] 573b; PS 741; Audo [1897] 148a; Sākā [1963] 8f.; Barsom-Samuel [1991] preface, 8th para.). - Cf. BH *Horr*. ad loc. (ed. Tullberg [1842] 6.19-7.1): ܟܐܡܪܗ ܠܩܘܪܒܢܐ : ܗ̅ ܐܢܐ ܦܚܡܗ ܕܢܘܪܐ . ܢܣܒ ܡܢ ܡܕܒܚܐ ܒܨܒܬܐ ܗ̅ ܐܢܐ ܒܫܡܐ ܕܡܫܬܡܫܢܘܬܗ ܡܫܬܡܫܢܐ ܠܩܘܪܒܢܐ (cf. also Tullberg's comm. ad loc.). - The "coal of the Seraphim" also features in the proem to BH *Rad*., Bodl. Or. 467, 1v 5f. (= ed. Istanbul 1.8f.): ܕܟܝܢܝ ܒܪ̈ܣܣܐ ܕܙܘܦܟ . ܙܕܩܝܢܝ ܒܓܘܡܪܬܐ ܕܣܪ̈ܦܝܟ . ("... cleanse me with the sprinkling of thy hyssop [cf. Ps. 51.7], and justify me with the coal of thy Seraphim").

[3] Cf. BH *Horr*. in Jes. 6.3: ܩܕܝܫ ܩܕܝܫ ܩܕܝܫ . ܗ̅ ܐܢܐ ܬܠܝܬܝܘܬ ܩܢܘ̈ܡܐ. - The phrase ܚܕ ܠܘܬ ܚܕ also echoes Jes. 6.2-3 (Peshitta: ... ܘܦܪܚܝܢ ܦܪܚܝܢ . ܘܩܥܝܢ ܚܕ ܠܚܕ . . .), as was evidently noticed by some copyist who inserted the words *w-qārēn*.

[4] Cf. also BH *Carmina* [Scebabi] 148.3-6, [Dolabani] p. 46, no. 3.8, incipit ܠܘܬܟ ܗܘ ܠܒܝ ܟܣܦ ܨܗܝܐ ܠܡܒܘܥܐ.

[5] Note the alliteration in the Syriac (six *pē*'s); perhaps: "Peripatetic phrases and apophantic aphorisms".

[6] Cf. the description in the list of BH's works in the continuation of *Chron. eccl.* (II.477.13-16): ܟܬܒܐ ܕܚܐܘܬ ܚܟܡܬܐ ܕܒܗ ܟܢܫ ܘܣܕܪ ܟܠܗܝܢ ܡܕ̈ܥܬܐ ܡܫܠܡܢܘܬܐ ܘܟܠܗ ܕܦܝܠܘܣܘܦܘܬܐ ܕܐܪܝܣܛܘܛܠܝܣ ("Liber butyri sapientiae, in quo collegit et digessit

had limited himself to the treatment of logic (*L. pupillarum*). In two others, one a relatively short work (*Sermo sapientiae*) and the other of an intermediate length (*Tractatus tractatuum*), he had dealt with a wider range of subjects, covering logic, the natural sciences and metaphysics. In his final and longest work on philosophy, he will now deal with these subjects more "fully" (*malyāʾīṯ*),[7] dealing with each subject at greater length.

The title of the work is given a rather cryptic explanation in the proem.[8] The allusion to Isaiah, chapter 6, at the beginning of the proem suggests that the author wishes us to associate the title "*ḥêwaṯ ḥeḵmtā*" with phrases which occur a little later in the Book of Isaiah, namely the "cream and honey" (*ḥêwtā w-ḏeḇšā*) of Jes. 7.15, on which the promised child, born of a virgin (Jes. 7.14), would be fed and the "cream and honey" of Jes. 7.22, eaten by those "who remain in the land". We might note further that, of these two verses, Jes. 7.15 is associated by Barhebraeus himself in his biblical commentary with Lc. 2.40, the verse in which we hear of the Holy Child growing up and being "filled with wisdom".[9]

Although we should perhaps beware of reading too much into these associations, if we are correct in making these associations, the implication is that the author is claiming his work to be an exposition of divine wisdom. This is somewhat surprising in view of the fact that

---

omnes partes sapientiales et sententias philosophiae Aristotelis").

[7] We note that in contrast, in the proem to the *Tractatus tractatuum*, BH talks of that work as an "introduction" (*maʿʿaltā*) to the sciences of logic, natural philosophy and metaphysics (ms. Cantab. Add. 2003, 3r 4).

[8] BH frequently mentions the titles of his works in his proems, i) sometimes simply stating that "such and such" is the name given by him to the work (so, for example, in *Eth.*, Bedjan [1898] 2.6, Teule [1993] textus 8.8; *Hieroth.*, Marsh [1927] 165*.18f.; *Splend.* [Moberg] 2.21f., cf. 2.9; *Tract.*, ms. Cantab. Add. 2003, 3r 5), ii) sometimes adding to such a statement a brief explanation as to why he has chosen that name (*Asc.* [Nau] 2.8-10; *Cand.* [Bakoš 1930] 27.11f.; *Serm. sap.* [Janssens] 46.8f.), and iii) sometimes, as in *But.* here, more allusively, mentioning the word/phrase used as the title, without explicitly stating that this is the title of the work (*Columb.* [Bedjan] 522.8, 19f., with reference to Noah's dove [Gen. 8.8] and the dove as the symbol of the Holy Spirit; *Nom.* [Bedjan] 2.2, [Çiçek] 1b.2; *Rad.* [Istanbul] 2.1).

[9] BH *Horr.* in Jes. 7.15 [Tullberg] 8.12f.: ܗ̄ ܐܢܫ ܗܘ ܕܛܠܝܐ ܕܝܢ ܪܒܐ ܗܘܐ ܘܡܬܚܝܠ ܒܪܘܚܐ - BH does not here quote the second half of Lc. 2.40, which reads: ܘܡܬܡܠܐ ܚܟܡܬܐ . ܘܛܝܒܘܬܐ ܕܐܠܗܐ ܐܝܬ ܗܘܐ ܥܠܘܗܝ (cf. BH *Horr.* in Lc. 2.40 [Steinhart] 10.6-8: «ܛܠܝܐ ܕܝܢ ܪܒܐ ܗܘܐ» ܗ̄ ܒܡܫܘܚܬܐ ܕܝܢ ܐܢܫܝܬܐ_ «ܘܡܬܚܝܠ ܒܪܘܚܐ ܘܡܬܡܠܐ ܚܟܡܬܐ» ܗ̄ ܒܡܬܓܠܝܢ ܗܘ̈ܝ ܒܗ ܐܬܘ̈ܬܐ ܐܠܗ̈ܝܬܐ .).

the actual contents of the work is what the author himself elsewhere calls the "wisdom of the Greeks" (*ḥek̠mat̠ yawnāyē*)[10] and in view of the contrast traditionally drawn between the "wisdom of the Greeks" (i.e. pagan philosophy) and the wisdom of God.[11]

Biblical allusion aside, it is probably not wrong also to understand the title "Cream of Wisdom" as meaning the "essence" or "best part"[12] of wisdom, and it was no doubt an understanding of the title in this sense that gave rise to the variant forms which are sometimes encountered. The title of the work is given as *ḥek̠mat̠ ḥek̠māt̠ā* ("wisdom of wisdoms") at Assemani, *BOCV* II.270. This is most probably based on the variant reading of the Vatican ms. (ܚܟܡܬ ܚܟܡܬܐ) in the list of Barhebraeus' works at *Chron. eccl.* II.477.13f., but otherwise has little manucript support.[13] One form which is frequently encountered in manuscripts is *ḥêwat̠ ḥek̠māt̠ā*, a hybrid form, as it were, between *ḥêwat̠ ḥek̠mtā* and *ḥek̠mat̠ ḥek̠māt̠ā*. This is, in fact, the form used in the great majority of the 19th century manuscripts accessible to me, which, it might be added, are mostly either themselves written in the East Syrian script or copied from East Syrian exemplars.[14] An interesting

---

[10] BH *Columb.* IV [Bedjan] 578.5-7: ܚܟܡܬ ܝܘ̈ܢܝܐ ... ܡܠܘܠܘܬܐ ܠܡ ܘܟܝܢܝܬܐ ܘܡܐܠܗܝܬܐ ܘܐܪܝܬܡܛܝܩܝ ܘܓܐܘܡܛܪܝܐ ܘܝܘܠܦܢܐ ܕܗܘ̈ܦܟܐ ܘܙܘ̈ܥܐ ܕܢܗܝܪ̈ܐ ("... wisdom of the Greeks, viz. logic, the natural sciences, metaphysics, arithmetic, geometry, and the teaching of the orbits and movements of the luminaries [i.e. astronomy]").

[11] Cf. the discussion on the phrase "*ḥek̠mat̠ yawnāyē*" in Ephrem, *Hymn. de fide* II.24 (ed. Beck 7.13) at Brock (1980) 19; Koonammakkal (1994) 170f.

[12] The use of the word *zubda* in this sense is common enough in Arabic (see, e.g. Wehr-Cowan, s.v.; cf. n. 19 below).

[13] See the discussion in the article "Crème de la science ou Science des sciences? ...", Janssens (1930). - In the list of BH's works in the Vatican ms. of *Chron. eccl.*, a contributing factor for the corruption into *ḥek̠mat̠ ḥek̠māt̠ā* may have been the title of the *Tractatus*, *Têḡrat̠ têḡrāt̠ā*, which stands at the beginning of the list.

[14] The form with the plural is used throughout the two Londinenses (both East Syrian), Or 4079 and 9380 (= Ll, dated 1808 and 1892), and the Vaticani (all E. Syr.), syr. 469 (1804), 603-4 (1826) and 613-615 (= V, 1887), as well as ms. Beirut, USJ syr. 48 (18/19th c., E. Syr.), Manchester, John Rylands syr. 56 (= R, 1887) and Ming. syr. 23 (= m, 1894; the last two, W. Syr., but copied from E. Syr. mss.). The undated Princeton manuscript (= P) also has the plural on the two occasions where the title is preserved, at fol. ٢٣a (= p.124).8 and fol. ١٥٤ (= p. 315).14 (on the likely date of this ms., see Chapter 2 below), while Sarau & Shedd (1898) also write ܚܐܘܬ ܚܟܡ̈ܬܐ in their catalogue (pp. 13-15), suggesting that the Urmia mss., too, may have had this form. The report, on the other hand, at Mingana (1933) 112 that ms. Ming. syr. 44 (dated 1573/4) had the form ܚܟܡ̈ܬܐ may be suspected, since Mingana also writes ܚܟܡ̈ܬܐ for Ming 310, where we read ܚܟܡܬܐ without *syāmē*.

case is ms. Mingana syr. 310 (dated 1865), a West Syrian manuscript copied in Mosul (i.e. in the "East") reportedly from an immediate copy of the autograph. In this manuscript, the copyist, while using the form ܚܟܡܬܐ without *syāmē* throughout, vocalises this on one occasion, at the very beginning of the manuscript, as ܚܟ̈ܡܳܬܐ. One suspects that our 19th c. copyist was used to hearing the work being referred to, in speech, as *ḥêwat ḥekmāṯā* and so duly added the vowel *-ā-* to the word in his text, but hesitated to add the *syāmē* which were not there in his exemplar. - That this form with the plural *ḥekmāṯā* cannot be the original form is, I believe, vouched for sufficiently by the fact that the singular form is used in the two oldest manuscripts of the work available to me (Florence, Laur. or. 69 & 83, = F, dated 1340 and Oxford, Bodl. Hunt. 1, dated 1498/9, both W. Syr.)[15] and by the proem above, where *ḥekmtāk* is parallel to *reḥmtāk*.[16]

As to the source of inspiration for the title, Margoliouth[17] has suggested that Barhebraeus might have modelled his title on the words "the cream of truth and choice morsels of the sciences" (*zubdat al-ḥaqq ... qafīy al-ḥikam*) found near the end of Ibn Sīnā's *K. al-išārāt wa-l-tanbīhāt*,[18] a work which Barhebraeus had translated into Syriac. We might mention in this connection that Barhebraeus also translated a philosophical work of Athīr al-Dīn al-Abharī entitled *Zubdat al-asrār* ("cream of secrets").[19] A further factor leading to the adoption of the title *ḥêwat ḥekmṯā* will have been the similarly alliterative titles the author had given to his two earlier compendia of philosophy, *Têḡraṯ têḡrāṯā* and *Swāḏ sōp̄iya*.

---

[15] Also, it appears, in the majority of manuscripts in the list of Barhebraeus' works in the *Chron. eccl*.

[16] In the proem, the singular *ḥekmtāk* is read in all six manuscripts collated above.

[17] Margoliouth (1887) 39; repeated by Janssens (1930) 372; cf. also Tkatsch (1928) I.88.

[18] Ed. Dunyā (1947) 256.4f., (1957-60) (IV).903.4f.; tr. Goichon 525.11. - BH translates the phrases as ܚܐܘܬܐ ܕܫܪܪܐ ... ܦܠܓ̈ܐ ܕܚܟ̈ܡܬܐ (BH *Ind*., ms. Florence, Laur. or. 86, 132r b12f.).

[19] Cf. Takahashi, *Bio-Biblio.*, I.2.2, no. 14. - "*Zubda*", of course, is a word commonly used in titles of Arabic works, as one may see, for example, by glancing at the index of Brockelmann, *GAL* (Suppl. III.1165f.), where we count 63 works whose titles begin with this word. One might mention among them, in view of BH's knowledge of Ṭūsī, the astronomical works of the latter, *Zubdat al-hai'a* and *Zubdat al-idrāk fī hai'at al-aflāk*.

1.3. *Date of Composition*

The date of composition of the *Butyrum* can be gathered from the following notes in our manuscripts.[20]

Mingana syr. 310 (M), fol. 216r 7-20:
"Here ends the Book on the Soul ... Thus was written in the copy [*ktāḇā*] of our illustrious father - may God preserve his life - on 22nd Āb, the year 1596 of the Greeks [= Aug. 1285 A.D.] in the fortified city of Mosul. Glory to God who has given strength and help. The second copy [*ktāḇā*] was finished on 11th Teshrīn I, the year 1597 of the Greeks [= Oct. 1285 A.D.]. [18-20] This our own third [copy] was finished on 17th Tammūz, the year 2176 of the Greeks [= July 1865] in the School of the Mother of God. Amen."

Laur. or. 83 (F), fol. 191v a 5-13:
"Here ends the Book of Theology. With its completion [*šullāmā*] comes to end the theoretical part of philosophy in the book of Cream of Wisdom, except for the mathematical [parts] which belong to another treatise. End of Kānōn I, the year 1597 of the Greeks [= Dec. 1285 A.D.]. Glory to God who has given strength and help in His goodness. - Thus was written in the copy of our blessed and deceased father."

Laur. or. 83, fol. 227r 5-13:
"Here ends the Book of Politics. With its conclusion [*šumlāyā*] comes to end the book of Cream of Wisdom. On 8th Shḇāṭ, the year 1597 of the Greeks [= Feb. 1286]. Glory to God, who has given strength and help in His goodness and abundance of His manifold mercies. Thus was written in a copy of our father, the holy Maphrian of the East, the late Mar Gregory."

cf. Laur. or. 83, fol. 227r, between columns:
"This date for the conclusion of the composition of this book is in the hand of the author, the late maphrian."

According to these notes, Barhebraeus finished the second part (natural sciences) of the *Butyrum* on 22nd Aug. 1285, the third part (metaphysics) at the end of Dec. 1285 and the whole work on 8th Feb. 1286. We also learn that the second part, at least, was written in Mosul.

[20] For the Syriac text, see Appendix I below.

1.4. *Divisions of the Work*

While it is clear that there are four major parts to this work 1) logic, 2) natural sciences, 3) metaphysics and 4) practical philosophy, the terminology used by the author to designate these divisions is not always consistent. It appears firstly that the author intended to divide the work into two parts (*pālgwāṯā*),[21] dealing with A) logic and B) philosophy proper. The parts on logic, natural sciences and metaphysics are also designated as the first, second and third "teachings" (*yulpānē*), but not the section on practical philosophy. This last section is given instead the heading "part" (*mnāṯā*), while it is also stated at the end of the preceding book (Book of Theology) that this is the end of the "part" (*mnāṯā*) on "theoretical philosophy".[22] Although there is nowhere an indication of the beginning of the "part" on theoretical philosophy, that must mean the natural sciences and metaphysics. The hierarchy of the divisions is therefore probably to be understood to be as follows.

- Pālgūṯā A. Logic
  - Logic = Yulpānā I
- Pālgūṯā B. Philosophy
  - Mnāṯā 1. Theoretical philosophy
    - Natural sciences = Yulpānā II
    - Metaphysics = Yulpānā III
  - Mnāṯā 2. Practical philosophy
    - Practical philosophy (= Yulpānā IV?)

For the sake of simplicity, the four major divisions of the work will be referred to below simply as Parts 1-4.

The four parts are further divided into a total of 22 "books" (*kṯāḇē*), as given below. Each book is divided into "chapters" (*qep̄ale'a*, κεφάλαια), "sections" (*pāsōqē*) and "theories" (*te'ōriyas*, θεωρίαι), so that the Book of Minerals edited below, for example, consists of 5 chapters, 17 sections and 72 theories and the Book of Meteorology of 5 chapters, 17 sections and 77 theories.

---

[21] Vocalised thus in the manuscripts, rather than *pelgwāṯā* ("halves"), which might also have been considered possible given the bipartite division.

[22] See the note quoted above from Laur. or. 83, fol. 191v a 5-13.

I. Logic (ܡܠܝܠܘܬܐ): 1. Isagoge (ܟܬܒܐ ܕܐܝܣܓܘܓܝ; 4 chapters); 2. Categoriae (ܟܬܒܐ ܕܩܛܝܓܘܪܝܐܣ; 3 chapters); 3. Perihermenias (ܟܬܒܐ ܕܦܪܝܗܪܡܢܝܐܣ; 9 chapters); 4. Analytica [priora] (ܟܬܒܐ ܕܐܢܠܘܛܝܩܐ; 7 chapters); 5. Apodeictica (ܟܬܒܐ ܕܐܦܘܕܝܩܛܝܩܐ; 9 chapters); 6. Topica/Dialectica (ܟܬܒܐ ܕܛܘܦܝܩܐ ܗܘ ܕܕܝܠܩܛܝܩܐ; 8 chapters); 7. Sophistici elenchi (ܟܬܒܐ ܕܣܘܦܝܣܛܝܩܐ; 5 chapters); 8. Rhetorica (ܟܬܒܐ ܕܪܝܛܘܪܝܩܐ; 7 chapters); 9. Poetica (ܟܬܒܐ ܕܦܘܐܝܛܝܩܐ; 3 chapters).
II. Natural sciences (ܟܝܢܝ̈ܬܐ): 1. Auscultatio physica (ܟܬܒܐ ܕܫܡܥܐ ܟܝܢܝܐ; 5 chapters); 2. De caelo (ܟܬܒܐ ܕܫܡܝܐ; 5 chapters); 3. De generatione et corruptione (ܟܬܒܐ ܕܗܘ̈ܝܐ ܘܚܘܒ̈ܠܐ; 4 chapters); 4. De mineralibus (ܟܬܒܐ ܕܡܛܠܘܣ; 5 chapters); 5. Meteorologica (ܟܬܒܐ ܕܦܘܚ̈ܬܐ; 5 chapters); 6. De plantis (ܟܬܒܐ ܕܢܨܒ̈ܬܐ; 4 chapters); 7. De animalibus (ܟܬܒܐ ܕܚܝ̈ܘܬܐ; 6 chapters); 8. De anima (ܟܬܒܐ ܕܥܠ ܢܦܫܐ; 4 chapters).
III. Metaphysics (ܒܬܪ ܟܝܢܝ̈ܬܐ): 1. Philosophia prima (ܟܬܒܐ ܕܦܝܠܣܘܦܝܐ ܩܕܡܝܬܐ; 8 chapters); 2. Theologia (ܟܬܒܐ ܕܬܐܘܠܘܓܝܐ; 6 chapters).
IV. Practical philosophy (ܦܝܠܣܘܦܝܐ ܣܘܥܪܢܝܬܐ): 1. Ethica (ܟܬܒܐ ܕܐܝܬܝܩܘܢ; 4 chapters); 2. Oeconomica (ܟܬܒܐ ܕܐܝܩܘܢܘܡܝܩܘܢ; 3 chapters); Politica (ܟܬܒܐ ܕܦܘܠܝܛܝܩܘܢ; 3 chapters).

An idea of the weight carried by each of these four parts may be gained from the number of folios they occupy in the following manuscripts.

Laur. or. 69 & 83 (F): a) Logic: **281** fols. (ms. 69 + 9 lines in ms. 83); b) Nat. sc.: **128** fols. (ms. 83, 2r-130r); c) Metaph.: **61** fols. (-191v); d) Pract.: **36** fols. (-227r).
Brit. Lib. Or. 4079 (L): a) Logic: **155** fols. (2v-157r); b) Nat. sc.: **81** fols. (-238v); c) Metaph.: **47** fols. (-285r); d) Pract.: **29** fols. (-314r).

1.5. *Models of Butyrum sapientiae*

The resemblance of the overall structure of the *Butyrum* and much of its contents to those of Ibn Sīnā's *K. al-šifāʾ* has long been known and D.S. Margoliouth, who edited the Book of Poetics over a century ago, aptly talks of the "cream of wisdom" as being drawn from the 'milk' that is the *Šifāʾ*.[23] At the same time, as Margoliouth was aware, the *Butyrum* is not simply a translation or summary of the *Šifāʾ*, and this is obvious even from a cursory look at the major divisions of the two works.

[23] Margoliouth (1887) 43: "... equidem scire gerens utrum totum illud butyrum lacte Avicennaeo expressum esset, ..."

*Šifā*ʾ	*Butyrum*
1: Logic (*manṭiq*)	Logic (*mlīlūṯā*)
2: Natural sciences (*ṭabīʿīyāt*)	Natural sciences (*kyānāyāṯā*)
3: Mathematics (*riyāḍīyāt*)	Metaphysics (*bāṯar kyānāyāṯā*)
4: Metaphys. (*baʿda al-ṭabīʿīyāt*)	Practical phil. (*pīlōsōpiya praqṭīqāytā*)

In the *Butyrum*, Barhebraeus has suppressed the section on the mathematical sciences (geometry, algebra, music and astronomy)[24] and has expanded the discussion of ethics, economics and politics (treated only briefly in the *Šifā*ʾ in the 10th *maqāla* of the Ilāhīyāt) into an independent section, which, as M. Zonta has recently shown, has as its principal source Naṣīr ad-Dīn aṭ-Ṭūsī's *Aḫlāq-i Nāṣirī*.[25] It may be noted that the alteration made here by Barhebraeus, in fact, brings the *Butyrum* more in line with the contents of the Aristotelian corpus.

As has been made known mainly through studies made by H.J. Drossaart Lulofs,[26] another major source used by Barhebraeus for the section on natural sciences in his *Butyrum* was the compendium of Aristotelian philosophy composed by Nicolaus Damascenus, the 1st century B.C. Peripatetic philosopher, historian and adviser to Herod the Great. It was long believed that Nicolaus' Περὶ τῆς Ἀριστοτέλους φιλοσοφίας was lost except for a few fragments preserved in the writings of later authors, until it was shown that one of the texts preserved in ms. Cantab. Gg. 2.14 was a Syriac version (or, to be more precise, long excerpts from a Syriac version) of this work. As far as can be made out from this manuscript and other available evidence, Nicolaus' compendium covered only the works of Aristotle dealing with the natural sciences and the *Metaphysica* (treated immediately after *Ausc. phys.*). Drossaart Lulofs has published what remains of the first five books of this Syriac version and the fragment (1 damaged folio) of the section *de plantis*.[27] Nicolaus' *De plantis*, however, survives

[24] For BH's reason for omitting the mathematics, see the note at ms. F, fol. 191v, which is quoted above; cf. with this Ibn Sīnā's note concerning the omission of the practical sciences in his work, *Šifā*ʾ, al-madḫal, ed. Madkour et al. (Cairo, 1952), 11.12f. (cf. ibid. intro. ١٢, 13).

[25] Zonta (1992); id. (1998).

[26] Drossaart Lulofs (1985); Drossaart Lulofs-Poortman (1989); cf. Drossaart Lulofs (1965); Takahashi (2002a); id. (2002b) 219-228; id. (2002d).

[27] Drossaart Lulofs (1965) and Drossaart Lolofs-Poortman (1989).

in a more complete form in an Arabic version made by Isḥāq b. Ḥunain and revised by Thābit b. Qurra,[28] and from a comparison with this Arabic version Drossaart Lulofs was able to show that large parts of the sections on plants in Barhebraeus' *Butyrum* and *Candelabrum* were based on Nicolaus' work.

The Cambridge manuscript in which the Syriac version is preserved is unfortunately badly damaged and has been rebound in the wrong order.[29] In addition, what the manuscript contained, even in its original state, was not the whole of the Syriac version, but excerpts from it. Some parts were, in fact, so severely curtailed that the part corresponding to Aristotle's *De generatione et corruptione*, for example, occupies only 7 lines in our manuscript. On the other hand, it appears that the excerptor took care to copy at least something out of each "book" (*mêmrā*) of his exemplar, so that what we have remaining is sufficient to allow us to reconstruct the order in which the materials were arranged.

It may be instructive here to compare the order in which the "books" were arranged in the parts of the *Butyrum* and the *Šifā᾿* dealing with the natural sciences and in Nicolaus' *Compendium*.

Butyrum	*Šifā᾿*	Nicolaus
I. Auscultatio physica	I. Auscultatio physica	I. Auscultatio physica
		II.-III. Metaphysica
II. De caelo	II. De caelo	IV. De caelo
III. De gen. et corr.	III. De gen. et corr.	V. De caelo (contd.), De gen. et corr.
	IV. De action. et passion.	
IV. De mineralibus	Va. De mineralibus	VI. Meteorologica I-III
V. Meteorologica	Vb. Meteorologica	VII. De mineralibus, De plantis?
VI. De plantis	VI. De anima	
VII. De animalibus	VII. De plantis	VIII.-IX. De animalibus (*HA* & *PA*)
VIII. De anima	VIII. De animalibus	X. De anima

[28] This Arabic version was translated into Latin by Alfred of Sarashel, and the work known as Ps.-Aristotle, *De plantis* is, in fact, a 13th c. Greek version (probably by Maximus Planudes [c. 1255-1305] or Manuel Holobolos) of this Latin version of Nicolaus' work.

[29] In what follows, references to this work will be according to the following page numbers. - pp. 1-2: fol. 328rv; 3-4: 369rv; 5-6: 329rv; 7-8: 330rv; 9-10: 331rv; 11-12: 332rv; 13-14: 333rv; 15-16: 334rv17-18: 370rv; 19-20: 335rv; 21-22: 336rv; 23-24: 337rv; 25-26: 338rv; 27-28: 339rv; 29-30: 340rv; 31-32: 341rv; 33-34: 343rv; 35-36: 342rv; 37-38: 367rv; 39-40: 368rv; 41-42: 344rv; 43-44: 345rv; 45-46: 346rv; 47-48: 347rv; 49-50: 348rv; 51-52: 349rv; 53-54: 365rv; 55-56: 366**vr**; 57-58: 371rv; 59-60: 350rv; 61-62: 383**vr**; 63-64: 382rv; 65-66: 378rv; 67-68: 379rv; 69-70: 380rv; 71-72: 381rv; 73-74: 384rv; 75-76: 385**vr**.

XI. (*Sens.*, *Somn.Vig.*, *Insomn.*)
XII. (*GA* I-IV)
XIII. (*GA* V).[30]

It may be seen that in suppressing the book on "actions and passions" (*al-afᶜāl wa-l-infiᶜālāt*) and placing the De plantis and De animalibus before the De anima, Barhebraeus is in agreement with Nicolaus against Ibn Sīnā and, more generally, with the traditional order of the Aristotelian corpus. On the other hand, in placing mineralogy before meteorology, Barhebraeus follows Ibn Sīnā, and the materials from Ibn Sīnā's book on "actions and passions", in fact, survive in the *Butyrum* by being incorporated into Books III. De gen. et corr. and IV. De mineralibus.

In addition to Ibn Sīnā's *Šifāʾ*, Ṭūsī's *Aḫlāq-i nāṣirī* and Nicolaus' compendium Barhebraeus has clearly used a number of further sources in the *Byturum*. The sources used in those parts of the work dealing with mineralogy and meteorology will be discussed in Section 3.2 below.

[30] Further fragments corresponding to Arist. *De longaevitate* and *Hist. An.* I are found at the end of the manuscript; the placement of these passages here is evidently due to the copyist/excerptor and not the original order.

CHAPTER TWO

MANUSCRIPTS OF *BUTYRUM SAPIENTIAE*

Given below is a list of manuscripts containing either the whole or part of the *Buyrum sapientae*, which are known to exist today or to have existed in the near past.[1]

The manuscripts which I have been able to examine are marked with a single asterisk (*). The manuscripts used in the edition of *But.* Min. and Mete. below are marked with two asterisks (**).

The colophons of the manuscripts to which I have had access are given in Appendix I below.

A. *Manuscripts which Contain the Whole Work or whose Exact Contents are Unknown*

1. Florence, Bibl. Medicea Laurenziana, or. 69 (olim Palat. 186)* and **83** (olim 187)** = **F**

Literature: S.E. Assemani (1742) 328f., Drossaart Lulofs-Poortman (1989) 43, 730.

1340 A.D. West Syrian. Copied by Najm b. Shams (al-Daula) b. Abū al-Faraj of Mardin.[2] 281 + 227 fols., 25 to 30 lines (double columns).

The two manuscripts originally formed a single whole, or. 69 containing the logical part of *But.*, except for the very last part of the Book of Poetics, which is found on the first page of the text in

---

[1] The manuscripts used by Drossaart Lulofs in his edition of *But.* Plant. (mss. FMmVLl below) were discussed at Drossaart Lulofs-Poortman (1989) 42-48. A further analysis of mss. FMVLlP may be found in Joosse (1999).

[2] The name ܢܓܡ of the copyist has been rendered as Negemus (S.E. Assemani [1742] 328f.) and Neğem (Margoliouth [1887] 42). In the 'light' of the names of the copyist's father, Shams ("sun", fol. 227r b17, v 16), and his brother, Badr ("full moon", fol. 227v 15), the name should no doubt be read as Najm ("star"). - Najm is wrongly made the son of Abū al-Faraj by Assemani (followed by Drossaart Lulofs-Poortman [1989] 43). - He is probably to be identified with the copyist, "the monk Najm", of a manuscript of BH *Splend.* (olim) in New Jersey dated 1336 (Barṣaum, *Luʾluʾ* 426 n.1).

or. 83 (fol. 3r). - The colophons suggest that the manuscript is an immediate copy of the autograph.[3] - The manuscript was purchased, according to a note on or. 69, fol. 1r., by Rabban Daniel (= Daniel of Mardin) from the priest Rabban Joseph b. Cyriacus, for 400 *zūzē*, in 1678 A.Gr. (1367 A.D.). - On the marginal notes in this manuscript, see Appendix II below.

2. London, British Library, Or. 4079 = L**

Lit.: G. Margoliouth (1899) 24f., Drossaart Lulofs-Poortman (1989) 46f.

1809 A.D. East Syrian, vocalised. Copied by George (Gīwargīs) b. Yaqo b. Dusho of the family of Yuḥanna,[4] of Alqosh.[5] 322 fols, ca. 35 lines (double columns).

The manuscript was copied, according to the colophon on fol. 314r, from an unvocalised exemplar which was difficult to read. - It was purchased by John Elijah Mellus from the great-grandson of the copyist George, and sold by him, "after he had made a copy of it", to E.A. Wallis Budge in 1889 (cf. no. 7 below).[6] - The manuscript was used along with Ming. syr. 310 (ms. M., no. 27 below) by Mellus in copying Vat. 613-615 (ms. V, no. 29 below). - There is a long lacuna in *But.* Gen. et corr. between I.ii.4. and III.iii.3., and in *But.* Min. the whole of chapter III. is missing. These lacunae have been supplied at the end of the manuscript in the hand of Mgr. Mellus (= λ). - The text is preceded, on fol. 1r, by the same drawings (relating to logic) as those found at the beginning of Vat. 603 (no. 16 below) and, on fol. 1v, by a glossary of Greek technical terms identical to that found at the beginning of John Rylands 56 (no. 17 below).

---

[3] See Appendix I (the notes on fol. 191v a 12f.; 227r b 10ff.; also the note between the columns on 227r.).

[4] Written ܝܘܚܐܢܐ by the copyist.

[5] Known to us also as the copyist of mss. Ming. syr. 109-111 (dated 1793-95) and Telkeph 16, 87, 88, 94 (dated 1791-1826; catal. Ḥabbī in Ḥabbī et al. [1977] 21-49). - For a manuscript copied by his son Yaqo b. George b. Yaqo, see Vosté (1928) 186 (N.D. des Semences 165 = Baghdad, Mar Antonius syr. 495 Ḥaddad-Isaac). - Yaqo (ܝܩܘ): vernacular form of Jacob (MacLean [1901] 121a).

[6] See the note of sale found on fol. 314v, reproduced at Drossaart Lulofs-Poortman (1989) 46.

3. Olim Urmia, Lib. of the Museum Association of Oroomiah College 64-68

Lit.: Sarau-Shedd (1898) 13-15.

These five manuscripts, all in the West Syrian script, measuring 6 x 9 cm and, it would seem, by the same copyist (or, possibly, father and son?), together covered the whole of the *Butyrum*. Total: ca. 3,000 pp.

64: 1826 A.D. Copied by Behnām Āṯōrāyā. 641 pp. *But*. De anima-Polit.
65: 1825 A.D. Copied by Behnān Āṯōrāyā. 448 pp. *But*. Isag.-Periherm.
66: 1830 A.D. Copied by Gabriel Behnān Māwṣlāyā. 624 or 632 pp. *But*. Ausc. phys.-Animal.
67: 1815 A.D. (?)[7] Copied by Gabriel Behnān Māwṣlāyā. 944 pp. *But*. Apod.-Topic.
68: 1830 A.D. Copied by Gabriel Māwṣlāyā. 663 (?)[8] pp. *But*. Soph.-Poet.

The fragmentary Princeton manuscript (no. 34 below) may well be a fragment of olim Urmia 66.

4. Baghdad, Chaldean Monastery of Mar Antonius, Syr. 177 (olim Notre-Dame des Semences 47 Scher = 60 Vosté)

Lit.: Scher (1906) VII.497; Vosté (1928) 25; Haddad-Isaac (1988) 86.[9]

1818 A.D. Copied at the monastery of Rabban Hormizd by Joseph Audo (the future Chaldean patriarch 1848-78). 373 fols. (38 quires), 35.5 x 23 cm (32 lines, double columns). Judging from its reported size, this manuscript may well contain the whole work (cf. no. 2 above).

5. Baghdad, Mar Antonius, Syr. 178 (olim Notre-Dame des Semences 61 Vosté)

Lit.: Vosté (1928) 25; Haddad-Isaac (1988) 87.

---

[7] Perhaps a printing error for "1825" in the light of the dates of the other four mss.

[8] Or 463 pp. - The numeral in question is smudged and unclear in the copy of the catalogue accessible to me.

[9] This is one of a large number of manuscripts from N.-D. des Semences which have been transferred to Baghdad via the Monastery of St. George by Mosul where Macomber found them in 1966 (Macomber [1969] 476 n.22; cf. Kaufhold [1990] 264f.). - On the Monastery of N.-D. des Semences by Alqosh, see, e.g., Scher (1906) 479-481; Leroy (1957) 221-228; Fiey (1965) 548f. On its name, better translated as "N.-D. des Moissons", see Fiey (1965) 548 n.3 ("Our Lady of the Harvest", Mingana [1933] 1189).

Undated. Copied by a certain Elijah, thought by Haddad-Isaac to be Elijah Shēr (Scher) of Shaqlawa (1860-1949).[10] 21 quires, 22 x 15 cm (20 lines). Unlikely to contain the whole work (cf. no. 4 above).

6. Olim Notre-Dame des Semences 62 Vosté

Lit.: Vosté (1928) 25.

Undated but of recent date. Copied by the monk Vincent. 21 quires, 22 x 16 cm (16 lines). Unlikely to contain the whole work (cf. no. 4 above).

7. London, British Library, Or. 9380 = l**

Lit.: Drossaart Lulofs-Poortman (1989) 47.

1892 A.D. East Syrian, vocalised. Copied in Alqosh by ᶜIsa b. Isaiah (Ishaᶜya) b. Cyriacus,[11] apparently at Budge's request.[12] 563 fols, 28 lines.

The readings closely resemble those of Brit. Lib. Or. 4079 (ms. L, no. 2 above). With L it shares the lacuna in *But.* Gen. et corr. The greater part of Min. III. too is also missing, but unlike in L, the beginning and the end of this chapter have survived, so that this manuscript cannot be derived, at least solely, from L. - The manuscript was later lent by Budge to J.P. Margoliouth who used it in preparing her *Supplement to the Thesaurus Syriacus.*[13]

8. Mosul, Syrian Orthodox Episcopal Residence 1.63, nos. 6, 7, 9.

Lit.: Ibrahim in Ḥabbī et al. (1981) 202.

Undated. West Syrian. - Items nos. 6 and 7 in the manuscript are described by Ibrahim as the third and fourth parts (*qism*) of "*Kitāb*

---

[10] A relative of Addai Scher and author of several works, including a continuation of BH's *Mêmrā zawgānāyā* composed in 1883 (Baumstark [1936] 99, ms. Paderborn syr. 3; Macuch [1976] 413f., 402 n.7; ᶜAwwād [1978] 86f.; Fiey [1993] 16); known also as a copyist from 1880 onwards (Haddad-Isaac [1988] index 464, 476; cf. also Fiey [1965] 171 n.2, 754).

[11] Known to us as the copyist of a large number of manuscripts written at Alqosh, dated, as far as I am aware, between 1853 (Dohuk 34) and 1898 (CSCO 4), including Berol. 60, 63, 75, 96, 99 Sachau; Cantab. Add 2811; Ming. syr. 58, 61, 149, 576, 579; Vat. syr. 598, 599; a *Gazzā* in ᶜAmadiya (Fiey [1965] 315); Batnaya 7; ᶜAqra 35, 89; Dohuk 14, 34; CSCO 4; Baghdad, Mar Antonius, multi (see Haddad-Isaac [1988] index 483); Strasbourg 4132, 4133 (Briquel-Chatonnet [1997] 218, 220).

[12] See the mention of Budge at fol. 453v 1 (in the colophon reproduced in Appendix I below); cf. no. 2 above and Ebied (1974) 510-512 et passim.

[13] See the preface to the *Supplement* (p. vii) and the list of abbreviations (p. x, s.v. "But. Sap."); cf. Drossaart Lulofs-Poortman (1989) 47.

zubdat al-ḥikma fī al-falsafa", and item no. 9 as the second part of "*Kitāb al-samāᶜ al-ṭabīᶜī*" (i.e. Auscultatio physica; whether of the *But.* is not stated).

B. *Manuscripts Containing Whole or Part of the Section on Logic*

9. Oxford, Bodleian Library, Huntingdon 1* no. 7 (p.347-419)
Lit.: Uri (1787) 5f., no. XXV; Payne Smith (1864) 368-398, no. 122; cf. Bruns (1780) 3f.; Bernstein (1832) ix-x; Martin (1872) I. intro. 13; Göttsberger (1900) 66f.; Carr (1925) xciii f.; Budge (1932) viii-ix and Plate II; Sauma (2003) 144-146.
ca. 1498 A.D.[14] West Syrian.
The huge manuscript,[15] occupying the pride of place in the collection of Robert Huntingdon (resident in Aleppo 1671-1681),[16] contains various works of Barhebraeus: 1. *Horr.* (p.1-147), 2. *Gramm.* (p.148-162, followed by *Aequ.*, p.162-171), 3. *Splend.* (p.172-237), 4. *Carm.* (p.238-254, followed by poems of John b. Maᶜdani, p.254-259), 5. *Nom.* (p.260-335), 6. *Columb.* (p.335-347), 7. *But.* (p.347-419), 8. *Chron.* & *Chron. eccl.* (p.420-638).
There seems to have been at least three copyists responsible for the manuscript. One of them, calling himself Joseph or "Joseph the Iberian" in the colophons to *Splend.* (bottom of p. 237), *Mêmrā zawgānāyā* (p.241) and *Carm. de divina sapientia* (p.243), may be identified with the later Metropolitan of Jerusalem, Gregory Joseph ᶜAbd Allah al-Kurjī (1510-1537).[17] Another calls himself Denḥā[18]

---

[14] At the beginning of the entry on this manuscript, Payne Smith (1864) 368 gives the date of the manuscript as 1491 A.D. (this is followed by editors who have used this manuscript, such as Moberg [1922] xv and Sprengling-Graham [1931] xi, as well as at Baumstark, *GSL* 314 n.1, 316 n.4, 317 n.5). This is presumably a misprint, since I am unable to find in the manuscript itself or Payne Smith's description of it any dates other than that of 23rd Teshri I 1810 (= 1498 A.D., not 1499 as Payne Smith [1864] 390 has it) at the end of the text of *Nom.*

[15] Measuring 22 x 15 in. sec. Budge (ca. 55 x 38 cm); most pages in four columns of up to 70 lines each. - "Mole insana" as Uri puts it, or "ein ungeheurer Foliant" Ewald (1838) 109; "forma maxima, quam imperialem dicimus" Bernstein (1832) ix-x.

[16] See Budge (1932) II.xviii.

[17] On Joseph the Iberian, see Barṣaum, *Luᵓluᵓ* 458f.; Macuch (1976) 22; Fiey (1993) 219f.; Kiraz (1988) 49f., 64 n.5, the last of whom cites an article by I.E. Barṣaum, "Mār Ġrīġūriyūs al-rābiᶜ al-Kurǧī", al-Maǧalla al-baṭriyarkīya al-suryānīya

in the colophon to *Nom.* and tells us furthermore that the manuscript was copied at the Monastery of Mar Ḥananiya and Mar Augin (= Dair al-Zaᶜfarān) at the request of Bp. Dionysius (Dionysius Abraham of Mardin?) and that this part was finished at the 8th hour on 23rd Teshri I 1810 (= Oct. 1498). A third, whose hand I find difficult to distinguish from Denḥā's, has left us a note in Arabic at the end of *Horr.* in Daniel. (end of p. 102) referring to himself as "al-nāqil Sulaimān".

The part of the manuscript containing *But.* is, I believe (as does Barṣaum, *Luʾluʾ* 419 n.2), in the hand of Joseph the Iberian.[19] The text of *But.* is preceded by an introduction, perhaps composed by Joseph himself.[20] The text breaks off in mid-passage at Apod. IV.i.1, but the manuscript may originally have contained the whole of the section on logic.[21] Blank spaces are frequently left beside and between parts of the text, where the copyist may have intended to insert the scholia, giving rise to the possibility that the manuscript was copied from an exemplar containing extensive scholia, such as Laur. or. 69 & 83.

10. Birmingham, Orchard Learning Resources Centre, coll. Mingana,[22] Syr. 44A (fols. 1-30)
Lit.: Mingana (1933) 112.

(Jerusalem) I (1933) no. 5, p. 145-156 (non vidi).

[18] Denḥā Ṣaifī al-Ṣalaḥī, according to Barṣaum, *Luʾluʾ* 424 n.5, 453 n.6.

[19] From an examination of the manuscript on microfilm, I would tentatively allocate the responsibilities for its different parts as follows: Denḥā/Sulaimān (with more angular hands): *Horr.*, *Nom.*, *Columb.* (up to p.340), *Chron./Chron. eccl.*; Joseph: *Gramm.*, *Splend.*, *Carmina*, *Columb.* (p.341ff.), *But.* - There appears to be a change also between pp. 243 and 244 (halfway through *Carmina*), but the change there seems to be of the pen rather than of the hand.

[20] Published with Latin translation at Payne Smith (1864) 392-5; cf. Barṣaum, *Luʾluʾ* 458.18 (The passage of *Luʾluʾ* is rendered inaccurately at Macuch [1976] 22. The date 1533 does not belong to this introduction). - The same introduction is found in Ming. syr. 326 (no. 15 below).

[21] Quires no. 12-14 are missing at this point. Since the surviving part of *But.* occupies 72 pages and Apod. IV.i.1. is roughly half way through the section on logic, 3 quires (60 pages) would have been about right for the rest of the section. Since, however, quires numbered 13 and 14 do occur later on in the manuscript between quires 20 and 21, it may be that there is only one quire missing here (see Payne Smith [1864] 395 and the Latin note to the same effect in the manuscript on p.419). The quires are numbered in Armenian numerals, not Coptic as we are told at Payne Smith (1864) 395.

1574 A.D. (Tammuz 1885 A.Gr.). West Syrian. Copied by a certain Ephrem at Dair al-Zaᶜfarān. The first book (Isagoge) only; followed by a number of other works related to Aristotelian logic and, at the end of the manuscript, by Barhebraeus' *Tractatus tractatuum* (incomplete). The manuscript was transferred from Mardin to Mosul in 1829 A.D. (cf. no. 19 below).

11. Florence, Bibl. Medicea Laurenziana, or. 37, 10, 6 and 8 (olim Palat. 176-9)*

Lit.: S.E. Assemani (1742) 322; Renan (1852) 59, 66; Margoliouth (1887) 40; Janssens (1930) 367 n.5.

16-17th c. The manuscripts, said by S.E. Assemani to be a translation of Aristotle by Ḥunain b. Isḥāq, in fact, contain the first part of *But.*, evidently copied from Laur. or. 69 (no. 1 above) by Antonius Sionita.[23] - The text is divided between the four manuscripts as follows: 37 (olim 176): Isag.-Anal. pr. (350 fols.); 10 (olim 177): Apod. (208 fols.); 6 (olim 178): Top. (200 fols.); 8 (olim 179): Soph.-Poet. (395 fols.).

12. Olim Diyarbakır/Amid, l'Archevêché chaldéen 32

Lit.: Scher (1907) 344f.

Completed June 1638 A.D. [2nd Ḥaziran 1949 A.Gr.]. Copied in Diyarbakır by Simeon, Metropolitan of Diyarbakır.[24] First five books of section on logic. 174 fols., 28 x 19 cm (23 lines).

The manuscript was located by Macomber together with Diyarbakır 33 below in Diyarbakır and later transferred to the Chaldean episcopal residence in Mardin.[25]

---

[22] Collection formerly at Selly Oak Colleges.

[23] Antonius Sionita (al-Ṣiḥyūnī) filius Euphimiani: a pupil of the Maronite College in Rome, archpriest and monk, copyist of several manuscripts in Bibl. Med. Laur. dated 1607-1635 (?) and a compiler of a "dictionarium arabico-turcicum" (mss. olim Palat. 335-6). See Assemani (1742) 51, 410, 385 (cf. Gemayel [1997] 280-283 nos. 528-537); Cheikho (1924) 136f.; Armalet (1996) 408; Graf, *GCAL* I.138, III.342 (the place and date of death given by Graf, Paris 1648, are those of Gabriel Sionita; the confusion also occurs elsewhere, e.g. at Leroy [1971] 139 n.40).

[24] Metropolitan 1638-1657 (Fiey [1993] 50).

[25] Macomber (1969) 480 n.45 (cf. Desreumaux [1991] 130, 183). Prof. W. Baars (in a personal communication to Dr. P. Joosse) also reports that he found most parts of the Diyarbakır manuscripts of *But.* in 1965 in the attic of the Chaldean church (former cathedral) in Diyarbakır. Cf. also Vööbus (1978) 189 n.15 (found the bulk of the Chaldean collection at Diyarbakır in the attic of the Chaldean Church of Mar Petion in 1964).

13. Olim Diyarbakır, l'Archevêché chaldéen 33
Lit.: Scher (1907) 345.
Copied in Dec. 1706 A.D. by Patr. Joseph II.[26] Last four books of section on logic. 147 fols., 31 x 21 cm (25 lines).

14. Beirut, Université Saint Joseph, Bibliothèque Orientale, syr. 48*
Lit.: Khalifé-Baissari (1964) 272f.
18/19th c. East Syrian. Copied by Rabban John Rokos b. Simeon b. ܡܩܪܠܩܘ of Mangīsh (Mangēshē).[27] First four books (to the end of Prior Analytics). 202 fols.
The manuscript was presented to Louis Cheikho by John Elijah Mellus (cf. no. 29 below) in 1895.[28]

15. Birmingham, Mingana, Syr. 326*
Lit.: Mingana (1933) 605f.
Undated, but ca. 1810 sec. Mingana. West Syrian. Containing the first three books (up to the beginning of Periherm. II.v.7.). 90 fols.
The text of *But.* is preceded (on fol. 1v 1-2v 6) by the same introduction as in Bodl. Hunt. 1 (no. 9 above).

16. Vatican, Bibl. Apostolica, syr. 603-604 (olim Mardin 56-57)*
Lit.: Scher (1908) 81; Lantschoot (1965) 135f.; cf. Macomber (1969) 481 n.52.
Completed on 11th Sept. 1826 A.D. (or 1886?).[29] East Syrian. 145 + 152 fols.

---

[26]= Joseph Sliwa/Ṣlībā, n. 1667, Chaldean Patriarch (Diyarbakır) 1696-1712 (Murre-van den Berg [1999] §35), a prolific writer (Macuch [1976] 42-44) and, among others, one of the continuators of BH *Mêmrā zawgānāyā* and commentator of BH *De divina sapientia.*

[27] Khalifé-Baissari give the date as "probablement XVIIIe s.", but make no mention of the copyist, who is otherwise unknown to me. Perhaps a nephew of David b. سفر of Mangīsh (near ᶜAmadiya), attested as a copyist in 1819 (Haddad-Isaac [1988] 142, no. 331). Any connection with Thomas Rokos Khanjarkhān? (titular bp. of Basra and patriarchal visitor to Malabar in 1860; for the family name of Thomas Rokos, see Ḥabbī [1980] 95, Ḥabbī et al. [1977] 98, 168, 174f., and for the connection of this name with Mangīsh, Haddad-Isaac [1988] nos. 413, 596, 627, "Eusebius b. Yonan of the family of خنجرو/خنجريان of Mangīsh").

[28] See Cheikho (1898) 26.7f. and the note of dedication on the last page (fol. 202r) of the manuscript.

[29] According to Scher, the manuscripts were copied by J.E. Mellus himself and finished on 11th Sept. 1886 (i.e. ܐܦܦܘ). Van Lantschoot read the year as 1826 (ܐܦܟܘ). The handwriting is, I believe, different from that of Mgr. Mellus. In the colophon at Vat. syr. 604, fol. 152r, the date "Ilul 11" is clear, but, unfortunately, not the year in the photocopy of the manuscript available to me.

In the possession of J.E. Mellus (cf. no. 29 below) in 1889. Vat. 603 covers *But.* Isag.-Apod., Vat. 604 Top.-Poet.

17. Manchester, John Rylands University Library, Syr. 56 (Mingana 17)* (=R)
Lit.: Coakley (1993) 181f.
Completed on 27th Sept. (Ilul) 1887.[30] West Syrian. Copied in Mosul by Matthew b. Paul (ob. 1947).[31] 183 fols.
The manuscript was copied, according to the Garshuni notes on fol. 183r, from a "Chaldean" (i.e. East Syrian) manuscript. Coakley (p. 182 n.113) suggests that this exemplar might be Vat. syr. 603-4 (no. 16 above), but Brit. Lib. Or. 4079 (= L, no. 2 above) or another manuscript closely related to it is a more likely candidate in view of the glossary of terms at the beginning of this manuscript which is identical to that in ms. L (but which is not found in Vat. 603) and the resemblance of the Syriac colophon on fol. 182v to that in ms. L (fol. 157r).[32]

18. Pampakuda (Kerala), Library of the Konat Family 229
Lit.: van der Ploeg (1983) 176; cf. Hambye (1977) 38.
1896 A.D. Copied in Mosul by the deacon Matthew (= Matthew b. Paul; cf. no. 17 above and no. 21 below).

19. Mosul (?), (olim) coll. Yaᶜqūb Sākā 28/1
Lit.: Dāniyāl (1981/2) 336f.; cf. I. Sākā (1999) 12.[33]
The manuscript as a whole finished on 23rd Dec. 1909. Copied by Yaᶜqūb b. Buṭrus Sākā. - *But.* Isag. only, followed by several other

---

[30] So the date of the Syriac colophon on fol. 182v. Coakley, following the Garshuni colophon on fol. 183r, gives the date as 27th May (Ayyār), but this must be a mistake (on the copyist's part) since we are given the following dates in the earlier parts of the manuscript: Bk. I (Isag.) completed on 25th Tammuz (July) 1887, fol. 12v; Bk. III (Perherm.) on 10th Ab (Aug.), fol. 46v; Bk. V (Apod.) in Ilul (Sept., the day of the month omitted), fol. 96r.

[31] The copyist also of mss. Konat 229, Ming. syr. 23 and Syr. Orth. Patr. 6/5 (nos. 18, 28, 30 below), as well as of a further 40 manuscripts in the Mingana collection alone dated between 1872 (Ming. syr. 161) and 1933. See Mingana (1933), s.v. "Matthew, son of Paul" in General Index; Fiey (1959) 30, 31, 33; Coakley (1993) 113 n.24.

[32] As stated by Lantschoot, there is only a "simple note de datation" at the corresponding place in Vat. syr. 604 (fol. 152r).

[33] Metr. Severius Isḥāq Sākā reports that Yaᶜqūb Sākā's manuscript collection is in his possession (in the introduction, written in 1998, to the 2nd edition of Yaᶜqūb Sākā's poems). Whether that collection still includes the manuscript in question here is not indicated.

works on Aristotelian logic, viz. 2) Arist. *Periherm.* with marginal commentary, 3) Proba's comm. on *Periherm.*, 4) Arist. *Anal.*, 5) Severus Sebokht's treatise on *Anal.* - These contents agree closely with Ming. syr. 44A-G (no. 10 above). - Whole manuscript: 85 fols., 27 x 20 cm.

20. Location unknown

1955 A.D. West Syrian. Begun by Philoxenus Yōḥannān [Dolabani] (to p. 45 of the Syriac numeration), continued by Ap̄rēm Gurǧo (ܓܘܪܓܘ), priest in Ḥabsus (to p. 422), and Alexander Özmen (to end, p. 810).[34] - This is probably to be identified with a manuscript, which Mr. Gabriel Afram kindly informs me as being in **Mardin** (787 pp.) and which was copied by Iskender Özmen from an exemplar located at the time in Amid.

21-22. (olim) Kandanad (Kerala) and Aleppo

Barṣaum, *Luʾluʾ* 419, reports two further recent manuscripts of the first part of the *Butyrum* in Kandanad and Aleppo. The first of these *could* be identical with Konat 229 (no. 18 above).[35] The latter *may* be the same as the manuscript of the *Butyrum* mentioned (without specification of which part) as being at the residence of the Syr. Cath. metropolitan of Aleppo at Armalet (1937) 206.[36]

---

[34] A photocopy of this manuscript in his possession was kindly shown to me by dayroyo Mattias Nayis (then in St. Ephrem's Monastery, Maʿarrat Saidnāyā).

[35] Since Barṣaum knew of a manuscript of BH *Ind.* which travelled to Kandanad with Mar Timothy Eugene (1884-1975, Catholicos Baselios Augen 1964-) in 1927 (see Barṣaum, *Luʾluʾ* n.6 and Mingana [1933] 1031-34, colophon of Ming. syr. 558), the reference is likely to be to another manuscript taken by Mar Timothy. - Some of the manuscripts once in the possession of Mar Basil Eugene are reported now to be in Pampakuda, others at the Catholicate in Kottayam-Devalokam (information of Fr. B. Varghese). - No manuscript of BH *But.* appears among the manuscripts formerly in the possession of members of the Karot/Karawat family of Kandanad which have now reached SEERI in Kottayam via Tiruvalla (van der Ploeg [1983] 109-115; Briquel-Chatonnet, Desreumaux & Thekeparampil [1997] passim).

[36] I have not seen the catalogue by Isḥāq Armala, *Fihris maḫṭūṭāt Maktabat Maṭrānīyat al-Suryān al-Kāṯūlīk fī Ḥalab*, sine loco, 1927 (mentioned by Y.M. as-Sawwas apud Roper [1992-4] III.188). A new catalogue of the collection, which contains some 300 manuscripts, is reportedly under preparation by Fr. Sabri Sabra (infomation of Mr. J. Fathi-Chelhod).

C. *Manuscripts of the Sections on Natural Sciences, Metaphysics and Practical Philosophy*

23. Damascus, Syrian Orthodox Patriarchate 6/2

Lit.: Dolabani/Lavenant et al. (1994) 587; cf. Barṣaum, *Luʾluʾ* 419, lines 4f.

1286 A.D. An immediate copy of the autograph, made within the lifetime of the author.

According to Dolabani/Lavenant et al., the manuscript contains "Tome II" of *But.*, with eight books beginning with the *Auscultatio physica*, i.e. the section on the natural sciences only. Barṣaum, *Luʾluʾ* 419, on the other hand, merely says that it contains the second volume (*muğallad*) of *But.* and tells us furthermore that it was copied from the autograph, completed by the author "at the end of 1285 and at the beginning of 1286".[37] Since we know from elsewhere that *But.* as a whole was indeed finished by the author at the beginning of 1286 (in February), whereas the section on the natural sciences had been finished earlier in August 1285 (see Section 1.3 above), the mention of the fact that the autograph from which this manuscript was copied was finished at the "beginning of 1286" suggests that it also contains the metaphysics and practical philosophy. If that is the case, this manuscript could be the exemplar of Ming. syr. 310 (no. 27 below).[38] - Cf. nos. 26, 31 below.

24. Diyarbakır/Amid, Chaldean Library (*al-ḫizāna al-kaldānīya*) 33 (?)

Lit.: Barṣaum, *Luʾluʾ* 419 n.5.

1379 A.D. Copied by Dāwud b. Abī al-Munā al-Qillithī. - The manuscript number is the same as that of no. 13 above, but the reported content, date and copyist are different. The reference *could* be to no. 25 below (Diyarbakır 34).

---

[37] Barṣaum, *Luʾluʾ* 419.4f.: وللثانى نسختان قديمتان اولاهما في خزانتنا انجزت على عهد المؤلف وهي اول مصحف نقل عن نسخته التي فرغ منها في اواخر سنة ١٢٨٥ واوائل سنة ١٢٨٦ ... - The other manuscripts of the "second volume" mentioned are "Amid, Chaldean 33" (no. 24 below), Ming. syr. 23 (no. 30 below) and another in his own library (presumably either no. 26 or 28 below).

[38] If so, this manuscript must still have been in Mosul in 1865 when Ming. syr. 310 was copied.

25. Olim Diyarbakır, l'Archevêché chaldéen 34
Lit.: Scher (1907) 345.
Date and copyist unknown, but dated to 17th c. by Scher. 21 quires, 25 x 18 cm (27 lines, double columns).
Natural sciences only. - Unlike Diyarbakır 32-33 (nos. 12-13 above), Macomber was unable to locate this manuscript at the time of his visit to Diyarbakır in the 1960's.

26. Damascus, Syrian Orthodox Patriarchate 6/3
Lit.: Dolabani/Lavenant et al. (1994) 587; cf. Barṣaum, *Luʾluʾ* 419, line 6.
Mostly 1748 A.D. (2059 A.Gr.). Copied in Quṭrabbul (near Diyarbakır). 30 quires, 32 x 22 cm.
"Identique à l'ouvrage ci-dessus" (i.e. Syr. Orth. Patr. 6/2, no. 23 above) sec. Dolabani et al. - Formerly in the possession of ʿAbd al-Nūr of Edessa, Metr. of Amid (cf. no. 31 below).[39]

27. Birmingham, Mingana, Syr. 310B (fols. 1-380)** = **M**
Lit.: Mingana (1933) 588f., Drossaart Lulofs-Poortman (1989) 43-45.
1865 A.D.[40] West Syrian. Copied at the School of the Mother of God (ܐܣܟܘܠ ܕܝܠܕܬ ܐܠܗܐ) in Mosul.[41]
According to the note on fol. 216r (at the end of *But.* De anima, = end of the part on natural sciences), the manuscript was transcribed from an immediate copy of the autograph, dated (for the part in question) 11th Teshri I 1597 (Oct. 1285),[42] i.e. only two months

---

[39] Dionysius ʿAbd al-Nūr Aṣlān of Edessa, Metr. of Kharput 1896-1913, Metr. of Amid 1917-33 (Fiey [1993] 163, 217).

[40] Mingana read the date, given on fols. 216r and 316v, as ܒܩܠܘ (2136 A.Gr. = 1825 A.D.). A comparison with the *lāmad*'s and *ʿē*'s elsewhere in the manuscript, however, shows that the third letter must in fact be *ʿē*, i.e. ܒܩܥܘ (2176 A. Gr. = 1865 A.D.), a reading which is also supported by J.E. Mellus (see no. 29 below).

[41] Mingana identifies the anonymous copyist with the equally anonymous copyist of Ming. syr. 306, a manuscript containing BH *Asc.* and, like Ming. syr. 310, with corrections in the margin in an East Syrian hand. - The "School of the Mother of God" is also mentioned in the colophon of Ming. syr. 309, a manuscript containing BH *Tract.* and dated Dec. 1865 A.D. (i.e. the same year as Ming. 310). In Ming. 309, the school is said to be in the Quarter of the Carpenters (*mašrīṯā d-naggārē*), so that it was probably attached to the Church of the Mother of God in the Citadel/in the Quarter of the Carpenters, where Matthew b. Paul, the copyist of John Rylands 56, Konat 229, Ming. syr. 23 and Syr. Orth. Patr. 6/5, later worked.

[42] Not 1286, as Mingana has it. The autograph of this part was completed on 22nd

after the autograph and before the work as a whole was completed by the author (cf. no. 23 above). - The manuscript was used by J.E. Mellus as one of the exemplars for Vat. 613-615 (no. 29 below) in 1887, but must have remained in the Syrian Orthodox School/Church of Our Lady in Mosul since Ming. syr. 23 (dated 1894, no. 29 below), too, is a copy of this manuscript. - The East Syrian hand which made the corrections in the margin has been identified by Drossaart Lulofs as that of Mgr. Mellus (= M^V).[43]

28. Damascus, Syrian Orthodox Patriarchate 6/5

Lit.: Dolabani/Lavenant et al. (1994) 587; cf. Barṣaum, *Luʾluʾ* 419, line 6.

1881 A.D. Copied by Matthew b. Paul of Mosul (cf. no. 17, 30). 215 pp., 25.5 x 20 cm.

Containing "Tome II" of *But.* sec. Dolabani et al. (cf. on no. 23 above).

29. Vatican, Bibl. Apost., syr. 613-615 (olim Mardin 58-60)** = **V**

Lit.: Scher (1908) 81f.; van Lantschoot (1965) 143-145, Drossaart Lulofs (1989) 45f, 48.

1887 A.D. East Syrian, vocalised. Copied in Mosul by Mar John Elijah Mellus (Mīlōs).[44] 237 + 160 + 88 fols.

The note at Vat. 614, fol. 39v (see Appendix I below) was understood

---

Aug. 1285.

[43] See Drossaart Lulofs-Poortman (1989) 44f. with the plates on pp. 66f. The identification is confirmed by the constant agreement of these corrections with the readings of ms. V.

[44] Born in Mardin 1831; monk of Rabban Hormizd and later of Notre-Dame des Semences; bishop of ᶜAqra 1864; in Rome as patriarchal envoy in 1868 and during the 1st Vatican Council (1869-70); in Malabar (resident in Trichur) 1874-1882, where he headed the "Mellusian schism", which later led to a group of the Malabar Christians joining the Church of the East in 1907, although Mellus himself was reconciled with Rome in 1889; metropolitan of Mardin 1890-1908 (Vosté [1939] 368f.; Graf *GCAL* IV.112f.; Macuch [1976] 402; Fiey [1993] 52, 108; in connection with the schism in Malabar, Mingana [1933] 1085f.; Tisserant [1941] 3135-3138; Spuler [1961b] 237; Ḥabbī [1980] 101-105; van der Ploeg [1983] 43, 134). - Mgr. Mellus was evidently particularly interested in BH and his works in 1887, since he reports himself in a note attached to Vat. syr. 612 (BH *Rad.*) that he copied eleven "books" of BH in that year (van Lantschoot [1965] 143) and has also left us a biography of BH based largely on *Chron. eccl.* but also including additional materials in a manuscript dated 1887 (Leeds syr. [coll. Budge] 3, fols. 114v-137v, cf. Ebied [1974] 516), while mss. Mardin, Chald. 47 (BH *Eth.*) and 52 (*Nom.*) were copied *for* Mellus in 1887. His interest in BH *But.* in particular is shown by his association with at least four other manuscripts in the list here (nos. 2, 14, 16, 27).

by van Lantschoot and Drossaart Lulofs as meaning that Mellus had access to three manuscripts dated 1285, 1286 and 1865. This note, however, is based on the note at the corresponding place in Ming. syr. 310 (= M, no. 27 above; fol. 216r), itself dated 1865. The manuscript, in fact, appears to be a product of the collation by Mellus of mss. Brit. Lib. Or. 4079 (= L, no. 2) and M. - Along with Vat. 603-604 (no. 16 above), these three volumes must have been among the manuscripts donated to the Vatican by Mellus' successor in the see of Mardin, Israel Audo (bishop 1910-1941).[45]

30. Birmingham, Mingana, Syr. 23* = m

Lit.: Mingana (1933) 68f., Drossaart Lulofs-Poortman (1989) 45.

Sept.-Oct. 1894. West Syrian. Copied in Mosul by the deacon Matthew b. Paul (cf. no. 17, 28). 129 fols.

The manuscript has been judged to be a copy of Ming. syr. 310 (= M, no. 27 above) by Drossaart Lulofs and, as noted by Drossaart Lulofs, the copyist has taken account of the corrections made by J.E. Mellus in the margin of ms. M.

31. Midyat, Mar Barṣawma

Information of Mr. Gabriel Afram.

20th c. - West Syrian. Copied by John b. Joseph of Zaz from an exemplar copied in 1923 by Mar Philoxenus [Dolabani] of Mardin. The exemplar itself was reportedly copied from two exemplars, one of them copied in 1286 in Mosul and in the possession at the time of Mar Dionysius ᶜAbd al-Nūr of Edessa, Metr. of Amid (cf. nos. 23 & 26 above).[46]

D. *Manuscripts Containing Excerpts or Fragments of the Section on Natural Sciences*

32. Manchester, John Rylands University Library, Syr. 44 (olim Harris 165) B/1 (fols. 97r-133v)

Lit.: Coakley (1993) 171.

17th-18th c. sec. Coakley. East Syrian. - John Rylands syr. 44

---

[45] So Macomber (1969) 481 n. 52; cf. van Lantschoot (1965) p. 131, 133.

[46] One suspects that the exemplar actually used by Dolabani was no. 26 above (dated 1748), the date 1286 being, as often in such cases, the date of an ancestor mentioned in the exemplar rather than the date of the immediate exemplar itself.

consists of two originally separate manuscripts, the second of which now makes up fols. 97-134 (= 44B). The part containing *But.* (97r 1-133v 11) begins in mid-sentence in *But.* Gen. et corr. I.i.2 (ܠܐ ... ܡܫܘܚܠܦܐ ܠܐ ܕܟܝܢܐ ܘܠܐ ܡܫܡܠܝܐ) and ends in mid-sentence in *But.* Min. IV.iii.6 (ܟܪ̈ܝܐ ܕܠܐ ܡܬܚܒܪܢܐ ܗܘܐ ... = end of IV.iii.5, followed by ܘܦܠ̈ܓܐ ܕܡܬܬܪܝܡܢܘܬܐ ܠܐ ܡܬܚܫܒܝܢ = beg. of IV.iii.6.4f.).

33. Vatican, Bibl. Apost., syr. 469 no. 3 (fols. 132v-151v)** = **W**

Lit.: van Lantschoot (1965) 8.

Book of Mineralogy. East Syrian, dated 1804 A.D., copied in Alqosh by Gabriel b. Ḥadbšabbā, perhaps to be identified with Gabriel b. Khaushaba b. Joseph,[47] known to us as the copyist of several other manuscripts written in Alqosh between 1803-1829.[48] The piece on fols. 124r-132r, immediately preceding the excerpt of *But.* (a piece said to be on "Merveilles de la création" by van Lantschoot), is an excerpt from BH *Eth.* (*Eth.* IV.13.3-8, corr. ed. Bedjan 450.1-465.7).[49]

As in mss. LlP (nos. 2, 7, 34) the greater part of *But.* Min. III is missing.

34. Princeton Theological Seminary, Speer Library, Nestorian 25 = P**

Lit.: Degen (1977); cf. Clemons (1966) 500 (no. 334); Desreumaux (1991) 223.

Date and copyist unknown. West Syrian. A fragmentary manuscript. The surviving folios contain the following parts of *But.* Ausc. phys., De caelo, Min. and Mete.

fols. 61r-68v: *But*. Ausc. phys. V.i.4.-De caelo I.vi.3.
fols. 100r-105v: *But*. De caelo V.iv.1-V.vi.3.
fols. 136r-157v: *But*. Min. III.iii.3-Mete. I.ii.6.
fols. 167r-178v: *But*. Mete. II.iii.2-III.iv.2.

---

47 ܚܘܫܒܐ, of course, is not "Gaspar" as van Lantschoot (1965) 99, following PS 1716, believed, but the vernacular equivalent of Class. Syr. *ḥad-b-šabbā* (MacLean [1901] 92b; cf. Abbeloos-Lamy [1872-7] I.785/6 n.2; Nöldeke [1868] 156 n.1).

48 Mss. Ming 94, 427, 519A, Berol. or. fol. 3122 (3 Aßfalg), Vat. syr. 573, ᶜAqra 26, Dohuk 2, 6 (cf. Fiey [1965] 680 n.1), Baghdad, Mar Antonius syr. 22, 713; cf. also Briquel-Chatonnet (1997) 165 (on the exemplar of Paris syr. 425). - Patriarch John mentioned in the colophon is the Chaldean patriarch John VIII Hormizd, patriarch officially 1830-1838, but *de facto* from much earlier on (Vosté [1939] 380 n.4; Murre-van den Berg [1999] §20).

49 Cf. Section 3.2.2g below.

The readings are usually identical with those of mss. LlW (no. 2, 7, 33), and like LlW it must have been copied from an exemplar in which most of *But*. Min. III. was already lost, since there is a note at 136r 3-4 (at the end of what remains of Min. III) telling us that "the rest of chapter III" was missing.[50]

Degen dated this manuscript to the 16th century on the evidence of the similarity of the style of writing to that of a manuscript dated 1575/6 in Hatch (1946) Plate CLVIII (opposite p. 209).[51] Since a number of manuscripts in Princeton are known to originate from Urmia,[52] this manuscript may well be a fragment of Urmia 66 (dated 1830 A.D., cf. no. 3 above), especially as the catalogue of Sarau & Shedd tells us that much of *But*. Min. III was also missing in that manuscript.

35. Baghdad, Mar Antonius, Syr. 173 (olim N.D. des Semences 54 Vosté), no. 7

Lit.: Vosté (1928) 23; Haddad-Isaac (1988) 84f.

Undated but assigned to the period of the restoration of the Monastery of Rabban Hormizd (i.e. 1808-1869) by Vosté. "The book of natural sciences (*kyānāyāṯā*) out of the *Butyrum sapientiae*", occupying, as far as can be made out from Vosté's description, the last six of the 14 quires of the manuscript (17 x 11 cm, 16 lines to a page). According to the catalogues, six chapters are announced at quire 9, page 1, but the manuscript ends in the middle of chapter 3, section 3. - If the announcement of "six chapters" is correct, the fragment may begin with the Book of Animals, since that is the only book in the section on natural sciences in the *Butyrum* which has six chapters.

36. Louvain, CSCO, syr. 22, no. 4, fols. 121r-142r

Lit.: de Halleux (1987) 45f.

Meteorology. The rest of the East Syrian manuscript, dated 1903 A.D., consists mainly of excerpts from medical works.[53] Copied in

---

[50] Due to the loss of the folios immediately preceding, only the last two and a half lines of Min. III survive in the manuscript as we have it today, but there may originally have been a little more of it there (as in ms. l).

[51] It may be remembered here that Hatch's *Album* does not cover manuscripts after the end of 16th c. (cf. Goshen-Gottstein [1979] 25 n.54).

[52] Macomber (1966) 335 n.2; Desreumaux (1991) 223f. - Besides the items mentioned there, Princeton 34a can be identified with olim Urmia 177 (BH *Serm. sap.*).

[53] Of the other texts in the manuscript left unidentified by de Halleux, nos. 1-2 and 6 correspond to Paris syr. 423 (olim coll. Pognon), nos. 1-3 (Briquel-Chatonnet [1997]

Alqosh by Elijah Homo,[54] ordered by Jacob Eugene Manna, presumably for J.-B. Chabot.

E. *Further Excerpts etc.*

37. Charfeh, fonds Rahmani 563 [766 Sony], fol. 16r-17v
Lit.: Sony (1997) 270.
On the division of languages and the descendants of Japheth, Ham and Shem (BH *Chron*. rather than *But.*?) and on the life-spans of animals, reptiles and birds (probably *But.* Animal.). In a manuscript dated by Sony to 13-16th c.

38. Birmingham, Mingana, Syr. 460K (fol. 23r-24r)
Lit.: Mingana (1933) 820f.
1796/7 (2108 A.Gr.). West Syrian. "On ethnological characteristics of the peoples of the earth" from the *Kitāb zubdat al-ḥikma*, i.e. probably *But.* Econ. II.iii.3[55] (in Garshuni like the rest of the manuscript?).

39. Charfeh, fonds Rahmani 540 [404 Sony], fol. 9v
Lit.: Sony (1997) 157
On the characteristics of the different nations, i.e. probably *But.* Econ. II.iii.3. In a manuscript copied by ᶜAbd Allāh Fāᶜūr[56] in 2119 A.Gr. (1807/8).

40. Charfeh, fonds Rahmani 541 [835 Sony], fol. 114v-116r
Lit.: Sony (1997) 328.

164f., copied in 1901 by Elijah Homo); nos. 1-2 also correspond to Ming. syr. 594A-B (copied in 1932 from a manuscript copied by Elijah Homo in 1921), a text which has been identified as the Syriac counterpart of Ḥunain b. Isḥāq's *Bk. of Nourishment* (*K. al-aġḏiya*) by Degen (1978) 67-71.

[54] Elijah b. Homo b. Isaiah b. Homo b. Ḥanna b. Homo b. Daniel b. Elijah b. Daniel d-Bēṯ Naṣro, 1856-1932 (see Aßfalg [1963] 21; Ḥaddad [1985] 188; cf. Mingana [1933] 1133), belonged to a well-known of family of copyists at Alqosh (on which, see Ḥaddad [1985]; cf. Fiey [1965] 393f., with 394 n.1) and himself evidently possessed a significant library (Mingana [1933] 526; Aßfalg [1963] xii, 1f., 4f., 19-21, 77-83, 141). For manuscripts copied by Elijah Homo see, Ḥaddad (1985) 188, 190f., and add to the list there Berol. Sachau 226 (89 Sachau), Ming. syr. 246A, Paris syr. 404, 423, 424-5, CSCO syr. 6, 21. - "Homo" (ܗܘܡܐ): a shortened form of Hormizd (Sachau [1899] 338; Ḥaddad [1985] 168; cf. Krotkoff [1982] 115; Fiey [1965] 680); latinised as "Humus" at Assemani, *BOCV* III/1.352 n.2 in the translation of the colophon of Vat. syr. 175, a manuscript copied in 1713 by the great-gr.-gr.-grandfather our Elijah.

[55] Ed. Zonta (1992) 82f., tr. 100.

[56] Deacon ᶜAbd Allāh b. Fāᶜūr b. Barṣaum of Ṣadad (?): cf. Dolabani (1994) 311.

Econ. II.iii.3. In a manuscript copied ca. 2152 A.Gr. (1840/1) at D. al-Zaᶜfarān.

41. Paris, Bibl. Nationale, syr. 384/7 (fol. 284-290)
Lit.: Briquel-Chatonnet (1997) 86.
Notes on the *Butyrum* by François Nau (notes left unpublished after Nau's discovery that the work had been described in the Florence manuscript catalogue).

In his catalogue of manuscripts at Charfeh, Armalet mentions a manuscript of the *Butyrum* dated 1284 at the "Chaldean library" (*maktabat al-kaldān*) in Diyarbakır.[57] In the absence of references elsewhere to such a manuscript, however, one is led suspect that the date given by Armalet is due to misunderstanding of the date of composition (1596 A.Gr. = 1284/5) given in the manuscript as the date of the copy and that the manuscript mentioned by Armalet should be identified with one of the Diyarbakır manuscripts listed above.

The text in ms. Florence, Laur. or. 298 (olim Palat. 62) identified as a part of the *Butyrum* by S.E. Assemani in his catalogue[58] is in reality the last part of the anonymous *Causa causarum* (*ᶜEllaṯ kul ellān*).[59]

F. *Manuscript Families*

The manuscripts used in the edition of *But.* Min. and Mete. below are the following.

1. F = Florence, Laur. or. 83 (olim Palat. 187), 1340 A.D.
2. L = London, British Library, Or. 4079, 1808 A.D.
3. l = London, British Library, Or. 9380, 1892 A.D.
4. M = Birmingham, Mingana syr. 310, 1865 A.D.

---

[57] Armalet (1937) 206.

[58] Assemani (1742) 109.

[59] Fols. 85r-139r, in the part of the manuscript copied in the Monastery of Mar Abel and Abraham (Midyat) in 1488. The part of the text covered in this manuscript, *Caus. caus.* V.vi-fin. (corr. tr. Kayser [1893] 280-352 and intro. xxi-xxiii [the part translated by Ryssel]), agrees exactly with that covered in Vat. syr. 191 (see Furlani [1948]; cf. Furlani [1946]; Furlani's dating of Vat. 191 to the end of 15th c., based on the identification of "Patriarch Peter" mentioned in the colophon with Peter b. Ḥasan al-Ḥadathī, 1458-92, might be questioned in view of the customary adoption of the name "Peter" by Maronite patriarchs).

5. V = Vatican, syr. 613-615 (olim Mardin 58-60), 1887 A.D.
6. P = Princeton Theological Seminary, Nestorian 25, undated (fragmentary).
7. W= Vatican, syr. 469, 1804 A.D. (*But*. Min. only)

These manuscripts can be classified into the following groups.

A. Ms. **F**, according to its colophon, is an immediate copy of the autograph and generally exhibits the best readings, although the manuscript seems to have been copied in some hurry, so that there are occasional lacunae, as well as other errors due to carelessness.
Among other manuscripts, Laur. 37, 10, 6 & 8 (no. 11) has been judged to be a copy of F. For the reason given above, Bodl. Hunt. 1 (no. 9), along with Ming. syr. 326 (no. 15), might also be suspected to belong to this group, a suspicion that is given some support by the general agreement of the readings in the proem (see Section 1.2 above). More generally, other manuscripts originating from D. al-Za^cfarān and environs *could* belong to this group.[60]

B. Ms. **M**, if the subscription on fol. 216r can be trusted, is a grandchild of the autograph, its parent possibly being the manuscript said to be at the patriarchate in Damascus (Syr. Orth. Patr. 6/2, no. 23). The manuscript is unfortunately marred by nonsensical readings due to the incompetence of the copyist,[61] but its readings, where they make sense, need to be treated with due respect.
Among other manuscripts, Ming. syr. 23 is a copy of M, while Syr. Orth. Patr 6/5, copied by Matthew b. Paul like Ming. syr. 23, is also likely to belong to this family, although John Rylands syr. 56, a copy of the logical section of *But*. (which is missing in both M and Syr. Orth. Patr. 6/2) by the same Matthew, belongs to group C below.

C. The East Syrian manuscripts **LIW** and the West Syrian **P** constantly exhibit the same variant readings and may be considered to form a close family. The statement that ms. L was copied from an unvocalised manuscript rules out the possibility that it was copied from the

[60] This would include Ming. syr. 44, along with Saka 28 for the reason given above. Perhaps also Barṣaum's "Diyarbakır 33" (no. 24), Syr. Orth. Patr. 6/3, Ming. syr. 460 and the Charfeh excerpts.

[61] Cf. the judgement of Drossaart Lulofs (1989) 48 that the copyist of ms. M was "hardly qualified to copy an old manuscript".

undated, but vocalised P.[62] The survival of a part of *But.* Min. III. in mss. P and l, on the other hand, rules out the possibility that these depend on L, in which the whole of the chapter is missing. Although ms. P is written in the West Syrian script, the East Syrian script and origin in Alqosh of mss. LlW suggest that this family represents a tradition of manuscripts handed down, perhaps from an early date, in East Syrian circles.

Among other manuscripts, the West Syrian John Rylands 56 (logic only), copied by Matthew b. Paul, is stated to have been copied from a manuscript in the "Chaldean" script and, in the proem, has the same variant readings as those of L (as well as Vat. syr. 603, Beirut, USJ syr. 48), so that it may be regarded as belonging to this family. The same probably applies to Konat 229, another manuscript of the logical part by Matthew b. Paul. Besides these, those 19th c. East Syrian manuscripts originating from Alqosh (incl. the monasteries of Rabban Hormizd and Notre-Dame de Semences) are all likely to belong here. Whether the earlier East Syrian manuscripts olim Diyarbakır 32-34 (nos. 12, 13, 25) also belong to this group is unknown.

D [< B & C]. Ms. V, as has been stated, may be considered to represent a collation of L and M, along, occasionally, with conjectures (more often infelicitous than not) made by the copyist, J.E. Mellus.

A conspectus of the manuscripts discussed above is given on the following page. The legends are as follows.

#1: number in the list above; **#2**: script; **#I**: section of *But.* on logic; **#II**: section on natural sciences; **#III**: section on metaphysics; **#IV**: section on practical philosophy; **x**: contains the whole of the section named; **p**: contains a part of the section named; **e**: contains an excerpt (a single book or less) of the section named; **f**: contains a fragment of the section named; **?**: may contain the section named; **#3**: family.

[62] The absence, in fact, of any reference to the script in which the exemplar of ms. L was written (unless this is what is meant by "the obscurity of the letters") would seem to favour the view that the exemplar was in the East Syrian script.

Date	Manuscript	#1	#2	Copyist	Place	#I	#II	#III	#IV	#3
1286	SOPatr 6/2	(23)	WS		Mosul?	-	x	?	?	B?
1340	Laur 69, 83	(1)	WS	Najm b. Shams		x	x	x	x	A
1379	"Diyarb 33"	(24)	WS?	David of Killit		-	x	?	?	
1498	Bodl Hunt 1	(9)	WS	Joseph Iberian	D. Za^{c}farān	p	-	-	-	A?
1574	Ming 44A	(10)	WS	Ephrem	D. Za^{c}farān	e	-	-	-	A?
16 c.?	CharfR 563	(37)	WS			-	e	-	-	
17 c.	Laur 37 etc.	(11)	WS	Antonius Sion.	Florence?	x	-	-	-	A
1638	Diyarb 32	(12)	ES	Metr. Simeon		p	-	-	-	
17 c.	Diyarb 34	(25)	ES?			-	x	-	-	
17/8 c.	MancJR 44B	(32)	ES			-	f	-	-	
1706	Diyarb 33	(13)	ES	Patr. Joseph II		p	-	-	-	
1748	SOPatr 6/3	(26)	WS?		Quṭrabbul	-	x	?	?	B?
1797	Ming 460K	(38)	WS			-	-	-	e	
1804	Vat 469/3	(33)	ES	Gabriel b. Ḥadb.	Alqosh	-	e	-	-	C
1808	CharfR 540	(39)	WS	^{c}Abd-Allāh Fā^{c}ūr		-	-	-	e	
1809	BL Or 4079	(2)	ES	George b. Yaqo	Alqosh	x	x	x	x	C
1810?	Ming 326	(15)	WS			p	-	-	-	A?
1815	Urmia 67	(3)	WS	G. Behn. of Mos.	Mosul?	p	-	-	-	C?
1818	BaghMA 177	(4)	ES	Joseph Audo	R. Hormizd	?	?	?	?	C?
1825	Urmia 65	(3)	WS	Behn. of Athor	Mosul?	p	-	-	-	C?
1826	Urmia 64	(3)	WS	Behn. of Athor	Mosul?	-	p	x	x	C?
1826?	Vat 603-604	(16)	ES			x	-	-	-	C
1830	Urmia 68	(3)	WS	Gabr. of Mosul	Mosul?	p	-	-	-	C?
1830	Urmia 66	(3)	WS	G. Behn. of Mos.	Mosul?	-	p	-	-	C?
1830?	Princeton 25	(34)	WS			-	f	-	-	C
1841	CharfR 541	(40)	WS		D. Za^{c}farān	-	-	-	e	
1865	Ming 310B	(27)	WS		Mosul	-	x	x	x	B
1881	SOPatr 6/5	(28)	WS	Matthew b. Paul	Mosul	-	x	?	?	B?
1887	MancJR 56	(17)	WS	Matthew b. Paul	Mosul	x	-	-	-	C
1887	Vat 613-615	(29)	ES	J.E. Mellus	Mosul	-	x	x	x	BC
1892	BL Or 9380	(7)	ES	^{c}Isa b. Isaiah	Alqosh	x	x	x	x	C
1894	Ming 23	(30)	WS	Matthew b. Paul	Mosul	-	x	x	x	B
1896	Konat 229	(18)	WS	Matthew b. Paul	Mosul	x	-	-	-	C?
19 c.?	BeirUSJ 48	(14)	ES	John Rokos		p	-	-	-	C
19 c.	BaghMA 178	(5)	ES	Elijah Scher?	ND Sem.?					C?
19 c.	NDSem 62	(6)	ES	Vincent	ND Sem.?					C?
19 c.	BaghMA 173	(35)	ES			-	p	-	-	C?
1903	CSCO 22	(36)	ES	Elijah Homo	Alqosh	-	e	-	-	C?
1909	Saka 28/1	(19)	WS	Ya^{c}qūb Saka		e	-	-	-	A?
1955	unknown	(20)	WS	Dolabani et. al.		x				
20 c.	Midyat	(31)	WS	John b. Joseph		-	?	?	?	B?
?	MosSO 1.63	(8)	WS		Mosul?					
?	Kandanad	(21)				x	-	-	-	
?	Aleppo	(22)				x	-	-	-	

CHAPTER THREE

BOOKS OF MINERALOGY AND METEOROLOGY

3.1. *Overview of the Books*

The two books of the *Butyrum* edited below correspond largely to the fifth "book", or *fann*, of the section (*ǧumla*) on the natural sciences (*ṭabīʿīyāt*) in Ibn Sīnā's *K. al-šifāʾ*, which is itself divided into two "treatises" (*maqāla*) dealing, respectively, with V/1. "minerals (*al-maʿādin*) and V/2. "meteorology" (*al-āṯār al-ʿulwīya*).

At one further remove, most of the themes treated in these two books go back to Aristotle's *Meteorologica*, although some of the subjects treated, especially in the Book of Minerals, are touched upon only briefly in that work.

The correspondences of the chapters in the *Butyrum* to the *Šifāʾ* and *Meteorologica* may be summarised as follows.

Butyrum	Ibn Sīnā, *Šifāʾ*	Arist. *Mete*.
Min. I. Rocks, springs and mountains	V/1.1-3	[I.13]
Min. II. Earthquakes	V/1.4	II.7-8
Min. III. Minerals	V/1.5	[III.6B, IV]
Min. IV. Habitable world	V/1.6	[II.5]
Min. V. Sea	[IV.2]	[I.13], II.1-3
Mete. I. Clouds, rain etc.	V/2.1	I.9-12
Mete. II. Illusions (rainbow etc.)	V/2.2-3	III.2-6
Mete. III. Winds	V/2.4	II.4-6
Mete. IV. Thunder, lightning etc.	V/2.5	II.9-III.1, I.4-8
Mete. V. Deluges etc.	V/2.6	[I.14]

It will be seen that the order of the material in the *Butyrum* is essentially the same as that in the *Šifāʾ*, which, in turn, differs from that of *Meteorologica* (and Nicolaus' *Compendium*, where the order in Aristotle is generally followed). On two occasions, however, Barhebraeus has reduced two or three chapters (*faṣl*) of the *Šifāʾ* into one chapter (*qep̄aleʾon*) in the *Butyrum* and, on one occasion, in Min. V, he has added a chapter which has no counterpart in the fifth *fann* of the *Šifāʾ*, incorporating into it some of the material from the preceding *fann*.

3.2. *Sources of Books of Mineralogy and Meteorology*

3.2.1. *Aristotelian Meteorology in Syriac*

A useful survey of the Arabic material related to Aristotle's *Meteorologica*, including summaries of the more important works, has recently been provided by Lettinck in his *Aristotle's Meteorology and its Reception in the Arab World* (Leiden 1999). Before entering a discussion of the sources used by Barhebraeus in the books dealing with mineralogy and meteorology in the *Butyrum sapientiae*, it will be helpful to provide a brief note here on the comparative material in Syriac.

a) *Syriac versions of Greek works*

Although it is usually assumed that Ibn al-Biṭrīq's Arabic version of Aristotle's *Meteorologica* was made from a Syriac intermediary,[1] nothing seems to survive of this Syriac version. Of the Syriac version of Theophrastus' *Meteorology*, on the other hand, we do have a fragment in ms. Cantab. Gg. 2.14,[2] while some further excerpts from the same work have been detected in Moses b. Kepha's *Hexaemeron*.[3]

A further Aristotelian, or rather - as most scholars today would agree - Pseudo-Aristotelian, work that is of concern to us here is the *De mundo* (*περὶ τοῦ κόσμου*), the Syriac version of which by Sergius of Rēsh-ᶜAynā (ob. 546) survives intact in a 7th c. manuscript (Brit. Lib. Add. 14658 [987/8 Wright]) and was published by de Lagarde in his *Analecta syriaca*.[4]

Another Syriac version of a Greek work is that of Nicolaus' *Compendium* of Aristotelian philosophy (see Section 1.5 above). Although this work is only preserved in fragments, the parts of it dealing with meteorology and mineralogy have fared much better than

[1] See Lettinck (1999) 7f.

[2] Fol. 351v-353v. - Published by Wagner-Steinmetz (1964) and again by Daiber (1992), the latter together with the Arabic versions made from the Syriac by Ibn al-Khammār and Bar Bahlūl.

[3] Daiber (1992) 188-191.

[4] De Lagarde (1858) 134-158; cf. the German translation by Ryssel (1880) and (1881) and the corrections on the text and analysis by Baumstark (1894) 405-438. - Three Arabic versions of the *De mundo*, two of which, at least, appear to be based on the Syriac, have been edited in an unpublished dissertation by Brafman (1985).

others and make up 52 pages out of the total of 76 occupied by this work in ms. Cantab. Gg. 2.14, while some further excerpts from the meteorological part are found together with excerpts from Barhebraeus' *Candelabrum* and Bar Kepha's *Hexaemeron* in ms. Paris syr. 346.[5] As far as can be made out from what survives in these manuscripts, the work as composed by Nicolaus appears generally to have been a faithful summary of Aristotle. The Syriac version as we have it in the Cambridge manuscript, however, also contains a large amount of later scholial material, most of which seems to derive from Olympiodorus' commentary on Aristotle's *Meteorologica.* The translator of the Syriac version remains uncertain. Although Barhebraeus, who made an extensive use of this work, ascribes the translation to Ḥunain b. Isḥāq (808-873)[6] and there are further grounds for associating the work with Ḥunain, in view of instances where the same Greek terms are translated differently in the main text and in the scholia, I am now less inclined than I was[7] to regard Ḥunain b. Isḥāq as the translator of the main text, although most of the scholia probably belong to him.[8]

This Syriac version of Nicolaus (or rather of Nicolaus-Olympiodorus) is of significance also for the transmission of Aristotelian meteorology in Arabic as the immediate source of at least two Arabic works on the subject. Firstly, it was apparently one of the sources used by the East Syrian Ibn al-Khammār (Ibn Suwār, fl. second half of 10th c.) in composing his *Treatise on Meteorological Phenomena* (*Maqāla fī al-āṯār al-mutaḫayyila fī al-ğaww*).[9] More importantly, the work that has come down to us as the Arabic version of Olympiodorus' commentary on the *Meteorologica* (*Tafsīr Ulimfīdūrūs li-kitāb*

---

[5] See Nau (1910) 228, 230f., Schlimme (1977) 863-879.

[6] BH *Hist. dyn.* 82.15f.

[7] Takahashi (2002a) 193f.

[8] See Takahashi (2002d) 026f.; cf. comm. on *But*. Mete. III.iii.4.7-9 below.

[9] Published by Lettinck (1999), Supplement 1, p. 313-379. The relationship of this work to Olympiodorus' commentary and the so-called Arabic version thereof ("Pseudo-Olympiodorus", Olymp. arab.) is discussed briefly by Lettinck at id. (1999) 277. The relationship is more easily explained if we assume that Ibn al-Khammār made use of the Syriac Nicolaus with its Olympiodorean scholia. Nicolaus is mentioned as a source by Ibn al-Khammār at *Mutaḫayyila* [Lettinck] 320.4 (cf. Sezgin, *GAS* VII.284). That Ibn al-Khammār used the Syriac version of the work is made likely by the references to Syriac terms at *Mutaḫayyila* 332.3; 378.10. - Cf. comm. on *But*. Mete. II.i.5.5-6 and II.ii.2.4-6 below.

Arisṭāṭālīs fī al-āṯār al-ʿulwīya, translated, according to the *codex unicus*, by Ḥunain b. Isḥāq and revised by Isḥāq b. Ḥunain),[10] is, in fact, largely a translation of the Syriac text we have in the Cambridge manuscript, not only of the Olympiodorean scholia but also of the main Nicolean text.[11] As has been noted by Lettinck,[12] that Arabic version of [Pseudo]-Olympiodorus was, in turn, one of the main sources, if not *the* main source, used by Ibn Sīnā in his discussions of meteorology in the *Šifāʾ*.

b) *Original Syriac Works*

It is not so easy to detect influences of specifically Aristotelian meteorology in the surviving original Syriac works from the pre-Islamic period which deal with germane matters, such as the meteorological-astronomical treatises of Ps.-Dionysius and Ps.-Berosus,[13] even if these "anti-Hellenistic" works have been found to betray significant influence of Greek learning in spite of themselves.[14]

The influence of Peripatetic meteorology becomes much more clearly recognisable in the *Hexaemeron* of Jacob of Edessa (640-708), even if the question as to whether such elements are taken directly from Aristotelian sources or indirectly, via other hexaemeral and related literature, is a matter that has still to be examined.[15]

---

[10] Preserved in ms. Tashkent, al-Biruni Institute of Oriental Studies 2385, fol. 347r-368r; published by Badawī (1971) 83-190 (cf. reviews by H. Gätje, OLZ 71 [1976] 382-385; F.W. Zimmerman & H.V.B. Brown, Der Islam 50 [1973], 313-324); cf. Lettinck (1999) ix, 3, 9 et passim.

[11] Cf. Takahashi (2002a) 191-193, 195 n.21, 196 n.24, 199 n.35; id. (2002d).

[12] Lettinck (1999) 10 et passim.

[13] Published, respectively, by Kugener (1907) and Furlani (1917); and Levi della Vida (1910).

[14] Levi della Vida (1910) 42f.: "Se insisto su questa ignoranza (della quale fa fede quasi ogni riga dello scritto pubblicato sopra), si è perchè ad essa fa contrasto la presenza, che sono andato notando, di elementi greci, più o meno male interpretati, nel corso di tutta l'esposizione della dottrina del Beroso. Ciò prova, a mio vedere, quanto profonda e durevole sia stata la penetrazione ellenistica nell'Oriente."

[15] Meteorological and mineralogical matters are frequently treated in the second and third treatises (*mêmrē*) of Jacob's *Hex.* (ed. Chabot p. 36-142; tr. Vaschalde 36-118). On the names of the twelve winds as given by Jacob, see Takahashi, (2002b) 230f. (the names evidently go back to the Greek, rather than the Syriac version, of the *De mundo*, but probably through an intermediary, as the displacement of two of the winds suggests); cf. comm. on *But*. Mete. III.ii.3.3-8 below. For a recent study on the manner in which material derived from Ptolemy's *Geog*. is used in Jacob's *Hex.*, see Schmidt (1999) 57-66.

Clearly Aristotelian are also the meteorological accounts in the *Book of Treasures* of the East Syrian Job of Edessa (c. 769-835). While some echoes also of Theophrastean *Meteorology* have been noted there, here again the identification of the immediate sources used by Job is a work still to be done and by this period we need to bear in mind the possibility of Syriac works being influenced by Arabic (Muᶜtazilite) sources.[16]

We are better informed, on the other hand, concerning the sources used in the *Hexaemeron* of Moses bar Kepha (833-903).[17] Much of the meteorological material in this work, besides those parts going back to Jacob of Edessa, can be traced back to two sources, the Syriac version of Theophrastus' *Meteorology* and the work we have in Arabic as Ḥunain b. Isḥāq's (803-873) *Compendium* of Aristotelian meteorology (*Ǧawāmiᶜ li-kitāb Arisṭāṭālīs fī al-āṯār al-ᶜulwīya*),[18] probably used by Bar Kepha in its Syriac version.[19]

Less well-known again are the sources used in the metrical *Hexaemeron* of the East Syrian Emmanuel bar Shahhārē (ob. 980), although one passage, at least, has been identified by Daiber as deriving from Theophrastus via Bar Kepha.[20]

Bar Kepha's *Hexaemeron* was probably also known to the anonymous author of the *Causa causarum* (11th c.?),[21] but that work generally betrays a greater influence of Arabic sources and this applies also to those parts dealing with meteorology.

It is unfortunate that the hexaemeral treatise of Dionysius bar Ṣalībī (ob. 1171) appears to be lost, since, given his proximity to Barhebraeus in time, such a treatise is likely to have been known to Barhebraeus.[22]

---

16 See Daiber (1992) 173f. with the literature cited there. Daiber suggests that Job may have been the translator of Theophrastus' *Meteorology* into Syriac.

17 I follow Reller (1994) 33 here in giving the year of Bar Kepha's birth as 833 rather than 813.

18 Published by Daiber (1975); also by Y. Ḥabbī, Mosul 1976 (sec. Sezgin, *GAS* VII.267).

19 See Daiber (1991) 46-49; cf. id. (1992) 171-173.

20 See ten Napel (1980) and Daiber (1992) 174. To the published excerpts of Emmanuel's *Hexaemeron* mentioned at Baumstark, *GSL* 238 n.4, add Manna (1901) II.143-207, which includes those parts of Mêmrā V dealing with meteorology (p. 171-178).

21 See Reinink (1997) 287 with n.51.

22 A mêmrā "on heaven, and on sun, moon and stars and other items according to

The discussions of meteorological matters in the theological work of Severus Jacob bar Shakko (ob. 1240/1), the *Book of Treasures*, is largely a summary of Bar Kepha's *Hexaemeron*, although he seems also to have had direct access to Jacob of Edessa's work.[23]

The situation is quite different with Bar Shakko's work devoted to the secular sciences, the *Book of Dialogues*, where three sources are used for the discussion of meteorology,[24] namely the Syriac version of the Pseudo-Aristotelian *De mundo*, used briefly at the beginning,[25] Job of Edessa's *Book of Treasures*, used for the discussion of the rainbow and the halo, and most importantly a hitherto unidentified (lost?) source in Arabic, which was itself closely based on the *K. al-mabāḥit̠ al-mašriqīya* of Fakhr al-Dīn al-Rāzī (1149-1209) and which was also used (along with the *Rasāʾil* of the Ikhwān al-Ṣafāʾ) by Zakarīyāʾ b. Muḥammad al-Qazwīnī (ca. 1203-1283) in his *ʿAǧāʾib al-maḫlūqāt*.[26] The existence of this last source has to be postulated on the ground of those instances where Bar Shakko and Qazwīnī agree with each other against Fakhr al-Dīn al-Rāzī,[27] since Bar Shakko, who died some forty years before Qazwīnī, cannot have used the latter's work, while Qazwīnī is most unlikely to have used a Syriac source.[28]

---

the hexaemeral order" is mentioned in a list of Bar Ṣalibi's works in ms. Vat. syr. 37 (olim Scandar 32; see Assemani, *BOCV* II.210b 32-41, 211a 22-30, cf. Baumstark, *GSL* 296 w. nn. 9,10; Blum, TRE IX.7.40-42). It has been suggested by Koffler (1932) 206 that this and other theological treatises listed there may have been a major source for BH's *Cand.* as a whole. The likelihood that Bar Ṣalibi, in turn, will have made use of Bar Kepha's *Hex.* in his hexaemeral treatise is indicated by his use of that work in his commentary on Genesis (see Schlimme [1977] 756-816). On the relationship between Bar Kepha, Bar Ṣalibi and Barhebraeus in their works in the field of biblical exegesis, see Reller (1994) 148, 155 n.305, 160f.

[23] See the summary of the work in Nau (1896); cf. Takahashi (2003a). For an instance where Bar Shakko has information which is in Jacob but not in Bar Kepha, see Takahashi, op. cit., n.107.

[24] Bar Shakko, *Dial*. II.2.3, Questions 13-19, ms. Göttingen, orient. 18c, 300r-304r; cf. Ruska (1897) 152.

[25] As was already noted by Ruska (1897) 154 n.1.

[26] Cf. Takahashi (2002b) 231-236 and comm. on *But.* Min. III.i.5, III.ii.3 below.

[27] For an instance of such divergence of the two from Fakhr al-Dīn al-Rāzī (an "error" shared by Bar Shakko and Qazwīnī), see comm. on *But.* Min. III.i.5.1-6 below.

[28] The discovery of this source is also of interest in connection with the question of the sources used by Qazwīnī. Sersen had been led by the divergences between Qazwīnī and the Ikhwān al-Ṣafāʾ to posit the use of an earlier common source by these two (Sersen [1976] 160-194; cf. Daiber [1992] 222-225, Lettinck [1999] 53).

3.2.2. *Mineralogy-Meteorology in Earlier Works of Barhebraeus*

a) *Tractatus tractatuum* (*Têḡrat̲ têḡrāt̲ā*)

Meteorological and mineralogical matters are treated in this work in the 5th and 6th sections of Treatise II, chapter 2.[29] The main source of the discussions there appears to be Ghazālī's *Maqāṣid al-falāsifa* (itself a reworking of Ibn Sīnā's *Dāniš-nāma-i ᶜAlāʾī*) and the order of the material too basically follows that of the *Maqāṣid*. Among the additional sources used, we may count the Syriac version of the *De mundo* and probably also Fakhr al-Dīn al-Rāzī's *Mabāḥit̲ al-mašriqīya*.[30]

b) *Sermo sapientiae* (*Swād̲ sōp̄iya*)

In this work of introduction to philosophy, there is a brief passage at Chapter II, Section 17 (ed. Janssens 78.5-79.3, tr./comm. p. 232f.) where Barhebraeus talks of the generation of cloud, rain and snow from vapour and of wind from smoke in the atmosphere, followed by that of springs from vapour and of earthquakes by smoke in the earth. One point of note here is that, as in *Tract.*,[31] the term *eṭrā* is still used in this passage in the sense of "vapour" (78.8, 79.1), as opposed to that of "exhalation" (ἀναθυμίασις, including both vapour and smoke) as is the case in *Cand.* and *But.*

c) Translation of *K. al-išārāt wa-l-tanbīhāt*

As its translator, Barhebraeus must have been familiar with the contents of Ibn Sīnā's *Išārāt* and we might therefore expect to find some echoes of that work in the *Butyrum*.[32] "Mineralogical-meteorological" matters, however, are touched upon only briefly in the *Išārāt* within the framework of more general discussions on the physical theories concerning the four elements,[33] so that the examination

---

We now see, however, that, if we take away from Qazwīnī's meteorology what he shares with Fakhr al-Dīn al-Rāzī and Bar Shakko, what remains turns out largely to be excerpts from the Ikhwān al-Ṣafāʾ.

[29] Ms. Cantab. Add. 2003, fol. 50r-57v.

[30] For a more detailed discussion, see Takahashi (2002c).

[31] Cf. Takahashi (2002c) 167 n.43.

[32] For the suggestion that the very title of the *Butyrum* might have been inspired by a phrase occurring in the *Išārāt*, see Section 1.2. above.

[33] IS *Išārāt* II.2.20-21, ed. Dunyā (1957-60) II.286ff.; corr. BH *Ind.*, ms. Laur. or. 86 (olim 185), fol. 64v-65r.

as to whether Barhebraeus did indeed make use of the *Išārāt* as a source will have to be left to the editions of other parts of the *Butyrum*.

d) *Candelabrum sanctuarii* (*Mnārat̠ qud̠šē*)

The most extensive discussion of mineralogical-meteorological matters by Barhebraeus prior to that in the *Butyrum* is to be found in the second book, or "base", of his major theological work, the *Candelabrum sanctuarii*, where the Creation of the world is discussed in a hexaemeral framework.[34]

Despite what the theological nature of the work as a whole and the hexaemeral scheme used in Base II might lead us to expect, the principal source for the discussions of mineralogy and meteorology in the *Candelabrum* is not a work of a Christian author, but the *K. al-mabāḥit̠ al-mašriqīya* of Fakhr al-Dīn al-Rāzī, the work which, as we have noted above, was also the indirect source of Bar Shakko's *Dialogues* and which is itself largely a summary of Ibn Sīnā's *K. al-šifāʾ* in those parts which concern us here. At a rough estimate, somewhat over a half of the passages dealing with mineralogy and meteorology in the *Candelabrum* may be considered either as translations or paraphrases of the corresponding parts of the *Mabāḥit̠*, while the order of the discussion too generally follows that of the *Mabāḥit̠*. Among the additional sources used, the most important is the Syriac version of Nicolaus Damascenus' *Compendium*, while there are also a number of instances where the Syriac version of the *De mundo* has been used, especially in the choice of the Syriac technical terms.

Those parts of the *Candelabrum* dealing with geography (especially climatology) is also of concern to us here, since this subject plays a rôle in the part of the *Butyrum* edited below (in Min. IV-V). The main source of the *Candelabrum* in this area is the *K. al-tafhīm li-awāʾil ṣināʿat al-tanǧīm*, a handbook of astrology by Abū al-Raiḥān Muḥammad al-Bīrūnī (973-after 1050), which includes an extensive discussion of geometry, geography and astronomy as prerequisites for the study of astrology. Here again, however, a number of other sources have been used. It is to be noted, for example, that the values used in

[34] To be more precise, in Base II, chapter (*qep̄aleʾōn*) 3, section (*pāsōqā*) 1, which deals with the first day of Creation, and within that section, in "topics" (*nīšē*) 2 and 4, which deal, respectively, with the elements earth and air. - Ed. Bakoš (1930-33) 82-104, 109-130; ed. Çiçek (1997) 65-80, 84-98.

the *Candelabrum* for the latitudes of the seven climes are the traditional values as given in Ptolemy's *Almagest*, rather than the values given by Bīrūnī or the values given by Naṣīr al-Dīn al-Ṭūsī in his *Tadkira fī ᶜilm al-haiᵓa* (these values are used by Barhebraeus in his *Ascensus mentis*), while there are also the odd phrases here and there which may go back to the hexaemeral works of Bar Kepha and Jacob of Edessa.[35]

e) *Liber radiorum* (*K. d-zalgē*)

Barhebraeus' shorter work on theology is basically an abridgement of the *Candelabrum* and this description also applies by and large to those parts of the work dealing with mineralogy and meteorology.[36] A closer look at the text, however, reveals instances where the *L. radiorum* contains materials which are not in the *Candelabrum* and many of these passages, in the parts dealing with meteorology at least, show an affinity with the *Tractatus tractatuum*,[37] suggesting that Barhebraeus made use of this earlier work of his as an additional source in composing these passages. There is also at least one instance where the *L. radiorum* is in agreement with the *Butyrum* against the *Candelabrum* (and *Tractatus*), namely in quoting Psalm 104.32 in connection with earthquakes (see comm. on *But.* Min. II.iii.4. 4-7 below). It just so happens that that verse is also quoted by Bar Kepha in his discussion of earthquakes in the *Hexaemeron*. Barhebraeus may therefore have picked up that verse from Bar Kepha when composing the *L. radiorum*

---

35 For a detailed discussion of the sources of mineralogy-meteorology in *Cand.*, see Takahashi (2002b).

36 Mainly in Mēmrā I, *qep̄aleᵓōn* 1, *pāsōqā* 2 (on element earth, ed. Istanbul p. 11-18) and I.1.4 (on air, p. 19-27).

37 E.g. the comparison of snow to carded cotton (ܐܝܟ ܥܡܪܐ ܕܢܕܝܦ) found in *Rad.* [Istanbul (1997)] 21.7f. and *Tract.*, ms. Cantab. Add. 2003, 55v 8 (cf. Ghazālī, *Maqāṣid* [Dunyā] 340.25 كالقطن المندوف), but not in *Cand.* - To give a more general idea, the correspondences with *Cand.* (ed. Bakoš) and *Tract.* (ms. Cantab. Add. 2003) in the part of *Rad.* dealing with evaporation and precipitation might be summarised as follows: *Rad.* [ed. Istanbul] **22.16-18** [evaporation]: corr. *Tract.* 55r 26-29; *Cand.* 111.11-112.2 (wording here closer to *Tract.*); **22.18-23.4** [cloud and rain]: corr. *Cand.* 112.4-6; *Tract.* 55v 4f.; **23.4-6** [fog]: < *Cand.* 123.6f.; **23.6-9** [snow]: corr. *Cand.* 123.8-10; *Tract.* 55v 6-9 (incl. comparison with carded cotton, see above); **23.9-13** [hail]: corr. *Cand.* 123.10-124.3; *Tract.* 55v 9-15 (*Rad.* here closer to *Tract.* than to *Cand.*, where, for example, there is no mention of the rôle played by heat and the occurrence of hail in spring and autumn); **23.13f.** [dew]: cf. *Cand.* 124.3-6; **23.14f.** [frost]: < *Cand.* 124.7f.; **23.15f.** [rainstorm]: < *Cand.* 113.8-10; **23.17** [snowstorm]: < *Cand.* 113.10.

and thought it worthwhile to reuse this verse when composing the *Butyrum* later on.[38]

In the treatment of geography, too, the *L. radiorum* shows a number of divergences from the *Candelabrum*. Noteworthy among these is the reference to works of Ptolemy called "ܟܬܒܗ ܦܪܘܟܝܪܘܣ" (i.e. the *Handy Tables*, πρόχειροι κανόνες) and "ܓܐܘܓܪܐܦܝܩܘܢ", respectively, at ed. Istanbul 16.6f. and 17.2, and the mention of a place called "ܛܐܦܪܘܒܢ" (Taprobane) at 16.9. These three words are also found in close proximity to each other and in forms very similar to these in Severus Sebokht's treatise "on the Constellations".[39] It is likely, in other words, that that work of Severus Sebokht (ca. 575-666/7) was used as an additional source by Barhebraeus in composing these parts of the *L. radiorum*.[40]

f) *Ascensus mentis* (*Sullāqā hawnānāyā*)

Mineralogy and meteorology as such do not play any rôle in Barhebraeus' work on astronomy, but this work needs also to be mentioned here because of its overlap with those parts of the *Butyrum* edited below dealing with geography and climatology.

The *Ascensus* as a whole is modelled on Naṣīr al-Dīn al-Ṭūsī's *Tadkira fī ᶜilm al-haiʾa*.[41] A comparison of the *Ascensus* with the *Tadkira*, however, shows that Barhebraeus must also have used a number of additional sources, as was his wont. In particular, for the great number of Syriac (and Greek) technical terms which we encounter in this work, he must have had a Syriac source. The mention of Severus

---

[38] Barhebraeus must, of course, have been thoroughly familiar with the Psalms, so that for the mere quotation of Psalm 104.32 we need not necessarily infer a use of Bar Kepha, but there are further correspondences with Bar Kepha in the passage of *But*. in question (though not in the shorter passage of *Rad.*).

[39] See Nau (1930-1) XXVII.407; cf. id. (1910) 237, l. 3; 240, l. 11f. (ܦܪܘܟܝܪܘܣ, ܓܐܘܓܪܐܦܝܩܘܢ, ܛܐܦܪܘܒܢ). - In *Cand.*, Ptolemy's *Geography* is called ܟܬܒܐ ܙܘܓܪܐܦܝܐ (*sic* with *zain*; ed. Bakoš 103.12), a form which suggests an Arabic Vorlage (جُغرافيا). - Cf. also comm. on *But*. Mete. III.iii.1.6 below.

[40] It may also be remembered here that the manuscript in which we have the most complete text of Severus' treatise (Paris syr. 346, copied in 1309 A.D. in Dair al-Zaᶜfarān) also contains excerpts from Nic. syr. and BH *Cand.*, which gives rise to the suspicion that much of the content of this manuscript may have been copied from manuscripts once in BH's possession.

[41] See Takahashi, *Bio-Biblio.*, Section I.2.2.H.

Sebokht' name on one occasion in the *Ascensus* itself (ed. Nau 106.22f.)[42] and the correspondences with Severus' treatise "on the Constellations" we have found in the *L. radiorum* lead us to suspect that that work was at least one of the major sources used for that purpose, even if a closer comparison of the vocabulary used in the *Ascensus* and that still unpublished treatise of Severus' is a work that has yet to be carried out.[43] Some observations will be made in the commentary to *But.* Min. IV below on the manner in which Ṭūsī's *Tadkira* is used in the *Ascensus* and on the manner in which some of these elements derived from the *Tadkira* are taken up again in the *Butyrum*. From the commentary to *But.* Min. V.i.,[44] it will be seen that the geographical overview of the world as found in the *Ascensus*, where Barhebraeus greatly expands what Ṭūsī had to say in the *Tadkira*, shares much in common with the overview in the *L. radiorum*.

g) Others

Passages dealing with mineralogical and meteorological matters are also occasionally to be found in other works of Barhebraeus.

In the *Ethicon*, for example, subjects normally treated under the heading of meteorology appear in the chapter concerned with the contemplation of Divine Creation (Book IV, chapter 13).[45] The account of the meteorological phenomena there (ed. Bedjan 450.15-451.9) is close to the account in the *L. radiorum* and is probably best seen as a summary of the account in that work, while the geographical overview of the world (ed. Bedjan 452.18-454.10), too, has much in common with the *L. radiorum*, as well as with the account in the *Ascensus*, a work which was composed in the same year as the *Ethicon*.[46]

Among the miscellaneous matters mentioned in the "sententiae" (*petgāmē*) in the fourth chapter of the *L. columbae*, the colours of the rainbow discussed in Sentence 5 (ed. Bedjan 580.20-581.3) falls within the scope of meteorology and the vocabulary used there is at least

---

[42] Cf. Nau (1910) 228, 234f.

[43] Cf. comm. on *But.* Mete. II.iii.3.7 (ἔξαρμα) below.

[44] See also Takahashi (2003a).

[45] Ed. Bedjan 447-466, Çiçek 233-242, especially in those parts of the section dealing with things created on the first day (meteorological phenomena) and on the third day (seas etc.).

[46] See comm. on Min. V.i. below.

reminiscent of that used in the discussion of the matter in the *Butyrum*.

A handy overview of places where Barhebraeus refers to philosophical matters (including the natural sciences) in his biblical commentary, *Horreum mysteriorum*, has been provided by Göttsberger.[47] Longer passages where mineralogical-meteorological subjects are dealt with include the discussion of the sea in connection with Gen. 1.9 (ed. Sprengling-Graham 10.19-28; cf. *But*. Min. V) and of the formation of rainbows in connection with Gen. 9.13 (Sprengling-Graham 40.17-23; cf. *But*. Mete. II).

One correction is due in the summary provided by Göttsberger, namely where he tells us that Barhebraeus denied the possibility of stones being formed in water in his comment on Job 14.19 (p. 162). If true, this would contradict the views expressed at *But*. Min. I.i.-ii,[48] but it transpires that Göttsberger's statement is based on a misreading. The Peshitta version of Job 14.19 reads: ܟܐܦܐ ܫܚܩܐ ܡܝܐ : ܐܦ ܟܠܬܐ ܥܦܪܗ ܕܐܪܥܐ.[49] Barhebraeus rightly realised that stones do not wear away water and so, having quoted Job 14.19a, comments:

> ܡܫܐܠܢܐܝܬ ܩܪܝ . ܗ̄ ܟܐܦܐ ܠܐ ܡܥܒܕܐ ܒܡܝܐ : ܐܦܠܐ ܡܢ ܟܠܬܐ ܠܒܨܘܪܬܐ
> ܕܡܬܚܝܒܐ ܗܝ ܥܦܪܗ ܕܐܪܥܐ .
>
> "Read as a question [i.e. read: 'Does stone wear away water, or a clod the earth's soil?'], [because] stone does not act upon [*ma*ᶜ*bdā* Aphel active + *b*-] water, nor is earth's soil diminished by a clod which is formed [from it]."[50]

It seems that Göttsberger read the verb *ma*ᶜ*bdā* here in the passive ("is not formed in").[51]

---

[47] Göttsberger (1900) 161-163.

[48] And in *Cand*. as noted by Koffler (1932) 159. Koffler, who evidently did not check the text of *Horr*. was thrown off at a tangent by Göttsberger's misreading here.

[49] The Peshitta here is, in fact, a poor rendition of the Hebrew "ᵃḇānīm šāḥᵃqū mayim, ...", where either of the plural nouns could in theory be the subject. - BH might have reached a more satisfactory solution than he did had he had access to the Syro-Hexapla (ܠܟܐ̈ܦܐ ܐܫܥܘ ܡܝܐ, "water made smooth the stones"; cf. Septuagint: λίθους ἐλέαναν ὕδατα; cf. further Bernstein [1836] 514, s.v. ܫܥܐ), but he probably did not have the Syro-Hexapla here, since the Bk. of Job (along with Ruth and Ezekiel) is one of the few books for which BH gives no citations from the Syro-Hexapla in *Horr*. (Göttsberger [1900] 127).

[50] Ed. Bernstein (1858) 5b.

[51] Or, perhaps, misunderstood Bernstein's Latin translation: "lapis non operatur in aquis".

3.2.3. *Sources of Mineralogy-Meteorology in Butyrum sapientiae*

As has been stated in Section 1.5. above, the model and principal source of the *Butyrum sapientiae* is Ibn Sīnā's *K. al-šifāʾ*. The use of the Syriac version of Nicolaus' *Compendium* in those parts, at least, of this work dealing with the natural sciences has also been mentioned. An analysis of the text edited below, however, makes it clear that these two were not the only sources used by Barhebraeus in the books on mineralogy and meteorology in the *Butyrum*. Listed below are the sources which have been identified so far, together with some brief comments on the manner in which these sources are used.

a) Ibn Sīnā, *K. al-šifāʾ*

In contrast to his exegetical and theological works, such as the *Horreum mysteriorum* and the *Candelabrum*, where the authorities cited (usually the Church Fathers, but occasionally also secular Greek authors) are frequently mentioned by name,[52] Barhebraeus rarely names his authorities in his philosophical works. With the exception of the "alchemists" and Theo of Alexandria mentioned, respectively, at Min. III.i.5 and IV.ii.2,[53] the only authorities mentioned in the part of *But*. studied here are Ibn Sīnā and Aristotle-Nicolaus and even these are never referred to by their personal names. In the case of Ibn Sīnā, it is not by that name (or such names as Abū ʿAlī, as might also have been expected) that he is referred to, but by the honorific title *sāḇā rēšānā*,[54]

---

[52] For a survey of the authors cited in *Horr.*, see Göttsberger (1900) 170-181; for those in *Cand.*, see the indices to the editions of the individual bases. Whether Barhebraeus cites these works directly or via intermediaries in each case is, of course, a matter that has in most cases still to be explored.

[53] We ignore here such names as those of Empedocles, Anaxagoras and Democritus, who are mentioned by Aristotle in his works.

[54] Used five times in the two books edited below, three times on its own (Min. I.i.2, IV.ii.3, Mete. II.ii.4) and twice accompanied by *myattrā da-ḥrāyē*, "the best of the moderns" (corr. Arab. *afḍal al-mutaʾaḫḫirīn*; at Min. I.i.3 and Mete. II.iii.7). At Min. I.i.4 we find Ibn Sīnā referred to simply as the *sāḇā*, "the doctor/shaikh", at Min. IV.iii.1 as the *sāḇā mšabbḥā* and at IV.iii.6 simply as the *mšabbḥā*. In the section heading of Min. IV.iii., we encounter the adjective *sāḇāyā* in the sense of "Avicennian" (not "of the elders", as the word was understood by J.P. Margoliouth, PS Suppl. 220a). - Elsewhere in Barhebraeus' works, the title *sāḇā rēšānā* is used of Ibn Sīnā at *Chron.* [Bedjan] 219.15, 220.23 and the Arabic title *al-šaiḫ al-raʾīs* at *Hist. dyn.* [ed. 1958] 77.12, 187.1f., 189.25. - Interestingly, the title *al-šaiḫ al-raʾīs* is found applied to BH himself on the title page (written in Arabic in the hand, I believe, of Daniel of Mardin) preceding the text of BH *Gramm*. in ms. Laur. or. 298 (fol. 3r).

which is of course a translation of the title by which Ibn Sīnā is commonly known in Arabic, *al-šaiḫ al-raʾīs*.

As has been mentioned above, the order in which the chapters are arranged in the two books of the *Butyrum* studied here generally follows the order in the *Šifāʾ*. The same usually applies for the arrangement of the material within each chapter. As for the actual contents, somewhere between one half and two thirds of Barhebraeus' books on minerals and meteorology may be considered to consist of translations and summaries of the corresponding parts of the *Šifāʾ*. Some of these passages are literal translations of the *Šifāʾ*, but more usually Barhebraeus' renditions are free paraphrases of Ibn Sīnā, in which Barhebraeus frequently combines materials taken from different parts of the *Šifāʾ* or inserts into a sentence of the *Šifāʾ* words and phrases taken from other sources.

This free rendition on the whole diminishes the worth of the *Butyrum* as a tool in the textual criticism of the *Šifāʾ*, but there are cases where the *Butyrum* serves as an indirect 13th c. withness for particular readings, allowing us to correct or confirm the readings of the received text of the *Šifāʾ*. Listed below are those instances which have been noticed so far, where the *Butyrum* might be considered to have a bearing on the readings of the *Šifāʾ* (usually in relation to the Cairo edition).

N.B.: F marg. = Quotations from the *Šifāʾ* in the margin of Laur. or. 83 (= ms. F).[55]

Šifāʾ, al-afʿāl wa-l-infiʿālāt, ed. Cair. 208.1: المادة المحترقة. - ܡܪܬܐ ܡܬܩܕܢܝܬܐ *But*. Min. V.ii.4.5, = المرة المحترقة? (so also *Mabāḥit* II.142.9).

Šifāʾ, al-maʿādin, ed. Cair. 3.11: التفجر: التفخير BDS et ed. Ṭ ‖ also 4.18 لتفجير: لتفخير ed. Ṭ (et, suspicor, codd.). - ܦܘܚܪܐ *But*. Min. I.i.2.3 (= تفخير).

5.20: جوزجان edd. Cair. et Ṭ: جوزجانان ed. Holmyard 75.10; جورجانان (lege جوزجانان) F marg. - ܓܘܙܓܢܝܐ (pl.) *But*. Min. I.i.4.4. - *But*. might be considered here to support the reading جوزجانان, with the extra "-ān" recognised as a Persian plural ending.

6.1: فنقر: فنفذ BM, edd. Ṭ et Holmyard, et F marg. - ܘܥܡܕ *But*. Min. I.i.4.5. - BH's *ʿmad* is closer to *nafada*.

6.10: الجميلة: الجليلة BDS, edd. Ṭ et Holmyard, et F marg. - ܓܒܝܐ *But*. Min. I.i.4.11 ("finest, choice").

[55] See Appendix II below. - The other sigla are those of the Cairo edition: B = ب = al-Azhar 331; D = د = Dār al-Kutub (falsafa 894?); S = سا = Dāmād; M = م = Brit. Lib. (Or. 7500?); Ṭ = ط (lithograph ed. Tehran).

12.12: تفصيلها:نفصلها BM et ed. Ṭ. - ܦܘܪܫܢܗܘܢ *But*. Min. I.ii.6.8.
14.2: متيسر: منتشر DSM: منتشرة ed. Ṭ. - ܕܒܕܝܪ *But*. Min. I.iii.2.4 (closer to منتشر).
13.18: الأرضية المتولدة من اختلاطاتها بعفونات : اختلاطاتها بها العفونات ... DM et ed. Ṭ. - ܐܪܥܢܝܘܬܐ ܕܡܬܝܠܕܐ ܡܚܘܠܛܢܐ ܒܗܘܢ ܡܣܪܝܘܬܐ *But*. Min. I.iii.4.2-3 (favours the reading of DMṬ).
17.14: خسف الأرض edd. Cair. et Ṭ. - ܠܐܪܥܐ ܢܬܪܐ *But*. Min. II.ii.2.8. - BH may have read خرق or فشق.
19.13: الهواء الكائن edd. Cair. et Ṭ. - ܐܐܪ ܡܠܝܠܐ *But*. Min. II.ii.3.7. - BH seems to have read الهواء اللطيف vel sim.; cf. 19.14: الأرض الكثيفة, corr. ܐܪܥܐ ܕܥܒܝܬܐ *But*. II.ii.3.8.
20.11: بالجبلة : بالحيلة edd. Ṭ. et Holmyard; بالجملة M. - ܒܡܘܪܣܐ *But*. Min. III.i.2.1 (with Ṭ & Holmyard).
21.6: جُلالة edd. Cair. et Ṭ (et *Mabāḥit* II.214.13): حلالة Holmyard. - ܫܪܝܐ *But*. Min. III.i.4.2 ("dissolution").
22.10: ثقيلة طينته ; ثقيلة طينية Holmyard; ثقيلا طينيًا Garbers-Weyer. - ܒܛܝܢܘܬܐ ܡܩܛܠ *But*. Min. III.ii.3.8-9. - BH probably read تقتله طينية.
22.17: حسية edd. Cair. et Ṭ: حسنة Holmyard. - ܡܬܪܓܫܢܘܬܐ *But*. Min. III.iii.1.6 (not the same, but closer to حسنة).
Šifāʾ, al-āṯār al-ᶜulwīya, ed. Cair. 35.14: فى الصحو edd. Cair. et Ṭ: فى الشمس B; فى الصبح DSM. - ܒܫܦܝܐ *But*. Mete. I.i.2.8 (= فى الصحو).
52.1: ارجاء edd. Cair. et Ṭ. - ܪܦܝܘܬܐ *But*. Mete. II.iii.1.6 (= ارخاء).
60.16: إلى فوق :إلى أسفل DSM et ed. Ṭ. - ܠܬܚܬ *But*. Mete. III.iv.2.1(i.e. with DSMṬ).
61.4: ترى :يرى ed. Ṭ. - ܡܬܚܙܝܐ *But*. Mete. III.iv.3.5.- IS probably wrote *tarā* (cf. Olymp. arab. 143.13 رأيتَ), but BH may have read *yurā*.
61.4-5: 4 التشكل edd. Cair. et Ṭ: الشكل SM ‖ 5 التشكل ed. Cair.: الشكل ed. Ṭ et M - ܣܘܟܡܐ . . . ܐܣܟܝܡܐ *But*. III.iv.3.6-8. - If *sukkāmā* (verbal noun, Pael) = *tašakkul* (verbal noun, V), and *eskēmā* = *šakl* (simple nouns), then *But*. supports the readings of ed. Ṭ.
70.6: مدبر : مدير ed. Ṭ. - See comm. on *But*. Mete. IV.i.5.5-7 below.
72.6-7, 15: كانت متعلقة, تتعلق. - ܕܡܫܬܠܗܒܢ, ܢܗܪܐ *But*. Mete. IV.iii.2.5, 7. - Did BH read مشتعلة, تشتعل?
72.15: لم يبق : لم يكن S. - ܠܐ ܦܫܐ *But*. Mete. IV.iii.2.6. - Cf. 72.14: حتى لم يبق: ܐܝܟܢܐ ܕܠܐ ܢܦܫ *But*. ibid.
72.18: الشهب والكواكب وذوات الأذناب : الشهب والكواكب ذوات الأذناب BDSM. - ܠܡܩ̈ܕܚܬܐ ܘܟ ܘܟܘ̈ܟܒܐ ܕܢܘ̈ܦܝܐ *But*. Mete. IV.iii.3.1 (with BDSM).
74.16: الحارة: الحادة S et F marg. ad *But*. Min. IV.iii.4.10-11 (the text of *But*. has ܚܡܝܡܐ).
75.14: لكيفية تسيّل أرضية باردة مجمدة - تسيّل ed. Cair.: شديدة ed. Ṭ; تشتد DSM. - ܒܐܝܟܝܘܬܐ ܕܪܕܝܘܬܐ ܡܫܝܚܢܝܬ ܩܪܝܪܘܬܐ ܘܡܓܠܕܢܝܬܐ *But*. Mete. V.i.1.10-11. - *But*. here, like ed. Ṭ and codd. DSM, talks of the severity of coldness.
76.5: البخار : البحار ed. Ṭ. - ܝܡܐ *But*. Mete. V.i.3.1 (with Ṭ).
77.3: التبن edd. Cair. et Ṭ, sed التين *Mabāḥit* II.219.1, et ܬܒܢܐ *But*. Mete. V.ii.1.6.

b) Nicolaus Damascenus, *Compendium of Aristotelian Philosophy* (= Nic. syr.)

Aristotle is cited 18 times by name in Base II of the *Candelabrum* and, where the relevant portions of Nicolaus' *Compendium* survive, the statements ascribed by Barhebraeus to Aristotle can almost invariably be traced back to that work. In the two books of the *Butyrum* studied here, the name "Aristotle" is never mentioned, but we find passages ascribed to "the Master" (*rabban*) on fifteen occasions[56] and such passages are almost invariably found to be quotations or paraphrases of passages out of the *Compendium*. Since Barhebraeus attributes the *Compendium* he had to Nicolaus in his *Historia dynastiarum*, he must have known that the author of this work was not Aristotle himself but Nicolaus. The fact that he quotes from this work under the name of Aristotle/"the master" indicates that he believed, rightly on the whole, that Nicolaus' *Compendium* was an accurate summary of Aristotle and, as Drossaart Lulofs put it, considered Nicolaus a "legitimate locum tenens" of Aristotle.[57]

As in the case of passages derived from the *Šifāʾ*, passages of Nicolaus are sometimes found quoted almost word for word in the *Butyrum*, sometimes more freely. In comparison, however, with passages taken from the *Šifāʾ* the quotations from Nicolaus tend to be more literal. This can be explained partly by the simple fact that in quoting Nicolaus Barhebraeus did not have to translate from Arabic into Syriac, partly by the fact that Nicolaus' work, itself conceived of as a summary of Aristotle, is on the whole less verbose than the *Šifāʾ* and did not require further summarisation, and partly perhaps also by the reverence on the part of Barhebraeus for what he believed to come close to the *ipsissima verba* of "the Master".

In comparison, on the other hand, with quotations from Nicolaus in the *Candelabrum*, the quotations in the *Butyrum* tend to be freer and there are a greater number of instances where materials taken from Nicolaus are combined with materials taken from other sources. This is probably to be explained in the first place by the greater maturity and greater skill as an epitomist acquired by Barhebraeus by the time he came to write the *Butyrum*. Another factor that would have contributed

---

56 *But*. Min. I.i.4, i.5; II.i.4 (bis), ii.1, ii.2; V.ii.4, iii.1, iii.2, iv.1; Mete. II.i.3, ii.2, iii.2; III.i.2, iii.1.

57 Drossaart Lulofs-Poortman (1989) 35f.

to this freer use of Nicolaus in the *Butyrum* and in particular to those instances where materials taken from Nicolaus and from the *Šifāʾ* are found combined with each other in a single passage is the affinity there already existed between the *Šifāʾ* and the Syriac version of Nicolaus. As has been noted in Section 3.2.1 above, one of the major sources used by Ibn Sīnā for his discussions of meteorology in the *Šifāʾ* was the so-called Arabic version of Olympiodorus' commentary, which, in turn, is basically a translation of the Syriac Nicolaus with its Olympiodorean scholia. This means that there is a great number of passages which resemble each other in the *Šifāʾ* and in the Syriac Nicolaus-Olympiodorus text used by Barhebraeus. These instances provided an opportunity for Barhebraeus to produce those passages in his work where he combined elements taken from the two in such a way that it is now often difficult for us to decide which part of a given passage should be considered as being based on the *Šifāʾ* and which on Nicolaus.

As has also been stated above, the Nicolaus text we have in the Cambridge manuscript is far from complete and, even if the parts covering meteorology and mineralogy have fared better than others, there are still a great number of lacunae there, including a long one in the discussion of the sea. We know at the same time that Barhebraeus must have had a more complete text of the *Compendium*, so that the *Butyrum* (along with the *Candelabrum*) can be used in our attempt to reconstruct these lost passages. This is especially the case where a passage is explicitly attributed to "the master" in the *Butyrum* but a corresponding passage cannot be found in the Cambridge manuscript.[58] There are also a significant number of cases where lost passages of the Syriac Nicolaus-Olympiodorus may be suspected to lie behind particular passages of the *Butyrum* for various reasons, the commonest being the fact that Barhebraeus stands closer in wording than Ibn Sīnā to the original text of Aristole or to the Greek Olympiodorus, especially where this is supported by parallel passages in other works derived from the Syriac Nicolaus-Olympiodorus, usually the Arabic version of Olympiodorus, but occasionally also Ibn al-Khammār's *Mutaḫayyila* and Barhebraeus' own *Candelabrum*.[59] The reconstruction of the exact

---

[58] This applies to *But.* Min. V.iii.1.8-11, iii.2.1-3, iv.1.1-5 below.

[59] See in particular the comm. on *But.* Min. V.ii-iv below (on the sea, corresponding

wording of the Syriac Nicolaus-Olympiodorus is often made difficult by the manner in which Barhebraeus combines his sources. What has been said concerning the relatively free use of Nicolaus in the *Butyrum* means that in this respect the *Candelabrum* is usually a more reliable tool than the *Butyrum*, although, at the same time, the very fact that the *Butyrum* is a more detailed work means that it provides a greater amount of material for such reconstruction work than the *Candelabrum*.

It is also most likely that Barhebraeus had a better text of Nicolaus' *Compendium* than is preserved in the Cambridge manuscript, which has been judged to date from the 15/16th century and which could therefore itself be a descendant of the very manuscript used by Barhebraeus.[60] This means that the passages taken from the *Compendium* in the *Butyrum* and the *Candelabrum* can often be used for correcting the text of the Syriac version of the *Compendium*.

For the correction of the original Greek texts that lie behind the Syriac the works of Barhebraeus are usually a little too far removed, but one instance has been noticed in the part of the *Butyrum* studied here which allows us to correct the Greek text of Olympiodorus' commentary: Olymp., *in Mete.* 116.13: χελύνων Stüve: γελλίνων codd.; ܛܠܠܝܢܘܣ *But.* Min. V.iii.2.3 (i.e. lege τελλίνων).

c) Abū al-Barakāt al-Baghdādī, *K. al-mu*ᶜ*tabar*

The work that is most frequently used as the source after the *Šifā*ᵓ and Nicolaus' *Compendium* in the parts of the *Butyrum* studied here[61] is the *K. al-mu*ᶜ*tabar* of Abū al-Barakāt Hibat Allāh al-Baghdādī (ob. 1165), an author who, though following in the footsteps of Ibn Sīnā, is known to have developed original lines of thought.[62]

---

to the long lacuna in the Cambridge manuscript); also on Min. I.ii.1.7-10, iii.1.1-4; II.ii.1.7-8, ii.2.2-4, ii.4.8-9; IV.ii.2.4-9; Mete. I.ii.6.6, iii.2.3-5; II.i.5.5-6, ii.2.4-6, iii.4.5-7, iii.5.7-8, iii.6.1-6; IV.iii.4.7. - It is needless to say that the proposed reconstruction in such cases need to be handled with care.

[60] We see, for example, that Barhebraeus was able to give the Greek names of the twelve winds in forms which are more faithful to the Greek than the forms found both in the Cambridge manuscript and in the excerpt in ms. Paris syr. 346 (see comm. on Mete. III.ii.3).

[61] Besides in the two books studied here, the *Mu*ᶜ*tabar* is certainly used as the source in the refutations of astrology and alchemy, respectively, in *But.* Gen. et corr. II.v and IV.iii. It may be expected that a closer study of other parts of the *Butyrum* will frequently reveal further instances where the *Mu*ᶜ*tabar* is used as the main source.

[62] On Abū al-Barakāt's originality, see in particular the works of S. Pines gathered together in the *Collected Works of Shlomo Pines*, vol. I.: Studies in Abu'l-Barakāt

Within the two books studied here, those parts where the *Muʿtabar* is used extensively include the dicussion of geology and hydrology in Min. I (esp. sections I.ii-iii), the discussion of minerals in Min. III, the discussion of climatology in Min. IV (where the whole of section IV.iv. is based almost entirely on the *Muʿtabar*) and the discussion of precipitation in Mete. I (where section I.ii, on rain, is again based almost entirely on the *Muʿtabar*). It may be noted that these parts, perhaps with the exception of Mete. I, deal with subjects which had not been treated in great detail by Aristotle and, consequently, by Nicolaus. Put another way, a part, at least, of the reason why Barhebraeus chose to use the *Muʿtabar* in these parts may have been that, in the absence of the relevant Nicolean material, he needed another source to supplement the materials taken from the *Šifāʾ*. It should also be noted, on the other hand, that in most of the parts where the *Muʿtabar* is used in the *Butyrum*, Abū al-Barakāt does have something different to say from the *Šifāʾ*, so that it may also be this originality that attracted Barhebraeus' attention, although we see that little or no use is made of the *Muʿtabar* in the discussions of those topics where Abū al-Barakāt's view diverged more significantly from the traditional Aristotelian-Avicennian position (e.g. on winds, on rainbow, halo etc.).

The discursive, not to say long-winded, style of Abū al-Barakāt means that in using passages taken from the *Muʿtabar* Barhebraeus is usually forced to shorten and summarise what he found there, although there are also instances, as in Min. III.iii.3, where the passage of the *Muʿtabar* is copied almost word for word. An interesting case is found in Min. IV.iv.4, where Barhebraeus, who disagrees with what Abū al-Barakāt was saying, follows the wording of the *Muʿatabar* for much of the passage but then reverses the whole purport of the passage by altering the wording at the end of the passage.

Following instances have been noticed, where the *Butyrum* suggests alternative readings to the Hyderabad edition of Abū al-Barakāt's work.

Muʿtabar II.216.13 مزجية: ܡܬܚܒܝܢܐ *But*. Mete. I.ii.5.7 (suggesting مرجية)

Muʿtabar II.231.17: نمزج: ܠܡܦܩܘ *But*. Min. III.iii.3.9 (suggesting نخرج).

---

al-Baghdādī, Physics and Metaphysics, Jerusalem-Leiden 1979.

d) Fakhr al-Dīn al-Rāzī, *K. al-mabāḥiṯ al-mašriqīya*

As has been noted above, the principal source of the discussions of mineralogy-meteorology in Barhebraeus' *Candelabrum* was the parts dealing with these subjects in Fakhr al-Dīn al-Rāzī's *Mabāḥiṯ al-mašriqīya*, which are themselves closely based on the discussions in Ibn Sīnā's *Šifāʾ*.

It is clear that in the *Butyrum* it is usually the *Šifāʾ* itself rather than the *Mabāḥiṯ* that Barhebraeus is following, but there are also a significant number of instances where Barhebraeus chooses to follow the *Mabāḥiṯ* rather than the *Šifāʾ*. This most often happens where the discussions in the *Šifāʾ* are long-winded or inconclusive and where Fakhr al-Dīn al-Rāzī instead gives a more succint or more definitive account of the subject. Examples of such instances include the explanations of the mechanisms of earthquakes (Min. II.ii.1), winds (Mete. III.i.1) and thunder and lightning (Mete. IV.i.1-2). Extensive use of the *Mabāḥiṯ* is also made in Section Min. IV.iii, where Barhebraeus argues against the Avicennian view that the climate at the equator is moderate and found a supporter for his view in Fakhr al-Dīn al-Rāzī (along with Naṣīr al-Dīn al-Ṭūsī, who was unusually in agreement with Rāzī on this occasion).

e) Ps.-Aristotle, *De mundo* (Syriac version)

Another work which had already been used by Barhebraeus (as well as by Bar Shakko) in his earlier discussions of mineralogy-meteorology was the Syriac version of the *De mundo*. The relative brevity of the *De mundo* in comparison, for example, with Nicolaus' *Compendium* naturally meant that less material could be extracted from it, but this work still plays a significant rôle as a source also in the *Butyrum*, most notably in the description of the Mediterranean Sea at Min. V.i.1. Influence of the *De mundo* can also be detected in the rest of Section Min. V.i. and in scattered places throughout the two books studied here, often in those places where the *De mundo* had already been used in the earlier *Candelabrum*.[63]

f) Moses b. Kepha (and Severus b. Shakko)

The only passage in the two books studied here where there is some positive indication of the use of Bar Kepha's *Hexaemeron* is the

[63] See comm. on *But.* Min. II.ii.2.2-4, ii.3.5-6, ii.4.6; Mete. I.ii.3.6-9, iii.2.3-5, iii.2.8; II.iv.

one at Min. II.iii.4.4-7, where we find the quotation of Ps. 104.32 (see Section 3.2.2e above). Elsewhere any similarities there are with Bar Kepha's work can usually be explained either as the result of the use of common sources or as indirect influence resulting from Barhebraeus' use of Bar Kepha's work in the earlier *Candelabrum*.[64]

It is to be assumed that Barhebraeus knew the works of Bar Shakko, who had worked only a generation before him in the very monastery where he usually resided during his maphrianate.[65] It is all the more surprising therefore that it has not been possible to find any positive indication of the influence of Bar Shakko's *Treasures* or *Dialogues* in the parts of Barhebraeus' work studied here. We find, in fact, that Bar Shakko and Barhebraeus frequently use different vocabulary in rendering into Syriac the same Arabic material.[66]

g) Severus Sebokht

It has been noted above that Severus' treatise "on the Constellations" is likely to have been used as a source by Barhebraeus in the *L. radiorum* and that Severus is also likely to be an important source in the *Ascensus mentis*. In the part of the *Butyrum* edited here, there are some possible echoes of Severus' treatise in the discussion of the habitable zones of the earth at Min. IV.ii.2.1-4.[67]

h) Theo of Alexandria

The account of the theory of trepidation at *But.* Min. IV.i.2 goes back to a passage in Theo's *Small Commentary* on Ptolemy's *Handy*

---

[64] See comm. on *But*. Min. II.ii.1.5, ii.3.5-6, iii.2.3, iii.3.1; IV.iii.6.3-11; Min. I.iii.2.3-5; III.ii.3.3-8; IV.i.3.2f.; V.i.1.1.

[65] Even if Barhebraeus tells us that the many books Bar Shakko had been taken to the public treasury (δημόσιον) in Mosul (*Chron. eccl*. II.411.13-15), this presumably does not mean that copies of Bar Shakko's own works were not available in Mar Mattai (or indeed in Mosul).

[66] See comm. on *But*. Min. III.i.5.6, 9; III.ii.3.1 below; cf. also III.ii.3.6. - More material for the comparison of the manner in which Bar Shakko and Barhebraeus translate the same (or almost the same) Arabic words and sentences is provided by those passages where Barhebraeus depends on Fakhr al-Dīn al-Rāzī's *Mabāḥiṯ* in his *Candelabrum* and Bar Shakko depends on the same passages of the *Mabāḥiṯ* via the unidentified source mentioned in Section 3.2.1b above. Such a comparison frequently shows that Bar Shakko tends to translate more literally, usually retaining the Arabic word order, whereas Barhebraeus' renditions are freer and less influenced by the Arabic syntax.

[67] See comm. ad loc.; cf. also on Min. IV.ii.1.1.

Tables, a passage that was also used by Barhebraeus in his *Chronicon* and *Ascentis mentis*. The vocabulary used by Barhebraeus shows that he must have known this passage in Syriac and not in Arabic. It is unclear whether his immediate source was a Syriac translation of the *Small Commentary* or some secondary source in which the passage was quoted, but the latter of these two possibilities seems more likely.

i) Alchemical Literature of Abū Bakr al-Rāzī (Rhazes) etc.

"Alchemists" are named as the source for the lists of "spirits" and "bodies" (metals) at Min. III.i.5 and the same source probably stands behind the examples of "stones" at Min. III.i.1.3.[68] The brevity of these passages makes it difficult to identify the immediate source, or to determine whether the information was taken from some alchemical work (e.g. the works of Abū Bakr al-Rāzī) or some encyclopedic work (such as Khwarizmī's *Mafātīḥ al-ʿulūm*).

j) Geographers

The sources used for the geographical account of the seas in Section Min. V.i. are clearly different from those used in the rest of the two books studied here. While the account in the *Butyrum* seems to consist largely of materials drawn from Barhebraeus' own earlier works, we can recognise behind the accounts in those earlier works a number of different sources, including, besides the Syriac *De mundo* and Bar Kepha's *Hexaemeron*, works of such Islamic authors as Bīrūnī and the geographers of the so-called Balkhī school. These sources will be discussed in greater detail in the introductory comments on that section.

k) Scripture

The only literal quotation from the Bible in the two books of the *Butyrum* studied here is that of Ps. 104.32 at Min. II.iii.4.6f. While the idea of using that verse of the Psalms is probably due to Bar Kepha (see above), the passage of the *Butyrum* where the quotation occurs itself corresponds to one of the few passages in the *Šifāʾ* where Ibn Sīnā uses the word "God". The mention, in other words, of God by Ibn Sīnā may have prompted Barhebraeus to make his own, Christian addition here.[69]

---

[68] The other reference to "Alchemists" at Min. III.iii.3 is taken from the *Šifāʾ*.

[69] Min. II.iii.4.6 is, incidentally, the only place where the word "God" occurs in

Further more or less recognisable biblical allusions are found in Mete. V, a chapter in which we come close to discussing matters such as Creation and eschatology. Barhebraeus' intentions in making these allusions are not altogether clear, but they are probably best seen as a part of his efforts towards a synthesis of theology and philosophy.[70]

l) Personal Observation

There are a number of occasions in the two books of the *Butyrum* studied here where Barhebraeus reports his own personal observation, namely at Min. I.i.2.10-12 (formation of stone; observed in his childhood in a monastery near Melitene), Mete. I.i.2.6-10 (rain; on a mountain between Gubos and Claudia), Mete. I.iii.4.9-10 (hailstone; between Aleppo and Antioch), Mete. IV.iii.3.7-10 (comet of 1264 A.D.), and possibly also Min. IV.iii.6.10-11 (fruit in Antioch).

In two out of these five instances, Barhebraeus' observations are offered as additions following on reports of observations made by Ibn Sīnā (Min. I.i.2, stone; Mete. I.iii.4, hail). In two others, Barhebraeus' observations replace accounts of observations reported in the *Šifā*ᵓ (Mete. I.i.2, rain; IV.iii.3, replacing a report of the comet of 1006/7 A.D.). This leaves Min. IV.iii.6 (if it is indeed a case of personal observation, see comm. ad loc.) as the only place where Barhebraeus makes a personal observation without being prompted to do so by the *sāḇā rēšānā*.

It is also interesting to note that the localities specified (except for the comet of 1264 A.D.) all belong to the "western" part of the Syrian Orthodox world, near those places where Barhebraeus is known to have spent his youth (Melitene and Antioch) rather than those places in the "east" where he later worked as maphrian. Although the observations could in theory belong to those occasions when Barhebraeus visited the "west" during his maphrianate,[71] this is less

---

the two books of the *Butyrum* studied here. The adjective "divine" (*allāhāyā*) occurs at Min. IV.i.2.9 ("divine wisdom", representing Ibn Sīnā's *ḥikam ilāhīya*) and Mete. V.iii.1.2, 8 ("divine revelation"; *ilhām allāh* Ibn Sīnā).

[70] See the introductory comments to *But.* Mete. V, together with comments on V.i.3.6-7; V.ii.2.10-11; V.iii.

[71] i.e. in 1268 and 1272, when Barhebraeus is known to have been in the region around Melitene, so that the observation at Mete. I.i.2 could in theory belong to one of those occasions. There are no records of his visits to Antioch and Aleppo during his maphrianate and a visit to Antioch becomes less likely after its fall to the Mamluks in 1268.

likely and the impression one receives is that of someone reminiscing about his youth in his old age.[72]

An attempt is made below to illustrate, inevitably in a simplified form, the relationships between some of the works mentioned above.

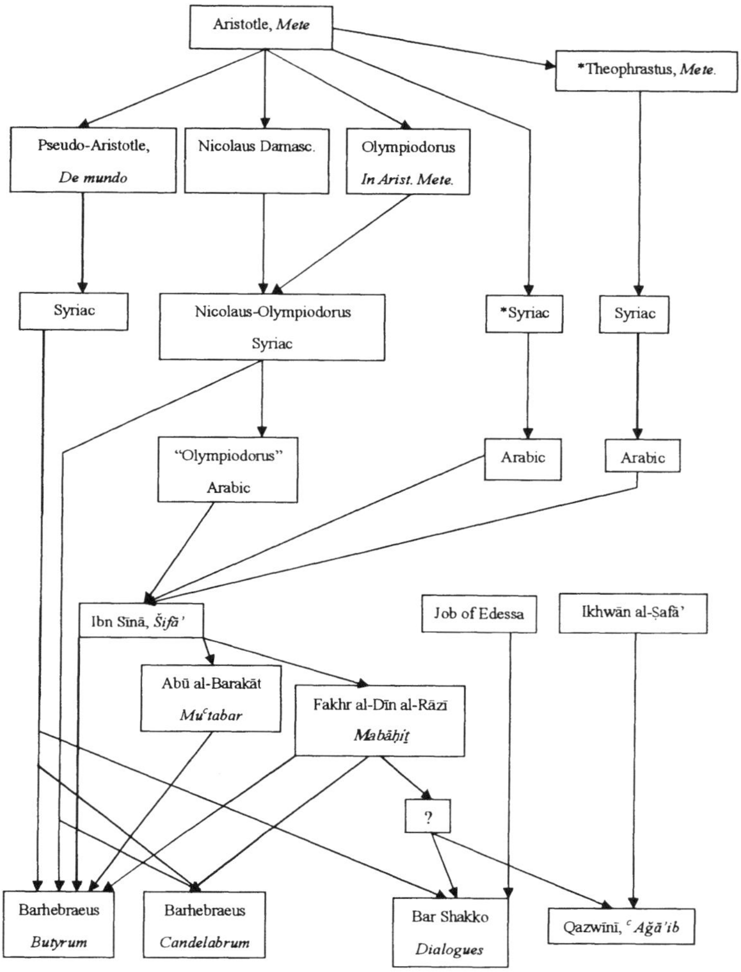

[72] If we can trust the well-known report that Barhebraeus prophesied his own death (*Chron. eccl.* II.465.20-467.8), this means that he may have had a foreboding of his death by the time he came to write this part of the *Butyrum* in 1285.

3.3. *Language and Style*

a) *General*

Few would deny Barhebraeus' skills as a prose writer.[73] His writings are usually marked by clarity and conciseness, and this applies by and large also to the work under study here. Even if the combination of concision and complexity of thought can occasionally result in obscurity, verbosity at any rate is not a vice that can be predicated of our author.

Barhebraeus' skills as an epitomiser becomes all the more evident when we make a close comparison of his writings with his sources and observe the manner in which he omits any redundant words and phrases.[74] A short example is given here. Further evidence of his skills in this respect will be provided in plenty in the commentary below.

Nic. syr. 39.7f.: ܠܐܐܪ ܕܡܢ ܟܒܪ ܚܫ ܡܢܢ ܘܐܫܬܚܢ
the air which has already been affected by us and has been warmed
But. Min. II.ii.1.10: ܠܐܐܪ ܕܐܫܬܚܢ ܡܢ ܦܓܪ̈ܝܢ
the air which has been warmed by our bodies

b) *Vocabulary*

For his vocabulary and especially for the technical terminology, Barhebraeus depends in the first instance on his Syriac sources. The terms used for the meteorological phenomena in the two books of the *Butyrum* studied here can almost invariably be traced either to the Syriac version of Nicolaus' *Compendium* or to the Syriac *De mundo*. This applies in particular to those Greek terms used in transliteration, most of which can be shown to have been taken from the two works just named, with the exception 1) of some astronomical terms (e.g. "ἔξαρμα" in Mete. II.iii.3) which are also used by Barhebraeus in his *Ascensus* and were probably taken from Severus Sebokht, and 2) of

---

[73] See, for example, the judgement of Manna (1901) II.358: ܘܡܕܡ ܠܒ ܡܢ ܕܪܐ ܐܬܩܢܘܗܝ ܐܪܝܐ : ܡܛܠ ܦܘܪܫܢ ܡܠܘܗܝ ܘܫܦܝܪܘܬܗ ܐܪܝܘܬ ܘܡܬܒܪܟܢܘܬ : ܕܒܠܝܫܘܬܗ ܩܘܕܫܐ . ܡܒܝܪܐ ܕܦܣ̈ܘܩܝܐ ܕܗܒܝܢܐ . . . .

[74] A related feature is BH's constant preference for shorter forms, such as the apocopate form of the verb *hwā* (e.g. Min. II.i.2.5 ܬܗ̈ܐ, where Nic. syr. has ܬܗܘܐ) and the construct construction instead of the perisphrastic construction with *d-* (e.g. Min. II.i.3.2 ܪ̈ܘܚܝ ܐܐܪ: ܪ̈ܘܚܐ ܕܐܐܪ Nic. syr.; Min. II.i.4.2 ܡܘܬܒ̈ ܚܫܐ: ܡܘܬܒܐ ܕܚܫܐ Nic. syr.).

the terms used in the passage on trepidation derived from Theo of Alexandria in Min. IV.i.2 ("ἀποτελεσματικοί", "ἔφοδος"), where the immediate source is still to be identified.[75]

It is clear from the great number of Greek terms used both in the *Butyrum* and in his other works that Barhebraeus had no problem in regarding these loanwords as a part of the Syriac language. His attitude, on the other hand, to the use of Arabic terms is quite different. There are only two instances in the two books studied here where Arabic words are given in transliteration, *ḫārṣīnī* (name of metal) at Min. III.i.5 and *haramān* (pyramids [dual]) at Min. V.iii.2. Elsewhere, the use of Arabic terms is avoided and, seeing that a large part of the work studied here is based on Arabic sources, this meant that Barhebraeus had constantly to find appropriate Syriac terms to represent the vocabulary he found in his Arabic sources. Where possible he makes use of his Syriac sources, but these Syriac sources could not always be relied upon to provide the right words for some of the newer concepts that appear in the works of Ibn Sīnā and others. The two books of the *Butyrum* studied here contain a significant number of hapax legomena, as well as words unknown to the dictionaries in the meanings in which they are used there. Many of these may be neologisms invented by Barhebraeus, even if our imperfect knowledge of Syriac lexicography means that we cannot be certain that they are. A selection of such instances is given below.[76]

puḥḥārā = *tafḫīr* (Min. I.i.2, baking, hardening)
metgallᵉlānūṯā = *tamawwuǧ* (Min. I.ii.1, fluctuation)
abbūḇānāyā < *unbūba*? (Min. I.ii.5, tubular)
qāᶜōyā = *muṣawwit* (Min. II.ii.3, sound-producing)
maḥmᶜānūṯā = *taḫmīr* (Min. III.i.3, leavening)
margānāyā = *marǧī* (Mete. I.ii.5, meadowy)
metgammānūṯā = *inqiṭāᶜ* (Mete. V.ii.3.10, extinction)

---

[75] As an adverb which may have been coined by Barhebraeus on the basis of a noun found in Nicolaus, we have *diyaqlasāʾīṯ*, "by διάκλασις", at Mete. II.iii.2.9. Another hybrid Greco-Syriac word known to the dictionaries only from Barhebraeus is *qaṯeḥṭīqāyūṯā* (vertical position, coordinate), which occurs in the part of the *Butyrum* studied here at Min. IV.i.2, answering to Ibn Sīnā's *musāmata*.

[76] It is needless to say that it is often only by comparing these words with the sources that we can ascertain the intended meanings of these coinages. - It is difficult to resist mentioning one example from elsewhere in the *Butyrum*: *mullāyā w-puḥḥāmā* (But. *Eth*. I.i.2) = *al-ǧabr wa-l-muqābala* (Ṭūsī, *Aḫlāq-i nāṣirī* [Mīnowī] 39.12; i.e. "algebra"; cf. Lane [1863-93] 374a, s.v. "ǧabr").

The avoidance of Arabic terms also applies to proper names and the consequent use of Syriac (and Greco-Syriac) appellations for names of peoples and places often results in the appearance of some archaic-sounding and anachronistic names in Barhebraeus' works, a factor which may lead (and has led) the unwary to suppose the use of much older sources than is in fact the case.[77] In the part of the *Butyrum* edited here, an example of this is found in Min. IV.ii.6, where Barhebraeus' "Scythian" in fact stands for Ibn Sīnā's "Turk", the "Cushite" for the "Abyssinian" (*ḥabaša*) and the "Midianite" for the "Arab/Bedouin". A good number of further examples are found in the geographical section at Min. V.i, whereby it may be noted that this tendency to "Syriacise" proper names is less marked in the corresponding part of the *Candelabrum* than in Barhebraeus' later works.[78]

c) *Syntactical Features*

One marked feature of Barhebraeus' syntactical style, at least in the kind of writing examined here,[79] is the constant placement of the verb at the end of the clause. In the parts of the *Butyrum sapientiae* studied here, where we are mostly dealing with things that generally happen, the verbs are found in the vast majority of cases in their participial form, the form used in describing general occurrences, and these are placed as a rule at the ends of sentences and clauses, with the result that the commonest word order is "subject-object-verb".

This feature is particularly noticeable when Barhebraeus is translating from Arabic, a language which adheres more strictly than Syriac to the Semitic rule of initial placement of the verb, as well as when he is otherwise following a Syriac source word for word, but makes this one alteration of the placement of the verb, as happens in the following sentence (an instance with a finite, perfect, verb).

Nic. syr. 38.22: ܐܬܦܣܩܬ ܡܢ ܪܘܚܐ ܘܠܥܠܐ ܕܩܘܪܣܐ ܒܗ ܢܘܪܐ ܐܝܟ ܒܠܝܓܘܢܝܬܐ .
But. Min. II.ii.2.8f.: ܘܡܢ ܪܘܚܐ ܘܠܥܠܐ ܕܩܘܪܣܐ ܒܗ ܢܘܪܐ ܐܝܟ ܒܠܝܓܘܢܝܬܐ ܐܬܦܣܩܬ .

[77] As happened with the "*Laughable Stories*", where this "Syriacisation" of names (and more generally the "Christianisation" of Islamic elements) was one of the factors that delayed the discovery of the fact this work was in fact largely based on an Arabic collection of short stories (see Marzolph [1985] 103-109).

[78] See Takahashi (2003a), para. 58, with n. 118.

[79] The situation seems to be somewhat different in his historical writings, for example, where the finite forms of the verb prevail.

The wind raised ash containing [lit. in which stood] fire in the form of sparks

This deliberate postponement of the verb can occasionally lead to some awkwardness, as in the following passage where the main verb "is" (*īṯēh*) is placed at the very end of a complex sentence.

But. Mete. I.iii.3.3-5:

ܘܥܠܬܐ ܕܚܘܠܠܘܬ ܦܪ̈ܕܝܬܐ ܕܒܪܕܐ . ܚܘܒܪܐ ܕܬܪ̈ܬܝܢ ܡܬܬܙܝ̈ܥܢܘܬܐ . ܗܝ ܠܬܚܬ
ܕܒܐܘܪܬܐ ܕܒܗ ܢܚܬ̈ܝܢ . ܘܗܝ ܕܒܦܬܝܐ ܕܒܗ ܡܢ ܪ̈ܘܚܐ ܡܬܬܪ̈ܕܝܢ . ܘܥܡ ܕܟܢ̈ܝܪܐ
ܡܬܬܪ̈ܥܥܢ ܘܡܬܚܘܠܠܝܢ ܡܬܬܚܟܟܝܢ ܘܡܬܚܫܟܠܠ ܐܝܬܝܗ .

The reason for the roundness of hailstones is the combination of two movements, namely the vertical in which they fall and the horizontal in which they are driven by winds, along with [the fact that] their corners become broken and rounded as they are rubbed against each other.

Once, however, the reader has become aware of it as a "rule",[80] this regular placement of the verb at the end is a factor that usually serves to clarify the syntactical connection, especially in some of Barhebraeus' longer and more complex sentences.

Although one does occasionally come across formulations untypical of Syriac which reflect the Arabic Vorlage,[81] given that a great deal of what is in the *Butyrum* consists of translations and paraphrases from Arabic works, the number of such "Arabisms" is surprisingly small.[82] What has been said concerning the placement of the verb, of course, means that the word order in the *Butyrum* is usually quite different from that in the Arabic source works, and as a result the translations, even when the vocabulary of the source is closely followed, are often quite different from their sources in their syntactical structure. The impression one receives from comparing the text of the *Butyrum* with its Arabic sources is that it is the work of someone who has a clear understanding of what he is reading and is able to reproduce its contents accurately in his own language.

---

[80] That the copyists did not always grasp this point can be gathered from the punctuation of some of the manuscripts.

[81] See comm. on Min. II.iii.1.9.

[82] The situation is somewhat different in Bahebraeus' translation of the *Išārāt*, which is intended specifically as a translation and in which we find a greater number of instances that might be considered as Arabisms, although even there it is difficult to find any major violation of the Syriac idiom. The impression that one receives from looking at the comparable material in the *Candelabrum* is that Barhebraeus was still less skilled in the avoidance of Arabism when composing that earlier work of his (see comm. on Min. V.ii.1.6-9 below; cf. Takahashi [2002b] 237 n.44).

One further syntactical feature, or rather absense thereof, which is worth mentioning because of the help this might provide in dating some of Barhebraeus' works is the absence of the construction "infinitive + participle" (*meᶜbaḏ ᶜāḇeḏ*). This construction occurs frequently, even in places where there is not much emphasis on the verb, in Barhebraeus' earlier works, such as the *Tractatus tractatuum* and the *Candelabrum*. It occurs only once in the parts of the *Butyrum* studied here in the form *law mehwā hāwē* at Mete. IV.i.6.1, where there is considerable emphasis on the verb "to be/exist". A similar observation applies to the use of the conjunction *men qeṣ* ("because"). This occurs in the *Tractatus* and the *Candelabrum*, but not in the parts of the *Butyrum* edited below.

d) *Rhetorical Features*

The subject matter treated in the part of the work studied here was not one that lent itself easily to rhetorical embellishment, but it appears that Barhebraeus could not always resist the temptation to exercise his skills in this area. - We know, of course, from his poems about his skills in versification. We would also assume that he was a practised preacher.

We see, for example, that the long opening sentence of the book on mineralogy (Min. I.i.1) is written in rhymed prose. Barhebraeus has also clearly made a conscious effort to find a variety of verbs to replace the one verb "completed" used by Ibn Sīnā. The result is that this opening sentence stands in quite a contrast to the more pedestrian passage of the *Šifāʾ* which, as far as the content is concerned, is its source.

Rhyming verbs are fairly common. At Min. IV.iii.6 we find three pairs within one 'theory' (*eṯraᶜᶜī*, *eṯrannī*; *maggrā*, *mahhrā*; *hāwēn*, *māṭēn*) and we find furthermore that the clauses ending in this last pair of verbs are isosyllabic. As an example of rhyme designed to produce an onomatopoeic effect, we might mention "*qālē dḥīlē wa-zmāmē mḥīlē*" describing the droning noise accompanying earthquakes (Min. II.ii.3).

As a clear instance of wordplay we have the collocation "*hānā kul law hānā kulleh*" in Min, IV.i.4. We might also consider in a similar light the use of the verb *eškaḥ* in the two senses of "to find" and "to be able" within the same sentence at Mete. III.iii.4.3-4. Much more common, however, are the opposite cases where different words are

used in the same senses,[83] while as an instance of such *variatio* on a larger scale we find the different formulae used to introduce four refutations at Min. IV.iii.1-4.

There are also instances where one can recognise a balanced, chiastic structure in a sentence, as in the case of the following, where this rhetorical consideration leads to an infraction of the "rule" of the final placement of the verb discussed above (verb-subject, verb-subject, *ᶜd̲ammā d-* subject-verb, subject-verb).

Min. V.i.3.6f.: ܘܡܛܠܠ ܠܐܪܥܐ ܗܘܐ ܘܡܩܪܪ ܐܣܛܘܟܣܗ. ܥܕܡܐ ܕܗܘܐ ܕܡܬܘܢܬܐ
ܕܗܘܐ ܡܬܦܫܛ ܘܥܠܠܐ ܢܚܬܐ ܢܬܪܐ

Such rhetorical embellishents are not to be found with any regularity, but they do occur, and this point is worth bearing in mind, both for the purpose of interpreting Barhebraeus' intentions in a particular passage and in the work of identifying the sources. When, for example, an unexpected word is used in a passage that might at first suggest the use of an additional source, the reason for the use of that word may turn out to be a rhetorical one.[84]

3.4. *Barhebraeus' Meteorology: an Assessment*

a) *Butyrum sapientiae as a Compilation*

With reference to the subtitle he gave to his edition of Barhebraeus' *Sermo sapientiae*, "Introduction aux oeuvres philosophiques de Bar Hebraeus", Janssens remarked that it was not possible to speak of the "philosophy of Barhebraeus" as opposed to the "philosophical works of Barhebraeus", since his works consisted of compilations of the works of others and it was not possible to discern a coherent system of philosophy there.[85]

As will be quite clear by now, *Butyrum sapientiae*, the work of which a part is edited here, is also very much a work of compilation,

---

[83] E.g. *lā māṭē-lā qāreb̲* (Min. IV.i.3.10, 11); *net̲maṭṭyān-net̲waᶜᶜad̲* (Mete. I.ii.1.3, 4); *met̲qrē-meštammah-met̲kannī* (Mete. I.ii.3.7, 8, 9).

[84] This applies in particular to instances of *variatio*, as in the opening sentence of the Book of Mineralogy mentioned above. Further examples include: *mšawdaᶜ-mbaddeq* at Min. II.iii.2.6, 8, both corresponding to Ibn Sīnā's *dalla*; and *qānē-mettaggar* at Min. III.i.4.4, 5, both answering to Ibn Sīnā's *istafāda*.

[85] Janssens (1937) 8f.

based, in this case, on Ibn Sīnā's *K. al-šifāʾ*, but with materials derived from a significant number of other sources. This compilatory character of the work occasionally results in minor contradictions within the two books studied here, as happens in the chapter on winds, where the use of Fakhr al-Dīn al-Rāzī's *Mabāḥiṯ* at the beginning of the chapter results in formulation suggestive of the definition of wind as "moving air", a definition clearly rejected by Aristotle, whose views are then followed in much of the rest of the chapter.[86] On the whole, however, such contradictions are rare and the *Butyrum sapientiae*, or at least those parts of it which have been studied here, may be considered a generally successful product of harmonisation. In the case of materials derived from Nicolaus' *Compendium* and the *Šifāʾ*, such harmonisation was made easier, as has been mentioned, by the fact that Ibn Sīnā himself depends to a large extent on the Nicolaus-Olympiodorus text via the Arabic Olympiodorus.

Because Ibn Sīnā's *Šifāʾ* is used as the main source, the views presented in the two books studied here are basically Avicennian. There are, however, a number of instances where Barhebraeus' access to the Syriac version of Nicolaus allowed him to come closer to Aristotle, as, for example, in the chapter on earthquakes where Barhebraeus like Aristotle considers only wind to be the cause of earthquakes, whereas Ibn Sīnā, probably influenced by Theophrastus, had allowed other causes.[87] We also see that in the same chapter access to Nicolaus enabled Barhebraeus to provide us with a more detailed account of the theories of Anaxagoras, Democritus and Anaximenes on earthquakes than is the case in the *Šifāʾ*.

One place where Barhebraeus differed more clearly with Ibn Sīnā is found in Min. IV.iii, where a whole section is devoted to the refutation of the Avicennian view concerning the climate at the Equator. While this view had already been refuted before Barhebraeus by Fakhr al-Dīn al-Rāzī, as well as by Naṣīr al-Dīn al-Ṭūsī, and the main points that Barhebraeus makes are already to be found in Rāzī's *Mabāḥiṯ*, it is possible to recognise a certain amount of originality in the manner in which Barhebraeus presents these arguments.

What originality there is in the *Butyrum* needs to be sought in the

---

[86] See comm. on Mete. III.i.

[87] Cf. comm. on Min. II.ii.1.7-8.

small details of the way in which Barhebraeus combines and alters the materials he found in his sources. Among the more notable instances of such alteration, one might count the "biblical" additions found in Mete. V, whereby Barhebraeus moves one step further towards a synthesis of philosophy and theology.[88]

b) *Assessment in Historical Context*

In making an assessment of the work edited here, we should probably begin by considering whether it meaures up in the first place to the author's own intentions. As has been stated elsewhere,[89] one has reasons to believe that the main purpose of Barhebraeus' literary activity was the revival of scientific learning in Syriac and the composition of the *Butyrum* is also best seen as a part of that project. What Barhebraeus needed to do for that purpose was not so much to produce a highly original piece of writing as to present a) the best scientific materials available in his day b) in clear and readable language in Syriac.

The choice of using the *K. al-šifāʾ* as the model of the *Butyrum* was probably an obvious one, since that work of the "Best of the Moderns" was undoubtedly the best systematic presentation of philosophy available to Barhebraeus. In using Nicolaus's *Compendium* and the Pseudo-Aristotelian *De mundo*, on the other hand, Barhebraeus was going back as close as he could to the "Master". At the same time, by using Fakhr al-Dīn al-Rāzī's *Mabāḥiṯ* and the more original work of Abū al-Barakāt, he also incoporated into his work the more recent developments since the time of Ibn Sīnā. The selection of the works used, in another words, is an impeccable one. Whether Barhebraeus then made the right selection out of the contents of these works is another question, but here again he is generally successful in selecting the essential points from each work. We have seen, for example, that in using Abū al-Barakāt's *Muʿtabar*, he chose those passages where Abū Barakāt had developed on Ibn Sīnā's ideas, while rejecting those parts where Abū Barakāt had diverged too far from the traditional Aristotelian-Avicennian positions.

The clarity and readability of the work under consideration is, I believe, beyond question (even if the English translation of the work

---

[88] Cf. Takahashi, *Bio-Biblio.*, Section I.1.10.

[89] See Takahashi (2001), para. 42-47; id., *Bio-Biblio.*, Section I.2.6.

presented below suffers from the inadequacies of the translator).[90] A related task was the creation of a new scientific vocabulary in Syriac, a point which has been touched upon in Section 3.3 above. Barhebraeus' achievement in this respect, as far as can be gathered from the materials edited below, is a sound and commendable one.

The catastrophic decline in the fortunes of the communities using the Syriac language in the period following that of Barhebraeus means that it is not so easy to judge the *Butyrum* by the success it enjoyed among its intended audience. While there are not many manuscripts today covering the whole of this long work and the majority of the manuscripts that we do have date from the 19th century, the existence at least of manuscripts from each of the centuries between the 14th and the 18th provides an indication of the continued interest enjoyed by this work. For the fact that the *Butyrum* was still considered a standard text on philosophy among Syriac speakers (or rather Syriac readers) six centuries after its composition in the 19th century, we have, besides the manuscript witness, the witness of Ernest Renan, who reports that a Chaldean priest from Mosul, upon being questioned by him on the philosophical textbooks used in the schools of his community, named the *Isagoge* and the *Butyrum sapientiae*.[91] The particular popularity enjoyed by the section of the *Butyrum* dealing with the natural sciences is witnessed to by the manuscripts containing excerpts of this section.

c) *Significance of the Work Today*

The question as to what value the *Butyrum* as a whole (and in particular the two books studied here) has for us today is, of course, a different

---

[90] There is, admittedly, also some evidence for the lack of editing on the part of the author, who died not longer after composing this work and may not have had time to properly revise it (see comm. on Min. III.i.1.3, III.iii.3.9-10, Mete. I.iii.3.3-5 below; also the heading of the Book of Meteorology, where four chapters are announced instead of five in all manuscripts).

[91] Renan (1852) 66, 71; cf. Göttsberger [1900] 31 n.2 and Drossaart Lulofs [1989] 38. - The identity of the "Isagogi" mentioned here will probably have to remain a matter of conjecture, but assuming that a Syriac text is meant, the likely candidates are the commentary on Porphyry's *Isagoge* by Proba (of which there are several 19th c. copies in the Chaldean monastery of Mar Antonius, mss. Baghdad, Mar Antonius syr. 169-172) and the *K. d-īsāgōgī* of the Chaldean patriarch Joseph II (ms. Ming. syr. 433BC); more generally on the tradition of Aristotelian logic in Syriac, see Brock (1993).

one from that of the value the author originally intended it to have and one that can be judged from a variety of angles.

Simply as a medieval presentation of Aristotelian philosophy (or meteorology in particular), its value will have to be assessed as not much above mediocre, seeing that it offers us little in the way of spectacular new developments, even if the clarity of the presentation helps place it above a good number of other works in that domain.

For the classicist, as well as being of general interest (if only as a curiosity) as a late piece of Aristotelica, it has a more concrete value as a tool for the reconstruction of the lost *Compendium* of Nicolaus Damascenus, whereby it will be remembered that the parts of the *Butyrum* edited here correspond to the portion where the Syriac Nicolaus is best preserved, whereas the remaining books correspond to those parts where the Nicolaus text is almost completely lost, so that the edition of the remaining books of the *Butyrum* may be expected to offer a great deal more for that reconstruction work.

For the Avicennian scholar and more generally for the student of Islamic philosophy, the work will be of interest as a relatively rare instance of the reception of Ibn Sīnā (as well as of Abū al-Barakāt and Fakhr al-Dīn al-Rāzī) in Syriac. It may also be of service both as an indirect witness for the textual criticism of the *Šifāʾ* and, though not intended to be such, as a commentary on the *Šifāʾ* by someone who was, if not a great philosopher, at least a competent one.[92]

For the Syriac lexicographer, it offers a great amount of fresh material where the meanings of the words intended by the author, who happens also to be an authority on Syriac grammar and an acknowledged master of Syriac prose, can be ascertained through comparison with the Arabic source passages. At the same time, for those interested in the preservation of Syriac as a living language, the neologisms created by Barhebaeus offer a ready-made stock of new vocabulary, as well as providing models for their work of creating further new words.

For the student of Syriac language and culture - and more generally for anyone interested in intellectual history - it is an important document from the period of so-called Syriac Renaissance, in which the one who is largely acknowledged to be the greatest author of that period made a valiant attempt to create a new type of philosophical-scientific writing

---

[92] See further Takahashi (2003b).

in Syriac by transferring from Arabic into Syriac the contents of the best philosophical-scientific literature of his day and combining this with the materials which had been made available in Syriac centuries earlier in translations from Greek. For members of Syriac communities today, it stands as an example given by one of their greatest scholars, showing the way in which they might go about reviving and fostering their language and culture in the third millenium.

It will be seen that, when all these considerations are put together, the value of the work, of which a part is edited below, is not altogether an insignificant one.

TEXT AND TRANSLATION

NOTES ON THE TEXT AND TRANSLATION

1. *On the Sigla etc.*

Given below are the sigla used for the manuscripts in the edition below.

F = Florence, Laur. or. 83 (olim Palat. 187), 1340 A.D.
M = Birmingham, Mingana syr. 310, 1865 A.D.
W= Vatican, syr. 469, 1804 A.D. (*But*. Min. only)
P = Princeton Theological Seminary, Nestorian 25, undated (fragmentary).
L = British Library, Or. 4079, 1808 A.D.
l = British Library, Or. 9380, 1892 A.D.
V = Vatican, syr. 613-615, 1887 A.D.
Λ: Agreement of all the manuscripts of the "eastern" group (WPLl) available at a given point.
λ: Additions at the end of ms. L (fol. 316r-322v),[1] corresponding to the lacunae in that manuscript, by the same copyist as that of V.
-^{1}: Original reading of the manuscript.
-$^{corr.}$: Correction judged to be by the original copyist.
-^{2}: Correction judged to be by a second hand.
M^{V}: Correction in the margin of M by the copyist of V.

The relevant chapters are found at the following places in each of these manuscripts.[2]

	F	L	l	M	V	P	W
Min. I	51v	182r	355v	70v	95r		132v
Min. II	54r	183v	359r	74v	100r		136v
Min. III	56r	-	361r	78r	104v	(136r)	140r
Min. IV	57v	184v	362r	81r	108r	136r	141r
Min. V	60v	187r	366r	87r	115r	147r	147r
Mete. I	63r	188v	369r	91v	120r	154r	(151v)
Mete. II	65r	190r	372r	95r	124r	(167r)	-
Mete. III	68r	192r	376v	101r	131r	171r	-
Mete. IV	70v	194r	380r	106r	137r	(178v)	-
Mete. V	72v	195v	382v	110r	141v		-
End	74r	196r	384v	112v	144v		-

---

[1] According to the European foliation. No original foliation in this part.

[2] The foliations followed in the table below are as follows: F: European; L. European (182r-196r: corr. 208b-222b of the original Syriac numeration); l: European = original Syriac numeration; M: European = original Syriac numeration; V: European (95r-144r: corr. 60r-112v of the original Syriac numeration); P: original Arabic numeration; W: European.

In the translation, square brackets [] indicate words and phrases supplied by the translator to clarify the meaning, while circular brackets () indicate words rephrased to clarify the meaning (e.g. pronouns replaced by their referents).

2. *General Remarks*

It will be seen from the apparatus criticus to the text that the readings most often followed are those of ms. F, the oldest manuscript accessible and most probably an immediate copy of the autograph.[3] At the same time, this manuscript betrays signs of having been copied in some hurry and contains a fair number of omissions, as well as instances where one is led for various reasons to prefer the readings found in the other manuscripts.

Ms. M shows signs of having been copied by someone with little undertanding of what he was copying in the form of the significant number of its nonsensical readings (many of which were subsequently corrected in the magin by John Elijah Mellus, the copyist of V). In this light, the fact that this manuscript is not more corrupt than it actually is lends support to the pedigree it claims of being a copy of an immediate copy of the autograph. Where they make sense, therefore, the readings of this manuscript have been treated with due respect.

The manuscripts of the "eastern" group (WPLl) share a great number of errors in common, meaning that these errors must be blamed on their common ancestor rather than on the copyists of each of these manuscripts, who seem on the whole to have executed their work with care. Within the group, W stands somewhat apart from the others and not infrequently exhibits the correct readings where PLl go wrong but at the same contains errors of its own. Of the other three (PLl) none can be a direct descendant of another, although L and l show a certain affinity to each other. That the copyist of l made use of more than one exemplar is suggested by the occasional reference to variants in its margin. It is also worth recording that the manuscripts of this group frequently offer us better transcriptions of Greek proper names and technical terms than F and M.

[3] It is a matter of some regret that the efforts to gain access to another even older manuscript reported to be an immediate copy of the autograph (Syr. Orth. Patr. 6/2) have so far remained unsuccessful.

Ms. V represents a collation of mss. M and L. As such, its readings could have been omitted from the apparatus. These have, however, been retained partly for the demonstration of the relationship of this manuscript to M and L, and partly because they may be regarded as conjectures made by a scholar.

In a small number of instances I have been forced to adopt readings which are not found in any of the manuscripts. The most glaring instance occurs at the beginning of the Book of Meteorology, where four chapters are announced, whereas the book, in fact, contains five. The other instances mostly relate to congruence of grammatical gender and number. If one is correct in the view that the different families of manuscripts descend independently of each other from the autograph, these instances must be attributed to slips of the pen on the part of the author and so suggest that the author never had the opportunity to properly revise this work, which was, as we have seen, composed not so long before his depature from this world.

3. *On the Orthography of the Manuscripts*

Of the manuscripts used, Ll are fully pointed and vocalised, while WPV, too, are almost fully vocalised with the exception of some obvious words. In F and M vowel signs are occasionally used to avoid ambiguities and on some rarer, usually Greek, words.

In matters of orthography, ms. F has usually been taken as the guide, being the oldest manuscript accessible and one written in the script also used by the author.[4]

In view of the fact that the present edition is intended to open a series of partial editions of the *Butyrum sapientiae* using the same manuscripts, it was judged best to retain in the apparatus criticus those manuscript readings which may be of little value for the establishment of the text but are of interest for the establishment of the relationship between the manuscripts, as well as for our judgement of the quality of each manuscript. Not registered, however, in the apparatus are such instances as obvious cases of dittography, abbreviations such as ܐܝܬܘ̄

[4] The signs used for vocalisation in F are a mixture of the system using points and the "Greek" vowel signs (ܰ, ܳ etc.). As these latter signs could not be combined with the Estrangelo script in the computer font used, such signs have been "converted" into the point signs.

for ܐܢܬܘܢ and ܡܪ̄ for ܡܪܢܝܬܐ, and regularly occurring orthographical variants such as the following.

- ܪܒܝܐ and derivatives: invariably ܪܒܐ etc. in the East Syrian mss.
- ܐܣܛܪܐ (so F) and derivatives: usually ܐܣܛܝܪܐ etc. in other mss.
- ܐܣܟܡܐ (so F): usually ܐܣܟܝܡܐ in other mss.
- ܟ̈ܠܬܐ (so F): ܟܠܠ̈ܬܐ in other mss.
- ܐܘܪܫܠܐܡ: constantly written ܐܘܪܫܠܡ in lV (also λ)
- Final *yōḏ* in feminine imperfect verbs: usually omitted in the East Syrian mss.

4. *On the Lacuna in Min., Chapter III*

As has been noted in Introduction, Chapter 2 above, the manuscripts of the group WPLl share the major lacuna in Chapter Min. III. - In ms. L the whole of chapter III is missing and the following note is added at the end of chapter II (in rubric and partly illegible on microfilm):

ܕܟ ܕܥܠܝܗܘܢ ܡܠܬܐ ܒܬܪ ܦܘ̈ܠܚܢܐ ܡܠܬܐ {--} ܒܐܢܫܘܬܐ ܘܐܢܬܘܢ ܡܢ ܠܒܪ
ܠܡܫ̈ܒܚܢܐ ܡܫܘܕܥܢܐ ܐܢܫ ܕܟܠ ܡܫܠܡܢܘܬܐ

The extent of the lacuna is the same in mss. W and l, namely from III.i.1.2 ܐܝܟܢܐ ܗܘ to III.iii.1.1 ܡܪܢܝܬܐ. - There is no note explaining this in W. - In l, the following note is found in the margin after III.i.1.2 (fol. 361r):

ܕܟ ܠܝ ܕܡܢ ܡܢܝܢܐ ܕܦܘܠܚܢܐ ܐܝܟ ܡܢܗ ܩܕܡ ܕܐܝܬ ܐܝܬܘܗܝ ܐܝܟ ܕܐܢܫܘܬܐ ܩܕܡ

and the following note at the end of chapter III:

ܗܕܐ ܕܢܦܩܬ. ܕܟ ܕܐܝܬܝܗ ܕܡܢܗ ܡܩܒܠܝܢ ܠܐ ܐܫܟܚܬ ܒܐܢܫܘܬܐ ܘܚܠܦܝܗ
ܠܐ ܫܠܡܬ

The original extent of the lacuna in the fragmentary manuscript P is unknown, since the part where the lacuna occurred is itself lost. The surviving part of the manuscript resumes at III.iii.3.10 ܐܣܟܠܝܡܐ and the following note (same as in l, except for the omission of the word ܒܐܢܫܘܬܐ) is found at the end of chapter III:

ܗܕܐ ܕܢܦܩܬ. ܕܟ ܕܐܝܬܝܗ ܕܡܢܗ ܡܩܒܠܝܢ ܠܐ ܐܫܟܚܬ ܘܚܠܦܝܗ ܠܐ ܫܠܡܬ.

[tit.]

ܟܬܒܐ ܕܡܝܛܐܠ̈ܝܣܘ .
ܕܐܬܦܠܓ ܐܡܝܢܐ ܕܒܢ̈ܝܬܐ ܡܢ ܟܬܒܐ ܕܚܐܘܬ ܚܟܡܬܐ . ܘܗܘ ܐܝܬ
ܦܣܘ̈ܩܐ ܬܠܬܐ .
ܦܣܘܩܐ ܩܕܡܝܐ .
ܡܛܠ ܦܘܪܫܐ ܘܡܬܚܝܐ . ܘܗܘ ܐܝܬ ܦܣܘܩ̈ܐ ܬܠܬܬܐ .
ܦܣܘܩܐ ܩܕܡܝܐ .
ܡܛܠ ܗܘܝܐ ܕܟܐܦ̈ܐ . ܘܬܪܐܘܢܘܗܝ ܫܥܐ .

[I.i.1]
ܡܬܝܕܥܬܐ . ܠܡܪܟܒܐ ܚܝܠܢܐ ܕܒܢ̈ܝܬܐ ܟܕ ܐܝܟ ܕܗܘܐ ܢܟܠܐ ܕܡܬܬܚܕܬܢܘܬܐ ܡܬܝܕܥܬܐ ܒܟܠܡܕܡ ܕܒܢܝܐ ܐܝܟܢܐ . ܘܠܡܪܟܒ ܒܥܘܡܩܐ ܠܐ ܡܬܚܫܟܢܐ ܘܡܬܚܫܟܢ ܒܝܬܘܬܐ ܕܪܕܝܢ . ܐܝܟܢ ܡܟܠܬܐ ܕܟܠ ܗܘܢܐ ܘܣܘܟܠܐ ܘܒܡܬܪܬܐ ܕܐܝܢܘ̈ܬܐ ܡܬܝܕܥܬܐ ܘܫܥܒܘܕܝܗܝܢ ܘܡܥܒܕܢܐ ܕܡܬܬܠܝܢ ܡܢܗܝܢ ܦܪܝܢ . ܡܬܚܦܝܢܘܬܐ ܦܐܝܐܬܐ ܕܡܬܬܠܡܢܘܬܐ ܠܘܬ ܢܟܠܐ ܕܡܪܟܒ ܕܡܝܛܐܠ̈ܝܣܘ ܐܝܟܢܐ ܕܒܥܘܡܩܐ ܡܬܚܫܟܢܐ ܢܟܐܐ ܠܝ . ܟܕ ܚܝܘܢ ܡܬܝܕܥ ܘܚܝܐ ܡܬܚܝܕܥ ܡܢ ܐܝܬܘܗܝ ܗܟܝܠ . ܘܗܘ ܟܬܒܐ ܗܢܐ . ܟܕ ܟܠܝܢ ܘܟܠ ܦܘܪܫܐ ܘܡܬܚܝܐ ܘܐܪ̈ܥܐ ܕܐܪܥܐ ܘܡܢܝܢܐ ܕܟܬܒܬܐ ܡܠܘ . ܒܥܘܡܩܘܢ ܐܬܚܕܝ . ܐܟܡܢܐ ܕܒܠܐ ܟܐܦܐ ܕܡܬܘܗܝ ܐܝܬ ܐܝܟܢܐ ܕܡܬܚܝܢܐ

[I.i.2]
ܕܬܬܝܕܥ . ܐܪܥܐ ܟܕ ܝܒܝܫܬܐ ܐܝܟ ܕܡܣܬܒܪܐ ܠܐ ܨܘܪܬܐ ܗܘܝܐ ܒܐܦܐ . ܟܕ ܗܘܟܢܐ ܕܡܬܩܒܠܐ ܠܐ ܗܘܐ ܚܪܝܐ . ܐܠܐ ܦܘܪܟܢܐ ܢܗܘܐ ܠܡܪܟܒܘܬܐ . ܗܘܝܢ ܕܝܢ ܒܟܐܦܐ ܟܕ ܙܢܝܢ . ܚܕ ܡܢ ܗܢܘܢ ܕܦܘܬܝ ܠܛܝܢܐ ܕܠܐܘܓܐ ܟܡܝܟܐ . ܐܟܡܢܐ ܕܒܟܐܦܐ ܕܦܬܝܐ ܝܒܝܫ . ܕܠܗܘܕ ܡܢ ܒܗ ܒܐܦܐ ܠܛܝܢܐ ܡܬܩܪܒ . ܡ ܡܬܬܚܬܡ . ܐܝܟܢܐ ܕܝܢ ܗܢܘܢ ܕܦܠܓܐ ܕܡܬܪܐ ܟܕ ܢܦܠܝܢ ܐܘ ܢܕܝܢ . ܦܠܓܝܢ ܕܝܢ ܡܪܐ ܐܘ ܒܝܒܫܐ ܡܬܦܠܓܝܢܘܬܐ ܕܐܪܥܐ ܗܘ ܕܟܠܝܗ ܢܦܠܝܢ ܐܘ ܒܟܝܢܐ ܐܪܥܢܐ ܕܢܡܘܣܢܘܬܐ ܟܠܒܐ ܒܗ . ܐܟܡܢܐ ܕܦܠܓܐ ܡܠܝܢܐ . ܐܘ ܡܬܬܪܘܬܐ . ܐܘ ܐܢܫܘܬܐ ܕܠܐ ܠܗ ܝܒܝܫܐ . ܗܢܘ ܗܘܝܐ ܪܒܝܢܐ ܕܟܕ ܦܠܝܓ ܝܬܝܪ ܒܗܘܢܘܢ ܒܝܒܘ ܠܛܝܢܐ ܕܝܒܝܫܘ ܕܪܟܝܢܐ . ܘܒܬܪ ܒܗܘܢ ܘܗܠܝܢ ܥܕܢܝܢ ܐܟܡܢܐ ܐܝܟ ܒܐܦܐ ܒܬܪ ܦܝܫܬܐ . ܘܐܝܟܢܐ ܡܢ ܒܫܘܠܬܐ ܠܗܘܬ ܒܗܝ ܐܟܡܢ ܠܪܟܢܐ ܫܪܬܐ ܕܟܠ ܚܝܕ ܡܠܝܠܝܢ ܡܬܚܬܢܐ ܗܘܬ . ܘܣܢܝܬ ܡܬܡ ܐܪܕܟܠܐ ܦܘܠܬܐ ܣܢܐ ܕܒܐܦܐ ܡܣܝܒܐ ܕܒܚܝܠܝ ܒܗ ܣܘܦܐ ܘܡܪܟܝܢܐ .

____

FMWLIV: tit.: 1 post ܕܡܝܛܐܠ̈ܝܣܘ: add. ܕܐܬܦܠܓ ܠܡܪܟܒܘܬܐ ܠܚܕܐ V || 2 ܕܡܬܚܝܐ ΛV || **I.i.1.**: 3 ܒܝܬܘܬܐ: ܝܬܘܬ Λ || 5 ܒ: ܒܗ ܒ W || ܕܡܬܠܡܢܘܬܐ: in marg. V || 6 ܡܬܝܕܥ (ܐ supra lin.) Λ || 7 ܡܢ: om. l || 7 ܟܬܒܐ: ܟܬܒܘܬܐ l || ܟܕ: ܟܕ l || 9 ܕܟܠܝܗ: ܟܠܗ lV || **I.i.2.**: 2 ܠܐ ܗܘܝܐ : ܠܐ ܗܘܐ MV; ܠܐ ܗܘ̈ܝܐ WL; ܠܐ ܗܘ̈ܝܢܐ l || 5 ܐܘ: ܐܘ l || 7 ܕܡܠܝܢܐ V || 11 ܐܪܕܟܠܐ F

THE BOOK OF MINERALOGY

Which is the fourth [book] on the natural sciences in the Book of the Cream of Wisdom. It contains five chapters.

FIRST CHAPTER

ON MOUNTAINS AND SPRINGS

It contains three sections.

First Section: On Formation of Stones. Five theories.

[I.i.1.]
First [theory]: Since to [the discussion of] matters common to [all] natural things we have duly added the science of the primary movements in the world of nature, accurately set down the matters of incorruptible and corruptible bodies and then completed the discourse on generation and corruption, the actions and passions of the primary qualities and the mixtures generated from them, the proper order [for the exposition] of the teaching draws us on to the science of the μεταλλικοι, i.e. mineral bodies. For these come first and closest to the elements. Although this book will teach about them [i.e. minerals] and [also] about mountains, springs, earthquakes and the position of the habitable world, it is called by their name, just as there are [other] cases where the whole is called by the name of its part.

[I.i.2.]
Second [theory]: Genuine earth does not usually turn into stone, because the abundance of its dryness imparts not cohesion but crumbling to its parts. Stones are formed in two ways: (i) firstly, by way of baking of glutinous clay in the sun - as we see in the case of soft stones, which are intermediate between stone and clay at first, and later become hard - and (ii) secondly, by way of congelation of water, as it drips or pours down. Water congeals either through the mineralising power of the earth on which it falls, or through an earthy nature in which heat predominates - this is how salt congeals - or [in which] coldness [predominates], or a property which is unknown to us. The Princely Doctor [i.e. Ibn Sīnā] has recounted how, when [he was] a child, he saw on the banks of the [river] Gihon clay [used for] washing the head. Twenty-three years later he found it in the form of a stone, though porous. I [too] in my childhood went with my parents to a new monastery which was being built beside Melitene, and saw in front of the builders a single block of hard stone in which (pieces of) pottery and coal were mixed.

[I.i.3.]

ܕܬܠܬ. ܐܝܬ ܠܗ ܕܡܟܝܟܐ ܠܗܘ ܕܢܣܒܬܐ ܡܟܝܟ ܘܡܬܡܟܟܬܐ ܟܕܒܘ ܕܐܝܟ ܕܐܝܟ ܒܐܦܐ
ܦܩܕ. ܟܕ ܡܬܒܟܝܢܐ ܕܡܬܒܟܝܐ ܠܟܐܦܘܫܘܬܐ ܟܕ ܐܬܟܢܫܘܬܐ ܕܐܝܬ ܒܗܘܢ
ܠܐ ܗܘܝܐ. ܐܠܐ ܡܠܘܗܝ ܦܬܓܡܐ ܡܢ ܡܟܢܫ ܡܬܐ ܦܥܠܬܐ ܠܟܐܦܘܫܘܬܐ.
ܘܡܬܪܐ ܗܢ ܗܘܐ ܠܗ ܒܐܢܫܐ ܟܕ ܡܡܫܐ ܕܒܠܐ ܡܬܢܠܐܦܐ ܟܡܕܡ ܡܢ ܦܢܝܬܐ. ܐܬ
ܟܕ ܚܘܪܐ ܕܡܬܟܪܟܘܬܐ ܕܬܐܟܐ ܐܬܚܐ ܡܬܥܠ ܒܝܬܐ ܚܟܝܡܐ. ܡܬܦܫܛ
ܠܟܝܒܟܐ ܕܚܠܝܐ ܡܫܬܡܚܝܢ. ܗܘ ܕܝܢ ܟܕ ܗܘ ܡܬܬܐ ܕܐܬܬܐ ܗܘܟܐ ܕܬܚܐ
ܐܡܪ ܕܒܬܪܐ ܠܗ ܚܣܝܪܘܬܐ ܗܘܐ ܐܝܟ ܚܣܝܪܘܬܐ ܗܠܝܢ ܕܐܝܬܝܗ ܡܬܦܐܣ ܡܬܚܫܚܝܢ
ܘܡܬܝܕܥ ܚܣܝܪܘܬܐ ܠܚܒܝܒ ܠܗܘܢ. ܕܐܝܟ ܟܐܦܐ ܗܘܬ. ܘܚܘܬܗ ܠܐ ܐܬܚܙܝܢ.
ܘܟܕ ܡܢ ܚܒܝܒܗ ܫܠܡܐ ܕܡܢ ܬܘܕܝܬܐ ܒܗܝ ܗܘܝ.

[I.i.4]

ܕܐܝܬܒܟ. ܐܡܪ ܟܕ ܕܝܢ ܕܐܘ ܒܡܫܡܠܝܘܬܐ ܕܡܬܐܡܪܝܢ ܡܬܦܫܛܐ ܡܛܠ
ܕܐܬܚܘܝ ܒܐܦܝܐ. ܘܠܐ ܕܗܘ ܒܒܝܬܗ. ܘܡܬܪܐ ܗܢ ܗܘܐ ܟܕ ܢܦܫܘܬܐ
ܕܡܬܥܠ ܕܟܒܐ. ܘܡܬܚܬܐ ܘܡܫܬܚܐ ܗܘܝܐ. ܘܡܫܟܚܐ ܗܢ ܗܠܝܢ ܕܒܬܪܐ ܠܗ
ܬܠܬܝܢ ܒܐܦܬܐ ܕܡܬܒܢܝܐ ܕܢܦܠ ܡܢ ܡܠܟܘܬ ܦܪܝܫܬܐ ܕܬܚܘܬ ܡܟܢܐ ܘܡܬܡܪ
ܡܠܝܢ ܘܡܬܒܪ ܒܐܬܪܐ. ܟ ܥܒܕ ܡܬܢܗ ܗܬܬܢ ܠܒܝܬ ܐܒܟܝܢܐ ܕܫܘܪܐ ܐܘܡܢܐ
ܡܢ ܐܘܡܬܐ ܕܡܬܬܠܐ ܒܗ. ܐܝܟܐ ܢܦܠ ܟܠ ܐܬܪܐ. ܘܠܐ ܕܢܠܐ ܡܢܗ
ܐܬܬܡܟܟ. ܘܠܡܢܘܢ ܐܬܦܢܝܘ ܩܕܡ ܐܢܬܐ ܠܦܘܩܕܢܐ ܡܢܗ ܟܠܦܪܝܘܗܝ ܡܢܝܬܐ
ܠܚܪܬܐ. ܘܐܘܦܠܐ ܒܝܘܡܗ ܠܡܬܟܝܢ. ܘܦܥܡ ܠܡܬܚܦܟܐ ܡܢܗ ܡܘܡܐ. ܘܟܕ
ܡܢܘܬܗ ܠܐ ܐܬܬܟܝܘ ܐܦܩܢܐ ܠܡܬܟܠܠܘܬܗ. ܘܗܘ ܦܬܐܠܐ ܡܢ ܦܬܓܡܐ
ܟܠܠܬܐ ܠܚܪܬܐ ܐܝܟ ܕܡܫܟܐ ܕܡܛܠ ܫܒܝܚܘܬ ܕܟܝܢ ܒܡܬܐ ܡܠܟܐ
ܗܘܐ. ܘܐܬܡܢܝܐ ܕܡܬܦܢܐ ܕܢܟܝܐ ܒܡܬܢܝܐ ܡܢ ܕܐܝܬ ܗܢܐ ܗܘܐ ܐܘܡܢܐ
ܡܬܬܗܦܟ. ܘܐܬܕܟܪ ܐܢܘܢ ܦܩܐܢܠܐ ܒܦܩܐܢܟܠܝܘܢ.

[I.i.5.]

ܕܡܬܟܢܫ. ܐܝܟ ܠܟܫܦܪܐ ܢܗܪܐ ܐܘܦܐ ܐܡܪ ܕܝܢ ܕܡܦܩ ܕܝܢ ܒܟܐܦܐ ܘܐܦ
ܒܚܟܡܬܐ ܕܐܠܗܐ ܘܡܫܪܝܬܐ ܕܐܝܠܝܢ ܕܐܝܬ ܠܗܘܢ ܢܟܝܐ ܕܦܘܪܥܢܐ ܘܐܡܪ
ܫܪܝܬܐ. ܘܡܫܬܠܡܘܬܐ ܕܐܠܗܐ. ܘܡܬܟܪܝܢ ܚܘܪܐ ܕܚܝܠܐ ܡܢ ܥܒܕܐ
ܕܐܠܗܘܬܐ ܕܡܠܬܐ ܢܦܩܢ ܟܫܝܪܘܬܐ. ܡܫܬܠܡܐ ܕܝܢ ܠܟܐܦܐ ܐܦ
ܠܚܒܝܒܘܬܐ ܘܡܬܐܠܝܢ ܐܘܪܚܬ ܦܩܕܢܐ. ܘܫܠܡܐ ܠܡܬܕܡܪܘܢ ܐܘܪܚܬ

FMWLlV: **I.i.3**: 2 ܒܕ: ܒܕ ܟܕ L; ܟܕ 1 || ܠܟܐܦܘܫܘܬܐ: ܠܟܐܦܘܫܘܬܐ L || 2-3 ܟܕ ܐܬܟܢܫܘܬܐ . . . ܠܟܐܦܘܫܘܬܐ: om. Λ || 4 ܗܘܐ F || 5 ܒܝܬܐ: ܒܝܬܐ Ll || 6 ܕܐܬܐܬܐ: ܕܐܬܐܬܐ 1 || 7 ܕܒܬܪܐ: ܕܒܬܪܐ Λ || ܕܐܘܟܐ L || 8 ܘܡܬܝܕܥ ܚܣܝܪܘܬܐ F; ܘܡܬܝܕܥ ܚܣܝܪܘܬܐ W; ܘܡܬܝܕܥ ܚܣܝܪܘܬܐ Ll || **I.i.4**: 2 ܘܠܐ: ܠܐ ܘܠܐ MΛV; ܕܠܐ Nic. || 4 ܕܡܬܒܢܝܐ F; ܕܡܬܒܢܝܐ (sine *syāmē*) M; ܕܡܬܒܢܝܐ V || ܢܦܠ: ܢܦܠ V || 5 ܡܢܗ: om. Λ || ܗܬܬܢ Λ || 6 ܒܗ: ܒܗ Λ || 7 ܟܠܦܪܝܘܗܝ Λ || 8 ܘܐܘܦܠܐ Ll || ܠܡܬܟܝܢ: ܠܡܬܟܝܢ Λ || ܡܢܗ: ܡܢܗ MV || 9 ܡܢܘܬܗ MΛV || ܐܬܬܟܝܘ: ܡܬܟ F; ܐܬܟܝܘ V || ܠܡܬܟܠܠܘܬܗ MΛV || 10 ܒܡܬܐ: ܠܡܬܐ Λ || 11 ܗܢܐ: ܗܠܝܢ Λ || 12 ܦܩܐܢܠܐ V || ܒܦܩܐܢܟܠܝܘܢ V || **I.i.5**: ܕܡܦܩ ܕܝܢ MΛV || 2-3 ܘܐܡܪ ܫܪܝܬܐ . . . ܕܐܠܗܐ: in marg V || 3 post ܫܪܝܬܐ: add. ܗܘܐ V || 3 ܚܘܪܐ: ܚܘܪ MΛV || 4 ܕܐܠܗܘܬܐ F ||

[I.i.3.]
Third [theory]: It is probably true that certain animals and plants have turned into stone in various places, because the transformation of compounds into a stony state through the agency of the earthiness which they contain is not difficult but is in fact easier than the transformation of pure water into a stony state. This happens through the severity of the mineralising force in certain regions, or [it is] through the intensity of the heat which the earth suddenly experiences during a severe earthquake [that] bodies which are found on it become hardened. The same [teacher], the most excellent of the moderns, the Princely Doctor [Ibn Sīnā] has said that he [once] saw a loaf of bread, like those loaves whose middle the bakers make thin and which they imprint with images of animals. It was like a stone, but its colour had not been altered and on one of its sides were lines [of the kind] which are formed on them by the oven.

[I.i.4.]
Fourth [theory]: The Master [Aristotle] has said that in the fall of a thunderbolt, too, something which is stony and not oily in its nature is observed. This occurs due to a fieriness which is suddenly extinguished and becomes cold and dry. The Doctor [Ibn Sīnā] has recounted that his disciple saw in the land of the Gūzgānāyē [a piece of] iron which fell from the αἰθήρ [and] which weighed a hundred and twenty minae. It [first] sank into the earth, then rebounded twice from it in the way a ball rebounds off a wall which it strikes. It then fell to the earth and a frightening sound was heard because of it. Only with difficulty could the inhabitants cut off a small piece from it with a chisel. They brought and showed it to the governor and he ordered a sword to be made from it, but because of its hardness the smiths could not beat it. This iron was made of small round grains like millet which stuck very closely to each other. It is said that the finest swords in Yemen are made from such a substance and poets have commemorated them in their poems.

[I.i.5.]
Fifth [theory]: For further corroboration, the Master [Aristotle] has added that stones are also formed in the tendons of legs and joints of those who are suffering from gout and arthritis, in the bladders of children and in the eyelids [?] through the filtration of moistures which undergo solidification with heat. Also transformed into stone are glass and κοράλλιον, i.e. corals, for which reason (the latter) are called λιθόδενδρον,

ܐܝܠܢܐ ܕܟܐܦܐ ܡܬܩܪܝܢ . ܘܒܠܘܪܐܝܢܐ ܒܗ ܡܬܒܟܠ ܠܒܢܐ ܘܡܬܬܒܟܠ
ܒܢܝܐ ܚܒܝܪܐܝܬ ܗܘ̈ܝܢ ܫܘ̈ܒܐ . ܘܒܡܠܚܘܗܝ ܒܗ ܫܒܝܢ ܡܬ̈ܠܐ ܒܡܒ̈ܥܐ .
ܚܒܝܢ ܠܗܘܢ ܒܐܬ̈ܪܐ . ܘܒܟܐ̈ܦܐ ܕܡܢ ܗܠܝܢ ܡܬܩܢܝ̈ܢ . ܢܦ̈ܩܢ ܘܡܬܦܫܪ̈ܢ .
ܗܢܐ ܕܐܦ ܬܘܠܕܬܐ . ܕܐܝܠܢܐ ܒܗ ܒܢܘܪܐ ܗܘ̈ܝ . ܠܐ ܡܢ ܫ̈ܟܝܢ . ܘܡܢ ܒܬܪ
ܕܐܬܦܫܚ ܡܢܗ ܡܬܦܫܪ̈ܢ . ܒܟܠ ܚܠܦ ܕܠܐ ܩ̈ܠܝܢ ܗܘ̈ܝ ܡܢ ܡܕܡ ܠܚܪܬܐ ܩ̈ܠܝܢ .

[tit.]

ܦܣܘܩܐ ܬܪܝܢܐ .

ܡܛܠ ܗܘܝܐ ܕܛܘܪ̈ܐ ܘܡܒܘܥܘܬܗܘܢ . ܘܐܬܪ̈ܘܬܗܘܢ ܫܬܐ .

[I.ii.1.]

ܡܕܡܝܢܬܐ . ܒܟܐ̈ܦܐ ܕܝܢ ܪܘܪ̈ܒܬܐ ܕܛܘܪ̈ܐ . ܐܦ ܡܬܥܒܕ ܗܘܝܐ ܡܢ ܝܒܝܫܘܬܐ
ܕܡܬܚܠܛܐ ܒܠܝܫܐ ܡܬܥܒܕܐ ܘܛܝܢܐ . ܐܦ ܒܐܝܕܐ ܒܐܝܕܐ ܗܘܝܐ ܒܡܬܘܬܐ
ܕܛܝܢܐ ܡܬܩܫܝܢܘܬܐ ܕܫܡ̈ܫܐ ܠܚܝܠܝܐ . ܒܗܘ ܕܐܬܪܐ ܒܗ ܡܬܝܒܫ ܛܝܢܐ
ܡܬܐܡܪܐ . ܘܠܟܐ̈ܦܐ ܪ̈ܘܚܐ ܒܐܝܕܝܘ ܡܬܝܒܫ . ܒܡܬܚܠܠܘܬ ܛܝܢܐ ܡܬܥܒܕܐ ܡܢ
ܐܬܪܐ ܡܬܒܩܝܒ . ܘܒܗ ܛܝܢܐ ܡܬܩܫܝܐ . ܘܡܛܘ ܛܝܢܐ ܒܗ ܦܠܐܪܝܢ
ܒܥܠܘܗܝ ܕܐܬܪܐ ܕܒܗ . ܘܒܗ ܙܒܢܐ ܗܘ ܡܬܠܐܐ ܒܗ ܡܬܦܬܬܐ ܕܝܪ̈ܬܐ
ܗܘܝܢ . ܐܦ ܒܬܪ ܘܒܡܣܝܢܐ ܗܘ ܗܢܐ ܡܬܩܫܝܐ . ܐܡܟܢܐ ܕܐܝܬܝܗ ܒܟܠܝܗ ܒܗ
ܒܟܐ̈ܦܐ ܠܚܝܠܐ ܫܒܝܢ ܘܠܐ ܡܬܚܣܢ . ܒܪ̈ܘܚܐ ܕܬܘܡܪܐ ܒܡܛܪܐ ܕܝܘܡܐ ܪܒܝܢ .
ܘܒܗ ܙܒܢܐ ܡܬܩܫܝܐ ܐܪ̈ܒܥܬܐ ܒܢܝ̈ܢ ܪܒܝܢ . ܘܒܐܝܕܐ ܒܐܝܕܐ ܪܒܝܢ .
ܒܝܘܡܐ ܕܒܟܐ̈ܦܐ ܕܫܟܝܢ ܠܚܝܠܐ ܗܘ̈ܝܢ .

[I.ii.2.]

ܕܬܬܗܘܝܢ . ܒܠܒܬ ܪܘܡܐ ܕܛܘܪܐ . ܐܦ ܐܝܬܘܬܗ ܗܘ . ܐܡܟܢܐ ܕܒܢ̈ܝܢܐ ܡܬܝܢܐ
ܪܘܚܐ ܚܪ̈ܝܬܐ ܒܟܐ̈ܦܘܬܗܘܢ ܠܒܢܝܢܬܐ ܡܢ ܐܬܪܐ ܗܠܝܢ . ܘܡܬܥܒܕܐ ܒܡܛܪܬܗ
ܢܦܩܐ ܒܟ̈ܦܐ ܒܡܥܝܢܐ . ܐܦ ܒܪܥܝܢܬܐ ܗܘ . ܐܡܟܢܐ ܕܡܬܝ̈ܢܬ ܐܬܪܐ
ܡܬܩܢܝܢܬ ܘܡܢܥܐ ܢܦܩܝܢ . ܒܪܓ̈ܠܐ ܗܢܝܢ ܕܒܢܝܬܐ ܒܡܫ̈ܡܫܢܐ ܕܪܘܚܐ ܕܢܦܩܝܢ ܐܦ
ܡܬܝܐ ܕܢܦܩܢ ܡܬܩܫܝܢ . ܘܒܚܝܠܗܘܢ ܢܦܠܐ ܗܘܢܐ . ܘܗܢܝܢ ܒܟܐ̈ܦܐܝܢܝܬܐ
ܒܪܘܡܗܘܢ ܦܬܚ . ܘܕܒܝܠܗܘܢ ܛܘܪܐ ܗܘܢܐ . ܘܪ̈ܠܠܬܐ ܕܡܬܝܐ ܠܗܘܢ ܕܢܦܩܝܢܐ
ܒܗܘܢܐ ܡܬܩܫܝܢ . ܘܠܝܬ ܢܦܩܝܢܐ ܪܘܡܐ .

FMWLIV: **I.i.5**: 6 ܐܝܠ̈ܢܐ ܕܟܐܦܐ Λ; ܐܝܠܢܐ ܕܟܐ̈ܦܐ V || ܡܬܩܪܝܢ (f.) LlV || ܘܒܠܘܪܐ ܢܒܥܐ M; ܘܒܠܘܪ ܢ̈ܒܥܐ V || 7 ܫܘܒܐ Λ || ܘܒܡܠܚܘܗܝ F; ܘܒܡܠܚܗ W; ܘܒܡܠܚܗ Ll; ܘܒܡܠܚܘܗܝ V || 8 ܗܠܝܢ: ܐܝܠܝܢ Λ; ܐܝܠܝܢ V || ܢܩ̈ܫܐ Λ || 9 ܒܗ: ܒܝܕ W; ܒ Ll; ܕܒܗ V || 10 ܕܐܬܦܫܚ L; ܕܐܬܦܫܚ l || ܘܚܟ̈ F || ܡܢ ܡܕܡ: ܡܨܠܝܐ M[1], ܡܬܥܒܕ M[corr.] || **tit.**: 2 ܘܡܒܘܥܘܬܗܘܢ FLl (cf. IS منابع) || **I.ii.1**: 2 ܡܬܥܒܕܐ Λ (cf. IS مدد) || 3 ܛܝܢܐ: ܛܝܢܐ̈ MV || 4 ܒܡܬܚܠܠܘܬ l || 6 ܕܒܗܝܢ MV; ܕܒܗܝܢ Λ || 8 ܠܚܝܠܐ: in marg. V || ܒܪ̈ܘܚܐ: ܒܪܘܚܐ Ll || 9 ܒܗܘܢ: ܕܠܗܘܢ Λ || ܘܒܐܝܕܐ: ܘܐܝܕܐ V || 10 ܕܫܟܝܢ F || **I.ii.2**: 1 ܪܘܡܐ M || ܗܘ: ܗܝ M; om. V || ܕܒܢ̈ܝܢܐ: ܕܒܢܝܢܐ V || 2 ܘܡܬܥܒܕܐ: ܘܡܬܥܒܕ ΛV || 4 ܡܬܩܢܝܢܬ Λ || ܒܡܫ̈ܡܫܢܐ MV || 5 ܕܢܦܩܢ: ܕܢܦܩܝܢ Ll; ܕܢܦܩܢ W et l marg. || 5-6 ܢܦܠܐ . . . ܦܬܚ: om. Λ; in marg. V || 7 ܪܘܡܐ: ܒܗܘܢܐ Λ ||

i.e. tree of stone. In Tyrrhenia when bricks are formed and are immersed in the sea, they quickly turn into rocks. On Cnidus when they throw lumps of earth into springs, (the springs) turn them into stones. Stones which are discharged from Etna burn and melt. This is in fact surprising. How [is it possible] that when they were in [the midst of] fire, they were not affected at all, but melt after they have been driven out from it? Perhaps they later gain a property which they did not possess before.

Second Section: On Formation of Mountains and their Uses. Six theories

[I.ii.1.]
First [theory]: Large stones (which make up) mountains are either formed suddenly by heat which comes upon a large quantity of glutinous clay, or are formed gradually over intervals of countless years. [This happens] because - since the earth is situated amid water and winds are constantly causing the water to move - particles of earth are made to move with the fluctuation of the water and become mixed with water. When the water [then] solidifies, (the particles) adhere to the surface of the earth. As this mass grows with time, islands are formed. This [process] can in fact also be observed in an experiment. For example, when Babylonians require stones for building but cannot find [them], they throw seeds of dates in running water. With time earthy particles adhere to them and gradually grow in size, until stones suitable for building work are formed.

[I.ii.2.]
Second [theory]: The cause of the elevation of mountains is either (i) essential: as when in severe earthquakes the strong winds which cause them lift a part of the earth and suddenly create a deep valley in its side - or (ii) accidental: as when parts of the earth differ in their constitution, so that the soft (parts) are excavated by blasts of tearing winds or [by] furrowing waters, and the place they [originally occupied] becomes a valley, while the stony (parts) remain in their elevated position and the place they [occupy] becomes a mountain. Torrents of water [then] increase the depth of that (part) which has been excavated and the elevation of that (part) which has not been excavated.

[I.ii.3.]

ܕܬܬܠܬ. ܢܘܪܐ ܗܿܝ ܚܙܝܢ ܠܘܬ ܟܠܢܫܐ ܕܠܫ̈ܡܐ ܢܘܪܢܝܐ ܗܘܘ ܕܟܕ ܡܬܚܫܒܝܢ
ܬܚܫܒܬܗܘܢ ܡܢ ܚܫܝܫܘܬܐ ܡܣܬܠܩܝܢ. ܘܡܫܪܪܘܬ ܐܚܪܬ ܢܘܪܢܝܐ ܟܕ
ܡܬܠܒܟܝܢ. ܗܠܝܢ ܠܟܝ̈ܢܐ ܗܘܝܢ. ܘܟܠܡܕܡ ܗܟܢܐ ܕܟܝ̈ܢܐ ܠܟܠ ܡܢ ܢܘܪܐ
ܡܬܦܠ̈ܓܢ. ܘܡܬܢܣܒܢ ܥܠ ܦܘ̈ܚܡܐ ܡܬܩܫܬܢ.

[I.ii.4.]

ܕܐܬܒܪܟ. ܢܦܫܢ ܕܝܢ ܢܘܪܐ ܘܠܘܬ ܝ̈ܕܪܐ ܕܬ̈ܚܠܬܐ ܕܡܢ ܟܝ̈ܢܬܐ ܗܘܝܢ ܕܟܠܢܘܪܐ
ܡܬܦܠ̈ܓܢ. ܗܘ ܟܕ ܠܝܡܐ. ܗܢܐ ܕܟܠܗ ܢܘܪܐ ܡܬܚܫܒ ܘܟܕ ܡܬܪܒܝܢܘܬܗ
ܒܐܦܝ̈ܬܐ ܡܬܬܪܨܐ. ܘܗܕܝܐ ܗܘܐ. ܡܢܐ ܟܕ ܡܬܟܢܫܝܢ ܒܐܬܪܐ ܚܕ
ܘܦܠܝܓ ܗܘ ܚܫܝܫܘܬܐ ܕܐܝܟ ܗܘ ܟܠܗ ܢܘܪܐ ܡܬܚܫܒܐ ܘܠܐ ܡܬܦܫܛܐ
ܡܕܪܟܝܢ. ܢܦܫ ܠܗ ܠܐܪܥܐ ܘܟܝܢܐ ܕܢܦܫܗ ܠܟܠܗܝܢ. ܘܥܠ ܐܝܟܢܐ ܦܕܝܢ.
ܘܪ̈ܚܠܬܐ ܘܢܘܪܬܐ ܘܦܪܕܐ ܒܝ̈ܕܥܝܗܘܢ ܟ̈ܠܗܝܢ.

[I.ii.5.]

ܕܫܒܥ. ܗܟܢܐ ܕܟܝܢܬܐ ܡܢ ܢܘܪܐ ܘܡܫܩ̈ܠܝܗܘܢ ܘܡܢ ܐܪ̈ܒܥܬܐ ܡܬܝ̈ܬܐ ܦܪܝܢ.
ܟܕ ܒܗܘܢ ܦܫܝܬܐ ܠܐ ܡܬܚܫܒܝܢ. ܐܠܐ ܡܠܘ ܡܬܦܫܩܝܢ ܠܫ̈ܡܐ.
ܘܢܘܪ̈ܬܐ ܠܟܝ̈ܢܐ ܒܟܠܗܘܢ ܡܢ ܟܝ̈ܢܬܐ ܢܘܪܢܝܬܐ ܢܬ̈ܝܢ. ܘܡܢ ܗܪܟܐ
ܡܠܬܐ ܚܠܝܨܘܬܐ ܐܝܬܝܗ. ܕܐܬܐܡܪ ܢܘܪܐ ܥܠ ܡܕ̈ܝܢܬܐ ܕܡܝ̈ܐ ܩܕܝܡ. ܘܐܝܟ
ܕܐܡܪ ܠܡܐܡܪ. ܢܘܪܐ ܗܿܝ ܒܐܦܝܐ ܐܝܟ ܐܢܒܕܩܠܝܣ ܡܢܝܐ ܕܐܪ̈ܒܥܐ ܐܘ
ܠܡܘܡܢܘܬܐ. ܟܕ ܦܫܝܛܐ ܐܝܟ ܣܘܦܐ ܘܦܫܝܛܐ ܠܐ ܢܦܩ ܠܫܡܐ ܗܘܝܢܐ.
ܘܡܘܣܘܣܗ ܕܐܪ̈ܒܥܐ ܗܘ ܕܝܕܥܬ ܐܬܬܠܝܢ ܐܝܟ ܢܘܪܐ ܕܒܗ ܗܘܝܐ.
ܣܚܝܦܘܬ ܢܘܪܐ. ܘܚܠܝܨܘܬܐ ܐܝܟ ܕܩܕܡܝܬܐ ܐܣܛܘܟܣܐ ܕܐܝܬܝܗܘܢ ܗܘܝܢ
ܕܡܬܝܩܝܢ ܢܦܫ ܠܦܩܐ. ܘܠܡܬܚܡܐ ܘܢܘܪܬܐ ܐܝܟ ܒܢ̈ܝܐ ܡܡܠܠܢܘܬ
ܢܩ̈ܦܬܐ.

[I.ii.6.]

ܕܫܬ. ܒܦܫܩܐ ܡܬܦܫܩܝܢ ܦܘܫܩܐ ܥܠ ܠܫܡܐ ܕܡܬܦܠܠܝܢ ܗܘܠܝܐܬ
ܟܐܪ̈ܣܛܘܬܐ. ܘܡܬܚܙܝܢ ܢܘܗܪܐ ܠܟܝܢܐ ܒܩ̈ܢܘܢܐ ܕܒܗܘܢ ܗܘܐ ܢܘܦܠܐܝܘܢ
ܚܝܘܬܐ ܟܠܗܝܢ ܦܘܪܣܐ ܘܦܝܠܣܘܦܐ. ܠܝܬ ܠܗܘܢ ܐܝܟ ܢܘܪܐ. ܘܟܠܡܕܡ ܒܗܘܢ
ܡܬܠܟܝܢ ܗܘܠܟܗܘܢ. ܟܕ ܒܐܪ̈ܒܥܬܐ ܦܫܝܬܐ ܘܦܫܝܛܘܬܐ ܠܡܬܦܫܩܘ ܦܪܫ
ܪܒܘ ܕܟܝ̈ܢܐ ܗܘܝܢ ܕܒܗܝܢ ܡܬܡܠܠܝܢ ܡܬܦܠܓܢܘܬܗܘܢ ܗܘ ܕܒܗܝܢܘܬ
ܗܘܟܠܐ ܠܦܪ̈ܣܝܗܘܢ ܡܢ ܢܗܘܐ ܦܪܬܐ ܢܗܘܐ ܠܐ ܡܬܚܝܢ. ܐܝܬ ܕܝܢ ܐܦ
ܫܪ̈ܫܬܐ ܐܢܫܝܬܐ ܡܢܝܢܬܐ ܕܢܘܪܐ ܕܠܢܩ̈ܠܝܐ ܒܢ̈ܝܐ ܡܢܝܬܐ ܐܝܟ
ܐܝܣܘܬܐ ܘܦܘܪܫ ܠܝܬ ܦܪܣܝܗܘܢ.

FMWLlV: **I.ii.3**: 1-2 ܡܬܚܫܒܝܢ ܬܚܫܒܬܗܘܢ: om. F || **I.ii.4**: 1 ܗܕ̈ܪܐ F || ܕܡܢ: ܡܢ Λ || 4 ܘܡܢ: ܡܢ Ll || **I.ii.5**: 5 ܐܢܒܕܩܠܝܣ F; ܐܢܒܕܩܠܝܣ V || ܠܡܘܡܢܘܬܐ ܐܘ: ܘܠܡܘܡܢܘܬܐ ΛV || 6 ܘܡܘܣܘܣܐ V || 7 ܐܬܬܠܝܢ V || 8 ܐܣܛܘܟܣܝܐ: ܐܣܛܘܟܣܐ W; ܐܣܛܘܟܣܐ ܐܝܬ L; ܐܣܛܘܟܣܝܐ l; ܐܣܛܘܟܣܝܬܐ V || ܐܢܒܕܩܠܝܣ V || 9 ܒܢܩ̈ܦܬܐ Ll; ܒܢܩܦܬܐ W || ܡܡܠܠܢܘܬ MlV || **I.ii.6**: 3 ܠܗܘܢ: ܒܗܘܢ V || 4 ܗܘܠܟܗܘܢ V || ܘܦܝܠܣܘܦܐ: om. L; in marg. V || 7 ܡܢܝܢܬܐ V || 8 ܦܪܣܝܗܘܢ Fl ||

[I.ii.3.]
Third [theory]: Mountains are useful for detention of those moist vapours, which, being [at first] confined below them, are made to rise by heat and when condensed by the coldness of the mountain air become the material for clouds. For this reason the majority of clouds are generated above mountains and spread out from them on to the plains.

[I.ii.4.]
Fourth [theory]: Mountains are useful for the flow of streams which are born of springs on mountains. For vapour, when it is confined inside a mountain and is condensed by its stony coldness, turns into water. This water, when it gathers in one place, grows in quantity and is compressed by the heat - which is also confined inside the mountain and is not dissipated - tears the earth, gushes forth from it with force, flows over its surface and forms streams, rivers and [other forms of] flowing waters as it proceeds.

[I.ii.5.]
Fifth [theory]: The majority of springs flow out of mountains and their foothills, and from hard grounds - because on soft (grounds) vapours are not retained but are rather dissipated - and the best-known rivers in the world descend from mountains springs. Hence the old saying that the foundations of mountains are placed on cisterns of water. To use an illustration, a stony mountain is like a hard alembic [made] of iron or glass. - For porous [material] such as earthenware or wood would not retain much vapour. - The depth of the earth below the mountain is like the cucurbit, which the heat of the fire touches. Springs are like the tubular tails of the alembic from which the droplets fall. The seas and rivers are like the flasks which receive the drops.

[I.ii.6.]
Sixth [theory]: For mineral substances requiring [for their formation] vapours which are mixed to a high degree with an earthy substance and remain for a longer period in places where their thorough mingling takes place without [them being subjected to] scattering and dissipation, there is no (place) like mountains. For this reason the majority of them are formed in mountains. For in loose and porous terrains they cannot be confined over myriads of years during which that mixture of theirs which makes them ready for the reception of their forms from the giver of forms is completed.

There are also other particular uses of mountains, the exposition of which belongs to [the discussion of] particular [branches of] natural sciences, such as medicine.

[tit.]

ܦܘܣܩܐ ܬܠܝܬܝܐ.

ܡܛܠ ܡܫܒ̈ܚܐ. ܬܐܘܪ̈ܝܐܘܗ ܐܪܒܥ.

[I.iii.1.]

ܡܫܡܫܢܘܬܐ. ܗܟܝܠ ܕܝܢ ܡ̈ܢ ܛܝܒܘܬܐ ܕܪ̈ܕܝܢ ܡܢ ܐܐܪ ܗܘ ܕܟܫܘܪܘ ܕܐܪܥܐ ܠܦܐ̈ܪܐ ܘܫ̈ܠܝܐ ܕܥܒܘܕܗ ܚܐܠ. ܘܡܢ ܡܬܪܘܬܗ ܒܗ ܡܬܠܒܒ ܘܡܬܠܒܒܢ ܡܪ̈ܢܐ ܗܘܐ. ܘܦܠܓ ܡܪ̈ܢܐ ܒܚܐܦܐ ܡܬܕܒܝܢ ܘܠܐܪܥܐ ܢܦܩܝܢ. ܘܡܢܗ ܠܥܠܡ. ܘܒܥܠܡܐܐ ܢܦܩܬܐ ܕܦܝܫ ܠܗ ܡܬܚܠܦܐ ܡܬܕܒܪܢܘܬܗܘܢ. ܘܡܪ̈ܢ ܛܝܒܘܬܐ ܡܫܡ̈ܫܬܐ ܡܢ ܠܥ̈ܠܡܐ ܕܢܦܠ ܕܙ̈ܕܩ ܐܪܥܐ ܠܗ ܩܝܡ ܗܘܝܐ. ܘܟܢܫ ܕܡܠܐܐ ܢܦܩܬܐ ܠܘܬ ܠܗܘܢ. ܦܫܡ. ܘܠܡܕܪܐ ܥܠ ܐܦ̈ܝ ܐܪܥܐ ܠܗ ܡܩܒܠܝܢ. ܘܡܢ ܒܪ̈ܐ ܡܬܟ̈ܢܫ ܒܐܒ̈ܕܢܐ ܐܢܘܢ. ܘܡܢ ܠܐܪܥܐ ܠܗ ܡܢܝܢ ܢܒܠܐ ܠܓܝܪ̈ܐ. ܠܡܕܝܢܐ ܢܘܪܐ ܕܚܡܐ ܡܢ ܐܩ̈ܢܘܡܝܗܘܢ ܐܩܢܘܡܝܐ ܡܬܚܠܦܝܢ. ܘܠܡܕܒܪܐ ܡܢܗܘܢ ܡܬܕܒܪܝܢ.

[I.iii.2.]

ܕܬܬܚܝܢ. ܒܡܫ̈ܚܬܐ ܕܫܡܫܐ ܡܢ ܒܪ̈ܐ ܠܒܪ̈ܐ ܥܒܕܢܐ ܒܗ ܐܪܥܐ ܬܬܒܝܢ ܐܝܟܢܐ. ܘܥܠ ܐܦ̈ܝ ܐܪܥܬܐ ܡܬܦܬܚܬܐ ܒܡܥܠܢ̈ܐ ܠܡܛܪ̈ܐ ܡܦܩܝܢ. ܘܡܢ̈ܐ ܬܪ̈ܝܢܐ ܡܢ ܠܡ̈ܥܠܐ ܗܟܝܢ ܕܡܠܐܐ ܡܫܟܚܐ ܩܝܡ. ܘܡܢ ܗܘ ܕܒܚܐܦܐ ܠܬܪ̈ܥܐ ܠܐܪܥܐ ܘܢܕܪܘܢ ܗܦܟ ܒܢܠܗܘܢ. ܬܦܠܐ ܗܘ ܒܢ̈ܐ ܐܪ̈ܒܥܬܗܘܢ. ܘܒܠܡܙܐ ܙܙܝܗ ܕ̈ܪܢܗܘܢ. ܘܦܫܟܐܐ ܕܗܘ ܡܢܐ ܕܡܬܠܒܟ ܡܬܦܫܫ ܡܢܗܘܢ. ܘܒ̈ܪܐ ܒܐܪ̈ܥܬܐ ܝܠܝ̈ܘܬܐ ܐܘ ܢܠܝ̈ܘܬܐ ܕܝܠܝܢ ܡܬܕ̈ܚܝܢ. ܘܡܫܩ̈ܠܢܘܬܐ ܠܗ ܡܬܕ̈ܚܝܢ. ܐܠܐ ܡܢ ܒܬܪ ܕܠܝܠܝܘܬܐ ܥܒܪ. ܒܗ ܢܦܫ ܡܬܕ̈ܟܪܝܢ. ܘܒܠܡܙܐ ܢܦܘܪܐ ܒܕܘܟ ܡܢ ܡܩܒܠܝܢ ܡܪ̈ܐ. ܒܕܘܟ ܕܝܢ ܠܐ. ܘܒܕܘܟ ܡܢ ܢܦܫ ܕܡܒܝܢ ܒ̈ܪܐ. ܒܕܘܟ ܕܝܢ ܡܠܝܠ.

FMWLIV: **I.iii.1**: 1 ܡ̈ܢ: om. W || ܕܪܕܝܢ FMl; ܕܪ̈ܕܝܢ V || 2 ܚܐܠ scripsi: ܚܐܠܝ codd. || ܡܬܠܒܒ (Ethpe) V || ܘܡܬܠܒܒܢ: ܘܡܬܠܒܒܝܢ M || 3 ܘܒܥܠܡܐܐ V || 4 ܕܦܝܫ: ܕܦܪܫܐ M^{1} ܕܦܝܫ M^{V} || ܡܫܡ̈ܫܬܐ: ܡܫܪܪܐ M || ܠܥ̈ܠܡܐ W || 5 ܩܝܡ: ܦܝܡ M; ܡܢܝ V || ܕܡܠܐܐ F || 6 ܒܪܐ F || 7 post ܠܓܝܪ̈ܐ: add. ܐܦ in marg. M || 8 ܐܩ̈ܢܘܡܝܗܘܢ: ܐܩܢܘܡܝܗܘܢ M^{1} ܐܩ̈ܢܘܡܝܗܘܢ M^{V} || **I.iii.2**: 1-2 . . . ܥܒܕܢܐ ܐܝܟܢܐ: ܒܗ ܐܪܥܐ ܬܬܒܝܢ ܥܒܕܢܐ ܐܝܟܢܐ V || 2 ܐܝܟܢܐ FWl || 2 ܒܡܥܠܢ̈ܐ: ܘܡܥܠܢ̈ܐ MV; ܒܡܥܠܢܐ l || 2-3 ܘܡܢ̈ܐ ܬܪ̈ܝܢܐ Λ; ܘܡܢ ܬܪ̈ܝܢܐ V || 5 ܡܢܗܘܢ: fort. ܡܢܗ scribend. || ܘܒܪܐ F || 6 ܝܠܝ̈ܘܬܐ ΛV || 6-7 ܘܡܫܩ̈ܠܢܘܬܐ ܠܗ ܡܬܕ̈ܚܝܢ: in marg. l || 6 ܘܡܫܩ̈ܠܢܘܬܐ FV || 7 ܒܗ: ܒܗ Λ || 8 ܒܕܘܟ (sec.): ܕܘܟ V || ܕܝܢ: om. M^{1} add. M^{V} ||

Third Section: On Springs. Four theories

[I.iii.1.]
First [theory]: The water of flowing springs is generated from air which enters through openings in the earth into the clefts and channels inside (the earth) and, when it is condensed and made into vapour by (the earth's) coldness, turns into water. This water is propelled violently, tears the earth and bursts out from it. Because of the additional material which it acquires its flow is not obstructed. The water of stationary springs is generated from vapours which do not have the power to tear the earth. Because they do not have any additional material, it stands still and is unable to flow over the surface of the earth. The water of wells is procured artificially. Because (such water) is unable to tear the earth, engineers use their skills to remove the weight of the soil from the surface of (the water) and devise means to draw from it.

[I.iii.2.]
Second [theory]: People create passages inside the earth from well to well in high places and cause the water to come out on the surface of low grounds in canals. Seepage water is generated from vapours which have a large quantity of material, but whose power is too weak for them to tear the earth with force and to flow. The ground [from which] they [emerge] is soft, and for this reason their flow is dispersed and much of what is turned into vapour is dissipated away from them. Wells are found in clayey grounds or sandy (grounds) which are clayey, but are not found in rocky (grounds) except after they have been made clayey by being dug to a greater depth. For this reason diggers find water in one place but not in another; and in one place the wells are deep, in another less so.

[I.iii.3.]

ܕܬܠܬ . ܕܠܩܘܒܠܐ ܕܒܪ̈ܝܬܐ ܕܚܫ̈ܠ ܫܡܫ ܡܝ̈ܐ . ܘܠܐ ܡܢ ܪܢܝܬܐ ܒܡܝܢܬ
ܐܪܥܐ ܣܒܝܥܝܢ ܡܝ̈ܐ ܐܟܡܐ ܕܦܫܝܛܐ ܦܩܕܝܢ . ܢܒܥܐ ܡܢ ܗܘ ܕܐܡܝܢ
ܕܒܫܘܠܐܐ ܒܐܪܥܐ ܪܡܬܐ ܒܪ ܡܠܝܠ ܡܬܩܫܪܐ ܡܬܬܚܫܝܢ ܡܝ̈ܐ .
ܘܒܐܪܥܐ ܡܬܓܠܝܬܐ ܒܪ ܦܓܪ ܡܬܩܫܪܐ ܠܐ ܡܬܬܚܫܝܢ ܡܝ̈ܐ . ܘܐܠܐ ܐܬܪܐ
ܚܠܝܐ ܕܡܝ̈ܐ ܟܠܗ ܐܪܥܐ ܗܘܐ . ܒܪ ܒܘܣܡܐ ܛܘܒܐ ܡܬܬܚܫܝܢ ܗܘܘ ܒܗ̇ .
ܓܦܢܝܢ ܕܝܢ ܡܢ ܒܪ̈ܐ ܒܬܪܐܐ ܡܢ ܡܛܪ̈ܐ ܘܬܠ̈ܓܐ ܕܕ̈ܝܢ . ܘܒܡܛܪ̈ܐ ܘܩܘܪܘ
ܕܐܪܥܐ ܠܒܘܣܡܗ̇ ܢܣܒܝܢ .

[I.iii.4.]

ܕܐܪܒܥ . ܡܝ̈ܐ ܕܪ̈ܕܝܢ ܒܥܘܫ ܕܒܡܬܬܚܝܕܘܬܐ ܡܬܦܠܓܝܢ . ܢܦܫ ܡܢ ܟܠܗܘܢ
ܡܝ̈ܐ ܡܫܬܚܠܦܝܢ . ܐܦ ܒܢܝ̈ ܘܟܝ̈ܢܐ ܕܫܘܠܛܢܗܘܢ ܒܐܪܒܥܘܬܐ ܕܟܝܢܗ̇ ܡܬܦܠܓܐ
ܒܗܘܢ ܡܫܘܬܐ ܕܝ̈ܐ . ܘܡܢ̈ ܐ̈ܝܢܐ ܢܦܫ ܡܢ ܟܠ ܡܢ̈ܝ ܡܟܐܟܝܢ . ܒܗ̇
ܕܟܡܘܫ ܪ̈ܕܝܘܗܝ ܠܘܬ ܘܟܝܢܐ ܕܫܘܠܛܢܗܘܢ ܒܐܪܒܥܘܬܐ ܚܘܠܛܢܝܬ ܡܫܘܬܐ .
ܘܡܝ̈ܐ ܕܦܫܝܛܝܢ ܡܪ̈ܣܝܢ ܒܡܝܐ ܕܡܬܬ̈ܠܝܢ ܡܝ̈ܐ ܐܚ̈ܪܢܐ ܒܪܘܚܬܗܘܢ ܡܬܚܠܛܝܢ .
ܒܪ ܠܗܠܝܢ ܡܠܘܐܐ ܕܡܬܝܒܠܐ ܗܘ ܒܪ ܢܒܠܐ ܕܡܬܝܒܠܘܬ ܡܝ̈ܐ ܒܪܘܡܐ
ܠܬܚܘܡܐ ܡܕܡ ܡܬܚܫܡܐ ܡܝܐ . ܒܡܐ ܕܟ̈ܝܢܝܢ ܘܡܬܚܠܦܝܢ ܡܢ ܗܘ ܪܘܡܐ .
ܠܬܚܘܬ ܬܘܒ ܡܟܝܠ ܠܡܟܠܘܬܗܘܢ .

FMWLlV: I.iii.3: 1 ܕܠܩܘܒܠܐ scripsi: ܠܩܘܒܠܐ codd. (v. comm.) || ܚܫܡ Λ || 4 ܡܝ̈ܐ: om. V || 5 ܒܪ: ܒܝܕ ΛV || ܡܬܬܚܫܝܢ: ܡܬܚܫܚܝܢ F || **I.iii.4**: 1 ܕܪ̈ܕܝܢ V (cf. supra I.iii.1.1) || ܡܬܦܠܓܝܢ Ll || 2 ܘܟܝܢܐ Λ || 3 ܐ̈ܝܢܐ: ܐܝܢܐ MLlV (cf. supra I.iii.2.3) || ܡܟܐ̈ܟܝܢ W; ܡܟܐܟܝܢ Ll; ܡܟ̇ܐܟܝܢ V || 4 ܕܟܫܘܫܝܗ WM || 5 ܡܪ̈ܣܘܢ F || ܒܡܝܐ: ܘܡܝܐ M || ܕܡܬܬܠܝܢ MΛV || ܒܪ̈ܘܚܬܗܘܢ L; ܒܪ̈ܘܚܬܗܘܢ l || 6-7 ܡܝ̈ܐ . . . ܡܝܐ: in marg. M || 7 ܡܬܚܫܡܐ: ܡܬܚܫܡܐ LlV ||

[I.iii.3.]
Third [theory]: That vapours become water in places which are suitable and [that] water was not enclosed in the middle of the earth from the beginning as the simple-minded think is known from the fact that on high ground water is found when (the ground) has been dug a little, whereas on low ground water is not found [even] when it has been dug much. If the natural place of water were inside the earth, it would be found in it at the same depth. Well water overflows in spring because of rain and melting snow and descends through clefts and holes in the earth to the depth of (the earth).

[I.iii.4.]
Fourth [theory]: Running water, because it is refined by movement, is the best of all waters. Furthermore, the duration of its mingling with earthiness, from which decay is generated in it, is short. Seepage water is the worst of all waters, because due to the slow speed of its flow the duration of its mingling with the decay-generating earthiness is long. Whenever wells whose water is stationary are drawn upon, other water is attracted into its place. [This is] because the vapour [which is] the material of that spring - since it has the power to lift water up to a certain predetermined level - is able, whenever (the water) decreases and falls below that height, to raise it again to (that level).

[tit.]

ܩܦܠܐܘܢ ܩܕܡܝܐ .

ܡܛܠ ܙܘܥܐ ܕܐܪܥܐ . ܘܒܗ ܐܪܒܥ ܬܐܘܪ̈ܝܐ .

ܬܐܘܪܝܐ ܩܕܡܝܬܐ .

ܡܛܠ ܥܠ̈ܬܐ ܕܡܥܠ̈ܝܢܐ ܥܠ ܙܘܥܐ ܕܐܪܥܐ . ܘܐܝܬܝܗܝܢ ܐܪܒܥ .

[II.i.1.]

ܡܬܚܙܝܐ . ܙܘܥܐ ܐܝܬܘܗܝ ܙܘܥܐ ܕܓܙܪ ܠܡܢܬܐ ܡܢ ܐܪܥܐ ܒܝܕ ܥܠܬܐ ܕܡܬܬܚܒܫܐ ܒܘܠܗ̇ ܗܝ ܕܒܠܥܗ̇ ܡܩܒܠܐ . ܘܐܢܟܣܝܡܢܝܣ ܬܢܝ ܠܐܐܪ ܦܣܩ ܥܠܬܐ ܙܘܥܐ ܕܐܪܥܐ . ܘܐܡܪ ܕܥܠ ܕܦܩܘܥܐ ܐܪܥܐ ܡܢ ܐܐܪ ܡܬܬܠܝܢܐ . ܐܝܟ ܠܘܚܐ ܕܛܐܦ ܥܠ ܡܝ̈ܐ . ܘܡܛܠ ܕܡܬܚܙܝܐ ܡܫܬܚܠܦܘܬܐ ܕܐܪܥܐ ܦܫܝܫܐ ܘܡܬܩܪܩܪܐ . ܘܡܛܠܗ ܗܘܐ ܥܠܬܐ ܕܒܗ ܚܙܝܢܢ ܟܕ ܠܒܝܐ ܕܡܢ ܫܡܫܐ ܡܬܦܫܐ ܠܟܝܢܐ ܘܪܘܚܐ . ܡܢܐ ܕܡܛܠܬܐ ܡܢ ܐܐܪ ܒܗ ܢܦܠܐ . ܘܡܬܗܦܟ ܠܐܐܪ ܠܟܠ ܙܗܝܐ . ܟܕ ܪܗܛܐ ܗܢܘܢ ܕܠܬܚܬ ܟܐܦ̈ܐ . ܘܟܕ ܪܗܛܐ ܕܠܟܠ ܠܡܥܒܕ ܠܗ ܡܬܟܪܟܝܢ . ܘܟܠܗܝܢ ܡܬܦܫܐ ܠܬܚܬ ܘܡܬܬܙܝܥܐ .

[II.i.2.]

ܕܬܚܬܝܢ . ܕܝܡܘܩܪܛܝܣ ܕܝܢ ܠܡܝ̈ܐ ܦܣܩ ܥܠܬܗ ܙܘܥܐ ܕܐܪܥܐ . ܘܐܡܪ ܕܟܕ ܟܐܦ̈ܐ ܕܐܪܥܐ ܡܝ̈ܐ ܡܠܝܢ . ܘܟܕ ܢܦܩܝܢ ܡܛܪ̈ܐ ܐܘ ܡܢܗܘܢ ܠܡܥܡܩܘܬܗ ܡܬܦܬܚܝܢ . ܘܟܕ ܢܘܚܝܢ ܡܝ̈ܐ ܡܫܬܠܚܝܢ . ܘܡܛܠ ܕܠܐ ܟܬܝܒ ܠܗܘܢ ܒܟܐܦ̈ܐ ܕܐܪܥܐ . ܡܫܬܚܝܢ . ܘܟܕ ܐܝܟܢ ܠܡܛܪ̈ܐ . ܘܒܗ ܘܢܝܚܐ ܕܐܪܥܐ ܟܐܦ̈ܝܢ . ܘܐܡܪ ܗܘܐ ܕܐܦ ܟܕ ܡܬܢܓܕܐ ܗܘܐ ܐܪܥܐ ܗܘܐ ܢܘܚܐ . ܒܡܬܢܓܕܘܬܗ ܟܕ ܡܬܪ̈ܓܫܢܐ ܠܬܚܬܝܬܐ . ܘܠܘܬܗ ܢܦܩܐ ܠܡܝ̈ܐ . ܘܡܢܘܢ ܡܝ̈ܐ ܕܡܬܚܫܒܝܢ . ܡܢܐ ܕܒܟܐܦ̈ܝܢ ܢܦܠܝܢ . ܠܗ ܡܙܝܥܝܢ .

[II.i.3.]

ܕܬܠܬ . ܘܐܢܟܣܓܘܪܐܣ ܠܗ ܠܐܪܥܐ ܦܣܩ ܥܠܬܗ ܙܘܥܐ ܕܢܘܪܐ . ܘܐܡܪ ܕܢܘܪܐ ܠܬܚܬ ܟܕ ܢܦܠܐ ܐܘ ܡܬܚܬܐ ܡܢ ܡܛܪ̈ܐ ܐܘ ܡܬܬܟܒܫܐ ܡܢ ܒܢܝܢܐ ܡܬܠܐܝܐ ܕܐܪܥܐ . ܟܡܫܘܠܛܘܬ ܗܘ ܕܢܘܪܐ ܡܙܝܥ ܠܐܪܥܐ . ܘܟܠܗܝܢ ܬܪ̈ܬܝܢ ܡܠ̈ܝܢ ܘܬܪ̈ܬܝܢ ܕܒܠܚܘܕ ܡܬܟܐܪܐ ܗܟܢ ܙܘܥܐ .

FMWLIV: **tit.**: 4 ܕܡܥܠ̈ܝܢܐ: ܕܡܥܠܝܢܐ Λ || **II.i.1:** 1 ܐܝܬܘܗܝ: ܕܐܪܥܐ ΛV[1]; ܐܝܬܘܗܝ V[corr.] || 2 ܘܐܢܐܟܣܝܡܢܘܣ M; ܘܐܢܐܟܣܝܡܢܝܣ Λ; ܘܐܢܟܣܝܡܢܝܣ V || 3 ܕܦܩܘܥܐ: ܕܦܩܘܥܐ FMW || ܐܪܥܐ: ܕܐܪܥܐ FMLW || 4 ܦܫܝܫܐ: ܦܫܝܫܐ MV || 8 ܘܟܠܗܝܢ V || **II.i.2:** 1 ܕܐܪܥܐ: om. Λ || 2 post ܕܐܪܥܐ: add. ܡܠܝܢ V || ܡܛܪ̈ܐ F || ܠܡܥܡܩܘܬܗ W; ܠܡܥܡܩܘܬܗ L; ܠܡܥܡܩܘܬܗ I || 5 ܒܡܬܢܓܕܘܬܗ F || 6 ܡܬܪ̈ܓܫܢܐ F || ܘܡܢܘܢ ܡܝ̈ܐ: in marg. V || **II.i.3:** 1 ܘܐܢܟܣܓܘܪܐܣ V || 2 ܐܘ sec.: ܗܘ M; ܗܘ V ||

SECOND CHAPTER

ON EARTHQUAKES

It contains three sections.

First Section:

On the Opinions of the Ancients on Earthquakes. Four theories

[II.i.1.]

First [theory]: Earthquake is a movement which occurs to a part of the earth due to a cause which [itself] moves and causes what is on it to move.

Anaxagoras attributes the cause of the movement of the earth to air, and says: Because the earth is flat it is borne by air, like a leaf that is spread on water. [Furthermore] because the lower part of the earth is porous and loose, while this upper part of it, on which we live, is thick and dense on account of the clay which is hardened by the sun, when a part of the air falls into (the earth) and strives to rise upwards, it passes through those openings which are (in the part) below but cannot pass through the openings which are (in the part) above, and for this reason moves and shakes the earth.

[II.i.2.]

Second [theory]: Democritus attributes the cause of the movement of the earth to water and says that the bowels of the earth are filled with water. When rains descend, their waters too reach its depth and mingle with that water [already there]; (then), because the bowels of the earth cannot contain (the two bodies of water), they become compressed and as they squeeze each other they cause the movement and shaking of the earth. He says furthermore that earthquake also occurs when the earth is dry. For when it becomes dry, (the earth) yearns for moisture and attracts water towards itself. This water which is attracted, when it falls into the bowels of (the earth), moves (the earth).

[II.i.3.]

Third [theory]: Anaximenes attributes to the earth itself the cause of its own movement, and says that when tops of mountains fall, either loosened by rains or broken up by excessive dryness of the earth, they cause the earth to move by that mighty fall of theirs. For this reason earthquakes occur in rainy seasons and seasons of absence of rain.

[II.i.4.]

ܕܐܒܪܗܡ . ܠܡܬܚܫܒܬܐ ܡܢ ܡܕܒܪܢܘܬܐ ܡܦܝܣ ܗܘܐ ܘܐܡܪ . ܐܠܐ ܫܪܝܪܐ ܗܘܬ . ܠܐ
ܗܘܐ ܒܪ̈ܗܒܢܘܬܐ ܕܐܝܬ ܠܗܝܢ ܒܡܫܘܚܬܐ ܠܘܬ ܡܣܒܠ ܣܓܝܐ ܗܢܐ ܗܘܐ ܗܘ ܚܝܐ
ܕܠܐܝܬ . ܐܠܐ ܒܚܠܡܘܗܝ . ܒܪ ܚܘܒܝܬ ܢܣܒܗ ܠܡܠܬܐ ܥܠ ܠܒܗ ܐܒܪܗܡ .
ܘܬܘܒ ܐܦ ܠܒܗ ܐܡܪ ܒܪ ܠܚܝܝ ܠܐܒܪܗܡ ܠܐ ܡܫܟܚܐ ܕܢܚܫܘܗ . ܐܒܗܬܐ ܡܬܚܫܒܬܐ
ܡܢܗ ܡܬܬܚܝܗ . ܘܬܘܒ ܕܠܐ ܫܠܝܚܐ ܐܒܪܗܡ ܒܡܫܬܐ . ܐܠܐ ܐܣܛܪܛܝܐ ܡܕܡ
ܐܬܚܙܝ . ܘܠܬܪܝܢܘܬܐ ܗܘܬ ܒܪ ܫܪܝܐ ܐܡܪ . ܕܠܐ ܢܚܫܚܘܢ ܡܢܗ ܒܠܒܐܝܬ
ܒܐܒܪܗܡ . ܐܠܐ ܫܒܝܩ ܠܗ ܡܢܟܝܢ ܒܢܝܐ . ܘܠܡܫܡܠܝܘ ܕܬܠܝܬܝܘܬܐ ܗܘܬ ܒܪ ܐܡܪ
ܗܘ . ܐܠܐ ܡܘܢ ܗܘܐ ܫܪܒܐ . ܝܕܥ ܗܘܐ ܕܫܡܥܬܐ ܗܘܬ ܐܒܪܗܡ . ܘܡ̈ܠܝܐ
ܕܡܢܗ ܩܪܝܒ ܗܘܘ ܡܢ ܟܠܗ ܕܗܘܘ . ܘܠܣܘܪܝܐ ܩܠܝܠ ܗܘܘ . ܒܪ ܪܝܫܐ ܠܐܒܪܗܡ
ܒܡ̈ܠܐ ܕܡܢ ܚܠܡ ܦܪܨܘ ܢܦܠܘ . ܘܡܢܘ ܗܘܐ ܗܘܬ ܕܒܐܬܪܘܬܐ ܕܠܝܬ ܒܗܘܢ
ܠܐܒܪܗܡ ܐܢܫ ܡܢܕܥܢ ܠܗ ܗܘܐ ܗܘܐ ܚܝܐ .

[tit.]

ܡܐܡܪܐ ܬܪܝܢܐ .
ܡܛܠ ܛܠܝܘܬܗ ܫܚܝܡܬܐ ܕܚܝܐ . ܡܬܪܒܝܢܘܬܗ ܘܐܒܪܗܡ .

[II.ii.1.]

ܡܕܒܪܢܘܬܐ . ܗܢܘ ܕܝܢ ܐܡܪܢ ܕܐܒܪܗܡ ܡܢ ܒܪ ܡܬܬܠܡܕܐ ܘܫܐܪܝܐ ܡܢ ܫܡܫܬܐ ܡܕܡ
ܡܫܡܫܢܘܬܐ ܕܥܠܝܗ . ܚܠܦ ܐܢܫ ܗܢܝܐ ܕܡܬܚܫܒܐ ܒܫܘܐܐ ܡܬܚܙܐ ܒܗ
ܘܬܚܘܝܐ ܗܢܐ ܗܘܝܘ ܡܫܠܡܐ ܕܟܘܢܐ . ܘܗܘ ܗܢܐ ܒܪ ܠܐ ܡܫܟܚܐ ܠܗ ܐܣܟܝܡܐ
ܕܬܗܘܐ ܒܗ ܠܟܝܢܘܬܐ ܕܐܒܪܗܡ . ܡܝܬܪܐ ܠܗ . ܘܒܚܝ̈ܠܘܗܝ ܕܟܝܢܗ
ܚܒܪܐ ܡܬܚܫܒܢܐ ܕܟܝܢܗ ܕܢܦܫܐ ܕܡܬܚܫܒܐ ܒܡܐܪܢܘܬܐ ܘܠܡܫܟܢܗ ܢܦܫܐ
ܘܩܘܡܐ . ܘܠܐ ܡܕܡ ܒܪ ܐܝܬ ܕܡܫܟܚܐ ܕܠܐ ܠܡܟܠܘ ܠܐܒܪܗܡ ܒܪ ܠܡ̈ܫܡܫܢܐ
ܢܦܫܐ ܕܒܠܝܗ ܐܝܟ ܕܘܟܐ . ܠܐ ܡܢܐ ܘܠܐ ܐܚܪ ܐܒܠܐ ܡܬܝܢܬܐ ܕܡܢ ܐܒܪܗܡ
ܛܠܝܐ . ܐܠܐ ܢܦܫܐ ܒܠܚܘܕ . ܡܕܘܪܐ ܐܠܐ ܒܢܝܐ ܛܠܝܐ ܠܡܬܠܡܕܘ ܒܗ
ܐܒܪܗܡ ܠܐ ܡܫܟܚܐ ܥܠܝܐ . ܘܐܝܟܐ ܡܢ ܒܪ ܛܠܝܐ ܗܘ ܢܦܫܐ ܥܠܡܗ . ܡܬܕܪܐ
ܠܗ . ܒܡܫ ܕܥܝܢܐ ܡܢ ܠܐܡܪ ܕܐܬܚܙܝ ܡܢ ܦܘܪܫܢ ܘܐܝܠܝܢ ܡܬܕܪܐ ܠܗ .

[II.ii.2.]

ܕܬܬܚܙܐ . ܒܪ ܙܒܢܐ ܕܐܒܪܗܡ ܡܢ ܢܘܗܪܝܐ ܠܚܪܢ . ܟܕ ܕܝܢ ܫܡܥܬܐ .
ܢܘܗܪܝܐ ܡܢ ܐܒܪܗܡ ܕܒܪ ܦܠܚܐ ܗܘܐ ܒܡܕܝܢܬܐ ܠܚ̈ܒܪܘܗܝ ܒܡܫܡܫܢܘܬܗ
ܕܐܒܪܗܡ . ܒܝܢܐ ܗܘܐ ܪܘܚܐ ܕܡܫܡܫܐ ܠܗܝܢ ܟܠ ܐܝܟܘܬܗ . ܘܡܬܚܫܒܐ ܕܠܐ

FMWLΛV: II.i.4: 1 ܡܢ: om. F; ܕܡܢ MV || ܫܪܝܪܐ ܗܘ M^1; ܫܪܝܪܐ ܗܘܬ M^V || 2 ܠܗܝܢ: ܠܗܘܢ F || ܗܘ: ܗܘܐ $M^{corr.}$ (in marg); om. ΛV || 2-3 ܕܠܐܝܬ ܚܝܐ V || 3 ܕܠܐܝܬ: om. Λ || ܢܣܒܗ (impf.) F || 4 ܠܒܗ V || 5 ܡܬܬܚܝܗ M || 6 ܘܠܬܪܝܢܘܬܐ V || 7 ܕܬܠܝܬܝܘܬ W || ܒܪ om. F || **II.ii.1**: 1 ܡܕܡ: ܘ V || 5 ܒܚܒܪܐ: ܚܒܪܐ F || ܡܬܚܫܒܢܐ F || 7 ܡܕܡ: ܡܢ F || 9 ܗܘܐ M^1; ܘܩܘܡܐ M^V || ܡܬܕܪܐ: ܡܬܕܪܢܐ ut vid. F || 10 ܢܘܗܪܝܐ M^1; ܢܘܗܪܝܐ M^V || ܡܬܕܪܐ scripsi: ܡܬܕܪܐ codd. || **II.ii.2**: 2 ܒܡܫܡܫܢܘܬܗ W ||

[II.i.4.]
Fourth [theory]: The Master [Aristotle] refutes the first opinion and says: If this were true, earthquake would not occur under particular conditions in places which were liable to be affected in this way, but [would occur] everywhere. For (Anaxagoras) apportioned the cause to all the earth. Furthermore, if all the air, though it bears the earth, cannot move it, how does a part of it move it? Furthermore, it has already been demonstrated that the earth is not flat along [its] latitude but is spherical. Dismissing also the second [opinion], he says: We have already demonstrated that water is not enclosed *en masse* in the earth but surrounds it. Against the third [opinion] the Master says: If the matter were so, the earth ought to be level, and movements which occur [ought to] be lessening in comparison with those which occurred [in the past], and [ought to] be ceasing in the end, since the tops of mountains would already have fallen in movements since the beginning. Furthermore, earthquake ought not to occur in regions where there are no mountains, such as Egypt.

Second Section: On the True Cause of the Earthquake. Four theories.

[II.ii.1.]
First [theory]: The Master [Aristotle] says: When the earth is moistened and is [then] heated by the sun and by the heat which is inside (the earth), smoky exhalation, which is driven violently, is generated in it. This smoke is the material of wind and this (wind), when it cannot find a way [by which] to escape because of the density of the surface of the earth, moves (the earth), and is dissipated through the tears which it makes in (the earth), like the wind, which is generated in must, tears its cask and escapes. For there is nothing which has the power to stir the earth together with the heavy objects which are upon it like wind, neither water nor air nor a part which is torn from the earth, except fire alone; and it is clear that fire cannot be generated inside the earth except in a smoky form. Although the cause of wind is fiery smoke, it cools us because it pushes away from us the air which has been warmed by our bodies and makes other [air] approach us.

[II.ii.2.]
Second [theory]: With a movement of the earth there sometimes occur beneficial effects, sometimes harmful effects: beneficial, as when the wind, meeting with vapoury materials in the depth of the earth, pushes

ܐܬܬܣܝܡ ܗܘܐ ܡܟܒܫܐ . ܫܘܘܕܝܐ ܕܝܢ ܐܝܟܢܐ ܕܒܗ ܗܘ ܦܘܫܐ ܗܟܢ ܒܝܫܬܐ
ܬܚܡܐ . ܡܢ ܡܬܬܦܪܢܣܢܘܬܐ ܟܝܢܝܬܐ ܡܬܡܠܝܢܐ . ܐܝܟܢܐ ܕܡܬܕܒܪܢ ܬܢܝܢܐ
ܒܡܦܣܢܘܬܐ ܡܟܝܢܬܐ ܘܡܣܟܢܐܬܐ . ܘܗܟܢܐ ܢܘܪܐ ܡܘܡܪܢܝܬܐ ܡܢ ܠܗ ܐܦ ܐܝܬܝܐ
ܒܪܢܫܐ . ܡܠܦ ܕܝܢ ܒܗ ܟܠܗ ܙܘܥܐ ܗܟܢ ܡܠܦܢܘܬܐ ܗܘܝܐ . ܘܒܡܦܣܢܘܬܗ
ܠܐܪܥܐ ܢܦܠܐ . ܗ̇ܠܟ ܒܗܝ ܕܝܢ . ܕܗܘܬ ܡܛܐ ܒܗܢ . ܘܗ̇ܘ ܦܘܫܐ ܡܠܟܐ
ܕܦܫܛܐ ܒܗ ܢܘܪܐ ܐܝܟ ܟܠܝܪܢܝܬܐ ܐܣܘܬ . ܘܡܠܬܗ̇ ܠܡܕܢܚܬܐ
ܕܠܦܐܪ̈ܘܗܝ . ܡܚܝܐ ܕܝܢ ܗܘ̇ܐ ܘܐܟܠܢܐ ܕܢܘܪܐ ܒܡܠܦܢ .

[II.ii.3.]

ܕܬܠܬ . ܗܠܝܢ ܡܢ ܡ̈ܠܐ ܕܣ̈ܠܩܐ ܘܪܘܡܡܐ ܡܣܠ̈ܩܐ ܦܘܫܢ ܠܢܦܫܐ . ܐܦ ܒܗܘܢ
ܗ̇ܘܝܢ . ܗܢܘܢ ܕܝܢ ܐܝܟ ܥܠ ܡܬܩܢܢܘܬܐ . ܗܢܘܢ ܗܦܟ ܐܝܟ ܒܟܬܒܐ ܕܗܘܬܪܐ .
ܗܘܝܐ ܕܝܢ ܗܕܐ ܡܛܠ ܒܫܘܠܐ ܐܣܟܡܐ ܕܫܘܬܦܘܬܗ ܕܐܪܒܥܐ . ܐܦ ܚܕ ܐܝܟ ܚܕ ܒܗ
ܕܡܬܢܐ ܐܝܬܘܗܝ ܗܢܐ ܕܒܟܝܘܢܐ ܡܫܬܠܡܢܐ ܢܩܫ ܡ̈ܠܐ ܡܫܬܠܡܢܐ ܚܒܪ .
ܘܐܝܬ ܐܝܬܘܗܝ ܕܡܬܚܙܐ ܕܠܐ ܢܦܫܐ ܗ̇ܘܝܐ . ܒܗܝ ܕܒܗ ܦܠܓܐ ܕܦܘܫܐ ܡܢ ܫܪܪܐ
ܡܚܘܐ . ܦܢܝܐ ܦ̈ܘܫܐ ܡܡܫܚܐ . ܘܒܗܘܢ ܠܢܦܫܐ . ܘܠܐܝܬܐ ܠܗ ܡܘܬܪܐ .
ܒܟܠܗ ܕܝܢ ܠܗܘܢ ܡܬܒܥܝܢ ܒ ܒܢܦܫܐ . ܟܕ ܙܘܥܐ ܕܐܝܬ ܡܠܝܠܐ ܠܦܓܪܐ
ܕܐܝܬܐ ܕܢܦܫܐ ܦܪܘܩ . ܡܢ ܚܕ ܒܥܘܬܐ ܡܫܬܟܚܬܐ ܫܪܝܪܐ . ܐܝܟܢܐ ܕܟܢ
ܡܫܬܟܚܬܐ ܫܪܝܬܐ . ܘܟܠܡܐ ܒܗ ܕܟܢܐ ܦܪܩ ܒܝܬܐ ܗܘܐ . ܒܒܝܬܐ ܠܦܘܪܩ
ܡܬܒܥܝܢ . ܐܝܟܐ ܒܝܬܢܐ .

[II.ii.4.]

ܕܐܝܟܢ . ܕܦܘܫܐ ܢܒܝܥܬ ܒܐܝܬܐ ܐܝܬܘܗܝ ܟܠܗ ܙܘܥܐ ܕܐܝܬܐ . ܢܒܥܐ
ܡܢ ܗ̇ܝ ܕܒܐܝܬܘܬܐ ܗ̇ܠܝܢ ܕܒܗܘܢ ܢܦܫܐܝܬ ܗܘܝܢ ܙ̈ܘܥܐ . ܒܗ ܒܪܝܐ
ܡܫܬܟܚܬܐ ܡܬܚܫܚܢ . ܦܪ̈ܝܢ ܢܦ̈ܫܐ ܘܠܐ ܒܟܝܢܝܬ ܗ̇ܘܝܢ . ܒܗ ܡܬܩܢܝܐ
ܡܫܬܟܚܐ ܡܬܚܫܚܢܐ ܠܗ̇ ܦܘܫܐ . ܘܗܢܘܢ ܙ̈ܘܥܐ ܕܐܝܬܐ ܗܢܘܢ ܡܢ ܕܟܠܢܝܬ
ܘܐܬܚܫܚܢܝܬ ܗܘܝܢ . ܒܐܝܠܝܢ ܐܝܟ ܕܠܒܫܢܐ . ܗܢܘܢ ܕܝܢ ܢܦܫܢܝܬ
ܘܡܫܬܚܠܦܢܝܬ ܡܢ ܦܘܫܐ ܠܚܘܡܐ . ܚܫܝܫܝܢ ܕܝܢ ܗܠܝܢ ܡܢ ܗܢܘܢ ܕܠܒܫܢܐ .
ܟܕ ܡܢ ܐܘܣܝܗ ܕܐܝܬܐ ܕܠܒܫܢܝܬ ܗܘܐ ܚܠܝܐ . ܡܢ ܒܫܘܡܠܝܗ ܕܝܢ ܒܣܠܩܐ ܢܦܫ
ܬܫܘܒܚܬܐ ܘܒܢܝܢܐ ܗܟܢ ܒܪܢܫܐ . ܘܡܠܦܐ ܚܕܐ ܠܒܪܢܫܐ ܗܘܝܢ . ܘܒܡܫܘܚܬܗܘܢ
ܒܩܘ̈ܢܐ ܕܐܝܬܐ ܦܬܝܢ .

FMWLlV: II.ii.2: 5 ܗܢ: ܗܘܢ LlV || 9 ܒܗ: ܒܗ̇ F || ܟܠܝܪ̈ܢܝܬܐ FW || ܐܣܘܬ Ll || ܠܡܕܢܚܬܐ Ll || 10 ܘܐܦܪܘܗܝ ܕܠܒ̣ܘܟܝ F; ܘܐܦܪܘܗܝ ܕܠܦܐ M; ܘܐܦܪܘܗܝ ܕܠܦ̈ܐ V || ܐܟܠܢܐ ΛV || **II.ii.3**: 2 ܡܬܩܘܢܢܘܬܐ M || 3 ܕܫܘܬܦܘܬܗ M || 4 ܕܡܬܢܐ: ܗܘ ܡܬܢܐ L; ܡܬܢܐ Wl || 5 ܐܝܬ ΛV || 6 ܡܚܘܐ: ܕܐܝܬܐ Ll; ܕܐܝܬܐ ܡܚܘܐ V || ܦ̈ܘܫܐ: ܦܘܫܐ F || 8 ܕܢܦܫܐ: ܢܦܫܐ ΛV || ܡܢ ܒܥܘܬܐ ܒܢܝܢ VM^V || ܕܟܢ: ܕܝܢ Λ || **II.ii.4**: 4 ܐܝܬ ܒܠܚܘܕ Ll || 5 ܘܐܬܚܫܚܢܝܬ scripsi cum Nic. syr. 41.15 et *Cand.* 128.10: ܘܐܬܚܫܚܢܝܬ MΛV; ܘܚܫܚܢܝܬܐ F || ܕܠܒܫܢܐ M; ܠܒܫܢܐ Λ || ܗܢܘܢ M || ܢܦܫܢܝܬ W; ܢܦܫܢܝܬ LlV || 6 ܡܫܬܚܠܦܢܝܬ F; ܘܡܫܬܚܠܦܢܝܬ MWl; ܘܡܫܬܚܠܦܢܝܬ L || ܗܢܘܢ: om. Λ || ܠܒܫܢܐ W; ܠܒܫܢܐ Ll; ܕܠܒܫܢܐ V || 9 ܒܩܘ̈ܢܐ W; ܒܢ̈ܘܢܐ Ll ||

them with a great force, causes them to come out on its surface and causes springs which did not exist (before) to gush out; and harmful, as when the wind, being very dry, is ignited by a strong movement - just as smoke is ignited with strong and repeated blowing - and so inflaming fire rises from inside the earth, especially when the cause of the movement is very severe and tears the earth by its exit. The Master [Aristotle] relates that this happened once: Wind raised ash containing fire in the form of sparks and filled the city of the Liparians. In this way the craters of fire on Etna were also formed.

[II.ii.3.]
Third [theory]: Sometimes terrifying sounds and weak humming noises precede the earthquake or occur with it, sometimes like the sound of piping, and sometimes also like the bellowing of a bull. This occurs because of the variation in the shapes of the openings of the earth. For air too, being similar, when it strikes different objects, makes different sounds. There are times when the noise occurs without an earthquake, because when the wind escapes from the noise-making holes, they find wide cracks and come out through them but does not cause the earth to move. We sense the sound first, then the earthquake, since the movement of light air precedes the movement of earth which is heavy; for hearing is swifter than [the sense of] touch, just as vision [is swifter] than hearing, and for this reason, although thunder occurs before lightning, we sense the lightning first, then the thunder.

[II.ii.4]
Fourth [theory]: That wind confined in the earth is the cause of the movement of the earth is known from the fact that in places in which movements occur frequently, when many wells are dug, earthquakes become fewer and occur not severely, since the wind can (then) find many escape routes. These movements of the earth occur sometimes with a shuddering and trembling motion, that is to say horizontally, sometimes with a pulsating motion and a sliding motion from a height to a depth. These are severer than the horizontal ones, since exhalation rises weakly from the surface of the earth, whereas from its depths [it rises] with greater strength and with a very severe force. For this reason they occur rarely and when they occur they cause great calamities [?].

[tit.]

ܦܣܘܩܐ ܬܠܝܬܝܐ . ܡܛܠ ܙܒ̈ܢܐ ܕܒܗܘܢ ܫܪܝܪ ܗܘܐ ܙܒܢܐ .

ܬܐܘܪ̈ܝܐ ܘܐܪܒܥ .

[II.iii.1.]

ܩܕܡܝܬܐ . ܒܟܠܗܘܢ ܕܪ̈ܐ ܫܪܝܪ ܗܘܐ ܙܒܢܐ ܕܐܬܐ . ܒܗ̇ ܕܡܬܚܫܒ
ܚܠܦܐ ܬܠܝܢܐ ܒܐܬܪܐ . ܘܒܐܬܪܐ ܗܢܐ ܗܟܢܐ ܐܝܟ ܕܒܣܘܓܐܐ ܚܝܐ
ܬܘܩܢܐ ܘܐܬܪ̈ܘܬܐ ܡܬܦܪܫܐ ܥܠ ܥܠܡ ܐܢܘܫ . ܗܐ ܗܟܢ ܦܘܫܐ
ܡܣܬܟܠܢܘܬܐ ܢ̈ܩܦ . ܙܪܝܐ ܒܚܝܠܬܐ ܠܡܣܠܬܐ . ܘܗܘ ܡܣܠܬܐ ܠܗܘ ܐܬܪܐ
ܡܬܚܫܒܢܐ . ܘܠܐܬܪܐ ܡܬܝܕܥܐ . ܡܣܬܟܠܢܘܬ ܒܪܫ ܟܕ ܗܘܠܝܢ ܚܠܦܐ ܬܠܝܢܐ
ܠܟܠ ܘܠܬܚܬ . ܘܗܘ ܕܒܚܝܠܐ ܠܐܦܝ ܠܟܠ ܒܬܪܝܗ ܡܬܚܫܒܢܘܬܐ . ܘܡܢܗ ܒܗ ܐܬܪ
ܦܐܪ . ܦܠܓ ܘܒܚܝܠܐ ܬܪܝܨܘܬܐ ܗܘܝܐ . ܒܐܦܠܐ ܕܝܢ ܗܘܐ ܐܬܪ ܒܚܝܠܐ . ܒܗܘ
ܕܟܠܗ ܐܬܪܐ ܢܦܫܐ ܚܣܝܢ ܦܘܫܐ . ܘܟܠܠܐ ܡܫܟܚܝܢ ܘܦܪܘܫܝܢ ܙܒ̈ܢܐ
ܐܬܪܐ . ܟܕ ܡܬܝܩܢܘܬ ܦܘܪܐ ܐܢܫ̈ܘܗܝ . ܘܡܫܬܪܐ ܗܘܒ . ܒܗܘ ܕܦܩܕ ܣܒܟܐ
ܠܦܘܫܐ . ܘܠܐ ܡܫܟܚ ܕܢܦܫܣ . ܒܐܢܫ̈ܟܐ ܕܝܢ ܗܘܘ ܙܒ̈ܢܐ ܒܠܗܘܢܐ
ܫܪܝܪܐܝܬ . ܒܗܘ ܕܒܡܫܟܢܐ ܢܦܫܐ ܡܬܚܫܒܢܘܬܐ ܠܚܠܦܐ . ܘܠܐܦܝ ܐܬܪܐ
ܦܢܝܐ ܘܡܢܝܢܐ . ܘܪܓܫܘ ܠܐ ܥܒܕܗܐ .

[II.iii.2.]

ܕܬܠܬ . ܒܙ̈ܒܢܘܬܐ ܕܒܥܘܡܪܐ ܗܪ̈ܟܢܐ ܘܦܫܝܛܐ ܩܕܡ ܙܒ̈ܢܐ ܒܚ̈ܝܠܐ ܗܘܘ .
ܫܪܝܪܐܝܬ ܐܢ ܡܛܠ ܡܫܟܚܐ ܒܠܚܘܕ ܐܕܡ . ܒܐܝܕܝܗܘܢ ܗܘܐ ܦܘܫܐ
ܡܫܟܚܬܐ ܡܬܚܫܒܢܐ ܘܡܬܚܫܒܢܘܬܐ ܕܦܫܝܛܘܬܐ ܠܐ ܢܦܫܟܐ ܐܬܪܐ ܠܚܝܠܐ .
ܒܟܠܗ ܡܬܚܫܒܢܐ ܦܘܫܐ ܘܡܢܝܢܐ ܠܐܬܪܐ . ܒܡܫܬܡܥܢܘܬܐ ܕܝܢ ܒܚܝܠ ܗܘܐ ܢܦܫܐ .
ܒܗ ܦܘܫܐ ܫܪܝܪܐ ܠܦܘܠܚܢܐ ܢܦܫܝܐ ܕܚܠܦܐ ܬܠܝܢܐ ܡܬܚܫܒܢܐ . ܘܐܦ ܚܠܦ ܥܠ
ܫܪܝܪܘܬ ܦܠܓܘܬܐ ܕܗܘ ܡܫܬܡܥܢܐ . ܘܦܫܝܛܘܬ ܡܬܪܘܢܘܬܗ ܡܫܟܚܐ . ܘܒܫܠܡܐ
ܡܬܚܫܒܬ ܒܚܝܠ ܗܘܐ ܙܒ̈ܢܐ ܒܗ ܦܘܫܝܐ ܕܚܠܦܐ . ܘܐܦ ܚܠܦ . ܥܠ ܫܪܝܪܘܬ
ܒܢܝܫܘܬܐ ܕܡܬܕܪܟܐ ܐܦܣܗ ܕܐܬܪܐ ܡܫܟܚܐ . ܒܬܪܐ ܘܡܬܚܫܒܢܘܬܐ ܫܪܝܪ ܗܘܐ
ܙܒ̈ܢܐ . ܒܗ ܒܟܠܗܘܢ ܫܪܝܪ ܗܘ ܕܒܡܫܬܐ ܕܒ̈ܢܝܐ ܡܬܚܫܒܢ ܦܘܫܐ .

FMWLΛV: II.iii.1: 1 ܙܒ̈ܢܐ M || 2 ܚ̈ܝܠܐ FMV || 3 ܪܘܚܢܘܬܐ F || ܘܐܪܥܢܝܘܬܐ: ܘܪܥܢܝܘܬܐ F || 5 ܡܣܬܟܠܢܘܬ: ܡܣܬܟܠܢܘܬ F || 6 ܒܬܪܝܗ W || ܘܡܢܗ F || 7 ܘܒܐܦܠܐ Λ || ܕܝܢ: om. F || 8 ܦܘܪܫܐ WL || 9 ܐܦܣܗ: in marg. V; ܕܐܦܣܗ l || 10 ܗܘܘ: om. Λ || 12 ܘܠܐ Λ || **II.iii.2:** 1 ܘܦܬܚܝܢܐ Λ; ܦܬܚܝܢܐ V || 3 ܡܬܚܫܒܢܐ: ܡܬܚܫܒܢܘܬܐ Λ || 3-4 ܒܟܠܗ ... ܡܬܚܫܒܢܐ: om. F || 4 ܘܡܢܝܢܐ W || 5 ܠܚܠܦܐ ܚ̈ܝܠܐ F || 8 ܕܡܬܕܪܟܐ F || 9 ܕܒ̈ܢܝܐ MΛV ||

Third Section: On Times when Movements Occur More.
Four theories.

[II.iii.1.]
First [theory]: Movement of earth occurs more in absence of winds, because the smoky exhalation is [then] confined in the earth. At such a time a thin, long cloud is usually seen beside the horizon. For when opposing winds blow, the stronger (wind) drives the weaker, so that the weaker (wind) is confined inside the earth and causes the earth to quake, though weakly because of the division of the smoky exhalation above and below. The stronger (wind) extends upwards in a straight line, and when the air is cooled by it, (this air) is condensed and a straight cloud is formed. The air becomes foggy during an earthquake, because wind is then confined inside the earth. Movements of the earth are frequent and great at night, because [its] coldness condenses the earth's surface, and also at dawn, because the sun causes the wind to move but is unable to dissipate (it). During the day, the movements occur most at noon, because [at noon] the heat attracts the exhalation with vigour, [and at the same time] causes the surface of the earth to contract and dry up, and [so] does not allow (the exhalation) to escape.

[II.iii.2.]
Second [theory]: Severe movements occur in places which have cavernous and porous depths, especially if [there is] much water flowing over them. For in such (places) much wind is observed, but the rapid movement [of the water] does not give room for penetration, so that the wind is confined [beneath the water] and causes the earth to quake. In winter earthquake occurs less, since excessive coldness retards the violent movement of the smoky exhalation. If it does occur, it indicates the excess of the moisture of that winter and the deficiency of its coldness. Likewise in summer movement occurs less because of the dissipation of the exhalation. If it does occur, it signifies the excess of dryness which thickens the surface of the earth. In spring and autumn movements occur more, since in them winds are generated more than in the rest of the seasons.

[II.iii.3.]

ܕܬܠܬ. ܗܘܢܐ ܕܝܢ ܐܦ ܦܝܠܣܘܦܘܣ ܕܝܘܢܝܐ ܘܡܣܗܕܐ ܥܠܝܗ ܘܗܝ ܢܦܫܐ ܕܐܝܬܝܐ ܒܠܚܘܕܘܬ ܡܬܒܣܡܢܘܬܐ ܕܡܢ ܝܠܘܦܐ ܡܢܗܠ. ܘܡܢܗ ܕܪܘܚܢܝܘܬܐ ܡܕܪܝ ܐܬܪܐ ܡܬܬܪܝܡ. ܘܒܠܥܕ ܬܘܕܝܬܐ ܒܠܒܗ ܡܢܗܠ ܡܬܚܡܫܢ. ܘܡܢ ܡܪܝܘܬܐ ܕܟܪܝܐ ܡܢܗܠ ܗܘܢܐ ܡܫܪܐ. ܘܡܢ ܡܪܝܘܬܐ ܕܟܪܝܐ ܒܡܘܬܒܐ ܠܐ ܗܘܢܐ. ܡܢ ܦܓܪܐ ܘܠܢܦܫܐ ܡܬܘܬܐ ܕܐܣܝܘܬܐ ܡܫܬܠܡܐ.

[II.iii.4.]

ܕܐܪܒܥ. ܐܢ ܐܡܪ ܠܢܦܫ. ܗܘ ܡܢܐ ܕܡܢ ܦܓܪܐ ܗܘܢܐ ܡܢܗܠ ܠܚܝܐ. ܘܪܘܚܐ ܕܐܪܥܐ ܒܡܢܐ ܐܬܚܙܝ ܐܡܝܢܐ ܘܡܫܬܡܫܐ ܗܘܢܐ. ܐܡܪܝܢ. ܕܒܗ ܡܬܒܪܝܢܘܬ ܦܓܪܐ ܘܒܠܚܘܕܘܬܐ ܕܢܦܫܗ ܕܐܪܥܐ ܘܥܡܗ ܥܠܬܐ ܕܡܠܟܐ ܠܒܪܢܫܐ ܐܝܬ ܠܒܪܢܫܐ ܗܘܢܐ ܡܫܪܐ. ܘܠܚܘܒܐ ܕܐܪܥܐ ܗܘ ܕܝܢ ܒܝܬ ܡܢ ܒܢܝܢܐ. ܚܕ ܡܢ ܦܘܠܚܢ ܦܓܪܐ ܕܐܪܥܐ ܠܡܚܫܘ ܚܝܘܬܐ. ܐܝܟܢܐ ܕܝܢ ܗܘܝܐ ܠܝܩܝܪܬܐ ܕܡܬܡܫܚܐ ܡܠܐ ܡܢ ܦܓܪܗ ܕܐܠܗܐ ܘܢܦܫܗ ܗܘ ܕܡܬܚܙܝܐ ܠܗܘܢ ܘܡܬܚܙܝܐ. ܢܦܫ ܠܗ ܒܐܪܥܐ ܘܪܘܚܐ ܘܒܢܝܐ ܟܠܗܝܢ ܘܗܘܐ. ܒܐܝܕܐ ܡܠܠ ܚܘܠܐ ܕܡܠܟܘܬܐ.

FMWLIV: II.iii.3: 1 ܐܦܝܠܣܘܦܘܣ lV || 2 ܝܘܢܝܐ Λ || 3 ܘܡܢ: ܡܢ W; ܕܡܢ Ll || 4 ܡܢܗܠ ܗܘܢܐ ... ܕܟܪܝܐ : om. F; in marg. V || ܒܡܘܬܒܐ: ܒܡܘܬܒܐ ܒܡܘܬܒܘܬܐ W; ܒܡܘܬܒܐ Ll || 5 ܡܢ: ܘܡܢ Λ || ܢܦܫܐ Λ || **II.ii.4:** 1 ܗܘ: ܗܘ V || 2 ܕܡܢ: ܕܡܢ F || 3 ܠܒܪܢܫܐ L || 7 ܕܡܠܟܘܬܐ: ܕܡܠܟܘܬܐ L; ܕܡܠܟܘܬܐ ܘܡܠܟܘܬܐ ܕܐܪܥܐ l ||

[II.iii.3.]
Third theory: The eclipse of the sun and the moon also becomes the cause of movement of the earth through the sudden deprivation of the heat which [arises] from radiation; because of the [resulting] coldness the openings of the earth become closed and the smoky exhalations are suddenly confined in (the earth). That this occurs due to coldness which arises suddenly, but does not occur due to coldness which arises gradually, one learns from the (human) body and particular experiments in medicine.

[II.iii.4.]
Fourth [theory]: If someone says: "what arises from winds [normally] subsides suddenly, but the movement of the earth is often constant and continuous", we say that this happens through the large quantity of the wind, the tortuosity of the openings of the earth and the persistence of the cause which generates the exhalation or retains [it]. The wise acknowledge two benefits of the movement of the earth: one is [its function of] opening up holes in the earth so as to cause springs to gush out; the other is [its function of] frightening the hearts of wicked peoples before the anger of God and His fury which disturbs them and terrifies [them]. "He looks on the earth and it quakes, and He rebukes the mountains and they smoke" [Ps. 104.32], that is to say, on account of the wickedness of the inhabitants.

[tit.]

ܡܐܡܪܐ ܬܠܝܬܝܐ .

ܡܛܠ ܒܥܢ̈ܢܐ ܡܬܚܘܝܢܐ . ܘܒܗ ܐܝܬ ܦܣܘ̈ܩܐ ܬܠܬܐ .

ܦܣܘܩܐ ܩܕܡܝܐ .

ܡܛܠ ܓܘܫܡܐ ܡܬܚܙܝܢܐ . ܬܐܘܪ̈ܝܗ ܫܒܥ .

[III.i.1.]

ܡܬܚܙܝܬܐ . ܐܝܟ ܟܠܐ ܘܐܝܟ ܓܘܫܡ ܡܬܚܙܝܢܐ ܐܘܟܝܬ ܒܥܢ̈ܢܐ ܡܬܚܘܝܢܐ . ܟܬܒܐ . ܐܢܘܢ ܡܬܚܙܝܢܝܬܐ . ܒܟܝܢܬܐ . ܡܠܝܐ . ܘܟܬܒܐ ܡܢ ܐܝܟ ܚܫܝܟܐ ܘܫܦܝܪܘܬܗ ܘܡܢܗܝܪܘܬܐ ܡܢ ܡܬܚܙܝܢ ܘܫܦܝ̈ܢ . ܘܠܐ ܡܬܚܙܝ̈ܢ ܐܠܐ ܡܬܚܙܝܢ . ܒܗܢ ܡܢܗܪܘܬܐ ܕܠܘ ܡܢ ܦܪܨܘܦܐ ܒܠܗܘܢ ܦܠܓܘܬ . ܐܠܐ ܡܢܗܝܪܘܬܐ ܕܠܐܦ ܒܥܢܢܐ ܥܠܝܗ ܠܚܙܝܬܐ ܡܢ ܒܫܢ . ܘܒܠܥܕ ܠܗ ܡܬܚܙ̈ܝܢ ܗܘܝܐܢ ܐܠܐ ܒܦܪܨܘܦܐ ܒܢܝܢܐ ܒܒܝ .

[III.i.2.]

ܕܬܬܚܡ . ܒܠܚܘܕ ܒܥܢ̈ܢܐ ܡܬܚܙܝܢܐ ܡܬܚܙܝܢ . ܘܐܝܟ ܒܦܪܨܘܦܐ . ܘܗܟܢܐ ܕܠܐ ܡܬܚܙܝܢܐ ܠܗ ܡܬܚܙܝܢ . ܐܠܐ ܡܬܚܙܝܢ ܒܠܚܘܕ ܒܠܚܘܕ . ܗܟܢ ܡܢ ܐܢܫܐ ܗܘ ܕܠܐ ܡܬܚܙܝܢܐ ܐܝܟ ܡܬܚܙܝܢܐ . ܐܠܐ ܒܡ ܡܬܚܙܝܢܐ ܥܠ ܗܘܢܐ ܒܨܒܘ ܥܒܝܐ . ܘܗܘ ܦܬܝܐ ܠܗ ܠܐ ܥܒܝܐ . ܘܗܘܠܐ ܕܡܬܚܙܝܢܐ ܐܢܘܢܐ ܡܢܗܝܪܘܬܐ ܕܐܢܘܢܐ ܐܝܟ ܚܙܝܬܐ ܢܗܝܪܐ ܕܠܐ ܦܪܨܘܦܐ ܡܢܗ ܐܝܟܢܐ . ܦܪܨܘܦܐ ܕܝܢ ܗܘ ܡܢܗܪܘܬܐ ܒܦܪܨܘܦܬܐ . ܘܥܒܝܐ ܡܢܗ ܡܢܗܪܐ ܢܗܝܪܐ ܕܠܐ ܒܥܒܠ ܦܪܨܘܦܐ ܡܛܠ ܕܢܗܝܪܘܬܗ . ܘܒܠܚܘܕ ܡܬܚܙܝܢ ܗܘ ܒܥܢ̈ܢܐ .

[III.i.3.]

ܕܬܬܠ . ܘܒܟܝܢܬܐ ܡܢ ܡܢܗܪܘܬܐ ܕܢܘܗܪܗ ܢܘܚܕܐ ܒܐܬܪܘܬܐ ܘܐܬܪܘܬܐ ܡܬܚܙܝܢ . ܘܕܢܗܪܘܬܐ ܡܬܚܙܝܢܬܐ ܒܫܘܚܠܦܐ ܢܗܝܪܘܬܐ ܡܢܝ . ܐܝܟܢ ܒܦܪܨܘܦܐ ܦܠܓܐ . ܘܐܝܟܐ ܘܐܬܪܘܬܗܘܢ ‹ܡܬܚܙܝܢ ܐܝܟܢܐ› . ܘܢܗܪܘܬܐ ܕܐܬܪܘܬܗܘܢ› ܢܗܘܐ ܡܢ ܐܬܪܘܬܐ ܡܬܚܙܝܐ . ܘܒܠܚܘܕ ܒܢܗܝܪܘܬܗ ܡܬܚܙܝܐ . ܐܝܟܢܐ ܕܝܢ ܗܘܠܐ ܕܠܗܘܢ ܡܬܚܙܝܢ ܕܢܗܪܐ ܢܗܝܪܐ ܕܦܠܓ ܘܢܘܗܪܐ ܡܢ ܢܚܝܢ ܒܗܘܢ ܘܒܢܗܪܘܬܐ ܠܗܘܢ ܦܠܓܐ .

FMWlVλ: tit.: 2 ܡܬܚܘܝܢܐ (*sic*) F || 4 ܒܓܘܫܡܐ ܡܬܚܙܝܢܐ λ || **III.i.1**: 1 ܓܘܫܡ: ܓܘܫܡܐ V || ܒܥܢ̈ܢܐ: ܓܘܫܡܐ V || 2 ܐܢܘܢ: **hinc FMλV** || 2 ܐܢܘܢ Vλ || 3 ܘܫܦܝܪܘܬܗ ut vid. F; ܘܫܦܝܪܘܬܗ λ || ܘܡܢܗܝܪܘܬܐ λ || 4 ܡܬܚܙܝܢ λ || ܕܠܘ: ܘܠܘ V || 5 ܕܠܐܦ ܒܥܢܢܐ: ܕܠܐ ܒܥܢܢܐ F || 6 ܗܘܝܐܢ Vλ || 7 ܒܦܪܨܘܦܐ ܒܢܝܢܐ Fλ (cf. IS حِيَل pl.) || **III.i.2**: 1 ܡܬܚܙܝܢܐ λ || 3 ܐܢܘܢܐ F; ܐܢܘܢ V; ܐܢܘܢܐ λ | || 4 ܒܨܒܘ scripsi: ܒܨܒܘܬ codd. || ܥܒܝܐ λ || 7 ܦܪܨܘܦܐ: ܦܪܨܘܦܐ F || **III.i.3:** 2 ܒܫܘܚܠܦܐ: ܒܫܘܚܠܦܐ F (cf. IS تخنير) || 3-4 ‹ܡܬܚܙܝܢ . . . ܕܐܬܪܘܬܗܘܢ› supplevi || 4 ܢܗܘܐ: om. V || ܐܬܪܘܬܐ: ܐܬܪܘܬܐ M; ܐܬܪܘܬܐ V; ܐܬܪܘܬܐ λ || ܡܬܚܙܝܐ V || ܒܢܗܝܪܘܬܗ F; ܒܢܗܝܪܘܬܗ M; ܒܢܗܝܪܘܬܐ λ || 5 ܢܗܪܐ: ܢܗܪ F || 6 ܒܢܗܪܘܬܐ F ||

THIRD CHAPTER

ON MINERAL BODIES

It contains three sections.

First Section: On the Categories of Minerals. Five theories.

[III.i.1.]
First [theory]: There are four categories of μεταλλικοι, i.e. minerals: stones, fusible substances, sulphurs, salts. Stones like vitriol, πυρίτης and μαγνησία are hard, are shatterable and burn, but are neither fusible nor malleable, since they are composed from wateriness which has been solidified not by coldness alone, but through dryness which has changed the wateriness into earthiness. For this reason, the majority of them are not fusible except perhaps by [certain] physical techniques.

[III.i.2.]
Second [theory]: All malleable bodies are fusible albeit with [special] technique and the majority of what are not malleable are not fusible, but are merely softened [and that only] with difficulty, except ἀδάμας [diamond], which is neither malleable nor are softened, but when beaten on an anvil penetrates inside it and is not affected by iron. The material of malleable (bodies) is a watery substance which has been mixed strongly with an earthy substance and cannot be separated from it. That wateriness solidifies with coldness, but there remains out of it a fluid part which has not yet been solidified because of its oiliness and for this reason (this) body is malleable.

[III.i.3.]
Third [theory]: Sulphurs are composed from wateriness which has undergone a vigorous leavening with earthiness and airiness; they have acquired much oiliness through leavening by heat, and have then solidified due to coldness. Alum and sal ammoniac <are salts, but the fieriness of sal ammoniac> is greater than [its] earthiness, and for this reason it is sublimated in its entirety. Their material is water, in which hot smoke, very fine and fiery, is mixed, and which dryness solidifies.

[III.i.4.]

ܕܐܬܩܪܒ. ܚܣܝܪܐ ܕܝܢ ܡܢ ܡܫܠܡܘܬܗ ܘܡܫܬܘܕܥܘܬܐ ܘܟܪ̈ܘܒܐ ܡܬܒܟܝܢ. ܘܣܠܩ
ܡܕܡ ܡܢ ܒܥ̈ܒܕܐ ܡܬܦܫܪܢܐ ܡܝܢ. ܘܚܠܡܘܬܗ ܘܚܠܡܝܢܗ ܡܢ ܟܪܝܐ
ܕܡܫܠܡܘܬܗ ܕܚܣܝܪܐ ܘܡܫܬܘܕܥܘܬܗ ܫܡܝܢ. ܘܟܕ ܣܠܩ ܕܢܚ ܡܢ ܒܥ̈ܒܕܐ
ܡܬܦܫܪܢܐ ܒܡܡ̈ܠܠܐ ܦܢܝ ܦܠܓܝ. ܒܕܡܘ ܗܘ ܡܢܐ ܕܣܠ ܦܪܠܐ ܦܢܐ
ܡܡܡܡ ܘܡܬܒܪܒܪ ܐܝܟ ܒܐܠܦܝܢܗ. ܘܗܘ ܡܢܐ ܕܣܠܩ ܚܣܝܪܐ ܡܬܬܒܪ
ܡܢܐ ܐܝܟ ܒܐܠܦܝܢܗܘܢ.

[III.i.5.]

ܕܢܣܒ. ܒܥ̈ܒܕܐ ܡܬܩܠܠܢܐ ܐܦ ܣܠܩܘܢ ܐܘܣܝܐ ܐܢܬܘܢ. ܡܬܦܣܩܢܐ
ܐܝܟ ܦܪܠܐ ܘܕܡܥܐ. ܐܦ ܡܬܦܪ̈ܣܢܐ ܐܝܟ ܢܡܘܣܐ ܘܣܘܪܠܐ. ܐܦ ܡܣܠܩ
ܐܘܣܝܐ ܘܡܠܦܢܐ ܘܡܣܬܦܩܢ ܘܦܫܝܩܐ ܘܒܝܬܢܝܐ ܐܝܟ ܠܒܠܒܘܬܐ. ܐܦ
ܡܢܝܢܐ ܕܓܪܐ ܠܗܘܢ ܘܦܠܓܘܬܐ ܘܐܘܣܝܐ. ܐܝܟ ܕܝܢܐ ܘܢܣܒܐ
ܘܐܪܡܘܣܝܢ ܘܒܐܠܦܝܢܗ. ܐܦ ܕܗܢܐ ܕܒܦܠܓܘܬܐ ܠܒܢܝܗ ܠܐ ܡܬܬܢܝ
ܘܐܘܣܝܐ. ܐܝܟ ܒܒܪܝܬܐ ܘܐܪܡܘܣܝܢ. ܘܦܠܓ ܒܡܠܬܐ ܐܪܒܥ ܦܢܝ
ܕܘܝܬܐ. ܐܪܡܘܣܝܢ ܒܒܪܝܬܐ ܢܦܠ. ܥܒܕܐ ܕܝܢ ܦܠܓܪܐ.
ܕܗܒܐ ܘܡܐܡܐ ܢܩܒܐ ܦܪܠܐ ܐܟܬܐ ܐܢܟܐ ܘܐܪܓܐ ܐܚܪܝܐ ܕܗܘܐ ܒܢܝ
ܘܒܐܪܟܢ ܡܬܬܡܡܐ.

FMVλ: **III.i.4**: 1 ܚܣܝܪܐ FλV || ܡܫܠܡܘܬܐ F || ܡܬܒܟܝܢ V || 2 ܡܢ prim.: om. λ || ܘܚܠܡܘܬܗܘܢ F || ܘܚܠܡܝܢܗܘܢ F; ܚܠܡܝܢܗܘܢ λ || 3 ܕܡܫܠܡܘܬܗܘܢ λ || ܘܡܫܬܘܕܥܘܬܐ λ || ܟܪܝܐ: ܟܪܝܐ M || 3 ܕܡܫܠܡܘܬܗ F || 4 ܒܡܡ̈ܠܠܐ scripsi: ܒܡܡܠܠܐ (sine *syāmē*) FM; ܒܡܡ̈ܠܠܢܐ V; ܒܡܡ̈ܠܠܢܐ ܠܠܐ λ || 5 ܒܐܠܦܝܢܗܘܢ F || 6 ܒܐܠܦܝܢܗܘܢ F || **III.i.5**: 1 ܡܬܩܠܠܢܐ λ || ܣܠܩܘܢ: ܣܠܩܘ V || 2 ܐܦ prim.: ܘ λ || ܢܡܘܣܐ F || ܘܣܘܪܠܐ F || 4 ܡܢܝܢܐ F || 5 ܘܒܐܠܦܝܢܗܘܢ F || ܕܠܒܢܝܗ Vλ || 6 ܒܡܠܬܐ M || 7 ܘܐܪܡܘܣܝܢ F || ܦܠܓܪܐ λ ||

[III.i.4.]
Fourth [theory]: Vitriols are composed of saltiness, sulphureity and stones, and have the power of some of the fusible bodies. Χάλκανθος and χαλκῖτις are generated through the dissolution of the saltiness and sulphureity of vitriol. When they acquire the power of one of the fusible bodies in the mines, they solidify, so that what acquires the power of iron turns red and yellow, like χαλκῖτις, and what gains the power of copper turns green, like χάλκανθος.

[III.i.5.]
Fifth [theory]: Mineral bodies are either (1) strong in substance [and] (a) malleable, like iron and gold, or (b) shatterable like corundum and rock crystal, or (2) weak in substance and mixture, and loosely composed, and (a) stony like glass, or (b) saline - moisture easily dissolves these - like alum, vitriol, sal ammoniac and χαλκῖτις, or (c) oily and - these are not easily dissolved by moisture alone - like sulphur and ἀρσενικόν. Alchemists name four "spirits": sal ammoniac, ἀρσενικόν, sulphur, mercury; and seven "bodies": gold, silver, copper, iron, lead, tin, and another species which occurs in China and is called *khārṣīnī*.

[tit.]

ܡܐܡܪܐ ܬܠܝܬܝܐ .

ܡܛܠ ܗ̇ܝ ܕܐܣܛܠܒܣܐ ܕܡܬ̈ܟܫܦܢܘܬܐ ܡܢܘ ܗܘ . ܘܐܝܬܪܘܬܗ̇ ܕܝܠܗ̇ .

[III.ii.1.]

ܡܕܒܪܢܘܬܐ . ܒܗܘܢ ܡ̇ܢ ܡܣܒܪܢܘܬܐ ܕܝܕܥܬܐ ܒܐܪܥܢܘܬܐ ܘܠܒܝܫܘܬܐ ܐܝܟ
ܫܪܝܐ . ܘܐܝܪܐܢܘܬܐ ܒܚܝܪܐ . ܘܒܝܪܢܘܬܐ ܥܠܝܐ . ܥܠܬ ܢܘܗܪܗ ܐܪܐ
ܥܠܝܡܘܬܗ̇ ܗܝ ܕܒܝܪܢܘܬܐ . ܘܟܝܪܘܬܗ̇ ܕܐܝܪܐܢܘܬܐ . ܘܐܝܟܢܘܬܗ ܡܛܠ
ܡܣܝܢܘܬܗ ܦܫܝܛܐ . ܘܡܫܘܬܦܘܬܐ ܡܬܦܩܕܢܘܬܐ ܕܐܝܪܐܢܘܬܗ ܒܚܝܪܬܐ
ܘܐܪܝܢܘܬܗ ܡܫܘܚܬܐ ܒܡܣܝܢܘܬܗ ܢܗܝܪܬܐ . ܥܠܬ ܢܘܗܪܗ . ܘܡܫܪܪ ܡܘܠܟܗ
ܗ̇ܘ ܕܒܪܝܗ ܠܐ ܡܬܦܪܫ ܡܢ ܚ̈ܕܪܐ ܡܢܘ̈ܬܗ . ܐܠܐ ܐܝܟܢܐ ܦܠܓ̈ܐ . ܥܠܬ
ܡܫܘܠܡܘܬܗ ܡܢ ܫܡܫܘܬܐ . ܘܡܣܝܢܘܬܗ ܥܠܬ ܠܪܘܫܗ ܡܢ ܢܘܪܐ . ܢܚ
ܒܒܪܝܬܐ ܕܝܢ ܦܠܓ ܠܗ . ܘܐܝܟ ܐܝܟܢܐ ܚܒܪ ܠܗ . ܒܗ̇ܝ ܕܐܝܟܢܐ ܢܗ̇ܪܐ ܐܝܟ
ܢܗܘܐ ܗܘ . ܘܢܗܘܐ ܕܦܠܓ ܐܝܟ ܐܝܟܢܐ ܗܘ .

[III.ii.2.]

ܕܬܬܠ . ܕܢܗܘܐ ܗܘ ܐܣܟܡܐ ܕܟܠܗܘܢ ܡܬܩܪܒܝܢܐ ܕܡܬܦܫܪܝܢ . ܢܗܝܪܐ ܡܢ
ܗܝ ܕܒܡܬܦܫܪܢܘܬܗܘܢ ܠܗ ܡ̇ܩܪܒܝܢ . ܘܟܠܗܘܢ ܗ̇ܢܘܢ ܕܒܫܡܫܘܬܐ ܡܫܝܢܐ
ܡܬܦܫܪܝܢ ܡ̇ܩܒܠ ܢܗܘܘܢ . ܐܦ ܥܒܕ ܘܐܝܟܢܐ ܕܒܫܡܫܘܬܐ ܡܫܠܡܬܐ
ܡܬܦܫܪ . ܒܗ ܢܗܝܪ ܡܬܟܠܦܐ ܢܘܪܐ ܬܫܘܬܦܬܐ . ܐܦ ܗ̇ܘ ܕܢܗܘܐ ܒܡܫܡܫܢܘܬܐ
ܢܘܗܪܢܐ ܡ̇ܩܒܠܝܢ . ܘܡܢ ܗ̇ܝ ܬܘܒ ܕܒܟܠܗܘܢ ܡܦܩܝܢ ܢܗܝܪܐ ܕܡܫܝܢ ܐܝܟܢ .
ܒܬܪ ܕܝܢ ܒܡܫܡܫܢܘܬܗ ܕܝܠܗ ܘܒܒܪܝܬܐ ܕܦܠܓܐ ܠܗ ܡܬܦܫܪ ܡܣܝܢܝܢ .
ܐܝܟܢܐ ܕܡܫܝܢܐ . ܐܝܟܢܐ ܡܬܦܠܓܝܢܘ ܡܬܦܫܪܝܢ ܐܪܐ ܢܗܘܐ ܘܒܒܪܝܬܐ
ܐܝܟܢ .

FMVλ: III.ii.1: 2 ܐܪܐ ... ܘܐܝܪܐܢܘܬܐ: ܐܪܐ ܢܘܗܪܗ ܥܠܬ ܥܠܝܐ ܕܒܝܪܢܘܬܐ ܗܝ ܥܠܝܡܘܬܗ̇ MV [ܐܪܐ ... ܗܝ in marg. V]; om. λ || ܥܠܬ: ܥܠܬܐ F || 3 ܥܠܝܡܘܬܗ̇ in marg. V || ܗ̇ܝ V || ܘܟܝܪܘܬܗ F || 4 ܕܐܝܪܐܢܘܬܗ ܡܬܦܩܕܢܘܬܐ ܒܚܝܪܬܐ Vλ || 5 ܡܣܝܢܘܬܗ: ܒܡܣܝܢܘܬܗ M || 6 ܕܒܪܝܗ: ܕܒܪ̈ܝܗ Vλ || ܘܥܠܬ Vλ || 7 ܥܠܬܐ ܕܥܠܬ ܠܪܘܫܗ Vλ || 8 ܠܗ ܚܒܪ: om. λ || 9 ܦܠܓ Vλ || III.ii.2: 2 ܕܒܡܬܦܫܪܢܘܬܗܘܢ F; ܕܒܡܬܦܫܪܢܝܗܘܢ λ || 6 ܦܠܓܐ V || 7 ܐܝܟܢܐ: ܕ ܐܝܟܢ λ; ܐ supra lin. V ||

Second Section: On the Fact that the Element of the Malleable [Minerals] is Mercury. Three theories.

[III.ii.1.]
First [theory]: In mercury, there predominates a wateriness which is mixed with fine earthiness; there is little airiness and no fieriness. The cause of its weight, therefore, is the absence of fieriness and the paucity of airiness, while its fluidity is due to its abundant wateriness. The sound mixture of its little airiness and its clear earthiness with its predominant wateriness is the cause of its whiteness. The firmness of its mixture on account of which its parts are not separated from each other but rise together is the cause of its ability to be sublimated by heat, its wateriness being the cause of its flight from fire. The vapour of sulphur solidifies (mercury), and makes it like lead, because molten lead is like mercury and solidified mercury is like lead.

[III.ii.2.]
Second [theory]: That mercury is the element of all fusible minerals is known from the fact that they revert to it upon being fused. The mercury in all those (fusible minerals) which are fused by a strong heat turns red. [In the case of] lead too, which is fused by a weak heat, when the fire is made to burn more [strongly] below it, its mercury turns fiery red. It is also known from the fact that (mercury) clings to all of them that (fusible minerals) are made of (mercury). Their formation, however, varies according to differences in (the mercury) itself and in the sulphur which solidifies it, as will be shown. The "parents" of the fusible μεταλλικοι, therefore, are mercury and sulphur.

[III.ii.3.]

ܕܬܠܬ݂ . ܡܢ ܚܢ̈ܪ ܙܘܓ ܕܟܝܢܐ ܘܡܟܝܢܘܬܐ ܣܘܪ̈ܬܐ ܘܕܟܝܢܬܐ ܘܠܐ ܢܩܝܪܬܐ
ܩܐܡܐ ܚܘܪ̈ܐ ܟܝܢܐ . ܘܐܝܟ ܒܟܝܢܘܬܐ ܟܡ ܕܟܝܢܘܬܗ̇ ܢܗܝܪ ܡܟܝܪܬܐ . ܘܣܠܩܐ
ܢܘܪ̈ܢܐ ܡܠܝܢܐ ܒ̇ܝܢܐ . ܘܡܣܘܦܐ ܠܐ ܢܟܝܪܐ . ܡܟܐ ܕܒܘܓ ܕܟܝܢܐ ܦܓܠܐ .
ܕܗܒܐ ܠܗ ܚܒܪܐ . ܘܐܝܟ ܐܘܣܟܐ ܕܙܘܓ ܠܒܢܐ . ܟܕܡ ܒܟܝܢܘܬܐ ܟܝܢܐ
ܘܢܩܒܐ . ܢܚܫܐ ܗ̇ܘܐ . ܘܐܝܟ ܙܘܓ ܟܝܢ ܘܠܒܢܐ ܘܐܪܟܢ ܘܒܟܝܢܘܬܐ ܗ̇ܘܟ
ܠܓܒܐܐ . ܦܪ̈ܙܠܐ ܗ̇ܘܐ . ܕܐܟܒܪܐ ܕܝܢ ܙܘܓܗ ܠܒ ܘܒܟܝܢܘܬܐ ܟܝܢܐ . ܘܟܡܗ̇
ܕܐܐܪܘܬܐ ܕܠܐ ܡܩܠܟܐ ܟܡܡܘܣ ܣܝܐܘܬ ܐܘܬ ܒܗ . ܟܡܝܪ̈ܐ ܠܢܩܐ .
ܟܡܦܡܗ̇ ܘܠܐ ܠܚܒܢܐ ܢ̇ܦܩܐ . ܘܕܐܢܟܐ ܙܘܓܗ ܟܝܢ ܘܟܠܝܢܘܬܐ
ܡܠܝܠ .ܘܒܟܝܢܘܬܗ ܗ̇ܘܟ ܟܝܢܐ ܘܗܘܪܝܐ ܘܡܣܝܠܐ . ܘܠܟܠܗܐ ܢ̇ܒܝܢ ܘܪܦܢܐ
ܡܠܝܗ .

[tit.]

ܦܣܘܩܐ ܬܠܝܬܝܐ . ܡܛܠ ܡܬܥܒܕܢܘܬ ܡܠܚ ܒܣ̈ܝܡܐ ܟܝܢ̈ܝܐ . ܬܐܘܪ̈ܝܣ
. ܬܠܬ݂

[III.iii.1.]

ܡܬܥܒܕܬܐ . ܠܘ ܠܐ ܡܚܝܢ ܡܟܝ ܦܘܪܣܐ ܕܟܒܪܝܬܐ ܠܙܘܓ ܒܐܘܡܢܘ ܢܦܠܛܘܢ .
ܘܐܦ ܠܟܒܪܐ ܕܟܝܢܐ ܚܒܪ ܐܘܡܢܘܬܐ ܠܐ ܡܫܠܐ . ܐܠܐ ܡܬܕܡܝܐ ܠܗ
ܟܝܢܐܝܬ ܟܠܗܘܢ . ܟܕ ܟܠܗܘܢ ܡܫܬܟܚܐ ܠܡܫܘ̈ܚܐ ܕܟܝܢܐ ܘܡܣܬܟܠܝܢ ܡܚ̈ܝܢܐ
ܕܗܘܝ ܟܐܒ̈ܟܐ ܗܠܝܢ ܡܢ ܟܝܢܐ . ܗܘܘ ܚܘܫܒܘܢ ܗܢܐ ܐܘܡܢܝܐ . ܘܟܬܪ̈ܐ
ܕܦܠܚ ܟܐܠܐ ܠܟܢܝܢܘܗܝ ܐܘܡܢܘܬ ܐܝܟܢܐ ܟܕܡ ܘܐܘܡܢܘܬܐ ܐܢܫܝܬܐ
ܠܫܘܠܡܘܬܗ ܠܐ ܡܫܠܝܐ ܟܐܢܝܫܘܢ . ܐܠܐ ܡܬܢܟܝܬܐ ܗܘܦܢܝܬܐ ܟܠܗܘܢ
ܡܬܚܫܚ . ܐܝܟܢܐ ܕܠܢܫܐ ܡܘܡܝܐ ܟܝܘܟܒܐ ܣܘܪܐ ܕܟܘܟܒ ܙܪܐ ܠܡܐܟܐ
ܢܟܚܝܢ . ܘܠܡܐܡܟܐ ܝܘܟܒܐ ܣܘܟܐ ܕܟܘܟܒ ܙܪܐ ܠܕܡܟܐ ܢܡܣܝܢ . ܘܢܕܟܘܢ
ܐܦ ܠܐܒܪܐ ܡܢ ܡܣܟܢܬ ܢܬܒܝܫܘܢ . ܘܢܘܫܘܢ ܡܟܝܡܘܢ .

FMVλ: III.ii.3: 1 ܘܕܟܝܢܬܐ: om. λ; post ܘܠܐ scripsit et ordinem postea correxit V ‖ 3-4 ܠܐ ܢܟܝܪܐ . . . ܚܒܪܐ: om., et in eius loco praebet ܗ̇ܘܐ ܕܗܒܐ in marg. F ‖ 4 ܠܗ: om. FM ‖ 5 ܘܢܩܒܐ: ܘܢܩܝܐ F ‖ ܘܠܒܢܐ: ܘܐܠܒܢܐ FV ‖ 6 ܠܓܒܐܐ: ܠܓܒܐ F ‖ ܕܐܟܒܪܐ F ‖ 8 ܟܡܦܡܗ̇: ܟܡܦܡܐ F; ܘܟܠܝܢܘܬܗ Vλ ‖ ܠܚܒܢܐ: ܠܚܒܢܐ V ‖ ܢ̇ܦܩ: ܢܦܩ MV; om. et spatium reliquit λ ‖ ܟܝܢܐ scripsi: ܟܝܢܐ codd. (IS ردى) ‖ **tit.**: 2 ܡܬܥܒܕܢܘܬ ܡܠܚ M; ܡܬܥܒܕܢܘܬ ܕܡܠܚ Vλ ‖ ܒܣ̈ܝܡܐ M ‖ **III.iii.1**: 1 ܠܘ: **hinc FMWIVλ** ‖ 1 ܡܚܝܢ scripsi: ܡܚܝܢ codd. ‖ ܦܘܪܣܐ λ ‖ ܕܟܒܟܒܝܬܐ MIVλ (cf. IS بالكبريت) ‖ ܙܘܓ V ‖ ܒܐܘܡܢܘܬܐ [ܒܐܘܡܢܘ l]: ܒܐܘܡܢܘ FMWVλ (cf. IS بالصناعة) ‖ 2 ܡܫܠܐ F ‖ 3 ܟܝܢܐܝܬ: ܟܝܢܐܝܬ MVλ ‖ 4 ܚܘܫܒ λ ‖ 5 ܦܠܚ V ‖ ܐܘܡܢܘܬ Vλ ‖ ܘܐܘܡܢܘܬܐ Vλ ‖ 6 ܡܬܢܟܝܬܐ Vλ ‖ 8 ܕܝܘܟܒܐ F ‖ ܡܟܝ V ‖ 9 ܠܐܒܪܐ Λ ‖ ܢܬܒܝܫܘܢ MIVλ ‖

[III.ii.3.]
Third [theory]: From pure mercury and white and pure sulphur which is not burnt, nature creates silver. If the sulphur, as well as being pure, is even better, has a subtle, fiery power and [is] clear-coloured and not burnt, when it meets pure mercury, it turns it into gold. If the substance of the mercury is good but the sulphur is bad and burnt, copper is formed. If the mercury is bad, impure and earthy, and the sulphur too is impure, iron is formed. The mercury in lead [tin] is good, but the sulphur is bad, and because it contains airiness which is not strongly mixed with them, when it is pressed, it escapes and as it escapes it lets out a tingling sound. The mercury in tin [lead] is bad and clotted by clayeyness, and its sulphur too is bad, fetid and weak, and for this reason its solidification is feeble and slack.

Third Section:
On Imitation of these Bodies by Means of Each Other.
Three theories.

[III.iii.1.]
First [theory]: Technicians are not unable by artificial means to solidify mercury using sulphurs - even if the action of [human] artifice does not attain to [the level of] the action of nature, but merely resembles it in outward appearance - so that it appears to those lacking in intelligence, and many believe, that the generation of these bodies by nature is the same as this artificial generation of them. In reality it is not in the power of the alchemists to remove the specific character of a certain species and to provide it with another specific character, but they are able merely [to effect] wonderful imitations, as when they dye red copper with a white colour which very much resembles [that of] silver, give silver a yellow colour which very much resembles [that of] gold, and cleanse leads of most of their impurity and obscure their blemishes.

[III.iii.2.]

ܕܬܬܬܠܡ . ܒܡܬܕܡܪܢܘܬܐ ܡܢ ܗܠܝܢ ܦܘܪ̈ܫܢܐ ܐܘ̈ܡܢܐ ܕܒܗܘܢ ܠܒ̈ܣܡܐ ܗܠܝܢ
ܡܬܐܡܪܝܢ ܠܗܠܝܢ . ܘܒ̈ܣܡܐ ܘܣܡ̈ܡܢܐ ܠܟܪ̈ܝܗܐ ܕܡܣܟܢ̈ܘܬܗܘܢ ܠܐ ܚܐܠܝܢ
ܡܣܬܟܠܝܢ . ܒܪܡ ܠܐ ܐܢܫ ܡܢܐ ܗܘܢܐ ܡܬܬܪܐ ܘܠܐ ܡܩܒܠ ܡܬܬܪܝܢܘܬ
ܕܡܣܟܢܐ ܗܘ ܕܒܡ̈ܐܟܠܐ ܐܫ̈ܝܢܐ ܒܡܠܝܘܬ ܐܣܝܘܬܐ ܒܣܝܪܬܐ ܘܡܣܟܢܐܝܬ
ܢܣܒܘܢܐ ܘܬܬܒܢܬ ܢܦܫܢܐ ܗܘܐ . ܐܢܫܢܐ ܕܠܘ ܒܠܒܘܫ ܦܫܝܛܐ . ܐܠܐ ܐܦ
ܡܬܬܥܝܢܐ ܚ̈ܫܘܟܐ ܡܢ ܦܘܪܫܢܐ ܕܒܝܢ ܫܪ̈ܝܪܐ ܠܐܠܗܐ ܢܦܩ ܒܦܘܪܫܢܗܘܢ .
ܘܗܘ ܡܢܐ ܕܡܢ ܠܒ̈ܣܡܐ ܗܠܝܢ ܡܬܬܪܐ ܝܟܬܒ ܟܘܢܐ ܗܘ ܐܦ ܣܘܡܟܐ ܕܗܘܝܢ
ܘܫܠܝܢ . ܘܗܠܝ ܡܢ ܢܦܫܐ ܕܗܘ ܕܡܣܒܪ . ܘܦܘܪܫܢܐ ܡܬܬܪܝܢܐ ܕܐܣܘܣܝܘܬܗܘܢ
ܡܢ ܚܡܝܢ . ܘܐܢܫܢܐ ܗܘ ܡܢܐ ܕܠܐ ܢܕܚܠ ܢܬܡܢܐ ܠܡܚܒܒ .

[III.iii.3.]

ܕܬܠܬ . ܦܠܘܚ ܒܡܠܐܟܐ ܒܥ̈ܒܝܕܘܬܗܘܢ ܫܦܝܪ ܕܠܡܒܚܪܢܘܬܐ ܠܐ ܢܩܘܡ ܐܢܫ̈ܝܢ .
ܐܠܐ ܒܩܕܝܫܘܬܐ ܕܒܫܝܢܐ ܫܘܒܚܐ ܐܦ ܡܩܕܫܢܘܬܐ ܐܝܟ ܗܘ ܕܒܚܢܐ ܒܗ ܡܬܚܙܝܢ
ܐܡܝܢܝܢ . ܘܠܡܬܠܠܝܬܗ ܒܚܘܒ ܫܠܝܠܐ ܡܟܝܢܐ ܐܬܚܪܝܢ . ܘܡܛܠ ܗܕܐ ܐܦ
ܕܡܢܐ ܒܝܕܝܢ . ܡܟܝܪܝܢ ܕܠܐ ܐܝܟܢܐ ܐܦܠܐ ܒܝܕܗ . ܒܗ ܫܠܐ ܒܝܕܥܐ
ܒܗ ܕܡܢ ܠܡܬܚܙܝܘ ܒܠܘܩܒܠ ܡܬܚܙܝܢ . ܘܟܠܗܝܢ ܡܬܝܕܥܢ ܕܗܠܝܢ ܒܫ̈ܡܝܢܐ ܠܐ
ܢܦܚ ܠܝ . ܐܦܠܐ ܡܬܚܙܝܢܘܬܐ ܗܝ ܠܝ ܕܥܒܕܝܢ . ܐܟܡܢܐ ܕܠܐ ܡܟܝܢ ܠܝܒܪ
ܡܦܠܓܐ ܘܫܠܐ ܒܚܘܠܐ ܕܒܗ ܘܒܐܝܕ̈ܝܗܘܢ ܐܬܡܠܠ ܒܗܘܢܐ . ܘܬܘܒ ܠܥܠܡܐ .
ܘܐܬܚܙܝ ܒܗܘܢ ܐܦܪܘܛܐ . ܘܐܢܫܝܢ ܐܝܬܝܗܘܢ . ܐܠܐ ܠܒܣܬܪܐ ܢܚܘܬ ܒܝܕܥܐ
ܒܝܕܥܬܐ ܡܬܕܒܪܝܢ . ܘܠܡܥܒܕܢܘܬܐ ܒܢܝܫܢܐ . ܒܪܡ ܐܟܡܢܐ ܕܠܦܫܘܩ ܡܢ
ܐܘܠ̈ܨܢܐ ܡܦܠܓܐ ܕܒܗ ܢܒܝܥ ܐܦܪܘܛܐ ܐܦ ܘܡܢܐ ܠܐ ܡܟܝܢ . ܐܢܫܢܐ
ܕܢܦܠ ܡܢܗܘܢ ܕܡܢܐ ܐܦ ܡܢܐ ܡܟܝܢ .

FMWlVλ: III.iii.2: 1 ܒܡܬܕܡܪܢܘܬܐ W || 2 ܕܡܣܟܝ̈ܢܘܬܗܘܢ F || 3 ܐܢܫ ܠܐ : ܠܐܢܫ Λ || ܘܠܐ: ܠܐ Λ || 4 ܐܫ̈ܝܢܐ: ܐܫܝܢܐ MVλ || 5 ܢܩ̈ܘܦܐ F || ܕܠܘ: ܠܘ Λ || 6 ܡܢ: ܡ̇ܢ M; ܡ̇ܢ ܡܢ Vλ || 7 ܗܘ [post ܟܘܢܐ]: om. V || 8 ܡܬܬܪܝܢܐ W || ܕܐܣܘܣ̈ܝܘܬܗܘܢ Vλ || 9 ܕܡܚܡ Vλ || **III.ii.3:** 2 ܫܘܦܪܬܐ Λ || ܡܩܕܫܢܘܬܐ Λ || 3 ܒܚܘܒ W || 4 ܕܠܐ: ܠܐ V || ܚܒ̈ܘܒܐ W || 6 ܠ sec.: supra lin. W; om. l || 7 ܕܚܝܠܐ FΛ || ܘܒܐܝܕ̈ܝܗܘܢ: ܘܐܢܫܢܐ l || 8 ܐܦܪܘܛܐ V || ܒܠܘܩܒܠ ΛVλ || ܘܠܡܥܒܕܢܘܬܐ Λ || 10 ܐܘܠ̈ܨܢܐ: hinc **FMWPlVλ** || ܪܘܡܢܐ ܐܦ: ܘܪܘܡܢܐ Λ || ܐܢܫܢܐ: ܐܢܫܐ Λ ||

[III.iii.2.]
Second [theory]: In these imitations, the essential distinctions whereby these bodies are differentiated into species are retained, while the accidental colours and consistencies which are not a part of their quiddity are altered. Therefore no intelligent person would dispute, and would not accept, the possibility of an imitation which reaches an utmost level [of perfection] through subtle and proven art and [through] much practice and assiduous experimentation, so that not only the layman but also the best money-changers fail to distinguish between the genuine and counterfeit coins. What our sense perceives in these bodies is [accidents such as] colour or consistency which undergo generation and cessation, while [their] substance remains uncorrupted, and the distinctions which constitute their essential character are concealed from us. [This being the case] how can we produce something which we do not know?

[III.iii.3.]
Third [theory]: Alchemists say in those fables of theirs which are never realised: "if we were to find pure, white or red sulphur like (the sulphur) which nature uses and were able to mix it firmly with mercury, we would [be able to] produce silver or gold." In reality, they have neither found [such sulphur] nor produced [silver and gold]. For the efficient power is found only where it can be found, while the proximate causes of these bodies are not known to us, nor is it possible for us to know them, just as we are unable to know the mixture and the formative power by and through which a walnut has been made round, an almond has been made elongated, the fleshy part of a citron has been made sweet and its pulp has become sour, but we merely perceive the outward appearance through sensation, and the effects through experience. Therefore, seeing that we are unable to extract from the elements a mixture with which to make a citron or a pomegranate, how can we mix [and produce] gold and silver from them?

[tit.]

ܡܠܐܘܢ ܪܒܝܥܝܐ . ܡܛܠ ܥܒ̈ܕ ܚܒܝܪܬܐ . ܘܗܘ ܐܝܬ ܦܣܘ̈ܩܐ ܐܪܒܥܐ .
ܦܣܘܩܐ ܩܕܡܝܐ . ܡܛܠ ܡܫ̈ܬܡܗܐ ܚܝ̈ܢܐ ܕܐܪ̈ܥܐ . ܬܐܘܪ̈ܝܣ ܐܪܒܥ .

[IV.i.1.]

ܡܙܡܢܬܐ . ܟܕ ܗ̇ܘ ܕܐܪܥܐ ܟܠܗ ܡܠܝܐ ܗ̇ܘܬ . ܘܡܠܝܐ ܡܢ ܟܠܗܘܢ ܟ̈ܝܢܗ
ܢܗܝܪܐܝܬ . ܟܕ ܟܠܗ ܡܟܝܠ ܩܢܘܡܐ . ܠܐ ܠܘܬ ܗ̇ܘ ܡܟܐ ܕܟܝܢ ܠܐܪܥܐ
ܘܠܡܠܝܐ ܐܫܬܟܚܘ . ܐܠܐ ܠܘܬ ܗ̇ܘ ܡܟܐ ܕܟܝܢ ܠܡܫܬܡܫܢܘܬ ܚܝܐ ܟܠ . ܟܕ ܗ̇ܘ
ܕܐܣܛܘܟܣܐ ܡܟܝܢܬ ܫܘܠܛܢܐ ܕܠܫܪܪܐ ܚܝܢ . ܟܕ ܡܢ ܡܢ̈ܘܬܐ ܡܢ ܐܪܥܐ
ܟܕ ܡܠܝܐ ܐܦ ܡܟܝܠ ܐܚܝܢ ܗܦܟ . ܘܗ ܐܣܛܘܟܣܘܬܗ ܡܬܬܚܬܝܐ . ܘܡܠ̈ܟܐ
ܘܫܡ̈ܝܐ ܘܦܩ̈ܕܬܐ ܘܚܝܠܐ ܒܗ ܗܘܝܢ . ܘܡܟܐ ܕܡܢ̈ܘܬܐ ܕܡܢ ܡܠܝܐ ܐܦ ܡܢ
ܥܠܡܐ ܕܐܣܛܘܟܣܐ ܐܪܥܐ ܗܦܟ . ܥܠܝܬܐ ܘܡܠܝܐ ܒܡܫܬܡܫܢܘܬܗ ܐܣܛܘܟܣܝܬܐ
ܕܟܢ . ܘܗܟܝܠ ܕܦܫ̈ܛܬܐ ܡܢ ܐܪܥܐ ܡܢ ܥܠܡܐ ܡܠܝܐ ܘܐܣܛܘܟܣܐ ܕܩܢܘܡܐ
ܘܡܟܢܫܢܐ ܡܬܚܙܝܢ .

[IV.i.2.]

ܕܬܬܬܝܢ . ܐܦ ܟܕ ܘܠܩ̈ܒܒܐ ܠܘܬ ܡܫܬܡܫܢܘܬܐ ܗ̇ܢܝܢ ܕܥܠ̈ܝܟܠܗܘܢ
ܡܫܬܠܡ ܡܬܟܪܝܢܘܬܐ ܐܝܟ ܕܟܠܗ ܒܟܘܠܗ ܐܣܛܘܟܣܐ ܠܫܪܪܐ . ܡܛܠ ܕܡ
ܠܡܬ̈ܝܪܢ ܗܘܝܢ ܕܐܝܟ ܕܟܡ ܕܠܘܬܐ ܕܐܣܛܘܟܣܘܬܗ ܒܡܬܟܪܝܢܘܬܐ
ܘܐܪܟܟܢ ܥܠܝܢ ܡܟܝܠ ܡܬܠ ܠܗܪܟܐ ܡܟܝܢ . ܘܒܐܝܟܢܘܬܗܘܢ ܗܦܟ ܡܟܝܢܐ
ܠܗܘܢܝܢ . ܚܕܐ ܡܬܪܐ ܟܬܒܝܢ ܥܠܝܢ . ܐܟܡܐ ܕܐܬܘ ܐܠܗܝܘܬܐ ܟܝܢܝܢ
ܠܐܣܛܘܟ ܕܣܘܟܠܐ ܗܢܐ ܦܩܕ . ܟܠ ܫܘܪܝܐ ܕܥܠܝ ܡܠܟܘܬܗ ܕܐܝܟܢܘܬܗ .
ܟܡ ܟܝܢܐ ܐܦܩܕܟܝܢ ܘܐܣܛܘܟܣܝܢ ܡܟܝܠܝܢ ܠܐܦܩܕܢܘܗ ܕܠܗܘܢ ܘܟܠܗܝܢ
ܗܘܝܢ ܬܘܪܟܬܐ ܒܡܬܩܢܢܘܬܐ ܕܡܟܝܠ ܒܟܝܢܐ ܡܟܝܢ ܘܡܬܩܝܡܝܢ ܒܐܝܬܘܬܐ .
ܘܡܟܝܢܐ ܒܡܫܡܫܢܘܬ ܢܟܝܠ ܢܗܘܐ ܗܘܐ . ܐܦ ܟܕ ܗܘ ܡܫܡܫܢܘܬܐ ܐܠܗܝܬܐ
ܡܟܝܢܐ ܕܗܘܬ ܕܠܫܡܝܐ ܡܫܟܚܬܐ ܐܠܗܝܬܐ ܟܝܢܐ ܢܗܘܐ .

FMWPLIV: **IV.i.1**: 2-3 ܠܐ ܠܘܬ ... ܐܫܬܟܚܘ: om. F || 3 ܠܡܫܬܡܫܢܘܬܗ ܕܚܝܐ M || 4 ܒܫܘܠܛܢܐ WPl || ܠܫܪܪܐ V || 5 ܗܦܟ FM || 7 ܐܣܛܘܟܣܝܬܐ L || **IV.i.2**: 1 ܡܫܬܡܫܢܘܬܐ F; ܡܫܬܡܫܢܘܬܐ Λ; ܡܫܬܡܫܢܘܬܐ V || 2 ܡܬܟܪܝܢܘܬܐ ܐܝܟ ܕܟܠܗ F || 3 ܕܐܣܛܘܟܣܘܬܗ F; ܕܐܣܛܘܟܣܘܬܗ Λ || 4 ܘܐܪܟܟܢ FM || 4 ܘܒܐܝܟܢܘܬܗܘܢ F || ܡܟܝܢܐ ܗܦܟ: add. ܘܐܪܟܟܢ F || 5 ܠܗܘܢܝܢ: add. ܠܗܘܢܝܢ ܡܬܪܐ ܘܡܢ ܡܟܝܢ F || ܟܝܢܝܢ: ܟܝܢܘܢ F; ܟܝܢܝܢ P || 6 ܠܐܣܛܘܟ F; ܠܐܣܛܘܟ MV || 7 ܟܝܢܐ: ܟܝܢܐ F || 8 ܡܟܝܢ F || ܘܡܬܩܝܡܝܢ F || 9 ܢܟܝܠ F || ܗܘܐ: ܗܘܐ M || 10 ܐܠܗܝܬܐ: ܐܠܗܝܬܐ V ||

FOURTH CHAPTER

ON MATTERS CONCERNING THE HABITABLE WORLD

It contains four sections.

First Section: On the Natural Position of the Earth. Four theories.

[IV.i.1.]

First [theory]: Although the nature of (earth and water) requires that earth be inside water and that water encircle it on all sides, [in reality] they have been found not [to be] in accordance with what is natural to earth and water, but in accordance with what is natural to the order of the universe. [That is] because the elements have a capacity for transformation into each other, so that when particles of earth turn into water or something else, the sphericity of (the earth) is breached and defiles, gorges, plains and valleys are formed in it. When [on the other hand] particles of water or other elements turn into earth, peaks and high grounds [are formed and] become attached to its spherical surface and in this way parts of the earth are freed from the rule and grasp of water which confines and conceals [the earth].

[IV.i.2.]

Second [theory]: Stars too have a major effect on the transformation of elements into one another in accordance with those coordinates which change with their movements - especially those fixed (stars) which, according to the opinion of the ancient astrologers, move eight degrees to the north in six hundred and forty years and [then] also eight degrees to the south in the same (number of years), [i.e.] one degree in eighty years, as Theo of Alexandria stated in his *Canon* as the method for this computation, based on the beginning of the reign of Augustus. In the same way the apogees and perigees change their positions and they become major causes of increase of water in one region and its decrease in another. In this way in the middle of the sea dry land is formed. For divine wisdom too has thus deemed it just that there be a natural place for animals that breathe [the air].

[IV.i.3.]

ܕܬܠܬ. ܠܐܘܣܛ ܐܪܥܐ ܡܢ ܚܘܕܪܐ ܗܘ ܕܐܘܪܝܙܘܢ ܠܬܪܝܢ ܦܠܓܝܢ ܓܠܝܐ
ܘܡܫܘܚܬܐ ܦܫܛ ܠܗ. ܘܚܘܕܪܐ ܗܘ ܕܡܢ ܡܕܢܚܐ ܠܡܥܪܒܐ ܠܐܘܩܝܢܘܣ.
ܦܠܠܠ ܠܚܠܣܪ ܡܢ ܦܠܓܐ ܓܠܝܐ ܡܫܘܚܬܐ ܠܬܪܝܢ ܦܠܓܝܢ ܓܪܒܝܐ ܘܬܝܡܢܝܐ
ܦܫܛ. ܘܡܕܢܚܐ ܡܢ ܐܘܣܛ ܐܪܥܐ ܠܐܪܒܥ ܡܢܘܢ ܡܬܦܠܓܐ. ܘܡܥܪܒܐ ܡܢ
ܬܘܒܒܐ ܓܠܝܐ ܓܪܒܝܐ ܘܠܐ ܒܟܠܗ. ܥܕ ܚܘܫܒܘܢ ܦܠܝܓܐ ܡܫܬܟܚ. ܥܕ ܟܠ
ܦܠܝܓܐ ܒܐܘܪܟܐ ܗܢ ܡܢ ܡܕܢܚܐ ܠܡܥܪܒܐ ܗܢܘ. ܒܦܬܝܐ ܕܝܢ ܡܢ ܬܝܡܢܐ
ܠܓܪܒܝܐ. ܘܐܪܒܥܐ ܕܡܢܝܬܐ ܦܠܓܗ ܕܚܘܕܪܐ ܐܪܒܥܬܗܘܢ. ܘܟܠܗ ܡܬܥܡܪܢܐ
ܗܘ. ܒܡܫܚ ܕܪܘܚܗ ܡܢ ܬܬܚܘܬܝܗܘܢ ܠܐ ܡܬܡܙܓܢܘܬܐ ܕܚܘܡܐ ܘܩܘܪܐ ܡܬܥܡܪܢܘܬܐ.
ܦܢܝܬܗ ܕܝܢ ܓܒܝܬ ܡܢ ܬܘܒܒܐ ܕܚܘܕܪܐ. ܥܕ ܡܢ ܬܝܡܢܐ ܥܒܪ ܚܘܡܐ ܡܘܡܪܢܐ
ܠܐܪܥܐ ܕܢܘܒܝܐ ܠܗܢܘܢ ܕܡܫܘܚܬܐ ܠܐ ܡܬܠܝܐ. ܘܡܢ ܓܪܒܝܐ ܥܒܪ ܩܘܪܐ
ܡܬܠܝܐ ܠܐܪܥܐ ܕܬܘܠܬ ܗܘܠܘ ܓܪܒܝܐ ܠܐ ܦܪܚ.

[IV.i.4.]

ܕܐܪܒܥ. ܟܠܗܝܢ ܕܝܢ ܡܕܢܐ ܠܬܘܒܒܐ ܐܣܝܐ ܓܠܝܐ ܘܬܪܝܢܘܬܗܘܢ ܬܫܥܝܬܐ
ܓܫܡܐ ܐܝܬ ܠܗ. ܘܠܐ ܚܕܐ ܡܫܘܚܘܬܐ ܦܫܝܛܬ ܗܘܬ ܡܢ ܡܫܘܚܬܢܘܬܗ ܐܠܝܨܘܬܐ
ܕܫܘܚܬܐ ܡܕܢܐ ܠܐܪܥܐ. ܘܟܪܟ ܦܘܪܢܣܗ ܟܟܒܐ ܠܐܪܥܐ ܕܡܫܘܚܬܐ. ܡܛܠ
ܕܟܢܝܫ ܒܗܢܐ ܗܘ ܐܩܠܝܡܐ ܕܠܗ. ܒܟܠܗ ܚܕܐ ܚܘܟܡܐ ܡܫܬܟܚܐ ܬܡܢ.
ܘܟܠܗܝܢ ܠܐ ܡܬܡܢܝܢܐ. ܠܐ ܓܘܐ ܠܡܘܟܠܐ. ܟܕ ܕܡܢܝܢܐ ܕܡܫܟܚܐ ܡܢ
ܡܠܦܢܐ ܕܝܘܢܝܐ ܟܠ ܠܗ ܗܘܝܐ ܟܠܗ ܢܦܫ. ܐܝܟܢܐ ܕܠܐ ܢܫܬܟܚ ܐܬܪܐ ܗܘ.
ܘܟܠ ܐܝܟ ܕܐܡܪܝܢ ܚܕ ܗܘܝܐ ܬܚܘܬ ܒܟܠܗ ܘܫܪܟܗ. ܠܬܝܡܢܐ ܡܟܝܠ
ܐܝܟܢܐ ܕܠܟܠ ܡܢ ܪܒܥܐ ܠܗ ܚܕܐ ܟܢܫܐ ܡܬܡܠܝܐ ܦܠܐ ܠܡܥܡܪܐ. ܘܒܗܢܐ
ܠܗܘܢܐ ܐܢܫܝܐ. ܐܦ ܠܐ ܡܫܬܟܚ ܒܗܢܐ. ܒܗܘܢܐ ܕܝܢ ܕܗܘܪܐ ܕܒܥܘܬܐ
ܘܠܟܪܝܗܐ ܡܢ ܦܠܝܓܐ ܓܒܝܬܐ. ܐܝܟ ܐܬܪܘܬܐ ܕܟܠܝܐ. ܟܪܟ ܐܢܫܐ ܗܢܘܢ
ܠܗ ܐܢܫܐ ܕܟܠܗܘܢ ܐܝܟ. ܘܠܫܢܬܐ ܒܟܬܝܒܬܐ ܕܟܕ ܗܘܝܢ ܡܠܟܐ ܡܫܠܝܢ.

FMWPLIV: **IV.i.3**: 1 ܗܘ: om. F || 3 ܦܠܠܐܠ F || 11 ܗܘܠܘ: ܗܘܠܘܢ V || **IV.i.4**: 4 ܗܘܐ ܦܘܪܢܣܗ V || 5 ܡܬܡܢܝܢ FMV || 6 ܢܫܬܟܚ V || ܕܠܐ ܢܫܬܟܚ: ܕܢܫܬܟܚ MV || 11 ܡܠܟܐ: ܡܠܟ V ||

[IV.i.3.]
Third [theory]: The circle of the horizon [*sic*] divides the terrestrial sphere into two halves, the upper and the lower [i.e. eastern and western], and that circle which is parallel to the isemeric (circle) from east to west divides each of the halves, the upper and the lower, into [further] two halves, the northern and the southern. Thus the terrestrial sphere is divided into four parts. The seven climes are situated in a part of the upper northern quarter, not the whole of it. For every clime extends from east to west in length and from south to north in breadth; and the length of the habitable world is a semi-circle, and all of it is habitable because it is far from the two extremes of deadly heat and cold, whereas its breadth is less than a quarter of the circle, since in the south it fails to reach the land near the equator because of the burning heat and in the north it fails to come near the land which is below the north pole because of the freezing cold.

[IV.i.4.]
Fourth [theory]: As to whether water covered the other upper quarter and the two lower (quarters) or not, there was not even one proof [that it did], except the notion of the necessity of the confinement of earth by water. The argument from the proximity of the sun to the earth in the south, [namely that] because its perigee is at the head of Capricorn, scorching heat prevails there and (it) is uninhabited for this reason, is not acceptable, because the deviation of the sun from the centre of the universe is not so very great as to make that place uninhabitable. Even if, as they say, the heat is strong below and around the [Tropic of] Capricorn, to the south of it, where the sun does not pass the zenith, it ought to be temperate and suitable for human habitation, provided it is not covered with water. In the proximity of the equator and to the north of the seventh clime, there are a few [inhabited] places, but those men [living there] are not men of understanding and are akin to the savage animals which are numerous there.

[tit.]

ܦܘܣܩܐ ܬܪܝܢܐ. ܡܛܠ ܡܩ̈ܦܠܐ ܐܬܪ̈ܘܬܐ. ܬܐܘܪ̈ܝܣ ܫܬ.

[IV.ii.1]

ܩܕܡܝܬܐ. ܩܕܡܝܐ ܕܡܫܬܘܕܥܢܘ ܚܡܫܐ ܦ̈ܠܓܐ ܦܠܓܗ̇ ܠܐܣܛܪ ܐܪܥܐ
ܒܐܪ̈ܒܥܐ ܚܘܕ̈ܪܐ ܕܦܠܠܝܢ ܠܐܘܩܝܢܘܣ. ܬܪܝܢ ܡܢ ܚܘܕ̈ܪܐ ܬܪܝܢ ܦ̈ܠܓܐ
ܓܪ̈ܒܝܝܐ ܒܪ̈ܝܬܐ ܘܡܬܝܒܢܐ ܦܪܝܫܝܢ ܡܢ ܐܪ̈ܥܐ. ܘܠܒܟܠܗ ܡܢܗܘܢ ܡܬܚܬܐ ܡܢ
ܣܦ̈ܝܬܗ̇ ܘܡܬܝܒܘܬܐ ܬܬܝܒܘܬܐ ܬܟܝܠ. ܘܬܪܝܢ ܚܘܕ̈ܪܐ ܐܚܪ̈ܢܐ ܬܠܬܐ ܦ̈ܠܓܐ
ܕ̈ܝܡܢܝܐ ܦܪܝܫܝܢ ܡܢ ܐܪ̈ܥܐ. ܘܠܒܟܠܗ ܡܢܗܘܢ ܬܬܝܢ ܡܬܝܒܘ̈ܬܐ ܬܬܝܒܘܬܐ ܬܟܝܠ.
ܕܦܠܓܐ ܡܬܚܬܐ ܚܕ ܕܡܬܢܒܝܐ ܕܒܘܬܐ. ܕܦܠܓܐ ܐܚܪܢܐ ܒܪܬ ܠܗ ܠܓܪܒܝܐ
ܒܪ̈ܝܬܐ. ܐܚܪܢܐ ܒܪܬ ܠܗ ܠܓܪܒܝܐ ܬܬܝܒܘܬܐ.

[IV.ii.2.]

ܕܬܪܬܝܢ. ܫܡܫ ܕܡܫܚܐ ܦܠܓܗ ܐܡܬܝ ܕܬܪܝܢܘܬ ܦ̈ܠܓܐ ܓܪ̈ܒܝܝܐ ܒܗ ܗܘܐ
ܡܥܢܐ ܕܡܢ ܪܘܚܐ ܫܡܫܐ ܠܐ ܡܬܚܙܝܐ. ܘܫܡܫܐ ܕܦܠܓܐ ܡܬܚܬܐ ܒܗ ܚܘܫܒܐ
ܦܘܪܫܐ ܕܡܢ ܗܘܐ ܫܡܫܐ. ܘܦ̈ܠܓܐ ܬܪܝܢ ܐܚܪ̈ܢܐ ܒܗ ܪܘܚܐ ܡܢ ܬܬܝܒܘܬܐ
ܠܐ ܡܫܬܟܚܢܘܬܐ ܠܕܘܟܐ ܐܝܟܢܐ ܕܫܡܫ. ܒܬܪ ܕܦܠܓܐ ܡܢ ܒܪ̈ܝܬܐ ܢܕܥܢ ܠܗ
ܒܥܡܘܪ̈ܐ ܘܚܙܝܐ ܕܒܥܡܘܪ̈ܐ. ܘܒܗ ܫܡܫܐ ܡܠܝܐ ܣܠܩ ܠܓܪܒܝܐ
ܡܚܝܢ. ܕܦܠܓܐ ܕܝܢ ܡܬܚܬܐ ܠܐ ܢܕܥ ܠܗ ܒܗ ܒܥܡܘܪܐ ܗܘܢ. ܐܦܠܐ ܒܐܝܕ
ܡܢ ܕܪ̈ܘܚܐ ܦܓܥܢ. ܐܠܐ ܒܠܚܘܕ ܡܬܚܙܝܢܘܬܐ ܕܘܟܐ ܕܬܡܢ ܘܒܚܘܢܬܐ
ܒܥܡܘܪܐ ܒܡܬܚܙܝܢܘܬ ܐܝܟ ܢܕܥܝܢ. ܐܢ ܠܝܬ ܒܠܗܐ ܐܚܪܬܐ ܕܒܠܚܘܕ ܐܝܟ
ܕܡܬܚܙܐ ܘܦܠܓ.

[IV.ii.3.]

ܕܬܠܬ. ܗܟܢܐ ܪܡܙܢܐ ܒܥܠܡܘܬܐ ܕܦܬܝܟܘܬܐ ܡܬܚܘܝܢ ܡܬܬܪܝܢ ܘܐܡܬܝ.
ܕܢܗܘܐ ܡܢ ܟܠ ܐܝܟܐ ܗܘܝܐ ܕܚܘܒܬܐ ܒܥܡܘܪܘܬܐ ܘܡܬܝܪܬܐ ܡܬܬܓܪ.
ܘܠܬܪܬܝܗܘܢ ܗܢܘܢ ܒܐܪ̈ܒܥܐ ܦ̈ܢܝܢ ܡܬܬܪ. ܗܘܝܐ ܡܕܢܚܐ. ܡܥܪܒܐ.
ܒܐܝܕܐ ܘܡܬܗܘܝܢܘܬܐ ܠܟܠ ܡܢ ܪܘܚܐ ܚܕܐ ܬܗܘܐ. ܘܒܪܘܚܐ ܡܬܢܝܐ ܒܗ
ܬܬܗܘܝܘܬܐ ܕܝܠܗ. ܘܒܠܚܘܕ ܠܗ ܗܘܐ ܚܘܫܒ. ܘܠܐ ܗܘܝܐ ܒܥܠܡܐ ܬܬܝܒܘܬܐ.
ܒܪ ܕܒܝܬ ܗܘܐ ܠܟܠ ܡܢ ܪܘܚܐ ܚܕܐ ܥܡܡܐ. ܘܦܘܬܝܐ ܗܘܝܐ ܐܝܬܘܬܐ
ܒܗܘܢ ܡܢܗ ܫܘܐ ܒܗ ܒܐܪܒܐ ܕܝܠܗ. ܘܒܠܚܘܕ ܗܘܐ ܚܘܫܒ. ܒܗ
ܢܗܘܐ ܒܪ ܗܘܝܐ ܡܢ ܚܕܐ ܐܣܛܘܟܣܝܐ ܡܬܝܒܐ ܝܠܝܐ. ܗܘ ܕܝܢ ܕܗܘܐ ܒܗܘܝܐ
ܘܐܪ̈ܒܥ. ܚܕܚܕܐ ܡܢܗ ܒܥܡܘܪܐ ܬܬܝܒܘܬ ܢܗܘܐ ܡܢ ܗܘ ܕܗܘܝܐ ܕܚܘܒܬܐ
ܒܪܝܬܐ.

FMWPLIV: **IV.ii.1**: 1 ܕܡܫܬܘܕܥܢܘ V || ܦܠܓܗ̇ ܦ̈ܠܓܐ V || 2 ܕܦܠܠܝܢ F || ܡܢ: ܡܢ̈ MPLI || ܘܪܒܐ: om. F || 3 ܒܪ̈ܝܬܐ ܘܡܬܝܒܢܐ PLI || ܦܪܝܫܝܢ MΛV || 4 ܬܟܝܠ: ܢܗܘܐ Λ || ܚܕܘܪ FMV || 5 ܦܪܝܫܝܢ MΛV || **IV.ii.2**: 1 ܒܗ: ܒܗ Λ || ܡܬܚܙܝܢ: ܡܬܚܙܝܢ MΛV || 3 ܕܬܪܝܢ PLI || 4 ܡܫܬܟܚܢܘܬܐ: ܠܐ PLI || 5 ܘܚܙܝܐ ܕܒܥܡܘܪܐ: om. F || 6 ܗܘܐ FMV || 7 ܦܓܥܢ Λ || **IV.ii.3**: 1 ܒܥܠܡܘܬܐ ܒܗ V || 5 ܒܠܚܘܕܐ V || 7 ܒܗܘܢ ܡܢܗ ܠܒܪ̈ܝܬܐ PLI; ܒܗܘܢ ܒܪ̈ܝܬܐ M; ܡܢܗ ܒܪܝܬܐ V || ܚܘܫܒ ܒܠܚܘܕܐ V || 10 ܒܪܝܬܐ F ||

Second Section: On Climates of [Various] Regions. Six theories.

[IV.i.1.]
First [theory]: The Peripatetics of old divided the terrestrial sphere into five divisions with four circles parallel to the isemeric. Two circles separate two tympanic divisions, the northern and the southern, out of the earth and each of these is confined by a part of the circumference (of the earth) and a flat surface. Two other circles separate three trapezial sections out of the earth and each of them is confined by two flat surfaces. The middle trapezial (division) [is located] where the equator (is), another trapezial (division) between it and the northern tympanic (division), and the other (trapezial division) between it and the southern tympanic (division).

[IV.ii.2.]
Second [theory]: Those who made such a division say that the two tympanic division are uninhabitable because of the severe coldness arising from the distance of the sun, and similarly the middle trapezial (division) because of the burning heat arising from the proximity of the sun. The other two trapezial (divisions), because they are far from the two extremes, are useable for human settlement. However, while the inhabitants of the northern trapezial (division) and the customs of its inhabitants are known to us - and here are situated these celebrated seven climes - we know of no inhabitant in the southern trapezial (division) at all, nor have we met anyone from its settlements. Rather, we only know of the possibility of settlement there and (its) suitability for habitation due to the temperateness of the air, provided there is no other cause preventing [habitation], such as [the presence of] seas.

[IV.ii.3.]
Third [theory]: The Princely Doctor [Ibn Sīnā] thinks the opposite of the Peripatetics, his teachers, and says that the equator is more temperate as regards heat and cold than all [other] places and confirms this opinion of his with four arguments.

First argument: The sun passes the zenith in Aries and Libra there and changes position with speed because of the increase [in the rate] of declination. For this reason it does not create much heat. [This is] not so in the second clime, where the sun passes the zenith at the head of Cancer and remains for many days in its proximity because of the decrease [in the rate] of declination. For this reason it creates much heat. For the declination changes by one minute in one day (in the second clime), but by twenty-four (minutes) there [at the equator]. Therefore, the heat in the second clime is stronger than that at the equator.

[IV.ii.4.]

ܕܐܬܪܒܒ . ܗܟܢܐ ܬܪܝܢܐ . ܒܣܘܛܠܐ ܕܫܘܚܬܐ ܐܣܦܩܢܝܐ ܠܠܠܝܐ ܒܟܠܗ ܫܢܬܐ ܥܝܢܐ . ܬܘܒ ܡܕܡ ܣܡܩܡܩܘܬܐ ܐܣܦܩܢܝܐ ܒܡܫܪܬܐ ܠܠܝܐ ܡܬܚܙܝܐ ܒܟܠܗ ܫܢܬܐ . ܘܫܪܐ ܡܢܗܘܢ ܠܐܣܝܘܬܐ ܠܐ ܚܠܒܐ . ܘܡܫܬܘܚܐ ܫܒܥܐ ܒܐܬܪܘܬܗ ܐ̈ܣܦܩܢܝܐ ܡܢܗܝܢ ܣܡܩܡܩܘܬܐ ܚܠܒܐ . ܘܒܐܬܪܘܬܗ ܠܠܐܬܐ ܡܫܘܚܬܐ ܡܪܘܬܐ . ܘܡܫܬܘܚܠܦܐ ܒܐܬܪܐ ܘܡܬܚܙܝ ܒܠܘܩܒܠ ܗܘܐ . ܒܣܘܛܠܐ ܕܫܘܚܬܐ ܐܪܐ ܕܐܬܚܙܝܘܬܐ ܗܝ ܒܠܗ ܥܕܡܐ . ܘܒܐܬܬܘܬܐ ܐܠܗܝܐ ܒܢܝܢ ܡܢ ܣܡܩܡܩܘܬܐ ܚܘܪܐ . ܒܢܝܢ ܕܝܢ ܡܪܘܬܐ .

[IV.ii.5.]

ܕܣܡܟܐ . ܗܟܢܐ ܬܠܝܬܝܐ . ܒܢܝܚܘܬ ܗܘܪܡܐ ܕܫܡܫܐ ܠܟܠ ܡܢ ܪܫܐ ܢܗܘܐ ܡܢ ܗܘܪܡܐ ܢܦܫܗ . ܘܟܠܡܐ ܒܪܫ ܗܘܦܠܐ ܒܝܬ ܡܢ ܕܒܪܫ ܐܪܥܐ ܢܦܫܗ ܫܡܫܐ . ܘܐܦ ܒܗܘܦܠܐ ܢܗܘܐ ܡܢܒܥ ܠܢܦܫ . ܐܦ ܟܕ ܘܒܬܪܐ ܘܬܘܪܡܐ ܒܝܬ ܡܢ ܡܢܐ ܕܒܐܪܥܐ ܘܫܒܠܬܐ ܢܦܫܗ . ܘܡܢܗ ܦܠܓܐ ܒܝܬ ܡܢ ܒܬܪ ܦܠܓܐ . ܘܐܦ ܒܬܪܝܗܘܢ ܗܘܪܡܐ ܒܫܡܐ ܡܢܒܥ ܠܝ . ܘܡܠܝ ܡܢ ܒܝܬ ܕܢܦܫܡܝܢܐ ܒܪܢܐ ܡܪܡܢܐ ܒܡܫܡܫܢܘܬܐ ܡܪܡ ܢܦܫܗ ܠܐܪܥ . ܘܒܪܗ ܢܗܘܐ ܒܫܡܝܐ ܠܗܘܠ ܫܡܫܡܘܬܐ ܒܪܢܐ ܦܬܝܢܐ ܗܘܐ . ܘܟܠܡܐ ܒܬܪܝܢܐ ܢܗܘܐ ܡܢ ܕܒܡܪܡܢܐ ܚܙܝܢ ܫܡܝܐ . ܒܣܘܛܠܐ ܕܫܘܚܬܐ ܐܪܐ ܒܪ ܗܘܪܡܐ ܕܫܡܫܐ ܗܘܦܠܐ ܡܢ ܢܚܝܬܘܬ ܗܘܪܡܐ ܫܡܫ . ܠܐ ܒܪܝ ܫܡܝܐ .

[IV.ii.6.]

ܕܫܬ . ܗܟܢܐ ܪܒܝܥܝܐ . ܒܣܘܛܠܐ ܕܫܘܚܬܐ . ܟܕ ܠܘ ܗܟܢ ܡܬܚܙܝܐ ܫܒܝܚܐ ܡܢ ܓܠܝܐ ܪܫܐ . ܠܘ ܡܢ ܗܘܪܐ ܗܘܟܢܐ ܠܘܬ ܫܡܫܐ ܡܢܗܝܢ ܒܡܫܘܪܗܝܢ . ܐܝܟܢܐ ܕܡܫܘܚܬܢܐ ܠܪܟܢܘ . ܟܕ ܗܘܦܠܐ ܘܒܪܢܐ ܡܠܟܐ ܡܬܚܙܝܢܘܢ ܡܢ ܐܡܪܐ ܘܡܫܘܚܬܐ ܡܠܟܐ ܡܢܝܢܘܢ ܬܘܒܪܐ ܕܫܡܪܐ ܒܠܘܩܒܠ ܪܫܡܝܢ . ܘܒܪܢܐ ܡܠܟܐ ܡܬܘܢ ܡܢ ܗܘܦܠܐ ܡܠܟܐ ܡܢܟ ܦܠܓܘܬ ܫܡܪܐ ܪܫܘ . ܘܐܟܢ ܡܟܣ ܐܝܢܐ . ܐܡܪ . ܠܡܢ ܕܐܡܪܝܢ ܫܡܝܐ ܠܒܪ . ܐܠܐ ܦܠܓܐ ܕܒܗ ܐܬܦܠܓܘ ܘܐܬܬܟܝܢܘ ܠܐ ܡܬܬܚܙܝܢ ܒܗ . ܐܟܢܐ ܕܡܫܘܚܬܐ ܕܡܬܐܪܐ ܠܗ ܡܬܟܪܟ . ܐܠܦܐ ܒܫܒܝ̈ܐ ܒܫܡܝܐ ܕܐܬܬܘܗ . ܘܐܡܪܝܢ ܕܫܒܢܐ ܒܡܟܐܪܐ ܡܫܝܚܘܬܗ ܒܪܢܐ ܐܝܟ ܡܫܒܚܢܐ ܕܡܬܪܒܠ ܡܢ ܗܘܐ ܒܐܪܥ ܢܘܪܐ . ܘܒܪ ܒܡܟܐܪܐ ܡܢ ܫܡܝܐ ܡܬܚܒܢ ܗܘܘ . ܗܘ ܠܒܢܝܐ ܡܫܘܬܐ ܠܒܢܝ ܗܘܐ .

FMWPLIV: IV.ii.4: 1 ܘܠܠܝܐ PL1 || 3 ܘܫܪܐ ... ܠܐ ܚܠܒܐ: om. Λ || 4 ܘܒܐܬܪܘܬܗܘܢ PL; ܘܒܐܬܪܘܬܗܘܢ W1 || 5 ܘܡܬܚܙܝ V || 6 ܕܐܬܚܙܝܐܝܬ F || **IV.ii.5**: 3 ܘܟܠܡܐ ... ܢܦܫܗ: om. F || 3 ܠܢܦܫܗ W; ܠܢܦܫܗ PL1 || 4 ܡܢܐ: om. Λ || 5 ܕܢܦܫܡܝܢܐ Λ || 6 ܒܡܫܡܫܢܗ: ܒܡܫܡܫܢܘܬܗ V || 7 ܠܗܘܠ Λ || 9 ܒܪܝ: ܒܪܢ F || **IV.ii.6**: 1 ܗܘܦܠܐ F || ܠܘ ܡܢ: ܘܡܢ MV || 6 ܐܡܪ: add. ܐܢܐ V || 7 ܕܡܫܘܚܬܢܐ M; ܕܡܫܘܚܬܢܐ V || 9 ܒܡܟܐܪܐ: ܒܡܟܐܪܐ F; ܒܡܟܐܪܐ M || 10 ܗܘ: om. FM ||

[IV.ii.4.]
Fourth [theory]: Second argument: At the equator day is equal [in length] to night throughout the year. There, therefore, the heat of the day is tempered by the coldness of the night throughout the year, and neither [heat nor cold] prevails over the other. In the seven climes, heat prevails during the long days of the summer heat prevails, while coldness (prevails) during the long nights of the winter, and temperate weather occurs only in spring and autumn. At the equator, then the whole year is spring-like, whereas in other places sometimes heat is strong and sometimes coldness.

[IV.ii.5.]
Fifth [theory]: Third argument: Prolonged proximity of the sun at the zenith is more heat-generating than [its mere] proximity. For this reason the sun generates less heat at the head of Cancer than at the head of Leo, even though it is closer to our zenith in Cancer. It also generates less heat in Taurus and Gemini than when it is in Leo and Virgo, and less before noon than after noon, even though it is equally close to us in both positions. This is because during the first period the heat-source warms the air with certain moderation; because of this (the air) becomes more suitable for reception of heat in the second period, and for this reason the heat is stronger in the second (period) than in the first. At the equator, then, since the proximity of the sun occurs without the prolongation of the proximity, the heat is not strong.

[IV.ii.6.]
Sixth [theory]: Fourth argument: Since at the equator the sun is does not move very far from the zenith, its inhabitants do not suffer a change from great coldness to heat in such a way that they feel the hurt. [This is] because Cancer and Capricorn, the signs of their winters, are separated only by a quarter of a circle from Aries and Libra, the signs of their summers, whereas Capricorn, the sign of our winter, is separated by a half of a circle from Cancer, the sign of our summer. "Even if I concede, he [Ibn Sīnā] says, "that the heat is constant there, nevertheless the bodies (of the people who) were born and have grown up in (the heat) do not feel it, just as the Scythian does not feel the cold of his region, and the Cushite does not (feel) the heat of his region". He says that he (once) saw in Bukhara, his city, a certain Midianite who was trembling with cold in the month of Īyār [May] and, while the people of Bukhara were complaining about the heat, he was wearing winter clothes.

[tit.]

ܦܣܘܩܐ ܬܠܝܬܝܐ . ܡܛܠ ܫܒܝܠܐ ܕܬܒ̈ܢܐ . ܘܐܝܬ ܒܗ̇ ܫܬ .

[IV.iii.1]

ܡܫܡܠܝܘܬܐ . ܗܘ ܚܘܪܐ ܗܘܢܐ ܡܫܡܠܝܐ . ܐܦ ܡܢ ܟܠܗ̇ ܓܘܐ ܕܣܦܝܪܬܐ ܕܢܦܫ . ܐܠܐ ܕܢܬܒܣܡܘܢ ܠܐܕܪܐ ܗܢܝܐ . ܘܟܠܡܕܡ ܠܡܦܪܐ ܡܬܝܘܡܐ ܠܗ ܡܬܦܠܓܢܝܢ . ܘܟܕ ܥܒܝܪ ܠܡܕܡ ܡܕܡ ܐܚܪܢܝ . ܕܡܫܬܡܠܝܐ ܕܢܦܫܘܬܐ . ܟܕ ܫܪܟܐ ܒܐܝܕܐ ܘܡܫܡܠܝܘܬܐ ܠܟܠ ܡܢ ܕܘܟܐ ܚܕܐ . ܐܦ ܒܝܘܡܐ ܡܫܢܝܐ ܘܠܐ ܦܫܛ . ܒܬܪ ܕܝܢ ܠܗ ܡܕܡ ܡܬܝܠܕ ܡܢ ܓܠܝܐ ܕܘܟܐ . ܒܟܠܗ̇ ܥܠܡܐ ܡܕܝܢ ܐܝܟ ܠܟܠ ܡܢ ܕܘܟܐ ܐܝܟ ܠܗܘܢܐ ܡܕܡ ܗ̇ܘܐ ܫܪܟܐ . ܘܐܠܝܢ ܕܒܟܠܗ̇ ܥܠܡܐ ܢܦܫ ܦܫܛ ܢܦܫܐ .

[IV.iii.2.]

ܕܬܬܝܬܝܢ . ܘܗܟܢܐ ܚܙܝܐ ܗܘܐ ܡܬܚܙܝܢܐ ܒܡܕܝܢܬܐ ܕܐܝܟ ܠܠܝܐ ܕܡܫܬܡܠܝܐ ܕܢܘܗܪܐ . ܡܛܠ ܕܡܬܚܙܐ ܡܕܡ . ܠܐ ܡܢܝܢܐ ܣܓܝܐ ܕܡܬܚܙܝܢܘܬܐ ܣܓܝܐܘܬܐ ܕܐܣܬܟܠܘ ܗܘ ܕܡܬܚܙܝܢܐ ܡܬܬܝܐ ܕܫܡܫܐ ܠܟܠ ܡܢ ܕܘܟܐ ܚܕܐ . ܐܝܟ ܕܐܠܐ ܡܬܚܙܝܢܐ ܫܦܝܪܐ ܕܡܬܠܩܚܐ ܡܬܚܙܝܢܐ ܠܣܓܝܐܘܬܗ ܠܡܬܚܙܝܢܘܬܐ .

[IV.iii.3.]

ܕܬܠܬ . ܘܟܕ ܐܝܬ ܕܡܬܚܙܝܢܐ ܬܠܝܬܝܐ ܡܬܚܙܝܢܐ ܕܢܘܗܪܐ ܣܓܝܐܘܬܐ ܡܢ ܬܠܝܬܝܐ ܕܢܘܗܪܐ ܒܢܘܗܪܐ ܡܬܚܙܝܢܐ ܕܐܝܟ ܠܟܠ ܡܢ ܕܘܟܐ . ܒܐܬܪܐ ܗܘܢ ܕܒܥܒܕܐ ܣܓܝܐܐ ܕܡܬܚܙܝܢܐ . ܟܕ ܢܘܗܪܐ ܬܠܝܬܝܐ ܣܓܝܐܘܬܐ ܕܐܬܚܙܝܬ ܘܐܬܚܙܝܢܐ ܕܢܘܗܪܐ ܐܬܪܐ ܗܘܐ . ܘܟܠܗ ܕܡܬܚܙܝܢܘܬܐ ܐܪܥܐ ܕܡܬܚܙܝܢܐ ܡܕܡ ܢܘܗܪܐ ܡܕܡ ܣܓܝܐܐ ܡܢ ܡܬܚܙܝܢܘܬܐ ܐܪܥܐ ܕܡܬܚܙܝܢܐ ܠܗ ܡܕܡ ܢܘܗܪܐ . ܒܟܠܗ ܕܒܗ ܫܪܟ ܢܘܗܪܐ ܢܦܫ ܣܓܝܐܐ . ܢܘܗܪܐ ܕܢܘܗܪܐ ܕܡܬܚܙܝܢܐ ܡܕܝܢ ܒܡܬܚܙܝܢܐ ܒܢܘܗܪܐ ܡܢ ܢܘܗܪܐ ܕܡܬܚܙܝܢܐ ܕܫܡܫܐ ܡܕܡ .

[IV.iii.4.]

ܕܐܬܪܐ . ܘܠܗܘܢ ܗܟܢܐ ܕܒܥܒܕܐ ܡܬܚܙܝܢܐ ܕܢܘܗܪܐ ܢܘܗܪܐ ܠܗ ܟܠ ܡܬܚܙܝܢܘܬܐ ܕܢܘܗܪܐ ܡܬܚܙܝܢܐ ܕܢܘܗܪܐ ܒܣܓܝܐܐ ܐܝܟ ܠܐ ܡܬܚܙܝܢܘܬܗܘܢ ܗܘ . ܐܠܐ ܕܟܝ ܐܪܐ ܡܕܡ ܐܝܟ ܠܗ ܡܕܡ ܢܘܗܪܐ . ܘܐܝܬܘܗܝ ܕܟܠܗ ܡܬܚܙܝܢܘܬܐ ܡܬܚܙܝܢܐ ܒܣܓܝܐܐ ܘܒܢܘܗܪܐ ܡܬܚܙܝܢܘܬܐ ܡܬܚܙܝܢܐ . ܟܠ ܐܬܪܐ ܡܕܝܢ ܡܬܚܙܝܢܐ ܗܘ . ܒܗܕܐ ܕܠܐ ܒܣܓܝܐ ܒܡܬܚܙܝܢܘܬܐ ܢܦܫܗ ܡܬܚܙܝܢ . ܐܠܐ ܡܬܚܙܝܢܐ ܒܢܘܗܪܬܐ ܗܘܢܘ .

FMWPLIV: **IV.iii.1**: 1 ܕܡܫܡܠܝܘܬܐ P ‖ 2 ܕܢܬܒܣܡܘܢ Λ ‖ ܠܗ: add. ܒܬܪ V ‖ ܡܬܦܠܓܢܝܢ: ܢ supra lin. P; ܡܬܦܠܓܢܝ I ‖ 5 ܒܟܠܗ̇ ܥܠܡܐ: ܒܥܠܡܐ V ‖ **IV.iii.2**: 3 ܕܐܣܬܟܠܘ MV ‖ ܘܗܘ MV ‖ 4 ܡܬܚܙܝܢܐ: ܡܬܚܙܝܐ Λ ‖ ܠܣܓܝܐܘܬܗ: ܠܣܓܝܐܘܬܐ ܕܠܗ PI ‖ **IV.iii.3**: 4 ܕܒܥܒܕܐ Λ ‖ ܕܡܬܚܙܝܢܐ MV ‖ (5 ܡܕܡ ܣܓܝܐܐ ... ܠܗ ܡܕܡ ܢܘܗܪܐ: om. Λ ‖ 7 ܕܡܬܚܙܝܢܐ ܒܢܘܗܪܐ V ‖ **IV.iii.4**: 1 ܕܒܥܒܕܐ Λ; ܕܒܥܒܕܠܐ M[1], ܕܒܥܒܕܐ M[V] ‖ 2 ܠܐ ܐܝܟ: ܘܠܐ F ‖ ܡܬܚܙܝܢܘܬܗܘܢ V ‖ ܕܟܝ ܐܪܐ: ܬܠܠܐ M[1]; ܟܝ ܐܪܐ VM[V] ‖ 4 ܡܬܚܙܝܢܐ Λ ‖ 5 ܕܠܐ: ܕܐܘܠܐ F ‖

Third Section: Refutation of the Avicennian Arguments. Six theories.

[IV.iii.1.]
First [theory]: Even though the illustrious Doctor [Ibn Sīnā] is held dear by the entire community of excellent men, their love for the truth is greater and for this reason they do not hesitate to refute his arguments. Refuting his first argument, they say: At the equator, even though the sun changes position quickly and does not linger as it passes the zenith in Aries and Libra, it nevertheless does not [then] move very far from the zenith. Throughout the year, therefore, the sun is either at the zenith or in the proximity of it and the heat must be strong there throughout the year.

[IV.iii.2.]
Second [theory]: The second argument too is refuted by the fact that the coldness of the night at the equator, because it is very weak, does not have the power to mitigate the heat of its day, which is strong due to the great proximity of the sun to the zenith, just as the morning coldness of the second clime also [does] not [have the power to mitigate] its noonday heat.

[IV.iii.3.]
Third [theory]: In refutation of the third argument, it is said: The excess of the summer heat, which arises due to the prolonged proximity of the sun to the zenith, occurs in those regions to the north of the equator after the excess of the wintry cold which has preceded [it] and has greatly cooled the air. It is clear that the incalescence of a land which has undergone much cooling during the winter is much less than the incalescence of a land which has not undergone much cooling during the winter. - For at a time when a cold [object] grows hot, a hot [object] will burn. - The prolonged proximity of the sun in the climes is therefore less heat-generating than [its mere] proximity at the equator.

[IV.iii.4.]
Fourth [theory]: Against the fourth argument it is said: The dispute here is not about whether the inhabitants of the equator feel the heat or not, but [about] whether it is [in fact] very [hot] or not very hot. If temperateness is established by the fact that a sentient being does not feel the heat and the cold, every place will be temperate, because neither does the Cushite feel the greatness of the heat of (his locality) nor the Scythian the excess of the coldness of (his).

[IV.iii.5.]

ܕܫܡܝܐ . ܡܬܐܡܪܬ ܗܘܬ ܕܦܬܚ ܫܡܝܢܗ ܕܡܠܟܘܬܐ ܫܡܝܢܝܬܐ ܐܬܟܒܫ ܘܫܒܩ
ܡܬܐܡܪ ܗܘ . ܐܝܟܢܐ ܕܠܝܬ ܟܠܢܫ ܕܫܡܝܐ ܡܢ ܐܢܛܝܟܪܝܣܛܘܣ . ܒܝܘܡܘܗܝ ܕܝܢ
ܫܡܝܐ ܒܝܫ ܡܬܓܠܝܐ ܓܠܝܐ . ܒܗܘܢܝܐ ܕܫܘܠܛܢܐ ܡܢ ܡܬܬܐ ܚܙܐ ܗܘܐ . ܒܡܠܟܘܬܐ
ܫܡܝܢܝܬܐ ܕܝܢ ܢܦܠܐ . ܘܟܠܢܫ ܕܢܦܠ ܫܡܝܐ ܚܙܝܢ ܡܢܗ ܕܡܠܟܘܬܐ ܫܡܝܢܝܬܐ
ܠܥܠܡܐ ܡܢ ܡܫܝܚܐ ܕܫܘܠܛܢܐ ܢܦܫܐ ܚܙܝܐ . ܘܐܠܗܐ ܕܡܢܗ ܘܟܠܢܐ ܠܡܬܝܗܘܢ
ܒܟܠ ܟܘܚܒܐ ܘܡܬܦܢܝܢܐ ܒܫܡܐ ܠܫܡ ܫܡܝܐ . ܒܝܘܡܘܗܝ ܕܝܢ ܢܦܩ ܡܢܗ ܡܠܟܐ
ܕܡܠܟܐ ܕܫܘܠܛܢܐ ܠܢܦܩ ܢܦܠܐ ܕܡܠܟܘܬܐ ܚܙܝܐ . ܢܦܩ ܢܦܠܐ ܐܝܟܢܐ ܢܦܫܐ .
ܒܐܪܥܐ ܕܠܐ ܡܬܡܠܝܢܐ ܗܘ .

[IV.iii.6.]

ܕܢܦܠ . ܟܕ ܗܘܝܢ ܗܘܝܢܐ ܐܝܟ . ܠܝܘܡܬܐ ܚܙܝܐ ܡܬܚܙܝܢܐ . ܘܐܝܟ ܡܪܘܢܐ ܡܠܟܐ .
ܐܝܟܢܐ ܗܘܝܢ ܕܠܥܠܡܐ ܐܬܐ ܟܕ . ܘܚܙܝܬ ܐܢܐ ܡܢ ܕܝܢ ܕܐܝܟ ܕܐܡܪ .
ܐܝܟ ܠܐ ܐܬܐ ܕܫܘܠܛܢܗ ܕܥܠܡܐ ܟܠ ܫܘܠܛܢܐ ܘܢܝܫܐ ܢܦܫ ܡܢ
ܫܘܠܛܢܐ ܕܟܠܗ ܡܠܟܘܬܐ ܐܝܬܝܗ ܚܙܝܐ . ܘܡܠܟܐ ܕܡܠܟܘܬܐ ܡܬܚܙܝܐ ܠܐ
ܡܬܦܠܓܢ ܘܟܠܗ ܒܫܡܝܐ ܫܘܠܛܢܐ ܢܦܠܐ ܠܐ ܠܥܠܡ ܐܠܗܐ . ܡܢܗ
ܡܬܚܙܝܢ ܗܘܝܢ ܟܕ ܕܥܠܡܐ ܚܙܐ . ܘܫܘܠܛܢܗ ܢܦܠܐ ܒܡܠܟܘܬܐ ܡܠܟܐ
ܐܠܗܐ ܕܬܬܓܠܐ . ܘܒܫܡܝܐ . ܐܝܟܢܐ ܕܡܬܚܙܝܐ ܡܢ ܕܟܠ ܢܦܫܐ
ܘܢܦܠܐ ܠܐ ܚܙܝܐ ܒܟܠܗܘܢ ܕܡܠܟܘܬܐ ܚܙܝܬܗ . ܘܐܬܐ ܕܒܐܝܕܘܗܝ ܡܬܚܙܝܢܐ
ܚܙܝܐ ܠܗ . ܠܥܠܡܐ ܕܢܦܠ ܟܠܗ ܚܙܝܐ . ܘܢܦܠܐ ܘܠܐ ܡܬܚܙܝܢܐ ܐܝܟ
ܦܐܪܘܣ . ܒܐܝܟ ܟܕ ܚܕ ܡܢܗܘܢ ܚܙܝܐ . ܐܝܟܢܐ ܡܬܚܙܝܐ . ܐܟܚܕܐ
ܕܟܠܗܘܢ ܒܫܡܐ ܠܥܠܡܐ ܒܫܡ ܐܝܟܢܐ ܟܠܗܝܢ ܘܒܫܡܝܐ ܘܫܡܝܐ ܫܬܝܢ .

FMWPLIV: IV.iii.5: 2 ܡܬܐܡܪܘܢ FM; ܡܬܐܡܪ V || 2 ܐܢܛܝܟܪܝܣܛܘܣ WL || 5 ܕܒܗܘܢ: ܒܗܘܢ MV || 6 ܠܫܡ: ܠܫܡ W || 7 ܕܡܠܟܘܬܐ: add ܫܡܝܢܝܬܐ in marg. F || ܢܦܠܘܗ: ܢܦܠܐ F || 8 ܗܘ: ܗܘܐ PLIM; ܗܘܝܐ V[1], ܗܘܐ V[corr.] || IV.iii.6.1 ܚܙܐ: add ܗܘ V || ܡܬܚܙܝܢܐ: add. ܗܘܐ supra lin. F || ܡܪܘܢܐ: ܡܪܘܢܐ FM[1]; ܡܪܘܢܐ M[V] || 3 ܠܐ: supra lin. F; om. M || 3-4 ܫܘܠܛܢܐ . . . ܕܡܠܟܐ: om. MV || 4-11 ܫܬܝܢ . . . ܘܡܠܟܐ: ponunt ante IV.iii.6 FM (ordinem codicis M correxit M[V]) || 5 ܚܙܝܬ V || 6 ܡܬܚܙܝܢܘܢ Pl; ܡܬܚܙܝܢ L; ܡܬܚܙܝܢܘܢ V || ܘܫܘܠܛܢܗ: ܘܫܘܠܛܢ L || 7 ܕܬܬܓܠܐ F || 9 ܚܙܝܬܗ M|| 10 ܟܕ: om. PLl || ܡܬܚܙܝܐ: ܡܬܚܙܝܢ F || 11 ܘܫܡܝܐ M[1], ܘܒܫܡܝܐ M[V] || ܫܬܝܢ: بتين WPL
||

[IV.iii.5.]
Fifth [theory]: It is also said that the latitude at the beginning the seventh clime is forty-seven degrees, [i.e.] double the total declination of the sun from the isemeric, so that when the sun stays at the head of Cancer, it is winter at the equator and summer in the seventh clime. It is clear that the distance of the sun at that time from the seventh clime will be equal to its distance from the equator. The sun must at that time be heating the two sides, the northern and the southern, equally. Therefore, seeing that the winter heat at the equator is equal to the summer heat in the [seventh] clime, what will the summer heat [at the equator] be like? It will surely be intolerable.

[IV.iii.6.]
Sixth [theory]: This being the case, [the opinion of] the illustrious [Doctor] is astonishing, though it [may seem] insolent [of me] to say so. How did he come to hold these perverse opinions? With all due respect to his great status, I say: Why did he not realise that continuous heat is more harmful to animals and plants than heat which comes after coldness? Bodies which are not cooled by the coldness of winter and within whose bellies the implanted heat is not reversed are in danger when they reach summer. The heat of the summer must be tempered by the coldness of the winter, and vice versa, so that the moderation, which is not found within each day and month, is found during the course of the year. In a place where moderation is always found, the whole year is like spring and its fruit are monthly, not yearly, that is to say, when one of them is maturing, another is [just] starting, just as in Antioch we see ripe grapes, unripe grapes and blossoms on a single vine at a single time.

|tit.|

ܦܘܣܩܐ ܕܪܒܝܥܝܐ . ܡܛܠ ܥܠܬܐ ܫܡܫܢܘܬܐ ܘܬܪܥܝܬܐ ܪ̈ܘܚܢܝܬܐ ܘܐܝܬ̈ܝܬܐ .
ܬܐܘܪ̈ܝܗ ܫܢܝܢ .

|IV.iv.1.|

ܡܫܡܫܢܘܬܐ . ܕܡܫܡܫܐ ܠܐ ܣܢܝܩ ܢܒܝܚܐ ܡܢ ܗ̇ܝ ܕܦܘ̈ܩܕܢܐ ܕܪܫܝܡ ܡܢܗ ܠܘܬ ܡܢ
ܬܪ̈ܥܝ ܓܘ̈ܫܡܐ ܕܡܬܚܒܝܢ ܠܗ ܣܓ̈ܝܐܢ . ܟܕ ܕܝܢ ܡܫܡܫܢܘܬܐ ܡܢ ܐܠܗܐ ܠܒܫܐ .
ܘܐܠܗܐ ܡܢ ܡܫܡܫܐ ܕܝܠܗ . ܘܥܠ ܥܠܝܡܘܬܐ ܕܡܫ̈ܡܫܢܐ ܪ̈ܝܘܢܐ ܘܐܪ̈ܚܝܢܐ
ܕܡܫܠܝܢ ܡܛܫܝܢܐ . ܘܠܐ ܥܠ ܡܫܡܫܢܐ ܐܪ̈ܝܬܝܢܐ ܘܩ̈ܕܝܫܢܐ ܡܩܕܫܢܐ . ܐܝܟܢܐ
ܕܡܫܬܪܝܢܘܬܐ ܫܪܝܢ . ܕܟܡܐ ܕܢܗܘܐ ܡܫܡܫܢܐ ܢܗܘܐ ܠܕܝܢ . ܘܠܡܫܡܫܢܘܬܐ ܐܠܗܐ
ܢܗܘܘܢ ܡܫܡܫܢܘܬܐ ܗܝ ܕܟܝܢܐ ܡܠܝܠ ܕܡܫܡܫܐ ܠܒܝܫܐ . ܘܠܗܝ ܗܝ ܕܡܠܝܠܘܬܐ
ܗܢܐ ܡܫܡܫܢܘܬܐ . ܗܕܝܢ ܕܝܢ ܐܠܗܐ ܡܡܫܝܫ ܠܡܫܡܫܢܘܬܗ ܟܠܗܘܢ ܘܠܡܫܡܫܢܐ ܕܟܗ̇
ܡܫܡܫܝܢ .

|IV.iv.2.|

ܕܬܠܬܝܢ . ܒܡܫܡܫܢܐ ܠܐ ܡܫܟܚܐ ܠܐ ܡܬܚܫܒܐ ܢܦܫܐ . ܒܗ̇ܝ ܕܒܪܘܚܢܘܬܐ
ܘܩܪ̈ܝܐ ܘܦܘ̈ܪܫܐ ܕܒܗ ܡܬܦܫܩܢ ܘܡܬܦܠܓܝܢ ܘܠܐ ܡܬܦܠܓܝܢ . ܐܠܐ ܥܠ ܒܟܠܡܫܡܫܢܐ
ܡܫܡܫܢܐ ܕܝܢ ܐܠܗܐ . ܘܟܠܡܐ ܕܡܫܬܡܥܐ ܟܠܗܝܢ ܠܐ ܢܦܫܐ ܐܠܐ ܡܫܡܫܢܐ .
ܘܠܐ ܡܫܟܚܐ ܠܡܫܡܫܘ ܕܐܬܪܐ ܡܫܬܡܥܐ ܠܐܠܗܐ ܕܡܫܡܫܐ ܟܠܗܝܢ ܘܠܐ ܠܐ ܡܫܡܫܐ
ܕܢܦܠܘܢ ܥܠ ܦܠܓܐ ܕܡܫܡܫܐ ܗܘ ܕܐܠܗ ܡܢ ܐܬܪܘܗܝ . ܒܗ̇ܝ ܕܡܫܡܫܢܐ
ܕܡܫܡܫܐ ܠܘܬ ܕܒܗܬܐ ܠܐܬܪܐ . ܐܠܐ ܐܝܟ ܢܦܫܬܐ ܒܐܦ̈ܝܗܘܢ ܦܢܝܐ .

|IV.iv.3.|

ܕܬܠܬ . ܒܟܠ ܐܬܪ ܘܩܕܡ ܒܪܢܫܐ ܕܒܗ ܢܦܫ ܐܫܬܡܫܐ ܚܙܝ ܫܡܫܐ . ܘܡܫܡܫܢܐ
ܕܒܗ ܡܫܬܘܐ ܐܫܬܡܫܐ ܚܙܝ ܗܘܝܐ . ܒܗ̇ܝ ܕܠܒܪܢܫܐ ܕܡܫܡܫܐ ܐܠܗܐ ܠܗܘ .
ܘܠܐܠܗܐ ܡܫܡܫܢܘܬܐ ܢܦܫܐ . ܒܪܓܘ ܢܬܚܙܐ ܐܠܗܐ ܬܫܡܫܬܐ ܕܡܫܡܫܢܘܬܐ
ܚܙܝܐ . ܐܝܟܢܐ ܕܡܫܠܡ . ܘܠܐ ܢܬܚܙܐ ܒܡܫܝܚܘܬܗ . ܐܝܟܢܐ ܕܡܫܡܫܬܐ .
ܘܒܝܬ ܐܠܗܐ ܕܡܫܡܫܢܘܬܐ ܐܠܗܐ ܠܒܪܢܫܐ ܕܡܫܡܫܢܘܬܗ ܢܦܫܐ ܚܙܝܐ . ܐܝܟܢܐ
ܕܬܫܡܫܬܐ ܘܡܫܬܪܝܢܘܬܐ . ܠܐ ܡܫܡܫܢܘܬܐ ܡܬܒܥܝܐ ܒܗ ܐܠܗܐ ܘܫܪܝܪܘܬܐ .
ܐܠܐ ܡܬܡܫܚܢ . ܘܡܠܝܢ ܗܢ ܠܗܘܬ ܦܘܫܩ ܐܫܬܡܫܐ ܕܥܠ ܐܬܪܐ ܟܠܗܝܢ ܕܠܗ .
ܠܗܘܬ ܕܝܢ ܦܘܫܩ ܐܫܬܡܫܐ ܘܟܠܗܝܢ ܕܐܬܪܐ ܡܫܡܫ ܠܗܘܬ ܐܫܬܡܫܐ ܘܟܠܗܝܢ ܕܐܬܪܐ
ܐܚܪܢܐ . ܠܘ ܗܟܢ ܡܫܬܡܥܢܐ ܡܫܡܫܢܘܬܐ ܕܐܬܪܝܢ ܘܒܝܬ ܕܟܠܢ ܐܠܗܐ
ܘܫܦܝܪ . ܐܠܐ ܒܡܫܡܫܐ ܐܝܟܢܐ ܕܡܫܡܫܢܐ .

FMWPLIV: tit.: 1 ܥܠܬܐ: om. ΛV / ܪ̈ܘܚܢܝܬܐ: ܪܘܚܢܝܬܐ MΛ || IV.iv.1: 1 ܕܡܫܡܫܢܘܬܐ V || 2 ܣܓ̈ܝܐܢ scripsi: ܣܓܝܐܢ codd. || ܐ̈ܠܗܐ M || 4 ܘܩ̈ܕܝܫܢܐ: ܘܩܕܝܫܢܐ M || 5 ܘܟܡܐ MV || ܡܬܦܠܓܝܢ W || 7 ܕܝܢ: om. F || ܠܫܡܫܐ F || ܕܟܗ̇: ܕܟܗ L || IV.iv.2: 1 ܒܡܫܡܫܢܐ ܠܐ ܡܫܟܚܐ Λ || ܕܒܪܘܚܢܘܬܐ Λ || 4 ܡܫܬܡܥܐ: om. F || ܟܠܗܝܢ: ܟܠܗ MV || ܘܠܐ: ܠܐ F || 5 ܕܡܫܡܫܐ: ܕܡܫܡܫܢܐ Λ || 6 ܒܐܦ̈ܝܗܘܢ: add ܕܡܫܡܫܢܐ L || IV.iv.3: 1 ܚܙܝ: ܚܙܐ MV; ܚܙܝ L || 2 ܚܙܝ: ܚܙܐ MV || ܕܡܫܡܫܢܐ: ܘܡܫܡܫܢܐ Ll || 4 ܘܠܐ: ܘܐܠܐ MV || ܡܬܒܥܝܐ FMV || ܒܡܫܝܚܘܬܗ V || 5 ܘܒܝܬ ܐܠܗܐ: ܘܒܝܬܐ Λ || ܕܡܫܡܫܢܘܬܗ: ܕܡܫܡܫܢܘܬܗ Λ || ܢܦܫܐ: ܢܦܫ MV || ܚܙܝܐ: ܚܙܝܟ PLl || 6 ܘܫܪܝܪܘܬܐ Λ || 8 ܕܝܢ: om. L || 8 ܕܐܬܪܐ . . . ܡܫܡܫ: rasa in M; om. V || 9 ܗܟܢ: ܡܫܬܡܥܢܐ MWPV || ܐ̈ܠܗܐ PMV ||

Fourth Section: On the Cause of Heat and Cold according to Time and Place. Five theories.

[IV.iv.1.]
First [theory]: That the sun is not hot is known from the fact that plains, which are far from it, are hotter than tops of mountains, which are close to it. Rather, heat proceeds from radiation, and radiation emanates from the sun and becomes visible on surfaces of dense and earthy bodies which are polished - not on loose airy bodies and transparent celestial bodies - as we see on mirrors, which shine the more, the more polished they are. Increase of radiation results in increase of heat which almost sets [objects] on fire, especially when the mirror is concave. In such a case the rays from its edges are reflected to its middle and sets the kindling there on fire.

[IV.iv.2.]
Second [theory]: Light does not become visible on an unpolished body because it is divided and scattered by the coarsenesses, bumps and holes on it and loses its coherence. Neither does the ray manifest itself on transparent bodies and for this reason the celestial sphere does not shine or give light at night. One cannot say that the earth conceals the rays of the sun at night and does not allow them to fall on that half of the sky above the horizon, because in relation to the sun the earth has no great size and stands like a dot before it.

[IV.iv.3.]
Third [theory]: In every place and region heat grows strong in the season in which the day becomes long and coldness grows strong in the season in which the day becomes short, because the appearance of the sun results in radiation, and radiation results in heat. As a result, prolongation of radiation causes an increase of heat, as [happens] in summer, while its curtailment [brings about] its decrease, as [happens] in winter. When the time during which radiation is observed is equal to the time during which it is not observed, as [happens] in spring and autumn, neither heat nor coldness increases, but [the two] temper each other. These [changes occur] according to the proportion of the day in each place to the night in the same [place]. [As regards what happens] according to the proportion of [the length of] the day and night in a given place to [the length of] the day and night in another place, the effect of the length of the time during which radiation is manifest or not manifest is found not [to be] thus, but to the contrary, as will be shown.

[IV.iv.4.]

ܕܐܪܒܥ . ܗܢ ܒܒܢܝܐ ܠܐܪܒܥܬܗܘܢ ܐܣܟܝܡܐ ܡܬܩܪܝܬ ܡܣܬܟܠܢܘܬܐ ܠܢܦܫܐ . ܚܠܦ ܗܕܐ
ܐܪܥܢܝܬܐ ܠܢܦܫܐ ܗܢܐ ܕܒܢܝܐ ܥܬ̈ܝܕ . ܘܫܡܫܐ ܕܡܠܟܐ ܟܘܟܒܝܐ
ܕܐܬܚܙܝܬܐ ܙ̈ܒܢܝܢ ܗܘ ܐܣܟܝܡܗ ܪܒܐ . ܒܫ̈ܚܩܐ ܡܓ̈ܘܫܐ ܒܝܪ ܡܢ ܚܡܝܐ
ܕܡܠܟܐ ܡܕܡܐ ܕܬܬܚܙܝܬܐ ܙ̈ܒܢܝܢ ܗܘ ܐܣܟܝܡܗ ܗܟܢ ܪܒܐ . ܘܗܟܢܐ ܠܗ
ܡܠܟܐ ܕܐܣܟܝܡܗ ܗܟܢ ܚܡܝܐ ܒܝܪ . ܘܗܪܐ ܡܢ ܒܡܐ ܕܐܬܚܙܝܐ ܗܘ ܕܐܣܟܝܡܗ
ܡܠܟܐ ܗܟܢ ܪܒ . ܠܠܢܗ ܡܬܚܙܝܐ ܬܘܒ ܗܟܢ ܪܒ . ܒܚܠܘ ܬܘܒܐ ܒܗ
ܡܪܝܬܐ ܒܡܬܚܙܐ . ܘܡܚ̈ܙܝܢ ܬܠ̈ܬܐ ܥܠ ܐܦܝ ܐܪܥܐ . ܘܡܛܠ ܕܐܪܥܐ
ܒܟܝܢܐ ܡܪܝܐ . ܠܡܪܝܬܐ ܡܬܚܙܝܢܘܬܐ ܗܟܢ ܡܢ ܕܠܣܟܝܡܗܘܬܐ ܡܠܟܬܐ ܐܣܝܪܐ .
ܘܚܠܡܪܐ ܣܟܝܡܗܬ ܡܠܟܐ ܡܢ ܡܪܝܬܗ ܡܬܚܙܝܐ ܬܘܒܝܐ . ܘܚܙܝ ܒܗ ܥܘܪܐ .
ܘܐܝܟܢܐ ܕܐܣܟܝܡܐ ܡܠܟܐ ܘܠܠܢܐ ܡܬܚܙܝܐ ܒܝܪ . ܠܐ ܬܘܒܐ ܒܗ ܡܪܝܬܐ .
ܘܚܠܡܪܐ ܣܟܝܡܗܬ ܡܠܟܗ ܡܢ ܡܪܝܬܗ ܡܬܚܙܝܐ ܠܐ ܫܒܝܐ . ܘܚܙܝ ܒܗ ܣܗܪܐ .

[IV.iv.5.]

ܕܣܟܝܡ . ܗܢ ܟܝܡܐ ܠܟܠ ܡܢ ܪܘܚܐ ܡܬܚܠܦܢܘܬܐ ܚܕܕ . ܗܟܢ ܡܟܣܝ . ܗܢ
ܡܕܝܢ ܡܬܠܬܠܝܢ ܠܠܠܐ ܠܦܪܐ ܘܐܪ̈ܦܐ . ܘܠܟܠ ܕܗܘ ܡܠܝܢ ܐܠܦܐ ܡܟܣܝܢܐ .
ܘܒܗܝ ܬܘܒ ܕܐܣܟܝܡܘܬܗ ܐܠܦܐ ܟܝܡܐ ܥܠ ܦܠܓܗ ܕܐܣܦܪ ܐܪܥܐ ܗܘܝܢ .
ܒܟܠ ܐܬܪܐ ܡܕܝܢ ܕܡܠܟܬܐ ܠܟܠ ܡܢ ܪܘܚܐ ܚܕܕ ܟܝܡܐ . ܒܡܬܚܙܝܢܗ ܚܙܪܐ
ܕܐܠܦܐ ܗܘܝܢ ܒܒܣܪ̈ܐ . ܘܐܝܟܢܐ ܕܠܟܠ ܡܢ ܪܘܚܐ ܠܐ ܚܕܕ ܟܝܡܐ .
ܒܣܘܦܐ ܕܫܘܪ ܐܠܦܐ ܗܘܝܢ ܒܒܣܪ̈ܐ ܒܠܡܪܐ . ܘܚܠܝܐ ܕܡܬܚܙܝܢܘܬܐ ܕܣܟܝܡܐ
ܗܟܢ ܡܢ ܡܩܦܣ ܣܟܝܡܐ . ܘܚܠܡܪܐ ܐܬܪܐ ܕܒܗ ܚܕܕ ܟܝܡܐ ܠܟܠ ܡܢ
ܪܘܚܐ ܗܟܢ ܡܢ ܐܬܪܐ ܕܠܘ ܗܒܝܢܐ ܣܟܝܡ . ܘܦܠܡܪܐ ܒܟܠ ܐܬܪ ܗܟܢ ܡܢ ܢܦܪܐ
ܘܪܟܝܢܐ ܣܟܝܡ .

FMWPLIV: **IV.iv.4**: 2 ܥܬ̈ܝܕ Λ || 3 ܕܐܬܚܙܝܬܐ PL1 || ܗܘ: ܗܘܐ WP1 || 4 ܗܘ: ܗܘܐ Λ || 7 ܐܦ̈ܝ: om. F || 8 ܡܪܝܐ: ܡܪܘܪܐ M || 9 ܚܠܡܪܐ V || ܒܗ MV || 10 ܒܬܪ MV || 11 ܡܠܟܗ: ܡܠܟܐ MV || **IV.iv.5**: 1 ܪܡܝ M || ܡܬܚܠܦܢܘܬܐ ܚܕܕ: ܚܠܡܪܐ ܚܕܕ F || 2 ܡܬܠܬܠܝܢ: ܡܬܠܬܠܝܢ Λ || ܕܗܘ: ܡܟܣܝ M^1V^1, ܕܗܘ $M^VV^{corr.}$ || 3 ܘܒܗܝ: ܘܒܗ FM || ܐܣܟܝܡܘܬܗ F; ܕܐܣܟܝܡܐ M || ܕܐܣܦܪ ܐܪܥܐ M; ܕܐܣܦܪ ܐܪܥܐ V || 5 ܕܐܠܦܐ: ܗܘ ܕܐܠܦܐ MV || 6 ܕܣܟܝܡܐ: ܕܐܣܟܝܡܐ $ΛVM^V$ || 7 ܣܟܝܡܐ: ܕܐܣܟܝܡܐ Λ; ܕܐܣܟܝܡܐ ܣܟܝܡܐ VM^V || 8 ܕܠܘ: ܘܠܘ Λ || 8-9 ܘܦܠܡܪܐ . . . ܣܟܝܡ om. FMV ||

[IV.iv.4.]
Fourth [theory]: Although by nature lengthening of the day results in increase of heat, other causes invalidate this law of nature. The heat in the seventh clime whose longest day is sixteen hours is many times weaker than the heat in the first clime whose longest day is thirteen hours and, in this way, in every clime the longer the day, the weaker the heat. This is because [in] that place whose summer day is longer, its winter night is also longer. As a result, coldness grows strong there in winter and snow lingers on the surface of the earth. Because earth is cold by nature, it retains the winter coldness more than the summer heat. For this reason the heat of the summer is subdued by the coldness of its winter and coldness prevails there. Where the summer day and winter night are short, coldness does not grow strong. For this reason, the heat of its summer is not overcome by the coldness of its winter and heat prevails there.

[IV.iv.5.]
Fifth [theory]: When the sun stand vertically at the zenith, it generates more heat, because at that time the shades of mountains and walls disappear and the heat-generating rays reach everywhere. Because, furthermore, the rays of the sun are constantly on a half of the terrestrial sphere, in every place where the sun passes the zenith at noon, its inhabitants are in the middle of the circle of its rays, whereas where the sun does not pass the zenith, the inhabitants are on the edge of the circle of the ray at noon - and it is clear that the middle of a hot thing is hotter than its edges. - For this reason, a place where the sun passes the zenith is hotter than a place where this is not the case and noon is hotter everywhere than morning and evening.

[tit.]

ܩܠܐܘܦ ܚܡܝܫܝܐ . ܡܛܠ ܫܪܟܐ ܕܟܝܢܐ . ܘܗܢܘܢ ܐܝܟ ܦܫܝܛܐ ܐܪܥܢܝܐ .

ܦܘܣܩܐ ܩܕܡܝܐ . ܡܛܠ ܡܥܕܢܐ ܕܬܚܘܡܐ . ܬܐܘܪܝܣ ܬܠܬ .

[V.i.1.]

ܡܕܒܪܢܘܬܐ . ܢܟܝܐ ܗܘ ܕܟܠܗ ܐܪܥܐ ܐܝܟ ܕܠܦܪܘܬܐ ܚܝܐ ܢܦܫ . ܐܠܐ ܐܝܬܝܗܘܢ
ܡܬܬܕܝܢ . ܘܡܬܥܒܕܢܐ ܒܗ ܡܛܠܬܗܘܢ ܐܪܒܥ ܡܬܚܫܒܐ ܕܐܬܪܘܬܗܘܢ ܦܘܫܟܐ ܐܠܦܝܐ
ܦܬܝܟ ܠܗ . ܘܗܢܐ ܚܒܠ ܠܡܬܪܬܐ . ܐܝܟ ܕܠܠܒܝܢܐ ܡܦܩ . ܘܠܒܝܢܐ ܚܝܐ
ܠܡܥܢܐ ܕܡܢ ܡܬܚܝܠܐ ܐܪܥܢܝܐ ܡܬܝܬܪܐ ܚܒܪ . ܘܡܬܚܫܚܢܘ ܡܢ ܐܪܥܐ
ܐܬܬܪܝܢ . ܡܬܢܬܝܢܐ ܕܝܢ ܡܛܠܬܗ . ܘܡܬܢܟܝܢܐ ܐܪܒܥ . ܘܡܬܚܫܚܢܘ
ܠܚܡܝܐ . ܘܗܢܐ ܐܝܬ ܒܪܬܐ ܦܪܕܬܐ . ܡܩܝܠܢܐ ܡܬܩܪܐ ܒܗܘܢ ܦܪܕܘܢ
ܡܦܪܕܘܢ . ܘܠܒܪܬܐ ܗܘܝ ܡܬܚܝܠܬܐ . ܘܚܠܘܦ ܒܫܡܝܐ ܒܐܝܩܪܐ ܘܓܠܠܬܐ .

[V.i.2]

ܕܬܘܬܝܢ . ܒܬܘܩܦ ܢܟܝܐ ܚܝܐ ܬܒܪ ܐܝܬ ܒܩܠܒܐ . ܘܒܗܘܢ ܬܬܝܢ ܒܪܬܐ .
ܕܡܬܦܬܝܢ ܗܘܦܠܝܢ ܬܟܬܐ ܘܠܒܪܬܐ . ܘܡܬܢܟܝܢܐ ܬܠܬܐ ܒܩܠܒܝܢ .
ܡܬܕܘܝܢܘܢ ܒܓܠܐܦܝܢܘܢ ܐܕܕܝܐܢܘܢ . ܘܒܬܪ ܗܠܝܢ ܒܗܘܟܐ ܠܢܟܝܐ ܕܡܬܬܪܐ
ܡܬܚܠܝܢܘܢ . ܘܡܢ ܐܕܕܝܐܢܘܢ ܠܢܟܝܐ ܡܠܝܢܐ ܢܗܘ ܠܐܘܪ ܒܪܬܐ . ܘܚܒܪ ܒܠ
ܒܡܠܝܢܐ . ܘܓܙܪܐ ܒܢܟܝܐ ܕܡܬܚܠܦܘܢ . ܗܘ ܕܐܝܟ ܡܢܗ ܠܢܟܝܐ ܢܛܪܢܐ ܢܗܘ ܠܐܘܪ
ܒܪܒ ܡܬܚܫ . ܘܡܬܚܠܦܝܢ ܕܐܬܬܝܗ ܢܟܝܐ ܕܦܘܫܟܐ ܘܡܬܦܪܣܢܐ
ܘܕܐܬܚܪܝܢ ܓܙܪܐ .

[V.i.3.]

ܕܬܠܬ . ܐܠܠܐܝܬܝܗܘܢ ܡܢ ܠܒܪ ܡܬܚܫܝܢ ܕܡܛܠܬܗ ܠܡܬܚܪܬ ܬܚܝܢ ܪܕܐ . ܘܚܒܪ
ܒܠ ܠܦܘܪܐ ܡܬܡܬܚܠܝܢܐ ܕܡܬܚܫ ܒܐܝܟ ܢܠܗܘ . ܘܡܢ ܪܕܐ ܠܡܬܚܫܢܐ ܫܪܪ ܠܐܪܒܥ
ܕܬܒܪܬܐ ܘܒܓܙܪܐ . ܘܡܬܐܪܬܒܥܐ ܚܒܪ ܒܗܘܟܐ ܕܢܟܝܐ ܡܦܬܚܢܐ . ܘܡܢܐ ܒܪܬ
ܐܝܟ ܠܢܟܝܐ ܡܠܝܢܐ ܡܬܚܡܬܢ ܒܐܝܟܢܦܠܘܢ . ܘܚܒܪ ܠܢܟܝܐ ܕܦܘܫ . ܡ ܚܒܪ
ܒܠ ܐܪܒܥ ܕܟܒܝܐ ܘܦܘܫܟܐ . ܘܚܒܪ ܠܢܟܝܐ ܕܟܠܗܘ . ܐܝܠܐ ܠܢܟܝܐ

FMWPLIV: **V.i.1**: 1 ܐܠܐܝܬܝܗܘܢ Λ; ܐܠܝܬܝܗܘܢ V ‖ 2 ܡܛܠܬܗ F ‖ ܕܡܬܦܬܠܝܢ M ‖ 4 ܐܬܚܝܐܘܢ: ܗܘܢ- F[1]; ܐܬܪܬܐܘܢ F[corr.] super rasura; ܐܬܚܝܐܘܢ V ‖ ܘܡܬܚܫܚܢܘ M; ܡܬܚܫܚܢܘ V ‖ 5 ܡܬܢܬܝܢܘ MV ‖ 5 ܕܝܢ: ܕ MV ‖ ܡܛܠܬܗ F ‖ ܐܪܒܥ: ܠܚܡܝܐ MΛV ‖ 6 ܠܚܡܝܐ: ܐܪܒܥ MΛV ‖ 7 ܡܦܪܕܘܢ: ܡܦܪܠܘܢ P ‖ **V.i.2**: 1 ܒܩܠܒܐ: ܒܩܠܒܐ MV ‖ 2 ܕܡܬܦܬܝܢ: om. Λ ‖ ܗܘܦܠܝܢ Λ ‖ ܘܡܬܢܟܝܢܐ WPl ‖ ܬܠܬܐ Λ ‖ ܒܩܠܒܝܢ: ܒܩܠܒܝܢ MV ‖ 3 ܡܬܕܘܝܢܘܢ F; ܡܬܕܘܝܢܘܢ MV ‖ ܒܓܠܐܦܝܢܘܢ PLl ‖ ܐܕܕܝܐܢܘܢ PLIV ‖ 3-4 ܘܒܬܪ ܗܠܝܢ . . . ܡܬܚܠܝܢܘܢ: om. Λ ‖ 3 ܒܗܘܟܐ: ܒܗܘܟܐ MV ‖ 4 ܐܕܕܝܐܢܘܢ PLIV ‖ 5 ܢܛܪܢܐ PLl; ܢܛܪܢܐ ܝܫܘܥ ܒܪܚܡܬܐ 1 marg. ‖ 6 ܕܦܘܫܟܐ: ܕܦܘܫܟܐ F ‖ ܘܡܬܦܪܣܢܐ V‖ 7 ܓܙܪܐ: ܡܬܓܙܪܐ ut vid M[1]; ܓܙܪܐ M[V] ‖ **V.i.3**: 1 ܐܠܠܐܝܬܝܗܘܢ Λ; ܐܠܠܝܬܝܗܘܢ V ‖ ܡܢ: ܡܢ M ‖ ܕܡܛܠܬܗ F ‖ 2 ܠܦܘܪܐ: ܦܘܪܐ ut vid. M[1], ܠܦܘܪܐ M[V] ‖ ܫܪܪ: om. Λ ‖ 3 ܘܡܬܐܪܬܒܥܐ ΛV ‖ ܚܒܪ: ܚܒܪ MV ‖ ܡܢܐ: ܡܢ MV ‖ 4 ܡܢ ܠܢܟܝܐ M ‖ ܒܐܝܟܢܦܠܘܢ V ‖ ܘܚܒܪ: ܘܚܒܪ MΛV ‖ 5 ܢܟܝܐ ܕܟܒܝܐ W ‖ ܘܦܘܫܟܐ: ܘܦܘܫܟܐ ΛV ‖ ܘܚܒܪ: ܘܚܒܪ MΛV ‖ 5-6 ܐܝܠܐ ܠܢܟܝܐ ܕܦܘܫ: om. Λ ‖

FIFTH CHAPTER

ON MATTERS CONCERNING THE SEA

It contains four sections.

First Section: On the Position of the Seas. Three theories.

[V.i.1.]

First [theory]: That sea which surrounds the whole earth like a single island is called the Atlantic. In the west a narrow mouth is open to it at the στήλαι, i.e. Pillars, of Hercules. Through it it enters into the habitable world as if into some harbour and forms this well-known sea which is called the Oceanus [*sic*] by many. To its east is Asia, to its west are the στήλαι, to its north is Europe and to its south is Libya. In it are the large islands Sicily, Samos, Chios, Rhodes and Cyprus, as well as many small ones. They all abound in people and crops.

[V.i.2.]

Second [theory]: In the south of this sea there are two gulfs and in them are two islands called the Greater and the Lesser Syrtes. In its northern (side) are three gulfs, the Sardinian, the Galatian and the Adriatic and after these a slanting gulf called the Sicilian. From the Adriatic a thin tongue goes out towards the north and passes by Byzantium and flows into the sea of Pontus, from which in turn a river-like tongue goes out towards the northeast and flows into the Maeotis, which is the lake of Caspia, Hyrcania and Iberia.

[V.i.3.]

Third [theory]: Outside the στήλαι the Atlantic flows to the southwest and passes by the Mountain of Silver, from which the Nile rises. As it flows eastwards, it encircles the land of the Berbers and the Cushites. In Arabia it forms the gulf of the Red Sea. This (gulf) in fact extends like a thin tongue into Egypt and forms the Sea of Reed. Then, it passes by the land of Sheba and of Saba and forms the Sea of Elam,

ܕܗܢܘ . ܘܡܟܬܒܝܢ ܬܘܒ ܟܬܒ ܥܠ ܗܢܘ ܒܡܠܘܬܐ . ܘܫܪܝ ܠܡܟܬܒܢܐ ܘܟܬܒ ܥܠ
ܬ̈ܠܝܬܝܐ . ܘܡܟܬܒܝܢ ܟܬܒܢ ܟܬܒ ܥܠ ܐܪܒܥܐ ܕܟܠܗܝܢ ܘܡܛܠܗܝܢ . ܘܡܟܬܒܢܐ
ܥܠ ܐܪܒܥܐ ܕܦܠܛܘܢܘܣ . ܘܐܪܒܥܬܐ ܫܡ̈ܗܬܐ ܘܦܘܪܐ ܠܐ ܡܬܦܠܓܢܐ . ܘܟܬܒ
ܠܢܦܫܐ ܕܒܪܝܠܝܢܐ . ܘܗܘ ܬܘܒܠ ܒܬܪܬܐ . ܡ ܟܬܒ ܥܠ ܐܪܒܥܐ ܘܬܠܝܬܝܐ
ܘܣܘܦܣܛܐ ܘܒܠܘܚܝܬܐ . ܘܡܟܬܒܢܐ ܟܬܒ ܥܠ ܐܢܠܘܛܝܩܐ ܘܥܠܬܗ
ܦܪܝܡܝܢܝܐ . ܘܡܟܬܒܢܐ ܥܠ ܐܦܘܕܝܩܛܝܩܐ . ܘܐܡܝܢܐ ܕܒܬܪ ܡܟܠܐ ܢܝܪ ܦܘܠܝܐ .
ܘܡܛܘܒܐ ܕܐܘܦܪܐܢܣ ܕܡܛܝܬܝܩܐ ܡܬܚܫܒ ܗܘܐ .

[tit.]

ܦܘܣܩܐ ܬܢܝܢܐ . ܡܛܠ ܡܠܟܝܙܕܩ ܒܪ ܢܘܚܐ . ܘܬܫܥܝܬܗ ܕܐܒܪܗܡ .

[V.ii.1.]

ܡܪܕܘܬܐ . ܡܠܬܐ ܟܕ ܡܢ ܫܘܠܛ ܡܬܡ ܐܬܝܐ . ܘܠܘ ܡܢ ܢܛܘܪܘܬܗܘܢ ܡܬܕܒܪܢܐ .
ܘܫܘܠܛ ܐܝܬ ܠܗ ܡܠܬܐ ܐܠܗܐ ܢܗܘܐ ܡܠܝܐ ܘܫܠܝܐ ܗܘܐ . ܥܬܝܕܐ ܡܕܡ
ܕܫܘܠܛ ܐܪܒܥܬܐ ܒܕܝܢܘܬܐ ܘܡܬܪܬܐ ܐܘܟܝܬ ܟܠܝܐ ܗܘܢܝܐ ܡܠܝܐ ܗܘܐ ܟܠܗ
ܕܡܠܝܘܬ ܢܦܫܐ . ܘܟܠܗܝܢ ܒܗ ܡܬܡܠܝܐ ܡܠܬܐ ܒܡܠܬܐ . ܘܢܦܫܬܐ ܗܢܘܢ
ܡܠܬܐ . ܘܦܪܘܫܬܐ ܢܬܡܠܝܐ . ܘܡܥܡܩܐ ܢܬܦܫܩܘܢ . ܦܠܓܝ ܘܗܢܘܢ ܡܠܝܐ .
ܘܡܠܝܐ ܬܘܒ ܒܗ ܫܡܥܝ ܠܗܘܢ ܡܠܬܐ ܡܬܬܦܠܓܢܐ ܒܢܚܬܐ . ܠܒܪ ܢܚܬܐ ܒܡܕܡܐ
ܪܒܝ . ܘܡܬܕܒܠܝܢ . ܘܥܠ ܡܢ ܡܕܡܐ ܐܘܩܝܢܘܣ ܦܪܫܝܢ . ܘܡܦܩܠܝܢ ܒܗܝܢ
ܠܠܫܢܐ . ܡ ܟܝܪܝ ܠܗܝܢ . ܘܡܬܡܠܝܢ ܡܠܬܐ ܫܠܝܐ . ܘܡܠܬܐ ܕܟܬܒܝܢ ܢܗܘܐ ܒܡܠܬܐ
ܘܡܠܝܐ ܡܬܡܠܝܢ . ܘܗܘܐ ܟܝ ܡܢ ܡܬܪܬܐ ܐܪܒܥܬܐ ܫܠܝܠܬܐ ܢܚܬܐ ܐܝܟ .
ܘܢܦܫܗ ܒܢܚܬܐ . ܡܠܬܐ ܕܐܬܝܢ ܒܗ ܦܘܪܫ ܕܡܠܬܐ ܘܠܗܠܝܢ ܕܐܘܪܫܐ
ܟܠܝ . ܫܠܝܐ ܡܬܡܠܝܢ . ܒܗܘ ܕܫܠܝܐ ܡܠܝܢ . ܘܒܦܘܪܫ ܕܡܠܬܐ ܐܝܟ
ܕܢܝܫܐ ܟܬܒܝ . ܘܟܠܗܝܢ ܗܢܝܢܐ ܕܒܝܬ ܠܗܝ ܡܦܩܐ .

[V.ii.2.]

ܕܬܬܬܐܡ . ܒܕܡܘܬܐ ܡܢ ܢܚܬܐ ܐܦ ܡܠܬܐ ܫܠܝܐ ܫܟܝܚܝܢ . ܘܢܦܪܘܫܬܐ ܒܗ ܥܒܕܝܢ .
ܐܠܐ ܒܗܝ ܕܫܠܝܐ ܡܠܝܢ . ܡܬܬܠܡܝܢ ܘܟܐܢܝܐ ܗܘܝܢ . ܘܡܫܠܝܐ ܟܠܝܢ . ܘܐܝܟ
ܬܠܬܐ ܦܣܩܝܢ . ܐܝܟ ܕܐܡܬܠ ܗܘܡܝܪܘܣ ܘܐܚܪܢܐ . ܕܢܚܬܐ ܡܢ ܥܦܪܐ ܠܐܪܥܐ
ܢܚܬ . ܘܫܠܝܐ ܠܗ ܢܦܠ . ܠܐ ܕܝܢ ܡܬܬܦܫܩ ܢܚܬܐ . ܘܐܦ ܢܦܪܘܫܬܐ ܕܪܒܬܐ
ܒܗ ܥܒܕ . ܒܗܝ ܕܟܕ ܦܘܫܩܗ ܡܬܚܙܝܬܐ ܒܟܢܘܫܬܐ ܡܬܚܙܝܬܐ ܡܢ ܡܠܘܗܝ
ܦܠܓ ܫܡܫܐ . ܘܠܗܢܐ ܡܬܚܙܝܐ ܡܢܗܘܢ ܚܫܫܐ . ܐܚܪܢܐ ܕܒܢܝܫܐ ܫܪܝܪ .

FMWPLIV: **V.i.3**: 6 ܘܡܟܬܒܝܢ V || 6-7 ܘܡܟܬܒܝܢ ܟܬܒܢ . . . ܟܬܒ: om. F || 6 ܘܟܬܒ: ܘܟܬܒܘ L || 7 ܟܬܒ: om. Λ || 8 ܕܦܠܛܘܢܘܣ: sine *syāmē* FM; ܕܦܠܛܘܢܝܣ Λ || ܘܐܪܒܥܬܐ: ܘܐܪܒܥܝܬܐ M[1], ܘܐܪܒܥܬܐ M[V] || ܘܟܬܒ: ܘܟܬܒܘ MΛV || 9 ܬܘܒܠ: ܬܘܒ Λ; ܬܘܒܠ MV || 11 ܕܒܬܪ: add ܗܘܐ: F || ܦܘܠܝܐ: ܦܘܠܝܘ ut vid. F || 12 ܕܐܘܦܪܐܢܣ: ܕܐܘܦܪܢܣ M || ܡܬܚܫܒܝܢ Λ || **V.ii.1**: 1 ܢܛܘܪܝܢ V || 3 ܢܦܫܬܐ V || 6 ܘܡܬܠܝܢܐ M || ܡܬܬܦܠܓܢܐ: ܡܬܦܠܓܢܐ W; ܡܬܬܦܠܓܢܐ PL1 || ܒܡܕܡܐ Λ || 7 ܐܘܩܝܢܘܣ FMV || ܒܗܝܢ V || 8 ܠܗܝܢ Λ || 9 ܟܝ ܡܢ: ܘܟܝ V || ܘܗܘܐ MΛV || 10 ܦܘܪܫ W || **V.ii.2**: 2 ܒܗ: ܒܗܝܢ Λ || 3 ܬܠܬܠܬܐ M[1]; ܬܠܬܐ ΛVM[V] || ܗܘܡܝܪܘܣ F; ܗܘܡܪܘܣ ܗܘ M || ܘܐܚܪܢܐ Λ || 5 ܦܘܫܩܗ W ||

and then the Sea of India. In the southeast it passes by Inner India. It (then) encircles the East and passes by the Chinese. In the northeast it passes by the land of Gog and Magog. In the north [it passes] by the land of the Turks and the desolate lands and impassible mountains and forms the Sea of Britain, in which is the island of Thule. Then, it passes by the Iberians, Alans, Scythians and Bulgars. In the northwest it passes by Italy and the whole land of the Franks. In the west [it passes] by Spain. It ends where it began, by the στήλαι. These things are presented plainly in books on geography.

Second Section: On the Salinity of Seawater. Four theories.

[V.ii.1.]
First [theory]: Water is transformed by admixture of something else and not by itself. Through admixture of air, it becomes not bitter but finer and sweeter. The remaining alternative therefore is that admixture of a burning and bitter earthy substance, i.e. smoky exhalation, is the cause of the salinity of the sea. For this reason, when ash is boiled in water and this water is [then] strained, boiled again and placed in the sun, it hardens and turns into salt. Sailors, too, when they run short of drinking water at sea, pour seawater into a caldron, and boil it. Above the caldron they place sponges and collect the vapours with them. They then squeeze them and obtain sweet water. The water that remains is found to be denser and saltier. Furthermore, if someone makes a hollow ball out of wax and throws it in the sea, the water which trickles through the pores in the wax and enters the inside of the ball is found to be sweet, because sweet (water) is fine, and passes through the pores in the wax as if by filtration, while the smoky exhalation, which is dense, remains outside.

[V.ii.2.]
Second [theory]: In (certain) parts of the sea sweet water is also enclosed and rivers flow into (the sea), but because sweet (water) is fine it is vaporised and turns into clouds, whereas salty (water) is dense and is left behind like dregs. As Aesop said in a fable: "the sea gives what is clear to another and keeps what is turbid for itself." The sea does not grow larger even though large rivers flow into it, because on account of its large area the sun meets many parts of its water and causes much vapour to rise from it, just as we see in an experiment that the same

ܕܫܪܐ ܒܫܘܒܚܬܐ ܕܡܪܝܐ ܒܗ ܡܬܦܫܩܐ ܒܟܠ ܡܬܛܠܠܐ . ܐܦ ܒܗ ܡܬܒܪܟܐ .
ܐܠܗܐ ܗܘܐ ܒܪ ܢܫܐ . ܘܐܝܟ ܠܫܒ̈ܚܐ ܡܫܝܚܐ ܩܠܣܘ ܡܢܗ . ܒܗܘ ܕܐܡܪ
ܡܢܝ ܕܒܠܫܒܚܐ ܩܠܣ ܡܢܗ ܡܢ ܢܨܚܬܐ ܚܙܐ ܠܗ .

[V.ii.3.]
ܕܬܠܬ . ܡܛܠ ܢܒܝܐ ܒܝܕ ܡܠܝܘܬܗܘܢ ܘܐܪܒܥܬܗܘܢ ܢܦܩ ܡܢ ܡܪܝܐ ܐܫܬܐܠ
ܢܫܪܐ ܘܕܒܫ . ܘܒܠܫܪܐ ܒܪܬܐ ܒܗܘܢ ܠܢܦܫܐ . ܘܚܫܠܝܐ ܠܗ ܠܢܦܫܐ .
ܘܐܠܦܐ ܕܢܒܝܐ ܢܫ̈ܝܐ . ܕܢܨܝܚܘܬܐ ܘܡܫܒܚܬܐ ܩܠܝܠ . ܘܐܝܠܝܢ ܕܒܫܒܚܐ ܕܡܝܬܘ
ܐ̈ܕܝܢ . ܒܗܘ ܕܒܬܫܒܚܬܐ ܕܠܢܒܝܐ ܡܫ̈ܝܚܐ ܐ̈ܕܝܢ . ܒܗ ܡܢ ܢܒܝܐ ܠܢܨܚܬܐ ܡܚܢܝܢ .
ܕܠܐܝܟ ܡܬܠܒܒܝܢ . ܐܝܟ ܕܝܢ ܢܨܚܬܐ ܡܫܝܚܐ ܒܟܐܠܗܝܢ ܘܡܫܝܚܐ ܠܢܒܝܐ
ܩܕܝܫܐ . ܘܒܝܢܐ ܚܠܦ ܩܠܝܠ ܘܒܟܝܢ ܡܫܒܚ . ܐܝܟܢܐ ܕܢܨܝܚܬܐ ܕܠܒ ܢܫ̈ܝܐ
ܠܝܗ ܒܗܘܢ ܘܠܐ ܠܒܒܝ . ܘܐܝܟ ܢܬܐܡܪܘܢ . ܘܢܝܫܐ ܗܘܐ ܠܗ ܡܬܠܒܒܝܢ
ܒܗܘܢ . ܘܒܐܪܒܥܐ ܕܒܐܘܢܐ ܕܡܫܝܚܐ ܠܡܕܘܢ ܡܫܝܚܐ ܕܡܪܝܐ ܩܠܝܠܐ
ܒܢܫܪܐ ܚܠܝܐ ܥܒܕܐ . ܘܠܐ ܡܬܠܒܒܝܢ ܒܗ ܢܝܫܐ . ܘܡܪ̈ܝܗ ܕܡܫܝܚܐ ܗܘܐ ܒܗ
ܢܬܒܠܡܘܢ ܘܢܬܗܕܪܘܢ ܦܠܓܝܢ ܘܩܕܡ ܩܠܝܠܐ .

[V.ii.4.]
ܕܐܪܒܥ . ܡܦܣܩܢܘܬܗ ܐܡܪ̈ . ܕܕܘܒܪܬܐ ܕܐܪܥܐ ܐܬܚܙܝܬ ܢܒܝܐ .
ܘܒܠܫܪܐ ܩܠܝܠ . ܘܪ̈ܒܢ ܐܡܪ̈ . ܕܫܪܐ ܡܠܬܐ ܒܐܣܟܡ ܡܣܬܝܟܪܐ ܐܬܚܬܢ
ܟܢܫ ܕܡܫܝܚܐ ܡܬܚܫܚܐ . ܘܠܦܪܘܫܘܬܐ ܠܗ ܠܡܠܘܬܐ ܦܫܝܛܐ . ܐܠܐ ܒܟܪ
ܒܢܝܐ ܕܐܢܫܐܠܗ̈ܐ ܡܬܪܥܝܢܐ . ܒܐܘܪܚ ܐܒܗܬܐ ܕܕܘܒܪܬܐ ܪ̈ܚܡܢܘܬܐ
ܕܒܫܘܠܛ ܡܬ̈ܬܐ ܢܒܙܬܐ ܐܬܚܙܝܬ ܐܬܒܪܝܬܗ . ܢܒܝܐ ܡܪܝܐ ܕܒܫܘܠܛ ܥܠܡܐ
ܬܘܒܢܐ ܐܬܚܙܝܘ ܐܬܚܙܝܬܗ . ܒܠܘܬܐ ܕܝܢ ܫܡܠܝܘܬܐ ܕܚܙܝܘܬ ܡܛܠ ܢܒܝܐ . ܗܘ
ܕܠܐ ܒܗ ܢܬܚܙܐ ܒܒܠܗ ܐܪܥܐ ܢܦܫ ܪ̈ܫܝܗܘܢ . ܘܫܡܠܝܐ ܠܥܠܡܐ ܒܪ̈ܝܬܐ
ܕܠܪ̈ܟܬܐ ܘܡܫܬܡܠܝܐ ܕܢ̈ܚܬܐ ܠܚܟܡ ܐܬܚܙܝ .

FMWPLIV: **V.ii.2**: 9 ܕܡܕܝܢ V || ܒܠܫܒܚܐ Λ || **V.ii.3**: 3 ܘܡܫܒܚܬܐ: om. V || 4 ܐ̈ܕܝܢ prim: ܐ̈ܕܝܢ Λ || ܐ̈ܕܝܢ sec.: ܐ̈ܕܝܢ PL || 5 ܒܟܐܠܗܝܢ M; ܒܟܠܗܝܢ PLlV || 7 ܡܬܠܒܒܝܢ M; ܡܬܠܒ̈ܒܝܢ V || 8 ܘܒܐܪܒܥܐ: ܘܒܐܪܒܥܬܐ PLl || ܕܒܐܘܢܐ: ܕܒܐܘܢܐ F; ܕܒܢܝܬܐ M; ܕܒܢܝܬܐ V || ܡܫܝܚܐ: add. ܒܢܫܪܐ ܕܡܪܝܐ ܩܠܝ̈ܠܐ Λ (dittogr.) || 10 ܢܬܗܕܪܘܢ F; ܡܬܗܕܪܘܢ MV || ܘܩܕܡ: ܘܩܡ Λ || ܩܠ̈ܝܠܐ M; ܩܠܝ̈ܠܐ V || **V.ii.4**: 1 ܡܦܣܩܢܘܬܗ V || 2 ܡܣܬܝܟܐ ܦܘܪܫܐ M; ܡܣܬܝܟܦܘܪܫܐ PLl || 3 ܡܬܚܫܚܐ V || ܘܠܦܪܘܫܘܬܐ W || ܠܡܠܘܬܐ: add. ܢܫܝܐ L || 5 ܐܬܒܪܝܬܗ scripsi: ܐܬܒܪܝܬ codd. || 6 ܒܠܘܬܗ P || ܫܡܠܝܘܬܐ PLlV || ܡܛܠ ܢܒܝܐ W || 7 ܕܠܐ ܒܗ: ܠܐ ܕܒܗ Λ || 8 ܘܡܫܬܡܠܝܐ P ||

quantity of water disappears more quickly when it is spread out than when is concentrated. Neither does the sea grow smaller, even though large quantities of vapour rise from it, because the same amount as that which rises from it in [the form of] vapour enters it from rivers.

[V.ii.3.]
Third [theory]: Seawater, because of its salinity and earthiness, is heavier and denser than other [types of] water and for this reason an egg will float on it, whereas it will not float on sweet [water]. Ships on the sea are heavy, whereas those on rivers and lakes are light, and those who sail on the Sea of India, because they sail on lakes which are connected to the sea, when they transfer from the sea to a lake, easily go under. There is a putrid lake in Palestine near the Red Sea. Its water is extremely salty and dense, so that very heavy animals will float on it and will not sink even if they are bound up and fish do not breed in it. In the land of Chaonia near Dodona, a spring of salty water flows into a fresh(water) river and fish do not breed in (this river). When the water of this spring is boiled and [then] is cooled, it hardens and becomes salt.

[V.ii.4.]
Fourth [theory]: Empedocles said that the sea is the sweat of the earth and is salty for this reason. The Master [Aristotle] said that this statement uses the figure of a metaphor - i.e. transfer of an image - and is appropriate for poets, not for philosophers, unless it is taken as an analogy. Indeed, just as sweat is moisture which has been made salty by admixture of yellow bile, sea is water which has been made salty by admixture of smoky exhalation.

The final cause of the salinity of seawater is to prevent its fumes, when it becomes putrid, from blowing all over the earth, and bringing about a universal corruption which causes plants to wither and animals to die.

[tit.]

ܦܣܘܩܐ ܬܠܝܬܝܐ . ܡܛܠ ܚܙܘܝ ܢܘܪܐ . ܬܐܘܪ̈ܝܣ ܕܝܠܗ ܬܠܬ .

[V.iii.1]

ܡܕܡܝܘܬܐ . ܕܢܘܪܐ ܒܪܘܡܬܐ ܕܠܗܠܬܐ ܢܗܘܐ ܒܚܢܢܐ ܠܐ ܐܠܝܨܐ . ܐܠܐ ܡܢ
ܠܡܥܒܕܘ ܡܢ ܕܗܘܬܐ ܠܐܫܬܘܬܐ . ܒܪ̈ܬܐ ܒܬܪ ܐܪ̈ܝܚܐ . ܕܠܡܫܬܡܫܘܬܗܘܢ ܚܕܐ
ܐܪ̈ܝܚܐ ܠܗ ܦܦܘܫ . ܡܥܒܕܐ ܕܝܢ ܢܘܪܐ . ܒܗܘ ܕܡܠܐܐ ܡܢ ܢܗܝܪܘܬܐ ܢܦܫܗ .
ܘܢܗܝܪܘܬܐ ܡܢ ܚܪ̈ܝܬܐ ܕܠܝܠ . ܘܡܢ ܡܛܠ ܡܠܐܐ ܓܘܫܡ . ܘܐܦܠܝܢ ܡܫܚ̈ܬܐ ܡܢ ܚܕ
ܚܒܠܐ ܚܪ̈ܝܬܐ ܢܓܕܥ . ܘܢܗܝܪܘܬܐ ܚܢܝܢ . ܘܡܫܟܚܐ ܠܗ ܡܚܝܢܐ . ܘܡܚܒܠܐ
ܐܫܝܢܐ ܚܪ̈ܝܬܐ ܢܓܕܥ . ܘܢܗܝܪܘܬܐ ܚܫܝܢ . ܘܡܫܟܚܐ ܡܬܦܠܓܐ . ܘܡܚܝܢܐ ܡܢ
ܣܘ ܚܒܠܐ ܒܗ ܡܦܘܫ ܡܬܪܐ ܢܘܪܐ ܗܘܐ ܢܒܥܐ . ܘܚܒܠܐ ܐܫܝܢܐ ܒܗ ܦܘܫܝܢ
ܢܒܥܐ ܗܘܐ ܢܘܪܐ . ܐܝܟ ܕܐܡܪ ܒܗ . ܕܡܫܬܟܚܝܢ ܒܡܥܒܕܐ ܠܚܒܠܐ . ܡܢ
ܕܚܒܠܐ ܡܬܚܠܝܢ . ܡܫܘܐ ܕܒܐ ܘܠܘܦܢܐ ܗܘܐ . ܘܡܢ ܪ̈ܒܝܐ ܠܦܪܐ
ܡܬܬܠܝܢ . ܘܡܢ ܕܒܡܠܟܐ ܡܬܚܠܝܢ . ܡܢܠܐ ܕܒܐ ܘܡܫܪܝܐ ܗܘܐ .
ܘܡܬܬܚܒܝܢ ܠܦܪܐ ܘܠܝܠܝܐ . ܘܡܚܒ̈ܠܐ ܡܬܚܒܝܢ ܘܒܚܒܘܫܝܢ ܐܦ ܢܗܝܪܘܬܐ .

[V.iii.2]

ܕܬܬܚܝܢ . ܡܬܕܢ ܒܗ . ܐܝܟ ܕܐܡܪܢ ܒܗ . ܢܘܪܐ ܗܘܐ ܡܫܬܘܚܐ . ܘܐܬܚܒܠܐ
ܘܗܘܐ ܐܠܦܐ . ܘܒܠܗܒܐ ܒܒܗܘܒܐ ܗܘ ܕܐܪܣܛܘܛܠܝܣ ܒܪܒܐ ܕܣܘ̈ܟܠܐ
ܡܕܡܘ̈ܬܐ ܡܩܒܠܢܘ ܘܦܠܠܝܣܘ ܡܬܬܚܝܢ . ܐܦ ܒܗ ܘܢܗܘܐ ܡܬܚܠܦܢܘܣ ܒܗ
ܢܗܘܐ ܐܦܘܢܐ ܗܘܐ . ܢܗܘܐ ܗܘܐ ܒܚܒܘܫܬܐ . ܘܐܬܘܒ ܓܪܘ ܡܬܘܗ . ܐܝܟܢܐ
ܕܐܠܦܐ ܕܐܪܟܬܐ ܠܗ ܢܚܒܟ ܒܗ . ܘܗܘܐ ܡܢ ܗܘܐ ܒܒܢܐ ܒܬܪ ܢܚܝܬܐ
ܕܠܐ ܡܬܬܚܝܫ ܡܢܝܢܐ ܕܚܝܠܘܬ ܒܚܫ̈ܒܐ . ܒܥܒܐ ܙܒܢܝܢ ܒܗ ܙܒܢܐ
ܒܠ̈ܦܘܢܐ ܐܚܪܢ . ܘܒܡܪ̈ܟܒܐ ܡܬܚܪܟܝܢ . ܘܒܪ̈ܦܝܢܐ ܡܬܟܪܝܢ . ܘܒܫܘܠܛܐ
ܕܐܐܪ ܦܫܝܛ . ܘܡܢ ܐܬܬܐ ܠܐܫܝܢܐ ܡܓܠܝܢ . ܘܠܡ̈ܪܝܢܘܢ ܡܬܚܫܠܝܢ .
ܘܒܬܪ̈ܬܗܘܢ ܡܬܚܫܠܝܢ . ܘܡܢ ܐܫܝܢܐ ܠܗ ܡܬܡܪܝܢ . ܘܗܘܐ ܒܠܦܪܐ ܡܫܝܐܐ
ܡܬܩܒܠ . ܘܒܬܪ ܡܬܬܒܠܐ ܐܬܠܝܢܐ ܢܘܦܐ ܐܡܨܡܨܐ ܒܫܓܝܪ . ܕܡܢ ܒܬܪ̈ܘܐ
ܦܪܐܡܢܣܘܣ ܐܘܒܢܬ ܡܬܦ̈ܬܐ ܕܠܐ ܡܬܩ̈ܫܝܢ ܡܬܦܬܪ . ܘܒܡܠܝܢ ܙ̈ܒܢܐ ܐܫܝܢܐ
ܡܙ̈ܒܢܐ ܡܬܚܡܨܝܢ . ܓܡܚܝܢ ܕܒܬܚ̈ܬܐ ܐܝܟ ܕܡܚܝܢ ܡܢ ܠܐ ܡܬܦܪܝܢ .
ܡܢܝܢ ܕܝܢ ܠܐ ܡܬܚܫܠ̈ܝܢ .

FMWPLlV: V.iii.1: 2 ܐܪ̈ܝܚܐ ܒܬܪ F || 3 ܕܡܠܐܐܐ PLl || 4-5 ܚܪ̈ܝܬܐ . . . ܕܠܝܠ: om. W || 5 ܡܚܝܢܐ: ܡܚܝܢܐ ܡܚܝܢܐ MΛV || 6 ܡܬܦܠܓܐ (Pael) V || 8 ܢܒܥܐ ܗܘܐ ΛV || ܐܡܪ: ܐܡܪܝܢ V || ܡܫܬܟܚܝܢ V || 11 ܡܬܬܚܒܝܢ V || ܘܒܚܒܘܫܝܢ ܐܦ ܢܗܝܪܘܬܐ: om. Λ || **V.iii.2**: 2 ܘܗܘܐ: ܘܗܘܐ V || ܒܒܗܘܒܐ M || ܕܐܪܣܛܘܛܠܝܣ F; ܕܐܪܣܛܘܛܠܝܣ Λ; ܕܐܪܣܛܘܛܠܝܣ V || 3 ܡܩܒܠܢܘ: ܡܩܒܠܢܘ M; ܡܩܒܠܢܘ Λ (una cum voce sequenti ܡܩܒܠܢܘܦܠܠܝܣܘ) || ܦܠܠܝܣܘ F || ܡܬܚܠܦܢܘܣ Λ; ܡܬܚܠܦܢܘܣ V || 5 ܒܒܢܐ ܒܬܪ ܢܚܝܬܐ ΛV || 6 ܒܗ ܙܒܢܐ ΛV || 10 ܡܬܩܒܠ F || ܐܬܠܝܢܐ: ܐܬܠܝܢܐ P; om. L || ܕܡܢ: ܡܢ ΛV || 11 ܦܪܐܡܢܣܘܣ F; ܦܪܐܡܢܣܘܣ MPLl || ܡܬܩ̈ܫܝܢ WPL[1]l (ܡܬܩ̈ܫܝܢ $L^{corr.}$) || ܡܬܦܬܪ FV || ܐܫܝܢܐ MW || 12 ܡܙ̈ܒܢܐ Λ || ܡܬܚܡܨܝܢ V || ܐܝܟ: om. FMV || 13 ܡܬܚܫܠ̈ܝܢ: add ܐܝܟ V ||

Third Section: On the Migration of the Sea. Three theories.

[V.iii.1.]
First [theory]: It is not necessary that the sea should be in its particular place by nature; rather, it is able to migrate from one place to another, albeit over long periods of time, which the lives of humans are not long enough to delimit. The sea migrates because it takes its material from rivers, while rivers draw their water from springs and grow copious with rainwater. Often in one area springs dry up, rivers fail, and the sky is not clouded over, while in another area springs emerge, rivers gush forth and the sky causes rain to fall. Thus, in one area, as water vanishes, sea becomes dry land, while in another area, as (water) becomes copious, dry land becomes sea. The Master [Aristotle], in fact, said: When the five planets gather in [the sign of] Capricorn, great winter and cataclysm occur, and mountains are torn apart by thunder. When they gather in [the sign of] Cancer, great summer and conflagration occur, mountains crumble and fall, and springs dry up and together with them the rivers.

[V.iii.2.]
Second [theory]: Egypt, as the Master [Aristotle] said, was sea in former times, but [then] dried up and became dry land. For this reason bones of marine animals, κογχύλια and τελλίναι, are found in the depression of Arsinoitis. Lake Maeotis, too, although it used to be sea, became [first] a deep lake and [then] its water decreased further, so that large ships cannot pass through it. Such a process does take place, but over a long time, the number of whose years cannot be recorded in books. For often nations are annihilated by floods, are exterminated by wars, are destroyed by famine, perish due to corruption of air and migrate from one place to another; their languages change, and their writings are replaced and cannot be read by others. In fact, (such writings) are found on many mountains. On two solid, flame-shaped towers in Egypt, called πυραμίδες - or indivisible particles - by the ancients and named "haramān" in modern times, we hear that there are writings, some of which cannot be read and some of which cannot be understood.

[V.iii.3.]

ܕܬܠܬ. ܢܚܬ ܥܠܝܗ ܒܚܝܠܐ. ܘܐܝܟ ܕܒܓܒܪܐ ܡܬܬܦܝܣ ܡܢ ܪܘܚܐ ܕܝܢ ܒܗܘܢ ܡܬܬܦܝܣܝܢ. ܐܦ ܟܠ ܐܦܣܩܘܦ ܠܢܟܝܢ. ܡܬܬܦܝܣ ܕܝܢ ܗܘܝܢ ܢܨܝܚܘܬܐ ܡܫ̈ܠܠܢܐ ܕܡܢܐܦܐ ܙܝܢ ܠܗ. ܡܫܠܡܢܐ ܗܘܐ ܕܒܪܟܬܐ ܐܠܗܝܬܐ ܐܝܟ ܗܢܘܢ ܕܒܒܘܠܠܝܐ ܡܬܪܕܝܢ ܡܩܢܝܢ. ܘܡܛܠ ܕܒܫܘܪܝܗ ܠܬܪܝܢܘܬܐ ܓܒܝܐ ܙܕܝܢ. ܐܝܟ ܒܗ̈ܠܐ ܕܡܫܡܫܢܘܬܐ ܕܠܘܬ ܢܘܗܪܐ ܒܝܬ ܡܢ ܗܝܐ ܒܢܝܢ ܕܝܢ ܗܘ ܠܗܠܟܐ. ܗܘܪܦܘܣ ܕܝܢ ܗܘܐ ܡܟܪܙܢܐ ܕܒܝܬ ܬܠܬ ܪܝܫܐ ܐܢܬܘܢ. ܘܒܪ ܠܫܡܫܐ ܪܘܚܐ ܡܬܪܢܝܢ ܡܢܝܢܐ ܒܗ ܒܡܟܪܙܢܐ ܘܡܬܬܬܬܝܢ ܒܪܡܐ ܠܝܡܐ. ܘܒܪ ܠܗ ܡܨܝܢ ܠܡܪܢܐ ܠܟܪ̈ܝ ܢܝܚܐ ܠܒܣܡܬܐ ܟܠܦܝܢ. ܘܡܫܡܫܢܘܬ ܪ̈ܘܚܐ ܩܕܝܫܝܢ. ܘܡܢܝܢܐ ܠܬܪܝܢܘܗܝ ܕܟܪܐ ܡܬܬܦܝܣܝܢ. ܘܟܠܗ ܬܫ̈ܒܚܬܐ ܘܫܠܝܡܬܐ ܒܚܝܠܐ ܕܟܠܗ ܘܡܢܘܙ ܠܪ̈ܝܫܐ ܕܗܘܝܐ ܘܟܪܝܗܘܢ ܢܘܢܐ ܘܦܝܫܐ ܢܪܟܝܢ ܦܘܠܚܢܐ.

[tit.]

ܦܘܣܩܐ ܪܒܝܥܝܐ. ܡܛܠ ܐܝܪܐܪܟܘܣ. ܘܐܝܪܐܪܟܝܘܬܗ ܬܠܬ.

[V.iv.1]

ܡܪܟܒܢܘܬܐ. ܠܟܠܢܫ ܡܢ ܗܘ ܕܒܚܝܠܐ ܒܗܘܢ ܕܐܠܗܝܘܗܝ. ܡܛܠ ܐܝܪܐܪܟܘܣ. ܕܐܢܫܘܬܐ ܠܗ ܒܡܫܡܠܝܘܢ ܕܐܪܟܐ. ܘܒܗ ܗܘ ܟܠܝܠܬ ܡܢܝܢܐ. ܘܡܛܘܠ ܡܢܝܢܐ ܠܗ ܡܫܬܘܬܐ ܐܝܟ. ܘܡܫܬܡܫܢܝܢ. ܘܟܠܝܠܬ ܕܢܝܐ ܢܗܝܪܝܢ. ܒܗܘ ܕܒܝܬ ܒܡܫܡܫܢܘܬܗܘܢ ܠܫܠܝܠܬ ܐܪܟܐ ܡܫܡܫܢܘܬܐ ܐܝܬ. ܡܢܝܢ ܕܝܢ ܘܐܝܟܢ ܕܠܝܐ ܡܬܡܢܝܢܘܬܐ ܐܝܬܘܗܝ. ܒܗܘ ܕܝܢ ܟܠܝܠܬ ܡܢܝܢܐ ܐܢܫܘܬܐ ܐܝܪܐܪܟܘܣ. ܡܫܬܡܫܢ ܡܢܝܢܐ ܕܠܟܠ ܡܬܬܦܝܣܝܢ ܒܚܝܠܐ. ܒܗܘ ܕܒܡܫܡܠܝܘܢ ܕܐܪܟܐ ܡܫܬܝܢ. ܘܒܐܪܡܘܬܐ ܟܠܝܐ ܕܠܟܪ ܡܢ ܡܫܡܠܝܘܢ ܠܟܠ ܐܢܫܘܬܐ. ܡܢܝܢ ܡܢܝܢ ܐܝܪܐܪܟܘܣ ܒܪ ܡܫܬܡܫܢܝܢ. ܠܟܠ ܡܬܬܦܝܣܝܢ ܒܢܝܢܬܗ. ܘܡܢܐ ܕܡܫܡܠܝܘܬ ܢܝܚܐ ܢܗܝܪܝܢ. ܠܬܠܬ ܡܬܬܦܝܣܝܢ ܡܠܝܐܝܬ. ܗܘ ܕܒܟܠܝܐ.

[V.iv.2.]

ܕܬܬܬܦܝܢ. ܐܝܟܢܐ ܗܘܝ ܡܠܬ ܦܘܩܕܢ. ܕܐܝܟ ܒܡܫܡܫܢܘܬܐ ܕܡܢܝܢܐ ܕܠܟܠܝܢ ܡܢ ܐܝܪܐܪܟܘܣ ܒܝܕ ܡܢܝܢܐ ܕܢܚܝܢ ܡܢ ܐܪܟܐ. ܫܡܥܝܢ ܚܟܝܡܝܢ ܠܗ ܒܝܕ ܦܘܠܘ ܕܐܪܟܐ. ܘܕܪܝܫ ܕܒܝܠܐ ܗܢܐ ܢܗܝܪܐ ܡܢ ܗܘ ܕܒܡܫܡܠܐ ܡܢ ܡܢܝܢܐ ܒܝܬܝܐ ܚܟܝܡܝܢ ܠܗ ܘܡܫܬܡܠܝܐ ܠܡܫܬܡܠܝܢ ܡܛܠ ܡܫܡܠܝܘܬ ܠܡܫܝܚܐ ܕܗܘܐ ܡܢ ܚܟܝܡܐ. ܒܡܬܪܐ ܕܝܢ ܡܢܝܢܐ ܒܝܬܝܐ ܠܡܫܡܠ ܡܛܠ ܡܫܬܡܠܝܐ ܚܟܝܡܝܢ ܠܗ ܡܛܠ ܡܫܡܠܝܘܬ ܡܠܝܐ. ܠܐ ܡܢܝܢ ܚܟܝܡܐ ܒܚܝܠܘܬ ܡܢܝܢܐ ܕܪܝܫܐ ܡܛܠ ܠܗܢ ܕܪܝܫܐ ܠܗ.

FMWPLIV: V.iii.3: 3 ܕܒܝܫܘܦܐ: add. ܥܕܡ Λ || 8 ܘܡܫܡܫܢܘܬ ܪ̈ܘܚܐ ܩܕܝܫܝܢ: om. F || 9 ܬܫܒܚܬܐ: ܬܫܒܚܬܐ W; ܬܫܒܚܬܗܘܢ PLIVM^V || tit.: ܐܝܪܪܟܘܣ P; ܐܝܪܐܪܟܘܣ LIV || V.iv.1: 1 ܒܗܘܢܗܘܢ M; ܒܗܘܢ V || ܕܐܠܗܝܘܗܝ Λ || 2 ܐܝܪܐܪܟܘܣ PLI; ܐܝܪܪܟܘܣ V || ܒܡܫܡܠܝܘܢ Λ || ܕܐܪܟܐ: om. L || 4 ܘܡܫܡܫܢܘܬܐ FM || ܡܢܝܢ: ܡܢܝܢܢ M || 5 ܡܢܝܢܐ: add ܡܢܝܢܐ Λ (dittogr.) || 6 ܐܝܪܐܪܟܘܣ PLI; ܐܝܪܪܟܘܣ V || ܕܒܡܫܡܠܝܘܢ MΛV || 7 ܘܐܪܡܘܬܐ V || ܡܫܡܠܝܘܢ scripsi: ܡܫܡܠܝܘܢ codd. || ܐܢܫܘܬܐ: add. ܢܦܫ MV || 8 ܡܢܝܢܐ ܡܢܝܢ: ܡܢܝܢ M (om. ܡܢܝܢܐ); ܡܢܝܢܐ ܕܝܢ Λ; ܡܢܝܢ ܡܢܝܢܐ V || ܐܝܪܐܪܟܐ Pl; ܐܝܪܪܟܐ LV; ܐܝܪܐܪܟܘܣ W || ܒܢܝܢܬܗ: ܒܠܢܝܢܬܗ Λ || ܘܡܢܐ: ܘܡܢܢܐ F[1] (ܘܡܢܐ F[corr.] ut vid.); ܡܢܢܐ M || V.iv.2: 1 ܐܝܟܢܐ: add. ܕܝܢ Λ || 2 ܐܝܪܐܪܟܘܣ l; ܐܝܪܪܟܘܣ V || ܫܡܥܝܢ P || ܠܗܘ ΛV || 3 illeg. M[1] (ܒܝܕ ܦܘܠܘ?); ܘܦܘܠܘܣ M[V] || 6 ܠܗܢ: in marg. M ||

[V.iii.3.]
Third [theory]: The sea is calm by nature, but is made to move accidentally by winds which are released from its depth or which blow on its surface. It is also made to move by wave-causing rivers, which push it violently, especially when its water is squeezed into narrow places such as those at Byzantium. The water in εὔριποι flows in both directions, just as [of] the scales on a balance sometimes the one and sometimes the other goes down according to weight. The εὔριπος is a passage between two seas. When wind blows, water is pushed in the passage and is brought to the sea. When (this water) is unable to push the water of the sea, it turns backwards, undergoes a reversal of flow and in this way moves in both directions. Many acknowledge as the cause of the ebb and flow in the Sea of Elam and of India the daily and monthly rising and setting of the moon.

Fourth Section: On Tartarus. Three theories.

[V.iv.1.]
First [theory]: Refuting the statement written in Plato's *Phaedo* about Tartarus - [namely] that it is at the centre of the earth and contains the entirety of water and [that] this water is unstable, fluctuates and undergoes reversal of flow because there is a void between its convex surface and the concave underside of the earth - the Master [Aristotle] says that this is impossible because, if Tartarus is the entirety of water, water will be found to move upwards 'by nature' - because (this water) is placed at the centre of the earth and in a sphere whatever is outside the centre is above [it]. - As it fluctuates, therefore, Tartarian water will move upwards 'by nature' and, when it undergoes reversal of flow, it will move downwards 'by force', which is absurd.

[V.iv.2.]
Second [theory]: The passage of *Phaedo* also says that the same quantity of water as that which exits from Tartarus through springs gushing out of the earth reenters it through the pores in the earth. That this opinion is defective is known from the fact that in summer little water enters it and much (water) exits from it because of the large quantity of vapour generated by the sun, whereas in winter little water exits from it and much water enters it because of the large quantity of rain. The quantity, therefore, of water which flows out of it is not equal to that which flows into it.

[V.iv.3.]

ܕܬܠܬܬ . ܐܦܠܐ ܒܠܚܘܕܝܬܐ ܕܡܢܝܢܐ ܐܪܝܬܡܛܝܩܘܣ ܗܘ ܗܘܐ ܦܝܠܘܣܘܦܘܣ . ܒܠܚܘܕ ܡܘܣܝܩܝܬܐ
ܒܗ ܓܝܪ ܗܘܘ . ܘܗܘܐ ܠܗ ܡܟܝܠܐ . ܟܕ ܕܐܦ ܡܘܣܝܩܝܬܐ ܡܢܝܢ ܡܬܚܫܒܝܢ .
ܒܬܪ ܕܝܢ ܠܐ ܡܫܟܚܝܢ ܐܢܫ ܐܢܫܐ . ܐܠܐ ܒܡܕܥܘܬܐ ܐܪܝܬܡܛܝܩܝܬܐ ܡܬܚܫܒܝܢ . ܠܐ ܠܐ
ܡܕܝܢ ܦܝܠܘܣܘܦܘܣ ܐܠܐ ܢܣܒܐ ܐܪܝܬܡܛܝܩܘܣ ܒܠܚܘܕܝܬܐ ܕܐܣܛܪܘܢܘܡܝܐ ܡܢܝܢܐ .
ܘܗܟܘܬ ܡܕܝܢܬܐ ܕܒܠܚܘܕ ܗܘܝܐ . ܘܡܛܠ ܕܐܣܛܪܘܢܘܡܝܐ ܘܠܐ ܗܘܝܐ ܗܘܐ ܐܪܝܬܡܛܝܩܘܣ
ܒܠܚܘܕܐ . ܐܦ ܢܣܒܐ ܡܢܝܢܬܗ ܠܐ ܗܘܝܐ ܗܘ ܘܡܬܚܫܒܢܐ ܐܪܝܬܡܛܝܩܘܣ . ܠܐ ܕܝܢ ܡܟܝܠ
ܒܐܢ̈ܫܐ ܟܕ ܕܠܝܬ ܐܢܫܐ ܐܢܫܐ ܚܕܢܐ ܗܘ ܕܒܡܢܝܢܐ ܕܐܝܟ ܗܟܢܐ ܕܠܝܬ
ܐܢܫ ܕܐܬܚܫܒ ܠܡܚܫܒܘ ܢܗܘܐ . ܐܠܐ ܡܬܚܫܒ ܡܢ ܢܣܒܗ ܗܘ . ܟܕ ܗܘ ܡܢܐ
ܕܡܢ ܢܣܒܬ ܢܦܫܗ ܗܘܐ ܐܚܪܢ ܘܐܦ ܐܢܫܐ ܠܡܢܝܢܐ ܗܘܐ ܘܠܐ ܡܢܝܢܐ . ܘܒܗ ܠܐ
ܡܬܚܫܒ ܒܐܢ̈ܫܐ ܐܝܟܐ ܡܢ ܢܣܒ . ܥܢܝܢܐ ܕܐܝܢܐ ܡܢܝܢܐ ܠܦܠܛܘܢ .

ܫܠܡ ܡܐܡܪܐ ܕܦܝܠܘܣܘܦܘܣ .

FMWPLIV: V.iv.3: 1 ܡܢܝܢܘܬܐ M[1]; ܒܠܚܘܕܝܬܐ M[V] || ܦܝܠܘܣܘܦܘܣ PL; ܦܝܠܘܦܘܣ IV || 2 ܕܐܦ scripsi: ܕܐܦ codd. || 4 ܦܝܠܘܣܘܦܘܣ LI; ܦܝܠܘܦܘܣ V || ܒܠܚܘܕܝܬܐ M || ܕܐܣܛܪܘܢܘܡܝܐ M || ܡܢܝܢܐ PLI; ܡܢ̈ܝܢܐ V || 5 ܡܕܝܢܬܐ: ܡܠܬܐ L || ܕܐܣܛܪܘܢܘܡܝܐ: add. ܫܠܡ ܡܐܡܪܐ ܕܦܝܠܘܣܘܦܘܣ ܘܫܘܪܝܐ ܫܘܡܠܝܐ ܐܡܝܢ W || ܘܠܐ: **hinc FMPLIV** || 7 ܟܕ: ܘܟܕ L || ܗܘ: ܗܘ Λ || 9 ܘܠܐ ܗܘܐ: ܠܐ ܐܦ ܗܘ Λ || ܠܐ ܒܗ: ܘܠܐ F || 10 ܥܢܝܢܐ: add. ܡܢܝܢ Pl || ܥܢܝܢܐ ܕܡܢܝܢܐ V || ܡܢܝܢܐ: om. P || ܠܦܠܛܘܢ: ܠܦܠܛܘܢ ΛV || **fin.**: ܫܠܡ ܡܐܡܪܐ ܕܦܝܠܘܣܘܦܘܣ: om. M || ܕܦܝܠܘܣܘܦܘܣ PV; illeg. L[1]; ܕܦܝܠܘܣܘܦܘܣ L[corr.] || add. ܘܠܐܠܗܐ ܫܘܒܚܐ ܕܝܗܒ ܘܥܕܪ ܠܟܠܢ ܒܟܠܙܒܢ ܐܡܝܢ Pl; add. ܘܠܐܠܗܐ ܫܘܒܚܐ ܕܝܗܒ ܘܥܕܪ ܐܡܝܢ ܘܠܡܠܟܐ ܘܡܫܝܚܐ ܒܫܘܠܡܗ ܕܡܢ ܒܬܪ ܢܦܫܗ ܕܡܠܬ ܫܘܒܚܐ ܕܠܐܠܗܐ L; add. ܘܠܐܠܗܐ ܫܘܒܚܐ V ||

[V.iv.3.]
Third [theory]: If Tartarus were the entirety of water, all rivers would flow into it. This, however, is not the case because, even though some rivers are swallowed up [into the earth], they nevertheless do not remain below the earth but appear in other places. Therefore, not Tartarus but the sea is the entirety of the watery element.

(The sea) is a part of this world, and because the world is without beginning and not created, likewise the sea, too, [which is] a part of it, is not created and [is] without beginning. (The sea) is not artificial, because it is beyond human power to excavate such a depth, which no one has been able to fathom. Neither is it spontaneous, since that which arises of its own accord is small, rarely arises and does not persist. Since it is neither artificial nor spontaneous, the remaining alternative is that nature has formed it in this way.

Here ends the Book of Minerals.

[tit.]

ܟܬܒܐ ܕܡܥ̈ܕܢܝܬܐ.

ܕܐܝܬܘܗܝ ܚܡܝܫܝܐ ܕܟܝܢ̈ܝܬܐ ܡܢ ܟܬܒܐ ܕܚܐܘܬ ܚܟܡܬܐ. ܘܗܘ ܐܝܬ

ܡܐܡܪ̈ܐ ܚܡܫܐ.

ܡܐܡܪܐ ܩܕܡܝܐ.

ܡܛܠ ܟ̈ܝܢܐ ܘܡܠܝܢ ܕܠܘܬܗ ܡܬܝܚܣܢ. ܘܗܘ ܐܝܬ ܦܣܘ̈ܩܐ ܬܠܬܐ.

ܦܣܘܩܐ ܩܕܡܝܐ.

ܡܛܠ ܐܝܟܢܝܘܬ ܗܘ̈ܝܐ ܕܟ̈ܝܢܐ. ܬܐܘܪ̈ܝܡܐ ܬܠܬ.

[I.i.1]

ܡܩܕܡܬܐ ⁘ ܠܟ̈ܝܢܐ ܡܠܝܢ ܕܟܠܗ ܐܝܟܢܐ ܡܬܩܠܝܢ ܦܝܠܘܣܘܦܐ ܗܢ ܕܩܕܡ ܟܬܒܐ ܡܛܐܘܪܘܠܘܓܝܐ ܐܘܕܥܬ ܡܛܠܠܝܬܐ ܦܫܝܛܬܐ ܡܬܦܪܫܢܝܬܐ ܠܢܦܫܐ. ܗܫܐ ܕܝܢ ܠܡܦܪܫ ܥܠ ܟ̈ܝܢܐ ܙܢܐ ܕܐܬܡܗܒ. ܘܐܡܪܢ. ܝܠܗܘ ܕܡܫܟܚܐ ܡܢ ܟܕ ܙ̈ܢܝ ܥܠ ܐܪܥܐ ܘܡܬܐ. ܡܬܚܙܝܢܘܬܐ ܕܡܛܠܝܬܐ ܡܕܡ. ܡܢ ܐܪܥܐ ܡܢ ܢܦܩܐ ܡܫܟܚܐ. ܡܢ ܡܬ̈ܐ ܕܝܢ ܠܡܥܒܐ. ܘܡܢܗܘܢ ܡܠܝܢ ܟܕ ܡܫܬܠܡܝܢ ܒܙܒ̈ܢܐ ܡܬܡܠܝܢ. ܘܐܝܟܐ ܕܡܬܡܠܝܢ ܡܢܗ ܡܢ ܐܘܩܝܢܘܣ ܕܝܡܐ. ܘܠܐ ܡܬܠܝܐ ܠܗ ܡܢ ܡܢܬܗ ܝܠܗܐ. ܡܬܩܢܝܢ. ܘܗܘܐ ܡܛܐ ܡܬܐܪ. ܘܡܬܬܝܩܢ. ܘܚܢܐ ܡܛܪ. ܗܘ ܕܝܢ ܗܘ ܐܐܪ ܡܢ ܥܪܒܐ ܕܐܐܪܐ ܡܪܢ. ܠܘ ܡܢ ܚܝܘܗ. ܐܠܐ ܡܢ ܡܪܢܬ ܐܪܥܐ ܘܡܬ̈ܐ ܘܗܘܐ ܠܟܢܐ †ܒܡܠܝܘܬ† ܗܘܐ ܘܪܥܘܬ ܡܫܡܥܐ ܥܠܝܗ ܠܟܠܐ ܘܬܘܬܐ.

[I.i.2]

ܕܬܠܬܝܢ. ܡܛܠ ܕܗܘ ܕܐܠܗܐ ܡܢܝܢܐ ܡܕܡ ܟܕ ܡܢܬܐ ܠܐܐܪ ܗܘ. ܡܢ ܬܠܬܝܗܘܢ ܡܢܝܢܐ ܕܗܘܢܐ ܚܝܢܐ. ܡܢ ܡܢܬܐ ܡܢ ܟܕ ܢܬܚܠܡܘܢ ܘܢܬܦܫܛܘܢ. ܡܢ ܐܐܪ ܕܝܢ ܟܕ ܢܬܦܚܚ ܘܢܬܦܝܚ. ܐܟܡܐ ܕܒܢܝܫ ܠܦܪܐ ܡܢܬܐ ܙܒܢܝܢ ܩܕ̈ܡܝܢ ܫܢܝܢ ܕܦܬܐܪ ܐܐܪ. ܘܒܬܪ ܝܘܡܐ ܡܬܬܪܝܨ. ܘܫܡܢܐ ܚܝܢܐ ܡܬܥܒܕ. ܘܡܬܠܐܐ ܐܦ ܬܠܝܬܐ ܠܝܬ. ܘܠܫܠܚܐ ܠܗ ܒܟܠܒܢ ܠܐܬܪܐ ܕܡܥܒܕ ܡܢܝܢ ܒܐܐܪ ܡܬܟܠܐ. ܒ ܫܘܐ ܚܝܢܐ. ܐܝܟܢܐ ܕܝܢ ܟܕ ܢܣܒܐ ܒܪܝܫ ܠܦܪܐ ܕܒܝܫ ܒܒܢܘ ܠܦܠܘܪܐ ܒܐ ܐܬܦܠܛܬ. ܢܦܩܬ ܚܝܢܐ ܕܒܥܒܐ ܠܦܠܓܐ ܕܠܦܪܐ ܐܬܟܠܝܬ. ܘܥܠ ܒܒܦܘ ܒܦܬ. ܘܐܝܟܢܐ ܘܕܒܥܒܕ ܒܝܣܐ ܠܒܠ ܡܪܢܗ. ܘܒܬܪ ܒܪܢܐ ܡܢܝܢ ܝܐܕܝܢ ܠܬܟܬ̈ܢܝܢ. ܟܕ ܚܝܢܝܢ ܡܬܪ̈ܝܢܘܬ ܡܢ ܡܢ ܡܬܢ̈ܝܐ ܕܢܒܥܬ ܥܠ ܐܬܪܐ ܘܒܠܝܢܘܬ.

FMPLIV: **tit.**: 1 ܕܡܥܕܢܝܬܐ (sg.) M || 2 ܚܡܝ̈ܫܝܬܐ (pl.) ΛV || 3 ܚܡܫܐ scripsi: ܐܪܒܥܐ codd. || 7 ܕܟܝܢܐ M || **I.i.1**: 1 ܦܝܠܘܣܘܦܐ ΛV || 2 ܡܛܐܘܪܘܠܘܓܝܐ ΛV; ܡܛܐܘܪ̈ܘܠܘܓܝܐ M || ܢܦܫ ܢܦܫܐ: M; ܢܦܫܐ تيققابة V || ܘܡܢ: ܡܢ M || 3 ܥܠ: ܕܥܠ ΛV || ܚܝܢܐ P || ܕܡܫܟܚܐ: ܡܫܟܚܐ ܕܡܢ ΛV || 4 ܡܬܐ: ܡܬܗ F; om. L || ܡܢ ܐܪܥܐ ܡܢ: om. F || 6 ܡܬܚܙܝܢܘܬܐ V || 8 ܕܐܐܪܐܗܘ M || 9 +ܒܡܠܝܘܬ†: sic ΛVM[v]; om. FM[1]; fort. ܒܡܠܝܘܬ ܚܝܠܐ ܘܡܫܡܥܬܐ || **I.i.2**: 1 ܡܛܠ ܕܒܝܫ ܠܐܠܗܐ V || ܗܘ: om. ΛV || 2 ܬܬܚܠܡܘܢ ΛV || ܘܢܬܦܫܛܘܢ P || ܡܢ [post ܘܢܬܦܫܛܘܢ]: ܘܡܢ V || 4 ܐܐܪ: om. Λ || 5 ܐܦ ܬܠܝܬܐ: ܘܬܠܝܬܐ F || ܡܥܒܕ PL1 || 6 ܫܘܐ Λ || ܕܒܝܫ: ܕܒܝܫܬ Λ || 7 ܠܦܠܓܗ V || 8 جيفبة Λ || ܝܣܐ Λ || 9 ܒܣܒܪ V || ܡܬܪܝܢܘܬ M ||

THE BOOK OF METEOROLOGY

Which is the fifth [book] on natural sciences in the Book of the Cream of Wisdom. It contains five chapters.

FIRST CHAPTER

ON CLOUDS AND THINGS WHICH DESCEND FROM THEM

It contains three sections.

First Section: On How Clouds are Formed. Three theories.

[I.i.1.]
First [theory]: The [discussion of] things generated inside the earth is followed by the [division of] philosophy which is called μετεωρολογία - viz. the science of the wandering things - by the ancients. Of these things, we ought first to investigate the clouds.

We say: When the sun's rays shine on earth and water, the heat which is generated by them raises dust from earth and vapour from water. These [i.e. vapour and dust] mingle with each other as they rise and they come to the [stratum of] air which is far below the element fire and which the heat of the rays [reflected from the earth] does not reach. There they grow cold and are condensed, and turn into cloud. This [stratum of] air is colder than the other [strata of] air, not by its nature but because of the coldness of earth and water, the chill of the night with [its] absence <of radiation> and of winter, and the remoteness of the heat-causing [zones/matter?] above and below.

[I.i.2.]
Second [theory]: Because vapour is, as it were, intermediate between water and air, it is possible for a cloud to form from the two: from water, when it is rarefied and dissipated; from air, when it is thickened and condensed. Just so, we often see on the peaks of cold mountains how the air grows cold; where there had been clear sky, [the air] becomes condensed and suddenly turns into a cloud; and there falls rain or snow.

Vapour does not always ascend to the very cold region of the air before a cloud is formed. For I myself, when I was on top of a mountain between Gubos and Claudia with others, saw a cloud which ascended half way up the mountain and lay over Gubos, while I and those with me were in the clear sky above it. After a while, our disciples came to us squeezing out of their clothes the water of the rain which had fallen on the place and on them.

[I.i.3]

ܕܬܠܬ. ܡܬܬܠܝܐ ܕܝܢ ܠܡܠܟܐ ܡܢ ܗܘ ܕܠܐܘܡܢܐ ܡܫܒܚܐܐ ܢܗܘܐ. ܐܦ ܡܢ ܢܘܗܪܐ
ܗܢܐ ܕܡܠܐܟܐ ܢܗܝܪܐ ܢܦܩܐ ܠܗ. ܐܦ ܡܢ ܢܘܗܪܐ ܕܐܝܟܢܐ ܠܟܠ ܡܕܡ. ܘܡܢ
ܬܫܒܘܚܬܐ ܟܠܝܠܐ ܠܗ. ܐܦ ܡܢ ܛܘܒܐ ܕܦܐܝܐ ܒܐܘܡܢܘܬܗ. ܐܦ ܡܢ ܢܘܗܪܐ
ܕܠܡܫܠܡܘܬܐ ܕܡܬܝܕܥܢܘܬܗ ܣܓܝܥ ܠܗ. ܐܦ ܒܗ ܠܡܠܟܐ ܡܕܡ ܦܕܡ ܦܐܝܡ.
ܐܝܟܢܐ ܠܗ ܢܗܘ. ܘܟܡ ܢܟܡ. ܐܦ ܡܟܠܠ ܬܘܩܢܐ ܕܡܬܝܕܥܢܘܬܐ ܡܫܟܠܠܢܘܬܗ
ܘܠܗܢܐ ܕܫܘܠܡܗ ܘܡܫܡܠܝܢܘܬܗ.

[tit.]

ܡܐܡܪܐ ܬܢܝܢܐ. ܡܛܠ ܡܠܐܟܐ. ܬܐܘܪܝܐ ܩܕܡܝܬܐ.

[I.ii.1]

ܡܕܡܝܘܬܐ. ܠܡܠܟܐ ܡܢ ܚܕ ܗܘ ܕܡܫܡܫܢܘܬܐ ܕܢܗܘܐ ܡܢ ܐܪܒܥܐ ܘܡܪܒܥܐ ܫܠܡ.
ܒܫܡܫܘܬܗ ܡܬܚܙܝܐ. ܒܗܘܢ ܡܫܬܠܚܝܢ ܕܐܠܝܢ ܠܐܡܬܐ ܚܡܫܐ ܕܐܚܝܕ ܡܬܚܙܐ.
ܘܒܦܘܪܫܢܗ ܐܚܪ̈ܢ ܡ̈ܠܬܗ. ܘܐܝܟܢܐ ܕܠܗܢܐ ܡܪܝܐ ܕܐܝܬ ܠܐ ܢܬܚܝܠ.
ܡܬܦܩܕܝܢ ܘܐܝܬ ܫܡ̈ܝܐ. ܘܐܦ ܠܐܢܫ ܐܝܬ ܡܪܝܐ ܢܬܚܙܐ ܠܡܠܟܐ. ܦܐܝܬ.
ܘܒܡܪܝܘܬܗ ܢܚܬ. ܘܒܫܡܫܘܬܗ ܡܬܚܙܝܢ. ܒܗܘܢ ܣܓܝܐܐ ܕܐܚܝܕ ܠܐܡܬ̈ܐ
ܡܫܬܠܚܝܢ ܕܐܠܝܢ ܡܬܚܙܝܢ. ܘܡܬܦܢܝܢ ܕܒܝ ܡܫܡܫܢܘܬܗ. ܘܢܦܩܝܢ. ܘܡܠܐܟܐ
ܡܬܠܡ. ܘܠܡܠܟܐ ܕܒܐܝܬ ܦܐܝܬ ܘܢܦܩ. ܐܝܟ ܡܢ ܢܘܗܪܐ ܕܟܠ ܦܩܕ ܢܚܬ.
ܟܠܗܘܢ ܐܪܒܥܐ ܗܘܝܢ ܡܬܚܠܦܝܢ ܒܫܪ̈ܪܐ ܡܬܚܙܝܢ. ܘܟܠܗܝܢ ܗܘܝܢ ܕܪ̈ܒܬܐ
ܡܬܚܙܝܢ ܒܗܠܝܢ.

[I.ii.2]

ܕܬܠܬܝܢ. ܚܝܠܐ ܠܗ ܐܘܣܝܐ ܐܝܬܝܗ ܕܠܚܝܘܬܐ ܘܢܘܗܪܐ ܡܠܐܟܐ ܐܝܬܝܗ. ܐܟܚܕܐ
ܕܦܫܝܛܝܢ ܡܦܠܓܐ. ܐܠܐ ܚܕ ܒܗ ܚܕ ܐܘܣܝܐ ܕܡܠܐܟܐ ܐܝܬܝܗ ܐܘܣܝܗ.
ܘܠܗܢܐ ܗܘ ܕܠܝܬܝܗ ܡܢ ܪܝܫ ܛܘܒܐ ܫܘܝܐ ܚܝܠܐ. ܒܗ ܢܗܝܬ ܐܝܟ ܠܐܢܫ
ܠܡܬܚܙܝܢܘܬܗ. ܘܢܗܝܪ ܡܢ ܗܢܐ ܕܒܫܐ ܒܗܘܢ ܡܠܐܟܐ ܘܒܬܠܝܬܐ ܠܐ ܡܕܡ ܫܘܐ.
ܚܝܠܐ ܡܢܗ ܠܡܠܟܐ ܕܢܗܘܐ ܘܠܚܝܘܬܐ ܐܝܬܝܗ. ܗܘ ܕܐܝܟ ܡܫܬܠܚܢܘܬܐ ܦܐܝܬ.
ܡܠܐܟܐ ܢܚܬ. ܘܐܝܟ ܢܗܝܪܘܬܐ. ܠܡܠܟܐ. ܘܐܝܟ ܫܡܫܐܝܬ. ܕܠܝ ܒܠܚܘܕ ܠܐܝܬ.
ܘܡܢ ܕܠܚܝܘܬܐ ܐܝܬ ܒܚܝܠܐ ܠܚܝܘܬܐ ܡܬܠܐ ܒܡܠܐܟܐ ܕܝܢܐ. ܘܡܫܡܫܢܘܬܐ
ܢܘܗܪܐ ܡܢ ܕܒܗܐ ܠܗܢܐ ܡܫܡܫܢܐ ܚܝܠܐ.

FMPLIV: **I.i.3**: 1 ܡܫܒܚܐܐ V || 2 ܕܡܠܐܟܐ F || 5 ܘܡܬܝܕܥܢܘܬܐ P || ܡܫܡܠܝܢܘܬܗ PI; ܡܫܡܠܝܢܘܬܗ L || 6 ܘܠܗܢܐ: ܘܠܗܢܐ Λ || ܕܡܫܡܠܝܢܘܬܗ Λ || **I.ii.1**: 1 ܡܢ: ܡ̈ܢ P || 2 ܡܫܬܠܚܝܢ V || 3 ܐܚܪܝܢ M[1], ܐܚܪܢ M[v] || ܠܐ: om. M; scripsit et delevit V || 4 ܐܝܬ ܠܐܢܫ: ܐܝܬ ܐܢܫܐ M; ܐܢܫܐ Λ; ܠܐܢܫܐ V || ܠܡܠܟܐ: ܠܗ ΛV || 5 ܘܒܫܡܫܘܬܗ ΛV || 6 ܡܫܬܠܚܝܢ V || ܢܦܩ̈ܝ M; ܡܫܦܩܝ Λ || 7 ܘܐܝܟ Λ || 8 ܐܪܒܥܐ: ܐܪܒܥ F || ܒܫܪ̈ܪܐ: in marg. F || ܡܬܚܠܦܝܢ M; ܡܬܚܠܦ̈ܢ V || **I.ii.2**: 1 || ܕܠܚܝܘܬܐ: ܠܚܝܘܬܐ ̈ܕ in marg. M (prima manu) || 2 fin. ܐܘܣܝܗ MV || 5 ܕܐܝܟ: ܐܝܟ Λ || ܡܫܬܠܚܢܘܬܐ l || 7 ܕܝܢܐ MΛV || ܡܫܡܫܢܘܬܐ Λ || 8 ܢܘܗܪܐ Λ || ܡܢ: ܡ̈ܢ M ||

[I.i.3.]
Third [theory]: Vapour is prevented from rising to a great height: either by [its own] weight, when it has an excessive volume; or by a wind that blows above it and prevents its ascent; or by a mountain which stands in front of it; or by opposing winds which confine it in their midst; or when a [body of] vapour precedes and is stationary, another joins it and clings to it; or because of the severity of the coldness which annuls the impulse of its ascent and makes it heavy.

Second Section: On Rain. Six theories

[I.ii.1.]
First [theory]: The vapour, which rises from earth and water with the heat of the rays, becomes scattered in its ascent, because it is moves from a centre which is narrow to a circle which is wide, and as it is scattered, its particles grow small. If (the particles) do not reach the cold region of the air, they are dissipated and turn into air. If it arrives at the cold place/region/part of the air, the vapour grows cold, and through its coldness, it falls. As it descends, it becomes concentrated because it moves from a circle that is wide to a centre that is narrow. As it becomes concentrated, its particles grow large and heavy and causes drops to fall as rain. If the vapour which grows cold and heavy in the air falls from a very elevated height, in the long duration of its [fall] its particles become bound with each other more, and for that reason its drops become larger.

[I.ii.2.]
Second [theory]: A cloud is not a distinct substance which bears and gives birth to rain as the simple think, but its substance is the same as the substance of rain. For this reason when a cloud forms below the top of a mountain, as one descends, one enters into its midst and sees nothing more than what one sees on a rainy and foggy day. A cloud is therefore condensed and thickened vapour.

If (the vapour) grows cold completely, rain falls; if excessively, snow; and if inadequately, it merely makes the air turbid. This turbidity of the air due to the cloud resembles the turbidity of water due to dust. A cloud is transferred from one place to another through movement by winds.

[I.ii.3]

ܕܬܠܬ . ܡܫܡܠܝܢܘܬܗ ܕܪܘܚܐ ܕܫܠܝܚܘܬܐ . ܠܡܫܡܠܝܢܘܬܗ ܡܫܡܫܢܘܬܐ ܕܡܠܦܢܐ .
ܕܡܫܡܠܝܐ ܟܠܗ . ܘܠܐ ܠܝܬܗ ܡܠܦܢܐ ܒܚܕܐ ܕܡܫܡܫܢܘܬܐ ܕܪܘܚܐ . ܒܟܠܗܝܢ
ܐܡܬܝ ܐܢܫ̈ܐ . ܕܡܠܦܢܐ ܠܪܘܚܐ ܫܡܥ . ܘܒܐܝܕܐ ܕܪܘܚܐ ܡܫܡܫܢܘܬܐ . ܡ
ܡܠܦܢܐ ܠܝܬ . ܐܠܐ ܐܢ ܡܢ ܪܘܚܐ ܪܒܬ ܦܐܪܐ ܒܢܝܐ . ܘܬܪܒܝܬܗ ܦܐܪܐ .
ܘܦܘܩܕܢܗ ܢܩܦ . ܣܡܝܢ ܘܢܒܐ ܡܠܦܢܐ ܠܦܘܪܫܢܐ ܕܪܘܚܐ ܪܒܐ . ܘܥܡ ܪܘܚܐ
ܡܠܦܢܐ ܠܝܬ . ܟܕ ܠܦܘܫܬ ܡܫܡܫܢܘܬܗ ܢ̈ܦܠܐ . ܘܡܠܦܢܐ ܕܪܘܚܐ ܢܚܠܦܘܗܝ . ܘܡܫܐܠܐ
ܐܝܟ ܕܐܡܫܡܫܢܘܬܐ ܠܝܬ . ܐܢܫܐ ܡܫܡܫܢܐ . ܘܗܘ ܕܪܘܚܢ ܢܚܠܦܘܗܝ . ܘܡܫܡܫܢܘܬܗ
ܘܥܡ ܪܘܚܐ ܠܗ . ܫܡܫܐ ܘܒܘܚܢܐ ܡܫܡܫܢܘܬܗ . ܘܐܢܫܐ ܕܡܠܦ ܪܘܚܢ
ܢܚܠܦܘܗܝ . ܘܠܐ ܡܫܬܡܥ ܒܐܪܐ . ܕܫܡܫܐ ܡܫܡܠܝܐ .

[I.ii.4]

ܕܐܪܒܥ . ܠܡܠܦܢܐ ܕܦܠܚ ܡܢ ܪܘܚܐ ܢܦܫܗ . ܘܒܗ ܡܢ ܪܘܚܐ ܦܐܪܐ . ܘܟܠ ܬ̈ܪܥܐ
ܡܢ̈ܐܬܗ ܡܫܬܡܠܝܢ . ܦܚܡܐ ܠܚܘܒܬܗ . ܘܠܝܬ ܡܠܦܢܐ ܟܠ ܐܪܥܐ ܕܡܬܪܥܐ
ܠܥܠܡܐ . ܐܦ ܟܕ ܘܠܐܪܥܐ ܕܪܝܫ ܗܘܐ ܡܢ ܪܘܚܐ ܕܪܘܚܐ ܠܥܠܡܐ ܪܘܚܐ .
ܘܡܠܦܢܐ ܡܫܡܫ . ܘܟܠܗܝܢ ܡܫܡܫܢܐ ܪܘܚܐ ܡܫܡܫܢܘܬܐ . ܕܠܐܪܥܐ ܡܢܗ
ܡܫܡܠܝܐ . ܘܟܢܫܘܬܐ ܠܐܪܥܐ . ܘܡܫܡܫܢܘܬܐ ܠܫܡܝܐ ܠܐܪܥܐ
ܘܐܪܥܐ . ܠܫܘܬ ܡܢܗܘܢ ܕܪܘܚܐ ܒܫܘܬ ܪܘܚܐ ܘܡܫܡܫܢܗ ܡܢ ܟܠܗܘܢ .

[I.ii.5]

ܕܚܡܫ . ܡܠܦܢܐ ܡܫܡܠܝܐ ܡܢ ܠܥܠ ܕܫܡܝܐ ܘܡܫܡܠܝܢܘܬܐ ܘܐܝܟܢܐ ܟܠܗܘܢ ܗܢܘܢ .
ܘܠܐ ܡܢ ܠܚܘܒܗ ܐܢܫ ܘܪܒܐ . ܘܐܪܥܘܬܐ ܟܠܗܝܢ ܡܠܦܢܐ . ܗܢܘܢ ܦܚܡܐ
ܕܡܫܡܫܐ ܕܪܘܚܐ ܡܢ ܪܘܚܐ . ܕܐܪܥܘܢ ܡܫܡܫܢܘܬܐ . ܘܟܠܗܘܢ ܒܡܫܡܫܢܘܬܐ
ܐܝܬ ܡܫܡܠܝܢܘܬܐ ܡܠܝܢܘܬܐ ܘܟܢܫܘܬܐ . ܕܡܫܡܫ ܐܢܫ ܘܡܢܗ ܐܪܥܐ . ܘܡܢ
ܟܕ ܪܘܚܐ ܒܝܬ ܢܦܫ̈ܐ ܟܠܗܘܢ ܪܘܚܐ . ܐܢܫܘܬܗܘܢ . ܘܐܪܥܘܬܐ ܡܫܡܫܢܐ
ܡܠܦܢܐ . ܗܢܘܢ ܕܠܡܫܡܫܢܐ ܡܫܡܫܝܢ . ܘܪܘܚܐ ܕܡܫܡܫܢܘܬܐ ܢܦܫ ܢܦܫ̈ܐ ܟܠܗܘܢ .
ܘܠܐܪܥܐ ܠܡܫܡܫܢܐ ܡܫܡܫܝܢ . ܘܡܫܡܫܢܐ ܠܡܫܡܫ ܬܠܬܘܬܐ ܐܢܘܢ . ܐܢܫܘܬܗܘܢ .

FMPLIV: I.ii.3: 1 ܫܠܝܚܘܬܐ Λ || ܡܫܡܫܢܘܬܐ: ܡܫܡܫܢܘܬܗ Λ || 3 ܐܢܫܐ l || 4 ܐܢ: om. F || ܘܐܪܐ PL; ܘܦܐܪܐ l || ܘܬܪܒܝܬܗ M || ܘܦܘܩܕܢܐ V || 5 ܕܪܘܚܐ FMΛ || 6 ܡܫܡܫܢܗ F || 7 ܐܢܫܐ: ܐܢܫ ΛV; ܐܢܫܐ M^1 || 8 ܕܫܡܫܐ FM || 9 ܫܡܫܐ: ܫܡܫܐ Λ || **I.ii.4**: 3 ܠܐܪܥܐ ܕܪܝܫ ΛV || ܠܥܠܡܐ ΛV || 5 ܡܫܡܫܢܘܬܐ Pl || 6 ܘܡܫܡܫܢܗ F; ܡܫܡܫܢܗ Λ; ܡܫܡܫܢܐ V || ܟܠܗܘܢ FMLIV || **I.ii.5**: 1 ܕܫܡܝܐ: ܫܡܝܐ ut vid. F; ܕܫܡܝܐ Λ || 4 ܡܠܝܢܘܬܐ LV || ܘܟܢܫܘܬܐ L || ܕܡܫܡܫ V || 6 ܡܠܦܢܐ F || ܢܦܫ̈ܐ: ܢܦܫ̈ܐ L ||

[I.ii.3.]
Third [theory]: The strong movement of wind blocks the weak movement of rain's descent, and rain does not fall until the wind has been interrupted. Consequently men say that rain interrupts wind, but the truth is that the wind is [first] interrupted and then the rain falls, unless the cloud is cooled greatly by the wind, its density increases and its raindrops become heavy; then, the motion of the rain overcomes the motion of the wind and rain falls with the wind inclining in the direction in which it is blowing.

Rain, whose drops are large and which falls with violent force and incessantly, is called *zīqā*. (Rain), whose drops are small and which falls gently and for a short time, is named *rsīsā* and *rzāpā*. (Rain), whose drops are very small and cannot be seen in the air, is termed *rsāmā*.

[I.ii.4.]
Fourth [theory]: Vapour that rises from the sea is hot, but when it is cooled by the wind and its particles are pressed against each other, its density increases and rain falls on land which is near the sea. In fact, the wind drives the cloud also to lands which are far from the sea and causes rain to fall. For this reason the south wind is found to cause rain to fall on one area, the north wind on another and east and west winds on others in accordance as to whether the sea is situated in the [source]-direction of the wind and whether it is blowing from its side.

[I.ii.5.]
Fifth [theory]: Summer rains are generated only from the vapours of seas, rivers and marshes, and not from condensation of cold air. Places which are deficient in rain are those that are very hot and far from the sea, [those] whose land is low-lying, [those] in whose soil there is mixed, salty and sulphureous heat - because of which the air becomes hot and dry - and [those] into which winds blow little from the coast. Places which abound in rain are those that are near the sea, [those] to which sea winds blow more, [those] which are adjacent to snowy mountains and [those] which are watery and retentive of moisture.

[I.ii.6]

ܕܐܬ. ܟܣܪܐ ܠܡܐܡܪ. ܠܟܠ ܡܢ ܓܘܪܐ. ܐܝܟܢܐ ܢܦܫ ܡܢܝ ܐܚܪܝ. ܦܓܪܐ
ܒܢܝܐ. ܘܡܢ ܓܘܫܐ ܠܕܘܟܬܐ ܕܢܘܫܐ ܡܬܦܪܫܝܢܐ. ܘܟܠܗܝܢ ܟܠ ܝܘܡܐ ܟܝܪ
ܒܫܦܝܪܘܬ ܐܚܪܝ ܠܗ ܦܩܕ ܢܦܫ ܡܠܟܐ. ܘܐܝܟ ܡܫܟܚܝܢ ܗܘܝܢ ܠܡܠܟܐ. ܘܐܠܗܐ
ܡܚܝܠܐ ܠܝ ܠܝ ܡܠܟܐ ܕܝܗܒܐ ܗܘܐ. ܘܟܐܪܟܐ ܕܒܩܝܐ. ܐܝܟ ܢܦܫ ܒܫܦܝܪܐ.
ܦܩܕ ܡܠܟܐ. ܟܕ ܕܠܠܡܠܟ ܝܗܒܐ ܠܝܘܚܢܢ ܕܢܫܩ ܓܘܫܐ. ܘܟܠܗܘܢ ܙܕܩ
ܠܗܘܢ. ܘܗܘܐ ܗܘܝܢ ܦܐܪܐ ܘܦܠܚܝܢ. ܘܟܢܝܫܐ ܗܘܘ. ܘܗܘܐ ܕܚܝܠܐ ܡܚܝܠܐ ܦܠܚ.
ܢܘܚܝ. ܘܡܬܘܠܢܝܢ.

[tit.]

ܫܘܠܡܐ ܕܡܠܘܫܬܐ. ܡܛܠ ܕܠܐ ܘܡܢܝܢܐ ܘܕܡܘܬܐ ܘܡܕܝܢܬܐ ܘܕܘܟܬܐ ܘܟܘܪܗܢܐ.
ܐܪܥܢܘܬܐ ܫܡܝ.

[I.iii.1]

ܡܬܚܫܒܐ. ܠܐ ܡܢ ܠܗ ܡܢ ܟܢܝܫܐ. ܐܠܐ ܡܢ ܠܡܠܟܐ ܢܘܚܐ. ܕܡܫܬܘܕܝܢ
ܘܡܠܘܢ. ܘܟܝܢܐ ܕܡܠܟܘܬܗ. ܟܕ ܚܙܝܐ ܠܗ ܡܢܗܪܬ ܠܐܢܫܐ. ܘܡܬܝܕܥ ܘܡܫܡܠܝܐ
ܘܠܥܡܐ ܡܫܬܠܡܐ ܠܗ. ܘܫܡܥܘ ܫܠܡ ܡܢܐ. ܟܕ ܡܠܝܐ ܘܒܪܘܚܬܐ
ܕܡܬܩܢܝܢ. ܒܝܕ ܐܝܟ ܕܝܠܗ ܡܬܬܪܥܝܢܐ ܢܘܚܝ. ܘܡܢܐ ܕܡܠܟܐ ܐܡܠܟܐ
ܡܬܟܢܫ ܡܢܢܫ ܕܡܬܬܚܐ ܫܠܡ ܡܫܝܚܐ. ܫܡܥܝ ܡܢ ܟܪܝܗܐ ܒܠܠܐ. ܟܕ
ܢܦܫ ܒܠܠܐ. ܘܡܢ ܡܬܟܠܘܢܬ ܠܐ ܟܕ ܡܬܪܘܬܐ ܡܬܝܕܥܐ ܡܕܝܢܐ ܐܘܟܝܬ
ܘܡܬܗܘܐ ܗܘܐ. ܘܡܢ ܗܘܟܢܐ ܠܐ ܠܘܬ ܡܕܝܢܐ. ܗܢܟܢܐ ܕܡܠܟܐ ܠܘܬ ܡܠܟܐ
ܡܢܐ.

[I.iii.2]

ܕܐܬܬܝܢ. ܡܠܟܐ ܕܝܢ ܗܘܐ ܡܢܐ ܕܡܕܡ ܗܘ ܕܡܬܟܢܫ ܘܡܠܟܘܬܐ ܟܢܝܫܐ ܡܬܩܢܢܗ
ܕܟܢܝܫܐ. ܘܡܬܟܠܠ ܡܢܗܝܢ ܢܦܠܬܐ. ܡܠܟܐ ܠܗܝܢ ܡܢܘܬܐ ܕܡܠܟܘܬܐ. ܘܟܕ
ܡܬܟܠܠ ܢܫܐ. ܘܒܠܗܘܢ ܕܢܘܪܘܬ ܡܠܟܐ. ܟܠܠ ܐܚܪܝ ܡܪܝܟܬ ܡܬܩܢܬܐ
ܡܬܢܝܬܐ. ܕܡܫܬܡܫ ܡܢ ܬܕܪܐ. ܘܡܫܬܪܫܝܢ ܘܡܬܢܝܚ. ܘܡܬܕܪܐ ܠܐ ܡܬܠܡܕܝܢ
ܐܢܬܘܢ. ܐܝܟܢܐ ܕܟܬܒܘܬܐ. ܘܗܘ ܡܠܟܐ ܠܠܡܠܟܐ ܕܡܠܟ ܡܢ ܐܪܥܐ
ܟܡܘܪܟܐ ܕܐܪܥܐ ܫܟܢ. ܘܟܠܡܪܐ ܟܝܘܡ ܡܠܟܐ ܡܬܬܟܪ ܗܘܪܐ. ܟܕܡܐ
ܕܦܩܕ ܡܠܟܐ ܟܠ ܐܪܥܐ. ܘܫܘܩܐ ܡܢܘܬܐ ܟܠ ܐܝܩܪܗ. ܘܟܠܠܝܢ
ܠܡܠܟܐ. ܘܗܘܝܢ ܟܢܫ ܗܘܪܐ. ܘܡܠܟܐ ܕܡܬܢܟܠ ܘܡܬܒܪܟܐ ܢܦܫ. ܒܡܠܟܘܬܐ
ܡܬܬܪܐ.

FMPLIV: **I.ii.6**: 1 ܓܘܪܐ F || 2 ܓܘܫܐ MΛV || ܠܕܘܟܬܐ: ܐܬܪܐ L || 3 ܒܫܦܝܪܘܬ: hinc FMLIV || 3 ܡܠܟܐ: ܡܠܟܐ M || 4 ܕܠܠܡܠܟܐ ܕܒܩܝܐ V || 5 ܠܝܘܚܢܢ MV || 6 ܘܗܘܐ: ܘܡܢ F || ܡܚܝܠܐ LV || **I.iii.1**: 1 ܠܗ ܡܢ: om. F || 4 ܘܠܐ: ܠܐ ΛV || ܡܫܟܚܐ V || **I.iii.2**: 2 ܡܬܩܢܢܗ F || 3 ܕܡܠܟܘܬܐ V || ܢܫܐ 1 || 6 ܫܟܢ: ܫܟܢ V || 8 ܡܬܬܪܐ F ||

[I.ii.6.]
Sixth [theory]: In a word, a cloud condenses above mountains where the air is very cold and is driven by winds to a distant place. For this reason rain does not fall much on the sea because of the heat of the air, although vapours are abundant there. If this were not the case, the rain on the sea would never cease. Rains are plentiful in Ethiopia, although it is very hot, because winds drive the vapours of the sea towards it and press them against its mountains. There (the vapours) grow cold, condense, turn into clouds and, once they have acquired a watery nature, they grow heavy and are discharged.

Third Section: On Dew, Frost, Snow, Hail, Fog and Mist. Six theories.

[I.iii.1.]
First [theory]: Dew [is formed] not from a cloud, but from diurnal vapour, whose ascent has been retarded and whose quantity is small, when the coldness of the night strikes it, condenses and thickens it, and turns it into water. This water because of the fineness and smallness of its particles falls gently and imperceptibly. When a large quantity has been gathered of these watery particles, then one senses the dew as it falls during the night. From the freezing of dew due to severe coldness, frost [*qarṣānā*], also called *zmāytā*, is formed. Hence dew has in relation to frost the relationship of rain to snow.

[I.iii.2.]
Second [theory]: Snow is generated when, before the particles of a cloud assemble and coalesce with each other and drops are formed from them, there comes upon them a powerful coldness, and they fall while freezing. The reason for the whiteness of snow is the intrusion of air in between the watery particles which are separated from each other, loose and rarefied and not connected to each other, as [is the case] with foam. Snow confines the vapour that rises from the earth in the proximity of the earth. For this reason, on a snowy day the chill is interrupted until there is much snow on the earth, the coldness grows strong over its surface and the vapours cease; then [at last] the chill gains strength. Snow which falls suddenly and heavily is called "snowstorm" [*kōḵīṯā*].

[I.iii.3]

ܕܬܠܬ. ܘܐܦ ܒܗܿܝ ܡܬܦܠܓܢܘܬ ܡܢܘ̈ܬܐ ܕܚܝܠܐ. ܘܡܬܚܒܠܢܘܬ ܢܦ̈ܠܬܐ ܕܐܬܒ̈ܪܬܐ. ܡܫܠܝܐ ܠܗܘܢ ܡܫܝܪܘܬܐ. ܘܡܫܬܠܡܐ ܠܗܘܢ. ܒܝܕܐ ܡܫܬܡܠܝܢ. ܘܢܦܩܬ. ܘܚܠܦܬܐ ܕܚܠܠܘܬ ܦ̈ܪܝܫܘܬܐ ܕܚܝܠܐ. ܒܚܘܫܒܐ ܕܐܬܬܗܝܢ ܡܬܬܦܩܚܢܘܬܐ. ܗܿܘ ܒܗ̇ ܕܒܐܘܪܚܐ ܕܒܗ̇ ܠܢܬܗܝ. ܘܗܘ̇ ܕܒܦܬܟܐ ܕܒܗ̇ ܡܢ ܕܘܟܐ ܡܬܬܕܢܝܢ. ܘܒܗ ܒ̈ܚܝܠܐ ܡܬܬܬܝܒܢ ܒܚܘ̈ܬܡܐ ܡܬܬܬܒ̈ܢܐ. ܘܡܬܚܠܠܢ ܐܝܬܘܗܝ. ܘܐܝܟ ܕܒܣܘܟܠܐ ܒܬܪܐܐ ܘܒܬܫܪܝܬܐ ܗܘܐ ܚܝܠܐ. ܒܣܘܟܠܐ ܕܝܢ ܠܐ ܗܘܐ. ܐܦܠܐ ܒܣܘܟܠܐ. ܒܣܘܟܠܐ ܗܿ ܒܗܿ ܕܡܬܝܪܘܬ ܣܘܟܠܐ. ܐܦ ܒܚܝܠܐ. ܘܡܢ ܕܦ̈ܪܝܫܐ ܬܘܟܠܐ ܚܝܠܐ. ܡܬܚܠܝܐ ܠܗ. ܘܬܠܝܬܐ ܚܒܪܐ. ܘܐܦ ܡܫܠܝܐ. ܠܐ ܡܢܡ ܚܒܪܐ. ܒܣܘܟܠܐ ܕܝܢ. ܒܗܿ ܕܠܡܠܬܐ ܢܦܠܝܐ ܘܢܫܪܐ. ܕܐܬܬܗܘܢ ܗܘܠܐ ܕܚܝܠܐ. ܦܩܕ ܒܝܬ ܒܗ.

[I.iii.4]

ܕܐܬܒܚܪ. ܒܬܪܐܐ ܕܝܢ ܘܒܬܫܪܝܬܐ. ܠܘ ܒܟܠ ܡܬܚܠܐ ܚܝܠܐ. ܐܠܐ ܒܗ ܡܬܦܠܓܢ ܡܢܘܬܗ. ܘܡܫܬܡܠܝܢܘܬ ܡܬܬܝܒܢ. ܗܘܝܢ ܡܬܚܠܝܢ. ܘܦ̈ܪܝܕ ܒܝܕܐ ܠܢܬܗܝ. ܒܬܫܪܝܢ ܕܝܢ ܢܦܫ ܗܘܐ ܒܝܕܐ. ܒܗܿ ܕܒܣܘܟܠܐ ܒܗ ܡܬܬܕܡܪ ܠܡܠܟܐ. ܕܠܠܐܝܬ ܡܬܚܠܐ. ܘܡܢܝܢܐ ܡܢܐ ܡܢ ܗܿ ܕܡܢܐ ܫܘܢܝܐ ܡܢܡ ܡܬܝܪܐ ܡܬܚܠܝܢ. ܘܒܝܕܐ ܕܡܢ ܚܘܠܝܐ ܕܐܬܝܬ ܦܩܕ ܡܢ ܐܬܪܐ ܠܢܬܗ. ܦܫܛ ܦ̈ܪܝܫܘܬܗ ܘܐܬܝ. ܘܫܚܠܐ ܢܦܝܐ ܕܒܐܬܪ ܡܬܬܒܢ ܒܚܘ̈ܬܡܝܢ ܘܡܬܚܠܠܢ. ܘܦ̈ܪܝܕܐ ܒܬܪܘܬܐ ܕܐܘܪܚ ܘܠܐ ܐܬܡܠܝܢ. ܡܢ ܚܝܠܐ ܕܬܘܪܒܐ ܠܐܬܪܐ ܠܢܬܗܝ. ܘܒܫܪ ܡܢ ܠܦܬܪ ܠܐܬܪܐ. ܫܒܝܠ ܕܫܪܐ ܦ̈ܪܝܫܬܐ ܕܒܝܕܐ ܢܦܫ ܡܢ ܡܢܝܢܐ ܬܫܠܡܬ. ܘܒܠܦܬܐ ܕܒܚܝܢ ܒܬܘܐܐ ܠܐܬܠܝܬܘܒܐ. ܐܝܟܐ ܡܘܫܚ ܫܝܢܬ ܦ̈ܪܝܫܐ ܕܐܬܒܪܬܐ ܕܒܝܕܐ ܕܢܦܠ ܒܐܘܣܝܐ ܡܬܝܪܘܬܐ ܘܦܪܫ ܗܘܪܐ ܘܚܝܠܐ ܘܦܪܫܐ.

[I.iii.5]

ܕܫܡܫܐ. ܒܪܘܠܐ ܗܿܝ ܡܢ ܐܘܣܝܐ ܕܚܝܠܐ ܐܠܗܝܐ. ܒܪܡ ܕܝܢ ܡܘܫܚܐ ܕܚܝܠܐ ܠܐ ܡܫܚܐ. ܡܠܝܐ ܗܘ ܡܢ ܚܝܠܐ. ܘܒܒܝܐ ܡܢ ܒܪܘܪܐ. ܐܟܡܐ ܕܒܪܘܪܐ ܡܠܝܠ ܡܢ ܒܪܘܠܐ. ܘܒܒܝܐ ܡܢ ܐܬܪ. ܘܒܪܘܠܐ ܕܡܢ ܬܘܡܟܐ ܠܢܬܐ. ܡܠܘܢ ܕܝܢ ܒܗܿ ܡܠܝܐ. ܒܠ ܢܦܫܐ ܡܡܠܝܐ. ܘܐܝܟܢܐ ܕܡܢ ܠܡܠܟܬ ܡܫܬܠܐ ܕܬܫܡܫ. ܘܠܐ ܡܬܚܠܠܢܐ ܒܡܫܬܚܬܐ. ܒܠ ܡܠܬܐ ܡܡܠܝܐ.

FMLIV: I.iii.3: 1 ܘܐܦ: ܡܢ ΛV ‖ ܡܬܒܪܟܢܘܬ V ‖ 3 ܕܚܠܠܘܬ Λ ‖ ܦ̈ܪܝܫܘܬܐ F ‖ 4 ܗܿܘ ΛV ‖ ܘܐܦ ܡܫܠܝܐ ܡܢܡ ܠܐ 8-9 ‖ Λ ܬܘܟܠܐ ‖ Λ ܦ̈ܪܝܕܐ 8 ‖ M ܕܡܬܝܪܘܬܐ 7 ‖ ΛV ܐܝܟܢ 5 ‖ L ܕܘܫܐ ‖ ܚܒܪܐ: om. L; in marg. V ‖ 8-9 ܠܐ ܡܢܡ: ܡܢܡ ܠܐ V ‖ 9 ܐܬܬܗܘܢ ΛV ‖ I.iii.4: 2 ܡܬܦܠܓܢ 1 ‖ 5 ܫܘܢܝܐ 1 ‖ ܦ̈ܪܝܕ F ‖ 3 ܠܢܬܗܝ: ܒܬܫܪܝܢ ܕܝܢ ܢܦܫ ܗܘܐ ܒܝܕܐ: om. F ‖ 4 ܡܢ ܗܿ: om. ΛV ‖ 5 ܚܝܠܐ M ‖ ܕܚܝܡ V ‖ ܦ̈ܪܝܫܘܬܗ F; ܦ̈ܪܝܫܘܬܗ M ‖ 6 ܒܚܘ̈ܬܡܝܢ M ‖ 7 ܒܬܪܘܬܐ M ‖ ܡܬܡܠܝܢ M ‖ ܐܬܪܐ M ‖ 8 ܒܬܘܪܬܐ V ‖ 9 ܠܐܬܠܝܬܘܒܐ V ‖ ܡܘܫܚ M[1]; ܒܡܘܫܚ ΛVM[V] ‖

[I.iii.3.]
Third [theory]: If the coldness comes to the particles of the cloud after their coalescence and formation of large drops, hail takes shape and falls. The reason for the roundness of hailstones is the combination of two movements, namely the vertical in which they fall and the horizontal in which they are driven by winds, along with [the fact that] their corners become broken and rounded as they are rubbed against each other. Hail usually occurs in spring and autumn. It does not occur in winter or in summer, in winter, because if the coldness of the winter is severe it causes the cloud to freeze before it turns into drops and causes snow, whereas if it is weak it causes nothing; in summer, because there is then very little moist and heavy vapour which is the material of hail.

[I.iii.4.]
Fourth [theory]: In spring and autumn, on the other hand, the cloud does not freeze immediately, but when its particles have coalesced and condensed completely, [only] then they freeze and hailstones fall. Hail occurs more in autumn, because the vapour freezes readily, since it has been rarefied during the summer. This [i.e. that rarefied vapour freezes more easily] is known from the fact that hot water freezes before the cold. Hailstones which fall from clouds that are very far from the earth melt and grow small and through the prolonged attrition in the air their corners become broken and rounded. Hailstones which are large and not spherical fall from a cloud that is close to the earth. We have heard that on one of the mountains in the north there was a single hailstone which weighed more than a mina. On a mountain between Beroea [Aleppo] and Antioch, I myself have seen large stones of hail which fell in the shape of horns and heads of oxen, sheep and horses.

[I.iii.5.]
Fifth [theory]: Fog is of the [same] substance as cloud, but it does not have the consistency of a cloud. For it is finer than cloud and denser than mist, just as mist is finer than fog and denser than air. Fog which descends from the height, especially after rain, announces clear weather, while that which begins to rise from below and is does not vanish through evaporation announces rain.

[I.iii.6]

ܕܫܬ . ܟܒܝܫܘ ܠܐܠܗܘܬ . ܒܫܥܒܘܕܘܬܐ ܫܥܒܘܕܝܬܐ . ܡܛܠ ܕܝܢ ܫܥܒܘܕܝܬܐ . ܡܢ
ܡܬܐ ܘܐܝܟܢܐ ܗܘܬܐ ܬܠܝܠܬܐ . ܡܢܗܘܬܐ ܐܝܟܢܝܘܬܐ . ܕܫܡܫܢܘܬܐ ܡܕܝܢܝܘܬܐ ܠܫܘܪܝܘܬܐ
ܡܪܝܟ . ܥܡܡܐ . ܘܡܕܒܪܢܐ ܕܡܕܝܢܝܢ ܠܡܠܟܐ ܡܫܡܠܝܐ . ܘܡܢ ܐܝܟܢܐ ܗܘ ܐܝܟܢܘܬܐ
ܒܢܝܫܘܬܐ ܡܢܗܘܬܐ ܢܦܝܫܘܬܐ ܕܫܡܫܢܘܬܐ ܐܝܟܢܝܘܬܐ ܡܪܝܟ . ܡܬܒܪܢܐ .
ܘܡܕܒܪܢܐ ܕܡܕܝܢܝܢ ܦܫܝܛܐ ܡܫܡܠܝܐ . ܘܠܡܠܟܐ ܡܢ ܡܠܐܟܐ ܗܘ ܕܒܝܫܐ ܘܡܠܝܐ
ܘܠܡܠܟܐ ܘܠܡܠܐ ܘܡܕܝܪܐ ܘܡܪܝܢܐ ܘܒܪܓܠܐ ܘܒܪܘܪܐ . ܘܟܠ ܡܫܡܫܝܢ
ܡܠܟܘܬܐ ܫܠܝܛܐ ܘܡܫܠܘܬܐ ܕܡܠܟܐ ܘܡܠܟܐ ܫܠܝܛܐ . ܦܫܝܛܐ ܕܝܢ ܡܠܐܟܐ
ܗܘ ܕܪܘܚܐ ܘܫܘܠܛܢܐ ܘܢܫܪܐ ܕܪܡܝܢ ܘܡܕܝܢܝܢ ܦܘܫܩܐ ܘܐܝܟܢܝܘܬܐ ܕܫܠܝܛܘܬܐ
ܕܐܝܟܢܐ ܡܫܡܫܝܢ . ܐܝܟ ܕܡܫܡܫܝܢ .

FMLIV: I.iii.6: 3 ܠܡܠܟܐ M || 6 ܡܫܡܫܝܢ MΛV || 7 ܡܠܟܘܬܐ V || 8 ܗܘ: ܗܘ̈ M || ܘܦܘܫܩܐ ܢܫܪܐ Λ ||

[I.iii.6.]
Sixth [Theory]: To sum up, celestial, especially solar, heat raises airy particles mixed with small watery particles from water and moist grounds and what is composed from them is called vapour. From dry earth, it lifts up fiery particles mixed with earthy particles and what is composed from them is called smoke. Vapour is the material of cloud, rain, snow, dew, hail, frost, fog and mist, and on it the halo, rainbow and mock suns appear as illusions. Smoke, on the other hand, is the material of wind, firewind, shooting stars, comets and terrible signs which are seen in the air, as we shall show.

[tit.]

ܡܛܐܘܪܘܢ ܬܪܝܢܐ . ܡܛܠ ܐܬܪܐ ܡܫܬܟܢܐ ܕܡܬܝܬܒ ܒܒ̈ܢܝܢܫܐ . ܘܗܘ ܐܝܬ
ܦܣܘ̈ܩܐ ܐܪܒܥܐ .

ܦܣܘܩܐ ܩܕܡܝܐ . ܡܛܠ ܒܘܚܪܐ ܕܡܫܬܟܢܘܬ ܟܠܗ̇ ܐܪܥܐ ܕܡܫܬܟܢܐ
ܙ̈ܢܝܐ . ܬܐܘܪ̈ܝܐ ܘܬܫܥ .

[II.i.1]

ܡܕܡܢܬܐ . ܣܘܟܠܐ ܘܡܫܘܚܬܐ ܘܟܘ̈ܢܬ ܫܩܫܐ ܘܦܢ̈ܝܬܐ ܡܫܬܟܢܐ ܐܝܟܢ . ܘܡܫܬܟܢܐ ܐܝܬܘܗܝ ܡܢ ܐܪ̈ܒܥܘܬ ܪ̈ܒܥܐ ܢܘܦܠܐ ܕܡܢ ܒܗܘܢܐ ܕܐܪܥܢܐ . ܒܪܐ ܕܡܫܬܟܠܢܘܬ ܬܪܝܢܐ ܐܝܟ ܡܫܘܚܬܐ ܠܡܕܡܢܐ ܐܝܟ ܢܘܡܐ ܕܒܪܝܬܐ ܠܘܬ ܣܬܘܐ . ܒܗ ܠܐ ܡܛܝܒ ܫܘܝܘܐܝܬ ܒܬܪ̈ܝܢܐ ܡܕܡܢܐ . ܘܥܠ ܐܢܟܝܢܘܬ ܣܬܘܐ . ܬܠܬ ܡܬܩܘܡܝܢ ܗܟܢ . ܫܒܐ ܕܝܢ ܗܘ ܕܐܝܬܘܗܝ ܕܡܢ ܣܬܘܐ ܠܗܘ ܝܠܝܦܐ . ܘܡܬܚܫܚܢܘܬܐ ܦܠܓ . ܘܠܗܬܘܗܝ ܕܣܬܘܐ ܡܢܚܠ ܢܘܡܐ . ܘܗܘ ܝܠܝܦܐ ܗܘ ܠܗܬ ܡܩܒܠܐ ܐܝܟ ܡܫܘܚܬܐ ܡܬܬܚܡܐ . ܘܫܠܝ ܡܢܗ . ܘܡܬܚܫܒ ܒܪܗܛܐ ܕܦܠܓ ܒܡܬܚܫܚܢܘܬܐ ܕܠܡܘܒܠ ܡܫܘܚܬܐ ܡܩܒܪ . ܘܠܗ ܘܠܡܫܘܚܬܐ ܐܚܪܢܐ ܡܕܪܟ . ܘܐܝܟ ܗܘ ܕܒܦܠܓ ܡܫܘܚܬܐ ܣܬܘܐ ܠܗ ܡܬܕܪ .

[II.i.2]

ܕܬܐܬܐܝܢ . ܡܬܚܘܝܢ ܕܬܐܬܐܝܢ ܡܥܪ̈ܒܐ ܕܢܘܡܐ ܕܡܬܚܫܚܢܐ . ܐܝܟ ܕܐܝܬܘܗܝ ܡܬܦܠܓ ܒܫܒܥܬܐ . ܘܡܬܚܫܚܢܐ . ܘܡܬܦܠܓ ܗܘܬ ܒܡܫܘܚܬܐ . ܘܗܝ ܒܟܒܗ ܒܫܒܥܬܐ ܡܬܦܠܓܝܢ . ܘܐܪܒܥܐ ܡܬܚܫܝܢ . ܘܕܕܠܟܐ ܬܘܕܝܬܐ ܡܢܐ . ܢܒܥܐ ܡܢ ܗܘ ܕܐܠܗ ܢܘܡܐ ܕܡܬܚܫܚܢܐ ܒܡܫܘܚܬܐ ܡܬܦܠܓ ܗܘܐ . ܒܫܒܥܬܐ ܕܠܢܝܬܐ ܡܟܝܢ ܗܘܐ . ܘܗܘܐ ܠܗ ܡܚܝܢܐ . ܐܠܐ ܒܗܘܪܒ ܫܥܬܐ ܦܪܒ . ܘܒܪܘܫܗ ܡܪܝܫ .

[II.i.3]

ܕܬܠܬ . ܡܬܚܘܝܢ ܕܬܠܬ . ܗܘ ܕܒܬܪ̈ܝܢܐ ܬܠܬܐ ܒܗܘܢܐ . ܕܐܢܬܬܐ . ܗܢ ܡܬܚܫܚܢܐ ܠܗܘܒܠ ܣܬܘܐ ܢܦܩܐ . ܘܟܠܗܘܢ ܡܫܘܚܬܐ ܢܦܩܐ ܒܫܒܥܬܗܘܢ . ܕܡܢܗ ܡܬܚܫܚܢܐ ܒܫܒܥܬܐ ܡܬܗܘܝܐ . ܘܟܠܬܐ ܕܡܬܬܫܡܫܢܘܬ ܢܘܡܐ ܒܫܒܥܬܐ . ܕܢܦܫܗ ܢܘܡܪܐ ܟܠܗܘܢ ܠܐ ܟܠܝܗ ܐܝܟܢܗ . ܘܒܪ ܠܗܘܒܠ ܒܫܒܥܬܐ ܡܫܘܚܬܐ ܗܘܬܐ . ܘܠܗܘܒܠ ܡܫܘܚܬܐ ܢܘܡܐ . ܢܘܡܐ ܒܡܫܘܚܬܐ ܡܬܬܫܡܫܐ . ܘܬܪܝܢ ܗܘܢ ܐܚܪܢܐ ܒܫܒܥܬܐ . ܬܪܝܢ ܕܝܢ ܒܡܬܚܘܝܢ ܡܕܡܢܬܐ . ܠܗ ܡܛܠ ܒܫܬܐܬܗ . ܐܠܐ ܡܛܠ ܠܒܝܬܗ . ܡܬܩܢܝܢ ܡܪܒܐ . ܘܐܢܬܐ . ܘܗܘܠܐ ܬܠܝܬܝܐ ܢܦܝܫܐ ܕܠܗܘܢ ܡܢ ܣܬܘܐ . ܘܐܬܐ ܠܗܬ ܡܬܚܫܚܢܐ ܕܢܘܡܪ .

FMLIV: **tit.**: 1 ܡܫܬܟܢܐ: ܠܡܫܬܟܢܐ MΛV || 3 ܕܡܫܬܟܢܘܬ ܟܠܗ: om. ΛV || 4 ܙ̈ܢܝܐ M || II.i.1: 2 ܪ̈ܒܥܐ MV || 4 ܠܡܛܝܒ: ܠܡܛܒ Λ || ܥܠ MV || ܣܬܘܐ: ܣܬܘܐ M[1] || 5 ܡܬܚܘܝܢ L || ܐܢܬܬܐ M || ܘܡܬܚܫܚܢܘܬܐ: ܡܬܚܫܚܢܘܬܐ Λ || 6 ܠܗܬܘ M || ܐܝܟ: om. Λ || 8-9 ܡܫܘܚܬܐ ܒܦܠܓ: ܒܡܫܘܚܬܐ Λ || 9 ܡܬܕܪ: ܡܬܕܪ F || **II.i.2**: 1 ܡܬܚܘܝܢ L || ܕܡܬܚܫܚܢܢܐ M || 3 ܕܕܠܟܐ V || 4 ܡܬܚܫܚܢܐ F; ܕܡܬܚܫܚܢܢܐ M || ܒܡܫܘܚܬܐ M || 5 ܦܪܒ: om. Λ || **II.i.3**: 1 ܡܬܚܫܚܢܐ V || 2 ܡܫܘܚܬܐ: ܡܫܘܚܬܐ L || 4 ܢܘܡܪܐ: ܢܘܡܐ V || 6 ܒܫܬܐܬܗ MV || 7 ܠܒܝܬܗ MV || ܢܦܫ 1 || ܣܬܘܐ: ܣܬܘܬܐ ΛM; ܒܫܒܥܬܐ ܣܬܘܬܐ V ||

SECOND CHAPTER

ON ILLUSORY MATTERS WHICH ARE SEEN ON CLOUDS

It contains four sections.

First Section: Introduction to the Teaching on the Efficient Cause of Cloud-Related Illusions. Nine theories.

[II.i.1.]

First [theory]: Halo, rainbow, mock suns and lances are illusions. An illusion is the perception by the sense of an image of something in the image of something else by way of conveyance by the latter - e.g. a mirror - of the first - e.g. the image of a man - towards vision, when the first is not imprinted in reality in the second.

Concerning how vision works there have been three opinions. One is that which says that a ray goes out from the visual organ and meets the object seen and conveys its image to the visual organ. This ray is also directed towards a smooth object, such as a mirror, is reflected from it, is conveyed until it meets the object seen which is placed opposite the mirror, perceives it and the mirror together, and believes that it is seeing it inside the mirror.

[II.i.2.]

Second [theory]: The second opinion asserts that the image of the object seen is imprinted as it is in the eye and is seen. It is also imprinted in the mirror and this [mirror] is imprinted together with (the image of the object seen) in the eye and they are seen together. That this assertion is false is known from the fact that, if the image of the object seen were imprinted in the mirror, it would be fixed to a particular place. In reality it is not so, but as the viewer comes near it it [too] comes nearer and as he moves away from it it [too] moves away.

[II.i.3.]

Third [theory]: The third opinion is that of the natural philosophers who are correct in [their] inquiry, who say: When the object seen is [placed] opposite the visual organ and there is a transparent body between them, a likeness of the object seen is formed in the eye. The reason for the depiction of the image in the eye is the shining of light on (the object) and not on (the eye). When the mirror is opposite the eye and the image opposite the mirror, the image is depicted in the mirror and the two of them [are depicted] together in the eye. The Master [i.e. Aristotle] adopts the first opinion here not because of its veracity but because of its renown and says: "We imagine a straight line which goes out from the visual organ and comes to the object seen which is bright."

[II.i.4]

ܕܐܬܟܬܒ . ܡܫܡܠܝܘܬ ܢ̈ܦܫܐ . ܐܦ ܗܢܐ ܕܐܢܛܝܪܝܛܝܩܘܣ . ܗܘ ܚܪ̈ܝܢܘܬ
ܚܠܘܦܢܐ . ܫܦܝܐ . ܐܝܟܢܐ ܕܢܦܩܘ ܠܝܘܡܐ ܡܢ ܟܪܣܐ . ܘܡܬܟܣܝܢ ܠܘܬ ܡܪ̈ܥܐ .
ܘܒܚܠܘܦ ܡܫܡܠܝܢ ܠܘܬ ܐܢܫܐ . ܘܡܫܡܠܝܢܐ ܡܫܡܠܝܘܬ ܡܬܚܫܒܢܐ ܒܡܪ̈ܥܐ ܐܢܫܐ .
ܐܦ ܗܢܐ ܕܐܢܛܝܪܝܛܝܩܘܣ . ܗܘ ܚܪ̈ܝܢܘܬ ܚܠܘܦܢܐ . ܐܝܟܢܐ ܕܢܦܩܘ ܠܝܘܡܐ ܡܢ
ܟܪܣܐ . ܘܡܬܟܣܝܢ ܠܘܬ ܐܣܘܬܐ . ܘܡܢܗ ܒܚܠܘܦ ܠܘܬ ܡܪ̈ܥܐ . ܘܡܢ ܡܪ̈ܥܐ ܠܘܬ
ܐܢܫܐ . ܘܡܫܡܠܝܢܐ ܡܫܡܠܝܘܬ ܡܬܚܫܒܢܐ ܒܐܣܘܬܐ ܕܢܦܫܬܐ ܠܦܘ̈ܪܐ .
ܘܒܐܢܛܝܪܝܛܝܩܘܣ ܡܢ ܒܘܠܒܐ ܕܫܘܥܐ ܡܬܚܫܒܢܐ ܕܡܬܚܫܒ ܒܚܫܚܘܬܐ . ܘܐܟܚܕ
ܡܢ ܡܢܐ ܕܗܠܝܢ ܡܬܚܫܒܐ . ܘܒܬܪ ܐܟܚܕܐ ܘܒܬܪ ܚܘܝܐ . ܒܐܢܛܝܪܝܛܝܩܘܣ ܕܝܢ
ܒܘܠܒܐ ܕܫܘܥܐ ܠܐ ܡܬܚܠܦܝܢ ܡܬܚܫܒܢܐ . ܘܐܦ ܡܢ ܡܢܐ ܕܗܠܝܢ ܡܬܚܫܒܐ .
ܘܒܬܪ ܐܟܚܕܐ ܘܒܬܪ ܚܘܝܐ .

[II.i.5]

ܕܫܡܫ . ܦܐܪ ܕܝܢ ܫ̈ܢܝܐ . ܐܦ ܒܡܫܡܠܝܘܬܐ ܕܡܬܚܫܒܢܐ . ܟܕ ܗܘܝܢ ܡܢ ܬܪܬܐ .
ܗܘܝܢ ܕܝܢ ܙܒܢܬܐ . ܫܢܐ ܠܗ . ܐܦ ܒܐܣܡܟܬܗ . ܟܕ ܕܐܝܢܐ ܕܚܣܝ ܠܗ
ܚܘܝܐ ܡܬܬܪܓܡܐ ܠܗ . ܘܠܐ ܒܡܣܬܟܠܢܘܬܐ . ܐܠܐ ܫܝܠܘܠܘܬܐ . ܐܠܐ ܦܠܘܣܐ ܐܦ
ܐܣܟܡܝܢܐ ܡܬܚܫܒܐ . ܐܦ ܒܡܫܡ̈ܠܬܐ ܕܒܦܠܘܣܘܦܘܬܗ . ܟܕ ܕܐܝܢܐ ܕܚܣܝ . ܠܐ
ܡܬܬܪܓܡܐ ܫܪܫܘܬܗ . ܐܝܟܢܐ ܕܦܠܓܘܬܐ ܕܒܚܝܠ ܒܐܝܠܝܢ ܘܢܛܪܐ . ܡܢ ܬܘܒܐ
ܒܝܬܝܐ ܡܬܚܫܒܐ . ܐܦ ܒܝܠܗ . ܟܕ ܗܘܝܢ ܡܢ ܒܝܬܗ ܝܘܡܢܐ . ܗܘܝܢ ܕܝܢ
ܦܫܝܛܐ ܫܢܐ ܠܗ . ܐܦ ܒܡܬܚܫܒܐ ܕܒܬܗ ܠܗ ܠܐܢܫܝܢܐ . ܟܕ ܕܐܝܢܐ ܕܦܠܓ
ܢܫܐ . ܠܐ ܡܬܚܫܒܐ ܒܒܝܬܗ ܬܘܒܗ ܡܢ ܫ̈ܘܥܐ . ܐܝܠܐ ܬܘܒܗ ܕܐܢܫܝܢܐ
ܡܢܗ . ܐܝܟ ܬܘܒܐ ܕܒܝܬܐ ܡܢ . ܘܕ̈ܘܒܪܐ ܡܢܗ .

[II.i.6]

ܕܗܟܢ . ܠܡ̈ܫܒܚܐ ܠܒܢ̈ܝܐ . ܟܕ ܗܟܝܠ ܠܒܢܝ̈ܗܘܢ ܡܢ ܚܘܒܬܐ ܕܡܪܝܐ ܠܗܘܢ .
ܚܘܝܐ ܠܒܢܝܢܐ ܡܫܡܠܝܢ . ܘܐܦ ܡܫܬܡܥ ܫܡܥܝܢ . ܟܕ ܗܘ ܡܢ ܠܒܢܝܐ
ܘܫܡܥܐ ܚܘ̈ܝܐ ܐܢܫܝܐ ܡܬܬܪܒܝܢ . ܐܝܟܢܐ ܕܟܕ ܢܦܠ ܠܒܢܝܐ ܥܠ ܚܘܝܐ
ܐܘܟܡܬܐ . ܘܡܫܡܠܝܘܬܐ ܡܬܚܫܒܢܐ . ܘܒܚܘܠܝܐ ܙܢܐ ܗܘܐ ܠܒܢܝܐ ܡܫܡܠܝܢܐ
ܒܡܕܝܢ ܬܘܒܐ ܘܐܘܟܡܐ ܐܢܫܝܐ .

FMLIV: II.i.4: 1 ܢܦܫܐ V || ܕܐܢܛܝܪܝܛܝܩܘܣ M; ܕܐܢܛܝܠܝܛܝܩܘܣ V || 2 ܫܦܝܐ: om. L || 3 ܡܫܡܠܝܘܬ V || ܡܬܚܫܒܢܐ MV || 4 ܕܐܢܛܝܠܝܛܝܩܘܣ V || ܚܠܘ̈ܦܢܐ F || 5 ܘܡܬܟܣܝܢ: ܘܡܬܟܣܣܝܢ MV || ܘܡܢܗ MV || ܠܘܬ ܡܪ̈ܥܐ ܘܡܢ ܡܪ̈ܥܐ ܠܘܬ: ܠܘܬ ܒܥܠܬܐ ܘܡܬܟܣܝܢ MV || 6 ܘܡܫܡܠܝܢܐ: ܘܡܫܡܠܝܢ MV || ܡܫܡܠܝܘܬ F; ܡܬܡܫܠܝܘܬ MV || 7 ܘܒܐܢܛܝܪܝܛܝܩܘܣ M; ܘܒܐܢܛܝܠܝܛܝܩܘܣ V || 8 ܒܐܢܛܝܠܝܛܝܩܘܣ V || **II.i.5**: 3 ܐܠܐ: ܐܦ Λ || 4 ܕܒܦܠܘܣܘܦܘܬܗ ΛV || 5 ܒܐܝܠܝܢ F || 6 ܡܢ: om. F || ܘܗܘܝܢ ܕܝܢ V || 7 ܕܒܬܗ: ܕܒܝܬܗ F || 9 ܘܕ̈ܘܒܪܐ: ܘܕ̈ܘܒܪܐ Λ || **II.i.6**: 1 ܠܒܢܝ̈ܗܘܢ V || 2 ܫܡܥܝܢ: ܡܫܬܡܥܝܢ Λ || 3 ܘܫܡܥܐ: ܘܠܫܡܥܐ V || 4 ܙܢܐ ܗܘܐ: ܗܘܐ ܗܢܐ Λ || 5 ܬܘܒܐ scripsi: ܡܬܚܒܐ codd. ||

[II.i.4.]
Fourth [theory]: Illusory perception of images occurs either [1] by means of ἀνάκλασις, i.e. single reflection - as when a ray goes out from the eye, is directed towards water, is reflected from it towards a tree, and thus the tree is seen as an illusion in water - or [2] by means of διάκλασις, i.e. double reflection - as when a ray goes out from the eye, is directed towards a wall, is reflected from it towards water and from water towards a tree, and thus the green colour of leaves is seen on the wall. In ἀνάκλασις, the object of vision in the mirror moves with the movement of the viewer and appears smaller, blacker and less white than it ought to. In διάκλασις, the object of vision is not disturbed with the movement of the viewer and appears larger, less black and whiter than it ought to.

[II.i.5.]
Fifth [theory]: Vision may err [1] concerning the size of the object seen, since it sees it sometimes as larger and sometimes as smaller [than in reality]; [2] concerning its shape, because with something that is far away, one cannot perceive its corner or its convexity and concavity, and it appears flat or round; [3] concerning the parts on its surface, because with something that is far away, one cannot perceive its roughness, as when a mountain which is thick with trees and crags appears smooth from a distance; [4] concerning its colour, since it sees it sometimes as deeply and sometimes as lightly tinged; or [5] concerning the distance between it and another object, because with something that is very far away, one cannot know the extent of its height above the viewer, or the height of another object from it, like the height of the moon from us and that of the fixed stars from (the moon).

[II.i.6.]
Sixth [theory]: Bright bodies cause an illusion of luminous colour when their light is reflected from a mirror which is close to them. If they move away, they grow dark, so that different colours are composed out of the light and the darkness, just as, when light falls on a black cloud, (the cloud) appears red. In this way illusory light is generated on a thing which is far away and black at the same time.

[II.i.7]

ܕܫܒܥ . ܓܘܫܡܐ ܡܫܬܐܠܐ . ܐܦ ܚܕ ܓܘܫܡܐ ܪܒܐ ܡܢ ܗܢܘܢ ܕܡܬܟܬܫܝܢ
ܠܡܬܟܬܫܘ ܒܩܘܒܠܗܘܢ ܒܡܬܕܒܪܢܘܬܐ . ܠܐ ܐܝܠܝܢ ܕܡܢܟܪܝܢ ܢܦܫܗܘܢ . ܒܡܕܒܪܝܐ
ܕܡܢܘܬܐ ܡܢܗܘܢ ܠܬܐܪܝܟܘܬ ܡܬܝܩܪܝܢ . ܐܦ ܢܣܒ . ܘܡܣܒܪܐ ܥܠܝܡܘܬܐ ܠܗ ܠܐ ܡܬܝܩܪܝܢ
ܐܚܝܕܐ ܚܠܡܝܢ . ܐܦ ܓܘܢܐ ܡܫܒܠܐ ܦܢܝܢ . ܡܟܝܟܝܢ ܕܝܢ ܢܦܫܗܘܢ ܠܬܐܪܝܟܘܬ ܠܗܘܬ
ܡܫܝܢܘܬܐ . ܐܝܟܢܐ ܕܢܬܚܕܬ ܢܫܪܐ . ܐܝܟܢܐ ܕܠܝܘܢܐ ܕܒܡܥܠܬܐ ܢܫܘܐ
ܒܡܫܝܢܘܬܐ .

[II.i.8]

ܕܬܡܢܝܐ . ܒܗ ܓܘܫܡܐ ܡܫܡܠܝܐ ܡܬܕܒܪܢܐ ܢܗܘܐ . ܘܡܬܕܒܪܢܘܬܐ ܢܦܫܢܝܬܐ
ܒܡܬܕܒܪܢܘܬܐ . ܡܫܬܥܠ ܒܗ ܚܕܝܐ . ܘܗܓܓܐ ܡܫܝܢܘܬܐ ܠܗ ܚܕܐ . ܘܐܦ
ܡܬܚܒܪܐ ܒܗ ܗܓܓܐ ܡܝܬܐ . ܠܘ ܡܬܕܒܪܢܐ ܦܐܫ ܒܡܬܕܒܪܢܘܬܐ . ܘܡܫܪܐ ܡܢ
ܗܘܢܐ ܒܗ ܓܘܫܡܐ ܕܡܬܓܠܐ ܒܥܒܕܘܗܝ ܢܗܘܐ . ܐܝܟ ܒܥܘܠܐ . ܗܘ ܕܐܦ
ܓܘܢܐ ܡܕܡ ܢܗܘܐ ܒܥܒܕܘܗܝ . ܐܝܟ ܡܫܝܢܘܬܐ . ܡܟܝܟܐ ܠܗܘܢܐ . ܘܐܠܐ ܠܐ .

[II.i.9]

ܕܬܫܥ . ܡܫܝܢܘܬܐ ܡܢ ܪܒܬܐ ܠܓܘܫܡܐ ܘܐܣܟܡܐ ܡܫܘܚܐ . ܠܚܘܪܬܐ ܕܝܢ
ܠܓܘܢܐ ܟܠܗ ܐܣܟܡܐ . ܒܗܝ ܕܐܣܟܡܐ ܒܣܘܦ ܩܕܡܐ ܡܬܬܚܡ .
ܒܓܘ ܐܝܟܢܐ ܕܐܓܪܐ ܠܗ ܡܦܪܫ ܘܡܬܚܡ ܡܚܘܐ . ܐܝܟܢܐ ܠܢܦܫܐ
ܡܦܪܫܐ . ܘܡܠܝ ܡܢ ܦܘܠܚܢܗ ܡܬܬܟܠܢܘܬܐ ܐܝܟ ܗܪܟܐ .
ܘܒܐܣܟܡܝܬܐ ܘܒܠܒܘܫܐ ܕܪܘܚܐ ܐܦܪܘܣܘܦܐ ܗܘ̈ܝܢ .

FMLIV: **II.i.7**: 2 ܠܐ: ܐܠܐ Λ || 3 ܠܬܐܪܝܟܘܬ: in marg. V || ܠܐ ܡܬܝܩܪܝܢ MV || 4 ܐܦ ܓܘܢܐ: ܘܓܘܢܐ F || 5 ܕܒܡܥܠܬܐ: ܡܥܠܬܐ ut vid. M[1]; ܒܡܥܠܬܐ IVM[V] || **II.i.8**: 1 ܒܗ ܓܘܫܡܐ: ܒܓܘܫܡܗ Λ || ܡܫܡܠܝܐ: ܡܫܠܡܢܐ MV || ܘܡܬܕܒܪܢܘܬܐ: ܘܡܬܕܒܪܢܐ MΛV || 4 ܗܘ: om. V || 5 ܢܗܘܐ: ܗܘܐ V || ܒܥܒܕܘܗܝ M || ܠܗ̈ܘܢܐ MΛV || **II.i.9**: 2 ܡܫܚܐ MΛV || 4 ܡܬܬܟܝܠܢܘܬܐ V || ܘܒܐܣܟܡܝܬܐ IV || 5 ܐܦܪܘܣܘܦܐ FMΛ (sed voc. ܐܦܪܘܣܘܦܐ F) || ܗܘܝܢ MΛV ||

[II.i.7.]
Seventh [theory]: A plural number of bodies - or a single body which is too large to be seen in a mirror - may not necessarily appear in a like manner, because parts of them may be positioned in a straight alignment, or they may be far away and not all of them are positioned together on a single plane, or they may have a weak colour. What is brighter casts its image more strongly towards the mirror, so that it obscures the image of a gloomy thing in the mirror.

[II.i.8.]
Eighth [theory]: When a smooth body is transparent and appears transparent in actuality, vision is conveyed through it and does not see an illusion of the type seen in mirrors. If such an illusion is seen on it, it is no longer transparent in actuality. This occurs when there is a coloured body behind it, as is the case with rock crystal, which shows an image just like a mirror if there is some colour behind it, but if not, does not.

[II.i.9.]
Ninth [theory]: A large mirror shows colour and shape, a small mirror colour but not shape, because shape is defined by division at edges. How, then, can something appear shaped which sense does not divide and whose edges it does not define?

These are the traditional premises here and the proofs are provided through geometry and optics.

[tit.]

ܦܘܫܩܐ ܬܢܝܢܐ . ܡܛܠ ܣܘܟܠܬܐ . ܬܐܘܪ̈ܝܐ ܘ ܐܪܒܥ .

[II.ii.1]

ܡܪܝܡܘܬܐ . ܣܘܟܠܬܐ ܗܝ ܘܡܫܠܡܘܬܐ ܘܫܠܝܘܬ ܥܒܕܐ ܘܩܢ̈ܝܢܐ ܡܫܟܚܐ ܐܢܬܘܢ .
ܘܠܐ ܗܘܝܢܐ ܕܡܫܬܚܠܦ ܠܢܦܫܟܘܢ . ܘܣܘܟܠܬܐ ܗܝ ܐܢܬܘܢ ܚܙܝܐ ܕܡܬܚܙܝܐ
ܚܙ̈ܝܢ ܗܘܝܐ . ܡܢ ܕܒܢܝܢܐ ܦܠܚܘܬܐ ܘܦܘܠܚܢܐ ܡܬܦܢܝܢܐ ܒܗ ܚܘܝܐ
ܘܗܘܝܐ . ܒܕܘܟ ܡܬܬܘܗ ܗܘ ܕܠܘܩܒܠ ܗܘܝܐ ܠܐ ܡܬܦܢܝܐ ܠܗܘܝܐ . ܐܝܟܢܐ
ܡܬܒܪܝܐ ܒܗ ܡܫܟܚܐ ܕܗܘܝܐ . ܒܫܠܡܘܬܐ ܬܘܒܐ ܗܘ ܡܬܦܢܝܢܐ ܒܢܦܫܗ
ܡܬܚܙܝܐ . ܘܠܐ ܡܫܟܚܐ . ܘܡܬܚܙܝܢܐ ܗܘܝܢ ܕܬܘܒܐܝܬ ܠܐ ܡܬܦܢܐ ܠܘܩܒܠ ܗܘܝܐ .
ܚܠܦܐ ܡܬܚܙܐ ܠܡܫܟܚܐ ܕܗܘܝܐ ܠܘܬ ܚܙܝܐ ܡܬܚܙܝܐ . ܘܡܛܠ ܕܐܝܬܘܗܝ ܚܠܦܐ
ܡܬܚܙܐ . ܠܗܘܝܘ ܒܠܚܘܕ ܐܝܬܘܗܝ ܡܬܚܙܝܐ . ܘܚܠܦܝܢ ܐܚܪܢܐ ܦܠܓܘܬܐ
ܕܗܘܝܐ ܚܙܝܢܝܐ ܒܐܝܬܘܗܝ ܗܘܝܐ ܚܙܝ ܗܘܝܐ ܐܚܪܢ .

[II.ii.2]

ܕܬܬܚܙܐ . ܡܬܚܙܝܐ ܕܣܘܟܠܬܐ ܐܝܬܝܬܐ ܡܬܚܙܝܢܐ . ܒܗܝ ܕܡܫܬܡܠܝܐ ܡܢ
ܚܙܝ ܙܒܢܐ . ܘܚܠܦܝܐ ܡܬܚܙܐ ܠܚܝܠܐ ܦܘܠܚܢܐ ܕܬܗܘܐ . ܘܒܗ ܠܐ ܡܬܚܙܝܢܐ .
ܠܐ ܡܫܬܡܠܝܐ ܐܝܟ ܗܘܝܐ ܕܚܙܝܐ . ܘܠܐ ܡܫܬܡܠܝܐ ܥܠ ܝܕ ܡܫܬܡܠܝܐ
ܐܝܬܝܐ ܡܬܚܙܝܐ . ܘܐܝܟ ܕܐܡܪܢ ܦܢ . ܐܝܬܝܘܬܐ ܠܗ ܚܙܝܢ ܗܘ ܕܠܐ ܢܗܘܐ
ܡܕܡ . ܗܘ ܚܕ ܒܗ ܠܐ ܚܙܝܢ ܠܗܘܦ ܥܒܕܐ ܒܠܚܘܕ . ܡܫܬܡܠܝܢ ܕܐܝܬܝܘܬܐ
ܚܙܝܢ . ܘܗܝܕܝܢ ܒܐܝܬܝܘܬܐ ܒܗ ܚܝܠܐ ܡܫܬܡܠܝܢ .

[II.ii.3]

ܕܬܬܠܬ . ܦܢ̈ܝܢܐ ܕܝܢ ܣܘܟܠܬܐ ܡܢ ܡܥܠܬܐ . ܒܗܝ ܕܒܣܘܟܠܬܗ ܡܢ ܚܕ ܗܘܝܐ
ܕܗܘܝܐ ܚܙܝܐ ܐܝܬܘܗܝ . ܘܐܝܬܝܐ ܗܘܝܐ . ܘܗܝ ܣܘܟܠܬܐ ܙܕܩ ܕܗܝ ܗܘܝܐ
ܐܝܬܘܗ . ܘܡܬܚܙܝܢ ܕܚܙܝܬܗ ܚܠܦܘܗܝ ܕܗܝ ܗܘܦܟܐ ܒܗܘ ܚܘܝܐ ܠܡܬܚܙܝܢܐ .
ܒܡܥܠܬܐ ܕܝܢ ܚܘܝܐ ܘܡܫܬܡܠܝܐ ܥܠ ܗܘܦܟܐ ܕܗܘܝܐ ܐܝܟܢ . ܒܗܝ ܡܬܦܟܢ
ܕܥܒܕ ܙܕܩ ܠܗ ܒܐܝܬܘܗܝ ܗܘ . ܘܣܘܟܠܬܐ ܡܢ ܚܙܝܐ ܗܝ ܦܠܚܢܐ . ܡܥܠܬܐ ܕܝܢ
ܦܠܚܘܬ ܚܙܝܐ

FMLΛV: **II.ii.1**: 1 ܡܢ: ܡܢ M ‖ ܘܡܫܠܡܘܬ M ‖ 2 ܠܢܦܫܟܘܢ F ‖ ܕܡܬܚܙܝܢܐ MV ‖ 4 ܒܕܘܟ . . . ܠܗܘܝܐ in marg. V ‖ ܡܬܚܙܝܐ MV ‖ 5 ܗܘ: ܘܗܘ M ‖ 6 ܘܡܬܚܙܝܢܗ MΛV **II.ii.2**: 2 ܡܬܚܙܝܐ F ‖ 3 ܠܐ: om. Λ ‖ 4 ܐܝܟ Λ ‖ 5 ܚܙܝܢ ut vid. M ‖ **II.ii.3**: 1 ܦ̈ܢܝܐ V ‖ 1-2 ܡܢ ܚܕ . . . ܘܗܝ ܣܘܟܠܬܐ: om. L ‖ 2 ܗܘܝܐ: ܗܘܝܢܐ MΛV (om. L) ‖ ܙܕܩ L; ܙܕ̈ܩܐ I ‖ 3 ܒܗܘ: illeg. M[1]; ܒܗܘ M[V] ‖ 4 ܙܕܩ ΛV ‖ 5 ܗܘ: om. M ‖ ܣܘܟܠܬܐ Λ ‖ ܡܢ M ‖ ܗܝ: ܗܝ M; om. V ‖ ܘܡܥܠܬܐ FMV ‖

Second Section: On Halo. Four theories.

[II.ii.1.]
First [theory]: The halo, rainbow, mock suns and lances are illusions and are not realities which exist by themselves. The halo is a circle which appears around the moon when a moist and thin cloud is interposed between the viewer and the moon, so that that part of (the cloud) which is in front of the moon does not conceal the moon and no illusion of the moon appears on it - because in a straight alignment the object of vision appears in itself and not its illusion - whereas each one of those parts of (the cloud) which are not positioned in a straight line in front of the moon conveys an illusion of the moon to vision. Because each of them is small, it conveys (the moon)'s light but not its shape, and all of them together form a circular phantasm of light in the shape of the moon around the moon.

[II.ii.2.]
Second [theory]: The middle of the halo appears black because in the middle the ray is strong and for that reason obscures the thin cloud which is there. Since it is not visible it does not glitter like those [parts] which are on the circle, and what does not glitter appears black beside what glitters. As the Master [Aristotle] said: "we see blackness when we do not see anything". For indeed, when we do not see the light of the sun at night, we think that we are seeing blackness; similarly in daytime when we close our eyes.

[II.ii.3.]
Third [theory]: Halo is distinguished from rainbow in that with the halo one end of the axis is the visual organ and the other is the moon. The halo is the belt around that axis and the centre of its circle is on this line between the viewer and the object seen. In the rainbow, on the other hand, the viewer and the sun are on the line of the axis, but the centre of the circular belt is not between them. The halo is a complete circle, while the rainbow is a semi-circle.

[II.ii.4]

ܕܐܪ̈ܒܥ . ܗܟܢܐ ܕܚܢܢܐ ܬܫܒܘܚܬ ܚܢܢܐ ܐܫܬܪܝܬܐ ܠܥܠܐ . ܡܬܝܕܥܐ ܕܬܡܘܬܐ ܫܘܒܚܬܐ
ܬܫܒܘܚܬ ܫܘܒܚܐ . ܘܬܫܒܘܚܬܐ ܪܒܬܐ ܗܘܝܐ . ܒܗܝ ܕܡܬܪܒܐ . ܘܠܫܘܒܚܐ
ܠܡܪܝܐ ܒܡܥ̈ܠܬܐ ܕܪ̈ܘܚܡ ܡܢ ܡܪܚܬܐ ܡܬܟܠܐ . ܘܐܬܥ̈ܬܕ ܐܦܪܘ ܕܐܒܒ
ܫܘܒ̈ܚܬܐ ܐܚܝܕܐ ܬܫܒܘܚܬ ܫܦ̈ܝܪܐ ܒܗܘ . ܘܡܢܗܝܢ ܕܡܬܪܐ . ܘܐܬܚܙܝ ܐܦܪܘ ܕܒܢܝܐ
ܫܘܒܚܬܐ ܗ̇ܝ ܕܒܗ ܐܬܡܢܝܬ ܒܡܥ̈ܠܬܐ ܕܠܘܬܗ ܩ̈ܢܘܡܗ ܐܬܒܪܝ ܘܫܟܝܢܐ
ܐܡܠܟ̈ܘܬܐ ܗ̇ܘܬ ܠܡܘܪܒܐ . ܫܪ̈ܝܪ ܥܡ̈ܡܐ ܕܝܢ ܫܡܐ ܠܥܡܐ ܗܘܝܐ ܫܘܒܚܬܐ .
ܒܗ̇ܝ ܕܐܝܟ ܕܒܫܘܒܚܐ ܠܡ̈ܠܐܟܐ ܫܘܒܚܬܐ ܕܠܐ ܡܟܝܟ̈ ܠܡܫܘܬܦܘܬܗ ܡܫܦܝܐ .
ܘܗܟܢܐ ܪ̈ܥܝܢܐ ܐܟܘܬ ܕܫܡܝܐ ܫܪ̈ܝܪ ܥܡ̈ܡܐ ܒܡܫܟܢܗ . ܫܘܒܚܬܐ ܕܒܓ̈ܠܝܢ ܡܬܝܐ
ܡܓܠܝܢܐ .

[tit.]

ܦܣܘܩܐ ܬܠܝܬܝܐ . ܡܛܠ ܡܣܟܢܘܬܐ . ܬܘܪܐܝܬܐ ܘܡܫܝܚܝܬܐ .

[II.iii.1]

ܡܕܡܝܬܐ . ܠܫܘܒܚܐ ܡܬܚܙܝܢܐ ܐܝܟ ܕܠܒܟܐ . ܕܡܢ̈ܝܬܐ ܡܢ ܡܪ̈ܝܐ ܕܡܫܡܫܢܐ
ܡܟܬܫܢܐ ܘܡ̈ܠܠܐ ܕܐܝܬ ܒܗ . ܠܫܘܐ ܠܡܫܡܫܢܐ ܡܫܡܫܢܘܬܐ . ܘܡܫܡܫܝܢ ܐܝܟ ܗܘܢܐ
ܕܡܫܡܫܢܐ ܠܡܫܡܫܢܐ ܕܡܫܡܫܝܢ ܗܘܘ ܠܡܫܡܫܢܐ . ܐܝܟ ܠܘܬܐ . ܐܦ ܚܢܢܐ
ܫܡܥܬܐ . ܐܡܪܐ ܕܗܘܬ ܒܗܠܝܢ ܡܫܡܫܢܐ ܘܚܠܡܘܬܐ ܠܡܫܢܐ ܪܡܫܬܐ . ܘܠܐ
ܦܪܝܫ ܒܫܢ ܕܗܘܬ ܡܫܡܫܢܐ ܠܚܢܢܐ ܕܠܒܫܬܐ ܘܡܫܡܫܢܘܬܐ ܕܟܡܢܐ ܡܫܬܐ .
ܘܡܬܢܝܢܬܐ ܗܘܬ . ܒܗ ܡܢ ܒܢܝ̈ ܚܝ̈ܠܐ ܕܫܪܝܪܐ ܒܡܪܝܐ ܕܡܫܡܫܢܐ ܡܠܬܝܢܐ
ܠܡܫܠ ܫܡܫܐ ܡܬܪܡܫܝܢ . ܒܗܢ ܡܬܚܙܝܢܐ ܒܫܡܐ ܡܬܚܙܝܢ . ܘܡܫܢܐ ܒܗ ܦܠܓܐ
ܐܝܟ ܡܪܝܐ ܒܫܡܫܐ ܘܠܡܫܠ ܫܡܫܐ ܐܦ ܫܠܡܐ ܠܗܘ . ܘܫܪ̈ܝܪ ܡܢܗܘܢܐ
ܒܓܠܝܢܐ . ܘܒܦܪ̈ܘܫܬܐ . ܒܗ ܡܬܬܚܬܝܢ ܐܝܟ ܡܢ ܫܡܝܐ . ܒܗ ܕܠܒܟܬ ܚܝܢܗ .
ܡܢ ܕܠܒܦܪܢܗ . ܫܘܒܚܐ ܡܬܚܙܝܢܐ ܫܢܐ .

FMLIV: **II.ii.4**: 2 ܬܫܒܘܚܬ ܫܘܒܚܬܐ: om. L || ܘܠܫܘܒܚܐ: ܘܠܫܘܒܚܐ F || 4 ܐܚܝܕܐ: om. Λ || ܬܫܒܘܚܬ ΛV || 7 ܐܝܟ V || ܠܡܫܘܬܦܘܬܗ V || 9 ܡܓܠܝܢܐ MV || **II.iii.1**: 1 ܕܡܢ̈ܝܬܐ: ܕܡ̈ܢܬܐ Λ || 2 ܡܟܬܫܢܐ M || 4 ܕܬܫܒܘܚܬ ΛV || 5 ܦܪܝܫܐ Λ || ܕܡܫܡܫܢܘܬ: ܕܡܫܡܫܬ F || 6 ܘܡܬܢܝܢܬܐ: add. ܒܗ ܡܢ V || ܒܗ ܡܢ M || 7 ܡܬܪܡܫܝܢ ut vid. M; ܡܬܪܡܫܝܢ V || ܘܡܫܢܐ V || 8 ܫܪ̈ܝܪ ܐܦ F || ܒܗ: ܒܗܢ M ||

[II.ii.4.]
Fourth [theory]: When a cloud occurs below another cloud it is possible for a halo to form below a halo. The lower [halo] will be larger because it is closer and conveys the luminous illusion through parts which are further out from the centre. Some have said that they saw seven haloes at one time below each other, and this is surprising. Another has said that he saw a halo which, when measured against the stars which were in front of its limits, was approximately forty-five stades [in diameter]. Around the sun halo occurs rarely, because (the sun) usually dissipates thin clouds which are unable to conceal it. The Princely Doctor [i.e. Ibn Sīnā] has said that he saw a halo with the colours of the rainbow around the sun in Hamadan.

Third Section: On Rainbow. Eight theories.

[II.iii.1.]
First [theory]: For [the appearance of] the illusion of a rainbow, moist air sprayed with particles of dewy, transparent and clear water should serve as the mirror. Behind this dewy air there should be a coloured body, such as a mountain or a dark cloud, as the example of the rock crystal confirms. As a result, sense errs and does not distinguish between the place (occupied by) the rainbow and the murky and dark cloud behind the rainbow. In mills too, when fine waterdrops are sprayed opposite the sun from the edges of the wheel rotating in water, the colours of the rainbow are seen in them. The same happens when one gathers water in one's mouth and blows in front of the sun or a lamp, and around a candle in a bath. In the morning when one awakes from sleep, because of the moistness of one's eye, before wiping it, one sees a rainbow-like illusion.

[II.iii.2]

ܕܬܪܬܝܢ . ܐܡܪ ܠܟ ܪܒ . ܕܒܡܫܝܚܐ ܕܐܬܦܠܚ ܡܢ ܠܒܩܐ ܒܫܡܐ . ܠܒ̈ܢܝܐ ܡܫ̈ܬܠܝܢܐ ܡܬܛܝܒܝܢ . ܘܐܝܠܝܢ ܕܡܫܬܠܛܝܢ ܫܪܝܘ . ܡܬܬܠܡܕܝܢ ܘܠܐ ܡܬܒܛܠܝܢ ܒܐܝܕܐ . ܐܝܟܢܐ ܕܒܪܫܝܬ ܠܐܪܝܣܛܘܛܠܝܣ ܠܦܝܠܣܘܦܘܬܐ . ܗܘ ܟܕ ܥܠܠ ܚܣܝܪ ܡܐܢܐ ܢܦܫܗ . ܦܬܚ ܗܘܐ ܕܝܠܝܕܘܬܐ ܕܒ̈ܢܝܢܫܐ ܫܪܝܐ ܗܘܐ . ܥܠ ܕܡܬܬܠܡܕܐ ܗܘܬ ܫܪܝܬ ܡܢ ܐܝܕܐ . ܒܐܝܕܐ ܓܝܪ ܕܡܬܝܠܕܐ ܡܢ ܐܝܕܐ ܟܝܢܝܐ ܗܘܬ ܠܗ ܡܬܬܪܝܐ . ܘܗܘ ܒܐܝܕܐ ܫܪܝܐ ܗܘܬ ܕܡܬܬܠܡܕܘܢ . ܐܝܟ ܕܐܦ ܡܢ ܡܬ̈ܠܐ ܟܝܢܝܐ ܠܗ ܠܒ̈ܢܝܢܫܐ ܘܒܗܘܢ ܒܡ̈ܠܐ ܫܪܝܐ ܕܡܬܬܠܡܕܘܢ . ܘܠܝܬ ܐܢܫܐ . ܗܘ ܟܠ ܠܗܠܝܢܬܐ ܕܢܦܫܗ ܟܠܝܐ ܢܫܠ . ܒܗ ܐܝܕܐ ܕܦܫܝܩܐ ܠܗ ܐܘܣܛܐ ܕܠܡܠܦܘܬܗ ܡܬܚܒܠ . ܘܡܢܗ ܠܗ ܐܘܣܬܐ ܐܘܪܚܐ ܕܐܝܠܘܦܘܬܗ ܥܠܡ . ܘܠܒ̈ܢܝܐ ܡܫܬܠܝܢܐ ܒܗ ܡܬܚܒܠ .

[II.iii.3]

ܕܬܠܬ . ܫܪܪܢܐ ܗܘ ܕܝܢ ܐܘܡܢ ܡܬܠܐ ܗܢܐ . ܒܗܿ ܕܒܝܢܬܐ ܕܦܣܩܢܘܬܐ ܕܡܢܗܝܢ ܟܠܗ ܐܠܗܐ ܕܫܪܝܐ ܠܗ ܫܡܫܐ . ܗܟܢܐ ܗܘܝܢ . ܐܝܟܢܐ ܕܐܠܗ ܠܫܡܫܐ ܡܫܠܝܢ ܕܫܪܪܐ ܒܒܪܝܢ . ܒܗܘܐ ܕܠܐ ܗܘܐ ܡܢ ܗܘ ܫܪܪܐ ܠܟܠ ܡܢ ܐܪܥܐ ܒܠܚܘܕ ܕܒܝܢܬܐ ܗܢܝܢ ܒܟܪ ܗܘܐ . ܒܒܪܗ ܐܠܐ ܫܡܫܐ ܒܠ ܐܘܪܚܘܢ ܢܗܘܐ . ܗܘ ܟܠܢܦܫܬܐ ܐܘܪܚܘܢ ܒܬܪܝܢ ܦܠܓܝܢ ܦܣܩܐ ܗܘܬ ܠܗ ܠܫܪܪܐ . ܦܠܓܐ ܠܟܠ ܡܢ ܐܪܥܐ ܘܦܠܓܐ ܠܫܡܝܐ ܡܢܗ . ܘܒܝܬܐ ܕܦܠܓܐ ܐܣܛܘܟܣܐ ܐܘܒܢܬ ܪܡܝܐ ܕܫܡܫܐ . ܐܪܥܐ ܒܒܝܬ ܡܬܠܐ . ܘܡܢܐ ܕܠܟܠ ܡܢ ܪܒܐ ܗܘܐ . ܟܠܠ . ܘܒܝܬܪܬܐ ܒܬܪܒܢܐ . ܒܠܡܐ ܡܬܘܝܐ ܗܘܝܐ ܡܬܠܐ . ܘܒܡܠܐ ܠܐ ܗܘܝܐ . ܒܗܿ ܕܒܡܬܘܐ ܒܝܬ ܪܐܡ ܫܡܫܐ . ܘܒܡܠܐ ܢܗܪ .

[II.iii.4]

ܕܐܪܒܥ . ܒܗܘܢ ܕܒܡܬܠܐ ܠܐ ܢܗܪ . ܐܠܐ ܢܗܘܪ . ܐܝܟ ܒܗܘܢ ܕܫܡܫܐ . ܒܗܿ ܕܒܡܬܠܐ ܡܢ ܡܬܘܐ ܕܪܘܚܢܝܬܐ ܕܦܣܩܢܘܬܐ ܡܢ ܫܡܫܐ . ܒܒܪܗ ܡܢ ܫܠܠ ܢܗܝܪܐ ܡܟܠܝܢܐ ܒܒܝܬ ܡܢ ܒܠܚܘܕ ܕܫܡܫܐ . ܠܒ̈ܢܝܐ ܡܬܠܡܕܐ . ܡܫܡܫܐ ܘܒܪ̈ܬܐ ܘܐܪܥܐ . ܡܬܠܡܕܝܢ . ܒܫܠܝܐ ܕܝܢ ܢܗܪ ܡܢܗܪܐ ܡܢܗܪܬܐ ܕܦܣܩܢܘܬܐ ܡܢ ܡܬܠܐ . ܘܒܠܚܘܕ ܢܗܪ ܡܗܪ ܒܗܘܢ . ܘܡܟܠ ܫܡܫܬܐ ܠܠܝܠܝܐ ܗܘܐ . ܐܝܟ ܕܐܦ ܡܬܠܐ ܣܓܝܐܐ . ܗܘ ܟܠ ܡܬܠܐ ܕܢܗܪ ܣܓܝܐܝܢ ܡܬܦܣܩܝܢ . ܢܗܪ ܫܪܪܐ ܡܬܦܫܩܝܢ .

FMLIV: **II.iii.2**: 1 ܐܡܪ ܟܕ Λ || ܒܫܡܝܐ F || ܠܒܩܐ: ܠܒܩܐ Λ; ܠܒ̈ܩܐ V || 3 ܠܐܪܝܣܛܘܛܠܝܣ Λ || ܠܦܝܠܣܘܦܘܬܐ L; ܠܦܝܠܣܘܦܘܬܐ V || 4 ܢܦܫܗ: om. Λ || ܕܝܠܝܕܘܬܐ MV || 5 ܡܢ ܐܝܕܐ: om. Λ || ܟܝܢܝܐ ܗܘܬ: add. ܡܢ V || 6 ܐܝܟ: **hinc FMPLIV** || 6-7 ܕܡܬܬܠܡܕܘܢ . . . ܐܝܟ: om. F || 8 ܟܠܝܐ PL || ܠܗ: ܠܘ M^1 || ܕܠܡܠܦܘܬܗ Λ; ܕܠܡܠܦܘܬܗ V || 9 ܐܘܪܚܬܐ M || ܕܐܝܠܘܦܘܬܗ V || ܡܬܠܡܕܢܐ M || 10 ܒܗ MV || **II.iii.3**: 1 ܕܝܢ: ܟܕ V || ܕܒܝܢ̈ܬܐ V || ܕܦܣܩܢܘܬܐ M || 3 ܡܫܠܝܢ Λ; ܡܫܠܝܢ V || 4 ܕܒܝܢ̈ܬܐ V || 5 ܟܠܢܦܫܬܐ ܐܘܪܚܘܢ M; ܟܠܢܦܫܬܐ ܕܐܘܪܚܘܢ V || **II.iii.4**: 1 ܘܢܗܪܬܐ P || 6 ܣܓܝܐܝܢ M; ܣܓܝ̈ܐܝܢ ΛV ||

[II.iii.2.]
Second [theory]: The Master [Aristotle] has said that rainbow colours are observed in water which is sprinkled from oars on the sea. With someone whose vision is weak, (the vision) is broken and is not conveyed through the air, as happened to Antipheron of Tarentum. For when he fixed his gaze a little, he thought he was seeing forms of men, because his vision was broken by the air. Indeed, because of its weakness (the vision) was reflected from the air towards visible objects and it saw their likenesses in the air, just as (vision) is also reflected from water towards objects and sees their likenesses in water. When the ray of the sun falls on the glass in the skylight of a bath, it is conveyed towards the wall opposite (the skylight) through the spray-filled air, is reflected from (the wall) towards another wall by διάκλασις and causes an illusion of the rainbow colours on it.

[II.iii.3.]
Third [theory]: The shape of this rainbow is circular because the sprayey particles from which the visual ray is reflected towards the sun are placed in such a way that, if we were to make the sun the centre of the circle, the segment of the circle which falls above the earth would pass along those particles. Therefore, if the sun is on the horizon, the plane of the horizon will divide the circle in two halves, a half above the earth and a half below it. As the ἔξαρμα - i.e. elevation - of the sun increases, the size of the rainbow decreases. When (the sun) is at the zenith it disappears. In northerly regions rainbow occurs at midday in winter, but not in summer, because the sun is less elevated in winter and more in summer.

[II.iii.4.]
Fourth [theory]: The colour of the rainbow is neither luminous nor white like the colour of the halo, because with the rainbow the sprayey moisture is far from the sun, so that various colours, [namely] red, green and purple, are generated from the mingling of the illusory luminous with something falling in the genus of the dark. With the halo, on the other hand, the sprayey moisture is closer to the moon and for this reason its colour is luminous and white - also on account of the darkness of night, just as dirty clothes appear whiter when they are compared with clothes which are [even] dirtier.

[II.iii.5]

ܕܢܟܝܐ . ܒܓܠܝܘܬܐ ܡܬܚܘܝܐ . ܥܠܬܐ ܕܡܬܚܫܒܢܘܬ ܟܝ̈ܢܐ ܡܬܝܕܥܢܝܬܐ ܕܡܬܚܘܝܐ .
ܒܗܝܠ ܡܫܝܚܐ ܕܐܬܐܬܝ ܒ̈ܢܝܐ ܢܫܒܚܐ . ܘܚܠܦܬܐ ܡܢ ܒܗܘ ܕܡܬܝܒܐ ܠܥܠܡܐ .
ܫܠܝܛܐܝܬ ܚ̇ܠܦ ܡܢܗ ܬܪ̈ܝܐ ܠܘܬ ܥܡܡܐ . ܒܓܘ ܩܢܘܡܐ ܢܗܘܐ ܡܬܚܙܝܐ
ܠܗܘܢ . ܒܗܘ ܕܢܣܒܐ . ܟܕ ܡܬܚܙܝܐ ܒܐܘܣܝܐ ܐܦ ܒܪ ܡܬܚܙܝܢܘܬ ܐܘܣܝܐ .
ܩܢܘܡܐ ܡܬܚܙܝܐ . ܐܠܗܘܬܐ ܕܝܢ ܒܗܘ ܕܐܢܫܐ ܡܢ ܥܡܡܐ . ܡܫܒܠܐܝܬ ܚ̇ܠܦ
ܡܢܗ ܬܪ̈ܝܐ ܠܘܬ ܥܡܡܐ . ܒܓܘ ܩܢܘܡܐ ܕܝܠܗ ܠܐܘܣܝܘܬܐ ܡܬܚܙܝܐ .
ܘܒܗܢܐ ܕܐܝܟ ܗܢܐ ܡܘܗܘ ܐܘܪܓܢܐ . ܘܡܢ ܬܪܝܨܘܬܗ ܒܩ̈ܢܝܐ ܗܠܝܢ ܡܩ̈ܦܠܢܐ .
ܠܗܢܐ ܒܐܬܘܬܐ ܡܫܠܡ . ܘܡܫܡܫܢܘܬܗ ܡܬܚܙܝܐ .

[II.iii.6]

ܕܟܠܗ . ܐܝܟܢܐ ܗܘܐ ܟܕ ܐܬܬܢܝ ܩ̈ܢܘܡܐ ܚܕܐ ܒܓܘ ܡܬܚܙܝܢ . ܘܗܕܪܐ
ܟܝܢܐ ܕܐܠܗܘܬܐ ܘܗܕܪܐ ܠܗܘܢ ܕܐܠܗܘܬܐ ܟܝܢܝܬܐ . ܩ̈ܢܘܡܐ ܡܬܚܙܝܢ
ܟܕ ܢܬܦܠܛܘܢ ܠܬܪܝܢ . ܘܗܕܪܐ ܠܗܘܢ ܕܐܠܗܘܬܐ ܠܗܘܢ ܘܗܕܪܐ ܟܝܢܐ
ܕܐܠܗܘܬܐ ܟܝܢܝܬܐ . ܒܗܘ ܕܩ̈ܢܘܡܐ ܐܝܟ . ܐܬܟܢܫܢ . ܘܡܬܝܕܥܘܢ ܗܕܪܐ ܡܬܚܙܝܢ
ܕܐܬܬܠܝܢ ܩ̈ܢܘܡܐ . ܗܘ ܟܕ ܠܗܘܢ ܘܗܘ ܟܝܢܝܐ . ܦܪ̈ܝܫܝܢ . ܒܗܢ ܩ̈ܢܘܡܐ
ܡܬܚܙܝܢ .

[II.iii.7]

ܕܟܝܢܐ . ܗܪܟܐ ܐܡܪ ܡܫܬܡܥܐ ܕܐܢܫܝܐ ܩܢܘܡܐ ܚܕܢܝܐ . ܕܚܠܝܡ ܗܘܐ ܡ̈ܠܐ
ܕܡܬܩܢܝܢܘܢ ܡܬܝ̈ܢ ܦܪ̈ܫܝܢ . ܦܢܝܐ ܠܗ ܚܕܢܝܐ ܠܗ . ܘܗܕܪܐ ܒܪ ܟܝܢܐ .
ܒܗܘ ܕܐܢܫܝܐ ܡܢܝܢ . ܟܕ ܩܢܘܡܐ ܗܘ ܢܗܘܐ . ܘܗܕܪܐ ܡܬܚܙܝܐ ܕܠܗ ܢܣܒ .
ܐܘܪܓܢܐ ܕܝܠܗ ܠܡܫܡܫܢܘܬܐ ܐܠܐ ܕܢܣܒܐ . ܘܓܒܝܐ . ܐܘܪܓܢܐ ܕܝܠܗ
ܠܐܘܡܢܘܬܐ . ‹ܘܓܒܝܐ ܐܘܪܓܢܐ ܕܝܠܗ ܠܡܫܡܫܢܘܬܐ› . ܟܕ ܡܬܚܙܝܐ ܒܗܘ
ܩܢܘܡܐ ܠܐܘܪܓܢܐ . ܩܢܘܡܗ ܡܢ ܐܘܪܓܢܐ . ܘܐܘܪܓܢܐ ܡܢ ܩܢܘܡܐ ܐܠܐ
ܕܢܣܒܐ . ܐܠܐ ܒܪܬܐ . ܕܐܝܬ ܡܬܚܙܝܢ ܠܗ ܡܢܝܢ . ܐܠܐ ܡܢ ܚܘܒܢܐ ܒܐܘܡܢܘܬܐ
ܡܬܬܠܝ . ܘܗܘܝܬ ܒܓ̈ܢܐ ܗܠܝܢ . ܐܝܟܢܐ ܕܗܘܬܐ ܡܢ ܩܢܘܡܐ ܕܡܬܚܙܝܐ ܢܣܒܐ .
ܘܐܝܟܢܐ ܐܘܪܓܢܐ ܢܗܘܐ . ܘܒܬܪܗ ܩ̈ܢܘܡܝܢ . ܘܐܢܬ ܐܡܪܬ ܕܓܒܝܐ ܚܘܒܢܐ
ܡܬܚܫܒ ܒܩܢܘܡܝܢ . ܚܠܦ ܟܠܗ . ܗܘ ܕܐܝܬܘܗܝ ܒܟܠ ܠܗ ܢܬܚܫܒ . ܐܡܪ .
ܘܟܠ ܗܘ ܡܢܐ ܕܠܐ ܡܬܚܙܝܐ ܠܗ ܠܘܬܟܠ ܐܢܫܐ .

FMPLIV: **II.iii.5**: 2 ܘܚܠܦܬܐ Λ (ܕ supra lin. V) || 3 ܡܢܗܘܢ MΛV || 4 ܕܢܣܒܬ F || ܡܬܚܙܝܐ ΛV || 6 post ܡܬܚܙܝܐ: fort. ܠܗܘܢ supplendum (cf. v. 4 supra) || 7 ܡܩ̈ܦܠܢܐ Λ || **II.iii.6**: 1 ܩ̈ܢܘܡܐ V || ܡܬܚܙ̈ܝܢ PL || 4 ܐܬܟܢܫܢ 1[corr] || 5 ܦܪ̈ܝܫܝܢ V || ܡܬܚܙܝܢ PLV || **II.iii.7**: 1 ܐܡܪ: ܐܡܪ F || ܕܐܢܫܝܐ PlM || ܗܘܐ: om. Λ || 2 ܡܬܝܢ ΛV || ܦܢܝܐ Λ || 3 ܩܢܘܡܐ: ܩܢܘܡܗ l || ܕܠܗ: ܕܠܗ V || 4 ܐܘܪܓܢܐ Pl || 4 ܐܘܪܓܢܐ ܕܝܠܗ ... ܠܡܫܡܫܢܘܬܐ: om. V || ܘܓܒܝܐ: ܘܓܒܝܐ MΛ (om. V) || 5 ܟܕ: ܟܕ Λ || 7 ܕܐܝܬ ܡܬܚܙܝܢ F || 8 ܗܠܝܢ: ܠܗܝܢ M || 9 ܘܠܐܘܪܓܢܐ l || ܒܬܪܗܘܢ ΛV || ܐܡܪܬ ܕܓܒܝܐ: ܓܒܝܐ ܗܘ M ||

[II.iii.5.]
Fifth [theory]: In the teaching of the ancients, the difference in the position of the two clouds is given as the reason for the variety of the threefold colours of the rainbow. From the upper [cloud], because it is close to the sun, vision is reflected strongly towards the sun, so that its colour appears a clear red, because a bright object appears red when seen amid a black [thing] or through a black medium. From the lower, on the other hand, because it is far from the sun, vision is reflected weakly towards the sun, so that it appears as red tending towards blackness, and such a colour is purple. From these two outer colours, a green colour is generated and appears between them.

[II.iii.6.]
Sixth [theory]: It is also said: When two rainbows are seen one inside another, the outer circle of the inner rainbow and the inner circle of the outer rainbow appear red on account of their juxtaposition with each other. The outer circle of the outer rainbow and the inner circle of the inner rainbow, because they are the extremities, are purple. The two middle circles of the two rainbows, the outer and the inner, are green, since they are compounded from the extremities.

[II.iii.7.]
Seventh [theory]: Here, the best of the moderns, the Princely Doctor [Ibn Sīnā] says: All these words which our friends the Peripatetics have put forward do not convince me. For, seeing that the outer circle is pure red because it is close to the light, the middle circle, which adjoins it ought to be a purple tending towards redness, and the inner [circle] a purple tending towards blackness. A purple colour tending towards redness, since it is intermediate between red and purple, ought to be redder than purple and more purple than red, not green, which is not related to either of them but is generated from yellow and black. The separation of these colours, whereby a certain band resembling red and another similar to purple arises, with green in between them and sometimes with a yellow colour accompanying them, requires an explanation, which - (Ibn Sīnā) says - "I have not yet found out and I am not convinced by what I have read".

[II.iii.8]

ܕܬܡܝܗܐ . ܡܢ ܕܝܢ ܡܘܣܪܐ ܗܘܢܐ ܡܫܬܐ ܒܠܠܝܐ . ܚܙܐ ܠܚܟܡܐ ܒܪܫ . ܘܡܠܟ ܚܐܪܐܝܬ . ܒܗܝ ܕܡܫܬܐ ܥܠ ܢܘܗܪܐ ܚܠܦ ܐܢ̈ܫܐ ܡܣܝܒܐ ܒܣܘܟܡܗ . ܘܐܠܐ ܠܬܘܬܗ ܕܒܢܝܐ ܡܫܠܛܐ ܠܐ ܫܠܘ ܡܢܗ ܡܠܟܘܬܗ . ܘܒܝܕ ܗܕܐ ܠܚ̈ܟܝܡܐ ܒܡܘܪܐ ܒܡܫܬܝܢܘܬܐ ܠܗ ܠܒܝܢܬܐ ܡܬܬܕܪܝܢ ܘܡܬܬܩܢܝܢ . ܒܕܠܐ ܫܠܘ ܠܬܘܬܗܘܢ ܕܡܬܘܬܗܘܢ . ܡܫܟܚܐ ܕܝܢ ܬܘܒ ܡܫܬܐ ܒܣܘܟܡܗ ܐܦ ܫܠ ܒܢܝܐ ܕܡܠܟ ܡܕܝܢܬܐ ܘܡܫܠܛܐ . ܘܒܣܘܟܡܐ ܡܠܟܘܬܐ ܗܢܝܐ . ܐܡܪܝܢ ܕܬܘܣܦܬ ܠܡܫܟܚܘ ܠܘܬ ܣܘܓܐܐ ܠܡܫܟܚܐ ܕܢܘܗܪܐ ܒܡܫܬܐ . ܘܫܠܝܐ ܕܠܘ ܒܐܡܝܢܘ . ܐܠܐ ܒܡܫܠܡܢܘܬܗ ܒܠܝܢܘܕ ܗܘܐ ܢܗܝܪ ܢܘܗܪܐ ܡܫܬܐ . ܘܒܗܢܐ ܕܒܢܝܐ ܫܠܝܛܘܬܐ ܒܒܝܬ ܡܫܠܡܢܘܬܐ ܕܡܫܬܐ ܚܙܐ ܠܚܟܡܐ ܗܘܐ . ܘܒܠܗܘܢ ܡܫܬܐ ܡܢ ܡܘܣܪܐ ܒܠܠܝܐ ܡܠܟ ܚܐܪܐܝܬ ܗܘܐ .

[tit.]

ܦܘܣܩܐ ܪܒܝܥܝܐ . ܡܛܠ ܟ̈ܝܢܬ ܫܡܫܐ ܘܣܗܪܐ . ܬܐܘܪܝܐ ܬܠܬ .

[II.iv.1]

ܡܕܡܢܝܬܐ . ܗܠܝܢ ܕܝܢ ܡܫܟܚܢܘܬܐ ܐܝܢ . ܕܐܝܟ ܫ̈ܡܫܐ ܒܟܝܢܐ ܕܬܪܝܨܐ ܘܡܩܒܠܐ ܒܡܘܪܒܐ ܕܫܡܫܐ ܡܬܚܙܝܢ . ܘܗܟܢ ܗܕܐ ܚܝܠܐ ܒܕܡܘܬܐ ܕܡܘܣܪܐ ܠܢܘܗܪ ܫܡܫܐ ܒܢܬܗ ܡܩܒܠܐ . ܘܒܬܪ ܢܘܗܪܐ ܠܡܫܟܚܐ ܗܘܐ ܕܐܡܣܒܪ ܫܡܫܐ ܠܘܬ ܣܦܪ̈ܐ ܡܢܒܥܐ . ܒܗܝ ܕܡܫܬܝܢܐ ܒܗܘܐ ܒܬܪ ܒܘܢܐ ܐܦ ܠܐܣܡܟܬܐ ܡܢܘܪܐ . ܘܠܡܫܟܚܐ ܗܘܐ ܠܐܣܘܪܐ . ܒܕ ܒܡܫܬܢܐ ܡܣܬܟܠ . ܘܒܐܣܡܟܬܐ ܚܕܪܢܝܐ ܡܣܬܡܟܢ . ܐܝܟ ܫܒܪܐ ܕܒܫ̈ܡܫܐ ܕܠܝܠܝܐ ܒܐܝܪ . ܘܠܚܘܒܬ ܢܘܪܐ ܡܬܚܫܟܐ . ܘܢܘܪܐ ܒܗ ܚܪܝܐ . ܐܝܟ ܫܡܫܐ ܡܬܚܙܝܢܐ . ܘܐܦ ܠܡܫܟܚܐ ܦܘܠܓ ܠܒܪܐ ܢܩܐ . ܘܡܬܐܪܢܝܬ ܡܥܒܕܢܐ ܒܦ̈ܪܨܐ ܡܬܚ̈ܘܝܢ . ܐܬܡܫܡܫܐ ܘܠܚܘܠܦܐ ܡܣܬܒܪܐ ܫܡܫܐ . ܘܒܬܪ ܡܫܠܡܢܘܬܐ ܚܕܪܢܝܬ ܒܐ̈ܪܥܐ .

[II.iv.2]

ܕܬܪ̈ܬܝܢ . ܟ̈ܝܢܬ ܫܡܫܐ ܥܠ ܡܠܦ̈ܢܐ ܡܣܬܒܪܢ . ܒܗܝ ܕܥܠ ܦܘܪܢܣܘܬ ܥܠܡܐ ܬ̈ܠܝܬܐ ܘܒܪ̈ܝܐ . ܕܡܬܬܩܢܘܬ ܫܡܫܐ ܫܪܝܐ ܡܢ ܦܘܫܩܘܢ ܘܫܪܝܢܘܢ . ܡܣܬܒܪܢ . ܘܗܟܢ ܕܐܡܪܝܢ ܐܢ̈ܫܝܢ . ܒܕ ܡܢ ܠܥܠ ܫܡܫܐ ܗܘ̈ܝܢ . ܟܝ̈ܢܢ ܡܣܬܒܪܢ ܥܠ ܡܠܦ̈ܢܐ . ܘܡܢܐ ܕܠܫܡܫܐ ܡܢܗ ܗܘ̈ܝܢ . ܗܟܝܠ . ܠܐ ܥܒܝܕܐ ܠܡܣܒܠܐ . ܒܗܝ ܕܫ̈ܡܫܐ ܕܒܗܝܢ ܗܘ̈ܝܢ ܡܫܟܚܢܘܬܐ ܗܠܝܢ . ܠܐ ܦ̈ܝܫܢ ܡܢܗ ܦܘܚܐ . ܐܝܟܢܐ ܕܒܫܡܫܐ ܕܫܡܫܐ ܢܬܦܪܫ ܡܢ ܬ̈ܕܪܐ ܦܠܛܘܗܝ . ܐܠܐ ܐܝܟܢܐ ܕܒܟܘܟ̈ܒܐ ܫܡܫܐ ܡܬܚܙܝܢܐ ܠܝ . ܒܕ ܐܣܛܘܟܣܐ ܡܬܒܪܝܐ ܡܬܚܕܝܢ . ܠܬܚܘܒܐ ܡܢ ܫܡܫܐ ܗܘܝܐ . ܘܡܬܚܙܝܐ ܐܦ ܬܫܒܘܚܬܐ . ܒܐܘܟܠܐ ܕܫܡܫܐ ܕܝܘܡܢ . ܒܪ̈ܝܫܝܬܐ ܗܘܝܐ .

FMPLIV: **II.iii.8**: 1 ܘܡܠܟ ΛV || 2 ܒܣܘܟܡܗ MΛ (sed ܒܣܘܟܡܗ PL) || 3 ܠܒܢܝܐ V || ܡܢܗ ΛV || ܡܫܠܛܢܗ F; ܡܫܠܛܢܐ ܡܫܠܛܢ ΛV || 4-5 ܠܬܘܬܗܘܢ ܕܡܬܘܬܐ F; ܠܬܘܬܗܘܢ ܕܡܬܘܬܗܘܢ M || 7 ܠܡܫܟܚܐ: ܠܡܫܟܚܐ Λ || ܠܒܝܢܬܗܘܢ: ܠܒܝܢܬܗܘܢ Λ || 8 ܗܘܐ ܢܗܝܪ: ܢܗܝܪܐ Λ; ܢܗܝܪ V || 9 ܡܠܟ: om. ΛV || **tit.** ܪܒܝܥܝܐ: ܐܪܒܥܝܬܝܐ LV || ܟܝܢܬ M || **II.iv.1**: 3 ܢܘܗܪܗ ΛV || ܒܡܫܬܢܐ: ܒܡܫܬܢܐ V || 7 ܐܝܟܢܐ V || **II.iv.2**: 1 ܡܣܬܒܪܝܢ V || 2 ܘܫܪܝܢܘܢ M; ܘܫܪܝܢܘܢ V || 3 ܠܥܠܐ PL || 4 ܕܠܫܡܫܐ F || 4-5 ܗܟܝܠ . . . ܗܘ̈ܝܢ: om. Λ || 6 ܕܒܟܘܟܒܐ L || 7 ܫܡܫܐ ܕܒܟܘܟܒܐ: add. ܗܘܝܐ V || ܡܬܚܙܝܢܐ Λ || 9 ܕܝܘܡܢ L ||

[II.iii.8.]
Eighth [theory]: From the moon the rainbow is formed at night, but seldom and very rarely, because the rainbow requires light with plenty of rays for its formation. Otherwise, the illusion (of the light) will not be reflected from it towards the smooth cloud. - Thus, images of pale objects appear indistinctly in mirrors, because their likenesses are not reflected towards them. - For its Formation, the rainbow also requires a cloud which is very compact and smooth and has a high level of suitability, so that it is capable of conveying to the visual organ the illusion of the pale luminary. It is obvious that the moon is not always very bright, but only when it is full, and the occurrence of a suitable cloud at the time of full moon happens seldom. For this reason, the rainbow is formed very rarely from the moon at night.

Fourth Section: On Mock Suns and Lances. Three theories.

[II.iv.1.]
First [theory]: These [mock suns] are illusions which appear like suns on a compact and smooth cloud near the sun. This cloud, like the moon, receives the light of the sun in itself and conveys together with the light also an illusion of the shape of the sun towards vision, because a large mirror shows together with the colour also the shape. Furthermore, when viscous vapour rises to the αἰθήρ, is formed into a circular shape like other moist bodies in air, reaches the sphere of fire and catches fire, it appears like the sun. If the vapour is very compressed and its particles are mixed firmly with each other, the mock sun persists day and night and goes round in circles with the celestial sphere.

[II.iv.2.]
Second [theory]: Mock suns indicate [coming of] rain because they indicate an abundance of moist and thick exhalations which the heat of the sun is unable to dissipate and dissolve. The view of some that they indicate rain less when they occur to the north of the sun and more when they occur to the south of it is unacceptable, because the clouds in which these illusions occur are not so very far from us that their directions can be distinguished from each other in relation to the sun, but the one which appears to us [to be] to the north of the sun is found to the south of the sun when we move by a few stades and, in the same way, the one in the south becomes one in the north with a small change of our position.

[II.iv.3]

ܕܬܠܬ. ܠܬܪ̈ܬܐ ܫܢܝ̈ܢ. ܕܐܦ ܥܒܕ̈ܐ ܘܥܒܕ̈ܘܬܐ ܡܬܚܫܒܢ. ܐܦ ܗܢܘܢ ܡܫܬܟܚܐ
ܕܪ̈ܚܡܝܢ ܠܡܫܘܚܬܐ ܒܓܢܣܗܘܢ ܐܝܬܝܗܘܢ. ܒܪܡ ܕܝܢ ܟܠ ܚܕ ܫܡܫܐ ܡܢ ܢܗܝܪܐ
ܘܡܢ ܡܡܠܠܐ ܗܘܝܢ. ܬܪ̈ܝܢܐ ܕܝܢ ܘܠܐ ܣܘܕܪ̈ܢܐ ܡܬܚܫܒܝܢ. ܐܦ ܡܬܡܠܠܝܢ ܕܒܩ̈ܘܦܐ
ܐܝܟ ܐܚܪ̈ܢܐ ܡܢ ܣܘܕܪ̈ܐ ܦܘܪܫܢܐ. ܘܡܠܘܫܐ ܠܐ ܡܬܬܪܝܨ ܕܒܥܐ
ܒܡܫܘܚܬܗܘܢ. ܐܠܐ ܐܝܟ ܡܘܪ̈ܓܠܐ ܬܪ̈ܝܢܝܬܐ ܫܪܐ ܠܗܘܢ. ܐܦ ܡܬܡܠܠܝܢ ܕܡܘܣܝܩܐ
ܕܥܘܢܝܐ ܠܡܫܘ̈ܚܬܐ ܬܪ̈ܝܢܝܬܐ ܡܢܗܘܢ. ܒܐܝܢܐ ܗܘ ܕܐܠܝܨܘܬܗ ܕܡܫܘܚܬܐ ܠܐ
ܬܫܡܫܬܐ ܠܡܦܠܚܘ. ܐܠܐ ܗܘ ܒܚܕܐ ܚܠܝܠܐ ܐܦ ܒܡܫܘܚܬܐ. ܘܠܬܪ̈ܬܐ ܒܚܕ ܠܡܐܪܐ
ܒܝܢ̈ܝ ܗܘܝ. ܐܠܐ ܒܡܫܘܚܬܐ ܘܐܝܬܝܐܝܬ ܒܡܢܝܢܐ. ܒܗ̇ ܕܒܠܚܘܕܝܗ̇ ܐ
ܡܢ ܣܦܝܩܘܬ ܡܫܡܫܐ ܠܘܬ ܡܬܡܫܚܢܐ ܒܟܠ ܡܢܝܢܐ ܬܫܡܫܬܐ.

FMPLIV: **II.iv.3**: 1 ܫܢܝ̈ܢ: ܫܢܝܢ V|| 3 ܬܪ̈ܝܢܬܐ Λ || ܐܦ ܡܬܡܠܠܝܢ: ܘܡܬܡܠܠܝܢ Λ || 5 ܒܡܫܘ̈ܚܬܗܘܢ FM || ܬܪ̈ܝܢܝܬܐ P || 6 ܠܡܫܘ̈ܚܬܐ: ܠܡܫܘܚܬܐ PL1 || ܬܪ̈ܝܢܝܬܐ add. ܠܐ Pl || 7 ܒܚܕܗ ΛV ||

[II.iv.3.]
Third [theory]: Those lances, which are also called rods and shoots, are also illusions which resemble the rainbow in their colour. They occur, however, to the side of the sun, to the right and to the left. They appear straight and not circular: either because they are small segments of large circles so that the sense does not perceive their curvature but sees them as straight lines, or because the position of the viewer makes the curved object appear straight, i.e. when he stands opposite not its curved face, but its concave or convex side. Lances occur rarely at midday, but [usually occur] in the morning and evening, especially in the evening, because at midday a thin cloud is quickly dissipated by the heat of the sun.

[tit.]

ܡܐܡܪܐ ܬܠܝܬܝܐ . ܡܛܠ ܪܘܚܐ . ܘܒܗ ܐܝܬ ܦܣ̈ܘܩܐ ܐܪܒܥܐ .

ܦܣܘܩܐ ܩܕܡܝܐ . ܡܛܠ ܡܘܠܕܐ ܕܪܘܚܐ . ܬܐܘܪܝܐܣ ܐܪܒܥ .

[III.i.1]

ܩܕܡܝܬܐ . ܗܕܝܢ ܐܢܘܢ ܚܠܝ̈ܬܐ ܕܡܫܬܡܠܝܢ ܠܐܐܪ . ܚܕ ܗܘ ܡܢ ܡܛܪ̈ܐ . ܘܐܝܬܘܗܝ ܐܝܟ ܕܐܡܒܝܪܝܢ ܬܠܝܩܐ ܘܠܡܥܝܢܐ . ܘܡܢܗ ܡܬܝܠܕܝܢ ܬܢܢܐ ܘܥܪ̈ܦܠܐ ܘܡܛܪ̈ܐ . ܘܗܠܝܢ ܕܠܗܘܢ ܪܓܝܢ . ܐܣܬܢܐ ܕܝܢ ܡܢ ܐܪܥܐ . ܘܐܝܬܘܗܝ ܢܒܥܐ ܘܦ̈ܝܢܢܐ . ܘܡܢܗ ܡܬܝܠܕܝܢ ܪܘܚܐ ܘܥܠܥܠܐ ܘܦܪܚܐ . ܘܗܠܝܢ ܕܠܗܘܢ ܪܓܝܢ . ܡܬܝܠܕܐ ܕܝܢ ܪܘܚܐ . ܟܕ ܦܠܚܝܢ ܬ̈ܢܢܐ . ܘܠܘܬ ܟܪܘܬܐ ܡܪܬܚܐ ܡܬܟܢܫܝܢ . ܘܐܟܢ ܗܟܢ ܦܐܪܝܢ . ܢܦܪܝܢ . ܘܡܣܬܗܟܐ ܠܐܦ̈ܝ ܠܗܘܬ ܢܦܫܗ . ܘܠܐܐܪ ܡܟܠܠܝܢ . ܘܪܘܚܐ ܚܕܒܝܢ . ܘܐܟܢ ܗܟܢ ܠܐ ܦܐܪܝܢ . ܡܬܟܠܝܢ . ܘܠܘܬ ܐܣܟܡ ܢܗܪܐ . ܕܡܒܘܥܐ ܕܡܫܬܘܠܛܐ ܡܬܦܪܫܢܐ . ܡܬܟܢܫܝܢ . ܘܟܕ ܡܢ ܙܘܥܐ ܗܘ ܫܘܪܝܐ ܘܫܪܝܐ ܕܠܘܬܐ ܡܬܟܢܫ ܡܬܠܗܒܝܢ . ܒܥܘܢܝܐ ܠܐܦ̈ܝ ܠܗܘܬ ܡܬܪܬܚܝܢ . ܘܗܘ ܗܟܢܐ ܪܘܚܐ ܚܕܒܝܢ .

[III.i.2]

ܕܬܪܬܝܢ . ܐܢܫܝܢ ܡܢ ܚܟܝ̈ܡܐ . ܟܕ ܡܬܦܫܩܝܢ . ܐܡܪܝܢ . ܐܝܟ ܕܠܐܪܥܐ ܡܬܢܦܫܐ ܐܝܬܘܗܝ ܗܘܠܐ ܕܪܘܚܐ . ܗܘܐ ܕܝܢ ܢܬܚܫܒ . ܢܬܚܫܒܐ ܕܝܢ ܠܐܦ̈ܝ ܠܟܠ ܟܪ̈ܝܐ . ܠܡܢܐ ܪܘܚܐ ܟܠ ܒܡܫܬܘܡܐ ܒܗܘܢܝܗ ܠܐܦ̈ܝ ܠܗܘܬ ܠܐ ܚܠܦ . ܐܠܐ ܐܠܢܦܫܗ ܡܬܦܬܚܐ . ܘܠܘܬܗܘܢ ܐܡܪ ܕܢ . ܕܟܠܗܝܢ ܡܬܢܦܫܐ . ܟܕ ܦܠܚ . ܦܠܓ ܒܐܐܪ ܐܝܬ ܒܠܘܬܐ ܕܫܪܘܝܗ . ܘܟܕ ܠܐ ܡܬܚܫܒ ܕܠܘܬܐ ܐܝܟ . ܒܗܕ ܕܦܪܫ ܢܦܫܗ ܥܡܗ ܒܗܝܢ ܠܘܬܗܘܢ . ܢܦܫ ܒܗܘܢ . ܘܗܘܐ ܚܠܦ . ܘܡܬܬܚܫܒܢܘܬܐ ܐܠܢܦܫܗ ܚܟܝܡ . ܐܝܟ ܕܐܦ ܟܪܝܐ ܕܒܥܝܬܐ ܒܢܦܫܐ ܡܬܚܫܒ . ܐܢ ܠܐ ܚܟܝܡ . ܠܘ ܒܫܪܫܗ ܠܐܦ̈ܝ ܠܗܘܬ . ܐܠܐ ܐܠܢܦܫܗ ܗܘܐ .

[III.i.3]

ܕܬܠܬ . ܐܢܫܐ ܕܫܡܥܘ ܐܢܫܝܢ . ܕܒܠܚܘܕ ܡܛܪܐ ܒܗ ܐܪܥܐ ܢܒܥܐ . ܫܡܥܘ ܐܣܛܘܢ . ܕܒܠܚܘܕ ܪܘܚܐ ܒܗ ܐܪܥܐ ܢܒܥܐ . ܘܡܢ ܠܗܒܬ ܦܠܚܐ . ܘܕܠܐ ܗܒܢܐ . ܐܠܐ ܡܢ ܠܟܠ ܐܝܬܘܗܝ ܚܒܪ ܡܬܟܢܫܐ ܕܪܘܚܐ . ܦܫܕܐ ܗܘ ܕܡܕܡ ܕܡܬܪܒܐ ܠܝ ܪܘܚܐ . ܡܬܦܫܩܝܢ ܚܝܠܐ . ܟܕ ܫܒܩ ܡܕܡܝܬ ܢܦܫ ܡܬܪܗ . ܠܐ ܕܝܢ ܡܬܚܒܫ ܒܟܠܗܐ ܡܬܢܦܫܐ ܟܕ ܦܠܚ . ܒܗܕ ܕܒܠܚܘܕ ܦܠܓ ܡܬܟܒܫ . ܘܠܐ ܐܝܟ ܕܡܢ ܐܘܪܝܐ ܡܕܡ ܡܬܦܠܚܝܢ . ܘܢܦܫܐ ܗܘܐ ܡܢ ܗܘ ܕܟܕ ܡܬܚܫܒ ܪܘܚܐ

FMPLIV: **III.i.1**: 2 ܕܐܡܒ̈ܝܪܝܢ P || ܬܠܝܩܐ M || 4 ܘܡܢܗ: ܘܡܢܗܘܢ LV || 5 ܬܢܢ̈ܐ: om. F || 6 ܡܟܠܠܝܢ L || 8 ܡܬܟܢܫܝܢ: ܡܬܟܢܫ Λ || **III.i.2**: 2 ܕܪܘܚܐ M || 3 ܒܡܫܬܘܡܗܘܢ F || ܒܗܘܢܝܬܐ ΛV || 4 ܐܠܢܦܫܗ: ܐܡܢܦܫܗ MV || 5 ܒܠܘܬܐ ܐܝܬ: ܘܒܠܘܬܐ Λ || 6 ܢܦܫܐ: ܢܦܫ MV || ܠܘܬܗܘܢ LM; ܠܘܬܗܘܢ V || 7 ܕܚܟܝܡ: ܡܬܚܟܡ MV || 8 ܐܢ: om. FMV || **III.i.3**: 1 ܕܫܡܥܘ MV || ܕܒܠܚܘܕ V || 2 ܐܣܛܘܢ: ܐܣܛܘܢ MV; ܐܬܐ Λ || 3 ܚܒܪ: om. PL1 || 4 ܚܝܠܐ: ܚܝܠ Λ || ܡܕܡܝܬ Λ (ܐ supra lin. L1) || ܡܕܡܝ P ||

THIRD CHAPTER

ON WINDS

It contains four sections.

First Section: On Generation of Winds. Four theories.

[III.i.1.]

First [theory]: There are two [kinds of] exhalations which rise into the air. One [rises] from water and is moist and vapoury, as we have said. From it are generated clouds, fog, rain and similar things. The other [rises] from earth and is dry and smoky. From it are generated winds, whirlwinds [^{c}*al^{c}ālē*], tornadoes [*qarḥē*] and similar things.

Wind is generated when [bodies of] smoke rise and come to the cold stratum. If they grow cold there, they become heavy, fall violently downwards, cause the air to surge and [thus] cause wind. If they do not grow cold there, they ascend [further] and reach the sphere of fire, which moves with the movement of the celestial sphere. When their ascent is obstructed by that circular and swift movement of the fire, they slide violently downwards and in this way cause wind.

[III.i.2.]

Second [theory]: Some of the ancients, doubting this, say: "If the material of wind is the smoky exhalation, which is hot, and what is hot rises upwards, why are winds not reflected vertically downwards retracing their paths, but move obliquely?" To them the Master [Aristotle] says: When the smoky exhalation rises, it meets the air or the fire which is around it. When it cannot penetrate them because their swift movement prevents their penetration, it strikes them and bounces back, and makes an oblique movement, just as an arrow that is shot at the ceiling of a house, if it does not stick, is reflected not vertically downwards but obliquely.

[III.i.3.]

Third [theory]: Just as some thought that the entirety of water was confined inside the earth, others thought that the entirety of wind was confined inside the earth and [that] it rose from below. That this is not so, but the winds begin to blow from above, is proven by the fact that clouds move before we become aware of the wind, because they are affected by it first. Nor do we perceive the smoky exhalation as it rises, because it gathers little by little and not as if it were discharged from some reservoir. This is known from the fact that when winds

ܡܫܡܠܝܘܬܐ ܢܓܕܢ . ܟܠܗܝܢ ܕܝܢ ܒܚܫܝܫܘܬܐ . ܘܐܦ ܗܠܝܢ ܕܐܠܝܨܝܢ ܥܠܘܗܝ ܢܣܒܢܢ
ܗܘܬ ܒܐܪܥܐ ܒܡܫܒܚܢܘܬܐ . ܗܢܘ ܗܘܐ ܕܒܚܫܝܫܘܬܐ ܢܓܕܝܢ . ܡ ܠܡܫܬܡܠܝܘ .
ܐܝܟ ܕܐܡܪ ܢܒܝܐ ܕܩܕܡ ܦܫܝܩ . ܗܐ ܡܬܦܬܚܝܢ . ܘܐܢܫܝܢ ܕܦܘܪܫܢܐ ܒܐܘܪܚܬܐ
ܕܠܒܝܫܢ ܦܘܪܫܢܐ ܠܐ ܐܢܬܘܢ . ܡܛܠ ܕܣܒܝܒܐ ܥܠܝܗ ܐܘܪܚܐ . ܡܛܝܒܐ ܠܗ
ܘܡܩܪܒܐ . ܐܦ ܠܐ ܥܠܝܗ ܗܘܠܐ ܕܦܘܪܫܢܐ ܥܠܝܗ ܐܘܪܚܐ ܣܒܝܒܐ ܗܘܬ . ܒܫܐܠܐ
ܪܒܐ ܦܠܓܘܬܗ . ܘܒܚܠܝܘܬܗ ܢܦܫܬ .

[III.i.4]

ܕܐܪ̈ܟܒ . ܗܘܠܐ ܕܦܘܪܫܢܐ ܠܗ ܡܢ ܗܘܠܐ ܕܡܠ̈ܦܢܐ ܐܢܬܘܢ . ܟܡܐ ܕܐܝܟ
ܕܒܡܣܟܢܐ ܒܡܫܘܚܬܐ ܡܫܚܢܐܝܬ ܡܠܦܢܐ . ܦܪ̈ܫܝ ܦܘܪܫܢܐ . ܘܒܡܫܘܚܬܐ ܡܫܚܢܐܝܬ
ܦܘܪܫܢܐ . ܦܪ̈ܫܝ ܡܠ̈ܦܢܐ . ܘܐܦ ܚܕܐ ܗܘܬ ܗܘܠܐ ܠܗܝܢ ܠܘܬܗܘܢ . ܠܐ ܐܬܕܠܠܘܗ
ܠܫ̈ܪܪܐ . ܚܕܝܢ ܕܝܢ ܡܠ̈ܦܢܐ ܠܦܘܪܫܢܐ . ܟܕ ܕܡܕܒܪܝܢ ܠܡܠ̈ܦܢܐ ܦܪ̈ܝܫܐ ܕܦܫܩܢ .
ܘܡܟܠܠ ܦܘܪܫܢܐ ܠܡܠ̈ܦܢܐ . ܟܕ ܕܡܫܡܠܝܢ ܘܡܦܩܕܝܢ ܘܡܕܒ̈ܪܝܢ ܠܠܡܝܕܐ . ܘܐܝܬ
ܐܝܟ ܕܡܫܬ̈ܪܐ ܡܠ̈ܦܢܐ ܠܗܘܢ ܗܘܐ ܕܦܘܪܫܢܐ . ܟܕ ܕܡܬܪ̈ܝܢ ܘܡܬܕܒܪܝܢ ܠܗ
ܠܐܪܚܐ . ܘܒܡܫܚܬܐ ܠܗܘܢ ܡܠܦܢܐ ܕܫܦ̈ܝܪܐ ܚܙܝ ܠܗ . ܐܝܟ ܕܐܡܪ ܦܘܠܘܣ
ܬ̈ܠܡܝܕܐ ܗܘ̈ܝܢ ܢܨܝ̈ܚܐ ܒܫܡܫܐ . ܘܦܘܪܫܢܐ ܗܘܐ ܒܡܠ̈ܦܢܐ ܡܠ̈ܦܢܐ . ܟܕ ܕܠܗܘܢ
ܫ̈ܪܪܐ ܢܫܐ ܠܟܪ̈ܝܐ . ܘܡܠܒܪܐ ܠܗܝܢ . ܘܟܕ ܕܡܬܦܫܩܢܐ ܡܬܪ̈ܬܐ ܕܚܝܝܐ
ܒܟܗܝܢ ܡܢ ܢܝܫܗܘܢ ܦܘܪܫܢܐ . ܘܟܕ ܕܡܫܬܟܚܐ ܠܡܠ̈ܦܢܐ ܦܪ̈ܝܫܐ ܡܢ ܠܠܡܝܕܐ .
ܘܦܐܪ ܘܡܬܬܒܟܐ . ܘܢܫܬܐ ܡܠ̈ܦܢܐ .

[tit.]

ܦܣܘܩܐ ܬܠܝܬܝܐ . ܡܛܠ ܦܘܫܩ ܩܘܫܬܐ ܕܡܫܒ̈ܚܝ ܦܘܪܫܢܐ . ܬܫܒܘܚܬܐ ܘܬܘܕܝܬܐ ܠܠܗ .

[III.ii.1]

ܡܫܡܠܝܘܬܐ . ܬܫܒܘܚܬܐ ܚܠܦ ܩܘܫܬܐ ܕܡܫܒܚܝܢ ܠܡ̈ܠܐ ܦܘܪܫܢܐ . ܟܡܐ
ܕܐܘܪ̈ܦܘ ܒܫܘ̈ܠܡܐ ܬܫܒܘܚܬܐ ܡܬܬܫܒܚ . ܬܠܬܐ ܡܫܝ̈ܚܢܐ . ܘܬܠܬܐ
ܡܫ̈ܒܚܢܐ . ܘܬܠܬܐ ܫ̈ܒܚܢܐ . ܘܬܠܬܐ ܬܫ̈ܒܚܢܐ . ܘܗܟ ܕܬܫܒܚܢܐ ܬܫܒܘܚܬܐ ܡܫܒܚܝܢ
ܟܠܗ ܘܠܟܠ ܢܦܫܗ . ܡܫܒܚܢܐ ܡܢ ܡܫܒܚܢܐ ܒܫܘܬܦܘܬܐ ܕܐܝܬ ܐܝܟܢܐ ܘܡܫܡܠܝܘܬܐ .
ܬܫܒܚܢܐ ܕܝܢ ܡܫܒܚܢܐ ܡܫܠܡܢܐ ܕܐܝܬ ܦܘܪܫܢܐ . ܬܫܒܘܚܬܐ ܡܫܒܚܢܐ ܡܫܬܘܕܝܢ ܕܐܝܬ
ܫܒܚܢܐ . ܫܒܚܢܐ ܡܫ̈ܒܚܢܐ ܒܫܘܬܦܘܬܐ ܕܐܝܬ ܐܝܟܢܐ ܘܡܫܡܠܝܘܬܐ . ܡܫܡܠܝܢܐ
ܡܫ̈ܒܚܢܐ ܡܫܠܡܢܐ ܕܐܝܬ ܦܘܪܫܢܐ . ܫܘܬܦܘܬܐ ܡܫ̈ܒܚܢܐ ܡܫܬܘܕܝܢ ܕܐܝܬ ܫܒܚܢܐ .
ܫܒܚܢܐ ܬܫܒܘܚܬܐ ܫܒ̈ܚܢܐ . ܕܝܢ ܢܫܒܚܘܬܐ ܗܝ ܕܟܠܗ ܦܫܘܫ ܡܫܪܪܐ ܠܡܫܪܪܐ
ܠܐܘܪ̈ܦܘ ܬܫܒܚܬ ܗܠܝܢ ܫ̈ܒܚܢܐ ܕܡܫܒܚܝܢ ܠܗ ܡܫܬܒܚܝܢ . ܬܫܒܘܚܬܐ ܬܫܒܘܚܬܐ
ܬܫܒܚܢܐ . ܕܝܢ ܢܫܒܚܘܬܐ ܗܝ ܕܟܠܗ ܦܫܘܫ ܡܫܪܪܐ ܠܡܫܪܪܐ ܠܐܘܪ̈ܦܘ ܠܟܠ ܡܢ
ܗܠܝܢ ܬܫܒܚܢܐ ܕܠܐ ܡܫܬܒܚܐ ܠܗ ܡܫܬܒܚܝܢ .

FMPLIV: **III.i.3**: 8 ܢܣܒܢܢ: ܢܣܒܘܢ MV || 9 ܕܩܕܡ: add. ܕܡܫܒܚܢܐ M || 10 ܐܦ: ܘܐܦ V || **III.i.4**: 1 ܗܘܠܐ sec.: om. Λ || ܐܝܟ V || 2 [ante ܡܠܦܢܐ]: ܡܫܚܢܐܝܬ V || ܡܠܦܢܐ F || 4 ܕܝܢ: om. F || 5 ܠܡܠ̈ܦܢܐ FMPL || ܘܡܦܩܕܝܢ: ܘܡܦܩܕܝܢ MV || 6 ܕܡܠ̈ܦܢܐ M || 8 ܢܨܝ̈ܚܐ: om. Pl || ܡܠ̈ܦܢܐ PL || 9 ܘܡܠܒܪܐ: ܘܡܠܒܪܐ F || 9 ܒܟܗܝܢ P || **III.ii.1**: 1 ܬܫܒܘܚܬܐ Pl || ܟܡܐ: ܟܡ MV || ܒܫܘ̈ܠܡܐ: ܒܫܘ̈ܠܡܐ ܬܫܒܘܚܬ MV || 3 ܘܬܠܬܐ bis V || ܕܬܫܒܚܢܐ: ܕܬܫܒܚܢܐ Λ || ܡܫܒܚܢܐ ΛV || 5 ܡܫܒܚܢܐ bis Λ || 6-7 ܡܫ̈ܒܚܢܐ ter Λ || 9 ܗܠܝܢ ΛV || 11 ܗܠܝܢ: ܗܠܝܢ V ||

start up they blow gently, later strongly. If all their material were really confined in the earth, they ought to start up strongly, then weaken, just as when the flood-gates of the water in a cistern are opened. If a little wind, which is not the entirety of wind, because it is confined inside the earth, causes (the earth) to quake and tremble, [then] if the entirety of the material of wind were confined inside the earth, it would open up (the earth) with great force and go out in its entirety.

[III.i.4.]
Fourth [theory]: The material of wind is not the material of rain, because, usually, in a year abounding in rains winds occur less and in a year abounding in winds rains occur less, but if the material for them both were the same, they would not oppose each other. Rains dismiss winds because they extinguish the smoky exhalation that rises. Winds stop rains because they rarefy, dissipate and scatter vapour. Sometimes rain helps towards the generation of wind because it moistens and waters the earth and makes it suitable for the production of fume, just as moist pieces of wood raise more smoke. Wind too produces rain because it drives clouds towards each other and compacts them, because the coldness of the cloud is made to withdraw inside (the cloud) by the heat of the wind and because (wind) dissipates the smoky exhalation out of the vapour, so that (the vapour) grows cold, thickens and falls as rain.

Second Section: Definition of the Directions of Winds. Three theories.

[III.ii.1.]
First [theory]: The ancients enumerate twelve directions from which winds blow, because the horizon is defined by twelve points, three easterly, three westerly, three northerly and three southerly. Know that eight of them are apparent and are known to all. The first is the equinoctial east which is [at the point of ascent of] the head of Aries and Libra. The second is the aestival east which is the head of Cancer. The third is the hibernal east which is the head of Capricorn. The fourth is the equinoctial west which is the head of Aries and Libra. The fifth is the aestival west which is the head of Cancer. The sixth is the hibernal west which is the point of Capricorn. The seventh is the northerly point which is fixed at the point at which the meridian intersects the horizon below the north pole which is visible to us. The eighth is the southerly point which is fixed at the point at which the meridian intersects the horizon above the south pole which is not visible to us.

[III.ii.2]

ܕܬܬܬܡ . ܐܪܒܥܐ ܬܫܡ̈ܫܬܐ ܐܚܪ̈ܢܐ ܡܢܗܘܢ ܕܐܝܬ ܒܗܘܢ . ܗܟܢܐ ܡܬܬܫܡܫܝܢ .
ܒܗ ܬܠܬܐ ܕܪ̈ܓܐ ܕܝܬܝܪ ܪܒ ܡܢ ܕܪ̈ܓܐ ܕܟܐܢܘܬܐ ܟܠܝܠ ܠܗ . ܘܐܬܚܫܒ ܗܘ
ܕܥܒܪ ܒܐܘܪܚܗܘܢ ܡܢ ܠܥܠ ܘܠܐ ܩܪܒ . ܘܕܪ̈ܓܐ ܐܚܪ̈ܢܐ ܕܝܬܝܪ ܪܒ ܡܢ
ܕܪ̈ܓܐ ܕܟܐܢܘܬܐ ܚܣܝܢ ܡܢܝܢ . ܘܐܬܚܫܒ ܗܘ ܕܥܒܪ ܒܐܘܪܚܗܘܢ ܡܢ ܠܬܚܬ ܘܠܐ
ܕܢܚ . ܒ ܬܠܬܐ ܗܠܝܢ ܕܪ̈ܓܐ ܦܐܠܠܝܐ ܠܕܪ̈ܓܐ ܓܡܝܪ̈ܐ . ܒܗܢܐ ܐܢܐ ܕܚܕ
ܡܢܗܘܢ ܟܚܕ ܠܥܠܐ ܕܕܪ̈ܓܐ ܡܬܚܫܢܐ ܢܥܒܪܘܢ ܠܟܠ ܡܢ ܐܘܪܚܗܘܢ . ܘܐܚܪ̈ܢܐ
ܒܥܒܕܗ ܐܚܪ̈ܢܐ ܡܣܩܬ ܠܟܠ ܡܢ ܐܘܪܚܗܘܢ . ܘܟܠܢܐ ܕܝܢ ܬܬܩܝܡܝܢ ܢܦܫ̈ܬܐ
ܕܥܒܪܐ ܬܪܝܢ ܬܫܡ̈ܫܬܐ ܡܬܬܫܡܫܝܢ ܡܢ ܬܪ̈ܝܗܘܢ ܠܥܠ ܠܒܪ̈ܝܐ . ܘܗܟܢܐ
ܒܥܠܐ ܡܬܒܪܝܢܐ . ܚܕ ܡܢ ܬܪ̈ܝܗܘܢ ܕܪ̈ܓܐ ܦܐܠܠܝܐ ܕܕܪ̈ܓܐ ܓܡܝܪ̈ܐ . ܟܚܕ
ܠܥܠܐ ܕܕܪ̈ܓܐ ܠܗ ܡܬܚܫܢܐ ܥܒܪ ܠܬܚܬ ܡܢ ܐܘܪܚܗܘܢ . ܘܐܚܪ̈ܢܐ ܒܥܒܕܗ
ܐܚܪ̈ܢܐ . ܡܣܩܬ ܠܬܚܬ ܡܢ ܐܘܪܚܗܘܢ . ܘܗܟܢܐ ܝܬܝܪ ܬܬܩܝܡܝܢ ܡܠܝܢ ܢܦܫ̈ܬܐ
ܕܥܒܪܐ ܬܪܝܢ ܬܫܡ̈ܫܬܐ ܐܚܪ̈ܢܐ ܡܬܬܫܡܫܝܢ ܡܢ ܬܪ̈ܝܗܘܢ ܥܒܪ ܡܬܒܪܝܢܐ .

[III.ii.3]

ܕܬܠܬ . ܐܪܒܥ ܦܘܪ̈ܣܐ ܠܟܪܝܗܘܬܐ . ܡܕܒܪܢܘܬܐ ܒܗ ܘܡܬܪܓܡܢܘܬܐ ܘܡܟܪܗܢܘܬܐ
ܘܡܫܡܫܢܘܬܐ . ܗܠܝܢ ܐܝܟ ܐܣܘ̈ܬܐ . ܘܡܬܟܪܗܐ ܐܪܝܟܘܬܐ ܡܢ ܟܝܢܬܗܘܢ ܡܬܦܠܛܝܢ .
ܘܡܢ ܢܦܫܐ ܒܫܘܡ̈ܗܐ ܕܚܠܬܐ ܡܬܐܣܝܢ . ܠܗܢ ܓܝܪ ܕܡܢ ܡܪ̈ܥܐ ܡܚܠܝܐ ܠܢܦܫܐ
ܦܘܝܣܗ ܦܪ̈ܐ . ܘܠܗܢ ܕܡܢ ܡܪ̈ܥܐ ܦܘܬܩܐ ܦܘܪܩܢܗ . ܘܠܗܢ ܕܡܢ ܡܟܪ̈ܗܐ
ܡܚܠܝܐ ܐܦܐܪܘܦܠܝܐܗ . ܘܠܗܢ ܕܡܢ ܡܟܪ̈ܗܐ ܦܘܬܩܐ ܠܒܫܗ . ܘܠܗܢ ܕܒܝܫ̈ܐ
ܡܟܪ̈ܗܐ ܡܚܠܝܐ ܠܟܪ̈ܝܗܐ ܡܫܡܫܝܢ . ܘܠܗܢ ܕܒܝܫ̈ܐ ܡܟܪ̈ܗܐ ܦܘܬܩܐ ܠܡܫܡܫܢܐ
ܦܘܢܩܐܗ . ܘܠܗܢ ܕܒܝܫ̈ܐ ܡܟܪ̈ܗܐ ܡܚܠܝܐ ܠܟܪ̈ܝܗܐ ܬܐܪܘܦܣܐܗ . ܘܠܗܢ
ܕܒܝܫ̈ܐ ܡܟܪ̈ܗܐ ܦܘܬܩܐ ܠܡܫܡܫܢܐ ܠܟܘܠܦܘܣܗ .

FMPLIV: III.ii.2: 2 ܕܪ̈ܓܐ: ܘܕܪ̈ܓܐ M || 3-4 ܡܢ ܠܥܠ . . . ܒܐܘܪܚܗܘܢ: om. Λ || 5 ܦܐܠܠܝܐ V || 6 ܢܥܒܪܘܢ: ܕܥܒܪ FMV || 6-7 ܡܢ ܠܟܠ ܡܢ ܐܘܪܚܗܘܢ ܒܥܒܕܗ ܐܚܪ̈ܢܐ: om. Λ; in marg. V || 7 ܬܬܩܝܡܝܢ L || 8 ܠܥܠ M || 9 ܦܐܠܠܝܐ V || 9 ܕܕܪ̈ܓܐ: ܕܪ̈ܓܐ M || 10 ܕܕܪ̈ܓܐ: ܕܪ̈ܓܐ V || 12 ܡܬܒܪܝܢܐ: ܡܬܒܪܝܢܐ MV || ܥܒܪ ΛV || **III.ii.3:** 2 ܗܠܝܢ: ܗܠܝܢ F; ܗ̈ܠܝܢ M || 3 ܕܚܠܬܐ: ܕܚ̈ܠܬܐ M || ܕܡܢ: ܡܢ V || 3-4 ܡܪ̈ܥܐ ܡܚܠܝܐ ܠܢܦܫܐ ܦܘܝܣܗ ܦܪ̈ܐ ܘܠܗܢ ܕܡܢ: om. Λ || ܦܘܝܣܗ: ܦܘܝ̈ܣܗ V (om. Λ) || 5 ܐܦܐܪܘܦܠܝܐܗ M; ܐܦܐܪ̈ܘܦܠܝܐܗ V || 6 ܡܪ̈ܥܐ ܡܚܠܝܐ ܠܟܪ̈ܝܗܐ ܡܫܡܫܝܢ ܘܠܗܢ ܕܒܝܫ̈ܐ: om. Λ || 7 ܦܘܢܩܐܗ F; ܦܘܢ̈ܩܐܗ V || 8 ܠܟܘܠܦܘܣܗ V; ܠܟܘܠܦܘܣܗ 1 ||

[III.ii.2.]
Second [theory]: The other four points out of the twelve are defined thus: When one imagines a circle that is greater than circles which are always apparent to us [i.e. the arctic circle] - this [circle] is one which touches the horizon from above and does not set - and another circle that is greater than circles which are always concealed from us [antarctic circle] - this [circle] is one which touches the horizon from below and does not rise - then, one imagines two circles parallel to the meridian in such a way that one of them touches one side of the visible circle above the horizon and the other its other side likewise above the horizon, it is apparent that at the two points of contact two points are defined on either side of the north. In the same way, on the southern side, one of the two circles parallel to the meridian touches one side of the invisible circle below the horizon and the other its other side likewise below the horizon, and thus at these two points of contact two other points are defined on either side of the south.

[III.ii.3.]
Third [theory]: The best-known winds are four, namely the east, the west, the north and the south. These are the mothers. The other eight are generated between them and are called by particular names by the Greeks. They call (the wind) which blows from the aestival east Caecias, that from the hibernal east Eurus, that from the aestival west Aparctias, that from the hibernal west Lips, that between aestival east and north Meses, that between hibernal east and south Phoenicias, that between aestival west and north Thrascias, and that between hibernal west and south Libonotus.

[tit.]

ܦܣܘܩܐ ܬܠܝܬܝܐ . ܡܛܠ ܣܘܠܩ ܕܗܘܢܐ ܒܣܘܠܩܐ ܐܬܪ̈ܘܬܐ ܘܪ̈ܢܝܐ .
ܬܐܘܪ̈ܝܐܗ ܫܬ .

[III.iii.1]

ܡܕܝܢܬܐ . ܕܪܘܚܐ ܒܪܘܚܢܘܬܐ ܡܝܬܪܐ . ܒܟܠ ܗܘܦܟܐ ܕܠܦܪܐ ܡܬܠܦܐ
ܚܒܝܪܐ . ܒ ܡܠܟܐ ܠܗܘܢ . ܘܐܝܟ ܡܬܡܨܝܐ ܠܡܬܡܨܝܢܘܬܗ ܠܗܝܢ ܡܫܝܚܐ . ܠܐ
ܕܡܬܝܪܐ ܐܝܟܢܐ ܕܡܫܬܒܚܢܘܬܗ ܥܠ ܐܬܪ̈ܘܬܐ ܫܡܝܢܝܐ ܡܫܬܡܥ . ܘܪܘܚܢܐ
ܡܫܬܘܕܥܬܐ ܢܫܡܬܐ . ܒܗܝ ܕܗܘ ܡܫܬܡܥܢܘܬܗ ܥܠ ܐܬܪ̈ܘܬܐ ܫܡܝܢܝܐ
ܡܬܚܙܝܢܐ ܝܐܝܬ . ܗܘܝܐ ܐܡܪ ܪܒ . ܐܢ ܢܐܡܪ ܐܝܟ . ܐܝܟܢܐ ܠܚܟܡܐ ܪܘܚܢܐ
ܡܢ ܩܕܝܫܬܐ ܗܝ ܢܗܝܪܬܐ ܕܠܒ ܢܗܝܪܐ . ܐܡܪܝܢ ܠܗ ܕܪܘܚܢܐ ܒܪܘܚܢܘܬܐ ܚܠܦܐ
ܠܡܚܝܢܐ ܕܫܐ ܠܗܘܢ . ܘܡܬܝܒܒ ܠܐܪܥܐ . ܘܩܠܝܠܐ ܡܫܬܐܐ ܡܫܡܫܐ . ܐܝܟ
ܕܐܦ ܡܫܡܫܐ ܪܠܒܐ . ܘܒܗܝ ܗܢܘܢ ܕܠܦܘܩܕܢܗ ܕܚܘܫܒܘܬܐ ܪܘܚܢ ܘܩܕܝܫܬܐ .
ܚܠܦܝܢ ܡܫܡܫܢܐ ܩܠܝܠ . ܘܡܛܠ ܠܐ ܒܢܚܬܘܬܗܘܢ ܡܚܝܠ ܕܪܘܚܐ ܕܡܢ ܠܗܘܢ
ܠܥܠ . ܕܐܢܬ ܒܢ ܒܪܘܚܐ ܘܫܬܠܝܢܐ ܠܐܝܕ ܦܝܫܘ . ܘܠܦܘܩܕܢܗ ܐܝܬܝܪܝܐ
ܠܡܫܡܫܢܘܬܗ .

[III.iii.2]

ܕܬܬܠܝ . ܪܘܚܐ ܡܪܚܡܢܘܬܐ ܘܗܝ ܡܬܒܪܝܢܬܐ ܡܬܝܒ ܠܡܬܦܫܩܢܘܬܐ . ܒܝܕ ܚܘܫܒܘܬ
ܪܘܚܐ ܘܡܪܒܝܢܐ ܕܚܘܫܒܐ ܡܢ ܦܠܚܘܬܗܝܢ ܘܠܘܬܗܝܢ . ܒܝܕ ܡܪܚܡܢܘܬܐ ܡܢ
ܠܡܫܬܡܥܢܘܬܐ ܝܠܝܐ . ܡܒܪܒܢܘܬܐ ܕܝܢ ܠܡܪܝܪܘܬܐ . ܡܛܠ ܕܡܪܚܡܢܘܬܐ ܥܠ ܢܚܫܐ
ܕܢܐܨ ܡܢ ܚܫܫܐ ܚܒܝܪܐ . ܒ ܡܠܟܐ ܠܗܘܢ . ܡܒܪܒܢܘܬܐ ܕܝܢ ܥܠ ܢܚܫܐ
ܕܟܠܗ ܒܚܒܪܬܐ ܚܒܠܐ ܚܒܝܪܐ ܘܦܐܪܐ . ܐܝܠܝܢ ܡܠܟܐ ܠܗܘܢ . ܘܡܪܚܡܢܐ
ܐܡܪܝܢ . ܕܠܡܫܬܡܥܢܘܬܐ ܕܢܗܒ ܫܡܫܐ ܠܡܪܝܫܝܐ . ܢܦܠ . ܥܠ ܕܠܥܠ ܡܢ ܐܪܥܐ
ܗܘ . ܘܠܗܝ ܕܠܡܪܒܝܢܐ ܢܗܒ . ܠܐ ܢܦܠ ܥܠ ܕܗܘܝܬ ܐܪܥܐ ܗܢܘܢ . ܘܐܡܪܝܢ
ܕܡܫܐ ܝܕܬ . ܐܝܟ ܝܦܪܐ ܘܪܡܫܐ ܕܡܫܝܚܐ ܘܡܪܒܝܢܐ ܠܐ ܡܫܬܠܡܝܢ ܗܘܘ .
ܘܗܘܐ ܠܗ ܡܚܝܢܐ . ܐܝܟ ܝܦܪܐ ܕܡܫܝܚܐ ܕܪܡܫܐ ܗܘ ܕܡܪܒܝܢܐ . ܘܝܦܪܐ
ܕܡܪܒܝܢܐ ܕܪܡܫܐ ܕܡܫܝܚܐ .

FMPLIV: III.iii.1: 9 ܒܡܫܡܫܐ F || ܘܒܗܝ: ܘܒܗ M || ܒܗܘܢ: om. F || ܕܚܘܫܒܬܐ M || 10 ܠܐܝܕ Λ; ܠܐܝܕܐ M || ܐܝܬܝܪܝܐ Λ || **III.iii.2**: 1 ܡܬܝܒ P; ܡܬܝܒܢܘܬܐ M || 4 ܚܒܝܪ Λ || ܚܒܝܪܐ: add. ܘܦܐܪܐ VM[V] || ܒ: ܐܝܠܝܢ V || ܡܠܟܐ Λ || 5 ܡܠܟܐ ܐܝܠܝܢ ܘܦܐܪܐ ܚܒܝܪ ܚܒܠ MΛ ||

Third Section: On the Variation of Winds according to Places and Seasons. Six theories.

[III.iii.1.]
First [theory]: The north wind is cold because it passes over a multitude of snow-covered mountains before coming to us. If it is able to proceed further in a southerly direction, it will no doubt grow hot through its passage over hot places. The south wind is hot because it reaches us after its passage over hot places. Here the Master [Aristotle] says: If someone says "how can wind blow from that torrid region which is extremely dry?", we say to him that the north wind drives vapoury exhalation there, moistens the earth and causes large quantities of smoke to rise, just like moist wood. Furthermore, because the equator is spacious and wide, large quantities of exhalations rise, but because of their lack of concentration, the winds which blow from there are weak. For the north and south are like the navels of a melon and the isemeric tropic its middle.

[III.iii.2.]
Second [theory]: The east and west winds are more temperate because of the equality of the distance and proximity of the sun to their regions. The east wind, however, is somewhat hotter and the west wind somewhat colder because the east wind passes over dry land which is heated by the sun before coming to us, whereas the west wind enters and passes over the sea inside the habitable world [i.e. the Mediterranean] and grows cold before coming to us. The ancients say that the sun retains the heat which it gives to those in the east because it is above the earth, but it does not retain that (heat) which it gives to those in the west because it is below the earth. We say that this would be true if the morning and evening of those in the east and the west were not different. This, however, is not the case, but [what is] morning for those in the east is evening for those in the west and the morning of those in the west the evening of those in the east.

[III.iii.3]

ܕܬܠܬ. ܒܗܠܝܢ ܡܢ ܚܘܒܐ ܕܠܒܪܝܬ ܢܦܫܐ. ܒܝܕ ܚܐܪܘܬ ܗܘܠܝܐ ܠܦܪܝܫܘܬܐ ܕܡܬܦܫܩܢܐ ܡܢ ܚܟܝܡܘܬܐ. ܐܝܟܢܐ ܕܗ̇ܝ ܕܠܘܬܐ ܕܐܪܝܟܐ ܡܢ ܫܠܡܘܬܐ. ܠܘ ܐܚܪ̈ܢܐ ܕܕܡܝܢܐ ܡܢ ܕܡܝܢܐ ܡܬܦܫܩܝܢ. ܐܠܐ ܐܝܟ ܢܒܘܥܐ ܡܢ ܡܥܝܢܐ. ܐܦ ܠܗܝܢ ܐܪܒܥܐ ܒܡܦܩܢܘܬ ܫܡܝܢܝܘܬܐ. ܢܦܫܐ ܘܠܐ ܡܬܦܠܓܢܐ. ܐܝܟ ܬܠܬܐ ܕܒܫܠܡܘܬܐ. ܡܕܡ ܕܥܠܝܗ̇. ܢܦܫ. ܒܡܘܬܐ ܕܝܢ ܡܬܦܠܓܐ ܢܦܫܐ ܟܕ ܚܐܪܘܬ ܬܠܬܝܗ̇ ܠܒܪܘܬܐ. ܕܝܢ ܫܠܡܘܬܐ. ܘܡܢ ܦܠܓܘܬܐ ܗܘܐ ܕܗ̇ܝ ܠܐܪܐ. ܘܒܐܬܪܐ ܡܢ ܒܬܪ ܢܦܫܐ. ܒܪܡ ܗܘܐ ܕܦܠܓܘܬܐ ܦܪܫ ܕܗ̇ܝ ܠܐܪܐ. ܠܡܥܒܕܘܬܐ ܕܝܢ. ܒܗܠܝܢ ܒܪܒܝܢ ܐܬܝܕܥ ܠܐܪܐ ܡܢ ܢܦܫܘܬܐ. ܒܐܬܪܐ ܕܗܘ ܐܝܟܢ ܕܒܪ ܢܫܐ ܚܘܒܐ ܠܒܪܝܬ ܢܦܫܐ. ܘܒܐܢܫܐ ܕܗܘ ܕܠܐ. ܒܪ ܠܒܬܐ ܕܡܬܚܫܒܝܢ ܠܒܪܝܢܘܬܗ ܦܠܓܝܢ. ܦܠܓܐ.

[III.iii.4]

ܕܐܪܒܥ. ܐܠܗܐ ܡܢ ܠܬܚܬܝܬܐ ܢܦܫܐ. ܒܪ ܫܡܝܢܐ ܠܡܫܬܘܬܗ ܡܠܐ ܢܦܫܐ. ܒܪܡ ܕܝܢ ܠܘ ܡܬܚܙܐ ܕܗܠܝܢ ܫܡܝܢܐ ܠܘܬܗ. ܐܝܟ ܕܟܬܝܒ ܡܬܚܙܝܢܘܬܗ. ܠܗܘ ܡܢ ܕܡܫܬܘܐ ܡܢ ܠܐܪܐ ܘܦܫܝܩܐ ܡܬܦܫܩ. ܘܠܘ ܠܒܝܫܘܬܐ ܕܐܪܓܝܠܝ ܐܠܐ ܒܬܪ ܐܝܢܐ ܠܢܦܠܬܗ ܠܗ ܡܬܒܥܐ. ܘܠܫܡܝܐ ܠܘ ܡܫܬܐ ܕܢܦܠܐ. ܐܠܐ ܡܢ ܒܬܪ ܒܗܝܢ ܢܘܟܪܝ ܠܢܦܫܐ ܢܦܫܐ. ܢܦܠ ܕܝܢ ܗܘ ܠܒܬܘܠܘܬܐ. ܕܡܢ ܐܝܟܢܐ ܠܢܦܫܐ ܕܒܗܘܢܬ ܡܬܦܫܩܢܐ ܢܦܫܐ. ܐܦ ܠܗܝܢ ܘܬܪܝܢ ܢܚܝܬ ܠܡܪܒ ܡܬܦܫܩܢܐ ܗܘ ܡܕܡ ܢܦܫܐ. ܕܡܢ ܐܚܪ̈ܝܢ ܡܢ ܚܘܪܬܐ ܡܬܦܫܩܢܐ. ܐܝܟ ܕܢܦܫܐ ܚܕܕܐ. ܡܢ ܐܚܪ̈ܢܐ ܕܝܢ ܒܬܚܬܝܬܐ. ܐܝܟ ܕܡܢ ܚܘܠܩ ܠܬܚܬܝܬܐ ܠܬܚܬܝܬܐ ܡܬܦܠܓܐ.

[III.iii.5]

ܕܚܡܫ. ܢܦܫܐ ܕܒܗ ܡܬܦܫܩܢܘܬ ܫܡܝܢܐ ܠܢܦܫ. ܫܡܝܢܝܘܬܐ ܡܬܦܫܩܝܢ. ܘܟܕ ܗܘ ܟܕ ܗܘ ܚܕܕܐ ܕܢܦܫܐ ܫܡܝܢܐ. ܘܗܘ ܟܕ ܗܘ ܗܘܠܐ ܕܠܗ ܠܬܠܬܐ. ܒܬܪܬܝܢ ܦܠܓܘܬܐ ܕܦܠܓܘ ܠܒܢܝܢܐ ܘܦܠܓܘ ܬܢܝܢܐ ܐܬܦܫܩܘ. ܢܦܫܐ ܠܒܢܝܢܘܬܐ ܫܡܝܢܝܘܬܐ ܗ̇ܝ. ܘܬܢܝܢܘܬܐ ܫܡܝܢܝܘܬܐ ܠܗ ܗ̇ܝ. ܒܐܝܕܐ ܒܗܠܝܢ ܡܢ ܟܕ ܫܡܝܢܐ ܒܗܢܘܢ ܦܠܓܘ ܠܒܢܝܢܐ ܗܘܐ. ܢܦܫܐ ܠܒܢܝܢܘܬܐ ܠܢܦܫ̈ܝܢ ܐܚܪ̈ܢ. ܒܗܠܝܢ ܕܝܢ ܟܕ ܫܡܝܢܐ ܒܗܘܢ ܬܢܝܢܐ ܗܘܐ. ܢܦܫܐ ܬܢܝܢܘܬܐ ܠܗ ܠܢܦܫ ܠܗܘܢ. ܐܝܟ ܕܠܒܢܝܢܐ ܚܕ. ܘܢܦܫܐ ܬܢܝܢܘܬܐ. ܟܕ ܡܢ

FMPLIV: **III.iii.3**: 2 ܕܠܘܬܐ: ܠܘܬܗ̇ ܕ ΛV || ܐܚܪ̈ܢܐ: ܐܝܟ F; ܐܝܟ ܡܢ ΛV || ܕܕܡܝܢܐ: ܕܡܝܢܐ V || ܐܝܟ sec.: om. V || 6 ܬܠܬܗܘܢ ΛV || 7 ܡܢ: ܡܢ P || ܦܠܓܘܬܐ FM || 8 ܠܡܥܒܕܘܬܐ V || 9 ܘܐܢܫܐ V || 10 ܠܡܬܚܫܒܢܘܬܐ V || ܦܠܓܝܢ: ܦܠܓܝܢ F; ܦܠܓܝܢ ΛV || ܦܠܓܐ: ܦܠܓܝܢ V; om. P || **III.iii.4**: 3 ܘܠܘ ܠܒܝܫܘܬܐ MΛV || ܕܐܪܓܝܠܝ ΛV || 4 ܠܢܦܠܬܗܘ V || 4-5 ܡܢ ܒܬܪ ܒܬܪ P || 7 ܗܘܐ: om. Λ; supra lin. V || 8 ܡܢ prim.: ܘܡܢ ΛV || ܒܬܚܬܝܬܐ: ܒܬܚܬܝܐ F; ܒܬܚܬܝܬܐ L || ܕ ܐܝܟ: om. ΛV || 8-9 [illegible] Λ || 9 [illegible] Λ; [illegible] M[1]; [illegible] VM[V] || **III.iii.5**: 1 ܒܗ̇ Λ || ܘܟܕ ܗܘ: ܘܗܘ ΛV || 2 ܕܢܦܫܐ FΛV || ܕܠܗ M || 7 ܟܕ ܡܢ: ܕܡܢ F ||

[III.iii.3.]
Third [theory]: In summer absence of winds occurs because of the shortage of the exhalationary material which is dissipated by the heat, just as the flame of a lamp is extinguished by a blaze, not as like is destroyed like, but as a lesser by a greater. For the earth too, in abundance of heat, dries up and does not produce smoke, just as a straw in a blaze is burned up before it produces smoke. In winter wind is made scarce because of the shortage of its efficient cause, namely heat, and the exhalation is extinguished by the abundance of the chill. In spring the winds are rare because the exhalation is extinguished beforehand by the chill of winter; in autumn, on the other hand, because the exhalation has not yet been generated because of drought. Absence of wind therefore occurs during [all] four seasons of the year, but it becomes abundant at any time when the causes which counter its hindrances abound.

[III.iii.4.]
Fourth [theory]: Each of the twelve winds blows when the sun inclines in its direction, but not as soon as the sun arrives in its area, because at the beginning of its visit it dissipates what exhalation and smoke it finds, but it can only turn the frozen moistures into exhalation after a while. For this reason, wind blows not as soon as (the sun) arrives, but after twenty days, especially the south wind that blows from dry land which is dissolved with difficulty. In fact, this wind can be delayed for as long as two months. It is called "white wind" by some because it brings clear weather and "egg wind" by others because it makes [hens] pregnant without mingling of cocks with hens.

[III.iii.5.]
Fifth [theory]: Winds that blow with the movement of the sun are called etesian. Although the same efficient cause of wind, the sun, and the same material for it, snow, are found in both the directions of the north pole and of the south pole, the north winds are etesian, whereas the south winds are not etesian. That is to say, in summer when the sun is in the direction of the north pole, north winds blow towards us. In winter, however, when the sun is in the direction of the south pole, south winds do not blow towards us because we are in the north and the south winds, since they blow from far away, are weakened and do

ܪ̈ܘܚܐ ܠܟ̈ܠܗܝܢ . ܡܬܬܚܝ̈ܠܢ . ܘܠܘܬ ܠܐ ܡܬܟܠܝܢ . ܒܫܡܝܐ ܕܪ̈ܘܚܐ ܒܪܘܚܢܝܘܬܐ
ܕܒܗܘܢܫ ܣܓܝܐܝܢ ܡܬܟܠܝܢ . ܘܗܢ ܡܪܟܐ ܠܒܪܝܐ ܕܠܫܡܫܘܬܗܘܢ ܫܘܝܢ ܕܒܥܒܕܘܬܐ
ܗܝ ܐܝܬܘܬܐ ܕܠܡܘܬܒ . ܐܢ ܐܝܬ ܗܘܐ ܒܒܪܝܬܐ . ܕܡܬܠܡܠܝܢܐ ܪ̈ܘܚܐ ܡܢ
ܡܫܬܪܝܬܐ ܒܡܬܪܝܬܐ ܗܘܝܢ . ܒܪܘܚܢܘܬܐ ܕܝܢ ܠܐ ܗܘܝܢ .

[III.iii.6]

ܕܝܠܗ . ܒܡܠܟܐ ܡܢ ܦܠܚܝܐ ܢܘܗܪܢܝܘܬܐ ܒܡܕܒܪܢܐ . ܘܦܪܘܫܐ ܡܢܗܪܐ ܠܟܠܗܘܢ .
ܒܡܠܟܐ ܕܝܢ ܡܬܢܨܚܢܐ ܢܘܗܪܢܝܘܬܐ . ܘܠܡܠܟܐ ܦܠܚܝܐ ܡܫܡܫܐ . ܘܡܣܬܒܪܐ
ܒܡܠܟܐ ܪܘܚܐ ܡܢܗܪܢܝܬܐ ܒܝܬ ܠܟܠ . ܘܡܬܕܒܪܢܘܬܐ ܢܦܫ . ܘܒܫܘܒܚܐ ܡܢ
ܘܪܘܚܐ ܢܘܗܪܢܝܘܬܐ ܒܡܠܟܐ . ܘܠܟܠܗܘܢ ܦܠܚܝܐ ܠܐ ܡܫܡܫܐ . ܒܡܕܒܪܢܐ
ܕܝܢ ܡܬܢܨܚܢܐ ܢܘܗܪܢܝܘܬܐ ܘܠܟܠܗܘܢ ܦܠܚܝܐ ܡܫܡܠܝܐ . ܘܡܣܬܒܪܐ ܒܫܘܒܚܐ
ܪܘܚܐ ܡܬܕܒܪܢܘܬܐ ܒܝܬ ܠܟܠ . ܘܡܬܢܨܚܢܘܬܐ ܢܦܫ . ܘܐܝܟ ܡܠܟܐ ܠܐ ܡܬܚܙܝܢܘܬܐ .
ܘܐܝܟ ܫܘܒܚܐ ܠܐܠܗܐ .

[tit.]

ܦܣܘܩܐ ܕܐܪܒܥܐ . ܡܛܠ ܪܘܚܐ ܚܢܢܝܬܐ ܘܚܠܠܝܬܐ ܘܢܘܪܢܝܐ ܘܙܠܝܩܐ .
ܘܬܐܘܪܝܐ ܘܗ ܐܝܟܢ .

[III.iv.1]

ܡܬܕܒܪܢܘܬܐ . ܪܘܚܐ ܢܘܗܪܢܝܬܐ ܠܪܘܚܐ ܕܡܬܠܒܫܐ ܒܢܝܐ ܟܠ ܕܟܠ ܦܪܝܢ . ܘܡܬܕܒܪܢܐ
ܠܗ ܠܡܠܟܐ . ܐܠܐ ܠܪܘܚܐ ܕܗܘ ܒܢܝܐ ܪܘܚܢܝܬܐ ܒܗ ܡܬܕܒܪܢܘܬܐ ܡܕܒܪܢܐ . ܪܘܚܐ
ܢܘܗܪܢܝܬܐ ܦܪܝܢ . ܘܐܡܝܢܝܢ . ܕܢܦܫ ܡܢ ܥܠܡܐ ܕܪܘܚܐ ܒܫܘܒܚܐ . ܒܗ ܫܡܝܢ ܡܢ ܒܗ
ܒܡܠܟܐ ܕܐܝܟ ܕܪܘܚܐ ܘܗܠܝܢ ܫܡܝܢ . ܗܕܐ ܕܝܢ ܐܝܟ ܕܗܘ ܐܘܢܓܐ ܡܕܒܪ ܡܬܠܒܫ
ܡܬܕܒܪܐ . ܘܫܡܝܢ ܡܢ ܒܗ ܠܗܘ ܟܠܢܐ ܐܦܝܢܐ ܒܐܝܪ ܡܬܕܒܪܢܘܬܐ ܘܟܠܗ
ܫܡܝܢ . ܗܕܐ ܕܝܢ ܒܗ ܡܬܕܒܪܢܐ ܡܢ ܒܢܝܐ ܒܡܠܟܐ ܫܘܪܝܐ . ܘܠܡܫܡܫܐ ܒܫܘܒܚܐ
ܠܐܠܗܐ .

FMPLIV: III.iii.5: 8 ܡ̈ܬܟܠܝܢ PL || 9 ܡ̈ܬܟܠܝܢ P || 10 ܡܢ: om. ΛV || **III.iii.6:** 1 ܦܠܚܝܐ M || 4-5 ܦܠܚܝܐ ܘܠܟܠܗܘܢ .... ܠܐ ܡܫܡܫܐ: om. I || 6 ܡܬܢܨܚܢܘܬܐ V || **tit.** ܪܘܚܐ ܚܢܢܝܬܐ ܘܚܠܠܝܬܐ ܘܢܘܪܢܝܐ F || ܘܙܠܝܩܐ: ܘܙܠܝܩܐ F; ܘܙܠܝܩܐ M (voc. ܘܙܠܝܩܐ) || **III.iv.1:** 1 ܕܟܠ PL || 2 ܡܬܕܒܪܢܘܬܐ ܡܕܒܪܢܐ Λ || 4 ܫܡܝܢ: om. F || 4-5 ܐܝܟ ܕܗܘ ܐܘܢܓܐ ܡܕܒܪ ܡܬܠܒܫ ܡܬܕܒܪܐ: om. Λ || 4 ܐܘܢܓܐ F (om. Λ) || 5 ܘܫܡܝܢ ܡܢ: ܘܕܡܐ ܡܢ F; om. Λ || 6 ܘܠܡܫܡܫܐ Λ ||

not reach us, unlike the north winds which reach us with their strength [intact]. Hence it happens that those in that other inhabited region opposite us, if there is an inhabited region there, wonder why the south winds are etesian but the north are not.

[III.iii.6.]
Sixth [theory]: In summer heat abounds in the east and dries up the exhalation in advance, whereas in the west heat is moderate and causes much vapour to rise. For this reason in summer east winds blow less and west winds more. In winter heat is small in the west and does not cause much exhalation to rise, whereas in the east heat is moderate and produces much exhalation. For this reason in winter west winds blow less and east winds more. Autumn is like summer and spring is like winter.

Fourth Section: On "Cloud Wind", Whirlwind, Tornado and Eddy Wind. Four theories.

[III.iv.1.]
First [theory]: People today call wind which generates cloud "cloud wind". The ancients call " cloud wind" not this but wind which is ejected from a dense cloud as it bursts. They say that it is severer than other winds because (other winds) occur with their material flowing little by little, whereas (the cloud wind) is released suddenly, as from some reservoir. (Other winds) occur when smoky exhalation strikes moving air and is reflected, whereas (the cloud wind) occurs when it is driven out of a cloud by force and for this reason blows violently.

[III.iv.2]

ܕܬܬܬܢܝܚ . ܢܘܗܪܐ ܚܫܘܟܬܐ ܕܢܗܝܪܐ ܘܥܪܦܠܐ ܘܡܬܟܣܝܢܐ ܠܬܚܘܬ ܡܫܬܟܠܐ .
ܐܠܝܨܘ ܕܒܚܝܠܐ ܠܒܣܬܪܬܐ ܢܫܟܚ . ܡܛܠ ܒܚܝܠܘܬܗ ܠܐ ܚܒܝܪܐ ܒܗ . ܘܡܛܠ
ܢܗܝܪܘܬܗ ܕܠܝܬ ܠܗ ܚܠܦܐ . ܐܠܐ ܠܚܒܐ ܡܬܬܚܬܝܐ . ܘܒܗ ܡܢ ܚܒܐ ܗܘܐ
ܒܢܘܗܪܐ ܡܫܝܚܐ ܠܢܦܫܐ . ܒܪܬܗ ܠܒܝܬ ܗܘܝܬܗ ܚܒܪܐ . ܘܗܒܝܢܐ ܡܢ ܢܘܗܪܐ
ܚܫܘܟܬܐ . ܟܠܟܠܗ ܕܚܙܝܘ ܡܢܗ ܗܘܐ . ܘܒܗ ܢܘܗܪܐ ܒܪܒܘܬܐ ܟܠܟܠܗ ܠܗ
ܗܘܐ . ܐܦܠܐ ܒܗ ܡܠܐܟܐ ܢܘܗܪܐ ܚܫܘܟܬܐ . ܘܢܘܗܪܐ ܚܫܘܟܬܐ ܢܗܝܪ ܡܠܝܐ .
ܒܟܒܫ ܟܠ ܡܕܝܢܬܐ ܠܐܝܬܘܬܐ ܡܫܝܚܐ ܕܠܐ ܬܒܥܐ . ܟܠܟܠܗ ܕܝܢ . ܒܗܝ
ܕܢܗܝܪ ܒܚܒܐ . ܟܠ ܡܕܝܢܬܐ ܡܫܬܐܠܬܐ . ܡܫܬܡܫܢܘܬܐ ܒܗ ܕܒܟܠܗܝܢ ܡܠܝܐ
ܢܗܝܪ ܡܢ ܗܘ ܕܒܒܒܐ ܕܐܝܬܘܗ ܡܬܦܪܫܐ .

[III.iv.3]

ܕܬܠܬ . ܗܘܐ ܕܝܢ ܟܠܟܠܗ ܐܦ ܡܢ ܗܘܠܐ ܕܘܝܫܬܐ ܕܡܬܕܒܪܐ ܠܬܚܘܬ . ܘܒܗ
ܢܦܫܐ ܒܐܪܥܐ ܚܠܦܐ . ܘܒܪܘܚܐ ܐܝܬܘܬܐ ܒܪܬܗ ܒܟܠܗܘܢ ܦܓܪܐ . ܘܒܪܬܗܘܬܐ
ܫܡܝܐ . ܘܐܝܬܗ ܟܠܟܠܗ ܕܢܦܫܗ ܐܝܬܘܬܗ ܗܝ ܕܒܒܪܬܗܘܢ . ܐܝܟ ܗܘ ܕܡܢܗܘܢ . ܘܥܠܘ
ܘܢܦܫ ܐܚܝܕܐ . ܘܫܠܡ ܠܚܘܫܒܐ ܕܦܓܪܐ ܒܗܘܢ . ܠܐ ܕܝܢ ܐܦ ܕܘܝܫܐ . ܘܐܝܬܗ
ܟܠܟܠܗ ܕܥܠܘ ܐܝܬܘܬܗ ܗܝ ܕܗܘܠܐ ܒܠܒܘܫ ܢܦܫܬܐ ܠܒܪܬܗܘܢ . ܘܡܬܒܥܝܬ ܕܘܝܫܐ
ܠܚܘܫܒܐ ܒ ܫܠܡ . ܫܪܝ ܒܗܘܢ ܕܝܢ ܗܘܟܢ ܡܬܒܪܝܢܘܬܐ ܡܛܠ ܢܗܝܪܘܬܐ
ܕܒܝܢܗ . ܘܡܬܢܗܪܢ ܘܢܦܩܢ ܡܛܠ ܕܠܒܘܫܗܘܢ . ܒܗܝ ܕܐܠܗ ܦܠܓ ܘܢܦܫܐ ܗܘܐ .
ܠܟ ܠܟ ܥܒܕܗ ܐܝܟܢܢܗ .

[III.iv.4]

ܕܐܪܒܥ . ܗܘܐ ܕܝܢ ܟܠܟܠܗ ܘܒܗ ܡܬܬܠܝܐ ܒܢܝܐ . ܘܒܝܢ ܚܘܠܡܢܘܬܐ
ܕܢܘܪܐ ܗܝܢܘܢ ܕܒܗܘܢ . ܦܠܓܐ ܢܘܗܪܐ . ܡܬܒܪܝܢܝܬ ܢܦܫܐ . ܐܝܟ ܗܝܕܝܢ
ܕܒܬܫܠܡܬ ܦܘܪܐ ܕܡܚܒܢܘܬܗ ܡܬܓܫܡܢ . ܗܘܐ ܗܘܝܬ ܟܠܟܠܗ ܘܡܢ ܦܓܪܐ
ܕܬܬܬܢܝܚ ܢܘܗܪܐ ܡܫܝܚܘܬܐ . ܘܢܗܝܪܐ ܡܬܚܒܫܢܘܗܝ . ܘܒܟܠܐ ܙܕܩܝܢ ܡܢ ܐܕܪܟܐ ܐܠܗܝܐ
ܚܙܝ . ܘܡܢ ܢܦܫܐ ܐܠܗܝܐ ܫܠܡ . ܘܒܗ ܒܟܝܢܘܬܐ ܚܕܝܢܘܬܐ ܦܫܝܛ . ܐܝܟ ܗܝܕܝܢܐ
ܕܠܘܬܗ ܒܐܪܐ ܡܬܚܫܒܐ . ܐܦ ܠܗ ܘܡܢ ܦܓܪܐ ܕܬܬܬܢܝܚ ܢܘܗܪܐ ܕܠܐ ܦܓܪ
ܚܝܠ ܟܠܟܠܗ ܗܘܐ . ܘܢܗܝܪܐ ܡܬܚܒܢܐ .

FMPLIV: **III.iv.2**: 3 ܢܗܝܪܘܬܐ MV || ܠܐ ܚܠܦܐ: **hinc FMLIV** || 4 ܘܗܒܝܢܐ: ܘܗܒܝܪܐ Λ || 5 ܚܙܝܘ Λ || 5 ܟܠܟܠܗ sec.: om. Λ || 8 ܡܫܬܐܠܬܐ: add. ܗܢܘܢ VM[V] || ܕܒܟܠܗܝܢ: ܕܟܠܗܝܢ MV || **III.iv.3**: 1 ܐܦ: ܐܦ F; om. Λ || 2 ܘܒܪܬܗܘܬܐ Λ || 3 ܕܢܦܫܬܐ MΛV || ܕܒܒܪܬܗܘܢ F; ܕܒܒܪܬܗܘܢ L; ܕܒܒܪܬܗܘܢ l; ܕܒܒܪܬܗܘܬܐ V || ܕܡܬܕܒܪ M || 3-4 ܘܢܦܫܬܐ ܥܠܘܐ Λ || 5 ܕܗܘܠܐ: ܕܥܠܘܐ F || 5-6 ܘܡܬܒܥܝܬ ܕܘܝܫܐ ܠܚܘܫܒܐ ܒ ܫܠܡ: om. Λ || 7 ܗܘܐ: om. l || ܥܒܕܗ M || **III.iv.4**: 4 ܡܬܚܒܫܢܘܗܝ V || 6 ܠܘܬܗ: ܠܘܬܗ V|| ܒܐܪܐ: om. MΛV || ܡܢ MΛV || 7 ܢܦܫܐ V || ܡܬܚܒܢܐ: ܡܬܚܒܢܘܗ V ||

[III.iv.2.]
Second [theory]: If a cloud wind, which is heavy and moist and is sliding downwards with force, collides with a thick cloud, because of (the cloud's) density it does not pass through it and because of its own weight does not turn back, but it is carried to the side. When on the side too it collides with a strong wind, it produces a downward spiral. In this way, from a cloud wind a whirlwind, which is severer than it, is generated. Whirlwind does not occur with the north wind, nor cloud wind with rain. Cloud wind is finer [than whirlwind], so that it requires little coldness for it not to occur, whirlwind much coldness because it is denser. For the heat which is in fine exhalation is cooled more easily than that in dense [exhalation].

[III.iv.3.]
Third [theory]: Whirlwind is also generated from windy material which is driven downwards and, colliding with the earth, turns back, meets another wind of the same kind and forms spirals. The sign of a descending whirlwind is that it ascends and descends simultaneously in its spirals like a dancer. It seizes the object it meets but does not push [it] away. The sign of an ascending whirlwind is that only ascent is observed in its spirals. It first pushes away the object, then seizes [it]. The configuration of the spiral perseveres in (the whirlwind) because of the weight of its nature and the solidity of its constitution due to its moisture, since if its constitution were fine its shape would quickly leave it.

[III.iv.4.]
Fourth [theory]: Whirlwind also arises when a cloud is torn and, because of the crookedness of those openings through which it escapes, the wind comes out in a spiral, like hair which is made curly by the tortuosity of the pores through which it grows. Whirlwind also arises from the meeting of two severe winds and [this] is called *qarḥā*. It often uproots trees from the earth and snatches ships from the sea. When it fastens onto a part of a cloud it appears like a sea monster flying through air. Also from the meeting of two winds which are not very strong a whirlwind arises, and [this] is called *zīqā*.

[tit.]

ܡܐܡܪܐ ܪܒܝܥܝܐ. ܡܛܠ ܫܪ̈ܒܐ ܕܠܥܠ ܡܢ ܐܪܥܐ ܗ̈ܘܝܐ. ܘܒܗ ܐܝܬ ܦܣܘ̈ܩܐ
ܬܠܬܐ.

ܦܣܘܩܐ ܩܕܡܝܐ ܡܛܠ ܥܢܢܐ ܘܡܛܪܐ. ܘܬܐܘܪ̈ܝܣ ܫܬ.

[IV.i.1]

ܡܕܡܢܝܬܐ. ܚܠܝܛ ܡܢ ܠܗܓܐ ܘܬܢܢܐ. ܠܐ ܡܢܗ ܘܠܗ. ܐܠܐ ܐܝܟ ܕܡܣܬܒܪܐ ܠܦ̈ܝܠܣܘܦܐ ܐܚܪ̈ܢܐ. ܒܪ ܕܒ̈ܪܬܐ ܣܠܩܢ ܦܠܓܝܢ. ܘܡܢܗ ܡܢ ܝܠܦܬܐ. ܡܢܐ ܕܠܓܘܢܝܬܐ ܡܕܝܪܬܐ ܕܐܐܪ ܡܬܥܒܕܐ. ܦܠܓ. ܘܥܢܢܐ ܡܢܗ ܫܡܝܐ. ܘܬܢܢܐ ܟܠܗ ܡܬܦܫܛ. ܒܪܡ ܐܢܫܐ ܕܡܣܬܒܪܢܘܬܗ ܦܐܫ ܡܢ ܚܡܝܢܐ ܗܘ ܬܢܢܐ. ܠܐܬܪ̈ ܠܥܠ ܡܬܬܚܬܬܐ. ܘܡܬܐܡܪܐ ܡܦܣܘ ܠܗ ܠܥܢܢܐ ܘܠܗܘ. ܘܡܢܗ ܕܡܛܪܐ ܥܠ ܐܪܥܐ ܗܘܐ. ܘܐܝܟ ܦܐܫ ܗܘ ܬܢܢܐ. ܢܦܠ ܘܠܐܬܪ̈ ܠܬܚܬ ܡܬܬܚܬܬܐ. ܘܡܦܣܘ ܠܥܢܢܐ ܘܠܗܘ. ܘܥܠ ܐܪܥܐ ܡܬܒܕܪ. ܘܐܝܟ ܕܗܘܐ ܕܠܐܐܪ ܕܗܘܝܐ ܘܡܛܪܐ ܒܡܒܥܝܢ ܙܪܝܐ. ܘܠܐ ܒܥܢܢܐ ܡܬܥܒܕܐ ܠܗ. ܒܪܡ ܡܬܐܡܪܐ ܠܐܪܥܐ ܕܠܥܢܢܐ ܠܒܪܬܐ ܡܬܦܣܩܐ. ܘܠܐ ܐܪܥܐ ܘܠܐ ܕܡܬܥܒܕ.

[IV.i.2]

ܕܬܬܚܬܐ. ܡܛܠ ܕܗܘ ܚܡܝܐ ܗܘ ܬܢܢܐ. ܒܡܫܟܢܐ ܗܘ ܡܠܝܢܐ ܕܡܬܩܢܘܬܐ ܘܐܬܪ̈ܚܝܢܘܬܐ ܢܦܩܝܢ ܒܗ. ܟܕ ܡܫܬܟܚܐ ܘܡܬܬܦܝܢܘܬܐ ܡܬܚܠܦܢܘܬܐ ܡܬܘܠ ܕܡܢܘܬܐ ܚܕ̈ܝܢ ܒܗ. ܒܪܡ ܡܢܗ ܕܙܘܥܐ ܗܘ ܒܥܢܢܐ ܘܢܦܫܐ ܬܡܢ ܕܡܢ̈ܘܬܐ ܒܢܝ̈ܬܐ ܕܡܬܩ̈ܦܢ ܒܬ̈ܪܕܐ. ܡܫܬܠܡܢ ܘܚܒܪ ܒܪܘܚܐ. ܘܒܪ ܬܪܓܡܐ ܠܡܪܒ ܕܒܬܪܐ ܡܠܝܐ ܡܢ ܡܬܚܠܦܢܘܬܐ ܒܝܢܬܐ ܦܫܘܩܝܬ ܡܬܒܥܪܝܢ. ܐܬܦܓܥܐ ܒܗܘ ܡܢܐ ܕܗܘܐ ܡܢ ܡܟܒܪܬܐ ܕܐܝܟ ܒܠ ܒܡ̈ܫܟܐ ܐܩ̈ܠܡܐ ܘܡܠܝܢ ܒܠܠܝܐ. ܐܝܟܢܐ ܙܠܝܓܐ ܘܢܫܒܐ ܕܡܠܝܢ ܡܬܠܒܫ. ܘܒܪ ܗܠܝܢ ܗܟܢܐ. ܡܢܐ ܐܝܬ ܠܡܐܡܪ ܡܛܠ ܚܡܝܐ ܕܫܡܝܐ ܕܡܬܚܠܦܢܘܬܐ ܡܠܝܢܘܬܐ ܕܐܝܟ ܗܘ ܢܫܐ.

[IV.i.3]

ܕܬܠܬ. ܐܝܬ ܐܡܝܪܐ ܕܒܐܪܥܐ ܒܠܗ ܐܬܪܐ ܗܘܐ. ܒܪ ܗܘ ܚܡܝܐ ܕܡܫܬܠܡܝܢ ܕܐܝܟ ܒܥܢܢܐ. ܘܡܢ ܕܗܘܒܠܗ ܘܠܐ ܐܬܪܢܐ ܢܬܥܒܕ. ܐܡܪܢܐ ܕܡܢ ܕܗܘܒܠܐ ܕܒܪܬܐ ܒܥܢܢܐ. ܡܢ ܒܪ ܢܘܗܪܐ ܒܬܪܘܢܗ ܡܢ ܡܢܢܐ ܒܥܢܢܐ ܡܬܚܠܦܐ. ܘܡܢܗܘܢ ܠܐܬܪ ܢܡܛܐ. ܘܠܐ ܚܕܐ. ܘܐܬܪܒܝܬܐ ܕܡܠܝܬܐ

FMLIV: tit. ܫܪ̈ܒܐ: ܫܪܒܐ L; ܫܪܒܐ l || ܕܠܥܠ Λ || ܗ̈ܘܝܐ M || **IV.i.1:** 4 ܕܡܣܬܒܪܢܘܬܐ Λ || 5 ܥܠ: ܥܠ MV || 6 ܢܦܠ: ܢܦܠ Λ || ܡܬܬܚܬܬܐ Λ || 8 ܘܠܐ Λ || ܕܒܥܢܢܐ Λ || ܠܥܢܢܐ Λ || ܠܐܪܥܐ Λ ||
IV.i.2: 1 ܘܡܛܠ MΛV || ܗܘ: ܗܝ MV[1]; ܗܘ V[corr.] || ܕܡܬܩܢܘܬܐ ΛV || 2 ܟܕ: ܘܟܕ Λ; ܕܟܕ V (ܕ supra lin.) || 3 ܡܬܘܠܐ M; ܡܬܘܠܐ V; ܡܬܘܠ ܣܘܟܠܐ V marg. || 4 ܘܡܫܬܠܡܢ V || 5 ܠܬܪܕܐ Λ ||
6 ܕܐܝܟ: ܕܐܝܬ M[1]; ܕܐܝܟ M[V] || 8 ܕܡܬܚܠܦܢܘܬܐ LV || **IV.i.3:** 4 ܕܡܠܝܬܐ F || 4-5 ܐܝܟ ܡܠܝܬܐ: ܘܡܠܝܬܐ F; ܕܡܠܝܬܐ ܐܝܟ V

FOURTH CHAPTER

ON THE REMAINING PRODUCTS OF THE SMOKY EXHALATIONS

It contains three sections.

First Section: On Thunder and Lightning. Six theories.

[IV.i.1.]
First [theory]: Vapour and smoke [are] not [found] by themselves but usually the two of them together, mixed with each other. When this mixture reaches the cold stratum of air, it thickens, a cloud is generated from it and smoke is confined inside it, so that, if the smoke retains its heat, it is carried upwards, and violently tears the cloud and goes out. From the tearing the sound of thunder is generated. If the smoke grows cold, it becomes heavy and is carried downwards, [also] tears the cloud and goes out, and causes the sound of thunder to be heard. If a loud sound is heard from wind tearing thin and fine air as it blows, how much more should a thundering sound be heard from (wind) tearing a thick cloud!

[IV.i.2.]
Second [theory]: Because this smoke is a fine body in which wateriness and earthiness are mingled, the heat and the mixing movement create an oily mixture in it, so that it is inflamed by that intense movement and the strong collision with each other of the parts of clouds which are torn, and causes lightning. If you wish to know that fine substances are easily lit by a little movement, observe what happens when you pass your hand over a black and smooth objects at night, [i.e.] how fine sparks and coruscations are generated. This being the case, what need is there to say about oily smoke which undergoes such a forceful movement.

[IV.i.3.]
Third [theory]: Lightning sometimes becomes the cause of thunder, when the inflamed smoke is extinguished in a cloud and from its extinction a thundery sound is heard, just like from the extinction of coal in water. For fire in its flight from water moves violently and strikes the air with force and makes a sound. From earths which are

ܐܦ ܠܡܫܡܫܘܬܐ ܕܡܫܝܚܘܬܐ ܐܝܟ ܕܒܐܘܪܚܐ ܕܠܗܘܢ. ܚܠܦ̈ܝܐ ܕܡ̈ܫܝܚܐ ܕܡܠܟܝܢ ܫܠܝܛܝܢ ܡܢܗܘܢ. ܘܡܢ ܫܘܬܦܘܬܐ ܟܝܢܝܬܐ ܡܫܡܫܢܘܬܐ ܐܝܟ ܕܒܗܝܡܢܘܬܐ ܡܫܬܠܡܝܢ. ܘܥܠ ܐܦ̈ܝܗ ܕܐܪܥܐ ܥܠܡ̈ܝܢܝܬܐ ܢܡܘ̈ܣܝܬܐ ܡܫܬ̈ܡܫܢ. ܘܟܕ ܡܢܝܚܘܬܗܘܢ ܠܗ ܦܩܕ ܡܫܩ̈ܠܐ. ܐܝܟ ܢܒܝܐ ܕܐ̈ܠܗܐ ܒܚܠܦ̈ܝܐ ܕܡܠܐ ܡܢ ܫܡܝܐ ܕܐܝܬ ܒܗ ܡܠܝܐ ܘܐܣܛܘܟܣܘܗܝ. ܒܗ ܡܬܡܫܚܢ ܫܠܝܡܘܬܐ ܟܠ ܒܪ̈ܝܬܐ.

[IV.i.4]
ܕܐܡܪܢܢ. ܠܐ ܓܝܪ ܗܘܐ ܒܪܘܝܐ. ܐܠܐ ܐܦ ܗܘܐ ܒܗܘܢ ܕܒܪܘܝܐ. ܒܗܘܢ ܪܘܚܐ ܕܩܘܕܫܐ ܒܒܢܝܢܐ ܘܡܬܩܢܐ ܠܗ ܘܡܬܪܒܝܢܐ ܘܫܡܠܝܐ ܡܢܗ ܒܗ ܢܦܫܐ ܗܘܐ ܒܪܘܝܐ. ܘܡܢܝܢ ܠܒܪܢܫܐ ܕܒܪܘܝܐ ܘܠܐ ܢܦܫܐ ܒܗܘܢ ܒܪܘܝܐ. ܒܗ ܪܘܚܐ ܠܗ ܦܩܕ ܚܐܪܐ. ܒܕܝܠܗ ܠܒܢܝܐ ܢܦܫܐ ܘܠܐܦܪܫ ܠܗ ܡܢ ܒܪܝܐ. ܘܟܕ ܪܒܪܘܝܐ ܡܢܗ ܒܪܘܝܐ ܗܘܐ. ܒܒܪܘܝܐ ܠܗܘܢ ܡܬܚܙܝܢ. ܐܝܟܐ ܕܒܪܘܝܐ. ܒܗܝ ܕܝܠܬܐ ܡܢ ܟܠ ܠܡܫܠܡܘܬܐ ܘܡܫܬܡܫܢܘܬܐ ܕܕܠܐ ܒܪܘܝܐ ܡܩܝܡ ܡܫܝܚܐ. ܒܡܫܚܐ ܕܝܢ ܟܠ ܡܫܬܡܫܢܘܬܐ ܠܠܘܝܬܐ ܕܐܡܪ ܡܫܝܚ. ܐܝܟܢܐ ܕܢܬܡܫܚܠܗ ܠܬܘܬܗ ܠܐ. ܘܟܠ ܡܫܬܡܫܢܘܬܐ ܒܪܘܝܐ. ܘܟܠܡܕܡ ܕܒ ܢܦܫܐ ܢܦܫܐ ܠܒܪܝܐ ܘܦܠܐ. ܘܗܘܬ ܢܦܫܐ ܘܦܠܐ. ܒܗܘܬ ܕܡܫܝܚܐ ܕܝܘܠܦܢܐ ܠܟܠܗ ܕܡܫܝܚܘܬܐ ܡܪܕܘܬܐ ܥܡܡܝܢ. ܘܠܡܫܝܚܘܬ ܢܦܠܐ ܗܘܒ ܡܢ ܕܘܟܬܐ ܪܘܚܢܝܬܐ. ܡܢܟ ܒܡܫܚܐ ܕܦܠܐ ܫܪܝܪ.

[IV.i.5]
ܕܝܠܗ. ܡܫܡܫܢܘܬܗܘܢ ܐܚܪܬ ܕܒܪܘܝܐ ܐܬܬܩܢܬ ܐܝܟܢܐ ܕܐܠܝܦܐ ܡܫܝ̈ܚܐ ܕܡܬܦܫܩܢ ܒܡ̈ܠܐ ܘܡܬܦܫܩܢ ܒܗܝܢ. ܘܟܕ ܪ̈ܓܫܝܢ ܒܪܘܝܐ ܗܘܐ. ܘܐܝܟ ܡܠܬܐ ܗܘܐ. ܐܦ ܗܘܐ ܕܒܐܪܥܐ ܘܒܫܡܝܐ ܕܟܠܗܘܢ ܡܫܝܚܝܢ ܐܠܗܐ ܗܘܐ ܗܘܐ ܒܪܘܝܐ ܘܐܦ ܒܪܘܝܐ. ܘܐܢܫܐ ܒܟܠܗܘܢ ܐܚܪܢܐ ܕܡܫܝܚܘܬܐ ܡܢ ܢܦܫܐ ܕܡܫܝܚ ܡܫܬܡܫܢܐ ܒܒܪܘܝܐ. ܘܒܪܘܝܐ ܘܐܦ ܒܪܘܝܐ ܚܕܬܐ. ܘܐܠܐ ܗܘܐ ܪܒܪܘܝܐ ܪܒܪܝܪ. ܒܗܝ ܕܢܦܫܐ ܒܐܝܬܘܗ ܒܡܫܚܐ ܚܕܪܘܝܐ ܕܡܫܝܚܘܬܐ ܫܪܝܪܐ ܘܡܬܦܫܩ ܠܬܘܬܗ ܕܒܪܘܝܐ ܠܐ ܡܬܕܪܟܐ.

[IV.i.6]
ܕܝܠܗ. ܘܠܡܐܫܬܘܕܥܘܬܗ ܐܚܪܬ ܕܠܘ ܡܫܝܚܐ ܗܘܐ ܒܪܘܝܐ. ܐܠܐ ܡܫܝܚܐ ܗܘ ܒܠܚܘܕ +ܕܡܢ ܡܫܡܫܐ ܕܒܪܘܝܐ ܒܠܚܘܕܝܗ ܗܘܐ ܠܫܡܝܐ ܡܫܡܫܢܐ ܠܡܬܬܡܫ+. ܐܝܟ ܕܐܦ ܡܢܐ ܕܡܫܬܠܚܝܢ ܡܢ ܫܡܝܐ ܡܬܡܫܚܝܢ ܒܠܚܘܕ. ܘܠܗܘܬ ܡܠܬܐ ܗܘܐ ܠܐ ܙܕܩ

FMLIV: **IV.i.3**: 5 ܚܠܦ̈ܝܐ : ܚܠܝܦܘܬܐ Λ || 6 ܟܝܢܝܬܐ ܐܝܟ : ܘܟܝܢܝܬܐ F || ܡܫܬܠܡܝܢ scripsi: ܡܫܬܠܡܢ codd. || 8 ܡܠܐ V || 9 ܘܐܣܛܘܟܣܘܗܝ Λ || IV.i.3: 4 ܢܦܫܐ: ܢܦܫܐ ut vid. M || 6 ܒܪܘܝܐ V || 7 ܠܠܘܝܬܐ F || ܠܐܦܪܫ Λ || 8 ܢܦܫܐ: ܠܢܦܫܐ M || 10 ܕܦܠܐ: ܘܦܠܐ F || **IV.i.5**: 1 ܡܫܡܫܢܘܬܗܘܢ F; ܡܫܡܫܢܘܬܗܘܢ ΛV || 3 ܡܫܝܚܝܢ V || 4 ܘܐܢܫܐ ܒܟܠܗܘܢ F; ܘܐܢܫܐ ܒܟܠܗܘܢ Λ; ܘܐܢܫܐ ܒܟܠܗܘܢ MV || ܕܡܫܝܚ: ܕܐܡܫܝܚ V || **IV.i.6**: 1 ܘܠܡܐܫܬܘܕܥܘܬܗ scripsi (cf. Nic. syr. 43.11; *Cand.* 119.4): ܘܠܡܐܫܬܘܕܥܘܬܗ FΛM (ܐ supra lin. 1); ܘܠܡܫܬܘܕܥܘܬܗ V || ܕܠܘ: ܕܠܐ L || ܡܫܝܚܐ M || 2 ܕܡܢ: ܡܢ V || ܕܒܪ̈ܘܝܐ M || ܒܠܚܘܕܝܗ: ܠܒܠܚܘܕܝܗ ΛV || 3 ܫܡܝܐ: ܫܡ̈ܝܐ F || ܘܠܗܘܬ: ܠܗܘܬ F ||

saline or contain oily grime fine oily exhalations rise and are ignited by a little heat of the sun or of lightning, and bright flames are seen on the surface of the earth. Because of their fineness they do not have much capacity for combustion, like the fire that burns in the exhalation rising from wine containing salt and sal ammoniac when the flask is placed on coals.

[IV.i.4.]
Fourth [theory]: Lightning only occurs when thunder occurs with it because [it is] from wind which is confined in a cloud, tears it, is ejected and escapes from it burning [that] lightning arises. Thunder can occur without lightning occurring with it when the wind is not very strong, so that it tears the cloud but does not inflame the air. Although thunder occurs before lightning, we sense the lightning first, then the thunder, because vision requires contraposition and transparency, which do not occur in time, whereas hearing requires undulating movement of the air for the sound to reach it, and all movement [takes place] in time. For this reason, when an oar strikes water and goes up, and strikes and goes up again, we hear the sound of the first stroke after [seeing] the second stroke. We also see the stroke of an axe from a distant place before hearing the sound.

[IV.i.5.]
Fifth [theory]: Empedocles says that lightning is a flash of the sun's rays which are confined in clouds and are ignited in them, and [that] thunder is generated when they are extinguished. According to this theory, lightning and thunder ought to occur during the day and in summer when the rays are abundant. Anaxagoras says that a part of the fire of the αἰθήρ is confined in a cloud and causes lightning and thunder. This view too is not correct, because fire circulates in its place with the circular motion of the celestial sphere and is not driven straight towards a cloud.

[IV.i.6.]
Sixth [theory]: Cleidemus says that lightning does not [really] exist, but is merely an illusion which arises from the division of a cloud, due to the reflection of the sun's rays from them, just as water sprinkled from oars flashes at night. According to this theory, lightning ought not to be seen at night, because the sun is [then] below the earth and

ܗܘܐ ܕܠܚܝܫܐ ܒܪܘܚܐ ܒܠܠܝܐ . ܒܗܝ ܕܠܚܝܫܬ ܐܪܥܐ ܗܘܝܐ ܫܟܝܟܐ . ܘܠܐ
ܡܬܝܢܐ ܕܡܢ ܚܡܝܡܐ ܠܘܬܗ ܡܬܟܠܦܬ ܢܓܝܪܬܐ . ܒܠܠܝܐ ܕܝܢ ܡܬܚܡܝܢ ܡܛܪ̈ܐ ܕܥܒܪܘܢ .
ܘܠܐ ܒܐܝܡܡܐ . ܒܗܝ ܕܡܫܘܚܐ ܢܘܗܪܗ ܕܫܡܫܐ ܒܐܝܡܡܐ ܠܚܡܝܡܘܬܗܘܢ .

|IV.ii.tit|

ܦܣܘܩܐ ܬܪܝܢܐ . ܥܠ ܦ̈ܝܬܐܓܘܪܝܘܣ ܘܐܢܟܣܓܘܪܣ . ܘܕܝܡܘܩܪܝܛܘܣ ܘܐ̈ܚܪܢܐ .

|IV.ii.1|

ܦܝܬܓܘܪ̈ܝܐ . ܘܦܝܬܐܓܘܪܝܘܣ ܡܢ ܐܝܬܘܗܝ ܪܘܚܐ ܚܕܢܝܬܐ ܕܟܘܟ̈ܒܐ ܕܫܒܝܢܐ . ܐܢܟܣܓܘܪܣ
ܕܝܢ ܪܘܚܐ ܓܠܓܠܝܬܐ ܕܟܘܟ̈ܒܐ ܕܫܒܝܢܐ . ܘܗܘܠܐ ܕܬܬܚܝܡ ܓܒܝܐ ܡܢ ܗܘܠܐ
ܕܒܪܘܐ . ܘܓܠܝܙܐ ܠܗ ܢܘܗܪܗ ܒܠܝܠܝܐ . ܐܠܐ ܗܘ ܒܥܝܢܗ ܒܪ ܙܠܘ ܡܢܠܐ
ܠܐܪܥܐ . ܒܝܕ ܪܦܝܬܘܬܗ ܘܒܝܢܝܬܗ ܢܘܗܪܗ ܐܪܥܢܐ . ܐܝܟ ܒܝܕ ܕܝܢܐ ܘܢܘܗܪܐ
ܕܒܪܐ ܠܗ . ܗܘܠܐ ܕܝܢ ܕܒܪܘܐ . ܒܝܕ ܡܠܝܘܬܗ ܠܘ ܗܘܐ ܡܬܚܙܝܐ . ܐܠܐ
ܒܓܠܠ ܡܬܚܙܝܢܐ ܘܪܒܐ . ܘܢܘܗܪܗ ܒܠܚܕ ܒܥܝܢܗ ܡܢܠܐ ܠܐܪܥܐ .

|IV.ii.2|

ܕܬܬܚܝܡ . ܦܝܬܐܓܘܪܝܘܣ ܕܗܘܐ ܡܢ ܗܘܠܐ ܕܢܦܫ ܡܠܝܐ . ܐܪܓܝܘܣ ܐܝܟܢܐ ܢܘܗܪܐ
ܡܬܚܙܝܐ . ܒܕܠܐ ܡܬܚܙܝܢܘܬ ܥܕܡܐ ܒܓܠܝܙܐ . ܐܝܟܢܐ ܕܐܝܠܐ ܓܒܝܐ
ܢܝܢܘܣ . ܗܘ ܕܝܢ ܕܡܢ ܗܘܠܐ ܕܢܦܫ ܒܒܝܐ . ܦܝܠܘܐܢܘܣ ܐܝܟܢܐ ܓܙܦܐ
ܡܬܚܙܐ . ܒܗܝ ܕܡܬܚܙܐ ܡܢ ܠܓܙܐ ܕܦܠܓ ܒܗ . ܠܐ ܕܝܢ ܡܬܚܙܐ ܠܗ .
ܐܠܐ ܒܒܪ ܢܦܫ ܢܘܗܪܐ ܐܚܝܕ . ܘܡܬܚܙܝܢܐ ܕܢܦܫ ܡܫܝܢ ܥܒܪ ܘܠܐ ܡܬܚܙܐ
ܠܗ . ܒܪܒܐ ܒܪ ܦܠܓ ܒܫܒܝܐ . ܠܗܘܢܐ ܘܢܝܣܐ ܕܒܗ ܡܦܫܪ . ܒܗܝ
ܕܠܓܒܝܢ ܘܟܠܝܢ ܠܗ ܡܢ ܡܬܚܙܝܢܘܬܐ . ܠܢܘܗܪܐ ܕܝܢ . ܐܝܟܢ ܕܢܦܫ ܡܫܝܢ . ܠܐ
ܡܬܚܙܐ . ܐܠܐ ܒܒܪ ܡܬܚܙܐ ܠܗ ܒܠܝܠܝܐ . ܘܠܢܘܗܪܐ ܕܒܝܢܝܐ ܡܦܫܪ . ܘܠܝܢܝܐ
ܠܐ ܡܬܚܙܐ .

|IV.ii.3|

ܕܬܠܬܐ . ܦܝܬܐܓܘܪܝܘܣ ܕܝܢ . ܒܗܝ ܕܡܢ ܗܘܠܐ ܕܢܦܫ ܒܒܝܐ ܗܘܐ . ܡܬܚܙܐ ܡܬܚܠܦܐ
ܠܐܪܥܐ . ܘܒܪ ܪܓܝܫ ܗܘ ܗܘܠܐ ܕܠܗ ܠܒܝܫܬܐ ܠܓܙܝܐ ܡܓܝܢܐ ܠܗܘܬ ܡܢܠܐ
ܕܢܘܢܐ ܡܬܚܙܝܢܐ . ܘܒܐܪܥܐ ܢܘܗܪܐ . ܘܡܢ ܪܘܚܐ ܗܘܝܢ ܗܠܝܢ . ܢܘܗܪܐ ܡܢ
ܡܬܚܙܝܢܘܬ ܓܠܝܙܘܬܐ ܕܠܓܝܢ ܒܐܪܥܐ . ܐܝܟܢܐ ܕܒܢܘܗܪܐ ܕܢܦܫ ܐܢܟܣܘܣ
ܐܬܚܙܝܢܬ . ܓܠܝܙܘܬܐ ܒܪ ܠܗ ܡܢܠܐ ܗܘ ܐܢܟܣ ܐܝܟܐ ܐܢ ܦܝܠܘܢܐ ܕܢܦܫ . ܘܡܢ

FMLIV: **IV.i.6**: 5 ܕܡܢ: ܡܢ V || 6 ܠܚܡܝܡܘܬܗܘܢ Λ (voc. ܠܚܡܝܡܘܬܗܘܢ L) || **tit.**: ܦܝܬܐܓܘܪܝܘܣ F || **IV.ii.1**: 5 ܡܠܝܘܬܗ ΛV || ܒܓܠܠ F || 6 ܘܢܘܗܪܗ F || **IV.ii.2**: 2 ܡܬܚܙܝܢܘܬ F || 3 ܕܡܢ: ܡܢ ΛV || ܦܝܠܘܐܢܘܣ: ܦܝܠܘܐܢܘܣ V || 4 ܠܓܙܐܗ M || 6 ܒܫܒܝܐ: ܒܫܒܐ F || 7 ܠܗ: ܡܢܗ M[1] || ܡܢ ܐܝܟ ΛV [orthog.] || **IV.ii.3**: 2 ܕܠܗ F || ܠܓܙܝܐ ܡܓܝܢܐ F || 4 ܐܢܟܣܘܣ V ||

vision cannot be reflected from the cloud towards it. Flashing water, however, is seen at night, not during the day, because the light of the sun obscures its brightness during the day.

Second Section: On Thunderbolt and Firewind. Three theories.

[IV.ii.1.]
First [theory]: Thunderbolt [κεραυνός] is inflamed cloud wind. Firewind [πρηστήρ] is inflamed whirlwindy wind. The material of both is denser than the material of lightning. For this reason, not only its light but [also] its body reaches the earth aflame, because of its density and concentration of its earthy weight, or because of the propulsion and compulsion which occur to it. The material of lightning, on the other hand, because of its fineness does not persist for long, but is quickly dissipated and extinguished, and its light reaches the earth without its body.

[IV.ii.2.]
Second [theory]: Thunderbolt which arises from finer material is called ἀργής, i.e. "white", because it passes imperceptibly through an object, so that it does not even impart colour to it. That [arising] from denser material is named ψολόεις, i.e. "touching", because it blackens the object it meets, but does not burn it, unless perchance (the object) is more compact; a looser object, it passes through and does not damage. Consequently, when it meets a shield it melts the silver and the copper in it because they are solid and prevent it from passing through. The wood, however, as being something more loose, it does not burn, although it may blacken it. [Similarly] it melts the gold in a purse, but it does not burn the purse.

[IV.ii.3.]
Third [theory]: Firewind, because it arises from denser material, burns and inflames the air. When it is extinguished, its thick material is transformed into hard bodies in accordance with the mixture it has and falls to the earth. That these things arise from winds is known from the observation of flames blowing through the air, as was observed in the fire in the temple in Ephesus. For flame is nothing other than burning

ܘܙܕܩ ܡܫܬܐܡܢܘܬܗܘܢ ܘܡܬܬܥܝܩܘܬܗ. ܘܗܟ ܒܬܪܗܘܢ. ܪܘܚܐ ܠܟܢܫܐ. ܒܗ ܕܒܫܘܪܝܐ
ܘܫܠܡܘܬܐ ܡܫܬܠܡܢܘܬ ܡܬܒܚܝܪܬܐ ܘܡܬܕܢܚܢܐ ܢܘܚܐ ܡܢ ܥܡܠܐ. ܘܫܠܡܬܐ ܠܐ
ܡܬܦܠܓܢܝܬܐ. ܒܚܘܒܐ ܕܝܢ ܕܢܘܦܩܗ ܡܬܦܠܓܢܝܬܐ. ܐܦ ܚܢܢ ܘܒܗܢ ܕܒܚܢܐ
ܗ̇ܘܢܐ ܪܘܚܐ. ܗܪܟܐ ܕܝܢ ܕܒܚܢܐ ܢܚܬܝܢ ܐܦ ܠܐܪܚܐ ܚܠܝܫܐ. ܘܡܠܟ ܡܢ ܟܠ
ܡܫ̈ܬܐܡܢܘܬܗ ܘܡܬܬܥܝܩܘܬܗܘܢ

[tit.]

ܦܘܣܩܐ ܬܠܝܬܝܐ. ܡܠܠ ܕܥܘܒܕ̈ܝܗܘܢ ܘܒܥ̈ܒܕܐ ܫܦܝܪܐ ܘܢܨ̈ܝܚܐ. ܬܫܡܫܬܗܘܢ
ܐܪܒܥ.

[IV.iii.1]

ܡܫܡܫܢܘܬܐ. ܘܡܠܐ ܚܢܢ ܕܕܘܒܪ̈ܝܗܘܢ ܐܘܪܚܬ ܠܡܩ̈ܕܫܐ ܕܡܬܚܫܚܝܢ ܒܡܫܡܫܢܘܬ ܘܫܪܝܐ
ܕܡܠܝܢ ܐܘܪܚܐ ܩܕܝܫܐ ܐܝܬܘܗܝ. ܒܗ ܠܡܣܒܐ ܚܝܐ ܢܗܘܪܐ ܫܒܚܬ. ܘܠܐ ܡܬܟܠܝܐ
ܠܬܡܢ. ܘܙܕܩ ܕܠܗܘܢ ܘܚܕܐ ܡܢ ܟܠܐ ܠܐܘܪ. ܘܗܘ ܚܕܐ ܐܝܠܢܐ ܩܕܝܫܐ. ܐܝܟ ܡܠܠ. ܚܕ
ܡܢܠܐ ܠܡܫܬ ܡܬܦܠܓܢܝܢ. ܘܡܠܠ ܦܠܘܚܘܬܗ ܒܪܘܚܐ ܕܥܠܡܐ ܒܗ ܫܠܡܘܬܐ.
ܘܡܫܕܪܐ ܕܬܠܬ ܡܬܦܩܫܫ. ܘܐܝܟ ܒܡܒܥܐ ܕܫܦܪ ܘܗܢ ܫܒܝܐ ܠܘܬ ܡܫܡܫܬ.
ܘܐܝܟܢܐ ܕܚܕܐ ܝܗܝܒܐ. ܘܐܘܦܣܩܘܦܘܣܐ ܐܘܪܚܬ ܡܠܘܐܐ ܡܕܢܚܐ ܡܝܐ.
ܘܟܗܢܐ ܘܡܫܡܫܢܐ ܡܫܡܫܝܢ. ܥܠܡܢܘܬܐ ܕܟܗܢܘܬܐ ܐܝܟ ܢܫܡܬܐ ܕܠܐܬܝܐ
ܡܬܚܫܚܢܐ ܕܢܡܘܣܐ. ܘܩܕܫܘܬܐ ܡܫܡܫܐ. ܘܐܦ ܒܟܗܢܐ ܡܫܡܫ. ܐܝܟ ܠܡܩ̈ܕܫܐ
ܡܬܚܫܚܢ. ܘܐܦ ܐܝܟ ܒܠܩ̈ܕܫܬܐ ܕܡܠܟܝܢ. ܘܐܝܟ ܡܫܡܫܐ ܡܢ ܚܕ ܕܢܟܐ ܥܩܝܢ
ܡܬܚܫܚܢ. ܚܙܐ ܡܬܚܫܚܢ. ܘܐܦ ܐܝܟ ܕܗܢܐ ܕܝܠܗ ܢܒܝܐ. ܕܐܠܗܐ ܡܬܬܢܝܐ.

[IV.iii.2]

ܕܬܬܚܫܚ. ܗܕܐ ܡܢ ܠܟܠܗ ܕܡܫܡ̈ܫܢܐ ܕܢܗܘܐ ܘܠܐ ܠܡܬܠܠ. ܒܗ ܐܝܟܢܐ ܕܢܗܘܐ
ܫܬܘܬܦܘܬܐ. ܗܢܐ ܕܡܠܟܘܬܐ ܐܝܠܝ̈ܢܐ ܕܫܒܚܐ ܕܠܗܘܢ. ܘܠܐ ܡܢ ܘܠܗ. ܐܝܟܢܐ.
ܕܙܕܩ ܐܬܬܢܣ. ܒܡܢܗ ܗܘ ܕܡܫ̈ܡܫܢܘܗ. ܐܦ ܚܢܢ ܗܘ ܕܡܫܡܫܢܝܢ ܐܢܫ. ܐܦ
ܡܬܠܐ. ܡܢ ܡܬܚܙܝܢ ܘܐܬܚܫܚܬ ܡܫܡܫܢܘܬܐ. ܘܗܘܝܢܐ ܐܢܫ. ܐܦ ܡܫܡ

FMLIV: **IV.ii.3**: 6 ܘܗܟ ܒܬܪܗܘܢ ܪܘܚܐ: om. L; in marg. V || ܒܗ: ܒܗܘ L || 8 ܕܢܘܦܩܗ LV || ܡܬܦܠܓܢܝܬ (sec.) LV || 9 ܕܝܢ: om. MΛV || ܘܡܠܟ: ܗܘܐ F; ܡܠܟ LV || ܘܡܫܬܐܡܢܘܬܗܘܢ FMV; ܡܫܬܐܡܢܘ l || 10 ܘܡܬܬܥܝܩܘܬܗܘܢ FMV || **tit.**: ܕܥܘܒܕܝܗܘܢ ut vid. FL; ܕܥܘܒܕܗܘܢ V || ܐܪܒܥܐ V || **IV.iii.1**: 1 ܕܕܘܒܪ̈ܝܗܘܢ scripsi: ܕܕܘܒܪܝܗܘܢ FMV; ܕܕܘܒܪ̈ܝܗܘܢܐ Λ || ܠܡܩ̈ܕܫܐ F || ܕܡܬܚܫܚܝܢ Λ || ܒܡܫܡܫܢܘܬ V || 3 ܐܝܟ ܚܕ ܡܠܠ: ܐܝܟ ܡܠܠ Λ || 4 ܠܡܫܬ: ܠܗ ܡܫܬ Λ; ܠܐܡܫܬ V || 5 ܕܫܦܪ: ܫܦܪ V || 6 ܘܐܦܝܣܩܘܦܘܣܐ F; ܘܐܦܣܩܘܦܘܣ Λ; ܘܐܦܝܣܩܘܦܘܣܐ V || 7 ܘܟܗܢܐ V || ܡܫܡܫ: ܫܡܫ V || 8 ܡܫܡܫ: ܡܫܡܫܢ Λ || 9 ܕܡܠܟܝܢ L || ܐܝܟ ΛV || ܥܩܝܢ M || 10 ܚܙܐ: ܚܙܝܐ F || **IV.iii.2**: 3 ܕܡܫ̈ܡܫܢܘܗ: ܡܫ̈ܡܫܢܘܗ MΛ; ܕܡܫ̈ܡܫܢܐ V || ܕܡܫܡܫܢܝܢ MV || ܐܦ ܡܬܠܐ: ܘܡܬܠܐ Λ || 4 ܡܢ: ܡܢ F ||

smoke. Before and after thunderbolt and firewind, wind blows, because at the beginning and at the end wind is squeezed and discharged weakly from the cloud and for this reason is not set on fire. When, on the other hand, it is ejected with force, it is set on fire. Wind in fact also occurs with thunder. This often tears even the earth. So much, then, on thunderbolts and firewinds.

Third Section: On Δοκίδες, Shooting Stars and Comets. Four theories.

[IV.iii.1.]
First [theory]: The material of δοκίδες, i.e. torches which are seen in the αἰθήρ, and the rest of these things is smoke, since vapour sinks because of its weight and does not rise there but grows cold before it reaches that place. The smoky exhalation, if fine, is ignited when it reaches the αἰθήρ, and because of its fineness, the flame travels through it with speed. As soon as it has caught fire it is dissipated, and it seems like a star which jumps and falls from the sky. If the smoke is dense, has much ὑπέκκαυμα, i.e. material, and is extensive in length and width, a large flame, like stubble in a field, is seen burning and persists for days. If it is extensive in length [only], things like torches are seen. If things like sparks, which are fine and sprout like hair from a single source, are seen, these are called "goats". If there occurs something like a stream of fire, it is named δαλός.

[IV.iii.2.]
Second [theory]: Here one ought to give the cause of extinction of fire. For we say that pure fire glows when it is burning with smoky material and not on its own, as has already been demonstrated. Therefore, its extinction occurs, either because in the region of air or water it is transformed by coldness and moisture and becomes air or something

ܐܝܟܢ . ܗ̇ܘܐ . ܐܦ ܟܕ ܗ̇ܘ ܕܡܢ ܗܘܠܐ ܕܒܗ̇ ܫܐܟܐ ܢܘܪܐ . ܠܢܘܪܢܝܘܬܐ
ܡܬܚܫܠܐ ܒܟܠܢܝܘܬܗ̇ . ܐܝܟܢܐ ܕܠܐ ܢܗܘܐ ܡܕܡ ܐܚܪܢܐ ܗܟܢ . ܘܗܟܢܐ
ܟܕ ܠܐ ܦܣܥܐ ܠܢܘܪܐ ܗܘܠܐ ܕܐܬܝܠܕ ܒܗ̇ ܘܐܬܓܠܝ . ܠܚܝܠܗ̇ ܡܚܒܠܐ ܦܣܥܐ .
ܘܠܐ ܡܬܚܪܝܢܘܬܐ ܗ̇ܘܝܐ . ܘܐܝܟܢ ܗ̇ܘ ܕܡܬܒܢܐ ܕܒܚܕ ܡܫܘܚܬܐ . ܟܕ ܟܬܝܪܐ
ܠܐ ܡܬܦܬܬ . ܐܠܐ ܫܪܝܬ .

[IV.iii.3]
ܕܬܠܬ . ܠܡܥܒܕܐ ܡܢ ܘܡܥܒܕܐ ܢܩܝܦܐ ܘܡܠܬ . ܠܘ ܡܢ ܟܠܗ̇ ܡܕܡܝܢܘܬܐ
ܕܒܝܬ . ܟܕ ܗܘ ܕܠܡܪܢܘܬܐ ܘܠܒܝܬܐ ܒܪܘܚܐ ܗ̇ܘ ܟܠܢܐ ܡܫܠܡܢܐ ܠܐ ܐܝܬ .
ܐܠܐ ܡܢ ܟܠܗ̇ ܡܬܝܠܕܢܘܬܐ . ܒܐܘܪܚ ܡܢܐ ܕܠܢܘܪܢܝܘܬܐ ܒܟܬܝܒܬܐ ܗܘܠܐ
ܕܠܗܘܢ ܡܬܚܫܠܐ . ܡܬܚܫܠܢܘܬܐ ܗ̇ܘܝܐ . ܘܡܢ ܫܪܟܐ ܕܠܗܘܢ ܡܬܚܫܠܐ . ܘܗ̇ܘ
ܗܘܠܐ ܐܝ ܦܩܕ ܡܠܝܢܐ . ܡܬܚܫܒ ܕܒܚܕܗ̇ ܡܠܝܢܐ . ܘܐܝ ܚܕܢܐ
ܡܬܚܫܒ . ܘܐܝܟܢ ܒܚܕܒܚܕܐ ܕܐܝܬܝܗ̇ ܠܗ ܢܩܝܦܘܬܐ ܐܦ ܕܗܘܢܐ ܐܦ ܕܘܠܐ ܡܬܒܢܝܐ
ܕܠܗܘܢ . ܘܐܝܟܢ ܫܘܬܐ ܬܘܒ ܡܬܝܠܕܢܘܬܐ . ܘܒܡܕܡܝܬ ܒܟܠܗ̇ ܢܘܪܢܝܐ . ܐܝܟܢ ܗܟܝܠ
ܕܐܬܚܫܒ ܒܟܢܝܬ ܐܠܗ ܘܫܡܫܡܫܐ ܘܟܘܟܒܐ ܘܫܡܫܐ ܕܩܠܡܩܘ ܫܪܒܐ ܡܠܘܫܐ
ܕܐܝܬܝܗ̇ ܟܕ ܢܗܪܝܢ ܒܗܘܢ . ܘܟܠܗ ܡܠܝܐ ܢܘܪܐ . ܘܟܐܝܢܐ ܒܐܝܢܐ ܟܕ ܐܬܦܠܓܘ
ܐܬܦܪܫܘ ܘܐܬܦܠܓܘ .

[IV.iii.4]
ܕܐܪܒܥ . ܫܪܐ ܠܡܥܒܕܐ ܕܝܢ ܘܠܘ ܒܐܡܪܝܢ ܗܟܝܠ ܐܬܩܢܬܐ ܗܠܝܢ . ܟܕ ܕܠܡܚܫܘ
ܡܬܚܫܒܐ ܗܘܠܐ ܡܢܝܢܘܬܐ ܕܗܘܝܐ ܟܠܗ ܓܒܝܐ ܘܦܪܝܫܐ . ܐܝܟܢܐ ܕܠܐ
ܬܬܩܒܥ ܒܐܘܪܚܗ̇ . ܘܢܟܪܐ ܢܦܘܐ ܕܠܗܘܢ . ܘܟܕ ܗܘܝܐ ܢܘܪܐ ܠܡܘܪܐ
ܟܠܢܐ ܕܗܘܝܬ ܬܬܩܒܥܐ ܠܡܬܩܒܠܘ . ܠܐ ܟܝܬ ܡܫܘ ܠܗ . ܐܠܐ ܫܠܐ ܡܠܝܐ ܕܘܪܐ
ܕܬܩܒܘ ܠܟ . ܩܕܡܝܐ ܕܝܢ ܦܩܕ ܠܟܝܢܐ ܘܐܬܝܠܕܬ ܕܡܬܚܫܠܝܢ . ܠܐ ܕܠܗܝܢ . ܐܠܐ
ܒܡܬܚܫܠܝܢ ܗܘܝܢ . ܘܐܬܩܢܬܐ ܕܡܠܬܐ ܕܡܩܘܡܝܢ ܘܐܩܒܠܝܢ ܒܗܘܝܬ ܡܬܚܫܒܝܢ .
ܡܬܩܒܠܝܢ ܕܝܢ ܬܘܒ ܐܦ ܩܢܝܬܐ ܒܐܝܟܐ . ܘܐܝܟܢܐ ܫܩܘܠܐ ܒܡܟܝܟܐ . ܘܐܝܬ
ܐܡܬܝ ܕܟܠܗ̇ܝܐ ܒܟܬܝܒܬܐ ܕܦܩܕ ܠܟܝܢ ܡܢ ܐܪܒܥܐ ܩܠܗ . ܘܠܬܘܢܐ ܟܠܢܐ
ܡܬܚܫܠܐ ܟܕ ܒܗܘܢ ܒܐܪܒܥܐ ܡܬܝܒܕ . ܘܡܢܐ ܕܢܠܗ ܡܢ ܗܘܝܬ . ܒܟܬܝܒܐ
ܕܢܘܪܐ ܡܢ ܐܪܒܥܐ ܠܥܡܝܢܐ ܡܬܚܫܠܐ ܕܡܬܚܫܒ . ܘܗܟܝܠ ܡܢ ܐܬܩܢܬܐ ܟܠ ܪܘܚܐ
ܘܟܠܢܝܬ ܡܠܬܐ ܘܒܘܪܕܢܐ ܡܬܚܬܝܬܐ ܫܚܝܢܐ ܘܝܒܝܫܐ ܡܬܩܕܪܝܢ .

FMLIV: IV.iii.2: 5 ܕܒܗ̇ ΛV || ܢܘܪܐ: ܠܢܘܪܐ Λ || 6 ܒܟܠܢܝܘܬܗ̇ Λ; ܒܟܠܢܝܘܬܗ̇ V || 7 ܒܗ̇ ܘܐܬܓܠܝ: ܐܬܓܠܝ ܒܗ̇ Λ || **IV.iii.3**: 3 ܡܢܐ: ܡܢ Λ || 4 ܡܬܚܫܠܐ: ܡܬܚܫܠ Λ || 6 ܒܚܕܒܚܕܐ l || ܘܕܘܠܐ ܡܬܒܢܝܐ MV || 7 ܕܠܗܘܢ FΛ || ܫܘܬܐ ܬܘܒ ܡܬܝܠܕܢܘܬܐ F || ܒܟܠܗ̇ ܢܘܪܢܝܐ M; ܒܟܠܗ̇ ܢܘܪܢܝܐ V || 8 ܕܩܠܡܩܘܘ MV || 9 ܐܬܦܠܓܘ M || **IV.iii.4**: 3 ܒܐܘܪܚܗ̇ M || 4 ܕܐܬܚܫܒ V || ܟܝܬ: om. F || 5 ܘܐܬܝܠܕܬ: ܘܐܬܝܠܕܬܐ V || 6 ܘܐܩܒܠܝܢ M || ܒܗܘܝܬ V || ܡܩܘܡܝܢ Λ || 7 ܡܬܚܫܒܝܢ V || ܒܡܟܝܟܐ M || 9 ܒܟܬܝܒܐ ܗܘܝܬ: ܒܟܬܝܒܐ ܘܗ̇ܢܘܢ ܚܝܠܐ Λ || ܡܢ: ܡܢ F || ܐܬܩܢܬܐ V || 10 ܠܥܡܝܢܐ M || 10-11 ... ܘܗܟܝܠ ܡܬܩܕܪܝܢ: om. Λ 10 ܪܘܚܐ: ܪܘܚܐ MV || 11 ܘܝܒܝܫܐ: ܘܝܒܝܫܬܐ F (om. Λ; cf. IS اليابسة) ||
ܡܬܩܕܪܝܢ scripsi: ܡܬܩܕܪܝܢ FMV (om. Λ) ||

else, or because the material with which the fire burns is transformed into fieriness in its entirety, so that there remains nothing earthy at all. Thus, since fire no longer has any material with which to burn and glow, it reverts to its transparent nature and becomes invisible, and it appears as if it is already extinguished, when in truth it is not dead but is [still] alive.

[IV.iii.3.]
Third [theory]: Torches, comets etc. are not extinguished by the first cause because coldness and moisture have no power in that upper height, but by the second cause, namely, when their material is transformed completely into fieriness, it becomes transparent and its glow is hidden from the eye. If the material is very fine its imaginary extinction is accelerated, but if it is thick it is retarded and its glow appears as a star with a lock of hair, a tail or a beard, and also as a horned beast and like fiery rods, such as those which appeared in the year 1575 of Seleucus around the sign of Leo and circulated with it. They persisted for the whole summer, but having thinned out little by little were dissipated and vanished.

[IV.iii.4.]
Fourth [theory]: These signs occur rarely and inconstantly because it is with difficulty that the smoky material of all this is formed [sufficiently] dense and compact so that it is not dissipated along its path and its glow persists for a long time, and that it is able to be lifted up to the upper height of the αἰθήρ despite being heavy in this way. For only a violent and very strong force will lift it. Very thick and moist bodies of smoke, which *are* lifted up, do not glow but become coaly and cause terrible red and black signs to appear in the αἰθήρ. They also cause illusions of chasms in the air and dark porches in the sky. Sometimes very thick sulphureous exhalation rises from the earth and reaches the upper height, while its root is fixed to the earth. When it catches fire from the αἰθήρ, there appears a column of fire extending from the earth to the sky.

These signs portend winds, drought and death-bringing, feverish and withering diseases.

[tit.]

ܩܦܠܐܘܢ ܚܡܝܫܝܐ. ܡܛܠ ܫ̈ܡܗܐ ܪ̈ܘܪܒܐ ܕܩܘܡ ܒܟܬ̈ܒܐ. ܘܒܗ ܐܝܬ

ܦܘܫܩ̈ܐ ܬܠܬܬܐ.

ܦܘܫܩܐ ܩܕܡܝܐ. ܡܛܠ ܛܘ̈ܒܐ. ܬܐܘܪܝܐܘܣ ܬܠܬ.

[V.i.1]

ܡܕܡܝܢܬܐ. ܢܦܩܐ ܕܝܢ ܠܐܝܠܝܢ ܕܐܬܐܡܪ̈ܝ ܕܥܠ ܫ̈ܡܗܐ ܪ̈ܘܪܒܐ. ܐܝܟ ܛܘ̈ܒܐ ܘܫ̈ܡܗܐ ܕܫܡܗܘܢ ܕܣܓܝܐ ܠܢܘܟܪܝܐ ܡܬܩܢܝܢ ܒܟܬ̈ܒܐ. ܢܫܠܠ. ܒܗ ܐܡܪܝܢ ܕܛܘܒܐ ܡܢ ܐܢܬܘܢ ܒܫܠܡ ܚܕ ܡܢ ܐܘܣܝܐ ܐܝܬܝܐ ܥܠ ܟܠܗ ܪܘܒܐ ܡܬܚܫܚܢܐ ܕܐܪܒܐ. ܐܘ ܥܠ ܡܕܡ ܡܢܗ. ܐܝܬܘܗܝ ܕܝܢ ܥܠܬܗ ܒܒܝܬܐ ܘܫܘܬܦܘ ܕܒ̈ܢܝܐ ܠܛܒܐ ܘܡܕܡ ܡܢ ܕܘܟܐ ܒܢ ܒܡܫܘܬܗ ܕܠܗ ܕܐܘܣܛܘܡܐ. ܘܛܘܒܐ ܡܚܫܚܢܐ ܡܢ ܫܡܐ. ܒܗ ܡܠܬܐ ܦܪܝܫܐ ܘܡܫܘܚܬܐ. ܐܦ ܒܗ ܡܠܬܐ ܬܒܥܬܐ ܘܐܡܪܝܢܐ. ܐܦ ܒܗ ܫܘܠܡܐ ܠܐ ܡܫܡܫܢܐ ܕܐܡܪ ܠܒܢܝܐ. ܡܫܬܠ. ܢܦܪܢܐ ܕܝܢ ܫܡܐ ܡܢ ܢܘܟܪܝܐ ܕܪܘܚܐ ܒܪ̈ܬܐ ܘܚܘܝܐ ܢܦܫ ܚܕܝܐ. ܘܐܡܪܝܢܐ ܡܢ ܡܠ̈ܟܐ ܒܒܢܝܐ ܘܡܩ̈ܕܡܢܐ ܕܪ̈ܘܚܐ. ܘܐܡܪܚܢܐ ܡܢ ܐܡܬܘܬ ܗܠܝܢ ܦܬܓܡܐ ܥܠ ܒܒܪܬܐ. ܐܦ ܒܡܫܝܢܐ ܕܡܕܪܫܬܐ ܡܫܡܫܢܘܬ ܚܝ̈ܬܐ ܘܢ̈ܣܒܬܐ. ܐܡܒܝܐ ܕܕܡ̈ܘܬܗܘܢ ܒܐܕ̈ܫܝܬܐ ܒܕ̈ܘܟܝܢ ܕܟ̈ܝܢ ܡܫܬܡܫ̈ܢ.

[V.i.2]

ܕܬܪܬܝܢ. ܥܠ ܫܪܒܐ ܡܦܟܠܢܐ ܕܢܛܪܘܬܐ ܘܒܝܬܪܘܬܐ. ܐܦ ܕܪܒܗ ܡܝܬܝܐ ܘܗܘ ܡܢܐ ܕܡܛܝܒ ܠܗ ܐܝܟ ܕܒܫܘܦܪܐ ܫܘܐ. ܒܬܪ ܕܝܢ ܫܘܝܐ ܕܬܪܝܨܘ ܡܬܩܘܡ. ܦܓܪ ܢܦܫܐ ܘܦܓܪ ܒܝܬܐ. ܠܘ ܠܐ ܡܡܬܚܫܚܐ ܗܘ. ܒܝܬܗ ܐܡܕܐ ܕܐܢܫ ܐܘܫܛ ܕܒܡܕܡ ܡܢ ܒܥܬܪܐ ܬܘܬܒܬܐ ܘܫܘܬܐܐ ܕܬܢܝܢܐ ܠܗ ܢܦܫܬ ܡܬܪܐ. ܡܝܬܐ ܦܘܩ ܕܡܩܒܠܘܬ ܡܬܪܐ ܒܦܩܕܐ ܐܘܪܚܐ ܬܘܩܢܐ ܡܫܬܠ. ܐܦ ܐܐܪ ܠܒܝ ܡܬܪܐ ܢܬܡܫܠܘ ܡܫܬܠ. ܘܛܘܒܐ ܕܡܬܪܐ ܢܦܫܐ. ܒܗ ܒܝܬܐ ܘܛܘܒܐ ܕܢܘܪܐ ܘܪܘܚܐ ܘܒܥܬܪܐ.

[V.i.3]

ܕܬܠܬ. ܝܢ ܒܠܗ ܫܘܝܐ ܕܢܫܒܐ ܕܡܬܘܬܐ ܡܕܡ ܗܘ ܫܡܫܝܢܬܐ. ܐܝܟ ܐܘܦܝܢܘܢ. ܐܦ ܐܘܦܪܐܘܢ. ܐܦ ܫܘܠܡ ܝܠܝܐ ܕܫܫܫܐ. ܐܝܬܝܐ ܕܟܠܝܬܗ ܐܣܛܪܟܝܘܢ ܢܡܘܣܐ ܘܦܪܐܘܢ. ܐܠܝܐ ܕܟܐܢܐ ܟܐܢܐ ܥܠ

FMLIV: tit.: ܩܦܠܐܘܢ: ܦܘܫܩܐ L || V.i.1: 1 ܕܐܬܐܡܪܘ Λ || ܥܠ F || 5 ܘܫܘܬܦܘ F || 6 ܡܚܫܚܢܐ ܡܢ: ܡܢ ܡܚܫܚܢܐ L; ܡܢ ܡܚܫܚܢܐ l; ܡܚܫܚܢܐ ܡܢ MV || ܘܡܫܘܚܬܐ: ܡܫܘܚܬܐ ܐܦ Λ || 7 ܐܦ ܒܗ ܡܠܬܐ ܬܒܥܬܐ ܘܐܡܪܝܢܐ: om. F || 8 ܕܠܒܢܝܐ F || 10 ܡܩܕܡܢܐ M || ܚܝ̈ܬܐ Ll[1]; ܚܝ̈ܬܐ l[corr.] || ܒܐܕܫܝܬܐ: ܒܐܕܫܝܬܐ Λ || ܡܫܡܫܢܘܬ: ܡܫܡܫܢܘܬ M[1]; ܡܫܡܫܢ M[V]; ܡܫܡܫܢܘܬ ܡܫܡܫܢ V || 11 ܒܐܕ̈ܫܝܬܐ: ܚܕ̈ܫܝܬܐ M || ܡܫܬܡܫ̈ܢ: ܡܫܬܡܫܝܢ M; ܡܫܬܡܫܝܢ V || V.i.2: 1 ܘܒܝܬܪܘܬܐ V || 5 ܐܘܪܚܐ MV || ܬܘܩܢܐ l || 6 ܢܬܡܫܠܘ ܠܒܝ ܡܬܪܐ Λ || ܠܒܝ: ܠܒ̈ܝ M || 7 ܘܪܘܚܐ: om. Λ || V.i.3: 1 ܝܢ: ܒܗ Λ; om. V || ܫܡܫܝܢܬܐ: ܫܡܫܝܢܬܐ Λ || 3 ܐܣܛܪܟܝܘܢ l || ܘܦܪܐܘܢ ΛV

FIFTH CHAPTER

ON GREAT EVENTS WHICH TAKE PLACE IN THE WORLD

It contains three sections.

First Section: On Deluges. Three theories.

[V.i.1.]

First [theory]: It accords with what has been said that we should speak about great events like deluges and the rest of those [phenomena] which are brought about once in a long while in the world. We say that a deluge is the predominance of one of the four elements over the whole or part of the habitable quarter of the earth. Its efficient cause is the conjunctions of the planets and certain fixed stars, along with the favourable disposition of the element itself. A deluge of water occurs suddenly, through repeated and continuous high tides, through heavy and incessant rains, or through immoderate transformation of air into water. That of fire arises from inflammation of strong winds, and this [type of deluge] is faster [to spread]. That of air [arises] from severe and destructive blast of winds. That of earth [arises] from the diffusion of a large quantity of sand over the habitable world, or severe coldness which hardens animals and plants, such that their petrified likenesses are found in various places.

[V.i.2.]

Second [theory]: [With] everything which is subject to increase and decrease - although its medium degree and that which is close to it occur in most cases - the occurrence of its two extremes, the excessive and the deficient, is not impossible. Consequently, just as there are times when in some of the great habitable worlds rain does not fall for many years, it is also possible for a profusion of rain to occur suddenly at the other extremes, or for air to be transformed suddenly into the nature of water, and [so] for a deluge of water to occur, [and] similarly also for a deluge of fire, wind and dust.

[V.i.3.]

Third [theory]: If the cause of the migration of the sea is a certain celestial configuration, like the apogee or the perigee, or the variation in the obliquity of the sun['s course] whereby the zodiac coincides with the plane of the equator, [then] it must be that the sea will gradually

ܐܬܪܘܬܐ ܡܬܡܪܡܪܢܐ ܢܓܠܠ ܢܦܫܐ . ܘܫܘܝܢ ܠܗ ܒܫ̈ܝܢܐ ܠܡܣܬܪܝܢܐ
ܢܬܟܠܠܘܢ . ܘܗܡܢܝܢܐ ܐܪܟܐ ܠܢܦܫܐ ܘܒܣܝܡܐ ܕܠܡܘܢܝܢ ܣܘܬܐ ܡ̈ܢܝܬܐ ܐܡܪ ܠܗ
ܢܦܫܝ ܡܬܟܠܠܐ . ܘܟܠܠ ܟܠܗܝܢ ܗܢܝܢ ܘܟܕ ܐܡܟܡܢܐ . ܟܕ ܡܟܐ ܕܡܢܐ
ܕܡܘܬܐ ܕܡܟܐ ܐܬܦܩܕ . ܘܟܠܗܝܢ ܒܪܬܐ ܢܦܫܐ . ܕܠܐ ܦܘܫܝܢ ܟܢ ܢܪܟܢܝܢ
ܕܗܢܝܬܐ ܗܢܐ ܒܪܟܢܘܬܐ ܟܡܪܐ ܡܢܝܘܬܐ ܗܘܬ . ܘܟܠܗܝܢ ܐܢܬܠܕܘ ܒܗ
ܠܦܪܐ . ܘܡܢܝܢܐ ܕܡܬܐ ܐܬܒܣܝܡ . ܢܦܫܐ ܡܪܝܡ ܡܫܝܢܐ ܗܘ . ܘܡܫܝܢܐ ܕܪܒܪܒ ܡܢ
ܣܘܫܡܐ ܟܠܗ ܟܐܠܠ ܒܢܚܬܐ ܢܦܫܐ . ܘܡܢܝܢܐ ܟܗܢ ܟܠ ܣܘܪܝܐ ܣܘܠܩܐ .
ܘܟܡܦܩܐ . ܟܫ̈ܝܢܐ ܟܕܡ ܒܩ̈ܘܬ ܡܘܟܐ .

[tit.]

ܩܘܡܘܩܐ ܬܪܝܢܐ . ܡܛܠ ܗܘ ܕܐܝܟ ܟܠܗܝܢ ܢܦܫܐ ܡܬܦܢܝܢ ܢܘܟܪܝܬܐ ܘܣܘܬܐ .
ܐܬܐܘܪܝܐܘܣ ܬܠܬ .

[V.ii.1]

ܡܕܡܢܬܐ . ܠܝܬ ܘܠܐ ܚܕܐ ܐܣܬܪܟܣܘ ܡܬܚܣܢܘܬ ܟܠ ܠܗ ܡܬܝܐ ܕܚܝ̈ܐ
ܟܕ ܣܘܟܠܗܘܢ ܒܢܝܐ ܕܒܪܝܬܐ ܢܗܪܐ . ܐܝܟ ܠܗ ܟܠ ܐܠܝܢܘܬ ܗܘܢܘܢ ܒܢܝܐ
ܘܒܢܝܐ ܘܡܟܠܐ ܡܢ ܕܒܪܐ ܘܢܦܩܬܐ . ܐܢܫܝܢ ܘܗܘܐ ܒܩܢܝܐ ܘܐܢܫ̈ܝܢ ܕܕܠܐ
ܘܒܪܐ ܡܬܟܠܠܝܢ . ܢܦܫ ܡܢ ܡܕܒܪܢܐ ܡܫܟܚܝܢ . ܘܒܣܘܟܬܐ ܗܘܐ ܕܒܗܪܬܐ
ܡܢ ܒܢܝܐ ܕܗܘܪܐ ܡܬܦܢܝܢ . ܘܣܘܟܬܐ ܡܢ ܡܟܪܐ ܕܡܬܬܪܬܐ ܒܢܝܐ .
ܘܒܡܪܟܐ ܡܢ ܐܘܦܝܢܘܢ ܐܝܬܘܗܝ ܣܘܟܐ ܘܐܬܢܝܐ . ܘܐܬܪܕܐ ܡܢ ܡܬܦܢܝܐ .
ܘܟܘܡܪܝܐ ܡܢ ܟܦܪܐ . ܟܠܗܘܢ ܟܢ ܗܠܝܢ . ܟܕ ܡܢ ܐܢܫܐ ܘܐܢܫܐ ܡܬܟܠܠܝܢ .
ܐܦ ܡܢ ܟܠܟܪܣܘܢ ܗܘܐ ܡܬܟܠܠܝܢ .

[V.ii.2]

ܕܬܬܦܢܝܢ . ܟܕ ܠܘ ܟܐܢܫܘ ܡܬܡܟܬܟܐ ܟܠܠܗܘܢ ܢܘܟܪܝܬ ܠܟܢܪܐ ܘܐܟܡܘܢܝܬܐ .
ܠܐ ܐܠܝܢܐ ܕܠܐ ܬܦܩܘܐ ܗܘܝ ܢܝܕ ܟܘܢ ܐܡܟܡܟܐ ܡܕܡ ܡܫܘܠܚܢܐ ܘܟܡܢܝܬܐ
ܐܣܦܠܟܡܢܝܬܐ ܕܟܡܬܘܢܐ ܐܢܫܐ ܠܚܫܘ ܡܬܟܣܝܢ . ܘܠܐ ܦܠܐ ܠܟܣܡܬܪܘ
ܕܡܘܫ ܣܘܬܐ ܐܠܐ ܟܬܪܡܢܬ ܘܒܪܐ ܟܡܪܟܪܟܐ ܠܗ ܡܬܟܠܠܐ . ܟܡܫ
ܕܡܬܟܪܐ ܟܠܗܘܢ ܡܢ ܡܬܦܢܝܢܘܬܐ ܕܐܣܦܠܟܡܢܐ ܦܢܝ ܕܢܘܬܐ .
ܘܡܬܦܢܝܢܘܬܐ ܟܢܘܫܘܬܐ ܩܘܫܐ . ܟܕ ܟܘܢ ܐܝܟܢܐ ܕܡܫ ܟܢܘܫܘܬܐ
ܟܡܪܟܪܟܐ ܩܘܫܐ . ܡܫܝܢ ܕܐܝܟ ܟܡܢܝܬܐ ܡܕܡ ܡܢ ܐܪܟܐ ܠܦܘܬܐ ܡܢ ܡܪܟܐ
ܐܐܪܢܐ ܬܬܟܢܝܫ . ܘܡܬܦܢܝܢܘܬܐ ܕܪܡܢܐ ܠܗܫ ܕܟܪܡܘܢܐ ܬܦܘܢܐ . ܘܫܢܝ
ܣܘܡܟܠ ܟܡܘܢܝܬܐ ܦܪܘܬܐ ܐܢܫܘܬܐ ܡܢ ܢܡܟ ܦܪܘܬܐ ܟܠܗ ܬܪܝܢ ܘܬܬܪܟܐ .

FMLIV: **V.i.3**: 4 ܡܬܡܪܡܪܢܐ M || ܘܫܘܝܢ MΛ || 5 ܢܬܟܠܠܘܢ: ܢܬܟܠܠܘܢ ΛV || ܩܘܡܬ ΛV || 6 ܘܟܠܠ: ܡܛܠ Λ || 8 ܟܡ M || 9 ܬܬܟܣܝܢ V || ܡܕܡ Λ || 11 ܟܡܘܬ ΛV || **V.ii.1**: 2 ܣܘܟܠܗܘܢ LV || 3 ܐܢܫܝܢ: add. ܠܐ Λ || ܒܡܫܟܢܐ Λ || ܕܠܐ ΛV || 5 ܕܗܘܪܐ M || ܕܡܬܬܪܬܐ Λ || 6 ܣܘܟܐ ΛV || **V.ii.2**: 3 ܐܣܦܠܟܡܢܝܬܐ FM || 4 ܟܬܪܡܢܬ: ܟܬܪ ܡܢܬ LM[1]; ܟܬܪ ܡܢܬ VM[V] || 5 ܡܬܦܢܝܢܘܬܐ: ܡܕܡܢܘܬܐ Λ || 5-6 ܘܡܬܦܢܝܢܘܬܐ . . . ܕܐܣܦܠܟܡܢܐ: om. F || 6-7 ܡܫܝܢ . . . ܟܕ ܟܘܢ: om. M || 8 ܐܐܪܢܐ F || 9 ܦܪܘܬܐ: add. ܟܠܝܢ Λ || ܬܦܢܝܢ: ܬܦܢܝܢ L; ܬܦܢܝܢ MIV ||

conquer the habitable places and those [places] unsuitable for habitation will be laid bare. In this way the earth will be divided into sea and dry land which are not suitable for the sustenance of animals which breathe the air, and this world of ours comes to an end and its shape passes away, until this present configuration recurs and a new world begins. For we know for certain that this northern region was covered by water and for that reason mountains were formed in it, whereas now the seas are in the south. The sea is therefore liable to migration and it may be that one or other of its migrations will be the cause of the disappearance of the habitable world. In this way after every beginning there is an end, and vice versa, though over an indeterminable number of years.

Second Section: On the Fact that Plants and Animals are also Generated without Reproduction. Three theories

[V.ii.1.]
First [theory]: There is not even one proof for the impossibility of the generation of species after their destruction in the manner of a new creation, nor for the necessity of their generation by means of seed and by reproduction from a male and a female. Consider how herbs and trees generated without seed are more numerous than those which are sown. Among animals, too, bees are generated from cow-dung, snakes from hair cast in water, scorpions from ὤκιμον - i.e. basil - and figs, frogs from rain and mice from dust. Although all of these are [usually] generated from a father and a mother, they are also formed without them.

[V.ii.2.]
Second [theory]: Although [such] extraordinary and strange formation is not effected constantly, it is not necessarily the case that it does not occur at all when there occur a certain configuration of the celestial sphere and a favourable disposition of the elements, which are realised on rare occasions in long periods of time. One must not think that the generation of an animal is only accomplished through the casting of the seed in the womb, because all compounds take their origin from mixture of elements and mixture occurs through assembly, so that, just as assembly occurs in the womb, it is also possible for a drop to be assembled out of airy water in some part of the earth and to acquire a mixture similar to that in the belly, and, when the conditions are favourable, for a human form to appear upon it and to grow through

ܫܡܥܠ ܒܫܘܠܡܐ ܕܦܬܓܡܐ ܐܢܫܘܬܐ ܡܢ ܢܣܒ ܕܦܬܓܡܐ ܠܠܘܬܗ ܡܬܚܒܐ ܘܐܬܬܟܝܐ.
ܘܐܡܟܢܐ ܕܐܒܕܘܗܝ. ܗܟܢܐ ܒܐܪܡܝܬܐ ܬܫܝܢܐ ܕܬܗܘܐ ܐܬܡܘܢܬܐ. ܘܐܟܢܐ
ܘܐܡܟܢܐ ܕܠܐ ܐܟܢܐ ܘܐܡܟܢܐ ܒܬܪ ܠܦܘܬܐ ܠܠܘܬܐ ܫܪܘܬܗ ܢܬܦܪܩ.

[V.ii.3]
ܕܬܠܬܐ. ܗܘ ܒܪ ܡܬܦܠܓܘܬܐ ܕܚܡܫܐ ܠܡܘܬܠ ܦܬܓܡܐ ܒܝܘܬܢܘܬܐ. ܐܝܟܢܐ
ܕܡܬܒܥܐ ܬܗܘܐ. ܬܘܒ ܕܝܫܘܬܐ ܘܡܬܦܠܓܘܬܐ ܗܘܝܐ. ܘܐܝܬܝܗ ܠܗ. ܠܘ ܠܗ
ܬܚܝܬܐ ܕܒܝܬ ܚܠܩܬܐ ܐܝܬܝܘܬܐ ܡܬܦܠܓܘܬܐ ܕܐܝܟܢܘܬܗ ܐܝܟ ܕܡܬܒܥܐ ܫܡܝܐ
ܡܢܗ ܬܬܒܥܐ. ܘܐܠܐ ܗܘܐ ܗܟܢܐ. ܐܬܦܠܓܘ ܐܪܒܥܐ ܠܒܪܝܬ. ܘܠܐ ܬܗܘܐ ܗܘܐ
ܘܒܗ. ܒܪܐ ܐܠܗܐ ܕܡܢ ܟܠ ܒܪܝܬܐ ܐܝܬܝܐ ܒܪܝܬܐ ܢܗܘܐ. ܐܝܟ ܡܢ ܟܠ
ܐܠܗܐ ܐܝܬܝܐ ܐܠܗܐ. ܒܬܪܬܝܢ ܢܦܠܘܬܐ ܒܬܪܬܝܢ ܘܬܪܬܝܢ ܒܬܪܬܝܢ.
ܝܚܝܕܝܘܬܐ ܗܝ ܘܡܬܦܪܫܢܘܬܐ. ܕܐܝܟ ܕܡܫܘܚܬܐ ܗܘܝܐ. ܘܠܐ ܐܝܟܢܘܬܐ.
ܘܗܝ ܕܐܝܟ ܗܘܝܐ. ܐܝܬ ܐܡܬܝ ܕܚܙܐ ܠܡܫܬܐ ܠܐ ܗܘܝܐ. ܘܐܠܗܐ
ܠܡܠܟܘܬܗ ܙܒܢܐ ܡܬܦܠܓܘܬܐ ܕܐܝܬܝܘܬܗ ܚܠܩܬܐ ܕܗܘܝܐ ܠܡܠܟܘܬܐ. ܐܝܬ
ܗܘܐ ܐܝܬܐ ܠܡܬܦܠܓܘܬܗ ܐܪܙܐ ܒܠܒܪ ܦܘܫܐ ܘܗܘܝܐ ܕܡܬܒܪܝܐ.

[tit.]
ܫܘܠܡܐ ܬܠܝܬܝܐ. ܡܫܠܡ ܕܝܢ ܐܡܘܪܘܬܐ ܘܠܒܝܫܐ ܘܫܘܪܝܐ ܡܬܚܫܒܢ.
ܬܫܬܐܝܬ ܘܡܐܡܪܐ ܬܢܝܢ.

[V.iii.1.]
ܡܬܒܥܝܐ. ܠܐ ܐܝܟܢܘܬܐ ܡܬܦܠܓܘܬܐ ܡܢ ܡܬܒܥܝܢܘܬܐ ܕܐܝܟܢܐ ܘܡܬܒܥܝܢܘܬܐ. ܐܝܟ
ܡܢ ܒܠܚܘܕ ܐܠܗܐ ܕܐܠܗܐ ܒܪܝܬܐ ܗܘܐ ܦܢܝܢ ܝܬܝܪܐ. ܒܗ ܒܪܝܬܐ ܒܠܚܘܕ
ܒܫܡܝܐ ܒܠܚܘܕ ܐܝܬܝܗ. ܘܠܐ ܒܗܘܝܢܐ. ܘܒܠ ܕܡܬܦܠܓ ܗܘ ܘܡܬܝܕܥܢܘܬܐ
ܠܡܬܗ. ܐܝܟ ܗܘ ܗܘܐ ܗܘ ܘܡܬܒܥܝܢܐ. ܢܬܝܕܥ ܗܟܢ ܘܐܝܟܢܘܬܐ ܡܬܝܕܥܢܘܬܐ
ܐܝܬ. ܘܒܠܚܘܕ ܒܝܬܝܐ ܒܝܬܝܐ ܡܬܦܠܓܘܢ. ܘܡܢ ܙܘ ܠܗܘܢ ܫܡܝܠܐ ܬܗܘܐ
ܦܪܩ. ܘܫܡܝܐ ܠܠ ܫܘܪܝܐ ܕܡܬܚܫܒܢ ܡܬܦܪܫ. ܘܡܫܘܐ ܐܝܟ ܕܐܝܬܝܗ.
ܒܝܬܝܐ ܡܬܝܕܥ ܕܬܗܘܐ ܐܝܬܝܐ ܒܠܚܘܕ ܕܐܝܬܝܐ ܘܦܘܫܐ ܠܗܘܢܐ ܒܚܝܘܬܐ ܡܬܝܕܥ.
ܘܗܘܐ ܗܘܐ ܒܒܠܚܘܕ ܐܠܗܐ ܐܝܟ ܒܫܘܠܐ ܒܠܚܘܕ ܚܝܘܬܐ ܫܪܝܪܬܐ ܕܩܕܝܡܐ
ܘܐܬܢܝܐ ܗܟܢ. ܘܡܢ ܗܘܝܢ ܕܐܝܬܝܗ ܒܬܪܗ ܒܠܚܘܕ ܘܠܚܘܕ ܡܬܝܠܕ ܒܗܘܝܐ.

FMLIV: **V.ii.2**: 10 ܐܟܢܐ ΛV || ܘܐܡܟܢܐ V || 11 ܕܠܐ ܐܟܢܐ ܘܐܡܟܢܐ: om. M[1] || ܠܠܘܬܐ: om. V || **V.ii.3**: 1 ܠܡܘܬܠ ΛV || 4 ܡܢܗ F || 6 ܘܬܪܬܝܢ F || ܒܬܪܬܝܢ: om. M[1] || 7 ܗܘ: ܗܝ M || 8 ܘܐܠܗܐ: ܘܐܠܗ MΛV || **tit.**: ܘܫܘܪܝܐ: illeg. M[1]; ܘܫܘܪܝܐ M[V] || **V.iii.1**: 1 ܐܝܟܢܘܬܐ M || ܘܡܬܒܥܝܢܘܬܐ Λ || 3 ܗܘ: om. FM || 4 ܗܟܢ: ܗܢ Λ || 8 ܚܝܘܬܐ V || 9 ܒܠܚܘܕ: ܠܚܘܕ LV ||

the agency of the giver of forms. Just as in the belly, so in the ground it is possible for a pair to be generated and for a father and a mother to be created at first without a father and a mother after a universal deluge.

[V.ii.3.]
Third [theory]: The mixture suitable for the reception of animal form is better and more complete if it occurs in the womb, but even if does not, it is not impossible for a mixture like it or somewhat inferior to it to be constituted through different causes. If this were not the case, species would be annihilated entirely and would not [then] be generated again and be restored, because it is not necessary that from every man another man be formed, or from every tree another tree, since the casting of the drop in the womb and of the seed on the earth is voluntary and possible, which usually takes place, but not necessary, and there are times, once in a long while, when something like this fails to take place. If it were not for the eternity of the motions of the celestial spheres which are the causes of eternal generations, there would be an opportunity for extinction of species without restoration and generation afresh.

Third Section: On the Fact that Skills, Languages and Writing too are Renewed. Two theories

[V.iii.1.]
First [theory]: Every acquired skill takes its origin from the deliberation of the individual soul or from the divine revelation which occurs to some men, since the universal man is something imaginary and not real and everything whose cause is created and has a beginning is itself created and has a beginning. The sciences and the skills, therefore, have beginnings and for this reason they grow gradually and acquire greater completeness from one period to another. Their renewal testifies to the renewal of their inventor, that is, as we have said, the first man who is created after the wholesale destruction of the species and the general deluge. This man lays the foundations of the teachings and crafts through divine revelation or natural power of reasoning and the "building" is perfected little by little by those who come after him.

[V.iii.2]

ܕܬܬܬܡ . ܕܒܟܘܠܠ ܦܐܪ̈ܐ ܘܢܓ̈ܠܐ ܣܦ̈ܪܐ ܘܡܠ̈ܐ ܚܠܠܝܢ ܚܠܝܐ . ܘܡܢܗܪܐ
ܐܢ̈ܫܝܐ ܕܗܘܝܢ ܒܗܘܬ ܣܘܟ̈ܠܐ ܡܗܝܡ̈ܢܐ . ܒܗ ܒܡܫܘܚܬ ܗܘ ܕܠܫ̈ܢܐ ܠܡܫܝ̈ܚܐ
ܘܙܘ̈ܚܐ ܗ̈ܘܬ ܚܝܘܬܢܐ ܠܒ̈ܢܝܐ ܢܫܘܕܥܘܢ ܘܡܢ ܢ̈ܒܝܐ ܢܫܬܘܕܥܘܢ ܒܟܝܢܐ ܡܝܢ .
ܙܡܝܢܐ ܕܒܡܫܡܫܢܐ ܒܬܦܩܐ ܒ̈ܢܝܢܐ ܘܡܫܡܫܢܐ ܫ̈ܡܝܢܐ ܡܫܬܘܕܥܝܢ
ܘܡܫܬܘܕܥܝܢ . ܐܝܟܢܐ ܠܘܬ ܗܘ ܕܒܫܡܝܐ ܕܗܘ̈ܝܢ ܘܡܫܬܘܕܥܘܬܐ ܫܒܝܚܬܐ
ܘܡܠ̈ܐ ܠܡܫܝܚܘ ܡܬܕܪܟܝܢ . ܒܗ ܡܢ ܐܠܗ̈ܝܐ ܡܫܬܪܝܢ . ܘܒܟܝܢ̈ܬܐ ܢܚܫܬܐ
ܡܫܠܡܝܢ . ܘܒܗ ܦܓܪܐ ܐܢܫܘܬܐ ܘܡܬܕܠܡܝܢ ܒ̈ܢܝܢܐ . ܘܗܪܢܐ ܘܡܬܒܢܝܢܬܐ
ܡܬܕܟܝܢ . ܘܡܢ ܗܪܟܐ ܡܫܘܕܥܢܘܬܐ ܡܠܝܘܬܐ ܒܪܘܚܢܘܬ ܩܘܕܫܐ ܦܪܝܐ .
ܘܠܬܪܒܒ ܕܡ̈ܢܬܐ ܡܬܟܬܒܢܘܬܐ ܒܚܘܫܒܐ ܕܥܡ̈ܡܐ ܘܡܠ̈ܐ ܦܪܝܐ ܒܠܬܐ .
ܘܗܢܐ ܐܦ ܦܐ̈ܪܐ ܡܬܝܗܒܝܢ . ܘܡܫܘܚܠܦ ܠܚܘܒܪܐ ܘܠܥܘܪܐ ܡܫܬܚܠܦܝܢ .

ܫܠܡ ܟܬܒܐ ܕܦܘܫܩܬܐ

FMLIV: V.iii.2: 1 ܕܬܬܬܡ: ܕܬܬܡ M || ܘܡܢܗܪܐ: ܘܡܢܗܝܐ F; ܘܡܢܗܪܢܐ Λ || 2 ܒܗ: ܒܗ Λ || ܕܠܫ̈ܢܐ: ܕܠܫܢܐ Λ || 3 ܘܡܢ ܢ̈ܒܝܐ ܢܫܬܘܕܥܘܢ: om. F || 5 ܘܡܫܬܘܕܥܘܬܐ V || 6 ܐܠܗ̈ܝܐ Λ || ܘܒܟܝܢ̈ܬܐ Λ || 8 ܡܬܕܟܝܢ L; ܡܬܕܟܝܢ M; ܡܬܕܟܝܢ V || 8 ܡܫܘܕܥܢܘܬܐ ܡܠܝܘܬܐ F || ܒܪܘܚܢܘܬ F || ܩܘܕܫܐ FM || 10 ܠܚܘܒܪܐ: ܚܘܒܪܐ M[1] || ܘܡܫܬܚܠܦܝܢ M || **fin.**: ܕܦܘܫܩܬܐ M || ܕܦܘܫܩܬܐ: add. ܘܠܐܠܗܐ ܫܘ̄ ܕܟܠ M; add. ܘܠܐܠܗܐ ܫܘܒܚܐ ܕܚܝܠ ܘܥܕܪ L; add. ܘܠܐܠܗܐ ܫܘܒܚܐ ܕܚܝܠ ܘܥܕܪ ܐܡܝܢ l; add. ܘܠܐܠܗܐ ܫܘܒܚܐ V ||

[V.iii.2.]

Second [theory]: It is obvious that writing and speech disappear with the disappearance of writers and speakers. Since the human individuals who arise after the annihilating destruction have by nature an aptitude for communicating to each other [their] emotions and thoughts, the likelihood is that they will at first communicate [them] with signs made with fingers and shapes made with limbs. They will then advance to a stage where instead of things and actions they will use nouns and verbs, beginning with necessary things and ending with less useful things. When the population increases and interests become opposed, villages and cities are built. Hence vocal communication becomes inadequate with the distance between persons and necessity calls for composition of written images instead of [spoken] nouns and verbs. Thus writing too is renewed and come to differ with the differences in modes of life and inhabitants.

Here ends the Book of Meteorology.

COMMENTARY

PRELIMINARY REMARKS

The emphasis has been placed in the following commentary on the identification of the immediate sources used by Barhebraeus (**BH**) and the examination of the manner in which he paraphrases and alters the material he found in his sources. For the purpose of comparing the text of *But.* with those of his sources, the relevant passages are usually quoted in extenso and underlining is used to indicate the correspondences between the passages (continuous lines for exact matches, dotted lines for less exact matches; italics are used when comparing a passage of *But.* with a plural number of passages).

The work most frequently cited in this commentary is the *Kitāb al-šifā* of Ibn Sīnā (**IS**), since that was the work used as the main source of *But.* by BH. Other Arabic works frequently cited include the *K. al-muʿtabar* (***Muʿtabar***, *Muʿt.*) of Abū al-Barakāt al-Baghdādī and the *K. al-mabāḥiṯ al-mašriqīya* (***Mabāḥiṯ***, *Mab.*) of Fakhr al-Dīn al-Rāzī. Both these works are themselves related in different ways to the *Šifāʾ*. Of the two, Abū al-Barakāt's work is usually more independent of the *Šifāʾ*, whereas Fakhr al-Dīn al-Rāzī's work, in those parts dealing with mineralogy and meteorology, is largely a summary of the *Šifāʾ*. While there are a number of instances where one can be certain that BH has used the *Mabāḥiṯ* as his source in preference to the *Šifāʾ*, there are also instances where the wording of the *Šifāʾ* and of the *Mabāḥiṯ* are so close to each that it becomes futile to decide which work BH is following as his source in a particular passage of *But.* In such cases the comparison is usually made only with the *Šifāʾ*.

The Syriac work most often quoted in the commentary is the Syriac version of Nicolaus Damascenus' *Compendium* of Aristotelian philosophy as preserved in ms. Cambridge, Gg. 2.14 (= **Nic. syr.**). As has been explained in the Introduction above, the text preserved in this Cambridge manuscript contains a great amount of scholial material which must be subsequent in date to Nicolaus and a large proportion of which can be linked to Olympiodorus' commentary on Aristotle's *Meteorologica* (**Olymp.** ***in Mete.***, or simply "Olymp."). Where there are positive indications in the manuscript that a particular passage is a scholion, this will be indicated in the commentary below. There are, however, also instances where one suspects a passage to be scholial but where there are no positive indications in the manuscript to that

effect and further study will be required to determine the exact provenance of the passage. In the commentary below, such passages too will simply be quoted under the designation "Nic. syr." without further qualification, a designation which should therefore be understood to stand, strictly-speaking, for "Nicolaus vel Olympiodorus syriacus".

Systematic examination of the relationships between the works used by BH as his sources and their relationships in turn to their Greek sources has been considered beyond the brief of the present commentary, although reference will be made as appropriate to such Greek works, in particular to Aristotle's *Meteorologica* (**Arist**. ***Mete***., or simply "Arist."). With quotations from Nic. syr. I limit myself largely to naming the related passages in related works without going into a discussion of the relationships, since that is a discussion which is better reserved for an edition of the text of Nic. syr. itself. The works named in such instances will normally include Arist. *Mete*. and Olymp. *in Mete*. (whereby the citation of passages of Olymp. does not necessarily indicate a judgement that the passage of Nic. syr. is a scholion derived from Olymp.), as well as the so-called Arabic version of Olympiodorus' commentary (**Olymp. arab**.), a work that is derived to a large extent from the text designated here as Nic. syr., and BH's own *Candelabrum sanctuarii* (***Cand***.) in those instances where BH had used the same passages of Nic. syr. as his source in that earlier work of his. References will also be made occasionally to the commentary on Arist. *Mete*. by Alexander of Aphrodisias (Alex. Aphr.), especially in those parts where the relevant portions of Olymp. *in Mete*. are lost in lacunae, as well as to the *Treatise on Meteorological Phenomena* (*Mutaḫayyila*) by Ibn Suwār Ibn al-Khammār, a work where Nicolaus (most probably in the Syriac version) is explicitly named as one of the sources.

The text of Nic. syr. as preserved in the Cambridge manuscript is incomplete and fragmentary, which means that there are likely to be passages in the parts of *But*. examined here which are based on lost passages of Nic. syr. This applies in particular to the discussion of the sea (*But*. Min. Chap. V), where there is a long lacuna in the Cambridge manuscript, but there are also instances elsewhere where one is led to suspect lost passages of Nic. syr. behind the text of *But*. In such instances passages from works related to Nic. syr. (Arist., Olymp., Olymp. arab.) are used to corroborate the suspicion.

BOOK OF MINERALOGY

The "book" (*ktā̱bā*) as a whole corresponds to the treatise (*maqāla*) on minerals (*al-ma*ᶜ*ādin*) in the *Šifā*ᵓ. For the overall correspondence of the chapters in this book to those in the *Šifā*ᵓ, see Introduction, Section 3.1 above.

1. 1: ܟܬܒܐ ܕܡܛܐܠܝܩܘܣ ("The Book of Mineralogy"): lit. "of minerals". - The word *mehṭalīqō* (< gr. μεταλλικοί) is used here as the equivalent of IS's *al-ma*ᶜ*ādin*, which is used in the title of the corresponding *maqāla* in the *Šifā*ᵓ (Ṭabīᶜbīyāt V.1). In *But*. the term is used as a synonym of *meṯhaprānē* (lit. "things which are dug/mined"), as may be seen from I.i.1.6 and III.i.1.1 below.

1. 2 ܟܬܒܐ ܕܚܐܘܬ ܚܟܡܬܐ ("the Book of the Cream of Wisdom"): On the title of the work, see Introduction, Section 1.2 above.

BOOK OF MINERALOGY, CHAPTER ONE

ON MOUNTAINS AND SPRINGS

The chapter is divided into three sections: I.i. Formation of stones; I.ii. Formation of mountains and their uses; I.iii. Springs. The corresponding part of the *Šifā*ᵓ (ed. Cairo p. 3-14) is also divided into three chapters (*fuṣūl*)[1]: I. on mountains and their formation; II. on the uses of mountains and formation of clouds and moistures; III. on springs. The dividing point between the first two sections, however, differs between *But*. and *Šifā*ᵓ The first *faṣl* in *Šifā*ᵓ is divided into (i) preliminary discussions (3.9-7.9) dealing with: a. "formations of stones, b. "formation of larges stones or stones in large numbers" and c. "formation of elevation",[2] and (ii) d. the discussions on the formation of mountains proper. The correspondences between *Šifā*ᵓ and *But.* may be summarised as follows.

---

[1] 1. فى الجبال وتكونها [وتكونها: om. DSM, Holmyard], p. 3-9; 2. فى منافع الجبال وتكون السحب والأنداء [والاندية Holmyard], p. 10-13; 3. فى منافع المياه [add. وفى الزلازل Holmyard, combining this with the following *faṣl*], p. 13-14.

[2] See *Šifā*ᵓ 3.6-8: a.: حال تكون الحجارة; b.: حال تكون الحجارة الكبيرة أو الكثيرة; c.: حال تكون ما يكون له ارتفاع وسمو.

But.	*Šifā*ʾ
I.i.: stones (formation)	I.a.
I.ii.: mountains (formation and uses)	I.bcd, II.
I.iii.: springs	III.

Much of the material in the chapter is taken from the *Šifā*ʾ. Abū al-Barakāt's *K. al-muᶜtabar* is used as an additional sources in Section I.ii. at I.ii.1.3-10 and again in Section I.iii. at I.iii.2.5-3.7. Nic. syr. is certainly used towards the end of Section I.i. There are also a number of passages where one is led to suspect the use of lost passages of Nic. syr. in Section I.ii. and I.iii.

Min. I. Section i.: *On Formation of Stones*

The first 'theory' of this section serves as a preamble to the whole of Book IV. In a number of other books in *But.*, BH devotes the first chapters of the books to what he calls the προθεωρία (ܩܕܡܬܐܘܪܝܐ), in which he lays down the principles which will be applied in the respective books and often also gives a list of the subjects treated there. - This happens in Part II (Naturalia) of *But.* in Books I (Ausc. phys.), III (De corr. et gen.), VII (De animalibus) and VIII (De anima). In Book II (De caelo) the προθεωρία occupies the first section of the first chapter. Similarly, the first chapters are entitled "on the προθεωρία" in Part III, Books I (metaphysics) and in all three books of Part IV (Eth., Econ. and Polit.). Of the remaining three books of Part II, the first chapter of Book VI (De plantis), though not entitled "προθεωρία" ("on the constitution [*quyyāmā*] of plants") deals with what applies in general to plants. There is no extensive προθεωρία, on the other hand, in Books IV (minerals) and V (meteorology). This is probably to be explained by the diversity of the subject matter treated in these books which make the kind of generalisation made elsewhere difficult and by the fact that many of the principles applied here have been laid down in earlier books, mainly in Books I and III of Part II (Ausc. phys. and De gen. et corr.).

The rest of the section (I.i.2-5) corresponds by and large to the first half of the chapter *(faṣl)* of the *Šifā*ʾ on the formation of mountains where IS discusses the formation of stones as a preliminary to his discussion of the formation of mountains. The source for much of I.i.2-4 is the *Šifā*ʾ and the order of the material too largely follows that of the *Šifā*ʾ, although BH has done some editing of the repetitive

material in I.i.2 and has combined a sentence taken from the *Šifā*ʾ with one taken from Nic. syr. at I.i.4.1-3. 'Theory' I.i.5, giving us further examples of formation of stone, is based not on the *Šifā*ʾ but on some passages in the mineralogical section of Nic. syr., whereby it is to be noted that that part of Nic. syr. apparently derives not from Aristotle but from Theophrastus, as has been argued elsewhere (see Takahashi [2002a]).

The correspondences of 'theories' I.i.2-5 to their sources may be summarised as follows.

	*Šifā*ʾ	Other sources	
I.i.2.1-2	3.9-10		Formation (1): not pure earth
I.i.2.2-5	3.10-15, 4.18-9		Formation (2): by baking
I.i.2.5-8	3.10-11, 4.1-13		Formation (3): congelation
I.i.2.8-10	3.15-17		IS's report
I.i.2.10-12	-		BH's report
I.i.3.1-6	5.1-7		Petrified plants and animals
I.i.3.6-9	5.10-13		Report of hardened bread
I.i.4.1-3	5.14-16	Nic. 58.25-26	Stones in thunderbolts
I.i.4.3-11	5.20-6.9		Meteorite in Jūzjān
I.i.4.11-12	6.10-11		Swords in Yemen
I.i.5.1-4	-	Nic. 58.27-29	Petrification in human body
I.i.5.4-6	-	Nic. 58.29-30	Glass and coral
I.i.5.6-8	-	Nic. 58.13-16	Artificial stone
I.i.5.8-10	-	Nic. 59.3-13	Volcanic stones on Etna

Min. I.i.1.1-7:

The first part of the preamble here, written in rhymed prose (in *-an*), is reminiscent of the opening words of Arist. *Mete*. (I.i, 338a 20-26), where Arist., after summarising the subjects of *Ausc. phys.*, *De caelo* and *De gen. et corr.*, introduces the subject of 'meteorology'. A similar preamble is found in the *Šifā*ʾ at the beginning of the fourth book (*fann*) on the natural sciences (De actionibus et passionibus, *al-afʿāl wa-l-infiʿālāt*). BH, who has no independent book corresponding to IS's fourth *fann* in the part of *But.* on the natural sciences but has incorporated much of the material taken from there into his Book III (De gen. et corr.), has transferred this preamble to the beginning of his Book IV (corresponding to IS's fifth *fann*), adding to the list of subjects already treated a summary of the subjects treated in IS's fourth *fann*.

But.: (1) Since to [the discussion of] matters common to [all] natural things (2) we have duly [*ak̠ zed̠qā*] added the science [*īd̠aʿtā*] of the primary

movements in the world of nature, (3) accurately set down the matters of incorruptible and corruptible bodies (4a) and then completed the discourse [*melltā*] on generation and corruption, (4b) the actions and passions of the primary qualities and the mixtures generated from them, (5) the proper order [for the exposition] of the teaching draws us on to the science of the μεταλλικοί, i.e. mineral bodies. (6) For these come first and closest to the elements.

Šifā, Act. et pass. [Qassem 1969] 201.4-10:

قد فرغنا من تعريف الأمور العامة للطبيعيات، ثم من تعريف الأجسام والصور والحركات الأولية فى العالم واختلافها فى طبائعها، ثم من تعريف أحوال الكون والفساد وعناصرها، فحقيق بنا أن نتكلم عن الأفعال والانفعالات الكلية التى تحصل عن الكيفيات العنصرية بمعاضدة من تأثيرات الأجرام السماوية، فإذا فرغنا من ذلك شرعنا حينئذ فى تفسير أحوال طبقات الكائنات، مبتدئين بالآثار العلوية والمعدنيات، ثم ننظر فى حال النفس. فإن النظر فى النفس أعم من النظر النبات والحيوانات، ثم ننظر فى النبات ثم فى الحيوانات.

"We have completed (1) the exposition of matters common to natural things, (2) then the exposition [*taʿrīf*] of bodies, shapes, the primary movements in the world and their differences in their natures, (4a) then the exposition of the conditions of generation and corruption, and their elements. (5) It is proper, therefore, (4b) that we should talk about the universal actions and passions which arise out of the elementary qualities with the help of the influences of the heavenly bodies. (5) When we have expounded these things, we shall begin the explanation of the conditions of the categories of things which come to be, starting with the meteorological phenomena and minerals. Then, we shall examine the condition of the soul. - For the examination of the soul is more general than the examination of plants and animals. - Then, we shall examine the plants, and then the animals.

l. 1. ܨܒ̈ܘܬܐ ܓܘܢܝ̈ܬܐ ܕܟܝܢܝ̈ܬܐ ("matters common to natural things", lit. "common matters of natural things"): IS الأمور العامة للطبيعيات. - The reference is to *But*. Part II, Book I (Ausc. phys.). - BH has altered IS's *li-* (syr. *l-*) to *d-*, although both constructions are possible with syr. *gawwānāyā* (see PS 668f.; cf. gr./lat. κοινός/communis + gen. or dat.).

l. 1. ܝܕܥܬܐ ("science"): IS تعريف. - Although *īdaʿtā* properly means "knowledge/science" as translated here (corr. arab. *maʿrifa*) rather than "exposition" (*taʿrīf*), the word may be intended here as an equivalent of *taʿrīf*, there being no noun derived from the Pael form of √*ydʿ* (corr. *ʿarrafa*) in Syriac and the nouns derived from its causative forms (*tawdīṯā*, *šuddāʿā*) having rather different meanings.

l. 1-2. "primary movements in the world of nature": i.e. the movement of the celestial bodies discussed in Book II (De caelo).

l. 2-3. "the matters of incorruptible and corruptible bodies": The phrase has no equivalent in the *Šifāʾ*. The "incorruptible bodies" are the celestial bodies treated in Book II, while the sublunar "corruptible bodies" were treated in Book III (De gen. et corr.).

l. 3. "generation and corruption": i.e. the subject matter of Book III (De gen. et corr.).

l. 4. "actions and passions of the primary qualities and the mixtures generated from them": These are the subjects treated in IS's fourth *fann* (De act. et pass.) and incorporated by BH in his Book III. "Mixtures" are not mentioned in IS's preamble quoted above, but are the subject of the second half (*maqāla* II) of the fourth *fann*.

I.i.1.7-9: Preamble (2)

But.: Although [*kad*] this book will teach about them [i.e. minerals] and [also] about mountains, springs, earthquakes and the position of the habitable world, it is called by their name, just as there are [other] cases where the whole is called by the name of its part.

The enumeration of the subjects treated in the book is reminiscent of Arist. *Mete*. I.i, 338a 26-339a 5, where Arist. gives us a rundown of the subjects treated in his book, although the content of the list naturally differs as BH follows IS in placing 'mineralogy' here instead of 'meteorology'.

The question as to why the book is called the "Book of Mineralogy" despite its treatment also of other subjects is a natural one to ask. An analogous question is posed in a passage marked off as a scholion near the beginning of the section of Nic. syr. on meteorology, although the answer there is naturally different.

Nic. syr. 10.17-20 [scholion[3]]:

ܥܠܬܐ ⁘ ܟܕ ܢܣܒܗ ܕܟܬܒܐ ܐܝܬܝܗܘܢ : ܟܠ ܐܝܠܝܢ ܕܗܘ̈ܝܢ ܡܢ ܬܪ̈ܬܝܗܘܢ
ܥܠܬܐ ⁘/18 ܐܝܟ ܡܬܩܪܐ ܣܟܘܠܬܢܘܬܐ ܦܚ̈ܗܝܬܐ . ܘܐܡܪܝܢܢ ܕܟܕ ܐܝܬܝܗܝܢ ܕܫܚܝܢܢ .
ܩܠܝܠܢ/19 ܥܠܬܐ . ܘܒܢܦܫܗܝܢ ܐܢܐܬܘܡܝܐܣܝܣ ܗ̄ ܕܝܢ ܪ̈ܗܛܢ ܠܥܠ ܐܢܝܢ .
ܐܦ/20 ܡܬܟܠܝܢ ܒܙܒܢ ܡܢ ܣܘܠܩܐ .

[19 ܐܢܐܬܘܡܝܐܣܝܣ scripsi : ܐܢܐܠܘܡܝܐܣܝܣ cod. || ܐܦ (cf. Olymp. πλήν εἰ): ܐܦ cod.]

Comment: Seeing [*kaḏ*] that the subject of the book is those things which arise from the two exhalations, why is it entitled "the science of the meteora [*paḥhāyāṯā*]"? We say: Because the exhalations are hot, they are light and are, in themselves, ἀναθυμιάσεις, i.e. things which run upwards, although they are sometimes prevented from ascending.

The answer given here in *But.* reminds one of IS's explanation as to why the science of metaphysics is called "theology" (*ʿilm al-ilāhī, ilāhīyāt*).

---

[3] Cf. Olymp. 3.30-33: ἡ αἰτία τῆς ἐπιγραφῆς. ἐπιγέγραπται Μετέωρα οὐ μάτην, ἀλλ' ἐκ τῆς ὑλικῆς αἰτίας, φημὶ δὲ διὰ τὰς ἀναθυμιάσεις. αὗται γὰρ ὅσον ἐφ' ἑαυτὰς ἐν τῷ μετεώρῳ φέρονται, πλὴν εἴ ποτε διὰ βίαν ἐν τῇ γῇ ἐπισχεθῶσιν.

Šifā', al-Ilāhīyāt, maqāla 1, faṣl 3 [Anawati-Zayed 1960] 23.5-6:

كثيراً ما تسمى الأشياء من جهة المعنى الأشرف، والجزء الأشرف، والجزء الذى هو كالغاية.

Often things are named after its noblest concept, its noblest part or the part which is its aim.

I.i.2.1-2: Formation of stones (1): not from pure earth

But.: Genuine earth does not usually turn into stone, because the abundance of its dryness imparts not cohesion but crumbling to its parts.

Šifā' 3.9-10:

ونقول أما فى الأكثر فإن الأرض الخالصة لا تتحجر لأن استسلاء اليبس عليها لا يفيدها استمساكا، بل تفتتا

We say: In most cases, unadulterated earth is not petrified, because the mastery of dryness over (the earth) does not favour its cohesion, but its crumbling.

l. 1. ܗܘܐ ܟܐܦܐ ("turn into stone"): تتحجر IS. - The Arabic verb *taḥaǧǧara* has constantly to be paraphrased as Syriac possesses no equivalent word.

l. 2. ܠܐ ܗܘܐ ("not"): = ܠܘ. - The variant readings here in mss. other than F indicate the failure of the later copyists to recognise this relatively rare and archaic phrase.[4]

I.i.2.2-5: Formation of stones (2): from clay by baking

But.: (1) Stones are formed in two ways: (2) firstly, by way of [*ba-znā d-*] baking [*puḫḫārā*] of glutinous [*ṭallūšā*] clay in the sun - (3a) as we see in the case of soft stones, (3b) which are intermediate between stone and clay at first, (3c) and later become hard.

Šifā' 3.10-15:

وإنما تتكون الحجارة فى الأكثر على وجهين من التكون: أحدهما على سبيل التفخير، والثانى على سبيل الجمود. فإن كثيرا من الأحجار يتكون من الجوهر الغالب فيه الأرضية، وكثير منها يتكون من الجوهر الغالب فيه المائية. فكثير من الطين يجف ويستحيل أولا شيئا بين الحجر والطين، وهو حجر رخو، ثم يستحيل حجرا. وأولى الطينات بذلك ما كان لزجا، فإن لم يكن لزجا فإنه يتفتت فى أكثر الأمر قبل أن يتحجر.

[11 التفخير : التفجر ed. Cair.]

(1) Stones are formed, usually, in two ways: (2) through [*ʿalā sabīl*] baking [*tafḫīr*] and through congelation. Many stones are formed from a substance in which earthiness predominates, while many of them are formed from a

---

[4] Modern grammarians agree in considering the particle *law* to be a contraction of *lā + hū* (PS 1898; Nöldeke [1898] §328B; Brockelmann [1925] §224; Duval [1881] §294 3; Joosten [1992] 588 n.41; Muraoka [1997] §93.11). BH himself, however, explains *law* as a shortened form of *lā (h)wā*, with which it is semantically equivalent (BH *Splend.* [Moberg] 174.31: ܠܘ ܡܢ ܠܐ ܗܘܐ ܡܬܚܒܪ). - For some other instances of *lā (h)wā* in BH, see *Denḥā* [Chabot] line 965; *Carmina* [Dolabani] 95.22, 97.10, 165.17 (all in verse works).

substance in which wateriness predominates. (3ab) Clay often dries and is transformed first into something [intermediate] between stone and clay, i.e. a soft stone; (3c) then, it is transformed into stone. (2) The most suitable (type of) clay for that (purpose) is that which is glutinous [*laziğ*].[5] For if it is not glutinous it usually crumbles before it is petrified."

Šifā' 4.18-19:

فتكون الأحجار إذن إما لتفخير الطين اللزج فى الشمس، وإما لانعقاد المائية من طبيعة ميبسة أرضية، أو سبب مجفف حار.

(1) Stones are formed, therefore, (2) either by baking of glutinous clay in the sun, or by the solidification of wateriness by a drying, earthy nature, or a hot, desiccating cause.

l. 3 ܦܘܚܪܐ ("baking"): التفخير IS (syr. √*pḥr*: corr. arab. √*fḫr*). - The noun *puḥḥārā* may be a coinage by BH in this sense.[6] The corresponding verb *paḥḥar* is used in the relevant sense (denominative, < *paḥḥārā*) at Bar Kepha, *Hex.*, Paris 311, 37v a5,[7] and Job of Edessa, *Book of Treasures* [Mingana] 405a 5f. (cf. also PS Comp. 441b).

I.i.2.5-8: Formation of stones (3): from water by congelation

The passage of *But.* here may be considered a summary of the passages of the *Šifā'* quoted below.

But.: (1) Secondly, by way of congelation [*qṭārā*] of water, (2) as it drips [*nāṭpīn*] or trickles [*nādrīn*]. (3) Water congeals either through the mineralising power of the earth on which it falls, (4a) or through an earthy nature [*kyānā arʿānāyā*] in which heat predominates - (4b) this being how salt congeals - (5) or [in which] coldness [predominates], (6) or a property which is unknown to us.

Šifā' 3.10-11 [see under I.i.2.2-5 above]: Stones are formed, usually, in two ways: (2) through baking (1) and through congelation [*ğumūd*].

Šifā' 4.1-13:

وقد تتكون الحجارة من الماء السيال على وجهين: أحدهما أن يجمد الماء كما يقطر أو كما يسيل برمته. والثانى يرست منه فى سيلانه شىء يلزم وجه مسيله وتتحجر. وقد شوهدت مياه تسيل، فما يقطر منها على موضع معلوم ينعقد حجرا أو حصى مختلفة الألوان. وقد شوهد ماء قاطر، إذا أخذ لم يجمد وإذا انصب على أرض حجرية تقرب من مسيله انعقد فى الحال حجرا. فعلمنا أيضا أن لتلك الأرض قوة معدنية، تحيل السيال إلى الجمود. فمبادئ تكوّن الحجارة، إما جوهر طينى لزج، وإما جوهر تغلب فيه المائية. وهذا القسم

[5] On the significance of the concept of glutinosity [*luzūğ*] here and its possible connection with the "unctuous moisture" of the alchemists, see Freudenthal (1991) 59-62.

[6] The word seems to be unknown to the lexica in this sense, except in the lexicon compiled by Adler, where, according to PS 3085, the word was explained as "*efformatio testae*", an explanation which is rejected by PS.

[7] Where we have ܡܦܚܪܐ, which is to be preferred against the reading ܡܦܚܕܐ of ms. Paris 241, 162r 5; cf. Bakoš (1930-3) 84, footnote.

يجوز أن يكون جموده من قوة معدنية مجمدة، ويجوز أن يكون قد غلبت عليه الأرضية على الوجه الذى ينعقد به الملح، بأن غلبت الأرضية فيه بالقوة دون المقدار؛ وإن لم يكن على نحو كيفية الأرض التى فى الملح، بل على كيفية أخرى، ولكن مشاركة لها فى أنها تتغلب بمعاونة الحرارة، فلما يصيبه الحر يعقده، أو قوة أخرى مجهولة عندنا. ويجوز أن يكون بالضد، فتكون أرضيته تتغلب بقوة باردة يابسة تعينه.

A stone may be formed from running water in two ways. (1) First: water congeals [*yağmudu*] in its entirety (2) as it drips [*yaqṭuru*] or as it flows [*yasīlu*]. Second: there is deposited from (the water) as it flows something which adheres to the surface of its channel and petrifies. It has been observed how a portion of running water, dripping on to a certain place, solidifies into stone or pebbles of different colours. (3) Dripping water has been seen, which when extracted does not congeal, but when poured on stony earth near its channel immediately solidifies into stone. We then also know that that earth has a mineralising power, which turns what is liquid into a solid. The bases for the formation of stones are either a glutinous clayey substance or a substance in which wateriness predominates. (3) The congelation of the latter category may occur because of a mineralising, solidifying power; (4a) it may occur (when) earthiness [*arḍīya*] has come to predominate over it (4b) in the (same) way as that in which salt is solidified [*yanʿaqidu*], (namely) in that the earthiness predominates in it by (its) power and not by (its) quantity - Even if it is not in the manner of the quality of the earth which is in salt, but of a different quality, nevertheless they are similar in that (earthiness) gains mastery through the help of heat, so that when heat comes upon it it solidifies them. - (6) or another power unknown to us; (5) it may occur through the opposite [of heat], so that earthiness predominates through a cold, dry power which helps it.

l. 6-7 ܐܘ ܒܟܝܢܐ ܐܪܥܢܝܐ ܕܚܡܝܡܘܬܐ ܫܠܝܛܐ ܒܗ ("or through an earthy nature [*kyānā arʿānāyā*] in which heat predominates"): This is not quite what IS says, who in fact talks of earthiness predominating w*ith the help of* heat.

I.i.2.8-10: IS's report of petrified clay

The corresponding passage in the *Šifāʾ* occurs between the passages corresponding to I.i.2.2-5 (stones formed from clay) and I.i.2.5-8 (from water) above and provides an example of how stones are formed from *clay*. The point becomes blurred in *But.* through its transfer to after the discussion of how stones are formed from *water*.

But.: (1) The princely doctor [i.e. Ibn Sīnā] has recounted how, (2) when [he was] a child, he saw on the banks of the [river] Gihon clay [used for] washing the head. (3) Twenty-three years later he found it in the form of stone, though porous [*paḥḥīḥā*].

Šifāʾ 3.15-17:

وقد شاهدنا فى طفولتنا مواضع كان فيها الطين الذى يغسل به الرأس، وذلك فى شط جيحون. ثم شاهدناه قد تحجر تحجرا رخوا، والمدة قريبة من ثلاث وعشرين سنة.

(2) I saw, in my childhood, places in which there was clay with which head was washed. - That was on the river Ǧaihūn. - (3) Later I saw that (the clay) had petrified into soft [*raḫw*] stone. The interval was approximately twenty-three years."

l. 8. ܐܣܝܐ ܪܝܫܢܝܐ ("the princely doctor"): = arab. *al-šaiḫ al-raʾīs*, the frequently used byname of Ibn Sīnā. This is the name by which BH usually refers to IS in *But*. See Introduction, Section 3.2.3a above.

l. 9. "[river] Gihon": شط جيحون IS. - i.e. the Oxus, Amū Daryā.

I.i.2.10-12: BH's report

Prompted by IS giving an example from his own experience, BH adds an observation of his own here. - For historians it will be a matter of regret that BH does not give the name or the exact location of the monastery mentioned here.

But.: I [too] in my childhood went with my parents to a new monastery which was being built beside Melitene and saw in front of the builders a single block of hard stone in which (pieces of) pottery and coal [*qarbōniya* < καρβώνια] were mixed.

I.i.3.1-6: Petrified plants and animals

But.: (1) It is probably true [lit. there is probability to the (assertion)] that certain animals and plants have turned into stone in various places, (2a) because the transformation [*šuḡnāyā*] of compounds into a stony state [*kēpānāyūṯā*] through the agency of [*b-yaḏ*] the earthiness which they contain is not difficult (2b) but is in fact easier than the transformation of pure [*pšīṭē*] water into a stony state. (3a) This happens through the severity [*ʿušnā*] of the mineralising force in certain regions, (3b) or [it is] through the intensity [*ʿuzzā*] of the heat which the earth suddenly experiences during a severe [*ʿaššīnā*] earthquake (3c) [that] bodies which are found on it become hardened.

Šifāʾ 5.1-7[8]:

وإن كان ما يحكي من تحجر حيوانات ونبات صحيحا، فسبب فيه شدة قوة معدنية محجّرة تحدث فى بعض البقاع الحجرية، أو تنفصل دفعة من الأرض فى الزلازل والخسوف، فتحجر ما تلقاه. فإنه ليس استحالة الأجسام النباتية والحيوانية إلى الحجرية، أبعد من استحالة المياه، ولا من الممتنع فى المركبات أن تغلب عليها قوة عنصر واحد يستحيل إليه. لأن كل واحد من العناصر التى فيها، مما ليس من جنس ذلك العنصر، من شأنه أن يستحيل إلى ذلك العنصر، ولهذا ما يستحيل الأجسام الواقعة فى الملاحات ألى الملح، والأجسام الواقعة فى الحريق إلى النار.

(1) If what is told concerning the petrification of animals and plant(s) is true, (3a) the cause of this is the intensity of the mineralising, petrifying

[8] Cf. Fakhr al-Dīn al-Rāzī, *Mabāḥiṯ* II.207.13-16.

power, which arises in certain stony places (3b) or is suddenly discharged from the earth during earthquakes and subsidences (3c) and petrifies whatever it comes into contact with. (2a) For the transformation [*istiḥāla*] of plant and animal bodies into a stony state [*ḥaǧarīya*] (2b) is no more improbable than the transformation of waters, (2a) and it is not impossible in compounds for the power of one element to predominate over them, into which (element) (the compound) is [then] transformed, because it is in the nature of every element in (such compounds) other than that element to be transformed into that element. For this reason, bodies falling into salt-pans are transformed into salt and bodies falling into flame into fire.

l. 2. ܐܪܥܢܝܘܬܐ ("earthiness"): In the latter part of the passage of the *Šifāʾ* quoted above, IS talks generally of "compounds" and "elements" and does not mention the word "earth/earthiness", but BH will have understood that the element which brings about petrification by its predominance can only be "earth".

l. 4-6. "or [it is] through the intensity of the heat ... [that] bodies which are found on it become hardened". - This is how the sentence has to be translated as the text stands, without any conjunction before the verb *metqaššēn* corresponding to IS's "*fa-*". The mention of "heat" is also an addition which is not in the *Šifāʾ*.

I.i.3.6-9: Report of hardened bread

But.: (1) The same [teacher], the most excellent of the moderns, the princely doctor [Ibn Sīnā] has said (2) that he [once] saw a loaf of bread [*ḥāḥurtā*], like those loaves whose middle the bakers make thin and which they imprint [*tābʿīn*] with images of animals. (3a) It was like a stone, (3b) but its colour had not been altered (3c) and on one of its sides were lines [of the kind] which are formed on them by [*men*] the oven.

Šifāʾ 5.10-13[9]:

وقد رأيت رغيفا على صورة الأرغفة المحرقة، المرققة الوسط، المرقومة بالسباع؛ قد تحجر، ولونه باق، وأحد وجهه عليه أثر التخطيط الذى يكون فى التنور. وجدته ملقى فى جبل قريب من بلدة من بلاد خراسان تسمى جاجرم، وحملته معى مدة.

(2) I [once] saw a loaf of bread [*raġīf*] in the form of baked loaves made thin in the middle [*muraqqaqat al-wasṭ*] and imprinted [*marqūm*] with animals.[10] (3a) It had been petrified (3b) but its colour remained. (3c) On one side of it was the trace of the line [*taḫṭīṭ*] which is formed in the oven. I found it discarded on a mountain near a town in Khurāsān called Jājarm and carried it with me for a while.

---

[9] Cf. Fakhr al-Dīn al-Rāzī, *Mabāḥiṯ* II.207.17-18; Qazwīnī, *Aǧāʾib* 209.19-23.

[10] IS's wording allows different interpretations; cf. tr. Holmyard: "showing the marks of a bite" (Holmyard-Mandeville [1927] 23).

l. 7. ܐܦܘܝܐ ("bakers"): BH makes the picture more concrete, as it were, by naming the agent where IS has a passive construction without naming the agent: cf. Min. I.i.4.9 ("smiths"); III.ii.3.2 ("nature") below.

l. 9. ܡܢ ܬܢܘܪܐ ("by the oven"): فى التنور ("in the oven") IS. - BH may have read من instead of فى (the two words are often indistinguishable in Arabic manuscripts).

<u>I.i.4.1-3</u>: Stones in thunderbolts (meteorites)

But. [underline: agreement with Nic.; italics: with *Šifāʾ*]:

(1) The master [Aristotle] has said (2) that <u>in the fall of a thunderbolt</u> [κεραυνός], <u>too, something which is stony</u> and <u>not oily in</u> its <u>nature</u> <u>is observed</u>. (3) This occurs due to a *fieriness* which *is* suddenly *extinguished and becomes cold and dry*.

Nic. syr. 58.25-26:

ܘܐܦ ܒܡܦܘܠܬܗ ܕܩܪܐܘܢܘܣ ܡܬܚܙܐ ܡܕܡ ܕܟܐܦܢܝܐ . ܕܠܐ ܕܗܝܢ ܒܟܝܢܐ .

(2) <u>In the fall of a thunderbolt, too, something which is stony</u>, which [*d-*] <u>is not oily in</u> <u>nature</u>, <u>is observed</u>.

Šifāʾ 5.14-16[11]:

وقد تتكون أنواع من الحجارة من النار إذا أطفئت. وكثيرا ما يحدث فى الصواعق أجسام حديدية وحجرية، بسبب ما يعرض للنارية أن تطفأ فتصير باردة يابسة.

[Certain] kinds of stone are formed from fire when they are extinguished. Ferreous and stony bodies are often formed in thunderbolts, due to an accidental cause by which <u>fieriness</u> <u>is extinguished and becomes cold and dry</u>.

<u>I.i.4.3-11</u>: Meteorite in Jūzjān

IS gives us four cases of stones falling from the sky (*Šifāʾ* 5.16-6.14). BH gives us a shortened version of the third and longest of these reports. - The meteorite reported here by IS is probably the same as that reported to have fallen in Jūzjān "some years ago" by Bīrūnī, *Ǧamāhir* 251.12-15 (cf. Belenickij-Lemmlejn [1963] 483f. n.24).

But.: (1) The doctor [Ibn Sīnā] has recounted (2a) that his disciple saw <u>in the land of the Gūzgānāyē</u> [a piece of] <u>iron</u> (2b) which <u>fell from</u> the αἰθήρ [*hetīr*] (2c) [and] <u>which weighed</u> a hundred and twenty <u>minae</u>. (3a) It [first] sank <u>into the earth</u>, (3b) <u>then rebounded</u> <u>twice</u> from it in the way <u>a ball</u> rebounds off <u>a wall</u> which it strikes. (3c) It <u>then</u> fell <u>to the earth</u> (3d) and <u>a frightening sound</u> <u>was heard</u> <u>because of it</u> [*menneh*]. (4) <u>Only</u> <u>with difficulty</u> could the inhabitants [*bnay aṯrā*] <u>cut off a</u> small <u>piece</u> [*mnāṯā*] <u>from it</u> with a <u>chisel</u> [τόρνος]. (5a) <u>They brought</u> and showed <u>it</u> to the

[11] Cf. *Mabāḥiṯ* II.207.19-21.

governor (6a) and he ordered a sword to be made from it, (6b) but because of its hardness the smiths could not beat it. (7) This iron was made of small round grains like millet which stuck very closely to each other.

Šifāʾ 5.20-6.9:

وقد صح عندى بالتواتر ما كان ببلاد جوزجان فى زماننا الذى أذركناه/ ١ من أمر حديد لعله يزن مائة وخمسين منًا، نزل من الهواء فنفذ فى الأرض، ثم نبا نبوة أو/ نبوتين نُبوّ الكرة التى يرمى بها الحائط، ثم عاد فنشب فى الأرض، وسمع الناس لذلك صوتا/ عظيما هائلا؛ فلما تفقدوا أمره، ظفروا به، وحملوه إلى والى جوزجان، ثم كاتبه سلطان/ خراسان فى عصرنا وهو امير يمين الدولة وأمين الملة أبو القاسم محمود بن سبكتكين المظفر/ ٥ المغلب، يرسم له إنفاذه أو إنفاذ قطعة منه، فتعذر نقله لثقله فحاولوا كسر قطعة منه، فما/ كانت الآلات تعمل فيه إلا بجهد، وكان كل مثقب وكل مقطع يعمل فيه ينكسر لكنهم/ فصلوا منه آخر الأمر شيئا فأنفذوه إليه؛ ورام أن يطبع منه سيفا، فتعذر عليه./ وحكى أن جملة ذلك الجوهر كان ملتئما من أجزاء جاورسيّة صغار مستديرة، التصق بعضها/ ببعض. وهذا الفقيه أبو عبيد عبد الواحد بن محمد الجوزجانى، صاحبى، شاهد هذا كله.

[5.20 جوزجان: جوزجانان Holmyard ‖ 1 فنفذ BṬM et Holmyard: فنقر ed. Cair. ‖ 6.2 يرمى Ṭ et Holmyard: ترمى ed. Cair. ‖ 3 وحملوه: وحمل BD et Holmyard; وحملوا Ṭ ‖ جوزجنان Holmyard ‖ 4 يمين الدولة Ṭ et Holmyard [v. EI² VI.65]: عين الدولة ed. Cair. ‖ 9 عبيد الله Ṭ et Holmyard]

I am convinced by the reliability of my source of the truth of (2a) what happened in the country of Jūzjān in our own time: [a piece of] iron, (2c) perhaps weighing 150 *mann*, (2b) fell from the air [*hawāʾ*] (3a) and penetrated into the earth. (3b) It then rebounded once or twice like a ball thrown against a wall. (3c) Then, it settled on the ground again. (3e) People heard a loud and terrifying sound caused by this [*li-ḏālika*]. When they had investigated the matter, they took possession of the object (5) and carried it to the governor of Jūzjān. He wrote about it to the Sultan of Khurāsān, (the one) in our time, the Amīr Yamīn al-Daula wa-Amīn al-Milla Abū al-Qāsim Maḥmūd ibn Sebüktigīn, the victorious conqueror, who ordered him to send him the object or a part of it. Its transportation, however, was impossible because of its weight. (4) So they tried to break a piece from it, but the tools would only affect it with difficulty, and every drill and chisel applied to it broke. In the end, however, they cut off a part of it (5) and brought it to (the Sultan), (6a) who ordered a sword to be struck from it. (6b) This was impossible. (7) It is said that the whole of this substance [*ğauhar*] was composed of small, round millet-like parts adhering to each other. (2) All this was witnessed by my friend, the *faqīh* Abū ᶜUbaid ᶜAbd al-Wāḥid ibn Muḥammad al-Jūzjānī.

l. 4. "his disciple": IS's source was his life-long companison and disciple Abū ᶜUbaid Allāh al-Jūzjānī, to whom he refers to as *ṣāḥibī* at *Šifāʾ* 6.9.

l. 4. ܐܬܪܐ ܕܓܘܙܓܢܝܐ ("the land of the Gūzgānāyē"): Jūzjān (pers. Gūzgān/Gūzgānān). To the northeast of Herat between the rivers Murghāb and Amū Daryā. Jūzjān was ruled by the Farīghūnids, who had become vassals of the Ghaznavid Maḥmūd, until 1010/1011 (401 h), when, upon the death the last Farīghūnid Abū Naṣr Aḥmad, it was placed under the governorship of Maḥmūd's son, Muḥammad.[12]

[12] Bosworth (1975) 172; EI² II.799. s.v. "Farīghūnids" [Dunlop]; Nāẓim (1931)

l. 4-5. ܡܐܐ ܘܥܣܪܝܢ ܡܢܝܢ ("120 minae"): مائة وخمسين منّا ("150 *mann*") IS. - The mina (gr. μνᾶ; arab. *mann*, also *manā*; syr. *manyā*, abs. pl. *mnīn*), a measure of weight which was especially widely used in Iran, was normally reckoned at 2 *raṭl* (corr. syr. *līṭrā* < gr. λίτρα), or just over 800 grams, in Islamic times.[13] As was noted by Holmyard (Holmyard-Mandevile [1927] 24 n.3), however, a weight just over 120 kg would not have been particularly difficult to transport, so that IS's *mann* here may be one of the larger units to which the term was also applied.[14] I am unable to find any ground on which BH might consciously have altered IS's to "150 *mann*" to his "120", so that the alteration is probably due either to corruption in BH's copy of the *Šifā'* or to carelessness on BH's part (misreading of ܢ = 50 as ܟ = 20 in his own notes?).[15]

l. 9. ܐܘܡܢ̈ܐ ("smiths", lit. "craftsmen"): cf. comm. on I.i.3.7 ("bakers") above.

l. 10. ܐܝܟ ܕܘܚܢܐ ("like millet"): جاورسي IS. - The Arabic *ǧāwars* (pers. *gāwars*) is millet (*Panicum miliaceum*, gr. κέγχρος), the usual Syriac term for which is *praggā*. *Duḫn*, with which Syr. *duḥnā* used here is cognate, is strictly a kind of sorghum (perh. *Andropogon sorgum* or *Pennisetum spicatum* sec. Dietrich; gr. ἔλυμος), but seems also to be a more general term, so that *ǧāwars* is said in some sources to be a kind of *duḫn* or is equated with it.[16]

I.i.4.11-12: Swords in Yemen

But.: (1) It is said that the finest swords in Yemen [*b-taymnā*] are made from such a substance [*ūsiya*] (2) and poets have commemorated them in their poems.

Šifā' 6.10-11:

وحُدّثت أن كثيرا من السيوف اليمنية الجليلة، انما اتخذ من مثل هذا الحديد. وشعراء العرب قد وصفوا ذلك في شعرهم.

[وحُدّثت: وحدث Ṭ || الجليلة BDSṬ et Holmyard (et F marg.): الجميلة ed. Cair. || اتخذ DSṬM et Holmyard (et F marg.): تتخذ ed. Cair.]

177f.

[13] Hinz (1955) 16-23; EI^2 VI.117-120 s.v. "Makāyil/Mawāzin"; PS 2164f. s.v. ܡܢܝܐ.

[14] E.g. the *mann* of Rayy, corresponding to 600 *dirham* (approx. 1920 g), which was widely used in northern Iran (Hinz, 19; EI^2 VI.120); this multiplied by 150 would give us a weight just under 300 kg.

[15] The quotation from the *Šifā'* in the margin of ms. F has "150", as does Fakhr al-Dīn al-Rāzī, *Mabāḥiṯ* II.208.1; "50" is read in Wüstenfeld's edition of Qazwīnī, *Aǧā'ib* 209.ult.

[16] For the Arabic terms, see EI^2 Suppl. 249 s.v. "Djāwars" [Dietrich]; Dietrich (1988) II.247-249; id. (1991) 114; also Lane 409 s.v. جاورس and 861 s.v. دخن; BH *Ġāfiqī* [Meyerhoff-Sobhy], p. 92 no. 201 [*ǧāwars*], p. 112 no. 238 [*duḫn*], with comm. ad loc. For the Syriac, Löw (1881) 101-103; PS 834 s.v. ܕܘܚܢܐ, 1016 s.v. ܗܠܘܡܘܣ and 3233 s.v. ܦܪܓܐ.

(1) I am told that many of the fine Yemeni swords are made from this kind of iron (2) and the poets of the Arabs have described that in their poems.

l. 11. ܐܘܣܝܐ ("substance"): IS has "iron" at the corresponding place, but cf. "substance" (*ǧauhar*) at *Šifāʾ* 6.9.

I.i.5.: Unusual petrification

In the last theory of the section BH turns to a part Nic. syr. dealing with unsual manners of petrification,[17] from which he selects five instances. The same part of Nic. syr. was also used by BH in *Cand*. The correspondences may be summarised as follows.

Nic. syr.	*But.*	*Cand.*	
58.3-4	-	89.5-6	Stalactites etc.
58.4-6	-	89.6	Wood petrified in water
58.6-7	-	89.7	In holes/caves [?]
58.7-8 [?]	-	89.8	Gypsum
58.9-13	-	-	Fossils
58.13-14	-	-	Artificial petrification
58.14-15	I.i.5.6-7		in Tyrrhenia
58.15-16	I.i.5.7-8	89.8-9	in Cnidus
58.16-23	-	-	Salt, soda etc.
58.23-25	-	-	Petrification by heat
58.25-26	I.i.4.1-2	-	in thunderbolts
58.27-29	I.i.5.1-4	89.9-10	in human body
58.29-59.3	I.i.5.4-6	-	Glass, coral etc.
59.3-13	I.i.5.8-10	-	Lava from Etna

I.i.5.1-4: Petrification in human body

But.: (1) For further corroboration [*ak̲ da-l-šurrārā yattīrā*], the master [Aristotle] has added [*awsep̄ emar*] (2a) that stones are also formed (2b) in the tendons of legs and joints of those who are suffering from [lit. have the affection (*ḥaššā*) of] gout [ποδάγρα] and arthritis [lit. disease of joints], (2c) in the bladders of children (2d) and in the eyelids [?], (2e) through the filtration of moistures which undergo [lit. take] solidification with heat.

Nic. syr. 58.27-29

27 ܗܟܝܢ ܕܝܢ ܟܐ̈ܦܐ : ܘܐܦ ܒܪ̈ܓܠܐ ܕܐܝܠܝܢ ܕܐܝܬ ܠܗܘܢ ܚܫܐ ܕܦܘܕܐܓܪܐ ܒܓܝ̈ܕܐ ./

28 [ܘܒ]ܦܪ̈ܩܬܐ . ܘܒܐܝܠܝܢ ܕܐܚܝܕܝܢ ܒܟܐܒܐ ܕܦܪ̈ܩܬܐ . ܘܒܫܠܘܚܝܬܐ ܕܛܠ̈ܝܐ . ܘܒܥܝ̈ܢܐ/

29 [-]ܟܢܐ ܡܢ ܥܒܪܐ ܕܪ̈ܛܝܒܘܬܐ : ܕܩܒ̈ܠܐ ܩܦܝܘܬܐ ܒܚܡܝܡܘܬܐ .

[27 ܕܦܘܕܐܓܪܐ (cum *syāmē*) cod. || 29 ܕܪܛܝܒܘܬܐ (sine *syāmē*) cod.]

(2a) Stones are also formed (2b) in the legs of those who have the affection of ποδάγρα in the tendons and joints, and those who have been seized by

[17] Cf. Takahashi (2002a) 209-210 (text), 217-219 (tr.).

<u>the disease of joints</u>, (2c) <u>in the bladders of children</u>, (2d) <u>and in the eyelids</u> [?], (2e) <u>through the filtration of moistures which undergo solidification with heat</u>.

l. 2. ܦܘܕܓܪ̈ܐ ("gout", gr. ποδάγρα): Written with *syāmē* in all mss. of *But.* other than F (and vocalised *-grē* in LIV) and in the ms. of Nic. syr., as seems often to be the case elsewhere in Syriac (see PS 3038).

l. 2-3. ܟܐܒ ܦܘܪ̈ܩܐ (lit. "disease of joints"): ܟܐܒܐ ܕܦܘܪ̈ܩܐ Nic. syr. - No doubt a translation of ἀρθρῖτις [νόσος] (cf. PS 1659 s.v. ܟܐܒ ܦܘܪ̈ܩܐ, 103 s.v. ܐܪܬܪܝܛܝܣ [*sic*]), which is often mentioned alongside gout in medical literature (e.g. Hipp. *Aff.* 30-31; Pliny, *NH* 20.9). For an instance where the bladder stone is associated with arthritis, see Hipp. *Nat. Hom.* 14.

l. 3. "in the bladders of children": The bladder stone (vesical calculus) is frequently mentioned in classical literature.[18] For its association in particular with children, see Hipp. *Nat. Hom.* 12 (cf. also Hipp. *Aër.* 9, Strabo 16.2.43).

l. 3. ܒܡܐܢ̈ܝ +ܓܘܙܐ ܕܥܝ̈ܢܐ sic F ("in the eyelids" [?], lit. "in the vessel of the nut of the eyes" [?]): ܒܡܐܢ̈ܝ ܓܘܐ ܕܥܝ̈ܢܐ MWLIV; ܒܡܐܢ̈ܝ [-] ܥܢܐ Nic. - We are likely to be talking about stone-like formations in the eyelid or the adjacent conjunctiva here,[19] since such formations were known to the ancients and were attributed by them to the solidification (λιθίασις, *taḥağğur*) of liquids entering them.[20] It is difficult to see, however, why, if eyelids are meant, this is not simply rendered by *tallīpā* or *temrā*. If we read *mānay gawwā d-ʿaynē* ("in the inwards of the eyes" [?]) with MWLIV, the reference may be to cataract, which was believed to be due to solidification of liquids entering the eye from the brain.[21] The reading *ʿānā* of Nic. syr. is tempting (*b-mānay gawwā d-ʿānā*: "in the intestines of sheep"), but BH clearly read *ʿaynē* and we seem elsewhere in this passage to be talking of petrification in the human body.

l. 3. ܡܢ ܫܚܠܐ ("by filtration"): This accords with Hippocrates' description of how bladder stones are formed (Hipp. *Aër.* 9; *Nat. Hom.* 12). The word

[18] Hipp. *Aër.* 9; *Nat. Hom.* 12, 14; *Aph.* 4.79; Rufus medicus, *De renum et vesicae affectionibus* 3, 13; Arist. *HA* III.15, 519b 19 (in animals); Ps.-Arist. *Prob.* 10.43; Pliny, *NH* 11.208, 28.212 etc. For instances in Arabic literature, see Schönfeld (1976) 178f.

[19] For modern clinical accounts of the pathologies involved, see Duke-Elder (1965) 139f. (*conjunctiva petrificans*), 585 (conjunctival concretions, lithiasis); Duke-Elder & MacFaul (1974) 34 (Meibomian lithiasis).

[20] Galen [Kühn] 14.771; cf. Ḥunain, *al-ʿAšr maqālāt fī al-ʿain* [Meyerhoff (1928)] 131.11, 132.7; *al-Masāʾil fī al-ʿain* [Sbath-Meyerhoff (1938)] §121, §137, §140; IS *Qānūn* III.3.3.14, 22, ed. Qassh III.988.17f., 990.18f. (tr. Hirschberg-Lippert, 109, 114).

[21] See Sbath-Meyerhoff (1938) 9. - "*Mānā*" could correspond to Galen's χιτῶνες (lit. tunics, i.e. the membranes within the eye; rendered as *ḥiğāb* by Ḥunain), but in *The Syriac Book of Medicines*, Galen's χιτών is rendered by *kottīnā* (ed. Budge [1913] I.68.9 etc.).

šeḥlā probably corresponds to gr. διήθησις, which is one of the processes by which stones are formed according to Theophrastus (Theoph. *Lap.* §3 [Wimmer] 341.2-6).

l. 4. ܢܣܒܢ ܩܛܪܐ ("undergo solidification", lit. take solidification): The phrase *nsab qṭārā* (< gr. πῆξιν λαμβάνω vel sim.?) is used four times in connection with petrification in Nic. syr. (58.4, 7, 9, 29).

I.i.5.4-6: Glass and coral

The sentence in *But.* is more complete here than the corresponding sentence in the Cambridge ms. of Nic. syr. and allows us to restore two Greek words which were omitted by the copyist of the Cambridge ms. (or one of his predecessors). On the other hand, BH is inaccurate in his paraphrase when he talks of glass and coral being transformed *into* stone, since what the original sentence must have meant is: "transformation into stone occurs/is observed in glass and coral", i.e. that they are the *results* of transformation into stone.

But.: Also transformed into stone are glass and κοράλλιον, i.e. corals [*kesnē*], for which reason (the latter) are called λιθόδενδρον, i.e. tree of stone.

Nic. syr. 58.29-30:

ܡܫܬܚܠܦܢܘܬܐ/30 ܕܝܢ ܠܟܐܦܐ ܗܘܝܐ ܒܙܓܘܓܝܬܐ ܘܒܩܘܪܠܝܐ .

[30 ܘܒܩܘܪܠܝܐ : ܘܒܩܘܪܐܠܝܐ ut vid. cod.]

Transformation into stone occurs in glass and in corals.

l. 4. ܡܫܬܚܠܦܐ ("are transformed"): fem. sg. - i.e. strictly speaking in agreement with *zgōgītā* ("glass") only.

l. 5. ܠܝܬܘܕܢܕܪܘܢ ("λιθόδενδρον"): The word is attested in Greek at Dioscorides [Wellmann] 5.121 and, given the possible use of Theophrastean material at Dioscorides 5.76ff. (Steinmetz [1964] 110-111), that instance too may go back to Theophrastus.

I.i.5.6-8: Artificial stones

But.: (1) In Tyrrhenia [?] when bricks are formed [*metgablān*] and are immersed [lit. are made to enter] in the sea, they quickly turn into rocks [*hāwyān šūᶜē*]. (2) On Cnidus when they throw lumps of earth [*qullāᶜē*] into springs, (the springs) turn them into stones.

Nic. syr. 58.13-16:

13 . ܟܐܦܐ ܕܝܢ ܡܢܗܝܢ ܡܢ ܟܝܢܐ ܡܬܥܒܕܢ :/

14 [ܡܢܗܝܢ ܕܝܢ] ܒܐܘܡܢܘܬܐ . ܐܝܟܢܐ ܕܐܝܟ ܗܠܝܢ ܕܡܬܓܒܠܢ ܒܛܘܪܗܢܝܐ ܒܝܡܐ :/

15 [.......] ܗܘܝܢ ܫܘܥܐ . ܐܡܪܝܢ ܕܝܢ ܕܐܦ ܒܩܢܝܕܘܣ ܟܕ ܫܕܝܢ ܩܘܠܥܐ/

16 [ܒ- -]ܐ ܠܡܒܘܥܐ ܗܦܟܝܢ ܠܗܘܢ ܟܐܦܐ .

For some stones are formed by nature, others by artifice, (1) like those which are formed in Tyrrhenia [?] in the sea <...> turn into rocks [or: rocks are formed?, *hāwēn šūᶜē*]. (2) They say that on Cnidus, too, when they

throw lumps of earth into <--> {in another copy: "into springs"} (the springs) turn them into stone.

l. 6. ܛܘܪܐܢܝܐ ("Tyrrhenia [?]"): ܛܘܪܐ ܢܫܝܐ M; ܛܘܪ ܢܝ̈ܫܝܐ V; ܛܘܪܝܢܝܐ Nic. - I take this as a transcription of Τυρρηνία, the *ālap̄* instead of the expected *yōḏ* possibly being due to confusion with the word τυραννία. The readings of MV are clearly later corruptions. - The process of petrification in the sea described here is reminiscent of what is said about the "dust of Puteoli" (pozzolan) at Pliny, *NH* 35.166: "quis enim satis miretur pessumam eius partem ideoque pulverem appellatum in Puteolanis collibus opponi maris fluctibus, mersumque protinus fieri lapidem unum inexpugnabilem undis et fortiorem cotidie"; cf. also Seneca, *QN* 3.20.3: "quemadmodum Puteolanus pulvis, si aquam attigit, saxum est, sic ..." - Puteoli, though not in Tyrrhenia/Etruria, is at least on the Tyrrhenian Sea.

l. 6. ܠܒ̈ܢܐ ("bricks"): There is no mention of "bricks" in the passage of Nic. syr. quoted above and it is difficult to fit in both this word and the verb *metta^clīn* (see below) in the short lacuna at the beginning of Nic. syr. 58.15,[22] so that we must suppose either that the notion of "bricks" is an addition by BH or that BH had a more complete text of Nic. syr. than that in the Cambridge ms.[23]

l. 6. ܡܬܬܥܠ̈ܝܢ ("are immersed", lit. "are made to enter", √ᶜ*ll* Ettaph.): The verb, though not in the passage of Nic. syr. quoted above, is used a little earlier at Nic. syr. 58.4, of wood being immersed in water. At the place corresponding to Nic. syr. 58.4 in *Cand.* (89.6), BH had turned the word into the Peal *ᶜā(ʾ)līn*.

l. 7. ܫܘ̈ܥܐ ("rocks"): Probably rendering gr. πέτρα as opposed to λίθος, since *šōᶜā* is used, strictly, of "rock" still anchored to the ground (*lapis vivus*) as opposed to detached stone (see PS 4098); cf. "saxum" in the passage of Seneca quoted in comm. on l. 6. "Tyrrhenia" above, also Pliny's "lapis unus inexpugnabilis undis".

l. 7. "on Cnidus": The petrifying spring on Cnidus is also mentioned by Pliny (*NH* 35.167: "... et in fonte Cnidio dulci intra octo menses terram lapidescere"; cf. RE XX.916 s.v. "Knidos" [Bürchner]).

---

[22] One *could* restore a sense approximating to that of *But.* by reading ܕܒܗ ܡܢ ܠܒ̈ܢܐ at the beginning of line 15 in Nic. syr. ("... the sea, in which from bricks rocks are formed").

[23] Cf. Pliny, *NH* 35.167: "eadem est terrae natura in Cyzicena regione, sed ibi non pulvis, verum ipsa terra qua libeat magnitudine excisa et demersa in mare lapidea extrahitur." - "Earth cut out in the size required" is not quite "brick", but is closer to it than "dust". - Cf. also Strabo 12.1.67 (= Posidonius, frag. 237 Edelstein-Kidd), mentioning the floating "bricks" made from volcanic pumiceous earth in Pitana (in Asia Minor) and comparing this with the floating "earth" in Tyrrhenia: φασὶ δ' ἐν τῇ Πιτάνῃ τὰς πλίνθας ἐπιπολάζειν ἐν τοῖς ὕδασι, καθάπερ καὶ ἐν τῇ Τυρρηνίᾳ γῆ τις πέπονθε.

I.i.5.8-10: Volcanic stones on Etna

The corresponding passage in Nic. syr., besides being interrupted by a scholion (59.5-9), has suffered from the damage to the manuscript, so that the passage of *But.* here has to be used to restore the lacunae.

> *But.*: (1) Stones which are discharged [*metnapṣān*] from Etna burn and melt. (2) This is in fact surprising [*tmīhā*]. (3a) How [is it possible] that when they were in [the midst of] fire, (3b) they were not affected at all, (3c) but melt after they have been driven out from it? (4) Perhaps they later gain a property [lit. power] which they did not possess before.

Nic. syr. 59.3-5, 9-13:

3 [...............]/
4 ܟܐܦܐ ܣܓܝܐܬܐ ܕܝܩܕܢ ܘܡܬܦܫܪܢ . ܐܝܟܢܐ ܕܗ̇[ܠܝܢ ܕܡܢ ܐܝܛܢܐ ܡܬܢܦܨܢ ....]/
5 [ܬܡܝܗ]ܐ ܗܝ .

9 [........]/
10 ܕܟܕ ܡܢ ܢܘܪܐ ܐܬܛܪܕܘ ܚܫܝܫܝܢ ܗܘܘ ܘܠܐ ܒܗ̇ ܒܙܒܢ ܒܐܚܪܝܢ . [....................]/
11 ܐܠܐ ܟܒܪ ܟܠܗ ܡܢ ܟܝܢܘܬܐ . ܚܝܠܐ ܣܓܝܐܐ ܘܡܫܚܠ[ܦܐ ܠܒܬܪܟܢ ܩܢܝܐ ......]/
12 ܕܠܐ ܩܢܝܐ ܗܘܐ ܠܗܘܢ ܡܢܩܕܡ . ܗܟܢܐ ܓܝܪ ܘܐܦ ܒܙܒܢܐ [ܢܓܝܪܐ ....]/
13 +ܡܬܝܠܕܢ ܡܕܡ ܕܠܘܩܕܡ ܠܝܬܝܗܘܢ ܗܘܘ . ܘܐܝܠܝܢ ܕܐܝܬܝܗܘܢ [ܗܘܘ ....]/

(1) <There are> many stones, which burn and melt, like those <which are discharged from Etna>. (2) This is surprising: (3ac) <why do those (stones)> which, when they were driven out from fire, (3b) were not affected at the time, (3c) <melt afterwards?>. (4) But perhaps every substance <later acquires> many different powers which it did not have before. For just so over a long period <of time there arise> certain -- which were not there before, and those which were there <disappear>.

Min. I. Section ii.: *On Formation of Mountains and their Uses*

The first half of the section here (I.ii.1-2) corresponds to the latter half of the chapter (*faṣl*) on the formation of mountains in the *Šifāʾ*, where IS deals with the formation of the mountains proper (as opposed to the formation of stones; cf. introductory comments to Section I.i. above). The latter half of the section (I.ii.3-6), on the other hand, corresponds to the chapter of the *Šifāʾ* on the "uses of mountains". BH has, in another words, brought together into this section those parts of the *Šifāʾ* dealing with "mountains".

Most parts of the section are based on the *Šifāʾ*. In 'theory' I.ii.1, however, BH has inserted a passage based on Abū al-Barakāt's *Muʿtabar*, a source which is used again in Section Min. I.iii. below. There may also be some echoes in this section of lost passages of Nic. syr.

	Šifāʾ	Other sources	
I.ii.1.1-3	6.15-16, 7.9-13		Large stones/mountains
I.ii.1.3-7		*Muʿt.* II.208.18-22 II.209.2-3	Islands
I.ii.1.7-10		*Muʿt.* II.208.18-209.2 + Nic.?	Artificial stones in water
I.ii.2.1-3	6.16-1		Differentiation (essential)
I.ii.2.3-7	6.18-7.8	(*Mab.* II.208.13-16)	Differentiation (accidental)
I.ii.3.1-3	10.4-7, 11.13-17		Utility (1): clouds
I.ii.3.3-4	12.7-8		Utility (1): clouds (contd.)
I.ii.4.1-6	10.8-11 10.18-11.2	+ Nic.?	Utility (2): rivers
I.ii.5.1-4	11.8-12 (10.13-11.2)		Utility (2): rivers (contd.)
I.ii.5.4-10	11.3-8		Comparison with alembic
I.ii.6.1-6	12.8-11		Utility (3): minerals
I.ii.6.6-8	12.12-13		Other uses

I.ii.1.1-3: Formation of large stones/mountains

BH has combined here two passages in the *Šifāʾ* dealing, respectively, with the formation of large stones (6.15-16) and the formation of mountains (i.e. the land or rock mass constituting the material of the mountains) (7.9-13). The result is a slightly inaccurate rendition of what IS had to say. IS had seen an analogy between the formation of (i) stones, (ii) "large stones" and (iii) the land mass. BH has reduced the three steps to two by equating the "large stones" with the land mass.

But. [underline: agreement with *Šifāʾ* 6.15-16; italics: with *Šifāʾ* 7.9-13]:
(1a) Large stones (which make up) [lit. of] *mountains* *are* either *formed* (1b) suddenly by heat which comes upon a large quantity of *glutinous clay*, (1c) or are formed *gradually* [*b-īḏā b-īḏā*] *over intervals of countless years* [lit. many years which pass beyond reckoning].
Šifāʾ 6.15-16:

وأما تكون حجر كبير فيكون إما دفعة، وذلك بسبب حر عظيم يغافص طينا كثيرا لزجا، وإما أن يكون قليلا قليلا على تواتر الأيام.

[يغافص: يعافص ed. Cair.]
(1a) The formation of large stones occurs (1b) either suddenly, because of a great heat which comes upon a large quantity of glutinous clay, (1c) or occurs little by little [*qalīlan qalīlan*] over a succession of days.
Šifāʾ 7.9-13:

فالجبال تكونها من أحد أسباب تكون الحجارة، والغالب أن تكونها من طين لزج جف على طول الزمان، تحجر فى مدد لا تضبط، فيشبه أن تكون هذه المعمورة قد كانت فى سالف الأيام غير معمورة؛ بل مغمورة

فى البحار، فتحجرت، إما بعد الانكشاف قليلا قليلا فى مدد لا تفى التأريخات بحفظ أطرافها، وإما تحت المياه لشدة الحرارة المنحقنة تحت البحر.

(1a) The formation of mountains is by one of the causes for the formation of the stone; (1b) as a rule [*al-ġālib*] their formation is from glutinous clay, which has dried up over a long period and has petrified over indeterminable intervals. It seems likely that this habitable world [*maᶜmūra*] was not formerly inhabited but was submerged [*maġmūra*] in the sea; then, it petrified either after being exposed, little by little over intervals whose limits the historical records cannot preserve, or [while still] under water, due to the intense heat confined beneath the sea.

l. 2 ܛܝܢܐ ܡܣܓܝܐܐ ܘܠܙܘܓܐ ("a large quantity of glutinous clay"): cf. I.i.2.2-5 above.

I.ii.1.3-7: Formation of islands

BH switches his sources here. The second part of this theory is based on the opening passage in the chapter on "mountains, seas, rivers, springs and wells" in Abū Barakāt's *K. al-Muᶜtabar*. Abū al-Barakāt's idea of stones being formed by the action of waves is reminiscent of Nicolaus, *De plantis* arab. [Drossaart Lulofs-Poortman] 181, §159 (corr. Ps.-Arist. *De plantis* 823b 11ff.),[24] a passage corresponding to which may also have been present in BH's copy of Nic. syr. It may be precisely because of the corroboration he found in Nic. syr. that BH decided to accept Abū al-Barakāt's theory here.

But.: [This happens] because - (1a) since the earth is situated amid water (1b) and winds are constantly causing the water to move - (2a) particles of earth are made to move with the fluctuation [*metgallᶜlānūtā*] of the water (2b) and become mixed with water. (3a) when the water [then] solidifies, (3b) (the particles) adhere to the surface of the earth. (4a) As this mass [*mlōʾā*] grows with time, (4b) islands are formed.

Abū al-Barakāt, *Muᶜtabar* II.208.18-22, 209.2-3:

لما كانت الارض يابسة ذات اجزاء لا تتجزأ وكان الماء يحيط بها والرياح تحرك الماء بالتمويج صارت الارض تتحرك اجزاؤها فى قعر الماء بحركته فتمتزج بالماء وتتصل به اجزاؤها ويبقى المتصل منها على شكل يتفق له فى حركته وامتزاجه بانعقاده وتنضاف اليه اجزاء بعد اجزاء من الاجزاء الارضية المختلطة بالماء فيزداد عظما بعد عظم ... فكذلك يعرض لما يعرض ان يتشكل من الاجزاء الارضية بالحركات الموجبة فى قعر الماء على طول الزمان ان تعظم ثم تعظم حتى تعلو على وجه الماء جبلا عظيما

(1a) Since the earth is dry and has inseparable parts, water surrounds it (1b) and winds cause the water to move by wave-causing action [*tamwīǧ*], (2a) particles of earth are made to move at the bottom of the water with (the water's) movement, (2b) so that (the earth) becomes mixed with the

---

[24] On the probable Theophrastean origin of the idea, see Steinmetz (1965) 265-266, 311.

water and its particles are joined to (water). (3b) The joined (particles) [lit. what is joined of them] remain in the shape which they acquire through their movement and mixture (3a) when they [or: water] solidify/ies [*bi-in*ᶜ*iqādihi*] (3b) and there is added to them particle after particle out of the earthy particles mixed with water and they continuously increase in size. ... [208.18-209.2: see below] ... (4a) Similarly, what has been formed out of the earthy particles by the necessitating [*mūğiba*][25] movements at the bottom of the water will in the course of time [ᶜ*alā ṭūl al-zamān*] grow larger and larger (4b) until they rise above the surface of the water as a large mountain.

l. 4. ܡܬܓܠܠܠܢܘܬܐ ("fluctuation"): The word is known to dictionaries only from the instances in BH *But*. (PS Suppl. 74). Here it answers to Abū al-Barakāt's *tamwīğ*. In Abū al-Barakāt the active *tamwīğ* is used of the wave-causing action of the wind (< II *mawwağa*, corr. Pael *gallel*), while here the passive *metgallᵉlānūṯā* is used of the waves caused by that wind in the water (< Ethpa. *etgallal*, corr. V *tamawwağa*). - Cf. *But*. Min. V.iii.3.3. ܡܓܠܠܐ (adj.): corr. *Šifā*ʾ مُمَوَّج; Mete. III.i.1.6 ܡܓܠܠ (Pael ptc.): corr. *Mabāḥiṯ* يُحصل تموّجًا; Mete. IV.i.4.7 ܡܬܬܙܝܥܢܘܬܐ ܡܓܠܠܬܐ: corr. *Šifā*ʾ تموّج (of the air-waves in connection with hearing).[26]

I.ii.1.7-10: Artificial stones formed in water

This passage too is closely based on the *Mu*ᶜ*tabar*, except that there is no mention of the "Babylonians" in the *Mu*ᶜ*tabar* (at least not in the Hyderabad edition). The artificial "stones" mentioned here are in fact probably solidified bitumen (ἄσφαλτος), whose use as mortar in Babylon was already known to Herodotus (I.179).[27] BH may therefore have had a Greco-Syriac source (a lost passage of Nic. syr.?) from which he picked up the reference to Babylon.

But. [underline: agreement with *Mu*ᶜ*tabar*; italics with Strabo]:
(1) This [process] can in fact also be observed [*meṯhazyā*] in an experiment. (2a) For example, when *Babylonians* [lit. some Babylonians, *nāšīn bāḇlāyē*] require [*bā*ᶜ*ēn*] stones for building (2b) but cannot find [them], (2c) they throw date pits in running water. (3a) With time earthy particles adhere to them (3b) and gradually grow in size, (3c) until stones [*kēpē*] *suitable for building work are formed*.

[25] Perhaps to be emended to *mumawwiğa*, "wave-causing".

[26] The noun ܡܬܓܠܠܠܢܘܬܐ is also used of the movement of air in connection with hearing at *But*. De anima III.i.2. (ms. F 118r b5) and Metaph. VI.iii.1. (F 154r b5, 7, 11).

[27] Cf. Pliny, *NH* 35.182; Vitruvius I.5.8, VIII.3.8.

Abū al-Barakāt, *Muʿtabar* II.208.18-209.2:

وترى هذا فى مياه وفى مواضع فان قوما اذا ارادوا احجارا لبنيانهم القوا فى الماء الجارى نوى التمر او ما يشبهه فيلتبس على كل واحدة اجزاء ارضية بعد اجزاء فتعظم كلما بقيت حتى تصير صخرا كبارا بقدر ما يريدون فيرفعونه من الماء ويبنون به بنيانهم ويبقى بقاء صالحا كغيره من الصخر

(1) You can see this in [various] waters and places. (2a) For people [*qaum*], when they require [*arādū*] stones for their buildings, (2c) throw date pits or something similar in running water. (3a) To each of them there adhere earthy particle after particle, (3b) so that they grow larger so long as they remain [in the water], (3c) until they turn into large boulders [*ṣaḫr*] of the size (the people) require. They then take (the boulder) out of the water and build their buildings with it and (such a boulder) is as enduring as other boulders.

Strabo 16.1.15[28]: γίνεται δὲ ἐν τῇ Βαβυλωνίᾳ καὶ ἄσφαλτος πολλή, περὶ ἧς Ἐρατοσθένης μὲν οὕτως εἴρηκεν, ὅτι ἡ μὲν ὑγρά, ἣν καλοῦσι νάφθαν, γίνεται ἐν τῇ Σουσίδι, ἡ δὲ ξηρά, δυναμένη πήττεσθαι, ἐν τῇ Βαβυλωνίᾳ· ταύτης δ᾿ ἐστὶν ἡ πήγη τοῦ Εὐφράτου πλησίον· πλημμύροντος δὲ τούτου κατὰ τὰς τῶν χιόνων τήξεις καὶ αὐτὴ πληροῦται καὶ ὑπέρχυσιν εἰς τὸν ποταμὸν λαμβάνει· ἐνταῦθα δὲ συνίστανται βῶλοι μεγάλαι πρὸς τὰς οἰκοδομὰς ἐπιτήδειαι τὰς διὰ τῆς ὀπτῆς πλίνθου. ἄλλοι δὲ καὶ τὴν ὑγρὰν ἐν τῇ Βαβυλωνίᾳ γίνεσθαί φασι. περὶ μὲν οὖν τῆς ξηρᾶς εἴρηται, πόσον τὸ χρήσιμον τὸ ἐκ τῶν οἰκοδομιῶν μάλιστα· φασὶ δὲ καὶ πλοῖα πλέκεσθαι, ἐμπλασθέντα δ᾿ ἀσφάλτῳ πυκνοῦσθαι.

I.ii.2.1-3: Height differentiation (1): essential cause

But.: (1a) The cause of the elevation [*rawmā*] of mountains is either essential [*ūsiyāytā*]: (2a) as when [*akmā d-*] in severe earthquakes the strong winds which cause them lift a part of the earth (2b) and suddenly create a deep valley [*naḥlā*] in its side - (3) or accidental [*gedšānāytā*] ...

Šifāʾ 6.16-18[29]:

وأما الارتفاع فقد يقع لذلك سبب بالذات، وقد يقع له سبب بالعرض. أما السبب بالذات، فكما يتفق عند كثير من الزلازل القوية أن ترفع الريح الفاعلةُ للزلزلة طائفة من الأرض، وتحدث رابية من الروابى دفعة.

(1) Elevation [*irtifāʿ*] may have an essential [*bi-l-ḏāt*] cause (3) or an accidental [*bi-l-ʿaraḍ*] cause. (2a) An essential cause [is involved when], as [*kamā*] happens with many strong earthquakes, the wind causing the earthquake causes a part of the earth to rise (2b) and suddenly creates [reading *tuḥdiṯa*] some hill [*rābiʾa min al-rawābī*].

l. 3. ܢܚܠܐ ("valley"): This is the opposite of IS's "hill" (*rābiʾa*). Did BH perhaps read أودية vel sim. instead of رابية and understand "creates valleys out of [*min*] the hills"?

---

[28] Although Strabo does not mention the use of date pits for collecting the bitumen, various uses of the date palm in Babylonia are discussed in the passage immediately preceding (16.1.14). He also talks of the scarcity of building materials in Babylonia (16.1.5).

[29] Cf. *Mabāḥiṯ* II.208.9-12.

l. 3. ܥܡܝܩܐ . . . ܢܚܠܐ ("deep valley"): The separation of the adjective from the noun is unsual.

I.ii.2.3-7: Height differentiation (2): accidental cause

The passage may be considered a summary of a longer passage in the *Šifāʾ*. In his rearrangement of the material, BH was probably guided by a similar summary in the *Mabāḥiṯ*, which he had earlier reproduced almost verbatim in *Cand*.

But. [underline: agreement with *Šifāʾ*, italics: with *Mabāḥiṯ*]:

(1) or accidental: (2a) as when parts of the earth *differ* in their constitution, (2b) so that *the soft (parts) are excavated* by blasts of tearing *winds* or [by] furrowing waters, (2c) and the place they [originally occupied] becomes a valley, (2d) *while the stony (parts) remain* in their elevated position [*b-rawm-hēn*] (2e) and the place they [occupy] becomes a mountain. (3a) Torrents [*reglāṯā*] of water [then] increase the depth of that (part) *which has been excavated* (3b) and the elevation of that (part) which has not been excavated.

Šifāʾ 6.18-7.8:

وأما الذى بالعرض، فأن يعرض لبعض الأجزاء من الأرض انحفارٌ دون بعض، بأن تكون رياح نسافة أو مياه حفارة تتفق لها حركة على جزء من الأرض دون جزء، فيتفحر ما تسيل عليه ويبقى ما لا تسيل عليه رابيا. ثم لا تزال السيول تغوص فى الحفر الأول إلى أن تغور غورا شديدا، ويبقى ما انحرف عنه شاهقا. وهذا كالمتحقق من أمور الجبال وما بينها من الحفور والمسالك. وربما كان الماء أو الريح متفق الفيضان، إلا أن أجزاء الأرض تكون مختلفة، فيكون بعضها لينة وبعضها حجرية، فينحفر الترابى اللين، ويبقى الحجرى مرتفعا. ثم لا يزال ذلك المسيل ينحفر وينحفر على الأيام، يتسع، ويبقى النتوء، وكلما انحفر عنه الأرض كان شهوقه أكثر.

(1) An accidental cause [is involved when] (2a) some parts of the earth are subjected to excavation [*inḥifār*] while others are not, (2b) because blasting [*nassāfa*] winds or excavating [*ḥaffāra*] waters happen to move over a part of the earth but not over another, (2b) so that the (part) over which they flow is excavated (2d) while (the part) over which they do not flow remain as a hill [*rābiʾa*]. (3a) Then, the streams continue to move deeper [*taġūṣu*] into the original cavities [*ḥufar*] until they form a deep trench [*taġūra ġauran šadīdan*], (3b) while (the part) from which they turned away [*inḥarafa*] remains as an eminence [*šāhiq*]. This is what may be regarded as certain concerning mountains and cavities and the passes between them. Or, the water or the wind may be uniform in its action [lit. flooding], (2a) but the parts of the earth differ [in their constitution], some parts being soft and others stony, so that the soft soily (part) is excavated, (2d) while the stony (part) remains elevated. (3a) Then, that channel [*masīl*] continues to be excavated and excavated for days and becomes widened, (3b) while the hillock [*nutūʾ*] remains and the more the earth is excavated from (its sides) the more lofty it becomes.

Fakhr al-Dīn al-Rāzī, *Mabāḥiṯ* II.208.13-16[30]:

واما الذى بالعرض فان الطين بعد تحجره تختلف اجزاؤه فى الصلابة والرخاوة فاذا وجدت مياه قوية
الجرى او رياح عظيمة الهبوب انفجرت (انحفرت) الاجزاء الرخوة وبقيت الصلبة ثم لا تزال السيول والرياح
تغوص فى تلك الحفرات الى ان تغور غورا شديدا فيبقى ما انحفر (انحرف) عنه شاهقا

(1) The accidental [cause]: (2a) After its petrification the parts of the clay differ in hardness and softness, (2b) so that when strongly-flowing rivers or violently-blowing winds occur [lit. are found], the soft parts are excavated [lege *inḥafarat*], (2d) while the hard (parts) remain. (3a) Then the streams and the winds continue to move deeper [*taġūṣu*] into those cavities [*ḥufrāt*] until they form a deep trench [*ġaur*], (3b) so that (the part) from which they turned away [lege *inḥarafa*] remains as an eminence [*šāhiq*]

I.ii.3.1-3: Utility of mountains (1): formation of clouds

BH now jumps to the second chapter (faṣl) of the *Šifā*ʾ, which is on the uses (*manāfi*ʿ) of mountains. The process whereby the mountains help towards the formation of clouds is touched on briefly near the beginning of this chapter at *Šifā*ʾ 10.4-7 and then explained more fully at 11.13-12.7,[31] after the explanation of how mountains help in the formation of springs. The passage of *But.* here is structured around the shorter of the two passages at *Šifā*ʾ 10.4-7, but BH has integrated into it some elements taken from the later, longer discussion.

But. [underline: agreement with *Šifā*ʾ 10.4-7; italics: with *Šifā*ʾ 11.13-17]:

(1) Mountains are useful for [*ḥāšḥīn lwāṯ*] detention [*kelyānā*] of those moist vapours, (2) which, being [at first] *confined* [*meṯḥabšīn*] below them, (3) are made to rise by heat (4) and when *condensed* [*meṯlabdān*] by the coldness of the mountain air (5) become the material for clouds.

*Šifā*ʾ 10.4-7:

منافع الجبال كثيرة، وذلك لأنه لا يشك شاك فى وفور المنافع المتصلة بالسحب، وبالأودية المنبعثة من
العيون، وبالجواهر المعدنية. فأما السحب إنها إنما تتولد، كما نتبين من بعد من الأبخرة الرطبة إذا
تصعدت بتصعيد الحرارة فوافت الطبقة الباردة من الهواء التى فرغنا من تقديم خبرها.

(1) Uses [*manāfi*ʿ] of mountains are many, and that is because noone doubts the abundance of uses connected with the clouds, with rivers originating from springs and with mineral substances. (1/5) Clouds are generated, as we shall explain later, from moist vapours (3) when they are made to rise [*taṣaʿʿadat*] by the raising-action [*taṣʿīd*] of heat (4) and they reach the cold stratum of air, which we have already discussed.

*Šifā*ʾ 11.13-17

وكما أن أكثر العيون والأودية من الجبال، فكذلك أكثر السحب تكون من الجبال، وتجتمع فى الجبال من

[30] Cf. *Cand*. 84.1-4, which is based on this passage of *Mabāḥiṯ*.

[31] Cf. also the summary of the latter passage at Fakhr al-Dīn al-Rāzī, *Mabāḥiṯ* II.210.8-14, which however is rather diferent from BH's summary here.

الأسباب ما لا تجتمع فى مواضع أخرى. من ذلك أنه يعرض للبخارات بها من الاحتقان والتقوّى ما يفجّر العيون، فكيف حالها إذا تصعّدت وهى بعد أبخرة. فإنها لقوتها فى اندفاعها ولكثافة جرمها لا تتحلل بسرعة، بل يكون لها أن تندفع إلى الحيز المبرِّد والعاقد للبخار من أحياز طبقات الهواء.

Just as most springs and rivers [*audiya*] [originate] from mountains, most clouds originate from mountains and gather on mountains for reasons for which they do not gather in other places, (2) namely that the vapours are subjected in them to confinement [*iḥtiqān*] and intensification [*taqawwā*] which cause springs to gush forth. But what happens when (the vapours) rise while still [in the form of] vapours? Because of their strength in their propulsion and the density of their mass [*ǧirm*] they will not be dissipated in a hurry, (4) but will be pushed towards that region among the strata of air which refrigerates [*mubarrid*] and condenses [*ʿāqid*] vapour.

l. 1. ܚܫܝܚܝܢ ܠܘܬ ("are useful for"): cf. the same phrase at I.ii.4.1 below, where it may go back to gr. "ἐπιτήδεια πρός".

l. 1-2. ܡܬܚܒܫܝܢ ܬܚܝܬܗܘܢ ("confined below them"): The notion of vapours being confined below (i.e. inside) the mountains is not easy to understand unless one has already followed the discussion of how springs are formed in mountains, which, due to rearrangement of the material, is placed after this point in *But*. at I.ii.4-5 below. This is probably why the phrase was omitted by the copyist of ms. F, who however left a blank space here.

l. 2. "the coldness of the mountain air": The coldness of mountains due to their elevation is discussed further in the *Šifāʾ* at 11.21ff.

I.ii.3.3-4: Utility of mountains (1): formation of clouds (contd.)

But.: For this reason the majority of clouds are generated above mountains, and spread out [*metpašṭān*] from them on to the plains.

Šifāʾ 12.7-8: فلذلك ما ترى أكثرُ السحب الماطرة إنما تتولد فى الجبال، ومنها تتوجه إلى السائر البلاد.

For this reason most rain-causing clouds are seen being generated on mountains and spreading [*tatawaǧǧahu*] from them to the rest of the land.

I.ii.4.1-6: Utility of mountains (2): formation of rivers

The agreement of certain phrases in this passage with the passage of Olymp. quoted below suggests that BH may have used a lost passage of Nic. syr. in combination with the *Šifāʾ* here.

But. [underline: agreement with *Šifāʾ*; italics: with Olymp.]:

(1) *Mountains are also* [w-] *useful for* the flow of streams which are born [*metyaldān*] of springs on mountains. (2a) For *vapour*, when it is confined inside a mountain (2b) and is condensed by its *stony coldness*, (2c) *turns into water*. (3a) This water, when it gathers [*metkannšīn*] in one place, (3b) grows in quantity (3c) and is compressed by the heat - (3d) which is also confined inside the mountain and is not dissipated - (3e) tears the earth, (3f) gushes forth from it with force, (3g) flows over its surface (3h)

and forms streams, rivers and [other forms of] flowing waters as it proceeds [lit. flows].

*Šifā*ʾ 10.8-11:

والعيون أيضا فإنها إنما تتولد باندفاع المياه إلى وجه الأرض بالعنف، ولن تندفع بالعنف إلا بسبب محرّك لها مصعّد إلى الفوق. والأسباب المصعدة للرطوبات إنما هى الحرارات المبخّرة للرطوبات، الملجئة إياها إلى الصعود. والعيون أيضا، فإن مبادئها من البخارات المندفعة صعدا عن تصعيد الحرارة المحتقنة فى الأرض من الشمس والكواكب.

(1) Springs too are formed [*tatawalladu*] by the propulsion [*indifāʿ*] of waters towards the surface of the earth with force [*bi-l-ʿunf*], and they can only be propelled with force by a cause moving it and raising it upwards. The causes lifting moistures can only be the heat vaporising the moistures, coercing them towards ascent. With springs too, their origins are from vapours propelled upwards by the raising action [*taṣʿīd*] of the heat from the sun and the stars confined [*muḥtaqina*] in the earth.

*Šifā*ʾ 10.18-11.2:

والجبال أقوى الأرضين على حقن الحرارة فى ضمنها، وحبس البخار المتصعد منها، حتى يقوى اجتماعه ويعد بقوته منفذا يندفع منه إلى خارج، وقد تكاثف واستحال مياها، وصار عيونا.

[Ṭ ونفذ :ويُعد]

... but mountains are the strongest [type] of grounds (3cd) for confinement [*ḥaqn*] of heat in their interior (2a/3c) and retention [*ḥabs*] of vapour raised by it, so that (3a) (vapour's) concentration [*iǧtimāʿ*] becomes strong (3ef) and it creates a passage with its strength whence it is propelled to the outside, (2b) after it has thickened (2c) and been transformed into water, (3gh) and becomes springs.

Olymp. 103.25-29: (1) τὰ γὰρ ὄρη ἐπιτήδεια καὶ πρὸς τὸ δέξασθαι πολλὴν ὑγρότητα διὰ τὸ σηραγγῶδες αὐτῶν, ἀλλὰ μὴν καὶ πρὸς τὸ φυλάξαι· ὡς πετρώδη γὰρ ὄντα οὐκ ἀπογαιοῦσι τὸ ὕδωρ, ἀλλὰ φυλάττουσιν. (1) ἔτι γε μὴν ἐπιτήδειά εἰσι καὶ πρὸς τὸ γεννῆσαι· (2) ψυχρὰ γὰρ ὄντα τὴν ἀναγομένην ἀτμίδα εἰς ὕδωρ μεταβάλλουσιν.

l. 2-3. ("stony coldness"): cf. *Šifā*ʾ 12.4: على أن جوهر الحجارة أشد قبولا للبرد من الأرض ("because the substance [*ǧauhar*] of stone is more susceptible to coldness than earth").

l. 5. ܨܳܪܶܐ ܠܗ̇ ܠܐܪܥܐ ("tears the earth"): cf. I.iii.1.3 below.

I.ii.5.1-4: Utility of mountains (2): formation of rivers (contd.)

The parts of the passage marked (1a) and (2) below closely follow the sentence at *Šifā*ʾ 11.8-12 (< Arist. *Mete*. 350a 2-7), which is found after the passage corresponding to I.ii.5.4-10 below. Part (1b) sums up the discussion concerning hard (*ṣulb*) and soft (*raḫw*) ground at *Šifā*ʾ 10.13-11.2.

But.: (1a) The majority of springs flow [*rāḏyān*] out of mountains and their foothills [*šp̄ūlē*], and from hard grounds - (1b) because on soft (grounds) vapours are not retained but are rather [*mālōn*] dissipated - (2) and the

best-known rivers in the world descend [*nāḥtīn*] from mountains springs. (3) Hence the old saying that the foundations of mountains are placed on cisterns [*qeḇyāṯā*] of water.

Šifā' 11.8-12:

فلذلك ما يرى من أن أكثر العيون إنما يتفجر من الجبال ونواحيها، وأقلها فى البرارى؛ وذلك الأقل لا يكون أيضا إلا حيث يكون الأرض صلبة، أو فى جوار الأرض صلبة. فإذا تتبعت الأودية المعروفة فى العالم، وجدتها منبعثة من عيون الجبلية

(1a) For this reason one sees that the majority of springs are discharged [*tatafağğaru*] from mountains and their sides [*nawāḥin*], and very few are on the plains; and even those few do not occur except where the ground is hard or in the neighbourhood of hard ground. (2) When you examine the well-known rivers [*audiya*] in the world, you will find that they originate [*munbaʿiṯa*] from mountain springs.

l. 3-4. "Hence the old saying ...": cf. *Šifā'* 11.2-3: فيكاد أن يكون ما تستقر عليه الجبال مملوءا ماء ("It is almost as if what the mountains rest [*yastaqirru*] on is filled with water").

I.ii.5.4-10: Comparison with alembic

Arist. had compared the mountain in its function of gathering and condensing vapour to a sponge at *Mete*. 350a 7-13. IS has replaced this simile with that of a still and this is followed by BH.[32]

But.: (1) To use an illustration [*ak da-ḇ-ṯaḥwī l-mēmar*], a stony mountain is like a hard alembic [made] of iron or glass. - (2) For porous [material] such as earthenware or wood would not retain much vapour. - (3) The depth of the earth below the mountain is like the cucurbit, which the heat of the fire touches. (4) Springs are like the tubular tails of the alembic from which the droplets fall [lit. drip]. (5) The seas and rivers are like the flasks which receive the drops.

Šifā' 11.3-8:

ويكون مَثَل الجبل فى حقنه الأبخرة وإلجائه إياها إلى فجر العيون، مَثَل الإنبيق الصلب من حديد أو زجاج أو غيره مما يعد للتقطير، فإنه إن كان سخيفا متخذا من خشب متخلخل أو خزف متخلخل لم يحقن بخارا كثيرا، ولم يقطر منه شيء يعتد به، وإذا كان من جوهر صلب لم يدع شيئا من البخار يتفشى ويتحلل، بل جمع كله ماء وقطّره. فالجبال كالإنبيق، وقعر الأرض التى تحته كالقرع، والعيون كالمثاعب، والأذناب التى فى الأنابيق والأودية والبحار كالقوابل.

(1) A model [*maṯal*] of the mountain in its [function] of confining vapours and forcing them to emerge as springs would be a hard alembic made of iron, glass etc. among (tools used) for distillation. (2) For if (the alembic) were flimsy and made of porous wood or earthenware, it would not retain much vapour and no significant amount would be distilled from it. When

[32] For diagrams illustrating the kind of distillation equipment envisaged here see, e.g., Berthelot-Duval (1893) 108; Wiedemann (1878) 578; id. (1909) 237; Hasan-Hill (1986) 134ff.; Mertens (1995) cxx.

it is made of a solid substance, it does not allow any vapour to diffuse and dissipate but gathers all of it as water and distils it. (1) Mountains, then, are like the alembic, (3) the depth of the earth below it is like the cucurbit, (4) springs are like the conduits [*maṯāʿib*] and tails [*aḏnāb*] on the alembics (5) and the rivers and the seas are like the receptacles [*qawābil*].

l. 5. ܐܢܒܐܩܝܢ ("alembic"): انبيق IS. - The Greek term ἄμβιξ, ἄμβικος usually referred either to the whole of the still or to what we now call the cucurbit[33] (i.e. the bottom part of the still in which the distilland was heated). In Arabic, however, *anbīq/inbīq* is usually the "cap" of the still fitted above the cucurbit. - The usual Syriac form of the word appears to be ܐܢܒܝܩܐ,[34] the form ܐܢܒܐܩܝܢ being known to dictionaries only from here (PS Suppl. 23), where the medial -*ā*- and final -*īn* may be due to a false analogy with *aṯālīn* (see below).

l. 7. ܐܬܐܠܝܢ ("cucurbit", properly "aludel"): قرع IS. - The word used by BH here, cognate with the Arabic *uṯāl/aṯāl*, is properly the term for the aludel, the instrument used for sublimation (arab. *taṣʿīd*) rather than for distillation (*taqṭīr*). The Syriac word (in the form ܐܬܠܐ) is used in the text edited by Berthelot-Duval in the sense both of "vapour, soot" (13.7, 16; 14.18 etc.) and "aludel" (19.2, 5; 23.16; 26.4). The corresponding Greek word in the first sense is no doubt αἰθάλη; for the latter an unattested form *αἰθάλιον has been suggested.[35]

l. 7, 8. "which the heat of the fire touches", "from which the droplets fall": These words appear to be explanatory additions by BH.

l. 8. ܕܘܢܒܬܐ ܐܒܘܒܝܬܐ ("tubular tails"): i.e. the delivery tube; المثاعب والأذناب IS. - Of the two terms used by IS, *miṯʿab* ("drain") may be related to gr. σωλήν and can be compared with *mīzāb* used by Abū Bakr al-Rāzī and Khwārizmī. The term "tail" (*ḏunāba/ḏināba*) is also used in Rāzī.[36] - BH

---

[33] Mertens (1995) cxxii f.; for the terms used for the "cap", see ibid., cxxi (βῖκος, φιάλη, χαλκεῖον, μαστάριον, μασθωτόν). - "Cucurbit": gr. λοπάς, also βῖκος, χύτρα, θηλυκόν, πατέλλιον, βωτάριον, ἄμβιξ (Mertens [1995] cxxi); syr. *qarʿā* (Berthelot-Duval [1893] 19.17f., 24; 21.23; 22.1, 3; 40.16, 20f.; 41.6; 43.21f.); arab. *qarʿ/qarʿa*, lit. "gourd", whence lat. *cucurbita*.

[34] So in Bar ʿAlī (see PS 254) and saepe in the text edited by Berthelot-Duval (21.23, 22.3, 33.20, 43.22 etc.); the form ܐܡܒܝܩܐ with *mīm* occurs in Bar ʿAlī (PS 224) and at Berthelot-Duval 33.11f.; ܐܡܒܝܩܘܣ at Berthelot-Duval 26.15, 28.21, 29.13 (cf. PS Suppl. 21).

[35] Ruska (1924) 23 n.3; id. (1937) 61 n.2; accepted by Ullmann (1972) 265 (also at EI² V.111). - For the Greeks terms used for the predecessor of what we now know as aludel, see Mertens (1995) cliii-clxi (φανός, rather than κηροτακίς); cf. Berthelot-Ruelle (1888) 145, 162; Berthelot-Duval (1893) 108.

[36] (a) *mīzāb*: Rāzī, *Asrār* [Dānishpazhūh] 9.18 (cf. Ruska [1937] 95 n.8); Khwārizmī, *Mafātīḥ al-ʿulūm* [van Vloten] 257.6 (cf. Stapleton et al. [1927] 363). - (b) *ḏunāba*: *Asrār* 9.13f., 16. - Other known terms include (c) *ḫaṭm* ("snout, beak"): *Asrār* 8.17, 9.1 (cf. Wiedemann [1878] 577); (d) *iḥlīl* ("urethra"): Garbers (1948) 18; and, possibly, (e) *ḏubāb* ("tip"): in Ibn al-ʿAwwām, but emended by Clément Mullet to *ḏanab* (see

has rendered IS's two nouns by an adjective-noun phrase (as if they constituted a hendiadys). - The adjective *abbūḇānāyā* is known to the dictionaries only from *But*.[37] - It may be noted that *unbūba* (corr. syr. *abbūḇā*) is found in one manuscript (Escorial) of Rāzī's *K. al-asrār* where others read *ḏunāba*.[38]

l. 9-10. ܒܝ̈ܣܬܐ ܡܩܒܠܢܝܬ ܢܛ̈ܘܦܬܐ ("flasks which receive the drops"): i.e. the receptacle for the distillate; قوابل IS. - The usual Arabic term is *qābila*, pl. *qawābil* used by IS here, while the Syriac text edited by Berthelot-Duval has *mqabblānā* (19.18, 21.23, 43.22ff.). BH has paraphrased IS's *qawābil* here, no doubt for the benefit of the layman (cf. comm. on ll. 7, 8 above). The word *bīstā*[39] chosen by BH may have a technical ring, since it features in the list of utensils at Berthelot-Duval 21.22.

I.ii.6.1-6: Utility of mountains (3): formation of minerals

But.: (1a) For mineral substances [*gušmē*] requiring [for their formation] vapours (1b) which are mixed to a high degree [*saggīyā'īṯ*] with an earthy substance [lit. earthiness] (1c) and remain for a longer period [*nuḡrā yattīrā*] in places where their thorough [*ḥattīṯā*] mingling takes place without [them being subjected to] scattering and dissipation, (1d) there is no (place) like mountains. (2) For this reason the majority of them are formed in mountains. (3a) For in loose and porous terrains [*arʿāṯā*] they cannot be confined over myriads [*rebbū rebbwān*] of years (3b) during which that mixture [*metmazzgānūṯā*] of theirs is completed, (3c) which makes them ready [which gives the readiness] for the reception of their forms from the giver of forms.

Šifā' 12.8-11:

وأما الأجسام المعدنية المحتاجة إلى أبخرة تكون أخلاطها بالأرضية أكثر، وإقامتها فى مواضع لا تتفرق عنها أطول، فلا شىء أطوع لها كالجبال، فلذلك تتولد أكثرها بها. وأما الأرضون السهلة، فكيف يكون فيها البقاء والاحتباس والاحتقان، الذى بسببه يتم لها الامتزاج المؤدى إلى استعدادها لصورتها.

(1a) For mineral bodies requiring vapours (1b) which are mixed more thoroughly [lit. whose mixings are more] with earthiness (1c) and which remain longer [lit. whose sojourn (*iqāma*) is longer] in places from which they are not scattered, there is nothing more congenial [*aṭwaʿ*] than [lit. like, *ka-*] mountains, (2) and for this reason most of them are formed in them. (3a) As for level [*sahil*] grounds, how can there occur in them the sojourn [*baqā'*], retention [*iḥtibās*] and confinement [*iḥtiqān*] (3b) through which their mixing [*imtizāǧ*] is completed, (3c) that leads to their readiness for [the reception of] their form.

Wiedemann [1909] 236 n.6; EI² I.486).

[37] See PS Suppl. 1. - The word occurs at *But*. De plant. II.ii.3 [Drossaart Lulofs-Poortman] line 169 ܣܦܝܩܘܬܐ ܐܒܘܒܢܝܬܐ ("tubular void" in plants); II.iii.5, line 202 ܢܨܒܬܐ ܐܒܘܒܢܝܬܐ ("tubular plant"); De animalibus II.iii.2, ms. F 89r a1 ܦܓܪܐ ܐܒܘܒܢܝܐ (tubular body of cuttlefish, σηπία and τευθις).

[38] Ruska (1937) 95 n.5, at the place corresponding to ed. Dānishpazhūh 9.13.

[39] Itself probably a word of Greek origin (< βῆσσα etc.; see Brockelmann 69; PS Suppl. 54).

l. 6. ܝܗܒ ܨܘܪ̈ܬܐ ("giver of forms"): cf. Mete. V.ii.2.9 below.

I.ii.6.6-8: Other uses

> *But.*: (1a) There are also other particular uses of mountains, (1b) the exposition [*purrāšā*] of which belongs [*lāḥem*] to [the discussion of] particular [branches of] natural sciences, such as medicine.
>
> *Šifā᾽* 12.12-13:
>
> فهذه منافع الجبال، ولها منافع أخرى جزئية، تفصيلها فى العلوم الطبيعية الجزئية، مثل الطب وغيره.
>
> [تفصيلها ṬBM: نفصلها ed. Cair.]
>
> (1a) Such are the uses of mountains. They also have other particular [*ǧuzʾīya*] uses, whose exposition [*tafṣīl*] are in the particular natural sciences, such as medicine.

Min. I. Section iii.: *On Springs*

The section corresponds to IS's short chapter (*faṣl* 3) on water sources (*fī manābiᶜ al-miyah*). In this chapter, IS first classifies water into fives types (*Šifā᾽* 13.4-5): i. flowing springs (*ᶜuyūn sayyāla*); ii. still springs (*ᶜuyūn rākida*); iii. wells (*ābār*); iv. canals (*qanan*); and v. seepage (*nazz*). IS goes on to discuss the origins of the types i, ii and iii/iv (13.6-16), then to talk of type i as the best type of water (13.16-18) and type v as the worst, combining this with the explanation of the origin of this type (14.1-4). The chapter ends with a discussion of how wells are replenished. The order of presentation is somewhat altered in *But.* as may be seen in the table below.

The classification of waters as given here goes back to Arist. *Mete.* 353b 17ff. (cf. Olymp. 127.17ff., Olymp. arab. 107.13-22). The wording of *But.* in these parts is not always very close to the *Šifā᾽* and the use in particular of terminology closer to that of Arist. suggests that BH is here combining the Avicennian material with material taken from lost passages of Nic. syr. The wording of *But.* becomes closer to that of the *Šifā᾽* in those parts dealing with water quality (itself probably of Theophrastean origin)[40] and replenishment of wells.

The middle part of the section dealing with additional matters relating to wells (I.iii.2.5-I.iii.3.7) is based on Abū al-Barakāt. There, BH has the problem that Abū al-Barakāt, although he agreed with BH (and

---

[40] See Theophrastus, Frag. 214A [Fortenbaugh et al.], 21-24, 26f. (= Athenaeus, 42C); cf. Gilbert (1907) 425 n.1.

Arist.) in denying the theory that there was a primordial mass of water below the earth (see I.iii.3.1-5 below), disagreed with BH (and Arist., IS) in denying that the waters of the rivers, springs and wells originated from condensation of air or vapour (a theory which was upheld by BH, see I.iii.1.1-4), attributing the origin of these waters rather to precipitation. In utilising the material borrowed from Abū al-Barakāt, therefore, BH has to make a certain amount of selection and rearrangement, and is forced to add a statement that "vapours are transformed into water" at I.iii.3.1. We also see that precipitation, which plays the central role in the supply of the waters of rivers, wells etc. in Abū al-Barakāt, is given a subsidiary role by BH (I.iii.3.6-7).

	*Šifā*ʾ	Other sources	
I.iii.1.1-4	13.6-7	+ Nic.?	Flowing springs (i)
I.iii.1.4-6	13.8-10	+ Nic.?	Still springs (ii)
I.iii.1.6-8	13.11-15	+ Nic.?	Wells (iii)
I.iii.2.1-2	13.15-16	+ ?	Canals (iv)
I.iii.2.2-5	14.1-3		Seepage
I.iii.2.5-9		*Mu*ᶜ*t.* II.211.13-17 II.211.21-22	Wells in clayey ground
I.iii.3.1-5		*Mu*ᶜ*t.* II.211.17-21	Primordial mass of water
I.iii.3.6-7		*Mu*ᶜ*t.* II.211.11-15	Precipitation
I.iii.4.1-4	13.16-14.1, 14.3-4		Water quality
I.iii.4.5-8	14.5-10		Replenishment

I.iii.1.1-4: Flowing springs

The passage corresponds in its position and in some of its contents to the passage of the *Šifā*ʾ quoted below, but does not follow it very closely. The idea of spring water originating from condensation inside mountains is what IS had been talking about in the preceding chapter (on uses of mountains), but what is condensed according to IS is "vapour" rather than "air". Arist., on the other hand, talks of river water originating from condensed "air" at *Mete.* 349b 19-27 (note 349b 22 ἐξ ἀέρος, 23 ὁ ἀτμίζων ἀήρ),[41] so that one is led to suspect

[41] Cf. Abū al-Barakāt, *Mu*ᶜ*tabar* II.210.8-10: قال قوم وهم الاكثرون من الحكماء المتقدمين والمتأخرين ان الهواء المحتقن فى باطن الجبل يبرد فيستحيل ماء ويسيل فيستمد هواء ويبرد فيستحيل ماء ... ("Some, that is to say the majority of the philosophers ancient and modern, have said that air confined in mountains grows cold, turns into water and flows, and then more air is drawn in, grows cold and turns into water ...."; Abū al-Barakāt goes on to refute this view).

the passage of *But*. here to be a conflation of Avicennian and Nicolean material. Further evidence for the use of lost passages of Nic. syr. in this part of *But*. is provided by the use of such terms as *qāyōmā* at I.iii.1.4 and *mᶜattḏay b-īḏayyā* at I.iii.1.6 (see comm. ad loc. below).

But.: (1a) The water of flowing springs is generated from air (1b) which enters through openings in the earth into the clefts and channels inside (the earth) (1c) and, when it is condensed and made into vapour by (the earth's) coldness, (1d) turns into water. (2a) This water is propelled violently [*b-ḥēpā*], (2b) tears the earth (2c) and bursts out [*nābᶜīn*] from it. (3a) Because of the additional material [*mlōʾā yattīrā*] which it acquires (3b) its flow is not obstructed.

Šifāʾ 13.6-7:

فأما مياه العيون السيالة، فإنها تنبعث من أبخرة كثيرة، قوية الاندفاع، كثيرة المادة، تفجر الأرض بقوة انفجارها، ثم لا تزال تفيض مستتبعة موادها، على ما تعلمه.

[7 تفجر: يتفجر ed. Ṭehr.]

(1a) The waters of flowing springs originate from large quantities of vapour, (2a) which are propelled strongly [lit. is strong of propulsion, *qawīyat al-indifāᶜ*] and contain much material [*kaṯīrat al-mādda*] (2bc) and which break up [*tufağğiru*] the earth by the force of its discharge [*infiğār*]. (3) Then, the subsequent (parts) [*mustatbiᶜa*] of their materials continue to pour forth, according to [the principle] you know.[42]

l. 1. ܪ̈ܕܘܝܢ ("flowing"): السيالة IS; corr. Arist., Olymp. ῥυτός; Olymp. arab. *ğārin*.
l. 2. ܥܐܠ ("enters"): ܥܐܠܝܢ codd. - The subject must be "air" (sg.).
l. 3. ܠܐܪܥܐ ܒ̈ܙܥܝܢ ("tears the earth"): cf. I.iii.1.5, I.iii.1.7 (corr. IS يشق الارض) and I.iii.2.3-4 (corr. IS يخرق الارض) below.

I.iii.1.4-6: Still springs

But. (1a) The water of still springs is generated from vapours (1b) which do not have the power to tear the earth. (2a) Because they do not have any additional material [*mlōʾā yattīrā*], (2b) it stands still (2c) and is unable to flow over the surface of the earth.

Šifāʾ 13.8-10:

وأما مياه العيون الراكدة، فإنها مياه حدثت من أبخرة بلغ من قوتها أن اندفعت إلى وجه الأرض، لكن لم يبلغ من قوتها وكثرة مادتها أن يطرد تاليها سابقها طرداً ويدفعه ويسيحه.

(1) The waters of stationary springs are waters that arise from vapours which are able because their strength to be propelled to the surface of the earth, (1b/2a) but [in which] it is not possible, because of [the lack of] their strength and the [small] quantity of their material, (2bc) for the following (part) of them to drive away the preceding (part) of them, and to push it and to cause it to flow.

---

[42] Rephrased by Bahmanyār, *Taḥṣīl* 716.5: على ضرورة الخلاء ("according to the necessity of the void", i.e. due to *horror vacui*).

l. 4. ܩܝܘܡܬܐ ("stationary"): الراكدة IS; corr. Arist., Olymp. στάσιμος; Olymp. arab. *qā'im* (107.19). The agreement of BH and Olymp. arab. in using words derived from the root √*qwm*, as opposed to IS's *rākid*, suggests that *qāyōmā* (or *qayyūmā*?)[43] was the term used in Nic. syr.

I.iii.1.6-8: Wells

But.: (1) The water of wells is procured artificially [lit. are prepared by hand]. (2a) Because (such water) is unable to tear the earth, (2b) engineers use their skills (2c) to remove the weight of the soil from the surface of (the water) (2d) and devise means to draw from it.

Šifā' 13.11-15:

وأما مياه الآبار والقنى، فإنها معانة فى ظهورها وبروزها بالصناعة. وذلك لأنها لما كانت ناقصة القوة عن أن تشق الأرض وتبرز، قصرت لها المسافة فأزيل عن وجهها ثقل التراب المتراكم، حتى يخلص الحفر إلى مستقر البخارات. فحينئذ تصادف منفذا تندفع إليه بأدنى حركة

(1) The waters of wells and canals are helped in their appearance [*ẓuhūr*] and emergence [*burūz*] [on to the earth's surface] by human skill [*ṣinā*ᶜ*a*]. (2a) This is because, since they lack the strength to tear the earth and to emerge, the distance [which they have to travel] is shortened, (2c) and the weight of the accumulated soil is removed from their surface until the digging reaches the stationary vapours. Then, they find a passage towards which they are pushed by the least movement.

l. 6. ܡܬܥܒܕܢ ܒܐܝ̈ܕܝܐ ("procured artificially", lit. prepared by hands): This must be a rendition of gr. χειρόκμητος (Arist. *Mete*. 353b 25, Olymp. 127 24 etc.; cf. also χειροποίητος Arist. 353b 33 etc.). - Cf. Min. V.iv.3.6f. below.

l. 8. ܡܬܛܟܢܝܢ ... ܡܬܦܪܣܝܢ ("use their skills ... devise means"): We might note the use of two Ethpaal denominative verbs derived from nouns of Greek origin (*teknā* < τεχνή; *pursā* < πόρος). *Etparras* occurs in Nic. syr. at 51.24, 26.

I.iii.2.1-2: Canals

IS mentions canals (*qanāt*) as the "running" counterpart of the "standing" wells (*bi'r*) at *Šifā'* 13.15-16, but does not discuss their mechanism. The sentence may be related (via a lost passage of Nic. syr.?) to the passage of Arist. quoted below, although what BH described here sounds like the underground "qanāts" of the type particularly widespread in Iran.[44]

---

[43] On adjectives of this form, see Duval (1881) §240; Nöldeke (1898) §119. The instances of the nomen agentis *qāyōmā* cited at PS 3532 all seem to mean "one who stands in charge" or "one who upholds" rather than "who/what stands still", although the lexica cited by PS give definitions such as قائم and ثابت, which would fit here.

[44] On these, see EI[2] VI.528-532 s.v. "Ḳanāt" [Lambton]; Hasan-Hill (1986) 84f.;

But.: People create passages [*šḇīlē*] inside the earth [*arʿā*] from well to well in high places and cause the water to come out on the surface of low grounds [*arʿāṯā*] in canals [*agōgē*].

Arist. *Mete*. 349b 35-350a 2: οἱ γὰρ τὰς ὑδραγωγίας ποιοῦντες ὑπονόμοις καὶ διώρυξι συνάγουσιν, ὥσπερ ἂν ἰδιούσης τῆς γῆς ἀπὸ τῶν ὑψηλῶν.

l. 2. ܐ̈ܢܫܐ ("people"): ܐܢܫܐ Fl. - If we may trust BH to have obeyed his own rules, he would have written *nāšā* used in the collective sense of "people" with *syāmē*; see BH *Splend*. 28.14f.[45]

l. 2. ܐܓܘ̈ܓܐ ("canals"): < ἀγωγός (or *agōgā d-mayyā* < ὑδραγωγός).[46] Elsewhere in BH the word is used of the aqueduct built by Hezekiah in Jerusalem at *Chron*. [Bedjan] 24.10 (< Michael, *Chron*. 51b 24; cf. Syro-Hexapla, II Reg. 18.17, 20.20).

I.iii.2.2-5: Seepage

But.: (1a) Seepage water [*may rāṣīnā*] is generated from vapours which have a large quantity of material, (1b) but whose power is too weak for them to tear the earth with force and to flow. (2a) The ground [from which] they [emerge] [lit. their ground] is soft, (2b) and for this reason their flow is dispersed [*zrīq*] (2c) and much of what is turned into vapour is dissipated away from them.

*Šifā*ʾ 14.1-3:

وأما النز فهو أردأ المياه، وإنما يتولد من بخارات لها مادة كثيرة، وليس لها من قوة الاندفاع ما يخرق الأرض بقوة؛ بل اندفاعها منتشر، وأرضها رخوة يتحلل عنها أكثر ما يتبخر ...

[2 منتشر DSM: منتشرة Ṭ; متيسر ed. Cair.]

(1a) Seepage [*nazz*] is the worst of water. It is generated from vapours which have a large quantity of material (1b) but does not have the force of propulsion to tear the earth with force; (2b) rather, their propulsion is diffuse [*muntašir*]. (2a) Their earth is soft [*raḫw*] (2c) (so that) most of what has become vapour is dissipated from it, ...

l. 4. ܐܪܥܗܘܢ (lit. "their ground"): The suffix here, as well as that in l. 5 ܪ̈ܕܝܗܘܢ, could refer back either to *may rāṣīnā* (seepage water) or to *lahgē* (vapours). The corresponding suffixes is IS (*arḍ-hā*, *indifāʿu-hā*), on other hand, can only refer back to *buḫārāt*.

l. 5. ܡܢܗܘܢ ("away from them"): The masc. pl. suffix must be taken to refer back to "vapours" (or "seepage water", see preceding comment), giving the sense: "away from the main body of the vapours". The suffix of *ʿan-hā* in the corresponding clause of the *Šifā*ʾ can also be taken as referring back to "earth".

Hill (1996) 755.

[45] Cf. Duval (1881) §275, p. 263; Nöldeke (1898) §146, p. 90.

[46] PS 23f. s.v.; add to the references there Jacob of Edessa, *Hex*. 53b 31.

I.iii.2.5-9: Wells found in clayey ground

But.: (1a) Wells are found in clayey grounds or sandy (grounds) which are clayey, (1b) but are not found in rocky (grounds) (1c) except after they have been made [*šārkān*] clayey [lit. into clayeyness] (1d) by being dug to a greater depth [lit. when they are deepened more]. (2a) For this reason diggers [*ḥāpōrē*] find water in one place but not in another; (2b) and in one place the wells are deep, in another less so [lit. little, *qallīl*].

Abū al-Barakāt, *Muᶜtabar* II.211.13-17, 211.21-22:

ومياه الآبار من مياه الثلوج والامطار تنزل وترشح من الاعلى الى المواضع الخالية والاغوار من الارض فيجدها المحتفرون فى ارض دون ارض وفى موضع اعمق واغور وفى موضع اعلى ولا يوجد فى الصخرية ويوجد فى الرملية والطينية وتنخرق الآبار الى اغوار عميقة كبيرة فيعتقد أن موضع الماء ابدا تحت الارض ... وانما توجد الآبار فى الارض الطينية او الرملية التى تنتهى الى طينية ولا توجد فى الصخرية ما لم تنته الى الطينية

The waters of wells originate from waters of snow and rain which descend and filters from the heights to the empty places and hollows [*aġwār*] in the earth, (2a) so that diggers [*muḥtafirūn*] find them in some grounds but not in others, (2b) and in deeper and lower places and in higher places. (1b) It is not found in rocky [places/grounds], (1a) but is found in sandy and clayey ones. (1d) Wells are bored into deep and great hollows [*aġwār*]. ... [211.17-21: see under I.iii.3.1-5 below] ... (1a) Wells are only found in clayey ground or sandy (ground) which has been turned into [*tantahī*] clayeyness [*ṭīnīya*]; (1b) they are not found in rocky (ground) (1c) so long as it has not been turned [*lam tantahi*] into clayeyness.

l. 7. "by being dug to a greater depth" [lit. when they are deepened more]: There is no obvious counterpart for this clause in the *Muᶜtabar*, but it may be connected to *Muᶜtabar* II.222.16-17 ("wells are bored ..."), which is best taken with what follows, but could be understood as a part of the preceding sentence by taking the conjunction "*wa*-" to be circumstantial ("when ...").

I.iii.3.1-5: Refutation of primordial mass of water

What is refuted here is in fact the Platonic theory of the Tartarian waters enclosed inside the earth, which is dealt with at greater length in Min. V.iv.1-3 below (cf. also Mete. III.1.3).

But.: (1) That vapours become water in places which are suitable (2) and [that] water was not enclosed in the middle of the earth from the beginning [*men rēšīṯā*] as the simple-minded [*pšīṭē*] think (3) is known (3) from the fact that on high ground water is found when (the ground) has been dug a little, (4) whereas on low ground water is not found [even] when it has been dug much. (5a) If the natural place of water were inside the earth, (5b) it would be found in it at the same depth.

Abū al-Barakāt, *Muᶜtabar* II.211.17-21:

فيعتقد أن موضع الماء ابدا تحت الارض ويوصل اليه بالحفر وليس كذلك فانك تجد أرضا عالية تحفر البئر فيها فتجد الماء قريبا ثم تنزل منها الى ارض مستفلة بقياسها استفالا كثيرا فتحفرها فلا تجد ماء او تجده فى عمق اعمق ولو كان ماء البئر هو الماء الذى تحت الارض لتساوى سطحه بالنسبة الى سطح الارض

... (2) As a result, it was believed that the place of water from the beginning [*abadan*] was below the earth and that it was reached by digging. This is not the case. (3) For [when] you find a high ground and dig a well in it, you may find water near [the surface]. Then, [when] you descend from it to a ground much lower down in relation to it, and dig it, you may not find any water or you may find it at a greater depth. (5a) If well water were the water which is below the earth, (5b) its level would be the same [everywhere] in relation to the level of the earth.

l. 1-2. ܕܠܗܓܐ ... ܝܕܝܥܐ ܡܢ ܗܝ ("that vapours ... is known from the fact that ..."): ܕܠܗܓܐ scripsi: ܠܗܓܐ codd. - For the construction "*d-* ... *yiḏīʿā men hāy d-*", see Min. II.iv.1, III.ii.2.1, IV.iv.1.1, V.iv.2.3; Mete. II.i.2.3. It is also possible to supply the "*d-*" instead at a later point, before "*law men* ..." (*wa-ḏ-law men* ...), thereby making the first part ("vapours ...") an independent sentence.

l. 1. "Vapours become ...": see the introductory comments on Section Min. I.iii. above.

I.iii.3.6-7: Well water supplemented by precipitation

But.: (1) Well water overflows [*šāpʿīn*] in spring because of [*men*] rain and melting snow (2) and descends through clefts and holes [*peʿrē w-pōrō*] in the earth to the depth of (the earth).

Abū al-Barakāt, *Muʿtabar* II.211.11-15:

... والانهار والعيون النزية والرشحية تزيد تارة بالامطار اذا كثرت وتارة بالثلوج اذا ذابت وتارة بهما ولا تزيد ببرد شديد مستول من غير مطر ولا ثلج ومياه الآبار من مياه الثلوج والامطار تنزل وترشح من الاعلى الى المواضع الخالية والاغوار من الارض

(1) Rivers and springs, be they of the trickling or oozing type [*nazzīya wa rašḥīya*], increase sometimes with the rains as they abound, sometimes with the snows as they melt, sometimes with both. They do not increase [merely] with intense, overpowering coldness without there being rain or snow. (1) The waters of wells originate from waters of snow and rain (2) which descend and filters from the heights to the empty places and hollows [*aġwār*] in the earth, ...

I.iii.4.1-4: Quality of waters

But.: (1) Running water, because it is refined by movement, is the best of all waters. (2) Furthermore, the duration [*zaḇnā*] of its mingling [*ḥulṭānā*] with earthiness, from which decay is generated in it, is short. (3) Seepage water is the worst [lit. humblest] of all waters, (4) because, due to the slow speed [*šuḥḥārā*] of its flow, the duration of its mingling with the decay-generating [*mawlḏānyat masyūṯā*] earthiness is long.

Šifāʾ 13.16-14.1, 14.3-4:

والسيالة أفضل، لأن هذه الحركة تلطفها. ومع ذلك فإن مدتها فى الاختلاط فى حركتها إلى البروز بالأرضية المتولدة من اختلاطاتها بها العفونات، تقصر. وأما النز فهو أردأ المياه،... والذى يبقى ويحتبس، تطول مدة مخالطته للأرض إلى أن يبرز؛ لأن حركته إلى البروز بطيئة، فيعفن ويتغير فى طريقه عند مخالطته للأرضية.

[13.18 بها العفونات DṬM: بعفونات ed. Cair.]

(1) Running [waters] are the best because this movement refines them. (2) Furthermore, their period [*madda*] in being mixed [*iḫtilāṭ*] - in their movement towards emergence [*burūz*] - with earthiness, from whose mixture with (the waters) decays [*ʿufūnāt*] are generated, is short. (3) Seepage [*nazz*] is the worst of waters ... [see under I.iii.2.2-5 above] ... (4) [For the portion of vapours] which [is not dissipated] but remains and becomes confined, the period of its mixing with the earth before it emerges is long, because its movement towards emergence is slow; as a result, it decays and is altered on the way by its mixture with earthiness.

l. 3. ܒܗ ("in it", i.e. in the water): بها IS. - The word *bihā* in IS is best taken with *iḫtilāṭ*, the whole phrase then meaning "from the mixture (of the earthiness) *with* (the waters)". It seems, however, that BH has taken *bihā* with *mutawallida*, giving the sense: "due to whose admixture decays are generated *in* (the waters)".

I.iii.4.5-8: Replenishment of wells

But.: (1a) Whenever wells whose water is stationary are drawn upon [*metdallyān*], (1b) other water is attracted into its place. (2a) [This is] because the vapour [which is] the material of that spring - (2b) since it has the power to lift water up to a certain predetermined level [*tḥōmā mtaḥḥmā*] - (2c) is able, (2d) whenever (the water) decreases and falls below that height, (2c) to raise it again to (that level).

Šifāʾ 14.5-10:

والعيون الراكدة والآبار الراكدة إذا نُزحت، يُجلب إليها بدل ما ينزح منها. وذلك لأنه إنما كان للبخار الذى هو مادة تلك العين أن يندفع إلى أن يبلغ المبلغ الذى كان استقر قديما عليه فقط، فإذا بلغ ذلك المبلغ صار فى الثقل بحيث لا يتمكن ما تحته أن يقلّه ويزيحه؛ بل يكون ما وقف من ذلك سدا، كما كانت الأرض قبل أن تحفر. فإذا نقص من ذلك الثقل، قَدَر البخار المندفعُ إلى جهته أن يتصعد ويُحرك ما يغمره من فوق إلى الحد المحدود.

(1a) When still springs and still wells are drawn upon [*nuziḥat*], (1b) an amount equivalent to what is drawn from them is attracted into them. (2a) This is because the vapour, which is the material of the spring, (2b) has the [property] of being propelled until it reaches the level [*mablaġ*] which it had before (and no further); when it reaches that level its weight becomes such that what is below it cannot raise it or remove it; rather, [the water/vapour] standing [above] becomes an obstruction, just as the earth was before it was dug. (2d) When something is [then] taken from that weight, (2bc) the vapour pushed in its direction is able to rise and move what is covering it from above [again] to the predetermined level [*al-ḥadd al-maḥdūd*].

l. 5. ܘܒܐܪ̈ܐ ("wells"): IS speaks of "still *springs and* still wells". The omission of "springs" by BH here results in a slight awkwardness when he goes on to say "the material of that *spring*" in the next line.

l. 5. ܡܬܕܠܝ̈ܢ sic F ("[wells] are drawn upon"): ܡܬܕܠܝܢ codd. ceteri. - The use of *dlā* in the sense of "*draw from* (a well)" rather than "*draw* (water)" seems unusual, but the reading of F, making "wells" the subject of its passive form,

provides an exact parallel of IS's *nuziḥat*. It is also possible to read *metdallēn* with the ceteri and take this with the following *mayyā*, but we would then need, in our minds at least, to supply *men-hēn* ("i.e. *kmā d-metdallēn mayyā [men-hēn], ḥrānē* ...).

BOOK OF MINERALOGY, CHAPTER TWO

ON EARTHQUAKES

Earthquakes, according to the Aristotelian theory, are caused by underground winds. In Arist. *Mete.*, therefore, the discussion of earthquakes (II.vii-viii) follows that of winds and precedes that of other phenomena produced, like winds and earthquakes, by the smoky exhalation (thunder, lightning etc.). IS, on the other hand, who accepts the smoky exhalation as the main, but not exclusive, cause of earthquakes, alters the order in his *Šifāʾ*, placing the chapter on earthquakes among those dealing with geological matters. This order is followed by BH in *But.*

l. 2. ܢܝ̈ܕܐ ܕܐܪܥܐ ("earthquakes"): The absence of a single word in Syriac (as in Latin and some modern European languages) corresponding to gr. σεισμός and arab. *zalzala* and referring specifically to the tremor of the earth results in various words meaning "trembling" or "movement" (*reᶜlā*, *nawdā*/*nyāḏā*, *zawᶜā*) being used in this sense, often qualified by "of the earth". In Nic. syr. the standard term is *zawᶜā d-arᶜā* (used 31 times), while *nyāḏā* is also found twice, on both occasions in the combined form *zawᶜā wa-nyāḏā d-arᶜā.*[1] These two are also the terms used by BH in the present chapter, *zawᶜā* usually being qualified by *d-arᶜā* (but without at II.i.1.1, 3.1; ii.2.7; iii.2.7) and *nyāḏā* being used on its own except here and at II.i.2.4 (*zawᶜā wa-nyāḏā d-arᶜā*, corr. Nic. syr. 36.15).[2] Of the two forms of the noun from √*nwd*, BH evidently prefers *nyāḏā*,[3] while Bar Kepha and Bar Shakko prefer *nawdā*.[4]

---

[1] Probably representing an original gr. σεισμός καὶ κίνησις γῆς, a combination encountered, for example, at Arist. 365a 14. - The verb *anīḏ* (√*nwd* Aph.) is also used twice in Nic. syr. (35.30, 37.10), similarly combined each time with *azīᶜ* (√*zwᶜ* Aph.).

[2] The verb *arᶜel*, related to *reᶜlā*, is used together with *anīḏ* at *But.* Mete. III.1.3.11 below.

[3] All the citations in the sense of "earthquake" under *nyāḏā* in PS (2310) and Brockelmann (418b) are from BH's works. The form *nawdā*, for which the lexica give no citations from BH, is found at BH *Serm. sap.* 79.2.

[4] Invariably combined with *reᶜlā* and *zawᶜā* in Bar Kepha, *Hex.* (Paris 241, 191r b12f., 191v b3f., 192r b4, 10, 20f., 192v a1f., 26f., 193r b12).

Min. II. Section i.: *On the Opinions of the Ancients on Earthquakes*

Arist. began his account of earthquakes by mentioning and refuting the theories of Anaxagoras, Democritus and Anaximenes (*Mete*. II.vii). IS, on the other hand, begins his discussion of earthquakes in the *Šifā*ʾ with an examination of the possible causes of earthquakes and reaches, like Arist., the conclusion that the smoky exhalation is the principal cause of seismic activity (15.5-16.3). In contrast to Arist., however, IS also accepts moving water (i.e. Democritus' theory, which is not refuted explicitly by Arist., see comm. on II.i.4.6-7 below) and "collapse of supports"[5] as possible causes (16.3-4).

The present section of *But*. depends almost entirely on Nic. syr., with the exception of the initial definition of earthquakes (II.i.1.1-2) and the odd phrase here and there, mostly taken from the *Šifā*ʾ. BH, however, reorders his material, placing the refutation of all three "ancient" views together in the final "theory" of the section (II.i.4), whereas as Nic. syr. (following Arist.) had refuted each theory immediately after describing them. The correspondences may be summarised as follows.

II.i.1.1-2	*Šifā*ʾ 15.4-5	Definition of earthquake
II.i.1.2-8	Nic. 35.12f., 20f., 26-31	Anaxagoras' theory
II.i.2	Nic. 36.13-19	Democritus' theory
II.i.3	Nic. 36.21-23, 30	Anaximenes' theory
II.i.4.1-6	Nic. 35.31-36.7, 36.11-13	Refutation of Anaxagoras
II.i.4.6-7	Nic. 36.20-21	Refutation of Democritus
II.i.4.7-11	Nic. 36.30-37.4	Refutation of Anaximenes

II.i.1.1-2: Definition of earthquake

But.: Earthquake is a movement which occurs to a part of the earth due to a cause which [itself] moves [*mettzīʿā*] and causes what is on it to move [*mzīʿā*].

Šifā 15.4-5:

وأما الزلزلة، فإنها حركة تعرض لجزء من أجزاء الأرض بسبب ما تحته، ولا محالة / ٥ أن ذلك السبب يعرض له أن يتحرك ثم يحرك ما فوقه.

Earthquake is a movement which occurs to one of the parts of the earth

[5] Probably to be seen as a modified version of the Anaximenean theory, received by IS via Theophrastus (see Theoph. arab. [Ibn al-Khammār] 15.2-7, and comments thereupon at Daiber [1992] 290).

due to a cause below it, and there is no doubt that that cause (itself) moves [*yataḥarraku*], then causes what is above it to move [*yuḥriku*].

II.i.1.2-3: Anaxagoras' theory (1).

In each of the "theories" II.i.1., II.i.2., II.i.3., BH begins the description of the theories of Anaxagoras, Democritus and Anaximenes with the formula "*x* attributes [*sā'em*] the cause of the movement of the earth to *y*". This may be compared in the first instance with the opening sentence of the section on earthquakes in Nic. syr., but the sentence structure is closer to that of a sentence occuring at a later point in the *Šifā'*.

But. [underline: agreement with Nic.; italics: with *Šifā'*]:
Anaxagoras attributes [*sā'em*] *the cause* of the movement of the earth *to* [*l*-] *air*
Nic. syr. 35.12-13:
ܢܐܡܪ ܥܠ ܙܘܥܐ ܕܐܪܥܐ. ܐܢܟܣܓܘܪܘܣ ܐܣܒܪ ܕܐܝܬܘܗܝ ܥܠܬܐ ܕܙܘܥܐ
13/ ܕܐܪܥܐ ܐܐܪ. ܕܡܘܩܪܝܛܘܣ ܡܝܐ. ܐܢܟܣܝܡܝܢܣ ܕܝܢ ܐܪܥܐ.
Let us talk about the earthquake. Anaxagoras thought that the cause of the movement of the earth was air, Democritus water, and Anaximenes earth.
Šifā' 17.1: وأما انكساغورس فإنه ينسب العلة إلى الهواء
Anaxagoras relates [*yansubu*] the cause to air

II.i.1.3-4: Anaxagoras (2).

But.: ... and says: Because the earth is flat it is borne by air, like a leaf [sg.] that is spread on water.
Nic. syr. 35.20-23[6]:
ܐܢܟܣܓܘܪܘܣ ܕܡܢ ܩܠܐܙܘܡܢܝ ܐܡܪ ܕܥܠ ܕܦܫܝܛܐ /21 ܐܪܥܐ ܡܬܛܥܢܐ ܡܢ
ܐܐܪ: ܐܝܟ ܛܪܦܐ ܕܦܪܝܣܝܢ ܥܠ ܡܝܐ ܘܡܬܛܥܢܝܢ/22 ܡܢܗ. ܗܠܝܢ ܕܟܕ
ܡܬܟܢܫܝܢ ܛܒܥܝܢ. ܘܐܝܟ ܛܪܦܐ ܕܕܗܒܐ ܕܪܡܝܢ/23 ܥܠ ܡܝܐ. ܗܠܝܢ ܐܦ ܟܕ
ܡܬܟܢܫܝܢ ܒܗ ܒܕܡܘܬܐ ܛܒܥܝܢ.
Anaxagoras of Clazomenae said: Because the earth is flat it is borne by air, like leaves [pl.] which are spread on water and are borne by it - these, when they are drawn together, sink - and like gold-leaves which are thrown on water - these too, when they are drawn together, sink in the same way.

l. 3. "because the earth is flat ...": The theory held by Anaxagoras and others that the earth is flat and *therefore* rests on air is not mentioned in Arist. *Mete.* but in Arist. *De caelo* (294b 13ff.). It is also brought into connection with Anaxagoras' theory of earthquakes by Alexander of Aphrodisias (*in Mete.* [Hayduck] 114.21-23; cf. Fontaine [1995] p. LV and 126-127).

[6] Cf. Olymp. arab. 133.12-13; *Cand.* 129.1-3.

l. 4. ܛܪܦܐ ("a leaf"): Of the two examples given by Nic. syr., BH uses that of the "leaves" in *But.* and that of "gold leaves" in *Cand.* (129.2). - Cf. Arist., *Mete.* 348a 9, where Arist. talks of "earth and gold" floating on water in connection with suspension of hailstones in air; cf. further Nic. *De plantis* arab. [Drossaart Lulofs-Poortman] §154 (corr. Ps.-Arist. *De plantis* 823a 17-27).

II.i.1. 4-8: Anaxagoras (3)

But. [underline: agreement with Nic.; italics: with *Šifā'*]:

[Furthermore] (1a) because *the lower part* of the earth is porous [*paḥḥīḥā*] and *loose* [*mp̄arsʿā*], (1b) while this upper part of it, on which we live, is thick [*lbīdā*] and dense [*rṣīpā*] on account of the clay which is hardened by the sun, (2a) when a part of *the air* falls into (the earth) (2b) and strives to rise upwards, (2c) it passes through those openings which are (in the part) below (2d) but cannot pass through the openings which are (in the part) above, (2e) and for this reason moves and *shakes the earth*.

Nic. syr. 35.26-31[7]:

ܘܐܡܪ ܕܟܠܗ̇ ܐܪܥܐ ܦܘܚܝܬܐ ܐܝܬܝܗ̇/27 ܐܠܐ ܡܫܘܚܬܐ ܕܠܥܠ ܕܥܠܝܗ̇ ܥܡܪܝܢܢ
ܟܬܝܪܐ ܦܘܚܝܐ. ܡܛܠ ܕܡܬܥܒܝܢ ܫܘܦܢܘܗ̇/28 ܡܢ ܡܛܪܐ. ܡܐ ܕܝܢ ܠܡ ܕܢܦܠ
ܡܢܬܐ ܡܢ ܐܐܪ ܒܗ̇. ܟܕ ܦܪܟܐ ܠܡܣܠܩ ܠܥܠ/29 ܠܥܠ. ܥܒܪܐ ܡܢ ܒܫܘܦܢܘܗ̇
ܕܠܬܚܬ ܥܠ ܕܦܘܚܝܐ ܐܝܬܝܗ̇ (ܗܘ) ܒܥܒܝܘܬܗ̇ ܠܬܚܬܝܬܐ: ܟܕ/30 ܠܐ ܕܝܢ ܡܫܟܚܐ:
ܕܢܥܒܪ ܠܥܠ ܒܗܢܘܢ ܡܛܠ ܥܒܝܘܬܗܘܢ. ܡܙܝܥܐ ܘܡܙܝܥܐ/31 ܠܐܪܥܐ.

[27 ܦܘܚܝܬܐ: ܦܘܚܝܐ cod. || 29 ܠܬܚܬܝܬܐ: ܠܬܚܬܐ cod.]

(1a) He says that the whole of the earth is porous, (1b) but its upper part, on which we live, is dense, because its openings are thickened by rain. (2a) When a part of the air falls into (the earth), (2b) as it strives to rise upwards, (2c) it passes through the openings which are (in the part) below because its lower side is porous; (2d) as, however, it cannot pass upwards through those (openings) above because of their density, (2e) it moves and shakes the earth.

Šifā' 17.2-4:

وأن الجنبة السافلة متخلخلة، والتى نحن عليها متكاثفة للأمطار التى تُعرّى وجهها. فإذا نفذ الهواء فى التخلخل الذى بتلك الجنبة، ثم لم يجد طريقا إلى الانفصال والصعود الطبيعى الذى له، وذلك من الجهة التى نحن عليها، زلزل الأرض.

... (1a) and the lower side is loose [*mutaḫalḫil*], (1b) while the (side) on which we are is thickened by the rains which bind together [*tuʿarrī*][8] its surface. (2a) When the air penetrates into the loose part which is in that side, (2d) then cannot find a way for the discharge and the ascent natural to it, namely from the side on which we are, (2e) it shakes the earth.

---

[7] Cf. Arist. 365a 19-25; Olymp. arab. 133.13-17; *Cand.* 129.3-5.

[8] The verb *ʿarrā* (√*ʿrw*) here needs to be understood in the light of *ʿurwa* in the sense of "tie, bond". - Cf. Arist. 365a 21f. συναληλεῖφθαι διὰ τοὺς ὄμβρους (Alex. 114.17 συναληλίφθαι καὶ ἡνῶσθαι ὑπὸ τῶν ὄμβρων).

l. 4. ܬܚܬܝܬܗ ܬܚܬܝܬܐ ܕܐܪܥܐ ("the lower part of the earth"): ܟܠܗ ܐܪܥܐ ("whole of the earth") Nic. - In talking of the "whole" of the earth as being porous, Nic. syr. has remained faithful to Arist. (365a 22f.). The alteration of this to "the lower part" is understandable and in making this alteration BH no doubt found support in IS's "the lower side" (*al-ğanba al-sāfila*). In paraphrasing the same passage of Nic. syr., BH uses a different solution at *Cand.* 129.3: "the whole of the earth is porous, *except for* [*sṭar men*] its surface which is dense."

l. 4-5, 6. ܦܚܝܚܐ ܘܡܦܪܣܥܐ ... ܠܒܝܕܐ ܘܪܨܝܦܐ ("porous and loose ... thick and dense"): ܦܚܝܚܬܐ ... ܪܨܝܦܐ Nic. - Where Nic. syr. has only one adjective on each occasion, BH has two each. The addition of "loose" (*mp̄arsʿā*) may be due to IS's *mutaḫalḫila*, while "dense" (*lḇīḏā*) may echo the verb *metlabbdīn* in Nic. syr. Since we find two adjectives each being used in the corresponding passage of Olymp. arab. (133.14: *saḫīf mutaḫalḫil ... mutakāṯif mutalabbid*), it is possible that BH has kept the original reading of Nic. syr., but against that possibility we have to reckon the fact that BH himself uses only the two adjectives used by Nic. syr. (*paḥḥīḥā* and *rṣīpā*) in the corresponding passage of *Cand.* (129.3), as well as the fact that *mp̄arsʿā* does not occur anywhere in the Cambridge manuscript of Nic. syr. (the noun *mp̄arsʿūṯā* is used once at Nic. syr. 5.5).[9]

l. 5-6. ܡܛܠ ܛܝܢܐ ܕܗܘ ܫܡܫܐ ܡܩܫܝܐ ("on account of the clay which is hardened by the sun"): It appears that BH found the attribution of the impermeability of the earth's surface to the action of "rain" (so Arist., Nic. syr. and IS) difficult to accept. - The fault here lies in part with the Syriac translation. Arist., in fact, talks of the earth's surface turning slimy and so being clogged up (συναληλεῖφθαι) under the influence of rain. The idea expressed by the verb συναλείφω has been lost in the rendition "its openings are thickened by rain" of Nic. syr.[10] BH's difficulty will have been increased by the notion expressed elsewhere that rain "loosens" the earth (in Anaximenes' theory; see II.i.3.2 below)[11] and that earth's surface is thickened

---

[9] Bar Kepha, in contrast, seems to have been more fond of *mp̄arsʿā*, since he uses this word and its cognates four times in his chapter on earthquakes, while not once using *paḥḥīḥā*: Bar Kepha, *Hex.*, Paris 241, 191v a15-18: ܐܪܥܐ ܓܝܪ ܐܝܬܝܗ ܩܪܝܪܬܐ ܘܝܒܝܫܬܐ ܘܩܫܝܬܐ . ܘܐܐܪ ܐܝܬܘܗܝ ܚܡܝܡܐ ܘܪܛܝܒܐ ܘܡܦܪܣܥܐ ("earth is cold, dry and solid, while air is hot, moist and loose" [passive sense and adjectival use as in *But.*]); 191v a12-13: ܢܪܥܐ ܠܗ ܘܡܦܪܣܥܐ ܠܗ ("[fire] tears and loosens [the earth]" [active/verbal]; 191v a26: ܡܦܪܣܥܘܬܐ ("looseness [of air]"); 192r a5-7: ܥܕܡܐ ܕܡܬܦܪܣܥܐ ܕܘܟܬܐ ܗܝ ("until that place is loosened").

[10] BH may also have failed to understand the sense of IS's *tuʿarrī* (see n. 8 above).

[11] Cf. Bar Kepha, *Hex*. Paris 214, 191r b20-22 [in what is also in fact a rendition of Anaximenes' theory, though he is not named]: ܘܟܕ ܡܬܝܒܫܐ ܐܪܥܐ ܕܠܥܠ ܘܡܬܩܫܝܐ . ܐܘ ܡܬܪܛܒܐ ܒܡܝܐ ܘܡܬܦܪܣܥܐ .: "when the earth above [the caves etc. in the earth] is dried and is hardened, or is moistened by water and is loosened"; also Theoph. arab. [Ibn al-Khammār] 15.4-5: (إما لأن الارض تجفّ وتفتّت وإما لأنها ترطب وتنحلّ). - In II.i.3., and in the corresponding passage of Nic. syr., what is loosened by rain is "the

by dryness (II.iii.2.8). - The solution adopted by BH may be explained by referring back to the preceding chapter of *But*. (Min. I.i.2.) where we hear of petrification through hardening "of glutinous clay by the sun".

II.i.2.1-7: Democritus' theory

But.: (1a) Democritus attributes the cause of the movement of the earth to water (1b) and says that the bowels of the earth are filled with water. (2a) When rains descend [*nāḥtīn*], (2b) their waters too reach its depth (2c) and mingle with that water [already there]; (3a) (then), because the bowels of the earth cannot contain (the two bodies of water) [*l-hōn*], (3b) they become compressed (3c) and as they squeeze each other [*zārbīn la-ḥḏāḏē*], (3d) they cause the movement and shaking of the earth. (4) He says furthermore [*tūḇ*] that earthquake also [*āp̄*] occurs when the earth is dry. (5) For when it become dry [lit. in its dryness] (the earth) yearns [*metragrgā*] for moisture and attracts water towards itself. (6a) This water which is attracted, (6b) when [*mā d-*] it falls into the bowels of (the earth), (6c) moves (the earth).

Nic. syr. 36.13-19[12]:

ܕܝܡܘܩܪܝܛ(ܘܣ) 14/ ܕܝܢ ܐܡܪ . ܕܟܪ̈ܣܬܗ ܕܐܪܥܐ ܡܝ̈ܐ ܡܠܝܢ . ܘܟܕ ܕܝܢ ܕܡܛܪ̈ܐ
ܢܗܘܘܢ . ܢܚܬܝܢ ܐܦ 15/ ܡܝ̈ܐ ܕܡܛܪ̈ܐ ܠܥܘܡܩܗ . ܘܟܕ ܣܓܝܢ ܡܝ̈ܐ ܘܠܐ
ܚܐܢ ܠܗܘܢ ܟܪ̈ܣܬܗ 16/ ܕܐܪܥܐ ܡܬܐܠܨܝܢ . ܘܟܕ ܡܙܕܪܒܝܢ ܥܒܕܝܢ ܙܘܥܐ
ܘܢܘܕܐ ܕܐܪܥܐ . ܐܡܪ ܕܝܢ/17 ܕܗܘܝܐ ܗܕܐ ܐܦ ܗܘ ܕܒܝܒܫܘܬܐ ܐܝܬܝܗ ܐܪܥܐ .
ܟܕ ܡܬܝܒܫܐ ܠܗ ܓܝܪ ܓܪܫܐ ܪܛܝܒܘܬܐ 18/ ܠܘܬ ܢܦܫܗ ܥܡ ܪܓܬܐ ܡܕܡ . ܗܘ ܗܟܐ
ܕܝܢ ܕܡܬܬܢܓܕ : ܟܕ ܢܦܠ ܒܟܪ̈ܣܬ{ܗ} 19/ ܡܙܝܥ ܠܗ .

(1a) Democritus (1b) says that the bowels of the earth are filled with water; (2a) when rains occur (2b) the water of the rain too descends [*nāḥtīn*] to its depth; (3a) and when the water increases and the bowels of the earth cannot contain it [*l-hōn*], (3b) it becomes compressed (3c) and as it is squeezed [*mezdarbīn*] (3d) it causes the movement and shaking of the earth. (4) He says that this occurs also [*w-*] when the earth is dry. (5) For as it becomes dry [*kaḏ metyabbšā*], (the earth) attracts moisture towards itself with a certain yearning [*reggṯā*], (6a) and that thing which is attracted, (6b) as [*kaḏ*] it falls into the bowels (of the earth), (6c) moves (the earth).

l. 1. "Democritus attributes ...": see comm. on II.i.1.2-3 above.

---

tops of mountains"; in Bar Kepha (and, it appears, in the Theophrastean tradition in general), it is "the earth above the caverns in the earth", which is more easily equated with "the upper part of the earth".

[12] Cf. Arist. 365b 1-6; Olymp. arab. 133.20-134.4; *Cand.* 129.5-7. - The second part of Democritus' theory, corresponding to Nic. syr. 36.16-19 and *But.* II.i.2.4-7, is omitted in *Cand*.

II.i.3.1-4: Anaximenes' theory

But. [underline: agreement with Nic.; italics: with *Šifā'*]:

(1a) *Anaximenes* attributes to the earth itself the cause of its own movement (1b) and says that when *tops* of *mountains fall*, (1c) either loosened by rains (1d) or broken up by excessive dryness of the earth, (1e) they cause the earth to move [*mzī'īn*] by that *mighty fall* of theirs. (2) For this reason ['*al hādē*] earthquakes [lit. movements] occur in rainy seasons and seasons of absence of rain.

Nic. syr. 36.21-23, 30[13]:

ܐܢܟܣܝܡܢܣ ܐܡܪ ܕܪ̈ܝܫܐ ܕܛܘܪ̈ܐ ܗ̄ ܩܘܠܘܢܘܝ 22/ ܩܪܐ ܠܗܘܢ. ܗܠܝܢ ܟܕ
ܢܦܠܝܢ ܐܘ ܟܕ ܡܬܦܫܪܝܢ ܡܢ ܡܛܪܐ ܐܘ ܟܕ ܡܢ ܝܒܝܫܘܬܐ /23 ܕܐܪܥܐ
ܡܬܚܬܟܝܢ: ܥܒܕܝܢ ܙܘܥܐ ܕܐܪܥܐ. ܡܛܠܗܢܐ ܠܡ ܒܙ̈ܒܢܐ ܡܛܪܝܐ .//30
ܘܒܙ̈ܒܢܐ ܕܠܘܩܒܠ ܡܛܪܐ ܗܘܝܢ ܙܘ̈ܥܐ ܕܐܪܥܐ .

(1a) Anaximenes (1b) says that when tops of mountains - he calls them κολωνοί - fall, (1c) either when loosened by rain (1d) or when broken up by the drynness of the earth, (1e) they cause a movement ['*ābdīn zaw'ā*] of the earth. (2) For this reason [*meṭṭul hādē*] movements of the earth occur in rainy seasons and seasons of absence of rain.

Šifā' 16.5-6:

وربما كانت للزلزلة أسباب فوق الأرض، كجبال يعرض لها أن تسقط قُللها أو أجزاء كبيرة منها سقوطا قويا فتتزلزل الأرض، على ما كان يراه رجل يقال له أراكيماس ...

Sometimes the earthquake has causes above the earth, such as mountains, whose tops or large parts of which fall strongly [lit. a strong fall], so that the earth is shaken, in the way a man called Arākīmās [Anaximenes] thought...

l. 1. "Anaximenes attributes": see comm. on II.i.1.2-3 above.

l. 3 ܝܒܝܫܘܬ ܡܣܓܝܐܬܐ ܕܐܪܥܐ ("excessive dryness of the earth"): ܝܒܝܫܘܬܐ ܕܐܪܥܐ Nic. - The addition of *saggī'tā* may simply have been prompted by the thought that a normal level of dryness would not cause such crumbling, but it is worth noting that Olymp. arab., too, mentions the severity of the dryness in the same connection (133.10: *šiddat al-yabs*).

II.i.4.1-6: Refutation of Anaxagoras

The four points made in Nic. syr. in refutation of Anaxagoras can all be traced back to Arist. 365a 25ff. (cf. Alex. 114.23-115.12), although the latter parts of the argument from the sphericity of the earth ("[this] is known from ...) are later additions. BH uses the last three of these four arguments in *But.* and there is little in our passage of *But.* which cannot be explained with reference to the passage of Nic. syr., although it is remarkable that the three arguments are presented in the reverse order of that in Nic. syr.

---

[13] Cf. Arist. 365b 6-12; Olymp. arab. 133.7-11; *Cand.* 129.7-9.

But.: (1a) The Master [Aristotle] refutes the first opinion and says: (1b) If this were true, (1c) earthquake [*nyādā*] would not <u>occur</u> under particular conditions [*dīlānāʾīt*] in <u>places</u> <u>which were liable to be affected in this way</u> [lit. suitable for the reception of this effect], (1d) but [would occur] everywhere. (1e) For <u>(Anaxagoras) apportioned the cause generally to all the earth</u>. (2a) <u>Furthermore, if all the air, though it bears the earth, cannot move it</u>, (2b) how does <u>a part of it</u> <u>move it</u>? (3) Furthermore, it has already been demonstrated that <u>the earth is not flat along [its] latitude but is spherical</u>.

Nic. syr. 35.31 [ult.]-36.7, 11-13[14]:

ܠܘ ܕܝܢ ܒܬܪܝܨܘܬܐ ܐܡܪ ܗܢܐ ܕܡܢܗ̇ ܬܚܝܬܐ ܗܝ ܡܢ ܐܪܥܐ ܠܥܠܝܬܐ ܐܚܪܢܝܬܐ ܡܢܗ̇ ‹ܕܝܢ› /
1 ܠܬܚܬ. ܟܠܗ̇ ܓܝܪ ܐܪܥܐ ܠܬܚܬ ܐܝܬܝܗ̇. ܐܝܟܢܐ ܕܡܚܘܝܢ ܟܠܗܘܢ ܓܘܫܡ̈ܐ
2/ ܝܩܝܪ̈ܐ ܡܬܩܠܐ ܕܠܐܪܥ ܠܬܚܬ ܢܨܝܢ. ܬܘܒ ܕܝܢ ܕܠܐ ܫܛܝܚܐ ܐܪܥܐ ܒܦܬܝܐ.
ܐܠܐ 3/ ܐܣܦܝܪܝܐ. ܘܡܬܝܕܥܐ ܡܢ ܗܝ ܕܟܕ ܫܡܫܐ ܗܘܐ ܠܥܠ ܡܢ ܐܪܥܐ. ܠܘ
ܟܠ ܡܢܢ 4/ ܡܩܒܠܝܢ ܢܘܗܪܗ. ܐܠܐ ܡܫܚܠܦܐܝܬ ܗܘܐ ܢܘܗܪܗ ܠܦܘܬ
ܫܘ̈ܚܠܦܐ 5/ ܕܐܘܪ̈ܝܙܘܢܣ. ܘܬܘܒ ܠܘ ܗܢܘܢ ܗܢܘܢ ܚܘܕܪ̈ܐ ܡܬܚܙܝܢ ܒܟܠܗܘܢ
ܩܠܝܡ̈ܐ 6/ ܐܠܐ ܡܫܚܠ̈ܦܐ. ܘܬܘܒ ܐܢ ܕܝܢ ܟܠܗ ܐܐܪ ܟܕ ܫܩܝܠ ܠܗ̇ ܠܐܪܥܐ
ܠܐ ܡܙܝܥܐ ܕܢܙܝܥܝܗ̇ 7/ ܚܕ ܟܡܐ ܝܬܝܪܐܝܬ ܡܢܘܬܐ ܡܢܗ ܕܐܐܪ ܠܐ ܡܙܝܥܐ
ܕܬܙܝܥܝܗ̇. //11 ܬܘܒ +ܕܢܠ ܠܘ ܒܟܠܙܒܢ ܐܝܢܐ ܕܗܘ ܐܦܠܐ ܒܟܠܕܘܟܐ ܐܝܢܐ ܕܗܘ
ܗܘܝܐ 12/ ܙܘܥܐ ܕܐܪܥܐ. ܐܠܐ ܒܐܝܠܝܢ ܕܐܝܬ ܠܗܘܢ ܚܫܝܚܘܬܐ ܠܘܬ ܩܘܒܠܐ
ܕܚܫܐ ܗܢܐ ܀ 13/ ܗܘ ܕܝܢ ܒܓܘܢܐܝܬ ܢܣܒ ܠܥܠܬܐ ܥܠ ‹ܟܠܗ› ܙܒܢܐ ܘܥܠ ܟܠܗ̇
ܐܪܥܐ ܀

[31 ‹ܕܝܢ›: supplevi || 6 ܠܐܪܥܐ: ܠܐܐܪܥܐ cod. || ܠܘ ܕܢܠ+: fort. ܕܝܢ ܕܠܝܬ ܠܘ vel sim. || 13 ‹ܟܠܗ› supplevi]

This [man] said not rightly that one part of the earth is below and another part above; for the whole of the earth is below, just as all heavy-weight bodies which incline downwards show. (3) <u>Furthermore</u>, <u>the earth is not flat along its latitude but is spherical</u> and [this] is known from the fact that when the sun is above the earth not all of us receive its light at the same time, but its light occurs differently according to the changes of the horizons. Besides, not the same but different circles appear in all the climes. (2) <u>Furthermore, if all the air, though it bears the earth, cannot move it</u>, how much more [is it the case that] <u>a part of the air</u> cannot <u>move it</u>. (1bc) Furthermore - [?] - the movement of the earth [*zawʿā d-arʿā*] <u>occurs</u> not at any time and in any <u>place</u>, but in those <u>which are liable to be affected in this way</u>; <u>(Anaxagoras)</u>, however, <u>apportioned the cause generraly</u> to all times and <u>to all the earth</u>.

1. 1. . . . ܐܠܘ ܫܪܝܪܐ ܗܘܬ ("if this were true ..."): The sentence is also turned into a conditional at the corresponding place in Olymp. arab., 135.13ff.: "If air were the cause of the earthquake, earthquakes ought to occur in all places in the same way ..."

[14] Cf. Arist. 365a 25-35; Olymp. arab. 135.1-19.

l. 3. ܕܝܠܢܐܝܬ ("under particular conditions"): The word does not occur in the text of Nic. syr. above as we have it in the Cambridge manuscript, but one suspects that the cluster of letters "*dyln*" (ܕܝܠܢ) in line 36.11 represents a remnant of some word related to *dīlānāyā*. - Cf. Olymp. arab. 135.14f.:

ونحن الآن نجدها ليس تكون فى جميع المواضع ولا فى جميع الأزمنة، لكن ذلك بحسب ملاءمة الوقت الخاص فى السنة والبلد الذى يكون (تكون) فيه.

We find, in fact, that [earthquakes] do not occur in all places and at all times, but [occur] according to the suitability of the particular [*ḫāṣṣ*] time of the year and of the country in which they occur.

l. 3. ܢܣܒܗ ܠܥܠܬܐ ܥܠ ܟܠܗ ܐܪܥܐ ("apportioned the cause to all the earth"): In the context, the verb *nsab̲* must mean "attribute, attach, associate". A similar usage is found at Ishoᶜdad, in Mt. 13.8 (ed. Gibson II.96.11-15):

ܐܘܪܝܓܢܣ ܕܝܢ ܠܗܕܐ ܕܫܬܝܢ ܥܠ ܐܝܠܝܢ ܕܒܐܪܡܠܘܬܐ ܡܬܕܒܪܝܢ ܢܣܒ ܠܗ . . .
ܗܕܐ ܕܝܢ ܕܬܠܬܝܢ ܥܠ ܡܙ̈ܘܓܐ . . . ܗܕܐ ܕܝܢ ܕܡܐܐ ܥܠ ܒܬ̈ܘܠܘܬܐ

Origen associates 'sixtyfold' with those who practise widowhood ... 'thirtyfold' with the married ... and 'a hundredfold' with the virgins.[15]

The unusual sense of the verb *nsab̲* (+ *ᶜal*) here, in fact, comes very close to the sense of arab. *nasaba* (+ *ilā*/*li*-).[16]

l. 5. ܦܬܝܐ ("latitude"): lit. "breadth". - For a proof of the sphericity of the earth, we strictly require the proofs of curvature along both its latitude *and* longitude. The two additional arguments given in Nic. syr. do, in fact, provide proofs for both, since the latter, which refers a little obscurely to the visibility of certain stars from different latitudes, proves its latitudinal curvature, while the former has to be understood as a proof of its longitudinal curvature in view of the word *meḥd̲ā* ("[not] at the same time").[17] Similar arguments are used at *But*. De caelo V.v.1-2 (ms. F 35v), where the earth's sphericity is demonstrated on the basis of its being convex (*kustānāyā*) along its longitude (*urkā*), as proven by the fact that the sun rises and lunar eclipses begin earlier in the east than in the west, and along its latitude (*pt̲āyā*), as proven by the difference in height and visibility of the polar and other stars from different regions of the earth.[18]

---

[15] Gibson (I.57.19ff.) translates: "Origen takes this of 'sixtyfold' about those who practise widowhood ...". In Ishoᶜdad, it is just possible to understand "*nsab̲ x* ᶜal *y*" to mean "take/assume *x* [to be] about *y*"; the meaning of *nsab̲* would, then, be close to that of Gk. ὑπο-λαμβάνω (usually + εἶναι, but sometimes with ellipse of εἶναι). The instance in our passages of Nic. syr. and *But*., however, gives support to Brockelmann's rendition of the verb as "rettulit ad" (432a, s.v. *nsab̲*, 7.).

[16] Cf. *Šifāʾ* 17.1 (*yansubu al-ᶜilla ilā al-hawāʾ*), quoted under Min. II.i.1. 2-3 above. - According to Nöldeke (1910) 188 (cf. Brockelmann 432a), the original sense of *nsab̲* was that of arab. *našiba* ("be attached, adhere"). Allowing for the correspondence of the transitive "attach" (*nsab̲*, *nasaba*) to the intransitive "be attached" (*našiba*), our *nsab̲ ᶜal* in the sense of "attach to" could be understood as a reflex of this earlier meaning of the verb.

[17] The implications are made clearer in the Arabic version of the passage (Olymp. arab. 135.7-9).

[18] A shorter version of the same proofs, without the arguments against the concavity

l. 5-6. ܡܩܕܡ ܐܬܚܘܝܬ ("has already been demonstrated"): The reference, as far as *But.* is concerned, will be to *But.* De Caelo V.v.1-2 (see preceding comment). - A clause similar to this may have dropped out in Nic. syr. before *w-īḏīʿā* in line 3, since the syntax of the text as we have it is awkward (lit.: "*that* the earth is ... spherical *and* is known ...") and the sphericity of the earth has indeed been demonstrated earlier in Nic. syr. at 8 [330v].17ff. (= Drossaart Lulofs [1965] p.86, frag. 37; corr. Arist. *De caelo* 297b 17ff.).

II.i.4.6-7: Refutation of Democritus

Arist. provides no explicit refutation of Democritus' theory of earthquakes at the corresponding place,[19] so that the refutation here is an addition by Nicolaus (or Nic. syr.).

> *But.*: Dismissing also the second [opinion], he says: We have already demonstrated that water is not enclosed en masse [*knīšāʾīt*] in the earth but surrounds it.
>
> Nic. syr. 36.20-21[20]:
>
> ܐܠܐ ܕܠܐ ܣܟܝܪܝܢ ܡܝ̈ܐ ܟܢܝܫܐܝܬ ܒܐܪܥܐ: ܒܡܐܡܪܐ/21 ܕܩܕܡ ܗܢܐ ܐܬܚܘܝܬ.
>
> ... but that water is not enclosed en masse in the earth has been demonstrated in the preceding book [*mêmrā*]."

It is not clear in what sense the word "book" (*mēmrā*) should be understood here, since the word is used in Nic. syr. both of the "books" within the works of Aristotle and of the divisions within Nicolaus' compendium. The reference here could be to the discussion of the relative positions of earth and water as found in Arist. *De caelo* II.iv,

---

of the earth's surface, is found at *Cand.* 82.4-13. Cf. further BH *Asc.* [Nau] 9.5ff. (tr. p. 8); Arist. *De caelo* 297b 31ff.

[19] The absence of the refutation is noted by Alex. Aphr. (116.10f.), who atttributed Aristotle's omission of the refutation to the insignificance of Democritus' theory (ἴσως διὰ τὸ ἐπιπόλαιον). There is a scholion on this point in our manuscript of Nic. syr., which, if we may trust our scholiast, is a new fragment of Olympiodorus (Nic. syr. 36.24ff.): "Aristotle does not here refute the opinion of Democritus, according to Alexander, out of reverence (*kuhhāḏā*). For (Democritus) was [still alive] in his time. ..." The scholiast then goes on to tell us that Olympiodorus denied this was the reason, pointing out that Arist. did refute Democritus' view that the sea was decreasing in volume (*Mete.* 356b 9ff.), as well as his atomic theory. - Arist. does explicitly deny that water is the cause of the earthquake in the following chapter (II.viii., 366a 3f.: οὐκ ἂν οὖν ὕδωρ οὐδὲ γῆ αἴτιον εἴη, ἀλλὰ πνεῦμα τῆς κινήσεως) and this is understood as a refutation of Democritus by Alex. (116.32f.: εἰ δὲ τοῦτο, οὔτ' ἂν ὕδωρ αἴτιον εἴη τῆς κινήσεως τῇ γῇ, ὡς Δημόκριτος ᾤετο, ...).

[20] Cf. Olymp. arab. 135.20-21.

IV.v,[21] or, perhaps more likely, to the refutation of the 'reservoir theory' of rivers at *Mete*. I.xiii, 349b 2ff. (corr. Nic. syr. 18.13ff.), which is in the preceding "book" of Arist. *Mete*.[22] This latter discussion, however, is not represented in *But*., so that in *But*., at least, the reference must be back to *But*. De caelo V.iv.6. (ms. F fol. 35r r15ff.)[23]:

> ܕܫܬ݂. ܝܡܐ ܪܒܐ ܕܚܕܝܪ ܠܐܪܥܐ ܗܘܝܘ ܐܣܛܘܟܣܐ ܟܠܢܝܐ ܕܡ̈ܝܐ. ܐܘ ܠܓܘ ܡܢ ܐܪܥܐ ܚܒܝܫ: ܐܘ ܠܒܪ ܡܢܗ ܣܝܡ. ... ܫܪܟܐ ܕܝܢ ܕܐܣܛܘܟܣܐ ܕܡ̈ܝܐ ܒܟܝܢܐܝܬ ܠܐܪܥܐ ܚܕܝܪ.
>
> Sixth [theory]: The great sea surrounding the earth is the universal element of water. It is either confined inside the earth or is placed outside it. ... The only remaining alternative is that the element water surrounds the earth by nature.

On the "position" of earth and water, see also *But*. Min. IV.i.1.1-2 and V.i.1.1-2 below.

l. 6. ܒܟܢܫܐܝܬ ("*en masse*"): If the reference in Nic. syr. is to the discussion in Arist. *Mete*. I.xiii. and Nic. syr. 18.13ff., this adverb could be related to the use of the verb *metkannaš* in that passage of Nic. syr.: Nic. syr. 18.14f.: ܗܢܐ ܡܬܒܪܐ ܟܕ ܢܚܬ ܡܬܟܢܫ ܬܘܒ ܬܚܝܬ ܐܪܥܐ ("[vapour] when it falls gathers again below the earth"; corr. Arist. 349b 4 ἀθροισθέν); ibid. 18.17f.: ܟܡܝܘܬܐ ܘܣܓܝܐܘܬܐ ܕܡ̈ܝܐ ܕܡܬܟܢܫܝܢ ܠܓܘ ("the quantity of the water which gathers inside [the earth]"; corr. Arist. 349b 10 συλλεγόμενον).

l. 7. ܐܠܐ ܚܕܝܪ ܠܗ ("but surrounds it"): This or a similar clause may have dropped out of Nic. syr., but it seems more likely that this is an addition by BH, probably on the basis of the passage of *But*. De caelo quoted above where the word *ḥḏīr* appears twice.

II.i.4.7-11: Refutation of Anaximenes

Here again, as in Min. II.i.4. 1-6 above, BH alters the order of the material he found in Nic. syr., but there is little in this passage which is not derived from Nic. syr.

> *But*.: (1a) Against the third [opinion] the Master says: (1b) If the matter [*šarbā*] were so, (1c) the earth ought to be level, (2a) and movements which occur [ought to] be lessening [*bāṣrīn*] in comparison with [*men*]

[21] The corresponding passages of Nic. syr., are lost, but, if they existed, they would have had their place either in the *mēmrā* immediately preceding or the one before (i.e. "book" IV or V, see Drossaart Lulofs [1965] 82-90).

[22] Another candidate would be the refutation of the Platonic theory of "Tartarus" at Arist. *Mete*. II.ii. 355b 32ff. (corr. Nic. Syr. 19.1.ff.; *But*. Min. V.iv.), but this is in the same "book" of Arist. *Mete*. as here.

[23] Cf. also BH *Cand*. 106.12f.

those which occurred [in the past], (2b) and [ought to] be ceasing in the end, (2c) since the tops of mountains would already have fallen in movements since the beginning. (3) Furthermore, earthquake [*nyāḏā*] ought not to occur [*hāwē*] in regions where there are no mountains, such as Egypt.

Nic. syr. 36.30-37.4[24]:

ܐܠܐ {ܐܠܘ} 31/ ܐܝܬܝܗ̇ ܗܘܬ ܗܕܐ ܗܟܢܐ. ܐܢܘ ܗܘܐ ܕܫܘܝܐ ܗܘܬ ܐܪܥܐ.
ܟܕ ܡܬܥܠܝܬܐ ܕܪ̈{ܡܢ} 32/ ܢܦܠ̈ ܒܡܟܝ̈ܟܢ ܕܬܚ̈ܬܝܢ. ܬܘܒ ܕܝܢ ܠܐ ܙܕܩ ܗܘܐ
ܕܒܐܬܪ̈ܘܬܐ ܕܐܪܥܐ ܕܠܝܬ 1/ ܒܗܘܢ ܛܘܪ̈ܐ ܐܝܟ ܡܨܪܝܢ ܘܙܘܥܐ ܕܐܪܥܐ ܐܡܪܢܐ
ܕܢܓܕܫ. ܬܘܒ ܕܝܢ ܐܝܟ ܗܕܐ ܗܘܐ ܠܡܠܬܐ 2/ ܗܕܐ. ܙܘܥ̈ܐ ܐܝܠܝܢ ܕܗܘܝܢ ܝܬܝܪ
ܐܡܝܢܐܝܬ ܒܨܝܪ̈ܝܢ ܢܗܘܘܢ ܡܢ ܗܠܝܢ ܕܩܕܡܘ ܗ{ܘܘ} 3/ ܡܛܠ ܗܐ ܕܪ̈ܡܝܢ ܢܦܠ̈
ܡܬܥܠܝܬܐ ܡܠܝ ܕܢܦܠܢ ܗܘܝ. ܘܠܫܘܠܡܐ ܢܬܒܛܠܘܢ ܡ{ܛܠ} 4/ ܕܢܦܠܝܢ ܠܗܘܢ ܪ̈ܝܫܝ
ܛܘܪ̈ܐ.

... (1b) but if this were so, (1b) the earth ought to be level, since the high parts fall on to the low (parts). (3) Furthermore [in that case] the movement of the earth [*zawʿā d-arʿā*] ought not to happen [*negdaš*] in regions of the earth where there are no mountains, as in Egypt. (2a) Furthermore, according to this [theory], those movements which occur more should constantly be becoming less [*bṣīrīn nehwōn*] than those which occurred before, (2c) because the highest parts would already have fallen, (2b) and should cease [impf.] in the end, [because] the tops of mountains would already have fallen [ptc.].

Min. II. Section ii.: *On the True Cause of the Earthquake*

The material of this and the following section (II.iii. "on times when earthquakes occur more") goes back ultimately to Arist. *Mete*. II.viii.

According to Aristotle, earthquakes are caused by winds (= dry/smoky exhalation) entrapped underground. IS, too, accepts this as the principal cause of earthquakes. He does not, however, give us a neat and tidy description of the mechanism by which such wind/exhalation causes the earthquake. This explains why for his description of the process in II.ii.1 BH turns to the summary of the *Šifāʾ* in Fakhr al-Dīn al-Rāzī's *al-Mabāḥiṯ al-mašriqīya*, a passage he had already used earlier in his *Cand*.[25]

The rest of Section II.ii. (II.ii.2-4) is concerned with various aspects of the earthquake, most of which, directly or indirectly serve to prove that earthquakes are caused by wind/exhalation. The subject of Section

24 Cf. Arist. 365b 12-20; Olymp. arab. 134.12-20.

25 Cf. Mete. III.1.i (on winds) below, where the *Mabāḥiṯ* is used for the same reason.

II.iii., according to its heading is the "times [*zabnē*]" when earthquakes are more frequent. Much of this section is concerned with the situations (both in time and place) under which earthquakes occur, although the last theory of the section, dealing with the reason for the persistence of earthquakes and benefits of earthquakes, does not really fit in under this description.

The order followed in these two sections is basically that of the corresponding chapter of the *Šifāʾ*, although some materials taken from near the end of that chapter are moved forward to II.ii.3 and II.ii.4.

Throughout the two sections, BH makes extensive use of Nic. syr. to supplement the materials taken from the *Šifāʾ*. This was made possible by the similarity between the *Šifāʾ* and Nic. syr., resulting not merely from the common Aristotelian origin of the materials in the two works, but from the fact that much of the material in the *Šifāʾ* in fact derives, via Olymp. arab., from the same Nicolean-Olympiodorean text as that preserved in the Cambridge manuscript of Nic. syr.

N.B.: *Mab.*: Fakhr al-Dīn al-Rāzī, *Mabāḥiṯ*; *DM*: Ps.-Arist. *De mundo*; BK: Bar Kepha, *Hex.*

	Šifāʾ	Others	
II.ii.1.1-6		Nic. 37.4-10 *Mab.* II.205.19-21	Aristotelian theory
II.ii.1.6-8	16.1-4	Nic. 37.11-19	contd.: motive force of wind
II.ii.1.8-9	15.8-9		contd.: fire as wind
II.ii.1.9-10		Nic. 39.6-16	contd.: why wind cools
II.ii.2.1-4	17.8-10		Benefits of earthquake
II.ii.2.4-8	17.10-14		Harms
II.ii.2.8-10		Nic. 38.22-23	Harm: example (Lipara & Etna)
II.ii.3.1-4	(17.14-15)	Nic. 40.7-12	Underground noise
II.ii.3.5-6	17.15-16	(DM & BK?)	contd.
II.ii.3.7-10	19.11-14	Nic. 40.13-17	contd.: hearing & touch
II.ii.4.1-4	17.17-19		Reduction of e'quakes by boring
II.ii.4.4-8	(19.4-9)	Nic. 41.14-17	Classification of earthquakes
II.ii.4.8-9		Nic. (lost)?	contd.
II.iii.1.1-2	(17.19-20)	Nic. 37.19-20	Relation to wind
II.iii.1.2-7	17.20-18-2	Nic. 37.20-24 Nic. 39.16-27	contd.: elongated cloud
II.iii.1.7-8	18.3-4	Nic. 38.25-39.2	contd.: nebulous air
II.iii.1.8-12	18.6-8	Nic. 37.25-30	Times of day
II.iii.2.1-4	18.9-11	Nic. 37.30-38.1	Earthquakes on porous land
II.iii.2.4-8	18.14-18	Nic. 38.6-10	Seasons: winter and summer
II.iii.2.8-9	(18.18)	Nic. 38.1-6	Seasons: spring and autumn
II.iii.3.1-5	18.19-19.2	Nic. 39.27-40.3	Eclipses
II.iii.4.1-4		Nic. 40.3-7	Persistence of earthquakes
II.iii.4.4-7	19.14-15	BK	Benefits of earthquakes

II.ii.1.1-6: The Aristotelian theory (a)

The beginning of the passage closely follows Nic. syr., but, then, for the explanation of the way in which smoke/wind causes the earth to shake, BH turns to a passage of Fakhr al-Dīn al-Rāzī's *Mabāḥit*, which he had used earlier in his discussion of earthquakes in *Cand*. BH also incorporates into this passage a number of elements taken from the longer discussion of earthquakes in the *Šifā*ᵓ.

But. [underline: agreement with Nic.; italics: with *Mabāḥit*]:
(1a) The Master [Aristotle] says: (1b) When the earth is moistened and is [then] heated by the sun and by the heat which is inside (the earth) [*da-ḇ-ğawwāh*], (1c) *smoky exhalation*, which is driven violently, *is generated* in it. (2a) This smoke is the material of wind (2b) and this (wind), *when* it *cannot* find a way [by which] to *escape because of the density of the surface of the earth*, (2c) moves (the earth) (2d) and is dissipated through the *tears* which it makes in *(the earth)*, (3a) like the wind, which is generated in must, (3b) tears its cask (3c) and escapes.

Nic. syr. 37.4-10[26]:

ܐܦ ܕܝܢ ܐܡܪܝܢ ܕܟܕ ܡܬܬܠܚܠܚܐ ܐܪܥܐ ܡܢ ܘ{ܫܚܢܐ} /5 ܡܢ ܚܡܝܡܘܬܐ ܗܘܝܐ
ܘܚܡܝܡܘܬܐ ܕܒܗ : ܬܢܢܐ ܣܓܝܐܐ ܥܒܕܐ . ܗܢܐ ܕܝܢ ܬܢܢܐ ܠ{ܕܡܬܝܠܕ} /6 ܟܠܗ ܗܘ
ܕܡܬܢܚܐ ܡܢ ܫܘܪܝܐ ܠܗ ܡܬܪܟܢܐ : ܐܢ ܒܠܥܠ ܘܐܢ ܠܬܚܬ ܠܗܢܐ ܐܬܪܐ ܠ{ܬܚܬ} /7
ܠܗ ܟܕ ܠܗ ܐܙܠ ܟܠܗ . ܘܝܕܝܥܐ ܗܢܐ ܘܓܠܝܐ ܗܘ : ܒܡܫܠܡܢܘܬܐ ܕܫܪܓܐ ܕܡܬܕܥܟܐ
/8 ܕܗܫܐ . ܗܢܐ ܬܢܢܐ ܟܠܗ ܕܢܦܩ ܡܢܗ . ܠܗܘ ܗܘ ܐܬܪܐ ܕܠܗ ܡܢ ܫܘܪܝܐ ܡܬܪܟܢܐ . ܠܗ /
9 ܟܕ ܠܗ ܢܩܦܐ ܬܢܢܐ ܟܠܗ . ܗܢܐ ܗܟܝܠ ܕܐܡܪܢ ܠܬܚܬ ܡܬܪܟܢܐ ܬܢܢܐ ܕܐܬܡܠܠ :
ܟܠܗ /10 ܐܙܠ ܠܬܡܢ ܘܟܕ ܡܬܥܨܪ ܡܙܝܥ ܠܗ ܠܐܪܥܐ ܘܡܙܝܥ ܠܗ .

(1a) We say: (1b) When the earth is moistened and {is heated} by the sun and by the heat which is in (the earth) [*d-ḇāh*], (1c) it produces much smoke. (2) This smoke all goes to that same place to which it immediately inclines from the beginning, be it upwards or downwards. - This is known and obvious in the wick of a lamp which has just been extinguished; for the whole of the smoke which comes out from it follows to that same place to which it inclines from the beginning. - When therefore the smoke mentioned inclines downwards, all of it goes there, and when it becomes compressed, (2c) it moves the earth and shakes it.

Fakhr al-Dīn al-Rāzī, *Mabāḥit* II.205.19-21[27]:

اذا تولد تحت الارض بخار دخانى حار كثير المادة وكان وجه الارض متكاثفا عديم المسام والمنافذ فاذا قصد ذلك البخار الخروج ولم يتمكن من ذلك بسبب كثافة وجه الارض فحينئذ يتحرك فى ذاته وتحرك الارض وربما بلغ فى قوته الى حيث تقوى على شق الارض ...

(1c) When smoky vapour, which is hot and rich in material, is generated under the earth, while the surface of the earth is dense and devoid of pores

---

[26] Cf. Arist. 365b 21-28, 366a 3-8; Alex. Aphr. 116.10-21, 116.32-117.2; Olymp. arab. 134.5-11.

[27] Cf. *Cand*. 127.6-10; also Qazwīnī, *Ağā*ᵓ*ib* 149.4-8, 12.

and passages, (2b) and when that vapour tries to escape but is unable to do so because of the density of the surface of the earth, (2c) then, it moves in itself and moves the earth. (2d) Sometimes its strength reaches such a level that it is able to tear the earth, ...

l. 2. ܕܡܬܕܚܝܐ ܒܥܘܫܢܐ ("which is driven violently"): cf. *Šifā*ʾ. 15.5-6: والجسم الذى يمكن أن يتحرك تحت الأرض، ويحرك الأرض، إما جسم بخارى دخانى قوى الاندفاع كالريح، ... ("And the body which can [itself] move under the earth and move the earth is either a smoky vapoury body which is driven violently [lit.: strong of propulsion] like wind, ...").

l. 3. ܘܬܢܢܐ ܗܢܐ ܗܘܝܘ ܗܝܘܠܐ ܕܪܘܚܐ ("This smoke is the material of wind"): For the equation of smoke and wind, see Mete. III (on winds) below.

l. 3-4. ܟܕ ܠܐ ܡܫܟܚܐ ܠܗ ܐܘܪܚܐ ܕܬܦܘܩ ("when it cannot find a way [by which] to escape"): cf. *Šifā*ʾ. 17.3f.: ... ثم لم تجد طريقا إلى الانفصال والصعود الطبيعى الذى له ("[when the air] ... and then cannot find a way for the discharge and the ascent which is natural to it"); cf. also *De mundo* syr. 146.5f.: ܟܕ ܒܥܐ ܕܢܫܟܚ ܠܗ ܡܦܩܢܐ ("when [the wind] seeks to find an escape route for itself").

l. 4-5. ܘܒܨܪܝܐ ܕܥܒܕܐ ܒܗ ܡܬܦܫܫܐ ("and is dissipated through the tears which it makes in (the earth)"): ܘܒܨܪܝܐ ܕܒܗ ܥܒܪܐ ܘܡܬܦܫܫܐ ("and passes through the tears which are in (the earth) and is dissipated.") cod. F. - The point of the comparison, that both the wind in the earth and the gas in the cask tear open what is confining it and escape, would be lost in the reading of ms. F.

l. 5. ܡܬܦܫܫܐ ("is dissipated"): cf. Bar Kepha, *Hex*., Paris 241, 192r a3-7: ܥܕܡܐ ܕܡܬܬܫܪܐ ܕܘܟܬܐ ܗܝ. ܘܢܦܩ ܘܡܬܦܫܫ ܀ ("... until that place is loosened, and [air] goes out and is dissipated"); also, in connection with the notion of "tearing" the earth, cf. Bar Kepha, Paris 241, 191v a11-13: ܘܟܕ ܢܦܩ ܡܢ ܓܘ ܐܪܥܐ ܨܪܐ ܠܗ ܘܡܫܪܐ ܠܗ. ("... as it goes out from inside the earth, (fire) tears and loosens the earth.").

l. 5-6. "like the wind ...": cf. *Šifā*ʾ 15.6-7: ... كالريح، كما يشق الخوابى إذا تولد فى العصير ("... like wind, just as it tears the casks when it is generated in must").[28]

II.ii.1.6-8: Motive force of wind

But. [underline: agreement with *Šifā*ʾ: italics: with Nic.]:

(1) For there is nothing which has the power to stir [*la-mdālū*] *the earth together with the heavy objects* which are upon it like *wind*, (1b) neither water nor air nor a part which is torn from the earth, except fire alone.

---

[28] Cf. a passage of Nic. syr. occurring a little later and marked off as a scholion (37.14-16): ܘܕܡܢ ܪܘܚܐ ܗܘܐ ܙܘܥܐ ܐܦ ܩܘܠܐ ܡܚܘܝܢ ܕܡܬܚܝܠ ܒܗܘܢ ܚܡܪܐ. ܗܠܝܢ ܟܕ ܡܬ ܡܬܥܨܪܝܢ ܒܣܘܓܐ ܕܪܘܚܐ ܕܡܫܬܪܝܢ ܒܗܘܢ: ܡܬܨܪܝܢ ܘܕܪܟܘ ܠܐ ܡܠܝܢ ܠܗܘܢ [ܡܬܟܐܒܢ: ܡܬܟܐܒܬ cod.] ("That earthquake arises from winds is also shown by pitchers [*qulē*] in which wine grows strong; these when they are pressed by the large quantity of winds which are released in them, are, torn [and] as a result are not filled [?]").

Šifā᾽ 16.1-4:

فهذه هى الوجوه التى يمكن أن تعرض معها الزلزلة، إما بخار ريحى أو نارى قوى يتحرك فيحرك
الأرض. وهذا هو وجه الأكثر، فإنه لا شيء أقوى على تحريك الأرض الحركة السريعة القوية التى للزلزلة
من الريح، وإما مياه تسيل دفعة، وهذا رأى ديمقراطيس، وإما انهدام بعض أركان القرار.

These then are the causes [*wuǧūh*] with which an earthquake can occur: either strong windy or fiery vapour which [itself] moves and moves the earth - this being the predominant cause, (1a) for there is nothing which has a greater power to move the earth with the swift and strong movement that occurs in an earthquake than wind - or water moving suddenly - this being the opinion of Democritus - or collapse of some of the pillars of support.

Nic. syr. 37.11-12, 16-19[29]:

ܕܐܝܬܝܗ̇ ܕܝܢ ܬܢܢܢܝܬܐ ܐܢܬܘܡܝܣܝܣ 11/ ܥܠܬܐ ܕܙܘܥܗ̇ ܕܐܪܥܐ. ܡܫܬܘܕܥܐ ܗܝ
ܕܡܬܝܠܕܐ ܗܘܬ ܡܫܠܝܐ ܘܫܠܝܐ ܠܥܓܠ. ... 16 ܬܘܒ ܕܝܢ ܙܘܥܐ ܕܗܠܝܢ ܕܗܘܐ ܠܐܪܥܐ (ܗܘ)
ܣܓܝܐ 17/ ܕܐܦ ܟܠܗ̇ ܡܕܝܢܬܐ. ܟܠܗ̇ ܓܝܪ ܡܬܬܙܝܥܐ ܒܥܕܢܐ ܚܕܐ ܡܬܬܙܝܥܐ. ܬܘܒ
ܕܝܢ ܘܩܫܝܢ ܐܢܘܢ 18/ ܒܡܐܢܐ ܓܝܪ ܝܩܝܪܐ ܡܬܬܙܝܥܝܢ ܥܡܗ̇. ܚܙܝܢܢ ܕܝܢ
ܕܠܪܘܚܐ ܒܠܚܘܕ ܐܝܬ ܚܝܠܐ 19/ ܕܡܬܬܙܝܥܢܘܬܐ ܕܐܝܟ ܗܢܐ ܠܡܥܒܕ.

That the smoky exhalation is the cause of the movement of the earth is proven by the fact that it arises suddenly and subsides quickly. ... Furthermore, this movement which occurs to the earth is very great; for often a whole city is moved. Furthermore, they are very severe; for heavy objects are moved with (the earth). We see that wind alone is able to cause such a movement.

l. 7-8. "neither water nor air ...": This addition at the end of the sentence, ruling out the elements water, air and earth as the possible causes of earthquakes amounts to a refutation of the theories of Democritus, Anaxagoras and Anaximenes. The wording is reminiscent of Arist. 366a 3-4[30]: οὐκ ἂν οὖν ὕδωρ οὐδὲ γῆ αἴτιον εἴη, ἀλλὰ πνεῦμα τῆς κινήσεως. As BH uses a similar formula at *Cand*. 129.9-10[31], this may go back to a lost passage of Nic. syr. Since IS, though regarding winds to be the main cause of earthquakes, also accepts water and collapse of "supports" as causes of earthquakes, the addition may also be meant to be a refutation of IS.

II.ii.1. 8-9: Underground fire as smoke

But.: and it is clear that fire cannot be generated inside the earth except in a smoky form [*ba-znā tennānāyā*].

---

[29] Cf. Arist. 365b 28-366a 5; Alex. Aphr. 116.21-34; Olymp. arab. 135..22-136.3.

[30] Cf. Alex. Aphr. 116.32f. - Neither Arist. nor Alex. Aphr. here explicitly rules out "air".

[31] *Cand*. 129.9-10, immediately following the accounts of the theories of Anaxagoras, Democritus and Anaximenes: ܕܪܘܚܐ ܐܢܝܢ ܥܠ̈ܠܬܐ ܕܙܘܥܐ. ܘܠܐ ܐܐܪ ܐܘ ܡ̈ܝܐ / ܐܘ ܪܝ̈ܫܝ ܛܘ̈ܪܐ ܡܬܝܕܥ ܡܢ ܗܝ ܕ ... ("That winds are the causes of the earthquake, and not air or water or tops of mountains, is known from the fact that ...").

*Šifā*ʾ 15.8-9:

والجسم الناري لا يحدث تحت الأرض، وهو نار صرفة؛ بل يكون لا محالة في حكم الدخان القوي وفي حكم الريح المشتعلة

But a fiery body does not occur below the earth as pure fire; rather, it occurs, without doubt, in the form of [*fī ḥukmi*] strong smoke and in the form of burning wind.

II.ii.1.9-10: Why wind cools

BH has inserted here a paraphrase of a passage which occurs at a later point in Nic. syr.

But.: (1a) Although the cause of wind is fiery smoke, (1b) it cools us (1c) because it pushes away from us the air which has been warmed by our bodies (1d) and makes other [air] approach us.

Nic. syr. 39.6-8, 16[32]:

ܚܙܝܢ ܕܝܢ ܕܟܕ ܢܫܒܢ ܪ̈ܘܚܐ ܡܬܬܩܪܪܝܢܢ . 7/ ܘܐܡܪܝܢ ܕܪ̈ܘܚܐ ܡܢ ܟܝܢܗܝܢ ܚܡܝܡܢ
ܐܠܐ ܡܩܪܪܢ ܠܢ . ܡܛܠ ܕܡܕܚܝܢ ܡܢܢ 8/ ܠܐܐܪ ܕܟܒܪ ܚܫ ܡܢܢ ܘܐܬܚܡܝ
ܘܐܚܪܢܐ .16// ܡܩܪܒܢ ܠܢ .

We see that when winds blow we are cooled. We say: Winds are hot by nature, (1b) but cool us (1c) because they push away from us the air which has already been affected by us and has been warmed, (1d) and make other [air] approach us.

II.ii.2.1: Benefits and harms of earthquakes

But.: With a movement of the earth there sometimes occur beneficial effects [*yutrānē*], sometimes harmful effects [*ḥusrānē*].

*Šifā*ʾ. 17.7-8: وقد تعرض مع الزلازل أحوال، فربما كانت نافعة، وربما كانت ضارة.

[Various] conditions occur with an earthquake: sometimes they are beneficial and sometimes they are harmful.

II.ii.2.2-4: Benefits

Although the content of the passage corresponds to that of the passage of the *Šifā*ʾ quoted below, the wording is different from that of the *Šifā*ʾ. The passage mentioning emergence of water with earthquakes at Nic. syr. 40.18ff.[33] has little resemblance with our passage, but there is

[32] Cf. Arist. 367a 33; Olymp. arab. 137.21-24. - The explanation given by Nic. syr. here is different from that given by Arist., according to whom the coldness is due to the cold vapour contained in the air. The same explanation as that given here is also found in the section on winds at Nic. syr. at 25.10-15, and the passage there, in turn, closely resembles Olymp. 98.19-22.

[33] Although Nic. syr. also talks of earthquakes causing water to gush out, the

a passage in Nic. *De plant.* arab., which suggests that there may have been another, lost passage of Nic. syr. which lies behind the passage of *But*. here.

> *But.*: (1a) beneficial, (1b) as when the wind, meeting [*pāǧʿā*] with vapoury materials in the depth of the earth, (1c) pushes them with a great force, (1d) causes them to come out on its surface (1d) and causes springs which did not exist (before) to gush out
>
> *Šifāʾ* 17.8-10:
>
> أما النافعة، فإن اتفق أن تشتمل تلك الرياح على مواد بخارية توجهها وتسوقها إلى جهة من الأرض، أو تجذبها إليها مستتبعة، فتعينها على التفجير للأرض، فتتفجر عيونا.
>
> (1a) Beneficial: (1b) if those winds [pl.] happen to implicate [*taštamila*] vapoury materials, (1c) then direct them and drive them towards an area of the earth, or attract them towards themselves, making them follow, (1d) and help them break through the earth, (1d) so that they gush forth as springs.

l. 3-4. ܡܒܘܥܬܐ ܕܠܐ ܐܫܬܟܚܘܢ ܩܕܡ. ("springs which did not exist"): cf. Nic. *De plant.* arab. §150 [corr. Ps.-Arist. *De plant.* 822b 31-35]: وقد قدمنا العلة ظهور الانهار والعيون فى الكون العلوى بأن الزلازل قد تظهر أنهارا وعيونا لم تكن قبل ذلك ... ("We have already discussed the cause for the appearance of rivers and springs in [the book on] Meteorology, saying that earthquakes sometimes cause rivers and springs to appear, which did not exist [*lam takun*] before that ..."); cf. also *De mundo* syr. 146.13-14 [< gr. 396a 6-7]: ܡܢܗܘܢ ܕܝܢ ܐܦ ܡܒܘ̈ܥܐ ܕܡܝ̈ܐ ܦܬܚܝܢ ܕܠܐ ܐܫܬܟܚܘ ܗܘܘ ܡܢ ܩܕܝܡ. ("Some [earthquakes] also open up sources which did not exist before"); cf. further *Cand*. 128.8-9.

II.ii.2.4-8: Harms

> *But.*: (1a) and harmful, (1b) as when the wind, being very dry, (1c) is ignited by a strong movement - (1d) just as smoke is ignited with strong and repeated blowing [*mappuḥtā*] - (1e) and so inflaming fire [*nūrā mawqdānītā*] rises from inside the earth, (1f) especially when the cause of the movement is very severe (1g) and tears the earth by its exit.
>
> *Šifāʾ* 17.10-14:
>
> وأما الضارة، فما يعرض من أن لا تكون المادة الريحية بهذه الصفة، بل تكون يابسة مائلة إلى طبيعة النارية، فتشتعل نارا عند الحركة القوية، فإن من شأن الحركة القوية أن تحيل الدخان والبخار والهواء نارا، فكثيرا ما تشتعل المنافخ والكيران إذا ألحّ عليها بالنفخ نارا. فإذا كان سبب الزلزلة قويا جدا، خسف الأرض باندفاعه وخروجه. وربما خلص نارا محرقة.

---

content as well as the context of the passage is quite different from here. - Nic. syr. 40.18-19, 21-23 [cf. Arist. 368a 26ff.]: ܢܚܫܘܒ ܕܝܢ ܡܚܫܒܢܐ / ܐܢܫ ܕܡܝ̈ܐ // ܐܫܬܟܚܘܢ ܥܠܬܐ ܕܙܘܥܐ / ܕܐܪܥܐ . ܡܢ ܗܝ ܕܒܙܒܢ ܙܘ̈ܥܐ ܣܓܝܐܐ ܙܒܢܝ̈ܢ ܡܝ̈ܐ ܠܒܪ ܢܦܩܝܢ . ܐܠܐ ܠܐ ܡܫܟܚܐ ܕܡܝ̈ܐ / ܢܗܘܘܢ ܥܠܬܐ ܕܙܘܥܐ . ("Someone might perhaps think that water is the cause of the movement of the earth on the basis of the fact that when movements occur, water often bursts out. But it is not possible for water to be the cause of the movement."). Cf. Arist. 368a 26-34; cf. Olymp. arab. 139.18-140.6.

(1a) Harmful: (1b) this results from the fact that the windy material is not of this type, but is dry, tending towards a fiery nature, (1c) so that it is inflamed [and becomes] fire in the strong movement - (1d) since it is a characteristic of a strong movement [*ḥaraka*] to turn smoke and vapour and air into fire, so that often bellows are inflamed [into] fire when they are subjected to a persistent blowing [*nafḫ*]. (1f) When the cause of the earthquake is very strong, (1g) it causes the earth to sink [*ḫasafa*] by its propulsion and its exit. (1e) Sometimes it is discharged as burning fire [*nār muḥriqa*].

l. 6. ܡܢ ܠܓܘ ܐܪܥܐ ܢܒܥܐ ("rises from inside the earth"): cf. *De mundo* syr. 145.17-20 [< gr. 395b 18-21]: ܡܠܝܐ ܓܝܪ ܐܦ ܐܪܥܐ ܒܓܘܗ̇. ܐܝܟܢܐ ܕܡܒܘ̈ܥܐ ܕܡܝ̈ܐ. ܗܟܢܐ ܐܦ ܡܒܘܥܐ ܕܢܘܪܐ ܘܪܘܚܐ. ܕܡܢܗܘܢ ܗܠܝܢ ܕܝܢ ܕܬܚܝܬ ܐܪܥܐ ܐܝܬܝܗܘܢ ܠܐ ܡܬܚܙܝܢ ܠܢ. ܗܠܝܢ ܕܝܢ ܕܠܥܠ ܓܕܝܢ ܘܢܦܫܝܢ. ܡܬܚܙܝܢ ܠܢ. ("For the earth has inside itself [*b-ḡawwāh*], just as [it has] springs of water, so also source of fire and wind. Of these, those which are below the earth are invisible to us; those which rise [*gāḏēn*] upwards and blow are visible to us").

l. 8. ܠܐܪܥܐ ܨܪܝܐ ("tears the earth"): خسف الأرض ("causes the earth to sink") IS. - BH may have read *ḫaraqa* (خرق) or *fa-šaqqa* (فشق) instead of *ḫasafa*.

II.ii.2.8-10: Lipara and Etna:

The mention of "burning fire" discharged from the earth in II.ii.2.4-8 has prompted BH to insert here a passage taken from another context in Nic. syr.[34]

But.: (1a) The Master [Aristotle] relates that this happened once: (1b) Wind raised ash containing [lit. in which stood] fire in the form of sparks and filled the city of the Liparians. (2) In this way the craters of fire on Etna were also formed.

Nic. syr. 38.22-23[35]:

ܘܗܕܐ ܗܘܬ ܒܚܕ ܙܒܢ: ܐܣܩܬ ܗܘ ܪܘܚܐ ܩܛܡܐ ܕܩܐܡܐ ܒܗ ܢܘܪܐ ܐܝܟ ܩܠܝܦܘܬܐ
23/ ܘܡܠܬܗ̇ ܠܡܕܝܢܬܐ ܕܠܝܦܪܐܘܬܐ. ܗܟܢܐ ܕܝܢ ܗܘܘ̈ ܘܐܩܪ̈ܛܪܐ ܕܢܘܪܐ
ܕܒܐܛܢܐ.

[ܕܠܝܦܪܐܘܬܐ: ܕܠܝܦܪܐܘܬܐ cod.]

(1a) Once it happened: (1b) wind raised ash containing fire in the form of sparks and filled the city of the Liparians. (2) In this way the craters of fire on Etna were also formed.

l. 10. ܐܩܪܛܪܐ ("craters"): lit. "bowls", = gr. κρατήρ.

---

[34] If one is correct in assuming an influence of *De mundo* in II.ii.2.4-8, the connection would have been facilitated by the mention of Lipara and Etna in *De mundo* immediately following the passage quoted above.

[35] Cf. Arist. 366b 31-367a 11; Olymp. arab. 137.9-10; *Cand.* 128.5-7. - There is no mention of the craters on Etna in Arist. *Mete.*, but Etna and Lipara are mentioned together in *De mundo* at gr. 395b 21 (*De mundo* syr. 145.21). The craters of Etna are mentioned later on in *De mundo* at 400a 33 (syr. 155.24; the word "crater" is not in the Syriac version). Cf. Nic. syr. 59.7f. (Takahashi [2002a] text 210, tr. 219).

II.ii.3.: Sounds accompanying earthquakes

Noises caused by subterranean winds are mentioned in two passages in Arist. *Mete*. Of these the first is found closely following on the passage describing the eruption on Lipara (367 a 17-20). The other, a longer passage dealing specifically with the subject of these noises, occurs at a later point at 368a 14-25. The passages corresponding to these are found at the following points in Nic. syr. and *Šifā*ʾ.

(a) Arist. 367 a 17-20: Nic. syr. 38.24-25; *Šifā*ʾ 17.14-16
(b) Arist. 368 a 14-25: Nic. syr. 40.7-8, 9-17; *Šifā*ʾ 19.9-14

The materials originating from these two Aristotelian passages have been combined into one theory in *But.*, with the material of II.ii.3.1-4 and II.ii.3.7-10 deriving mainly from the second of the Aristotelian passages and that of II.ii.3.5-6 from the first.

II.ii.3.1-4: Noises accompanying earthquakes

But. [underline: agreement with Nic.; italics: with *Šifā*ʾ]:
(1a) *Sometimes terrifying* sounds [*qālē*] and weak *humming noises* [*zmāmē*] precede the earthquake [*nyādā*] or *occur* with it, (1b) sometimes like the sound of piping [*mašrūqītā*], (1c) and sometimes also like the bellowing [*gʿāṯā*] of a bull. (2) This occurs because of the variation in the shapes of the openings of the earth. (3) For air too, being similar, when it strikes different objects, makes different sounds.

Nic. syr. 40.7-8, 9-12[36]:

ܐܠܐ ܘܩܠܐ ܬܘܒ ܡܫܚܠܦܐ ܡܩܕܡܝܢ ܠܙܘ̈ܥܐ ܕܐܪܥܐ /8 ܐܘ ܥܡܗܘܢ //9
ܓܕܫܝܢ. ܒܙܒܢ ܡܢ ܐܝܟ ܩܠ /10 ܡܫܪܘܩܝܬܐ. ܒܙܒܢ ܕܝܢ ܐܝܟ ܓܥܬܐ
ܕܬܘܪܐ. ܗܕܐ ܕܝܢ ܗܘܝܐ ܡܛܠ ܫܘܚܠܦ /11 ܐܣܟܡܐ ܕܦܬܘ̈ܚܝܗ̇ ܕܐܪܥܐ.
ܗܟܢܐ ܓܝܪ ܐܦ ܐܐܪ. ܟܕ ܕܡܝܐ ܐܝܬܘ(ܗܝ) ܗܘ /12 ܕܢܩܫ ܒܨܒ̈ܘܬܐ
ܡܫܚܠܦܝ ܐܣܟܡܐ ܩܠܐ ܡܫܚܠܦܐ ܥܒܕ ܀

[9 ܒܙܒܢ: ܒܙܠܒܢ cod.]

(1a) But different sounds also precede the movements of the earth [*zawʿē d-arʿā*] or occur with them; (1b) sometimes like the sound of piping, (1c) sometimes like the bellowing of a bull. (3) This occurs because of the variation in the shape [sg.] of the openings of the earth. (3) For thus too, air, being similar, when it strikes objects of different shapes, makes different sounds.

*Šifā*ʾ 17.14-15: وربما حدثت أصوات هائلة ودوى يدل على شدة الريح.
(1a) Sometimes terrifying sounds occur, and a drone [*dawīy*] which shows the severity of the wind.

[36] Cf. Arist. 368a 14-25; Alex. Aphr. 122.14-123.1; Olymp. 12.11-13; Olymp. arab. 139.3-9; *Cand.* 128.3-5.

l. 1. ܡܚܝܠܐ ("weak"): Perhaps to be understood in the sense of "low-pitched". The addition of the adjective, which has no counterpart in Nic. syr. or *Šifā'*, may have been prompted by a desire to produce an onomatopoeic effect through the assonance of "*dḥīlē... mḥīlē*".

II.ii.3.5-6: Noises unaccompanied by earthquakes

BH's explanation as to why noises sometimes occur without an earthquake resembles the explanation in Arist. 367a 17-20, rather than that in Arist. 368a 21-22. Nic. syr. gives no explanation at the corresponding place.[37] The explanation given here in *But.* is that of *Šifā'* 17.15-16, but the wording of the passage has a certain affinity with that of the passage of *Cand.* quoted below, which, in turn, has affinities with *De mundo* syr. and Bar Kepha, *Hex*.

But. [underline: agreement with *Šifā'*: italics: with *Cand.*]:
(1a) *There are times when* [*īt emat d-*] the noise [*q'āṯā*] *occurs without* [*dlā*] *earthquake*, (1b) because when the wind escapes from the noise-making [*qā'ōyē*] holes [*ḥrūrē*], (1c) they find wide *cracks* [*srāyē*] (1d) and *come out* [*nāp̄qā*] through them (1e) but does not cause the earth to move.

Šifā' 17.15-16:

فإن وجدت هذه الريح المصوتة منفذا واسعا بعد المنفذ الذى تصوت فيه، حدث عن اندفاعها صوت ولم تزلزل.

(1b) If this sound-making [*muṣawwita*] wind (1c) finds a wide passage [*manfaḏ*] (1b) after the passage in which it makes the sound, a sound arises from its propulsion, (1e) but it does not cause an earthquake [*tuzalzil*].

Cand. 128.3-5 [underline: agreement with *But.*]:

ܒܙܒܢ ܡܢ ܣܠܩܐ ܡܢ ܐܪܥܐ ܓܥܬܐ ܣܛܪ ܡܢ ܙܘܥܬܐ. ܟܕ ܠܐ ܡܣܦܩܐ
ܪܘܚܐ ܠܡܙܥܙܥܘܬܗ ܠܐܪܥܐ. ܐܠܐ ܨܪܝܐ ܠܗ ܘܢܦܩܐ.

(1a) Sometimes [*ba-zḇan man*] the bellowing [*g'āṯā*] arises from the earth without [*sṭar men*] movement [*zaw'ā*], [which happens] when the wind is insufficient to move the earth, (1c) but [nevertheless] tears [*ṣāryā*] it (1d) and comes out [*nāp̄qā*].

De mundo syr. 146.18-22 [< gr. 396a 11-15; underline: agreement w. *Cand.*]:

ܗܘܝܢ ܕܝܢ ܙܒܢ ܐܦ ܓܥܝܐ ܙܘܥܐ ܕܡܙܝܥܝܢ ܠܗ ܠܐܪܥܐ ܥܡ ܚܕܐ ܓܥܬܐ
ܡܣܩ. ܒܙܒܢܐ ܕܝܢ ܣܓܝܐܐ ܐܦ ܣܠܩܐ ܡܢ ܓܥܬܐ ܗܘܝܐ ܒܠܥܕ ܙܘܥܐ ܡܢ ܐܪܥܐ.
ܟܕ ܪܘܚܐ ܗܝ ܕܡܬܚܒܫܐ ܒܗ ܠܐ ܢܣܒܐ ܗܘܐ ܠܡܙܥܙܥܘܬܗ. ܐܠܐ ܟܕ ܢܣܒܐ
ܡܬܚܒܛܐ ܒܗ ܘܡܬܦܠܓܐ ܒܦܠܓܘܬܐ ܣܓܝܐܐ. ܢܦܩܐ ܠܒܪ ܕܐܝܟ ܗܢܐ.

There sometimes occur bellowing [*gā'ōyē*] earthquakes which move the earth with a certain bellowing [*g'āṯā*]. (1a) Often a certain bellowing also arises from the earth without movements, when the wind which is enclosed

[37] Nic. syr. 38.24-25: ܗܘܬ ܕܝܢ ܡܢ ܟܠܦܪܘܣ ܩܠܐ ܩܕܡ ܠܙܘܥܐ. ܐܘ ܗܘܐ ܥܡܗܘܢ. ܩܠܐ ܕܝܢ/ ܗܘܝܢ ܐܦ ܒܠܥܕ ܙܘܥܐ.: "Furthermore, a sound always [*men kul pros*] precedes the movements, or occurs with them; sounds also occur without movements."

in (the earth) is insufficient to move the earth, but as it turns in it and is divided by a strong force, gives out such a sound.

Bar Kepha, *Hex.*, ms. Paris 241, 192v b 12-16; Paris 311, 59b a 10-13 [underline: agreement with *Cand.*]:

ܐܝܬ ܐܡܬܝ̈ ܕܟܕ ܨܪܝܐ ܐܪܥܐ ܘܢܦܩܐ ܘܗܘܝܐ ܙܘܥܐ ܘܬܙܝܥܬܐ. ܗܘܐ ܩܠܐ ܥܫܝܢܐ ܘܚܣܝܢܐ ܕܫܡܥܝܢ ܠܗ ܒܢܝ̈ܢܫܐ ܘܪܥܠܝܢ.

[ܗܘܐ 311: ܘܗܘܝܐ 241]

Sometimes, when (the wind) tears [*ṣāryā*] the earth and comes out [*nāp̄qā*] and there occur tremor and movement, there occur a strong and mighty sound which people hear and tremble.

l. 5, 6. ܩܥܬܐ, ܩܥ̈ܘܝܐ ("noise", "noise-making"): The choice of the word *qᶜāṯā*, replacing *qālā* used at II.ii.3.1 above, is probably due to the *gᶜaṯā* of *Cand.* and *De mundo* syr. The rare word *qāᶜōyā* in this sense may be a coinage by BH,[38] formed on the analogy of *gāᶜōyā* in *De mundo* syr.[39] and intended to render IS's *muṣawwita, allaḏī tuṣawwitu fīhi*.

II.ii.3.7-10: Sensation of sound and movement

But. [underline: agreement with Nic.; italics: with *Šifāʾ*]:

(1) We sense [*margšīnan*] *the sound* [*qālā*] first, then *the earthquake* [*nyāḏā*], (2) *since the movement* [*zawᶜā*] *of* light *air precedes the movement of earth* which is heavy [*yaqqīrā*]; (3) for hearing is swifter than [the sense of] touch, (4a) just as *vision* [is swifter] than *hearing*, (4b) and for this reason, although thunder occurs before lightning, (4c) we sense the lightning first, then the thunder.

Nic. syr. 40.13-17[40]:

ܡܛܠ ܡܢܐ ܕܝܢ ܟܕ ܐܟܚܕܐ ܗܘܐ ܩܠܐ ܘܡܬܬܙܝܥܢܘܬܐ ܩܕܡܐܝܬ ܡܪܓܫܝܢܢ/14 ܒܩܠܐ ܘܠܒܬܪܟܢ ܒܡܬܬܙܝܥܢܘܬܐ. ܘܐܡܪܝܢܢ ܕܝܢ ܗܕܐ ܗܘܝܐ. ܡܛܠ ܕܫܡܥܐ/15 ܚܪܝܦܐ ‹ܡܡܫܘܫܢܘܬܐ› ܡܢ ܓܫܫܬܐ. ܗܟܢܐ ܕܝܢ ܘܟܕ ܪܥܡܐ ܗܘܐ ܠܩܕܡ: ܘܡܚܪ/16 ܒܬܪܗ ܗܘܐ ܒܪܩܐ. ܒܒܪܩܐ ܠܩܕܡ ܡܪܓܫܝܢܢ: ܘܡܚܪ/17 ܒܪܥܡܐ. ܡܛܠ ܕܚܙܬܐ ܚܪܝܦܐ ܝܬܝܪ ܡܢ ܡܫܡܥܢܘܬܐ.

[15 ܡܡܫܘܫܢܘܬܐ supplevi]

(1) Why, seeing that the sound and movement [*mettziᶜānūṯā*] occur together, do we sense the sound first and later the movement? (3) We say: This happens, because hearing is swifter than [the sense of] touch. (4a) Just so, (4b) although thunder occurs first and then after it the lightning, (4c) we sense the lightning first and then the thunder, (4a) because vision is swifter than hearing.

---

[38] The only citation in PS (3682) is from *Cand.* 43r (= ed. Bakoš 203[357].1), where it is used as the name of the constellation Boötes.

[39] Also a rare word, the passage of the *De Mundo* syr. here being the only instance cited in PS (760) and Brockelmann (127b).

[40] Cf. Arist. 368a 19-21; Alex. Aphr. 122.23-28; Olymp. arab. 139.10-17.

Šifā' 19.11-14:

وكما أن البصر يستبق السمع، فإنه إذا اتفق أن قرع إنسان من بعد جسما على جسم، رأيت القرع قبل أن تسمع الصوت. لأن الإبصار ليس فى زمان، والاستماع يحتاج فيه إلى أن يتأدى تموج الهواء الكائن إلى السمع، وذلك فى زمان. كذلك الصوت فى الزلازل يسمع قبل الزلزلة، وذلك لأن تموج الهواء أسرع وأسبق من تموج الأرض الكثيفة.

[الهواء الكائن ed. Cair. et Ṭ: fort. الهواء اللطيف legendum, cf. 19.14 الأرض الكثيفة]

(4a) Just as vision precedes audition, - when someone strikes one object after another, you see the stroke before you hear the sound, because seeing does not occur in time, whereas hearing requires that the vibration of the existent [*kā'in*] air [or: fine air?] be carried to the hearing and that occurs in time - (1) just so the sound in earthquakes are heard before the earthquake, (2) and that is because the vibration [*tamawwuǧ*] of air is faster than and precedent to [*asbaq*] the vibration of the thick [*katīfa*] earth.

l. 8-10. "just as ...": On the reason why we sense lightning before thunder, cf. Mete. IV.i.4. below.

II.ii.4.1-4: Reduction of earthquakes through excavation of wells

But.: (1) That wind confined in the earth is the cause of the movement of the earth is known from (2) the fact that in places in which movements occur frequently, (3) when many wells are dug, (4) earthquakes become fewer and occur not severely, (5) since the wind can (then) find many escape routes [*mappqānē*].

Šifā' 17.17-19:

ومن الدليل على أن أكثر أسباب الزلزلة هى الرياح المحتقنة، أن البلاد التى تكثر فيها الزلزلة إذا حفرت فيها آبار وقنى كثيرة حتى كثرت مخالص الرياح والأبخرة، قلت الزلازل بها.

(1) One indication that the commonest of the causes of the earthquake is confined winds [pl.] is (2) the fact that (in) places in which earthquakes are frequent, (3) when many wells and channels are dug in them (5) so that the escape routes [*maḫāliṣ*] for winds and vapours abound, (4) the earthquakes become fewer.

l. 3. ܠܐ ܡܬܚܣܢܢܐܝܬ ܗܘܝܢ ("occur not severely"): This seems to be an addition by BH; cf. II.iii.1.8 "more frequent and *greater*".

II.ii.4.4-8: horizontal and vertical tremors

Aristotle (*Mete*. 366b 18f., 368b 22-32) classified earthquakes into the horizontal "τρόμοι" and the vertical "σφυγμοί". Theophrastus, on the other hand, seems to have made a threefold classification.[41] In our manuscript of Nic. syr. we find a twofold classification in a passage

[41] Theoph. *Mete*. arab. [Daiber] tr. Ibn al-Khammār 15.26-35 (p. 244f.), p. 271 (tr.), 282f. (comm.), tr. Bar Bahlūl 15.16-21 (p. 210); cf. Sharples (1998) 164f.; Lettinck (1999) 217 n.19, 220.

which probably belongs to genuine Nicolaus (Nic. syr. 41.14-17) and a threefold classification in a passage marked off as a scholion (41.8-12).[42] The two passages are conflated into one at Olymp. arab. 140.17-141.2 to give a new threefold division and this is followed, in turn, by IS at *Šifā*ʾ 19.4-9. BH follows the twofold classification of Nic. syr. 41.14-17 both here in *But.* and at *Cand.* 128.9-129.1.

> *But.*: (1a) These movements of the earth occur sometimes with a shuddering and trembling motion [*reʿlānāʾīṯ wa-rṯīṯānāʾīṯ*], that is to say horizontally, (1b) sometimes with a pulsating motion [*nqāšānāʾīṯ*] and a sliding motion [*meštarglānāʾīṯ*] from a height to a depth. (2a) These are severer than the horizontal ones, (2b) since [*baḏ*] exhalation rises weakly from the surface of the earth, (2c) whereas from its depths [it rises] with greater strength and with a very severe force.
>
> Nic. syr. 41.14-17[43]:
>
> ܡܫܚܠܦܐܝܬ ܕܝܢ ܗܘ̈ܝܢ ܙܘ̈ܥܐ ܕܐܪܥܐ . ܡܢܗܘܢ ܓܝܪ ܠܦܬܝܐ /15 ܗܘܝܢ :
> ܪܥܠܢܐܝܬ ܘܪܬܝܬܢܐܝܬ . ܡܢܗܘܢ ܕܝܢ ܠܥܘܡܩܐ ܢܩܫܢܐܝܬ . ܥܫܝܢ /16 ܕܝܢ ܡܢ
> ܡܢ ܗܢܘܢ ܕܠܦܬܝܐ ܐܝܟ ܕܥܝܕܐ ܗܘܝܢ ܡܛܠ ܗܝ ܕܣܓܝ ܐܬܡܝܕܐ /17 ܚܠܝܫܐܝܬ
> ܕܐܪܥܐ ܕܠܒܪܝܬܐ ܣܠܩܐ ܠܦܐܬܐ ܡܢ ܥܘܡܩܐ . ܡܢ ܥܘܡܩܗ ܕܝܢ ܥܫܝܢܐܝܬ ܀
>
> (1a) Movements of the earth happen differently. Some of them occur horizontally with a shuddering and trembling motion; (1b) some of them towards the depth with a pulsating motion. (2a) These are usually severer than the horizontal ones, (2b) because [*meṭṭul-hāy d-*] much exhalation rises weakly from the outer face of the earth, (2c) but strongly from the depth.

l. 4-5. ܪܥܠܢܐܝܬ ܘܪܬܝܬܢܐܝܬ ("with a shuddering and trembling motion"): The two words together correspond to Arist. ὥσπερ ὁ τρόμος. In their adverbial forms they are known to PS (3953, 3992) only from BH *Cand.* 31r (= ed. Bakoš 128.10), where we have *rʿeltānāʾīṯ* instead of *reʿlānāʾīṯ*.

l. 5. ܐܝܟ ܕܠܦܬܝܐ ("horizontally"): lit. "to the width", "breadthwise"; cf. Arist. ἐπὶ πλάτος.

l. 5. ܢܩܫܢܐܝܬ ("with a pulsating motion"): corr. Arist. οἷον σφυγμός. *Neqbāʾīṯ* (ܢܩܒܐܝܬ) read at the corresponding place in *Cand.* (128.10; also PS 2449) should be corrected to *nqāšānāʾīṯ*, which is found in the Vaticanus of *Cand.*, as well as in *But.* and Nic. syr.

l. 6. ܡܫܬܪܓܠܢܐܝܬ ("(with) a sliding motion"): < *eštargal* "to slip, slide down" (√*rgl*, ܪܓܠ). - The word is known to the lexica only from here (PS Suppl. 315, where Margoliouth read ܡܣܬܪܓܠܢܐܝܬ with ms. l). Since it has no counterpart in Nic. syr., *Cand.* or Olymp. arab., it is likely to be an addition by BH. The origin of the word is uncertain, but it may be related to the

---

[42] Takahashi (2002a) 198f.

[43] Cf. Arist. 366b 18f., 368b 22-32; Alex. Aphr. 125.17-35; Olymp. arab. 140.17-141.2; *Cand.* 128.9-129.1.

name given to the "sinking" earthquakes at *De mundo* 396a 3-4 (ἰζηματίαι), for which the Syriac version, as published by de Lagarde, has *re*ʿ*lē* (ܪܥܠܐ; "shudderings"),[44] a word which has no connotation of a downward motion and is too general a term to be the name of a specific type of earthquake. It may be that BH, coming across this word and realising that it made no sense, postulated a corruption of some word derived from the root √*rgl* (ܪܓܠ) instead of √*r*ʿ*l* (ܪܥܠ); or it may be that his copy of *De mundo* syr. in fact contained some word derived from √*rgl* (*reḡlāyā*?). The latter possibility is given some support by one of the Arabic versions of the *De mundo* (which were made from the Syriac), in which the term is translated as "*ḏāt al-arğul*" ("possessing feet, legged"?).[45]

l. 6. ܡܢ ܪܘܡܐ ܠܥܘܡܩܐ ("from a height to a depth"), i.e. vertically downwards: ܠܥܘܡܩܐ Nic. (and *Cand.*). - Arist. talks of the motion of the vertical quakes as being *upwards* (368b 25 ἄνω κάτωθεν; cf. Alex 125.24 κάτωθεν ἄνω).[46] The phrase *l-umqā* in Nic. syr. is probably intended simply to mean "vertical" (as opposed to *la-p̄ṭāyā*: "horizontal"), without necessarily indicating a "downward" movement (cf. Olymp. arab. 140.21: *fī l-*ʿ*amqi*, rather than *ila l-*ʿ*amqi*). In adding the phrase *men rawmā*, BH specifies the direction as being downwards and this allowed him further to identify this type of earthquake as the "sliding" one (*meštarglānā*ʾ*iṯ*; see preceding note).

II.ii.4.8-9: Vertical tremors (contd.)

The close agreement of those parts of *But.* marked (1a) and (1b) below with Arist., Olymp. arab. and *Cand.* suggests that these parts are based on a lost passage of Nic. syr. What we have in (1c), however, is different from what is said in the corresponding parts of Arist. and Olymp. arab., so that this is probably a substitution made by BH. The ambiguity of the word *gunḥā* unfortunately makes it difficult to determine what BH meant.

---

44 *De mundo* gr. 396 a 3-4: οἱ δὲ συνιζήσεις ποιοῦντες εἰς τὰ κοῖλα ἰζηματίαι· [ἰζηματίαι Lorimer cum Paris. 2381: χασματίαι codd. ceteri et Bekker: χωματίαι Stobaeus]; *De mundo* syr. 146.10: ܗܠܝܢ ܕܝܢ ܕܡܢ ܗܕܐ ܘܡܢ ܗܕܐ ܡܙܝܥܝܢ ܒܟܢܫܐ ܠܥܘܡܩܐ. ܪܥܠܐ ܡܬܩܪܝܢ. ("Those (earthquakes) which cause to move from this direction and that [cf. gr. ***syn****-izēseis*] down to a depth [gr. εἰς τὰ κοῖλα] are called *re*ʿ*lē*").

45 Brafman (1985) 149, ms. Yehuda 308, 299v 10-11: ومنها ما يتحرك الى الجانبين جميعا في عمق الارض تسمى ذات الارجل. - The other two versions have رجفة, which would correspond to *re*ʿ*lē* (Brafman p.98, F 96v 12; p.130, K187r 15).

46 So also IS, *Šifā*ʾ 19.4, 6, 9 *ilā fauqin*. - According to the Aristotelian theory such vertical quakes are caused by exhalation gathering in the depth of the earth and thence travelling *upwards* (see Arist. 368b 26-30; cf. Alex. 125.25-29) and in vertical tremors it is indeed the initial upward motion which usually leaves the greatest impression. In reality, of course, the oscillatory nature of seismic waves means that the movement is both up and down (hence, no doubt: Ibn al-Biṭrīq 80.9: ʿ*ulūwan wa-suflan*; Ibn Tibbon II.511: *l-ma*ʿ*ᵃlāh ū-l-maṭṭāh*; Ḥunain, *Comp*. 237: *fauq wa-asfal*).

But.: (1a) For this reason they occur rarely (1b) and when they occur [lit. in their occurrence] (1c) they cause great calamities [?: *gunḥē*].

Arist. 368b 26-27: (1a) διὸ καὶ ἐλαττονάκις σείει τοῦτον τὸν τρόπον· ... (1b) ὅπου δ' ἂν γένηται τοιοῦτος σεισμός, (1c) ἐπιπολάζει πλῆθος λίθων, ὥσπερ τῶν ἐν τοῖς λίκνοις ἀναβαττομένων·

Olymp. arab. 140.22:

وحدوثها يكون فى الفرط، إلا أنها إذا حدثت ارتفع مع الريح التى تخرج حجارة.

(1a) Their occurrence is rare, (1b) but when they do occur (1c) stones are raised with the wind which exits.

Cand.: 129.1: ܡܛܠܗܕܐ ܚܕܐ ܠܟܡܐ ܗܘܝܢ: (1a) For this reason they occur rarely.

l. 8-9. ܓܘܢܚܐ ܪܘܪܒܐ ܡܣܥܪܝܢ ("they cause great calamities [?]"): *Gunḥā* is a word with many possible meanings[47] and unfortunately for our purpose there are several which would fit here: i. "atrocity, calamity",[48] ii. "crack, fissure" or iii. "clamour, terrifying noise". - The situation is not helped greatly by the verse and the scholion dealing with this word in BH's list of aequilitteral words[49] and one is tempted to think that BH is playing on words here in *But*. and is being deliberately ambiguous. - I incline towards the meaning "calamities" in view of the context where it is stated that these

[47] See PS 752f.; Brockelmann 125a; Cardahi, *Lobab* I.193f. - The problem caused by the ambiguity of this word is well illustrated by the different interpretations of the following sentence (from Isaac of Antioch, *De signo quod apparuit in coelo*): ܒܠܫܢܐ ܕܙܘܥܐ ܡܡܠܠ ܐܠܗܐ ܥܡ ܚܛܝܐ. ܘܐܪܝܡ ܩܠܗ ܒܦܩܥܬܐ: ܒܓܘܢܚܐ ܘܦܬܝܚܘܬܐ ܘܠܐ ܫܡܥܘ ܒܗ . . . This is rendered by J.S. Assemani, *BOCV* I.219, as: "Terraemotûs linguâ peccatores Deus alloquitur. Per *scissuras*, terraeque hiatus vocem extulit; nec tamen eum audierunt ...". Citing the same passage, Brockelmann translates the word as "horrendum, calamitas" (loc. cit., no. 4), Cardahi as "clamour" [*ḍağğa, ğalaba*] (I.193b 18ff.; reading ܘܓܘܢܚܐ and attributing the passage to Ephrem). - The basic meaning of the root √*gnḥ* appears to be that of causing "commotion, shock", both in Syriac and in other Aramaic idioms (see Levy [1867/8] I.149; Nöldeke [1875] 52; Drowser-Macuch [1963] 82 s.v. "guha"; Macuch [1965] 50.5, 85.25, 176.2).

[48] The word is often used by BH in this sense in his *Chron.*, usually of atrocities committed in warfare: ed. Bedjan 306.18; 378.8; 402.27 (with the verb *sʿar* as here); 432.4; 598.8 (in the continuation; with *sʿar*; and of natural disasters, wrought by "Justice").

[49] BH *Aequ.*, ed. Martin (1872) vol. ii, p. 89, l. 1049: . ܚܙܘ ܓܘܢܚܐ ܪܒܐ ܒܓܘܢܚܐ ܕܚܬܗ ܕܫܠܝܚܐ ܫܠܝܚܐ ("They saw the great *gunḥā* [calamity (?)] in *gunḥā* [astonishment], that the sister of the *šlīḥā* [messenger] was *šlīḥā* [naked]"). - Scholion ad loc.: ܓܘܢܚܐ ܒܬܪܬܝܢ ܦܘܫܩܐ. ܬܗܪܐ ܘܙܘܥܬܐ ܘܚܒܠܐ. ܘܡܬܐܡܪ ܓܘܢܚܐ ܐܦ ܠܦܩܥܐ ܕܗܘܐ ܒܐܪܥܐ. ܘܡܬܒܙܥܐ ܒܗ ܪܘܚܐ [ܕܡܬܒܙܥܐ ܒܗ Bar Bahlūl (Duval) 468.12f.] ("[To be vocalised] *gunḥā* in both meanings. [The word can mean] astonishment [*tehrā*], trembling [*zawʿṯā*] and destruction [*ḥḇālā*]. *Gunḥā* is also said of the fissure [*paqʿā*] which is made in the earth and which the wind tears through (?)"; or, reading ܕܡܬܒܙܥܐ ܡܢ ܪܘܚܐ/ܒܗ ܒܪܘܚܐ: "(in the earth), which is torn open by/with the wind"?).

vertical tremors are severer than the horizontal ones[50] and the fact that the verb *sʿar* is more commonly used with abstract objects than with concrete, physical objects (PS 2686). - The interpretation of the word as "fissures", on the other hand, would find some support if the word *meštarglānāʾīṯ* of II.ii.4.6 can be related to *reʿlē* of *De mundo* syr. 146.10 (see above), since the sentence there in the *De mundo* is immediately followed by one dealing with earthquakes that create openings in the ground.[51]

Min. II. Section iii.: *On Times when Movements Occur more*

For the overview of the section, see the introductory comments to Min. II. Section ii. above.

II.iii.1.1-2: Relation to wind

This is closer to Nic. syr. than to *Šifāʾ*, but the phrase "in absence of winds" betrays an influence of the latter.

But. [underline: agreement with Nic.; italics: with *Šifāʾ*]:

(1a) *Movement of earth* occurs more [*yattīr hāwē*] *in* [*b-*] *absence of winds* [pl.], (1b) because the smoky exhalation is [then] *confined* in the earth.

Nic. syr. 37.19-20[52]:

ܬܘܒ ܕܝܢ ܗܘܐ ܝܬܝܪܐܝܬ ܙܘܥܐ ܕܐܪܥܐ . ܟܕ ܗܘܝܐ/20 ܚܣܝܪܘܬܐ ܕܪܘܚܐ .
ܡܛܠܬ ܕܡܬܚܒܫܐ ܠܓܘ ܬܢܢܝܬܐ ܒܐܪܥܐ .

(1a) Furthermore, movement of earth occurs more [*hāwē yattīrāʾīṯ*] when there is an absence of wind [sg.] (1b) because the smoky exhalation is confined in the earth.

Šifāʾ 17.19-20:

وأكثر ما تكون الزلازل إنما تكون عند فقدان الرياح، لأن مواد الرياح يعرض لها الاحتباس

(1a) Usually, earthquakes only occur in [*ʿinda*] absence of winds [pl.], (1b) because the materials of winds are [then] subject to confinement.

---

[50] One tends in reality too to associate great disasters with the vertical shocks caused by the transverse seismic waves (S-waves), since these do not travel as far as the longitudinal waves (P-waves) and are therefore only felt in areas nearer the epicentre.

[51] *De mundo* syr. 146.11-12 [< gr. 396a 4-6]: ܘܗܢܘܢ ܬܘܒ ܕܒܕܩܝܢ ܦܬܚܐ ܘܦܬܚܝܢ ܠܗ ܠܐܪܥܐ . ܦܬܘܚܬܐ ܡܬܩܪܝܢ . ܘܡܢܗܘܢ ܕܝܢ ܗܠܝܢ ܡܬܥܝܪܝܢ ܪܘܚܐ ܚܬܝܪ . ("Those (earthquakes) also which create openings and open up the earth are called 'cleaving' (earthquakes) [gr. ῥῆκται]; and from these winds rise"). - Unlike the preceding sentences in *De mundo* syr. which all begin *hānōn dēn*, the sentence here begins *w-hānōn tūḇ*. This could have led BH to suppose that the sentence here refers not to a different type of earthquake but to the same type of earthquake as described in the preceding sentence (especially if BH read ܒܕܩܝܢ rather than ܕܒܕܩܝܢ).

[52] Cf. Arist. 366a 5-8; Alex. Aphr. 116.34-117.2; Olymp. arab. 136.3-6.

II.iii.1.2-7: Relation to wind (contd.): elongated cloud

BH combines here two different passages of Nic. syr., namely 37.20-24, which immediately follows the passage corresponding to *But.* II.iii.1.1-2, and 39.16ff. (the passages of Nic. syr. in turn correspond to two different passages in Arist.), as well as elements taken from a passage of *Šifā᾿*.

But. [underline: agreement with Nic. (both passages); italics: with *Šifā᾿*]:
(1a) *At such a time* [*wa-ḇ-ḏ-aḵ hānā zaḇnā*] a thin, long cloud is usually [*aḵ da-ḇ-suggā*] *seen* beside the horizon. (2a) For when [*kaḏ*] *opposing winds* blow, (2b) the stronger (wind) drives the weaker (2c) so that [lit.: and] the weaker (wind) is *confined* [*meṯhaḇšā*] *inside the earth* (2d) and causes the earth to quake [*l-arʿā mnīḏā*], (2e) though [*bram*] weakly because of the division of the smoky exhalation above and below. (3a) The stronger (wind) *extends* upwards in a straight line [*ba-trīṣū*] (3b) and when the air is cooled [*qā᾿ar*] by it [*menāh*] (3c) (this air) is condensed (3d) and a straight [*trīṣtā*] cloud is formed.

Nic. syr. 37.20-24[53]:

ܗܘܐ ܕܝܢ ܘܟܕ ܢܫ̈ܒܢ/21 ܪ̈ܘܚܐ. ܡܐ ܓܝܪ ܕܪ̈ܘܚܐ ܡܣܬܩܒܠܢܝܬܐ ܢܫ̈ܒܢ: ܘܕܚܝܐ
ܗܝ ܕܚܣܝܢܐ ܠܗܝ ܕܡܚܝܠܐ/22 ܟܕ ܥܡܕܐ ܗܝ ܕܡܚܝܠܐ ܬܚܘܬ ܐܪܥܐ. ܗܘܐ
ܙܘܥܐ ܕܐܪܥܐ. ܒܪܡ ܐܦ ܗܘܐ ܙܘܥܐ/23 ܕܐܪܥܐ ܟܕ ܢܫ̈ܒܢ ܪ̈ܘܚܐ. ܐܠܐ
ܝܬܝܪ ܡܚܝܠܐܝܬ ܗܘܐ ܡܛܠܬܗ ܕܡܬܦܠܓ ܚܠܦܐ ܬܢܢܐ/24 ܠܥܠ ܘܠܬܚܬ.
ܘܡܬܚܘܝܐ ܗܝ ܗܕܐ ܥܠܬܗ ܕܪ̈ܘܚܐ ܐܝܬܝܗ ܥܠܬܐ ܕܙܘܥܐ ܕܐܪܥܐ.

[24 ܥܠܬܗ: ܕܥܠܬܗ cod.]

(Earthquake) also occurs when winds are blowing; (2a) for whenever [*mā d-*] opposing winds blow (2b) and that (wind) which is stronger drives that which is weaker [and] (2c) when that which is weaker sinks [*ʿāmdā*] below the earth, (2d) a movement of the earth occurs. (2e) However [*bram*], even though movement of earth occurs when winds are blowing, it occurs more weakly, because the smoky exhalation is divided above and below; and this proves that the wind is the cause of the movement of the earth.

Nic. syr. 39.16-17, 22-27[54]:

ܘܡܚܘܐ ܐܘܟܝܬ ܥܠܬܐ ܕܐܬܝܐ ܗܝ ܕܡܬܢܫܡܢܘܬܐ/17 ܚܢܢܐ ܕܐܠܝܨ ܡܢ ܡܕܡ ܘܚܠܐ
ܕܐܪܥܐ. ܐܟܘܬ ܐܝܢܐ ܕܝܢ ܚܢܢܐ//22 ܦܫܝܛܘܬܐ ܘܐܪܝܟܘܬܐ ܕܡܫܘܚܬܐ ܒܐܘܪܟܐ
ܬܪܝܨܐܝܬ/23. ܟܕ ܓܝܪ ܪ̈ܘܚܐ ܡܣܬܩܒܠܢܝܬܐ ܢܫ̈ܒܢ: ܗ ܡܚܣܢܐ ܗܝ ܕܡܚܝܠܐ
ܠܗܘ ܐܪܥܐ. ܘܥܡܕܐ/24 ܘܚܠܐ. ܚܢܢܐ ܡܕܡ ܐܪܝܟܬܐ ܘܦܫܝܛܬܐ ܡܬܚܙܝܢܐ
ܕܡܬܚܙܝܢܐ ܥܠ ܟܠ ܐܘܪܚܐ/25 ܬܪܝܨܐܝܬ. ܟܕ ܓܝܪ ܡܬܚܣܢܐ ܪܘܚܐ ܗܝ
ܡܚܝܠܬܐ ܠܗܘ ܐܪܥܐ ܕܗܝ ܥܡܕܐ ܘܚܠܐ/26 ܕܐܪܥܐ. ܦܐܪ ܐܐܪ ܟܕ ܠܐ
ܡܟܝܠ ܡܬܚܙܝܢ ܡܢ ܪ̈ܘܚܐ. ܘܦܐܪ ܚܘܫܒܐ ܚܢܢܐ./27 ܘܡܛܠܬܗ ܕܟܝܢܐ ܪ̈ܘܚܐ
ܢܫ̈ܒܐ ܬܪܝܨܐܝܬ ܡܬܚܙܝܢܐ.

[17 ܚܢܢܐ: ܚܢܢܐ ܦ̄ ܚܢܢܘܬܐ cod. || 22 et 25 ܬܪܝܨܐܝܬ: ܬܪܝܨܝܬ cod. || 26 ܦܐܪ: ܦܐܐܪ cod.]

[53] Cf. Arist. 366a 8-12; Alex.Aphr. 117.2-9; Olymp. arab. 136.6-8.

[54] Cf. Arist. 367b 7-19; Alex. Aphr. 120.13-19; Olymp. arab. 138.6-13.

(1a) This is the cause of the sign which is often [*kmā zabnīn*] seen before a movement of the earth, I mean, the thin and long cloud which extends lengthwise [and] straight [*b-urkā trīṣā'īt*]; (2a) for when [*kad*] opposing winds blow (2bc) and then that (wind) which is weaker is driven inside the earth, and causes a movement, (1a/3d) a certain long and thin cloud is seen, which extends straight beside the horizon; (2c) for when that weaker wind is driven inside the earth, (2d) which causes the movement of the earth [*ʿābdā zawʿā d-arʿā*], (3b) the air grows cold [*qā'ar*], since it is no longer heated by the wind (3c) and is condensed, (3d) and a cloud is formed; (3a) because the wind is one, it blows extending straight.

Šifā' 17.20-18.2:

وفى مثل هذه الحال كثيرا ما تُرى فى الجو سُحب مستطيلة استطالة توجبها الرياح المختلفة إذا تهابّت وغلب منها واحد فامتد وحبس المغلوب فى قعر الأرض.

(1a) Under such a condition [*wa-fī mitli hādihi l-ḥāli*] elongated clouds are often [*katīran mā*] seen in the air, (2a) an elongation which differing winds bring about when they blow against each other and one of them prevails, (3a) so that it becomes extended (2c) and confines [*ḥabasa*] the (wind) which is overcome in the depth of the earth.

l. 3-4. ܪ̈ܘܚܐ ܡܩܒ̈ܠܢܝܬܐ ("opposing winds"): cf. Mete. I.i.3.3f. ܪ̈ܘܚܐ ܕܠܩܘܒܠܝܬܐ.

l. 6, 7. ܠܐܦ̈ܝ ܠܥܠ ܒܬܪܝܨܘܬܐ ܡܬܡܬܚܐ، ܥܢܢܐ ܬܪܝܨܬܐ ("extends upwards in a straight line", "a straight cloud"): There is no mention in Nic. syr. or IS of the "stronger wind" or the resulting "cloud" extending "upwards" (*l-appay l-ʿel*), even if there is a mention of the exhalation being divided "above and below" (*l-ʿel wa-l-taḥt*) in the first passage of Nic. syr. quoted above. - Arist., in fact, probably envisaged a straight cloud extending horizontally (see *Mete*. 367b 9-11) and this is also implied in Nic. syr. by the words "beside [*ʿal geb*] the horizon" (39.24). The word *trīṣā*, however, which is used three times in Nic. syr. 39-16-27 in the adverbial form *trīṣā'īt*, means not only "straight" but also "upright, vertical" and it seems that BH was led by this to envisage a cloud extending vertically.[55]

II.iii.1.7-8: Relation to wind (contd.): nebulous air

This is based on the *Šifā'*. The alteration made by BH in clause (1b) may be explained with reference to II.iii.1.1-2 above, although the passage Nic. syr. quoted below may also have played a part.

---

[55] Curiously the word "upwards" also occurs in a paraphrase of the passage of the *Šifā'* quoted above in Ibn Tibbon's *Otot ha-Shamayim*, ed. Fontaine II.489-492: "But Ibn Sina explains here that this is because of the opposition of winds, for one moves the vapour upwards [*l-maʿalāh*] and the other downwards and thus they form a long stretched cloud. According to Ibn Sina's reason it looks as if this is a long vertical cloud [*šᵉ-ôrek hā-ʿāb hû mi-l-maʿalāh l-maṭṭāh*]" (tr. Fontaine). - Although this may simply be a "free rendering" of IS's words on Ibn Tibbon's part (see Fontaine, intro. p. LXXI), the coincidence is worth noting.

But. [underline: agreement with *Šifā*ʾ]:
(1a) The air becomes foggy [*ʿarpellāyā*] during [*b*-] an earthquake, (1b) because wind is then confined inside the earth.

*Šifā*ʾ 18.3-4:

وكثيرا ما يكون فى وقت الزلازل غمامات راكدة فى الجو، ويكون الجو ضبابيا، / ٤ وذلك لفقدان الرياح فى ذلك الوقت.

Often at time of [*fī waqti*] earthquakes [pl.] still clouds occur in the air, and the air becomes foggy [*ḍabābī*]; (1b) that is due to the absence of winds at that time.

Nic. syr. 38.25-39.2[56]:

ܬܘܒ ܕܝܢ ܟܕ ܥܬܝܕ ܙܘܥܐ ܕܐܪܥܐ ܠܡܗܘܐ ܩܕܡ ܫܡܫܐ ܣܓܝܐ/26 ܩܐܡ ܥܡܛܢܐ ܘܐܦ
ܟܕ ܠܐ ܥܢܢܐ . ܟܕ ܠܘܬ ܥܘܡܩܐ . ܡܬܢܚܬܐ ܕܬܢܢܐ ܥܠܬܐ ܕܐܚܫܘܢܐ/27 ܠܡܗܘܐ .
ܘܠܐ ܡܥܒܪܐ//1 ܡܢ ܐܐܪ ܠܪܛܝܒܘܬܐ/2 {ܕܚܠܝܛܬ}ܠܝܛܐ ܒܗ .

[25 ܠܡܗܘܐ: ܠܡܗܘܝܐ cod.]

(1a) Furthermore, when a movement of earth is about to occur, a darkness [*ʿamṭānā*] often stands before the sun, although there is no cloud, (1b) because the smoky exhalation sinks to the depth [cf. Arist. ὑπονοστεῖν] and does not remove from the air the moisture which is mixed in it.

II.iii.1.8-12: Times of day

But. [underline: agreement with Nic.; italics: with *Šifā*ʾ]:
(1a) Movements of the earth [*zawʿay arʿā*] are frequent and great at night (1b) because [its] coldness condenses the earth's surface [lit. *because of the condensation by coldness of (the earth's) surface*]; (2a) and *also at dawn* [*šaprā*], (2b) because the sun causes the wind to move, but is unable to dissipate (it). (3a) During the day, the movements occur most at noon, (3b) *because* [at noon] *the heat attracts the exhalation with vigour*, (3c) [and at the same time] *causes* [fem., subj. 'heat'] *the surface* [*appē*] *of the earth to* contract and *dry up*, (3d) and [so] does not allow [fem.] (the exhalation) to escape.

Nic. syr. 37.25-30[57]:

ܬܘܒ ܕܝܢ ܒܠܠܝܐ ܗܘܝܢ ܣܓܝܐܝܢ ܘܪ̈ܘܪܒܝܢ ܙܘ̈ܥܐ ܕܐܪܥܐ . ܝܬܝܪ ܡܢ
ܕܒܐܝܡܡܐ/26 . ܡܛܠ ܕܒܠܠܝܐ ܗܘܐ ܒܠܚܘܕ ܦܘܪܐ ܝܬܝܪ ܡܢ ܕܒܐܝܡܡܐ .
ܒܐ{ܨܦܪܐ}//27 ܕܝܢ ܗܘܝܢ ܒܠܚܘܕ . ܡܛܠ ܕܫܡܫܐ ܢܫܒܐ ܗܘܐ ܒܠܚܘܕ
ܪܘܚܐ . ܫܡܫܐ ܕܝܢ ܒ{--}/28 ܡܛܠ ܕܫܪܝܐ ܡܫܟܚ ܠܡܥܒܪܘܬܗ ܕܐܪܥܐ
ܡܕܝܢ ܠܐ ܫܒܩ ܠܕܠܐ ܕܢܦܩ{ܠܐܦ̈ܐ}/29 ܠܟܠ ܢܦܫ . ܗܘܝܢ ܕܝܢ ܙܘ̈ܥܐ
ܕܐܪܥܐ ܡܣܬܟܝܐ . ܡܛܠ ܕܚܡܝܡܘܬܐ ܗܘܐ ܐܢܫ +ܡܢ/30 ܫܡܫܐ ܢܓܕܐ ܠܦܘܚܐ .
ܠܐ ܕܝܢ ܡܫܒܩ ܕܢܦܫܬ .

[27 --ܒ: fort. ܒܫܡܫܐ, cf. *But*. ‖ 29 ܐܫܪ: fort. ܐܫܬ legend., cf. Olymp. arab. 136.12 تحل]

---

[56] Cf. Arist. 367a 20-367b 7; Alex. Aphr. 119.23-28; Olymp. arab. 137.17-19.

[57] Cf. Arist. 366a 13-23; Alex. Aphr. 117.9-22; Olymp. arab. 136.8-15; *Cand.* 129.10-130.2.

(1a) Furthermore, movements of the earth [*zaw*ᶜ*ē d-ar*ᶜ*ā*] are frequent and great at night, more [so] than during the day, (1b) because the absence of wind occurs more at night than during the day. (3a) During the day, they occur at noon; for then the absence of wind occurs more; (3c) for the sun then causes [masc., subj. 'sun'] the outside [*barrāyūṯā*] of the earth to dry up and contract {...}; (3d) and therefore does not allow [masc.] the smoky exhalation to rise upwards. (2a) Movements of the earth also occur at dawn, (2b) because at this time the sun sheds [?] and causes wind to move, but is unable to dissipate (it).

*Šifā*ʾ 18.6-8:

وذلك يكون فى الأكثر ليلا لتخصيف البرد وجه الأرض، وبالغدوات أيضا وقد يكون فى أنصاف النهار بسبب شدة جذب الحر للبخار، مع تجفيف وجه الأرض وإعادة البرد إلى داخلها على سبيل التعاقب.

(1a) That occurs mostly at night (1b) because of the closure [*taḫṣīf*] by the coldness of the surface of the earth, (2a) and also in early morning [*ġadawāt*]; (3a) it sometimes occurs at noon (3b) due to the severity of the attraction by the heat of vapour, along with the desiccation of the surface of the earth and the return of the coldness towards its interior by mutual repulsion [of heat and cold] [*ta*ᶜ*āqub*].[58]

l. 9. ܒܝܕ ܡܥܒܝܢܘܬ ܩܘܪܐ ܐܦ̈ܝܗ (lit. "because of the condensation of/by the coldness of (the earth's) surface"): The construction of the phrase mirrors that of IS "*li-taḫṣīfi l-bardi* (gen.) *wajha* (acc.) *l-*ʾ*arḍi*".

l. 12. ܡܩܦܕܐ ܘܡܝܒܫܐ ("causes to contract and dry up"): ܡܝܒܫ ܘܡܩܦܕ ("causes to dry up and contract) Nic. - The more logical order of Nic. syr. is retained at *Cand*. 130.1.

II.iii.2.1-4: Earthquakes on porous land

Another passage where BH combines Nic. and *Šifā*ʾ. It may be noted that the association of the phenomenon described here with the "sea" in Arist. (366a 25, 31; cf. Ibn al-Biṭrīq [Schoonheim] l. 768) is lost in the main text of in Nic. syr. as well as in the *Šifā*ʾ, though not on the scholiast of Nic. syr.[59]

But. [underline: agreement with Nic.; italics: with *Šifā*ʾ]:

(1a) Severe *movements* *occur* *in* *places* which have cavernous and *porous* [*pe*ᶜ*rānāyē w-p̄aḥḥīḥē*] *depths* [ᶜ*umqē*], (1b) *especially if* [there is] *much water flowing* over them; (2a) for in such (places) much wind is observed [*meṯhawwyā*], (2b) but the rapid *movement* [of the water] *does not give room for penetration*, (2c) so that the wind is confined [*meṯabšā*] [beneath the water] (2d) and causes the earth to quake."

---

[58] i.e. by ἀντιπερίστασις, see comm. on Mete.I.iii.4.1-4 below.

[59] Nic. syr. 38.3-4: ܥܠ ܡ̈ܝܐ ܕܝܢ ܕܢܬܦܪܫ ܗ̈ܢܘܢ ܕܝܡܐ ܟܕ {ܝܩܝܪܝܢ}/ {ܝܩܝܪܝܢ} ܘܠܐ ܡܫܟܚܐ ܠܗܠܓܐ ܕܛܝܦܐܝܬ ܐܝܟ ܠܡܥܠܬܗܘܢ. ܐܠܐ ܥܡ ܙܘܥܐ. ("[By 'water'] we should understand 'seawater', since it is heavier and exhalation is unable to penetrate it in its upward movement [*tāypā*ʾ*īṯ*] except with movement [i.e. without causing an earthquake].")

Nic. syr. 37.30-38.1[60]:

ܬܘܒ ܕܝܢ ܐܬܪ̈ܘܬܐ ܕܒܗܘܢ/31 ܗܘܝܢ ܙ̈ܘܥܐ ܥܙܝ̈ܙܐ. ܡܥܪ̈ܢܝܬܐ ܘܦܣܝ̈ܣܝܬܐ
ܐܝܬܝܗܝܢ. ܡܛܠ ܕܗ̇ܘ ܪܘܚܐ/32 ܣܓܝܐܐ ܡܫܬܟܚܐ ܚܒܝܫܐ ܒܗܘܢ.
ܘܝܬܝܪܐܝܬ ܐܢ ܡ̈ܝܐ ܪܕܝܢ ܒܐܦ̈ܘܗܝ ܒܪ̈ܝܐ/1 ܕܐ{ܪ}ܥܐ ܘܡܛܒܥܝܢ
ܠܗ̇.

(1a) Furthermore, the places in which severe movements occur are cavernous and porous, (2ac) because much wind is found confined [*ḥbīšā*] in them, (1b) especially if water is flowing on the outer face of the earth and submerges it.

Šifāʾ 18.9-11:

وأكثر ما تكون الزلزلة فى بلاد متخلخلة غور الأرض، متكاثفة وجهها، أو مغمورة الوجه بماء يجرى، أو ماء غمر كثير لا يقدر ريح على خرقه. وخصوصا إذا كان متحركا، فإن المتحرك أشد ممانعة لأنه تسبق بحركته خرق الخارق إياه

(1a) Usually, earthquakes occur in countries where the depth [*ġaur*] of the earth is loose [*mutaḫalḫil*] and its surface is thick, (1b) either flooded on the surface with flowing water or much flood water, (2c) which the wind cannot penetrate [*ḫarq*]: (1b) especially when (the water) is moving, (2b) since moving [water] provides the greatest hindrance [against penetration], because it forestalls by its movement the penetration of the [wind] penetrating it.

l. 1. ܡܥܪܢܝܐ ܘܦܣܝܣܝܐ ("cavernous and porous"): ܡܥܪ̈ܢܝܬܐ ܘܦܣܝ̈ܣܝܬܐ (in agr. w. "places") Nic. - The two adjectives no doubt correspond to Arist. 366a 25 σομφὴ καὶ ὕπαντρος,[61] though in the reverse order. - Cf. the scholion on these two words at Nic. syr. 38.2-3: ܦܪܝܫܐ ܡܥܪܢܝܬܐ ܡܢ ܦܣܝܣܝܬܐ. ܒܕ ܗܝ ܒܡܥܪ̈ܬܐ /{--} {ܘܗܕܐ} ܒܢܩ̈ܒܐ ܐܝܟ ܦܣܝܣܘܬܐ. ("'Cavernous' is distinguishded from 'porous', since the former {--} with hollows, the latter with perforations.")

l. 3. ܠܐ ܝܗܒܐ ܐܬܪܐ ܠܢܦܩܐ ("does not give room for penetration"): For the expression, cf. Bar Kepha, *Hex.*, Paris 241, 191v b21 [corr. tr. Schlimme, p. 627, line 55ff.]: *kad lā yāhbīn l-āʾār atrā d-neppoq* ("without giving the air room to escape").

II.iii.2.4-8: Seasons of year: winter and summer

In Nic. syr., as in Arist., the discussion of the frequency of earthquakes in spring and autumn precedes the discussion of their rarity in winter and summer. In the *Šifāʾ* the order is reversed. In *But.*, though not in *Cand.* (130.2-3), BH follows the order in the *Šifāʾ*, and in its content and internal arrangement too the discussion of earthquakes in winter and summer in *But.* is closer to IS than to Nic. syr., although the

---

[60] Cf. Arist. 366a 23-b 2; Alex. Aphr. 117.23-118.14; Olymp. arab. 136.15-22.

[61] Cf. Alex. 117.24f.: καὶ ἔχουσι χώραν ἀραιάν τε καὶ πολλὰς ἔχουσιν κοιλότητας.

similarity of the material in Nic. syr. and IS makes it difficult to determine the exact extent to which *But.* is dependent on each.

But. [underline: agreement with *Šifā'*; italics: with Nic.]:

(1a) *In winter earthquake* [*nyādā*] *occurs less*, (1b) *since* [*bad*] excessive coldness [*qurrā*] retards [*mtāhhē*] the violent movement of *the smoky exhalation*. (2a) If it *does occur* [*gādeš*], (2b) *it indicates* [*mšawda*ᶜ] the *excess* [*yattīrūtā*] of the *moisture* of that *winter* and the *deficiency* [*bṣīrūtā*] of its *coldness*. (3a) Likewise *in summer movement occurs less* (3b) because of the *dissipation* [*puššāšā*] of the exhalation. (4a) If it *does occur*, (4b) *it signifies* [*mbaddeq*] *the excess of dryness* which thickens [*mlazz*ᶜ*zā*] the surface of the earth.

Šifā' 18.14-18:

وقلما تكون الزلزلة فى الشتاء، لشدة إجماد برده للبخار الدخانى. فإن عرض دل على أن رطوبة ذلك الشتاء أشد من برودته، فيولد ببلته وقلة برده بخارا كثيرا. وقلما تعرض الزلزلة أيضا فى الصيف، لشدة تحليله، فإن حدثت فى الصيف، دلت على أن السنة يابسة فيكثف وجه الأرض باليبس، وتخصف مسامها فتحتبس فيها الرياح ولا تخرج، حتى تجتمع لها مادة كثيرة تقوى على الزلازل

(1a) Earthquake rarely occurs in winter, (1b) due to the severity of the freezing [*iğmād*] by its coldness of the smoky vapour. (2a) If it does occur [*in* ᶜ*araḍa*],[62] it indicates [*dalla*] that the moisture of that winter is stronger than its coldness, so that it generates much vapour with its dampness and the deficiency [*qilla*] of its coldness. (3a) Earthquake also occurs rarely in summer, (3b) due to the severity of its dissipation [*taḥlīl*]. (4a) If it does occur in summer, (4a) it indicates that the year is dry, so that the surface of the earth is thickened by the dryness, and its openings are closed, so that winds are confined in it and cannot go out until a large enough amount of material has collected to cause earthquakes.

Nic. syr. 38.6-10[63]:

ܒܣܬܘܐ ܓܝܪ ܘܒܩܝܛܐ ܒܨܝܪ ܗܘܝܢ ܙܘܥܐ ܕܐܪܥܐ. ܡܛܠ/7 ܕܒܣܬܘܐ ܩܛ
ܪ ܠܗ ܗܠܓܐ ܬܢܢܢܝܬܐ ܡܢ ܩܪܝܪܘܬܐ. ܒܩܝܛܐ ܕܝܢ ܡܬܦܫܫ/8 ܡܛܠ ܫܚܝܢܘܬܐ.
ܟܕ ܗܟܝܠ ܕܓܕܫܐ ܕܢܗܘܘܢ ܙܘܥܐ ܒܣܬܘܐ. ܡܚܘܝܢ/9 ܕܣܬܘܐ ܝܬܝܪ ܪܛܝܒ ܗܘ
ܘܒܨܝܪ ܩܪܝܪ. ܘܟܕ ܓܕܫܐ ܕܢܗܘܘܢ ܒܩܝܛܐ. ܡܚܘܝܢ/10
ܕܝܬܝܪ ܝܒܝܫ ܘܒܨܝܪ ܚܡܝܡ.

(1a/3a) For in winter and in summer movements of the earth [*zaw*ᶜ*ē d-ar*ᶜ*ā*] occur less, (1b) because [*meṭṭul d-*] in winter the smoky exhalation is condensed [*qāṭar*] by the coldness [*qarrīrūtā*], (3b) and in summer it is dissipated [*metpawšaš*] on account of the heat. (2a) When therefore it happens [*gādšā*, impers. fem.] that earthquakes occur in winter, (2b) they indicate that the winter is more moist [*yattīr raṭṭīb*] and less cold [*bṣīr qarrīr*]; (4a) and when it happens that they occur in summer, (4b) they indicate that it is more dry and less hot.

---

[62] Either understand *ḏālika* with the masculine verbs or read ᶜ*araḍat dallat*; cf. 18.16 فإن حدثت ... دلت.

[63] Cf. Arist. 366b 4-7; Alex. Aphr. 118.18ff.; Olymp. arab. 136.23-137.3; *Cand.* 130.2-3.

l. 5. ܟܕ ܩܘܪܫܐ ܝܬܝܪܐ ܡܘܚܪ ܠܙܘܥܐ ܚܪܝܦܐ ܕܠܓܘܪܐ ܬܢܢܝܐ ("since excessive coldness retards the violent movement of the smoky exhalation"): The notion of 'retardation' of the exhalation is not found in the corresponding passages of either IS or Nic. syr., who both talk of the exhalation 'freezing' (*iğmād*, *qāṭar*). The alteration made by BH is perhaps to be explained with reference to Mete. III.iii.4. below.

l. 8. ܡܠܙܙܐ ("thickens [the surface of the earth]") voc. ܡܠܰܙܶܙܐ LIV: يكثّف (with وجه الأرض as subject rather than object) IS. - This is the only instance of the word in this sense known to the dictionaries (PS Suppl. 179a, also citing Ethpe. *etlzez* from *But*. Gen. et corr. III.iv). As a witness for the correspondence with arab. √*ktf*, we have a definition of *lzīzūṯā* as *takāṯuf* in Bar Bahlūl [Duval] 960.16.

II.iii.2.8-9: Spring and autumn

IS tells us that earthquakes occur mostly in spring and autumn but does not say why (*Šifāʾ* 18.18). Here, BH turns to Nic. syr.

But.: (1a) In spring and autumn movements [*zawʿē*] occur more [*yattīr*], (1b) since in them winds are generated more than in the rest of the seasons.

Nic. syr. 38.1-2, 5-6[64]:

ܬܘܒ ܕܝܢ ܙܘܥܐ ܕܐܪܥܐ ܗܘܝܢ ܝܬܝܪܐܝܬ ܒܪܒܝܥܐ ܘܒܬܫܪܝܢܐ/2 {ܒܗ}ܠܝܢ ܓܝܪ
ܙܒܢܐ//5 ܝܬܝܪ ܡܢ ܕܒܫܪܟܐ ܕܙܒܢܐ ܗܢܘܢ/6 {--}ܐ ܡܬܝܠܕܢ ܪܘܚܐ.

(1a) Furthermore, movements of the earth occur most [*yattīrāʾīṯ*] in spring and autumn; (1b) for in these seasons winds are generated more than in the rest of those {--} seasons.

II.iii.3.1-5: Solar and lunar eclipse

This passage seems to be based primarily on IS, but in the middle part of the passage BH closely follows the wording of Nic. syr.

But. [underline: agreement with *Šifāʾ*; italics: with Nic.]:

(1a) *The eclipse* of the sun and *the moon* also becomes the cause of *movement of the earth* (1b) through the sudden deprivation [*glīzūṯā*] of the heat which [arises] from radiation [*zallīqā*]; (2a) because of the [resulting] *coldness* (2b) *the openings* [*ḥrūrē*] *of the earth become closed* [*metraṣpīn*] (2c) and *the smoky exhalations are* suddenly *confined in (the earth)* [*b-ḡawwāh*]. (3a) That this occurs due to coldness which arises suddenly, (3b) but does not occur due to coldness which arises gradually [*b-ḏurgā*], (3c) one learns [impersonal, *metyallpā*] from the (human) body and particular experiments in medicine.

---

[64] Cf. Arist. 366b 2-4; Alex. Aphr. 118.15-18; Olymp. arab. 136.23-24; *Cand*. 130.2-3.

Šifā' 18.19-19.2:

والكسوفات ربما كانت سببا للزلازل، لفقدان الحرارة الكائنة عن الشعاع دفعة، ويعقب/ ٢٠ البرد الحاقن للرياح فى تجاويف الأرض بالتخصيف بغتة. والبرد الذى يعرض دفعة يفعل/ ١ من ذلك ما لا يفعله العارض بالتدريج. تأمل ذلك فى الأبدان وفى جزئيات تجارب صناعة/ ٢ الطب وغيرها.

(1a) Eclipses sometimes become the cause for [*li*-] earthquakes, (1b) through [*li*-] the sudden deprivation [*fuqdān*] of the heat which arises from rays; (2a) and there ensues coldness (2bc) which suddenly confines winds in the cavities of the earth through the closure [*taḫṣīf*] [of the cavities]. (3a) Coldness which occurs suddenly causes (3b) that which (coldness) occurring in degrees [*bi-l-tadrīǧ*] does not cause. (3c) Observe that in bodies and particulars of experiments in the art of medicine etc.

Nic. syr. 39.27-30, 39.32-40.3[65]:

ܬܘܒ ܕܝܢ ܗܘܝܢ ܝܬܝܪܐܝܬ/28 ܙ̈ܘܥܐ ܕܐܪܥܐ. ܟܕ ܗܘܐ ܐܩܠܝܦܣܝܣ ܕܣܗܪܐ.
ܟܕ ܠܐ ܥܕܟܝܠ ܠܓܡܪ ܛܠܩ/29 ܢܘܗܪܐ. ܟܕ ܓܝܪ ܢܘܗܪܗ ܕܣܗܪܐ ܠܐ ܡܬܬܚܬܐ
ܠܐܐܪ ‹ܐܪܥܐ› ܡܬܩܪܪܐ ܗܘܬ. ܘܟܕ ܡܬܚܒܫܝܢ/30 ܡܦܩܢܗ ܕܬܪ̈ܥܝܬܗ ܒܩܘܪܫܗ
ܕܐܪܥܐ ܡܬܚܒܫܝܢ ܠܓܘ ܡܢܗ//32 ܡܦܩܢܗ ܕܠܬ̈ܐ ܬܢܢܢܝܬܐ. ܩܕܡ ܡܢ ܫܘܪܝܗ
ܕܐܩܠܝܦܣܝܣ ܪ̈ܘܚܐ ܗܘܝܢ/1 ܟܕ ܗܘ ܓܝܪ ܕܒܦܠܓܗ ܕܠܠܝܐ ܥܬܝܕܐ ܕܬܗܘܐ
ܐܩܠܝܦܣܝܣ: ܪ̈ܘܚܐ/2 ܗܘܝܢ ܒܫܘܪܝܗ ܕܠܠܝܐ. ܗܘ ܕܝܢ ܕܒܨܦܪܐ ܥܬܝܕܐ ܕܬܗܘܐ
ܐܩܠܝܦܣܝܣ/3 ܒܦܠܓܗ ܕܠܠܝܐ.

[29 ‹ܐܪܥܐ› supplevi || 32 ܡܦܩܢܗ : ܡܦܩܗ cod.]

(1a) Furthermore, movements of the earth occur more when an eclipse of the moon occurs, when the light has not yet completely disappeared; for when the light of the moon does not reach the air, (2a) <the earth> becomes cold; (2ab) and when the openings of the earth are closed by the coldness, (2c) smoky exhalations are confined inside it [*l-gaw menāh*]. Before the beginning of the eclipse winds occur, since, when the eclipse is about to occur at mid-night, winds occur at the beginning of the night, whereas when the eclipse is about to occur in the morning, winds occur at mid-night.

l. 1. ܩܠܝܦܣܝܣ ܕܫܡܫܐ ܘܣܗܪܐ ("the eclipse of the sun and the moon"): BH here talks of the eclipse of the sun *and* the moon, whereas Nic., following Arist., talks only of the lunar eclipse and IS simply says "eclipses" without further specification. Ibn al-Biṭrīq talks of the "eclipse of the sun by the moon"[66] while Ḥunain (*Comp*. 224-229) and Bar Kepha, (*Hex*. Paris 241, 193r a1-11) mention the eclipses both of the sun and the moon as causes of earthquakes. The addition of the "sun" here in *But*. may be due to Bar Kepha (note also the spelling ܩܠܝܦܣܝܣ *But*. and Bar Kepha; ܐܩܠܝܦܣܝܣ Nic. syr.).

l. 5. "from the (human) body": cf. Arist. *Mete*. 366b 14-30.

[65] Cf. Arist. 367b 19-32; Alex. Aphr. 120.20-121.28; Olymp. arab. 138.13-17.

[66] Ibn al-Biṭrīq [Schoonheim] l. 802: كسوف الشمس بالقمر [والقمر cod. Istanbul].

II.iii.4.1-4: Persistence of earthquakes

But.: (1a) If someone says: (1b) "what [*haw mā d-*] arises from winds [normally] subsides suddenly [*menšel*], (1c) but [*w-*] the movement of the earth is often constant [*amīnā*] and continuous [*sbīsā*]", (2a) we say that this happens through [*b-yad*] the large quantity of the wind, (2b) the tortuosity [*ʿqalqlūṯā*] of the openings of the earth (2c) and the persistence of the cause which generates the exhalation or retains [it].

Nic. syr. 40.3-7[67]:

ܐܠܐ ܟܒܪ ܐܡܪ ܐܢܫ ܕܐܝܠܝܢ ܕܗܘܝܢ ܡܢ ܪܘܚܐ ܚܝܣܐܝܬ/4 {ܠܐ}ܢܬܢܝܚܢ. ܘܙܘܥܐ ܕܝܢ
ܕܐܪܥܐ ܙܒܢܐ ܣܓܝܐܢ ܡܬܡܣܟܢܐ ܘܐܡܝܢܐ ܗܘܐ ܥܕܡܐ/5 ܠܙܒܢܐ ܣܓܝܐܐ.
ܐܠܐ ܗܢܐ ܗܘܐ. ܡܛܠ ܣܘܓܐܐ ܕܪܘܚܐ ܘܡܛܠ/6 ܥܩܠܩܠܘܬܐ ܕܦܬܚܘܗ̈ܝ
ܕܐܪܥܐ. ܘܡܛܠ ܡܩܘܝܢܘܬܐ ܕܥܠܬܐ ܕܡܘܠܕܐ ܠܗ/7 ܠܠܚܝܬܐ ܐܘ ܐܚܝܕܐ ܠܗ.

But someone will perhaps say: (1b) "Those things which [*aylēn d-*] arise from winds subside quickly [*ḥīṣāʾīṯ*]; (1c) movement of the earth, however [*dēn*], is often continuous and constant, lasting [lit. until] a long time." (2a) This happens because of [*meṭṭul*] the large quantity of the wind, (2b) because of the tortuosity of the opening of the earth (2c) and because of the persistence of the cause which generates the exhalation or retains it.

l. 1. ܕܡܢܫܠܝ ("suddenly"): The substitution of this word here for Nic.'s "quickly" (*ḥīṣāʾīṯ*) suggests that BH may also have been looking at another passage which occurs at an earlier point in Nic. syr. (37.10-12) : ܕܠܚܝܬܐ ܬܢܢܢܝܬܐ ܐܚܘܝܬܗ/11 ܥܠܬܐ ܕܙܘܥܗ ܕܐܪܥܐ. ܡܬܚܘܝܐ ܗܝ ܕܡܢܫܠܝ ܗܘܐ ܘܚܝܣܐܝܬ{ ܢܬܢܝܚ.} ܐܝܠܝܢ/12 ܓܝܪ ܕܗܘܝܢ ܡܢ ܪܘܚܐ ܡܢܫܠܝ ܗܘܝܢ. ܘܚܝܣܐܝܬ ܢܬܢܝܚܢ. ("That the cause of the movement of the earth is the smoky exhalation is proven by the fact that it occurs suddenly and {subsides} quickly; for those things which arise from winds occur suddenly and subside quickly").

II.iii.4.4-7: Benefits of earthquakes

The passage corresponds to the last sentence in the chapter on earthquakes in the *Šifāʾ*. We see that the mention of God by IS prompted BH to make his own, Christian addition. BH had made a similar addition at the end of his discussion of earthquakes in *Rad.*[68] The contents of both the passage here in *But.* and in *Rad.* point to Bar Kepha as a possible source of inspiration.

But. [underline: agreement with *Šifāʾ*; italics: with Bar Kepha]:
(1) The wise [*ḥakkīmē*] acknowledge two benefits of the movement of the earth: (2a) one is [its function of] opening up [*puttāḥā*] holes in the earth so as to cause springs to gush out [*la-mgāḥū*]; (3a) the other is [its function of] frightening [*surrāḏā*] the hearts of *wicked* peoples [*ʿammē ʿawwālē*]

[67] Cf. Arist. 367b 32-378a 14; Alex. Aphr. 121.29-122.14; Olymp. arab. 138.18-139.2; *Cand.* 128.1-3.

before the anger of God and His fury which *disturbs* [*mdallḥā*] and terrifies [*mdaḥḥlā*] them. (3b) "*He looks on the earth and it quakes, and He rebukes the mountains and they smoke*" [Ps. 104.32], (3c) that is to say, on account of the wickedness of the inhabitants.

Šifā' 19.14-15:

ومن منافع الزلازل تفتيح مسام الأرض للعيون، وإشعار قلوب فسقة العامة رعب الله تعالى.

(1) Benefits of earthquakes (2a) include [its function of] opening up [*taftīḥ*] holes in the earth for springs; (3a) and [its function of] teaching [*iš'ār*] the hearts of the wicked among the people the fear [*ru'b*] of God.

Bar Kepha, *Hex.*, ms. Paris 241, 192r b 5-7, 192r b 18-192v a 12 [tr. Schlimme 628f.]:

ܐܢܫܝܢ ܕܝܢ ܡܢ/6 ܡܠܦܢܐ ܕܥܕܬܐ ܟܕ ܠܐ ܡܩܒܠܝܢ/7 ܠܗܠܝܢ ܐܡܪܝܢ . . . 18 ܡܟܝܠ
ܠܘ ܡܢ ܟܠ ܙܒܢ/19 ܪ̈ܘܚܐ ܘܡ̈ܝܐ ܘܢܘܪܐ ܘܠܗ̈ܓܐ/20 ܐܝܬܝܗܘܢ ܥܠ̈ܬܐ
ܕܙܘܥܐ/21 ܘܫܘܓܫܐ ܘܢܘܕ̈ܐ ܕܐܪܥܐ. ܐܠܐ/22 ܐܠܗܐ ܒܪܘܝܐ. ܗܘ ܕܨܒܐ/23
ܕܢܕܚܠ. ܗܘ ܕܡܦܪܥ/24 ܠܪ̈ܫܝܥܐ ܘܡܟܐܣ ܠܥܘ̈ܠܐ. /25 ܐܝܟ ܗܘ ܕܡܙܡܘܪܐ ܕܘܝܕ/
26 ܕܐܡܪ. ܕܚܐܪ ܒܐܪܥܐ ܘܙܝܥܐ. /27 ܘܡܟܐ ܒܛܘ̈ܪܐ ܘܬܢܝܢ. ܘܐܬܕܠܚܬ/28
ܘܙܥܬ ܐܪܥܐ. ܘܐܙܝܥܬܗ ܠܐܪܥܐ/29 ܘܦܬܚܬܗ. ܚܢܢ ܕܝܢ/30 ܐܡܪܝܢ. ܕܐܢ
ܥܠ̈ܬܐ ܟܝܢܝ̈ܬܐ/1 ܐܝܬܝܗܝܢ ܕܡܥܒ̈ܕܢ ܙܘܥܐ/2 ܘܫܘܓܫܐ ܘܢܘܕ̈ܐ ܕܐܪܥܐ. /3
ܐܢ ܥܠ̈ܬܐ ܐܚܪ̈ܢܝܬܐ. ܟܠܗܝܢ/4 ܒܡܕܒܪܢܘܬܐ ܐܠܗܝܬܐ ܩܝ̈ܡܢ. /5 ܘܐܦ ܒܨܒܝܢܗ
ܕܒܪܘܝܐ/6 ܡܫܬܡܠ̈ܝܢ. ܐܢ ܡܛܠ ܕܘܚܠܐ/7 ܕܚܛ̈ܝܐ. ܘܐܢ ܡܛܠ/8
ܡܣܬܟܣܢܘܬܐ ܕܪ̈ܫܝܥܐ. /9 ܘܐܢ ܡܛܠ ܙܘܗܪܐ ܕܐܚܪ̈ܢܐ. /10 ܘܐܢ ܡܛܠ
ܥܠ̈ܬܐ ܐܚܪ̈ܢܝܬܐ. /11 ܕܗܘ ܝܕܥ ܠܗܝܢ. ܐܝܟ ܕܝܢ̈ܘܗܝ/12 ܠܐ ܡܬܒܨܝ̈ܢܐ .

Some doctors [*malpānē*] of the Church, not accepting these [theories] say: ... Therefore, not always are winds, water, fire and exhalations the cause of tremors, movements and quakings of the earth, but God the Creator, who wishes to terrify [*ndaḥḥel*], who punishes the impious and the wicked ['*awwālē*], as [stated by] the Psalmist David who says: "He looks on the earth and it quakes, and He rebukes the mountains and they smoke" [Ps. 104.32], and "the earth was disturbed [*etdallḥat*] and it moved" [Ps. 18.8], and "Thou hast caused the earth to move and hast opened [*ptaḥtāh*] it" [Ps. 60.4]. We say: whether they be natural causes which cause tremors, movements and quakings of the earth, [or] other causes, they all arise with divine providence and are brought about with the assent of the Creator, be it for terrifying [*duḥḥālā*] of sinners, or for punishment of the impious, or for warning [*zuhhārā*] to others, or for other causes which He [alone] knows according to His inscrutable judgements."

BH *Rad.* [Istanbul] 27.16-19 [corr. Bodl. Or. 467, 13r 17-19]:

ܐܠܐ. ܘܠܗܠܝܢ ܗܟܝܠ ܥܠ̈ܬܐ ܟܝܢܝ̈ܬܐ ܟܠܗܝܢ. ܥܠܬܐ ܠܗܘ ܡܣܓܕܐ ܥܠܬ ܟܠ
ܥܠ̈ܝܠܢ ܙܕܩ ܠܡܫܬܘܕܥܘ. ܗܘ ܕܚܐܪ ܒܐܪܥܐ ܘܙܝܥܐ. ܘܡܟܐ ܒܛܘ̈ܪܐ
ܘܬܢܝܢ.

Ray: For all these natural causes, one should acknowledge as the cause the One who is worshipped, the cause of all things caused. For "He looks on the earth and it quakes, and He rebukes the mountains and they smoke".

[68] The rest of the discussion of earthquakes in *Rad.* (ed. Istanbul 27.3-15) closely follows the account in *Cand.* This last part, however, has no counterpart there.

l. 4. ܘܠܙܘܥܐ ܕܐܪܥܐ ܠܬܪܝܢ ܡܘܬܪܢܝܢ ܡܘܕܝܢ ܚܟܝ̈ܡܐ ("the wise acknowledge two benefits of the movement of the earth"): cf. arab. *ʿarafa* + *li-*: concede, acknowledge; cf. Min. V.iii.3.9-10 below.

l. 5. ܦܘܬܚܝܢ ("opening up"): This corresponds to IS's تفتيح, but cf. also ܦܬܚܬܝܗ in the quotation from Ps 60.4 in Bar Kepha.

l. 5. ܠܡܓܝܚܘ ܥܝܢ̈ܬܐ ("so as to cause springs to gush out"): cf. Min. II.ii.2.3-4 above (*ʿaynāṯā ... mgīḥā*).

l. 5. ܣܘܪܕ ("frightening"): This word itself is not in the passage of Bar Kepha quoted above, but we find there *duḥḥālā* ("terrifying", 192v a6) and *zuhhārā* ("warning", a9), both, like *surrāḏā*, verbal nouns of the form "CuCCāCā".

l. 6. ܕܡܕܠܚܐ ܠܗܘܢ ܘܡܕܗܠܐ ("which disturbs them and terrifies [them]"): Note alliteration of *mḏallḥā*... *w-mḏaḥḥlā*; and cf. Bar Kepha 192r 23 *nḏaḥḥel* and 27 *eṯdallḥaṯ*

BOOK OF MINERALOGY, CHAPTER THREE

ON MINERALS

The chapter corresponds to the fifth chapter ("on the formation of minerals") of the treatise on minerals in the *K. al-Šifāʾ*, and is based largely on the *Šifāʾ*. The correspondences are:

Section i.	*Šifāʾ*.20.4-21.9
Section ii.	*Šifāʾ* 21.10-22.11
Section iii.	*Šifāʾ* 22.11-23.11

There are, however, a number of passages which depend on other sources. In most of these the immediate source is the mineralogical section of Abū al-Barakāt's *K. al-muᶜtabar* (II.227-231).[1] For III.i.5.6-9, BH names the "alchemists" as his source. The brevity of the passage makes it difficult to determine who exactly these "alchemists" are, who are probably also the source of the names of the three 'stones' at III.i.1.3.

The use of Nic. Syr. can only be confirmed for III.i.2.3-4. - It is worth noting that the Aristotelian theory of the formation of minerals from the earthy and watery "exhalations" (Arist. *Mete*. III.vi.(b) 378a 15ff.) is not mentioned in this chapter or the corresponding part of the *Šifāʾ*, although it is reflected in an earlier discussion on the "uses of mountains" (see *But.* Min. I.ii.6. above).[2] Although Nic. syr. follows Arist. in discussing the formation of minerals from "exhalations" at 55.23-56.6, BH, following IS, makes no use of this in this chapter or in the corresponding parts of *Cand*. (86.1-89.10).

---

[1] See III.i.1.2-4; III.i.5.1-3; III.ii.1.; III.ii.3.6-8; III.iii.3.

[2] The formation of minerals from the two exhalations is mentioned in other works of IS: *Naǧāt* [Dānishpazhūh] 314.7ff., 317.6ff.; *Dāniš-nāma* [Mishkāt] III.73.9f. (tr. Achena-Massé II.51); *ᶜUyūn* [Ülken] 29.19-30.4, [Cairo (1908)] 24.5-12 (tr. Wiedemann [1907] 79).

Min. III. Section i.: *On the Categories of Minerals*[3]

After listing the four categories of minerals (*Šifāʾ* 20.4-5: corr. *But.* III.i.1.1-2), IS attempts a classification of minerals according to what we might call their "physical" characteristics (*Šifāʾ* 20.5-9) before going on to discuss the four categories, along with a fifth not named in the initial list (vitriols), in terms of the elements from which they are composed (*Šifāʾ* 20.9-21.9). - We could consider the latter an attempt at a "chemical" classification.

BH alters this order and deals with the "chemistry" of each of the five categories before going on to the classification according to physical characteristics in III.i.5. BH also alters the order in which the four categories are discussed at *Šifāʾ* 20.9-21.4, interchanging the positions of "metals" and "stones", and of "salts" and "sulphurs". The resulting order in *But.* (i.e. stones, metals, sulphurs, salts) is that of the initial list at *Šifāʾ* 20.5 and *But.* III.i.1.1-2.

III.i.1.1-2	a. *Šifāʾ* 20.4-5	
III.i.1.4-6	d. *Šifāʾ* 20.15-18	Stones
III.i.2.	c. *Šifāʾ* 20.9-14	Metals
III.i.3.1-3	f. *Šifāʾ* 21.3-4	Sulphurs
III.i.3.3-6	e. *Šifāʾ* 21.1-2	Salts
III.i.4.	g. *Šifāʾ* 21.5-9	Vitriols
III.i.5.1-6	b. *Šifāʾ* 20.5-9	"Physical" classification

Attempts to classify various substances (including organic matter, as well as minerals, in the case of Arist.) according to their "chemical" composition and "physical" properties goes back in the Peripatetic tradition, in the first place, to the fourth book of Aristotle's *Meteorologica*, where, in "chemical" terms, most substances are explained as being composed of the elements "earth" and "water", while "air", too, occasionally plays a part as in the case of "oil" (see, e.g., 383b 24f., 388a 31). The "physical" classification comes to the fore in *Mete*. IV.viii-ix (384b 24ff.), where the eighteen pairs of opposing qualities are discussed.

---

[3] For an overview of the different schemes for classification of minerals found in Islamic literature, see Ullmann (1972) 140-144.

In his "chemistry", in talking of stones and metals as being formed of earth and water, IS follows in the tradition of Aristotle, but particularly in the recognition of such substances as sulphurs, salts and vitriols as categories distinct from stones[4] and in his discussion of the manner in which they are formed, IS incorporates newer material which he evidently owes to the tradition of alchemy.[5] The debt to the alchemical tradition is especially evident in the discussion of the formation of metals from mercury and sulphur (*Šifā᾽* 21.10-22.11), which BH uses in Section ii. below.

In his "physical" classification, too, IS was probably inspired, in part at least, by Aristotle. Among the criteria he uses at *Šifā᾽* 20.5ff., malleability and solubility (also fusibility) feature among the eighteen qualities listed by Aristotle.[6] IS's classification of minerals into those which are strong and weak "in substance", however, is new.

tit.: ܡܛܠ ܓܢܣܐ ܡܬܐܠܝܩܝܐ ("on the Categories of Minerals"): lit. "on the mineral (μεταλλικά) genera".

III.i.1.1-2: Classification of minerals

But.: There are four categories [lit. genera] of μεταλλικοί, i.e. mineral bodies [*gušmē methaprānē*]: stones, fusible substances, sulphurs, salts.
Sifā᾽ 20.4-5:

إن الأجسام المعدنية تكاد أن تكون أقسامها أربعة: الأحجار، والذائبات، والكباريت، والأملاح.

The divisions of mineral bodies [*ağsām al-ma῾dinīya*] are, largely speakingh, four: stones, fusibles, sulphurs and salts.

l. 1. ܓܢ̈ܣܘܗܝ ("categories", lit. genera, < gr. γένος): أقسام IS. - Although IS has *qism/aqsām*, in the correspnding passage here, he also uses *ğins/ağnās*, in the same sense (e.g. *Šifā᾽* 21.1).

l. 1. ܡܬܐܠܝܩܘ ܐܘܟܝܬ ܓܘܫܡ̈ܐ ܡܬܗܦܪ̈ܢܐ ("μεταλλικοί, or mineral bodies"): cf. Min. I.i.1.5-6 above.

l. 2. ܐܘ̈ܣܝܐ ܡܬܦܫܪ̈ܢܝܬܐ ("fusible substances"): الذائبات IS. - i.e. metals. - cf. III.i.4.2 *gušmē metpaššrānē* (corr. IS *al-ağsād al-dā᾽iba*); III.i.4.3-4: id. (corr. IS *al-ağsād*); III.ii.21 *methaprānē d-metpaššrān* (corr. IS *dā᾽ibāt*); III.ii.2.7 *mehṭalīqō metpaššrānē*.

---

[4] On the history of the recognition and classification of "salts" (in the modern, chemical sense of the word), see the dissertation by Dittberner (1971).

[5] See on III.i.4.2-6 below.

[6] "Shatterable" which is added by BH at III.i.5.2 below would also correspond to Aristotle's θραυστός (385a 14).

III.i.1.2-4: Stones

The five qualities predicated of the "stones" are taken from Abū al-Barakāt.

> *But.*: Stones like vitriol, πυρίτης and μαγνησία are hard, are shatterable and burn, but are neither fusible nor malleable
> *Muᶜtabar* II. 229.21:
> وبالجملة فان المعدنيات منها احجار صلبة تتفتت وتحترق ولا يذوب ولا تنطرق
> Generally speaking, among minerals there are the hard stones, which are shatterable and combustible, but are not fusible or malleable.

Neither Abū al-Barakāt here nor IṢ in the passage of *Šifā᾽* used by BH in lines 4-6 below give any concrete examples of "stones".[7] The choice of the three 'stones' named here suggests that the source is to be found among alchemical literature or literature related to it (cf. III.i.5.6-9 below)[8]. Of the three items, "pyrite" (corr. Arab. *marqašīṯā*) and "magnesia" appear as the first two items in the list of thirteen 'stones' in Abū Bakr al-Rāzī's *K. al-asrār*, which is then followed by a list of the seven "vitriols", while in his earlier work, *al-Madḫal al-taᶜlīmī*, we find the vitriols listed before the stones. The three items appear, in the same order as here, as the fourth, firth and sixth items in the list of 29 'drugs' (*ᶜaqāqīr*) used by alchemists in al-Khwārizmī's *Mafātīḥ al-ᶜulūm.*[9]

---

[7] An idea of the kind of substances IS might have had in mind may be gathered from the section of the *Dāniš-nāma*, Ṭabīᶜīyāt, on minerals (ed. Mishkāt [1951/2] 73-78), where the qualities of infusibility (strictly, the difficulty of fusion, *dušḫwār gudāzand*) and non-malleability (*zaḫm na-paḏīrad*) are applied, in the first place, to substances such as corundum (*yāqūt*) and rock crystal (*billaur*), while infusibility is also applied a little later to substances such as iron filings (*sūniš-i āhan*), 'marcassite' (*mārqašīšā*) and talc (*ṭalq*). Both these types would probably be considered 'stony' in IS's scheme.

[8] At *Cand.* 84.6-85.3, BH gives a completely different list of stones, apparently based on Dioscorides' *De materia medica* (cf. Takahashi [2002b] 240).

[9] Rāzī, *Asrār* [Dānishpazhūh] 2.14-17 (and 3.19-21, 4.20-5 ult.; tr. Ruska 84, 85f.; Garbers-Weyer [1980] 8. - id., *Madḫal*, tr. Stapleton et al. (1927) 348f. - Khwārizmī, *Mafātīḥ* [van Vloten] 260.3, 260.7, 261.1. (tr. Wiedemann [1911] 709f.; Stapleton et al. [1927] 364. - "*Maqšīṯā*" and "magnesia" are also found, corresponding to Aries and Taurus, at the beginning of the list of twelve minerals corresponding to the signs of the Zodiac at Berthelot-Duval [1893], Syr. text. 6.14f. (cf. Ruska [1923] 12-14). - The two also appear, though in the reverse order and at the end, in a list of stones in the *K. al-ḫamsīn* attributed to Jābir b. Ḥayyān (Kraus [1942/3] II.22 n.7). - A list of stones which begins with 'arsenics' (*zarānīḫ*), 'marcassite' and 'magnesia' is found at Ikhwān al-Ṣafā᾽ [Bustānī] II.121.15ff. (vitriols are listed in the preceding paragraph

l. 3. ܙܐܓܐ ("vitriol"): "Vitriols" are treated at greater length in III.i.4. below, where they are said, in contradiction of what is said here, to be "composed of saltiness, sulphureity and stones". The double appearance and the contradiction is no doubt due to the use of different sources and may be regarded as a sign of incomplete editing by the author. - The possibility that "vitriol" here represents a misreading of *zuǧāǧ* ("glass") as *zāǧ* might also be considered.[10]

l. 3. ܦܘܪܝܛܝܣ ܘܡܓܢܝܣܝܐ ("πυρίτης and μαγνησία"): Syr. *pūrīṭīs* corresponds to arab. *marqašīṯā*, for which Syriac also has the word *maqšīṯā*,[11] while *maḡnīsiya* is arab. *maġnīsiyā*.[12] - *Marqašīṯā* is not only our "marcassite" (FeS_2), but the sulphides of various heavy metals, while *maġnīsiyā*, too, seems to cover various types of manganese ores, as well as our "magnesia" (MgO).[13] - In *But*. *pūrīṭīs* is also used in Gen. et corr. III.i.8. (ms. F 46r b12), where the word corresponds to *Šifā'* De act. et pass. 232.17 *mārqašīṯā*. - The word *pūrīṭīs* is glossed, wrongly, as "alum" (*al-šabb*) in mss. FM.

l. 3. ܡܬܬܪܥܥܢ ("are shatterable"): cf. III.i.5.2 below.

l. 3. ܘܝܩܕܢ ("and burn"): Aristotle, who included combustibility among his eighteen pairs of qualities (*Mete*. 385a 18: καυστόν ἄκαυστον), defined stones as incombustible (387a 18f.). Theophrastus, however, accepts that some stones burn and, in fact, makes combustibility and, more generally, the effects of fire on stones an important criterion in his attempt at classification of stones (Theoph. *De lap*. §9-22 [Wimmer] 342.7-343.47).[14]

[II.121.7ff.] along with salts, alums and boraxes).

[10] The mistake is common enough in Arabic manuscripts (so, for example, "Glas" in Dieterici's translation of the *Rasā'il Iḫwān al-ṣafā'* needs in many cases be corrected to "Vitriol", e.g. Dieterici [1876] 116.9f., corr. ed. Bustānī II.108.9f.; 130.17, corr. 121.7). - "Glass" appears as the last item in the list of thirteen stones in Rāzī's *K. al-asrār* (see preceding note).

[11] This form without *-r-* is no doubt due to a false derivation of the word (which appears, in fact, to be of Persian origin, Vullers [1855-64] 1167a) from the Syr. root √*qšy* (cf. PS 1666, s.v. ܟܐܦܐ ܡܩܫܝܬܐ, "lapis induratus"). For the identification of *pūrīṭīs* with *maqšīṯā*/*marqašīṯā* in Syriac, see Berthelot-Duval (1893) Syr. text, 3.10; 4.9 (tr. p. 7, 9); Bar Bahlūl [Duval] 970.15f. s.v. ܦܘܪܝܛܝܣ ܦܘܪܝܛܝܣ; 883.3 s.v. ܦܘܪܝܛܝܣ (*sic*; cf. 970.3 s.v. ܦܘܪܝܛܝܣ: حجر النار ܟܐܦܐ ܕܢܘܪܐ); PS 1714, 1946, 3073.

[12] For the use of the word in Syriac, see PS 2007; PS Suppl. 184 and 154 s.v. ܟܐܦܐ ܡܓܢܝܣܝܐ (All the citations in PS Suppl. are from the alchemical text edited by Berthelot-Duval).

[13] On the identity of the substances designated by these two terms in Arabic, see Wiedemann (1911) 97-99; Ruska (1937) 43; Siggel (1950) 88; Garbers-Weyer (1980) 112; on *marqašīṯā*, also Schönfeld (1976) 176f.

[14] Cf. Steinmetz (1964) 87-91. This passage of Theoph. is represented at Nic. syr. 57.19-26 (Takahashi [2002a] 208 [text], 216 [tr.]). - Cf. *Rasā'il Iḫwān al-ṣafā'*, where fire is said to be "like a judge of mineral substances" (ed. Bustānī II.109.1: واعلم أخي أن النار هي كالقاضي بين الجواهر المعدنية; cf. tr. Dieterici [1876] 116).

IS, on the other hand, while making fusibility an important criterion, makes no mention of combustibility in his classification of minerals in the *Šifāʾ*.[15] - By "burning" with reference to "stones", we should probably understand not only carbonisation but also other forms of chemical change under the effect of fire, such as calcination.[16]

III.i.1.4-6: Stones (contd.)

Cf. *But.* Min. I.i.2 above (on formation of stones).

But.: ... (1a) since they are composed from wateriness [*mayyanūṯā*] (1b) which has been solidified [*qeṭrat*] not by coldness alone (1c) but through dryness which has changed the wateriness into earthiness. (2a) For this reason, the majority of them are not fusible (2b) except perhaps [*ellā kḇar*] by [certain] physical techniques.

Šifāʾ 20.15-18:

وأما الحجريات من الجواهر المعدنية الجبلية، فمادتها أيضا مائية، ولكن ليس جمودها بالبرد وحده؛ بل جمودها باليبس المحيل للمائية إلى الأرضية. وليس فيها رطوبة حية دهنية، فلذلك لا ينطرق. ولأجل أن أكثر انعقادها باليبس، فلذلك لا يذوب أكثرها إلا أن يُحتال عليه بالحيل الطبيعية المذيبة.

(1a) The material of the stony (substances) among the natural mineral substances is also watery, (1b) but their congelation [*ǧumūd*] is due not only to coldness; (1c) rather [*bal*], their congelation is by dryness which transforms wateriness into earthiness.[17] They contain no quick, oily moisture, so that they are not malleable. (1) In most cases their solidification is by dryness, (2a) and for this reason most of them cannot be fused (2b) unless [*illā an*] they are subjected to some physical techniques which facilitate fusion.[18]

l. 6. ܦܘܪܣܐ ܟܝܢܝܐ ("physical techniques"): الحيل الطبيعية IS. - The word *pursā* (< gr. πόρος) is used for IS's *ḥīla*, pl. *ḥiyal*, in this chapter, besides here, at III.i.2.1 (*w-āp̄en b-p̄ursā*: corr. IS *wa-lau bi-l-ḥīla*) and at III.iii.1.1 (*māray*

---

[15] IS does make combustibility a criterion in his *ʿUyūn al-ḥikma* [Ülken] 30.1-3, [Cairo (1908)] 24.8-12.

[16] See, for example, Theoph. *De lap.* [Wimmer] 342.14f., where marble is said to burn and become ash (κατακαίεσθαι καὶ κονίαν ἐξ αὐτοῦ γίνεσθαι).

[17] The words *māʾīya* and *arḍīya* here could stand for *mādda māʾīya/arḍīya* (so Dittberner [1971] 129 for *milḥīya wa-kibrītīya wa-ḥaǧarīya* at *Šifāʾ* 21.5). In translating them as abstract nouns, I follow the interpretation of BH (and Holmyard-Mandeville). They are evidently synonymous with *ǧauhar māʾī/arḍī* of *Šifāʾ* 20.12, rendered as *ūsiya mayyānāytā/arʿānāytā* by BH (III.i.2.5).

[18] For the notion of solidification either by coldness or heat/dryness, cf. Arist. *Mete.* 382b 31-33: πήγνυται δὲ ὅσα πήγνυται ἢ ὕδατος ὄντα ἢ γῆς καὶ ὕδατος, καὶ ταῦτα ἢ θερμῷ ξηρῷ ἢ ψυχρῷ.; 383a 13f.; 385a 22-24. - For the difficulty of fusing substances solidified by dry heat, *Mete.* 383b 9f.; 388b 24ff. - Cf. IS, *Dāniš-nāma* [Mishkāt] III.77.3-78.3.

pursē: corr. IS *aṣḥāb al-ḥiyal*). - An idea as to what IS may mean by "physical techniques" can be gathered from *Dāniš-nāma* [Mishkāt] III.76.6-77.4, where it is said that substances such as 'marcassite' are made fusible through mixture with 'arsenic' (*zarnīḫ*) or sulphur.[19] - Cf. also BH *Serm. sap.* [Janssens] 78.4 ܐܟܢܐ ܕܚܙܝܢ ܠܟܝܡܝܩܐ ܕܫܪܝܢ ܠܓܘܫܡܐ ܡܩܫܝܐ ("... just so we see alchemists dissolve hard bodies").

III.i.2.1-3: Metals

But.: (1) All malleable bodies are fusible albeit [*w-āpen*] with [special] technique [*b-pursā*] (2a) and the majority of what are not malleable are not fusible, (2b) but are merely softened [and that only] with difficulty [*ʿatlāʾīt*]

Šifāʾ 20.10-12:

وجميع المنطرقات ذائبة ولو بالحيلة، وأكثر ما لا ينطرق ولا يذوب بالإذابة الرسمية وإنما يلين بعسر.

[11 بالحيلة Holmyard: بالجبلة Cair.; بالجملة M]

(1) All malleable (bodies) are fusible albeit [*wa-lau*] with [special] technique [*bi-l-ḥīlati*]. (2a) Most non-malleable (bodies) cannot be fused in the usual way, and are only softened with difficulty [*bi-ʿusrin*].

l. 1. ܒܦܘܪܣܐ ("with [special] technique"): بالحيلة IS. - Cf. III.i.1.6 above.

III.i.2.3-4: Metals (contd.): ἀδάμας

The 'theory', which is otherwise a paraphrase of the *Šifāʾ*, is interrupted in the middle by an insertion of a passage taken from Nic. syr. (The passage of *But.* here, along with another passage in *Cand*, helps us to restore the mutilated passage of Nic. syr.).

But.: ... (1a) except [*sṭar men*] ἀδάμας, (1b) which is neither malleable nor are softened, (1c) but when beaten on an anvil penetrates inside it (1d) and is not affected by iron.

Nic. syr. 60.2, 5-7[20]:

{ܡܬ}ܦܫܪܝܢ ܕܝܢ ܟܠܗܘܢ ܗܠܝܢ ܡܢ ܢܘܪܐ .// 2

{..........................} ܘܟܠܗܘܢ ܡܬܦܫܛܝܢ. ܣܛܪ ܡܢ ܐܕܡ[ܐܣ]/ 5

{....................ܕ}ܐ: ܐܠܐ ܟܕ ܡܬܢܩܫܐ ܥܠ ܣܕܢܐ ܥܒܪܐ ܠܓܘ[ܗ]/ 6

{...................} ܡܢ ܦܪܙܠܐ . 7

[6 ܕܐ- . . . : fort. ܠܐ ܓܝܪ ܡܬܦܫܛܐ ܘܐܦܠܐ ܡܬܪܟܟܐ ܗܘܐ || 7 init.: fort. ܘܠܐ ܡܬܓܠܦ ܘܐܦܠܐ ܡܬܦܠܚ]

All these [i.e. metals] are melted by fire {...} and all of them are malleable, (1a) except ἀδάμας. (1b) {This is not malleable and cannot be softened [?]} but when it is beaten on an anvil penetrates inside it {and cannot be carved or worked [?]} by iron.

---

[19] Cf. Theoph. *De lap.* §9 [Wimmer] 342.11f.: ὡσαύτως δὲ καὶ οἱ πυριμάχοι καὶ οἱ μυλίαι ῥέουσιν οἷς ἐπιτιθέασιν οἱ καίοντες.

[20] Takahashi (2002a) 211, 221; cf. *Cand.* 86.8-87.1.

l. 3. "ἀδάμας" (ܐܕܐܡܐܣܣ F; ܐܕܐܡܐܐܣ M; ܐܕܡܐܣ V): - In Nic. syr. "adamas" is also counted among the metals at 59.17 and 60.12, although at 59.19 it is referred to as the "stone of adamas" (ܟܐܦܐ ܕܐܕܡܘܣ). - It is difficult to decide how BH is likely to have written this word. Three variants similar to those in *But*. are found at *Cand.* [Bakoš] 86.8: ܐܕܡܐܣ B et ed. Çiçek (68.32); ܐܕܐܡܐܣ V; ܐܕܡܘܣ P. The ms. of Nic. syr. has ܐܕܡܘܣ (59.19), ܐܕܡܣ (59.17, 28) and ܐܕܡܐܣ (60.12). Cf. PS 38f. - The gender of ܐܕܡܐܣ, according to dictionaries (PS, Brockelmann) is masculine (as is gr. ἀδάμας). BH treats it as feminine here, as does Nic. syr. in the corresponding passage and elsewhere. At *Cand*. 86.8, it is in apposition to the fem. *kēpā* (ܟܐܦܐ ܐܕܡܐܣ), and the related pronoun and verbs are consequently also feminine.

l. 4. "when beaten on an anvil ...": cf. Strato, Frag. 56 Wehrli [Wehrli, 20.3]: [τὸν ἀδάμαντα] τυπτόμενον δὲ εἰς τοὺς ἄκμονας καὶ τὰς σφύρας ὅλον ἐνδύεσθαι. - Cf. further RAC III.955-963 s.v. "Diamant" [Hermann-Stumpf].

l. 4. "is not affected by iron": cf. Nic. syr. 57.17-19: "There are those [stones] which are easily sawn apart and carved and worked with a τορνός; there are those which are hardly affected in this way; and there are those which iron does not touch at all" (< Theoph. *De lap.* §5, 341.34-36). - Nic. syr. 59.14-15: "Why some stones cannot be carved with iron nor <worked> with a τορνός, but can be carved and worked with other stones, is not <known>" (cf. Theoph. *De lap*. §41-44, esp. §41 [Wimmer] 346.16f).

III.i.2.4-8: Metals (contd.)

But.: (1a) The material of malleable (bodies) is a watery substance [*ūsiya mayyānāytā*] (1b) which has been mixed [pf.] strongly with an earthy substance (1c) and cannot be separated from it. (2a) The wateriness [*mayyānūṯā*] solidifies [*qāṭrā*] with coldness, (2b) but there remains [*šārkā*] out of it a fluid [lit. alive] part (2c) which has not yet been solidified because of its oiliness (2d) and for this reason (this) body is malleable.

*Šifā*ʾ 20.12-14:

ومادة المنطرقات جوهر مائى يخالط جوهرا أرضيا مخالطة شديدة لا يبرأ منه، ويجمد الجوهر المائى منه بالبرد بعد فعل الحر فيه وإنضاجه، ويكون فى جملة ما هو حى بعد لم يجمد لدهنيته، ولذلك ينطرق.

[13 الحر فيه : الحارة B || جملة ما : جملتها ما Ṭ; جملته مما M]

(1a) The material of malleable (bodies) is a watery substance [*ǧauhar māʾī*] (1b) which is mixed [impf.] strongly with an earthy substance (1c) so that it cannot be separated from it. (2a) The watery substance [*al-ǧauhar al-māʾī*] is congealed [*yaǧmudu*] by coldness after heat has acted upon it and matured it. (2b) There is in the whole [a part] which is still fluid [lit. alive] (2c) and has not congealed on account of its oiliness, (2d) and for this reason it is malleable.

l. 6-7. ܡܫܬܪܟܐ ܡܢܗ ܡܢܬܐ ("there remains out of it a part"): ويكون فى جملة ما هو ... بعد IS. - Holmyard-Mandeville (p. 34) interpret IS differently: "Included in the group [of malleable bodies], however, are some ...". BH's interpretation is supported by *Šifā*ʾ 20.16f. (in the passage quoted under III.i.4-6 above),

where the non-malleability of 'stones' is attributed to the absence of the "fluid, oily moisture" in them.[21]

l. 7. ܚܝܬܐ ("fluid"): حى IS. - lit. "alive", or "quick" as in "quicksilver" (see Holmyard-Mandeville 34 n.11). Similarly at *Šifā'* 20.16 in the passage quoted under III.i.1.4-6 above, though not in the corresponding passage of *But*.

III.i.3.1-3: Sulphurs

But.: (1) Sulphurs [*kebryāṯā*] are composed from wateriness which has undergone a vigorous leavening [*ḥīṣā'īṯ ḥemʿaṯ*] with earthiness and airiness; (2a) they have acquired much oiliness through leavening [*maḥmʿānūṯā*] by [lit. of] heat, (2b) and have then solidified due to coldness.

Šifā' 21.3-4:

وأما الكباريت فإنها قد عرض لمائيتها أن تخمرت بالأرضية والهوائية تخمرا شديدا بتخمير الحرارة حتى صارت دهنية، ثم انعقدت بالبرد.

(1a) In the case of sulphurs [*kabārīt*], (1b) their wateriness has happened to be leavened strongly with earthiness and airiness (2a) by the leavening action [*taḫmīr*] of heat (2a) so that it has become oily; (2b) it has then been solidified by coldness.

l. 2. ܡܚܡܥܢܘܬܐ ("leavening"): ܡܚܠܡܥܢܘܬܐ F: تخمير IS. - Possibly a coinage by BH, since the word is unknown to the lexica (< √ḥmʿ, more likely from the commoner Aph. *aḥmaʿ* than Pa. *ḥammaʿ*).

III.i.3.3-6: Salts

But.: (1a) Alum [*ṣrāpā*] and sal ammoniac [*armehnīqōn*] <are salts, (1b) but the fieriness of sal ammoniac> is greater than [its] earthiness, (1c) and for this reason it is sublimated [*mestallaq*] in its entirety [*b-ḵullāyūṯeh*]. (2) Their material is water, in which hot smoke, very fine and fiery, is mixed, and which dryness solidifies.

Šifā' 21.1-2:

وأما الشب والنوشادر فمن جنس الأملاح، إلا أن نارية النوشادر أكثر من أرضية، فلذلك يتصعد بكليته، فهو ماء خالطه دخان حار لطيف جدا كثير النارية، وانعقد باليبس.

[2 فهو ماء: فهما B marg.; cf. Bahmanyār 719.9 فهو كأنه ماء/فكأنه ماء]

(1a) Alum [*šabb*] and sal-ammoniac [*nūšādir*] belong to the category of salts, (1b) except that the fieriness of sal-ammoniac is greater than the earthiness, (1c) so that for this reason it is sublimated [*yataṣaʿʿadu*] in its entirety. (2) It is water which has been mixed with hot, very fine smoke containing much fieriness, and has been solidified by dryness.

---

[21] Cf. also *Dāniš-nāma* [Mishkāt] III.76.5f.: وهر گروهی که اندر وی روغن بود تمام نفسرد پس زخم پذیرد ("Every category [of minerals] which contains oil does not become solidified completely, so that they are malleable").

l. 3. ܐܪܡܗܢܝܩܘܢ ("sal ammoniac"): نوشادر IS. - The identification of *nūšādir* with the "salt of Ammon" (τὸ ἀμμωνιακόν) has been attributed by Ruska to the Syriac physicians and translators.[22] The corruption of ἀμμωνιακόν into *ἀρμονιακόν/ἀρμενιακόν[23] may have begun early. We find *armeniyaqōn* defined as *melḥā ammōnīqōn* at Bar Bahlūl [Duval] 297.1.[24] - In *But*. the word is written ܐܪܡܘܢܝܩܘܢ at Gen. et corr. III.i.5,[25] but the form ܐܪܡܗܢܝܩܘܢ, with *-eh-*, prevails elsewhere.[26] This latter form may be due to an association with ܐܪܣܢܝܩܘܢ (ἀρσενικόν), since the two substances are found listed together at Gen. et corr. III.i.8 and at Min. III.i.5.7.[27]

l. 3. ‹ܡܠܚܐ ܐܢܘܢ ܐܠܐ ܢܘܪܢܝܘܬܐ ܕܐܪܡܗܢܝܩܘܢ› ("are salts, but the fieriness of sal ammoniac") scripsi: om. codd.: من جنس الأملاح إلا أن نارية النوشادر IS. - The text as it stands in the mss. (here available only in FMV) does not make sense. The sense, if not the exact wording, can be reconstructed from the corresponding passage of IS. The loss will be due to haplography of *armenīqōn*.

l. 3, 4. "fieriness", "earthiness": It is taken for granted that "salts" contain fierines and earthiness. This may be understood in the light of the Aristotelian theory that the salinity of the sea is due to the "dry, smoky exhalation" which is fiery and earthy (see Min. V.ii. below).

l. 5. . . . ܐܝܬܝܗ̇ ܕܝܢ ܗܘܠܐ ܕܝܠܗܘܢ ("their material is ..."): فهو IS. - It is more natural to understand the latter part of the passage of the *Šifā*ʾ quoted above as referring to sal ammoniac only, although we then have no explanation of how alum is formed. BH has evidently understood it as referring to both sal ammoniac and alum, perhaps following a reading such as that found in the margin of ms. B (*fa-humā*).

l. 5-6. ܩܛܝܢܬܐ ܘܢܘܪܢܝܬܐ ܣܓܝ ("very fine and fiery"): لطيف جدا كثير النارية IS. - I have taken"very" (*saggī*) with both adjectives in view of the corresponding phrase in IS.

---

[22] See Ruska (1923) 18-21 (cf. id., EI(D)[1] III.1045, = EI[2] VIII.148f.), who was of the opinion that sal ammoniac was not known in the Middle East until the 9th century. For an argument that it was already known at least in the late Antiquity, see Dittberner (1971) 98-123.

[23] Which later produced forms such as "salarmoniac" and "sal armoniaco" in Europe (Ruska [1923] 3, 23).

[24] See also Bar Bahlūl 1090.5 s.v. ܡܠܚܐ ܐܡܘܢܝܩܘܢ. See further PS 227, 390, 392; and Ruska (1923) 14-18. - The alteration of *-m-* (μ) into *-r-* (ρ) is easier in Greek, while that of *-o/ō-* (ܘ) into *-e/eh-* (ܗ) can easily occur in Syriac.

[25] So ms. F 45v b 22; ܐܪܡܘܢܝܩܘܢ with -r- later erased M; ܐܡܘܢܝܩܘܢ V; lacuna Ll.

[26] So in all three mss. available here and at III.i.5.5, 7 below; also *But*. Gen. et corr. III.i.8. (ms. F 46r b 15: ܐܪܡܗܢܝܩܘܢ MV, lacuna Ll) and Mete. IV.i.3.9 (ܐܪܡܢܝܩܘܢ Ll).

[27] At *Cand*. 87.3 the word for sal ammoniac is written *ammōnīqōn* (ܐܡܘܢܝܩܘܢ; ܐܡܘܢܝܐܩܘܢ ms. B), while *armāniyaqōn* appears a little earlier as lapis lazuli (*lāzward*) at *Cand*. 84.6 (ܐܪܡܐܢܝܐܩܘܢ ܐܒܢܐ ܠܐܙܘܪܕ).

III.i.4.1-2: Vitriols

But.: (1a) Vitriols are composed of saltiness, sulphureity and stones, (1b) and have [*qnēn*] the power of some of the fusible bodies.
Šifā᾽ 21-5-6: وأما الزاجات فإنها مركبة من ملحية وكبريتية وحجارة، وفيها قوة بعض الأجساد الذائبة.
(1a) Vitriols are composed of saltiness, sulphureity and stones, (1b) and have in them [*wa-fīhā*] the power of some of the fusible bodies.

l. 1. ܡܠܚܝܘܬܐ MV ("saltiness"): ܡܠܝܚܘܬܐ F. - F has *mallīḥūṯā* again in l. 3 below, and what appears like *mlīḥānē* (ܡܠܝܚܢܐ), a form unknown to dictionaries, for *melḥānāyē* at III.i.5.4. I accept the readings of MV here, since *melḥānāyā*, though a rare form, is also found in Bar Shakko,[28] and since it is also quite likely that BH differentiates between *melḥānāyā* / *melḥānāyūṯā*, corresponding to IS's *milḥī* /*milḥīya*, and *mallīḥ* /*mallīḥūṯā*, corresponding to IS's *māliḥ* / *mulūḥa*.[29]

III.i.4.2-6: Vitriols (contd.): χάλκανθος and χαλκῖτις

But.: (1) Χάλκανθος and χαλκῖτις are generated through the dissolution of the saltiness and sulphureity of vitriol. (2a) When they acquire the power of one of the fusible bodies in the mines [*b-mehṭalla*], (2b) they solidify, (2c) so that what acquires [*qānē*] the power of iron turns red and yellow, like χαλκῖτις, (2d) and what gains [*mettaggar*] the power of copper turns green, like χάλκανθος.
Šifā᾽ 21.6-9:

وما كان منها مثل القلقند والقلقطار فكونها من حلالة الزاجات، وإنما تنحل منها الملحية مع ما فيها من الكبريتية، ثم تنعقد وقد استفادت قوة معدن أحد الأجساد؛ فما استفاد من قوة الحديد احمر واصفر كالقلقطار وما استفاد من قوة النحاس اخضر، ولذلك ما أمكن أن تُعمل هذه بالصناعة.

[6 فكونها: فتكونها Ṭ (et *Mabāḥiṯ* II.214.13) ‖ حلالة Holmyard: جُلالة Cair. (et *Mabāḥiṯ*, ibid.) ‖ 9 بهذه الصناعة: هذه بالصناعة B, Holmyard]
(1) Those of them such as qalqand and qalqatār are formed by dissolution of the vitriols, the saltiness alone dissolving, together with whatever sulphureity they may contain. (2b) Then, they solidify, (2a) after they have acquired the power of the ore [*maᶜdin*] of one of the [metallic] bodies, (2c) so that that which acquires [*istafāda*] the power of iron turns red and yellow, like *qalqatār*, (2d) and that which acquires the power of copper turns green. For this reason, these can be prepared through the art [of alchemy].[30]

---

[28] Bar Shakko, *Dial.*, Göttingen, Or. 18c, 303v b4, where it is clearly vocalised ܡܠܚܢܝܐ (cf. Ruska [1897] 158.5; Brockelmann 390b).

[29] For the latter correspondence, see Min. V.ii.2.2, ii.3.1, ii.4.6 below.

[30] The idea here of the formation of *qalqand* and *qalqaṭār* through dissolution of vitriol and acquisition of a metallic "power" is based on chemical experiment, as exemplified by the following method given by Abū Bakr al-Rāzī for preparation of *qalqand*: *K. al-asrār* [Dānishpazhūh] 5.10f.: تدبير القلقند يؤخذ الزاج فيحلّه بماء ويصفّيه ويلقى عليه

l. 2. ܟܠܩܢܬܘܣ ܘܟܠܩܝܛܝܣ ("χάλκανθος and χαλκῖτις"): القلقند والقلقطار IS. - IS names the following four "vitriols" with their colours under "*zāğ*" in the list of simple drugs (*al-adwiya al-mufrada*) in his *al-Qānūn fī al-ṭibb* [ed. Cairo (1987) 494][31]: 1. *qalqadīs* (< χαλκῖτις): white; 2. *qalqand* (= *qalqant*, < χάλκανθος): green; 3. *sūrīn* (< σῶρυ): red; 4. *qalqaṭār* (< *χαλκητάριον [?][32]): yellow. - Only two of these are named in the passage of the *Šifā'* quoted above, where *qalqaṭār* is said to be "red" as well as "yellow".[33] - One would have expected BH to render IS's *qalqaṭār* by *χαλκητάριον (*klqṭryn*). The identification of *qalqaṭār* with "χαλκῖτις" may be explained partly, at least, by such passages as the entry in Bar Bahlūl (ed. Duval 859.5f.) which equates ܟܐܠܩܝܛܝܣ with ܟܠܩܛܪܝܢ.[34] - The two words are written without *alaph* (ܟܠ-) in all three mss. here, but with alaph (ܟܐܠ-) in ll. 5-6 and at III.i.5.5 below.[35] - The word *kalqīṭīs* occurs again at *But*. De animalibus III.i.6. (ms. F 92r a 24),[36] where the word can be traced back to the χαλκῖτις λίθος ("copper ore") of Arist. *Hist. An*. 552b 10.

l. 4. ܒܡܬܟ̈ܠܬܐ ("in the mines"): cf. وقد استفادت قوة معدن أحد الأجساد IS. - The word *ma'din* here in IS appears to mean "ore" rather than "mine".[37] BH

برادة النحاس ويطبخ حتى يخضر ويصفّيه ويتركه ينعقد. ("Preparation of *qalqand*: Take vitriol and dissolve it in water, then purify [filter] it and pour copper filings on it. Let it undergo coction until it turns green. Leave it to solidify"); cf. tr. Stapleton et al. (1927) 373; Ruska (1937) 88; the latter follows a different text here.

[31] The same four are found with the same colours in Rāzī, *K. al-asrār* 4.24 (ult.) f., tr. Ruska (1937) 87; see ibid. 47. See further Holmyard-Mandeville (1927) 36f. n. 3-4; Dittberner (1971) 80-82, 128.

[32] So Ruska (1937) 47 and Brockelmann 330a for the Syriac *klqṭryn*/*klqyṭryn*. Berthelot-Duval (1893) tr. 75 n.2, 123f. etc. write *χαλκητάριν; J.P. Margoliouth at PS Suppl. 166 *χαλκιτάριον. The attempts to derive the word from χαλκάνθη (Dozy I.399) or χαλκοκρᾶτον (Meyerhoff [1940] 68) seem more far-fetched.

[33] *Qalqaṭār* is also said to be "yellow and red" in Rāzī's *al-Madḫal al-ta'līmī*, where *sūrīn* is missing (Stapleton et al. [1927] 349).

[34] Bar Bahlūl [Duval] 859.5f.: ܟܐܠܩܝܛܝܣ . ܟܠܩܛܪܝܢ قلقطار . ܗܟܢ ܪܒܢ وكذا جبريل ("χαλκῖτις: χαλκητάριον, *qalqaṭār*. So the Master [i.e. Ḥunain]. So Gabriel [bar Bōkhtīshō']"). - Elsewhere in Bar Bahlūl, we are told about χαλκῖτις being transformed into *χαλκητάριον: 898.26ff. s.v. ܟܠܩܕܝܣ (χαλκῖτις); 899.2f. s.v. ܟܠܩܛܪܝܢ; 1797.8ff. s.v. ܩܠܩܕܝܣ; cf. IS *Qānūn* [Cairo (1987)] 709.2: قال جالينوس: إنّ قلقديس قد يستحيل قلقطاراً.

[35] In *Cand.*, we find ܩܠܩܢܬܘܣ (with *qōp* as in Arabic, ܩܠ- = قل-) at 87.3 and 89.5 (corr. Fakhr al-Dīn al-Rāzī, *Mabāḥit*, II.211.4, 214.13 قلقند) and ܩܠܩܕܝܣ at 247.11. - See further Bar Bahlūl [Duval] 859 (ܟܐܠ-), 898f. (ܟܠ-) and 1797f. (ܩܠ-); cf. PS, Brockelmann ad locc.; *Syriac Book of Medicines*, ed. Budge: ܟܠܩܢܛܪܝܢ 63.20 and ܟܠܩܝܛܝܣ 90.20, but ܩܠܩܕܝܣ 63.21, 64.11, 87.7; Jacob of Edessa, *Hex*. [Chabot] 53b 17f. ܚܣܝܢܐ ܟܠܩܛܪܝܢ ܘܟܐܠܩܢܬܘܣ (ܟܐܠܩܛܪܝܢ cod. Leiden.).

[36] Written ܟܐܠܩܝܛܝܣ M; ܟܐܠܩܝܛܝܣ F; ܟܐܠܩܝܛܝܛܣ LIV and, thence, PS Suppl. 158.

[37] So Holmyard-Mandeville (p. 38): "after a virtue has been acquired from a metallic ore."; cf. Baffioni (1980) 95: "quando ha acquistato la forza di un corpo minerale".

omits the word at the corresponding place, but adds "in the mines" at the end of the clause. BH *may* be correct in a way in thinking of mines here, since we hear of the solidification of the constituents of *qalqand*, *qalqaṭār* etc. in "mines" in Rāzī, *K. al-asrār* 5.3-4: واصلها زاجات وشبوب يغسلها السيل، وينزل بها الى حفرة المعادن، فيقع عليها الشمس فيعقّدها. ("They originate from vitriols and alums which streamwater washes and carries down into the cavities of the mines. The sun then shines on them and solidifies them").[38]

l. 4-5. ܩܢܐ ... ܡܬܬܓܪ ("acquires ... gains"): استفاد ... استفاد IS. - An instance of *variatio* where BH uses two different words for rendering the same word in IS. We note also the variation of the construction in *ḥayl parzlā* ... *ḥaylā da-nḥāšā* ("power of iron ... power of copper") in the same sentence.

l. 5. ܡܣܡܩ ("turns red"): Intransitive both here and at III.ii.2.3, 5 below (similarly, *mawreq*, "turns green" in the following line).

l. 7. ܐܝܟ ܟܠܩܢܬܘܣ ("like χάλκανθος"): IS does not say "like *qalqand*" (*ka-l-qalqand*) at the corresponding place, although this is no doubt intended. The same addition as that made by BH here is made by Fakhr al-Dīn al-Rāzī (*Mabāḥiṯ* II.214.16).

III.i.5.1-6: "Physical" classification

The passage of *But.* here combines materials taken from the *Šifāʾ* and Abū al-Barakāt's *K. al-muʿtabar*. - A classification similar to the one here is given by BH at *Cand.* 86.1-87.3. The source there is Fakhr al-Dīn al-Rāzī, *Mabāḥiṯ* II.210.5-211.5, which is itself based on the classification in the *Šifāʾ*, but differs from it in a number of respects.[39]

But. [underline: agreement with *Šifāʾ*; italics: with *Muʿtabar*]:
(0) Mineral bodies are either (1) *strong in* [lit. of] *substance* [*ḥayltānay ūsiya*] [and] (1a) *malleable, like iron and gold*, or (1b) *shatterable like corundum* [*yaqqundā*] *and rock crystal* [*berūlā*], or (2) weak *in substance* and *mixture* [*mḥīlay ūsiya w-muzzāgā*] *and loosely composed* [*mp̄arsʿay rukkāḇā*] and (2a) stony *like glass*, or (2b) saline - moisture easily dissolves these - like alum, vitriol, sal ammoniac and χαλκῖτις, or (2c) oily and - these are not easily dissolved by moisture alone - like sulphur and ἀρσενικόν.

Šifāʾ 20.5-9:

وذلك /٦ أن من الأجسام المعدنية ما هو سخيف الجوهر، ضعيف التركيب والمزاج. ومنها ما هو /٧ قوى الجوهر؛ وما هو قوى الجوهر، فمنه ما ينطرق، ومنه ما لا ينطرق. وما هو ضعيف /٨ الجوهر، فمنه ما هو ملحى تحله الرطوبة بسهولة مثل الشب والزاج والنوشادر والقلقند. /٩ ومنه ما هو دهنى لا ينحل بالرطوبة وحدها بسهولة مثل الكبريت والزرنيخ.

[6 أن: لأن ST || ومنها: ومنه BDSṬ, Holmyard || 7 وما هو قوى الجوهر: om. M, Holmyard [sed habet C] || ومنه ما لا ينطرق: om. D || 8 ما هو ملحى: ملحى D || والزاج: om. BDC]

---

[38] Cf. tr. Stapleton et al. (1927) 373; Ruska (1937) 87.

[39] For an overview of the various systems for classification of minerals in Islamic literature, see Ullmann (1972) 140-144.

(0) That is to say, there are mineral bodies which (2) are feeble in substance [*saḫīf al-ǧauhar*] and weak in composition and mixture [*daʿīf al-tarkīb wa-l-mizāǧ*], and (1) there are those which are strong in substance. (1) Among those which are strong in substance [*qawīy al-ǧauhar*], (1a) some are malleable, (1b) while others are not malleable. (2) Among those which are weak in substance [*daʿīf al-ǧauhar*], (2b) some are saline - which moisture easily dissolves -, e.g. alum, vitriol, sal ammoniac and *qalqand*. (2c) Others are oily - which are not easily dissolved by moisture alone -, e.g. sulphur and *zarnīḫ*.

Muʿtabar II.230.1-3:

ومنها ما هو سخيف الجوهر متخلخل التركيب والمزاج كالزجاج ومنها ما هو قوى /٢ الجوهر والقوى
الجوهر منه منطرق كالحديد والذهب ومنه ما ينكسر ولا ينطرق /٣ كالياقوت والبلور

(2) There are those which are feeble in substance and loose in composition [*mutaḫalḫil al-tarkīb*] and mixture, (2a) such as glass. (1) There are those which are strong in substance. Among those which are strong in substance, (1a) some are malleable like iron and gold. (1b) Others are breakable and non-malleable like corundum [*yāqūt*] and rock crystal [*billaur*].

Cand. 86.1-87.3 [underline: agreement with *Mabāḥit*]:

(0) These [i.e. minerals] are either firmly-composed [*ḥlīmay rukkāḇā*] or weakly-composed [*mḥīlay rukkāḇā*].[40] (1) The firmly-composed are either (1a) malleable, like gold ... or (1b) not malleable and these are not malleable either (1b.i) because of their great softness, like mercury, or (1b.ii) because of their great hardness, and these [latter] either (1b.ii.a) lack oiliness and for this reason are shatterable [*metparkkīn*], like corundums [*yūqandē*], or (1b.ii.b) have oiliness which has become strongly attached to the stony nature, has solidified its parts and has become dry; from them are formed, for example, the ἀδάμας stone; for this [stone] is not only not malleable, but when beaten on an anvil penetrates inside it [< Nic. syr. 60.5-7] and, although in this way it conquers and overcomes every hard body, it is shattered [*metparkkā*] and is destroyed by the blood of goats [< Nic. syr. 59.19-23]. (2) The weakly-composed are either (2a) dissolved by moisture, like all the salts, I mean soda [*nītrōn*, cf. Nic. syr. 58.16-23, 60.18-20], alum, sal ammoniac [*ammōnīqōn*] and *qalqantīs*, or (2b) not dissolved, like *zarnīkā* and sulphur.

Mabāḥit II.210.5-211.5 [underline: agreement with *Cand.*]:

(0) Mineral substances are either strongly-composed or are weakly-composed. (1) If they are strongly-composed, they are either (1a) malleable - and these are the seven 'bodies' - or (1b) are not malleable, (1b.i) either due to its extreme moistness, like mercury, or (1b.ii) due to its extreme dryness, like corundum etc. (2) If they are weakly-composed, they are either (2a) dissolved by moisture - i.e. those which are salty - like vitriol, sal ammoniac, alum and *qalqand*, or (2b) are not dissolved by moisture, i.e. those which are oily - like sulphur and *zarnīḫ*

Provided below is a summary of the classifications encountered in the five passages quoted above and in two passages closely resembling each other in Bar Shakko, *Dial.* II.2.3.19 (Göttingen or. 18c, 303v

[40] Note sound play of "*ḥlīmay ... mḥīlay*".

a6-b9; = Ruska [1897] 157.22-158.6) and Qazwīnī, *ʿAǧāʾib* [Wüstenfeld] 203.29-204.4. These last resemble the passage in the *Mabāḥiṯ*, but at the same time share a number of features where they differ from the *Mabāḥiṯ*, indicating - since Bar Shakko's work predates that of Qazwīnī's and Qazwīnī is unlikely to have used a Syriac source - that they must depend on an unknown common source, which, in turn, was probably based on the *Mabāḥiṯ*.[41]

Šifāʾ	*Muʿtabar*	*But.*
1. Strong in substance	1. Strong in substance	1. Strong in substance
1a. Malleable	1a. Malleable	1a. Malleable
1b. Not malleable	1b. Breakable/non-mall. (corundum)	1b. Shatterable (corundum etc.)
2. Weak in substance	2. Weak: glass	2. Weak in substance
		2a. Stony: glass
2a. Saline (soluble) (alum etc.)		2b. Saline: alum etc.
2b. Oily (insoluble) (sulphur etc.)		2c. Oily: sulphur etc.

Mabāḥiṯ	*Cand.*	*Dial. & ʿAǧāʾib*
1. Strongly-composed	1. Strongly-composed	1. Strongly-composed
1a. Malleable (the 7 metals)	1a. Malleable (metals)	1a. Malleable: (the 7 metals)
1b. Not malleable	1b. Non-malleable	1b. Not malleable
1b.i. Moist (mercury)	1b.i.. Soft (mercury)	1b.i.. Soft (mercury)
1b.ii. Dry (corundum etc.)	1b.ii. Hard	1b.ii. Hard (corundum)
	1b.ii.a. Shatterable (corundum)	
	1b.ii.b. adamas	
2. Weakly-composed	2. Weakly-composed	
2a. Soluble/salty (vitriol etc.)	2a. Soluble/salty (vitriol etc.)	1b.ii.a. Soluble/salty: (vitriol etc.)
2b. Insoluble/oily: (sulphur etc.)	2b. Insoluble/oily (sulphur etc.)	1b.ii.b. Insol./oily (sulphur etc.)
		2. Weakly-composed

[41] The similarity between Qazwīnī and Bar Shakko was noted by Ruska (1897) 157-161, that between Qazwīnī and *Cand.* by Bakoš (*Cand.* 86 n.1); the similarity between Fakhr al-Dīn al-Rāzī and Qazwīnī is mentioned by Ullmann (1972) 143. - The scheme here in Bar Shakko and Qazwīnī, whereby substances such as vitriol and sulphur are classed under non-malleable-hard substances, is clearly a nonsense (the problem results from the repetition of the phrase "*kul kulleh qšayyā*" and "*fī ġāyat al-ṣalāba*", respectively, at *Dial.* [Ruska] 158.4 and *ʿAǧāʾib* [Wüstenfeld] 204.3). The agreement of the two, however, indicates that this must be what the respective authors wrote and that the error goes back to their common source. The 'error' in Qazwīnī was already noted by de Chézy (in the chrestomathy of de Sacy [1826/7] vol. III.), who retains the manuscript reading in his Arabic text (١٧٠.3) but gives the 'required sense' in his translation (p. 390) and explains his correction in his comments (p. 466f., n. 8). Ethé (1868) in his translation of Qazwīnī circumvented the problem by using a free rendering (p. 417, l. 26: "Letztere"). Garbers-Weyer (1980) 42 and Giese (1986) 118, on the other hand, translate as the text of Qazwīnī stands without any comment and Garbers-Weyer go so far as to provide us with a diagram of the classification of minerals based on the nonsensical text of Qazwīnī in their comments (p. 93).

l. 2. ܡܬܦܪܟܟܢܐ ("shatterable"): ينكسر ("breakable") A. al-Barakāt. - Although BH's *metparkkānā* is close to Abū al-Barakāt's *yankasiru*, it is not quite the same, since BH usually uses √*tbr* for arab. √*ksr*, while using *parkeḵ* for arab. *tafattata*.[42] - In looking for the source for the word *metparkkānā* here, we need to turn to the passage of *Cand*. quoted above, where the verb *metparkeḵ* is used twice and in one of these two instances is applied, as here in *But*., to the corundums. There in *Cand*., in turn, this verb, along with the part concerning ἀδάμας, is an addition by BH, which has no counterpart in the corresponding passage of the *Mabāḥiṯ*, but was taken from Nic. syr. 59.19-23: "Also, the stone of ἀδάμας is said in truth to be a branch of gold. It is to be wondered why fire does not heat it and nothing damages it at all. Perhaps its density is the reason why it does not receive anything inside it. Why the blood of goats shatters [*mp̄arkeḵ*] it and destroys it, we must investigate in the [book of] *Problems*."[43]

l. 3. ܝܩܘܢܕܐ ("corundum"): ياقوت Abū al-Barakāt. - The Syriac *yaqqundā*, also written *yūqandā* etc.,[44] corresponds to gr. ὑάκινθος and arab. *yāqūt*. In Arabic, the term is applied to various corundums, such as ruby, emerald and sapphire, whereby *yāqūt* without further qualification will often mean "ruby", the corundum par excellence.[45] - In *But*., the word is also used at Gen. et corr. III.i.5. (ms. F 45r b 13, corr. *Šifāʾ* Act et pass. 230.17 *yāqūt*).

l. 3. ܒܠܘܪܐ ("rock crystal"): بلور Abū al-Barakāt. - The word which appears as *billaur*/*ballūr* ("rock crystal") in Arabic and βήρυλλος ("beryl"[46]) in Greek is believed to be of Indian origin.[47] If this is correct, *bellūrā* may be the

---

[42] So, for example, at Min. I.i.2.2 (corr. IS *tafattut*) and III.i.1.3 (of "stones", corr. IS *tafattata*) above and, more significantly, in *But*. Gen. et corr. III.iii.3. (ms. F 47v b23-48r a12), which is related to the list of 18 pairs of physical qualities at Arist. *Mete*. 385a 12-18. Where Arist. had distinguished between objects "broken" into large pieces (κατακτός) and "shattered" into small pieces (θραυστός), IS (*Šifāʾ* Act. et pass. 245.3ff.) made a tripartite division into fragmentation into large pieces (*inkisār*), into small pieces by a strong force (*inriḍāḍ*) and into small pieces by a weak force (*tafattut*). The three corresponding terms used by BH as *tḇārā*, *rṣāṣā* and *purkāḵā*. - On *purkāḵā*, see further *But*. Gen. et corr. III.iii.1. (47v a 18, 22) *metparkkḵānūṯā*; III.iii.6. (48v a 8) *metparkek*.

[43] See Takahashi (2002a) 210 (text), 220 (tr.).

[44] For the various forms of the word, see Brockelmann 307 s.v. ܝܩܘܢܕܐ, also 175b ܗܘܩܝܢܬܘܣ.

[45] See, for example, Ps.-Arist. *K. al-aḥǧār* [Ruska] 99f. (tr. 135f.); al-Tamīmī, *Muršid* XIV [Schönfeld] 39 (comm. p. 143); de Sacy (1826-7) III.464f. [de Chézy]; Wiedemann (1912b) 212 n.1; Siggel (1950) 89; EI² XI.262 s.v. "Yāḳūt" [al-Qaddumi].

[46] See RE III/I 320f. s.v. "Beryllus 3." [Blümner]; RAC II (1954) 157ff. s.v. "Beryll" [H.E. Killy]

[47] From the Sanskrit *vaiḍūrya*, Pali *veḷuriya*, whence also the Chinese *bìliúli, liúli* (璧流璃, 琉璃, 瑠璃) etc. - See de Lagarde (1866) 22 §48; Hirth (1890) 62f.; Franke (1893) 600; Meissner (1903) 248-250; Brockelmann 78 s.v. ܒܠܘܪܐ; Needham et al. (1962) 104-106; cf. Dozy I.100; EI² I.1220 s.v. "al-Billawr" (Ruska-[Lamm]); Kahle

older form in Syriac, the commoner *bērūlā* having arisen under the influence of Greek.[48] Despite the correspondence of the latter to the gr. βήρυλλος in form, it seems reasonable to assume that BH uses *bērūlā* here as the equivalent of *billaur* with the sense of "rock crystal". - At *But*. Mete. II.i.8.4 and II.iii.1.4 below, *bērūlā* corresponds to IS *billaura*. - *Yāqūt* and *billaur*, though not in the above passage of the *Šifāʾ*, are given together as examples of minerals formed through solidification of vapour by IS in his *Dāniš-nāma* [Mishkāt] III.74.3 (tr. Achena-Massé II.52).

l. 3. ܘܟܐܦܢܝܐ ܐܝܟ ܠܙܓܘܓܝܬܐ ("and stony like glass"): Abū al-Barakāt does not subdivide the "weak-substance" minerals and does not say that glass is "stony". So the definition of this category as "stony" to distinguish it from "saline" and "oily" is BH's own. - For the treatment of glass as a type of stone, see Min. I.i.5.5 above.

l. 5. ܟܠܩܝܛܝܣ ("χαλκῖτις"): القلقند IS. - The word used as the equivalent of IS's *qalqand* at III.i.4.2, 6 above was ܟܠܩܢܬܘܣ (χάλκανθος), but the manuscript evidence is in favour of BH having written "χαλκῖτις" here. Although F has the word ending in *-ws*, this is how the word is also written at III.i.4.2, 5 in that manuscript.

l. 6. ܐܪܣܢܝܩܘܢ ("ἀρσενικόν"): الزرنيخ IS. - Not our "arsenic" (the element, As, and arsenic trioxide), but the sulphides of arsenic, realgar (As_4S_4) and orpiment (As_2S_3).[49] The Greek ἀρσενικόν/ἀρρενικόν and, accordingly, the Syriac *arsenīqōn*, are strictly yellow orpiment (syr. *zarnīḵā yurrāqā*, arab. *zarnīḫ aṣfar*), the word for realgar being σανδαράκη/σανδαράχη (Syr. ܣܢܕܪܟܐ etc., also *zarnīḵā summāqā*; arab. *zarnīḫ aḥmar*).[50] At the corresponding place in *Cand*. (87.3) BH uses the 'Arabic' form *zarnīḵā*, while Bar Shakko (Ruska [1897] 158.6) uses the 'Greek' form (ܐܪܣܢܝܩܘܢ).[51]

III.i.5.6-9: "Spirits" and "bodies"

> *But*.: Alchemists name four "spirits" [*rūḥāṯā*]: sal ammoniac, ἀρσενικόν, sulphur, mercury; and seven "bodies" [*pagrē*]: gold, silver, copper, iron, lead, tin, and another species which occurs in China [*ṣīn*] and is called *khārṣīnī*.

The direct source is probably the same as that used at III.i.1.2-4 above. The four "spirits" and the seven "bodies" (i.e. metals) listed here are the same as those given in such works as Rāzī's *K. al-asrār* and *K.*

(1936) 325 n.1; Nedospassowa (1993).

[48] So de Lagarde, loc. cit.

[49] Garbers-Weyer (1980) 107.

[50] See LSJ s.vv.; PS 2674, 371, s.vv. ܣܢܕܪܟܐ, ܐܪܣܢܝܩܘܢ.

[51] Nic. syr. uses the form *zarnīḵā* throughout, either alone (56.28) or qualified by *summāqā* (59.16; 60.7, 24) and *yurrāqā* (60.24).

al-madḫal al-taᶜlīmī, and Khwārizmī's *Mafātīḥ al-ᶜulūm*.[52] - The order in which BH lists the four "spirits" is quite different from the order found in these three works,[53] and for the metals, too, it is difficult to find a list which exactly matches the order here, although it is worth noting that arrangements similar to that in *But.* are found in works close to it in date.[54] For the metals, the order of the list here is, in fact, the same as the order in which they are dealt with in III.ii.3. below, except for the reversal of the position of "gold" and "silver", and for *kharṣīnī* which is not dealt with in III.ii.3.

l. 6. ܦܠܚܝ ܟܝܡܝܐ ("alchemists"): lit. "cultivators/practicians of alchemy". This phrase, along with *pālḥay ummānūtā*, is regularly used as a designation for alchemists in Bar Bahlūl's lexicon.[55] - Cf. III.iii.3.1 below.

l. 7. ܪܘܚܐ ("spirits"): corr. Arab. *arwāḥ*. - The use of the word in this technical, alchemical sense seems to be unregistered in Syriac dictionaries,

---

[52] Rāzī, *Asrār* [Dānishpazhūh] 2.11-13; cf. Holmyard-Mandeville (1927) 33 n.2; tr. Stapleton et al. (1927) 370; tr. Ruska (1937) 84. - *Madḫal*: Stapleton et al (1927) 412 (tr. p. 345f.); cf. Garbers-Weyer (1980) 3. - Khwārizmī, *Mafātīḥ* [van Vloten] 258.6f., 12f. - BH is in agreement with the *K. al-asrār*, against the other two works mentioned, in listing the "spirits" before the metals (on this point, see Dittberner [1971] 74f.).

[53] ***But.***: 1. sal ammoniac, 2. *arsenikon/zarnīḫ*, 3. sulphur, 4. mercury - ***Asrār***: **4132** (4123 in Stapleton et al. [1927] 370, whose translation is based mainly ms. Escorial 700 [Derembourg]). - ***Madḫal***: **3421** - ***Mafātīḥ***: **3241** - While the number of the "spirits" varies in the Jabirian corpus, with camphor and "oils" (*adhān*) sometimes being included, a list with the same four items as here is found in the *Seventy Books* (Kraus [1942/3] II.19, 22 nn. 4, 5, 7).

[54] ***But.***: 1. Au, 2. Ag, 3. Cu, 4. Fe, 5. Pb, 6. Sn, 7. *khārṣīnī* - **Jābir**, *K. al-ḫawāṣṣ al-kabīr* (Kraus [1942/3] II.19): **5641327** [Pb: *usrub*; Sn: *qalaᶜī*] - Rāzī, ***Asrār***: **1243657** (Au and Ag reversed in ed. Danīshpazhūh; *al-ḥadīd al-ṣīnī* instead of *ḫārṣīnī* in the Leipzig and Göttingen mss.) - ***Madḫal***: **1265437** (or **1235467**, see Stapleton et al. [1927] 345 n.5, 363 n.4) - **Khwārizmī**, *Mafātīḥ*: **1243567** [Sn: *al-raṣāṣ al-qalaᶜī*] - **Fakhr al-Dīn al-Rāzī**, *Mabāḥiṯ* II.211.8f.: **1264375** [Sn: *raṣāṣ*; Pb: *ānuk*] - **Qazwīnī**, *ᶜAǧāʾib* [Wüstenfeld] 203 ult. f.: **1236457** [Sn: *raṣāṣ*; Pb: *usrub*] - **Bar Shakko**, *Dial.* [Ruska (1897) 158.2] **1234657** [*parzlā ṣīnāyā* for *khārṣīnī*] - BH ***Cand.***: **12**-electrum-**3456** - **Abū al-Qāsim al-ᶜIrāqī (al-Sīmāwī)** [mid-13th c.] *Muktasab* [Holmyard (1923) 7.7] **123456** [no seventh; Sn: *qaṣdīr*] - **Dimashqī,** *Nuḫba* [Mehren] 48.17f. (tr. p.53): **1234765** [Sn: *qalaᶜī*; Pb: *raṣāṣ*]. - The books of Jābir's "Seven Books on Metals" are found arranged in the order 1234657 in two late mss. (Paris 2606, 16th c. A.D., and Ṭalᶜat kīmiyā 187, 10/11th c. A.H., see Kraus (1942/3) I.111; on the date of the mss. in question, see id. I.184f. and de Slane [1883-95] 471). - On the various appellations of lead and tin, see Wiedemann (1911) 85, Stapleton et al. (1927) 345 n.6. - Cf. also **Jacob of Edessa**, *Hex.* [Chabot] 53a 34-b 1: **4365**-glass (!)-**21**-electrum.

[55] Bar Bahlūl [Duval] 19.22, 210.16, 267.23, 377.21 etc. (cf. Duval [1901] proemium p. xv; PS 3148).

although the word probably carries a more technical sense than is brought out by its translation simply as "vapour" (so PS Suppl. 317) at Berthelot-Duval (1893) text. 36.4 (tr. 63.23).[56]

l. 7. ܦܓܪܐ ("bodies"): The term normally used in Arabic is *ǧasad*, *aǧsād*.[57] At III.i.4.2 and 3-4 above, IS's *al-aǧsād al-ḏāʾiba* and *al-aǧsād* are rendered by *gušmē metpaššrānē*. The choice of the word *pagrā* here, besides being necessitated by the need to use a word other than *gušmā* to render *ǧasad* unqualified by an adjective, was probably influenced by the contrast with the "spirits", since *pagrā* is the word used in the sense of "flesh" (i.e. σάρξ) as opposed to "spirit" (see PS 3033 s.v. ܦܓܪܐ and 793 s.v. ܓܘܫܡܐ).[58]

l. 9. ܚܐܪܨܝܢܝ ("*khārṣīnī*"): = خارصينى - The older identification of this reportedly rare Chinese metal with zinc[59] has generally been rejected in favour of a view which regards it as an alloy of some kind.[60] Needham et al. (1980) 429-432 identify it with the copper-zinc-nickel alloy, *paktong*.[61] - This "metal", also known as "Chinese iron" (*al-ḥadīd al-ṣīnī*), came to replace mercury in the list of seven metals, especially where mercury was counted among the "spirits".[62] - It is worth noting that both BH and Bar Shakko accept this "*kharṣīnī*/Chinese iron" as the seventh metal in their secular 'philsophical' works, while in their 'theological' works, they follow the tradition of the Syriac hexaemeral literature in giving "electrum" (ἤλεκτρον, properly, the gold-silver alloy) in its place.[63]

---

[56] Berthelot-Duval (1893) text 36.3f.: ܟܕ ܚܙܝܬܝܗ ܫܪܝܐ ܘܕܡܝܐ ܠܪܘܚܬܐ . ܘܦܪܚܐ ܘܕܡܝܐ ܠܙܝܘܓܐ ("when you see it [sc. the mixture of sublimated '*zarnīḵā*', 'water of sulphur' and whitened 'magnesia'] melt/be loosened and resemble the spirits (*rūḥāṯā*), and fly and resemble mercury").

[57] The terminology fluctuates in the Jabirian corpus, where we find *ǧism* applied to metals and *ǧasad* to other minerals, and vice versa (see Kraus [1942/3] II.19 n.1.). - Cf. Stapleton et al. (1927) 339.

[58] Bar Shakko calls the metals the *šaḇʿā gušmē kyānāyē* in the passage of the *Dial.* (Göttingen, Or. 18c, 303v a14f., Ruska [1897] 158.2) where the parallel passages of Fakhr al-Dīn al-Rāzī and al-Qazwīnī have *al-aǧsād al-sabʿa*.

[59] This seems to go back to de Sacy (1826/7) III. 452-464, and is accepted by Dozy I.346 (also Wehr), Stapleton et al. (1927) 405-410.

[60] See Kraus (1942/3) II.22 n.3; Siggel (1950) 79; EI² IV.1084 "khārṣīnī" [Dietrich], together with the literature cited there.

[61] See also Needham et al. (1974) 238. - *Paktong*: lit. "white copper", 白銅 [báitóng].

[62] Mercury still has its place among the seven metals in Pseudo-Apollonius of Tyana (Weisser [1980] 200f, §4.8 and nn. 17, 20). Its replacement by *khārṣīnī* occurs in the later works of the Jabirian corpus (Kraus [1942/3] II.22, with n.6).

[63] Bar Shakko has "Chinese iron" (*parzlā ṣīnāyā*) in *Dial.* (Göttingen, Or. 18c, 304r a3; cf. 303v a 16, Ruska [1897] 158.2, 11), corresponding to *ḫārṣīnī* in Fakhr al-Dīn al-Rāzī, *Mabāḥiṯ* II.211.8, 213.15 and Qazwīnī, *ʿAǧāʾib* 204.1, 205.4 (cf. PS Suppl. 285 s.v. ܨܝܢܝܐ), but "electrum" in *Thes.* (Paris 316, 170r 3, Nau [1896] 308).

Min. III. Section ii.: *On the Fact that the Element of the Malleable [Minerals] is Mercury*

In this section, BH deals firstly with the manner of formation and the properties of mercury (III.ii.1). This is followed by the argument that all metals are, in fact, formed from mercury with the aid of sulphur (III.ii.2) and by the discussion of the formation of different metals from various combinations of mercury and sulphur (III.ii.3.). In the second and third theories BH closely follows IS. In the first theory, the main source is Abū al-Barakāt, although certain elements have also been taken from the *Šifāʾ*.

III.ii.1.	*Muʿtabar* II.230.17-21 [& *Šifāʾ* 21.10-15]
III.ii.2.	*Šifāʾ* 21.15-22.2
III.ii.3.	*Šifāʾ* 22.2-11

On the "sulphur-mercury theory" ("Schwefel-Quecksilber Theorie") of formation of metals, first attested in its fully-fledged form in Pseudo-Apollonius of Tyana (Balīnūs),[64] see Wiedemann (1907) 66-71; Ullmann (1972) 260; Garbers-Weyer (1980) 73-74.

It is to be noted that IS, while rejecting the claim of the alchemists to be able to produce gold and other metals (see section III.iii. below), accepts this "sulphur-mercury" theory. Abū al-Barakāt rejects both, while Fakhr al-Dīn al-Rāzī accepts both. BH is in agreement with IS in rejecting the former but accepting the latter.

III.ii.1.1-7: Mercury: its composition and properties

But. [underline: agreement with *Muʿtabar*; italics with *Šifāʾ*]:

(1a) In *mercury*, there predominates [lit. is much more, *ṭāḇ yattīrā*] wateriness *which is mixed with fine earthiness*; (1b) there is little airiness [lit. is less/little] (1c) and no fieriness [lit. is absent]. (2) The cause of its weight, therefore [*ara*], is the absence [*glīzūṯā*] of fieriness and the paucity [*bṣīrūṯā*] of airiness, (3) while its fluidity is due to its abundant wateriness. (4) The sound *mixture* [*šappīrūṯ metmazzgānūṯā*] of its little *airiness* and

---

Similarly, BH has *ḵārṣīnī* here in *But.* but "electrum" in *Cand*. 86.3, 88.7, and *Rad*. [Istanbul] 12.8. - Cf. Jacob of Edessa, *Hex*. 53b 1 (tr. 43.13); Bar Kepha, *Hex*. IV/2.12, Paris 241, 164r a30 ܐܠܩܛܪ; Paris 311, 38b b 24 ܐܠܩܛܪܘܢ (infeliciter "Bernstein" tr. Schlimme 549); *Causa causarum*, tr. Kayser 348-351.

[64] Ps.-Apollonius, *K. sirr al-ḫalīqa* III.4-5 [ed. Weisser] 246ff.; cf. Kraus (1942/3) II.1; Ruska (1926) 151; Weisser (1980) 199f.

its *clear earthiness* with its predominant wateriness is the cause of its whiteness. (5a) The firmness of its mixture [*šurrār muzzāḡeh*] (5b) on account of which its parts are not separated from each other but rise together (5c) is the cause of its ability to be sublimated [*mestallqānūṯā*] by heat, (5c) its wateriness being the cause of its flight from fire.

Muʿtabar II.230.17-19:

وفى الزئبق مائية اغلب ونارية قليلة جدا وكذلك هوائيته فثقله لعدم النارية والهوائية وميعانه للمائية وصعوده بالحر لمائيته ولجودة امتزاجه يعسر انحلال مزاجه وبياضه لهوائيته القليلة الجيدة الامتزاج بالمائية

(1a) In mercury wateriness is predominant [*aġlab*], (1c) [there is] very little fieriness (1b) and so its airiness. (2) Its weight is due to the absence [*ʿadam*] of fieriness and airiness. (3) Its fluidity is due to the wateriness. (5c) Its rising [*suʿūd*] with heat is (5c) due to its wateriness, (5a) and due to the goodness of its mixture [*ǧūdat imtizāǧihi*] (5b) the dissolution of its mixture [*mizāǧ*] is difficult. (4) Its whiteness is due to its little airiness which is mixed well with [*al-ǧayyidat al-imtizāǧ bi-*] the wateriness.

Šifāʾ 21.10, 13-4:

وأما الزئبق فكأنه ماء خالطته أرضية لطيفة جدا كبريتية مخالطة شديدة ... وبياضه من صفاء تلك المائية، وبياض الأرضية اللطيفة التى فيه وبممازجة الهوائية إياه.

(1a) Mercury appears to be water, with which very fine, sulphureous earthiness is strongly mixed. ... (4) Its whiteness [arises] from the clarity [*ṣafāʾ*] of that wateriness and the whiteness of the fine earthiness which it contains, and [is realised] through [*bi*] the mixture of the airiness with it.

l. 2, 3. "there is little airiness and no fieriness", "the absence of fieriness and the paucity of airiness": Abū al-Barakāt says at first that there is "very little" (*qalīl ǧiddan*) fieriness and airiness in mercury, but in the following sentence attributes the weight of mercury to their "absence" (*ʿadam*). BH alters this and says that there is *little* airiness and *no* fieriness. This may be based on BH's own deduction, since the presence of airiness in mercury is confirmed a little later by the attribution of the whiteness of mercury to its airy content, while there is nothing to confirm the presence of fieriness, and the presence of fieriness would, in fact, be made difficult by the predominance of the wateriness.[65]

l. 4. ܫܦܝܪܘܬ ܡܙܝܓܘܬܐ ("the sound mixture"): lit. "the goodliness of the mixture, well-mixedness" (*šappīr*/*šappīrūṯā* + noun often corresponds to gr. compounds in εὐ-). The phrase here renders *Muʿtabar* II.230.19 *al-ǧayyidat al-imtizāǧi*, but, in its nominal form, it is closer to II.230.18 *ǧūdat imtizāǧi-hi*; cf. also *Muʿtabar* II.230.5 *ǧūdat mizāǧi-hi*.

l. 5. ܐܪܥܢܝܘܬܗ ܫܦܝܬܐ ("its clear earthiness"): صفاء تلك المائية وبياض الأرضية اللطيفة ("the clarity of that wateriness and the whiteness of the fine earthiness") IS. - One suspects some carelessness on BH's part here, since *špē* ("clear,

[65] On the airy content of mercury, see Arist. at *Mete*. 385b 4-5: ... ἔστιν δὲ πλέον ἀέρος, ὥσπερ τὸ ἔλαιον καὶ ὁ ἄργυρος ὁ χυτός.

limpid", corr. *ṣāfin*) is rather awkward as an attribute of "earth", although it may be noted that Qazwīnī talks of the "purity" (*naqāʾ*) of the earth at the corresponding place, *ʿAǧāʾib* 243.8: فأما بياضه فبسبب صفاء ذلك الماء ونقاء التراب الكبريتى الذى ذكرناه ("Its whiteness is due tó the´clarity of that water and the purity of the sulphury earth mentioned above").[66]

l. 7. ܡܝܢܝܘܬܗ ܥܠܬܐ ܕܥܪܘܩܝܗ ܡܢ ܢܘܪܐ ("its wateriness being the cause of its flight from fire"): cf. *Muʿtabar* II.230.4: لأنه يهرب من النار ويتصعد بسرعة كالماء ("... because (mercury) flees from fire and rises quickly like water").

l. 7. ܡܣܬܠܩܢܘܬܗ ("its ability to be sublimated"): lit. "its being made to rise" (cf. PS Suppl. 234). - In connection with mercury, the reference will be to sublimation (arab. *tasʿīd*); cf. III.i.3.5 *mestallaq* above.

III.ii.1.7-9: Mercury (contd.): solidification by sulphur

But. [underline: agreement with *Muʿtabar*; italics with *Šifāʾ*]:

(1a) *The vapour of sulphur* [*rīḥ kebrīṯā*] *solidifies* (mercury), (1b) and makes it like lead [*abbārā*], (1c) because [*b-hāy d-*] *molten* [*šaryā*] *lead* is like *mercury* (1d) and solidified mercury is like lead.

Muʿtabar II.230.19-21:

ويعقده الكبريت بما يحل من مائيته فيجعله كالرصاص فان الرصاص الذائب كالزئبق والزئبق المنعقد كالرصاص الجامد.

(1a) Sulphur solidifies it with what is dissipated out of its wateriness (1b) and makes it like lead [*raṣāṣ*]; (1c) for [*fa-*] molten [*ḏāʾib*] lead is like mercury (1d) and solidified mercury is like frozen [*ǧāmid*] lead.

Šifāʾ 21.14-15, 17:

ومن شأن الزئبق أن ينعقد بروائح الكباريت، ولذلك يمكن أن يعقد بالرصاص أو رائحة الكبريت بسرعة ... وأما الرصاص فلا يشك مشاهدُه إذا ذاب أنه زئبق

(1ab) A characterstic of mercury is that it is solidified by the vapours of sulphurs [*rawāʾiḥ al-kabārīt*], and for this reason it is possible for it to be solidified quickly with lead [*bi-l-raṣāṣ*] or the vapour of sulphur. ... (1cd) In the case of lead, the observer has no doubt, when it melts, that it is mercury.

Why mercury is said to solidify "with lead" [*bi-l-raṣāṣ*] in the above passage of the *Šifāʾ* has confounded commentators.[67] While the text of the *Šifāʾ* as we have it is supported by Bahmanyār, *Taḥṣīl* [Mutahhari] 720.5-6, Abū al-Barakāt has evidently interpreted IS differently and

[66] Fakhr al-Dīn al-Rāzī simply follows IS at *Mabāḥiṯ* II.213.1f.: وبياض الزيبق من بياض الارضية اللطيفة وصفاء المائية من ممازجة الهوائية ("The whiteness of mercury [arises] from the whiteness of the fine earth and the clarity of the wateriness [and] from the mixture of airiness").

[67] Holmyard-Mandeville (1927) 38 n.5, suggested that IS might have confused lead with galena (PbS), or that *raṣāṣ* should be corrected to *zarnīḫ*. Cf. Garbers-Weyer (1980) 90f.

talks of mercury being solidified *into* something like lead by the vapour of sulphur.[68] The same interpretation is found in Fakhr al-Dīn al-Rāzī.[69] - This notion of "sulphur" or the "vapour of sulphur" solidifying "mercury" and making it "like lead" is what provides the basis of the theory that all metals are formed from mercury and sulphur, which is developed in theories III.ii.2-3 below and in the corresponding parts of the *Šifā᾿*.

l. 8. ܐܒܪܐ ("lead"): رصاص Abū al-Barakāt/IS. - In view of the low fusion point ascribed to the metal at *Šifā᾿* 21.17, it is best to take *raṣāṣ* in the passages quoted above as meaning "lead". - See comm. on *But*. III.ii.3.6 below.

<u>III.ii.2.1-5</u>: Mercury as basis of metals: first proof

The line of argument in the the passage of the *Šifā᾿* quoted below is a somewhat complicated. IS is trying to prove that mercury is the 'element' of all metals on the ground that they turn into mercury upon fusion, but he has a problem since, while this may apply to lead, metals other than lead, when fused, turn red, which is not the colour of mercury. As a result, IS is forced to say (i) that because of the strong heat involved the mercury in these metals turns red. (ii) He then gives the case of lead and (iii) goes on to argue that lead, too, though it is fused without turning red, does turn red like the others if heated more intensely. BH simplifies this in *But.* by omitting the second step of the argument, which he has, at it were, transferred to III.ii.1.7-9 above.[70]

---

[68] If, however, solidification by the "vapour of sulphur" means conversion into sulphides as Homyard-Mandeville understand it (38 n.4), the resulting product would be cinnabar (HgS), which is not exactly "like lead".

[69] *Mabāḥiṯ* II.213.6f.: وثالثها ان الزيبق يمكن لن يعقد برائحة الكبريت حتى يكون مثل الرصاص ("The third [proof that mercury is the element of metals] is that mercury can be solidified with the vapour of sulphur, so that it becomes like lead"); similarly, *Cand*. 88.3.

[70] Fakhr al-Dīn al-Rāzī (*Mabāḥiṯ* II.213.4f.) simplifies the argument in a different way, reversing the order of steps (i) and (ii) and omitting step (iii): اما رصاص فلا شك عند ذوبه انه زيبق واما سائر الاجساد فانها عند الذوب تكون زيبقا محمرا. ("As for lead, there is no doubt, when it is fused, that it is mercury. The remaining metals turn into *reddened mercury* upon fusion"). The argument is simplified further in *Cand.* (88.1-2: "[metals] when molten resemble mercury in constitution"), although BH's awareness of the problem is reflected in the phrase "in constitution" (*b-quyyāmā*; i.e. "in constitution, though not in colour").

But.: (1a) That mercury is the element of all fusible minerals is known (1b) from the fact that they revert [*hāp̄kīn*] to it upon being fused [*b-metpaššrānūt-hōn*]. (2) The mercury in all those (fusible minerals) which are fused by a strong heat [*ḥammīmūtā qšītā*] turns red. (3a) [In the case of] lead too, which is fused by a weak heat, (3b) when the fire is made to burn more [strongly] below it, (3d) its mercury turns fiery red [lit. is reddened with a fiery redness].

*Šifā*ʾ 21.15-18:

فيشبه أن يكون الزئبق أو ما يشبهه هو عنصر جميع الذائبات، /١٦ فإنما كلها عند الذوب تصير إليه لكن أكثر ما يكون ذوبه بعد الحمى، فيرى زئبقه /١٧ محمرا. وأما الرصاص فلا يشك مشاهدُه إذا ذاب أنه زئبق، لأنه يذوب قبل الحمى، /١٨ وإذا حمى فى الذوب كان لونه كلون سائر الذائبات، أعنى فى الحمرة النارية.

[18 كلون : لون Ṭ, Holmyard]

(1a) It seems that mecury or something resembling it is the element of all fusible [minerals]; (1b) for all of them, upon fusion [*ʿinda al-ḏaub*], turn into (mercury). (2) But its fusion usually occurs after intense heating [*ḥummā*], so that its mercury appears reddened. (3a) In the case of lead, the observer has no doubt, when it is fused, that it is mercury, because it is fused before intense heating. (3b) When it is heated intensely in the process of fusion, (3d) its colour becomes like the colour of other fusibles, i.e. fiery red.

III.ii.2.5: Second proof

But.: (1a) It is also known from the fact that (mercury) clings to all of them (1b) that (fusible minerals) are made of (mercury).

*Šifā*ʾ 21.18-19: ولذلك ما يعلق الزئبق بهذه الأجساد كلها، لأنه من جوهرها.

For this reason, mercury adheres to all these bodies [i.e. metals], [namely] because it is of their substance.

l. 5. ܡܬܕܒܩ ("clings"): يعلق IS. - What is meant is the amalgamation of mercury with other metals (so Holmyard-Mandeville [1927] 39 n. 4).

III.ii.2.6-7: Variation

But.: (1a) Their formation [*hwāyā*], however, varies according to [lit. with, *b*-] differences in (the mercury) itself [*dīleh*] (1b) and in the sulphur which solidifies it, (1c) as will be shown.

*Šifā*ʾ 21.19-22.2:

لكن هذه الأجساد يختلف تكونها عنه بسبب اختلاف الزئبق وما يجرى مجراه فى نفسه؛ وبسبب اختلاف ما يخالطه حتى يعقده

(1a) But the formation [*takawwun*] of these bodies differs by reason of [*bi-sababi*] the difference of the mercury itself, or whatever takes its place, (1b) and by reason of the difference of what is mixed with (the mercury) so as to cause its solidification.

l. 6. ܟܒܪܝܬܐ ("sulphur"): ما يخالطه ("what is mixed with it") IS. - That IS means "sulphur" by this phrase is clear from what follows.

l. 7. ܐܝܟܢܐ ܕܡܬܚܘܐ ("as will be shown"): i.e. in the following theory.

III.ii.2.7-8: "Parents" of metals

But.: The "parents" [*abāhē*] of the fusible μεταλλικοί, therefore [*ara*], are mercury and sulphur.

Abū al-Barakāt, who rejects the "sulphur-mercury" theory makes a passing reference to the idea given here at *Muᶜtabar* II.230.8: لانه الام والمادة ("... because (mercury) is [supposedly] the mother and the material [of metals]"). The idea is found in at least three earlier works of BH and elsewhere (cf. Wiedemann [1907] 71).

BH, *Tract*., Cantab. Add 2003, 57v 4: ܘܐܒܗ̈ܝܗܘܢ ܕܟܠܗܘܢ ܙܝܘܓ ܘܟܒܪܝܬܐ
The parents of them all are mercury and sulphur.
BH, *Cand*. 87.4-88.1:
ܟܝܢܝ̈ܐ ܕܝܢ ܘܐܟܝܡܝ̈ܐ ܡܡܠܠܝܢ ܐܡܪܝܢ ܕܐܒܗ̈ܝ ܡܛܠܠܝܩܘ ܙܝܘܓ ܘܟܒܪܝܬܐ ܐܝܬܝܗܘܢ.
Natural philosophers [*kyānāyē*] and alchemists say that the "parents" of the μεταλλικοί are mercury and sulphur.
BH, *Rad*. [Istanbul] 12.1: ܐܒܗ̈ܝ ܡܛܠܠܝܩܘ ܠܙܝܘܓ ܢܕܥܝܢ ܘܠܟܒܪܝܬܐ
We acknowledge mercury and sulphur as the parents of the μεταλλικοί.
Bīrūnī, *Ǧamāhir* 229.17-18: وقد ذكر الطبيعيون ان الكبريت ابو الأجساد الذائبة والزئبق أمها
Natural philosophers [*tabīᶜīyūn*] have said that sulphur is the father of fusible bodies and mercury their mother.
Dimashqī, *Nuḫba* [Mehren] 56.7 (tr. Mehren 62):
فالزيبق أصل المعادن وأمّها كما أن الكبريت أصلها أيضا وأبوها
Mercury is the root [*aṣl*] and mother of the metals, just as sulphur is also their root and father.[71]

III.ii.3.: Formation of metals from mercury and sulphur

The formation of metals from mercury and sulphur is also discussed by BH at *Cand*. 88.3-11. As with the classification of minerals (see III.i.5.1-6), that passage of *Cand*. is dependent on Fakhr al-Dīn al-Rāzī, *Mabāḥiṯ* II.213.11-20, and closely related passages are also found in Bar Shakko, *Dial*., Göttingen, Or. 18c, 303v b9-304a 11 (Ruska [1897] 158.7-16) and Qazwīnī, *ᶜAǧāʾib* 204.29-205.8. References will occasionally be made in the following notes to these passages, often simply as "Fakhr al-Dīn al-Rāzī et al."[72]

[71] The idea of mercury and sulphur as the "roots" (*aṣlāni*) of metals is found in Ikhwān al-Ṣafāʾ [Bustānī] II.121.3 (tr. Dieterici [1876] 130): واعلم أن الكبريت والزئبق أصلان للجواهر المعدنية الذائبة ("Know that sulphur and mercury are the 'roots' of fusible mineral substances").

[72] Fakhr al-Dīn al-Rāzī et al. differ from IS in introducing the notions of "coction"

III.ii.3.1-6: Formation of metals (1): silver, gold, copper, iron

But.: (1a) From pure [*dakyā*] mercury (1b) and white and pure sulphur which is not burnt [*lā yaqqīdtā*], (1c) nature creates silver. (2a) If the sulphur, as well as being pure [lit. together with its purity], is even better, has [*qanyā*] a subtle, fiery power and [is] clear-coloured [*naṣṣūpā*] and not burnt, (2b) when it meets pure mercury, (2c) it turns it into gold. (3a) If the substance of the mercury is good (3b) but the sulphur is bad [*bīšā*] and burnt, (3c) copper is formed. (4a) If the mercury is bad, impure [*ṭmaʾ*] and earthy, (4b) and the sulphur too is impure, (4c) iron is formed.

Šifāʾ 22.2-7:

فإن كان الزئبق نقيا وكان ما يخالطه فيعقده قوة كبريت أبيض/٣ غير محرق ولا درن، بل هو أفضل مما يتخذه أهل الحيلة، كان منه الفضة. وإن كان/٤ الكبريت مع نقائه أفضل من ذلك وأنصع، وكان فيه قوة صباغية نارية لطيفة غير محرقة/٥ أفضل من الذى يتخذه أهل الحيلة، عقده ذهبا. ثم إن كان الزئبق جيد الجوهر، ولكن/٦ الكبريت الذى يعقده غير نقى، بل فيه قوة احتراقية، كان منه مثل النحاس. وإن/٧ كان الزئبق رديئا دنسا متخلخلا أرضيا، وكان كبريته نجسا أيضا، كان منه الحديد.

(1a) If the mercury is pure [*naqīy*], (1b) and what is mixed with it and solidifies it is the power of white sulphur, which is unburnt [*ġair muḥraq*] and is not filthy, but is better than what the alchemists [*ahl al-ḥīla*] use, (1c) silver is produced from it. (2a) If the sulphur, besides being pure, is [even] better and clearer [*anṣaʿ*] [in colour] than that [just described], and contains [*kāna fīhi*] a tinctorial, fiery, subtle, unburnt power, [and is] better than that which the alchemists use, (2bc) it solidifies (the mercury) into gold. (3a) Next, if the mercury is good in substance, (3b) but the sulphur which solidifies it is not pure but has in it a power of burning [*qūwa iḥtirāqīya*], the likes of copper are formed from it. (4a) If the mercury is bad, unclean [*danis*], loose and earthy, (4b) and its sulphur too is impure [*nağis*], (4c) iron is formed from it.

l. 1. ܠܐ ܝܩܝܕܬܐ ("unburnt"): غير محرق IS. - IS talks of the sulphur in silver as being غير محرق (22.3), that in gold as having a قوة غير محرقة (22.4) and that in copper as having a "*qūwa iḥtirāqīya*" (22.6). The word "محرق" in the first two cases could be read as active (*muḥriq*) or passive (*muḥraq*), but the adjective *iḥtirāqīya* in the last must have an intransitive sense, and this would suggest that "محرق", too, should be taken as passive.[73] This, at any

---

and "ripening" (*tabḫ*, *nadğ*) and the role played by heat and cold. In this, they are in agreement with the Ikhwān al-Ṣafāʾ (ed. Bustānī, II.106.14ff.; tr. Dieterici [1876] 114; cf. also II.116.13ff.; tr. Dieterici 125.), who mention the purity of the mercury and sulphur required for formation of gold, but otherwise attribute the differences of metals largely to degrees of coction and ripening rather than to the qualities of the mercury and sulphur. Qazwīnī differs further from the others in placing gold before silver and giving a number of additional conditions required for formation of gold. In these respects, he is in greater agreement with the Ikhwān al-Ṣafāʾ.

[73] Holmyard-Mandeville (p. 39) evidently take "محرق" as active ("sulphur which neither induces combustion ..."; "sulphur ... possesses a ... non-combustive virtue"), although in *qūwa iḥtirāqīya* the verb is understood intransitively, as it must ("...

rate, is how BH must have taken the word, since the participial adjective *yaqqīd*, which he uses here and in ll. 3 and 5, as well as in the corresponding passage of *Cand.* (88.5, 8, 10), cannot be transitive. - As to what is then meant by *muḥraq* a number of interpretations are possible (i. "set on fire, [actually] burning"; ii. "burnt"; iii. "combustible"), but if the phrase "*qūwa muḥraqa*" can be equated with "*qūwa iḥtirāqīya*", it would seem easiest to understand it in the sense of "combustible".[74] - It is more difficult, however, to assign the meaning "combustible" to BH's *yaqqīḏā* ("*mawqdānā*" used by Bar Shakko would fit this meaning better). In considering what BH means by *yaqqīḏā*, it is worth noting that he avoids saying "sulphur has a *yaqqīḏā* power" and says instead "sulphur is *yaqqīḏā*" (see on ll. 2-3 below). It is also to be noted that in line 3 below BH couples *lā yaqqīḏā* with *naṣṣūpā*, which must refer to the colour of the sulphur. Bakoš, in fact, in his translation of *Cand.*, renders *yaqqīd* simply as "brun foncé" (tawny, dark brown). This is probably essentially correct. It seems that BH understood IS's *muḥraq* in the sense of "burnt, parched, charred".[75]

---

possessing ... a property of combustibility"). Holmyard-Mandeville may have been influenced in this by the Latin version of the *Šifā'* (Holmyard-Mandeville, pp. 52f.: "sulphuris ... non urentis"; "vis igneitatis ... non urentis"; "sulphur ... quod non sit in eo vis adurens"; the negative "non" in the last of these is erroneous). - Translators are equally divided over *muḥraq*/*muḥriq* at Qazwīnī, *ʿAǧāʾib* 205.7 (of sulphur in iron), with Ethé taking it as passive ("verbrannt") while Garbers-Weyer, Giese and Bakoš (*Cand.* p. 87 n.1) apparently take it as active ("brennend, brûlant"). There, the passive reading is supported by Fakhr al-Dīn al-Rāzī, who has *muḥtariq* at the corresponding place. - The situation, of course, is not quite so clear cut as I suggest above, since, fire being contagious, something which is being burnt will cause other things to burn.

[74] This could, then, perhaps be brought into connection with such passages as the following: Jābir, *K. al-īḍāḥ*, Holmyard (1928) 54.11-14: ... فكان الطف تلك الكباريت واصفاها واعدلها الكبريت الذهبى فلذلك انعقد به الزيبق عقدا محكما معتدلا ولاعتداله قاوم النار وثبت فيها فلم تقدر على احراقه كقدرتها على احراق سائر الاجساد ("... the most subtle, clean and even of those sulphurs is the sulphur of gold. For this reason, the mercury is solidified by it firmly and evenly and because of its evenness (gold) resists fire and remains unaffected in it. For (fire) cannot burn [i.e. calcine, so Garbers-Weyer 88] it in the way it can burn other bodies [i.e. metals]"; cf. Kraus [1942/3] II.1; Garbers-Weyer [1980] 34f., 88; Ikhwān al-Ṣafāʾ [Bustānī] II.116.5-9; tr. Dieterici [1876] 124). - Although what is "incombustible" here is gold rather than sulphur, it is an easy step to transfer this incombustibility from gold to its sulphureous content. - The passage quoted above, in fact, allows a different translation: "... and because of its evenness (this sulphur) resists fire and remains unaffected in it, so that (fire) cannot burn (gold) in the way it it can burn other metals." The 'incombustibility" (i.e. incalcinability) of gold could, then, be attributed to the incombustibility of its sulphur.

[75] How sulphur might have become "burnt, charred" may be understood, if we remember the role played by heat in the process of its formation as described in III.i.3. above. - We might also note the mention of the use of sulphur "untouched by fire" for treatment of leprosy in IS's *al-Qānūn fī al-ṭibb* [ed. Cairo 1986], 559.23, s.v. كبريت: من أدوية البرص خصوصًا ما لم تمسّه النار.

l. 2. ܒܪܐ ܟܝܢܐ ("nature creates"): That is to say, "nature" in contrast to human artifice as practised by the alchemists. Although IS does not use the word "nature" here, this is implied in the phrase "better than [the sulphur which] the alchemists use" (cf. III.iii.1.1-4 below). - The contrast is underlined here by the use of the verb *brā*, the word used of divine creation at Gen. 1.1 (cf. the use of *etbrī* at *But*. Mete. V.ii.2.11, V.iii.1.7 below). - Here again, we observe a case where BH has supplied the implied subject and substituted an active verb for a passive or intransitive verb in IS (here, *kāna*; see on Min. I.i.3.7 *appāyē* above).

l. 2-3. ܚܝܠܐ ܢܘܪܢܝܐ ܘܩܠܝܠܐ ܩܢܝܐ ܘܢܨܘܦܐ ܠܐ ܝܩܝܕܐ ("has a subtle, fiery power and [is] clear-coloured and not burnt"): فيه قوة صباغية نارية لطيفة غير محرقة IS. - In IS, the word "power" (*qūwa*) is accompanied by four adjectives. In BH too, it is possible to take all four adjectives attributively with *ḥaylā* (masc.), but the position of the verb *qanyā* suggests that the last two adjectives, *naṣṣūpā* (on which, see below) and *lā yaqqīḏā*, should rather be taken as predicates of *keḇrīṯā* (fem.). - The thinking behind the alteration made by BH is not difficult to understand. While it is natural to speak of a "subtle, fiery power", an "unburnt power" is more awkward. BH similarly avoids talking of sulphur as having a "combustible power" (*qūwa iḥtirāqīya*) at lines 4-5 below by simply saying "sulphur is burnt".

l. 3. ܢܘܪܢܝܐ ("fiery"): نارية IS. - *But*. omits IS's "tinctorial" (*ṣibāġīya*), while Fakhr al-Dīn al-Rāzī et al. omit "fiery" (*nārīya*) at the corresponding place. Although the omission may simply be due to oversight in both cases or, in the case of "fiery", due the difficulty of reconciling this with "incombustibility", it is possible that these two words were felt to be synonymous, since the property of "tincture" (*ṣibġ*) seems to have been closely associated with fire by alchemists.[76]

l. 3. ܢܨܘܦܐ ("clear-coloured"): أنصع IS. - The word is probably to be read as *naṣṣūpā* (so Brockelmann 443b., cf. Min. I.ii.1.4 *qayyūmāṯā* above). - Since the word *naṣṣūpā/nāṣōpā* is defined in Bar Bahlūl and other lexica as *nāṣiᶜ* (also *ṣāfin*)[77] and the phrase "*summāqā naṣṣūpā*" at *But*. Mete. II.iii.5.3 and II.iii.7.3 corresponds to IS "*ḥumra nāṣiᶜa*", "*nāṣiᶜ al-ḥumra*", it seems safe to assume that the word here renders IS's *anṣaᶜ* (elative of *nāṣiᶜ*), rather than *ṣibāġīya* which stands in the corresponding place. - The Syriac lexica also define *naṣṣūpā/nāṣōpā* as "red" (*aḥmar*),[78] and this is of note since Fakhr al-Dīn et al. do not mention the clear colour of the sulphur in gold but say that it is "red" (*aḥmar*; *Cand*. 88.5 *summāqtā*).[79] I hesitate, however,

---

[76] See Kraus (1942/3) II.5.

[77] See Bar Bahlūl 1268.10-14, with n. 15; PS 2441. On *nāṣiᶜ*, see Freytag 288a, Kazimirski 1273a.

[78] See preceding note; cf. the application of *naṣṣūp* to the colour of blood in *The Syriac Book of Medicines*, ed. Budge, 201.ult.; also *nuṣpā*: "twilight, afterglow".

[79] See also the Latin version of the *Šifāʾ*, where IS's *anṣaᶜ* seems to have become "cum rubore clarum" (Holmyard-Mandeville 53.2).

to translate *naṣṣūpā* here as "red", since the elative form of the word *anṣa*ᶜ implies that the sulphur in silver is also *nāṣi*ᶜ, but the sulphur in silver is not "red" but "white".[80]

III.ii.3.6-9: Formation of metals (2): tin

But. [underline: agreement with *Šifā*ᵓ; italics: with *Mu*ᶜ*tabar*]:

(5a) The *mercury* in lead [*abbārā*] is good, (5b) but the sulphur is *bad*, (5c) and because it contains *airiness* (5d) which *is not strongly mixed* [*d-lā mzīgā ḥīṣāᵓīt*] with them, (5e) when it is *pressed*, (5f) it *escapes* (5g) and as it escapes it lets out a tingling sound [*qālā zgāgānāyā*].

*Šifā*ᵓ 22.8-9:

وأما الرصاص القلعي فيشبه أن يكون زئبقه جيدا ، إلا أن كبريته ردئ وغير شديد / ٩ المخالطة، وكأنه مداخل إياه سافا فسافا ، فلذلك يصر.

[9 يصرّ Holmyard: يضر Cair.]

(5a) As for tin [*al-raṣāṣ al-qala*ᶜ*ī*], it appears that its mercury is good, (5b) but its sulphur is bad; (5d) it is not strongly mixed, and it is as if it has penetrated it layer by layer. (5g) For this reason it shrieks [*yaṣirru*].

*Mu*ᶜ*tabar* II.231.8-10:

ومن طاهر الزئبق وردئ الكبريت الرصاص قالوا ولرداءة مزاجه وقلة امتزاجه يصر وانما يصر لهوائية مخالطة غير ممتزجة يخرجها العصر

[يصر bis: يضر ed. Hyderabad]

(5ab) From clean mercury and bad sulphur [alchemists claim to make] lead [tin, *raṣāṣ*]. (5g) They say that it shrieks [*yaṣirru*] (5d) because of the badness of the mixture [*mizāǧ*] and the lack of mixing [*imtizāǧ*], (5g) but [in reality] it shrieks (5c) because of airiness that is [physically/mechanically] mingled [*muḫāliṭ*] but not [chemically] mixed [*mumtaziǧ*] [with the lead/tin], (5e) which squeezing (5f) causes to escape."

Abū al-Barakāt, who does not accept the "sulphur-mercury theory", here also rejects the attribution of the "cry" to "weak mixture" and "stratification" and provides his own explanation of the phenomenon. BH, while accepting the "sulphur-mercury theory", rejects IS's explanation of the "cry" in favour of Abū al-Barakāt's. In doing so, he ignores the distinction between mechanical and chemical mixture (*ḫalṭ* and *imtizāǧ*).[81]

---

[80] Hence, presumably, Holmyard-Mandeville's translation of *anṣa*ᶜ as "whiter". - The ground for the substitution of "red" for "more clear-coloured" in Fakhr al-Dīn al-Rāzī et al. might be sought, besides in the fact that in comparison with silver the colour of gold is closer to red, in the rarity and the consequent high value of red sulphur; see Abū Bakr al-Rāzī, *K. al-asrār* 3.13, tr. Stapleton et al. 371, tr. Ruska 85 (cf. p. 41); Ps.-Arist. *K. al-aḥǧār* 112.10 (cf. Qazwīnī, ᶜ*Aǧā*ᵓ*ib* 243. penult.).

[81] On "chemical" and "mechanical" mixture, see Garbers-Weyer (1980) 73. - The following passage of the Ikhwān al-Ṣafāᵓ is of some interest here, since it may well be

l. 6. ܐܒܪܐ ("lead", understand "tin"): الرصاص القلعى IS; الرصاص Abū al-Barakāt. - Since syr. *abbārā* is usually "lead" and *abbārā* (corr. IS *raṣāṣ*) at III.ii.1.8 above, with its low fusion point, must mean "lead", I translate the word here too as "lead", although we must be talking about "tin" here. Having done so, I translate *ānkā* in l. 8 below (*al-ānuk* IS, *al-usrub* Abū al-Barakāt), which usually means "tin" in Syriac, as "tin", although we are in fact talking there about "lead". - The usage in *Cand*. is the other way round (88.3 *ānkā*, "lead", corr. Fakhr al-Dīn al-Rāzī *raṣāṣ*; 88.9 *ānkā*, "tin", corr. *raṣāṣ*; 88.11 *abbārā*, "lead", corr. *ānuk*); that in Bar Shakko, *Dial*. the same as in *But*. (Ruska [1897] 158.13 *abbārā*, "tin", corr. Qazwīnī *raṣāṣ*;158.16 *ānkā*, "lead", corr. *usrub*).

l. 6. ܕܐܒܪܐ ܕܝܢ ܙܝܘܗ ("The mercury in lead is ..."): lit. "Of lead, its mercury is ...". - The construction is analogous to that of the Arabic (*ammā* x, *fa*-y-*hu*), except that the first noun is preceded by the particle *d*-. The same construction is used in l. 8 below.

l. 8. ܩܠܐ ܙܓܓܢܝܐ ("a tingling sound"): The reference is to the so-called "cry" of tin (*Zinngeschrei*), the sound caused by the friction of the crystalline particles when tin is subjected to bending.[82] - The word normally used to describe the sound in Arabic is *ṣarīr* ("chirping, stridulation, screeching"), appearing in the verbal form *yaṣirru* in the passage of IS quoted above. - The adjective *zgāgānāyā* used by BH is not given in dictionaries. *Zgāgā*[83] is explained in lexica as "ringing (in the ear)" (*dawīy al-āḏān*, *ṭanīn*)[84] and

---

the source of IS, but at the same time also mentions "airiness" as the cause of the "laxity" (*raḫāwa*) of tin: Ikhwān al-Ṣafā᾽ [Bustānī] II.119.17-21 (tr. Dieterici [1876] 128f.): وأما القلعى فهو قريب من الفضة فى لونه ولكن يباينها بثلاث صفات: الرائحة والرخاوة والصرير؛ وهذه الآفات دخلت عليه وهو فى معدنه كما تدخل الآفات على الجنين وهو فى بطن أمه. فرخاوته لكثر هوائيته، وصريره لغلظ كبريته وقلّة مزاجه بزئبقه وهو سافٌ فوق ساف فلذلك يصر، وتنتن رائحته لقلة نضجه ("Tin is close to silver in its colour but is distinguished from it by three characteristics, smell, laxity and shriek. These defects have entered them while they were in the mine, just as defects enter the fetus while it is in its mother's womb. Its laxity is due to the large quantity of its airiness [*hawā᾽īya*]. Its shriek is due to the thickness of its sulphur and the inadequacy of its mixture with the mercury; [it is arranged] layer upon layer and for this reason it shrieks. Its foul smell is due to the inadequacy of its coction."; Dieterici translates *hawā᾽īya* as "mercury": "Das Zinn ist so weich wegen der Menge des darin enthaltenen Quecksilbers"). - Cf. further Ps.-Arist., *K. al-aḥǧār* [Ruska] 123.1-2; also Qazwīnī, *ʿAǧā᾽ib* [Wüstenfeld] 207.20-22, where the same three defects are mentioned and we also find the comparison of the "mine" to the mother's womb, but no details are given as to the cause of the defects. - Ps.-Apollonius [Balīnūs] attributes the "cry" to "moisture": *K. sirr al-ḫalīqa* III.2.3 [Weisser] 229.11-13: لأن رطوبته كثيرة، فلذلك صار له صَرِيرٌ وصريره فى رطوبته ونَتْنه فى سواده وسِواده فى يبسه ويبسه فى برده وبرده مع روحه، لا مع جسده. (see also III.4.4, 252.4).

[82] See Holmyard-Mandeville (1927) 40 n.1; Garbers-Weyer (1980) 91.

[83] Itself apparently derived from *zaggā*, "bell", the form perhaps being due to analogy with *zmāmā*.

[84] See Bar Bahlūl [Duval] 675.3 s.v. ܙܓܓܐ ܕܐܕܢܐ (*sic*); Bar Ali [Hoffman] 123

is used in the form *zḡāḡē d-eḏnē*, for one of the symptoms preceding vertigo, in the *Syriac Book of Medicines* [Budge] 31.13.[85]

III.ii.3.8-10: Formation of metals (3): lead

But.: (6a) The mercury in tin [*ānkā*] is bad and clotted [lit. killed] by clayeyness [*b-ṭīnāyūṯā qṭīl*], (6b) and its sulphur too is bad, fetid [*saryā*] and weak, (6c) and for this reason its solidification is feeble and slack [*naššīš wa-rp̄ē*].

Šifā' 22.9-11:

وأما الآنك فيشبه أن يكون/ ١٠ ردئ الزئبق، تقتله طينية، ويكون كبريته رديئا منتنا ضعيفا، فلذلك لم يستحكم/ ١١ انعقاده.

[10 تقتله طينية scripsi (v. comm. infra): ثقيلة طينته Cair.[86]; منتنة طينية B[87]; ثقيلة طينية Holmyard[88]; ثقيلا طينيًا Garbers-Weyer]

(6a) As for lead [*ānuk*], it seems that it is bad in [its] mercury - which clayeyness is killing -, (6b) while its sulphur is bad, fetid [*munattan*] and weak; (6c) and for this reason its solidification is not firm [*lam yastaḥkim inʿiqāduhu*].

l. 8. ܘܒܛܝܢܝܘܬܐ ܩܛܝܠ ("and clotted by clayeyness"): lit. "and killed by clayeyness" - The corresponding phrase in the *Šifā'* has caused editors difficulties. BH seems to have read تقتله طينية, and this may well be the correct reading. How mercury can be 'killed' may be understood if we remember the use of the word "alive" (Arab. *ḥayy*, as in "quicksilver", Fr. "argent vive") with reference to its fluidity (see comm. on III.i.2.7 *ḥayyṯā* above). For the use of the word 'kill' and 'dead' with reference to mercury, see, for example, Rāzī, *K. al-asrār* [Dānishpazhūh] 16.19 (tr. Ruska 105.16): تأخذ آبقا حيا فتقتله بمرقشيثا بوزنه ("Take live mercury and 'kill' it with marcassite of the same weight")[89]

l. 9. ܒܝܫܐ ("bad") scripsi: ܝܒܝܫܐ ("dry") codd.; IS رديا IS. - I correct the reading of the manuscripts here, since the alteration of *bīšā* into *yabbīšā* is more like than that of *radī'an* into *yabisan/yābisan*, although the latter is not inconceivable since a variant reading *yābisan* (يابسا) instead of *danisan* (دنسا) is reported at *Šifā'* 22.7 (in the sentence on "iron").

---

no. 3421; PS 1080.

[85] The phrase here in *But*. is rendered as *ṣaut rannān* by Behnam (1975) 33.18 in his paraphrase.

[86] Followed by Anawati (1996) 877 ("having heavy clay").

[87] Cf. *Muʿtabar* II.231.10f.: وردئ الزئبق ومنتنه مع ردئ الكبريت يكون منه الاسرب ("From bad and fetid mercury, [mixed] with bad sulphur, lead is formed").

[88] Followed by Baffioni (1980) 96.6 ("pesante, argilloso").

[89] Cf. further, ibid. 18.21 (tr. Ruska 108.8); 16.5 (tr. 104.24); 15.15 (where the word "dead" is missing in ed. Dānishpazhūh, but is included in tr. Stapleton et al. [1927] 386.24 and tr. Ruska 103. fin.); Abū al-Barakāt, *Muʿtabar* II.230.6.

Min. III. Section iii.: *On Imitation of these Bodies by Means of Each Other*

The veracity of alchemy or, in more concrete terms, the possibility of converting one metal into another, was, along with the related problem of the veracity of astrology, the subject of an ongoing dispute among scholars in the Islamic world.[90]

BH has already dealt with these subjects in the preceding book of *But.*, in sections entitled "on destiny and refutation of astrologers" (Gen. et corr. II.v. *meṭṭul ḥelqā w-makksānūṯ asṭrōlōḡō*, ms. F fol. 44v-45r) and "on the falsity of the art of alchemy" (Gen. et corr. IV.iii. *meṭṭul lā ḥattīṯūṯā d-ummānūṯā d-ḵemele'a*, fol. 50r-51r). Both these sections are based closely on Abū al-Barakāt (respectively, *Muᶜtabar* II.232.24ff., 231.20-232.23).[91]

The first two 'theories' of the present section may be regarded as a paraphrase of the well-known passage refuting alchemy in the *Šifā'* (p. 22-23)[92], although BH makes certain departures from the text of the *Sifā'* especially towards the end of the second theory. The last 'theory' of the section, on the other hand, is based on Abū al-Barakāt.

l. 2. ܡܬܕܡܝܢܘܬܐ ("imitation"): more literally, "the being imitated", or "the ability to be imitated".

III.iii.1.1-4: Possibility of imitation

But.: (1a) Technicians [*māray pursē*] are not unable [*law lā mṣēn*] by artificial means [*b-ummānūṯā*] to solidify mercury using sulphurs - (1b) even if [*āp̄en*] the action of [human] artifice [*ᶜḇāḏ ummānūṯā*] does not attain [*lā māṭē*] to [the level of] the action of nature, (1c) but merely resembles it in outward appearance [*barrānā'īt*] - (1d) so that [*badḡūn*] it appears to those lacking in intelligence, and many believe [*haymnīn*], (1e) that the generation [*hwāyā*] of these bodies by nature is the same as this artificial generation of them.

---

[90] For a survey of the subject, see Ullmann (1972) 249-255; id. (1979) 113f.; on the debate concerning astrology, Ullmann (1972) 274-277.

[91] For criticism of astrology in BH, see also *Eth.* IV.1.2 [Bedjan] 318.13-16.

[92] Translations: Holmyard-Mandeville (1927) 40-42; Garbers-Weyer (1980) 38-40; cf. also the comments in Garbers-Weyer 91f.; cf. further Ruska (1934); Anawati (1996) 875-879; Sezgin, *GAS* IV.8f.

Šifā' 22.11-15:

وليس يبعد أن يحاول أصحاب الحيل حيلا تصير بها أحوال انعقادات الزئبق / ١٢ بالكباريت انعقادات محسوسة بالصناعة، وإن لم تكن الأحوال الصناعية على حكم / ١٣ الطبيعية وعلى صحتها، بل تكون مشابهة أو مقاربة لذلك، فيقع التصديق بأن جهة كونها / ١٤ فى الطبيعة هذه الجهة، أو مقاربة لها، إلا أن الصناعة تقصر فى ذلك عن الطبيعة / ١٥ ولا تلحقها وإن اجتهدت.

[13 الطبيعية : الطبيعة STṬM, Holmyard et F marg. ‖ مشابهة: متشابهة B et F marg. ‖ أو مقاربة: ومقارنة Holmyard]

(1a) It is not unlikely that experts [*aṣḥāb al-ḥiyal*] are able to devise means [*yuḥāwila ḥīlan*] whereby the solidifications of mercury by sulphur become solidifications perceptible to the senses by artifice [*bi-l-ṣināʿa*] - (1b) even if [*wa-in*] the artificial conditions [*al-aḥwāl al-ṣināʿīya*] do not attain [*lam takun*] the level and accuracy of the natural (conditions), (1c) but are [merely] similar [*mušābiha*] or close to it - (1d) so that [*fa-*] the belief [*taṣdīq*] arises (1e) that the manner of their generation [*kaun*] in nature is [the same as] this [artificial] manner [of generation] or close to it, except that art falls short and cannot overtake nature in that respect, even if it tries.

l. 1. ܠܘ ܠܐ ܡܨܝܢ scripsi ("are not unable"): ܠܘ ܠܐ ܡܨܢ codd. - On *law lā mṣēn* as a set phrase see PS 1898 (s.v. ܠܘ).

l. 1. ܡܪ̈ܝ ܦܘܪ̈ܣܐ ("technicians"): lit. "masters/possessors of techniques"; أصحاب الحيل IS. - Cf. on III.i.1.5 above.

l. 1. ܒܐܘܡܢܘܬܐ l ("by artificial means"): ܒܐܡܝܢܘܬܐ FMV ("always"): The reading of ms. l is supported by IS's *bi-l-ṣināʿa*.

l. 3. ܠܒܨܝܪ̈ܐ ܪܥܝܢܐ ("to those lacking in intelligence"): lit. "to defective minds". - Cf. III.iii.2.3 below.

III.iii.1.5-9: Impossibility of altering species

The passage is based mainly on *Šifā'* 22.16-23.2, but BH incorporates into it the notion of "specific difference" taken from a passage found a few lines later in the *Šifā'*.

But. [underline: agreement with *Šifā'* 22.16-23.2; italics: with *Šifā'* 23.5-7]: (1a) In reality it is not in the power of [*lā māṭyā b-īḏayhōn*] the alchemists [*pālḥay kemele'a*] to remove [*la-mšallāhū*] *the specific character* [*adšānāyūṯā*] of a certain species and to *provide* [*l-maʿṭāpū*] it with another specific character, (1b) but they are able [*meškḥīn*] merely [to effect] wonderful [*tmīhāṯā*] imitations [*mdammyānwāṯā*], (1c) as when they dye [*ṣābʿīn*] red copper with a white colour [*ṣubʿā*] which very much resembles [that of] silver, (1d) give silver a yellow colour which very much resembles [that of] gold, (1e) and cleanse leads [*abbārē*] of most of their impurity [*ṣā'ūṯā*] and obscure their blemishes [*mōmē*].

Šifā' 22.16-23.2:

وأما ما يدعيه أصحاب الكيمياء، فيجب أن تعلم أنه ليس فى أيديهم أن يقلبوا الأنواع / ١٧ قلبا حقيقيا، لكن فى أيديهم تشبيهات حسية، حتى يصبغوا الأحمر صبغا أبيض شديد ١٨ الشبه بالفضة،

ويصبغوه صبغا أصفر شديد الشبه بالذهب؛ وأن يصبغوا الأبيض أيضا /١ أى صبغ شاءوا، حتى يشتد شبهه بالذهب أو النحاس؛ وأن يسلبوا الرصاصات أكثر /٢ ما فيها من النقص والعيوب

[17 حسنة : حسية Holmyard]

(1a) Concerning what the alchemists [*aṣḥāb al-kīmiyāʾ*] claim, you ought to know that it is not in their power [*laisa bi-aidīhim*] to effect a true change [*yaqlibū*] of species [*anwāʿ*], (1b) but it is in their power to produce sensory [*ḥissīya*] imitations [*tašbīhāt*], (1c) so that they dye [*yaṣbuġū*] the red [metal[93]] with a white colour [*ṣibġ*] which closely resembles [that of] silver, (1d) and with a yellow colour which closely resembles [that of] gold. (1d) [They are also able] to dye the white [metal] with whatever colour they wish, so that it closely resembles gold or copper, (1e) and to cleanse leads [*raṣāṣāt*][94] of most of the defects [*naqṣ*] and impurities [*ʿuyūb*] which are in them.

Šifāʾ 23.5-7:

وأما أن يكون الفصل المنوع يسلب أو يكسى، فلم يتبين لى إمكانه؛ بل بعيد عندى جوازه، إذ لا سبيل إلى حل المزاج إلى المزاج الآخر

(1a) As regards [the notion] that the specific difference [*al-faṣl al-munawwiʿ*] can be removed [*yuslabu*] or imparted [*yuksā*], the possibility of this has not been clear to me; rather, its possibility seems remote to me, since there is no way of dissolving a mixture [and converting it] into another

l. 5. ܒܩܘܫܬܐ ܕ ("in reality"): The expression occurs again at Min. III.iii.3.4 and Mete. I.ii.3.3.

l. 5-6. ܠܡܫܠܚܘܬܗ . . . ܠܡܠܒܫܘ ("to remove ... to impart"): The metaphor is that of clothing, both these verbs, like IS's *salaba* and *kasā*, being words used of removing and putting on clothes.[95]

l. 5. ܐܕܫܝܘܬܐ ("specific character"): See PS 46; PS Suppl. 5; Brockelmann 7a; cf. *But*. Gen. et corr. IV.iii.1, ms. F 49r a21: ܡܬܚܪܝܢ ܐܝܬܝܗ ܐܕܫܝܘܬܗ ܐܕܫܝܘܬܐ; 49r a fin.-b 1: ܐܕܫܝܘܬܗ ܐܝܬܝܗ.

l. 6. ܠܐ ܒܐܝܕܝܐ ܐܝܬܝܗܘܢ ("it is not in their power to"): ليس فى أيديهم IS. - This may be a genuine Syriac expression (see PS 1548, s.v. ܝܕ, ܒܐܝܕܐ ܕ), although here it reflects the similar Arabic expression used by IS. In the following line the same expression in IS is rendered by *meškḥīn*.

l. 6-7. ܡܫܟܚܝܢ . . . ܬܕܡܝܬܐ ("are able [to effect] imitations"): The absence of the verb is slightly awkward, but the phrase may be understood as a translation of IS *fī aidīhim tašbīhāt*.

l. 6. ܬܡܝܗܬܐ ("wonderful"): Though not quite the same as *ḥasana* ("beautiful, excellent"), the word here is at least closer to this than to *ḥissīya* ("perceived") of the Cairo and Tehran editions.

---

[93] i.e. copper; cf. Garbers-Weyer (1980) 92.

[94] i.e. lead and tin; see Holmyard-Mandeville (1927) 41 n.4; Garbers-Weyer (1980) 92.

[95] Cf. the same metaphor at Ikhwān al-Ṣafāʾ [Bustānī] II.59.2: انخلاعها, 59.3 إذا انتزعت منها صورة ألبست أخرى (generation and corruption as clothing and divesting of form; tr. Dieterici [1876] 62.27, 30).

l. 7-8. ܐܟܡܐ ܕܠܚܫܝܐ ... ܢܚܫܡܢ ("as when ... gold"): IS gives four examples: imitation (a) with red metal (i.e. copper) of (i) silver and (ii) gold, and (b) with white metal (interpreted by BH as silver) of (iii) gold and (iv) copper. BH has reduced these to two (copper into silver; silver into gold).

III.iii.2.1-9: Impossibility of altering species (contd.)

The line of argument is essentially the same as that of the *Šifāʾ*, although the material has been rearranged (especially through incorporation into his sentence (1) of materials taken from that corresponding to sentence (3)). For sentence (2), where BH expands on what IS had to say, one suspects an additional source.

But.: (1a) In these imitations, the essential distinctions [*puršānē ūsiyāyē*] whereby these bodies are differentiated into species [*metaddšīn*] are retained, (1b) while the accidental colours and consistencies which are not a part of [lit. do not enter] their quiddity are altered. (2a) Therefore no intelligent person would dispute, and would not accept, the possibility of an imitation (2b) which reaches an utmost level [of perfection] [*sākē ḥrāyē*] through subtle and proven art and [through] much practice and assiduous experimentation, (2c) so that not only the layman but also the best money-changers fail to distinguish between the genuine and counterfeit coins. (3a) What our sense perceives in these bodies is (3b) [accidents such as] colour or consistency which undergo generation and cessation [*d-hāwēn w-ḇāṭlīn*], (3c) while [their] substance [*haw d-sīm*] remains uncorrupted [lit. without (*star men*) corruption of the substance], (3d) and the distinctions which constitute their essential character [*ūsiyāyūṯā*] are concealed from us. (4) [This being the case] how can we produce something which we do not know?

Šifāʾ 23.2-5, 7-11:

إلا أن جواهرها تكون محفوظة، وإنما يغلب عليها كيفيات/٣ مستفادة بحيث يغلط فى أمرها، كما أن للناس أن يتخذوا الملح والقلقند والنوشادر/٤ وغيره./٥ ولا أمنع أن يبلغ فى التدقيق مبلغا يخفى الأمر فيه على الفُرْهَة. ... فإن هذه الأحوال المحسوسة يشبه أن لا تكون هى/٨ الفصول االتى بها تصير هذه الأجسادُ أنواعا، بل هى عوارض ولوازم وفصولها مجحولة،/٩ وإذا كان الشىء مجهولا كيف يمكن أن يقصد قصد ايجاده أو إفقاده. وأما سلخ هذه/١٠ الأصباغ والأعراض من الروائح والأوزان أو كسوها فهذه مما لا يجب أن يُصر على/١١ جحده، لفقدان العلم به، فليس يقوم البتة برهان على امتناعه.

(1a) ... except that their essence [*ǧauhar*] is retained, (1b) while the acquired qualities merely prevail over them in such a way that one is deceived concerning the matter - just as people take salt, *qalqand*, sal ammoniac etc. [to be genuine]. (2a) I do not deny that one can reach through meticuous work [*fī al-tadqīq*] such a level [*mablaġ*] (2c) that the matter remains concealed [even] to the experts. ... (1a/3a) For it seems that these perceived properties are not the distinctions [*fuṣūl*] whereby these bodies are differentiated into [lit. become] species [*taṣīr anwāʿan*], (1b/3b) but are accidents and attributes, (3d) while the distinctions are unknown. (4) And if the thing is unknown, how can anyone endeavour to produce [*īǧād*] it or

to destroy [*ifqād*] it? (1b/3b) As for the removal and imparting of these colours and the accidents such as smells and weights, (2) these are things which one ought not to persist in denying merely because of lack of knowledge concerning them, for there is no proof whatever of their impossibility

l. 1. ܦܘܪ̈ܫܢܐ ܐܘܣܝ̈ܐ ("essential distinctions"): The phrase, as it were, combines *Šifā'* 23.2 *ğauhar* and 23.8 *fuṣūl*.

l. 2. ܡܬܐܕܫܝܢ ("are differentiated into species"): On this denominative verb, see PS Suppl. 5 with citation from *But*. Ausc. I.v.1. (ms. F 5v ult.). It should probably be regarded as Ethpa., rather than Ethpe. as J.P. Margoliouth has it. - Here, it corresponds to IS *taṣīr anwāʿan*.

l. 2. ܓܘ̈ܢܐ ܘܩܘܝ̈ܡܐ ("colours and consistencies"): cf. l. 7 below.

l. 2. ܡܢܘܬܗܘܢ ("their quiddity"): The word, though not in the corresponding passages of IS, may be used here as the antonym of IS's "qualities" (*kaifīyāt* 23.2).

l. 3. ܠܐ ܐܢܫ ܡܪܐ ܗܘܢܐ ("no intelligent person"): lit. "noone in possession of intelligence"; cf. III.iii.1.3 "those lacking in intelligence" (*hawnē rʿīʿē*).

l. 4. ܘܠܛܝܦܘܬ ܐܘܡܢܘܬܐ ܒܚܝܪܬܐ ("subtle and proven art"): lit. "subtlety of tested art/skill"

l. 4-5. ܣܓܝܐܘܬ ܕܘܪܫܐ ܘܚܦܝܛܘܬ ܢܣܝܘܢܐ ("much practice and assiduous experimentation"): lit. "abundance of training and assiduity of experiment".

l. 5. ܦܫܝ̈ܛܐ ("the layman"): lit. "the simple", contrasted here with the professional money-changers.

l. 6. ܥܘܪܦܢܐ . . . ܥܪܘܦܝܗܘܢ ("the best money-changers fail to distinguish between the genuine and counterfeit coins"): lit. "the money (*ʿurpānā*) of the best money-changers fall short of the distinction between the genuine and the counterfeit". - *ʿUrpānā* could also mean "(the act of) money-changing" (corr. arab. *ṣarf*).[96] It could, just possibly, mean "investigation", but the evidence for this is rather weak.[97]

---

[96] Neither PS (2994f.) nor Brockelmann (549b) gives this latter meaning for *ʿurpānā*, although PS does quote lexica where the word is rendered simply as *ṣarf*. Cardahi, *Lobab* II.296b, who has the Peal *ʿrap̄* (as opposed to the Pael *ʿarrep̄* as in PS and Brockelmann) as the verb "*to change* money", gives *ʿurpānā* as the verbal noun (*maṣdar*) of this verb.

[97] The word *ʿurrāp̄ā* seems to have this meaning, if the reading is correct, in the phrase *ʿurrāp̄ē w-ḇuḥḥānē* in Bar Kepha, *Hex.*, Paris 241, 1v b17 (cf. ibid. b5f. *buḥḥānē w-ʿuqqāḇē*, b21 *ʿuqqāḇē w-ḇuḥḥānē*; PS 2994). - Brockelmann, 549b, distinguishes between two roots for syr. √*ʿrp*, one corresponding to arab. √*ṣrf* and the other to √*ʿrf* (the third root given by Brockelmann, *ʿrapṯā* "elm", corr. arab. √*ḍrf*, does not concern us here). The correspondence with arab. √*ʿrf*, however, is dubious. The phrase *ʿurpānā d-renyā* of *Acta martyrum et sanctorum* [Bedjan] V.73.10, which is rendered as "reputatio" by Brockelmann (cf. also PS Suppl. 256a, whose rendition of this phrase is even more dubious) can be understood as meaning, literally, "turning of thought", and is probably the same as (or corruption of) the phrase *ʿurrāp̄ā d-renyā*,

l. 7-8. ܕܗܘܝܢ ܘܒܛܠܝܢ ("which undergo generation and cessation"): The two verbs are in plural, although the two singular nouns to which the relative clause refers back to are joined by "or" (*aw*) rather than "and".

III.iii.3.: Further refutation

BH, now following Abū al-Barakāt, raises two more points in refutation of the claim of the alchemists in this 'theory', namely the fact that the "efficient force" required for the production of metals and other substances are only found in certain localities and our ignorance of the "proximate causes" of these substances. - These two points are also mentioned, albeit in a somewhat different form, among the five additional points Fakhr al-Dīn al-Rāzī gives as arguments against the claims of alchemy at *Mabāḥiṯ* II.215.4-216.4 after summarising the point made by IS in the *Šifāʾ* (Rāzī, a supporter of alchemy, then goes on to refute these).[98]

III.iii.3.1-4: Alchemists' claim

But.: (1) Alchemists say in those fables of theirs which are never realised: (2a) "if we were to find pure, white or red sulphur like (the sulphur) which nature uses (2b) and were able to mix it firmly with mercury, (2c) we would [be able to] produce silver or gold."

---

rendered as *taṣrīf al-fikr* at Bar Bahlūl [Duval] 1423.17f. (cf. PS 2994). Furthermore, in at least one of the two passages cited by Brockelmann for *ʿurrāpā* in the sense of "disputatio", the meaning is "*exchange* of words" (*ʿurrāpā d-mellē*, *Chron. ad 1234 pertinens* [Chabot] II.284.18, cf. tr. Abouna, 214.8). The use of the verb *ʿrap̄* in the sense of "disputavit" can, then, be explained as an ellipsis for *ʿrap̄ melltā* vel sim. (so *ʿrap̄(w) ʿamman*, "they disputed with us", Michael, *Chron.* [Chabot] IV.532a 7, cf. tr. Chabot III.92b 15f., "quand ils échangèrent leurs vues avec nous"; perhaps also *ʿrap̄nan ʿal sīgīlīyōn d-law šarrīrtā īṯēh*, ibid. IV.511b 1, tr. III.57.21 "nous prétendîmes que le diplôme n'était pas authentique"). The word *ʿurrāpē* in the sense of "investigations, disputations" in Bar Kepha (not cited by Brockelmann), too, could be explained in a similar manner.

[98] The first point made by Abū al-Barakāt/BH resembles Fakhr al-Dīn al-Rāzī's fourth (II. 215.18f.; see comm. on III.iii.3.2 below). The second may be compared with Rāzī's first, II. 215.4-8: اولها ان الطبيعة انما تعمل هذه الاجساد من عناصر مجهولة عندنا ولتلك العناصر مقادير معينة مجهولة عندنا ولكيفيات تلك العناصر مراتب معلومة وهى مجهولة عندنا ولتمام الفعل والانفعال بينها زمان معين هو مجهول عندنا ومع الجهل بكل ذلك كيف يمكننا عمل هذه الاجساد. ("Firstly: Nature produces these bodies from elements unknown to us; these elements have fixed quantities unknown to us; for the qualities of these elements there are definite grades, and they are unknown to us; for the whole of the action and passion between them there is a fixed time, which is unknown to us. Given (our) ignorance concerning all of this, how can we produce these bodies?")

Muᶜtabar II.230.21, 230.24-231.3:

والذين يرون ان الزئبق هو العنصر للمنطرقات ... فطلبوا /١ الكبريت الاحمر واعتقدوا انهم اذا وجدوه معدنيا او صناعيا اصابوا الكيمياء /٢ وعملوا من الزئبق دهبا وكذلك اذا وجدوا الكبريت النقى الابيض المصفى /٣ وقدروا على خلطه بالزئبق عملوا فضة

Those who are of the opinion that mercury is the element of the malleables [i.e. metals] ... (2a) and have sought red sulphur. They believe that (2a) if they could find it, be it from mines [*maᶜdinī*] or artificial [*ṣināᶜī*], they would be able to realise alchemy (2c) and produce gold from mercury. (2a) Similarly, if they could find pure, white, clean sulphur (2b) and could mix it with mercury, (2c) they would produce silver.

l. 1. ܦܠ̈ܣܐ ܟܡܝܐܐ ("alchemists"): cf. III.i.5.6 above.

l. 1. ܫܘ̈ܥܝܬܗܘܢ ("their fables"): خرافاتهم F marg. - The same word, *šūᶜīṯā*, is used of the claim of the astrologers at *But*. Gen. et corr. II.v.3. (ms. F 45r a 17; the corresponding word in F marg. there is also *ḫurāfa*).

l. 1. ܕܠܡܥܒܕܢܘܬܐ ܠܐ ܢܦ̈ܩܢ ("which are never realised"): لم تخرج الى الفعل F marg. - lit. "which do not come out into operation/actuality" (cf. PS 2777).

l. 2. "which nature uses": i.e. in order to produce gold and silver. - Abū al-Barakāt does not quite say "natural" sulphur, but this is implied in his *maᶜdinī* ("from the mines"); cf. Fakhr al-Dīn al-Rāzī, *Mabāḥiṯ* II. 215.18f.: ورابعها ان لهذه الاجساد اماكن طبيعية وهى معادنها وهى لها بمنزلة الارحام للحيوان ... ("Fourthly: For these bodies there are natural places, namely their mines, and these are for them like what the wombs are for animals ...").

III.iii.3.4-5: Locality

But.: (3a) In reality, they have neither found [such sulphur] nor produced [silver and gold]. (3b) For the efficient power [*ḥaylā ᶜāḇōḏā*] is found only where it can be found [*mḵān l-meštḵāḥū*] ...

Muᶜtabar II.231.3-4:

وما وجدوا وما عملوا لما قيل من/٤ ان القوة الفعالة لا تعرض ولا توجد الا حيث يوجد وعما عنه يوجد

[4 تعرض: v.l. تعدم; تعزم F marg.]

(1) They have neither found [such sulphur] nor produced [silver and gold] (2) because of what has been said concerning the fact that the efficient power [*al-qūwa al-faᶜᶜāla*] occurs and is not found except where it is found and by the [agent] by which it is found.

l. 4. ܒܫܪܪ̈ܐ ܕ ("in reality"): وبالحقيقة F marg. - Cf. III.iii.1.5 above.

III.iii.3.5-11: Proximate causes

But.: ... (1a) while the proximate causes [*ᶜellāṯā qarrīḇāṯā*] of these bodies [*gušmē*] are not known to us, (1b) nor is it possible for us to know them [lit. their knowledge is impossible for us], (2a) just as we are unable to know the mixture and the formative power [*muzzāḡā w-ḥaylā gāḇōlā*] (2b) by and through which [*d-ḇeh w-ḇ-īḏaw*] a walnut has been made round, (2c) an almond has been made elongated, (2d) the fleshy part of a

citron has been made sweet (2e) and its pulp has become sour, (3) but we merely perceive the outward appearance [*dmūṯā barrāytā*] through sensation [*rgēštā*], and the effects [*ma*ᶜ*bḏānwāṯā*] through experience [*nesyānā*]. (4a) Therefore, seeing that [lit. just as] we are unable to extract from the elements a mixture with which to make a citron or a pomegranate, (4b) how can we mix [and produce] gold and silver from them?

*Mu*ᶜ*tabar* II.231.12-15, 16-18:

ومعرفة الاسباب القريبة المتوسطة فى هذه الاشياء /١٣ متعذرة علينا بل ممتنعة كما امتنع علينا وتعذر ان نعرف السبب المزاجى والفاعلى /١٤ الذى تدورت به النارنجة واحمرت وتطاولت به الاترجة واصفرت وحمضت /١٥ به الرمانة وحلت ... بل نعرف الصورة من جهة المشاهدة والافعال /١٧ بالتجربة وكما لا نقدر أن نمزج من العناصر ما نتخذ منه اترجا ولا رمانا كذلك /١٨ لا نقدر على ان نمزج منها ذهبا ولا فضة

[14-15 وحلت ... تدورت : تدورت به الجــوزا (الجــوزة) وتطاولت به اللوزة وحلي به لحم الاتروجــه (الاترجه) وحمض به شحمها F marg. (57v-58r; = *But.*)]

(1a) The knowledge of the immediate and intermediate causes [*al-asbāb al-qarība al-mutawassiṭa*] of these things [*ašyā*ᵓ] is difficult [*muta*ᶜ*aḏḏir*] for us; (1b) rather, it is impossible [*mumtani*ᶜ], (2a) just as it is impossible and difficult for us to know the mixing and efficient cause [*al-sabab al-mizāǧī wa-l-fā*ᶜ*ilī*] (2b) through which [*bihi*] an orange has been made round and red, (2c/d) a citron has been made elongated and yellow (2d/e) and a pomegranate has been made sour and sweet. ... (3) Rather, we know the shape [*ṣūra*] on the basis of observation [*mušāhada*] and the effects [*af*ᶜ*āl*] through experience [*taǧriba*]. (4) Just as we are unable to mix [and produce] out of the elements something from which we can make a citron or a pomegranate, (4b) so we are unable to mix [and produce] gold or silver from them.

l. 5. ܥܠܠܬܐ ܩܪܝܒܬܐ ("the proximate causes"): The term goes back to Aristotle, τὰ ἐγγύτερα/ἐγγύτατα αἴτια (*Metaph.* 1044b 1, 1014a 5; *Ausc. phys.* 195b 2, 197a 24); cf. Daiber (1993) 121.

l. 7. ܡܘܙܓܐ ܘܚܝܠܐ ܡܨܒܬܢܐ MV ("the mixture and the formative power"): ܡܘܙܓܐ ܕܚܝܠܐ ܡܨܒܬܢܐ Fl. - While the singular suffixes of *beh w-ḇ-īḏaw(hy)*, along with the relatively unusual agreement of mss. F and l, favours the reading ***d**-ḥaylā*, the sense, along with Abū al-Barakāt's *al-sabab al-mizāǧī wa-l-fā*ᶜ*ilī*, favours ***w**-ḥaylā*.

l. 7-8. "by and through which a walnut ...": There are a number of differences here from the *Mu*ᶜ*tabar*. The agreement of what is otherwise an excerpt from the *Mu*ᶜ*tabar* in the margin of ms. F with *But.* suggests the possibility that BH and the scholiast had a different recension of the *Mu*ᶜ*tabar* from the text in the Hyderabad edition, but it is more likely that the agreements are due to the scholiast translating these parts back from BH's Syriac into Arabic to make his excerpt agree with the text of *But.* This is betrayed, in a way, by the Syriacisms in the spelling of the words الجوزا (recte الجوزة, cf. syr ܓܘܙܐ) and الاتروجه (recte الاترجة, cf. syr. ܐܬܪܘܓܐ). It is also likely that Abū al-Barakāt did mention "pomegranate" in these parts (at II.231.15, omitted here by BH and the scholiast), since its mention at II.231.18 would otherwise be somewhat sudden and awkward.

l. 8, 10. ܐܬܪܘܓܐ ("citron"): اترجة Abū al-Barakāt. - The mention by Abū al-Barakāt of the fruits and of citron in particular may be related to the story told by Ibn al-Faqīh of how the alchemists imprisoned by a Persian king asked for citron in their rations, pointing out its various uses.[99] If an allusion to the story is intended, the implication on Abū al-Barakāt's part will be: "these clever alchemists may know about and make use of the various properties of the citron and other fruits, but it still remains the case that they are ignorant of the causes of these properties and so are unable to produce the fruits themselves."

l. 8. ܕܡܘܬܐ ܒܪܝܬܐ ("outward appearance"): الصورة Abū al-Barakāt. - Abū al-Barakāt's *ṣūra* is infelicitous, since the word is one often used for "form" (εἶδος) in its Aristotelian sense (and that is what determines a species). BH's paraphrase indicates his awareness of the problem.

l. 9-10. ܐܟܡܐ ܕ . . . ܗܟܢܐ FMV(lit. "just as ..., how ...?"): ܐܟܡܐ ܕ . . . ܐܦܠܐ PIW ("just as ..., neither ..."): The expected construction is: "*akmā d- ... hākannā ... lā ...*" (cf. Abū al-Barakāt *kamā ... ka-ḏālika lā ...*"). The reading of PIW is closer to it and syntactically more tolerable than that of FMV. Precisely for that reason, however, the reading of FMV is the *lectio difficilior*, whose origin is difficult to explain. I suspect here an instance of inadequate editing on the part of the author, who began the sentence with *akmā d-* following Abū al-Barakāt's *kamā*, but then decided to end this 'theory' with a rhetorical question as he had done in the preceding 'theory'.

l. 9. ܠܡܦܩܘ ("to extract"): أن نمزج ("to mix") Abū al-Barakāt. - BH probably read *nuḫriju* (نخرج) instead of *namzuju* (نمزج).

---

[99] Ibn al-Faqīh [de Goeje] 205.9-11: اما قشره الظاهر فطيبٌ نشتمّه واما داخله ففاكهة يُنتفع به واما حُمَاضه فانه خلّ نافع طاهر واما حبّه فدهن ينتفع به ("Its outer rind is good - we can enjoy its smell. - Its inside is a fruit which can be used [as such]. Its sourness is pure, useful vinegar. Its seed is oil which can be used"). For the whole story, ibid. 205.6-17 (and paraphrase by Wiedemann [1907] 123); the same story is in Yāqūt, *Muʿǧam al-buldān* [Wüstenfeld] IV.264.19-265.5 and Qazwīnī, *Āṯār al-bilād* [Wüstenfeld] 164.20-26 (both of whom name Ibn al-Faqīh as their authority).

BOOK OF MINERALOGY, CHAPTER FOUR

ON MATTERS CONCERNING THE HABITABLE WORLD

Arist. discussed the conditions of the habitable world (the οἰκουμένη) briefly in a digression from his discussion of winds at *Mete*. II.5, 362a 31-b 30. IS treated the subject at greater length in the *Šifā*ᵓ, where he devoted the seventh (last) chapter of the treatise (*maqāla*) on mineralogy (p. 24-32) to this topic, and it is to that chapter that this chapter of *But.* corresponds.

In the first section of the chapter and in the first two 'theories' of the second section BH is basically following the *Šifā*ᵓ (respectively, *Šifā*ᵓ 24.7-26.13 and 26.17-27.8), although the Avicennian material is supplemented by materials taken from other sources. - The rest of the second section (IV.ii.3-6) and the whole of Section iii. deal with the climate of the equatorial region. Here, BH disagrees with IS's view that the equatorial region is temperate and therefore first gives us an account of IS's arguments (corr. *Šifā*ᵓ 28.13-30-15) at IV.ii.3-6 and then goes on to refute these in Section iii. BH seems to have used a number of sources in composing these parts of the chapter, including works of Fakhr al-Dīn al-Rāzī and Ṭūsī, both of whom disagreed with IS's view on the subject, as well as his own earlier discussion of the subject in *Asc*. - The discussion of the causes of heat and coldness in Section iv. is based almost entirely on Abū al-Barakāt's *Mu*ᶜ*tabar* (II.202.6-205.3).

l. 1. ܡܛܠ ܨܒܘ̈ܬܐ ܕܡܬܝܒܠܬܐ: lit. "on the matters of the habitable world". - Cf. *Šifā*ᵓ 24.3: في أحوال المسكونة وأمزجة البلاد ("on the conditions of the habitable world and the climate of the countries"); the second part of IS's chapter heading is used by BH in the section heading of Min. IV.ii.

l. 1. ܡܬܝܒܠܬܐ ("habitable world"): here corr. IS المسكونة; more often المعمورة IS; corr. gr. ἡ οἰκουμένη.

Min. IV. Section i.: *On the Natural Position of the Earth*

The section, whose title really only applies to the content of its first 'theory', corresponds to the first part of IS's chapter on the habitable world (*Šifā*ᵓ 24.7-26.13). BH follows here the order of presentation in the *Šifā*ᵓ and much of the content too is taken from the *Šifā*ᵓ, although

the Avicennian material is supplemented as usual by materials taken from a number of other sources. Of particular note is the account of the theory of trepidation in IV.i.2, which goes back to Theo of Alexandria's *Small Commentary* on Ptolemy's *Handy Tables*.

But.	*Šifāʾ*	Other sources	
IV.i.1.1-3	24.7-10		Earth and water
IV.i.1.3-9	24.10-18		Unevenness of earth's surface
IV.i.2.1-2	24.18-19		Influence of stars
IV.i.2.2-6	24.19-25.1	Theo	contd. (trepidation)
IV.i.2.7-8	25.1		contd. (apogees & perigees)
IV.i.2.8-10	25.9-10		Formation of dry land
IV.i.3.1-4			Quarters of earth
IV.i.3.4-11	25.13-17	Nic., Ṭūsī	Extent of the habitable world
IV.i.4.1-3	25.17-18		Other quarters
IV.i.4.3-9	25.20-26.6		Southern hemisphere
IV.i.4.9-11	26.6-13	(*Cand.*)	Arctic and equatorial regions

1. "Natural position": cf. *But.* Ausc. phys. II.iv.: ܡܛܠ ܗܝ ܕܠܟܠ ܓܘܫܡܐ ܐܣܟܡܐ ܐܝܬ ܘܐܬܪܐ ܟܝܢܝܐ ("That for every object there is a natural shape and place").

IV.i.1.1-3: Earth and water

But.: (1a) Although the nature (of earth and water) requires [*bāʿē*] (1b) that earth be inside [*b-gāw*] water (1b) and that water encircle it on all sides, (2a) [in reality] they have been found [*eštkaḥ(w)*] not [to be] in accordance with [*lpūt*] what is natural to earth and water, (2b) but in accordance with what is natural to the order [*mṭakksūṯā*] of the universe [*hānā kul*].

Šifāʾ 24.7-10:

فنقول أولا: إنا كنا قد أشرنا فيما تقدم إلى أن الواجب بحكم طبيعة الماء والأرض أن تكون الأرض فى ضمن الماء، ويكون الماء محيطا بها من جميع الجوانب؛ ولكن الوجود ليس على ذلك، وليس على ما هو طبيعى للأرض والماء، بل ما هو طبيعى لنظام الكل.

We say firstly: We have indicated previously that (1a) the necessary condition of the nature of water and earth is (1b) that earth be inside [*fī ḍimn*] water (1c) and that water surround the earth on all sides; (2a) but the actual condition [*wuğūd*] is not like that [*ʿalā ḏālika*] and not in accordance with [*ʿalā*] what is natural to earth and water, (2b) but what is natural to the order [*niẓām*] of the universe [*kull*].

IV.i.1.3-9: Cause of unevenness of earth

BH here simplifies a longer passage of IS. With the passage here, cf. also the discussion of the formation of mountains at Min. I.ii.1-2 above.

But.: (1a) [That is] because the elements have a capacity [*mkānūṯā*] for transformation [*šuḥlāpā*] into each other, (2a) so that when parts [*mnawwāṯā*] of earth turn into water or something else, (2b) the sphericity [*espērānāyūṯā*]

of (the earth) is breached [*mettarʿā*] and defiles, gorges, plains and valleys are formed in it. (3a) When [on the other hand] particles of water or other elements turn into earth, (3b) peaks [*šnānātā*] and high grounds [*nḏāyē*] [are formed and] become attached [*dābqīn*] to its spherical surface (4) and in this way parts of the earth are freed from the rule [*šulṭānā*] and grasp [*uḥdānā*] of water which confines and conceals [the earth] [lit. the rule of water and its confining and concealing grasp].

Šifāʾ 24.10-18:

وذلك أنه لما كان من شأن العناصر أن يستحيل بعضها ألى بعض بأجزائها، كانت الأرض لو وجدت على ما هو طبيعى لها لم يثبت. لأن فى طبيعة الأرض أن تستحيل أجزاء منها ماء أو نارا، أو غيرهما من الجواهر الأخرى. وتلك الجواهر أيضا قد تستحيل أجزاء منها أرضا، فما يستحيل من الأرض إلي غيره ينقص من جملة حجم الأرض، فيلزم ضرورة أن يقع هناك ثُلمةٌ فى تدوير الأرض وغور إذا كانت الأرض يابسة لا تجتمع إلي شكلها الطبيعى، بل يبقى عليها الشكل المستفاد. وما يستحيل إلى الأرض يكون لا محالة زيادة ونتوا ملحقا بها، فلا ينبسط عليها انبساط الماء المهراق على ماء غيره، حتى يصير منهما حجم واحد مستديز؛ فيلزم ضرورة أن يتولد على كريّة الأرض تضريس من غور ونجد.

(1a) That is to say, since it is a characteristic [*šaʾn*] of the elements to be transformed into each other in their parts, even if the earth were found in the state natural to it, it would not remain so, (2a) because it is in the nature of earth that parts [*ağzāʾ*] of it be transformed into water or fire or some other substance [*ğauhar*], (3a) and parts of those substances too are sometimes transformed into earth. (2a) Therefore, the [part] of the earth that is transformed into something else (2b) is deducted [*yanquṣu*] from the overall mass [*ḥağm*] of the earth, (2b) so that by necessity there occur here gaps [*ṯulma*] and depressions [*ġaur*] in the roundness [*tadwīr*] of the earth when the earth becomes dry while it has not gathered into its natural shape but the acquired shape still remains on it. (3a) What is transformed into earth [on the other hand] (3b) undoubtedly becomes additions [*ziyāda*] and protrusions [*nutūw*] attached [*mulḥaq*] to the earth, so that water does not spread out over (the earth) in the way that water poured on other [bodies of] water does, such that a single round mass is formed out of the two. (2b/3b) By necessity, therefore, undulations [*taḍrīs*] will be formed on the sphericity [*kurīya*] of the earth (consisting of) hollows and highlands.

l. 8. "rule and grasp of water": cf. *Šifāʾ* 25.11-12: والأولى أن يكون المستولى على الأرض هو الماء الذى من حقه أن يفيض على كليتها. ("... whereas it is more appropriate that the one having mastery [*mustaulī*] over the earth be water whose property it is to flow over the whole of it").

IV.i.2.1-2: Influence of stars

But.: Stars too have a major effect [*maʿbdānūṯā*] on the transformation of elements into one another in accordance with those coordinates [lit. vertical positions, *qaṯehṭīqāywāṯā*] which change with their movements

Šifāʾ 24.18-19:

وخصوصا وللكواكب لا محالة تأثير فى إيجاب هذه الإحالة بحسب المُسامتات التى تتبدل بحسب حركتها

In particular, the stars undoubtedly have an influence [*taʾṯīr*] in bringing about this transformation according to their vertical positions [*musāmatāt*] which change according to their movements.

l. 1. ܡܬܡܠܢܝܘܬܐ ("coordinates", lit. verticalities, vertical positions): < ܡܬܡܠܢܝܐ, gr. καθετικός; المُسامتات IS. - The nominal form in *-ūṯā* is known to the lexica only from BH *But*. (PS Suppl. 314); cf. IV.iv.5.1 below (ܣܪܛܐ ܡܬܡܠܢܝܐ).

IV.i.2.2-8: Influence of stars (contd.)

BH here expands the passage of the *Šifā*ʾ by combining it with materials taken from the well-known passage dealing with the theory of trepidation[1] in Theo of Alexandria's *Small Commentary* on Ptolemy's *Handy Tables* (Πρόχειροι κανόνες), a passage BH had also used in his *Chron.* and *Asc*.

While the forms of the transliterated Greek terms (ἀποτελεσματικοί, ἔφοδος) leaves little doubt that the passage of Theo was known to BH in a Syriac, and not Arabic, version, it is less clear whether BH's immediate source was the Syriac version of Theo's work itself or some secondary source in which this passage was quoted. Three considerations would seem to speak in favour of the latter option, namely (1) the fact that BH seems to know relatively little about Theo's work besides his account of the trepidation theory,[2] (2) the fact

---

[1] "Trepidation of the equinoxes", or "of the fixed stars", depending on one's point of reference; also referred to as the theory of "accessio et recessio" (arab. *iqbāl wa-idbār*). - In its "simple linear oscillatory version" as reported by Theo, this was an alternative, as it were, to the theory of "precession", whereby it was thought that the equinoctial point not only moved forward (i.e. eastward) along the ecliptic, but also backward after a certain period. Although this theory became untenable when no such retrogression was found to take place after the predicted period, it survived in an altered form, combined with the theory of precession and used to account for the variation in the rate of precession, and in the form as found in (Ps.)-Thābit b. Qurra's *De motu octavae spherae* was to have a lasting influence on astronomical theories in the West until its final rejection by Tycho Brahe (see, e.g., Neugebauer [1975] 631-634, Ragep [1993] 397-400; on the theory as found in *De motu*, see Carmody [1960] 84-101, Goldstein [1964], Ragep [1993] 400-408).

[2] Theo's "Canon" is also mentioned by BH in his discussion of the different "eras" at *Asc*. 199.17-19, where we are told that Theo used the beginning of the reign of Philip Arrhidaeus as his epoch in "his renowned Book of the Canon [*ktāḇā d-qānōneh mšammhā*]" (see Theo, *Small Comm*. 200.2f.). There too the presence of the name "Arrhidaeus" (ܐܪܝܕܐܘܣ) suggests the use of a Syriac source (cf. Bīrūnī, *Āṯār* 28.5-9, where "Arrhidaeus" is omitted). - A further mention of Theo's "Canon" is found at BH, *Hist. dyn.* ([Beirut (1958)] 73.10: ولثاون من الكتاب الزيج المسمى بالقانون). There, however, BH gives us no details as to its contents and he is probably following Qifṭī (*Tārīḫ al-ḥukamā*ʾ [Lippert] 108.9), who equally gives no details (Qifṭī does mention Theo's theory of trepidation in another context, 170.9; Sāʿid al-Andalusī, another source used by BH mentions the trepidation theory as one of the noteworthy features of Theo's

that BH constantly refers to Theo's work as the "Book of the Canon" and (3) the combination of the fact (a) that BH is unaware that Theo himself was against the trepidation heory, (b) that BH knew Ptolemy was against the theory (*Chron.* [Bedjan] 54.23-25) and (c) that BH makes Theo a contemporary of Ptolemy (*Chron.* 54.17). This is best explained if the passage was known to BH only in quotation and BH believed Theo's comment that Ptolemy was against the theory not to be a part of the original text of Theo, but a comment made by the author of his immediate source.[3]

But. [underline: agreement with Theo; italics: with *Šifāʾ*]:
... (1a) *especially* those *fixed (stars) which*, (1b) according to the opinion [*reᶜyānā*] of the ancient astrologers, (1c) move [*mšannēn*] eight degrees [*mōras*] *to the north* in six hundred and forty years (1d) *and* [then] also eight degrees *to the south* in the same (number of years) [*b-d-akwāṯhēn*], (1e) [i.e.] one degree in eighty years, (1f) as Theo of Alexandria stated [*sām*] in his "Canon" [*b-qānōneh*] as the method [*epōdōs*] for this computation [*ḥušbānā*], based on the beginning of the reign [lit. of the years of the reign] of Augustus. (2a) In the same way *the apogees and perigees change their positions* [*epōḵiyas*] (2b) and they *become major causes of increase of water in one region and its decrease in another.*
Šifāʾ 24.19-25.1:

وخصوصا الثوابت الصائرة تارة إلى الجنوب وتارة إلى الشمال، والأوجات والحضيضات المتغيرة فى أمكنتها. فيشبه أن يكون هذه أسبابا عظاما فى إحداث المائية من جهة أو نقلها إليها، وإبطال المائية من جهة أو نقلها عنها

... (1a) especially the fixed stars (1d) which come [*ṣāʾira*] sometimes to the south (1c) and sometimes to the north, (2a) and the apogees and perigees which vary in their positions [*amkina*]. (2b) For it appears that these are the major causes in the production [*iḥdāṯ*] of wateriness in one region - or its conveyance [*naql*] towards it - and the annihilation [*ibṭāl*] of wateriness in another - or its conveyance away from it.
Theo, *Small Comm.*, Chapter 12 [Tihon] 236.4-237.2 (corr. Halma [1822] I.53): According to certain opinions [κατά τινας δόξας] (1b) the ancient astrologers [οἱ παλαιοὶ τῶν ἀποτελεσματικῶν] claim [βούλονται] (1c) that the solstitial points move 8 degrees in the forward direction [εἰς τὰ

"Canon", see *Ṭabaqāt* [Cheikho] 40.14, tr. Blachère 86).

[3] I am unaware of references to Theo's works elsewhere in Syriac. References to Ptolemy's *Handy Tables* themselves are found in Sergius of Rēshᶜainā (as ܟܬܒܐ ‹ܕ›ܩܢܘܢܐ, Sachau [1870] 225.17) and Severus Sebokht (as ܦܪܘܟܝܪܘܢ, see Nau [1910] 240, id. [1930-1] 343 [index]; and most probably thence in BH, *Rad.* [Istanbul] 16.9, see Introduction, Section 3.2.2e above; see also Neugebauer [1959]). The knowledge of the latter work in the Syriac world is also suggested by a Greek fragment of it in a Syriac palimpsest (Vat. gr. 623, 886 A.D.; see Tihon [1992] 68ff.).

ἑπόμενα μετακινεῖσθαι μοίρας η̅] from a certain initial moment on, then turn back again by the same amount [καὶ πάλιν τὰς αὐτὰς ὑποστρέφειν]; but this is not so in Ptolemy's opinion because without the additional term based on such calculation the said computations, made with his tables, agree with the instrumental data. Nor do we recommend such a correction; (1f) nevertheless we shall explain [ἐκθησόμεθα] the procedure of the calculation they (devised) concerning this [τὴν ἔφοδον τοῦ περὶ τούτου γινομένου αὐτοῖς ἐπιλογισμοῦ]. They assume that 128 years before the beginning of the reign of Augustus [πρὸ τῆς ἀρχῆς τῆς Αὐγούστου βασιλείας] the greatest shift occurred in the forward direction of these 8 degrees, and also the beginning of the return motion; then they add to it the 313 years from the beginning of the reign of Augustus to the beginning of Diocletian and to it the additional years since Diocletian; and they take the position corresponding to 1/80 of this total, (1e) such that their motion amounts to one degree in 80 years [ὡς κατὰ π̅ ἔτη μίαν μοῖραν αὐτῶν μετακινουμένων], and they substract from 8° the degrees which result from this division. They add the remainder as the shift of the solstices to the result obtained by the said computations of the positions of the sun and the moon and the five planets.[4]

BH, *Chron.* [Bruns-Kirsch] 59.9-15, [Bedjan] 54.17-25 [underline: agreement with *But.*; italics with Theo]: (1f) Theo, the geometrician, of Alexandria also lived at this time.[5] There are to (his credit) among those books which have become famous nearly throughout the whole earth: the "Book of the Working of the Brazen Spheres" [*k. d-sāʿōrūṯ ḥuḏrē nḥāšāyē*],[6] by means of which the observations of the motions [*zawʿē*] of the stars are performed, (1f) and the "Book of the Canon" [*k. d-qānōnā*], in which he stated [*sām*] *the method* [*epōdōs*] *for the calculation* [*ḥušbānā*] of the movement [*šunnāyā*], relative to them [i.e. the stars], *of the solstitial points*[7], (1c/e) *which move in the forward direction* [*nāqpīn*][8] *one degree* [*mōrā*]

---

[4] This is based on the translation of Neugebauer (1975) 632, modified to facilitate its comparison with the related passages of *But.*; cf. also the translation of Tihon (p. 319) and the older translations and discussions of the passage at Delambre (1817) II.625f., Nallino (1899-1907) I.298 etc.

[5] i.e. at the same time as Ptolemy, during the reign of Titus Antoninus (i.e. Antoninus Pius, 138-161 A.D.), which is, of course, wrong as far Theo (4th c.) is concerned.

[6] i.e. the *Book of the Armillary Sphere* (*K.* [*al-ʿamal bi-*]*ḏāt al-ḥalaq*); see BH, *Hist. dyn.* 73.11; Sezgin, *GAS* V.185f.

[7] [Syriac] [Syriac] Bruns-Kirsch; [Syriac] ([Syriac]) [Syriac] Bedjan. - The manuscript reading reproduced by Bruns-Kirsch is correct: cf. Theo 236.5 τὰ τροπικὰ σημεῖα. Nau's reading [Syriac] in *Asc.* (103.23) should be corrected.

[8] lit. "follow", i.e. move following the sequence of the signs of the Zodiac (< Theo 236.5). The word was completely misunderstood by Budge (1932) in his translation (p. 54). Cf. BH, *Asc.* 103.1: *zawʿā ... neqpāʾīṯ*; Bīrūnī, *Tafhīm* 101.5f.: *ḥaraka muqbila ilā tawālī al-burūǧ*; Biṭrūǧī [Goldstein] 174, 44v 1, 179, 47r 2: (*intiqāl/naqla*) *ilā*

every 80 years, up to 8 parts [*mnawwān*], (1d) and *also move back* [*hāp̄kīn*]. *This is not so in Ptolemy's opinion, because without this additional term the calculations agree with the measurements* [lit. preceptions, *metdarkānwātā*] *made with observation instruments.*[9] ...

BH, *Asc.* I.8.8 [Nau] 103.22-104.5 [underline: agreement with *But.*]: We say in addition [*ʿam hālēn*]: If those ancient Chaldeans [*kaldāyē qaššīšē*][10] gave the solstitial point (1e) a motion [*zawʿā*] of one degree every 80 years, in the forward direction [*neqpāʾīt*] up to 8 parts (1c) in 640 years - (1d) they also move back [*hāp̄kīn*] in the same (number of years) [*b-d-akwāthēn*], (1f) according to Theo of Alexandria in the "Book of the Canon" - and [if] the ancient astrologers made use of this movement, how could this motion of the fixed stars have been unnoticed by them?[11]

l. 3. ܩܕ̈ܡܝܐ ܕܐܦܘܛܠܣܡܛܝܩܘܣ ("ancient astrologers"): οἱ παλαιοὶ τῶν ἀποτελεσματικῶν Theo. - The construction of the phrase mirrors that of the Greek. The phrase is rendered in most Arabic sources based on Theo as *aṣḥāb al-ṭalismāt/ṭilismāt.*[12]

l. 3. ܐܦܘܛܠܣܡܛܝܩܘܣ (sic M): ܐܦܘܛܠܣܡܝܩܘܣ (without -*ṭ*-) F; ܐܦܘܛܠܣܡܛܝܩܣܘܣ (with extra -*s*-) PLIV. - I adopt the reading of M as being the closest to the Greek, although the two older manuscripts of *Asc.* used by Nau for his

---

tawālī al-burūǧ.

[9] Cf. Theo 236.6-9: ὅπερ τῷ Πτολεμαίῳ οὐ δοκεῖ, διὰ τὸ ἄνευ τῆς ἐκ τοῦ τοιούτου ἐπιλογισμοῦ προσθέσεως συμφωνεῖν τὰς εἰρημένας διὰ τῶν κανονογραφιῶν ψηφοφορίας, ταῖς διὰ τῶν ὀργάνων καταλήψεσιν.

[10] For the mention of Chaldeans in connection with precession and trepidation, cf. Bīrūnī, *Āṯār* 326.1ff. (tr. 322; cf. Ragep [1993] 398) and Biṭrūǧī [Goldstein] 173, 44r 13ff. (tr. lat. [Carmody] 101, X.11). The passage of Biṭrūǧī is particularly interesting, since, as in the immediately preceding passage of *Asc.* (103.19-22), we find Timocharis, Aristyllus, Menelaus, Hipparchus and Ptolemy (though not Agrippa) mentioned as those involved in the discovery of precession.

[11] This passage of *Asc.* occurs at the end of the section dealing with the invariability of the relative positions of the fixed stars and here BH is refuting the view that the fixed stars were so called not because of the invariability of their relative positions but because the Chaldeans found them to be immobile. - The precession of the fixed stars (or, of the eighth sphere) is treated elsewhere in *Asc.* (I.1.8 [Nau] 12f.), where he gives the Hipparchian/Ptolemaic value of 1° in 100 years, as he also does at *But.* De caelo II.i.4 (ms. F, 24r-v), although at *Cand.* 198.9f. we are given the value of 360° in 23,760 years (i.e. 1° in 66 years).

[12] So Battānī [Nallino] II.190.11, 15; Yaʿqūbī [Houtsma] 159.11f.; Ṣāʿid al-Andalusī, *Ṭabaqāt* [Cheikho] 40.14; Biṭrūǧī [Goldstein] 174 (44r) ult.; cf. *aṣḥāb ṣināʿat al-ṭalismāt*, Ps.-Maǧrīṭī, *Picatrix* [Ritter] 78.12; *ahl al-ṭalismāt*, Ṭūsī, *Taḏkira* [Ragep] 125.1; *aṣḥāb al-aḥkām*, Ibn Yūnus, in quotation from letter of Thābit b. Qurra to Isḥāq b. Ḥunain (Caussin [1804] 117 [101] 13); *aṣḥāb al-ṭalismāt wa-hum aṣḥāb al-aḥkām min qudamāʾ ahl bābil* Bīrūnī, *Tafhīm* 101.4f. (corr. pers. 132.11f. *ḫudāwandān-i ṭalismhā* ...); *aṣl bābil*, *Kifāyat al-taʿlīm* (arab., 12th c.; apud Sezgin, *GAS* VI.102; lege *ahl bābil*).

edition[13] too are reported to have the form without -*ṭ*- as in ms. F here at ed. Nau 104.4. The reduction, on the other hand, of -τελεσμ- (cf. Bar Bahlūl [Duval] 257.27 ܐܦܘܛܠܣܡܛܝܩܘܣ) to -*ṭlysm*- may be due to the Arabic *ṭalism-/ṭilism-* etc.

l. 3-4. "six hundred and forty years": The figure is not in the text of Theo, although it is found in the marginal scholia in several manuscripts of the Greek text of Theo (information of Prof. A. Tihon). It is, of course, a figure that can easily be worked out (80 x 8).

l. 4. ܡܘܪ̈ܣ ("degrees"): < μοῖρα, μοίρας. On the terms used for "degree" in Greek and Arabic, see Kunitzch (1974) 156-160 (μοῖρα, βαθμός, τμῆμα; *daraǧa, ǧuzʾ* etc.). BH, while using only *mōrā* in our passage of *But*., also uses *mnāṯā* in the same sense at *Chron*. 59.13 and *Asc*. 104.1.

l. 4, 5. ܠܬܝܡܢܐ . . . ܠܓܪܒܝܐ ("to the north ... to the south"): in l. 5, after ܠܬܝܡܢܐ, add. ܡܫܢܝܢ ܘܡܢ ܡܕܢܚܐ ܠܡܥܪܒܐ F. - These words are taken from the *Šifāʾ*, but BH errs when he incorporates these words into his account of trepidation and talks of the stars moving "8 degrees to the north/south", since in trepidation we are concerned with the movement along the ecliptic (as BH was aware, see the passages of *Chron*. and *Asc*. quoted above) and a movement of 8 degrees along the ecliptic would only entail a movement through a fraction of that distance in the north-south direction. - The reading of ms. F ("and [then] also move eight *degrees* to the south ... and *from east to west* one degree in eighty years") does not make sense as it stands, but one suspects that the phrase "from east to west" originated from a marginal correction by someone (the author himself?) who noticed the error concerning the direction of trepidation.

l. 4. ܘܒܕܐ ܟܘܬܗܝܢ ("and ... in the same (number of years)"): ܘܒܕܐ ܟܘܬܗܘܢ F. - The feminine suffix -*hēn* could refer back either to *šnīn* (years) or to *mōras* (degrees). Although the word "τὰς αὐτάς" at Theo 236.6 must mean "the same number of degrees" (sc. μοίρας), in *But*. the interpretation of the suffix as meaning "years" is favoured by the parallelism with the preceding clause (***b**-šeṯmā wa-arbᶜīn šnīn tmānē mōrās*), as well as by the consideration that otherwise we have no specification of the duration of the return movement. - There are no masculine nouns in sight to which the suffix -*hōn* of ms. F (and *Asc*. 104.2) could refer to.

l. 5. ܒܩܢܘܢܗ ("in his *Canon*"): BH constantly refers to the work in which the trepidation theory was mentioned as "Theo's [Book of the] Canon". He is not alone in this, since "Theo's Canon" is also cited as the source by Sāᶜid al-Andalusī (*Ṭabaqāt* [Cheikho] 40.13) and in the *Kifāyat al-taᶜlīm* (Sezgin, *GAS* VI.102).[14]

---

[13] Paris syr. 244 (= A, 14/15th c.) and Hunt. dxl (= C, 1548 A.D.; whence PS 333, "ἐπιλησματικοί").

[14] According to Yaᶜqūbī, the report of the trepidation theory occurs in the 3rd chapter of the "Canon", which, however, is counted among Ptolemy's works by Yaᶜqūbī. Similarly, Battānī (III.190.5) ascribes the account of the trepidation theory

l. 7. ܐܦܘܓܝܘܢ ܘܦܪܝܓܝܘܢ ("apogees and perigees"): < gr. ἀπόγειον, περίγειον; الأوجات والحضيضات IS. These are the forms in which the two words are constantly written by BH (sg. ܐܦܘܓܝܘܢ, ܦܪܝܓܝܘܢ); see Mete. V.i.3.2 below and *Asc.* 235 (index).

l. 7. ܐܦܘܟܐܣ ("positions"): أمكنة IS; < gr. ἐποχή; cf. *Asc.* 235 (index).

IV.i.2.8-10: Formation of dry land

But.: (1) In this way [*hākannā*] in the middle of the sea dry land [*yabšā*] is formed [*hāwē*]. (2a) For divine wisdom too has thus deemed it just [*zedqat*] (2b) that there be a natural place for animals that breathe [the air] [*ḥayywātā sāyōqātā*].

*Šifā*ʾ 25.9-10:

فإذا كان كذلك، لم يكن بد من أن يكون بر وبحر وفى ذلك حِكم إلهية لولاها لم يكن للحيوانات الأرضية التى تعيش بالنسيم مكان طبيعى

(1) This being so [*kadālika*], there inevitably arise [*yakūna*] dry land [*barr*] and sea; (2a) and in this there is divine wisdom [*ḥikam ilāhīya*], (2b) were it not for which there would have been no natural place for terrestrial animals which live by breathing.

l. 10. "animals that breathe the air": cf. Mete. V.i.3.5 below. The idea in this context goes back to the Stoics; see Strabo 17.1.36 (< Posidonius); cf. Freudenthal (1991) 52.

IV.i.3.1-4: Four quarters of the earth

IS talks about the habitable "quarter" (*rub*ᶜ) of the earth in the parts of the *Šifā*ʾ corresponding to IV.i.3.4ff. below, but does not give us a definition of the quarters there.[15] BH gives us here a definition of the quarters, which in fact does not work, since the "circle of the horizon", which he uses here, has the position of the observer as its point of reference and is of no use for the purpose of dividing the earth into its eastern and western hemispheres. A correct definition of the quarters is given by BH at *Cand.* 91.11-92.11 and *Asc.* 128.1-14, where he divides the earth (a) into the northern and southern hemispheres with

---

to Ptolemy rather than to Theo. - There is generally much confusion about the authorship of the *Handy Tables* in Arabic literature and this has led scholars to assume that the the tables were available to the Islamic world only in a version which had been reworked by Theo (see Klamroth [1888] 18-20; Honigmann [1929] 116-121, 186-188; Sezgin, *GAS* V.174, 180-184), but the existence of such a specifically Theonic version of the tables is doubted by Neugebauer (1975) 968 (so also Prof. A. Tihon, in a personal communication).

[15] The definition is given later in the *Šifā*ʾ, ᶜIlm al-haiʾa, ed. Cairo 83.4-5 (corr. Ptolemy, *Alm.* II.1 [Heiberg] 78.22-88.4).

the equator and (b) into the upper and lower (i.e. our eastern and western) hemispheres with the great circle perpendicular to it, (c) this latter circle, in turn, being at right angles to the circle that intersects the equator at the "cupola of the earth" ("qubbṯā d-arᶜā" *Cand.*, "ḥāṭōrtā d-arᶜā" *Asc.*, arab. *qubbat al-arḍ*, *qubbat arīn* etc.).[16]

> *But.*: The circle of the horizon [*sic*] divides the terrestrial sphere [*espēr arᶜā*] into two halves, the upper and the lower [i.e. eastern and western], and that circle, which is parallel [*paralīlāyā*] to the isemeric (circle) [*īsīmerīnōn*] from east to west, divides each of the halves, the upper and the lower, into [further] two halves, the northern and the southern. Thus the terrestrial sphere is divided into four parts.

l. 1. ܚܘܕܪܐ ܗܘ ܕܐܘܪܝܙܘܢ ("the circle of the horizon"): For BH's own definition of the term "horizon", see *Asc.* 19.11-20.3 (where he distinguishes between the "perceived" [*regšānāyā*] and "real/rational" [*ḥattīṯā*] horizons, i.e. one tangential to the surface of the earth and the other passing through the centre of the earth).

l. 2. ܐܝܣܝܡܪܝܢܘܢ ("isemeric (circle)"): < ἰσημερινός. - i.e. the celestial equator; cf. *Asc.* 234 (index).

l. 2-3. "that circle which is parallel to the isemeric (circle)": In *Cand.* the circle dividing the earth into its northern and southern hemispheres is simply called the "equator" (lit. line of equality, *surṭā d-šawyūtā*, 92.1). This is defined more precisely at *Asc.* 128.1-4: ܚܘܕܪܐ ܪܒܐ ܦܪܠܝܠܝܐ ܠܐܝܣܝܡܪܝܢܘܢ ... ܘܗ̇ܘ ܗܘ ܚܘܕܪܐ ܣܘܪܛܐ ܕܫܘܝܘܬܐ ܡܬܩܪܐ ܡܛܠ ܫܘܝܘܬ ܐܡܝܢܬܐ ܘܠܠܝܐ ܕܐܝܡܡܐ ܬܡܢ ܒܐܡܝܢܘ ("the great circle parallel to the isemeric ... and that circle is called the line of equality because of the constant equality of day and night there").

IV.i.3.4-11: Extent of the habitable world

The passage of *But.* here combines, as it were, two traditions concerning the extent of the habitable world. One is Aristotelian, going back to *Mete.* 362b 12-30, where we are told of the possibility of the habitable zone extending to the whole length (longitude) of the earth and the limits placed on its latutidinal extent by the cold in the north and the heat in the south. The other is Ptolemaic, going back to *Alm.* II.1 [Heiberg] 88.5-19, where we are simply told that the habitable world

---

[16] The third point, which defines the longitude of the second (north-south) circle, is not in the definition of the quarters in Ptolemy (*Alm.*, loc. cit.), *Šifāʾ* (ᶜIlm al-haiʾa, loc. cit.) or Ṭūsī, *Tadkira* 245.19-247.1 (the "cupola of the earth" is mentioned later by Ṭūsī, at 251.14); see further Nau (1899) tr. 113 n.1. - On the "cupola of the earth", see EI² V.297 s.v. "al-Ḳubba, Ḳubbat al-ᶜĀlam, Ḳ. al-Arḍ, Ḳ. Arīn" [Ch. Pellat].

falls within one of the quarters of the earth, which by definition has a longitude of 180° and a latitude of 90°.[17] - In the passage of the *Šifā᾿* quoted below, IS is concerned only with the "Ptolemaic" definition of the extent of the habitable world. BH modifies this by adding to it the "Aristotelian" considerations about the northern and southern limits of habitability (hence the alteration of IS's "quarter of a circle" to "less than a quarter"), expressed in vocabulary borrowed, in part at least, from Nic. syr. - The additions in parts (1) and (2) concerning the climes suggest that BH also had before him a copy of Ṭūsī's *Tadkira*.

But. [underline: agreement with *Šifā᾿*; italics: with Nic.]:

(1) The seven climes [*qlīmē*] are situated in a part [*meddem*] of the upper northern quarter [*rubʿā*], not the whole of it. (2) For every clime extends from east to west in length [*urkā*] and from south to north in breadth [*ptāyā*], (3a) and *the length* of the habitable world is a semi-circle [*pelgūt ḥudrā*] (3b) and *all of it is habitable* [*metʿamrānā*] (3c) because it is *far from the two extremes* [*lā mmaššḥwātā*] of deadly [*mmītānē*] *heat* [*ḥummā*] and *cold* [*qurrā*], (4a) whereas its breadth is less than a quarter of the circle, (4b) since in the south it fails to reach the land near the equator because of [*b-yad*] the *burning* heat [*ḥummā mawqdānā*] (4c) and in the north it fails to come near the land which is below the north pole because of the freezing *cold* [*qurrā maglḏānā*].

Šifā᾿ 25.13, 14-17:

ثم ان أصحاب الرصد وجدوا ربع الأرض برا ... ووجد هذا الربع آخذا فى طوله نصف دور الأرض، على ما سنوضح هذا فى الفن الذى نتكلم فيه على الحيئة، ووجد عرضه آخذا ربع دور الأرض إلى ناحية الشمال، حتى يكون الربع الشمالى بالتقريب منكشفا.

(1) Then, the astronomers [*aṣḥāb al-raṣad*] found a quarter of the earth [*rubʿ al-arḍ*] to be dry land [*barr*]. ... (3a) This quarter was found to encompass in its length a half of the circumference [*daur*] of the earth - as we shall explain in the section [*fann*] in which we shall talk about astronomy [*al-hai᾿a*][18] - (4a) and its breadth was found to encompass a quarter of the circle of the earth on the northern side, so that the northern quarter is more or less [*bi-l-taqrīb*] uncovered [by water].

Nic. syr. 31.8-10[19]:

ܝܬܝܪ ܕܝܢ ܐܘܪܟܗ ܕܥܡܪܬܐ/9 ܡܢ ܦܬܝܗ̇ . ܫܪ̈ܐ ܓܝܪ ܘܐܦ ܐܬܪܐ ܕܡܬܚܡܡܐ
ܥܡܘܪܬܐ ܠܐ ܡܬܥܡܪܝܢ . ܟܠܗ ܕܝܢ/10 ܐܘܪܟܐ ܡܬܥܡܪ ܡܛܠ ܕܪܚܝܩ ܡܢ
ܬܚܘܡܝܢ ܠܐ ܡܡܫܚܘ̈ܬܐ .

(3a/4a) The length [*urkā*] of the habitable world [*ʿāmartā*] is greater than its breadth [*ptāyā*]. For the navels [i.e. the arctic regions] and the torrid

[17] Cf. *Asc*. [Nau] 128.20-129.18.

[18] Cf. *Šifā᾿*, ʿIlm al-hai᾿a, ed. Cairo 83.5-12 (corr. Ptolemy, *Alm*. II.1 [Heiberg] 88.5-19).

[19] Cf. Arist. 362b 12-30; Olymp. arab. 124.2ff.

[region] [*yaqqīdtā*] are, as has been said, uninhabitable, whereas the whole length [of the earth] is habitable because it is far from the two extremes [*lā mmaššḥwāṯā*].

Nic. syr. 30.18-25 [see under IV.ii.2.1-4 below]: The navels [i.e. the arctic zones] and the equatorial tropic are uninhabitable [*lā meṯʿamrānē*], (4c) the navels because of [*meṭṭul*] the immoderate coldness [*qarrīrūṯā lā mmaššaḥtā*] - for the sun never comes to these - (4b) the isemeric tropic because of the immoderate heat - for the sun is always in it. - (3b) The drums are habitable because they are free [*mḥarrar*] from the two extremes [*lā mmaššḥwāṯā*] mentioned; neither is immoderate coldness there [lit. in them] because the sun sometimes comes there, nor is immoderate heat there because the sun is not always there.

Taḏkira 251.1-5: (1) Most of the inhabited world [*ʿimāra*] on the northern side falls between the [area] beyond 10 degrees in latitude up to 50 [degrees]. The practitioners of the science [*ahl al-ṣināʿa*] have divided it into seven climes [stretching] lengthwise so that each clime is beneath a day-circle [*madār*], the conditions of the places in it then being similar. (2) Thus, every clime extends [*yamtaddu*] from east to west [*baina l-ḫāfiqaini*] in longitude [*ṭūlan*], while its [extent in] latitude is a small amount, ... (tr. Ragep).

l. 5. ܫܒܥ ܟܠܝܡܘܣ ܩ̈ܠܝܡܐ ("seven climes"): The seven climes are discussed at some length by BH at *Cand*. 94-102 (cf. also 160-162), where for the values of the latitudes of the climes BH follows the traditonal values given by Ptolemy (*Alm*. II.6 [Heiberg] 101-117), and at *Asc*. II.i.8 [Nau] 141f., where he uses the "latest" values given by Ṭūsī in his *Taḏkira* III.1 [Ragep] 251-253.[20] - BH usually has *qlīmē* as the plural form (besides here, *But*. Min. IV.ii.2.5, IV.ii.4.3, IV.iii.3.7; *Asc*. 141.13, 142.22), but the "Greek" plural form ܩ̈ܠܝܡܛܐ is encountered at *Asc*. 77.14.

IV.i.4.1-3: Other quarters

But.: (1a) As to whether water covered [*kassī(w)*] the other upper quarter and the two lower (quarters) or not, (1b) there was [lit. stood] not even one proof [that it did], (1c) except the notion [*masbarnūṯā*] of the necessity [*ālṣāyūṯā*] of the confinement [*ḥāḇōšūṯā*] of earth by water .

Šifāʾ 25.17-18:

ثم لم يقم برهان واضح على أن الأرباع الأخرى مغمورة بالماء، إلا ما يوجبه أغلب الظن بسبب وجوب غمور الماء للأرض.

(1b) Then, there was [lit. stood] no patent proof (1a) that the other quarters are flooded [*maġmūra*] by water, (1c) except what probability [*aġlab al-ẓann*] necessitates on the ground of the necessity [*wuǧūb*] of the flooding [*ġumūr*] of the earth by water.

---

20 See Honigmann (1929) 163 n.2; Ragep (1994) II.469-471; Takahashi (2002b) 254f.

l. 3. ܚܒܘܫܘܬܐ ("confinement"): This is not quite the same as IS's *ġumūr*. The picture presented by IS is, so to speak, two-dimensional. BH's wording here indicates a greater consciousness of the sphericity of the earth and the Aristotelian theory of the concentric circles of elements. Cf. IV.i.1. above (esp. l. 8. *ḥāḇōšā*).

IV.i.4.3-9: Southern hemisphere

But.: (1a) The argument from [lit. the matter of] the proximity [*qarrīḇūṯā*] of the sun to the earth in [lit. of] the south, (1b) [namely that] because its perigee is at the head [*rēš*] of Capricorn scorching heat [*ḥummā mḥarrḵānā*] prevails [*ʿāʾez*] there (1c) and for this reason (it) is uninhabited, (2a) is not acceptable, (2b) because the deviation [*šnuzyā*] of the sun from the centre of the universe is not so very great (2c) as to make that place uninhabitable [lit. so that that place be not inhabited]. (3) Even if, as they say, the heat is strong [*ʿazzīz*] below and around the [Tropic of] Capricorn, (4a) to the south of it, where the sun does not pass the zenith [*lʿel men rēšā*], (4b) it ought to be temperate and suitable for human habitation [*ʿumrā nāšāyā*], (4c) provided it is not covered with water.

Šifāʾ 25.20-26.3:

وأما أمر كون الشمس فى ناحية الجنوب أقرب إلى الأرض، ووجوب تسخين قوى بسبب ذلك، فليس ذلك مما يقع به تفاوت بعيد. فإن خروج الشمس عن المركز ليس بالكثير، وليس مما يوجب جزم القول بأن العمارة لا تحتمل أن يكون عنده. ولنفرض أن ما تحت مدار نقطة الجدى قد يشتد حره، فليس يبعد أن يكون الإمعان إلى ناحية القطب الجنوبى يتدارك ذلك، فيكون إمكان العمارة هناك أوغل من إمكان العمارة فى قطب الشمال.

(1a) As for the matter of the sun being nearer to the earth in the region of the south (1b) and the necessity of strong heating [*tasḫīn*] on that account, (2a) this is not something due to which there occurs a great [*baʿīd*] difference [*tafāwut*]; (2b) for the deviation [*ḫurūǧ*] of the sun from the centre [*markaz*] is not great, (2c) and is no something which necessitates the judgement [*ǧazm al-qaul*] that habitation [*ʿimāra*] cannot [*lā taḥtamil*] exist there. (3) Let us decide that the heat of what is below the tropic of Cancer is intense. (4) It is not improbable then that the intensity [*imʿān*] will continue towards the south pole, so that the possibility [*imkān*] of habitation [*ʿimāra*] there will penetrate further [*auġal*] [towards the pole] than the possibility of habitation at the north pole.[21]

l. 3-4. "because its perigee is at the head of Capricorn": IS's statement that the sun is closer to the earth in the southern hemisphere requires some explanation for the layman and this is duly provided by BH. - For BH's own account of the eccentricity of the solar sphere, see *Asc.* I.2 [Nau] 22-27. - The solar apogee and perigee (or the terrestrial aphelion/perihelion in a heliocentric system) are not stationary. BH tells us that the apogee was at the "28th minute of the last degree of Gemini" in 1590 A.Gr./1279 A.D.

---

[21] IS's argument here is also used by Ṭūsī at *Taḏkira* 249.7-15.

(*Asc.* 26.4-7), close enough for our purposes to the beginning of Cancer. The perigee would accordingly have been close to the beginning of Capicorn.

l. 6. ܗܢܐ ܟܠ ܠܐ ܗܢܐ ܟܠܗ ("the universe, not so very great"): Note word play: *hānā kul ... hānā kulleh*.

l. 8. ܠܥܠ ܡܢ ܪܝܫܐ ܠܐ ܥܒܪ ("does not pass the zenith"): lit. "does not pass overhead", i.e. the point vertically above one's head, = the zenith. - The phrase *lᶜel men rēšā* recurs in this chapter at IV.ii.3, 4, 5; ii.5.1; iii.1.4, 5f.; iii.2.3; iii.3.2; iv.5.1, 4, 5, 7 (also Mete. II.iii.3.7f.; *men ᶜellāway rēšā* Min. IV.ii.6.1f., iii.1.5), where it usually corresponds either to *samt al-raʾs* ("zenith") or some form of the verb *sāmata* [*al-raʾsa*] ("to be parallel to, to be vertically above [the head of]", i.e. "to be at the zenith of").

l. 9. "provided it is not covered by water": cf. IV.i.4.1-3 above.

IV.i.4.9-11: Equator and arctic

The passage here is not very close to the passage at the corresponding place in the *Šifāʾ*, but there are enough points of contact to indicate that BH intends this to be a summary of that passage. - A number of elements suggest that BH may also have had before him what he had written earlier in his *Cand*.

But [underline: agreement with *Šifāʾ*; italics: with *Cand.*]:

(1a) In the proximity of the equator and *to the north of the seventh clime*, there are a *few* [*dallīlē*] [inhabited] places, (1b) but those *men* [*nāšē hānōn*] [living there] are not men of *understanding* [*buyyānā*] (1c) and are akin to the *savage* *anlmals* [*ḥayywāṯā baᶜrīrāyāṯā*] which are numerous there.

Šifāʾ 26.6-13:

فهذا الربع يشبه أن يكون حده الجنوبى وهو خط/٧ الاستواء مجتازا فى أكثر الموضع على البحر. ويشبه أن تكون عمارة التى تتعدى ذلك إلى/٨ الجنوب عمارة لا يعتد بها، ولا يكون أولئك الناس ناسا يعتد بهم وهم مع ذلك جزيريون/٩ ليسوا مقيمين على بر متصل بالبر الأعظم. ثم يشبه أن يكون حده الشمالى حيث ارتفاع/١٠ القطب مثل تمام الميل. ولم يتبين لنا بعد أن مثل ذلك الموضع موضع يصلح لتولد/١١ الناس فيه ولمقامهم الدائم فيه أو لا يصلح لذلك، بل يمكن أن يسافروا إليه فى الصيف/١٢ ولا تكثر هناك إقامهم. وعسى أن يكون ذلك الموضع أو ما وراءه إن لم يكن صالحا/١٣ لأن يتوالد فيه الناس، كان صالحا لأن يتولد فيه حيوانات مخصوصة.

[7 مجتازا Ṭ: مختارا Cair.]

(1a) It seems that the southern limit of this quarter, which is the equator, pass in most places over the sea. It seems that habitation [*ᶜimāra*] which is beyond that towards the south is insignificant [*lā yuᶜtadd bihā*]. (1b) Those people [*ūlāʾika al-nās*] [who live there] are insignificant [in number] and are in addition islanders not living on land connected to the mainland [*al-barr al-aᶜẓam*]. (1a) Then, it seems that its northern limit lies where the elevation of the pole corresponds to the completion of declination.[22] I

[22] i.e. at the latitude where the arc distance from the North Pole corresponds to the maximum solar declination (= obliquity of the ecliptic), i.e. the arctic circle (90° N -

am not yet sure whether such a place is suitable for people to be born in and for them to stay continuously in, or that it is not suitable for that but it is possible for them to travel to it in summer and not stay there for a long time. (1c) Perhaps that place or what is behind it, even if it is not suitable for men to reproduce [*yatawāladu*] in, is suitable for certain animals to be born in.

Cand. 101.6-102.1: (1a) Beyond this [sc. seventh] clime to the north are a few [*bṣīrātā*] tribes called Īsū, Warang and Yūra in Turkish, (1bc) bestial and very savage people [*nāšē bᶜīrāyē w-saggī baᶜrīrē*], who resemble animals and men.[23]

l. 10. "to the north of the seventh clime": IS, in setting the northern limit of the habitable zone at the arctic circle, is following Arist. *Mete*. 362b 2-3 (that is to say, as the phrase τὸν διὰ παντὸς φανερόν there was normally understood).[24] BH, on the other hand, does not set the northern limit of habitability, but merely says that there are only a few inhabited places beyond the northern limit of the seventh clime, which in fact lies some way to the south of the arctic circle (50;04° N sec. Ptolemy, *Alm*. [Heiberg] 111.6; "approx. 50°" sec. BH, *Cand*. 100.7).[25]

l. 11. "are not men of understanding ...": cf. further *Cand*. 95.2-7: "What has been said concerning the equatorial region, that it is temperate, is improbable. This is known from the fact that its inhabitants and all those who are near [them] are burnt, blackened and singed in colour, hair, customs [*ᶜyāḏē*] and intelligence [*hawnē*]" (< Bīrūnī, *Tafhīm* 125.5-7)[26]; also *But*. Eth. I.iii.4: "This is the highest degree of animals, which adjoins the lowest degree of Man, the degree (of Man) in which are placed those men who dwell at the edges of the habitable world, like the Cushites [*kuššāyē*] in the south and the Huns [*hunnāyē*] in the north, who in their habits resemble the animals."

---

ca. 24° = ca. 66° N).

[23] On "Īsū", "Yūra", see Jwaideh (1959) 50 n. 4-5; cf. Bīrūnī, *Tafhīm* 145.11 (pers. 200.11f.); Yāqūt, *Muᶜǧam al-buldān* I.34.21; Qazwīnī, *Āṯār* 410.11f. - The last part of the sentence has no counterpart in Bīrūnī; cf. Yāqūt I.35.9f.: هو آخر العمارة ليس وراءه الا قوم لا يُعْبَأ بهم وهم فى ضيق العيش وقلة الرياضة بالوحش أشبه.

[24] See Lee (1952) 179 n. c.; Strohm (1970) 185; Lettinck (1999) 194 n.1.

[25] The value given by Ṭūsī in his *Taḏkira* is 50;20°. This value is used, by oversight, as the value for the middle of the seventh clime in BH, *Asc*. (142.17f.), where no value is given for the northern limit of that clime. For the northern limit of habitability, BH gives the two possibilities of 66° or 63°, whereby he considers the latter to be more likely (*Asc*. 129.2-4, 144.8-11, cf. Nau's note ad loc., *Asc*. tr. 129 n.1).

[26] Cf. also BH, *Asc*. 146.20-147.2 (where, *Asc*. 146.20-22 < Ṭūsī, *Taḏkira* [Ragep] 259.3-5; *Asc*. 146.23-147.2 < Bīrūnī, *Tafhīm*).

Min. IV. Section ii.: *On Climates of [Various] Regions*

The first two 'theories' of the section deal with the division of the earth into five zones as discussed by Aristotle at *Mete.* 362a 31ff. For the explanation of the division (IV.ii.1), BH follows the text and the vocabulary of the *Šifā'*, but his account of the habitability of these zones (IV.ii.2.1-4) is closer to that given in Nic. syr., while the discussion of the analogous habitable world in the southern hemisphere (IV.ii.2.4-9) may be based on a lost passage of Nic. syr.

IV.ii.1.1-7	*Šifā'* 26.17-27.8 (cf. Nic. 30.11-18)	Division into five zones
IV.ii.2.1-4	Nic. 30.18-25	Habitability of zones
IV.ii.2.4-9	Nic.? (cf. Olymp. 184.10-18)	Southern hemisphere

The rest of this section (IV.iv.3-6) and the whole of the following section (IV.iii.1-6) are concerned with the question of the climate and habitability of the equatorial region. - Whereas Aristotle simply considered the region between the two tropics to be uninhabitable (*Mete*. 362b 6-9; followed by Nic. syr. 30.20-22), others, led in part no doubt by the discovery of habitations south of the Tropic of Cancer, later argued that the equator itself was more temperate than the two tropics, giving as one of the reasons the increase in the speed of solar declination at the equator, the argument represented here in IV.ii.3.3-10.[27] - Ibn Sīnā disagreed with the "Peripatetics" and maintained that the equatorial region was temperate both here in the *Šifā'* and in his *Qānūn fī al-ṭibb.*[28] - Among those subsequent to IS, Abū al-Barakāt agreed with IS in considering the equatorial region to be temperate, although his arguments for this are somewhat different from IS's (*Muᶜtabar* II.202-208). Fakhr al-Dīn al-Rāzī, on the other hand, disagreed with IS and refuted IS's arguments at some length (*Mabāḥiṯ* II.199-204),[29] while Ṭūsī briefly reports IS's argument and Rāzī's

[27] The argument is reported at Ptolemy, *Alm.* II.6 [Heiberg] 103.5-10. IS does not go into a detailed discussion of the matter at the corresponding place in the book (*fann*) on astronomy in the *Šifā'* but refers the reader to his "books on the natural sciences" (*Šifā'*, ᶜIlm al-hai'a, ed. Cairo [1980] 96.3-8). - In Syriac, we have a report of the argument in Severus Sebokht, *Constell.*, Sachau (1870) 128.10-21, tr. Nau (1930-1) XXVIII.95.

[28] *Qānūn* I.2.2.1.8, ed. New Delhi (1982) 151.1-20; Cairo (1987) 120.14-121.1; Būlāq (1294 h.) 88.6-22.

[29] I have not had access to Fakhr al-Dīn al-Rāzī's commentary on IS's *Qānūn*, where he is also reported to have refuted IS's view (see Ragep [1994] II.473).

refutation and agrees, on this occasion, with Rāzī against IS (*Taḏkira* III.2.2-4 [Ragep] 257-259).[30] - The subject is also treated by BH himself in at least two of his earlier works, at *Cand.* 95.2-7, where he follows Bīrūnī, *Tafhīm* 125.5-9 (pers. 171.7-12) in rejecting the theory that the equator is temperate, and at *Asc*. 144-147, where he basically follows Ṭūsī's *Taḏkira* but adds some elements which are not in Ṭūsī.

In his report of the four Avicennian arguments here in IV.ii.3-6, BH does not follow the text of the *Šifā*ᵓ closely, although the ideas explained by BH are there in the *Šifā*ᵓ, so that BH is not actually falsifying IS's views. We notice in fact that some of IS's arguments are shortened while others are expanded. This may be due in part to a desire on BH's part to make each of the four arguments fit neatly into a 'theory'. The wording of a number of these passages also show a resemblance with the corresponding parts of *Asc*., suggesting that BH made use of his earlier work in composing these passages, while the influence of Fakhr al-Dīn al-Rāzī can be detected in the passage at IV.ii.5.5-9. We also note that BH often makes IS's abstract arguments more concrete by giving us concrete astronomical and geographical values.[31]

A number of the points made in refutation of the Avicennian arguments in Section IV.iii. can be found in Fakhr al-Dīn al-Rāzī's *Mabāḥiṯ*, but here again BH's wording is not very close to Rāzī's, so that we are led to suspect the use of a source which has so far escaped our notice. In the first argument (IV.iii.1.3-7), BH's wording agrees more closely with that in Ṭūsī's *Taḏkira* than with Rāzī. In the last argument (IV.iii.6), which BH presents as his own, he in fact closely follows the text of Abū al-Barakāt's *Muᶜtabar*, but turns this argument, through the alteration he makes at the end against Abū al-Barakāt's view that the equator is temperate.

Section ii

But.	*Šifā*ᵓ	*Qānūn*	*Mabāḥiṯ*	*Asc.*	
3.1-3	(27.9ff.)				Introduction
3.3-10	29.4-15	151.11-16	200.20-201.9	144.20.145.4	(1) Solar declination
4.1-7	29.7-8, 13			145.7-9	(2) Equal. of day/night
5.1-5	28.13-29.3	151.4-10	199.16-200.9	145.3-7	(3) Prolonged heating
5.5-9	28.20-29.2		200.10-17		(contd.)
6.1-9	29.16-30.11				(4) Temp. variation

[30] Cf. also Ṭūsī, *Muᶜīnīya* III.2, [Dānishpazhūh] 64.11-65.10.

[31] e.g. IV.ii.3.7-9, the values for rate of solar declination; IV.ii.3.5, "in the second clime" instead of "these countries which are next to us".

Section iii

But.	*Mabāḥit̠*	*Tad̠kira*	*Muᶜtabar*	
1.1-2				Introduction
1.3-7	202.19-21	257.18-20		(1) Sun never far from equator
2.1-4	203.1-8			(2) Weakness of night coldness
3.1-7	201.21-202.17			(3) Infl. of prior heating/cooling
4.1-6	(203.18-204.14)			(4) Objective heat/coldness
5.1-8	201.18-202.18	257.21-24		(5) Dist. of sun from equat./7th clime
6.1-2				BH's apology
6.3-11			206.7-16	(6) Need for heat/cold balance

l. 1. ܡܛܠ ܡܘܙ̈ܓܐ ܐܬܪ̈ܘܬܐ: lit. "mixtures/constitutions of places/regions". - See comm. on the title of Min. IV. above; cf. Fakhr al-Dīn al-Rāzī, *Mabāḥit̠* II.199.2: في امزجة البلدان. - For *muzzāḡā* in this sense, cf. gr. κρᾶσις; arab. مزاج.

IV.ii.1.1-7: Division of the earth

This division of the earth into five zone goes back to Arist. *Mete*. 362a 31ff. There is a corresponding passage in Nic. syr. 30.11-18 (cf. Alex. Aphr. 101.12ff.; Olymp. 183.18ff.; Olymp. arab 122.17-21), where the text is accompanied by a simple diagram. BH will no doubt have had this text and diagram of Nic. syr. before him when composing this passage, but the wording of the passage here follows that of the *Šifāᵓ*.

But.: (1) The ancient Peripatetics divided [*palleḡ*] the terrestrial sphere [*espēr arᶜā*] into five divisions [*pullāḡē*] with four circles parallel to the isemeric (circle). (2a) Two circles separate [*pāršīn*] two tympanic [*tab̠lānāyē*] divisions, the northern and the southern, out of the earth, (2b) and each of them is confined by [*ḥāb̠šīn*] a part of the circumference (of the earth) [*ḥōḡtāh*] and a flat [lit. straight] surface. (3a) Two other circles separate three trapezial [*dappānāyē*] divisions out of the earth, (3b) and each of them is confined by two flat surfaces. (3c) The middle trapezial (division) [is located] where the equator (is), another trapezial (division) between it and the northern tympanic (division), and the other (trapezial division) between it and the southern tympanic (division).

Šifāᵓ 26.17-27.8:

فنقول: إن قوما جعلوا كرة الأرض مقسومة بخمسة أقسام، تفصلها دوائر موازية لمعدل النهار. فمن ذلك دائرتان تفصلان الغامر الخراب من العالم، بسبب القرب من القطب وشدة البرد، إحداهما شمالية والأخرى جنوبية. وهاتين تفصلان من الأرض قطعتين طبليتين تحيط بكل واحدة منهما طائفة من محيط الكرة وسطح مستقيم، والحد المشترك بينهما دائرة. وأما الحد بين الغامر والعامر من جهة الحر عندهم، فهو ما بين البلاد التى تكون خارجة عن مجاز الشمس إلى الأرض المحترقة التى تحاذيها الشمس بمدارها، فتسخنها تسخينا لا يحتمل عندهم الحيوان المقام فيه. وهو مُكشف بين العمارتين، فتكون الأرض المحترقة محدودة بدائرتين شمالية وجنوبية تليهما من جهة القطبين عمارتان، فتكون ثلاثة قطوع دُفِّيّة يحيط بكل واحد منها من الجانبين سطحا دائرتين، ويصل بينهما سطح دفى، وكذلك تكون هيئة العمارتين. لكن السطحين المحيطين بكل واحد منهما لا يكونان متساوين، بل الذى يلى القطب يكون أصغر. وأما سطحا دُفِّ الأرض المحترقة عندهم فمتساويان.

We say: (1) People [*qaum*] have made [*ǧaʿalū*] the terrestrial sphere [*kurat al-arḍ*] divided [*maqsūma*] into five divisions [*aqsām*], which are separated by [*tufaṣṣiluhā*] circles parallel to the equinoctial [*muʿaddal al-nahār*]. (2a) For out of them, two circles separate [*tufaṣṣilāni*] the desolate uninhabitable land [*al-ġāmir al-ḫarāb*] - [which is uninhabitable] due to the proximity to the pole and severity of the cold - one [circle] in the north and the other in the south. (2a) These two [circles] separate out of the earth two tympanic sections [*qiṭʿataini ṭablīyataini*], (2b) each of which is surrounded by [*tuḥīṭu*] a part [*ṭāʾifa*] of the circumference [*muḥīṭ*] of the sphere and a straight [*mustaqīm*] surface, the shared border [*ḥadd*] between the two being a circle. As for the boundary between the uninhabitable and habitable land [*al-ġāmir wa-l-ʿāmir*] on [what] according to them [*ʿindahum*] [is] the side of heat, this lies in the lands beyond the path [*maǧāz*] of the sun towards the torrid earth [*al-arḍ al-muḥtariqa*] which the sun stands directly above [*tuḥāḏīhā*] in its circulation [*madār*] and heats to such an extent that, according to them, an animal cannot remain in (that heat). It is found between the two habitable worlds [*ʿimāratāni*]. The torrid earth [*al-arḍ al-muḥtariqa*] is bordered by two circles, in the north and in the south, which are adjoined by the two habitable worlds on the sides of the two poles, (3a) so that there are three tambourine-shaped [*duffīya*] sections (3b) each of which is surrounded on its two sides by the two surfaces of two circles and there come between them a tambourine-shaped surface. Similarly the shape of the two habitable worlds, except that the two surfaces surrounding the two of them are not equal, but that which is nearer the pole is smaller. The two faces of the tambourine of what, according to them, is the torrid earth are equal.

l. 1. ܩܕܡܝܐ ܕܦܪܝܦܛܝܩܘܣ ("The ancient Peripatetics"): That this is the view of the "ancient Peipatetics" is mentioned later on by IS at *Šifāʾ* 27.9: فهذه هو قول قدماء المشائين. - Fakhr al-Dīn al-Rāzī tells us the view is held by "most Peripatetics and all astronomers" (II.199.3 اكثر المشائين وجمهور المنجمين). - The division is also said to be that of "the ancients" (*qaḏmāyē*) by Severus Sebokht, *Constell.*, Sachau (1870) 127.7 (see under IV.ii.2.1-4 below).

l. 2. ܦܘܠܓܐ ("divisions"): أقسام IS. - BH uses the term *pullāgā* throughout this passage. IS uses *qism* here, but switches later to *qiṭʿa* (*Šifāʾ* 26.20), which is in fact more faithful to Aristotle's τμῆμα (362a 32; also ἔκτμημα 362b 5, 363a 29).

l. 3. ܛܒܠܝܐ ("tympanic") طبلى IS. - The term goes back to Arist. (362a 35 οἷον τυμπάνου; cf. also *De caelo* 293b 34 τυμπανοειδής) and is also used by Nic. syr. (*ṭaḇlā*) and Olymp. arab. (*ṭabl*). In these works, however, the tympanic sections are the two habitable sectors of the earth (in other words, Arist. et al. envisage a drum with two flat surfaces above and below). IS alters this and applies the term *ṭablī* to the arctic and antarctic sectors (i.e. envisaging a drum with a round bottom) and this is followed here by BH. - For the arctic and antarctic sectors Nic. syr. uses the term "navel" (*šerrā*; cf. Olymp. 184.4 ὀμφαλός) and Olymp. arab. "edge" (*ṭaraf*); cf. Mete. III.iii.1.10 (ܫܪ̈ܐ ܗܘܦܘܣ) below.

l. 5. ܕܦܝܝܐ ("trapezial"): < *dappā* board, table, plank etc.; دفى IS. - The word in IS is vocalised *duffī* (< *duff*, tambourine) by the editors of the Cairo edition and this is probably correct in view of the contrast with the "drum". BH, however, seems to have derived the word from *daffa* (side, board etc.).

IV.ii.2.1-4: Habitability of the zones

If we ignore the terms "tympanic" and "trapezial", where BH is following IS's usage, the explanation of the habitability or uninhabitability of each zone here is closer to that given in Nic. syr. than that given in the *Šifā*ʾ. A number of correspondences with the passage of Severus Sebokht's *De constellationibus* (*Mēmrā ᶜal demwātā*) quoted below gives rise to the possibility that BH also made use of that passage.

But. [underline: agreement with Nic.; italics: with Severus Sebokht]:
(1a) Those who made such a *division* [lit. who divided (*palleḡ(w)*) thus] say (1b) that the two tympanic divisions are uninhabitable [*lā metᶜamrīn*] (1c) because of [*b-yaḏ*] the severe coldness [*qurrā qašyā*] arising from [*d-men*] *the distance* [*ruḥqā*] *of* the sun, (2a) and similarly the *middle* [*meṣᶜāyā*] trapezial (division) (2b) because of the burning heat [*ḥummā mawqḏānā*] arising from *the proximity* [*qurbā*] *of the sun*. (3a) [They say that] *the other two* trapezial (sections), (3b) because they are far from the two extremes, (3c) are useable for human settlement [*duyyārā nāšāyā*].

Nic. syr. 30.18-25[32]:

ܘܡܨܥܝܐ/19 ܕܝܢ ܘܛܘܪܦܘܣ ܕܫܘܝܘܬܐ: ܠܐ ܡܬܥܡܪܢܐ ܐܢܘܢ. ܫܪ̈ܐ ܡܢ ܡܛܠ
ܩܪܝܪܘܬܐ/20 ܠܐ ܡܡܫܚܬܐ. ܠܐ ܡܛܐ ܓܝܪ ܡܬܚܡܝܢ ܫܡܫܐ ܠܥܠܡ ܠܘܬܗܘܢ.
ܠܛܘܪܦܘܣ/21 ܕܝܢ ܐܝܣܡܪܝܢܘܣ. ܡܛܠ ܚܡܝܡܘܬܐ ܠܐ ܡܡܫܚܬܐ. ܒܐܡܝܢܘ
ܓܝܪ/22 ܒܗ ܐܝܬܘܗܝ ܫܡܫܐ. ܛܒ̈ܠܐ ܕܝܢ ܡܬܥܡܪ̈ܢ. ܡܛܠ ܕܡܚܪܪ̈ܝܢ ܡܢ
ܬܪ̈ܬܝܢ/23 ܠܐ ܡܡܫܚ̈ܘܬܐ ܕܐܬܐܡܪ̈ܝ. ܘܐܦܠܐ ܒܗܘܢ ܩܪܝܪܘܬܐ ܠܐ ܡܡܫܚܬܐ
ܐܝܬ ܒܗܘܢ/24 ܡܛܠ ܕܐܬܐ ܒܙܒܢ ܠܙܒܢܝܢ ܫܡܫܐ. ܐܦܠܐ ܚܡܝܡܘܬܐ ܠܐ
ܡܡܫܚܬܐ ܐܝܬ/25 ܒܗܘܢ. ܡܛܠ ܕܠܐ ܐܡܝܢܘ ܗܘܐ ܒܗܘܢ ܫܡܫܐ.

(1b/2a) The navels [i.e. the arctic zones] and the equatorial tropic are uninhabitable [*lā metᶜamrānē ennōn*], the navels (1c) because of the immoderate coldness [*qarrīrūṯā lā mmaššaḥtā*] - for the sun never comes to these - (2a) the isemeric tropic (2b) because of the immoderate heat - for the sun is always in it. - (3a/c) The drums are habitable (3b) because they are free [*mḥarrar*] from the two extremes [*lā mmaššḥwāṯā*] mentioned; neither is immoderate coldness there [lit. in them] because the sun sometimes comes there, nor is immoderate heat there because the sun is not always there.

Severus, *Constell.* XVIII, Sachau (1870) 127.7-128.1 (tr. Nau [1930-1] XXVIII.94f.): The ancients [*qaḏmāyē*] also said thus: Because the whole surface of the earth is divided [*etpallḡat*] into five zones [*zōnas*] like the

---

[32] Cf. Arist. 362a 31-362b 9; Olymp. 184.10-24; Olymp. arab. 122.21-123.12.

celestial ones, as has been shown, (1) the two zones situated opposite [each other] below the poles, I mean the northern and the southern, because they are cold [*qarrīrāṯā*] and intemperate [*lā mmazzḡāṯā*] because of [*meṭṭul*] the distance of the sun [*raḥḥīqūṯeh d-šemšā*] from them, they say, are completely uninhabitable [*lā sāḵ meṯʿamrān*]. (3) The other three [zones] in the middle, I mean that below the summer (tropic),[33] that below the winter (tropic) and that below the isemeric or the equality of day and night, because they are temperate [*mmazzḡāṯā*] because of the passage [*maʿbartā*] of the sun over them, they say, are habitable. However, the two below the tropics, i.e. the summer and the winter, are more temperate and for this reason are more habitable [*yattīr meṯʿamrānyāṯā*]. (2) That in the middle [*meṣʿāytā*], i.e. that below the isemeric, they say, is κεκαυμένη, i.e. burned [*mawqaḏtā*], because of its constant proximity [*qarrīḇūṯā ammīntā*] of the sun to it, or its passage [*maʿbartā*] over it as it ascends [*sāleq*] northwards and descends [*nāḥeṯ*] southwards, towards the summer (tropic) and the winter (tropic), and is for this reason less, and not usually, habitable [*bṣīrāʾīṯ w-law aḵ da-ḇ-suggā meṯʿamrā*].

IV.ii.2.4-9: Habitations in southern hemisphere

The possibility of there being a habitable world in the southern hemisphere analogous to that in the north is hinted at by Arist. at *Mete.* 362b 30-363a 1. - BH has already touched upon this at IV.i.4.3-6 above where he was following the *Šifāʾ*. Here, the correspondences between our passage and the passage of Olymp. below leads us to suspect that BH had before him a lost passage of Nic. syr. based on Olymp.

But.: (1a) However, while the inhabitants [*ʿamōrē*] of the northern trapezial (division) and the customs of its inhabitants are known to us - (1b) in it are situated these celebrated seven climes - (2a) we know of no inhabitant in the southern trapezial (division) at all, (2b) nor have we met anyone from its settlements. (3a) Rather, we only know of the possibility of settlement there and (its) suitability for habitation [*ʿumrā*] due to the temperateness of the air [*mmazzḡūṯ āʾār*], (3b) provided there is no other cause preventing [habitation], such as [the presence of] seas.

Olymp. 184.10-18 [cf. Olymp. arab. 123.14-124.2]: τούτων οὕτω προδιατυπωθέντων ἰστέον, ὅτι τούτων τῶν ἓξ τμημάτων δύο μόνον οἰκοῦνται, λέγω δὴ αἱ μεταξὺ τοῦ ἀρκτικοῦ πόλου καὶ τοῦ θερινοῦ τροπικοῦ, ἔνθα καὶ ἡμεῖς οἰκοῦμεν, (1b) ἥτις εἰς ἑπτὰ κλίματα τέμνεται, ἀλλὴ δὲ ἡ ταύτης ἀνάλογος αὕτη ἐστὶν ἡ μεταξὺ ἀνταρκτικοῦ πόλου καὶ χειμερινῆς τροπικῆς. (3a) ἀλλὰ αὕτη λόγῳ μόνον οἰκεῖται διὰ τὸ

[33] ܬܚܝܬ ܡܕܝܠܗܘܢ ed. Sachau (with BL Add. 14538): lege ܬܚܝܬ ܗܘ ܡܕܝܠܝܬܐ with Paris 346 (see Nau, loc. cit., 94 n.4).

ὑπαγορεύειν τὸν λόγον εὔκρατον εἶναι ὥσπερ καὶ τὴν καθ' ἡμᾶς ζώνην. (2a) οὐκ ἀκριβῶς δ ' ἴσμεν, τίνες οἰκοῦσιν ἐκεῖσε, (2b) διὰ τὸ μὴ δύνασθαι μήτε ἡμᾶς ἐκεῖσε πορεύεσθαι μήτ ' ἐκείνους πρὸς ἡμᾶς διὰ τὴν διακεκαυμένην ζώνην μεταξὺ οὖσαν.

l. 8-9. "such as [the presence of] seas": cf. IV.i.4.1-3, 9 above.

IV.ii.3.1-3: Ibn Sīnā's view

But.: (1a) The princely doctor [Ibn Sīnā] thinks the opposite of the Peripatetics, his teachers, (1b) and says that the equator is more temperate as regards heat and cold than all [other] places (1c) and confirms this opinion [*tar*ᶜ*ītā*] of his with four arguments [*s*ᶜ*āyē*].

*Šifā*ʾ 27.9, 12-13:

فهذا هو قول قدماء المشائين، وليس التحقيق والوجود على ما حكوه. ... والقياس يجوّز، بل يوحب أن تكون بقعة خط الاستواء أصلح المواضع للسكنى وأولاها بالاعتدال

(1a) These are the words of the ancient Peripatetics, but reality and fact are not as they thought ... (1b) Reasoning allows, rather, compels [one to conclude] that the region [*buq*ᶜ*a*] of the equator is the most salutary for habitation and most moderate [*aulā bi-l-i*ᶜ*tidāl*] of places.

IV.ii.3.3-10: First argument: rate of solar declination

The argument put forward here is that found in the passage of the *Šifā*ʾ quoted below, but BH's wording is rather different from IS's.[34] Of the alterations made by BH, the concrete figures for the rate of solar declination (sentence (3)) may be an addition based on his knowledge of astronomy. The agreement of the wording (especially in clauses (2b-c)) with the passage of *Asc.* quoted below suggests that BH also had before him that earlier work of his when composing this passage of *But.* The passage of *Asc.*, in turn, follows Ṭūsī's *Taḏkira*, but contains elements which are not found in the *Taḏkira*. - As noted by Nau in his translation of *Asc.* (tr. 129 n.2), the argument is one which is already reported by Ptolemy (*Alm.* II.6 [Heiberg] 103.5-10).

But. [underline: agreement with *Šifā*ʾ; italics with *Asc.*]:

First argument: (1a) *The sun* passes *the zenith* [lit. above the head, *l*ᶜ*el men rēšā*] in Aries and Libra there (1b) and changes position [*mšannē*] with speed [*b-rahhībū*] (1c) *because of* [*b-yad*] *the increase* [*tawseptā*] [in the rate] *of declination* [*ṣlāyā*]. (1d) For this reason it does not create much

34 The corresponding passage of Fakhr al-Dīn al-Rāzī's *Mabāḥiṯ* (II.200.20-201.9) is essentially a summary of the *Šifā*ʾ, but it may be noted that the phrase "for many days" which is not in the *Šifā*ʾ but is in *But.* also occurs there, while the mention of "the sign of Cancer" in *But.* may also be related to Rāzī's rephrasing of IS's "these countries which are next to us" as "the places which are on the path of the two Tropics".

heat [*maḥḥem*]. (2a) [This is] not so in the second clime, (2b) where the sun passes the zenith at *the head* [*rēš*] *of Cancer* (2c) *and remains* [*hāwē*] *for many days in its proximity* [*b-qurbā menneh*] (2d) because of the decrease [in the rate] of declination. (2e) For this reason it *creates much heat* [*saggī maḥḥem*]. (3a) For the declination changes by one minute in one day (in the second clime) [lit. here], (3b) but by twenty-four (minutes) there [at the equator]. (4) Therefore, the heat in the second clime is stronger than that at the equator.

Šifāʾ 29.4-15:

فهذه البلاد التى تلينا يعرض لها أن الشمس تقرب منها بتدريج يتقدمه تسخين بعد/ ٥ تسخين؛ ثم إذا وازاها وحاذاها، عرض أن يقيم عندها مدة كثيرة لا تتنحى عن رؤوسها،/ ٦ لأن الميول عند قرب من المنقلبين تقل وتصغر جدا؛ ثم إن كانت تسامت الرأس/ ٧ وتجاوزه، عاودت المسامتة عن قريب، ويكون النهار أيضا طويلا والليل قصيرا،/ ٨ فيدوم إلحاح الشمس عليها بالتسخين، لكون مددها متقاربة ومع ذلك طويلة، ومع/ ٩ ذلك حافظة لقرب واحد من الشمس، فيكون الحر متجاوزا للحد./ ١٠ وأما فى خط الاستواء، فإن الشمس تبلغ المسامتة دفعة، لأن الميول هناك تكثر/ ١١ وتتفاوت تفاوتا لا يؤثر إلا أثر المسامتة والمغافصة، ثم تبعد عن سمت الرؤوس بسرعة،/ ١٢ ولا تلح عليها، وتأخذ كل ساعة تزداد بعدا إلى أن يبعد الميل كله، غير ملحة ولا لجوج،/ ١٣ ويكون النهار مساويا لليل فى الطول والقصر، ثم لا تعود ألى سمت الرأس عن قرب،/ ١٤ بل إلى نصف السنة. ثم تكون المسامتة خفيفة على الجملة المذكورة. ثم تأخذ فى البعد،/ ١٥ فلا يشتد الحر جدا، لما قلناه، ولا يشتد أيضا البرد.

[4-5 تسخن بعد تسخن Ṭ || 5 كثيرة Ṭ: om. Cair.]

(2) The sun approaches these countries which are next to us [*talī-nā*] gradually and [periods of] heating after heating precede (its arrival). Then when it is parallel to and faces them, (2c) it remains by them for a long period [during which] it does not go away from their zenith [*ʿan ruʾūsihā*], (2d) because the [rates of] declination [*muyūl*] decrease and become very small in the proximity of the two tropics [*munqalabāni*]. Then, when it reaches the zenith [*tusāmitu al-raʾsa*] and goes beyond it, it soon comes back to the zenith. Furthermore, the day is [then] long and the night short, so that the persistence of the sun in heating them is prolonged [*yadūmu*] because of the periods (of the visits) [*mudaduhā*] being close to each other as well as long, as well as retaining the same proximity to the sun [?],[35] (2e) with the result that the heat will become excessive [*mutaǧāwiza lil-ḥadd*]. (1) At the equator, the sun reaches [*tabluġu*] the zenith [*musāmata*] suddenly (1b) because the [rates of] declination increase and differ in such a way that it does not have any effect other than that of [merely] being at the zenith [*musāmata*] and of a sudden arrival [*muġāfaṣa*]. Then, then it goes away from the zenith [*samt al-raʾs*] with speed [*bi-surʿa*] and does not persist, and its distance increases every hour until it is distant by the total declination, without persistence and insistence. The day is [then] equal

[35] Cf. Fakhr al-Dīn al-Rāzī, *Mabāḥiṯ* II.201.3-5: فيدوم الحاح الشمس عليها بالتسخين من وجهين، احدهما طول النهار وقصر الليالى، والثانى بقاؤها على موضع واحد وعلى ما يقرب منه مدة طويلة ("... so that the persistence of the sun in heating them is prolonged for two reasons, firstly the length of the day and the brevity of the night, and secondly its lingering [*baqāʾuhā*] in the same place or what is close to it for a long time [*mudda ṭawīla*]").

to night in length and shortness. Then, it does not return to the zenith soon, but only after half a year. Then, the presence at the zenith is fleeting for all the reasons mentioned. It then goes away. (1d) As a result the heat is not very intense, for the reasons stated, nor is the cold intense.

Qānūn [New Delhi (1982)] 151.11-16: (1ab) At a place [*al-buqʿa*] near [*muṣāqiba*] the equator, the sun is at the zenith [*tusāmitu ... al-raʾsa*] only for a few days [*ayyāman qalīlatan*], then distances itself [*tatabāʿadu*] with speed [*bi-surʿa*] (1c/2d) because the increase of the degrees of the declination [*tazāyud al-aǧzāʾ al-mail*] at the equinoxes [*ʿuqdatāni*] is much greater by far [*aʿẓam kaṯīran fāḥišan*] than their increase at the two tropics [*munqalabāni*]. (2) Rather, it may not have any perceptible effect at the two tropics for three or four days or more. (2c) Then, the sun remains [*tabqī*] there in one limited [lit. nearby] area [*ḥayyiz wāḥid mutaqārib*] for a long time [*mudda madīda*] (2e) and [*fa-*] assiduously generates heat [*yumʿinu* (*tumʿinu*) *fī al-isḫān*]. It must therefore be believed from this that the countries whose latitudes are close to the total declination are the hottest countries ...

Asc. 144.20-145.4 [underline: agreement with *But*.; italics: with *Taḏkira*]: (1) This [sc. that the equator is temperate] is known from *the non-lingering* [*lā mqawwyūṯā*] of the sun *at* the zenith [*lʿel men rēšā*] there, because of [*meṭṭul*] the greatness of the *speed* of its declination [*suggaṯ surhāḇā d-mesṭalyānūṯeh*], (2ab) in contrast to *the place* [*aṯrā*] *whose latitude is equal to the total declination* [*ṣlāyā kullānāyā*], namely that below the head [*rēšā*] of Cancer, (2c) where it *remains* [*hāwē*] for many days [*yawmāṯā saggīʾē*] *in the proximity* [*qarrībūṯā*] *of the zenith* [*nuqdṯā da-lʿel men rēšā*]. (2e) *Although the presence* [*hwāyā*] of the sun *at the zenith* [*lʿel men rēšā*] is [in itself] *heat-generating* [*maḥḥmānā*], heat [*ḥummā*] is greatly multiplied [*yattīr metʿappap̄*] by its *persistence* [*kuttārā*]. ...[36]

Ṭūsī, *Taḏkira* III.2.2 [Ragep] 257.6-11, 15-17 [underline: agreement with *Asc*.]: (1) He [Ibn Sīnā] stated [that the equator is temperate] because the sun does not linger [*talbaṯu*] there long at the zenith [*samt al-ruʾūs*], but rather it passes by it at the times of its crossing [*iǧtiyāz*] from one of the directions to the other and its motion in declination [*ḥarakatuhā fī al-mail*] will be at its fastest [*asraʿ mā takūna*], then the heat of their summer will therefore not be intense. (2e) This is because even though being directly overhead [*musāmata*] leads to heating [*muqtaḍīya li-l-tasḫīn*], nevertheless the duration of this [state] [*makṯ ʿalaihā*] is more effective [*ablaġ*] for that than the [state] itself ... (2) He also stated that the hottest localities [*buqāʾ*] in summer are those whose latitudes are equal to the obliquity [*al-mail al-kullī*]. For the sun will be directly overhead [*tasāmatu-hā*] and will linger near this alignment [*fī qurbi musāmatati-hā*] for nearly two months (tr. Ragep).

---

[36] The argument carries on into that corresponding to the one given in IV.ii.5. below.

l. 5. ܣܛܝܐ ("declination"): ميل IS. - Solar declination is the distance of the sun from the celestial equator, measured either perpendicular to the celestial equator or to the ecliptic.[37] Here in *But*. as in the *Šifāʾ*, however, the "increase/decrease of the declination" must mean not the increase/decrease of the declination itself, but of the rate of its change.

l. 5. "in the second clime": At the beginning of the passage of the *Šifāʾ* quoted above, IS merely says "these countries which are next to us", but it is clear that he is talking about the region below the Tropic of Cancer.[38] The Tropic of Cancer (ca. 23;27° N today; 23;51° sec. Ptolemy), in turn, is aligned by Ptolemy with the seventh parallel (*Alm*. II.6 [Heiberg] 107-108; cf. tr. Toomer [1984] 85 n. 34.), the parallel which passes through the middle of the second clime. The second clime itself lies between 20;14° and 27;12° according to the traditional Ptolemaic values and between 20;27° and 27;30° according to the figures given by BH himself in *Asc*. II.1.8 [Nau] 141 (where he is following Ṭūsī's *Taḏkira*).

l. 6. ܝܘܡܬܐ ܣܓܝܐܐ ("for many days"): cf. Fakhr al-Dīn al-Rāzī, *Mabāḥiṯ* II.201.2 *ayyāman kaṯīratan*.

l. 8. ܐܟܣܝܩܘܣܛܐ ("minute"): < gr. ἑξηκοστή, ἑξηκοστόν; cf. *Asc*. 15.16-18.

IV.ii.4.1-7: Second argument: equal length of day and night

Here again, BH greatly expands what IS had to say. The notion of heat and cold "tempering" each other may be taken from *Asc*.

But. [underline: agreement with *Šifāʾ*; italics with *Asc*.]:

Second arugument: (1a) At the equator day is equal [*šwē*] [in length] to night throughout the year. (1b) There, therefore, *the heat* of the day [*ḥammīmūṯ īmāmā*] *is tempered* [*meṯmazzḡā*] *by the coldness* of the night [*b-qarrīrūṯ lelyā*] throughout the year, (1c) and neither [heat nor cold] prevails over the other. (2a) In the seven climes, heat prevails [*ʿālbā*] during the long days of the summer [lit. in the length of summer days] heat prevails, (2b) while coldness [prevails] during the long nights of the winter [lit. in the length of the winter nights], (2c) and temperate weather occurs only in spring and autumn. (3a) At the equator, then, the whole year is spring-like [*taḏānāytā*], (3b) whereas in other places sometimes heat is strong [*ʿāšnā*] and sometimes coldness.

Šifāʾ 29.7-8, 13 [see under IV.ii.3.3-10 above]: (2) Furthermore, the day is [then] long [*yakūnu ṭawīlan*] and the night short [*qaṣīran*], so that the persistence of the sun in heating them is prolonged [*yadūmu*] ... (1a) The day is equal [*musāwī*] to night in length and shortness.

---

[37] See EI[2] VI.914f. s.v. "al-Mayl" [King]; for BH's own explanation of the term, *Asc*. I.i.11 [Nau] 17.20-18.13.

[38] Fakhr al-Dīn al-Rāzī rephrases those words of IS as "the places which are on the path of the two Tropics" (المواضع التي على مدار نقطتي الانقلابين, *Mabāḥiṯ* II.200.20).

Asc. 145.7-9 [cf. Ṭūsī, *Tadkira* 257.13-15]: (1a) For also through [*b-*] the equality [*šawyūtā*] of the days and nights (1b) the severity [*ʿuzzā*] of the heat [*ḥammīmūtā*] and the cold [*qarrīrūtā*] is broken [*mettbar*] by their interaction [lit. by each other, *ba-ḥbartāh*] and (the heat and cold) revert to temperateness [*l-mmazzgūtā hāpkān*].

l. 6. ܪܒܝܥܝܬܐ ("spring-like"): cf. *Šifā*ʾ 30.13-14: ويكون كأنه فى ربيع دائم ("and it is as if [the inhabitant of equator] is in a perpetual spring"); cf. also IV.iii.6.9 below.

IV.ii.5.1-5: Third argument: prolonged heating

Here again, BH seems to be conflating the material in the *Šifā*ʾ with what he had written earlier in *Asc.*[39]

But. [underline: agreement with *Šifā*ʾ; italics: with *Asc.*]:
Third argument: (1) *Prolonged* proximity [lit. prolongation of proximity, *naggīrūt qurbā*] *of the sun* at the zenith [*lʿel men rēšā*] is *more heat-generating* [lit. generates heat more, *yattīr maḥḥem*] than [its mere] proximity. (2a) *For this reason* the sun generates less heat at the head of Cancer than *at* the head of *Leo*, (2b) even though it *is closer to* our *zenith* [lit. to our heads] in Cancer. (2c) It also generates less heat in Taurus and Gemini [i.e. April-June] than when it is in Leo and Virgo [i.e. July-September], (2d) and less before *noon* than *after noon*, (2e) even though it is equally [*b-šawyū*] close to us in both positions.
*Šifā*ʾ 28.13-29.3:

لكن ليس كل ما يسخن الجو من الشمس إنما هو بهذه المسامتة وإلا لكان الحر والشمس فى نقطة السرطان أشد منه وهي فى نقطة الأسد؛ وليس كذلك، وإلا لكان الحر والشمس فى نقطة الجوزاء مساويا للحر وهى فى نقطة الأسد، والحر وهى فى نقطة الثور مساويا للحر وهى فى نقطة السنبلة، وليس الأمر كذلك، ولكانت البلدان التى هى أقرب إلى مجاز الشمس لا تكون البتة أبرد من البلاد النائية عنه، وقد يكون كثيرا. وبالجملة فإن الشمس لو كان يجوز لها أن تنتقل دفعة إلى نقطة السرطان، لكانت لا تسخن البلاد التى تحتها تسخينا شديدا مفرطا، بل كان يكون إلى حد ما. وهذا مثل النار التى تدخل بيتا ما دفعة، فإنها لا تؤثر تأثيرا كبيرا، وإنما تؤثر بالمداومة؛ فإن المداومة تزيد كل وقت حرا إلى حر، وتجعل الهواء أيضا شديد الاستعداد للتسخن. ولهذا ما تكون الحرارة بعد زوال الشمس فى صيف أشد منها قبله، والنسبة واحدة.

(1) But the air is not heated by the sun merely by its being at the zenith [*bi-hādihi al-musāmata*]. (2a) Otherwise, the heat would be more intense when the sun is at the point of Cancer than when it is at the point of Leo, but this is not so. (2c) Otherwise, the heat when the sun is at the point of Gemini would be equal to the heat when it is at the point of Leo, and the heat when it is at the point of Taurus would be equal to the heat when it is at the point of Virgo; but the fact is not so. The countries which are nearer

[39] Cf. also Fakhr al-Dīn al-Rāzī, *Mabāḥit* II.199.16-200.9, who, though he does not accept the view that the equator is temperate, accepts the point made by IS here and expands on what IS had to say.

the course of the sun would never [*lā ... al-battata*] be colder than the countries remote from it, but they often are. In short, if the sun could be conveyed suddenly to the point of Cancer, then it would not heat the countries below it intensely and excessively, but only to a certain extent. This is like fire which is suddenly brought into a house. (1/3) It does not have much effect [at first]. It has an effect only with prolonged presence [*mudāwama*]. For prolonged presence constantly adds heat to heat, and also makes the air intensely ready [*šadīd al-istiᶜdād*] for being heated. (2d) For this reason the heat after the withdrawal of the sun [or: afternoon, *baᶜda zawāl al-šams*][40] in summer is more intense than that before, (2e) although the [positional] relationship is the same [*wāḥida*] [before and after].

Qānūn [New Delhi (1982)] 151.4-10 (corr. ed. Cairo [1987] 120.16-21): That is because there is only one celestial heat-generating cause [*al-sabab al-samāʾī al-musaḫḫin*] there, namely the presence of the sun at the zenith [*musāmatat al-šams li-l-raʾs*] (1) and this presence at the zenith [*hāḏihi l-musāmata*] alone does not have much effect. Rather, only a prolongation of the presence [*mudāwamat al-musāmata*] has an effect. (2d) For this reason, the heat after midday prayer [*baᶜda al-ṣalāt al-wusṭā*] is more intense than that at noon [*fī waqt istiwāʾ al-nahār*] (2a) and for this reason the heat when the sun is at the end [*āḫir*] of Cancer and the beginning [*awāʾil*] of Leo is more intense than when the sun is at the limit of the declination [*ġāyat al-mail*]. (2c) For this reason the sun, when it has turned away [*inṣarafat*] from the head [*raʾs*] of Cancer towards some point [*ḥadd*] elsewhere in the [course of] declination, is more heat-generating [*ašadd tasḫīnan*] than when it is at the equivalent point [*miṯl ḏālika l-ḥadd*] in the declination but has not yet reached the head of Cancer.

Asc. 145.3-7 [underline: agreement with *But*.]: (1) Although the presence [*hwāyā*] of the sun at the zenith [*lᶜel men rēšā*] is [in itself] heat-generating [*maḥḥmānā*], heat [*ḥummā*] is greatly multiplied [*yattīr metᶜappap̄*] by its persistence [*kuttārā*]. (2a) For this reason the summer heat [*ḥummā qayṭāyā*] in Leo is stronger [*ᶜaššīn*] than at the head of Cancer, (2b) which is closer to the zenith [lit. the point above the head, *nuqdṯā da-lᶜel men rēšā*]. (2d) Similarly the heat after noon is greater than that at noon.

Taḏkira 257.10-13 [underline: agreement with *But.*]: (1) This is because even though being directly overead [*musāmata*] leads to heating [*muqtaḍīya li-l-tasḫīn*], nevertheless the duration of this [state] [*makṯ ᶜalaihā*] is more

[40] The phrase is ambiguous. Given the context, IS may be talking about the withdrawal [*zawāl*] of the sun from the Tropic of Cancer after the summer solstice. Ṭūsī, however, either understood the phrase to mean "afternoon" [*baᶜda al-zawāl*] or deliberately changed it to mean that (hence the omission of "*al-šams*" and "*fī al-ṣaif*" in the passage of the *Taḏkira* quoted below; the same alterations are made by Fakhr al-Dīn al-Rāzī (*Mabāḥiṯ* II.200.7), who therefore probably also intends the phrase "*baᶜda al-zawāl*" to mean "afternoon", pace Lettinck [1999] 203). This interpretation as "afternoon" is then also followed by BH.

effective [*ablaġ*] for that than the [state] itself. (2c) It is because of this that [*wa-li-ḏālika*] summer is warmer [*aḥarr*] than spring (2d) and the afternoon [*ba*ᶜ*da al-zawāl*] is warmer than before [noon] (2e) despite the equality of the alignment in each case [*ma*ᶜ*a tasāwī al-musāmata fīhumā*] (tr. Ragep).

IV.ii.5.5-9: Effect of prolonged heating: explanation

The explanation given here in sentence (3) as to why the prolonged presence of the sun causes more heat may be seen as an interpretation of the passage at *Šifā*ᵓ 28.20-29.2. In formulating this interpretation BH has made use of the explanation given in more abstract terms by Fakhr al-Dīn al-Rāzī.

But. [underline: agreement with *Šifā*ᵓ; italics: with *Mabāḥiṯ*]:
(3a) This [lit. these (pl.)] is because *during the first period* [ᶜ*eddānā qaḏmāyā*] the heat-source [*maḥḥmānā*] warms the air with certain moderation [*mšuḥṯā meddem*]; (3b) because of this [the air] becomes more suitable [ᶜ*āhnā*] for reception of heat *in the second period*, (3c) and for this reason the heat is *stronger* [*yattīr* ᶜ*ā*ᵓ*ez*] *in the second* [period] than *in the first*. (4a) At the equator, then, since the proximity of the sun occurs [lit. is found] without the prolongation of the proximity, (4b) the heat is not strong.

*Šifā*ᵓ 28.20-29.2 [see above]: This is like fire which suddenly brought into a house. (3a) It does not have much effect [at first]. (3b) It has an effect only with prolonged presence [*mudāwama*]. For prolonged presence all the time [*kull waqt*] adds heat to heat, and also makes the air intensely ready [*šadīd al-isti*ᶜ*dād*] for being heated [*li-l-tasaḫḫun*].

Mabāḥiṯ II.200.10-17:

واما اللمية فهى ان السبب يفيد في الوقت الاول اثرا فاذا بقى الى الوقت الثانى افاد اثرا جديدا ومتى كان ذلك السبب اطول بقاء كانت آثار المجتمعة اقوى فلا جرم كان الاثر اقوى وهاهنا شكوك قد مضى ذكرها. ومن وجه آخر وهو ان السبب فى الوقت الاول اذا افاد اثرا انضم ذلك الاثر الى السبب الاول وصار المجموع مقتضيا لاثر آخر ولا شك ان تاثير المجموع اقوى من تاثير السبب وحده وعلى هذا الطريق كلما كان السبب ابقى كانت المعلولات المعينة للعلة على تاثير اكثر فلا جرم كان الاثر اقوى وهذه مقدمة يقينية لا شك فيها

The "why's"[41]: The cause brings about an effect [*aṯar*] in the first period [*al-waqt al-awwal*]. Then, it persists into the second period and brings about a new effect. The longer that cause persists, the stronger [*aqwā*] the cumulative effects become, so that the [end] effect will surely become stronger. But here there are the doubts which have been mentioned before.
- From another perspective: When the cause brings about an effect in the first period, that effect is joined [*inḍamma*] to the first cause, and the sum [*maǧmū*ᶜ] becomes the necessitator [*muqtaḍī*] of another effect. There is

[41] اللمية < لمّا; cf. ibid. II.199.20 الانية < أن; 199.19 امور انية وامور لمية.

no doubt then that the effect [*ta'ṯīr*] of the combined thing will be stronger than the effect of the cause alone. In this way, the longer the cause persists, the greater the things caused [*ma'lūlāt*] stipulating the cause [*'illa*] to bring about the effect become. Then, the [end] effect will surely become greater. These are certain premises about which there is no doubt.

IV.ii.6.1-5: Variation of temperature

The main point of the argument presented in clauses (1b)-(1c) is Ibn Sīnā's. In expanding the argument, in sentence (2), BH has, at it were, replaced IS's argument based on the movement of the sun in the longitudinal direction (which is more to the point) by one based on the movement along the ecliptic (which is less to the point, though not altogether irrelevant).[42]

But.: Fourth argument: (1a) Since at the equator the sun does not move very far [*law saggī marḥeq*] from the zenith [*men 'ellaway rēšā*], (1b) its inhabitants do not suffer a change [*mšannēn*] from great coldness to heat (1c) in such a way that [*aykannā d-*] they feel the hurt [*sugpānā*]. (2a) [This is] because Cancer and Capricorn, the signs [*malwāšē*] of their winters, are separated [lit. are distant] only by a quarter of a circle from Aries and Libra, the signs of their summers, (2b) whereas Capricorn, the sign of our winter, is separated [lit. are distant] by a half of a circle from Cancer, the sign of our summer .

Šifā' 29.16-30.1:

وذلك لأن بلادنا وخصوصا حيث نحن، فقد يكون بُعد الشمس فيها عن سمت رؤوسنا ضعف الميل، وزيادة بعد سمت رؤوسنا عن مدار البروج. فيعرض برد شديد، ثم يتعقبه حر شديد، وتبتلى الأبدان بالانتقال من ضد إلى ضد. وأما هناك فلا يُنتقل من ضد إلى ضد، بل إنما يُنتقل من واسطة إلى حد غير بعيد.

(1a/2b) That is because in our lands, especially where we are, the distance [*bu'd*] of the sun from our zenith [*samt ru'ūsinā*] can be [as much as] double the declination plus the distance of our zenith from the ecliptic [*madār al-burūǧ*],[43] (1b) so that an intense cold occurs. Then, an intense heat may follow, (1c) and the bodies will be afflicted [*tubtalā*] by the change [*intiqāl*] from one extreme [lit. opposite] to another. (1b) There [at the equator], one does not undergo a change [*yuntaqalu*] from one extreme to another, but only undergoes a change from the medium [*wāsiṭa*] to a limit which is not far away.

---

[42] Fakhr al-Dīn al-Rāzī's summary at *Mabāḥiṯ* II.201.11-16 is more faithful to the *Šifā'* than BH's

[43] i.e. at winter solstice, the arc distance of our zenith from the sun will be equal to double the obliquity of the ecliptic (= the distance between the two tropics) plus the extra distance between the Tropic of Cancer and the zenith.

IV.ii.6.5-9: Temperature variation (contd.)

Here BH follows the *Šifā᾽* more closely than he has done for much of this section. - In his sentence (3), BH conflates two passages of IS, *Šifā᾽* 30.6-7 (a general observation; in Khurāsān) and 30.7-9 (a particular case witnessed by IS; in Bukhārā).

But.: (1a) "Even if I concede [*maššep̄*]", he [Ibn Sīnā] says, "that the heat [*ḥummā*] is constant [*ammīn*] there, (1b) nevertheless the bodies [*pag̱rē*] [of the people] who were born [*etīled*] and have grown up [*etrabbī*] in (the heat) (1c) do not feel it, (2a) just as the Scythian [*sqūṯāyē*] does not feel the cold of his region, (2b) and the Cushite [*kuššāyē*] does not [feel] the heat of his region". (3a) He says (3b) that he [once] saw in Bukhārā, his city, a certain Midianite [*medyannāyā*] man (3c) who was trembling with cold in the month of Īyār and, (3d) while the people of Bukhārā [*buk̲ārāyē*] were complaining about the heat, (3e) he was wearing winter clothes.

Šifā᾽ 30.1-11:

ولو كان هناك حر دائم وكانت الأبدان هنالك قد نشأت على مزاجه، لا تنفعل عنه كثيرا، ولا يعرض لها خروج بعيد عما نشأت عليه، لكانت لا تحس بأمر مغير، فكيف وليس هناك إفراط البتة. وللأبدان ملاءمة لما نشأت عليه، حتى لا تنفعل عنه كثيرا. تأمل ذلك فى حال أبدان الترك، فإنهم لا ينفعلون من برد بلادهم انفعالا شديدا، ولا الحبشة ينفعلون من حر بلادهم انفعالا شديدا. وربما كان البدوى بخراسان يشكو البرد، فى وقت ما يكون الخراسانى يشكو الحر فى وقت واحد. وقد شاهدت هذا ببخارا من حال بدوى حضرها فى ماه أردى بهشب أو خرداد وقد تسلط بها أكثر الحر وهو يرتعد ويتزمل ويستغيث من البرد، وأهل البلد يتأذون من الحر؛ لأن مزاج العربى ألف مزاجا حارا، وألف الآخر مزاجا باردا؛ فيكون ذلك المزاج باردا بالقياس إلى الأعرابى، حارا بالقياس إلى البخارى بحسب مزاجه الذى له فى ظاهر بشرته.

(1a) Even if there is constant [*dā᾽im*] heat there, (1b) the bodies there which have grown up [*naša᾽at*] in its climate [*mizāǧ*] (1c) are not affected [*tanfaᶜilu*] much by it, and no great deviation [*ḫurūǧ*] occurs from what they have grown up in, so that they do not feel the alteration. For there is no excess there. Bodies have an adaptability [*mulā᾽ama*] to what they grow up in, so that they are not affected much by it. (2a) Observe this in the case of the bodies of the Turks [*al-turk*]. For they are not affected very much by the coldness of their lands. (2b) Nor are the Abyssinians [*al-ḥabaša*] affected very much by the heat of their lands. (3b) There were once Bedouin [*al-badw*] in Khurāsān who were complaining about the cold (3d) when the Khurāsānīs were complaining about the heat at the same time. (3b) I once saw this in Bukhārā in the case of a Bedouin [*badawī*] who was there in the month of Urdī Bihišt or Khurdād. There prevailed [*tasallaṭa*] the greatest heat and [yet] he was trembling (3e) and wrapping himself up [*yatazammalu*] and calling for help because of the cold, (3d) while the natives were suffering from the heat; because the condition [*mizāǧ*] of the Arab was accustomed to [*alifa*] a hot climate [*mizāǧ*], while the other was accustomed to a cold climate, so that that climate was cold for [*bi-l-qiyāsi ilā*] the desert Arab [*aᶜrābī*] and hot for the Bukhārī in proportion to his temperature [*mizāǧ*] which he had on the outside of his skin.

l. 7, 8. ܣܩܘܬܝܐ, ܟܘܫܝܐ ("Scythian", "Cushite"): الترك, الحبشة ("Turk", "Abyssinian") IS. - The designation of the Ethiopian/Abyssinian (*ḥabaša*) as *kuššāyā* is common enough in Syriac (PS 1716). *Squṯāyā*, on the other hand, where it does not actually mean "Scythian", is more often used as the equivalent of Arabic *ṣaqlab* (Slav) than as that of *turk*, although the latter correspondence is not unknown to the lexica (PS 2715), so that the change of "Turk" to "Scythian" here is best considered as a deliberate alteration on BH's part. This may be seen, in the first place, as an instance of his tendency to "Syriacise"/"Christianise" Arabic/Islamic proper names,[44] but in this instance there may be a connection with the tradition going back to ancient times of regarding the Scythians as a typical representative of the barbaric peoples of the north. - The "Scythian" and the "Cushite" appear again as typical representatives of the northerly and southerly peoples at *But*. Eth. II.iii.2; at *But*. Eth. I.iii.4, we find the "Cushites" in the same role, but the "Scythians" are replaced there by the "Huns"; cf. further *But*. Polit. II.iii.3 (Zonta [1992] 83, tr. 100); and *Cand*., Base X [Zigmund-Cerbü] 44.22, where the "Scythian" and the "Cushite" are mentioned in connection with the colours of skin in the resurrected body (the reference to St. Paul there points to Col. 3.11, where the Scythians are mentioned, though not the Cushites/Ethiopians).

l. 8f. ܠܚܕ ܐܢܫ ܡܕܝܢܝܐ ("a certain Midianite"): IS 30.6 البدوي; 30.7 بدوي; 30.10 الأعرابي. - Another instance where an Arabic proper name is replaced by a Syriac (biblical) name.

l. 9. ܒܐܝܪ ܝܪܚܐ ("in the month of Īyār [May]"): في ماه أردي بهشت أو خرداد IS. - Within IS's lifetime (980-1037 A.D.), Urdī Bihišt and Khurdād, the second and third months of the Persian (Yazdegerdian) calendar, would have corresponded to the period between 20th-6th April and 18th-4th June. BH is therefore basically correct in rendering this as "May".[45]

Min. IV. Section iii.: *Refutation of the Avicennian Arguments*

See the introductory comments to IV.ii. above.

l. 1. ܫܒܪܝܐ ("Avicennian"): < *šāḇā rēšānā* (= *al-šaiḫ al-raʾīs*, Ibn Sīnā). The word was misunderstood by J.P. Margoliouth (PS Suppl. 220, "refutation of the conjectures of the elders").

---

[44] See Introduction, Section 3.3b above; cf. Takahashi (2003a), para. 58.

[45] The unintercalated 365-day Yazdegerdian year usually moves forward by a day every four years against the Julian year. Within IS's lifetime, 1st Farwardīn moved from 21st to 7th March (Spuler [1961] 38). - At *Asc*. 200.15-201.2, BH tells us that 1st Teshrī I, 1590 A.Gr. (Oct. 1278 A.D.) corresponded to 23rd Āḏur 647 Anno Yazdegerdi. Assuming a system where the 5-day epagomenae were still placed at the end of Ābān rather than after Isfand (cf. Taqizadeh [1939-40] IX.917f.), this would make 1st Farwardīn coincide with 7th Jan., which is one day off the value given at Spuler, loc. cit. (6th Jan.).

IV.iii.1.1-2: Introduction

But.: Even though the illustrious Doctor [*sāḇā mšabbḥā*] is held dear [*rḥīm*] by the entire community of excellent men [*kulleh gawwā da-myattrē*], their love for the truth is greater and for this reason they do not hesitate [*mesṭamʿrīn*] to refute his arguments.

l. 1. ܣܒܐ ܡܫܒܚܐ ("the illustrious Doctor"): So here instead of the usual "*sāḇā rēšānā*"; cf. IV.iii.6.1 below (*mšabbḥā*).

IV.iii.1.3-7: Constant proximity of the sun

The refutation of IS's first argument is found in Fakhr al-Dīn al-Rāzī and in Ṭūsī, who explicitly ascribes this refutation to Fakhr al-Dīn al-Rāzī. BH's wording here is closer to that of Ṭūsī, although there are elements which he shares with Fakhr al-Dīn al-Rāzī but not with Ṭūsī.

But. [underline: agreement with *Mabāḥiṯ*; italics with *Taḏkira*]: (1) *Refuting* [*kaḏ šārēn*] *his first argument* [*sʿāyeh qaḏmāyā*], they *say*: (2) *At the equator, even though the sun* changes position [*mšannē*] quickly and does not linger [*lā maggar*] as it passes the zenith [*lʿel men rēšā ʿāḇar*] in Aries and Libra, (3) it *nevertheless* [*bram dēn*] *does not* [then] move *very far* [*law saggī marḥeq*] *from the zenith* [*men ʿellaway rēšā*]. (4) *Throughout the year*, therefore [*mādēn*], the sun is either *at the zenith* or in the proximity of it [*l-qurbā meneh*] (5) and the heat [*ḥummā*] must [*ālṣā d-*] be strong [*neʿšan*] there throughout the year.

Mabāḥiṯ II.202.19-21: واما الذى ذكره الشيخ من ان المشامتة لا تبقى الا زمانا قليلا فهو مسلم ولكن بعد الشمس عن مسامتة رؤسهم ليس بعظيم فهو دائما اما فى المشامتة او فيما قريب من المسامتة فكيف لا يكون الحر هناك عظيما

(2) The Doctor [*al-šaiḫ*] said that the presence [of the sun] at the zenith [*al-musāmata*] persists [*tabqī*] only for a short time. This is correct, (3) but [*wa-lākinna*] the distancing [*buʿd*] of the sun from their zenith is not great, (4) since it is constantly either at the zenith or in the proximity of [*fīmā qarīb min*] the zenith. (5) How, then, can the heat [*ḥarr*] there not be great [*ʿaẓīm*].

Taḏkira 257.18-20: The eminent Imām Fakhr al-Dīn al-Rāzī rejected the first argument [*radda ... ʿalaihi l-ḥukma l-awwala*], saying [*bi-an qāla*]: (2) Even though the sun lingers on the equator only briefly [*labṯu al-šamsi fī ḫaṭṭi l-istiwāʾi wa-in kāna qalīlan*], (3) it nonetheless [*lākinnahā*] is never too far from being directly overhead [*lā tabʿudu kaṯīran ʿani l-musāmatati*]; (4) it is thus virtually overhead [*fī ḥukmi l-musāmatati*] for the length of the year [*ṭūla s-sanati*] (tr. Ragep).

l. 3. "Refuting the first argument they say": We note the variation in the way each refutation is introduced here and in the following theories: "The second argument too is refuted by the fact that ..."; "In refutation of the third argument, it is said ...", "Against the fourth argument it is said ..."

IV.iii.2.1-4: Weakness of nocturnal coldness

BH's refutation here is rather different from Fakhr al-Dīn al-Rāzī's, although it may be considered as taking up the point made at the end of the passage of the *Mabāḥiṯ* below about the length of the equatorial night not being such as to strengthen the coldness there.

But.: (1) The second argument too is refuted [*meštrē*] (2a) by the fact that [*b-hāy d-*] the coldness of the night at the equator, (2b) because it is very weak, (2c) does not have the power to mitigate [lit. power of mitigation of, *ḥaylā da-mšahhyānūt*] the heat of its day, (2d) which is strong due to the great proximity of the sun to the zenith, (3) just as the morning coldness of the second clime also [does] not [have the power to mitigate] its noonday heat.

Mabāḥiṯ II.203.1-8:

واما ما ذكره من ان النهر والليل هناك متساوية ونهار صيف الآفاق المائلة اطول فالجواب ان تأثير طول النهار فى التسخين قليل فان الموضع الذى يكون القطب فيه على سمت الرأس يكون النهار فيه ستة اشهر ومع ذلك فهو من البرد بحيث لا يعيش فيه الحيوان وايضاً فلان طول نهرهم فى الصيف مقابل لطول لياليهم فى الشتاء وذلك تقتضى استحكام البرد فى ذلك الهواء وهو مانع من التسخين التام فى الصيف واما فى خط الاستواء فكما لم يوجد هناك فى الصيف طول النهار المقوى للسخونة كذلك لم يوجد طول الليالى المقوى للبرودة

(1) As for what he said concerning the day and night being equal there and the summer day in declining horizons [*al-āfāq al-mā'ila*, i.e. at higher latitudes] being longer, the answer is as follows: The effect of the length of the day in causing heat is small, since at the place where the pole is at the zenith the day lasts for six months, but in spite of that, because of its cold, it is such that animals cannot live in it. Furthermore, because the length of the summer day is opposed by [*muqābil*] the length of the winter night, and that necessarily brings about the coldness of the air and that prevents complete heating in the summer. (2) At the equator, on the other hand, just as the length of day that strengthens [*muqawwī*] the heat is not found there in summer, (2ab) neither is the length of night that strengthens the coldness.

IV.iii.3.1-7: Effect of prior heating/cooling

The argument put forward by BH here and in IV.iii.5 below can be found in a long passage which forms the main part of Fakhr al-Dīn al-Rāzī's refutation of IS's view (*Mabāḥiṯ* II.201.18-202.18). There, Rāzī first points out that the sun is at an equal distance from the equator and the place at the latitude twice the obliquity of the ecliptic and then goes on to give a long explanation as to why even the winter heat at the equator will be much greater than the summer heat at the latter place using an argument he has put forward before concering the

effect of pre-heating. The two points of the argument are separated in *But.*, the point about pre-heating being used here and that about the equidistance of the sun to the equator and to the seventh clime being used in IV.iii.5.

But.: (0) In refutation of the third argument, it is said: (1a) The excess [*yattīrūṯā*] of the summer heat, which arises due to the prolonged proximity of the sun to the zenith, (1b) occurs in those regions to the north of the equator (1c) after the excess of the wintry cold [*yattīrūṯ qarrīrūṯā saṯwāytā*] which has preceded [it] [*etqaddmat*] (1d) and has greatly cooled the air. (2) It is clear that the incalescence [*meštaḥḥnānūṯā*] of a land which has undergone much cooling [*d-saggī qerrat*] during the winter is much less [*saggī bṣīrā*] than the incalescence of a land which has not undergone much cooling during the winter. - (3) For at a time when a cold [object] grows hot, a hot [object] will burn. - (4) The prolonged proximity of the sun in the climes is therefore less heat-generating than [its mere] proximity at the equator.

Mabāḥiṯ II.201.21-202.17:

وايضًا فالشمس عند كونها فى غاية الميل قد كانت قبل ذلك فى القرب من/١ سكان خط الاستواء وذلك سبب سخونة وفى البعد عن سكان بلدة/٢ المفروضة وذلك سبب اشتداد البرد فخط الاستواء لم يخل قبل ذلك فى جميع/٣ السنة من مثل هذا التسخين او مما هو اقوى منه بكثير اما ما هو مثل هذا/٤ التسخين فذلك عند كونها فى غاية الميل من الجانب الآخر واما ما هو اقوى/٥ من هذا التسخين فذلك عند ما لا تكون فى غاية الميل فانها تكون لا محالة/٦ اقرب الى خط الاستواء مما اذا كانت فى غاية الميل وحينئذ يكون تسخينها/٧ لخط الاستواء اقوى مما اذا كانت فى غاية الميل واما سكان ضعف الميل/٨ فاسباب البرد الشديد فى حقهم قد كانت موجودة فى كل السنة السابقة/٩ فالشمس حين ما تكون فى غاية الميل تكون كالمسخن المتوسط بين جسمين/١٠ احدهما كان المسخن العظيم ملاقيا له طول السنة السابقة والثانى كان/١١ البرد العظيم ملاقيا له طول السنة السابقة فمن المعلوم ان التسخن البارد من ذلك/١٢ للمسخن اضعف كثيرا من تسخن ذلك بل لا نسبة لاحدهما الى آخر فانا/١٣ قد بينا ان الآثار الحاصلة من المسخن فى سالف الزمان تنضم اليه ويصير/١٤ المجموع مؤثرا فى التسخن فيخرج مما قلنا ان حر سكان خط الاستواء فى صميم /١٥شتائهم لا نسبة له الى حر البلدة المفروضة فى صميم صيفهم ثم ان الحر الشديد/١٦ فى البلدة المفروضة حر عظيم لا يطيقه اهلها وحر شتاء خط الاستواء اعظم/١٧ كثيرا من ذلك الحر بل لا نسبة له اليه

(1) Furthermore, when the sun is at the limit of the declination [i.e. at the Tropic of Cancer], it has before that been in the proximity [*qurb*] of the inhabitants of the equator - and that is a cause of heating [*suḫūna*]. - It has been in the distance [*buʿd*] from the place under consideration [*al-balda al-mafrūḍa*] - and that is a cause of intensification of coldness [*ištidād al-bard*]. - The equator has never been free before that for the whole year from heating at the same level [*miṯl hāḏa t-tasḫīn*] or from [heating] which is much stronger than that: heating at the same level, when (the sun) was at the limit of the declination on the other side [i.e. at the Tropic of Capricorn]; stronger heating, when it was not at the limit of declination, since it was then inevitably nearer to the equator than when it is at the limit of declination and at such times its heating of the equator is stronger than when it is at the limit of declination. (1c) As for the inhabitants of [the place] double the declination, causes of intense coldness [*al-bard*

al-šadīd] were to be found concerning them for the whole of the preceding year [*al-sana al-sābiqa*]. When, therefore, the sun is at the limit of declination, it is like a heat-source [*musaḫḫin*] halfway between two bodies, one of which has been faced with a great heat-source for the length of the preceding year, while the other has been faced with a great coldness for the length of the preceding year. (2) Now, it is known that the incalescence [*tasaḫḫun*] of a cold object [*al-bārid*] by that heat-source is much weaker [*aḍʿaf kaṯīran*] than the incalescence of the latter, rather there is no comparison between the two [lit. the one has no relationship to the other]. - For we have explained before[46] that the effects arising from the heat-source in the preceding period is joined to it and the sum [*maǧmūʿ*] becomes the one causing the effect of heating. - (4) It results from what we have said that the heat [*ḥarr*] of the inhabitants of the equator at the height [*ṣamīm*] of their winter is incomparably greater than [lit. has no relation to, *lā nisba lahu ilā*] the heat of the place under consideration at the height of their summer, and that the intense heat of the place under consideration is a great heat which its inhabitants cannot endure, but the winter heat of the equator is much greater than that heat, rather it is incomparable.

l. 4. ܕܡܫܬܚܢܢܘܬܐ ("incalescence"): lit. "the being warmed"; corr. *tasaḫḫun*. - Cf. PS Suppl. 331, with a citation of the word from *But*. Isag. II.ix., where it is contrasted with the active *mšaḥḥnānūṯā* (corr. arab. *tasḫīn*).

IV.iii.4.1-6: Objective heat/coldness

IS's argument, represented in theory IV.iii.6 above, that there is not much difference between the seasons at the equator is anwered by Fakhr al-Dīn al-Rāzī at *Mabāḥiṯ* II.203.18-204.14, who argues that there are in fact eight seasons at the equator. The refutation provided by BH, who takes up a different aspect of IS's argument, is quite different.

But.: Against the fourth argument it is said: The dispute [*ḥeryānā*] here is not about whether the inhabitants of the equator feel the heat or not [lit. about the sensation (*margšānūṯā*) by (of) the inhabitants of the heat or their non-sensation], but [about] whether it is [in fact] very [hot] or not very hot. If temperateness [*mmazzgūṯā*] is established by the fact that a sentient being [*margšānā*] does not feel the heat and the cold [lit. by the non-sensation ...], every place will be temperate, because neither does the Cushite feel the greatness of the heat of (his locality) [lit. his heat] nor the Scythian the excess of the coldness of (his).

[46] i.e. at *Mabāḥiṯ* II.200.10-17, cf. IV.ii.5.5-9 above.

IV.iii.5.1-8: Further refutation: equidistance of the sun from equator and seventh clime at summer solstice

BH's argument here corresponds to the beginning and the end of Fakhr al-Dīn al-Rāzī's argument at *Mabāḥiṯ* II.201.18-202.18 (cf. IV.iii.3 above). There, Rāzī had explained why even the winter heat at the equator will be *much greater than* the summer heat at the latter place using his argument concering the effect of pre-heating. Having used the pre-heating argument elsewhere, BH here limits himself to saying that the winter heat at the equator will be *as great as* the summer heat in the seventh clime. - The same argument as here is found in BH, *Asc*. - Rāzī's argument is also used by Ṭūsī at *Taḏkira* 257.21-24, where Ṭūsī too greatly simplifies Rāzī's argument but still concludes like Rāzī that the winter heat at the equator is much greater than the summer heat elsewhere.

But. [underline: agreement with *Mabāḥiṯ*; italics: with *Asc.*]:

(1a) *It is also said* (1b) that *the latitude* [*pṭāyā*] at the beginning the seventh clime is forty-seven degrees, [i.e.] *double the total declination* [*aᶜpā da-ṣlāyā kullānāyā*] of the sun from the isemeric, (1c) so that *when the sun* stays [*šārē*] *at the head of Cancer*, (1d) it is *winter* at the equator (1e) and *summer* in the seventh clime. (2) It is clear that *the distance* of *the sun* at that time from the seventh clime will be *equal* to its distance *from* the equator. (3) The sun must at that time *be heating* [*ālṣā d- ... naḥḥem*] the two sides, the northern and the southern, *equally* [*b-šawyū*]. (4a) Therefore, *seeing that* [*kaḏ*] the winter heat at the equator is equal to the summer heat in the [seventh] clime, (4b) *what* will the summer heat [at the equator] be *like* [*aykannā nehwē*]? (5) It will *surely* [*ba-šrārā d-*] be intolerable.

Mabāḥiṯ II.201.18-202.2, 202.14-18:

... انا نفرض بلدة عرضها ضعف الميل كله فاذا وصلت الشمس الى غاية القرب من سمت الرؤس اهلها كان بعدها عن سمت رؤسهم كبعدها عن سمت رؤس سكان خط الاستواء وايضًا فالشمس عند كونها فى غاية الميل قد كانت قبل ذلك فى القرب من سكان خط الاستواء وذلك سبب سخونة وفى البعد عن سكان بلدة المفروضة وذلك سبب اشتداد البرد ... فيخرج مما قلنا ان حر سكان خط الاستواء فى صميم شتائهم لا نسبة له الى حر البلدة المفروضة فى صميم صيفهم ثم ان الحر الشديد فى البلدة المفروضة حر عظيم لا يطيقه اهلها وحر شتاء خط الاستواء اعظم كثيرا من ذلك الحر بل لا نسبة له اليه واذا بلغ حر غاية شتائهم الى هذا الحد العظيم فما ظنك بحر صيفهم فثبت بهذا ان الحرارة فى ذلك الموضع عظيما جدا

... (1b) Let us consider a place whose latitude [*ᶜarḍ*] is double the total declination [*ḍiᶜf al-mail kullihi*]. (1c) When the sun arrives [*waṣalat*] at its closest point [*ġāyat al-qurb*] to the zenith of its residents [*ahl*] [i.e. at the Tropic of Cancer], (2) its distance from their zenith is like [*ka-*] its distance from the zenith of the inhabitants [*sukkān*] of the equator. Furthermore, when the sun is at the limit of the declination, it has before that been in the proximity of the inhabitants of the equator - and that is a cause of heating. - It has been in the distance from the place under consideration [*al-balda*

al-mafrūḍa] - and that is a cause of intensification of coldness. - ... It results from what we have said (1d) that the heat [*ḥarr*] of the inhabitants of the equator at the height [*ṣamīm*] of their winter is incomparably greater than [lit. has no relation to, *lā nisba lahu ilā*] (1e) the heat of the place under consideration at the height of their summer, and that the intense heat of the place under consideration is a great heat which its inhabitants cannot endure, but the winter heat of the equator is much greater than that heat, rather it is incomparable. (4a) Seeing that [*iḏā*] the heat at the depth [*ġāya*] of their winter reaches such a great level, (4b) what do you think the heat in their summer [will be like]. (5) It is confirmed [*ṯabata*] by this that the heat in that place [the equator] is very great.

Asc. 145.13-22: (1a) They say that this [that the equator is hotter than other places and is uninhabitable] is known from the fact (1b) that at the place where the latitude is double the total declination, (1c) when the sun is [*hāwē*] at the head of Cancer, (2) the sun will be equally far [*šawyāʾīṯ raḥḥīq*] to the south from it as to the north from the equator. (1d) The sun is then at its nearest summer point [*qurbā yattīrā w-qayṭāyā*] to that place and at the furthest winter point [*ruḥḥāqā yattīrā w-saṯwāyā*] from the equator. (3) Therefore, the summer heat [*ḥummā qayṭāyā*] of that place is equal to the winter heat of the equator. (4a) Since [*baḏ* Nau, *kaḏ* legendum?] (the equator) is so hot in winter, (4b) it must be hotter in the rest of the seasons. How [*aykannā*] can [a place] which is in this condition be more temperate than other [places]?

l. 1. ܦܬܝܐ ܕܫܘܪܝܗ ܕܩܠܝܡܐ ܫܒܝܥܝܐ ("the latitude at the beginning the seventh clime"): The latitude at the 16th parallel, the parallel at the southern limit of the 7th clime, is 46;51° according to Ptolemy (*Alm*. II.6 [Heiberg] 110.15, followed by BH at *Cand*. 99.9f.) and 47;12° according to Ṭūsī's *Taḏkira* ([Ragep] 253.8, followed by BH at *Asc*. 142.16).

l. 2. ܨܠܝܐ ܟܠܢܝܐ ܕܫܡܫܐ ("the total declination of the sun"): غاية الميل, الميل كله Rāzī. - On "declination" (*ṣlāyā*), see IV.ii.3.5 above. The "total declination" or the "limit of the declination" is the furthest point/the greatest angle the sun reaches from the celestial equator (= the obliquity of the ecliptic; = the latitudes of the two tropics). The traditional Ptolemaic value for this was 23;51° (x 2 = 47;42°), the values given by Arabic astronomers a little lower.

IV.iii.6.1-2: Apologia

But.: This being the case, [the opinion of] the illustrious [Doctor] is astonishing [lit. worthy of wonder], though it [may seem] insolent [of me] to say so. How did he come to hold these perverse opinions? With all due respect to his great status [*rebbūṯ dargeh*], I say ...

l. 1. ܘܐܢ ܡܪܚܝܐ ܡܠܬܐ ("though it [may seem] insolent [of me] to say so"): lit. "although the word is insolent".

l. 2. ܐܝܟܢܐ ܗܠܝܢ ܕܠܩܘܒܠܐ ܐܬܪܥܝ ("How did he come to hold these perverse opinions?"): lit. "How did he think these things which are contrary?"

l. 2. ܐܬܪܒܝ: this rhymes with *etrannī* in l. 3. below (see further on l. 3. *maggrā* etc. below).

l. 2. ܘܒܕܚܠܬ ܐܢܐ ܡܢ . . . ܕܐܡܪܢ ("With all due respect ..., I say"): lit. "I stand in fear of ... that I say".

IV.iii.6.3-11: Need for balance of heat and cold

Despite what the opening passage of the theory might lead us to expect, the argument put forward here is not really an original contribution by BH but is closely based on Abū al-Barakāt's *Muᶜtabar*. The situation, however, is made complicated by the fact that Abū al-Barakāt is a supporter of the view that the climate at the equator is temperate. What Abū al-Barakāt is taking issue with in the passage quoted below is the view of those who claim that, even if the equator is constantly hot, the constancy of the temperature is salutary in itself (see IV.ii.6 and IV.iii.4 above). Abū al-Barakāt first points out here that continuous heat is in fact injurious since we require a balance of heat and cold and then goes on to say that the climate at the equator is indeed constantly temperate and not constantly hot. BH has no problem in accepting Abū al-Barakāt's view concerning the need for balance (*iᶜtidāl*) and moderation (*mmaššḥūṯā*) and so closely follows Abū al-Barakāt for much of this passage. Then, at the end of the passage, he suddenly alters the purport of Abū al-Barakāt's passage by forcefully applying all that has been said not to the equator, but to Antioch, a city in the fourth clime, the clime which happens to be best suited for human habitation because of its distance from the two extremities of heat and cold, as BH tells us in *Asc*. (147.6-10; cf. Ṭūsī, *Taḏkira* 257.24), and to be inhabited by the wisest and finest men, as he tells us in *Cand*. (98.10-11; prob. < Bar Kepha, *Hex*., Paris 241, 170r a2-6).

The part of the passage here from l. 3 *w-pāgrē* to the end is placed at the end of IV.iii.5 in mss. F and M.[47] That the order of WPLl, which has been adopted here, is correct is confirmed both by the internal logic of the passage and the corresponding passage of the *Muᶜtabar*.

But.: (1a) Why did he not realise [*eṯrannī*] (1b) that continuous heat [lit. heat that is prolonged, *ḥammīmūṯā d-maggrā*] is more harmful [*mahhrā*] to animals and plants [lit. animal and plant (sg.)] than heat which comes [*āṯyā*] after coldness? (2a) Bodies [*pagrē*] which are not cooled [*metpayyḡīn*]

[47] Mgr. Mellus, the copyist of V, who used L and M as his exemplars, has inserted a note in the margin of M correcting the order.

by the coldness of winter (2b) and within whose bellies [*karsāṯā*] the implanted heat [*ḥammīmūṯā nṣībṯā*] is not reversed [*ʿāṭpā*] (2c) are in danger [*qindūnōs*] when they reach summer. (3a) The heat of the summer must be tempered [*ālṣā d-teṯmazzaḡ*] by the coldness of the winter, (3b) and vice versa, (3c) so that the moderation [*mmaššḥūṯā*], which is not found [*šḵīḥā*] within each day and month, (3d) is found during the course [*klīlā*] of the year. (4a) In a place where moderation is always [*b-ammīnū*] found [*šḵīḥ*], (4b) the whole year is like [*dāmyā l-*] spring (4c) and its fruit are monthly, not yearly, (4d) that is to say [*kēmaṯ*], when one of them is maturing [lit. arriving], another is [just] starting, (5) just as in Antioch we see ripe grapes [*ʿennḇē*], unripe grapes [*besrē*] and blossoms [*smāḏrē*] on a single vine at a single time.

Muʿtabar II.206.7-16:

وما علموا ان الحر الدائم على الحيوان والنبات اضر من/٨ الوارد بعد البرد وان الابدان التى لم تأخذ حظها من البرد وانعكاس الحرارة/٩ الغريزية واعداد الرطوبة الصالحة فى بواطن الابدان شتاء لا تسلم صيفا/١٠ وان مضرة برد الشتاء يتلافاها حر الصيف والمضرة حر الصيف يتلافاها برد الشتاء/١١ حتى يكون الذين يفقدون الاعتدال فى كل الزمان يجدونه فى جملة الزمان/١٢ لان الاعتدال الذى لا يجدونه فى كل يوم وشهر من سنتهم يجدونه فى جملة سنتهم/١٣ والذين يجدونه فى كل زمان فحالهم احسن ومثل هولاء كمثل من يجوع فيشبع/١٤ ويمرض فيعافى وهؤلاء كمن لا يجوع ولايمرض وكذلك يكون زمانهم ابدا/١٥ كالربيع وثمارهم شهورية لا سنوية اذا ادرك منها شيء بدا غيره لتشابه الاحوال/١٦ فى الازمان

(1a) But they did not realise [*mā ʿalimū*] (1b) that continuous heat [*al-ḥarr al-dāʾim*] is more harmful [*aḍarr*] for the animal and the plant than [heat] coming [*wārid*] after coldness; (2a) that bodies [*abdān*] which do not receive their allotment [*ḥaẓẓ*] of coldness, (2b) the reversal [*inʿikās*] of the innate heat [*al-ḥarāra al-ġarīzīya*] and the preparation [*iʿdād*] of the beneficial moisture [*al-ruṭūba al-ṣāliḥa*] in the interiors of the bodies [*bawāṭin al-abdān*] in winter (2c) are not safe [*lā taslamu*] in summer; (3b) and that the harm [*maḍarra*] of the winter cold is corrected by [*yatalāfā-hā*] the summer heat (3b) and the harm of the summer heat is corrected by [*yatalāfā-hā*] the winter cold. (3cd) As a result, those who fail to find [*yafqidūna*] the balance [*iʿtidāl*] in every period [*kull al-zamān*] find it over the whole period [*ǧumlat al-zamān*], because they find the balance which they do not find [*lā yaǧidūna*] within each day and month of their year in [the course of] the whole of their year [*ǧumla sanatihim*]. (4a) The condition, however, of those who find a balance [*iʿtidāl*] at all times [*fī kull zamān*] is the best. The former are like those who hunger and are sated and who become ill and recover; the latter like those who never hunger or become ill. (4a) Similarly, their seasons are perpetually [*abadan*] like spring [*ka-l-rabīʿ*] (4c) and their fruit are monthly, not yearly; (4d) when one of them is maturing [or: arriving, *adraka*] another is appearing [*badā*; or: starting, *badaʾa*] because of the constancy [*tašābuh*] of the conditions throughout the seasons

l. 3, 4, 6. ܡܓܓܪܐ (*maggrā*), ܡܗܗܪܐ (*mahhrā*), ܗܘܝܢ (*hāwēn*), ܡܛܝܢ (*māṭēn*): The frequency of rhyming verbs in this theory is worth noting (cf. on l. 2 *etraʿʿī* above). The two clauses ending in *hāwēn* and *māṭēn* are also isosyllabic (six syllables each).

l. 5. ܚܡܝܡܘܬܐ ܢܨܝܒܬܐ ("implanted heat"): The concept goes back to Aristotle (ἡ ἔμφυτος θερμότης, τὸ σύμφυτον θερμόν etc., *Mete*. 355b 9, *GA* 784b 7 etc.); cf. Abū al-Barakāt, *Muʿtabar* II.197-202. The term recurs in *But*. De plant. III.ii.3, IV.ii.3 [Drossaaart Lulofs-Poortman] lines 305, 405).

l. 8. ܒܟܠܝܠܐ ܕܫܢܬܐ ("in the course of the year"): فى جملة سنتهم Abū al-Barakāt. - It is difficult to find *klīlā* (lit. [circular] crown) used in this sense in the lexica, although its metaphorical application to the zodiac circle is known (PS 1733) and the comprehension of the sense here is aided by its derivation from the root √*kll*.

l. 9. "its fruit are monthly, not yearly": cf. *But*. De plant. III.i.1 [Drossaart Lulofs-Poortman] line 240ff.: "Some [fruits] reach maturity quickly ... Some fruits come twice a year, others ripen only in their season ..."

l. 10. ܡܫܪܐ ("is starting"): ܡܫܢܐ ("is changing", "is going off (?)") F; بدا Abū al-Barakāt. - The corresponding passage of the *Muʿtabar* confirms the majority reading. BH's rendition suggests that he read *badaʾa* rather than *badā* as printed in the Hyderabad edition of the *Muʿtabar* (which edition is, admittedly, sparing in its use of the *hamza*).

l. 10-11. ܐܝܟܢܐ ܕ ... ܚܙܝܢܢ ("just as we see ..."): ܚܙܝܢ ("just as we have observed ...") PL. - Since BH is known to have spent a part of his youth in Antioch, the addition here may well be one based on his own observation. That being the case, the vocalisation of verb as perfect in mss. PL may be correct, although we note that in other reports of personal observation at Min. I.i.2.10-12 above and Mete. I.i.2.6-10, I.iiii.4.9-10 below BH uses the first person singular and not plural as would be the case here.

Min. IV. Section iv.: *On the Cause of Heat and Cold according to Time and Place*

Some of the material treated in this section was dealt with by IS in the *Šifāʾ*. The subject, however, was treated at greater length by Abū al-Barakāt in his *K. al-muʿtabar* and it is that work which BH uses as his source here. As has already been noted in the comm. on IV.iii.6.3-11 above, Abū al-Barakāt agreed with IS in maintaining that the equatorial region was temperate and his discussions of the causes of heat and coldness are geared to proving that point. On two occasions in this section, we observe how BH skillfully manipulates his paraphrase to make Abū al-Barakāt's argument say the opposite of what he intended (see IV.iv.3.1-7; IV.iv.4.5-11).

But.	*Muʿtabar*		
IV.iv.1.1-8	II.202.7-16	cf. *Šifāʾ* 27.16-28.1	Radiation as cause of heat
IV.iv.2.1-2	II.202.16-17		contd. (unpolished bodies)
IV.iv.2.2-6			contd. (celestial sphere)

IV.iv.3.1-7	II.202.22-203.6		Length of day and night
IV.iv.3.7-10	II.203.6-8, 10-11		contd.
IV.iv.4.1-5	II.203.8-10, 11-15		Other factors
IV.iv.4.5-11	II.203.15-24		Retention of coldness
IV.iv.5.1-2	II.204.2-11		Shades
IV.iv.5.3-9	II.204.11-205.5	cf. *Šifā᾽* 28.3-10[48]	Circle of radiation

l. 1. ܥܠ ܥܠܬܐ ܕܚܡܝܡܘܬܐ ܘܩܪܝܪܘܬܐ ܙܒܢܝܬܐ ܘܐܬܪܝܬܐ: lit. "on the cause of temporal and local heat and coldness". - Cf. the chapter heading at *Muᶜtabar* II.202.6: الفصل العاشر فى الحر والبرد الزمانيين واسبابهما ("on heat and cold in time and their causes").

IV.iv.1.1-8: Heat due to solar radiation

But.: (1a) That the sun is not hot is known (1b) from the fact that plains, which are far from it, are hotter than tops of mountains, which are close to it. (2a) Rather, heat proceeds [*nāḇhā*] from radiation [*zallīqā*], (2b) and radiation emanates [*dānaḥ*] from the sun (2c) and becomes visible [*meṯhawwē*] on surfaces of dense and earthy bodies which are polished - (2d) not on loose airy [bodies] and transparent celestial [bodies] - (3a) as we see on mirrors, (3b) which shine the more [*yattīr nāhrān*], the more polished they are [*kmā d-yattīr mestaqlān*]. (4a) Increase [*yattīrūṯā*] of radiation results in increase of heat [lit. increase of heat follows increase of radiation] (4b) which almost [*bṣīr qallīl*] sets [objects] on fire [*mawqḏā*], (4c) especially when the mirror is concave [*ḥlīltā*]. (5a) In such a case [*hāydēn*] the rays from [lit. of] its edges are reflected to its middle (5b) and sets the kindling [*ḥabbūḇā*] there on fire [*mawqḏīn*].

Muᶜtabar II.202.7-16:

قد سبق القول فى الحرارة الصادرة عن شعاع الشمس وانها انما تصدر عن الشمس/٨ فى الاجسام الكثيفة الارضية والمائية دون اللطيفة الهوائية والشفافة السمائية وان/٩ الشمس نفسها وباقي الكواكب ليست بحارة لما وجدناه من برد أعالى الارض/١٠ والجو الذى يليها ولوكانت الشمس حارة لأسخنت الاعلى فالأعلى لكونه اقرب/١١ اليها وان هذه الحرارة تصدر عن الشعاع والشعاع انما يصدر عن نور الشمس/١٢ ويظهر على سطوح الاجسام الكثيفة وخاصة الصقيلة منها فانه يتصل فيها/١٣ باتصال السطح كما نراه على المرايا الصقيلة فان ظهوره فيها بحسب صقالها يكون/١٤ اشد وبحسب شدته يوجب الحرارة حتى تبلغ حد الاحراق خصوصا اذا كانت/١٥ المرايا مقعرة ينعكس شعاع اطرافهتا على وسط واحد فيحرق عند مجتمع الشعاع/١٦ المنعكس

We have talked before (2a) about the heat issuing [*ṣādira*] from solar radiation [*šuᶜᶜāᶜ al-šams*] (2c) and the fact that it only issues from the sun on to dense earthy and watery bodies (2d) and not [*dūna*] loose airy [bodies] and transparent celestial [bodies].[49] (1a) The sun itself and the rest of the stars are not hot (1b) as evidenced by [lit. on account of] the cold we have found in the heights [*aᶜālī*] of the earth and the air around it;

---

[48] Cf. also IS, *Qānūn* I.2.2.1.3, ed. New Delhi (1982) 142.9-27; Cairo (1986) 113.18-25; Būlāq I.81.28-82.3.

[49] Cf. *Muᶜtabar* II.189-191.

if the sun were hot, it would heat the highest part [*al-aʿlā*] because it is nearest to it. (2a) This heat issues from radiation (2b) and radiation issues from the light of the sun (2c) and becomes manifest [*yaẓharu*] on surfaces of dense bodies, especially polished ones, since it coheres [*yattaṣilu*] to them through the coherence [*ittiṣāl*] of the surface, (3a) as we see on polished mirrors. (3b) For its manifestation [*ẓuhūr*] on them becomes more intense in accordance with [*bi-ḥasabi*] their polishedness, (4a) and in accordance with their polishedness (the radiation) produces [*yūǧibu*] heat, (4b) until (the heat) reaches the limit of combustion [*ḥadd al-iḥrāq*], (4c) especially when the mirrors are concave [*muqaʿʿara*], (5a) so that the rays from [lit. of] the edges are reflected to one middle [point] [*wasṭ wāḥid*] (5b) and sets [objects] on fire by the gathered reflected rays.

l. 1, 2. ܦܩ̈ܥܬܐ ... ܪ̈ܫܝ ܛܘܪ̈ܐ ("plains ... tops of mountains"): Abū al-Barakāt talks only of "heights" (*aʿālī*) in the passage above, but talks more concretely of the temperature difference between "mountains" (*ǧibāl*), "level ground" ([*arḍ*] *mustawīya*) and "hollows" (*ġaur*) later on in the same chapter (II.207.7-24).

IV.iv.2.1-6: Unpolished bodies and celestial sphere

The theory provides the explanation as to why radiation does not become visible on "unpolished bodies" and "transparent celestial bodies" (IV.iv.1.4 above). Only the explanation relating to the first is given at the corresponding place in the *Muʿtabar*. - The point made in part (3) of the passage concerning the earth being smaller than the sun and therefore being unable to shield the stars from the sun's light is one that goes back ultimately to Arist. *Mete*. 345b 2-9.[50]

But.: (1a) Light does not become visible [*methawwē*] on an unpolished [*lā sqīlā*] body (1b) because it is divided and scattered [*metpseq w-metpalhad*] by the coarsenesses, bumps and holes on it (1c) and loses its coherence [lit. does not cohere]. (2a) Neither does the ray manifest itself [*dānaḥ*] on transparent bodies (2b) and for this reason the celestial sphere [*mawzaltā*] does not shine or give light [*lā nāhrā āplā manhrā*] at night. (3a) One cannot say that the earth conceals the rays of the sun at night (3b) and does not allow them to fall on that half of the sky above the horizon, (3c)

[50] This is a part of Arist.'s refutation of the theory of Anaxagoras and Democritus concerning the Milky Way. The corresponding passage is missing in the Cambridge ms. of Nic. syr., although there is a scholion relating to it at Nic. syr. 13.20-24 (cf. Olymp. 68.13-28). That a passage representing Arist. *Mete*. 345b 2-9 was present in BH's copy of Nic. syr. is made likely by the fact that the theory of Anaxagoras and Democritus is itself also missing in the Cambridge ms. of Nic. syr. but is represented at *Cand*. 122.10-12.

because in relation to the sun the earth has no great size and stands like a dot [*nuqdṯā*] before it.

Muᶜtabar II.202.16-17:

وما ليس بصقيل لا يتصل فيه النور لانقطاعه وتفرقه بما فى السطح/ ١٧ الخشن من نتوات صاعدة ومسام نازلة

(1a/c) Light does not cohere [*lā yattaṣilu*] to what is not polished (1b) because of [*li-*] its division [*inqiṭāᶜ*] and dispersion [*tafarruq*] by the ascending protrusions and descending pores that are on the coarse surface.

l. 6. "like a dot": cf. BH, *Asc*. I.i.5. [Nau] 10.15f.: ܦܣܘܩܐ ܚܡܝܫܝܐ. ܡܛܠ ܕܐܝܟ ܢܘܩܕܬܐ ܗܝ ܐܪܥܐ ܠܘܬ ܫܡܝܐ ("Section 5. On the fact that the earth is like a dot in comparison with the heaven"; cf. Ptolemy, *Alm*. I.5).

IV.iv.3.1-7: Length of day and night

The passage is based on Abū al-Barakāt. Here again, however, BH has the problem that Abū al-Barakāt is a proponent of the view that the equator is temperate. In the part of the passage corresponding to sentence (3), Abū al-Barakāt tells us that neither heat nor cold grows intense in the *place* where the day and the night are equal, implying thereby that the equator where the day and night are constantly of a similar length must be temperate. BH alters this by talking about the *time* when the day and night are equal and applying this to spring and autumn, which are not mentioned by Abū al-Barakāt at all.

But.: (1a) In every place and region [*b-ḵul aṯar w-p̄enyān*] heat grows strong [*ᶜāšen*] in the season [*zaḇnā*] in which the day [*īmāmā*] becomes long [*yārek̠*] (1b) and coldness grows strong in the season in which the day becomes short [*meṯkrē*], (1c) because the appearance [*denḥā*] of the sun results in [lit. is followed by] radiation [*zallīqā*], (1d) and radiation results in heat. (2a) As a result, prolongation [*naggīrūṯā*] of radiation causes [*ᶜāḇdā*] an increase [*tawsep̄tā*] of heat, as [happens] in summer, (2b) while its curtailment [*lā naggīrūṯā*] [brings about] its decrease, as [happens] in winter. (3a) When the time [*zaḇnā*] during which radiation is observed [lit. the time of appearance (*meṯhawwyānūṯā*) of the radiation] is equal to the time during which it is not observed [lit. the time of its concealment (*meṯhappyānūṯā*)], as [happens] in spring and autumn, (3b) neither heat nor coldness increases, (3c) but [the two] temper each other [*meṯmazzḡān*].

Muᶜtabar II.202.22-203.6:

فنقول ان الحر يشتد فى كل موضع يطول نهاره الذى هو زمان طلوع الشمس/ ٢٣ فى ذلك الموضع وذلك هو الحر الصيفى ويقابله فى كل موضع البرد الشتوى الذى/ ٢٤ يوجبه قصر النهار فى كل موضع فان الطلوع الشمس فى كل موضع يوجب/ ١ الحرارة من شعاعها الواقع على ما يطلع عليه من الارض ودوام ذلك الطلوع/ ٢ يوجب زيادة فى ذلك الحر فحرارة النهار الأطول اشد واقوى وبهذا الاعتبار/ ٣ يكون الزمان الذى نهاره اطول اشد حرا وذلك هو زمان الصيف فى كل موضع/ ٤ والزمان الذى ليله اطول اشد بردا وذلك هو زمان الشتاء فى كل موضع والموضع الذى/ ٥ يساوى نهاره وليله ابدا تتشابه وتتقارب احوال زمانه فى الحر والبرد ولا يشتد/ ٦ فيه حر ولا برد

We say: (1a) heat becomes intense [*yaštaddu*] in every place [*fī kull mauḍiᶜ*] where the day [*nahār*], which is the period of the appearance [*ṭulūᶜ*] of the sun in that place, is long [*yaṭūl*]. That is the summer heat, (1b) and this is opposed in every place by the winter cold which is brought about by [*yūǧibuhu*] the shortness [*qaṣr*] of the day in every place. (1c) For in every place the appearance [*ṭulūᶜ*] of the sun brings about [*yūǧibu*] heat through its rays [*šuᶜᶜāᶜ*] which fall on the parts of the earth on which it appears [*taṭluᶜu*] (2a) and the prolongation [*dawām*] of that appearance brings about an increase [*ziyāda*] in that heat, so that the heat of the longest day is the most intense and strong. (1a) By that consideration [*iᶜtibār*] the season [*zamān*] whose day is the longest is the most intense in heat (2a) and that is the season of summer in every place. (1b) The season whose night is the longest is the most intense in coldness (2b) and that is the season of winter in every place. (3a) In every place where the day and night are always equal, the conditions the season in terms of heat and coldness are always similar and close to each other (3b) and neither does the heat grow intense there nor the cold.

IV.iv.3.7-10 & IV.iv.4.1-5:

The end of the third theory and the beginning of the fourth are best taken together, since BH has here made some rearrangement of the order in paraphrasing a somewhat repetitive passage of Abū al-Barakāt.

But.: (1) These [changes occur] according to the proportion [*lpūṯ puḥḥāmā*] of the day in each place to the night in the same [place]. (2a) [As regards what happens] according to the proportion [*lpūṯ puḥḥāmā*] of [the length of] the day and night in a given place to [the length of] the day and night in another place, (2b) the effect [*maᶜbḏānūṯā*] of the length [*arrīḵūṯā*] of the time during which radiation is manifest or not manifest is found not [to be] thus [*law hākan meštaḵḥā*], (2c) but to the contrary [*b-hep̄kā*], (2d) as will be shown. [IV.iv.4] (3a) Although by nature lengthening [*arrīḵūṯā*] of the day results in increase [*yattīrūṯā*] of heat, (3b) other causes invalidate [*šāryān*] this law of nature. (4a) The heat in the seventh clime whose longest day is sixteen hours is many times [*b-aᶜpē saggīʾē*] weaker (4b) than the heat in the first clime whose longest day is thirteen hours (4c) and, in this way, in every clime the longer the day, the weaker the heat.

Muᶜtabar II.203.6-15:

... والذى يتقارب يتقارب والذى يتفاوت يتفاوت وبحسب التقارب/٧ والتفاوت يخالف الصيف الشتاء فى شدة الحر والبرد الموجودين فى الشتاء والصيف/٨ فى كل مكان وتتفاوت بعد ذلك الاقاليم والاصقاع فى شدة الحر والبرد فالذين نهارهم/٩ الأطول اطول لا يكون حر صيفهم اشد من حر صيف الذين اطول نهارهم اقصر من/١٠ اطول نهار هؤلاء وكان القياس يقتضى ان تكون زيادة الحر على الحر مثل زيادة النهار/١١ على النهار ولا تجد الامر كذلك بل تجده بالضدة اذ يكون حر الصيف عند الذين نهارهم/١٢ اطول اضعف وبرد شتائهم اقوى واشد والذين نهارهم اقصر حر صيفهم اشد/١٣ وبرد شتائهم اقل والسبب فى ذلك هو أن الحال كذلك فى كل صقع بقياس/١٦ نهاره الاطول الى نهاره الاقصر واما فى مقايسة صقع الى صقع فتختلف لأسباب/١٧ اخرى ...

... (1) and [a place] where [the lengths of the day and night] are close to each other [the greatest heat and cold] are close to each other; where they are dissimilar, they are dissimilar; and according as to whether they are close or dissimilar [*bi-ḥasab al-taqārub wa-l-tafāwut*], the summer and winter differ in the intensity of the heat and cold found in the winter and summer in every place [*makān*]. (2) The climes and regions [*aqālīm wa-l-aṣqāᶜ*] are dissimilar after that in the intensity of heat and cold. (4a) For the summer heat of those whose longest day is the longer is not more intense (4b) than the summer heat of those whose longest day is the shorter than the longest day of the first. (2a) Proprotionality [*qiyās*] would have demanded that the increase of heat be proportional to [*mitl*] the increase [in the length] of day, (2b) but you find the matter not to be so [*ka-ḏālika*], (2c) but you find it to be to the contrary [*bi-l-ḍidda*]. (4a) For the summer heat among those whose day is longer is weaker and their winter cold stronger and more intense, (4b) while the summer heat of those whose day is shorter is more intense and their winter cold less. (1) The reason for that is that the situation is thus in every region [*ṣuqᶜ*] according to the proportion [*bi-qiyās*] of its longest day to its shortest day, (2b/3b) but in the comparison [*muqāyasa*] of one region to another (the situation) is different because of other causes ...

iv.3.9-10. ܙܒܢܐ ܕܓܠܝܢ ܙܠܝܩܐ ܘܟܣܝܘܬܗ ("the time during which radiation is manifest or not manifest"): lit. "the time of the manifestation of the ray and its concealment". - i.e. day and night.

IV.iv.4.5-11: Retention of coldness

Parts (1)-(4) of the passage may be seen as a summary of the passage of the *Muᶜtabar* quoted below. Then, in parts (5)-(6), in what looks at first like an innocuous continuation applying the priniciples explained in the first part of the passage to the equatorial region, BH turns Abū al-Barakāt's argument on its head in arguing that the equatorial region must be hot. In doing so, BH has to downplay the emphasis made by Abū al-Barakāt on the role of the earth in the retention of coldness.

But: (1a) This is because [in] that place whose summer day is longer, (1b) its winter night is also longer. (2a) As a result, coldness grows strong [*tāqpā*] there in winter (2b) and snow [lit. snows] lingers [*maggrīn*] on the surface of the earth. (3a) Because earth is cold by nature, (3b) it retains the winter coldness more than the summer heat. (4a) For this reason the heat of the summer is subdued [*tāḥbā*] by the coldness of its winter (4b) and coldness prevails there. (5a) Where the summer day and winter night are short, (5b) coldness does not grow strong. (6a) For this reason, the heat of its summer is not overcome [*ḥāybā*] by the coldness of its winter (6b) and heat prevails there.

Mu^ctabar II.203.15-24:

احدها ان الذين نهار صيفهم اطول من نهار صيف آخرين يكون ليل شتائهم/ ١٦ اطول من ليل شتائهم وطول الليل يوجب شدة البرد وبقاء الثلوج والبرودة/ ١٧ فى الارض فلا يعتدل حرهم وبرودهم فى اعتدال نهارهم بل يغلب البرد لما استقر/ ١٨ فى الارض من البرودة ولا يسخن الا فى زمن اطول ثم لا يدوم السخونة مدة/ ١٩ فى مثلها تعود اسباب البرودة من طول الليل ولا تبقى الحرارة فى الارض مثل/ ٢٠ بقاء البرودة لان البرودة للأرض بالطبع وتستقر ببقاء الثلج فتكون للبرودة/ ٢١ بعد انقضاء السبب الموجب اسباب حافظة وهى برودة الارض الطبيعية/ ٢٢ وما اكتسبته من برد الثلوج والحرارة تنقضى مع انقضاء اسبابها ولا تلبث الا قليلا/ ٢٣ لان طبيعة الارض تضادها وتبطلها وليس لها مدد يبقى كالثلج للبرودة ولذلك/ ٢٤ ترى البلاد التى يدوم بقاء الثلج فيها صيفها ابرد وحرها اضعف

(1a) One [cause] is that [with] those whose summer day is longer than the summer day of others, (1b) their winter night is longer than the winter night of [the latter]. (2a) The length of the night brings about an intensity of the cold (2b) and persistence [*buqā^ᵓ*] of snows and of coldness [*burūda*] on the the earth, (4a) so that their heat and coldness are not balanced [*ya^ctadilu*] at their equinox [*i^ctidāl al-nahār*], (4b) but the cold prevails because of the coldness that remains fixed [*istaqarra*] in the earth and is warmed [*yasḫunu*] only over a longer period of time. Then, the warming [*suḫūna*] does not last for any time, in the like of which the causes of the coldness returns because of the length of the night. Heat does not persist in the earth like the coldness (3a) because coldness belongs to the earth by nature and remains fixed with the persistence of the snow, (3b) so that [even] after the termination of the necessitating cause [*al-sabab al-mūǧibu*] the coldness has retentive causes [*asbāb ḥāfiẓa*], namely the natural coldness of the earth and what it has acquired from the coldness of the snow. Heat on the other hand comes to an end with the termination of its causes and only lingers for a short time because the nature of earth is opposed to it and annuls it, and because it has no persistent helper like the snow is for the coldness. For this reason, you see that [in] the countries where the persistence [*buqā^ᵓ*] of the snow is enduring [*yadūmu*] their summer is cold and their heat weak.

IV.iv.5.1-2: Influence of shades

But.: (1a) When the sun stand vertically at the zenith [*l^cel men rēšā*], (1b) it generates more heat [*yattīr maḥḥem*], (1c) because at that time the shades of mountains and walls [*essē*] disappear [*metlaytīn*] (1d) and the heat-generating rays [*zallīqē maḥḥmānē*] reach everywhere.

Mu^ctabar II.204.2, 6-11:

وسبب آخر وهو مسامتة الشمس لرؤوس سكان الاقاليم ولا مسامتتها ... ٦ وانما المسامتة توجب الحر من/ ٧ وجهين - احدهما يخص والآخر يعم - والذى يخص هو عدم الاظلال والافياء/ ٨ الحاصلة بالجبال والجدران فانها لا تبقى عند المسامتة ولا يوجد لها ظل فى اوقات/ ٩ الظهائر وما يقاربها بل يستولى الحر عليها كما يستولى على البرارى والاراضى/ ١٠ المستوية ومع عدم المسامتة توجد فيها اظلال وأفياء تسترها عن الشعاع فتبقى/ ١١ مواضع الاظلال باردة مبردة فهذا من اسباب البرودة فى البلاد الجبلية

(1a) Another cause is the presence of the sun at the zenith [*musāmatat al-šams l-ru^ᵓūs*] of the inhabitants of the climes and its non-presence at

the zenith [*lā musāmatatu-hā*]. ... (1a) The presence at the zenith brings about [*tūğibu*] heat in two ways, one particular and the other general. (1c) The particular way is the absence [*ʿadam*] of the shades and shadows [*al-aẓlāl wa-l-afyāʾ*] made by mountains and walls [*ğudrān*]. (1a/c) For they [shades] disappear [*lā tabqī*] when the sun is at the zenith [*ʿinda l-musāmata*] and they [mountains/walls] have no shade [*lā tūğadu la-hā ẓill*] at noon and times near it, but the heat prevails over them as it does over plains [*barārī*] and level grounds. When the sun is away from the zenith [*maʿa ʿadam al-musāmata*] shades and shadows are found in them, which shield them [*tasturu-hā*] from the radiation [*šuʿʿāʿ*], so that the places of the shades remain cold and refrigerated. This is one of the causes of coldness in mountainous countries.

l. 1. ܥܒܪ ܡܬܩܬܠܝܣܐ ("stands vertically"): ܡܬܩܬܠܝܣܐ, apparently, < gr. καθετίσαι; cf. IV.i.2.1 ܡܬܩܬܠܝܣܘܬܐ. - J.P. Margoliouth read the word here as plural, mistaking the flatly-written *zqāpā* (ܩ̈ܦܐ) in ms. 1 for *syāmē* (PS Suppl. 313b).

IV.iv.5.3-9: Circle of radiation

But.: (1a) Because, furthermore, the rays of the sun are constantly on a half of the terrestrial sphere, (1b) in every place where the sun passes the zenith [*lʿel men rēšā ʿāḇar*] at noon, (1c) its inhabitants are in the middle of the circle of its rays, (1d) whereas where the sun does not pass the zenith, (1e) the inhabitants are on the edge of the circle of the ray at noon - (2) and it is clear that the middle of a hot thing is hotter than its edges. - (3) For this reason, a place where the sun passes the zenith is hotter than a place where this is not the case [lit. a place which is not so] (4) and noon is hotter everywhere than morning and evening.

Muʿtabar II.204.11-19, 204.24-205.2, 205.3-5:

والذى يعم/ ١٢ هو أن الشمس اذا شرقت على الارض كان شعاعها على نصف كرة منها تدور/ ١٣
بدورانها فان كرة الشمس يحاذى كرة الارض منها نصف لنصف ابدا وهذا/ ١٤ النصف يدور على الارض/
١٥ فيكون الناس كل يوم فى كل صقع فى كل غداة وعشية منه فى طرفه وعند/ ١٦ محيطه ثم يتوسطونه
فى وسط نهارهم فيكونون فى تحته ووسطه من جهة/ ١٧ الطول فان كانت الشمس مسامتتة لرؤوسهم
كانوا فى وسط نهارهم فى/ ١٩ الوسط الحقيقى من دائرة الشعاع المذكورة وتحتها وان لم تكن
المسامتة/ ٢٠ لم يكونون فى الوسط من كل جهة على ما فى هذا الشكل ... ٢٤ وانت تعلم ان/ ١
الوسط احر من الاطراف لاحاطة الحرارة به من كل الجهات والطرف/ ٢ يكون اضعف ... ٣ فانك ترى
الوسط احر والتأثير/ ٤ فيه اشد فهذا هو سبب اشتداد الحر عند مسامتة لا القرب الذى تأثيره فى ذلك/ ٥
اقل مما يظنه الجاهلون بعلم الهيئة

The general way: (1a) When the sun shines [*šaraqat*] on the earth, its rays [*šuʿʿāʿ*] are on a half of (the earth's) sphere which it encircles in its circulation; for a half of the sphere of the sun always [*abadan*] faces a half of the sphere of the earth and this half circulates above the earth. (1b/4) Every day, in every region [*ṣuqʿ*], every morning and evening people are at its edge and its circumference and then come to the middle of it [*yatawassaṭūna-hu*] in the middle of their day [*fī wasṭ nahārihim*]. For

they are below and in the middle of it along the lengthwise direction. (1b) If the sun comes to their zenith [*kānat musāmita li-ruʾūsihim*], (1c) in the middle of the day they are in the true middle of that circle of rays [*dāʾirat al-šuʿʿāʿ*] and below it. (1d) If it does not come to the zenith, (1e) they are not in the middle in every direction that is in that shape. ... (2) You know that the middle is hotter than the edges because the heat surrounds it [*li-iḥāṭat al-ḥarāra bihi*] in all directions, whereas the edges are weaker ... You see therefore that the middle is hotter and the effect is more intense there. This is the cause of the intensity of heat when the sun is at the zenith, not the proximity [of the sun to the earth] which has less influence than those ignorant of astronomy think.

l. 8-9 ܘܠܛܗܪܐ . . . ܚܡܝܡܝܢ ("and noon ... evening"): om. FMV. - The clause looks somewhat like an afterthought and this along with its absence in mss. FMV raises the suspicion that it may be an interpolation, but the fact that Abū al-Barakāt too speaks of the situation in the "morning and evening" speaks in favour of its authenticity, while the omission in FMV can be explained through the homoeoteleuton "*ḥammīm* ... *ḥammīm*".

BOOK OF MINERALOGY, CHAPTER FIVE

ON MATTERS CONCERNING THE SEA

Various aspects of the sea were discussed by Arist. at *Mete*. I.xiii-II.iii. Some of these subjects were discussed in the *Šifāʾ* by IS not in his *fann* on "mineralogy-meteorology" but in the second chapter (*faṣl*) of the *fann* on "actions and passions" (*al-afʿāl wa-l-infiʿālāt*), a chapter entitled "on general conditions of the sea" (*fī aḥwāl kullīya min aḥwāl al-baḥr*, ed. Qassem, Cairo 1969, p. 205-210).

Of the four sections in this chapter of *But.*, Sections V.ii. (salinity of the sea) and V.iii. (migration of the sea) are based to a large extent on that chapter of the *Šifāʾ*. BH has at the same time supplemented the Avicennian material with materials taken from elsewhere. In our search for the sources of these supplementary materials we encounter the problem that there is a long lacuna in the Cambridge manuscript of Nic. syr. at the place corresponding to Arist. *Mete*. I.xiii-II.iii. The correspondence, however, of much of these supplementary materials in *But.* to materials found in other works related to Nic. syr. allows us to be reasonably certain that they are derived from lost passages of Nic. syr. in most cases.

Of the other two sections, Section V.i. gives us a geographical overview of the seas. The sources used in this section are rather different from those used in the rest of the two books of *But.* edited here (see the introductory comments on that section below). Section V.iv. deals with the Platonic theory of "Tartatus" (or, rather, Aristotle's refutation thereof). The likelihood is that most, if not all, of the material in this section is derived from lost passages of Nic. syr.

It is needless to say that this chapter of *But*. is of particular importance for the reconstruction of the lost portions of Nic. syr.

Min. V. Section i.: *On the Position of the Seas*

IS does not go into geographical details in his discussion of the sea in the *Šifāʾ*. One could argue that BH is being more faithful to Arist. in inserting a section here on geography, since Arist. does mention a

good number of geographical names in his discussion of the rivers and seas (*Mete*. I.xiii.-II.i.) and a geographical description of the world is also found in the Ps.-Aristotelian *De mundo*.

Geography, especially geography discussed in terms of the seas, lakes and rivers, is a subject BH had already dealt with in four of his earlier works, *Cand*., *Rad*., *Asc*. and *Eth*.[1] A comparison of the section of *But*. here with the corresponding parts of these four earlier works reveals that there is little here which is not already in the earlier works, so that the likelihood is that BH turned to his own earlier works rather than to external sources in composing this section (although he may have had a renewed look at the Syriac version of the *De mundo*) and an elucidation of our passage of *But*. consequently requires constant reference to the earlier works.

While the repetition of the same or similar materials in both Syriac and Islamic (Arabic or Persian) geographical works makes it difficult to identify the immediate sources used by BH in composing his geographical accounts, it is still possible to distinguish a number of sources which lie behind these accounts with varying degrees of certainty.

Among the Syriac sources certainly used is the Syriac version of the *De mundo*, which is the principal source of BH's information on the Mediterranean both in *But*. (Min. V.i.1-2) and in his earlier works. There seems, on the other hand, to be little in BH's geographical accounts which goes back to Nic. syr., even if the lacuna in our text of Nic. syr. makes it difficult for us to be certain that this is the case. There are certain points of contact between BH's geographical accounts and the account found in Jacob of Edessa's *Hexaemeron* [*Hex*.], which was later reproduced in abbreviated forms in Bar Kepha's *Hexaemeron* and Bar Shakko's *Book of Treasures* [*Thes*.]. There seems to be nothing in BH's accounts which requires us to assume a direct knowledge of the longer account found Jacob's work, so that BH's immediate source may be Bar Kepha (more likely) or Bar Shakko. BH's geography also betrays influences of historiographical and exegetical works in Syriac,

---

[1]*Cand*. II.3.3.1 (ed. Bakoš 150-166, ed. Çiçek 108.16-117.4); *Rad*. I.3.1-2 (ed. Istanbul 34.15-40.8); *Asc*. II.1.3-6 (ed. Nau 133.12-18); *Eth*. IV.13.5 (ed. Bedjan 452.18-454.10; Çiçek 236a 6-b 24). - There is also a brief list of the seas at *Horr*. in Gen. 1.9 (ed. Sprengling-Graham 10.21-23). - For a more detailed discussion of these passages than is provided below, see Takahashi (2003a).

which BH will have studied for the composition of his own works in these fields.

Among the Arabic sources already used in *Cand.*, the earliest of BH's works which come into question here, is the astrological-encyclopedic work of Bīrūnī (973-1048), *K. al-tafhīm li-awāʾil ṣināʿat al-tanǧīm.*[2] A number of passages in *Cand.* indicate a debt to the 10th c. geographers of the so-called Balkhī School (Iṣṭakhrī, Ibn Ḥauqal, Muqaddasī), although the brevity of these passages makes it difficult to determine whether these passages go back directly to the works of these geographers or to later compilations such as Yāqūt's *Muʿǧam al-buldān*. One change BH makes between the account in *Cand.* and that in *Rad.* and *Asc.* is the addition of the values for the length and breadths of the seas. These values, along with some further details in the accounts given in *Rad.* and *Asc.*, agree with those given by the astronomer Battānī (858-929) and in the related accounts of Ibn Rusta (fl. ca. 922) and Kharaqī (ob. 1138/9).

V.i.tit.: "on the Position of the Seas": cf. the section headings at *Cand.* 154.3: ܡܛܠ ܣܝܡܗܘܢ ܕܝܡܡܐ ܘܡ̈ܫܘܚܬܗܘܢ; and Bīrūnī, *Tafhīm* 121.5f.: كيف وضع البحار من المعمورة

V.i.1.1-4: Atlantic (Oceanus)

But.: (1) That sea which surrounds [*ḥdīr*] the whole earth like a single island is called the Atlantic. (2) In the west a narrow mouth is open to it at the στήλαι, i.e. Pillars, of Hercules. (3) Through it it enters into the habitable world as if into some harbour and forms this well-known sea which is called the Oceanus [*sic*] by many.

De mundo syr. 139.16-21 [< gr. 393a 16-21; > *Cand.* 154.3-9]:

ܝܡܐ ܕܝܢ ܗܘ ܗܢ ܕܐܝܬܘܗܝ ܠܒܪ ܡܢ ܟܠܗ ܡܬܥܡܪܢܘܬܐ ܐܛܠܢܛܝܩܘܣ ܘܐܘܩܝܢܘܣ ܡܬܩܪܐ. ܘܗܪܟܐ ܐܦ ܚܕܪܝܢ ܗܘܐ. ܟܕ ܕܒܡܥܪܒܐ ܡܬܦܬܚ ܡܢ ܠܓܘ ܦܘܡܐ ܐܠܝܨܐ ܦܬܝܚ ܠܗ. ܘܗܘ ܒܝܕ ܗܢܝܢ ܕܡܬܩܪܝܢ ܣܛܝܠܐ ܕܗܪܩܠܝܣ. ܠܗ ܪܕܝܗ ܠܝܡܐ ܗܢܐ ܕܠܘܬܢ. ܐܝܟ ܕܠܠܡܐܢܐ ܡܕܡ. ܘܡܬܦܫܛ ܒܠܝܠ ܒܠܝܠ ܡܬܦܬܐ ܗܪܟܐ. ܥܕ ܡܬܦܫܛ ܒܗܪܟܐ ܕܠܒܟ ܠܥܘ̈ܒܐ ܪ̈ܘܪܒܐ ܕܐܚܝܕܝܢ ܚܕ ܒܚܕ.

(1) That sea which is outside the whole habitable world is called the Atlantic and the Ocean. It also flows around us [*ḥdārayn*] here. (2) Because on the west a narrow mouth is open to it from the inside - at what are called the στήλαι of Hercules - (3) its flow proceeds into this sea by us, as if into some harbour, and thus widens out little by little here, being spread out until it embraces [*lābek* < περιλαμβάνω] of the large gulfs which adjoin each other.

---

[2] On the use of this work in *Cand.*, see besides Takahashi (2003a) also id. (2002b) 247-252, 255f.

BH had used this same passage of the *De mundo* in his earlier description of the Mediterranean at *Cand.* 154.5-9, although there the material taken from the *De mundo* is combined with material borrowed from Bīrūnī's *Tafhīm*.[3]

l. 1. "which surrounds [*ḥdīr*] the whole earth like a single island": cf. *Cand.* 154.2-3: "the Universal Sea which is outside the whole of the Habitable World and surrounds [*ḥdīr*] the whole earth" and *Cand.* 157.11-158.2: "Thus the whole Habitable World rests like a single island inside the Universal Sea, and [the latter] surrounds [*ḥdīr*] the earth like a crown the head and a belt the loins.[4] - The idea is already there, among BH's immediate sources, in the *De mundo*, some lines before the passage quoted above: syr. 138.18-19 (< gr. 392b 21-22):

ܥܠ ܕܠܐ ܣܟܠܘܗ̇. ܕܐܝܟ ܟܠܗ̇ ܠܝܬܒܬܐ ܚܕܐ ܓܙܪܬܐ ܐܝܬܝܗ̇. ܕܚܕܝܪܐ ܡܢ ܝܡܐ ܗܘ ܕܡܬܩܪܐ ܐܛܠܢܛܝܩܘܣ

... because they did not discern that the whole habitable world [here: *yāṯeḇtā*, but *ʿāmartā* 138.17] was a single island, surounded by that sea called the Atlantic.[5]

The comparison with the crown [*klīlā*] and belt [*qamrā*] in *Cand.* also allows us to connect the passage with Bar Kepha, *Hex.*, ms. Paris 311, 46r a 6-10:

ܐܘܩܝܢܘܣ ܗܘ ܐܝܬܘܗܝ ܝܡܐ ܪܒܐ. ܗܘ ܕܐܝܬ ܐܢܫܝܢ ܕܐܡܪܝܢ ܥܠܘܗܝ. ܕܚܕܪ ܠܗ̇ ܠܟܠܗ̇ ܐܪܥܐ ܐܝܟ ܟܠܝܠܐ ܠܪܫܐ ܘܐܝܟ ܩܡܪܐ ܠܚܨܐ.

The Oceanus is that great sea, concerning which some say that it encircles the whole earth like a crown the head and a belt the loins.[6]

---

[3] We note, for example, the mention there in *Cand.* of the "passage [*maʿbartā*] of Hercules" (154.6), corresponding to the "maʿbarat Hirqlis" of *Tafhīm* 123.5, alongside the "στῆλαι, i.e. pillars [*qāymāṯā*], of Hercules" (154.7f.), taken from the *De mundo*.

[4] Cf. further *Rad.* [Istanbul] 35.17-18, [Gottheil] 53.1: "The sea Oceanus surrounds the whole Habitable World [*meṯʿamrānīṯā*] like a single island"; *Asc.* 134.19-20: "The Universal Sea, while it surrounds the whole Habitable World [*ʿāmartā*] in a circle [*qīqlōsāʾīṯ*] like a single island, ...".

[5] We also find the idea expressed somewhat more obscurely in Nic. syr. 15.6-7: ܐܬܐܡܪ ܕܝܢ/ ܠܥܬܝܩܐ ܕܐܘܩܝܢܘܣ ܐܝܬ. ܕܚܕܪ ܐܪܥܐ ܘܡܬܚܙܝܐ ܠܝ ܕܗܢܐ /ܐܝܬܘܗܝ. ("It has also been said by the ancients that there is an Ocean which flows around the earth, and it seems to me that it is this [current, which they have in mind]") < Arist. *Mete*. I.ix, 347a 6-8: ὥστ' εἴπερ ᾐνίττοντο τὸν ὠκεανὸν οἱ πρότερον, τάχ' ἂν τοῦτον τὸν ποταμὸν λέγοιεν τὸν κύκλῳ ῥέοντα περὶ τὴν γῆν.

[6] The comparison with the belt is missing at the corresponding place in ms. Paris 241 (172r 29-v3) and hence in Schlimme's translation (p. 572). - Interestingly, it is also there in ms. Paris 316 (172v 15-18) of Bar Shakko, *Thes.*, but missing in BL Add. 7193 (70r b20-21). - This comparison with a crown is attributed to Aristotle in Kharaqī [Nallino] I.175.13f.: وقد حكى ارسطاطاليس انّ بحر اوقيانوس محيط الارض بمنزلة اكليل لها (also in Quṭb al-Dīn al-Shīrāzī, *Nihāyat al-idrāk*, partial tr. Wiedemann [1912a] 31; and id. *al-Tuḥfa al-šāhīya* sec. Nallino, I.173 n.(1)). - For further instances in Syriac talking of the Ocean surrounding the earth/habitable world, see PS 88, s.v. ܐܘܩܝܢܘܣ;

Cf. further comm. on Min. II.i.4. 6-7 above.

l. 4. ܐܘܩܝܐܢܘܣ ("the Oceanus"): ܐܘܩܝܢܘܣ V; ܐܕܪܝܐܛܣ F[corr.] super rasura. - The agreement of the manuscript readings suggests that "Oceanus" is what BH wrote. In *De mundo* no proper name is given for the whole of the Mediterranean. In Syriac the name Adriatic is often applied to the whole of the Mediterranean (PS 64 s.v. ܐܕܪܝܐܛܣ; cf. Nau [1899] tr. 119 n. 2) and this is followed by BH himself in *Cand*. 154.9, 155.1, *Rad*. [Istanbul] 17.5, 37.12, *Asc*. 135.4, 137.3, 140.13-15, *Eth*. [Bedjan] 453.12 and *Horr*. in Gen. 1.9, in Act. 2.10. That his rejection of the name "Adriatic" here and his substitution in its stead of the "Oceanus" are not simply a slip of the pen is suggested by the fact that he now applies the name "Atlantic" alone to the outer sea both at V.i.1.1 and V.i..3.1, a name he had, following the *De mundo*, given as an alternative name for the "Oceanus" at *Cand*. 154.4 ("Atlantic" not in *Rad*., *Asc*. and *Eth*.). For his rejection of "Adriatic" as the name for the Mediterranean, he may have had a reason in his realisation, perhaps on re-reading the *De mundo*, that the name Adriatic was applied there to a particular gulf within the Mediterranean (see V.i.2.3-4 below; BH had not given the names of the four Mediterranean gulfs in his earlier works). The reason for his erroneous application of the name "Oceanus" to the Mediterranean is more difficult to find, although his knowledge that "some call only the western side [of the universal sea] the 'Oceanus'" (*Cand*. 154.5) may have had a role to play here, since the name the "Sea of the West" (*baḥr al-maġrib*) is sometimes used by the Arabs of the Mediterranean.

V.i.1.4-6: the three continents[7]

But.: (1) To its east is Asia, (2) to its west are the στήλαι, (3) to its north is Europe (4) and to its south is Libya.

The sentence combines, as it were, the materials found in two different passages in BH's earlier works, 1) that naming the lands around the Mediterranean and 2) that dealing with the division of the habitable world into three continents. The first of these began its life in *Cand*. as a passage based on Bīrūnī, and was then expanded in *Rad*. with a number of additions, including the names of the three continents which we encounter here in *But*.

Bīrūnī, *Tafhīm* 123.2-4 [pers. 168.13-15]:
... the Sea of Syria [*baḥr al-šām*], (4) to the south of which are the land of Maghrib [*bilād al-maġrib*; add *wa-ifrīqīya* pers.] up to Alexandria and

also Ps.-Berosus [Levi della Vida (1910)] 15.8-9: ܚܠܡܗ ܚܕܪ ܝܡܡܐ ܒܐܘܩܝܢܘܣ ܪܒܐ ܗܘ ܕܢܛܪ ܠܗ̇ ܠܚܠܡܗ̇ ܐܪܥܐ ܟܠܗ̇. - The notion is, of course, also very much present in the term frequently used in Arabic for the Ocean, *al-baḥr al-muḥīṭ*, the "surrounding sea".

[7] Cf. Takahashi (2003a), para. 18-21.

Egypt [*miṣr*]. (3) Opposite them in the north are al-Andalus and al-Rūm [*rūmīya wa-rūm* pers.] up to Antioch. (1) Between the two are Syria [*bilād al-šām*] and Palestine.

Cand. 155.1-3, 4-5 [underline: agreeement with *But.*; italics: with *Tafhīm.*]: (4) Thus *to the south* of the Adriatic Sea lie [*pāyšā*] *Alexandria and Egypt* [*meṣrēn*] (3) and *to its north Constantinople* [*qōsṭanṭīnōpōlīs*], *Rome* and the whole land of the Franks [*kulleh aṯrā da-p̄rangiya*]. ... (1) To its east are the lands of *Syria*. So much on this sea of ours.

Rad. [Istanbul] 37.11-17, [Gottheil] 54.2-6 [underline: agreeement with *But.*; italics: places **not** in *Cand.*]:
... this sea of us westerners called the Adriatic. (3) To its north are Rome, land of the Franks [*p̄rangiya*], Byzantium [*būzanṭiya*] and *the whole of Europe*; (4) to its south, which is called *the Sicilian (gulf)* [lege: *sīqīlīqōn*], are *Abyssinians, Nubians, Berbers*, Egypt [*īguptōs*], Alexandria and *the whole of Libya*; (1) to its east, which is called *the Sea of the Syrians*, are *Tyre, Sidon* and *the whole of Asia*.

Asc. 135.3-6: ... this great sea called the Adriatic [*aḏriyanōs*], (2) to whose west are the Isles of the Blessed, (1) to whose east Palestine, (3) to whose north Great Rome and Byzantium (4) and to whose south Africa [*ap̄rīqī*].

Eth. 453.13-14: (1) To its east is Syria, (3) to its north are Byzantium and Rome, (4) and to its south is Africa.

V.i.1.6-7: islands in the Mediterranean[8]

But.: In it are the large islands Sicily, Samos, Chios, Rhodes and Cyprus, as well as many small ones. They all abound in people and crops.

The five islands named here are the same as those mentioned at *Cand.* 155.3-4.

l. 7. "They all abound [*kahhīnān*] in people [*nāšūṯā*] and crops [*ʿallāṯā*]": cf. *Eth.* [Bedjan] 453.17-18: "... these [islands], though not famous, are nevertheless not devoid of people [*ellā men nāšūṯā lā spīqān*]."

l. 7. ܥܠܠܬܐ ("crops"): I follow the manuscripts in vocalising the word as *ʿallāṯā*,[9] but the reading *ʿellāṯā*, "things, equipment, wares",[10] would also make good sense.[11]

---

[8] Cf. Takahashi (2003a), para. 22-27.

[9] The fully vocalised manuscripts LIPV all read *ʿallāṯā*, as does M which is not normally vocalised. There is, however, no pointing here in F.

[10] For this meaning of *ʿelltā* (reminiscent of fr. "chose", ital. "cosa"), see PS 2877, s.v., no. 4; Brockelmann 524a, s.v., no. 7; also, BH *Chron.* 414.7f. (< Juwainī, *Tārīḫ-i ǧahānguša* I.60.1f. *ǧāmah-hā*, "fabrics"); perhaps also ibid. 504 24.

[11] See further Takahashi (2003a) n. 41.

V.i.2.1-4: gulfs in the Mediterranean[12]

But.: (1) In the south of this sea there are two gulfs and in them are two islands called the Greater and Lesser Syrtes. (2) In its northern (side) are three gulfs, the Sardinian, the Galatian and the Adriatic, and after these a slanting gulf called the Sicilian.

De mundo syr. 139.23-140.1 [< gr. 393a 23-28]:

ܡܬܐܡܪܬ ܩܕܡܝܬ ܕܬܬܦܬܚ ܠܝܡܝܢܐ. ܡܢ ܒܬܪ ܕܢܦܩܬ ܡܢ ܩܘܡܠܣ
ܕܗܪܩܠܝܣ. ܘܡܬܦܠܓܐ ܠܬܪܝܢ ܥܘܒܝܢ ܘܥܒܪܐ ܠܓܙܪܬܐ: ܗܠܝܢ ܕܡܬܩܪܝܢ ܣܘܪܛܝܣ.
ܗܢܝܢ ܕܠܚܕܐ ܡܢܗܝܢ ܣܘܪܛܝܣ ܪܒܬܐ ܡܩܪܝܢ ܠܗ. ܠܚܕܐ ܕܝܢ ܐܚܪܬܐ. ܣܘܪܛܝܣ
ܙܥܘܪܬܐ. ܠܓܒܐ ܕܝܢ ܗܘ ܐܚܪܢܐ ܓܪܒܝܝܐ. ܠܘ ܒܗ ܒܕܡܘܬܐ. ܒܪܫܝܬܗ
ܡܬܦܬܚܐ. ܘܥܒܕܐ ܕܝܢ ܬܡܢ. ܐܦ ܗܝܕܝܢ ܬܠܬܐ ܥܘܒܝܢ. ܗܘ ܕܡܬܩܪܐ
ܣܪܕܘܢܝܘܣ. ܘܗܘ ܕܡܬܩܪܐ ܓܠܛܝܩܘܣ. ܘܐܕܪܝܘܣ ܪܒܐ. ܘܡܢ ܒܬܪ ܗܠܝܢ ܬܘܒ
ܗܘܐ ܥܘܒܐ ܐܚܪܢܐ ܠܣܛܪܢܐܝܬ. ܗܘ ܕܡܬܩܪܐ ܣܩܠܝܩܘܣ.

(1) It is said first to widen out to the right after proceeding from the Stelae of Hercules and is divided into two gulfs and passes the islands called the Syrtes, one of which they call the Greater Syrtis and the other the Lesser Syrtis. (2) On the other, northern, side it does not widen out immediately in the same way, but makes there too three gulfs, that called the Sardinian, that called the Galatian and the Great Adriatic. After these is another, slanting gulf which is called the Sicilian.

l. 1-2. "two islands called the Greater and Lesser Syrtes": The error of turning the Syrtes into islands, as may be seen from the passage quoted above, is due to the Syriac version of the *De mundo* (cf. Baumstark [1894] 412; Ryssel 27 note a.).[13]

l. 2. "three gulfs": In the Greek original of the *De Mundo*, the northern subdivisions of the Mediterranean are called "seas" (πελάγη) as opposed to "gulfs" (note 393a 26: οὐκέτι ὁμοίως ἀποκολπούμενος). The Syriac version, which BH is following, ignores this distinction.

l. 3, 4. ܐܕܪܝܐܢܘܣ "Adriatic": The original Greek form of the name applied, as we have seen, to the Mediterranean in Syriac is 'Αδρίας, although ἀδριανός is attested as an adjective. This is usually rendered as ܐܕܪܝܐܢܘܣ or ܐܕܪܝܢܘܣ in Syriac,[14] while in the above passage of *De mundo* syr. we have ܐܕܪܝܘܣ. BH, too, in his earlier works usually has ܐܕܪܝܢܘܣ,[15] but the form in "-nōs" does

---

[12] Cf. Takahashi (2003a), para. 8-13.

[13] The Syrtes are rightly counted among the gulfs of the Mediterranean by Jacob of Edessa (*Hex.* 100a 18f.) and, following him, by Bar Kepha (*Hex.*, P172v a14, Q46r a19, tr. 572) and Bar Shakko (*Thes.* P173r 7, L70v a27; cf. Nau [1896] 312).

[14] So, for example, Jacob of Edessa, *Hex.* 99b 3, 35, 100b 2, 7 etc. (ܐܕܪܝܐܢܘܣ); Bar Kepha, *Hex.* Q42r b32, 46r a 3 (ܐܕܪܝܐܢܘܣ), 42v a1, 6, 24 (ܐܕܪܝܢܘܣ); Michael, *Chron.* 149b 29 (ܐܕܪܝܢܘܣ, tr. I.292), 411b 6, 9 (ܐܕܪܝܐܢܘܣ, ܐܕܪܝܐܢܣ, tr. II.414), 515c 3 (ܐܕܪܝܐܢܣ, III.62). - The form with "-n-" is, however, found at *Chron. ad 724 pertinens* [Brooks] 351.7: ܝܡܐ ܪܒܐ ܕܐܕܪܝܢܘܣ.

[15] ܐܕܪܝܢܘܣ: *Cand.* 154.9, 155.1; *Rad.* 17.5, 38.10; *Asc.* 137.3, 140.13-15; *Horr.* in

occur at *Asc*. 135.4, *Eth*. [Bedjan] 453.12 (ܐܕܪܝܐܢܘܣ) and *Chron*. [Bedjan] 449.6 (ܐܕܪܝܢܘܣ). That BH distinguished the form with "-n-" as an adjective is unlikely, although it is noteworthy that he constantly writes with both names "yammā Adriyōs/Adriyānōs" without "d-" (as he does with "yammā Ōqiyānōs"; in contrast to "yammā *d*-Ponṭos", cf. V.i.2.5 below).

V.i.2.4-5: Hellespont/Pontus[16]

The channel between the Mediterranean and the Black Sea is mentioned by BH in three earlier works, although in *Rad*. and *Asc*. (but not *Cand*.) the description moves from the Black Sea to the Mediterranean. The earlier passages, in turn, exhibit features which allow one to relate them to Bīrūnī's *Tafhīm*.

But. [underline: agreement with *Cand*.; italics: with *Asc*.]:
(1) *From* the Adriatic *a thin tongue* goes out [*nāpeq*] towards the north (2) *and passes by Byzantium* (3) *and flows* [*šāḏē*] *into the Sea* of Pontus, ...

Cand. 154.9-155.1 [underline: agreement with *But*.; italics: with *Tafhīm*]:
(1) From this gulf [i.e. the Mediterranean] a certain tongue [*leššānā meddem*] is produced [*metyabbal*] and *thins out*, (2) *and passes by the wall* [*šūrā*] *of Constantinople*, and is called the *Sea* of *Pontus*.

Rad. [Istanbul] 38.6-10, [Gottheil] 54.11-13 [underline: agreement with *But*.; italics: with *Tafhīm*]:
Ray. *The Sea* of *Pontus* is *in the land of the Scythians*.[17] Its length is 1,300 miles up to *Trebizond* and its width 300 miles. (1) *From it* a narrow tongue [*leššānā alīṣā*] (2) *passes by the wall of Byzantium* (3) and is cast [*rmē*] *into the Adriatic Sea*.

Asc. 136.21-137.5 [underline: agreement with *But*.; italics: with *Tafhīm*]:
Among these is *the Sea* of *Pontus*, i.e. *of Trebizond*, but it is called a sea and not a lake because of its size. Its length is 1,300 miles and its width 300 miles. (1) *From it* a *thin* tongue (2) *passes* [*ʿāḇeḏ* ed. Nau: lege *ʿāḇar*] *by the wall of Byzantium* (3) and flows [*šāḏē*] *into the Adriatic Sea*, just as the Adriatic [flows] into the Oceanus. Some call this tongue the canal [*ṭūraʿtā*] of Alexander the son of Philip, as if he had opened it.

Bīrūnī, *Tafhīm* 122.15-123.2[18] [underline: correspondence with *Cand*.]:
Then, in the middle of the habitable world, in the lands of the Slavs [*ṣaqāliba*] and Russians [*rūs*] is the sea known as the Pontus among the Greeks and known among us as [the Sea of] Trebizond because (Trebizond) is a port on (its shore). From it a channel [*ḫalīǧ*] goes out, which passes by

---

Gen. 1.9, in Act. 2.10. - ܐܕܪܝܐܣ: *Cand.* 161.10 cum cod. P (sed ܐܕܪܝܘܣ B), *Rad.* [Istanbul] 37.12 cum cod. O (sed ܐܕܪܝܘܣ Gottheil cum PB).

[16] Cf. Takahashi (2003a), para. 14-17.

[17] Scythians as equivalent of *Ṣaqāliba* in BH; cf. Min. IV.ii.6 above.

[18] Corr. *Tafhīm* pers. 168.10-13; Yāqūt, *Muʿǧam* I.21.13-16.

the wall [*sūr*] of Constantinople and continually thins out [*lā yazāl yataḍāyuq*] until it falls into the Sea of Syria.

l. 4. ܠܫܢܐ ("tongue"): The word "leššānā (d-yammā)" is used several times of a narrow stretch of the sea in the Syriac Bible (see PS 1973 s.v. ܠܫܢܐ, 4)), while the Arabic "lisān" is also frequently used in that sense. As the direct source of the expression here, however, one might reckon the phrase "leššānā meddem alīṣā" at *De mundo* syr. 140.27, whence we have "leššānā meddem" in the passage of *Cand*. quoted above and "leššānā alīṣā", coupled with the verb "rmē" as in the *De mundo*, in *Rad*.

l. 5 ܝܡܐ ܕܦܘܢܛܘܣ ("the Sea of Pontus"): So constantly with "d-" in BH, evidently unaware of the meaning of the Greek word πόντος. The usual Arabic form "baḥr bunṭus" might have favoured the interpretation "Sea *of* Pontus" and BH knew furthermore that there was a region/Roman province called "Pontus" (see BH *Chron*. passim), so that he may have thought that the name of the sea was derived from the region, not vice versa. The error is avoided by Jacob of Edessa et al. (e.g. Jacob, *Hex*. 100a 14: ܝܡܐ ܗܘ ܕܡܬܩܪܐ ܦܘܢܛܘܣ; 100b 25: ܝܡܐ ܗܢܐ ܦܘܢܛܘܣ). Cf. V.i.2.3 above (*yammā adriyōs*).

V.i.2.5-7: Maeotis/Caspian[19]

But. [underline: agreement with *Rad*./*Asc*.]: ... from which in turn a river-like tongue goes out towards the northeast and flows into the Maeotis, which is the lake of Caspia, Hyrcania and Iberia [*īberīs*].

Cand. 158.3-12: In the land of the Iberians [*īberāyē*] there is a lake which is by itself [*d-mennāh w-lāh (h)ī*] and is not connected to the Universal Sea, so that someone setting out from a certain known place on its shores and travels around it would be able to reach the spot from which he started, if it were not for that great river, called the Ātil [Volga], which pours its waters into this lake. Because of its size and width, they call this lake a sea and not a lake in the books and in the language of common usage [*mamllā da-ʿyāḏā*], Ptolemy calls it the Sea of Hyrcania and in our day they call it "Khazar". To the west of this lake are the Gate of Iron [Darband], the plains of the Qipchaqs, Shirwān and Ṭabaristān, to its south Greater Armenia, to its east the land of the Iberians and to its north the great desolate black mountain which is at the extremity of the land in the northeast.

Rad. [Istanbul] 38.10-16, [Gottheil] 54.13-55.2: To the north of it is situated the Maeotis, the lake of Caspia, i.e. of rushes [*qnayyā*]; the ancients call it (the lake) of Hyrcania and of the Iberians [*īberāyē*], but in our day it is called "Khazar". From it a river-like tongue flows into the Sea of Pontus, just as from the Sea of Pontus into the Adriatic and [from] the Adriatic into the Oceanus.

---

[19] Cf. Takahashi (2003a), para. 37-52.

Asc. 137.5-9: To the north-east of this sea is the Maeotis, the lake of the Caspians, i.e. of rushes; Ptolemy calls it the Hyrcanian and others the Sea of the Iberians [*īberāyē*]. From it a river-like tongue goes out into the Sea of Pontus, just as from the Pontus into the Adriatic.

l. 5. ܠܫܢܐ ܢܗܪܝܐ ("river-like tongue"): The adjective "river-like", which is also found in *Rad*. and *Asc*., may reflect the association of this "strait" with the river Tanais (Don).[20] The idea can be traced back ultimately to Arist. *Mete*., where we are told that, due among others to the "large number of rivers (flowing into it)", the Maeotis flows into the Pontus, and the Pontus into the Aegean.[21]

l. 6-7. ܕܩܣܦܝܐ ܘܗܘܪܩܢܝܐ ܘܐܝܒܪܝܣ ("of Caspia, Hyrcania and Iberia"): Of the three countries/peoples, Caspia(ns) and Hyrcania(ns) are those after which the Caspian Sea is usually named in Greek, Greco-Syriac and Greco-Arabic sources (e.g. Arist. *Mete*. 354a 3, *De mundo* gr. 393b 5, Ptolemy, *Geog*. VII.5.9; Jacob of Edessa, *Hex*. 100b 32f., see further RE X.2275f., s.v. "Kaspisches Meer" [Herrmann]). - On the problem of the Caspian Sea being linked to "Iberia", see Takahashi (2003a), para. 39-47. - The three words are written without *syāmē* in the manuscripts. We also note that whereas BH had spoken of the "Sea of the Elamites" and "of Indians" in his earlier works, he speaks of the "Sea of Elam" and "of India" in *But*., two instances where the difference consists of more than the presence or absence of the *syāmē*. The two words ܩܣܦܝܐ and ܗܘܪܩܢܝܐ, at least, are therefore best understood as place-names rather than as names of peoples ("Caspians" etc.). The case is less clear for "īberīs". Greek plurals (and false Greek plurals) in "-s", "-īs" etc. are common in Syriac.[22] In two instances in BH *Chron*. (66.3, 6; 419.26; also *But*. Econ. II.3.3, [Zonta (1992)] p. 83, l. a12), "īberīs" is the name of a people. Similarly, "turqīs", which occurs at V.i.3.8 below and elsewhere in BH's works, seems usually to refer to the people and there is at least no instance where it cannot be construed as a plural.[23] It is also possible, however, to construe "īberīs" as a

---

[20] See Takahashi (2003a), para. 49-51. Note also the phrase "as if it were a river" in the passage of Battānī quoted there in para. 51 (*Zīğ*, ed. Nallino 27.9f.).

[21] Arist. *Mete*. 354a 13-14: ἡ δ' ἐντὸς Ἡρακλείων στηλῶν ἅπασα κατὰ τὴν τῆς γῆς κοιλότητα ῥεῖ, καὶ τῶν ποταμῶν τὸ πλῆθος· ἡ μὲν γὰρ Μαιῶτις εἰς τὸν Πόντον ῥεῖ, οὗτος δ' εἰς τὸν Αἰγαῖον (cf. Olymp. *in Mete*. 132.7-9; Olymp. arab. 108.23-24). The same idea thence, by a different route, also in Jacob of Edessa, *Hex*. 100b 18-28 (but not in Bar Kepha and Bar Shakko). See further RE III.744.20-48 (s.v. "Bosporos" [Oberhummer]).

[22] Nöldeke (1898) 59f. §89.

[23] 1) Written ܛܘܪܩܝܣ in published editions: *Cand*. 156.12, 162.6, *Asc*. 134.7, 190.20, 194.12, *But*. Econ. II.3.3 (Zonta [1992] p. 83, l. 7). - 2) Written ܛܘܪܩܘܝܣ with an extra *waw*: *Asc*. 140.6, 19. - 3) Without *syāmē*: *Cand*. 99.9 (ܐܬܪ ܛܘܪܩܝܣ); *Asc*. 138.17 (ܛܘܪ ܛܘܪܩܘܝܣ); but here too the "land/mountains of the Turks" would be more natural than "land/mountains of Turkey". - Cf. further PS 136, 1453, s.vv. ܐܝܒܪܝܣ,

(false) Greek place-name in "-ις" and I opt for this possibility here, though with some hesitation, in view of the absence of the *syāmē* in manuscripts and the parallelism with "Caspia" and "Hyrcania".

V.i.3.1-11: the Outer Sea ("Atlantic")

But. [underline: agreement with *Cand*.; italics: agreement with *Asc*.[24]]:
(1) Outside the στήλαι the Atlantic flows [*rāḏē*] to the southwest (2) and passes by [*ʿāḇar ʿal*] *the Silver Mountain*, from which the Nile rises [*gāʾaḥ*]. (3) As it flows eastwards, it encircles [*ḥāḏar*] *the land of the Berbers and the Cushites*. (4) In *Arabia* it forms [*ʿāḇeḏ*] the gulf [*ʿubbā*] of the Red Sea. (5) This (gulf) in fact *extends* [*metmtaḥ*] like a *thin tongue* into *Egypt* [*īguptos*] and forms *the Sea of Reed*. (6) Then, it passes by the land of Sheba and Saba (7) and forms *the Sea of Elam*, (8) and then *the Sea of India*. (9) *In the southeast it passes by Inner India*. (10) It (then) encircles the East and passes by *the Chinese*. (11) In *the northeast* it passes by *the land of Gog and Magog*. (12) In *the north* [it passes] by the land of the Turks [*ṭurqīs*] and the desolate lands and *impassible mountains* (13) and forms the Sea of *Britain*, in which is *the island* of *Thule*. (14) Then, it passes by the Iberians, Alans, *Scythians* and Bulgars. (15) *In the northwest it passes by* Italy and the whole *land of the Franks*. (16) In the west [it passes] by *Spain*. (17) It *ends* where it *began*, by the στήλαι.
Cand. 155.5-157.11: (1) The Sea Oceanus, which is outside the Herculean Pillars, with its flow [*reḏyā*] proceeding southwards, (2) passes by the lands of the Western Arabs and by the Silver Mountain and the Moon Mountains, from whose caves the waters of the River Nile spring [*nāḇʿīn*], (3) and by the lands of the Abyssinians and the lands of the Nubians, i.e. the whole land of the Cushites. (5) At the end [*ṣawpā*] of this land there extends [*metmtaḥ*] from it a certain small gulf [*ʿubbā*] towards the north opposite Egypt [*meṣrēn*], which is called the Sea of Reed [*yammā d-ṣūp̄*], as being the end and extremity [*ṣawpā w-sāḵā*] of the great sea. The sons of Israel crossed it on foot and in it the Pharaoh was drowned. Because of the great number of mountains and rocks in this gulf ships are unable to sail through it except during the day and along the shores. (4) That great sea from which this gulf proceeds is called the Red Sea (6) and this [sea], as it flows towards the east, passes by the lands of Sheba and Saba and the whole land which is simply called the South [*taymnā*] and where the trees of frankincense are. (7) At the extremity of this sea called the Red, it forms a great gulf towards the north called the Sea of Fārs. On the western side of this sea is the city called Baṣra and the whole land of Babel and

---

ܠܘܗܝܐ. - The same applies also for the ܢܘܒܐܕܐܣ (Nubians) at *Cand*. IV [Nau (1916)] 154.7 etc. (further references from BH at PS 2258).

[24] Cf. further *Rad*. [Istanbul] 35.18-37.6; *Eth*. [Bedjan] 453.3-10, [Çiçek] 236a 13-23; and Takahashi (2003a), para. 28-36.

Seleucia and Ctesiphon. On its northern side are all the lands of the Persians and on its eastern side are the lands of the Indians. (8) The universal sea which is outside this gulf, as it flows further towards the east, passes by the lands of the Indians and at their extremity forms a gulf towards the north called the Sea of the Indians. To the west of this sea are the lands of the Indians, to its east are the lands of the Tibetans and after them the lands of the Chinese, (12) and to its north are the lands of the Huns or the Turks [*ṭurqīs*], who are the Mongols, that is to say, their first land from which they proceeded [*npaq(w)*]. The universal sea which is outside this gulf, flowing further towards the east, passes by the famous islands of the Indians [of] which [one] is called Sarandīb and another Qumair, as well as the rest of the islands and mountains from which are brought and conveyed those hot, choice and aromatic spices [*sammānē*], cloves [*qaranful*], aloe [*alanǧūǧ*], pepper, camphor etc., as well as precious stones, corundum [*yaqqundā* = *yāqūt*] etc. (10) In the same way, extending from the east towards the north, it passes by the lands of the Tibetans and the Chinese, (11) by the lands of the Huns, which I have mentioned, (14) by the lands of the Iberians, (12) and by many lands that are desolate, and impassible mountains. It passes also by the great black mountain in the north and the plains of the Qipchaqs (14) and by the land of the Alans. (13) There it forms a gulf from the north to the south, which is called Warang in the language of that place. (14) In the same way, extending from the north towards the west, it passes by the land of the Scythians, by the cities of the Bulgars, (15) by the whole of France (16) and by the land of Andalus of the Arabs [*Ṭayyāyē*] which now in our day the Franks have conquered. (17) It ends by the Herculean Pillars, whence it began.

Asc. 134.4-16 [on the Ocean]:

(16) This [Ocean] extends [*māṭah*] northwards and encircles Spain. (6) It passes in the northwest by France. (14) It stretches [*metpšeṭ*] to the north outside the Scythians. (11) It reaches [*metmaṭṭē*] the northeast and confines [*ḥābeš*] the land of the Inner Turks [*ṭurqīs gawwāyē*], who are the Gog and Magog. (12) It is thought to pass behind the impassible mountains which have been mentioned [cf. 133.23-134.1]. (10) It joins [*meṯhayyaḏ*] the eastern sea of the Manzāyē, i.e. the Chinese. (9) It passes in the southeast by the whole of Inner India (4/6) and then extends to the south along Inner Arabia and the Desert of the Himyarites, outside of the Mountain of Pharan. It passes by Egypt [*īguptos*] (3) and the lands of the Berbers and the Cushites and the whole of Africa. (2) In the proximity of the Silver Moutain, which is also called the Moon [Mountain], it becomes imperceptible, (17) but it is thought to unite after it with the Western Sea, from which it began and with which it ends.

Asc. 135.16-136.11 [on seas projecting into the habitable world]

From the southern sea four tongues enter [*ʿāʾlīn*] into the Habitable World. The first is the Sea of the Berbers, which is near the west; its length is 500 miles and the breadth of its extremity 100 miles. (4) The second is the Red Sea, (5) which thins out [*metqaṭṭan*] and extends [*māṭah*] 400 miles in

length and the breadth of whose extremity, which is the Sea of Reed, is 200 miles. (7) The third is the Sea of the Elamites, into which the Euphrates and the Tigris pour their waters. Its length is 400 miles and its breadth 500 miles. (8) The fourth is the Sea of the Indians, whose length is 1,600 miles. In it are 1,370 islands, the best-known of which is Ṭirani (Taprobane), called Sarandīb in the language of that place. Its circumference is 1,300 miles and in it are tall mountains, from which flow many rivers. From it are brought the red corundum [ruby] and other precious stones. From the eastern sea there proceed many tongues whose sizes and properties have not been charted [*ršīmān*] in books. (13) From the northern sea that well-known tongue called the Galatian proceeds. In it are the nineteen islands of Britain, the best-known of which is Thule, the freezing island.

l. 2. "the Silver Moutain": This is, so to speak, the Aristotelian name, found at *Mete*. 350b 14 (τὸ ᾿Αργυροῦν καλούμενον ὄρος), while the other, the "Mountain(s) of the Moon" (Σελήνης ὄρος), mentioned in *Cand.*, *Rad.* and *Asc.* but omitted in *Eth.* and *But.*, is the Ptolemaic name (*Geog*. IV.8.3, 6) and is the one usually encountered in Islamic geographical literature (also in Jacob of Edessa, *Hex.* 113b 36f.).

l. 3. ܝܡܐ ܣܘܡܩܐ ("the Red Sea"): In Jacob of Edessa (*Hex.* 101b 2-102b 4), as well as in Bar Kepha (*Hex.*, Paris 311, 46r a5-6, b6-33; tr. 571, 573) and Bar Shakko (*Thes*. P173v 11-174r 11, L71r a15-v a6), "the Erythrian, or the Red, Sea" is what we know as the Indian Ocean, from which two gulf called by them the "Arabian" and "Elamitic or Persian" (i.e. respectively, our Red Sea and Arabian/Persian Gulf") branch off. With BH, too, this is still the case in *Cand*., where the "Sea of Reed" (corresponding there to our Red Sea) and the "Sea of Fārs" are said to branch off from the "Red Sea".[25] In his later works, *Rad*., *Asc*. and *But*., however, the "Red Sea" is placed on the same level as the sea of the "Berbers", "Elamites" and "Indians" and corresponds to our Red Sea, usually called "Baḥr al-Qulzum" by the Arabs (but "al-Ḫalīǧ al-Aḥmar", Ṭūsī, *Taḏkira* 249.1).[26]

l. 5. ܐܪܥܐ ܕܫܒܐ ܘܕܣܐܒܐ ("the land of Sheba and of Saba"): The combination of "Sheba" and "Saba", which is also found at the corresponding place in *Cand*. 156.1f. and *Eth*. 453.6 (ܐܬܪ̈ܘܬܐ ܕܫܒܐ ܘܕܣܐܒܐ) and in the list of localities in the second clime at *Cand*. 97.1 (ܫܒܐ ܘܣܒܐܐ; corr. Bīrūnī, *Tafhīm* 143.15 "Sabaᵓ"), is no doubt to be explained with reference to Psalm 72.10 and there is no indication that BH distinguished two different

[25] We find the same system, placing the "Sea of Reed" and the "Sea of Elam" on the same level as branches of the "Oceanus" (no mention of the "Red Sea") at BH *Horr.* in Gen. 1.9 (ed. Sprengling-Graham 10.22).

[26] Our Red Sea is also called by that name in the *Causa causarum* (*yammā summāqā*, F132r b 23, tr. 339). - In the *De mundo*, the ἐρυθρὰ θάλασσα of the Greek (393b 4) is rendered by *ᶜubbā haw d-metqrē d-sūp* in the Syriac version (140.8-9; the Arabic versions all have the "Sea of al-Qulzum", Brafman 88.13, 125.14, 143.3f.).

places. BH will have located "Sheba and Saba" in Yemen or more generally in the southern part of the Arabian Peninsula,[27] even though in the maps accompanying *Cand.* "Sheba and Saba" is found in the northeastern part of the Arabian peninsula, while the southernmost part of the peninsula is taken up by "Inner Yemen" (*taymnā gawwāyā*). This is to be explained by the fact that Bīrūnī (loc. cit.) places "Sabaʾ" in the second clime but "the parts of Yemen south of Ṣanʿa", "such as Dhofar, Ḥaḍramaut and Aden", in the first clime (*Tafhīm* 143.10f.).

l. 5. ܝܡܐ ܕܥܝܠܡ ("the Sea of Elam"): i.e. the Arabian/Persian Gulf. - This is called the "Sea of Fārs" in the text of *Cand.* 156.4, and the "Sea of the Elamites, i.e. of the Persians (*pārsāyē*)"[28] in the map accompanying *Cand.* (160.12, 2nd clime); "Sea of the Elamites" in *Rad.* [Istanbul] 36.10, *Asc.* 135.21 (also 138.15) and *Eth.* [Bedjan] 453.20.

l. 6 ܗܢܕܘ ܓܘܝܬܐ ("Inner India"): As will be clear from the context, as well as the map accompanying *Cand.*, BH's "Inner India" is the eastern parts of India (almost Indochina), rather than Ethiopia as is sometimes the case in Syriac.[29]

l. 7. "land of Gog and Magog": The Biblical "Gog and Magog" (Ezech. 38.2, also "Magog, son of Japheth" Gen. 10.2), who appear as "Yāǧūǧ and Māǧūǧ" in the Koran (XVIII.93-98, XXI.96) and play a significant role in Jewish, Christian and Muslim eschatology, were generally identified with the primitive peoples in the north, especially with the Turks and, after their appearance on the scene, Mongols.[30] At *Cand.* 99.2f. BH follows Bīrūnī in locating the "land of the Huns and those enclosed [*ḥbīšāyē*] Gog" at the eastern end of the fifth clime (< *Tafhīm* 144.14f. "land of the eastern Turks and the walled-in Gog [*yāǧūǧ al-musawwarīn*]).[31]

---

[27] See *Horr.* in Ps. 72.10 (ed. de Lagarde [1879] 182.16, tr. Siegel [1928] 144), where "Saba" is glossed as "the South/Yemen" (Taymnā). At BH *Chron.* [Bedjan] 155.17, "the South/Yemen and the land of Sheba and of Saba" (ܬܝܡܢܐ ܘܐܪܥܐ ܕܫܒܐ ܘܕܣܒܐ) are mentioned among the lands allotted by Caliph al-Mutawakkil to his son Muḥammad al-Muntaṣir in 235 A.H., where Ṭabarī, for example, has "Yemen, ʿAkk, Ḥadramaut, al-Yamāma, al-Bahrain ..." (Ṭabarī, III.1395, see partial tr. Kraemer [1989] 96 with the parallel passages cited there in 95f. n. 326).

[28] Cf. Jacob of Edessa, *Hex.* 101b 29f. ܝܡܐ ܗܘ ܕܥܝܠܡܝܐ ܐܘܟܝܬ ܕܦܪܣܝܐ; similarly Bar Kepha, *Hex.* P173r a4-7, Q46r b27f. (tr. 573); Bar Shakko, *Thes.* P174r 6f., L71r b20-23.

[29] See PS 1026f. s.vv. ܗܢܕܘ et ܗܢܕܘܝܐ; Michael, *Chron.* tr. I.258 with n.2. - At BH *Chron.* 8.3f., 5, India is divided into four, the "northern" included among the inheritance of Shem and the "inner, outer and southern" included in those of Ham (< Michael, *Chron.* 8a 34, 9b 3, tr. I.18, "outer" not in Michael).

[30] On the location of Gog and Magog among Koranic commentators and Islamic geographers and historians, see EI^2 11.232b [van Donzel-Ott]; on the association with the Mongols, Boyle (1980); cf. Takahashi (2003a) n. 86.

[31] For further references to Gog and Magog in BH, see *Chron.* [Bedjan] 30.9, where Holophernes is said to be "a Magogian, i.e. a Turk" (< Michael. *Chron.* 64a

l. 8. ܐܪ̈ܥܬܐ ܚܪ̈ܒܬܐ ܘܛܘ̈ܪܐ ܠܐ ܡܬܗܠܟ̈ܢܝܬܐ ("desolate lands and impassible mountains"): The phrase occurs in much the same form in *Cand*. 157.5 and in an abbreviated form in *Rad*. 37.2-3 and *Asc*. 134.8 (cf. also 133.23f.). It goes back to Bīrūnī, *Tafhīm* 121.11-12: ارضون وجبال مجهولة خربة غير مسلوكة ("unknown, desolate and impassible lands and mountains"; in the East, between the Ocean and the land of the Turks)

l. 9. ܝܡܐ ܕܒܪܛܢܝܐ. ܘܒܗ ܬܘܠܐ ܓܙܪܬܐ ("the Sea of Britain, in which is the island of Thule"): In *Cand*. the sea entering the habitable world from the north is called the "Sea of Warang". In *Rad*. and *Asc*. this is replaced by the "Galatian Sea/Gulf" of *De mundo* (syr. 140.15 < gr. 393b 9), in which are the Britannic islands of Albion and Ierne according to *De mundo*, "the 19 islands of Britain" according to *Rad*. and "the 19 islands of Britain, the best-known of which is Thule" according to *Asc*. The identification of the Varangian (Baltic) Sea with the Galatian Sea (Bay of Biscay/North Sea) will have been caused by the general uncertainties of Islamic geographers over these northern regions, and more particularly by the rarity of references to the Varangian Sea[32] and by the fact that Thule is sometimes said to be located near (or even "in")[33] Britain but is sometimes associated with the northern sea corresponding to the Baltic.[34] A further contributing factor to the eastward displacement of Britain may be the coupling of Britain with "Iberia" at *De mundo* syr. 140.20 (< gr. 393b 17), where the latter will have been understood to mean "Georgia" rather than as "Spain" by BH. - The alteration in *But*. whereby this northern sea itself is now called the "Sea of Britain" rather than "Galatian" may have been prompted in part by an unease on BH's part with the fact that a part of the Mediterranean is also called the "Galatian" (*De mundo* syr. 139.29 < gr. 393a 27; cf. V.i.1.3

30f., tr. I.103); *Chron*. 217.15, where a group of "Huns" called "Guzzāyē" (Ghuzz, Oghuz), who "went forth with the emirs of the Seljuks", are said to be children of Magog b. Japheth b. Noah (< Michael, *Chron*. 566a 31, tr. III.149, where simply "Turks" instead of "Huns" etc.); *Horr*. in Ezech. 38.2 (ed. Gugenheimer, 27.10-13; tr. Dean 58), where "Gog and Magog" is glossed as "the Scythians, the sons of Japheth" (this identification with the Scythians goes back at least to Josephus, *AJ* I.6.1 [123], cf. Czeglédy [1957] 233-237); and *But*. Polit. I.iii.6 ("Gōgāyē" in the northeast; mentioned along with "Arabs" in the southwest). The continuator of BH *Chron*., though probably not BH himself, refers several times to the Ilkhanids as the "House of Magog" (ed. Bedjan, 555.7, 560.20, 579.23, 585.2, 586.9; the passage at 555.7, though relating to events just before BH's death, probably stems from the pen of the continuator).

32 Abū al-Fidāʾ, *Taqwīm* 35.9-12 tells us that Bīrūnī and Ṭūsī's *Tadkira* are the only authorities to mention this sea, although this is not quite true as it is in fact mentioned by a number of other authors predating Abū al-Fidāʾ. See Takahashi (2003a), para. 35 with n. 65.

33 So Battānī, *Zīğ* 25.16 (جزيرة ثولي التى فى برطانية).

34 See, e.g., *Ḥudūd al-ᶜālam*, tr. 59, §4.25, with comm. p. 181 (on §3.8); Miquel [1967-88] III.240.

above). It is unclear whether (and how) the name "Sea of Britain" can be traced back to the Ptolemaic "Βρεταννικός ὠκεανός" (*Geog*. II.3.3, 8.2, 9.1, VII.3.2, 5.2).

l. 9-10. ܐܝܒܪ̈ܝܐ ܘܐܠܢ̈ܝܐ ܘܣܩܘܬ̈ܝܐ ܘܒܘܠܓܪ̈ܝܐ ("Iberians, Alans, Scythians and Bulgars"): All four peoples are mentioned in *Cand*. 134.5-9, in the same order as here but separated out from each other. In *Rad*. [Istanbul] 37.3 "Bulgars, Scythians and Alans" are mentioned together, in that order. Only the "Scythians" appear at *Asc*. 134.6.[35] - Of the four, only the Scythians (assuming that BH's Scythians = Slavs/Ṣaqāliba) and Bulgars are mentioned in Bīrūnī, *Tafhīm* (211.9, 10). One place where all four are mentioned together (assuming that BH's Iberians and Scythians are Avars and Slavs)[36] is a list of the peoples living beyond the seventh clime in Ibn Rusta 98.17f.: "Gog [*yāǧūǧ*], Tughuzghuz, Turks, Alans [*al-lān*], Avars [*al-abar*], Bulgars [*burǧān*], Slavs [*ṣaqāliba*]".[37]

l. 10. ܐܝܛܠܝܐ ("Italy"): Italy, which does not face the outer Ocean and which had not been mentioned in this context in BH's earlier works, is out of place here, but on a schematic map of the seas, where Europe is usually drawn as a thin wedge between the Mediterranean and the outer Ocean, Italy or its equivalent in BH's Vorlage ("Rūm"?)[38] will not have been very far from the Ocean.

V.i.3.12: concluding remarks

But.: These things are presented plainly in books on geography.

Cf. the similar references to detailed discussions in works of other disciplines at Min. I.ii.6-8 ("medicine etc.") above and Mete. II.i.9 below ("geometry and optics"). - Cf. further Ṭūsī, *Muʿīnīya* 60.22: ودر كتب مسالك وممالك بعضى از آن موصوف باشد; id. *Tadkira* 249.6: ... يعرّفها اهل العلم بالمسالك والسيّاح وغيرهم

[35] "Iberians, Scythians and Alans" are mentioned together again as warlike peoples similar to the "Turks" in a passage describing the characteristics of the different peoples at *But*. Econ. II.3.3 (Zonta [1992] p. 83).

[36] On the possible association in BH's mind of Avars with Iberians, see Takahashi (2003a) 43-45.

[37] References to Avars are relatively hard to come by in Islamic geographical works; where they are mentioned they tend to be mentioned in the same breath as Slavs and Bulgars (*burǧān* or *bulġar*): Ibn Khurdādbih 92.4 (corr. Ibn al-Faqīh 83.18), 119.6 (also with Alans, but not Bulgars); Masʿūdī, *Tanbīh* 32.17, 182.9, 191.3; cf. Minorsky (1937) 419, 429.

[38] Cf. *Ḥudūd al-ʿālam*, tr. 52, §2.2, where the Western Ocean is said to pass by the extreme limit of "Rūm".

Min. V. Section ii.: *On the Salinity of Seawater*

The salinity of the sea was discussed by Aristotle at *Mete*. II.ii-iii. This dicussion was taken up by Ibn Sīnā in the first half of the chapter on the sea (*faṣl* 2) in the *fann* on "actions and passions" in the *Šifāʾ* (ed. Qassem, Cairo 1969, p. 205-208).

The section of *But*. here corresponds in the first place to that passage of the *Šifāʾ* and the order of the discussion too follows the order in the *Šifāʾ*. At the same time, BH has, as usual, supplemented the materials taken from the *Šifāʾ* with materials from elsewhere. Some of this supplementary material can be located in the text of Nic. syr. as we have it in the Cambridge manuscript. Others cannot, but, as has been stated, there is clearly a long lacuna in the fragmentary Cambridge manuscript at this point. The likelihood therefore is that a part of the material found in this section of *But*. is based on those lost passages of Nic. syr. and our suspicion in this regard can usually be corroborated by the appearance of the same material in other works related to Nic. syr., namely Arist. *Mete.* itself, the Greek Olympiodorus (as the source of much of the scholial material in Nic. syr.), Olymp. arab. and BH's own *Cand*.[39]

	Šifāʾ	Others	
V.ii.1.1-4	205.10-12		Salinity due to earthiness
V.ii.1.4-5	205.14-206.1		Proof (1) ash
V.ii.1.6-9	-	Nic.?	Proof (2): sailors:
V.ii.1.9-12	206.15-17	Nic. 23.6-11	Proof (3): wax-ball
V.ii.2.1-3	207.1-4	(Nic.?)	Evaporation of freshwater
V.ii.2.3-4	-	?	Saying of Aesop (?)
V.ii.2.4-7	207.2-3	Nic.?	Why the sea does not increase
V.ii.2.8-9	-	Nic.?	Why the sea does not increase
V.ii.3.1-2	207.11-12	Nic. 21.6-7	Density of seawater (eggs)
V.ii.3.3-5		Nic. 23.13-18	Density of seawater (ships)
V.ii.3.5-8	207.12-13	Nic. 23.22-26	Dead Sea
V.ii.3.8-10		Nic. 24.1-4	Spring in Chaonia
V.ii.4.1-4	207.16-17	Nic. 20.6-21	Empedocles' theory
V.ii.4.4-6	207.17-208.1		Analogy with sweat
V.ii.4.6-8	208.2-4		Purpose of salinity

[39] In *Cand.* the discussion of the salinity of the sea is found at 151.6-152.7.

V.ii.1.1-4: Salinity due to earthiness

But.: (1) Water is transformed by admixture [*ḥulṭānā*] of something else and not by itself. (2) Through admixture of air, it becomes not bitter but finer and sweeter. (3) The remaining alternative therefore is that admixture of a burning and bitter earthy substance [lit. earthiness, *arʿānāyūṯā*], i.e. smoky exhalation, is the cause of the salinity of the sea.

Šifāʾ, Act. et pass. 205.10-12:

والماء لا يتغير التغيرات التى بعد الكيفيات الأول بنفسه، إنما يتغير لمخالطة شىء آخر. والهواء إذا خالطه جعله أرق وأعذب، ولم يجعله ملحا. إنما يصير ملحا بسبب الأرضية المحترقة المرة إذا خالطته.

(1) Water does not undergo any alterations after its first [acquisition of its] qualities by itself; it can only be altered by the admixture [*muḫālaṭa*] of something else. (2) Air, when it mingles with (water), makes it finer and sweeter, and does not make it salty. (3) (Water) can only turn into salt because of [*bi-sababi*] a burning and bitter earthiness [*arḍīya*], when it mingles with it.

l. 2-3. . . . ܫܪܝܟܐ ܗܟܝܠ ܕ ("the remaining alternative therefore is"): The expression *šarrīḵā d-* occurs again at Min. V.iv.3.10 below.

V.ii.1.4-5: Proof (1): salt obtained from ash

The passage here may be considered to be based on the *Šifāʾ*. The example of the Umbrians in the passage of Nic. syr. quoted below is in fact used by IS immediately following the passage of the *Šifāʾ* quoted below at Act. et pass. 206.2-3.

But.: (1) For this reason, when ash is boiled in water, (2) and this water is (then) strained, (3) boiled again (4) and placed in the sun, (5) it hardens and turns into salt.

Šifāʾ Act. et pass. 205.14-206.1:

وأنت فيمكنك أن تتخذ الملح من رماد كل محترق، ومن كل حجر يفيده التكليس حدة ومرارة، إذا طبخته فى الماء، وصفيته، ولم تزل تطبخ ذلك الماء أو تدعه فى الشمس، فإنه ينعقد ملحا.

(1) You can obtain salt from the ash of anything burned, from any stone which acquires acidity [lit. sharpness] and bitterness through calcination, when you boil it in water, (2) strain it, (3) and continue to boil that water (4) or leave it in the sun; (5) for it (then) solidifies into salt.

Cf. also Nic. syr. 24.4-6[40]:

ܬܘܒ ܕܝܢ ܒܐܘܡܒܪܝܐ ܕܡܬܩܪܝܐ ܐܬܪܐ ܐܝܬܝܗܘܢ ܢܚܝܢ / ܩܢܝ̈ܐ ܘܣܘܟܢܘܣ . ܥܡܘܪ̈ܘܗܝ ܕܝܢ ܕܐܬܪܐ ܗܘ . ܡܘܩܕܝܢ ܠܗܠܝܢ / ܘܪܡܝܢ ܩܛܡܐ ܕܝܠܗܘܢ ܒܡܝ̈ܐ ܘܡܒܫܠܝܢ ܠܗܘܢ . ܘܥܒܕܝܢ ܠܗܘܢ ܡܠܚܐ .

(1) Furthermore, in a place called Umbria, there grow reeds and rushes [σχοῖνοι]. (2) The inhabitants of this place burn them and throw their ashes in water and boil it, and make it salt.

---

40 Cf. Arist. 359a 35-359b 4; Olymp. 164.32-36; Olymp. arab. 116.1-4.

V.ii.1.6-9: Proof (2): sailors

The close agreement of the passage here with Olymp. suggests that the source used is a lost passage of Nic. syr. - The same passage was used by BH in *Cand.*, where, however, this is used as an illustration of the explanation of the salinity as being due to "evaporation" (cf. V.ii..2. below).

But.: (1) Sailors, too, when they run short of drinking water [lit. when drinking water becomes deficient for them] at sea, (2) pour seawater into a caldron, and boil it. (3) Above the caldron they place sponges, (4) and collect [lit. receive] the vapours with them. (5) They then squeeze them and obtain [lit. find] sweet water. (6) The water that remains is found to be denser and saltier.

Olymp. *in Mete*. 158.38-159.4 [part of 4th proof (ἐπιχείρημα) that salinity of seawater is due to smoky exhalation[41]]: ... ἐπειδὴ καὶ αὐτοὶ οἱ ναυτικοί, ἐπὰν ὕδατος λεῖψις γένηται αὐτοῖς ἐν τῇ θαλάττῃ, ἑψῶντες τὸ θαλάττιον ὕδωρ ἐξαρτῶσι σπόγγους μεγάλους τοῦ στόματος τοῦ χαλκείου, ἵνα δέχωνται τὸ ἀτμιζόμενον· εἶτα ἐκεῖνο τὸ ἐξατμισθὲν ἐκπυρηνίσαντες ἐκ τοῦ σπόγγου εὑρίσκουσι γλυκὺ μὴ μετέχον τῆς καπνώδους ἀναθυμιάσεως.
... (1) since sailors, too, when a shortage of water occurs to them at sea, (2) boil seawater (3) and hang large sponges to the mouth of the cauldron, (4) so that they [sponges] receive what is vaporised; (5) then, squeezing what has been vaporised out of the sponge, they find sweet [water], not having in it anything of the smoky exhalation.

Cand. [Bakoš] 151.8-11, [Çiçek] 109.6-12:

ܡܛܠܗܕܐ ܡܠ̈ܚܐ ܟܕ ܚܣܪ ܠܗܘܢ ܡܝ̈ܐ ܡܫܬܝܢܝ̈ܐ ܒܝܡܐ. ܡܒܫܠܝܢ ܠܡܝ̈ܐ ܕܝܡܐ
ܘܒܐܣܦܘܓܐ ܗܝܕܝܢ ܡܩܒܠܝܢ ܠܠܗ̈ܓܐ. ܘܡܝ̈ܐ ܚܠܝ̈ܐ ܡܥܨܪܝܢ ܡܢܗܘܢ. ܘܡܝ̈ܐ
ܗܢܘܢ ܕܫܪܝܢ ܝܬܝܪ ܥܒܝ̈ܐ ܘܡܠܝ̈ܚܐ ܡܫܬܟܚܝܢ.

For this reason, when sailors run short of drinking water [lit. when drinking water becomes deficient[42] for them] at sea, (2) they boil seawater, (3-4) collect [lit. receive] the vapours with a sponge (5) and squeeze sweet water out of them.[43] (6) The water that remains is found to be denser and saltier.

---

[41] Cf. also Alex. *in Mete.* 86.20-24, where there is however no mention of "sailors" (simply τινες), or "sponges" (instead: "lids", ἐν τοῖς ἐπικειμένοις αὐτῶν πώμασιν ἀθροίζοντες).

[42] The verb "ḥāsar" in the singular in both editions, disagreeing with the subject "mayyā" (Arabism?); cf. *But.* "ḥāsrīn".

[43] As the text stands, "men-hōn" must refer back to "lahgē (vapours)". It is tempting to read ܒܐܣܦܘ̈ܓܐ in the plural as in *But*.

V.ii.1.9-12: Proof (3): wax-ball

But. [underline: agreement with Nic. syr.; italics: with *Šifā'*]:

(1) Furthermore, if someone makes a hollow *ball* out of wax [*qērūṯā*] (2) and throws it in the sea, (3) the water which trickles through the pores in the wax [*š'ūṯā*] and enters *the inside* [*l-gawwāh*] *of the ball* is found to be sweet, (4) because *sweet* (water) is fine, and passes through the pores in the wax as if by filtration, (5) while the smoky exhalation, which is dense, remains outside.

Šifā' Act. et pass. 206.15-17:

والدليل على أن ماء البحر يتملح بمخالطة الأرضية، وليس ذلك طبيعيًا له، أنه يقطر ويرشح فيكون عذبا، وقد تتخذ كرة من شمع فترسل فيه، فيرشح العذب إلى باطنه رشحًا.

The proof that seawater is made salty by the admixture of earthiness and that (saltiness) is not natural to it, is the fact that when it it trickles and percolates it becomes sweet. (1) When you take a ball [made] out of wax (2) and let it fall in (seawater), (3-4) sweet [water] will percolate [*yaršaḥu*] into its inside [*ilā bāṭinihi*] as percolation [*rašḥ*].

Nic. syr. 23.6-11[44]:

ܐܢܗܘ ܕܬܥܒܕ ܡܐܢܐ/7 ܡܢ ܫܥܘܬܐ ܬܣܟܘܪ ܦܘܡܗ: ܘܬܫܕܝܘܗܝ ܒܝܡܐ ܗܢܘܢ ܕܡܬܢܛܦܝܢ ܒܩܪܘܬܐ:
ܡܝܐ ܕܐ̈ܝܠܝܢ/8 ܡܢ ܩܪܘܬܐ ܘܥܐܠܝܢ ܠܚܠܝܠܘܬܗ ܕܡܐܢܐ: ܚܠܝܐ ܡܫܬܟܚܝܢ.
ܗܕܐ ܕܝܢ ܗܘܝܐ܀/9 ܡܛܠ ܕܡܬܡܫܚܠܝܢ ܡܝܐ ܐܝܟ ܕܒܡܫܚܠܐ. ܘܡܛܠ ܕܩܠܝܠܝܢ
ܒܡܢܘ̈ܬܗܘܢ/10 ܥܒܪܝܢ ܒܢܩܘ̈ܒܘܗܝ ܕܫܥܘܬܐ. ܥܠܬܐ ܕܝܢ ܬܢܢܝܬܐ: ܐܟܡܢ ܕܥܒܝܐ
ܒܡܢܘ̈ܬܗ܀/11 ܠܒܪ ܡܩܘܝܐ.

(1) If you make a vessel out of wax [*š'ūṯā*], close its mouth (2) and throw it in the sea, (3) the water which trickles through the wax [*qērūṯā*] and enters the hollow [*ḥlīlūṯā*] of the vessel is found to be sweet. (4) This happens because water is strained as if by filtration, and because it is fine in its particles, it passes through the pores in the wax, (5) while smoky exhalation, being [*akman d-*] dense in its particles remains outside.

l. 9 ܐܣܦܝܪܐ ("ball"): The alteration of Aristotle's "jar, vessel" (ἀγγεῖον) to "ball, sphere" (*espērā* < *kura*) is due to Ibn Sīnā (ἀγγεῖον Olymp.; *mānā* Nic. syr.; *inā'* Olymp. arab. and Ibn al-Biṭrīq 62.6; *klī* Ibn Tibbon II.173).

l. 11-12 ܐܝܟ ܕܒܫܚܠܐ ("as if by filtration"): vocalised *šeḥlā* FL1; *šḥellā* V; no voc. MP; cf. also ܡܫܚܠܐ Nic. syr.; ترويق، تصفية F marg. - I follow the manuscripts in reading *šeḥlā* ("filtration", i.e. process), as opposed to *šaḥlā/šāḥlā* ("strainer, filter", i.e. instrument; see PS 4114f. s.vv.), although it should be noted that Arist. himself had talked of a "filter" (*Mete*. 359a 4 ὥσπερ γὰρ δι' ἠθμοῦ).

V.ii.2.1-3: Evaporation of freshwater

The passage is based in the first instance on the *Šifā'*. There may, however, be some echoes here of a lost passage of Nic. syr.

---

[44] Cf. Arist. 358b 35-359a 5; Olymp. 158.27-32; Olymp. arab. 114.17f.

But. [underline: agreement with *Šifā'*]:
(1) In (certain) parts [lit. places] of the sea sweet water is also enclosed [*ḥbīšīn*] (2) and rivers flow into (the sea), (3) but [*ellā*] because sweet (water) is fine (4) it is vaporised and turns into clouds, (5) whereas salty (water) is dense (6) and is left behind like dregs [*teṭrā*].

Šifā' Act. et pass. 207.1-4:

والبحر أيضًا قد تكون فى مواضع منه مياه عذبة، وقد تمده مياه عذبة، إلا أنها ألطف من ماء البحر المجتمع فيه قديمًا، فيسبق إليها التحلل. فإن اللطيف يسبق إليه، وخصوصًا فى حال الانتشار. فإن الانتشار، يعين على ذلك، كما لو بسط الماء على البر. وإذا كان كذلك صار العذب يتحلل بخاراً ويصير سحبًا وغير ذلك، والمالح الكثيف يبقى.

(1) In (certain) places of the sea there is sweet water, (2) and sweet water augments it, (3) except [*illā*] that (sweet water) is finer than the seawater gathered in (the sea) from the beginning [*qadīman*], (4) so that evaporation reaches it it first; for a fine body reaches (evaporation) first, especially in a state of diffusion; for diffusion favours (evaporation), as when water is spread out on the ground. This being the case, (4) sweet (water) evaporates as vapour and turns into clouds etc., (5-6) while the thick salty (water) is left behind [*pāyšīn*].

The content of the passage goes back ultimately to Arist. *Mete*. 355a 32-b 20. The part corresponding to that passage of Arist. is lost in the lacuna in the Cambridge manuscript of Nic. syr. The additions/alterations made by BH here ("is enclosed", "rivers flow into it" and "like dregs") may therefore be taken from a lost passage of Nic. syr. IS does not explicitly mention "**rivers**" in the sentence quoted above, although this is implied in clause (2)."Rivers" flowing into the sea are mentioned explicitly at Arist. 355b 16 and will no doubt have been mentioned by Nic. BH's "**dregs**" may be related to Aristotle's "ὑπόστασις καὶ περίττωμα" (355b 8).

A sentence which may also be related to the relevant passage of Nic. syr. is the following:

Olymp. arab. 114.18-21[45]:

والثالثة أنه قد تختلط به مياه عذبة كثيرة، لا عن الأنهار التى تصب إليه فقط، لكن من الأمطار أيضًا الذى تنحدر إليه فى الشتاء ولا تعذب مياهه.

Third proof [that the mixture of smoky vapour is the cause sea's salinity] is that much sweet water is mixed [*taḫtaliṭu*] in it, not only from the rivers which flow into it, but also from the rains which descend on it in winter, and yet its waters are not sweet. ...

---

[45] The discussions of the sea and its salinity in both Olymp. gr. and in Olymp. arab. involve considerable rearrangement of the Aristotelian material (see Lettinck [1999] 131-133, 138-141), with the result that the material derived from one passage of Arist. sometimes reappears at several points. Rivers flowing into the sea are also mentioned in these discussions at Olymp. arab. 106.24-107.3, 107.9f.

The verb *taḫtaliṭu* in this sentence of Olymp. arab. leads one to suspect that the word which was read as "**is enclosed**" (ܡܬܚܒܫ) by BH and which makes for a slightly awkward sense in our passage of *But.* should perhaps have been "is mixed" (ܡܬܚܠܛ).

Another sentence most probably based on the same passage of Nic. syr. is found at *Cand.* 151.6-8[46]:

ܡܠܝܚܐ ܕܝܢ ܗܘܘ ܡܝܐ ܕܝܡܐ ܡܛܠ ܕܩܛܝܢܘܬܐ ܕܒܗܘܢ ܐܬܥܛܪܬ ܡܢ
ܫܚܝܢܘܬܐ ܕܫܡܫܐ. ܥܒܝܘܬܗ ܕܝܢ ܘܫܘܫܠܬܗ ܘܛܝܛܪܗ ܩܘܘ.

Seawater has become salty (3-4) because the fine part in it has been vaporised by the heat of the sun, (5-6) while the dense part [lit. its density] and its sediments and dregs [*teṭrā*] have remained behind [*qawwī(w)*].

V.ii.2.3-4: Saying of Aesop (?)

But.: ܐܝܟ ܕܐܡܪ ܐܝܣܘܦܘܣ ܒܡܬܠܐ. ܕܝܡܐ ܗܘ ܫܦܝܐ ܠܐܚܪܢܐ ܝܗܒ. ܘܥܟܝܪܐ ܠܗ ܢܛܪ

As Aesop said in a fable: "the sea gives what is clear to another and keeps what is turbid for itself."

or, possibly: ... "a clear sea gives to another, while a turbid (sea) preserves itself."

There must be some error here, since what we have here is quite different from the story about Charybdis, which is attributed to Aesop at Arist. *Mete*. 356b 9-15 (cf. Alex. Aphr. 78.18-22, Olymp. 118.30-34) and is duly preserved in Nic. syr. in a passage designated as a scholion.[47] One is rather clueless as to the origin of the saying here, unless it can somehow be related to Heraclitus, Frag. 61 Diels-Kranz (I.164.6-8)[48]:

θάλασσα ὕδωρ καθαρώτατον καὶ μιαρώτατον, ἰχθύσι μὲν πότιμον καὶ σωτήριον, ἀνθρώποις δὲ ἄποτον καὶ ὀλέθριον.

---

46 The account of the salinity of the sea at *Cand.* 151.6-152.7 is based largely, if not exclusively, on Nic. syr., but BH attributes the salinity there to evaporation and makes no mention of the earthy exhalation (cf. Bakoš's footnote, 151 n.4).

47 Nic. syr. 19.24-28 [scholion]: ܢܘܗܪܐ. ܐܡܪܝܢ /25 ܦܘܐܝܛܐ ܕܡܬܡܠܝܐ ܟܠܗ ܐܪܥܐ ܡܝܐ ܡܠܝܐ ܗܘܬ: ܘܟܕ ܫܬܬ ܚܕܐ ܙܒܢ /26 ܐܚܙܝܬ ܛܘܪܐ ܟܐܪܘܒܕܝܣ ܚܙܝܬ ܕܬܪܬܝܢ ܓܙܪܬܐ. ܘܟܕ ܕܬܠܬܐ/27 ܙܒܢܝܢ ܬܫܬܐ ܐܪܥܐ ܬܗܘܐ ܝܒܝܫܬܐ ܠܓܡܪ. ܐܝܟ ܕܬܠܬܐ ܙܒܢܝܢ. ܠܓܡܪ ܐܪܥܐ ܡܬܝܒܫܐ. ܡܬܠܬܐ ܕܐܝܣܘܦܘܣ ܐܝܬܝܗ ÷ 28/. ܐܪܥܐ [26 ܠܟܐܪܘܒܕܝܣ: ܠܟܐܪܘܒܕܝܣ cod.] ("Comment: Poets say that the whole of the earth was formerly full of water, and when Charybdis drank once from it she made the mountains appear. When she drank for the second time, she revealed the islands. If she drinks a third time, the earth will be completely dried up. - This is the story [*mellṯā*] of Aesop.")

48 In making a link with Heraclitus here, it may be noted that Empedocles' theory that "the sea is the sweat of the earth" (Arist. 357a 24ff.) is for some reason found wrongly attributed to Heraclitus at Olymp. 151.4, 30.

This could have been summarised as: "The sea is clear for some, but dirty for others" (syriacè: "ܝܡܐ ܗ̇ܘ ܫܦܝܐ ܠܐܢܫܝܢ. ܘܕܠܝܚܐ ܠܐܢܫܝܢ" vel sim.), which could then have been further misunderstood and corrupted to produce what we have here in *But*.

l. 3. ܐܝܣܘܦܘܣ ("Aesop"): The name is written ܐܣܘܦܘܣ at Nic. syr. 19.28, but ܐܝܣܘܦܘܣ, as here, at Nic. syr. 36.28 in a scholion referring back to Nic. syr. 19.24ff.

V.ii.2.4-7: Why the sea does not increase

Although the acceleration of evaporation due to diffusion is mentioned at *Šifā*ʾ Act. et pass. 207.2-3 (in the passage quoted under V.ii.2.1-3 above), the rest of this sentence of *But*. is not represented in IS. The sentence must be based on a lost passage of Nic. syr. corresponding to Arist. *Mete*. 355b 20-32. The same passage of Nic. syr. is the source of Olymp. arab. 109.19-20. Although some features of the sentence in *But.* which are not found in Arist. (verb "increase", "the sun") are found in the Greek Olympiodorus, the correspondences between *But.* and Olymp. gr. are not so remarkable for the rest, so that these similarities can be attributed to the use of a common commentary tradition and the passage of Nic. syr. in question may be a part of genuine Nicolaus rather than a scholion based on Olympiodorus.

But. [underline: agreement with Arist.; italics: with Olymp. arab.]:
(1) The sea *does not grow larger* [lit. increase, *mettawsep̄*] (2) even though large rivers flow into it, (3) because *on account of its large area* [lit. width] *the sun meets many parts of* its water (4) and causes *much vapour to rise from it*, (5) just as we see in an experiment that the same quantity of water disappears more quickly when it is spread out than when is concentrated.

Arist. *Mete*. 355b 20-32: τὸ γὰρ ζητεῖν τὴν ἀρχαίαν ἀπορίαν, διὰ τί τοσοῦτον πλῆθος ὕδατος οὐδαμοῦ φαίνεται (2) (καθ ʼ ἑκάστην γὰρ ἡμέραν ποταμῶν ῥεόντων ἀναρίθμων καὶ τὸ μέγεθος ἀπλέτων (1) οὐδὲν ἡ θάλαττα γίγνεται πλείων), τοῦτο οὐδὲν μὲν ἄτοπον ἀπορῆσαί τινας, οὐ μὴν ἐπιβλέψαντά γε χαλεπὸν ἰδεῖν. (5) τὸ γὰρ αὐτὸ πλῆθος ὕδατος εἰς πλάτος τε διαταθὲν καὶ ἀθρόον οὐκ ἐν ἴσῳ χρόνῳ ἀναξηραίνεται, ἀλλὰ διαφέρει τοσοῦτον ὥστε τὸ μὲν διαμεῖναι ἂν ὅλην τὴν ἡμέραν, τὸ δ' ὥσπερ εἴ τις ἐπὶ τράπεζαν μεγάλην περιτείνειεν ὕδατος κύαθον, ἅμα διανοουμένοις ἂν ἀφανισθείη πᾶν. ὃ δὴ καὶ περὶ τοὺς ποταμοὺς συμβαίνει· συνεχῶς γὰρ ῥεόντων ἀθρόων ἀεὶ (3-4) τὸ ἀφικνούμενον εἰς ἀχανῆ καὶ πλατὺν τόπον ἀναξηραίνεται ταχὺ καὶ ἀδήλως.

Olymp. *in Mete*. 141.11-20: νῦν αὐτὸ τοῦτο ζητεῖ, (2) τί δήποτε πολλοῦ καὶ μεγίστου ποταμοῦ ἐμβάλλοντος ἐν τῇ θαλάττῃ (1) ὅμως αὕτη κατ' οὐδένα τρόπον φαίνεται αὐξανομένη παρὰ τὸ πρότερον. καὶ ταύτην τὴν

ἀπορίαν λύει τοιούτῳ λόγῳ κεχρημένος· ἐπειδὴ γάρ, φησί, (3) τὸ τῆς θαλάττης ὕδωρ κατὰ τὸ τῆς γῆς πλάτος φαίνεται ἐκτεταμένον, (5) ὥσπερ ἄν τις κύαθον ὕδατος ἐπὶ τραπέζης ἐξαπλώσῃ, τούτου χάριν ἐξηπλωμένου τοῦ ὕδατος αὐτῆς ἐπὶ μέγιστον τῆς γῆς τόπον συμβαίνει τὸ πολὺ αὐτοῦ μέρος ἀπόλλυσθαι, (3) πῇ μὲν ἐξατμιζόμενον ὑπὸ τῆς τοῦ ἡλίου περιφορᾶς, πῇ δ᾽ ὑπὸ τῆς ὑποκειμένης γῆς ἀναπινόμενον. (4) ἐπειδὴ οὖν πολύ ἐστιν ἐξ αὐτῆς ἀπολλύμενον ὕδωρ, (1-2) εἰκότως οὐ φαίνεται αὐξανομένη, εἰ καὶ πάντες ἐσβάλλοιεν οἱ ποταμοί.

Olymp. arab. 109.19-20: وأما مياه البحر فلكثرة عرضه تلقى شعاع الشمس أجزاء كثيرة منه. ولهذه العلة لا تزيد، أعنى لكثرة ما ينبخر منها.

As for the waters of the sea, (3) because of the greatness of its width the sun's rays meet many parts of it. (1) For this reason, they [waters] do not increase [*tazīdu*], (4) I mean, because of the large quantity of what is vaporised from them.

l. 4. ܡܬܬܘܣܦܐ ("grow larger", lit. increase): cf. Arist. γίγνεται πλείων; Olymp. αὐξανομένη; Olymp. arab. *tazīdu*. - The agreement of *But*. and Olymp. arab. suggests that Nic. gr., like Olymp. gr., used the verb αὐξάνομαι.

l. 7. ܡܢ ("than"): On this use of ܡܢ (cf. Greek ἤ), see PS 48, s.v. no. 3; Duval, *Gram. syr.* 347f. §366e.

V.ii.2.8-9: Why the sea does not diminish

Why the sea does not diminish is explained by Arist. at *Mete*. 356b 21-30 (cf. Olymp. 143.35-39) in his refutation of Democritus' theory concerning the sea. The corresponding passage of Nic. syr. is preserved shortly after the resumption of the text after the lacuna. The parts corresponding to (1) and (2) of *But*. is missing in the preserved text of Nic. syr., but is found in a passage of Olymp. arab. immediately following that corresponding to V.ii.4-7 above, so that something resembling these parts may have been there in the portion of Nic. syr. lost in the lacuna. The alteration in the last part of the sentence of "rain" (implied in Nic. syr., explicit in Olymp. arab.) to "rivers" appears to be due to BH and is perhaps to be explained with reference to the mention of rivers earlier on in the same theory (V.ii.2.1).

But. [underline: agreement with Nic.; italics: with Olymp. arab.]: (1) *Neither does* the sea *grow smaller* [*bāṣar*], (2) *even though large quantities of vapour* [lit. much vapours] rise *from* it, (3) *because* the same amount as [*ak meddem*] that which rises from it in [the form of] vapour [*b-lahgā*] (4) enters it from rivers.

Nic. syr. 19.28-20.2:
ܬܘܒ ܕܝܢ ܠܝܬܗ ܒܒܘܨܪܐ/29 ܕܡܝ̈ܐ ܕܥܐܠܐ ܡܢ ܢܗܪ̈ܐ. ܟܕ ܡܬܠܗܓܢܐ ܡܢ ܫܡܫܐ.
ܡܬܒܕܪܝܢ ܢܫܒ̈ܬܐ/30 ܠܗܘܢ. ܫܡܫܐ ܓܝܪ ܠܐ ܟܠܗܘܢ ܦܪܥ ܠܗ: ܐܠܐ ܡܬܒܕܪܝܢ ܡܢ./
[20] ܘܗܘ ܗܘܐ ܕܢܦܩܘ ܠܗ ܠܡܬܟܢܫܘ ܟܕ ܦܪܥ ܠܗ: ܡܫܘܚܬ/2 ܡܬܒܕܪܝܢ ܠܗܘܢ: ܟܕ
ܡܬܪܣܝܢ ܡܢ ܗܘ ܫܡܫܐ.

Furthermore, (3) the whole quantity of water which rises from the sea as it is vaporised by the sun (4) falls down again towards (the sea). For the sun does not only approach us, but also recedes from us, (3) and that which it causes to rise in the form of vapour [*lahgānā'it*] when it approaches us, (4) that (same) sun causes to fall again towards us when it recedes from us.

Olymp. arab. 109.20-110.1:

وأما السبب فى أنها لا تنقص، وإن كانت الشمس تخطف منها بخاراً كثيراً، فلأن ما ينحل فى الصيف يعود إليها بالأمطار فى الشتاء.

(1) The reason why they [waters of the sea] do not decrease, (2) even though the sun takes away much vapour from them (3) is because what evaporates in summer (4) returns to them with rain in winter.

l. 8f. ܐܝܟ ܡܫܘܚܬ ܕ ("the same amount as"): cf. V.iv.2.1 below: ܐܝܟ ܟܡܝܘܬܐ ܕܡ̈ܝܐ ܕ ("the same quantity of water as").

V.ii.3.1-2: Density of seawater (eggs)

But. [underline: agreement with *Šifā'*: italics: with Nic.]:

(1) *Seawater*, because of its salinity and earthiness, is heavier and *denser* [*ʿbēn*] than other [types of] water (2a) and for this reason *an egg* [*bartā*] *will float* [*ṭāypā*] *on* it, (2b) whereas it will not float *on sweet* [water].

Šifā' Act. et pass. 207.11-12:

والبحر لملوحة مائيته، وكثرة أرضيته أثقل من المياه الأخرى وزنا. ولذلك فقل ما يرسب فيه البيض.

(1) The sea, because of the salinity of its wateriness and the large amount of its earthiness, is heavier than other waters in weight. (2) For this reason, an egg will rarely sink in it.

Nic. syr. 21.6-7[49]:

ܡܕܥܒܝܢ ܕܝܢ ܡܠܝܚܐ: ܡܢ ܗܕܐ ܝܕܝܥ ܕܐܢ ܬܣܝܡ ܒܪܬܐ ܐܘܟܝܬ ܒܝܥܬܐ / ܒܡ̈ܝܐ ܕܝܡܐ ܛܐܦܐ. ܐܢ ܕܝܢ ܒܡ̈ܝܐ ܚܠܝܐ ܬܣܝܡܝܗ̇ ܛܒܥܐ.

(1) That it [salt water] is dense [*ʿbēn*] (2a) is known from the fact that if you place an egg [*bartā*] - i.e. an egg [*bīʿtā*][50] - in seawater, it floats [*ṭāypā*]; (2b) but if you place it in sweet water it sinks.

Cf. also Nic. syr. 23.18-20:

ܘܗܕܐ ܠܐ ܡܢ ܣܓܝܐܘܬܗܘܢ ܕܡ̈ܝܐ ܕܝܡܐ. ܡܬܝܕܥܐ ܕܝܢ ܕܐܢ ܬܣܒ 19/
ܟܡܝܘܬܐ ܕܢܗܪܐ ܡܢ ܡ̈ܝܐ ܕܝܡܐ ܘܡܢ ܕܢܗܪܐ ܘܬܣܝܡ ܒܟܠܚܕܐ ܒܝܥܬܐ 20/
ܒܚܕܐ ܛܐܦܐ. ܐܚܪܬܐ ܕܝܢ ܒܡ̈ܝܐ ܕܝܡܐ ܛܒܥܐ. ܘܚܕܐ ܕܝܢ ܠܚܕܐ ܗܝ ܐܚܪܬܐ.

This does not happen because of the quantity [*saggī'ūṯā*] of the seawater. It is known that if you take the same quantity [*kmāyūṯā*] of seawater and river (water) and place one egg in each portion, one floats on seawater, while the other sinks in sweet (water).

Cand. 152.3-4:

ܘܡ̈ܝܐ ܡܠܝܚܐ ܢܛܘܦ ܚܝܠ ܡܢ ܗܘܢ ܚܠܝܐ ܕܒܥܬܐ ܡܢ ܗܘ ܕܡܬܚܠܒܐ ܡܢ ܒܪܬܐ
ܛܒܥܐ. ܒܚܠܝܐ ܕܝܢ ܛܥܒܬܐ.

---

[49] Cf. Arist. 359a 11-14; Olymp. 159.18-21; Olymp. arab. 111.20-22

[50] Probably an interpolated gloss. The Syriac word *bartā* standing alone is of course ambiguous ("daughter"!).

(1) That salt water is denser than sweet (water) (2a) is known from the fact that an egg will float on salt (water) (2b) but will sink in sweet (water).

V.ii.3.3-5: Density of seawater (ships)

But. [underline: agreement with Nic. 23.13-18]:
(1a) Ships on [lit. of] the sea are heavy [*yaqqīrān*], (1b) whereas [those] on [lit. of] rivers and lakes are light [*qallīlān*], (2a) and those who sail on the Sea of India, (2b) because they sail on lakes which are connected to the sea, (2c) when they transfer from the sea to a lake [sg.], (2d) easily go under [*mettab̲ʿīn*].
Nic. syr. 23.13-18[51]:

ܬܘܒ ܕܝܢ ܘܝܩܝܪܝܢ ܡ̈ܝܐ ܕܝܡܐ ܡܢ ܡܫܬ̈ܝܢܐ ܘܥܒܝܢ ܝܬܝܪ ܡܢܗܘܢ. /14 ܡܛܠ ܕܐܠܦ̈ܐ
ܕܒܝܡܐ ܪ̈ܕܝܢ: ܝܬܝܪ ܝܩܝܪ̈ܢ. ܗ̇ܠܝܢ ܕܒܢܗܪ̈ܘܬܐ/15 ܝܬܝܪ ܩܠ̈ܝܠܢ. ܘܡܬܝܕܥܐ ܡܢ ܐܝܠܝܢ
ܕܪ̈ܕܝܢ ܒܝܡܐ ܕܗܢܕܘ. ܒܕܝܠ ܕܪ̈ܕܝܢ ܒܐܓܡ̈ܘܬܐ/16 ܕܡܣܬܡ̈ܟܢ ܠܝܡܐ. ܟܕ ܡܥܒܪܝܢ ܡܢ
ܝܡܐ ܠܐܓܡ̈ܘܬܐ. ܒܩܠܝܠܘܬ ܛܒܥܝܢ. ܡܛܠ/17 ܕܡ̈ܝܐ ܡܫܬ̈ܝܢܐ. ܠܐ ܡܨܝܢ ܠܡܫܩܠ
ܝܘܩܪܐ ܕܐܠܦ̈ܐ ܕܫܩܠܝܢ ܠܗܘܢ ܡ̈ܝܐ/18 ܕܝܡܐ.

Furthermore, seawater is heavier and denser than potable (water), (1a) because ships which sail on the sea are heavier [*yattīr yaqqīrān*], (1b) whereas those on rivers and lakes are lighter [*yattīr qallīlān*]. (2a) [This] is known from those who sail on the Sea of India, (2b) because they sail on lakes which are connected to the sea. (2c) When they transfer from the sea to lakes [pl.] (2d) they easily go under [*ṭābʿīn*] because the potable water cannote carry the weight of ships which the seawater carries.

l. 3-4. ܐܝܠܝܢ ܕܒܝܡܐ ܕܗܢܕܘ ܪ̈ܕܝܢ ("those who sail on the Sea of India"): ܪ̈ܕܝܢ (fem. pl.) PL[2]l
The reading "rād̲ēn" (masc. pl., i.e. people, not ships) of FMV is supported by Nic. syr., as well as by "hānōn" of *Cand*. 152.6.

l. 3. "Sea of India": Arist. does not specifically name the "Sea of India". Olymp. does not either at 159.10-16, but does in the passages at 163.2-5 and 81.20-32 (Olymp. 81.26, 163.3 "ἰνδικοπλεῦσται"). There is no mention of India in Olymp. arab.

V.ii.3.5-8: Dead Sea

But. [underline: agreement with Nic.; italics: with *Šifāʾ*]:
(1) There is a putrid *lake* in *Palestine* near the Red Sea. (2) Its water is extremely [*hānā kulleh*] salty and dense, (3) so that very heavy *animals* will float on it [*b-hōn*] and *will not sink* even if they are *bound up* (4) and fish do not breed [lit. *are not born*] in it [*b-hōn*].

---

[51] Immediately preceding the passage on "eggs" at 23.18-20. - Cf. Arist. 359a 7-11; Olymp. 159.10-16, 163.2-5, 81.20-32; Olymp. arab. 111.22-112.2; *Cand*. 152.5-7. - See also Nic. syr. 21.7-8 [following the passage on "eggs" at 21.6-7]: ܘܬܘܒ ܐܠܦ̈ܐ/ ܕܥܒܪ̈ܢ ܒܝܡ̈ܡܐ ܝܩܘܪ̈ܢ. ܗ̇ܠܝܢ ܕܝܢ ܕܒܐܓܡ̈ܘܬܐ ܐܘ ܒܢܗܪ̈ܘܬܐ ܩܠܘ̈ܠܢ ܀ ("(1a) Furthermore ships which pass through seas are heavy [*yaqqūrān*], (1b) whereas those [passing] through lakes or rivers are light [*qallūlān*]").

Nic. 23.22-26[52]:

ܐܝܬ ܓܝܪ/23 ܒܦܠܣܛܝܢܐ ܝܡܐ ܕܡܬܩܪܐ ܡܝܬܐ ܘܩܪܝܒ ܠܝܡܐ ܣܘܦܐ.
ܣܘܦ/24 ܕܡܝܘܗܝ ܥܒܝܢ ܘܡܠܝܚܝܢ. ܘܡܬܝܕܥܐ ܡܢ ܗܝ ܕܚܝܘܬܐ ܪܘܪܒܬܐ ܕܠܠܐܝܘܬ ܛܐܦܢ ܒܗ
ܒܗ./25 ܘܐܦܢ ܢܬܐܣܪܢ: ܠܐ ܛܒܥܢ ܡܛܠ ܣܘܓܐܐ ܕܡܠܝܚܘܬܗ. ܠܐ ܡܬܝܠܕܝܢ
ܒܗ/26 ܢܘܢܐ ܒܟܒܪ ܡܝܐ.

(1) For there is in Palestine a sea which is called "dead" and [which] is near the Red Sea, (2) whose water is dense and salty. (3) [This] is known from the fact that large animals easily float on it [*beh*] and, even if they are bound, do not sink because of the large quantity of its saltiness. (4) Fish are not born in it [beh]{perhaps [*kbar*] "in the water"}.

Šifā' Act. et pass. 207.12-13:

وأما بحيرة فلسطين فلا يرسب فيها شيء، حتى الحيوان المكتوف. ولا يتولد فيها الحيوان، ولا يعيش.

(1) In the lake of Palestine (3) nothing sinks, not even a bound animal. (4) No animals is born in it, nor do they survive.

l. 5. ܒܚܝܪܬܐ ("lake"): Arist. talks of the Dead Sea as a λίμνη, but Olymp. as θάλασσα. Nic. syr. has "sea", while IS, Olymp. arab. and *But*. have "lake".

l. 5. ܣܪܝܬܐ ("putrid"): The word is missing in the Cambridge manuscript of Nic. syr., but should probably be restored there, since it is also found in Olymp. arab. (115.3: البحيرة المنتنة التي بأرض فلسطين).

l. 7, 8 ܒܗܘܢ ("in/on it", i.e. in/on the water): ܒܗ (i.e. in the sea) Nic. syr. - It appears that BH took notice of the words "ܒܟܒܪ ܡܝܐ" (probably a gloss) at the end of the passage of Nic. syr. quoted above.

V.ii.3.8-10: Spring in Chaonia

But.: (1) In the land of Chaonia near Dodona, a spring of salty water flows into [*b*-] a fresh(water) [lit. sweet] river (2) and fish do not breed [lit. are not born] in (this river). When the water of this spring is boiled and [then] is cooled, it hardens and turns into salt.

Nic. syr. 24.1-4[53]:

ܒܟܐܘܢܝܐ. ܒܟܐܘܢܝܐ ܗܝ ܕܩܪܝܒܐ ܠܕܘܕܘܢܐ ܡܥܝܢܐ/2 ܚܕ ܕܡܝܐ ܡܠܝܚܐ ܢܒܥܐ
ܠܢܗܪܐ ܕܐܝܬ ܒܗ ܡܝܐ ܚܠܝܐ. ܘܠܐ ܡܬܝܠܕܝܢ/3 ܒܗ ܢܘܢܐ. ܡܝܐ ܕܝܢ ܕܠܗ
ܕܡܥܝܢܐ ܗܢܐ ܟܕ ܡܬܒܫܠܝܢ ܘܡܬܩܪܪܝܢ ܩܫܝܢ/4 ܘܗܘܝܢ ܡܠܚܐ.

Text: (1) In Chaonia, near Dodona, a certain spring of salty water flows into [*l*-] a river that contains sweet water (2) and fish are not born in (this river). (3) When the water of this spring is boiled and cooled, it hardens and turns into salt.

V.ii.4.1-4: Empedocles' theory

But. [underline: agreement with Nic.; italics: with *Šifā'*]:

(1) *Empedocles said* that *the sea is the sweat of the earth* and is *salty for this reason*. (2a) Our master [Aristotle] said that this *statement* uses the figure of a metaphor - i.e. transfer of an image - (2b) and is appropriate for *poets*, (2c) not for *philosophers*, (2d) unless it is taken as an *analogy*.

---

[52] Cf. Arist. 359a 16-22; Olymp. 163.17-22, 29-32; Olymp. arab. 115.2-6.

[53] Cf. Arist. 359a 24-35; Olymp. 164.7-32; Olymp. arab. 115.16-21.

Nic. syr. 20.6-7[54]:

ܕܐܪܒܥܬܐ / ܕܐܡܦܕܘܩܠܝܣ ܐܡܪ ܕܝܡܐ ܕܘܥܬܐ ܕܐܪܥܐ ܐܝܬܘܗܝ ܗܢܐ ܘܡܛܠܗܢܐ
ܡܠܝܚ ܀

Fourth [sc. theory proposed by ancients on the salinity of the sea]: (1) Empedocles says that the sea is the sweat of the earth and is salty for this reason.

Nic. syr. 20.16-21[55]:

ܗܢܘܢ ܕܐܡܪܝܢ ܕܕܘܥܬܐ ܕܐܪܥܐ ܐܝܬܘܗܝ/17 ܝܡܐ . ܒܐܣܟܝܡܐ ܕܡܛܐܦܘܪܐ
ܡܫܬܡܫܝܢ ܐܝܬܝܗ ܫܘܚܠܦ ܕܕܡܘܬܐ/18 ܗܘܝܘ ܕܝܢ ܫܘܚܠܦ ܫܡܗܐ ܐܝܬܘܗܝ . ܘܐܝܬܘܗܝ
ܠܡ ܫܘܚܠܦ ܫܡ̈ܗܐ : ܐܝܟ ܗܘ/19 ܕܩܪܐ ܦܘܐܝܛܐ ܦܫܝܛܐܝܬ ܠܡܠܟܐ ܪܥܝܐ
ܕܥܡ̈ܡܐ . ܟܕ ܦܝܠܣܘܦܐ/20 ܠܘ ܗܟܢܐ . ܐܠܐ ܒܐܢܐܠܘܓܝܐ ܗܝ ܫܘܝܘܬ ܦܘܚܡܐ
ܗܢܘܢ ܕܠܘܩܒܠ ܚܕܕܐ . ܐܝܟ ܕܪܥܝܐ/21 ܒܨܝܐ ܕܥܢܗ . ܗܟܢܐ ܘܡܠܟܐ ܕܐܝܠܝܢ ܕܡܫܬܥܒܕܝܢ .

[17 et 18bis ܫܘܚܠܦ (cf. *But*.): ܫܘ̈ܚܠܦ cod. ‖ 18 ܫܡܗܐ: lege ܫܡ̈ܗܐ sive ܫܡܗܐ ‖ ܐܝܬܘܗܝ: ܐܝܬ cod. ‖ ܘܐܝܬܘܗܝ: ܘܐܝܬܘܗܝܢ (ܘܗܝܢ sub lin.) cod. ‖ 19 et 21 ܪ̈ܥܝܐ cod. ‖ 20 ܒܐܢܐܠܘܓܝܐ cod. ‖ 21 ܒܨܝܐ: ms. ܕܒܨܝܐ cod.]

(2a-b) Those who say that the sea is the sweat of the earth are using the figure of metaphor - i.e. transfer of an image, i.e. transfer of a name, and "an example of a transfer of appellations is when a poet simply calls a king the shepherd of the peoples"[56] - (2c) whereas [*kad*] a philosopher should not [speak] thus, but [should] use analogy - i.e. equation of corresponding things: [e.g.] "just as a shepherd cares for his sheep, so a king those subject to him".

Šifā' Act. et pass. 207.16-17:

وقد قال أنبادقليس إن ملوحة البحر بسبب أن البحر عرق الأرض. وهذا كلام شعرى ليس بفلسفى. لكنه مع ذلك يحتمل التأويل.

(1) Empedocles said: the salinity of the sea is due to [*bi-sababi*] the fact the sea is the sweat of the earth. (2) This is poetic talk, not philosophical, but nevertheless it contains an analogy.

l. 3 ܫܘܚܠܦ ܕܕܡܘܬܐ ("transfer of an image"): This definition of "metaphor" is taken from Nic. syr. There, in turn, the phrase is probably an infelicitous rendition of "ἐπιφορὰ εἴδους" vel sim., whereby the word εἶδος ought to have been translated as *āḏšā* (species) rather than as *dumyā* (image, likeness). - Cf. Arist. *Poet.* 1457b 6ff.: μεταφορά ἐστιν ὀνόματος ἀλλοτρίου ἐπιφορὰ ἢ ἀπὸ τοῦ γένος ἐπὶ εἶδος ἢ ἀπὸ τοῦ εἴδους ἐπὶ γένος ἢ ἀπὸ τοῦ εἴδους ἐπὶ εἶδος ἢ κατὰ τὸ ἀνάλογον.

V.ii.4.4-6: Analogy with sweat

But.: Indeed, just as sweat is moisture which has been made salty by admixture of yellow bile, (2) sea is water which has been made salty by [*b*-] admixture of smoky exhalation.

---

[54] Cf. Arist. 357a 24-26; Olymp. 151.3-7; Olymp. arab. 112.8-9.

[55] Cf. Arist. 357a 25-28.

[56] ποιμὴν λαῶν, *Iliad* 2.243 etc.; cf. Arist. *Eth. Nic.* 1161a 15.

Šifā' Act. et pass. 207.17-208.1: فإن العرق رطوبة من البدن تملحت بما يخالطها من المادة المحترقة من البدن. وماء البحر قد يملح بقريب من ذلك.

(1) For sweat is moisture from the body, which has been made salty by the combustive matter [*mādda muḥtariqa*] in the body that has mingled with it. (2) Seawater has been made salty by [*bi*-] something similar to that.

l. 5. ܡܪܪܬܐ ܝܩܝܕܬܐ ("yellow bile"): lit. "combustive/tawny bile". - I take the term here to be a technical one. "Yellow bile", the humour corresponding to the element fire, is usually called ܡܪܪܬܐ ܣܘܡܩܬܐ or ܡܪܪܬܐ ܫܥܘܬܢܝܬܐ in Syriac (PS 2204, PS Suppl. 199a; arab. المرة الصفراء), but the term "fiery/tawny bile" is found in Greek ("χολὴ πυρρά" Galen [Kühn] 15.658; Hippocrates *Epid*. 6.1.5). - The text of the *Šifā'* has المادة المحترقة. The alteration to *merrtā yaqqīḏtā* (= المرة المحترقة) may be due 1) to the text of the *Šifā'* in BH's hands, 2) deduction by BH as a physician ("fiery matter in the body" = yellow bile), or 3) Fakhr al-Dīn al-Rāzī, *Mabāḥiṯ* II.142.8-9: وسبب ملوحة العرق والبول مخالطة المرة المحترقة للمائية ("The cause of the salinity of sweat and urine is the mixture of yellow bile [*mirra muḥtariqa*] with the wateriness"); cf. comm. on V.ii.4.6 ("causa finalis") below.

V.ii.4.6-8: Purpose of salinity

The sense here appears to be that because of its salinity and consequent weight, the seawater cannot blow all over the earth, even if it turns putrid. In the passage of *Cand*. quoted below, on other hand, the meaning seems to be that the salinity prevents it from turning putrid in the first place (so also Olymp. arab. 113.19-114.2). - In connection with the idea of the corruption of air, cf. V.ii.2.7-8 (ܘܚܘܒܠܐ ܕܐܐܪ ܡܣܩܒ) below.

But.: (1) The final cause of the salinity of seawater is (2a) to prevent [lit. ... is so that not] its fumes, when it becomes putrid, (2b) from blowing all over the earth, (2c) and bringing about a universal corruption (2d) which causes plants to wither and animals to die.

Šifā' Act. et pass. 208.2-4:

فإذاً كانت ملوحة البحر لهذه العلة ولغاية هى حفظ مائه عن الأجون، ولولاه لأجن، وانتشر فساد أجونه فى الأرض، وأحدث الوباء العالم. على أن ماء البحر يأجن إذا خرج من البحر أيضًا، وإنما ينحفظ بعضه بمجاورة بعض وبمدد التمليح الذى يصل إليه.

The salinity of the sea is therefore due to this cause. (1) The purpose [*ġāya*] is the preservation of its water from putridity [*uǧūn*]. (2a) Otherwise it would turn putrid, (2b) and the corruption of its putridity would spread over the earth (2c) and would cause a contamination of the world, although seawater too turns putrid when it is taken out from the sea and it is only preserved by its parts being juxtaposed[57] to each other and the salinating agent that reaches it.

---

[57] بمجاورة: cf. Olymp. arab. 114.4; Olymp. 153.3 κατὰ παράθεσιν (cf. Lettinck [1999] 144 n. 53).

Cf. also *Cand*. 152.1-3:

ܘܐܝܟ ܕܗܘ ܥܠܬܐ ܫܘܡܠܝܝܬܐ ܕܡܠܝܚܘܬܗܘܢ ܐܝܬܝܗ ܗܝ ܕܠܐ ܢܣܪܘܢ ܘܢܚܒܠܘܢ
ܠܐܐܪ ܘܠܢܫܡܘܢ ܕܢܫܡܝܢ ܠܗ .

Be that as it may, the *causa finalis*[58] of its salinity is so that (the water) should not turn putrid and corrupt the air and those who breathe it.

Cf. *Eth*. IV.13.5, ed. Bedjan 452.18-453.3; ed. Çiçek 236a.6-13:

ܒܗܢܐ ܝܘܡܐ . ܟܕ ܐܬܟܢܫܘ ܡ̈ܝܐ ܠܐܬܪܐ ܚܕ . ܗܘܐ ܝܡܐ ܗܘܐ ܟܠܢܝܐ . ܕܐܣܝܪ
ܠܐܪܥܐ ܐܝܟ ܐܣܪܐ . ܘܡܢܗ ܝ̈ܡܡܐ ܡܢܬܢܝܐ ܥܠ ܐܦ̈ܝ ܐܪܥܐ ܒܕ̈ܘܟܝܢ ܦ̈ܪܝܫܢ
ܐܫܬܠܛܘ . ܘܗܪܟܐ ܡܬܕܡܪܝܢܢ ܒܚܟܡܬܗ ܕܐܠܗܐ ܕܥܒܕܬ ܠܡ̈ܝܐ ܡܠܝ̈ܚܐ . ܕܐܠܘ
ܕܐܠܐ ܗܟܢܐ . ܒܥܓܠ ܣܪܝܘ ܘܐܬܚܒܠܘ . ܘܠܐܐܪ ܘܠܡܫܬܡܫܢܘܗܝ ܚܒܠܘ .

On this day, when the waters gathered in one place, there was formed the universal sea, which encircles the earth like a belt, and [branching out] from (this sea) the particular seas came to rule over the face of the earth in different places. Here we marvel at the Wisdom of God that it made the waters salty, because otherwise (the waters) would have quickly become putrid and rotten and would have destroyed the air and those who breathe it.

l. 6. ܥܠܬܐ ܫܘܡܠܝܝܬܐ ("final cause", *causa finalis*, i.e. purpose): IS simply says "purpose" (*ġāya*) but Fakhr al-Dīn al-Rāzī has *al-sabab al-ġāʾī* at the corresponding place (*Mabāḥiṯ* II.142.11).

l. 8. "animals": cf. Min. IV.i.2 above ("animals that breathe [the air]"); Mete. V.i.2 below ("animals which breathe the air").

Min. V. Section iii.: *On the Migration of the Sea*

The section of *But.* here corresponds in the first place to the latter half of the chapter on the sea in the *Šifāʾ* (ed. Qassem, Cairo 1969, p. 208-210; cf. the introductory comments on V.ii. above). Here, none of the supplementary materials can be traced back to the surviving text of Nic. syr., but the likelihood is that most of these materials go back to lost portions of that work.

Of the three 'theories' in this section, the first two deal with the migration of the sea proper. The third theory, though placed within the same section, deals not so much with its migration as its currents and tides. BH had earlier treated the two subjects under two different headings in his *Cand*. (*Cand*. 152.8-153.1 (ܡܛܠ ܕܡܫܬܢܝܢܐ ܗܘ ܝܡܐ ܡܢ ܕܘܟ ܠܕܘܟ; cf. Fakhr al-Dīn al-Rāzī, *Mabāḥiṯ* II.142.16-143.4; and *Cand*. 153.3-154.2 (ܡܛܠ ܪ̈ܘܡܐ ܕܬܗܘܡܐ; cf. *Mabāḥiṯ* II.143.5-9).

---

[58] Male "causa efficiens" PS 2129 (passage of *Cand.* here cited as "*Cand.* 35v"), "la cause efficiente" Bakoš.

	*Šifā*ʾ		
V.iii.1.1-3	208.11-13	(Nic.?)	Posibility of migration
V.iii.1.3-8	208.13-209.11		Process of migration
V.iii.1.8-11		Nic.?	Great winter and summer
V.iii.2.1-3	(209.13-14)	Nic.?	Fossil shells in Egypt
V.iii.2.3-6		Nic.?	Maeotis
V.iii.3.5-9	209.15-210.4		Duration of migration
V.iii.2.9-13	210.4-6		Writings on pyramids
V.iii.3.1-2	210.7-8		Agitation by winds
V.iii.3.4-9		Nic.?	Euripus
V.iii.3.9-10		?	Ebb and flow

V.iii.tit. ܥܠ ܫܘܢܝ ܝܡܐ: cf. *Cand.* 152.8: ܥܠܬܗ ܕܡܫܢܝܢܘܬܐ ܗܝ ܕܝܡܐ ܡܢ ܕܘܟ ܠܕܘܟ; Fakhr al-Dīn al-Rāzī, *Mabāḥit* II.142.16: المباحث الثالث عن اختصاص البحر بجانب من الأرض دون جانب

V.iii.1.1-3: Posibility of migration

But.: (1) It is not necessary that the sea should be in its particular [*dīlānāytā*] place by nature; (2) rather, it is able to migrate from one place to another, (3) albeit over long periods of time, which human lives are not [long] enough to delimit.

*Šifā*ʾ Act. et pass. 208.11-13:

وأما اختصاص البحر فى طباعه بموضع دون موضع فأمر غير واجب، بل الحق أن البحر ينتقل فى مدد لا يضبطها الأعمار، ولا تتوارث فيها التواريخ والآثار المنقولة من قرن إلى قرن إلا فى أطراف يسيرة وجزائر صغيرة

[13 يسيرة : بسيرة ed. Cair.]

(1) The exclusivity [*iḫtiṣāṣ*] of the sea by nature in one place and not in another is not something that is necessary. (2) Rather, the truth is that the sea migrates [*yantaqilu*] (3) over periods which life spans [*aʿmār*] cannot embrace and chronicles and transmitted records cannot inherit [as records] from century to century, except in [the case of the disappearance] tiny tips [of land] and small islands

l. 2. ܒܙܒܢܐ ܣܓܝ ܐܪܝܟܐ ("over long periods of time"): BH may be inserting into his paraphrase of the *Šifā*ʾ here elements taken from a lost passage of Nic. syr., since the word "long" is not in the *Šifā*ʾ but is found in Arist. (351b 9f.: καὶ ἐν χρόνοις παμμήκεσι πρὸς τὴν ἡμετέραν ζωήν; note also ζωή) and Olymp. arab. (104.13: فى مدة طويلة من الزمان).

V.iii.1.3-8: Process of migration

The content of the passage goes back ultimately to Arist. 351a 19ff. - The beginning of the passage of *But.* agrees closely with the passage of the *Šifā*ʾ quoted below, while the latter part, though less close, can still be understood as a systematised paraphrase of the *Šifā*ʾ.[59]

---

[59] Cf. a similar but briefer account in *Cand.* 152.8-11, which employs some of the

But.: (1) The sea migrates (2) because it takes [its] material from rivers, (3) and rivers draw [their] water from springs and grow copious [*šāp̄ʿīn*] with rainwater. (4a) Often in one area [lit. side, *gabbā*] springs dry up, (4b) rivers fail (4c) and the sky is not clouded over, (5a) while in another area springs emerge, (5b) rivers gush forth (5c) and the sky causes rain to fall. (6a) Thus, in one area, as water vanishes, (6b) sea becomes dry land, (7a) while in another area, as (water) becomes copious, (7b) dry land becomes sea.

Šifāʾ Act. et pass. 208.13-14, 209.1-5, 7-11:

لأن البحر لا محالة مستمد من أنهار وعيون تفيض إليه، وبها قوامه. ... ومن شأن الأنهار أن تستقى من عيون، ومن مياه السماء. ومعولها القريب إنما هو على العيون، فإن مياه السماء أكثر جدواها فى فصل بعينه دون فصل. ثم لا العيون ولا مياه السماء يجب أن تتشابه أحوالها فى بقاع واحدة بأعيانها تشابهاً مستمراً. فإن كثيراً من العيون يغور وينضب ماؤها. وكثيراً ما تقحط السماء فلا بد من أن تجف أودية وأنهار، وربما طمت الأنهار، بما يسيل من أجزاء الأرض، جوانب من النجاد ... وإذا كان كذلك فستنحسم مواد أودية وأنهار، ويعرض للجهة التى تليها من البحار أن تنضب، وتستجد عيون وأنهار من جهات أخرى، فتقوم بدل ما نضب. ويفيض الماء فى تلك الجهة على البر. فإذا مضت الأحقاب، بل الأدوار، يكون البحر قد انتقل عن حيز إلى حيز

[The sea migrates] (2) because the sea undoubtedly receives its material [*istamadda*] from rivers and springs, and its existence depends on them [lit. in them is its standing/support]. ... (3) It is the nature of rivers that they draw water from springs and from precipitation [lit. waters of the sky]. What it immediately depends upon is the springs. For precipitation is available mostly only during a particular season. There are no rivers or precipitation, whose conditions necessarily remain constant in the same places. (4a) For the water of many springs oozes way and peters out (4c) and the sky often remains rainless [*taqḥatu*], (4b) so that the wadis and rivers will inevitably dry up [*taǧiffu*]. Often the rivers may engulf, while it flows over parts of the earth, the sides of higher ground [*niǧād*]. ... This being the case (6a) the materials of the wadis and rivers will come to an end (6b) and the side [*ǧiha*] of the sea that adjoins them will be depleted [*tanḍubu*], (5/7) while the springs and rivers will be formed anew on other sides, so that they will arise instead of what has been depleted, and the water on that side will flood over dry land. Then, when times [*aḥqāb*], or rather ages [*adwār*], have passed, the sea will have migrated from one region to another.

V.iii.1.8-11: Great winter and summer

The introduction "the master said", along with the agreement with the passage of Olymp. quoted below, suggests that this was taken from Nic. syr. - Aristotle only mentions the "great winter and excess of rains" at *Mete*. 352a 31 (μέγας χειμὼν καὶ ὑπερβολὴ ὄμβρων; cf. Strohm [1970] 165, comm. ad. loc.).

---

same vocabulary as that used here, along with what is essentially a summary of the passage of the the *Šifāʾ* in Fakhr al-Dīn al-Rāzī, *Mabāḥiṯ* II.142.18-143.4.

But.: (1) The master [Aristotle], in fact, said: (2a) When the five planets gather in [the sign of] Capricorn, great winter and cataclysm occur, (2b) and mountains are torn apart by thunder. (3a) When they gather in [the sign of] Cancer, great summer and conflagration occur, (3b) mountains crumble and fall, (3c) and springs [*mabbūʿē*] dry up and together with them the rivers.

Olymp. 111.30-112.1, 112.9-11: (2a) μέγας δέ ἐστιν ὁ χειμῶν, ἡνίκα πάντες οἱ πλάνητες ἐν χειμερινῷ ζῳδίῳ γένωνται, ἢ Ὑδροχόῳ ἢ Ἰχθύσι· (3a) μέγα δέ ἐστι θέρος, ὅταν ἐν θερινῷ ζῳδίῳ γένωνται, ἢ Λέοντι ἢ Καρκίνῳ. ... (2a) ἐν οὖν τῷ μεγάλῳ χειμῶνι ἡ ἤπειρος θαλαττοῦται, (3a) ἐν δὲ τῷ μεγάλῳ θέρει τοὐναντίον διὰ τὸ ποῦ μὲν ἔκκαυσιν καὶ πολλὴν ξηρότητα, ποῦ δὲ ὑγρότητα.

l. 11. "mountains crumble and fall": cf. Min. II.i.3. above.

V.iii.2.1-3: Fossil shells in Egypt

Arist. mentions Egypt as a land that was formerly under the sea (*Mete.* 352b 20-353a 1), but does not mention the bones/fossils found there as evidence for this. The idea is found in Olymp., Olymp. arab. and *Šifāʾ*. Our passage of *But.*, when we subtract what is also in the *Šifāʾ*, closely resembles Olymp., so that it may be regarded as a conflation of the *Šifāʾ* and a lost passage of Nic. syr., which in turn may have been a scholion based on Olymp., except that the "Depression of Arsinoitis" (modern Fayyūm) is not mentioned in the text of Olymp. known to us. The passage of *Cand.* below, free of elements taken from the *Šifāʾ*, no doubt represents more faithfully than *But.* what must have stood in the text of Nic. syr. in BH's hands.

But. [underline: agreement with *Šifāʾ*; italics: with Olymp.]:
(1) *Egypt*, as the master [Aristotle] said, *was sea in former times*, (2) but [then] dried up and *became dry land.* (3) For this reason bones of marine animals, *κογχύλια and τελλίναι*, are found in the depression [*ʿumqā*] of Arsinoitis.

Šifāʾ Act. et pass. 209.13-14:

وقد يعرف من أمر النجف الذى بالكوفة أنه بحر ناضب، وقد قيل ان أرض مصر هذه سبيلها، ويوجد فيها رميم حيوان البحر.

[That the sea migrates] is known from the condition of al-Najaf by al-Kūfa, [namely] that it is dried-up sea. (1) It has been said: the land of Egypt is like this [i.e. dried up sea] and decaying bones [*ramīm*] of sea animals are found in it.

Olymp. 116.10-15: (1) ὅτι δὲ Αἴγυπτος πάλαι θάλαττα ἦν (2) καὶ ὕστερον ἠπειρώθη, ... (3) δῆλον δὲ τοῦτο καὶ ἐκ τοῦ τῶν κογκυλίων καὶ τελλίνων ἐκεῖ εἶναι ὄστρακα, ὅπερ ἐστὶν ἐγκαταλείμματα τῶν ὀστρακοδέρμων τῶν ἀπομεινάντων ἠπειρωθείσης τῆς θαλάττης.
[τελλίνων scripsi: γελλίνων codd. VG et editio Aldina; χελύνων Stüve]

Olymp. arab. 104.7-8:

ويدل على صحة ذلك دلالة عظيمة المواضعُ المعروفة إلى هذا الوقت أن البحر فيما تقدم كان يغمرها، بمنزلة بلاد مصر، فإنه توجد إلى هذه الغاية فى مواضع عميقة منها دلائل البحر، وهى أنواع من الأصداف وغيرها.

Major proof that this [theory] is correct is provided by places, concerning which it is known until now (1-2) that the sea used in former times [*fīmā taqaddama*] to cover them, e.g. Egypt. (3) For until this day, there are found in deep places [*mawāḍi*ᶜ ᶜ*amīqa*] of it evidence [*dalā*ʾ*il*] of the sea, i.e. species of shells [*aṣdāf*] etc.

Cand. [Bakoš] 152.11-152.2, [Çiçek] 110.3-13:

ܡܬܝܕܥܐ ܗܕܐ ܡܢ ܗܝ ܕܟܕ ܝܡܐ ܐܝܓܘܦܛܘܣ ܗܘܬ ܛܡܪܗ ܢܝܠܘܣ. ܘܐܪܝܡ ܒܗ ܠܐܪܥܐ. ܒܟܢ ܐܬܢܓܒ ܝܡܐ ܘܗܘܐ ܝܒܫܐ. ܘܕܝܡܐ ܐܝܓܘܦܛܘܣ ܗܘܬ ܡܩܕܡܬܝܢ ܡܬܚܘܝܐ ܡܢ ܓܘܡܨܐ ܗܘ ܕܐܪܣܝܢܘܐܝܛܝܣ ܗܘ ܕܥܕܡܐ ܠܝܘܡܢܐ ܡܬܚܙܝܢ ܒܗ ܫܘܕܥܐ ܕܐܪܥܐ ܝܡܝܬܐ. ܕܐܝܬܝܗܘܢ ܩܘܢܟܘܠܝܐ ܘܛܠܠܝܢܘ [illegible] ܕܡܫܬܟܚܝܢ ܬܡܢ.

[ܟܕ: ܟܪ Bakoš || ܕܐܪܣܝܢܐܘܐܝܛܝܣ Çiçek || ܘܛܠܠܝܢܘ sic B: ܘܛܠܠܐܝܢܘ Bakoš et Çiçek]

This [sc. that the sea migrates] is known from that fact (1) that although Egypt was [once] sea, the Nile has buried [*tmar*] it and raised land in it, (2) so that the sea dried up and became dry land.[60] (1) That Egypt was sea in former times (3) is proven by the depression of Arsinoitis, where to this day one can see evidences of [its having been] the seabed [*ar*ᶜ*ā yammāytā*], namely the κογχύλια and τελλίναι, i.e. shells, which are found there.

l. 2. ܓܘܡܨܐ ܗܘ ܕܐܪܣܝܢܘܐܝܛܝܣ ("the depression of Arsinoitis"): i.e. Fayyūm. - Cf. Strabo 17.1.38-39.

l. 3. ܩܘܢܟܘܠܝܐ ("κογχύλια"): κογχύλιον: "a small kind of mussel or cockle" LSJ; see also PS 3548.

l. 3. ܛܠܠܝܢܘܣ ("τελλίναι"): τελλίνη: "a small bivalve shell-fish = ξιφύδριον" LSJ; see also PS 1434. - The word is written ܘܛܠܠܝܢܘ/ܘܛܠܠܐܝܢܘ in *Cand.* What stood in Nic. syr. may have been be ܛܠܠܝܢܗ, with the final *hē* representing the sound of "-αι", which BH may then have construed as a corruption of the plural ending in "-ō" (= -οι) when writing the *Cand*., and then as the plural ending in "-ōs" (-ους) when writing *But*.

V.iii.2.3-6: Maeotis

This goes back ultimately to Arist. *Mete*. 353a 1-7, but the wording of the passage is again closer to Olymp. than to Arist., so that it is likely to be based on a lost scholion based on Olymp. in Nic. syr. Here, as with the example of Egypt above, BH omits any reference to the process of silting.

---

[60] The idea that Egypt was turned into land by silting due to the Nile is already in Arist. and this is followed by Olymp., as well as by Olymp. arab., where we note the use of the verb "inṭamara", cognate with "ṭmar" which is used in *Cand*. (Olymp. arab. 103.22, 104.5).

But. (1) Lake Maeotis, too, although it used to be sea, (2) became [first] a deep lake (3) and [then] its water decreased further, (4) so that large ships cannot pass through it.

Olymp. 123.2-5 [lemma comment on Arist. 353a 1]: (1) τρίτη πίστις ἐκ τῆς Μαιώτιδος λίμνης. αὕτη γὰρ πάλαι μὲν ἦν θάλαττα· (2) ἠιόνος δὲ γενομένης λίμνη γέγονε μεγάλη, ὡς καὶ ἀπροσδιορίστως λέγεσθαι λίμνην. (3) τῶν δὲ ποταμῶν ἐπιφερόντων ἰλὺν μικροτέρα γινομένη, (4) ὡς μὴ δύνασθαι τὰ αὐτὰ ἐν αὐτῇ πλοῖα φέρεσθαι.

l. 3. ܡܝܐܘܛܝܣ ("Maeotis"): sic FM; ܡܝܐܘܛܝܣ PL1; ܡܝܘܛܝܣ V. - The transcription here, which must have been taken from Nic. syr., represents the Greek "Μαιῶτις" more accurately than "ܡܘܐܛܝܣ" found at V.i.2.6 above, where the word was no doubt copied from *De mundo* syr.

V.iii.3.5-9: Long duration of migration

This goes back to Arist. *Mete.* 351b 8-27. The passage of *But.* is a conflation of materials taken from the *Šifāʾ* with what must have been taken from Nic. syr. The materials taken from Nic. syr., in turn, show greater affinity with Olymp. and Olymp. arab. than with Arist.[61]

But. [underline: agreement with *Šifāʾ*; italics: with Olymp. arab.]:
(1a) *Such a process* [lit. this] *does take place*, (1b) but *over a long time*, (1c) the number of whose years cannot be recorded [*metthed*] in books. (2a) For often nations are annihilated by floods, (2b) are exterminated *by wars*, (2c) are destroyed by *famine*, (2d) perish *due to corruption of air* (2e) and *migrate from one place to another*; (3a) their languages change, (3b) and their writings are replaced (3c) and cannot be read by others.

Šifāʾ Act. et pass. 209.15-18, 210.1-4.

... إلا أن أعمارنا لا تفى بضبط أمثال ذلك فى البحار الكبار، ولا التواريخ التى يمكن ضبطها، تفى بالدلالة على الانتقالات العظيمة فيها. وربما هلكت أمم من سكان ناحية دفعة بطوفان أو وباء، أو انتقلوا دفعة، فينسى ما يحدث بها بعدهم. ... ولكن التاريخ فيه لا يضبط. فإن الأمم يعرض لهم آفات من الطوفانات والأوبئة، وتتغير لغتهم وكتاباتهم فلم يدرى ما كتبوا وقالوا.

[18 فينسى: فتنوسى ed. Cair.]

... (1bc) except that our lives are not long enough [for us] to record [*ḍabaṭa*] such [changes] in large seas, nor are chronicles which can record them sufficient for proving large-scale migrations in them. (2a) Often nations [*umam*] inhabiting a [certain] region are suddenly annihilated by flood or plague, (2e) or migrate suddenly, so that what has happened there is forgotten after their [departure]. ... (1bc) but history cannot record them. (2a) For nations are subject to ruin [*āfāt*] by floods and plagues, (3ab) and their languages and writings change, (3c) so that what they wrote and said cannot be understood.

---

[61] Note in particular the omission of diseases (Arist. 351b 14 νόσοις) among the causes of annihilation of peoples, shared by *But.*, Olymp. and Olymp. arab.

Olymp. 113.25-33: (1a) αὗται μὲν οὖν αἱ μεταβολαὶ γίνονται, πλὴν λανθάνουσιν διὰ τρεῖς αἰτίας· (1c) ἢ γὰρ διὰ τὸ ἐν πολλῷ χρόνῳ γίνεσθαι κατὰ τὴν ἡμετέραν ζωὴν τὰς τοιαύτας μεταβολὰς οὐ δυνάμεθα αὐτὰς γνῶναι (2) ἢ διὰ τὰς ἀθρόας τῶν ἐθνῶν ἀπωλείας, αἵτινες γίνονται τριχῶς, (2b) ἢ διὰ πόλεμον (2c) ἢ διὰ λιμὸν ἢ διὰ ἀφορίαν καρπῶν. (2e) τρίτη αἰτία· καὶ διὰ τὸ τὰς κατὰ μέρος γινομένας οἰκίσεις τε καὶ μεταναστάσεις ἡ ἄγνοια τούτων ἕπεται· συμβαίνει γὰρ τὸν αὐτὸν ἄνθρωπον τὸν πάντα χρόνον αὐτοῦ τῆς ζωῆς μὴ ἐνὶ τόπῳ +ποιοῦντα [οἰκοῦντα Diels] ἀγνοεῖν τὰ ἐν πᾶσι τοῖς τόποις, ἐν οἷς ᾤκισεν.

Olymp. arab. 104.13-21:

وحدوث هذا العارض على نظام فى مدة طويلة من الزمان. والأسباب التى لا يتهيأ لنا معها رؤية هذا التغير إلا فى الفرط، ثلاثة: أحدها أن طول زمان كونه يفضل على زمان حياة واحد واحد منا، ولذلك يموت قرن بعد قرن قبل أن يروه. والثانى أن من الأمم الكبيرة ما يبيد دفعة واحدة إما بسبب الحروب، وإما بسبب فساد الهواء، وإما بسبب الجدب. الثالث أن الناس ينتقلون دائمًا من موضع ألى موضع بسبب تعذر الغذاء أو غيره من الأسباب التى تحفز إلى النقلة، ولا يعرف المنتقلون ما كان تقدم من الحادث فى الموضع الذى رحلوا عنه حتى يحفظوه ويعيدوا ذكره. والذين يسكنون أيضًا فى ذلك الموضع لا يعلمون ما كان فيما تقدم فينسى الخبر ويعفو ذكره.

(1a) This process takes place [lit. the occurrence of this happening] (1b) as a rule over a long period of time. The reasons why we are only able to observe such alteration rarely are three. (1c) Firstly, the length of time [over which] it occurs exceeds the lifespans [*zamān ḥayāt*] of individuals among us and for this reason generation after generation die before they observe it. (2a) Secondly, some large nations are suddenly annihilated [*yabīdu*] (2b) either by war, (2d) by corruption of air, (2c) or by drought. (2e/c) Thirdly, people constantly migrate from one place to another because of the difficulty of [obtaining] food [*taʿaddur al-ġidā*ʾ < Olymp. ἀφορία καρπῶν?] or some other reason which prompts migration and those who have migrated [away] no [longer] know what took place earlier in the place from which they have migrated, so that they cannot retain and perpetuate its memory, while those who have [newly] settled in that place too are ignorant of what took place earlier, so that its report is forgotten and its memory perishes.

l. 7f. ܚܘܒܠܐ ܕܐܐܪ ("corruption of air"): corr. Olymp. arab. فساد الهواء. - Cf. V.ii.4.6-8 above.

V.iii.2.9-13: Writings on pyramids

The passage is based on the *Šifā*ʾ. To this BH has added an explanation of the pyramids. BH rightly equates "haramāni" with the Greek "πυραμίδες" and rightly tells us that they are "flame-shaped", i.e. pyramidal, polyhedral. We see, however, that BH goes amiss in his explanation of the Greek word "πυραμίς".

But.: (1) In fact [lit. behold!], (such writings) are found on many a mountain.

(2a) On two solid, flame-shaped towers in Egypt, called πυραμίδες or indivisible particles by the ancients, and named "haramān" in modern [lit. these latter] times, (2b) We hear that (2c) there are writings, some of which cannot be read and some of which cannot be understood.

Šifā' Act. et pass. 210.4-6:

وهو ذا يوجد فى كثير من الجبال. وبالهرمين الذين بمصر، على ما بلغنى، كتابات منها ما لا يمكن إخراجه، ومنها ما لا تعرف لغته.

(1) And that [*wa-huwa dā*] is found on many mountains. (2a) On the two pyramids [*bi-l-haramaini*] which are in Egypt, (2a) according what I have heard, (2c) there are writings, some of which cannot be read [lit. (whose meaning) cannot be extracted] and [with] some of which the language is unknown.

l. 10-12. ܬܪܝܢ ܡܓܕܠܐ ... ܘܡܛܠ ܕܒܙܒܢ̈ܐ ܐܚܪ̈ܝܐ ܗܪܡܐܢ ܡܫܬܡܗܝܢ ("two ... towers, ... named 'haramān' in modern times"): i.e. the two largest and best-known pyramids of Cheops and Chephren in Giza (see EI[2] III.173, s.v. "Haram" [Graefe-Plessner]). - "ܗܪܡܐܢ" is a transcription of the Arabic nominative dual هَرَمَان.

l. 10. ܐܛܝܡܐ ("solid"): The adjective is also used in the description of the pyramids at BH *Chron. eccl.* I.379.10f. (corr. Michael, *Chron.* 526b 35f., tr. III.82), a passage which was taken from Dionysius of Tellmaḥrē's report of his journey to Egypt[62]: ܣܩܘܦܐ ܘܐܛܝܡܐ ܘܠܐ ܚܠܝܠܐ ܘܣܪܝܩܐ ("slanting and solid, not hollow and empty").

l. 10. ܢܘܪܢܝܐ ܐܣܟܡܐ ("flame-shaped"): i.e. pyramidal, polyhedral; cf. gr. πυραμίς, πυραμοειδής. - For BH's own definition of *nūrānāyā* as a geometrical term in the sense of "polyhedral", see *Asc*. 7.13-19. - Cf. Nic. syr. 9.11-15 (= ed. Drossaart-Lulofs [1965] 89.4-8)[63]:

ܢܘܪ̈ܐ . ܠܢܘܪܐ ܐܚܪ̈ܢܐ ܐܣܟܡܐ ܐܣܦܝܪܝܐ ܝܗ̈ܒܘ ܗܘܘ ./12 ܘܐܚܪ̈ܢܐ
ܦܘܪܡܐܝܣ ܗ̄ ܡܪܒܥܐ ܥܒܕ ܕܡܢ ܐܪܒܥܐ . ܦܘܪܡܐܝܣ ܐܡܪ /13
ܬܠܡܚܪܝܐ ܕܡܢ ܥܠܬ ܨܠܝܢܘܬܐ ܐܣܟܡܐ ܗܟܢܐ ܡܛܠ ܕܦܘܪ̈ܡܐܝܣ ܕܚܙܐ /14
ܒܡܨܪܝܢ . ܟܕ ܡܢ ܠܬܚܬ ܦܬܝܢ ܗܘܘ ܚܡܫܡܐܐ ܐܡ̈ܝܢ : ܥܠ ܐܡܬܐ ܚܕܐ
ܡܬܟܢܫܝܢ/15 ܗܘܘ ܠܥܠ ܐܠܐ ܗܟܢ ܟܬܒ ܡܟܬܒܢܐ .

[11 ܝܗ̈ܒܘ: ܝܗܒܘ Drossaart Lulofs || 12 ܥܒܕ: om. Drossart Lulofs]

Comment: Some have given fire a spherical shape, others a pyramid.[64] - i.e. a square, from four [?][65] - A pyramid, says [Dionysius] of Tellmaḥrē, is so called on account of its slanting shape, because the pyramids he saw in Egypt, while 500 cubits wide below, narrowed down to one cubit at the top.[66] - But thus the author writes [?]."

---

[62] This report by Dionysius, reproduced by BH in *Chron. eccl.*, is one which had come to the attention of de Sacy long before the publication of Abbeloos-Lamy's edition of *Chron. eccl.* (see de Sacy [1801; rep. 1905] I.254-261; id. [1810] 501-8, 552-7). On BH's use of Dionysius' chronicle in general, see Abramowski (1940) 18f. et pass.

[63] See also Drossaart Lulofs' comments on the passage, p. 165f.

[64] Cf. Arist. *De caelo* 306b 32f., also 304a 12.

[65] "i.e. four-planed, consisting of four triangles" tr. Drossaart Lulofs.

[66] Cf. Michael, *Chron.* 526b ult.-527a 4; BH *Chron. eccl.* I.379.14-17; see comm.

l. 11. ܡܢܘ̈ܬܐ ܕܠܐ ܡܬܦܠܓ̈ܢ ("indivisible particles"): The phrase ought to mean "atoms" (ἄ-τομοι).[67] - A clue, at least, as to the possible cause for this disconcerting confusion of "pyramids" and "atoms" is provided by a curious phrase encountered at *But*. De anima I.ii.1: "Democritus of the party of the ever-moving spherical pyramids" (ܕܝܡܩܪܝܛܘܣ ܗܘ ܕܡܢ ܟܢܘܫܝܐ ܕܦܘܪܐܡܝܕܣ ܐܣܦܐܪܘܐܝܕܝܬܐ ܡܬܬܙܝܥܢܝܬܐ ܐܡܝܢܐ, ms. F 113v a 16f.; see Furlani [1931] 32, with n.3). As noted by Furlani, this must be related to Democritus' notion of "fire" and "soul" as consisting of "spherical atoms".[68] The phrase "spherical pyramids" ought, in other words, to read "spherical *fiery* (atoms)". Corruption of πυρώδης or πυροειδής (ܦܘܪܘܕܝܣ, ܦܘܪܘܐܝܕܝܣ) into πυραμίδες (ܦܘܪܐܡܝܕܣ) is quite conceivable.[69]

V.iii.3.1-2: Movement of sea by winds

The sentence closely follows the *Šifāʾ*. - Cf. *Cand.* 153.3f., Fakhr al-Dīn al-Rāzī, *Mabāḥiṯ* II.143.5f.

But.: (1) The sea is calm by nature, (2) but is made to move [*mettzīᶜ*] accidentally [*aḵ da-ḇ-ḡeḏšā*] by winds which are released from its depth or which blow on its surface.

Šifāʾ Act. et pass. 210.7-8:

وأعلم أن البحر ساكن فى طباعه، وإنما يعرض ما يعرض من حركته بسبب رياح تنبعث من قعره، أو رياح تعصف فى وجهه، ...

(1) Know that the sea is calm by nature. (2) Whatever movement that occurs [*ᶜaraḍa*] to them is due to winds which are released from its depth, or winds which blow on its surface, ...

l. 1. ܐܝܟ ܕܒܓܕܫܐ ("accidentally"): i.e. "accident" in its technical, Aristotelian sense. The phrase reflects the verb *ᶜaraḍa* used by IS.

---

on "ܐܠܦܐ" above. - This scholion found in the Cambridge manuscript of Nic. syr. must, of course, be an interpolation by someone posterior to Dionysius of Tellmaḥrē. Michael begins the sentence with the words: "they are called pyramids because ..."; these words are omitted in BH *Chron. eccl.*, so that the interpolator is likely to be someone who had direct access to the chronicle of Dionysius itself or that of Michael rather than BH's *Chron. eccl*. One is tempted to ask whether the interpolator might not in fact be BH himself, which would then mean that the copy of Nic. syr. we have in the Cambridge manuscript is a descendant of the very copy used by BH.

[67] Cf. Nic. syr. 36.28f.: ܐܣܩܗ ܠܐ ܡܬܦܠܓܢܘܬܐ ܕܡܢܘܬܗ ("[Aristotle] refuted his [i.e. Democritus' theory of] atoms").

[68] Arist. *De anima* 405a 2 τὰ σφαιροειδῆ πῦρ καὶ ψυχὴν λέγει; see also 405a 12.

[69] A further candidate for the cause of the confusion would be some passage derived from Arist. *De caelo* III.v, 304a 14ff., lying before BH in a garbled form: "... so all solid figures are composed of pyramids; but the finest body is fire, while among figures the pyramid is primary and finest; ... For if, on the other hand, they make the primary body an atom, ... (tr. Stocks, in "revised Oxford translation" p.497).

V.iii.3.2-4: Current in narrow channels: Bosporus

The sentence is based primarily on the *Šifā'*. The mention of Byzantium, I suspect, is due to a lost passage of Nic. syr., which in turn is probably the source of the passage of Olymp. arab. quoted below.[70] - BH would, of course, have known about the sea flowing in a narrow channel by Byzantium-Constantinople also from his readings of the geographers (see comm. on V.i.2.4-5, 5-7 above).

But.: (1) It is also made to move by wave-causing rivers, which push it violently, (2) especially when its water is squeezed into narrow places such as those at Byzantium.

Šifā' Act. et pass. 210.10-11:

... أو لاندفاع أودية فيه مموجة له بقوة، وخصوصاً إذا ضاقت مداخلها وارتفعت وقل عمقها، فيعرض أن يتحرك إلى المغار.

... (1) or due to propulsion [*indifā*ʿ] by rivers [*audiya*] [flowing] into it and causing it to surge [*mumawwiğa lahu*] with force, (2) especially when its entrances are narrow and raised and their depth small, so that it moves into a hollow place [*maġār*].

Olymp. arab. 108.24-109.2:

وأما المواضع التى لا تستوى فيها أرض البحر، كالذى يوجد في المواضع الضيقة من البلاد المعروفة + ببوزانطر، فإن البحر هناك يضغط الأرض لضيقه فيصغر عظمه وتسرع حركة الماء فيه.

[109.1 ببوزانطر: sine punctis diacriticis cod.; proponit بطورانقس (Τυρρηνικός) Badawī ‖ يضغط الأرض sic. cod. et Badawī: fort. يُضغَط بالأرض?]

In places where the seabed is uneven, such as that found in the narrow places in the country called Byzantium [?], the sea squeezes the land [or: the sea is squeezed by the land?] because of its narrowness, so that its size decreases and the movement of the water in it becomes faster.

l. 3. ܡܓܠܠܢܐ ("wave-causing"): corr. IS مُمَوِّجة له; cf. *But*. Mete. III.i.1.6 below.

l. 4. ܡܬܐܠܨܝܢ ("is squeezed"): There is no corresponding word in the sentence of the *Šifā'* quoted above, but the verb "to squeeze" (يضغط) is found in the passage of Olymp. arab. quoted above, as well as at the place corresponding to *Šifā'* Act. et pass. 210.10-11 in Fakhr al-Dīn al-Rāzī, *Mabāḥiṯ* II.143.6 (ينضغط) and thence in *Cand*. 153.4 (ܒܥܠܬ ܐܘܠܨܢܗܘܢ).

[70] The reading "Byzantium" at Olymp. arab. 109.1 is uncertain. The passage immediately preceding in Olymp. arab. (108.16ff.) is related to Arist. 354a 13ff. and Olymp. 132.3ff., and Badawī proposed for this reason to read the word as "Tyrrhenian" which occurs at Arist. 354a 21 and Olymp. 132.10. As has been noted above, however, the account of the seas in Olymp. arab. is the product of a considerable rearrangement of the materials taken from Arist. (Nic.) and Olymp. gr. and the passage at Olymp. arab. 108.24ff. is better associated with Olymp. gr. 128.21ff., where Olymp. does not actually use the word "Byzantium" but does talk about the strait (πορθμός) by "Bosporus" (128.22, 34; on the "strait by Calchedon" mentioned at 128.23, 35, see n.71 below).

V.iii.3.4-9: Euripus

The Euripus is mentioned in passing (with reference to the reversal of the flow there) in Arist. *Mete.* not in the discussion of the sea but in the discussion of earthquakes (366a 23).

The latter part of the passage of *But.* here (parts 2-3) must be based on Nic. syr., since a passage closely resembling ours is found in Olymp. arab. as well as in *Cand.* The passage of Olymp. gr. quoted below, although it has certain similarities with both *But*/*Cand.* and Olymp. arab. (note, e.g., παλίρροια corr. *But*/*Cand.* *hpīkūt redyā*), is not very close and it is particularly noteworthy that the word "Euripus" does not occur there.[71] At the same time, we note that there are certain elements which are common to Olymp. gr. and Olymp. arab. but are absent in *But.* and *Cand.*,[72] a fact which suggests that the passage in Olymp. arab. *may* be a conflation of materials derived from Nicolaus and Olympiodorus, while the passages in *But.* and *Cand.* preserve the Nicolean material in a purer form.

Of the first sentence of the passage, which gives us a definition of "εὔριποι", part (1a) has a counterpart in *Cand.*, but the comparison with scales in (1b) is not in Olymp. arab. or *Cand.* The comparison in fact takes up the word "ταλαντεύεσθαι" used Arist. *Mete.* 354a 8 (also 354a 11 ταλάντωσις) and looks very much like a scholion based on the second passage of Olymp. quoted below.

> *But.* [underline: agreement with Olymp. arab.]:
> (1a) The water in εὔριποι flows in both directions, (1b) just as [of] the scales on a balance, sometimes the one and sometimes the other goes down according to weight. (2) The εὔριπος is a passage between two seas. (3a) When wind blows, (3b) water is pushed in the passage (3c) and is brought to the sea. (3d) When (this water) is unable to push the water of the sea, (3e) it turns backwards, (3f) undergoes a reversal of flow (3g) and in this way moves in both directions.

---

[71] The "Calchedonian (Chalcedonian) strait" in the text of Olymp. gr. 128.35ff. is, of course, a nonsense, since what we are talking about here is clearly *the* "Euripus", i.e. the strait by Chalcis in Euboea. Given that Olympiodorus is explaining the difference between the currents in this "Euripus" and the "strait by Bosporus" in this passage, the alteration of "Chalcidian/Chalcidic" into "Calchedonian" is more likely to be due to some ignorant copyist than to Olympiodorus himself (the error probably being prompted precisely by the mention of the "Bosporus"). - Whether Olymp. discussed the Euripus at the place corresponding to Arist. 366a 23 is unknown, since the section of Olymp. *in Mete.* on earthquakes is lost.

[72] e.g. Olymp. arab. *mawwağa* (cause to surge): gr. ἀνασοβεῖν, *But./Cand.* simply "push"; arab. "their force weakens": gr. μηκέτι ἰσχύοντος.

Olymp. 128.35-129.4: ... ἐν δὲ τῷ Καλχηδονίῳ πορθμῷ παλίρρυτον ὑπομένει τὸ ὕδωρ τῆς θαλάσσης· ῥεῖ γὰρ ἐκ τοῦ ἑνὸς μέρους καὶ πάλιν ἀναστρέφεται εἰς τὸ ἕτερον. τοῦτο δὲ γίνεται, (3a) ἐπειδὴ συμβαίνει τὸν ἄνεμον σφοδρότατον ὄντα πολλάκις, φησί, πνεῖν ἐκ τοῦ ἑνὸς μέρους τῆς θαλάσσης κατ᾽ ἐναντίωσιν τοῦ ῥεύματος (3b) καὶ ἀνασοβεῖν τὸ ὕδωρ ὡς ἐπὶ τὰ ὀπίσω, (3d) εἶτα αὐτοῦ μηκέτι ἰσχύοντος ἐπ᾽ ἄπειρον ἀποσοβεῖν τὸ ὕδωρ ἐνδίδωσι, (3e) καὶ λοιπὸν φέρεται τὴν οἰκείαν ῥύσιν, ὥσπερ καὶ πρώην. (3f) καὶ τούτου χάριν φαίνεται ὡς παλίρροιαν ἔχει τὸ ὕδωρ.

Olymp. 134.17-19 [lemma comment on Arist. 354a 7f.]: καλῶς δὲ παλίρροιαν ταλαντεύεσθαι ἐκάλεσε, (1b) τῷ δίκην ταλάντων εἰς τὰ ἐναντία περιάγεσθαι τὸ ὕδωρ. καὶ γὰρ τὸ τάλαντον ὅπου δ᾽ ἂν καταβαρήσῃ, ἐκτρέπεται, ὡς συμβαίνειν καὶ εἰς τὸν ἐναντίον ῥέπειν.

Olymp. arab. 109.2-7:

وأما المواضع التي تهب منها الرياح بمنزلة الموضع المسمى اوريفس، وهو مسلك فيما بين بحرين، فأن المياه هناك إذا موّجتها الرياح فى ذلك المسلك صادمت مياه البحر ولانها لا تقوى على قهر تلك المياه ودفعها، إذ كانت أكثر منها وأقوى، تضعف حميتها لذلك السبب فترجع إلى خلف، وتجرى إلى ضد الجهة التى كانت تجرى إليها. ولهذه العلة تسمى المياه المنكفئة.

[4 ولانها sic cod.: لانها Badawī]

As for places from which winds blow, like the place called Euripus - (2) which is a passage between two seas - (3ab) when winds cause the waters to surge in that passage, (3c) the waters collide with the waters of the sea [outside the passage], (3d) but because they are unable to overcome and push away those waters - the latter being greater and stronger - their force [*ḥamīya*] weakens for that reason, (3e) so that they turn backwards (3f) and flow in the opposite direction from that in which they were flowing. For this reason they are called reversed waters [*al-miyah al-munkafiʾa*].

Cand. 153.5-8:

ܡܛܠ ܡܝ̈ܐ ܐܘܪܝܦܝܩܝܐ ܕܪ̈ܕܝܢ ܠܬܪܝܗܘܢ ܓܒ̈ܐ ܐܡܪ ܐܪܝܣܛܘܛܠܝܣ ܕܐܘܪܝܦܘܣ ܐܢܬܘܗܝ ܡܥܒܪܢܐ ܕܒܝܬ ܬܪ̈ܝܢ ܝ̈ܡܡܐ ܘܡܐ ܕܢܫܒܐ ܪܘܚܐ ܡܬܕܚܝܢ ܡܝ̈ܐ ܕܒܗ ܒܡܥܒܪܢܐ ܗܘ ܘܡܬܬܝܒܠܝܢ ܒܝ̈ܡܡܐ ܠܝܡܐ. ܘܡܐ ܕܫܠܐ ܢܫܒܗ ܕܪܘܚܐ ܗܦܟܝܢ ܡܝ̈ܐ ܠܕܘܟܬܗܘܢ. ܘܡܩܒܠܝܢ ܗܦܟܬܐ ܕܪܕܝܐ ܫܚܠܦܐܝܬ.

(1) Concerning the Euripian waters which flow in both directions, Aristotle said: (2) the Euripus is a passage between two seas. (3a) When wind blows, (3b) the water in that passage is pushed (3c) and is brought to the sea. (3d) When the blowing of the wind ceases, (3e) the water returns to its place (3f) and undergoes a reversal of flow.

l. 4, 6. ܗܘܪܝܦܘ ("εὔριποι"), ܗܘܪܝܦܘܣ ("the εὔριπος"): In line 4, the word, standing in the plural, must be taken as a common noun ("*any strait* or *narrow sea, where the flux and reflux is violent*" LSJ 729). Whether BH understood and intends the word as a proper noun (i.e. *the* Euripus between Boeotia and Euboea) in l. 6 is unclear. - The word is written with an initial *ālap̄* in *Cand.* and in other instances cited in PS 94, 97 (s.vv. ܐܘܪܝܦܘܣ, ܐܘܪܝܦܘܣ). The source for the form with *hē* cited at PS Suppl. 99a is the passage of *But*. here.

l. 7. ܡܬܕܚܝܢ ܡܝܐ ܒܗ ܒܡܥܒܪܬܐ ("water is pushed in the passage"): cf. ܡܝܐ ܕܒܗ *Cand*. - The meaning is "the water in the passage is pushed", as is clear from the parallel passage of *Cand*., rather than "water is pushed *into* the passage".

l. 8. ܗܦܝܟܘܬ ܪܕܝܐ ("reversal of flow"): cf. V.iv.1.3, 9 below (ܚܘܠܦܘܬ ܪܕܝܐ, ܗܦܝܟܘܬ ܪܕܝܐ).

V.iii.3.9-10: Ebb and flow

IS does not talk about tides in the section on the sea in the *Šifā*ʾ. Fakhr al-Dīn al-Rāzī does in the corresponding section of his *Mabāḥit̠* but his explanation of "the cause of the ebb and flow [*al-madd wa-l-ǧazr*] of seas and rivers" is unfortunately lost in a lacuna (see *Mabāḥit̠* II.143.8-9, with footnote ad loc.). BH had already associated the movement of seawater with the moon in *Cand*., where he does not, however, distinguish between currents and tides and where there is no mention of the "Sea of Elam and of India". The attribution of the view mentioned here in *But*. to "many" suggests that BH may have come across this view in several sources, while the mention of specific localities suggests that these sources may include geographical works.[73]

But.: Many acknowledge as the cause of the ebb and flow [*tawbā w-meleʾā*] in the Sea of Elam and of India the daily and monthly rising and setting of the moon.

Cand. 153.8-11:

ܘܐܪܝܪܬܐ ܐܫܪܝܪܬܐ ܐܡܝܪܘ . ܘܗܝ ܕܢܣܝܢܝܬܐ ܗܘܝܐ ܦܣܩܐ ܠܗ ܗܕܐ ܗܝ . ܕܥܠܬܗ ܕܗܦܝܟܘܬܐ ܪܕܝܐ ܕܐܝܟ ܗܕܐ . ܫܘܚܠܦܗ ܗܘ ܝܘܡܝܐ ܘܝܪܚܝܐ ܕܣܗܪܐ ܐܝܬܘܗܝ ܒܡܠܝܘܬܗ ܘܒܨܝܪܘܬܗ .

The correct view, which is proven by experiment is this: the cause of the reversal of flow such as this is the daily and monthly change of the moon as it waxes and wanes.

l. 9-10. ܥܠܬܐ . . . ܠ- . . . ܝܘܡܢܝܐ ܣܓܝܐܐ ("many acknowledge as the cause ... the daily ..."): cf. Min. II.iii.4.4 above (also *Rad*. [Istanbul] 27.16-19, quoted under Min. II.iii.4.4-7).

l. 9-10. "the Sea of Elam and of India": "Sea of Elam": Arabian/Persian Gulf; "Sea of India": either the Sea of Bengal or the Indian Ocean in general (cf. V.i.3. above). - Tidal fluctuation is more pronounced in the Indian Ocean and its appendages than in the Mediterranean (cf. Miquel [1967-88] III.261).

---

[73] Ibn Khurdādhbih (70.1-6), for example, speaks of the tides in the Persian Gulf (*baḥr al-fārs*) and on the Indian Ocean (*al-baḥr al-aᶜẓam*), associatin the former with the rise of the moon (*maṭāliᶜ al-qamar*). For a brief survey of views concerning "ebb and flow" in Islamic literature, see EI² V.949f., s.v. "al-Madd wa 'l-Dj̲azr" [Martínez]; for views mainly in geographical works, Miquel (1967-88) III.261-266.

Min. V. Section iv.: *On Tartarus*

The greater part of this section (up to V.iv.3.4) corresponds to the discussion and refutation of the Platonic theory of "Tartarus" by Aristotle found towards the end of *Mete*. II.ii. This subject had not been discussed by IS in his *Šifā*ᵓ. Only the very end of the discussion of "Tartarus" survives in the Cambridge manuscript of Nic. syr.,[74] but the likelihood is that the whole discussion here in *But.* is based on lost portions of Nic. syr. and this applies also for the additional material at the end of the section (V.iv.5-10).

Indicated below are the passages of Arist. and Olymp. to which the passages in this section correspond.

	Arist.	Olymp.	
V.iv.1.1-5	355b 32-356a 14	141.21-34	Plato's theory
			142.14-18
			147.2-5
V.iv.1.5-9	356a 14-19	141.35-142.25	refutation (1): upward flow
V.iv.2.1-6	356a 19-22	142.26-31	refutation (2): imbalance
V.iv.3.1-4	356a 23-33	142.36-143.7	refutation (3): rivers
V.iv.3.5-6	356b 6-9	143.11-12	sea is eternal
V.iv.3.6-10	353b 30-35	127.37-128.8	sea not artificial or spontaneous

V.iv.1.1-5: Plato's theory

The actual content of the alleged Platonic theory in the passage of *But.* here (parts (2a-f) below) agrees more closely with Olymp. than with Arist., while the parts providing the overall framework of the sentence (parts (1) and (3) below) is closer to Arist. (note *But.* "written": Arist. γεγραμμένον; *But.* "is impossible": Arist. ἀδύνατον ἐστιν). These latter parts may therefore belong to the genuine Nicolean material. Nicolaus' summary of Arist. may, in fact, have been very brief and may have consisted of not much more than what is preserved in the passage of *Cand.* quoted below.

But. [underline: agreement with Arist.; italics: with Olymp.]:
(1) *Refuting*[75] that statement [lit. word] written *in the Phaedo* of *Plato* about *Tartarus* - (2a) [namely] that it is *at the centre of the earth* (2b) and

[74] Nic. syr. 19.1-24, evidently scholia, corresponding to the discussion of the "ethical" and "physical" (ἠθικῶς καὶ φυσικῶς) interpretation of "Tartarus" at Olymp. 144.11ff., 146.2ff.

[75] lit. "the Master refutes ... and says ..."

contains *the entirety of water* (2c) and [that] this water *is unstable*, (2d) *fluctuates* [*metmaḥšlīn*] (2e) and undergoes a reversal of flow (2f) because *there is a void* [*spīqūtā*] *between its convex surface* [lit. convexity, *kustānūtā*] *and the concave underside* [lit. concavity, *ḥlīlūtā*] *of the earth* - (3) the Master [Aristotle] says that this is impossible ...

Arist. 355b 32-356a 14: (1) τὸ δ᾽ ἐν τῷ Φαίδωνι γεγραμμένον περί τε τῶν ποταμῶν καὶ τῆς θαλάττης (3) ἀδύνατον ἐστιν. λέγεται γὰρ ὡς ἅπαντα μὲν εἰς ἄλληλα συντέτρηται ὑπὸ γῆν, ἀρχὴ δὲ πάντων εἴη καὶ πηγὴ τῶν ὑδάτων ὁ καλούμενος Τάρταρος, (2a) περὶ τὸ μέσον ὕδατός τι πλῆθος, (2b) ἐξ οὗ καὶ τὰ ῥέοντα καὶ τὰ μὴ ῥέοντα ἀναδίδωσιν πάντα· (2cd) τὴν δ᾽ ἐπίρρυσιν ποιεῖν ἐφ᾽ ἕκαστα τῶν ῥευμάτων διὰ τὸ σαλεύειν ἀεὶ τὸ πρῶτον καὶ τὴν ἀρχήν· οὐκ ἔχειν γὰρ ἕδραν, ἀλλ᾽ ἀεὶ περὶ τὸ μέσον εἰλεῖσθαι· κινούμενον δ᾽ ἄνω καὶ κάτω ποιεῖν τὴν ἐπίχυσιν τοῖς ῥεύμασιν. τὰ δὲ πολλαχοῦ μὲν λιμνάζειν, οἷον καὶ παρ᾽ ἡμῖν εἶναι θάλατταν, (2e) πάντα δὲ πάλιν κύκλῳ περιάγειν εἰς τὴν ἀρχήν, ὅθεν ἤρξαντο ῥεῖν, πολλὰ μὲν κατὰ τὸν αὐτὸν τόπον, τὰ δὲ καὶ καταντικρὺ τῇ θέσει τῆς ἐκροῆς, οἷον εἰ ῥεῖν ἤρξαντο κάτωθεν, ἄνωθεν εἰσβάλλειν. εἶναι δὲ μέχρι τοῦ μέσου τὴν κάθεσιν· τὸ γὰρ λοιπὸν πρὸς ἄναντες ἤδη πᾶσιν εἶναι τὴν φοράν. ...

Olymp. 141.21-34: οὕτω περὶ τῆς ὁλότητος τῆς θαλάττης διαλεχθεὶς ὁ φιλόσοφος, (1) ἐπεὶ μὴ συνῳδὰ αὐτῷ φθέγγεται ὁ Πλάτων - βούλεται γὰρ αὐτὸς ἐν τῷ Φαίδωνι μὴ τὴν θάλατταν εἶναι ὁλότητα τοῦ παντὸς ὕδατος, ἀλλὰ ποταμόν τινα καὶ τοῦτον ὑπὸ γῆν ὄντα, (1) φημὶ δὴ τὸν Τάρταρον. (2a) ἔλεγε γὰρ περὶ τὸ τῆς γῆς κέντρον εἰλεῖσθαι τοῦτον τὸν ποταμόν, (2b) ἀπ᾽ αὐτοῦ δ᾽ ἑτέρους τρεῖς προιέναι, λέγω δὴ Ἀχέροντα, Κωκυτόν, Πυριφλεγέθοντα· ἐκ δὲ τούτων πάντας προιέναι τοὺς ποταμοὺς καὶ τὴν θάλατταν. (2c) τοῦτον δ᾽ αὐτὸν τὸν Τάρταρον ἀνάστατον[76] ὄντα (2d) καὶ ὄχλου μεστὸν κυμαίνεσθαι περὶ τὸ κέντρον (2e) καὶ ποτὲ μὲν ἐάσαντα τὸ κέντρον γίνεσθαι περὶ τὸ διάμετρον, ποτὲ δὲ περὶ τὸ ἕτερον διάμετρον, καὶ οὕτως ἄλλοτε ἀλλαχόθεν φερόμενον ταράχου εἶναι μεστόν. (2b) καὶ τούτων οὕτως ἐχόντων ἔλεγεν ὁλότητα αὐτὸν εἶναι τοῦ παντὸς ὕδατος· πάντας δὲ τοὺς ποταμοὺς καὶ τὴν θάλατταν εἰς αὐτὸν ἔλεγε χωρεῖν ὡς αὐτῶν ἐφιεμένων τῆς οἰκείας αὐτῶν ὁλότητος. (1/3) ταύτην τὴν ὑπόθεσιν ἀναιρεῖ Ἀριστοτέλης διὰ τεττάρων ἐπιχειρημάτων, ...

Olymp. 142.14-18: ... ἐπειδὴ οὖν ὑπετίθετο ὁ Πλάτων (2a) περὶ τὸ κέντρον τῆς γῆς εἰλεῖσθαι τὸν Τάρταρον, (2f) μὴ +ψαύοντα [fort. ψαύειν Stüve] δὲ τὴν τούτου κυρτὴν ἐπιφάνειαν τῆς κοίλης ἐπιφανείας τῆς γῆς ὡς ὕδωρ ἐν ἀγγείῳ, ἀλλ᾽ εἶναι μεταξὺ τῶν δύο ἐπιφανειῶν τόπον κενόν, εἰς ὃν κυμαίνων ὁ Τάρταρος ἄλλοτε ἄλλον καταλαμβάνοι τόπον, ...

Olymp. 147.2-5 [lemma comment on Arist. 356a 4: οὐκ ἔχειν γὰρ ἕδραν, ἀλλ᾽ ἀεὶ περὶ τὸ μέσον εἰλεῖσθαι]: ἰδοὺ ἐνταῦθα σαφῶς ὑποτίθεται ὁ Πλάτων (2f) μὴ συνεχῆ εἶναι τὴν κυρτότητα τοῦ Ταρτάρου τῇ κοιλότητι

---

[76] Here "unstable" (= ἄστατος), see LSJ 121 s.v. ἀνάστατος, 5.

τῆς γῆς, (2c) ἀλλ' εἶναι μὲν ἀνερμάτιστον διὰ τὸ μὴ ἐρείδειν ἐν τῇ γῇ, ἔχειν δὲ βάσιν τὸ κέντρον.

Olymp. arab. 106.8-11:

الذين زعموا أن الهاوية تحيط بكلية الماء، وأن ذلك نحو مركز الأرض، وأن المياه تجرى منها وإليها دائمًا على التساوى بسبب الأمواج العارضة فيها من غير قرار ورجوع الحركة إليها، يُردّ عليهم قولهم من وجوه كثيرة

(1) The statement [*qaul*] of those who asserted (2b) that the abyss [*hāwiya*] embraces the entirety of water, (2a) that it is at the centre of the earth and that waters constantly and regularly flow from it and to it (2d) because of the waves that arise in it (2c) without stability (2e) and the reversal of the movement towards it (1) can be refuted in many ways.

Cand. 151.3-5:

ܐܦܠܛܘܢ ܕܝܢ ܒܦܐܕܘܢ ܕܝܠܗ ܐܡܪ ܕܒܛܪܛܪܘܣ ܠܟ ܗܘ ܕܒܟܢܛܪܘܢ ܐܘܟܝܬ
ܡܨܥܬܐ ܓܘܝܬܐ ܕܐܪܥܐ ܐܝܬܝܗܘܢ ܟܠܗܘܢ ܡ̈ܝܐ. ܘܠܗܕܐ ܕܝܢ ܬܪܥܝܬܐ
ܣܬܪܗ̇ ܒܡ̈ܠܐ ܣܓܝ̈ܐܬܐ ܐܪܝܣܛܘܛܠܝܣ.

(1) Plato said in his *Phaedo*: (2a) in Tartarus which is at the κέντρον, or inner midst, of the earth (2b) are the entire waters. (1/3) Aristotle has refuted this opinion with many [words].

l. 3. ܡܬܚܫܠ ("fluctuates"): The denominative verb derived from *maḥšōlā* ("agitatio undarum" PS 1404) corresponds to Olymp.'s "κυμαίνομαι". - Cf. l. 8 below.

V.iv.1.5-9: Refutation (1): absurdity of water flowing upwards

The argument here corresponds to the argument put forward by Arist. at *Mete*. 356a 14-19. The passage of Nic. syr. used by BH here is probably a part of the scholia based on Olymp. rather than genuine Nicolaus, since the wording of our passage has much greater affinity with Olymp. than with Arist., even though the passage here is much shorter than the rather involved argument put forward by Olymp. The same passage of Nic. syr. lies behind the passage of Olymp. arab. quoted below. In part (2) BH may have rephrased the argument he found in Nic. syr., since Olymp. arab. agrees here with Olymp. gr. against BH.

But. [underline: agreement with Olymp.]:

... because, (1a) if Tartarus is the entirety of water, (1b) water will be found to move upwards 'by nature' - (2a) because (this water) is placed at the centre of the earth (2b) and in a sphere whatever is outside the centre is above [it]. - (3a) As it fluctuates, (3b) therefore, Tartarian water will move upwards 'by nature' (4a) and, when it undergoes reversal of flow, (4b) it will move downwards 'by force', (5) which is absurd.

Olymp. 141.35-142.25: (1a) οὐ καλῶς, ὦ Πλάτων, τὸν Τάρταρον λέγεις τοῦ παντὸς ὕδατος εἶναι ὁλότητα· δυσὶ γὰρ ἀτόποις ὑποπίπτεις οὕτως

ὑποτιθέμενος· πρῶτον μέν, ὅτι εὑρεθήσεται ἡ τοῦ ὕδατος ὁλότης ὑπὸ τὴν γῆν οὖσα περὶ τὸ παντὸς κέντρον, ὅπερ ἄτοπον· (1b) ἔπειτα δὲ εὑρεθήσεται τὸ ὕδωρ μηκέτι τὴν κάτω φορὰν κινούμενον κατὰ φύσιν, ἀλλὰ τὴν ἄνω ἐφιέμενον τῆς οἰκείας ὁλότητος, (5) ὅπερ ἐστὶν ἀδύνατον. πᾶσι γὰρ τοῦτο γνωστόν, ὅτι ἡ μὲν ἐπὶ τὸ κάταντες φορὰ κατὰ φύσιν ἐστὶ τῷ ὕδατι, ἡ δ' ἐπὶ τὸ ἄναντες παρὰ φύσιν· ἐπειδὴ οὖν τῆς οἰκείας ὁλότητος ἐφιέμενον τὸ ὕδωρ, τουτέστι τοῦ Ταρτάρου, ὡς σὺ φῄς, μέχρι μὲν τοῦ κέντρου τὴν κάτω <φορὰν κατὰ> φύσιν κινηθήσεται, ὑπερβὰν δὲ τὸ κέντρον καὶ ὡς ἐπὶ τὴν ὁλότητα χωροῦν τὴν ἄνω φορὰν παρὰ φύσιν, (5) ὅπερ ἄτοπον (ἄτοπον γὰρ τὸ πρὸς τὴν οἰκείαν ὁλότητα κινούμενον παρὰ φύσιν κινεῖσθαι· (2) ὅτι δὲ ἐπὶ τὸ ἄνω κινεῖται τὸ ὕδωρ τὸ ἀπὸ τῆς ἐπιφανείας τῆς γῆς φερόμενον ἐπὶ τὸν Τάρταρον <εἰ> ὑπερβαίνει τὸ κέντρον, δῆλον, (2b) ἐπειδὴ πᾶσα μὲν ἡ ἐπὶ τὸ κέντρον κίνησις κατωφερὴς ὑπάρχει, ἡ δ' ἀπὸ τοῦ κέντρου ἐπὶ τὸν κύκλον ἀνωφερής), ... [see under V.iv.1.1-5 above] τόπον κενόν, (3a) εἰς ὃν κυμαίνων ὁ Τάρταρος ἄλλοτε ἄλλον καταλαμβάνοι τόπον, συμβαίνει ποτέ, ἐν τῷ ἑνὶ μέρει τοῦ κέντρου εἰ ἐγένετο, εἶτα τοῦ ὕδατος τοῦ κατὰ τὴν ἐπιφάνειαν τῆς γῆς πρὸς αὐτὸν ἐρχομένου, ἐπειδὴ πρῶτον διὰ τοῦ κέντρου παρερχομένου ἐπίγνυτο τῷ Ταρτάρῳ, (4b) μέχρι μὲν τοῦ κέντρου τὴν κάτω φορὰν ἐκινεῖτο, (3b) μετὰ δὲ τὸ κέντρον ἐπειγόμενον παρ' αὐτοῦ τὴν ὁλότητα ἀνάγκη τὴν ἄνω φορὰν κινηθῆναι, εἴπερ πᾶσαι αἱ ἀπὸ τοῦ κέντρου κίνησις ἀνωφερεῖς εἰσι· καὶ συνέβαινε παρὰ φύσιν ἐπὶ τὴν οἰκείαν ὁλότητα κινεῖσθαι, (5) ὅπερ ἄτοπον.

Olymp. arab. 106.11-13:

وأحدها أن الهاوية إن كانت تحيط بكلية الماء وكان موضعها فى مركز الأرض، وكان جميع المياه تجرى منها وإليها، فيجب أن تكون المياه متحركة إلى العلو، لأن كل ما يبتدئ بالحركة من مركز الأرض فحركته إلى فوق.

Firstly, (1a) if the abyss [*hāwiya*] embraced the entirety [*kullīya*] of water, its place were at the centre of the earth (3a) and the entirety [*ǧamīʿ*] of water flowed from it and to it, (1b/3b) then waters would necessarily [*yaǧibu an*] move upwards, (2b) because the movement of whatever begins moving from the centre of the earth is upward.

l. 6, 8, 9. ܒܟܝܢܐ, ܟܝܢܐܝܬ, ܩܛܝܪܐܝܬ ("by nature", "by force"): Water's "natural" (κατὰ φύσιν) movement is downwards, while its upward movement is "contrary to nature" (παρὰ φύσιν). That is the sense in which these two phrases are constantly used in the passage of Olymp. quoted above. When BH reverses this and talks of the upward movement as being "by nature" and the downward as "forced", this is, of course, ironic ("according to Plato's theory ..."). Since Olymp. arab. does not use these terms, it is unclear whether this reversal is due to BH or to the Syriac scholiast. The word *qṭīrāʾīṯ* (by force) may correspond, besides to "παρὰ φύσιν", to "ἐπειγόμενον" of Olymp. 142.22.

V.iv.2.1-6: Refutation (2): imbalance of inflow and outflow

BH's second argument, which corresponds to Arist. 356b 19-22, is again closer to Olymp. than to Arist., although in this instance BH's argument is longer than Olymp.'s.

But. [underline: agreement with Olymp.]:
(1a) The passage of *Phaedo* also says (1b) that the same quantity [*ak kmāyūṯā*] of water as that which exits from Tartarus through springs gushing out of the earth reenters [*hāp̄kīn ʿāʾlīn*] it through the pores in the earth. (2a) That this opinion is defective is known (2b) from the fact that in summer little water enters it (2c) and much (water) exits from it because of the large quantity of vapour generated by the sun, (2d) whereas in winter little water exits from it (2e) and much water enters it because of the large quantity of rain. (3) The quantity, therefore, of water which flows out of it is not equal to that which flows into it.

Olymp. 142.26-31[77]: δεύτερον ἐπιχείρημα· (1a) οὐ καλῶς, ὦ Πλάτων, ἔφης (1b) τοσοῦτον ὕδωρ εἰσρεῖν ἐν τῷ Ταρτάρῳ, ὅσον καὶ ἐκρεῖ. (2a) μᾶλλον γὰρ εἰκὸς τοῦτο λέγειν, (2de) ὅτι ἐν μὲν τῷ χειμῶνι πλεῖον εἰσρεῖ εἰς αὐτὸν ἤπερ ἐκρεῖ διὰ τὴν ἀπὸ τῶν ὑετῶν πλεονεξίαν τοῦ ὕδατος, (2bc) ἐν δὲ τῷ θέρει μᾶλλον ἐκρεῖν ἤπερ εἰσρεῖν διὰ τὸ μὴ ἐπιμίγνυσθαι τὸ κάτω φερόμενον ὕδωρ διὰ τῶν ὄμβρων.

Olymp. arab. 106.13-15:

والثانى أنه لا يصح أن تكون كمية ما يجرى من الهاوية مساويًا لما يجرى إليها دائمًا، وذلك أن ما يجرى منها فى الصيف أقل، وما يجرى إليها فى الشتاء أكثر.

[14f. منها فى الصيف: fort. فيها فى الصيف]

Secondly, (2a) it is incorrect (1b/3) that the quantity of what flows out from the abyss is constantly equal to what flows into it, (2bc) i.e. what flows out from it [or: into it?] in summer is less, (2e) and what flows into it in winter is more.

V.iv.3.1-4: Refutation (3): rivers flow into the sea

BH's argument here, corresponding to Arist. *Mete*. 356a 23-33, is again closer to Olymp. than to Arist.

But. [underline: agreement with Olymp.]:
(1a) If Tartarus were the entirety of water, (1b) all rivers would flow into it. (2a) This, however, is not the case (2b) because, even though some rivers are swallowed up [into the earth], (2c) they nevertheless do not remain below the earth (2d) but appear in other places. (3) Therefore, not Tartarus but the sea is the entirety of the watery element.

---

[77] Cf. also Olymp. 148.31-149.3 (lemma comment on Arist. 356a 19).

Olymp. 142.36-143.7: τέταρτον ἐπιχείρημα· ἐχρῆν, ὦ Πλάτων, (1a) εἴπερ ὡς σὺ φῄς ὁλότης ὁ Τάρταρος, (1b) πάντας τοὺς ποταμοὺς ὁρᾶν ἐν τῇ γῇ ἀναπινομένους καὶ εἰς αὐτὸν ἀφικνουμένους ὡς πρὸς οἰκείαν ὁλότητα. (2a) νῦν δ' οὐ τοῦτο ὁρῶμεν γινόμενον, ἀλλὰ πάντας τοὺς ποταμοὺς ἢ ἄντικρυς εἰσρέοντας ἐν τῇ θαλάσσῃ ἢ ἄλλοις ποταμοῖς μείζοσι μιγνυμένους καὶ εἰσβάλλοντας ἐν τῇ θαλάσσῃ, (2b) οὐδέποτε δὲ ἐν τῇ γῇ χαουμένους, ἀλλ' εἰ καὶ τοῦτο γένηται, (2d) πάλιν ἀναφαινομένους καὶ μηδέποτε πρὸς τὸν Τάρταρον ἀφικνουμένους· (3) οὐκ ἄρα ὁλότης ὁ Τάρταρος.

Olymp. arab. 106.17-19:

رالرابع أن الهاوية لو كانت موضع كلية الماء لوجب أيضًا أن تصب مياهه إليها، وذلك شىء لم يكن على وجه الدهر. فقد وجب إذاً أن لا تكون الهاوية موضع كلية الماء.

Fourthly, (1a) if the abyss were the place of the entirety of water, (1b) then its waters must flow into it, (2) but that is something which has never happened. (3) It must therefore be that the abyss is not the place of the entirety of water.

l. 2. ܐܦܢ ("even if") scripsi: ܐܦ codd. - For the construction *āpen... bram dēn*, cf. IV.iii.1.4-5 (also *āpen... ellā*, IV.iii.1.1-2); cf. further V.ii.2.4-5: *w-āpen nahrwāṯā... šādēn*.

V.iv.3.5-6: Sea is eternal

But. [underline: agreement with Olymp.]:

(1) (The sea) is a part of this world, (2) and because the world is without beginning and not created, (3) likewise the sea, too, [which is] a part of it, is not created and [is] without beginning.

Arist. 356b 6-9: τοῦτο μὲν οὖν ἐοίκασι πάντες ὁμολογεῖν, ὅτι γέγονεν, εἴπερ καὶ πᾶς ὁ κόσμος· ἅμα γὰρ αὐτῆς ποιοῦσι τὴν γένεσιν. ὥστε δῆλον (2) ὡς εἴπερ ἀΐδιον τὸ πᾶν, (3) καὶ περὶ τῆς θαλάττης οὕτως ὑποληπτέον.

Olymp. 143.11-12: ... πάντων γὰρ λεγόντων, (2) ὅτι εἰ ἀΐδιον τὸ πᾶν, (3) ἀΐδιός ἐστι καὶ ἡ θάλασσα (1/3) μέρος οὖσα τοῦ παντός, εἰ δὲ φθαρτὸν τοῦτο, φθαρτὴ καὶ ἡ θάλασσα ...

Olymp. arab. 105.4-6:

إن جميع مَن قال فى العالم إنه محدث قد لزمه فى البحر أن يكون محدثًا، لأن البحر أحد أستقسات العالم وجزء منه. وأما جميع الذين قالوا فى العالم أنه غير محدث فقد أثبتوا قدَمه، خلا ديمقراطيس ...

All those who said that the world is created necessarily [said] that the sea [too] is created, (1/3) because the sea is one of the elements of the world and a part of it. (2) All those who said that the world is not created have (3) confirmed its eternity [*qidam*], except Democritus ...

l. 5, 6 ܗܘܝܐ ("[not] created"): i.e. passive ptc. - Vocalised thus (*pṯāḥā*) in both instances in ms. F; ܗܘܝܐ in l. 5, but ܗܘܝܐ in l. 6 in M; ܗܘܝܐ in both instances in PLIV, all of which constantly write "-āw-" for "-aw-".

l. 5-6. ܡܛܠ ܕܠܐܫܘܪܝܐ ܘܠܐ ܗܘܝܐ ܐܝܬܘܗܝ ܥܠܡܐ ("because the world is without beginning and not created"): Surprising words to be found flowing

from the pen of a Christian bishop without comment, even if he is largely reproducing the words of Nic. syr. here. Cf. the introductory comments to Min. V. below; cf. also the attribution of the formation of the sea to "nature" (rather than "God") below.

V.iv.3.6-10: Sea not artificial or spontaneous

As the vocabulary used and the similarities in particular with the passage of Olymp. arab. quoted below show, this is related to the passage at Arist. *Mete*. 353b 30-35, where Aristotle is arguing that the sea has no sources on the ground that the sea is unlike waters of sources (cf. Min. I.iii above), which are either running, artificial or spontaneous. The whole point has been altered in *But*., where all mention of the "sources" disappears and the point being made is the formation of the sea by "nature".

But. [underline: agreement with Olymp. ; italics: with Olymp. arab.]: (1a) (The sea) is not artificial [lit. prepared by hand, *m^c^attad b-īdayyā*], (1b) because *it is beyond* [*^c^ab̲rā*] *human power* to excavate such a depth [*^c^umqā*], (1c) which noone has been able to fathom. (2a) *Neither* is it spontaneous [lit. prepared by self, *m^c^attad men yāt̲eh*], (2b) since *that which arises of its own accord is small*, (2c) *rarely arises* and (2d) does not persist. (3a) Since it is neither artificial nor spontaneous, (3b) the remaining alternative is that nature has formed it in this way.

Arist. 353b 30-35: τούτων δ' οὕτω διωρισμένων ἀδύνατον πηγάς εἶναι τῆς θαλάττης· ἐν οὐδετέρῳ γὰρ τούτων οἷόν τ' εἶναι τῶν γενῶν αὐτήν· οὔτε γὰρ ἀπόρρυτός ἐστιν (1a) οὔτε χειροποίητος, τὰ δὲ πηγαῖα πάντα τούτων θάτερον πέπονθεν· (2ab) αὐτόματον δὲ στάσιμον τοσοῦτον πλῆθος οὐδὲν ὁρῶμεν πηγαῖον γιγνόμενον.

Olymp. 127.37-128.8: ἐπειδὴ οὖν τριτταί εἰσιν αἱ πηγαὶ ἢ ῥυτὸν ἔχουσαι ὕδωρ ὡς ἐν τοῖς ὄρεσιν, ἢ στάσιμον ὡς ἐν τοῖς πεδίοις, καὶ αἱ στάσιμοι ἢ χειρόκμητοί εἰσιν ἢ αὐτόματοι, ἴδωμεν, εἰ κατὰ μίαν διαφορὰν τῶν πηγῶν τούτων εἰκός ἐστι γίνεσθαι τὴν θάλασσαν. ἀπὸ μὲν γὰρ τῶν ῥυτὸν ὕδωρ ἐχουσῶν πηγῶν οὐκ εἰσρεῖται ἡ θάλασσα· οὐδαμοῦ γὰρ τῆς γῆς ὁρῶμεν ἀπὸ ὑψηλῶν πηγὰς ῥεούσας ἁλμυρὸν ἐχούσας ὕδωρ. (1a) ἀλλ' ἆρα μὴ ἀπὸ τῶν χειροκμήτων εἰσρεῖται; οὐδὲ τοῦτο· (1b) οὐδὲ γὰρ ὑπὸ τέχνης ὑδροφαντικῆς θάλασσα γίνεται. (2a) ἀλλ' ἆρα μὴ ἀπὸ τῶν αὐτομάτων; οὐδὲ τοῦτο· οὐδὲ γὰρ ἀπὸ τῶν μερῶν τὸ ὅλον γίνεται. οὐκ ἠδύνατο οὖν ἀπὸ τοιούτων πηγῶν μερικῶν γίνεσθαι ἡ θάλασσα ὁλότης οὖσα τοῦ παντὸς ὕδατος· (3b) λείπεται ἄρα μὴ ἔχειν πηγὰς ἁλυκὰς τὴν θάλασσαν, ὑφ' ὧν εἰσρεῖται.

Olymp. arab. 108.2-15:

الدليل على أن البحر ليس لمياهه ينابيع يجرى منها أن جميع العيون لما كانت مياهها إما جارية وإما غير جارية: أما الجارية فمثل الأنهار، وأما غير الجارية فبعضها متخذة مثل الآبار، وبعضها حادثة من تلقاء أنفسها مثل المياه التى تحدث من زلازل الأرض. ولم توجد مياه البحر بمنزلة الجارية التى تنبع من العيون

وِلا من المياه التى تجرى الحادثة من تلقاء أنفسها، وذلك أنه لو كانت للبحر عيون يجرى منها لأمكن أن يوقَف على هذه فى موضع من المواضع، كما قد وُقف على العيون التى تجرى منها مياه الأنهار العظيمة، وخاصة فى البحار التى يحيط بها الناس ويسكنون حولها. وليس نجده فى وقت من الأوقات وُقف له على عيون تجرى مياهه منها إليه. وقد نعلم أيضًا أنه ليس للبحر عيون منها مياهه، بمنزلة المياه المتخذة التى لا تجرى. إن ما يتخذ منها له مقدار يستطيع الإنسان الوقوف عليه. ومساحة البحر تتجاوز فى الطول والعَرْض مقدار ما يمكن فى قوة الإنسان الوقوف عليه. وكذلك أيضًا نعلم أنه ليس بمنزلة ما يجرى مما حدث من تلقاء نفسه من أن الحادث من تلقاء نفسه يكون فى الفرط وقليلاً جداً. وحال البحر ضدة هذا الحال.

[14f. وحال البحر ضدة هذا الحال sic cod.: وحال البحر هذه الحال Badawī]

The proof that the waters of the sea do not have sources from which they flow is as follows: While waters of all springs are either running or not running, the running waters are, for example, rivers. Some non-running waters are artificial [*muttaḫida*], like wells, while others are spontaneous [*ḥādiṯa min tilqāʾi anfusihā*], like waters that arise from earthquakes. Waters of the sea have not been found [to be] like flowing waters issuing from springs or from spontaneous running waters. That is, if the sea had sources from which it flowed, it should be possible to discover them in some place, just as one has discovered the sources from which the waters of large rivers flow, especially in the case of seas which people surround and around which they live, but we find no seas where at some time a source from which its waters flow to it has been discovered. (1a) We also know that the sea has no sources from which its waters flow, which are like non-flowing artificial waters. (1b) What is artificial has a size which a human is able attain, but the extent of the sea exceeds [*tataǧāwazu*] both in length and width the size which human power can attain. (2a) Similarly, we know that it is not like spontaneous flowing water (2bc) from the fact that what is spontaneous occurs rarely and is very small, but the condition of the sea is the opposite of this.

l. 6. ܡܬܥܒܕ ܒܐܝ̈ܕܝܐ ("artificial"): cf. Min. I.iii.1.6 above (ܡܬܥܒ̈ܕܢ ܒܐܝ̈ܕܝܐ).
l. 10. ܫܪܟܐ ܕ ("the remaining alternative is"): cf. Min. V.ii.1.2-3 above.

BOOK OF METEOROLOGY

The "book" (*kṯāḇā*) as a whole corresponds to the treatise (*maqāla*) on meteorological phenomena (*al-āṯār al-ulwīya*) in the *K. al-šifā*ʾ. For the overall correspondence of the chapters in this book to those in the *Šifā*ʾ, see Intro. Section 2.3.1 above.

l. 1. ܦܗܗܝܬܐ ("meteorology/meteora"): lit. the roaming things; corr. arab. الآثار العلوية. - Aristotle's *Meteorologica* is constantly called *ʿal pahhāyāṯā* or *kṯāḇā d-pahhāyāṯā* in Nic. syr. (10.11, 13, 14, 16, 18; 14.20; 42.21, 47.18; 56.16, 17, 19). See further on I.i.1.1-2 below. - Cf. Wright (1870-2) 776a 17f.: ܟܕ ܐܝܟ ܐܪܣܛܘ ܐܘܕܥ ܒܡܠܠ ܦܗܗܝܬܐ (in the introduction to Olympiodorus' commentary on the Organon, BL Add. 18821; cf. PS 3041).

l. 2-3. "It contains five chapters": All manuscripts read "four chapters", although there are in fact five chapters in the book.

BOOK OF METEOROLOGY, CHAPTER ONE

ON CLOUDS AND THINGS WHICH DESCEND FROM THEM

The chapter as a whole corresponds to the first chapter (*faṣl*) of the treatise (*maqāla*) on meteorology in the *Šifā*ʾ (*Šifā*ʾ 35-39).

The very first 'theory' of the chapter (I.i.1), however, which deals with the process of evaporation and cloud-formation, is based not on the *Šifā*ʾ but on Abū al-Barakāt's *Muʿtabar*, with the result that solar radiation is given a more prominent role in the production of vapour in *But.* than in the *Šifā*ʾ, as was the case with the production of heat discussed in Section Min. IV.iv. above where the whole section is closely based on Abū al-Barakāt. - Abū al-Barakāt is also used as the main source for the whole of Section I.ii., which deals with rain. This is partly due to the fact that the discussion of rain as such in the *Šifā*ʾ is relatively short and BH had consequently to turn to a different source for a longer discussion.

Among other sources BH had access to, Nic. syr. is not used to any great extent in this chapter, while the relatively frequent use of the *De mundo* may be noted (I.ii.3.6-9, iii.2.3-5, iii.2.8). Fakhr al-Dīn al-Rāzī's *Mabāḥiṯ* too is used at least in two passages (I.iii.2.1-3, iii.3.1-2).

l. 2. ܡܛܠ ܥܢܢܐ ܘܡܐ ܕܢܚܬܝܢ ܡܢܗܘܢ: The chapter title is borrowed from the *Šifā*ʾ (35.3): في السحب وما ينزل منها وما يشبه ذلك ("on clouds, what descends from them and similar things"). IS's title is, strictly, more accurate, since the phenomena discussed in the chapter include frost and fog, which do not "descend" from clouds.

Mete. I. Section i.: *On How Clouds are Formed*

The first theory, after the introductory words, follows Abū al-Barakāt's *Muʿtabar*. The remaining two theories are based on the first part of the corresponding chapter of the *Šifā*ʾ. The sources used may be summarised as follows.

I.i.1.1-3	-	Introduction
I.i.1.3-5	*Muʿtabar* II.213.7f.	Generation of exhalations
I.i.1.5-7	*Muʿtabar* II.213.8-16, 214.6-7	Formation of cloud
I.i.1.7-10	*Muʿtabar* II.213.16-19	Cold stratum of air
I.i.2.1-5	*Šifā*ʾ 35.6-9	Formation of vapour
I.i.2.5-6	*Šifā*ʾ 35.11-12	Elevation of condensation
I.i.2.6-10	(*Šifā*ʾ 35.12-15)	contd.: Personal observation
I.i.3.1-5	*Šifā*ʾ 35.15-20	Causes preventing rise of vapour

l. 2. ܡܛܠ ܐܝܟܢܝܘܬܐ ܗܘܝܐ ܕܥܢܢܐ: cf. *Šifā*ʾ 35.4: في كيفية تولد السحاب.

<u>I.i.1.1-3</u>: Introduction

But.: The [discussion of] things generated inside the earth is followed by the [division of] philosophy which is called μετεωρολογία - viz. the science of the roaming things [*pahhāyātā*] - by the ancients. Of these things, we ought first to investigate the clouds.

l. 1. ܒܬܪ ܕܒܓܘ ܐܪܥܐ ܡܬܝܠܕܝܢ ("things generated inside the earth"): i.e. the minerals etc. discussed in the preceding book.

l. 1-2. ܗܝ ܕܡܬܩܪܝܐ ܡܛܐܘܪܘܠܘܓܝܐ ܐܝܟܘܬ ܝܕܥܬܐ ܕܦܗܝܬܐ ܡܢ ܩܕܡܝܐ ("which is called μετεωρολογία - viz. the science of the roaming things [*pahhāyātā*] - by the ancients"): cf. *Cand*. 111.9f.: ܗܝ ܕܡܬܩܪܝܐ ܡܛܐܘܪܘܠܘܓܝܐ ܡܢ ܩܕܡܝܐ ܕܝܕܥܬܗܘܢ. ܐܝܟܘܬ ܝܕܥܬܐ ܕܦܗܝܬܐ ܡܢ ܩܕܡܝܐ ("the knowledge of which is called μετεωρολογία by the ancients, viz. the science of the roaming things"). - This goes back to Arist. *Mete*. 338a 26: ὃ πάντες οἱ πρότεροι μετεωρολογίαν ἐκάλουν. - The phrase "ܡܛܐܘܪܘܠܘܓܝܐ ܐܝܟܘܬ ܝܕܥܬܐ ܕܦܗܝܬܐ" occurs in exactly the same form as here at Nic. syr. 10.16, where it does not connect with what immediately precedes and follows, so that it is likely to be a gloss which has crept into the text.

I.i.1.3-5: Generation of exhalations by radiational heat

But.: (1a) When the sun's rays [*zallīqā*, sg.] shine on earth and water, (1b) the heat which is generated by them (1c) raises dust from earth and vapour from water.
Muᶜtabar II.213.7f.:

اذا اشرق الشعاع على سطح الارض والماء احدث فيهما حرارة فيصعد بتلك الحرارة من الارض غبار ومن الماء بخار ومن الممتزجات ممتزج

(1a) When the rays [*šuᶜᶜāᶜ*, sg.] shine on the surface of earth and water, (1b) they cause heat in them (1c) and through that heat there rises dust from earth, vapour from water and a mixture [of dust and vapour] from mixtures [of earth and water].

l. 3. ܙܠܝܩܐ ("rays"): On the role of solar radiation in creating heat, see Min. IV.iv.1 above, where, as here, the source is Abū al-Barakāt.
l. 4. ܚܠܐ ("dust"): Elsewhere the exhalation rising from earth is called "smoke" (*tennānā*, corr. arab. *duḫān*) or "smoky exhalation". The word *ḥellā*, which occurs only here in *But.* Min. and Mete., renders Abū al-Barakāt's *ġubār*.

I.i.1.5-7: Formation of cloud

BH here greatly shortens a discursive passage of Abū al-Barakāt. - The reason given in the middle part of the passage here as to why the middle region of the air is cold goes back ultimately to Arist. *Mete.* 340a 24-32.[1]

But.: (1) These [i.e. vapour and dust] mingle [*methallṭīn*] with each other as they rise (2a) and they come to the [stratum of] air (2b) which is far below [*mtaḥtay saggī*] the element fire (2c) and which the heat of the rays [*ḥammīmūt zallīqā*] [reflected from the earth] does not reach [*lā māṭyā*]. (3a) There they grow cold [*qārīn*] (3b) and are condensed [*metraṣpīn*], (3c) and turn into cloud.
Muᶜtabar II.213.8-16, 214.6-7:

... الممتزجات ممتزج والصاعد/ ٩ بالحرارة من ذلك كما قيل يصعد من مضيق الى سعة ومن جهة مزكز الى/ ١٠ محيط فتصعد أجزاؤه على خطوط مستقيمة كلما امعنت فى الصعود تباعدت/ ١١ فتفرق مجتمعها وتباعد متقاربها وتشتت فى طريقها وتنتهى حركتها بعضها ببعض/ ١٢ فيتعلق الرطب بالرطب والرطب باليابس واليابس باليابس بواسطة الرطب/ ١٣ حتى ينتهى الى حد من الجو يقصر الحرارة الشعاعية المنعكسة من الارض الى/ ١٤ ما يليها من الجو عن الوصول اليه والحرارة النارية ايضا لبعد موضعها الطبيعى/ ١٥ عنه لا تنتهى اليه وذلك هو الجو الذى بين الجوين الأدنى المستخن تتسخن به/ ١٦ الارض والماء عن مشرق الشعاع ‹و›الاعلى المستخن بحر النار وهذا المتوسط العديم الحرارة من الجانبين ... ٦ فيبرد ما فى اعاليه من بخار أصعدته اليه الحرارة فاذا برد/ ٧ ذلك البخار عاد هابطا ولقى صاعدا فبرده فتراكم من ذلك سحاب كثيف ...

(1) ... and a mixture [of dust and vapour] from mixtures [of earth and water]. What rises with heat rises, as has been said, from a narrow to a

[1] Note in particular 340a 26-29: ψυχρότερος, διὰ τὸ μήθ ' οὔτω πλησίον εἶναι τῶν ἄστρων θερμῶν ὄντων μήτε τῶν ἀπὸ τῆς γῆς ἀνακλωμένων ἀκτίνων.

wide place and from the centre to the periphery, so that its parts rise in straight lines. The more (the parts) endeavour to rise, the more they distance themselves from each other, so that (the parts) joined together become separated, (the parts) close together distance themselves from each other, they are dispersed in their paths, and their movements end in each other; (1) then, a moist (particle) becomes attached to a moist, a moist to a dry, and a dry to a dry through the agency of the moist, (2) until they come to the limit [*ḥadd*] of the air [*ǧauw*], (2c) which the radiational heat [*al-ḥarāra al-šuᶜᶜāᶜīya*] reflected by the earth up to the (part of the) air adjacent to it fails to reach [*taqṣuru ᶜan al-wuṣūl*], (2b) and which the fiery heat [i.e. the heat from the fiery sphere], too, cannot come to because of the distance of its natural place from it - that is to say, (2a) the air between the two airs, (2c) [namely] the warm lower [air], where the earth and the water are warmed by the shining of the rays, (2b) and the upper, warmed by the heat of the fire. ... (3a) Then, [the coldness] refrigerates the vapour in its heights which have been raised to it by the heat. When that vapour grows cold, it descends again, meets [a body of] rising [vapour] and refrigerates it, (3c) and a thick cloud is accumulated from that [process].

l. 5. ܟܕ ܡܬܥܠܝܢ ܒܚܕ̈ܕܐ ܡܬܚܠܛܝܢ ("mingle with each other as they rise"): cf. IS, *Šifāʾ* 39.6: وفى أكثر الأمر فيصعدان من الأرض مختلطين ("[vapour and smoke] usually rise from the earth mixed together"); also Mete. IV.i.1.2 below (ܟܕ ܒܚܕ̈ܕܐ ܚܠܝܛܝܢ ܣܠܩܝܢ).

I.i.1.7-10: Cold stratum of air

But.: (1a) This [stratum of] air is colder than the other [strata of] air, (1b) not by nature (1c) but because of the coldness of earth and water, (1d) the chill [*qurrā*] of the night with [its] absence <of radiation> and of winter, (1e) and the remoteness of the heat-source [*maḥḥmānā*] above and below.

Muᶜtabar II.213.16-19:

وهذا المتوسط العديم الحرارة من الجانبين هو الى الارض اقرب ويدنو من رؤوس الجبال الشامخة والظهور العالية فيكون ابرد موضع فى الهواء وبرده انما يكون عن برد الارض والماء اذا كانا على بردهما بغيبة الشعاع فى الليل وضعف اشراقه فى نهار الشتاء

(1a) This middle air devoid of heat [is so] on two accounts; it is closer to the earth and it adjoins the summits of lofty mountains and high prominences, so that it becomes the coldest place in the air. Its coldness is (1c) due to the coldness of earth and water, (1d) when these two are at their coldest due to the absence [*ġaiba*] of the radiation [*šuᶜᶜāᶜ*] at night and weakening of its illumination [*išrāq*] during the day in winter.

l. 9. ܘܩܘܪܐ ܕܠܠܝܐ ܥܡ ܓܠܝܙܘܬ‹ܙܠܝܓܐ› ‹ܘ›ܣܬܘܐ ("the chill of the night with [its] absence <of radiation> and of winter"): ܓܠܝܙܘܬ om. FM[1]; ‹ܙܠܝܓܐ› ‹ܘ› supplevi. - Cf. *Muᶜtabar* II.213.19: بغيبة الشعاع فى الليل وضعف اشراقه فى نهار الشتاء.

l. 9. ܡܚܡܡܢܐ ("heat-source"): cf. Min. IV.ii.5.5, IV.iv.5.2 above.

l. 9-10. "and the remoteness of the heat-source above and below": cf. I.i.1.5-7 above.

I.i.2.1-5: Formation of vapour from air and water

But.: (1a) Because vapour is, as it were [*meddem*], intermediate between water and air, (1b) it is possible for [*maṣyā d-*] a cloud to form from the two: (1c) from water, when it is rarefied and dissipated; (1d) from air, when it is thickened and condensed. (2a) Just so, we often see on the peaks of cold mountains how the air grows cold; (2b) where there had been clear sky [lit. after clear sky], [the air] becomes condensed and suddenly turns into a cloud; (2c) and there falls rain or snow.

*Šifā*ʾ 35.6-9:

وهذا الجوهر البخارى كأنه متوسط بوجه ما بين الماء والهواء، فلا يخلو إما أن يكون ماء قد تحلّل وتصعّد، أو يكون هواء قد تقبض واجتمع. وقد يعرض تكون السحاب من كلا الوجهين جميعا. وذلك أنا كثيرا ما شاهدنا الهواء يبرد فى أعالى الجبال الباردة فينقبض بعد الصحو سحابا دفعة، ثم يثلج

(1a) This vapoury substance is, as it were [*ka*ʾ*annahū*], intermediate, in a way [*bi-wağhin mā*], between water and air, (1c) so that it is without doubt either water which has been dissipated and vaporised (1d) or air which has been thickened and condensed. (1b) The formation of clouds [*takawwun al-saḥāb*] may sometimes arise [*qad ya*ʿ*raḍu*] from both factors together. (2a) We often see how air grows cold on the peaks of cold mountains (2b) and - where there had been clear sky [lit. after clear sky] - suddenly contracts into a cloud; (2c) then it snows.

l. 4. ܒܬܪ ܨܚܘܐ ("where there had been clear sky"): lit. "after clear sky": بعد الصحو IS.

l. 5. "and there falls rain or snow": IS talks only of "snow".

I.i.2.5-6: Elevation at which condensation takes place

But.: Vapour does not always [*b-k̲ul-zb̲an*] ascend to the very cold region of the air before a cloud is formed.

*Šifā*ʾ 35.11-12 وهذا البخار ليس يحتاج كل مرة أن يبلغ الموضع البارد الشديد البرد فى الجو

This vapour does not need every time [*kull marra*] to reach the region of the air which is extremely cold.

l. 6. "before a cloud is formed": lit. "then a cloud is formed". - There is not equivalent clause in the *Šifā*ʾ, but the idea is implicit there.

I.i.2.6-10: Personal observation

IS had recounted an experience of his to illustrate the point made in the immediately preceding passage (*Šifā*ʾ 35.11-12, corr. *But.* I.i.2.5-6). BH has replaced this with a personal observation of his own. The connection with the preceding passage is somewhat blurred in *But.* due to the omission of the clause corresponding to IS's "the air was ... not that very cold".

But.: (1) For I myself, when I was on top of a mountain between Gubos and Claudia with others, saw a cloud which ascended half way up the moutain and lay [*kpāṯ*] over Gubos, (2) while I and those with me were in the clear sky above it. (3) After a while, our disciples came to us squeezing out of their clothes the water of the rain which had fallen on the place and on them.

Šifā' 35.12-15:

وقد شاهدنا البخار وقد صعد فى بعض الجبال صعودا يسيرا حتى كأنه مكبّة موضوعة على الوهدة تحتها قرية، إحاطة تلك الوهدة لا يبلغ نصف فرسخ. وكنا نحن فوق تلك الغمامة فى الصحو وكان الهواء خريفيا ليس بذلك البارد جدا، فكان أهل القرية يمطرون من تلك الغمامة.

(1) We once saw vapour ascending gently up a certain mountain until it looked as if it lay placed [*mukabba mauḍūʿa*] over a gorge, at the bottom of which was a village and whose circumference did not reach half a parasang [ca. 3 km].[2] (2) We ourselves were above that cloud in the clear sky and the air was autumnal and not that very cold. (3) The inhabitants of the village were rained upon by that cloud.

l. 7. "Gubos and Claudia": Townships on the Euphrates downstream of Melitene, Gubos probably to be identified with Kale, opposite İzolu, and Claudia some way downstream, perhaps at the confluence of the Širǒ Suyu with the Euphrates. Both were still seats of Syr. Orth. bishops in BH's time and BH was himself bishop of Gubos in his youth. There were several monasteries in the area (visits to which may have been the reason for BH's presence on these mountains), most notably Ss. Sergius and Bacchus of Sargīsiya; the patriarchal monastery of Mār Barṣawmā, if Honigmann's identification of its site with Borsun Kalesi is correct, will have lain to the southwest of Claudia and could not be described as being between Gubos and Claudia.[3] There seems to be no way of knowing whether the experience recounted here took place during BH's occupancy of the see of Gubos or during one of his later visits to the area, although he must at any rate have been old enough at the time to have his own "disciples" (*talmīḏayn* l. 9).

I.i.3.1-5: Causes preventing rise of vapour

But.: (0) Vapour is prevented from rising to a great height [*rawmā saggī'ā*]: (1) either by [its own] weight, when it gains additional material [*mlō'ā yattīrā nehwē leh*]; or (2) by a wind that blows above it and prevents its ascent; or (3a) by a mountain which stands in front of it; or (3b) by opposing winds which confine it in their midst; or (4) when a [body of] vapour precedes and is stationary, another joins it and clings to it; or (5a) because of the severity [*tuqpā*] of the coldness (5b) which annuls the impulse of its ascent [*zawʿā d-sullāqeh*] and makes it heavy.

---

[2] 1 *farsaḫ*: ca. 6 km (Hinz [1955] 62).

[3] On Gubos: see Honigmann (1954) 124f.; Fiey (1988); id. (1993) 202. - On Claudia: Honigmann (1954) 120; EI[2] IV.484 s.v. "Ḳalāwdhiya" [Cahen]; Fiey (1993) 256f.

*Šifā*ʾ 35.15-20:

فعلمنا أن البخار كثيرا ما يؤدى به تكاثفه وتواتر مدده وبطء حركته المصعدة إياه إلى فوق، فيحوج ألى أن يتكاثف ويقطر مثل المعصور، وربما أحوجته الرياح إلى ذلك إما مانعة إياه عن الصعود بحركتها فوق، وإما ضاغطة إياه إلى الاجتماع بسبب وقوف جبال حائلة قدام الريح أو بسبب اختلاف رياح متقابلة، وإما لإلحاق المتأخر بالمتقدم الواقف وإلصاقه به من غير أن يكون حاجز من قدام، وإما لشدة بردها فيكثف به السحاب.

(0) We know that often the vapour is subject to thickening, (1) successive flow of its materials [*tawātur madadihi*] (5b) and slowing down of the movement raising it upwards [*ḥarakatuhu al-muṣṣaʿʿida*]; then it is compelled to thicken and drip like something being squeezed. (2) Often it is compelled to do so by winds, either by [winds] preventing its ascent by their movements [in the region] above [the vapour]; or (3) [winds] forcing it to concentrate, (3a) by reason of mountains standing before the wind and barring the way, or (3b) by reason of conflict [*iḫtilāf*] of opposing winds; or (4) through its action of causing the back part [of the vapour] to join and cling to the stationary front part, without there being something blocking it from the front; or (5) through the intensity of their coldness.

l. 4 "which confine it in their midst": There is no equivalent concept in IS. Did BH perhaps read احتباس vel sim. for *Šifā*ʾ 35.18 اختلاف?

Mete. I. Section ii.: *On Rain*

The discussion of rain in the *Šifā*ʾ is relatively short. For a more detailed discussion BH had to turn elsewhere and, as may be seen from the summary below, the section of *But.* here is based almost entirely on Abū al-Barakāt.

But.	*Muʿtabar*	Other sources	
I.ii.1.1-9	II.213.8-11, 214.6-13		Generation of rain
I.ii.2.1-8	II.215.12-15, 17-21		Cloud and rain
I.ii.3.1-6	II.215.21-24		Rain and wind
I.ii.3.6-9		*De mundo*; Nic.	Classification of rain
I.ii.4.1-6	II.215.24-216.6		Evaporation from sea
I.ii.5.1-2	II.216.6-7		Summer rain
I.ii.5.2-7	II.216.8-13		Influence of topology
I.ii.6.1-4	II.216.14, 17-21		Summing up
I.ii.6.4-7		*Šifā*ʾ 36.1-5	Ethiopia

I.ii.1.1-9: Generation of rain

The passage may be considered a paraphrase of a somewhat involved discussion of how rain is generated by Abū al-Barakāt, parts of which have already been used in I.i.1 above. It may be noted that Abū al-Barakāt (and BH following him) has a rather exaggerated idea of the height to which the vapours ascend in relation to the size of the earth.

But.: (1a) The vapour, which rises from earth and water with the heat of the rays, (1b) becomes scattered [*metparraq*] in its ascent [*b-massaqteh*], (1c) because it moves from a centre [*qehnṭrōn*] which is narrow to a circle [*ḥudrā*] which is wide, (1d) and as it is scattered, (1e) its particles grow small. (2a) If (the particles) do not reach the cold place of the air [*waᶜdā qarrīrā d-āʾār*], (2b) they are dissipated [*metpawššān*] (2c) and turn into air. (3a) If it arrives [*netwaᶜᶜad*] at the cold region of the air [*aṯar āʾār qarrīrā*], (3b) the vapour grows cold, (3c) and because of its coldness [*b-qarrīrūṯeh*], it descends [*nāḥet*]. (3d) As it descends [*b-maḥḥtteh*], it becomes concentrated [*metkannaš*] (3e) because it moves from a circle that is wide to a centre that is narrow. (3f) As it becomes concentrated [*b-kunnāšeh*], its particles grow large and heavy (3g) and let drops fall [*maṭṭpān*] as rain. (4a) If the vapour which grows cold and heavy in the air falls from a very elevated height [*rawmā d-ᶜellāy saggī*], (4b) in the long duration of its [fall] [*b-ṭawḥeh arrīkā*] its particles become bound with each other more [*yattīr mesṭammdān*], (4c) and for that reason its drops become larger.

Muᶜtabar II.213.8-11, 214.6-13 [cf. under I.i.1.5-7 above]: (1a) What rises with the heat [*al-ṣāᶜid bi-l-ḥarāra*] rises, as has been said, (1c) from a narrow to a wide place and from the centre [*markaz*] to the periphery [*muḥīṭ*], so that its parts rise in straight lines. (1b) The more (the parts) endeavour to rise [*amᶜanat fī al-ṣuᶜūd*], the more they distance themselves from each other [*tabāᶜadat*], so that (the parts) joined together [*al-muǧtamiᶜ*] become scattered [*tafarraqa*], (the parts) close together [*al-mutaqārib*] distance themselves from each other [*tabāᶜada*], (2b) they are dispersed [*tašattata*] in their paths. ... Then, [the coldness] refrigerates what vapour there is in its heights, which has been raised to it by the heat. (3b) When that vapour grows cold, (3c) it descends again, meets [a body of] rising [vapour] and refrigerates it, and a visibly thick cloud [*saḥāb katīf fī al-marʾā*] is accumulated from that [process]. The whole of it then drips down [*qaṭara*] as rain, or a part of it drips down while a part is scattered. What is made to drip down by that coldness drips down [at first] as watery drizzle [*al-radād al-māʾī*], which is [then] heated and rises [again], rises and is scattered, is scattered and its particles [*aǧzāʾ*] grows small [*ṣaġurat*]. (3c) Then, it falls again with the coldness [*bi-l-bard*] (3e) from the wide periphery [*saᶜa muḥīṭ*] towards the narrow centre [*ḍīq markaz*], (3d) so that the lines between its particles [*ḫuṭūṭ masāfātihi*] approach each other [*taqārabat*] and the particles meet with the approaching (of the lines). Then, (the particles) becomes united [*ittaṣalat*] with each other; (3f) (particles) which were small becomes large [*kabura*]; those which were hot becomes cold; (3g) and they fall as rain. (4a) If they come from the high (part of the) air [*ǧauw ᶜālin*], (4b) the duration of their [fall] [*masāfatu-hā*] will be longer, so that [there will be] more [opportunity for] their joining together [*ittiṣāl*] in that duration and their drops will become larger.[4]

---

[4] *Muᶜtabar* II.214.6-13: فيبرد ما فى اعاليه من بخار أصعدته اليه الحرارة فاذا برد ذلك البخار عاد هابطا ولقى صاعدا فبرده فتراكم من ذلك سحاب كثيف فى المرأى فقطر اما كله مطرا او يقطر بعضه ويتفرق البعض وانما يقطر

l. 1. "The vapour which rises from earth and water with the heat of the rays": cf. I.i.1.3-5 above.

l. 3, 4. ܘܕܘܟܬܐ ܩܪܝܪܐ ܕܐܐܪ, ܐܬܪ ܐܐܪ ܩܪܝܪܐ ("the cold place of the air", "the cold region of the air"): i.e. the cold middle stratum of the air discussed at I.i.1.6-10 above. - *Wa^cdā* (also the verb *netwa^{cc}ad* in l. 4): cf. III.iii.4.2, IV.iii.1.3 below.

l. 6. ܘܢܬܝܩܪ ("and [grow] heavy"): Abū al-Barakāt does not talk about the vapour growing heavy, but IS does (*Šifā*ʾ 36.5 *fa-yatqulu*): cf. also I.ii.3.5 below.

I.ii.2.1-8: Cloud and rain

But.: (1a) A cloud is not a distinct substance [*ūsiya ḥrēṯā*] which bears and gives birth to [*t^cīnā w-yāldā*] rain (1b) as the simple [*hedyōṭē*] think, (1c) but its substance is the same as the substance of rain. (2a) For this reason when a cloud forms below the top of a mountain, (2b) as one descends, (2c) one enters into its midst (2d) and sees nothing more than what one sees on a rainy and foggy day. (3) A cloud is therefore condensed and thickened vapour [*lahgā lbīdā wa-rṣīpā*]. (4) If (the vapour) grows cold completely, rain falls; if excessively, snow; and if inadequately, it merely makes the air turbid [*dālaḥ*]. (5) This turbidity [*dlīḥūṯā*] of the air due to the cloud resembles the turbidity of water due to dust [*^caprā*]. (6) A cloud is transferred [*mšannyā*] from one place to another through movement by winds.

Mu^ctabar II.215.12-15, 17-21:

والسحاب ليس غير المطر والثلج فى الجو اذا رؤى من بعيد وليس شيئا يقطر منه المطر كما يظنه من لا يتأمل ويتفكر فان السحاب قد يكون تحت الجبل ويراه الانسان وهو فوق الجبل والسحاب تحته ويدخل الانسان فى السحاب فلا يرى الا ما يراه فى يوم المطر والضباب. ... وكدر الهواء بالسحاب ككدر الماء بالتراب وليس هناك شيء يحمل الماء كما تظنه الدهماء وانما السحاب هو المطر بعينه حيث يرى من بعيد والسحاب الذى لا يمطر يكون عن بخار تراكم فكدر ولم يبرد ولو برد لقطر وينجر السحاب بحركة الرياح من موضع الى موضع

(1ab) A cloud is nothing other than rain and snow in air when it seen from a distance. It is not something from which rain falls, (1c) as someone who does not examine and consider [things] might think. (2a) For a cloud is sometimes formed below a mountain (2b) and someone who is above the mountain and has the cloud below him looks at it. (2c) He [then] enters the cloud (2d) and sees nothing other than what he sees on a rainy and foggy day. ... (5) Turbidity [*kadar*] of air due to cloud is like turbidity of water due to dust [*turāb*]. (1a) There is nothing there which bears [*yaḥmilu*] water, (1b) as the simple [*al-dahmā*ʾ] think. A cloud is nothing other than rain itself where it is seen from a distance. (4) A cloud which does not

---

(يقطر) ما يقطر من ذلك البرد الرذاذ المائى الذى سخن فصعد فتفرق وتفرق فصغرت اجزاؤه وعاد بالبرد هابطا من سعة محيط الى ضيق مركز فتقاربت خطوط مسافاته فتلاقت الاجزاء فى تقاربها فاتصلت بعضها ببعض فكبر صغيرها وبرد سخينها فهبط مطرا فان وردت من جو عال كانت مسافتها اطول فكان اتصالها فى مسافتها اكثر وقطراتها اكبر

produce rain is formed from accumulated vapour which has grown turbid but has not grown cold. If it were to grow cold, it would fall in drops. (6) A cloud is brought [*yanğarru*] from one place to another by the movement of winds.

l. 1. ܕܛܥܝܢܐ ܘܝܠܕܐ ܡܛܪܐ ("which bears [*ṭ*ᶜ*īnā*] and gives birth to rain"): cf. *De mundo* syr. 141.29f.: . ܥܢܢܐ ܐܝܬܘܗܝ ܥܒܝܘܬܐ ܕܐܠܡܐ ܕܡܬܟܢܫ ܘܡܬܥܒܐ . ܘܝܠܕܐ ܡܛܪܐ (< gr. 394a 26f.: νέφος δέ ἐστι πάχος ἀτμιδῶδες συνεστραμμένον, γόνιμον ὕδατος); BH, *Tract.* ms. C55v 5f.: ܘܥܢܢܐ ܡܢܗ ܡܬܒܪܝܐ ܕܚܒܠܐ ܘܝܠܕܐ ܐܘܟܝܬ ܡܚܬܐ ܡܛܪܐ ("... and cloud is formed from it, which bears [*ḥāḇlā*] and gives birth to - i.e. lets fall - rain").[5]

I.ii.3.1-6: Rain and wind

But.: (1a) The strong movement of wind blocks [*kālyā*] the weak movement of rain's descent, (1b) and rain does not fall until the wind has been interrupted [*metpasqā*]. (2a) Consequently men say that rain interrupts the wind, (2b) but the truth is that the wind is [first] interrupted and then the rain falls. (3a) unless the cloud is cooled greatly by the wind, (3b) its density [*rṣīpūṯā*] increases (3c) and its raindrops become heavy; (3d) then, the motion of the rain overcomes the motion of the wind (3e) and rain falls with the wind inclining in the direction in which it is blowing.

*Mu*ᶜ*tabar* II.215.21-24:

فتقاوم الحركة الريحية لقوتها حركة نزوله لضعفها فلا يمطر حتى تكف الريح عنه فيقول الناس ان قطع المطر الريح وانما انقطع الريح فنزل المطر او حتى يتراكم وتتصل اجزاؤه فى حركته ويشتد برده وتكثر قطراته فتقاوم بنقلها الريح ويمطر مع هبوبها والى جهة مهبها

The movement of wind with its strength opposes [*tuqāwimu*] the movement of (rain's) descent with its weakness, (1b) so that it does not rain until the wind has stopped [*takuffa*]. - People therefore say that rain interrupts [*qaṭa*ᶜ*a*] wind, (2b) but in fact the wind is [first] interrupted, then the rain falls. - (3a) Or [it does not rain], until its particles have accumulated and been united in its movement, the coldness has grown intense (3bc) and the raindrops have increased; (3d) then, (the raindrops) oppose the wind with their impulse [*naql*] (3e) and rain falls while the wind is blowing [lit. together with its blowing] and in the direction in which it is blowing.

I.ii.3.6-9: Classification of rain:

Arist. classified rain into ψακάς and ὑετὸς (347a 10-12; cf. Olymp. 86.10-12; respectively, *nadan*, *maṭar* Ibn al-Biṭrīq). IS does not give us a classification, although he does talk, besides of *maṭar*, of *dīma* (continuous rain) and *wābil* (heavy downpour) at *Šifā*ᵓ 36.5. Abū al-Barakāt, too, does not have a classification as such, although he talks

[5] Cf. Takahashi (2002c) 161.

of *radād* (drizzle) alongside *maṭar*. - BH gives us a threefold classification here. Of the four terms used here, two, at least, can be traced back to definite sources.

> *But.* [underline: agreement with *De mundo*; italics: with Nic.]:
> (1) Rain, whose drops are *large* and which falls with violent force and incessantly [*b-ḥēpā ḥīṣā w-ammīnā'īt*], is called [*metqrē*] "zīqā". (2) (Rain), whose drops are small and which falls gently [*nīḥā'īt*] and for a short time, *is named* [*meštammah*] "*rsīsā*" and "rzāpā". (3) (Rain), whose drops are very small and cannot be seen in the air, is termed [*metkannē*] "rsāmā".
>
> *De mundo* syr. 142.2-5 [< gr. 394a 30-32]:
>
> ܗܘ ܟܕ ܕܢܝܚܐܝܬ ܡܬܥܨܪܐ ܡܢ ܠܒܘܟܗ. ܢܘܛܦܬܐ ܪܟܝܟܬܐ ܡܪܣܣܐ. ܗܘ ܕܝܢ
> ܕܬܬܥܒܝܢܘܬ ܬܬܠܒܟ. ܢܘܛܦܬܐ ܥܒܝܬܐ ܢܦܩܢ. ܘܠܗ ܠܗܢܐ ܐܝܟ ܕܡܛܪܐ ܩܪܝܢ
> ܙܝܩܐ. ܡܛܠ ܕܒܚܝܦܐ ܚܝܨܐ ܘܐܡܝܢܐ ܢܚܬ ܥܠ ܐܪܥܐ.
>
> (2) For when (the cloud) is squeezed gently [*nīḥā'īt*] by its condenser, it sprinkles soft drops [*nuṭpāṯā rakkīkāṯā*, gr. μαλακὰς ψακάδας]. (1) When the cloud is condensed firmly, it produces large drops [gr. ἀδροτέρας sc. ψακάδας]; they call [*qārēn*] this kind of rain *zīqā* [gr. ὑετός],[6] because it falls on the earth with violent and incessant force [*b-ḥēpā ḥīṣā w-ammīnā*].
>
> Nic. syr. 15.9-11 [< Arist. 347a 10-12; > *Cand.* 113.8-10]:
>
> ܘܟܕ ܪܟܝܟܐܝܬ ܡܢ ܩܪܝܪܘܬܐ ܠܐܬܠܒܘܫܬܐ ܕܒܗ /. ܪܛܝܒܘܬܐ ܡܫܬܡܗܐ ܗܘ ܚܫܐ.
> ܘܟܕ ܚܣܝܢܐܝܬ ܩܪܝܪܘܬܐ ܠܗ ܡܥܨܪܢܘܬܐ ܘܐܬܒܪܬܐ /. ܡܛܪܐ.
>
> When [coldness] squeezes the moisture in it softly [*rakkīkā'īt*], the result [*ḥaššā*, Arist. πάθος] is named [*meštammah*] drizzle [*rsīsā*, ψακάδες]; (1) when it squeezes it strongly and [it falls] in large particles, rain [*meṭrā*, ὑετός].

7. ܙܝܩܐ FM[V] ("*zīqā*"): ܙܢܩܐ PLlV; ܐܝܩܐ M[1]. - The passage of *De mundo* syr. above, as well as the corresponding passages of *Cand.* (ed. Bakoš 113.9, reporting no variants; Çiçek 86.24) and *Rad.* (ed. Istanbul 22.16) confirms the reading *zīqā*. The citation of the passage of *But.* here under "znāqā" at PS Suppl. 113a (from ms. l) should therefore be deleted. - It is curious that J.E. Mellus, while following L in writing *znāqā* in his own copy (V), wrote *zīqā* in the margin of M. - The term *zīqā* reappears as the name for a type of whirlwind at III.iv.tit., III.iv.4.7 below.[7]

I.ii.4.1-6: Evaporation from the sea

> *But.*: (1) Vapour that rises from the sea is hot, (2a) but when it is cooled [*qā'ar*] by the wind and (2b) its particles are pressed [*metkabšān*] against each other, (2c) its density increases [*sāgyā lbīdūteh*] and (2d) rain falls

---

[6] *De mundo* arab. F [Brafman] 92.4: *wābil*.

[7] The *zīqē* also appear elsewhere in Syriac as "shooting stars", e.g. Bar Kepha, *Hex.* 196v b4ff. (cf. Nau [1899] tr. 14 n.1); see further PS 1118 s.v.

on land which is near the sea. (2e) In fact, the wind drives the cloud also to lands which are far from the sea and causes rain to fall. (3a) For this reason the south wind is found to cause rain to fall on one area, (3b) the north wind on another (3c) and east and west winds on others (3d) in accordance as to whether the sea is situated in the [source]-direction of the wind and whether it is blowing from its side [lit. in accordance with (*lpūṯ*) the position of the sea in the direction (*pnīṯā*) of the wind and its blowing (*maššḇāh*) from its side].

Muᶜtabar II.215.24-216.6:

وكذلك يصعد/ ١ البخار من البحار وهو جار (حار) فيتراكم بمدده وبحركة الرياح ويكبس بعضه الى/ ٢
بعض فيشتد تراكمه ويبرد بالترويح من الريح فيمطر على ارض قريبة او بعيدة/ ٣ من البحر فى زمن البرد
والحر فان الرياح الحارة قد تتبرد كما علمته فترى اكثر/ ٤ السحب الممطرة بقرب البحار او برياح قوية
تحملها من جهتها فكذلك/ ٥ ترى الرياح الجنوبية تمطر بلادا والشرقية بلادا اخرى والشمالية والغربية
كلا (كذا)/ ٦ بحسب قربه من البحر وهبوبه من الجهة القريبة

(1) Similarly vapour rises from the seas and (that vapour) is hot. (2c) It accumulates [*yatarākumu*] due to [the continued supply of] its matter and the movement of the wind and (2b) a part [*baᶜḍ*] of it is pressed [*yukbasu*] against another, (2c) so that its accumulation intensifies [*yaštaddu tarākumuhu*] and (2a) it grows cold [*yabrudu*] due to ventilation [*tarwīḥ*] by the wind; (2d/e) it then causes rain to fall on the earth near to and far from the sea in [both] cold and hot seasons. For hot winds sometimes grow cold as you know. You therefore see most rain-causing clouds near the seas or with strong winds which bring them from their direction. (3) In this way you see south winds causing rain on one region, (3bc) the east wind on another, and the north and west winds similarly, (3d) in accordance with [*bi-ḥasabi*] (the region's) proximity to the sea and (the wind's) blowing [*hubūbu-hā*] from the nearby direction [*min al-ǧiha al-qarība*].

l. 3. "In fact, wind drives ...": This is not in the above passage of *Muᶜtabar*, but cf. I.ii.2.7f. above.

I.ii.5.1-2: Summer rain

But.: (1a) Summer rains are generated only from the vapours of seas, rivers and lakes, (1b) and not from condensation of cold air.

Muᶜtabar II.216.6f.:

ولا يكاد المطر (المطر؟) الصيفى يكون الا منَ ابخرة البحار التى تحملها الرياح لا من برد الجو
وسحب المنعقدة فيه

(1a) Summer rain-producing [clouds] [or simply: "summer rain"?] are rarely formed except from vapours of seas which are carried by winds, (1b) not from the coldness of the air and the clouds condensed in it.

l. 1. "rivers and lakes": Not in the above passage of *Muᶜtabar*, but see *Muᶜtabar* II.216.12 (الانهار العظيمة والبطائح; quoted under I.ii.5.2-7 below).

I.ii.5.2-7: Influence of topology

But.: (1a) Places which are deficient in rain are those that are very hot and far from the sea, (1b) [those] whose land is low-lying, (1c) [those] in

whose soil there is mixed, salty and sulphureous heat [*ḥammīmūṯā muzzāgāytā mallīḥtā wa-keḇrīṯānāytā*] - because of which the air becomes hot and dry - (1d) and [those] into which winds blow little from the direction of sea. (2a) Places which abound in rain are those that are near the sea, (2b) [those] to which sea winds blow more, (2c) [those] which are adjacent [*šḇāḇāyīn*] to snowy mountains (2d) and [those] which are meadowy [*margānāyē*] and retentive of moisture [*nāṭray tallīlūṯā*].

*Mu*ᶜ*tabar* II.216.8-13:

والبلاد التى لا تمطر هى التى جوها احر والبحر منها ابعد وارضها اخفض وفى تربتها حرارة مزاجية كالسبخة المالحة والحمائية والكبريتية التى ينعكس منها الى جوها حر اكثر ويكون هبوب الرياح التى من جهة البحر فيها اقل والارض الكثيرة المطر هى المقاربة للبحر والتى الرياح البحرية تهب فيها وبقرب الجبال الحاملة للثلوج وبقرب الانهار العظيمة والبطائح وتكون فى نفسها مزجية (مرجية؟) حافظة للانداء من مطر ومن شتاء الى شتاء.

(1a) Regions which are not rained upon are those whose air is very hot, [those] from which the sea is very far, (1b) [those] whose land is low-lying, (1c) [those] in whose soil is mixed heat [*ḥarāra mizāğīya*] like a salty, feverish [?, *ḥummāʾīya*] and sulphureous marsh/salina [*sabaḫa*],[8] from which heat is mostly reflected into their air, (1d) and [those] into which the winds from the direction of the sea blow very little. (2a) A land abounding in rain is one that is close to the sea, (2b) [one] into which the sea-winds blow, (2c) [one] in the proximity of snow-capped mountains, [one] in the proximity of great rivers and lakes (2d) and [one] which has in itself a mixture [*mizağīya* (?); or: "which is meadowy (*marğīya*) in itself"?] retentive of the moisture [*ḥāfiẓa li-l-andāʾ*] from the rain from one winter to another.

l. 7. ܡܪܓܢܝܐ ("meadowy"): < *margā*, "meadow". - I suspect BH read the word printed as مزجية in the Hyderabad edition of the *Mu*ᶜ*tabar* as مرجية (< *marğ*, meadow), although the adjective *margānāyā* appears to be unknown to the lexica in this sense.

I.ii.6.1-4: Summing up

But.: (1) In a word [*ba-ḥdā l-mēmar*], a cloud condenses [*qāṭar*] above mountains where the air is very cold, (2) and is driven by winds to a distant place. (3) For this reason rain does not fall much on the sea because of the heat of the air, although vapours are abundant there. (4) If this were not the case, the rain on the sea would never cease.

*Mu*ᶜ*tabar* II.216.14, 17-21:

١٤ وبالجملة فالمطر عن بخار سخن فصعد وتفرق ثم تراكم وبرد فاجتمع ونزل ... ١٧ ويتأدى من الجبل الى موضع بعيد كما يتأدى/ ١٨ من البحر الا ان جو البحر لحرارته لا تنعكس منه الابخرة على اكثر الامر اليه/ ١٩ بل الى حيث تحمل السحب الرياح من جو بارد فيبرد وينزل فيه والجبال على/ ٢٠ الاكثر لبرد جوها تمطر ابخرتها من موضع صعودها او ما يقاربه ولولا ذلك/ ٢١ لدام مطر البحر واتصل لاتصال بخاره الصاعد

[8] "Marsh" (*sabaḫa*): cf. comm. on Mete. IV.i.3.4-9 below.

(1) In short [*bi-l-ǧumla*], rain (arises) from vapour which has grown hot, has risen and has been dispersed, then has come together and grown cold, has become condensed and falls. ... (2) It is conveyed from the mountain to a distant place, just as it is conveyed from the sea, (3) except that because of the heat of the air (above) the sea, vapours are usually not returned from (the air) to (the sea), but to whatever place with cold air the winds carry the clouds; it then grows cold and falls there. The mountains, because of the coldness of their air, usually cause their vapours to fall as rain in the place where they have risen or in the vicinity. (4) If this were not the case, the rain on the sea would be incessant and continuous due to the continuous (flow) of the vapour [rising] from it.

I.ii.6.4-7: Ethiopia

Ethiopia as an example of a hot county abounding in rain goes back to Arist. *Mete*. 349a 5-9 (there along with Arabia). The passage there, however, is rather different from ours. The immediate source of our passage of *But.* is the *Šifā*ʾ. There, however, the latter half of the passage applies not to rain in Ethiopia but to rain in general.[9]

But.: (1) In the land of the Cushites [Ethiopia, *arʿā d-kuššāyē*], although it is very hot, rains are plentiful, because winds drive the vapours of the sea towards it [*l-ṣawbāh*] and press them against its mountains. (2) There (the vapours) grow cold, condense, turn into clouds and, once they have acquired a watery nature, they grow heavy and are discharged.

*Šifā*ʾ 36.1-5:

وإنما يكثر المطر بأرض الحبشة مع حرارتها لاندفاع الأبخرة إليها وانضغاطها فى جبالها وهى بين يدى رياحها. وأما فى اكثر الأمر فإن الأبخرة تتصعد وتعلو إلى الحيز البارد من الهواء فتبرد ويعين ذلك انفصال ما ينفصل عنها من الدخان الحار اليابس الذى نذكره. وقد شاهدنا ذلك الانفصال على بعض قلل الجبال. فإذا بردت بالسببين انعقدت هناك غماما، ثم يستحيل ماء فيثقل فينزل.

(1) Rain is plentiful in the land of the Abyssinians despite its heat because the vapours are driven to it and are pressed against its mountains; these [processes come about] through the agency of winds. In most cases, vapours rise and ascend to the cold region of the air and grow cold. - This is helped by the discharge [*infiṣāl*] from it of a portion [lit. what is discharged, *mā yanfaṣilu*] of the hot and dry smoke, which we have mentioned. We have seen such discharge on some mountain tops. - (2) Then, when they have grown cold for the two reasons, (the vapours) thicken there into a cloud, then (the cloud) turns into water, then grows heavy, and then falls.

l. 6. ܟܝܢܐ ܡܝܢܝܐ ("watery nature"): ماء IS. - The alteration may be due to a lost passage of Nic. syr., since the expression "is transformed into the

[9] On the rôle of mountains in generation of cloud and rain, cf. Min. II.ii.3 and Mete. I.i.3 above.

nature of water [*ṭabīʿat al-māʾ*]" is found in the corresponding passage of Olymp. arab. (100.1).

l. 7. ܡܬܦܨܝܢ ("is discharged"): ينزل IS. - BH's choice of the word may have been influenced by the verb *infaṣala* which occurs earlier on the passage of the *Šifāʾ*.

Mete. I. Section iii.: *On Dew, Frost, Snow, Hail, Fog and Mist*

Most of the material in the section is taken from the *Šifāʾ*, although BH has incorporated into it materials taken from other sources, especially in his discussion of snow and of hail.

	Šifāʾ	Other sources	
I.iii.1.1-6	36.7-9		Formation of dew
I.iii.1.6-7	36.9		Formation of frost
I.iii.1.7-8	38.9	(Nic. syr. 16.16-19)	Relationships
I.iii.2.1-3	36.10-11	*Mab.* II.173.5-8; (*Cand.*)	Formation of snow
I.iii.2.3-5		*De mundo* syr. (+ Nic.?)	Whiteness of Snow
I.iii.2.5-8		*Muʿtabar* II.214.13-17:	Confinement of vapour
I.iii.2.8		*De mundo* syr. 142.9f.	Snowstorm
I.iii.3.1-2	36.12-13	*Mabāḥiṯ* II.173.5-8	Formation of hail
I.iii.3.3-5		*Muʿtabar* II.214.19-22	Rotundity of hailstone
I.iii.3.5-10	36.13-15, 38.2f.		Hail in winter/summer
I.iii.4.1-4	36.16-37.5		Spring and autumn
I.iii.4.5-7	37.12-14	Nic. syr. 17.9-12	Size/rotundity of hail
I.iii.4.8-10	38.1-2	own observation	Unusual hailstones
I.iii.5.3-5	38.7-9		Fog & clear weather
I.iii.6.1-5	39.1-6		Summing up
I.iii.6.5-9	39.10-13		Vapour and smoke

l. 1. ܛܠܐ ("dew"): So in all mss. here, as well as I.iii.1.1, 5-7, I.iii.6.6, although the form ܛܠܠܐ seems to be preferred in *Cand*. (ed. Bakoš 111.11, 113.6).

<u>I.iii.1.1-6</u>: Formation of dew[10]

But.: (1a) Dew [*ṭallā*] [is formed] not from cloud, (1b) but from diurnal vapour [*lahgā yawmāyā*], (1c) whose ascent has been retarded [*da-mšawḥar sullāqeh*] (1d) and whose quantity is small, (1e) when the coldness of the night strikes [*māḥyā*] it, (1f) condenses and thickens it [*marṣpā w-maqṭrā*] (1g) and turns it into water. (2) This water because of the fineness and smallness of its particles falls gently [*niḥāʾīt*] and imperceptibly. (3) When a large quantity has been gathered of these watery particles, then one senses the dew as it falls during the night.

[10] Cf. Arist. *Mete*. 347a 13-16, 347b 20-22.

Šifā᾽ 36.7-9:

فإن الطل ليس يتكون من سحاب، بل من البخار اليومى المتباطئ الصعود القليل المادة إذا أصابه برد الليل وكثفه وعقده ماء ينزل نزولا ثقيلا فى أجزاء صغار جدا لا نحس بنزولها إلا عند اجتماع شىء يعتد به

(1a) Dew [*ṭall*] is formed not from cloud, (1b) but from diurnal vapour [*al-buḫār al-yawmī*], (1c) retarded in ascent [*al-mutabāṭiᶜ al-ṣuᶜūd*] (1d) and small in quantity, (1e) when the coldness of the night strikes [*aṣāba*] it, (1fg) thickens it and condenses it [*kaṯṯafahu wa-ᶜaqqadahu*] into water. (2) It falls heavily [*nuzūlan ṯaqīlan*, i.e. slowly (?)] in very small particles, [such that] you will not sense their descent (3) except when a considerable (quantity) has been gathered.

l. 1. "Dew [is formed]": The omission of the verb is a little awkward. IS has يتكون.

I.iii.1.6-7: Formation of frost[11]

But.: From the freezing [*maḡlḏānūṯā*] of dew due to severe coldness, frost [*qarṣānā*], also called *zmāytā*, is formed.
Šifā᾽ 36.9: فإن جمد كان صقيعا - If (the dew) freezes, it is frost [*ṣaqīᶜ*].

l. 6f. "frost [*qarṣānā*], also called *zmāytā*": The same two terms are used at *Cand.* 113.7f. (*zmāytā* only *Rad.* [Istanbul] 22.15). Of BH's Syriac sources, Nic. syr. has *zmāytā* throughout (15.14, 16, 20; 16.10, 11, 14, 15, 18; 21.28; almost invariably corr. gr. πάχνη), while *De mundo syr.* has *qarṣānā* at 141.20 (gr. πάγος) and ܐܘܪܣܢܐ, no doubt to be emended to ܩܪܣܢܐ (gr. πάχνη), at 141.28 (at 138.9 we find ܐܓܠܝܕܐ corresponding to gr. πάχνη).

I.iii.1.7-8: Relationship of dew, frost, rain and snow

But.: Hence dew has [*qnē*] to frost the relationship [*peḥmā*] of rain to snow.
Šifā᾽ 38.9: ويجب أن تعلم أن نسبة المطر إلى الثلج نسبة الطل إلى الصقيع.
You ought to know that the relationship [*nisba*] of rain to snow is the relationship of dew to frost.
Cf. also *Šifā᾽* 36.11-12: ونظيره من البخار الفاعل الطل هو الصقيع.
The counterpart [*naẓīr*] (of snow) that is formed from vapour that makes dew is frost.
Nic. syr. 16.16-19 [< Arist. *Mete*. 347b 29-31]:

ܐܠܐ ܒܠܚܘܕ ܒܦܘܚܡܐ ܗܘ ܕܡܛܪܐ ܡܬܥܒܕ ܠܬܚܬ ܛܠܐ. ܒܦܘܚܡܐ ܕܝܢ ܕܬܠܓܐ ܙܡܝܬܐ. ܒܦܘܚܡܐ ܕܝܢ ܕܒܪܕܐ ܠܝܬ ܡܕܡ ܠܘܬܢ.

But there is formed below only dew in place of [*b-p̄eḥmā d-*] rain and frost in place of snow. There is nothing by us in place of hail.

[11] Cf. Arist. *Mete*. 347a 16f.

I.iii.2.1-3: Formation of snow

The passage corresponds in the first place to that of the *Šifā*ʾ quoted below. As may be seen, however, clause (1d) agree more closely with the *Mabāḥit*, while (1b) has affinities with *Cand*. (The passage of *Cand*. is itself probably based, at least partly, on the *Mabāḥit*).

But. [underline: agreement with *Šifā*ʾ; italics: with *Mabāḥit*]:
(1a) Snow [*talgā*] is generated (1b) when, *before* [*qdām hāy d-*] *the particles* [*mnawwātā*] of a cloud *assemble* [*netkannšān*] and coalesce [*nesṭammdān*] with each other (1c) and drops [*nutpātā*] *are formed* [*metgablān*] from them, (1d) *there comes upon them* [*māṭyā l-hēn*] *a powerful coldness* [*qarrīrūtā d-ḥayltānyā*], (1e) and they fall while freezing [*kad maḡldān*].
*Šifā*ʾ 36.10-11:

وهذا السحاب يعرض له كثيرا أنه كما يأخذ فى التكاثف، وفى أن يجتمع فيه حب القطر، يجمد ولم تتخلق الحبات بحيث تحس فينزل جامدا فيكون ذلك هو الثلج.

It often happens to such a cloud that, (1b) when it is in the process of condensation and when the grains of raindrops [*ḥabb al-qaṭr*] are gathering [*yağtamiʿu*] in it, it freezes (1c) while the grains have not been formed [*yataḫallaqu*] to the extent that they can be perceived. (1e) Then, it falls while freezing [*ğāmidan*] (1a) and that is snow [*talğ*].
Mabāḥit II.173.5-8:

واما ان كان البرد شديدا فلا يخلو اما ان يصل البرد الى الاجزاء البخارية قبل اجتماعها وانخلاقها حبات كبارا او بعد صيرورتها كذلك فان كان على الوجه الاول نزل ثلجا وعلى وجه الثانى نزل بردا

(1d) If the coldness is intense, it must be either that the coldness comes to [*yaṣilu ilā*] (1bc) the vapoury particles before their assembly [*iğtimāʿ*] and formation [*inḫilāq*] into large grains or after their development in this way. (1ae) In the first case, (the vapour) falls as snow. In the second case it falls as hail.
Cand. 112.8-10: If the air is very cold [*saggī qarrīrā*] and, (1b) before [*men qdām d-*] the particles of a cloud are united [*metḥayydān*] and assemble [*metkannšān*] with each other, (1d) the coldness comes into contact with [*tegšōp(y)*] them, (1ae) there falls snow.

I.iii.2.3-5: Whiteness of Snow

Given the certain use of *De mundo* syr. at I.iii.2.8 below, as well as the agreement of the phrase "whiteness of snow" and the notion of "separation", *De mundo* syr. 142.5-7 is likely to be one of the sources of this passage. The role played by air, however, is not mentioned explicitly in the *De mundo* (though it may be implied there). The attribution of the whiteness of snow to air is made by Arist. not in his *Mete*. but at *GA* 735b 21 (in connection with the whiteness of semen). A passage where the attribution is made in a discussion of snow itself is found in Theoph. *Mete*. and in a passage copied thence in Bar

Kepha, but the wording there is not so close to ours. A passage which, though not concerned with the whiteness of snow, does talk about the confinement of air in snow in terms similar to ours (note phrase "between particles" and notion of "rarity") is found in Nicolaus' *De plantis*, and this leads us to suspect that Nicolaus discussed the whiteness of snow in such terms in a passage that is not preserved in the Cambridge manuscript and that the passage of *But.* here is to be seen as a conflation of *De mundo* syr. and that lost passage of Nic. syr.

But. [underliness: agreement with *De mundo*; italics: with Nic. *De plant.*]: The reason for the whiteness of snow [*ḥewwārūṯ talgā*] is the intrusion of *air* [*ʿellāl āʾār*] *in between the* watery *particles* [*meṣʿaṯ mnawwāṯā mayyānāyāṯā*] which are separated [*mp̄assqān*] from each other, loose and rarefied [*mp̄arsʿān w-saḥḥīḥān*] and not connected to each other, *as* [is the case] with foam [*akmā da-ḇ-ruʿtā*].

De mundo syr. 142.5-7 [< gr. 394a 32-35]:

ܬܠܓܐ ܕܝܢ ܗܘܐ ܡܐ ܕܥܢܢܐ ܕܡܬܚܫܠܢ ܩܕܡ ܠܡܛܪܐ. ܡܬܦܣܩܢ ܡܛܠ

ܥܒܝܘܬܗܝܢ ܘܡܬܦܪܕܕܢ. ܦܣܩܗܝܢ ܗܟܝܠ ܥܒܕ ܠܪܥܬܢܘܬܐ ܘܚܘܪܘܬܗ ܕܬܠܓܐ.

Snow is formed when the clouds are separated [*metpasqān*] because of their density and are made loose [*metparddān*] before they turn into rain. Their separation [*psāqā*] therefore causes the foaminess [*ruʿtānūṯā*] and whitness of snow [*ḥewwārūṯeh d-ṯalgā*].

Nicolaus, *De plantis* arab., ed. Drossaart Lulofs-Poortman (1989) p. 187f., §178 (corr. Ps.-Arist., *De plantis* 825a 5-8):

... وذلك ان الثلج ينزل شبيها بالزبد فتجمده الريح ويضغطه الهواء، فيكون بين اجزائه تخلخل ...

... The reason is that snow falls like foam; the wind congeals it and air presses it together. So there will be rarity [*taḫalḫul*] between its particles [*baina aǧzāʾihi*] ... (tr. Drossaart Lulofs).

Bar Kepha, *Hex.*, Paris 241, 182r b19-26[12]:

ܚܘܪܐ ܕܝܢ ܗܘܐ ܬܠܓܐ ܡܛܠ ܐܐܪ ܕܚܒܝܫ ܒܓܘܗ. ܐܦ ܓܝܪ ܚܙܝܢ ܠܟܠܗܘܢ

ܦܓܪ̈ܐ ܐܝܠܝܢ ܕܡܬܚܒܫ ܒܗܘܢ ܐܐܪ ܣܓܝܐܐ. ܡܬܚܙܝܢ ܚܘܪܐ

ܐܟܙܢܐ ܕܪܘܥܬܐ ܘܐܝܟ ܡܫܚܐ ܕܡܬܚܒܛ ܥܡ ܡܝ̈ܐ

Snow is white [*ḥewwārā*] because of the air confined inside it. For we also see that all bodies in which much air is confined appear white, such as foam [*ak-znā d-ruʿtā*] and like oil beaten with water.

l. 3. "intrusion of air": For the role of air in whiteness, cf. also Min. III.ii.1.4-5 above (whiteness of sulphur); Olymp. 307.13, 18-20 (on whiteness of oil): λευκαίνεται δὲ διὰ τὸ κατεργάζεσθαι τὸ ἐν αὐτῷ ἀερῶδες ὑπὸ τῆς κινήσεως. ... λευκαίνεται δ' ὑπο ψύξεως, ἐπειδὴ παχυνόμενος ὁ ἐν αὐτῷ ἀὴρ οἷον χιονώδης γίνεται λευκός· ἡ γὰρ χιὼν οὐδὲν ἕτερον ἐστιν ἢ ἀὴρ πεπηγὼς μεταβλητὸς εἰς ὕδατος γένεσιν.

---

[12] ≈ Theoph. *Mete.* syr., Cantab. Gg 2.14, 353r 24-26, ed. Daiber [1992] p. 186f. - Cf. Theoph. *Mete.* arab. (Ibn al-Khammār), ed. Daiber (1992) p. 238, 9.8-11, and Daiber's comm. ad loc. (p. 277).

I.iii.2.5-8: Confinement of vapour by snow

But.: (1) Snow confines [*ḥābeš*] the vapour that rises from the earth in the proximity of the earth. (2a) For this reason, on a snowy day the chill is broken [*mettḇar*] (2b) until there is much snow on the earth, the coldness grows strong over its surface and the [supply of] vapours ceases [*bāṭlīn*]; (2c) then [at last] the chill gains strength.

Muʿtabar II.214.13-17:

وان كان البرد اشد جمد الرذاذ ونزل ثلجا وحبس البخار الصاعد بقرب الارض فلم يتصعد ولذلك ترى الجو الادنى فى يوم الثلج اذفأ فاما اذا نزل الثلج واشتد برد وجه الارض انقطعت الابخرة فبرد الجو باسره وما علا منه وما دنا الى الحد الذى تنتهى اليه التبريد ...

If the coldness becomes more intense, the drizzle [*raḏāḏ*] freezes, (1) falls as snow, and confines [*ḥabasa*] the rising vapour in the proximity of the earth, so that it does not rise. (2a) For that reason, you see that the lower air is warmer on a snowy day. (2b) When, however, the snow has fallen and the coldness of the surface of the earth has intensified, the [supply of] vapours is interrupted [*inqaṭaʿat*], so that the air grows completely cold along with what is above and below it to the utmost limit of refrigeration ...

I.iii.2.8: Snowstorm

But.: ܘܬܠܓܐ ܕܢܚܬ ܘܬܟܝܒܐܝܬ ܫܠܝܬ . ܟܘܟܝܬܐ ܡܬܩܪܐ

Snow which falls suddenly and heavily [*menšel(y) wa-tkībāʾīt*] is called [*metqrē*] "snowstorm" [*kōkītā*] (ܘܬܠܓܐ ܕܢܚܬ ܘܬܟܝܒܐܝܬ ܫܠܝܬ . ܟܘܟܝܬܐ ܡܬܩܪܐ)

De mundo syr. 142.9f. [< gr. 394a 36-b 1]:

ܡܐ ܕܝܢ ܕܗܢ ܬܠܓ ܘܡܫܬܟܝܒܐܝܬ ܢܚܬ ܗܘ ܬܠܓܐ . ܟܘܟܝܬܐ ܡܫܬܡܗ

When that snow falls suddenly and heavily [*men šel(y) wa-tkībāʾīt*], it is named [*meštammah*] "snowstorm" [*kōkītā*, gr. νιφετός].

Cand. 113.10: ܘܬܠܓܐ ܗܘ ܕܡܫܠܝܐܝܬ ܘܬܟܝܒܐܝܬ ܢܚܬ ܟܘܟܝܬܐ ܡܫܬܡܗ

Rad. [Istanbul] 22.17: ܘܬܠܓܐ ܣܓܝܐܐ ܕܡܫܠܝܐܝܬ ܢܚܬ ܟܘܟܝܬܐ

l. 8. ܟܘܟܝܬܐ ("snowstorm"): The word *kōkītā* also means and is more often used in the sense of "whirlwind" (PS 1695f., Brockelmann 320b). BH himself uses the word in both senses in *Cand.*, in the sense of "snowstorm" at *Cand.* 113.10 and in that of "whirlwind" at 125.3, 8 (where the immediate source is *De mundo* syr. 143.26 ܥܠܥܠܐ ܕܝܢ ܘܟܘܟܝܬܐ, < gr. 395a 7 λαῖλαψ δὲ καὶ στρόβιλος).

I.iii.3.1-2: Formation of hail

But. [underline: agreement with *Šifāʾ*; italics: with *Mabāḥit*]:

(1a) If, *after the coalescence* [*mesṭammḏānūṯā*] of *the particles* of the cloud and *formation* of *large drops*, (1b) *coldness comes to them* and freezes them, (1c) *hail* [*bardā*] takes shape [*mestakkam*] and *falls*.

Šifāʾ 36.12-13: وأما إذا جمد بعد ما صار ماء وصار حبّا كبارا، فهو البَرَد.
When (the cloud) freezes (1a) after it has turned into water and has turned into large drops [lit. grains], then it is hail [*barad*].

Mabāḥit II.173.5-8[13]: If the coldness is intense, it must be either (1b) that the coldness come to [*yaṣilu ilā*] the vapoury particles (1a) before their assembly [*iğtimāʿ*] and formation [*inḫilāq*] into large grains or (1a) after their development in this way. In the first case, (the vapour) falls as snow. (1c) In the second case it falls as hail.

l. 1. ܡܬܟܢܫܢܘܬܐ, ܡܬܓܒܠܢܘܬܐ ("coalescence", "formation"): cf. I.iii.2.1-2 above (ܘܡܬܓܒܠ ... ܢܬܟܢܫܢ).

I.iii.3.3-5: Rotundity of hailstone

The roundness of hailstones is discussed again at I.iii.4.5-7 below.

But.: (1a) The reason for the roundness [*glīlūṯā*] of hailstones [*parṣnāṯā d-bardā*] is the combination [*knušyā*] of two movements, (1b) namely the vertical [lit. in the length, *da-ḇ-urkā*] in which they fall, (1c) and the horizontal [lit. in the breadth, *d-ḇa-p̄ṯāyā*] in which they are driven by winds, (2a) along with [the fact that] as their corners [*gōnawwāṯā*] are rubbed against each other [*mettšīp̄ān*], (2b) they become broken [*mettaḇrān*] and rounded [*metgalglān*].

Muʿtabar II.214.19-22:

فان نزل الثلج من عال ايضا وحركته رياح فى نزوله صدمت الاجزاء بعضها بعضا وتشبث بعضها ببعض ودارت بحركتين الطولية التى بها هبطت والعرضية التى بها تشبثت فيدور فيستدير شكل البرد النازل او يقارب الاستدارة

If the snow falls from the height and is moved about by the wind during its descent, the particles collide with each other and coalesce [*tašabbatat*]. (1a) They become round [*dārat*] through the two movements, (1b) the vertical [lit. lengthwise, *ṭūlīya*] through which they fall (1c) and the horizontal [lit. breadthwise, *ʿarḍīya*] through which they coalesce, so that the shape of the descending hail becomes round and spherical [*yadūru wa-yastadīru*] or approaches sphericity [*istidāra*].

l. 3-5. ܐܝܬܝܗ ... ܘܥܡ ... ܟܢܘܫܝܐ ... ܘܥܠܬܐ ("The reason ... is the combination ... along with [the fact that] ..."): The subject of the whole sentence is *ʿellṯā* in l. 3, the main verb *īṯēh* in l. 5. The syntax is awkward unless some correction is made (e.g. ܘܡܛܠ ܕܥܡ for ܘܥܡ in l. 4), but the text as it stands is (just) understandable and may well be what BH wrote. The lack of editing here on BH's part is also evident in the fact that the discussion of the roundness of hail is repeated below (I.iii.4) and especially in the fact that the three words *gōnawwāṯā*, *mettaḇrān*, *metgalglān* recur together there in line 6.

---

[13] See under I.iii.2.1-3 above; cf. also *Cand.* 112.10-113.2.

I.iii.3.5-10: Scarcity of hail in winter and summer

The main source of the passage is *Šifāʾ* 36.13-15. BH has joined to it the reason for the scarcity of hail in summer which is mentioned much later on at *Šifāʾ* 38.2f.

But.: (1) Hail usually occurs in spring and autumn. (2) It does not occur in winter or in summer, (3a) in winter, because if the coldness of the winter is severe (3b) it causes the cloud to freeze (3c) before it turns into drops and (3d) causes snow, (3e) and if it is weak it does not do anything; (4) in summer, because there is then [lit. "in it"] very little moist and heavy vapour which is the material of hail.

Šifāʾ 36.13-15:

وأكثر البرد إنما يكون فى الربيع والخريف، ولا يكون فى الشتاء. وذلك لأن البرد الشتوى إن كان شديدا، فعل الثلج، وأجمد السحاب، ولا يمهله ريثما ينعقد حبا؛ وإن كان ضعيفا لم يفعل شيئا.

(1) Most hail occurs in spring and autumn. (2) It does not occur in winter and (3a) that is because if the coldness of the winter is severe (3d) it causes snow, (3b) causes the cloud to freeze and (3c) does not allow it to condense; (3e) if it is weak it does not do anything.

Šifāʾ 38.2f.: ويقل البَرَد فى الصيف، لأن البخار الرطب الثقيل يقل فيه، وفى الشتاء ...

(4) Hail is rare in summer, because there is then [lit. "in it"] little moist and heavy vapour, and in winter ...

I.iii.4.1-4: Spring and autumn

In his summary of the *Šifāʾ* here BH omits IS's discussion of the flight of coldness from heat, which goes back to Aristotelian notion of the mutual repulsion of heat and cold (*Mete*. 348b 2 ἀντιπερίστασις), a notion that plays a key role in Aristotle's theory of the formation of hail.[14]

But.: (1a) In spring and autumn, on the other hand, the cloud does not freeze [*maḡlḏā*] immediately [*ba-ʿḡal*], (1b) but [only] when its particles have coalesced [*meṣtammḏān*] and condensed [*metraṣpān*] completely, (1c) then they freeze (1d) and hailstones fall. (2a) Hail occurs more in autumn, (2b) because the vapour freezes readily [*dlīlāʾīṯ*], (2c) since it has been rarefied [*metparsaʿ*] during the summer. (3) This [i.e. that rarefied vapour freezes more easily] is known from the fact that hot water freezes before the cold.

Šifāʾ 36.16-37.5:

وأما فى الربيع والخريف فإن السحاب ما دام لم يتكاثف بعد تكاثفا يعتد به يكون الحر مكتنفًا إياه فلا يجمد ثلجا؛ حتى إذا استحكم استحصافه وأحاط به الهواء الحار والرياح القوية الحارة، هربت البرودة

[14] On ἀντιπερίστασις, see Lee (1952) 82f. n. b; Lettinck (1999) 97-99, 112f. etc. (see index 498 s.v. *antiperistasis*). - Cf. Mete. III.i.4.8-11 below (also Mete. IV.i.3.3, Min. III.ii.1.7).

دفعة إلى باطن السحاب، واستحصف السحاب دفعة على ما علمت من التعاقب المشروح فيما سلف صورته. ويكون الاستحصاف قد جمع البخار قطراً، قد عرض له استعداد شديد للجمود لخلخلة الحر إياه. كما أن الماء الحار أسرع جمودا من البارد، فيجمد وقد صار قطرا كبارا. ولذلك ما يكون البَرَد فى الخريف أكثر لأن الصيف يكون قد أفاد الأجسام زيادة تخلخل، والمتخلخل أقبل لتأثير البرد والحر جميعا.

(1a) In spring and autumn, the heat encloses [*muktanif*] the cloud while it has not undergone any appreciable condensation [*takāṯuf*], so that it does not freeze into snow. As a result, when it is firmly constituted [*istaḥkama istiḥṣāfuhu*] and the hot air and hot strong winds surround it, the coldness suddenly flees [*harabat*] inside the cloud and the cloud is suddenly constituted [*istaḥṣafa*] in accordance with what you know about the mutual repulsion [of heat and cold] [*taʿāqub*, ἀντιπερίστασις],[15] explained previously. The constitution [*istiḥṣāf*] [of the cloud] occurs (1b) when the vapour has already gathered into drops, [the vapour] having received a high degree of aptitude [*istiʿdād šadīd*] for freezing [*ǧumūd*] through rarefication by the heat, (3) in the way that [*kamā*] hot water freezes more quickly than cold, (1c) with the result that it freezes [*yaǧmudu*] (1b) when it has become large drops. (2a) For this reason hail [*barad*] occurs more in autumn, (2c) because the summer has already provided objects with an increase in rarefication [*taḫalḫul*], and a rarefied object is more receptive to the influence of both coldness and heat.

I.iii.4.5-7: Size and rotundity of hail

The passage is based mainly on IS, but BH has added to it some elements taken from Nic. syr. 17.9-12, a passage he had used earlier at *Cand.* 113.2f.

But. [underline: agreement with *Šifāʾ*; italics: with Nic.]:
(1a) Hailstones which fall from clouds that are very *far from the earth* (1b) melt and grow small (1c) and through the *prolonged* [*naggīrā*] attrition [*ḥkākā*] in the air their corners [*gōnawwāṯā*] become *broken* [*mettaḇrān*] and rounded [*metgalglān*]. (2a) Hailstones which are *large* and not *spherical* [*espērnāyān*] (2b) fall from a cloud that is close to the earth.

Šifāʾ 37.12-14:

وما كان من البَرَد نازلا من سحب بعيدة يكون قد صغر وذاب واستدار لذوبان زواياه بالاحتكاك فى الجو. وأما الكبار وخصوصا التى لا استدارة فيها، فهى التى تنزل من سحب دَوَان.

(1a) Hail falling from distant clouds (1b) have become small, molten (1c) and round [*istadāra*] because of the melting of their corners [*zawāyā*] through attrition [*iḥtikāk*] in the air. (2a) Large [hailstones], especially those which have no roundness [*istidāra*], (2b) are those which fall from low clouds.

---

[15] See Lettinck (1999) 112; cf. *Šifāʾ* 18.8, 60.4.

Nic. syr. 17.9-12[16]:

ܛܒ ܕܝܢ ܪܘܪܒܬܐ ܗܘܝܢ ܟܐܦܐ ܕܒܪܕܐ ܟܕ ܠܐ ܒܕܘܟܬܐ ܕܪܚܝܩܐ ܡܢ ܐܪܥܐ
ܡܬܩܢܝܢ. ܘܡܛܠܗܕܐ ܠܐ ܗܘܝܢ ܓܠܝܠܬܐ. ܢܓܝܪܘܬܗ ܓܝܪ ܕܡܫܘܚܬܐ ܡܬܒܪܐ
ܠܩܪܢܬܗܝܢ: ܘܥܒܕܐ ܠܗܝܢ ܐܣܦܪܢܝܬܐ.

Stones of hail [*kēpē d-bardā*] are very large when they are formed not in a place far from the earth and for this reason are not round [*glīlātā*]. For [when they are formed far away] the long duration [*naggīrūtā*] of [their] descent breaks [*mtabbrā*] their protrusions and makes them spherical [*espērnāyātā*].

l. 5. ܟܐܦܝܗ . . . ܒܪܕܐ ("hailstones"): lit. "hail ... its stones".

I.iii.4.8-10: Unusual hailstones

The first example is taken from the *Šifāʾ*.

But.: (1) We have heard that on one of the mountains in the north there was a single hailstone which weighed more than a mina. (2) On a mountain between Beroea [Aleppo] and Antioch, I myself have seen large stones of hail which fell in the shape of horns and heads of oxen, sheep and horses.

Šifāʾ 38.1-2:

وقد حُدِّثت أن ببلاد الجبل قطعة بردة وقعت من السماء فنقلت إلى بدر بن حسنويه تزن كذا منا

(1) It has been reported that in a mountain country a piece of hail fell from the sky and was brought to Badr b. Ḥasanwaih [979-1014]. It weighed about a mann.

l. 8. "mina": *mann* IS; approx. 830 g. (Hinz [1955] 17f.; cf. comm. on Min. I.i.4.5 above).

I.iii.5.1-3: Fog

But.: (1a) Fog [*ʿarpellā*] is of the substance [*ūsiya*] of the cloud, (1b) but it does not have the consistency [*quyyāmā*] of a cloud. (2) For it is finer [*qattīnā*] than cloud and denser [*ʿabyā*] than mist [*ʿrūrā*], just as mist is finer than fog and denser than air.

Šifāʾ 38.6-7 وأما الضباب فهو من جوهر الغمام إلا أنه ليس له قوام السحاب

(1a) Fog [*dabāb*] is of the substance [*ǧauhar*] of the cloud, (1b) except that it does not have the consistency [*qawām*] of a cloud.

l. 2-3. "For it is finer than ...": The definition of *ʿarpellā* as the denser and *ʿrūrā* as the finer type of fog is found in the lexica of Bar ʿAlī, Bar Bahūl etc. (see PS 2974). "Mist" (*ʿrūrā*) is not mentioned in the corresponding passage at *Cand*. 112.7-8: "Hence it is known that fog [*ʿarpellā*] is denser than air and finer than cloud."

---

[16] Cf. Arist. 348a 27-36.

I.iii.5.3-5: Fog announcing clear weather or rain

The content of the passage goes back to Arist. *Mete*. 346b 33-35, but the immediate source is IS.

But.: (1) Fog which descends from the height, especially after rain, announces [*mbaddqā ʿal*] clear weather, (2) while that which begins to rise from below and does not vanish through evaporation [*lā mettalqā b-pūššāšā*] announces rain.
Šifāʾ 38.7-9:

فما كان منه منحدرا من العلو خصوصا عقيب لأمطار، فإيه [فإنه] ينذر بالصحو. وما كان منه مبتدئا من الأسفل متصعدا إلى فوق ولا تتحلل فهو ينذر بالمطر.

(1) [Fog] which descends from the height, especially following on rains, announces [*yundiru bi-*] clear weather. (2) [Fog] which begins from below and rises upwards and is not dissipated [*lā tataḥallalu*] announces rain.

I.iii.6.1-5: Summing up

The passage may be seen as an attempt to present more systematically the content of the passage of the *Šifāʾ* quoted below. In paraphrasing the passage, BH will have had in mind what has already been said in I.i.1 above, as well as the discussions concerning the four elements and their compounds in earlier books of *But*. (materials corresponding to Arist. *Mete*. IV and *De gen. et corr.*).

But.: (1a) To sum up [*ba-knīšū l-mēmar*], celestial, especially solar, heat raises [*massqā*] airy particles mixed [*mzīgān*] with small watery particles from water and moist grounds [*arʿāṯā tallīlāṯā*] (1b) and what is composed from them [*mrakkbā d-menhēn*] is called vapour. (2a) From dry earth, it lifts up [*mrīmā*] fiery particles mixed with earthy particles (2b) and what is composed from them is called smoke.
Šifāʾ 39.1-6:

... فيجب أن نعلم أن جميع الآثار العلوية تابعة لتكون البخار والدخان، وذلك لأن الحرارة السمائية إذا أثرت فى البلة الأرضية أصعدت منها أبخرة، وخصوصا إذا أعانتها حرارة محتقنة فى الأرض، فما تصعد من جوهر الرطب فهو بخار وصعوده بطئ ثقيل، وما يصعد من جوهر اليابس فهو دخان وصعوده خفيف سريع. والبخار حار ورطب، والدخان حار ويابس، وقلما يتصعد بخار ساذج أو دخان ساذج، بل إنما يسمى الواحد منهما باسم الغالب، وفى أكثر الأمر فيصعدان من الأرض مختلطين.

We ought to know that all meteorological phenomena [*al-āṯār al-ʿulwīya*] depend on [*tābiʿan li-*] the generation of vapour and smoke. (1a) That is because when the celestial heat acts [*aṯṯarat*] upon earthy moisture [*al-billa al-arḍīya*], it raises [*asʿadat*] vapours [*abḫira*] from it, especially when it is aided by the heat confined [*muḥtaqina*] in the earth. (1b) The moist substance that rises is vapour [*buḫār*] and its ascent is slow and heavy. (1c) The dry substance that rises is smoke [*duḫān*] and its ascent is light and swift. Vapour is hot and moist, while smoke is hot and dry. Rarely does pure vapour or pure smoke rise, but [each] one is named after the

predominant [component], (1a/2a) and in most cases the two rise from the earth mixed with each other [*muḫtaliṭaini*].

l. 1. ܒܟܢܫܘ ܠܡܐܡܪ ("to sum up"): cf. I.ii.6.1 above (ܒܟܢܫܐ ܠܡܐܡܪ); cf. also *De mundo* syr. 144.22: ܘܒܟܢܫܘܬܐ ܠܡܐܡܪ (< gr. 395a 28 συλλήβδην; at the beginning of a list of meteorological phenomena similar to that given at III.6.5.5-9 below).

l. 1. ܫܡܫܢܝܬܐ ("solar [heat]"): on the role played by the sun in the process, see I.i.1.3 above.

I.iii.6.5-9: Vapour and smoke

What we have here is, in effect, a list of the subjects treated in chapters II-IV below.

But. : (1a) Vapour is the material of cloud, rain, snow, dew, hail, frost, fog and mist, (1b) and on it the halo, rainbow and mock suns appear as illusions. (2) Smoke, on the other hand, is the material of wind, firewind, shooting stars, comets and terrible signs which are seen in the air, as we shall show.

Šifā' 39.10-13:

فالبخار مادة السحاب والمطر والثلج/ ١١ والطل والصقيع والجليد، وعليه تتراءى الهالة وقوس قزح والشميسات والنيازك. والدخان مادة/ ١٢ الرياح والصواعق والشهب والرجوم وذوات الأذناب من الكوكاب والعلامات الهائلة./ ١٣ وسيرد عليك تفصيل جميع ذلك.

[11 والصقيع habent ed. Ṭ et cod. D; om. ed. Cair.]

(1a) Vapour is the material of cloud, rain, snow, dew, frost and ice [*ǧalīd*], (1b) and on it appear the halo, rainbow, mock suns and lances. (2) Smoke is the material of winds, thunderbolts, shooting stars [*šuhub*], meteorites [*ruǧūm*], comets and terrible signs. The explanation of all these will follow.

l. 7. ܡܚܙܝܢܝܬ ("as illusions"): cf. II.i.4.3, 6 below; also *Cand.* 114.4.

l. 7. ܩܫܬܐ ܕܒܥܢܢܐ ("rainbow"): lit. "[rain]bow in clouds" (i.e. to distinguish from other kinds of "bows").

BOOK OF METEOROLOGY, CHAPTER TWO

ON ILLUSORY MATTERS WHICH ARE SEEN ON CLOUDS

The chapter of *But.* here corresponds to the second and third chapters of the treatise on meteorology in the *Šifāʾ*, whereby the first section of the chapter in *But.* corresponds to IS's Chapter 2 (ed. Cairo p. 40-46), which is devoted to a discussion of optics as a preliminary to the discussion of illusory phenomena such as the rainbow and the halo, and the remaining three sections of *But.* correspond to IS's Chapter 3 (ed. Cairo p. 47-57), which actually deals with these phenomena.

Mete. II. Section i.: *Introduction to the Teaching on the Efficient Cause of Cloud-Related Illusions*

The section corresponds, as has been said, to the second chapter of the treatise on meteorology in the *Šifāʾ* (ed. Cairo p. 40-46) where IS discusses optics as a preliminary to his discussion of the rainbow, halo etc.,[1] and is based largely on that chapter of the *Šifāʾ*.

BH does, however, make a number of departures from the *Šifāʾ* with (*bāhet̲ nā d-emar*, to use BH's own expression)[2] not very felicitous results. In II.i.4, BH develops what looks like a novel theory of reflexion and double reflexion. This is, in fact, based on a misunderstanding of the two Greek terms "ἀνάκλασις" and "διάκλασις". The attempt to incorporate into II.i.7 materials taken from the discussion of the halo is equally unsatisfactory.

BH has also made some rearrangement of the material. In the discussion of the different theories of vision in II.i.1-3, BH brings forward the discussion of the "imprinting" theory, so as to have the erroneous theories discussed first before the correct theory.[3] The rest

---

[1] Optics are discussed again by Aristotle in his *De anima* and accordingly by IS in the corresponding *fann* of the *Šifāʾ* and by BH in the corresponding 'book' of *But*. (*But*. De anima III.iii-iv; see the summary by Furlani [1931] 40-42).

[2] Cf. Min. IV.iii.6.2 above.

[3] Cf. the order of presentation in Chapter Min. II. above, where the erroneous theories concerning the earthquake are treated first in Section II.i. and this is followed

of the section (II.i.4-9) deals with a number of other points relating to optics which need to be borne in mind in discussing the rainbow, halo etc. Here, BH first applies his theory of "anaklasis/diaklasis" to phenomena discussed at two different points in the *Šifā*ʾ, at 42.10-13 and 44.9-13. Having jumped as a result from *Šifā*ʾ p. 42 to p. 44, he continues, in II.i.5-7, to paraphrase the material in the latter part of IS's chapter. Then, in II.i.8-9, he comes back to the material he missed out on by jumping from p. 42 to p. 44.

	*Šifā*ʾ	Other sources	
II.i.1.1	40.5-6	Nic. 46.18	Definition
II.i.1.2-4	40.6-9		Definition of illusion
Erroneous theories			
II.i.1.4-5	40.11		Theories of vision
II.i.1.5-9	40.12-16		Visual-ray theory
II.i.2.1-3	42.6-7		Imprinting theory
II.i.2.3-6	42.8-10, 40.8-10		Refutation
Correct theory			
II.i.3.1-4	41.1-6		Correct theory
II.i.3.4-6	41.7-11		(contd.): reflexion
II.i.3.6-8	43.3-5	Nic. 49.17-20	Aristotle
Additional points			
II.i.4.1-6	42.10-13	Nic. 47.5-9	"Anaklasis/diaklasis"
II.i.4.7-10	44.9-13		(contd.)
II.i.5.1-9	44.14-45.2	Nic.?	Errors of vision
II.i.6.1-5	45.2-6		Illusory light
II.i.7.1-6	45.6-16		Disparity of images
II.i.8.1-5	43.16-44.2	*Šifā*ʾ 51.13-15	Transparency
II.i.9.1-4	44.3-6	(Nic. 48.23-26)	Shape and colour
Concluding remarks			
II.i.9.3-5	46.13-14		Concluding remarks

1. 1: "on the introduction [*ʿuttādā*] to the teaching on the efficient cause [*ʿelltā ʿābōdā*] of cloud-related illusions [*haggāgē ʿnānāyē*]": The title of the section is based on the corresponding chapter title in the *Šifā*ʾ (40.3f.): في المقدمات التى توطأ لتعليم السبب الفاعل للهالة وقوس قزح وسائر ما يشبهها ("on the preliminaries [*muqaddimāt*] which prepare [*tuwaṭṭiʾu*] for the teaching of the efficient cause [*al-sabab al-fāʿil*] of the halo, rainbow and other similar things.

by the presentation of the correct theory (II.ii.1) and the discussion of the various accidental features relating to earthquakes in the rest of the chapter.

II.i.1.1: Definition

But. [underline: agreement with *Šifā'*; italics: with Nic.; cf. II.ii.1.1-2 below]: *Halo* [*ḥugtā*], *rainbow* [*qeštā*], *mock suns* [*bnāt šemšā*] and lances [*nayzkā*] are illusions [*haggāgē*].

Šifā' 40.5f.:

ولنعرف حال الخيالات التى تتكون فى الجو، مثل الهالة وقوس قزح والنيازك والشميسات فإن هذه كلها تشترك فى أنها خيالات.

Let us learn about the condition of the illusions [*ḫaiyālāt*] which are formed in the air, like the halo [*hāla*], rainbow [*qaus quzaḥa*], lances [*nayāzik*] and mock suns [*šumaisāt*]. For these all have it in common that they are illusions.

Nic. syr. 46.18 [< Arist. 371b 18f.]:

ܢܡܠܠ ܕܝܢ ܥܠ ܗܘܓܬܐ ‹ܘܩܫܬܐ› ܘܒܢܬ ܫܡܫܐ ܘܫܒܘ̈ܩܐ.

[18 ܘܩܫܬܐ: supplevi]

Let us then speak about the halo [ἅλως], <rainbow> [ἶρις], mock suns [παρήλιοι] and shoots [*šabbūqē*, ῥάβδοι].

1\. 1. ܢܝ̈ܙܟܐ ("lances"): نيازك ("lances") IS; ܫܒܘ̈ܩܐ ("shoots") Nic. - For the other three phenomena mentioned, BH's terminology is in agreement with Nic. syr.; for the "rod"/"lance" he follows IS here. - At II.iv.3.1 below, BH gives us three terms for this phenomenon: *nayzkē* (lances), *šabṭē* (rods) and *šabbūqē* (shoots). Of these, *nayzkā*/*naizak* and *šabbūqā* are the terms used, repectively, by IS and Nic. syr., while *šabṭē* is the term used in *De mundo* syr. (144.24, 28; corr. gr. 395a 30, 35 ῥάβδος). - The same three terms, along with a fourth, *gērē* (arrows), are mentioned by BH in *Cand.*[4] - Elsewhere in Syriac, *nayzkā* appears as a type of shooting star, as, for example, in *De mundo* syr. (138.3, 145.11; corr. gr. 392b 4, 395b 12 δοκίδες) and BH, *Asc.* 14.21: "inflamed lances and fiery arrows" (ܢܝ̈ܙܟܐ ܕܡܫܬܠܗܒܢ ܘܓܐܪ̈ܐ ܢܘܪ̈ܢܝܐ). The last instance is interesting as this goes some way towards explaining the "arrows" also mentioned by BH in *Cand*.

II.i.1.2-4: Definition of illusion

But.: (1a) An illusion [*haggāgā*] is the perception by the sense [*madrkānūt regšā*] of the image [*yuqnā*] of something in the image [*yuqnā*] of something else (1b) by way of conveyance [*ba-znā d-mawblānūt*] by the latter [lit. the second], e.g. a mirror, of the first, e.g. the image of a man, towards vision [*ḥzāyā*], when the first is not imprinted [*lā ṭbī'*] in reality in the latter [lit. the second].

---

[4] *Cand*. [Bakoš] 117.2; [Çiçek] 89.7f.: ܢܝ̈ܙܟܐ ܘܫ̈ ܘܓܐܪ̈ܐ ܕܡܬܚܫܚܝܢ ܐܘ ܫܒܛ̈ܐ ܘܫܒܘ̈ܩܐ [ܫܒܛ̈ܐ ܕܫܒܘ̈ܩܐ Bakoš]. - Fakhr al-Dīn al-Rāzī, whom BH is following in *Cand*., has "lances and rods" (*al-nayāzik wa-l-'aṣā*) at *Mabāḥit* II.186.16. - The same four terms as in *Cand*. are mentioned by BH at *Rad*. [Istanbul] 24.9.

Šifā' 40.6-9:

ومعنى الخيال هو أن يجد الحس شبح شيء مع صورة شيء آخر، كما نجد صورة الإنسان مع صورة المرآة، ثم لا يكون لتلك الصورة انطباع حقيقي في ذلك الشيء الثاني الذي يؤديها ويرى معها.

(1a) The meaning of illusion is that the sense [*ḥiss*] finds the likeness [*šabaḥ*] of something together with the image [*ṣūra*] of something else, (1b) as [when] we find the image of a man with the image of a mirror, then [we find that] there is no true imprint [*inṭibā*ᶜ *ḥaqīqī*] of that image in that second thing which conveys [*yu'addī*] it and with which it is seen.

II.i.1.4f.: Theories of vision

But.: Concerning how vision works [lit. the manner of vision, *aykannāyūṯ ḥzāyā*] there have been [*hway*] three opinions [*heresiyas*].

Šifā' 40.11: والمذاهب المعتد بها في إدراك البصر لهذه الأشباح ثلاثة مذاهب.

The opinions [*maḏāhib*] of note concerning the perception by the vision of these likenesses are three.

II.i.1.5-9: First opinion: "visual-ray" theory

But.: One is that which says that (1) a ray [*zallīqā*] goes out from the visual organ [*ḥzāyā*] and (2) meets the object seen [*meṯhazyānā*] and conveys [*myabbel*] its image to the visual organ. (3) This ray is also directed towards a smooth object [*sqīlā*], such as a mirror, (4) is reflected [ᶜ*āṭep̄*] from it, (5) is conveyed until it meets the object seen, which is placed opposite the mirror, (6) perceives it and the mirror together, (7) and believes that it is seeing it inside the mirror.

Šifā' 40.12-16:

مذهب أصحاب الشعاعات، وهم يرون أنه يخرج من البصر شعاع فيمتد هو بنفسه إلى الصقيل الذي هو المرآة ويحيل ما يشوبه من الشعاع الذي في العالم إلى طبعه ويحعله (يجعله) كالآلة له، فيلقى الأملس، ثم ينعكس عنه مارا على الاستقامة، حتى يلقى شيئا يقابل ما انعكس عنه، فيدرك معا الأملس الذي هو المرآة وذلك الشيء، فيخيل عنده أنه يدرك صورة ذلك الشيء في المرآة.

The opinion of the 'rays-theorists' [*aṣḥāb al-šu*ᶜᶜ*ā*ᶜ*āt*]: They believe that (1) a ray [*šu*ᶜᶜ*ā*ᶜ] goes out from the visual organ [*baṣar*], (3) is itself directed towards a smooth object [*al-ṣaqīl*], i.e. the mirror, transforms any rays in the world which contaminates it to its own nature and makes it a tool for itself. It meets the most smooth object, (4) then is reflected [*yan*ᶜ*akisu*] from it, (5) travelling in a straight line until it meets something placed opposed the thing from which it was reflected. (6) It then perceives together the most smooth object, i.e. the mirror, and that thing. Thereby it is imagined that it is perceiving an image of that thing in the mirror.

l. 6. "meets the object seen and conveys its image to the visual organ": There is no equivalent of this part in the *Šifā'*. What BH is doing here is to insert first an explanation of how vision works according to the theory under consideration when no reflexion is involved. He makes the same addition in the following 'theory' (I.i.2.1-2).

II.i.2.1-3: Second opinion: "imprinting" theory

But.: (1a) The second opinion asserts (1b) that the image [*yuqnā*] of the object seen [*methazyānā*] is imprinted as it is [*ak d-ītaw*] in the eye [lit. the pupil of the eye, *bābtā*] and is seen. (2a) It is also imprinted [*mettbaʿ*] in the mirror (2b) and this [mirror] is imprinted together with (the image of the object seen) in the eye and they are seen together.

Šifāʾ 42.6-7:

والمذهب الثالث، مذهب من يقول: إن شبح المرئى يتصور كما هو فى المرآة، فإذا رؤيت المرآة بالمحاذاة رؤى أيضا الشبح المنطبع فيها.

(1a) The third opinion is the opinion of the one who says: (1b/2a) The image [*šabaḥ*] of the object seen [*marʾī*] is depicted [*yataṣawwaru*] as it is [*kamā huwa*] in the mirror, (2b) so that when the mirror is seen from opposite the image imprinted in it too is seen.

l. 1-2. "the image of the object seen is imprinted as it is in the eye and is seen": This is what happens when there is no reflexion involved; see comm. on I.i.1.6 above.

II.i.2.3-6: Refutation of "imprinting" theory

The refutation given by IS at the corresponding place (*Šifāʾ* 42.8-10) is worded somewhat obscurely, but IS has been rather repetitive in this chapter and has already said much the same thing near the beginning of the chapter (40.8-10) and in what is proffered as the refutation of the "visual-ray" theory (40.17-19), and BH makes use of these passages in his refutation of the "imprinting" theory.[5]

But. [underline: agreement with *Šifāʾ* 42.8-10; italics: with 40.8-10]:

(1) That this assertion [*tawdītā*] is false [*daggālā*] is known from the fact (2a) that, if *the image* [*yuqnā*] of the object seen *were imprinted in the mirror*, (2b) it would be *fixed* [*qbīʿ*] *to a particular place* [*dukktā dīlānāytā*]. (3a) In reality it is not so, (3b) but as *the viewer* [*ḥāzōyā*] comes near it it [too] comes nearer (3c) and as he moves away from it it [too] moves away.

Šifāʾ 42.8-10 [refutation of "imprinting" theory]:

وهذا الانطباع قول لا معنى له، لأن انطباع صورة شىء فى شىء يوجبه نوع من المحاذاة لا يتغير عن موضع إلى موضع بزوال شىء ثالث لا تأثير له فيه.

(1) This [theory of] imprinting is nonsense, (2a) because the imprinting of the picture [*ṣūra*] of a thing in another (2b) requires a kind of opposition [*muḥādāt*] which does not vary in position from one place [*mauḍiʿ*] to another with the withdrawal [*zawāl*] of the third party [i.e. the viewer] which has no influence on it.

[5] Cf. also Fakhr al-Dīn al-Rāzī, *Mabāḥit* II.177.13-14, which follows *Šifāʾ* 40.8-10.

*Šifā*ʾ 40.8-10:

... كما أن صورة الإنسان لا تكون منطبعة بالحقيقة ولا قائمة فى المرآة، وإلا لكان لها مقر معلوم، ولما كانت تنتقل بانتقال الناظر فيه، والمرئى ساكن.

... (2a) just as the picture [*ṣūra*] of the man is not truly imprinted [*muntabiʿa*] and does not stand in the mirror. (2a) Otherwise, (2b) (the picture) would have a fixed position [*maqarr maʿlūm*] (3) and would not move [*yantaqilu*] with the movement of the viewer [*nāẓir*] when the object seen remains stationary.

*Šifā*ʾ 40.17-19 [refutation of "visual-ray" theory]:

قالوا: وليس الأمر كذلك، وإلا لما كان المرئى ينتقل عن المرآة بانتقال الرائى، ولكان الرائى لا يرى بُعد ما بين المرآة وبين المرئى، والرائى يرى ذلك البعد وإن نظر فى المرآة.

[So] they say, but the matter is not so. (3) Otherwise the object seen would not move away [*yantaqilu*] from the mirror with the movement [*intiqāl*] of the viewer [*al-rāʾī*] and the viewer would not see the distance [*buʿd*] between the mirror and the object seen. [In reality] the viewer sees that distance even if he is looking in the mirror.

II.i.3.1-4: Correct theory: mechanism of vision

But.: (1) The third opinion is that of the natural philosophers [*kyānāyē*] who are correct in [their] inquiry [*ḥattītay buḥḥānā*], who say: (2a) When the object seen is [placed] opposite [*luqbal*] the visual organ (2b) and there is a transparent [*maḇrḥā*] body between them, (2c) a likeness [*dmūṯā*] of the object seen is formed [lit. depicted, *metyaqqan*] in the eye [*bāḇtā*]. (3) The reason for the depiction [*metrašmānūṯā*] of the image in the eye is the shining [*dnīḥūṯā*] of light on (the object) and not on (the eye).

*Šifā*ʾ 41.1-6:

ومذهب الطبيعين المحصلين؛ وهو أنه لا يخرج من البصر شعاعات البتة، بل من شأن المرئى إذا قابل البصر وبينهما مشف، والمرئى مضىء بالفعل، أن صورته تتشبح فى العين من غير أن يكون ذلك كشىء يخرج ويلاقى المشف المتوسط وينفذ فيه إلى البصر البتة، بل إنما يحدث الشبح فى العين نفسها، ويكون المشف المتوسط مؤديا بمعنى أنه يمكن تأثير ذى الشبح بشبحه فى العين. والعلة التى بها يمكن إلقاء الشبح، هو وقوع الضوء على ذى الشبح دون القابل.

(1) The opinion of perceptive natural philosophers [*al-ṭabīʿīyūna al-muḥaṣṣilūna*], namely: There go out no rays from the visual organ at all, (2a) but it is a characteristic of the object seen - when it stands opposite [*qābala*] the visual organ, (2b) there is a transparent [object] between the two and the object seen is luminous in effect [*muḍīʾ bi-l-fiʿl*] - (2c) that its picture [*ṣūra*] is formed [*tatašabbaḥu*] in the eye [*ʿain*], without that object being like something at all that goes out from the visual organ, meets the transparent medium and penetrates through it to the visual organ. Rather, the image [*šabaḥ*] arises in the eye itself. The transparent medium becomes a conveyer [*muʾaddī*] in the sense that it enables the influence of the image-source [*ḏū al-šabaḥ*] to form an image in the eye. (3) The cause which makes possible the projection [*ilqāʾ*] of the image is the falling [*wuqūʿ*] of light on the image-source [*ḏū al-šabaḥ*] but not on the recipient [*al-qābil*].

l. 3. ܡܬܪܫܡܢܘܬܐ ("depiction"): cf. the verb *metrašmā* in l. 5 below. The cognate verb *irtasama* is used by IS later on in the same chapter at *Šifā'* 46.3, 7-8 and in the following chapter at 51.4, 7, 16, 54.8.

l. 4. ܕܢܚܝܘܬܐ ܕܢܘܗܪܐ ܥܠܘܗܝ ܘܠܐ ܥܠܝܗ ("shining of the light on (the object) [masc.] and not on (the eye) [fem.]"): وقـوع الضـوء على ذى الشـبح دون القـابل IS. - Horten-Wiedemann (1913) 534 (followed by Lettinck [1999] 278) take IS's *al-qābil* to be the transparent medium ("das aufnehmende Substrat (das Medium)"). BH has taken it to mean the "eye" (*bābtā*), as the feminine ending of *'lēh* indicates.

II.i.3.4-6: Correct theory (contd.): reflexion

The sentence here may be seen as an attempt to express more simply what is said at the corresponding place in the *Šifā'*.

But.: (4a) When the mirror [*mahzītā*] is opposite [*luqbal*] the eye (4b) and the image [*yuqnā*] opposite the mirror, (4c) the image is depicted [*metrašmā*] in the mirror (4d) and the two of them [are depicted] together in the eye.

Šifā' 41.7-11:

وهذه من الأفعال الطبيعية التى لا يحتاج فيها إلى مماسة بين الفاعل والمفعول، بل تكفى فيها المحاذاة. وكذلك إيقاع الشعاع، فإن اتفق أن كان الجسم ذو الشبح صقيلا تأدى إلى العين أيضا صورة جسم آخر، نسبته من الصقيل نسبة الصقيل من العين، لا بأن يقبل الصقيل فى نفسه شيئا ينتبع فيه البتة، بل يكون تأدى صورته سببا لتأدى صورة ما يكون منه ومن العين على نسبة مخصوصة.

This is one of the actions of nature where there is no need for contact between the agent and the thing acted upon, but opposition [*muḥādāt*] alone suffices. It is the same with the projection of rays. (4) If, then, the image-source [*al-ğism dū al-šabaḥ*] happens to be smooth [*ṣaqīl*] (4cd) the picture [*ṣūra*] of another object too is conveyed to the eye, (4b) when the relationship [*nisba*] of (that other object) to the smooth object [*al-ṣaqīl*] is the same as (4a) the relationship of the smooth object to the eye, not because the smooth object receives in itself something imprinted on it. (4cd) Rather, the conveyance [*ta'addī*] of its picture becomes the cause of the conveyance of what is in a special relation to it and to the mirror.

II.i.3.6-8: Aristotle

The first part of the sentence is based on the *Šifā'*. The quotation in the second part of the sentence has been picked out by BH from a passage of Nic. syr., where Nic. is explaining why haloes are circular (corresponding, but not very close in wording, to Arist. *Mete*. 373a 9-16) and where he is, in fact, not concerned so much with the theory of visual rays as with a geometrical explanation of the circularity of the halo on a diagram.

But.: (1) The Master [i.e. Aristotle] adopts [*methaššaḥ*] the first opinion here not because of its veracity [*ḥattīṯūṯā*] but because of its renown [*ṯbīḇūṯā*] (2) and says: "We imagine a straight line which goes out from the visual organ and comes to the object seen which is bright [*meṯhazyānā d-nahhīr*]."

Šifā' 43.3-5:

فلهذا ما لم يشاق المعلم الأول فى هذا الموضع من كتابه، بل استعمل انعكاس البصر، إذ كان ذلك أشهر وأعرف؛ وإذ لم يكن بين القول فى الحس والمحسوس بعد، فجرى على المشهور.

For this reason the First Teacher [*al-mu'allim al-auwal*] did not go against the tide at this place in his book, but adopted [*ista'mala*] [the theory of] reflexion of vision, since that was more famous and well-known [*ašhar wa-a'raf*], and as he had not yet come to talk about the sense and the sensed [i.e. *De sensu et sensato*], he follows the well-known [theory].

Nic. syr. 49.17-20:

ܐܢ ܥܢ̈ܢܐ ܗܠܝܢ ܕܩܪܝܒܝܢ ܠܚܙܝܬܐ. ܘܐܝܟ ܕܒܩܐܬܛܘܣ ܐܝܬܝܗܘܢ ܠܥܠ ܐܝܬ ܐܢܘܢ. ܗ̈ܢܘܢ ܕܝܢ:
ܕܪ̈ܚܝܩܝܢ ܠܬܚܬ ܐܝܬ ܐܢܘܢ ܢܣܝܒܝܢ. ܘܡܣܬܟܠܝܢܢ ܣܪܛܐ ܬܪܝܨܐ ܕܢܦܩ ܡܢ ܚܙܝܬܐ
ܘܡܛܐ ܠܘܬ ܢܗܝܪܐ . . . .

If the clouds, which are closer to the visual organ and are as in a perpendicular, are higher up, while those further away are lower, (2) and we imagine a straight line which goes out from the visual organ and comes to the luminary [*nahhīrā*], ...

II.i.4.1-6: "ἀνάκλασις" and "διάκλασις"

The passage here is based on a misunderstanding. BH has picked up the two terms "ἀνάκλασις" and "διάκλασις" from a scholion in Nic. syr. (47.5-9). Thereby he seems to have misinterpreted the prefix "δια-" of "διάκλασις" as "δι-, δισ-" or "δυο-" and, possibly, the prefix -ܐܢܐ of ܐܢܐܩܠܐܣܝܣ (ἀνάκλασις) as "ἑνο-", and redefines these two terms, respectively, as "double reflexion" and "single reflexion". - BH then attempts to apply these two terms to the explanation of the phenomena described by IS at *Šifā'* 42.10-13, in what is a continuation of IS's refutation of the "imprinting" theory. - In doing so, BH seems also to have forgetten that he had just followed IS in rejecting the "visual-ray" theory in II.i.1 and II.i.3 (note l. 2, 4 here "as when the ray goes out from the eye ...").

But. [underline: agreement with *Šifā'*; italics: with Nic.]:
Illusory perception of images [*meṯhagg'gānūṯ yuqnē*] occurs *either* (1a) *by means of* [*ba-znā d-*] *ἀνάκλασις*, i.e. single reflection [*ḥdānāyūṯ 'ṭupyā*] - (1b) as when [*akmā d-*] a ray [*zallīqā*] goes out from the eye, (1c) is directed [*neṯmṭaḥ*] towards water, (1d) is reflected [*ne'ṭōp*] from it towards a tree, (1e) and thus the tree is seen [*methzē*] as an illusion [*haggāgānā'īṯ*] in water - *or* (2a) *by means of διάκλασις*, i.e. double reflection [*trayyānūṯ*

ᶜṭupyā] - (2b) as when [*akmā d-*] a ray goes out from the eye, (2c) is directed towards a wall, (2d) is reflected [*neᶜṭōp*] from it towards water (2e) and from water towards a tree, (2f) and thus the green colour[*yurrāqūṯā*] of leaves is seen on the wall.

*Šifā*ʾ 42.10-13 [in the refutation of the "imprinting" theory]

كما أن الضوء إذا نقل على الوجه المحاذى لوّن الشىء مع انتقاله عكسا، مثل ما يعرض للحائط أن يخضر بسبب انعكاس الضوء عن الخضرة إليه. فإن ذلك اللون يلزم موضعا واحدا بعينه ولا يختلف على المنتقلين. وأنت ترى صورة الشجرة فى الماء، ينتقل مكانها من الماء مع انتقالك.

... (2b) just as [*kamā anna*] light, when it is carried on to a facing surface [*al-waǧh al-muḥāḏī*], colours the object with its transportation in a reflexion [*maᶜa intiqālihi ᶜaksan*],[6] (2f) as [*miṯla mā*] happens to a wall that it turns green because of the reflexion [*inᶜikās*] of light from the greenness [*ḫuḍra*] towards it. That colour adheres to the one and the same place and does not move [*lā yaḫtalifu*] in the manner of movable objects. (1e) But you see the shape of a tree in water; the position [*makān*] of (the tree) in the water changes [*yantaqilu*] as you move.

Nic. syr. 47.5-9 [scholion][7]:

ܙܕܩ ܠܡܕܥ ܕܒܚܙܝܬܐ ܐܘ 6/ ܬܪܝܨܐܝܬ ܚܙܝܢܢ ܐܘ ܒܝܕ ܩܠܐܣܝܣ ܗ̄ ܬܒܪܐ.
ܘܒܗܕܐ. ܐܘ ܒܙܢܐ ܕܐܢܐܩܠܐܣܝܣ 7/ ܗ̄ ܥܛܝܦܘܬ ܗܦܘܟܝܐ: ܡܐ ܕܬܪܝܪܬܐ
ܗܘܬ ܡܚܙܝܬܐ ܗܝ ܕܡܛܠܬܗ ܗܘܐ ܬܒܪܐ ܕܚܙܝܬܐ 8/ . ܐܘ ܒܙܢܐ ܕܕܝܐܩܠܐܣܝܣ.
ܗ̄ ܡܬܬܒܪܢܘܬܐ. ܡܐ ܕܒܚܘܠܦܐ ܠܘܩܒܠܝܬ 9/ ܗܘܬ ܡܚܙܝܬܐ.

[6 ܕܕܝܐܩܠܐܣܝܣ: ܕܝܐܩܠܐܣܝܣ cod.]

One should know that with vision we see [things] either directly [*trīṣāʾīṯ*] or by means of κλάσις, i.e. breaking [*ṯḇārā*], and in the latter, either (1a) by means of [*ba-znā d-*] ἀνάκλασις - i.e. double turning [*ᶜṭīpūṯ hpūkyā*] - when the mirror due to which the breaking of vision occurs is solid, or (2a) by means of διάκλασις - i.e. refraction [*mettaḇrānūṯā*] - when on the contrary the mirror is moist.

II.i.4.7-10: "ἀνάκλασις" and "διάκλασις" (contd.)

For the notion of the image being movable in ἀνάκλασις and fixed in διάκλασις, see II.i.4.1-6 above.

But.: (1) In ἀνάκλασις, the object of vision in the mirror moves with the movement of the viewer and appears smaller, blacker and less white than it ought to. (2) In διάκλασις, the object of vision is not disturbed with the movement of the viewer and appears larger, less black and whiter than it ought to.

[6] For the various possible interpretations of the phrase, see Horten-Wiedemann (1913) 735, with n.1.

[7] The passage is a scholion on Nic. syr. 47.3-5 (see under II.ii.1.1-2 below). With the scholion here, cf. Olymp. 211.1f., 211.23-214.28; Olymp. arab. 145.15-22. On "ἀνάκλασις" and "διάκλασις", see further Gilbert (1907) 585f.

*Šifā*ʾ 44.9-13:

وإذا كان المرئى فى مشف ثان وراءه وبينهما سطح بالفعل، فإنه يؤدى مقدار الشيء أعضم مما ينبغى أن يؤديه، وخصوصا إذا كان سيالا مثل ما يرى الشىء فى الماء، إلا أنه يقصر فى تأدية لونه، فيُريه أقل سوادا وصبغا من صواده وصبغه. فإن كان ذلك الشىء خارجا عن ذلك السطح، وكان ذلك السطح يؤديه على أنه مرآة، رؤى ذلك الشىء أصغر حجما، وأشد سوادا من سواده، وأقل بياضا من بياضه.

(2) When the object seen is inside a second transparent (medium) and there is behind it and between the two a plane in actuality, then it conveys the size of the object as being larger than it ought to, especially when (the medium) is liquid, as when a thing is seen in water, except that it is defective in the conveyance of colour, so that it causes it to be seen as less black and tinted. (1) When that object is outside that plane and that plane conveys (the image of the object) in the manner of a mirror, that object appears smaller in size and blacker and less white than in reality [lit. ... than its blackness/whiteness].

II.i.5.1-9: Errors of vision[8]

But.: Fifth [theory]: Vision may err [*pāʾed*] [1] concerning the size [*mšuḥtā*] of the object seen [*meṯhazyānā*], since it sees it sometimes as larger and sometimes as smaller [than in reality]; [2] concerning its shape [*eskēmā*], because with something that is far away, one cannot perceive its corner [*gōniya*] or its convexity and concavity [*lā kustānūṯā āplā ḥlīlūṯā*], but it appears flat or round; [3] concerning the parts on its surface, because with something that is far away, one cannot perceive its roughness [*ḥarrūsūṯā*], as when a mountain which is thick [*ʿbīṯ*] with trees and crags [*qattārē*] appears smooth [*šaʿʿīʿā*] from a distance; [4] concerning its colour, since it sees it sometimes as deeply [*ʿammīq ṣubʿā*] and sometimes as lightly tinged [lit. open, *ptīḥā*]; or [5] concerning the distance [*metḥā*] between it and another object, because with something that is very far away, one cannot know the extent of its height [*kmāyūṯ rawmeh*] above [lit. from] the viewer, or the height of another object from it, like the height of the moon from us and that of the fixed stars [lit. decans, *daqensē*] from (the moon).

*Šifā*ʾ 44.14-45.2:

والبصر يعرض له الغلط فى الشىء من وجوه، منها فى مقدار الشىء كما ذكرناه من أنه تارة يراه أعظم وتارة أصغر؛ ومنها فى شكله، فإن البعيد لا يحس بزواياه ولا بتقبيبه، بل يرى مستديرا ومسطحا؛ ومنها فى وضع أجزائه، فإن البعيد لا يحس بخشونته؛ ومنها فى لونه، فإن تارة يرى الشىء أشد صبغا وتارة أقل صبغا؛ ومنها في وضعه من شىء آخر، فإن البعيد جدا لا يحس البعد الذى بين الرائى وبينه ولا الذى بينه وبين بعيد آخر مثله، كما لا يحس البعد بين القمر والثوابت فى جهة ارتفاعه.

Error [*ġalaṭ*] occurs to vision concening the object in various ways, [1] concerning the size [*miqdār*] of the object [*šaiʾ*], as we have said [cf. *Šifā*ʾ 42.9-14], namely that it sees it sometimes as larger and sometimes smaller [than in reality]; [2] concerning its shape [*šakl*], since with a distant object

[8] On related discussions of errors of vision, see Lettinck (1999) 258 n.31, with the literature cited there.

one cannot perceive its corners [*zawāyā*] and bulges [*taqbīb*], but it appears round and flat; [3] concerning the position of its parts, since with a distant object one cannot perceive roughness [*ḫušūna*]; [4] concerning its colour, since an object sometimes appears more tinted [*ašadd ṣibġan*] and sometimes less [*aqall ṣibġan*]; and [5] concerning its position in relation to another, since with a very distant object one cannot perceive the distance between it and the viewer or that between it and another distant object like it, as one cannot perceive the distance between the moon and the fixed stars [*ṯawābit*] in terms of their height [*fī ǧihat irtifāʿihi*].

l. 5-6. "as when a mountain which is thick with trees and crags appears smooth from a distance": This has no counterpart in the *Šifāʾ*. The addition is probably due to a lost passage of Nic. syr., since comparable passages occur in Olymp. arab. and Ibn al-Khammār, both of which here most probably depend on Nic. syr.[9]

Olymp. arab. 159.17-19:

والثالث أن يظن بالأشياء الخشنة أنها مُلس، وذلك أن ما ينتأ منها يفوته. ولهذه العلة ترى الأرض المختلفة الكثيرة الحجارة من بُعد متساوية ملساء.

The third [error of vision is] that it thinks that rough things are smooth [*muls*], i.e. it fails to notice what bulges out from them. For that reason variegated ground [*al-arḍ al-muḫtalifa*] abounding in stones [*al-kaṯīrat al-ḥiǧāra*] appears level [*mutasāwiya*] and smooth [*malsāʾ*] from a distance [*min buʿd*].

Ibn al-Khammār, *Mutaḫayyila* [Lettinck] 357.21-358.2:

This is also the reason why we see the earth, which is much jagged [*al-arḍ al-kaṯīrat al-zawāyā*], from far away [*min buʿd*], as if it were smooth [*mustawīya*], so we err in its roughness; due to the distance [*li-l-buʿd*] the holes [*ṯuqab*] in the earth are hidden. ... (tr. Lettinck).[10]

l. 6-7. ܥܡܝܩ ܨܒܥܐ ... ܦܬܝܚܐ ("deeply [tinged] ... lightly tinged"): أشد صبغا ... أقل صبغا IS. - The use of *pṯīḥā* (sc. *ṣuḇʿā*, i.e. *pṯīḥ ṣuḇʿā*) in this sense seems to be unregistered in dictionaries.

II.i.6.1-5: Illusory light

But.: (1) Bright bodies [*gušmē nahhīrē*], when their light is reflected [*hāp̄ek*] from a mirror which is close to them, cause an illusion of [*mhaggʿgīn*] luminous colour [*gawnā nuhrānāyā*]. (2a) If they move away, (2b) they grow dark [*ḥāškīn*], (2c) so that different colours [*gawnē ḥrānē*] are composed [*metrakkḇīn*] out of the light and the darkness, (2d) just as,

[9] The whole passage of the *Šifāʾ* above is itself most probably indebted to the discussion of the errors of vision in Olymp. arab. (159.13ff.) and possibly also on Ibn al-Khammār (cf. Lettinck [1999] 283).

[10] "Holes" (الثقب) here is Lettinck's conjecture; the reported manuscript readings are الشر (H) and الشق (T). BH's "trees" allows us to make a new suggestion (الشجر).

when light falls on a black cloud, (the cloud) appears red. (3) In this way illusory light [*nuhrā haggāgānāyā*] is generated on a thing which is far away [*raḥḥīqā* scripsi: *qarrībā* codd.] and black at the same time.

*Šifā*ʾ 45.2-6:

والأجسام المضيئة إذا انعكس ضوؤها عن المرايا القريبة منها، لم يبعد أن يخيل لون نير. فإن بعدت وكانت مظلمة لم يبعد أن تتركب من الضوء ومن الظلمة ألوان أخرى. كما أن الضوء إذا وقع على السحابة السوداء رؤيت حمراء، وكذلك يجوز أن يكون حال الضوء الخيالى فى شىء بعيد وأسود معا.

(1) When the light of luminous bodies [*al-ağsām al-muḍīʾa*] is reflected from mirrors that are close to them, it is likely that it causes an illusion of [*yuḫayyilu*] a bright colour [*laun nayyir*]. (2a) If they move away (2b) and grow dark [*kānat muẓlima*], (2c) it is likely that different colours [*alwān uḫrā*] are composed [*tatarakkabu*] out of the light and the darkness, (2d) just as, when light falls on a black cloud, (the cloud) appears red. (3) In this way it is possible for the condition of an illusory light [*al-ḍauʾ al-ḫayālī*] to occur in a thing that is far away [*baʿīd*] and black at the same time.

l. 2. ܡܚܝܠܝܢ: ("[bright bodies] cause illusions of"): يخيل IS. - In the above passage of IS, if we read يخيل with the editions (Cairo and Tehran), the subject will be either "light" (*ḍauʾ*, active *yuḫayyilu*) or "bright colour" (*laun nayyir*, passive, *yuḫayyalu*). BH may have read تخيل (*tuḫayyilu*), making "luminous bodies" (*al-ağsām al-muḍīʾa*) the subject.

l. 5. ܪܚܝܩܐ scripsi: ܩܪܝܒܐ codd. - The emendation is required by the sense, as well as the corresponding phrase in the *Šifā*ʾ.

II.i.7.1-6: Disparity of images in the case of several or large objects

The points made here are intended to serve as preliminaries for the explanation of the shape of the halo, which is discussed by IS in the following chapter and by BH in Section II.ii. below. In inserting points that are relevant in particular to the discussion of the halo (clauses (2) and (3b) here) into his paraphrase of the *Šifā*ʾ, BH has confused those points which apply to the "mirror" and those which apply to the "object" (it is the "mirrors" that are on a single plane in a halo and it is when the potential "mirror" is on the straight line between the viewer and the luminary that no image of the luminary occurs). The result is a rather confusing passage which in fact makes little sense.

But.: (1) [With] a plural number of bodies [*gušmē saggīʾē*] - or a single body which is too large to be seen in a mirror - it is not necessary that they appear in a like manner [*šawyāʾīṯ*], (2) because parts of them may be positioned in a straight alignment [*trīṣāʾīṯ sīmān*], or (3a) they may be far away (3b) and not all of them are positioned together on a single plane [*ba-ḥḏā šṭīḥūṯā*], or (4) they may have have a weak colour [*gawnā mḥīlā*]. (5a) Whatever is brighter [*yattīr nahhīr*] casts [*mšammar*] its image more strongly [*yattīrāʾīṯ*] towards the mirror, (5b) so that [*aykannā d.*] it obscures [*mḥappē*] the image of a gloomy thing [*ʿammūṯā*] in the mirror.

*Šifā*ʾ 45.6-16:

وإذا قام قائم وحاذى بصره أشياء كثيرة أو شيئا واحدا عظيما مما من شأنه أن يؤدى الشبح، فليس يجب أن تكون كل تلك الأشياء والشيء بحيث يؤدى شبح شيء واحد أو أشياء كثيرة، بل ربما كانت النسبة مع بعض تلك الأجزاء نسبة توجب أداء شبح ما، ومع أجزاء أخرى نسبة توجب أداء شبح آخر. وربما كانت الأجزاء الأخرى لا توازى ما يوجب تأدية شبحه، فتتعطل تلك الأجزاء ويبقى الفعل لما يوازى ذا الشبح الواحد الذى قد مر ذكره. وتلك الأجزاء تتعطل على وجهين: فإنها تتعطل إما لفقدان شيء من شأنه أن يؤدى شبحه، فإذا كانت لا مؤدى لها وللأجزاء المقدم ذكرها مؤدي اختلفا، وإما لأن ما نسبته أليه نسبة الأداء، ليس يبلغ من قوة إرساله الشبح وتمثيله إياه مثلا فى المرآة قوة الشيء الآخر، إما لبعد، وإما لضعب اللون. وأقوى ما يرسل شبحه هو الأقوى ضوءا، وكلما اشتد الضوء اشتد التأثير حتى يمنع أيضا من تأثير أشياء أخرى من شأنها أن تؤثر.

(1) When someone is standing and his vision faces many things [*ašyā*ʾ *kaṯīra*] or a single thing which is too large for its image to be conveyed, then it is not necessary that all those things or the thing be such as to convey the image of the single thing or the many things [as they are]. Rather, it may be that the relationship [of the vision] to some of those parts is a relationship which necessitates the conveyance of a certain image, while that to other parts is a relationship which necessitates the conveyance of a different image. It may be that the other parts do not face something which necessitates the conveyance of its image, so that those parts remain ineffective [*tataʿaṭṭalu*] and only that part retains the effect which faces the source of the one image mentioned. Those parts become ineffective in two ways. They become ineffective either [i] due to the absence of a thing, whose nature it is to convey its image, since when they do not have a conveyer, while the parts mentioned have a conveyer, the two will [appear] differently; or [ii] because its relationship to it is not a relationship of conveyance, its ability to convey an image and to create a likeness of it in a mirror falls short of the ability of the other thing, either (3b) because of the distance [*buʿd*] or (4) because of the weakness of the colour [*ḍuʿf al-laun*]. (5a) What casts its image most strongly [*aqwā mā yursilu šabaḥahu*] is the most strongly illuminated thing [*al-aqwā ḍauʾan*]. The more intense the light, the more intense its effect [*taʾṯīr*], (5c) until [*ḥattā*] it comes to hinder [*yamnaʿu*] the effect of other things, whose characteristic it is to cause an effect.

l. 3. ܬܪܝܨܐܝܬ ܡܬܬܣܝܡܝܢ ("are positioned in a straight alignment"): ܬܪܝܨܐܝܬ ܠܐ ܡܬܬܣܝܡܝܢ MV. - This is not very well expressed and the alteration made in mss. MV is understandable. The point is that no image will occur when the mirror is between the viewer and the object and the three are in a straight alignment. Cf. II.ii.1.5f. below: "because in a straight alignment the object of vision appears in itself and not its illusion".

l. 3-4. "and not all of them are positioned together on a single plane": The parts of the cloud that act as the "mirrors" in a halo are in a single plane. See Arist. *Mete*. 373a 14: καὶ ἐν ἑνὶ ἐπιπέδῳ πᾶσαι.

l. 5-6. "obscures the image of a gloomy thing": cf. II.ii.2. below.

l. 5. ܕܥܡܛܢܐ ("a gloomy thing"): The choice of the word is probably due to the following scholion of Nic. syr., where *ʿamṭānā* in fact stands for Aristotle's

ἀχλύς ("mist"). - Nic. syr. 49.6-8: ܗ̄ ܒܕ ܐܡܪ ܗܘ 7/ ܒܛܒܥܐ ܕܐܢܐܩܠܐܣܝܣ . ܚܙܝܐ ܣܒܠ . ܡܢ ܥܡܛܢܐ ܕܡܬܩܢ ܩܕܡ/8 ܫܡܫܐ ܐܘ ܣܗܪܐ . [7 ܕܐܢܐܩܠܐܣܝܣ : ܕܐܢܐܩܠܣܝܣ cod.] ("i.e. since he [Aristotle] says in the text [*b-ṭab̲ʿā*] that vision undergoes ἀνάκλασις due to the gloom [*ʿamṭānā*] which is formed in front of the sun or the moon").[11]

II.i.8.1-5: Transparency

The example of the rock crystal used here by BH does not occur in the passage at the corresponding place in the *Šifāʾ*, but later on in the discussion of the rainbow (in the passage corresponding to II.iii.1.1-5 below).

But. [underline: agreement with *Šifāʾ* 43.16-44.2; italics: with 51.13-15]:

(1) When a smooth body is transparent [*mab̲rḥā*] and appears transparent [*mab̲rḥāʾīt̲ net̲hzē*] in actuality [*b-maʿbdānūt̲ā*], (2a) vision [*ḥzāyā*] is conveyed [*metyabbal*] through it (2b) and does not see an illusion of the type seen in mirrors [lit. a specular illusion, *haggāgā mahzīt̲ānāyā*]. (3) If this illusion is seen on it, it is no longer transparent in actuality. (4a) This occurs when there is a coloured body [*gušmā da-mgawnan*] behind it [*bestreh*], (4b) as is the case with *rock crystal* [*berūlā*], (4c) which shows an image just like *a mirror* [*ak̲ maḥzīt̲ā*] if there is some colour [*gawnā meddem*] behind it [*bāt̲reh*], (4d) *but if not, does not* [*w-ellā lā*].

Šifāʾ 43.16-44.2:

وإذا كان/١٧ الجسم الصقيل مشفا، ورأى مشفا بالفعل، لم يمكن أن يُرى عليه هذا الخيال. فإذا رؤى عليه/١٨ الخيال لم يؤد ما وراءه ولم يكن مشفا بالفعل حينئذ بالقياس إلى ما وراءه. وإن كان وراء/١ الجسم الشفاف جسم ذو لون يحدده، أرى هذا الخيال؛ وإن لم يكن وراءه ما يحدده،/٢ نفذ فيه البصر، ولم ير هذا الخيال.

[17 ورأى: ورؤى DS[12] || رؤى: رأى BṬ || 1 أرى: لوّن DSṬ]

(1) When the smooth body is transparent and appears [*ruʾiya* (?)] transparent in actuality, it is not possible for this illusion to be seen on it. (3) When the illusion is seen on it, it does not convey what is behind it and is not transparent in actuality then in relation to what is behind it. (4a) If behind the transparent body there is a coloured body [*ǧismun d̲ū launin*] which limits it, it causes this illusion to be seen. (4d) If there is nothing behing it limiting it, (2a) vision penetrates [*nafad̲a*] through it (2b) and does not see this illusion.

Šifāʾ 51.13-15:

وذلك كالبلورة، فإنها إذا سترت من الجانب الآخر صارت مرآة فى الجهة التى تليك، وإن لم تستر وتركت وراءها فضاء مشف غير محصور لم تكن مرآة.

---

[11] Cf. Arist. 372b 34-373a 2: ἀνακλᾶται δ' ἀπὸ τῆς συνισταμένης ἀχλύος περὶ τὸν ἥλιον ἢ τὴν σελήνην ἡ ὄψις: ("Our vision is reflected from the mist which condenses round the sun and moon").

[12] Cf. *Mabāḥit̲* II.177.19: اذا كان الجسم الصقيل مشفا ويرى ما وراءه مشفا بالفعل

(4b) That is like rock crystal [*billaur*]. (4c) For when (rock crystal) is veiled [*sutirat*] on the other side it becomes a mirror on the side facing you, (4d) but if it is not veiled and behind it is an unlimited transparent open space it does not act as a mirror.

l. 4. ܒܠܘܪܐ ("rock crystal"): بلورة IS. - See comm. on Min. III.i.5.4 above.

II.i.9.1-4: Conveyance of shape and colour

This may be understood as a paraphrase of the passage of the *Šifā'* quoted below. BH will also have seen the passage of Nic. syr. quoted below, but there is no *positive* indication that BH has made use of that passages here.

But. [underline: agreement with *Šifā'*; italics: with Nic.]:

(1) A large mirror shows [*mḥawwyā*] colour and shape, (2) a small *mirror colour but not shape*, (3) because shape is defined [*mettaḥham*] by division at edges [*purrāš sāḵē*]. (4) How, then, can something appear shaped [*msakkmā*] which sense [*regšā*] does not divide and whose edges it does not define?

Šifā' 44.3-6:

إن المرايا إذا كانت بحيث لا يحدها الحس، لم يمكن أن يُؤدِّى اللون والشكل معا؛ فإن كانت صغارا، أدت اللون، ولم تف بأداء الشكل. لأن الجسم لا يمكن أن يرى مشكلا إلا وهو بحيث يقسّمه الحس، فكيف يرى ما لا ينقسم فى الحس مشكلا؟

When the mirrors are such that the sense cannot define [*yuḥaddidu*] them, colour and shape cannot be conveyed together. (2) If they are small, they will convey colour, but are not able to convey shape, (3) because a body cannot appear shaped unless it is such that the sense can divide [*yuqassimu*] it. (4) How, then, can something appear shaped [*mušakkal*] which is not divided according to [lit. "in"] sense [*ḥiss*] cannot divide?

Nic. syr. 48.23-26[13]:

ܐܢܗܘ ܕܝܢ/ ܕܙܥܘܪ̈ܢ ܡܚܙܝܢܝ̈ܬܐ. ܠܓܘܢܐ ܡܢ ܕܢܗܝܪܐ ܡܩܒܠܢ: ܠܐܣܟܡܝܗܘܢ/
ܕܝܢ ܠܐ ܡܩܒܠܢ. ܡܛܠ ܕܐܣܟ̈ܡܬܐ ܐܝܠܝܢ ܕܛܒ ܙܥܘܪܝܢ: ܠܐ ܡܬܚܙܝܢ ܒܗܝܢ/
ܐܣܟܡ̈ܐ.

[25 ܕܐܣܟ̈ܡܬܐ: ܕܐܣܟ̈ܡܬܐ ut vid. cod.]

(2) If the mirrors are small, they receive the colour of the luminary but do not receive their shape, because shapes are not seen in those things which are very small.[14]

---

[13] Cf. Arist. 372a 32-b 6; Olymp. 219.16-19, 226.16-21.

[14] The passage is accompanied by a scholion, Nic. syr. 48.30-49.2: ܟܕ ܐܦ ܒܟܝܢܐ ܐܝܟ/ ܕܒܡܘܠܝܐ ܡܬܬܦܠܓ ܐܣܟܝܡܐ ܐܝܢܐ ܕܗܘ ܠܓܠܐ ܗܘ ܘܓܠܐ ܣܘܦܠ. ܐܝܟ ܐܢܬ ܠܡܣܬܟܡܘ/ ܠܡܬܚܙܝܘ. ܗܢܐ ܗܘ ܡܬܚܙܝܢ ܕܠܐ ܗܘ ܢܦܠ ܠܐܣܟܡܐ ܒܟܝܢܐ. ܐܠܐ ܠܗ ܒܬܚܘܡܬܐ/ 1 ܟܕ ܐܝܬ ܠܡܬܚܙܝܘ ܕܠܐ ܡܬܦܠܓ ܕܐܚܝܕ ܡܢܗ ܢܦܠ ܬܚܘܡܐ ܬܚܘܡܬܐ. ܠܡܢܐ/ ܕܠܐ ܡܬܦܠܓ ܐܣܟܝܡܐ ܕܠܐ ܕܠܒܪ ܐܝܟ ܐܣܟܝܡܐ ܐܦ ܒܗܘܢ ܡܬܦܠܓܐ. [31 ܐܝܬ: ܐܝܬܐ cod.] ("Since even if in nature as well as in thought [i.e. φύσις and νοῦς] a given

II.i.9.3-5: Concluding remarks

But.: These are the traditional premises [*prōṭasīs mettašlmānāyāṯā*] here and the proofs [*apōdīksē*] are provided through geometry [*ge'ōmīṭriya*] and optics [*yulpānā da-ḥzāyē*].

Šifā' 46.13-14:

فهذه الأشياء كمقدمات وتوطئات، بعضها يعوّل فيه على صناعة الهندسة، وبعضها على علم البصر، ونحن نتكلم فيه في موضعه، وبعضها على الامتحان بالحس.

These things are as as preliminaries and preparations [*muqaddimāt wa-tauṭī'āt*], some of them depend on the art of geometry [*ṣinā'at al-handasa*], others on optics [*'ilm al-baṣar*] - we shall talk about this in its place - and others on the examination of sensation.

Mete. II. Section ii.: *On Halo*

The section on the halo is mostly a paraphrase of the *Šifā'*, although the actual explanation of how a halo is formed (II.ii.1.3-9) is based on Fakhr al-Dīn al-Rāzī. Nic. syr. is used in the definition of illusions at the beginning of the section, while the quotation from Aristotle in II.ii.2.4-6 is most probably taken from a lost passage of Nic. syr.

	Šifā'	Other sources	
II.ii.1.1-2		Nic. syr. 47.3-5[15]	Illusions
II.ii.1.3-9	47.4-5, (47.17-48.4)	*Mab.* II.178.17-179.4	Formation of halo
II.ii.2.1-4	48.9-17		Middle of halo
II.ii.2.4-6		Nic.?	Blackness
II.ii.3.1-5	49.17-50.2		Halo and rainbow
II.ii.4.1-6	49.8-12		Multiple haloes
II.ii.4.6-7	49.4-5		Solar halo
II.ii.4.8-9	49.13-15, 50.6		Ibn Sīnā's report

II.ii.1.1-2: Illusions

The first part of the sentence repeats what has been said at II.i.1.1 above (with the addition of the particle *man* and alteration of the copula *ennōn* to *īṯayhōn*). The second part is taken from Nic. syr.[16]

shape can be divided endlessly and unceasingly - let us say a triangle - [even if] this is thus reduced endlessly, it retains its shape. This, however, is not the case with sense-perception [*regšā*, αἴσθησις], since there is a triangle, one smaller than which cannot fall under perception. A very small mirror does not receive the shape of such [a triangle], even though it receives its colour"; cf. Olymp. arab. 146.15-19; also Arist. *EN* 1142a 28).

[15] See also Mete. II.i.1.1-2 above.

[16] The comparable definition of "realities" and "illusions" at *Cand.* 117.9-11

But.: (1a) The halo, rainbow, mock suns and lances are [*īṯayhōn*] illusions (1b) and not realities which exist by themselves [*law suʿrānē da-mqayymīn l-yāṯhōn*].
Nic. syr. 47.3-5:
ܟܠܗܘܢ ܡܟܝܠ ܗܠܝܢ܆ ܡܢ/4 ܗܠܓܬܐ ܠܡܓܚܝܢܐ ܗܘܝܢ. ܘܡܚܓܓܢܐ ܐܝܬܝܗܘܢ ܘܠܘ
ܣܘܥܪܢܐ ܕܡܩܝܡܝܢ/5 ܠܝܬܗܘܢ. ܘܒܦܢܝܐ ܕܓܠܦܐ ܕܚܙܝܬܐ ܡܬܚܙܝܢ.
(1a) All these then arise from the vapoury exhalation. They are illusions (1b) and not realities which exist by themselves, and they appear by way of reflection of vision.

l. 2. ܣܘܥܪܢܐ ܕܡܩܝܡܝܢ ܠܝܬܗܘܢ: ("realities which exist by themselves"). - In Nic. syr. *suʿrānā* usually corresponds to gr. πρᾶγμα, but in the scholia in the section on halo, rainbow etc., we frequently find the combination *suʿrānā mqayymā* (Nic. syr. 50.25, 28, 30, 31; 51.4, 12, 13; 52.13), which often answers to "ὑπόστασις" in Olymp. The phrase here with the further addition of *l-yāṯhōn* may go back to such an expression as found at Olymp. 210.33: οὐκ ἄρα ἰδίαν ἔχουσιν ὑπόστασιν.

II.ii.1.3-9: Formation of halo

The explanation here of how the halo is formed is based on Fakhr al-Dīn al-Rāzī's *Mabāḥiṯ* rather than on the *Šifāʾ*, although the opening line is taken from the *Šifāʾ*.

But. [underline: agreement with *Mabāḥiṯ*; italics: with *Šifāʾ*]:
(1) *The halo is a circle which appears around the moon* (2) when a moist and thin cloud [*ʿnānā raṭṭīḇtā w-raqqīqtā*] is interposed [*meṯmaṣṣʿā*] between the viewer and the moon, (3) so that that part of (the cloud) which is in front of the moon *does not conceal* [*msattrā*] *the moon* and no illusion of the moon appears on it - (4) *because in a straight alignment* [*ba-syāmā trīṣā*] the object of vision *appears in itself and not its illusion* [*haggāgā*] - (5) whereas each one of those parts of (the cloud) which are not positioned in a straight line in front of the moon conveys an illusion of the moon to vision. (6) Because each of them is small, it conveys (the moon)'s light but not its shape, (7) and all of them together form a *circular* phantasm [*panṭasiya*] of light in the shape of the moon around the moon.
Mabāḥiṯ II.178.17-179.4:
بل الحق ان الهالة خيال وذلك لان اذا توسط بين الرائى وبين القمر غيم رطب رقيق لطف بحيث لا يستر
القمر فالذى يقابل القمر من ذلك الغيم لا يستره ولا يرى ايضًا خيال القمر فيه فان الشىء انما يرى على
الاستقامة نفسه لا شبحه واما الاجزاء التى لا يقابل القمر وكانت لطيفة رقيقة ادى كل واحد من تلك
الاجزاء خيال القمر على الوجه الذى عرفت معنى الخيال ولما كان كل واحد من تلك الاجزاء السحابية

---

combines Nic. syr. with *De mundo* syr. 144.22-25 (gr. 395a 28-32): *Cand.* ܣܘܥܪܢܐ ܡܩܝܡܢܐ < *De mundo* syr. 144.24 ܡܩܝܡܝܢ ܡܩܘܡܝܢ ܐܝܬܝܗܝܢ ܗܘܝܢ; *Cand.* ܠܦܘܬ ܚܙܝܬܐ ܡܬܚܙܝܢ < 144.23 ܠܦܘܬ ܚܙܬ ܡܬܚܘܝܢ. - On meteorological phenomena which are καθ' ὑπόστασιν and κατ' ἔμφασιν, see further Gilbert (1907) 587f.

صغيرا لا جرم ما ادى شكل القمر بل ادى ضوءه فلا جرم ظهر الضوء فى كل واحد من تلك الاجزاء وان لم يظهر الشكل فى شىء منها ولما كانت النسبة الحاصلة بين الرائى وبين كل واحد من تلك الاجزاء وبين المرئى واحدة لا جرم كان شكل الهالة دائرة

But the truth is that the halo is an illusion [*ḫayāl*] and that is because (2) when a moist, thin and fine cloud [*ġaim raṭib raqqīq laṭif*] (3) such that it does not conceal the moon (2) is interposed [*tawassaṭa*] between the viewer and the moon, (3) that [part] of the cloud which is in front of the moon does not conceal [*yasturu*] it and no illusion of the moon appears on it. - (4) For in a straight line [*ʿala l-istiqāma*] the object itself is seen and not its image [*šabaḥ*]. - (5) Each one of those parts which are not in front of the cloud, even if they are fine and thin, conveys an illusion of the moon in the sense that you know. (6) Since each one of those cloudy parts is small, it will doubtless not convey its shape but [only] its light, so that the light appears on each one of those parts, although the shape does not appear on any of them. (7) When the relationship occurring between the viewer, each of those parts and the object seen is the same, the shape of the halo will without doubt be round.

Šifā' 47.4-5, 47.17-48.4:

وأما الهالة فإنها دائرة بيضاء تامة أو ناقصة ترى حول القمر وغيره، إذا قام دونه سحاب لطيف لايغطيه، لأنه يكون رقيقا... وذلك إذا كان السحاب مائيا لطيف الأجزاء رقيقا لا يغم القمر أو الكوكب، وأدى نفس الكوكب مع أداه شبح الكوكب، لا على استقامة ما بين الناظر والمنظور إليه. فإن الشىء إنما يرى على الاستقامة نفسُه لا شبحه، وإنما يؤدى شبحه زائلا عن محاذاة الاستقامة التى بينه وبين الرائى ضرورة. فإذا كان جميع أجزاء السحاب أو أكثره مستعدا لهذه التأدية، وكان نسبة كل مرآة فى وضعها من الرائى والكوكب يجب أن تكون نسبة واحدة من جميع جوانب الكوكب، وجب أن يكون ما يرى من الهالة مستديرا.

(1) The halo is a white circle, complete or incomplete, which appears around the moon etc. (2) when a fine cloud [*saḥāb laṭīf*] stands below it, (3) which does not conceal [*yuġaṭṭī*] it (2) because it is thin. ... That [happens] (2) when the cloud is watery, fine in its parts and thin, (3) so that it does not obscure [*yaġummu*] the moon or the luminary [lit. star, *kaukab*], (5) and conveys the luminary itself as well as the image of the luminary, not in a straight line between the viewer and the object viewed. - (4) For in a straight line the object itself is seen and not its image. - (5) (The cloud) necessarily conveys the image of (the luminary) away from the straight line between it and the viewer. (7) Then, when all or most parts of the cloud is in a position suitable for such conveyance and the positional relationship of all the mirrors to the viewer and the luminary is the same on all sides of the luminary, it necessarily results that the halo that appears is round.

l. 5. ܒܡܘܙܠܬܐ ܬܪܝܨܬܐ ("in a straight alignment"): cf. II.i.7.3 above.
l. 7-8. "Because each of them is small, ...": cf. II.i.9 above.
l. 8. ܦܢܛܣܝܐ ("phantasm"): < gr. φαντασία. - The word is used twice in Nic. syr. (48.28, 49.4).

II.ii.2.1-4: Why the middle of the halo is dark

The passage of *But.* here may be seen as an attempt at a systematic paraphrase of a somewhat obscure and long-winded passage of the *Šifā*ᵓ.[17]

But.: (1a) The middle [*msa*ᶜ*tā*] of the halo appears black [*ukkāmtā methawwyā*] (1b) because in the middle the ray is strong (1c) and for that reason obscures [*mḥappē*] the thin cloud which is there. (2a) Since it is not visible [*lā methawwyā*] (2b) it does not glitter [*maḇhqā*] like those [parts] which are on the circle, (2c) and what does not glitter appears black beside [ᶜ*al geb*] what glitters.

*Šifā*ᵓ 48.9-17:

ولأن ما يخرج عن المرآة وما يدخل فيها مما لا يخيل، لا يكون له اشراق ما يردّ الضوء ويعكسه إلي البصر، فيخيل أن خارجه وداخله اسود؛ فإن كل ما نقص من إشراقه عن الأبيض، ووضع فى جنب الأبيض يرى أسود. وداخل الهالة يعرض له سبب أخر، وهو أن قوة الشعاع الذى للكوكب تخفى حجم السحاب لا يستره، فكأنه ليس سحاب ولا شيء آخر لأن ما فيه من السحاب ليس يستر القمر، إذ كان هو سحابا رقيقا. ويعرض للصغير والرقيق أن لا يرى فى الضوء القوى خصوصا إذا كان بحيث لا يستر الشيء فيكون كأنه ليس موجودا، مثل ما لا ترى الهبات الجوية فى الصحراء، وإن رؤى لم ير مضيئا بل أسود مثل الشعلة فى النهار، وإذا لم ير أو رؤى أسود يتخيل كأن هناك منفذا أو مدخلا أو شيئا أسود.

Because what is outside and inside the mirror are among things that do not appear as illusions [*lā yuḫayyalu*], there being no illumunation [*išrāq*] that returns the light and reflects it to vision [?], (1a) as a result the outside and the inside [*dāḫil*] of (the halo) are imagined [*yaḫayyalu*] to be black. (2c) For everything which is not white because of [the lack of] its illumination [*min išrāqihi*] and is placed beside [*fī ǧanbi*] a white thing appears black. The inside [*dāḫil*] of the halo is subject to another cause, (1b) namely that the strength of the ray [*qūwat al-šu*ᶜᶜ*ā*ᶜ] of the luminary [*kawkab*] (1c) hides [*tuḫfī*] the bulk of the cloud which does not conceal [*yasturu*] it, so that it is as if there is no cloud or anything else there, because what cloud there is does not conceal the moon, since it is a thin cloud. It happens to a small, thin thing that it is not seen in a strong light, especially when it is such as to be unable to conceal the thing, so that it is as if it does not exist [*laisa mauǧūdan*], as happens to air-borne dust in the desert. Even if it is seen [*ru*ᵓ*iya*], it does not appear [*lam yura*] luminous but black, like a torch [*šu*ᶜ*la*] in daylight. (2a) When it is not seen [*lam yura*] (1a) or appears [*ru*ᵓ*iya*] black, it is imagined [*yataḫayyalu*] to be an [empty] passage [*manfaḏ*] or an entrance or something black.

II.ii.2.4-6: Blackness

The explicit attribution to Aristotle suggests that this is a lost passage of Nic. syr. and this is supported by Olymp. arab., as well as by Ibn al-Khammār, who explicitly tells us that he used Nicolaus (evidently

---

[17] The first part of the passage of the *Šifā*ᵓ quoted below is particularly problematic as the text stands (see the comment ad loc. in Horten-Wiedemann [1913] 538; Fakhr al-Dīn al-Rāzī omits this part in his paraphrase at *Mabāḥiṯ* II.179.5ff.).

in the Syriac version) as one of his sources in composing his treatise on the "illusory phenomena". - In comparing the four passages quoted below, we find that BH's sentence (1b) corresponds to what is in Arist., while (1c) corresponds to what is in Olymp. arab. and Ibn al-Khammār. BH may have preserved here what is closest to the complete text of Nic. syr.

But. [underline: agreement with Arist.; italics: with Ibn al-Khammār]: (1a) As the Master [Aristotle] said: (1b) "we see blackness [*ukkāmūṯā*] when we do not see anything". (1c) For indeed [*hā gēr*], *when we do not see* the light of *the sun* [*nhur šemšā*] at night, *we think* [*masbrīnan*] *that we are seeing* [*ḥāzēnan*] *blackness* [*ukkāmūṯā*]; (1d) similarly in daytime when we close our eyes.

Arist. *Mete*. 374b 12-14: τρίτον δ' ὅτι τὸ μέλαν οἷον ἀπόφασίς ἐστιν· (1b) τῷ γὰρ ἐκλείπειν τὴν ὄψιν φαίνεται μέλαν.

Ibn al-Khammār, *Mutaḫayyila* [Lettinck] 359.15-17: The third preliminary is that the colour black is equivalent to [*bi-manzila*] absence of vision, (1c) for [*li-anna*] if we do not see the sun or some (other) radiating object [*al-šams au al-muḍīʾ*], we think [*ẓanannā*] that we see something black [*annā narī šaiʾan aswad*]. (tr. Lettinck)

Olymp. arab. 160.18-20:

لأن الظلمة عدم النور عن الشیء الذی یکون له السواد. ولهذا السبب نجد البصر إذا لم یر الشمس توهم أنه رأی سواداً

... because darkness [*ẓulma*] is absence of light from the thing to which the blackness [*sawād*] occurs; (1c) and for this reason, we find that when the vision does not see the sun, it imagines [*tawahhama*] that it is seeing blackness.

l. 5. ܗܐ ܓܝܪ ("for indeed"): The combination of particles *hā gēr* (only here in *But*. Min., Mete.) occurs at least twice in Nic. syr. (3.1, 31.13) and may therefore be taken as it was from Nic. syr.

l. 5. ܐܘܟܡܘܬܐ ("blackness"): شیئا اسود ("something black") Ibn al-Khammār; سواداً ("blackness") Olymp. arab. - The agreement of BH and Olymp. arab. suggests that the abstract noun *ukkāmūṯā* was what was in Nic. syr.; this noun occurs once in the surviving parts of Nic. syr. at 54.31.

II.ii.3.1-5: Halo and rainbow

But.: (1) Halo is distinguished from [*prīšā men*] rainbow in that (2) with the halo one end [*sāḵā*] of the axis [*sarnā*] is the visual organ and the other is the moon. (3) The halo is the belt [*zōnī*] around that axis and (4) the centre of its circle is on this line between the viewer and the object seen. (5) In the rainbow, on the other hand, the viewer and the sun are on the line of the axis, (6) but the centre of the circular belt [lit. circle of the belt, *ḥḏur zōnī*] is not between them. (7a) The halo is a complete circle, (7b) while the rainbow is a semi-circle.

Šifāʾ 49.17-50.2:

وهالة الشمس تخالف قوس قزح فى أن محور هذه الدائرة ينتهى إلى البصر وإلى المرئى فى الجانبين جميعا. وتكون الهالة منطقة لذلك المحور، ويكون مركز دائرتها على هذا الخط بين الرائى والمرئى. وأما القوس فإن الرائى والشمس يكونان جميعا على خط المحور، لكن مركز دائرة المنطقة لا يكون واقعا بينهما. والقوس لا يزيد على نصف دائرة لكن الهالة قد تتم دائرة

(1) The halo of the sun differs from [*tuḫālifu*] the rainbow in that (2) the axis [*miḥwar*] of this circle terminates [*yantahī*] at the visual organ and the object seen at its two ends [lit. sides]. (3) The halo is the belt [*minṭaqa*] around that axis and (4) the centre of its circle is on this line between the viewer and the object seen. (5) With the rainbow, the viewer and the sun are together on that line of axis, (6) but the centre of the circular belt [*dāʾirat al-minṭaqa*] does not fall between them. (7b) The rainbow does not exceed a semi-circle, (7a) while the halo may be a complete circle.

II.ii.4.1-6: Two or more haloes

But.: (1) When a cloud occurs below another cloud it is possible for a halo to form below a halo. (2) The lower [halo] will be larger because it is closer and conveys the luminous illusion through parts which are further out from the centre. (3) Some have said that they saw seven haloes at one time below each other, and this is surprising. (4) Another has said that he saw a halo which, when measured against the stars which were in front of its limits [*d-luqbal sāḵēh*], was approximately forty-five stades [in diameter].

Šifāʾ 49.8-12:

وإذا وقعت سحابة بهذه الصفة تحت سحابة، أمكن أن تتولد هالة تحت هالة. والتحتانية تكون أعظم من الفوقانية، لأنها أقرب، فتكون تأديتها المرئى بأجزاء أبعد من الوسط. ومنهم من ذكر أنه رأى سبع هالة معا وهو بعيد. وقد حكى بعضهم أنه رأى هالة، فلما قدرت بالكواكب التى حاذت أقطارها كانت قريبة من خمسة وأربعين اسطاذيا.

(1) When a cloud with such a property occurs below [another] cloud, it is possible for a halo to form below a halo. (2) The lower [halo] will be larger than the upper, because it is closer, so that the conveyance of the object seen will be through parts further from the middle. (3) Someone has said that he saw seven haloes at one time, but this is improbable. (4) Another has recounted that he saw a halo and, when it was measured against the stars which were in front of its diagonals [*ḥāḏat aqṭārahā*], it was approximately forty-five stades [in diameter].

l. 6. ܐܣܛܕ̈ܝܐ ("stades"): اسطاذيا IS. - Olymp. (230.7) gives the diameter of the halo as "40 *degrees*" (μοῖρα). The blame for turning this into an absolute measure of length lies with Olymp. arab. ("45 *ġalwa*", 149.17). See Lettinck (1999) 257 n. 29, 268 n. 67, 281 n. 107.[18]

[18] The emendation of IS's "stades" (اسطاذيا) to "lines" (سطرا) made by Horten-Wiedemann (1913) 539 n. 1 is therefore unnecessary.

II.ii.4.6-7: Solar halo

But.: Around the sun halo occurs rarely, because (the sun) usually dissipates thin clouds which are unable to conceal it.
Šifā᾽ 49.4-5:

وقلما تكون حول الشمس هالة، لأن الشمس فى الأكثر تحلل السحب الرقيقة التى تبلغ من رقتها أن لا تستر الشمس.

Halo occurs rarely around the sun, because the sun usually dissipates thin clouds which because of their thinness are unable to conceal the sun.

II.ii.4.8-9: Ibn Sīnā's report

BH conflates here two passages of the *Šifā᾽*, where IS must be speaking about two different incidents.

But.: The princely doctor [i.e. Ibn Sīnā] has said that he saw a halo with the colours of the rainbow around the sun in Hamadan.
Šifā᾽ 49.13-15:

وقد رأيت حول الشمس فيما بين سنة تسعين وثلاث مائة وإحدى وتسعين هالة تامة فى ألوان قوح قزح

I once saw around the sun between the years 390 and 391 [999-1001 A.D.] a complete halo in the colours of a rainbow.
Šifā᾽ 50.6: وقد رأيت بهمذان هالة حول القمر قوسية اللون
I once saw in Hamadan a rainbow-coloured halo around the moon [!].

Mete. II. Section iii.: *On Rainbow*

The section corresponds by and large to IS's discussion of the rainbow in the middle part of his chapter on halo, rainbow etc. (ed. Cairo p. 50-54), although the passage on the rainbow around the moon (II.iii.8) corresponds to a passage placed at the end of that chapter in the *Šifā᾽*.

The first part of the section (II.iii.1-2) is concerned with the fact that rainbows are due to light being reflected not from clouds as is the case with the halo but from moist air. The material is taken mainly from the *Šifā᾽*, although BH combines this with materials that go back to the same Aristotelian source in Nic. syr., most notably in the passage on Antipheron of Tarentum (II.iii.2.2-7).

For much of the discussion of the shape of the rainbow (II.iii.3), BH's wording is closer to that of Fakhr al-Dīn al-Rāzī's paraphrase of the *Šifā᾽* than to the text of the *Šifā᾽* itself.

'Theories' III.iii.4-7 are concerned with the colours of the rainbow. Here, IS, who disagreed with the traditional Aristotelian theory, reported the Aristotelian theory and then gave his objections. BH does not

commit himself either way on the matter, but gives both the Aristotelian theory and IS's objections. In his report of the Aristotelian theory, he makes use of Nic. syr. in the explanation of how the red colour is generated, while his discussion of the colours of a double rainbow (II.iii.6) is probably based on a lost passage of Nic. syr.

	Šifā᾽	Other sources	
II.iii.1.1-5	50.16-19, 51.10-17	(*Mab.* II.180.19-181.5)	Required conditions
II.iii.1.6-10	52.1-6		Rainbow-like illusions
II.iii.2.1-2	52.10-11	Nic. 53.14-16	Spray from oars
II.iii.2.2-7	52.11-13	Nic. syr. 53.6-9, 13-14	Reflexion from air
II.iii.2.7-10	52.6-10		Bath
II.iii.3.1-4		*Mab.* II.182.10-12	Shape of the rainbow
II.iii.3.4-8	53.12-16	*Mab.* II.182.12-17	(contd.)
II.iii.3.8-10	53.16-18	(*Mab.* II.182.18-21)	Rainbows at midday
II.iii.4.1-5	53.6-8	(Nic. 53.32-54.3)	Colours (rainbow/halo)
II.iii.4.5-7	(57.4-11)	Nic.?	Colour of halo (contd.)
II.iii.5.1-2	54.4-7		Peripatetics on colour
II.iii.5.2-5	54.8-9	Nic. syr. 54.3-7	(contd.): red
II.iii.5.5-7	54.9-10		(contd.): purple
II.iii.5.7-8	54.11-12		(contd.): green
II.iii.6.1-6		Nic.?	(contd.): double rainbow
II.iii.7.1-2	55.8-9, 50.10-11		IS's objection
II.iii.7.2-8	54.12-13, 55.4-7		(contd.)
II.iii.7.8-11	54.14-15, 50.14-15		(contd.)
II.iii.8.1-10	57.11-16	Nic. 47.9-16	Lunar rainbow

II.iii.1.1-5: Conditions required for formation of rainbow

IS begins his discussions on the rainbow by telling us that he is less satisfied with the traditional doctrines concerning the rainbow than with those concerning the halo (*Šifā᾽* 50.8ff.) and launches into a lengthy discussion of the conditions required for the formation of a rainbow based on his own observations. In the passage of *But.* here BH summarises the concluding part of IS's discussion ((1)-(2)) and adds to it a sentence taken from an earlier part of IS's discussion ((3)). In his summarisation and in particular in naming the two required conditions first, BH may have been influenced by *Mabāḥiṯ* II.180.19-181.5, a passage he had used earlier at *Cand.* 114.11-115.4.

> *But.*: (1) For [the appearance of] the illusion of a rainbow, moist air sprayed with [lit. in which are sprayed (*zrīqān*)] particles of dewy [*rsīsānāyē*], transparent and clear water should serve as [lit. be/become] the mirror. (2a) Behind this dewy air there should be a coloured body, such as a mountain or a dark cloud, (2b) as the example of the rock crystal confirms. (3) As a result, sense errs and does not distinguish between the place (occupied by) [lit. of] the rainbow and the murky and dark cloud behind [*ᶜnānā dlīḥtā w-ᶜammūṭtā*] the rainbow.

*Šifā*ʾ 50.16-19, 51.10-17:

وأقول: أما أن هذا العارض لا بد من أن يكون وراءه فى أكثر الأمر سحاب مائى مستوى الأجزاء، فأمر توجبه المشاهدة لأن الأثر لا يكون فى نفس السحاب البتة، ولا نفس السحاب هو الذى يؤديه، لكن البصر يغلط فلا يميز بين مكان مرآته وبين السحاب الذى يكون وراءه. ... فظهر لى أن السحاب الكدر ليس يصلح أن يكون مرآة البتة لحدوث هذا الخيال، وإنما ينعكس للبصر منه (فيه) عن هواء رطب منتشر فيه أجزاء صغار من الماء مشفة صافية كالرش، وليست بحيث تكدر وتزيل الإشفاف، لكنها إذا لم يكن ورائها ملون لم تكن مرآة. وذلك كالبلورة، فإنها إذا سترت من الجانب الآخر صارت مرآة فى الجهة التى تليك، وإن لم تستر وتركت ووراءها فضاء مشف غير محصور لم تكن مرآة. فيجب أن يكون فى أكثر الأمر وراء هذا الهواء الرطب شىء لا يشف: إما جبل، وإما سحاب مظلم، حتى يرتسم هذا الأثر منعكسا عن الأجزاء المائية الشافة المنتشرة الواقعة فى الجو، دون البخارية الكدرة؛ فإنها إذا كانت بخارية كدرة لم تصلح لذلك.

[50.16ff.] I say: That behind this phenomenon [*ʿāriḍ*] there must in most cases be a watery cloud that is even in its parts is a fact confirmed by observation since this effect [*aṯar*] does not occur in the cloud itself at all, nor is the cloud itself what conveys the effect, (3) but vision errs and does not distinguish between the place (occupied by) [lit. of] its mirror and the cloud which is behind it. ... [51.10ff.] It then became clear to me that the murky [*kadir*] cloud was not suited to be the mirror for the occurrence of this illusion at all and (1) that (the illusion) was only reflected to the vision through it [*minhu* ed. Cair.: *fīhi* ed. Ṭ] from moist air in which were dispersed (*muntašir*) small, transparent, clear particles of water like dew [*rašš*], not such as to cause murkiness and to diminish the transparency, (2a) but [such that], when there is no coloured object behind it, it does not become a mirror. - (2b) That is like rock crystal. For when rock crystal is veiled on the other side it acts as a mirror on the side facing you, but when it is not veiled and behind it is an unlimited transparent open space it does not act as a mirror. - (2a) There must in most cases be behind this moist air something non-transparent, either a mountain or a dark cloud [*saḥāb muẓlim*], so that this effect [*aṯar*] may be formed [*yartasima*] by being reflected from the transparent watery particles diffused and situated in the air, unaccompanied by turbid vapoury particles, since when they are vapoury and murky they are unsuitable for that purpose.

l. 4. ܒܪܘܠܐ ("rock crystal"): بلورة IS. - Cf. II.i.8.4 above.

II.iii.1.6-10: Examples of rainbow-like illusions

But.: (1) In mills [*raḥwāṯā*] too, when fine waterdrops are sprayed opposite the sun from the edges of the wheel [*gīglā*] rotating in water, the colours of the rainbow are seen in them. (2) The same happens [lit. just so] when one gathers water in one's mouth and blows in front of the sun or a lamp [*šrāgā*], (3) and around a candle in a bath [*balanē*]. (4) In the morning [*ṣaprāyāṯā*] when one awakes from sleep, because of the moistness of one's eye, before wiping it, one sees a rainbow-like illusion.

*Šifā*ʾ 52.1-6:

ورأينا مثل هذا الخيال يتولد فى أرحاء الماء إذا انتضح عن أجنحة الآلة المنصوبة/ ٢ فى وجه الماء رَشُّ ماء صغير الأجزاء طلّى، توازيه الشمس، فيحدث دائرة بألوان القوس./٣ وكذلك إذا أخذ الإنسان الماءَ فى فمه، ونفخه فى الجو حذاء الشمس أو السراج. ورأينا/٣ الشمعة فى الحمام يتولد حواليها من رطوبة جو الحمام هذا الخيال؛ بل قد رأينا فى الغدوات/ ٥ حول الشمس خيالا هلالى الشكل قوسى اللون، والسبب فيه رطوبة المنتبه عن نومه، فكان/٦ إذا مسحت العين لم يظهر منه شىء.

[1 أرحاء : أرجاء edd. Cair. et Ṭ]

(1) We have seen a like of this illusion generated in water mills [*arḥā*ʾ] when a dewy spray of water with small particles is sprinkled from the sides of the wheel [*āla*] planted [*manṣūba*] in water and the sun stands opposite it. A circle with the colours of a rainbow is then formed. (3) Just so, when someone takes water in his mouth and blows it in the air opposite the sun or a lamp [*sirāǧ*]. (3) We have seen a candle in a bath [*ḥammām*], around which such an illusion was formed by the moisture of the air in the bath. (4) Indeed, we have seen in the morning around the sun an illusion crescent-like in shape and rainbow-like in colour. The reason for it is the moisture of the one awaking [*muntabih*] from his sleep; for when the eye is wiped nothing of the kind is seen.

l. 9. ܨܦܪ̈ܝܬܐ ("morning"): Probably to be understood with an ellipsis of *šāʿē*, "hours" (see Brockelmann 635b).

II.iii.2.1-2: Water spray from oars

BH has, as often, combined words taken from a passage of the *Šifā*ʾ with those taken from one in Nic. syr., both of which derive ultimately from the same place in Arist. (374a 29f., also 374b 6f.).

But. [underline: agreement with *Šifā*ʾ; italics: with Nic.]:

The Master [Aristotle] has said that *rainbow* colours [*gawnē qeštānāyē*] are observed *in water which is sprinkled* [*mezdalḥīn*] *from oars* [*līqē*] on the sea.

*Šifā*ʾ 52.10-11: وقد يحكى أن هذه الألوان تظهر من ماء ينتشر من مجاديف السفن فى البحر

It has been reported that these colours appear from water sprinkled [*yantaširu*] from oars of the boat on the sea

Nic. syr. 53.14-16[19]:

ܐܚܪܢܐ ܕܝܢ ܗܝ ܕܡܢ ܐܐܪ ܕܠܒܝܫ ܥܡܛܢܐ/15 ܐܝܟ ܕܒܫܪܓܐ. ܘܬܠܝܬܝܬܐ ܒܝܕ ܥܢܢܐ.
ܐܝܟ ܕܒܩܫܬܐ. ܘܪܒܝܥܝܬܐ ܒܝܕ ܡܝܐ ܘܒܝܕ ܗܕܐ/16 ܡܬܚܙܝܐ ܩܫܬܐ ܡܬܚܙܝܐ
ܒܡ̈ܝܐ ܕܡܙܕܠܚܝܢ ܡܢ ܠܝܩܐ.

Another [type of ἀνάκλασις] is that from murky air, as in a lamp, the third [type] is due to a cloud, as in a rainbow, and the fourth is due to water - Through this [last] type rainbow is seen in water which is sprinkled [*mezdalḥīn*] from oars [*līqē*].

---

19 With the passage of Nic. syr. as a whole here, cf. also Olymp. 230.11-13, 230.18-27; Olymp. arab. 145.22-146.10.

II.iii.2.2-7: Reflexion from air: Antipheron of Tarentum

But. [underline: agreement with *Šifā'*; italics: with Nic.]:

(1) With someone whose *vision* [*ḥzāṯā*] is *weak*, (2) (the vision) is broken [*mettaḇrā*] and is not conveyed through the air, (3) *as happened to Antipheron of Tarentum*. (4) *For when he fixed his gaze* [*ḥyārā*] *a little, he thought he was seeing forms* [*ṣūrāṯā*] *of men, because his vision was broken by the air*. (5a) Indeed, because of its *weakness* [*mḥīlūṯā*] (the vision) was reflected [*ʿāṭpā (h)wāṯ*] from *the air* towards visible objects [*meṯḥazyānē*] (5b) and it saw their likenesses [*demwāṯā*] in the air, (5c) just as (vision) is also reflected from water towards objects and sees their likenesses in water.

Šifā' 52.11-13:

ومن ضعف بصره حتى صار كأنه لا ينفذ فى الجو فقد يتخيل له ذلك، يتخيل له أشباح أشياء أخرى،
وربما يخيل له شبح نفسه أمامه، فإن الهواء يصير بالقياس إلى بصره محدودا منقطعا.

(1) Someone, whose vision [*baṣar*] is weak (2) so that it is, as it were, unable to penetrate through the air, sometimes sees such an illusion. (4) He sees illusory images of other things and he may even see an illusory image of himself before him; (5) for the air becomes limited and cut off [*maḥdūdan munqaṭiʿan*] in relation to his vision.

Nic. syr. 53.6-9, 13-14[20]:

6 ܐܪ̈ܒܥܐ ܕܝܢ ܐܕܫ̈ܝܗܘܢ ܐܢ̈ܝܢ ܕܐܢܐܩܠܐܣܝܣ. ܚܕ ܡ̇ܢ ܡܢ ܐܐܪ ܐܝܟܢܐ
ܕܓܕ[ܫ ܠ]ܐܢܛܝܦܪܘܢ 7/ ܠܛܪܐܢܛܝܢܐ. ܡܛܠ ܡܚܝܠܘܬܐ ܕܚܙܬܗ. ܐܝܟ ܕܐܡܪ
ܟܠܒܝܬܐ ܐܪܣܛܘܛܠܝܣ. 8/ ܗܟܢܐ ܓܝܪ ܡܚܝܠܐ ܗܘܬ. ܘܐܦ ܡܠܝܠܐ ܗܘܬ.
ܐܝܟܢܐ ܕܡܚܙܝܬܐ ܗܘܐ ܠܗ 9/ ܐܐܪ ܕܩܪܝܒ. ܘܠܐ ܡܫܟܚܐ ܗܘܬ ܕܬܕܚܘܝܗܝ
ܐܝܟ ܐܐܪ ܕܪܚܝܩ ܘܥ[ܒܐ]// 13 ܟܕ ܚܪ ܩܠܝܠ ܡܣܒܪ ܗܘܐ ܚܙܐ. ܨܘ̈ܪܬܐ
ܗܘܐ ܕܨܘ̈ܪܬܐ ܕܒ[ܢ̈ܝ ܐܢܫܐ ܚܙܐ] 14/ ܗܘܐ. ܥܠ ܕܡܬܬܒܪܐ ܗܘܬ ܚܙܬܗ ܡܢ
ܐܐܪ.

[7 ܠܛܪܐܢܛܝܢܐ : ܠܛܪܐܢܛܠܝܢܐ cod.]

There are four types of ἀνάκλασις: one is from air, (3) as happened to Antipheron of Tarentum (1) because of the weakness of his vision [*mḥīlūṯā da-ḥzāṯeh*], as Aristotle says in the text. (5) For it was so weak and very faint that the air nearby became for him a mirror and (his vision) was not able to push it as [it happens with] distant and dense air. (4) When he fixed his gaze a little, he thought he was seeing forms of men, because the vision was broken by the air.

l. 2. ܚܙܬܗ ("vision"): Hitherto in this section BH has used the word *ḥzāyā* in the sense of "vision" (corr. IS *baṣar*). Here he uses *ḥzāṯā*, which is the word used at the corresponding place in Nic. syr. This word recurs at II.iii.8.6, II.iv.1.4 below (also Mete. IV.i.4.6, IV.i.6.5; Min. II.ii.3.9). - It is difficult to distinguish between the sense of the two words. - BH himself seems to prefer *ḥzāṯā* in *Cand*. (117.11; 119.1, 2; 131.2) and *ḥzāṯā* is also the form constantly used in Nic. syr. (43 times), except on one occasion, at Nic. syr. 51.17, where *ḥzāyē* (pl.) is found, apparently in the sense of "lines of vision".

---

[20] Cf. Arist. 373a 35-b 10; Olymp. 230.11-18, 232.9-15; Olymp. arab. 145.22-146.5.

l. 3. ܐܢܛܝܦܪܘܢ ܛܪܐܢܛܝܢܐ ("Antipheron of Tarentum"): Arist. does not tell us the name of the person in question at *Mete*. 373a 35-b 10, but mentions him as Antipheron of Oreus (Ὠρείτης) at *Mem*. 451a 9. The name is given as Antipheron of Tarentum by Alex. Aphr. (147.32) and Olymp.; no name is given by Olymp. arab.

l. 4. ܟܕ ܚܝܪ ܩܠܝܠ ܡܚܝܪ ܗܘܐ ܚܝܪܗ ("when he fixed his gaze at little"): These words evidently correspond to Aristotle's "ἠρέμα καὶ οὐκ ὀξὺ βλέποντι". It seems that somewhere along the line these words were detached from what preceded them and taken with what followed. See Arist. 373b 4-7: οἷόν ποτε συνέβαινέ τινι πάθος ἠρέμα καὶ οὐκ ὀξὺ βλέποντι· ἀεὶ γὰρ εἴδωλον ἐδόκει προηγεῖσθαι βαδίζοντι αὐτῷ ... ("An example of this is what used to happen to a man *whose sight was weak and unclear*: he always used to see an image going before him as he walked ...", tr. Lee, my italics).

II.iii.2.7-10: Rainbow colours on the wall of a bath

But.: (1) When the ray of the sun falls on the glass [*zgōgīṯā*] in [lit. of] the skylight [*kawwṯā*] of a bath, (2) it is conveyed towards the wall opposite (the skylight) through the spray-filled air [*ā'ār rsīsānāyā*], (3) is reflected from (the wall) towards another wall by διάκλασις and causes an illusion of the rainbow colours on it.

Šifā' 52.6-10:

وقد رأينا فى بعض الحمامات هذا الخيال منطبعا تمام الانطباع فى حائط الحمام، ليس على سبيل الخيال، بل كان الشعاع يقع على جام الكوة فينفذ فى الرش المملوء منه هواء الحمام، ثم يقع على حائط الحمام وهو شعاع مضىء، ثم ينعكس عنه فى الهواء الرشى إلى الحائط الآخر ألوان قوس مستقرة ليس مما تبرح موقعة بانتقال الناظر.

We have seen in some baths such an illusion completely imprinted on the wall of the bath. That was not in the manner of an illusion. (1) Rather the a ray was falling on the [glass] cup [*ǧām*] of the skylight [*kūwa*] and (2) penetrated through the spray [*rašš*] which filled the air of the bath. It then fell on the wall of the bath in the form of a bright ray. (3) Then, it was reflected from it through the sprayey air [*al-hawā' al-raššī*] towards the other wall in the colours of a rainbow, which were fixed and did not alter their place with the movement of the viewer.

l. 9. ܕܝܐܟܠܐܣܐܝܬ ("by διάκλασις"): The process described here resulting in a stationary image is what BH decided to call "διάκλασις" in Mete. II.i.4 above. - The adverb seems to be a hapax legomenon.

II.iii.3.1-4: Shape of the rainbow

IS discusses the shape of the rainbow at *Šifā'* 53.9-54.4, but does not go into a detailed explanation of the circularity of the rainbow. For the first part of his explanation BH turns here instead to Fakhr al-Dīn al-Rāzī.

But.: (1) The shape of this rainbow is circular [*ḥudrānāyā*] (2) because the sprayey particles from which the visual ray is reflected towards the sun are placed [*sīmān*] in such a way that, if we were to make the sun the centre of the circle, the segment [*kpāpā*] of the circle which falls above the earth would pass along those particles.

Mabāḥit II.182.10-12:

البحث الخامس عن علة استدارة هذا القوس وهى ان الاجزاء التى ينعكس عنها شعاع البصر وقعت بحيث لو انا جعلنا الشمس مركز دائرة كان القدر الذى يقع من تلك الدائرة فوق الارض يمر على تلك الاجزاء

(1) Fifth examination: on the cause of the circularity [*istidāra*] of this rainbow. (2) The particles from which the visual ray is reflected occur [*waqaᶜat*] in such way that, if we were to make the sun the centre of the circle, the segment [*qadr*] of that circle which falls above the earth would pass along those particles.

l. 2. ܣܝܡܢ ("are placed"): وقعت Rāzī. - Did BH read وضعت?

II.iii.3.4-8: Shape of the rainbow (contd.)

In the latter half of the explanation, where the wording of the *Mabāḥit* becomes close to that of the *Šifā*ʾ, it becomes more difficult (and perhaps futile) to decide which of the two BH is following, but in sentence (3), at least, BH's wording is closer to the *Mabāḥit*.

But. [underline: agreement with *Mabāḥit*; italics: agreement with *Šifā*ʾ]
(3) *Therefore*, if *the sun is on the horizon*, the plane [*šṭīḥūṯā*] of *the horizon* will divide [*pāsqā*] the circle in two halves, a half above the earth and a half below it. (4) As the ἔξαρμα - i.e. *elevation* [*rawmā*] - of *the sun increases* [*sāgē*], the size of the *rainbow* *decreases* [*zāᶜrā*]. (5) When (the sun) is at the zenith [*lᶜel men rēšā*] it *disappears* [*bāṭlā*].

Mabāḥit II.182.12-17:

فان كانت الشمس على الافق كان الخط المار بالناظر والنير على بسيط الافق وهو المحور فيكون حينئذ سطح الافق يقسم المنطقة بنصفين ويرى القوس نصف دائرة فان ارتفعت الشمس انخفض الخط المذكور وصار الظاهر من المنطقة الموهومة اقل من نصف دائرة حتى اذا ارتفعت الشمس ارتفاعا كثيرا لم يكن قوس واما اذا كان ارتفاعها الى حد كان قوسا

(3) Therefore, if the sun is on the horizon, the line passing through the viewer and the luminary will be on the plane [*basīṭ*] of the horizon - that line being the axis [*miḥwar*]. - The plane [*saṭḥ*] of the horizon will then divide [*yaqsimu*] the belt [circle] in two halves and the arc [rainbow, *qaus*] will appear as a semi-circle. (4) If the sun rises, the above-mentioned line diminishes and the visible part of the imaginary belt becomes smaller than a semi-circle, until, (5) when the sun has risen to a great elevation [*irtifāʾan kaṯīran*], the arc disappears [*lam yakun*], whereas, when the sun has [only] risen to a limited elevation, the arc is there.

*Šifā*ʾ 53.12-16:

فإذا كانت الشمس على الأفق قطعت الأفق من الدائرة الموهومة له نصفها لا محالة، فإن ارتفعت الشمس ارتفع محور المنطقة، فانحطت المنطقة لا محالة، فنقصت القوس لا محالة. حتى إذا ارتفعت الشمس ارتفاعا كبيرا لم يكن قوس، وأما إذا كان ارتفاعها إلى حدٍّ كان قوس.

(3) Therefore, when the sun is on the horizon, it will undoubtedly cut [*qaṭaʿat*] from the imaginary circle a half of it. (4) If the sun rises, the axis of the belt will rise, and the arc [*qaus*] will undoubtedly diminish, so that when the sun has risen to a great elevation [*irtifāʾan kabīran*], the arc disappears, whereas, when the sun has [only] risen to a limited elevation, the arc is there.

l. 7. ܐܟܣܐܪܡܐ ("ἔξαρμα"): For BH's own definition of the term "ἔξαρμα", see *Asc*. 21.3-5: "The ninth circle is that of the ἔξαρμα, or elevation [*rawmā*]. It is the great circle which meets the two poles of the horizon, that is to say, it passes through the two points above the head and [below] the feet [i.e. the zenith and nadir] ..."; cf. further *Asc*. 21.9; 79.21; 80.5, 8; 82.23; 206.2. - Elsewhere in Syriac, the word is known from Severus Sebokht (see Nau [1910] 232; id. [1930-31] XXVII.344, 397).

l. 7f. ܠܥܠ ܡܢ ܪܝܫܐ ("at the zenith"): lit. "above the head". - See comm. on IV.i.4.8 above.

II.iii.3.8-10: Rainbows at midday

But.: (1) In northerly regions rainbow occurs at midday in winter, (2a) but does not occur in summer, (2b) because [*b-hāy d-*] the sun is less elevated [*bṣīr rāʾem*] in winter and more [*yattīr*] in summer.

Šifāʾ 53.16-18[21]:

فلذلك يجوز أن تحدث القوس فى بعض البلاد فى الشتاء فى أنصاف النهار، ولا تحدث فى الصيف، لقلة ارتفاع الشمس فى أنصاف نهار الشتاء وكثرته فى أنصاف نهار الصيف

For that reason, it is possible for the rainbow to occur in some regions at midday in winter, (2a) but it does not occur in summer (2b) because of [*li-*] the smallness of the elevation [*qillat irtifāʾ*] of the sun at midday in winter and its greatness [*kaṯratuhu*] at midday in summer.

II.iii.4.1-5: Colours of the rainbow and halo

The last part (sentence (5)) of the passage of *But.* here has no counterpart in the *Šifāʾ*, but contains nothing that cannot be deduced from parts (1)-(4). BH may have been encouraged in making the addition by the passage of Nic. syr. quoted below.

But. [underline: agreement with *Šifāʾ*; italics: with Nic.]:

(1) The colour of the rainbow is neither luminous [*nahhīr*] nor white [*ḥewwār*] (2) like the colour of the halo, (3) because with the rainbow the sprayey moisture is far from the sun, (4) so that from the mingling [*ḥulṭān*] of the illusory luminous [*nahhīrā haggāgānāyā*] with something falling in [lit. from, *min*] the genus of darkness [*gensā d-ḥeššōḵā*] various colours, [namely] red, green and purple, are generated. (5a) With *the halo*, on the other hand, the sprayey moisture is closer to the moon (5b) *and for this reason* its colour is luminous and *white*.

---

[21] Cf. *Mabāḥiṯ* II.182.18-21, where 19-21 ≈ *Šifāʾ* 53.16-18.

Šifā' 53.6-8:

وأما لونه فلعله إنما لا يكون منيرا أبيض، لأن مرآته بعيدة عن النير، ليس كما يرى فى الهالة. فلذلك يختلط الضوء الخيالى بشيء من جنس الظلمة، فتتولد حمرة وأرجوانية وغير ذلك.

(1) As for its colour, it is not luminous [*munīr*] and white [*abyaḍ*] (3) perhaps because its mirror is far from the luminary [*nayyir*], (2) unlike what is seen in the halo. (4) For this reason the illusory light [*al-ḍau' al-ḫayālī*] mixes with something of the genus of darkness [*ǧins al-ẓulma*] and red, purple and other (colours) are generated.

Nic. syr, 53.32-54.3[22]:

32 ܘܒܗܠܘܬܐ ܡܛܠ ܕܚܙܘܐ ܐܙܠܐ ܣܢܝܐ. ܠܐ ‡ܕܪܝܫܐ ܘܠܐ ܒܬܪܝܨܘܬܐ
/[54] [--ܐ‡ . ܐܝܟ] ܕܒܐܣܛܘܡܛܐ ܓܝܪ ܐܣܬܝܩܝܢ ܥܢܢܐ ܘܫܡܫܐ:
ܘܡܛܠܗܢܐ ܚܘܪܬܐ/2 [ܡܬܚ]ܙܝܐ ܗܠܘܬܐ. ܒܩܫܬܐ ܕܝܢ ܡܟܝܠ ܚܙܘܐ ܐܙܠܐ
ܐܙܠܐ ܣܢܝܐ: ܐܝܟ ܕܒܕܝܐܡܛܪܘܢ/3 ܓܝܪ ܦܪܝܫ ܫܡܫܐ ܡܢ ܥܢܢܐ.

[32 ܡܛܠ: ܡܢ cod. || 1 ܐܣܬܝܩܝܢ: ܐܣܬܝܩ cod. || 3 ܦܪܝܫ incert.]

(5a) In a halo vision travels a short distance, because -- [?] in a straight line - for the cloud and the sun are in a perpendicular [relationship] - (5b) and for this reason the halo appears white. In a rainbow, on the other hand, vision travels a long distance - for the sun is separated in a diametrical [position] from the cloud.

l. 2. ܪܛܝܒܘܬܐ ܪܣܝܣܝܬܐ ("sprayey moisture"): مرآته ("its mirror") IS. - As has been explained in II.iii.1-2, it is the particles of water sprayed in the air which act as the "mirror" for the purpose of producing a rainbow.

l. 4. ܣܘܡܩܐ ܘܟܪܬܢܝܐ ܘܐܪܓܘܢܝܐ ("red, green and purple"): The three colours of the rainbow according to Arist. are φοινικοῦς, πράσινος and ἁλουργός. In Nic. syr. these are rendered by *summāqā*, *kartānāyā* (lit. leek-coloured) and *argwānāyā* (corr. ἁλουργός at Nic. syr. 54.9, corr. πορφυροῦς at 54.21, 31). IS has *aḥmar*, *karrāṯī* and *urǧuwānī*.

II.iii.4.5-7: Colour of halo (contd.)

Arist. talks of apparent alteration of colours due to contrast with other nearby colours in his explanation of the fouth, "yellow" (ξανθός), colour seen in rainbows (375a 7-28) and mentions there the "moon rainbow", which appears white because it appears on a dark cloud at night (375a 17-22). IS takes up the discussion of the "moon rainbow" at *Šifā'* 57.4-16 (cf. II.iii.8. below) and follows Arist. in attributing its whiteness to the contrast with the darkness of the night. BH's attribution of the halo to the darkness of the night may be connected to that passage. - The comparison with dirty clothes can be traced back to Olymp. and may therefore be taken from a lost passage of Nic. syr.

---

[22] Cf. Arist. 374a 2f., Olymp. 234.14-28; Olymp. arab. 155.4-7.

But.: ... (5c) also on account of the darkness of night, (5d) just as dirty clothes appear whiter when they are compared [*metpaḥḥmīn*] with clothes which are [even] dirtier.

Olymp. 243.30-32: ἐὰν ἧττον μέλαν παρὰ μελάντερον παραβληθῇ, δόξει λευκὸν εἶναι τὸ ἧττον μέλαν διὰ τὸ τὰ ὅμοια ἐν τοῖς ὁμοίοις λανθάνειν. (5d) ἰδοὺ γοῦν τὸ ἧττον ῥυπαρὰ ἱμάτια παρὰ τὰ μᾶλλον ῥυπαρὰ παραβαλλόμενα καθαρὰ εἶναι δοκεῖ.

II.iii.5.1-2: Peripatetic theory on the colours of the rainbow

The Peripatetic theory which is described here in II.iii.5. and II.iii.6. is rejected by IS. BH does not commit himself either way, but gives IS's refutation in II.iii.7.

But.: In the teaching of the ancients [lit. the first teaching], the difference of the position of two clouds is given as the reason for the variety of the threefold colours of the rainbow.

*Šifā*ʾ 54.4-7:

وأما وجوب كون الألوان ثلاثة، ومرافقة لون أصفر إياها، وما يرى معها فى الأحيان بأعيانها، وترتيبها، فليس يمكننى أن أقف على السبب فيه. والذى يقال إن السبب فيه اختلاف وضع سحابتين وامتزاج لون ثالث منهما فشىء لا أصل له

Concerning the necessity of colours being three, the fact that a yellow colour accompanies them, what appears with them at certain times, and their arrangement, I am unable to settle on the reason for them. What has been said, [namely] that reason for it is the difference in the position of the two clouds and the mixture a third colour with them, has no basis.

II.iii.5.2-5: Peripatetic theory (contd.): red

But. [underline: agreement with *Šifā*ʾ; italics: with Nic.]:

(1a) From the upper [cloud], because it is close to the sun, (1b) vision is reflected strongly towards the sun, (1c) so that its colour appears [*metḥzē*] a clear red [*summāqā naṣṣūpā*], (2) because *a bright object* [*nahhīrā*] *appears red* when seen *amid a black* [thing] [*b-ukkāmā*] *or* through a black medium [lit. *through the mediation of a black* (substance), *b-yad meṣʿāyūt ukkāmā*].

*Šifā*ʾ 54.8-9:

ولا ما يقال إن الناحية العليا تكون أقرب إلى الشمس، وانعكاس البصر يكون أقوى فترى حمرة ناصعة

Neither [is there any ground for] the assertion that (1a) the upper side is nearer the sun (1b) and the reflection of vision is stronger, (1c) so that a pure red colour [*ḥumra nāṣiʿa*] is seen [*turā*].

Nic. syr. 54.3-7[23]:

ܢܗܝܪܐ ܕܝܢ ܟܕ ܚܘܪܐ ܐܝܬܘܗܝ. ܐܘ ܒܝܕ ܡܨܥܝܘ[ܬܐ] 4/[ܕ]ܐܘܟܡܐ. ܐܘ
ܒܐܘܟܡܐ ‹ܣܘܡܩܐ› ܡܬܚܙܝܐ. ܫܡܫܐ ܓܝܪ ܟܕ ܕܢܚ ܘܥܪܒ ܒܝܕ ܠܗܓܐ/5
ܐܘܟܡܐ ܗ̄ ܫܡܫܐ ܒܠܗܓܐ ܕܐܝܬܘܗܝ ܡܨܥܝܐ ܘܐܘܟܡܐ. ܘܠܐ ܒܗ ܒܥܛܪܐ
ܘܢܘܪܐ/6 ܒܗ ܒܬܢܢܐ ܐܘܟܡܐ. ܘܠܐ ܒܝܕ ܡܨܥܝܘܬܗ ܕܩܛܪܐ ܡܬܚܙܝܐ. ܢܘܪܐ
ܕܝܢ/7 ܟܕ ܕܠܩܐ ܒܩܝ̈ܣܐ ܪ̈ܛܝܒܐ ܐܘܟܡܐ.

[4 ‹ܣܘܡܩܐ›: supplevi || 7 ܐܘܟܡܐ: lege ܣܘܡܩܐ, cf. Olymp. 236.3-5]

(2) Although the luminary [*nahhīrā*] is white, [when seen] either through the mediation of (something) black [*b-yad meṣʿāyūṯā d-ukkāmā*] or in (something) black [*b-ukkāmā*] it appears <red>. For when the sun rises and sets, [it is seen] through black vapour - i.e. the sun [is seen] through [*b-yaḏ*] the vapour which is in the middle and black, and not in [*b-*] the exhalation [*ʿeṭrā*], whereas fire is seen in [*b-*] black smoke, and not through [*b-yaḏ*] the mediation of the soot. Fire, when it burns with moist (pieces of) wood, is black [read: "red"].

l. 2 ܘܥܠܝܬܐ ("upper [cloud]"): As the text stands, *ʿellāytā* here and *taḥtāytā* in l. 5 below must mean "upper/lower [cloud]" (fem.), whereas IS has *al-nāḥiya al-ʿulyā/al-suflā* ("upper/lower side"), meaning the upper/lower "band" of the rainbow. The word used below by BH for "band" is *ḥudrā* (masc.).

l. 3. ܣܘܡܩܐ ܢܗܝܪܐ ("clear red"): حمرة ناصعة IS. - See comm. on Min. III.ii.3.3 (ܢܗܝܪܐ) above; cf. also II.iii.7.3 below.

II.iii.5.5-7: Peripatetic theory (contd.): purple

But.: (1) From the lower [cloud], on the other hand, (2) because it is far from the sun, (3) vision is reflected weakly towards the sun, (4) so that it appears as red tending towards blackness, and such a colour is purple.

Šifāʾ 54.9-10:

وأن الناحية السفلى أبعد منها وقل لذلك إشراقا فيرى في الطوق الثاني حمرة إلى السواد وهو الأرجواني

... and that (1) the lower side (2) is further from it and (3) for that reason has less illumination, (4) so that there appears in the second circle [*tauq*] a redness [tending] towards blackness, and that is purple.

l. 5-6. "vision is reflected weakly towards the sun": Instead of following the wording of IS, BH here reproduces, *mutatis mutandis*, the formulation of l. 3 above.

l. 6. ܡܬܚܙܝܐ: It is tempting to supply ܠܘܩܒܠܗ after ܡܬܚܙܝܐ to produce the same phrasing as in l. 3-4 above, especially as this could have dropped out before the following ܘܓܘܢܐ.

---

[23] Cf. Arist. 374a 3-8, b 9-11; Olymp. 235.34-236.6, 240.10-18; Ibn al-Khammār, *Mutaḫayyila* [Lettinck] 356.3-13; *Cand.* 115.4-5.

II.iii.5.7-8: Peripatetic theory (contd.): green

But.: From these two outer colours, a green colour is generated and is seen between them.

Šifā' 54.11-12:

وأنه يتولد فيما بينهما لون كراثى كأنه مركب من إشراق حمرة الفوقانى وكدر ظلمة السفلانى

... and that between them a green colour is generated, as if it were compounded from the illumination of the redness above and the murkiness of the darkness below.

l. 7. ܣܘܦܢܝܐ ("outer"): The adjective *sawpānāyā* is used elsewhere by BH in the sense of "finite" (see PS 2579 s.v., < *sawpānā*, "end, destruction"). Here it must mean "at the edge" (< *sawpā*, "edge, extremity"). BH may have the word from a lost passage of Nic. syr., where it could have corresponded to gr. τελευταῖος (cf. Olymp. 237.14 "τελευταῖαι [sc. ζῶναι]", in the passage quoted under II.iii.6.1-6 below).[24] This sense "at the edge" of *sawpānāyā* seems to be unregistered in dictionaries (PS, Cardahi, Brockelmann, Audo).

II.iii.6.1-6: Peripatetic theory (contd.): colours of a double rainbow

Arist. discusses the colours of a double rainbow at *Mete*. 375a.30-b 9 (also 375a 1), where he mentions that the arrangement of the colours in the outer rainbow is reverse of that of the inner. IS does mention the double rainbow in passing at *Šifā'* 55.17, but does not discuss their colours. The passage of *But.* here is probably based on a lost passage of Nic. syr. which, in turn, may be traced back to Olymp. 237.12-16.[25]

But. [underline: agreement with Olymp.; italics: with Olymp. arab.]:

It is also said: (1) *When two rainbows* are seen one inside another, (2) *the* outer *circle* [*ḥuḏrā*] of *the inner rainbow* and *the* inner *circle* of *the outer rainbow* appear red on account of their *juxtaposition* [*naqqīpūṯā*] *with each other*. (3) *The* outer *circle* of *the outer rainbow* and *the* inner *circle* of *the inner rainbow*, because they are *the extremities* [*sāḵā*], are *purple*. (4) The two *middle circles* of the two rainbows, the outer and the inner, are *green*, since they are compounded from the extremities.

Olymp. *in Mete*. 237.12-16: ἕκτον καὶ τελευταῖον ἦν ζητῆσαι τὴν αἰτίαν τῆς τάξεως τῶν χρωμάτων, διὰ τί τῶν δύο ἰρίδων αἱ προσεχεῖς τελευταῖαι [sc. ζῶναι] μόναι φοινικοῦν ἔχουσι χρῶμα, τουτέστι τὸ πέρας τῆς μιᾶς καὶ ἡ ἀρχὴ τῆς ἐντός, αἱ δὲ παρ' ἑκάτερα αὐτῶν πράσινον ἔχουσι χρῶμα, αἱ δὲ κατὰ τὰ πέρατα ἁλουργόν.

---

[24] Nic. syr. uses the noun *sawpā* elsewhere as the equivalent of gr. πέρας, but for the πέρας at Olymp. 237.16 the word used seems to have been *sāḵā* (see II.iii.6.5 below).

[25] BH does not make use of the explanation of the arrangement given at Olymp. 237.16-239.28 (Olymp. arab. 157.13-158.9).

The sixth problem was to examine the cause of the arrangement of the colours, [namely] (1) why in a double rainbow [lit. of two rainbows] (2) only the adjoining [προσεχής] extreme [τελευταῖος] bands, i.e. the end [πέρας] of one and the beginning of the inner, have a red colour, (4) while those next to them have a green colour and (3) those at the end [πέρας] a purple [colour].

Olymp. arab. 157.4, 7-13:

متى حدثت قوسان معًا ألوانها تلك الثلاثة الألوان التى ذكرنا فقط. ... وأما ترتيب الألوان فيها فليس يوجد فيها جميعًا واحداً بعينه، لكن يوجد لون الدائرة العظمى فى القوس الداخلة: أحمر، ولون الدائرة الوسطى لون الكراث، ولون الدائرة الصغرى لون البنفسج. وأما قوس الخارجة ... فتوجد بخلاف ذلك، أعنى أن لون الدائرة الصغرى منهما (منها) يكون أحمر، ولون الدائرة العظمى بنفسجًا؛ فتكون عند ذلك الدائرتان الحمراوان إحداهما مجاورة للأخرى، والدائرتان التان لونهما لون الكراث فى الوسط، والدائرتان التان لونهما لون البنفسج فى الطرفين.

(1) When two rainbows occur together, their colours are those three colours only which we have mentioned. ... The arrangement of the colours in them is not the same. (2) The colour of the great circle [*dāʾira*] in the inner rainbow is red, (4) the colour of the middle circle green [*laun al-kurrāt*], and (3) the colour of the small circle violet [*banafsağ*]. As for the outer rainbow ... [the arrangement] is opposite of that, I mean, (2) the colour of the small circle is red and (3) the colour of the great circle is violet. The result is that (2) the two red circles are adjacent [*muğāwir*] to each other, (4) the circles whose colour is green are in the middle and (3) the circles whose colour is violet are at the extremities [*taraf*].

l. 5-6. "since they are compounded from the extremities": cf. *Šifāʾ* 54.11-12 [see under II.iii.5.7-8 above]: "... and that between them a green colour is generated, as if it were compounded [*murakkab*] from the illumination of the redness above and the murkiness of the darkness below".

II.iii.7.1-2: Ibn Sīnā's objection (init.)

The introductory comment here combines what IS says at the end of his objection to the Peripatetic theory (*Šifāʾ* 55.8-9) and what he had said earlier near the beginning of his discussion of the rainbow (50.10-11).[26]

But.: Here, the best of the moderns, the princely doctor [Ibn Sīnā] says: All these words which our friends the Peripatetics have put forward [*sādrīn*] do not convince me [*pīsā lā ʿābdān lī*].

Šifāʾ 55.8-9:

وبالجملة فإن أصحابنا من المشائين لم يأتوا فى أمر هذه الألوان وهذه الفصول بشىء فهمته

In sum, our friends among the Peripatetics have not brought forward anything I understand concerning these colours and these separations [*fuṣūl*, cf. under II.iii.7.8-11 below].

[26] Cf. also *Šifāʾ* 54.6: فليس يمكننى أن أقف على السبب فيه; 54.7: فشىء لا اصل له; 54.12: فكله ليس بشىء.

*Šifā*ʾ 50.10-11: وليس يقنعنى ما يقوله أصحابنا من المشائين فيها
What our friends among the Peripatetics have said about (the rainbow) does not convince me [*yuqniᶜu-nī*].

II.iii.7.2-8: Ibn Sīnā's objection (contd.)

But.: (1) For, seeing that the outer circle is pure red [*summāqā naṣṣūpā*] because it is close to the light, (2) the middle circle, which adjoins it ought [*wālē*, √*wly*] to be a purple tending towards redness, and the inner [circle] a purple tending towards blackness. (3) A purple colour tending towards redness, since it is intermediate between red and purple, ought to be redder than purple and more purple than red, (4) not green, which is not related to either of them but is generated from yellow and black.

*Šifā*ʾ 54.12-13[27]:

فكله ليس بشىء لأن الأولى هو أن يكون الأقرب ناصع الحمرة، ثم لا يزال كذلك على التدريج يضرب إلى الأرجوانية والقتمة، فيكون طرفه الآخر أقتم أرجوانيا.

... All this is nonsense, since the most appropriate case is that [*aulā an*, √*wly*] (1) the nearest [part] should be pure red [*nāṣiᶜ al-ḥumra*]; (2) then, it should continuously and gradually change towards purpleness and blackness [*qutma*], so that the other extremity will be the blackest purple.

*Šifā*ʾ 55.4-7:

وتولّد هذا الكراثى أيضا بين الأرجوانى والأحمر الناصع بديع. فإن اللون الممتزج منهما شىء هو أشد نصوعا من الأرجوانى وأشد أرجوانية من الناصع، لا لون كراثى لا مناسبة له مع واحد منهما. ولأن يتولد الكراثى بين الأصفر وبين الأسود والنيلى، أولى من أن يتولد بين أحمر ناصع وبين أرجوانى.

The generation also of this green (colour) (3) between purple and pure red is astonishing. (3) For the colour [produced by] mixing the two is something more intense in pure-redness [*nuṣūᶜ*] than purple and more intense in purpleness than pure red [*nāṣiᶜ*], (4) not a green colour which has no relationship with either of the two. [It is also astonishing] because it is more appropriate [*aulā*] that (4) green be generated from [lit. between] yellow and black-indigo, than that it be generated from [lit. between] pure red and purple.

l. 7. ܡܢ ܫܥܘܬܐ ܘܐܘܟܡܘܬܐ ("between yellow and black"): بين الأصفر وبين الأسود والنيلى ("between yellow and black-indigo") IS. - The "indigo" (*al-nīlī*) is also omitted at the corresponding place in the paraphrase by Fakhr al-Dīn Rāzī (*Mabāḥiṯ* II.182.9).

II.iii.7.8-11: Ibn Sīnā's objection (contd.)

At *Šifā*ʾ 54.14-15, IS is evidently objecting to the arbitrary division by the Peripatetics of the colours of the rainbow, whose gradation is in

[27] For *Šifā*ʾ 54.4-12 see under II.iii.4 above.

fact more continuous than the Peripatetic theory supposed.[28] BH seems, however, to have understood the passage to simply mean that IS himself did not know why different colours occurred and links this passage to the sentence where IS tells us that he has no satisfactory answers concerning the causes of the colours of the rainbow.

But.: (1) The separation of these colours, whereby [*aykannā d-*] a certain band resembling red and another similar to purple arises, with green in between them and sometimes with a yellow colour accompanying them, (2) requires an explanation [lit. cause], (3) which - (Ibn Sīnā) says - "I have not yet found out, and I am not convinced by what I have read [lit. seen]".

Šifā' 54.14-15:

وأما انفصال هذه الألوان بعضها عن بعض حتى يكون عرض واحد متشابه الحمرة وآخر متشابه الأرجوانية وبينهما قطع، فلا معنى له.

(1) The separation of these colours from each other, so that there arises one band resembling red and another resembling purple, there being a division [*qaṭ*ᶜ] between the two, makes no sense.

Šifā' 50.14-15:

وأما الألوان فلم يتحصل لي أمرها بالحقيقة، ولا عرفت سببها، ولا قنعت بما يقولون، فإن كله كذب وسخف.

(2/3) As for the colours [of the rainbow] I have no satisfactory answer. I have not yet found out their cause and I am not convinced by what they say, since it is all false and absurd.

l. 9-10. "and sometimes with a yellow colour accompanying them": This is mentioned elsewhere by IS: *Šifā'* 54.4-5: وأما وجوب كون الألوان ثلاثة، ومرافقة لون أصفر إياها ("As for the necessity of the colours being three, and the accompaniment of them by a yellow colour ..."). - BH does not mention elsewhere in his discussion of the rainbow the fourth, yellow (ξανθός), colour of the rainbow which the Peripatetics recognised (Arist. *Mete*. 375a 7-28).

II.iii.8.1-10: Lunar rainbow

The discussion of the rainbow formed by the moon and its rarity goes back to Arist. *Mete*, 372a 21-29. The passage of *But.* here is based mainly on *Šifā'* 57.11-16, but also contains some elements evidently taken from Nic. syr. 47.10-16, a passage BH had earlier copied almost verbatim at *Cand*. 115.9-116.3.[29]

---

[28] See Horten-Wiedemann (1913) 742 n.1.

[29] Cf. also *Mabāḥiṯ* II.184.10-14, which is a summary of *Šifā'* 57.11-16 and adds little that is not there, but we note: a) *Mabāḥiṯ* II.184.10 *nādira ǧiddan* (*Šifā'*: ᶜ*alā*

But.: (1) From *the moon* the rainbow is formed at night, but *seldom* [*ḥdā lakmā*] and very rarely, (2) because the rainbow requires light with plenty of rays for its formation. (3) Otherwise, the illusion (of the light) will not be reflected from it towards the smooth cloud. - (4) Thus, images of pale objects appear indistinctly in mirrors, because their likenesses are not reflected towards them. - (5) For its formation, the rainbow also requires a cloud which is very compact and smooth and has a high level of *suitability* [*ʿāhnūṯā*], (6) so that it is capable of conveying to the visual organ the illusion of the pale luminary. (7) It is obvious that *the moon* is not always very bright, but only when it *is full*, and (8) the occurrence of a suitable cloud at the time of *full moon* happens seldom. (9) *For this reason*, the *rainbow* is formed very rarely from *the moon* at night.

Šifāʾ 57.11-16:

وأما قوس الليل فإنه يقع فى الأحيان وعلى سبيل الندرة، فإنها يحتاج فى تكونها إلى أن يكون النير شديد الإضاءة حتي ينعكس منه خياله. فإن الأشياء الضعيفة اللون لا ينعكس عنها ضوؤها انعكاسا يظهر. وأن يكون أيضا الجو شديد الاستعداد، فإنه إن كان قاصرا لم يؤد خيال ما ليس بدلك البالغ فى كيفيته، وإنما يكون القمر شديد الإضاءة عندما يتبدر فى الشهر مرة، فيقل أن يجتمع تبدره والاستعداد التام من الجو، فلهذا لا تتولد قوسه إلا فى الندرة

(1) The night rainbow occurs [only] occasionally and as a rarity, (2) since it requires for its formation that the luminary be intense in its illumination, (3) so that its illusion can be reflected from it. - (4) For the light of objects which are weak in colour is not reflected from them in such a way as to be visible. - (5) [It] also [requires] that the air be intense in its suitability [*istiʿdād*], (6) since if is deficient it will not convey an illusion which is not so strong in its quality [i.e. brightness]. (7) The moon, however, is intense in its illumination only when it is full, once in a month, and (8) its being full rarely coincides with a complete suitability on the part of the air, (9) *so that* its rainbow is only formed rarely.

Nic. Syr. 47.9-16[30]:

ܘܩܫܬܐ ܡܢ ܣܗܪܐ ܡܣܬܥܪܐ ܗܘܬ ܐܝܟ ܕܒܡܣܟܠܢܘܬܐ. /10 ܣܗܪܐ ܥܒܝܐ ܕܝܢ
ܟܝܢܐܝܬ . ܣܗܪܐ ܚܘܪ̈ܒܐ ܕܝܢ ܣܓܝ ܠܚܒܐ . ܩܫܬܐ ܡܢ ܠܐ/11 ܡܬܚܙܝܐ ܕܝܢ
ܡܬܥܒܕܢܘܬܐ ܣܗܪܐ ܚܘ̈ܪܒܐ . ܐܝܟ ܕܒܡܣܟܠܢܘܬܐ ܕܝܢ ܡܬܥܒܕܢܘܬܐ ܣܗܪܐ/12 ܥܒܝܐ .
ܣܓܝ ܠܚܒܐ ܡܬܥܒܕܢܘܬܐ ܣܗܪܐ ܡܣܬܥܪܐ . ܡܛܠ ܕܟܠܠܐ ܠܓܘ ܚܘܪܒܝܬܗ/13
ܠܡܥܒܕܗ . ܘܡܛܠ ܕܗܕܐ ܕܠܗܘܐ ܡܠܝܐ ܡܣܬܥܪܐ . ܟܕ ܕܝܢ ܢܘܗܪܐ ܡܢ ܢܗܝܪܐ/14 ܗܘܐ
ܡܠܝܐ . ܘܟܕ ܕܠܗ ܗܠܘ ܒܗ ܚܕܐ . ܐܝܬ ܕܝܢ ܡܣܬܝܢܐ ܘܟܠ ܡܫܬܟܚܐ ܚܫܝܟܐ/15
ܕܥܢܢܐ . ܒܝܪܚܐ ܒܣܝܓܐܐ ܥܢܢܝ ܡܫܬܟܚ ܘܟܘܠ ܒܝܢܝ ܕܗܘܬ ܡܬܥܒܕܐ ܣܗܪܐ/16
. ܡܣܬܥܪܐ

[10-11 ܩܫܬܐ ܕܝܢ ܠܐ ܡܬܚܙܝܐ ܡܢ legendum?]

sabīl al-nādira), cf. *But.* saggī *bṣīrāʾīṯ*; b) 184.11 *ʿinda išdād nūr al-nayyir* (*Šifāʾ*: *an yakūna al-nayyir šadīd al-iḍāʾa*), cf. *But.* *nuhrā ʿattīr zallīqē*; and c) 184.11 *al-ašyāʾ al-ḍaʿīfat al-nūr* (*Šifāʾ*: *al-ašyāʾ al-ḍaʿīfat al-laun*), cf. *But.* *gušmē bahhūrē*.

30 Cf. Arist. 371b 22-24 (halo), 373a 27-31 (halo), 372a 21-29 (rainbow); Olymp. 218.10-13, 225.29-226.7; Olymp. arab. 145.4-5 (halo), 153.10-19 (rainbow); *Cand.* 114.5-8 (halo), 115.9-116.3 (rainbow).

Halo usually occurs around the moon, less around the sun, and occasionally around stars. (1) Rainbow is never formed around stars, but is usually formed around the sun, and is formed occasionally [*ḥdā lakmā*] around the moon, because at night its colours are lost to sensation, and (7-8) because the moon needs to be full, but it is full [only] one day in a month, and [it only occurs] when it is rising and not when setting. (5) There is also a need for a suitable positioning [*syāmā ʿāhnā*] of the cloud. (9) Consequently in fifty years we have [only] twice seen a rainbow occur around the moon.

l. 3, 5-6. "the smooth cloud", "a cloud which is very compact and smooth [*rṣīptā wa-sqīltā*]": For IS what acts as the "mirror" in the formation of rainbow is not the clouds as the Peripatetics thought, but the sprayey air (*ğaww*: so in the passage quoted above at *Šifā'* 57.13, 16; cf. under III.iii.1. above). - BH seems to have forgotten this. - The need for the cloud on which the rainbow forms to be "compact" and "smooth" is frequently mentioned in Nic. syr., e.g. Nic. syr. 48.11-12, 23: ܗܘ/12 ܕܝܢ ܕܡܚܙܝܢܝܬܐ ܕܡܦܢܝܐ ܠܚܙܝܬܐ ܐܝܟܢܐ ܕܬܚܙܐ 23// ܗܠܝܢ ܫܥܝܥܬܐ ܘܪܨܝܦܬܐ ܘܢܗܝܪܬܐ. ("The mirror which reflects the vision, so that you see these things, should be smooth [*šaʿʿīʿtā* < gr. λεῖος], compact [*rṣīptā* < πυκνός] and transparent [*nahhīrtā* < διαφανής]"; cf. Arist. 372a 29ff.; Olymp. 219.9-12; Olymp. arab. 146.11-15; *Cand*. 114.1-2).[31]

Mete. II. Section iv.: *On Mock Suns and Lances*

The section deals with the "mock suns" (παρήλιοι) and "rods" (ῥάβδοι) or "lances", which are discussed by Arist. at *Mete*. III.vi. (377a 29ff.). The section here is based mainly on the *Šifā'*, but the account of how "mock suns" are formed is taken from Fakhr al-Dīn al-Rāzī, while the

[31] See further Nic. syr. 52.13-14, 22-23: [ܩܫ]ܬܐ ܕܝܢ ܕܡܝܐ ܠܣܘܓܠܬܐ. ܒܗܝ ܕܡܚܙܝܢܘܬܐ ܐܝܬܝܗ ܘܠܐ ܫܪܪܐ ܡܘܡܝܢܐ. /14 ܘܒܗܝ ܕܒܝܕ ܐܢܐܩܠܐܣܝܣ ∴//22 ܗܘܝܐ /23 ܘܠܐ ܒܝܕ ܕܝܐܩܠܐܣܝܣ: ܘܒܗܝ ܕܡܢ ܥܢܢܐ ܕܪܨܝܦܐ ܝܬܝܪܐܝܬ ܗܘܝܐ ∴ [14 ܐܢܐܩܠܐܣܝܣ: ܐܢܐܠܐܩܣܝܣ cod.] ("Rainbow is similar to halo in that it is an illusion and not a reality, in that it occurs by way of ἀνάκλασις and not by way of διάκλασις, and in that it arises from a cloud which is very compact [*rṣīpā*]"). - Nic. syr. 53.16-17, 30-31: ܒܣܘܓܠܬܐ ܗܟܝܠ ܡܚܠ ܡܚܙܝܢܝܬܐ /17 ܡܛܠܗܕܐ //30 ܡܬܚܙܝܐ. ܘܡܛܠܗܕܐ ܚܘܪܬܐ ܡܚܙܝܢܝܬܐ. ܒܩܫܬܐ /31 ܕܝܢ ܥܒܝܐ ܥܢܢܐ ܘܐܘܟܡ ܝܬܝܪ. ܘܡܛܠܗܕܐ ܠܐ ܚܘܪܬܐ ܡܬܚܙܝܐ ܩܫܬܐ. /. ("In a halo, then, the mirror is finer, and for this reason (the halo) appears white. In a rainbow, on the other hand, the cloud is denser [*ʿabyā* < gr. παχύς] and more black, and for this reason the rainbow does not appear white"; cf. Olymp. 234.10-12). - Nic. syr. 52.18-19: ܫܩܝܠܘܬܐ ܕܝܢ ܕܩܫܬܐ ܕܠܒܪ ܡܢ ܥܢܢܐ. ܡܢ ܥܒܝܘܬܐ /19 ܘܪܨܝܦܘܬܐ ܕܥܢܢܐ ("The lustre [? *sqīlūṯā*] of the rainbow which is outside the cloud is due to thickness and density of the clouds"). - Nic. syr. 52.25-26: ܘܐܝܟ ܡܚܙܝܢܝܬܐ ܫܩܝܠܬܐ ∴ ܠܒܪ ܡܢ /26 ܥܢܢܐ ܡܬܚܙܝܐ ܩܫܬܐ. ("Rainbow appears like a shining [? *sqīlā*] mirror, outside the cloud").

two additional names given for the "lance", namely "rod" (*šabṯā*) and "shoot" (*šabbūqā*), are the names used, respectively, in *De mundo* syr. and Nic. syr.

	*Šifā*ʾ	Other sources	
II.iv.1.1-5	(56.3-5)	*Mab.* II.184.15-185.4	Mock suns (formation)
II.iv.2.1-8	56.13-18		Mock suns and rain
II.iv.3.1-7	56.5-8		Lances
II.iv.3.7-9	56.11-14		Lances: time of occurrence

II.iv.1.1-5: Formation of mock suns

The account of the mock suns given by IS is rather brief and obscure, but it appears that he accepted two causes for them. Fakhr al-Dīn al-Rāzī gives a longer account, where he adds a third cause. BH begins the passage below with phrases taken from the *Šifā*ʾ, but then closely follows Fakhr al-Dīn al-Rāzī for the rest of the theory. BH, however, combines the first two causes given by Fakhr al-Dīn al-Rāzī into one. - The third type of mock suns mentioned by Fakhr al-Dīn al-Rāzī (BH's second) is due to combustion of vapour, rather than reflection of light. As such it does not belong to the category of mock suns and other optical illusions discussed in this chapter, but falls in the same category as comets and shooting stars, which are discussed in chapter IV below.[32]

But. [underline: agreement with *Mabāḥiṯ*; italics: agreement with *Šifā*ʾ]:
These [mock suns] *are illusions* [*haggāḡwāṯā*] (1a) which appear *like suns* [*šemšē*] on a cloud, which is compact and smooth [*rṣīpā wa-sqīlā*], near [lit. in the proximity of] *the sun*. This cloud, like the moon, *receives* the light of the sun *in itself* (1b) and conveys together with the light also an illusion [*haggāḡā*] of the *shape of the sun* towards vision, because a large mirror shows together with the *colour* also the shape. (2a) Furthermore, when viscous [*ṭallūšā*] vapour rises to the αἰθήρ, (2b) is formed into a circular shape like other moist bodies in air, (2c) reaches the sphere of fire and (2d) catches fire [lit. and fire is kindled in it], (2e) it appears like the sun. (2f) If the vapour is very compressed [*lzīzā*] and its particles are mixed firmly with each other, the mock sun persists day and night and goes round in circles with the celestial sphere [*ʿam mawzaltā*].
Mabāḥiṯ. II.184.15-185.4:
الفصل الخامس فى الشميسات. ان لها اسبابا ثلاثة احدها انه يحصل بقرب الشمس غيم كثيف مندمج
الاجزاء صقيل فيقبل فى ذاته ضوء قبول الجرم الكثيف للضوء كما فى القمر. والثانيها ان لا يقبل الضوء
الشمس ولكنه يكون مؤديا خيال الشمس لان المرآة الكبيرة كما يؤدى اللون يؤدى الشكل ايضا وثالثها

[32] The third type of mock sun mentioned by Fakhr al-Dīn al-Rāzī in fact has a certain similarity with IS's solar halo formed by thickened smoky vapour (*Šifā*ʾ 49.5-6).

ان البخار اللزج اذا تصاعد وتشكل بشكل الاستدارة كما على ما هو طبيعة الاجرام الرطبه فى الهواء وبلغ فى صعوده الى كرة النار اشتعلت النار فيه وهو مستدير الشكل فلا جرم شكله شكل الشمس وربما كانت المادة كثيفة فتبقى اياماً وليالى بل شهورا وربما وصل الى موضع الذى يتحرك عليه بتبعية الفلك فهو ايضا تحركت على الاستدارة

Fifth Section: on Mock Suns. They have three causes. (1a) First: A cloud, [which is] thick, united [*mundamiğ*] in its parts and smooth, occurs in the proximity of the sun and receives in itself the light of the sun in the way that a thick body receives light, like in the moon. (1b) Second: It does not receive the light of the sun but becomes a coveyor for the illusion of the sun, because large mirrors, just as they convey colour, convey the shape also. (2a) Third: When viscous [*laziğ*] vapour rises, (2b) is shaped into a circular shape, according to what is natural for moist bodies in the air, and (2c) reaches in its ascent the sphere [*kura*] of fire, (2d) fire is kindled in it, while it is circular in shape, (2e) so that, undoubtedly, its shape will be the shape of the sun. (2f) Sometimes the material [*mādda*] is thick [*katīfa*], so that it remains for days and nights, even months. Sometimes it reaches the place in which things move [*yutaḥarraku ʿalaihi*] following the celestial sphere [*bi-tabaʿīyat al-falak*] and it too moves in circles.

Šifāʾ 56.3-5:

وأما الشميسات فإنها خيالات كالشموس عن مراى، شديدة الاتصال والصقالة،/ ٤ تكون فى جنبة الشمس، فتؤدى شكلها ولونها، أو تقبل ضوءا شديدا فى نفسها، وتشرق على/ ٥ غيرها بضوئها، وتعكسها أيضا

[56.3 مراى ed. Cair.: مراتى Ṭ; مرايا B]

Mock suns [*šumaisāt*] are illusions, (1a) [which appear] like suns [*šumūs*], [reflected] from mirrors [?] that are strongly united and smooth, which occur at the side of the sun, (1b) so that they convey (the sun's) shape and colour. (1a) Or, they receive an intense light in themselves, illuminate other objects with their light and also reflect them.

l. 1. ܗܠܝܢ ("These [mock suns]"): Following immediately after the title "on mock suns and lances", "these" [*hālēn*] would appear to mean both mock suns and lances, but it is clear that the reference here is only to mock suns.

l. 1. ܗܓܓ̈ܬܐ ("illusions"): This plural form is used here and at II.iv.2.5 below, with no appreciable difference in meaning from *haggāgē*.

l. 4. "because a large mirror ...": cf. II.i.9 above.

l. 5. ܐܬܐܪ ("the αἰθήρ"): see comm. on Min. I.i.4.4 above.

l. 8. ܠܙܝܙܐ ("compressed"): كثيف Rāzī. - Although *lzīzā* usually means "importunate, grievous" in older, "classical", Syriac and is used in that sense by BH himself (see PS, Brockelmann s.v.), lexica (BA, BB) also allow the meaning of "thickened, compressed", comparable to arab. *lazza*, and that must be the meaning of the word here. For the use of the root √*lzz* in the sense of "to compress, contract" elsewhere in *But*., see PS Suppl. 179a, s.vv. ܠܙ and ܠܙܝܙܘܬܐ (all the instances cited are from BH *But*.).

l. 8. ܘܡܫܪܪܐܝܬ ܡܬܚܠܛܝܢ ܒܚܕ̈ܕܐ ܡܢܘ̈ܬܗ ("and its particles are mixed firmly with each other"): This has no counterpart in the corresponding sentence of the *Mabāḥit*, but resembles the phrase *mundamiğ al-ağzāʾ* which occurs earlier on the passage of *Mabāḥit* quoted above (II.184.16-17).

II.iv.2.1-8: Mock suns indicating rain[33]

But.: (1a) Mock suns indicate [*mšawdʿān ʿal*] [coming of] rain (1b) because they indicate an abundance of moist and thick exhalations [*ʿeṭrē*] (1c) which the heat of the sun is unable to [*ḥāybā men*] dissipate and dissolve [*puššāš-hōn wa-šrāy-hōn*]. (2a) The view of some that [lit. what some say] (2b) when they occur to the north of the sun they indicate rain less, (2c) and when they occur to the south of it, more, (2d) is unacceptable [*lā šawyā l-qubbālā*], (3a) because the clouds in which these illusions [*haggāḡwāṯā*] occur are not so very far from us (3b) that their directions can be distinguished [*netparšān*] from each other in relation to the sun, (3c) but the one which appears to us [to be] to the north of the sun is found to the south of the sun when we move by a few stades [lit. nearby stades, *esṭaḏwāṯā qarrīḇē*] (3d) and, in the same way, the one in the south becomes one in the north with a small change of our position.

Šifāʾ 56.13-18:

... وذلك لأن الشمس فى هذا الوقت تحلل السحاب الرقيق فى الأكثر. وهذه الشميسات تدل على المطر، لأنها تدل على وفور أبخرة رطبة. قال بعضهم: إنها إن كانت شمالية عن الشمس قلت دلالتها هذه، وإن كانت جنوبية اشتدت. وقد غفل هذا عن أن السحاب التى عنها تتأدى هذه الخيالات لا يبلغ بعدها عنا أن يتميز ما بين شماليها عن جنوبيها؛ وأنه لا يبعد أن يكون كم هو شمالى عندنا يصير جنوبيا منا عن فراسخ قريبة، والجنوبى شماليا.

... (1c) That is because at such a time the sun dissolves [*tuḥallilu*] the thin cloud. (1a) These mock suns indicate [*tadullu*] rain (1b) because they indicate an abundance of moist vapours [*abḫira*]. (2a) Some have said: (2b) If they are to the north of the sun they indicate this less [lit. their indication of this is less], (2c) and if they are to the south, more. (2d) This [assertion] ignores (3a) the fact that the distance from us of the clouds from which these illusions are conveyed (3b) is not such that what is to the south (of the sun) can be distinguished [*yatamayyazu*] from what is to the north of it; (3c) and [the fact] that it is not unlikely that what is in the north with us becomes something in the south with [a movement of] few parasangs [lit. from nearby *farāsiḫ*], (3d) and what is in the south in the north.

l. 2. ܘܥܒܝܐ ("and thick"): For the addition of "thick" here, cf. the antonym "thin" [*raqīq*] in *Šifāʾ* 56.14 and the requirement of clouds on which mock suns are produced to be "dense/compact" mentioned in II.iv.1 above.

l. 2. ܦܘܫܫܗܘܢ ܘܫܪܝܗܘܢ ("dissipate and dissolve"): The verbal noun *šrāyā* answers here to IS's *ḥallala*. The verb *pawšeš* ("dissipate") is frequently used by Nic. syr., notably for our purposes in the following passage:

Nic. syr. 52.4-9:

4 [ܘܐ]ܦ ܡܬܐܠܦ ܕܢܚܘܬ ܥܡܛܐ ܡܢ ܐܐܪ ܕܡܬܥܒܝܐ ܗܘܝܐ܆ ܡܛܠ ܕܥܡܛܐ ܡܢ /
5 ܬ[ܠܬ] ܠܫܡܫܐ. ܡܛܠ ܚܘܠܫܗ ܕܝܢ ܠܐ ܡܫܟܚ ܕܢܦܘܫܫܝܘܗܝ. ܒܝܕ ܐܝܬܝܗ/
6 ܕܝܢ ܚ[ܕܬ] ܥܒܝܐ. ܡܛܠ ܕܠܘ ܒܠܚܘܕ ܡܚܒܠ ܠܗܘ ܥܒܝܐ ܠܫܡܫܐ. ܐܠܐ/7
ܡܦܫ[ܫ] ܠܗ܆ ܡܛܠ ܒܡܥܒܪܐ ܕܚܠܘܗ. ܚܕܬ ܒܥܒܝܐ ܕܝܢ ܚܝܐ/8 ܠܡܛܪܐ

[33] Cf. Arist. *Mete*. 377b 24-27.

[ܡܬܐ]ܡ[ܪܐ ܡܢ] ܠܐ ܕܒܟܘܟܒܐ ܡܫܟܚܐ ܠܗ ܡܟܝܟܘܬܐ ܡܟܝܟܘܬܐ ܡܫܝܢܐ/9 ܡܟܠ
[ܚܝܠ]ܠܘܬܗ ܀

We say: [Halo] usually occurs around the moon, because the moon attracts vapour, but because of its weakness cannot dissipate it. [It occurs] less around the sun, because the sun does not only attract vapour, but also dissipates it because of the strength of its power. Around stars it is formed rarely, because a star attracts weakly and dissipates weakly because of its weakness.

l. 7. ܐܣܛܕܘ̈ܬܐ ("stades"): فراسخ ("parasangs") IS. - A *farsaḫ* was reckoned at 3 *mīl* (miles), or ca. 6 km (Hinz [1955] 62), which is, of course, much longer than the classical stade of ca. 185 m, but BH also defines the "stade" as "3 miles" elsewhere.[34]

II.iv.3.1-7: Lances

But.: (1a) Those lances, which are also called rods [*šabṭē*] and shoots [*šabbūqē*], are also illusions which resemble [*dāmēn l-*] the rainbow in their colour. (2) They occur, however, to the side of the sun, to the right and to the left. (3) They appear straight and not circular: (4a) either because they are small segments of large circles (4b) so that the sense does not perceive their curvature [*kpīpūṯā*] but sees them as straight lines, (5a) or because the position [*qawmā*] of the viewer makes the curved object [*kpīpā*] appear straight, (5b) i.e. when he stands opposite not its curved face, but its concave or convex side [lit. when its face which is curved is not in front him, but its concave or convex side (*gabbeh ḥlīlā aw kustānā*)].

Šifā᾽ 56.5-8:

وأما النيازك فإنها أيضا خيالات فى لون قوس قزح إلا/٦ أنها ترى مستقيمة، لأنها تكون فى جنبة الشمس يمنة عنها أو يسرة لا تحتها ولا أمامها./٧ وسبب استقامتها أنها إما أن تكون قطعا صغارا من دوائر كبار فترى مستقيمة لا سيما إذا/٩ توالت من سحب، وإما لأن مقام الناظر وأوضاع السحب بحيث يرى المتحدب مستقيما.

[56.8 توالت من ed. Cair., ed. Ṭ[1]: توالت فى ed. Ṭ[corr.], DS ‖ المتحدب Ṭ: المنحدب Cair.]

(1) Lances are also illusions in the colour of a rainbow, (2) except that (3) they appear straight, (2) because they occur to the side of the sun, [to the] right of it and [to the] left, not below it or above it. (3) The reason for its being straight [*istiqāma*] is (4a) either that they are small segments [*qiṭaᶜ*] of large circles (4b) so that they appear straight, especially when they are continuous with [?] clouds, (5a) or because the position [*maqām*] of the viewer and the placements of the clouds are such that the convex thing [*al-mutaḥaddib*] appears straight.

l. 1. ܢܝ̈ܙܟܐ ܗܠܝܢ ܕܐܦ ܫܒܛ̈ܐ ܘܫܒܘ̈ܩܐ ܡܬܩܪܝܢ ("those lances which are also called rods and shoots"): On the terms, see comm. on II.i.1.1 above.

---

[34] See PS 294 s.v. ܐܣܛܕܐ. - All the instances cited there equating the stade with 3 miles are from BH's works: Hunt. dxl. 96v = *Asc*. [Nau] 201.ult.; Bod. Or. cccclxvii. 8v = *Rad*. [Istanbul 1997] 17.15; *Cand*. [Paris 210] 20v, 21r = ed. Bakoš 90.8, 9, 11.

l. 2. ܕܕܡܝܢ ܠܩܫܬܐ ("which resemble the rainbow"): The addition of the verb *dāmēn* may be connected with *De mundo* syr. 144.28-29 [< gr. 395a 35-36]: ܫܒܛܐ ܕܝܢ ܐܬܚܙܝܬܗ ܕܕܘܡܝܐ ܕܩܫܬܐ ܕܡܬܡܬܚܐ ܬܪܝܨܐܝܬ ("The rod [*šabṭā* < ῥάβδος] is a likeness [*dumyā* < ἔμφασις] of the rainbow stretched straight").

II.iv.3.7-9: Lances (contd.)

The last part of the passage of the *Šifāʾ* quoted below is awkward as the text stands, since it is difficult to see why the dissolution of the cloud should be the cause of the occurrence of lances at sunrise and sunset.[35] The sentence is more easily understood as the cause for their rarity at midday and that, at any rate, is how BH has understood the passage.

But.: (1) Lances occur rarely at midday [*ʿeddān ṭahrā*], (2) but [usually occur] in the morning [*ṣaprā*] and evening [*ramšā*], especially in the evening, (3) because at midday a thin cloud is quickly dissipated [*metpawššā*] by the heat of the sun.

Šifāʾ 56.11-14:

وقلما تكون هذه عند كون الشمس فى نصف النهار، بل عند الطلوع والغروب، لا سيما عند الغروب، ففى ذلك الوقت يكثر تمدد السحاب. وكثيرا ما تتفق لهذه أن تساير الشمس طالعة وغاربة، وذلك لأن الشمس فى هذا الوقت تحلل السحاب الرقيق فى الأكثر.

(1) These rarely occur when the sun is on the meridian [lit. is at midday], (2) but at sunrise [*ṭulūʿ*] and sunset [*ġurūb*], especially at sunset, since at that time the extension [*tamaddud*] of the cloud increases. It often happens that these [lances][36] accompany [*tusāyira*] the sun as it rises and sets. (3) That is because at that time the sun usually dissolves [*tuḥallilu*] the thin cloud.

[35] See Horten-Wiedemann [1913] 543 n.1

[36] Or "clouds" (so Horten-Wiedemann [1913] 543).

BOOK OF METEOROLOGY, CHAPTER THREE

ON WINDS

The chapter deals with winds, which were discussed by Arist. at *Mete*. II.iv-vi, as well with whirlwinds, discussed by Arist. somewhat later at III.i. - Both these subjects were treated together by IS in his fourth chapter of the treatise on meteorology in the *Šifāʾ* (ed. Cairo p. 58-66) and it is to that chapter that this chapter of *But*. as a whole corresponds.

BH has made one major alteration in the order of presentation in comparison with the *Šifāʾ* in that he has transferred the discussion of whirlwinds from the middle of the chapter to the end. For the rest, the order is basically as in the *Šifāʾ* and the correspondences may be summarised as follows.

	Šifāʾ	
Section i	58.4-60.9, (66.2-7)	Generation of wind
Section ii	61.12-62.4	Directions of wind
Section iii	62.4-66.2	Variation of wind
Section iv	60.12-61.8	Whirlwind

As regards the sources of the material in the chapter, it may be noted that a far more extensive use is made of Nic. syr. in this chapter (and the following chapter IV) than in the two preceding chapters.

l. 1. ܡܛܠ ܪ̈ܘܚܐ ("on winds"): cf. *Šifāʾ* 58.3 [chapter heading] في الرياح.

Mete. III. Section i.: *On Generation of Winds*

The section begins with a brief mention of the two exhalations and of the fact that winds etc. are generated from the dry, smoky exhalation. This is then followed in the rest of 'theory' III.i.1 by the explanation of the mechanism by which winds are generated by this smoky exhalation. The work BH follows in this explanation is not the *Šifāʾ* but Fakhr al-Dīn al-Rāzī's *Mabāḥit̲*. This results in wording sneaking into BH's explanation which suggests a view of winds as "moving air", a view that was refuted at length by Aristotle. In the rest of the chapter, however, the view followed is the Aristotelian one that makes the smoky exhalation the material as well as the cause of winds.

The remaining parts of the section are based either on the *Šifā᾽* or Nic. syr. and are concerned with points that are related to proving that the smoky exhalation is the cause and material of wind. - 'Theory' III.i.2 deals with the objection that the Aristotelian theory does not accord well with the sideways movement of the wind. - 'Theory' III.i.3 deals with the question as to whether winds originate from above or below. This point is discussed by IS much later on towards the end of his chapter on winds. The transfer of the discussion to here in *But*. may have been prompted by the order in Nic. syr., since those parts of Nic. syr. corresponding to III.i.3.2-9 (Nic. 28.22-29.3) immediately follow those corresponding to III.i.2.4-8 there (Nic. 28.18-22). - In the last theory of the section (III.i.4) the relationship between wind and rain is explained in terms of how each of the two either dissipate the other or help in the production of the other.[1]

	Šifā᾽	Other sources	
III.i.1.1-4	58.4-5	Nic. 25.17-20	Two exhalations
III.i.1.4-9	(58.6-9)	*Mab.* II.190.19-191.5	Generation of wind
III.i.2.1-4	58.9-11	Nic. 28.4-6	Oblique movement
III.i.2.4-8	(58.11-14)	Nic. 28.18-22	(contd.)
III.i.3.1-2	66.2-4	Nic. 24.22-24	Reservoir theory
III.i.3.2-4	(60.10)	Nic. 28.22-24	Origin of winds
III.i.3.4-9	(66.4-7)	Nic. 28.24-29.3	(contd.)
III.i.3.9-12	66.7-9		(contd.)
III.i.4.1-4	59.17-19		Wind and rain
III.i.4.4-5	(60.6-9)	Nic. 26.20-23	(contd.)
III.i.4.5-8	(59.19-60.2)	Nic. 26.28, 25.20-23	Rain assists wind
III.i.4.8-11	60.2-6	Nic. 26.23-27	Wind assists rain

1. 1. ܡܛܠ ܡܘܠܕܐ ܕܪܘܚܐ: cf. *Mabāḥiṯ* II.190.13: الفصل سابع فى حد الريح وكيفية تولدها ("on the definition of <u>wind</u> and the manner of their <u>generation</u>").

<u>III.i.1.1-4</u>: Two exhalations

But. [underline: agreement with Nic.; italics: with *Šifā᾽*]:

(1) There are <u>two</u> [kinds of] <u>exhalations</u> [*ʿeṭrē*] which rise into the air. (2a) One [rises] <u>from water</u> and is *moist* and <u>vapoury</u>, <u>as we have said</u>. (2b) *From it are generated* clouds, fog, *rain and similar things* [lit. those that

[1] The relationship of wind and rain is more clearly associated with the proof that wind is generated from the smoky exhalation in BH *Cand*. 124.2-5: "Hippocrates and the Stoics said that winds are a current and pouring of air [*redyā w-ašīḏūṯā d-ā᾽ār* < Nic. syr. 24.20f. < Hipp. ἠέρος ῥεῦμα καὶ χεῦμα]. That this is not true is known from the fact that in dry years when there is little rain winds are abundant ..."

resemble them]. (3a) The other [rises] from earth and is *dry* and *smoky*. (3b) *From* it *are generated winds*, whirlwinds [*ʿalʿālē*], tornadoes [*qarḥē*] *and similar things*.

Nic. syr. 25.17-20 [beginning of section on winds]:

ܒܡܐܡܪܐ ܩܕܡܝܐ ܚܘܝܢ ܕܬܪ̈ܝܢ ܐܝܬܝܗܘܢ ܥܛܪ̈ܐ . ܘܚܕ 18/ ܡܢܗܘܢ ܐܝܬܘܗܝ
ܬܢܢܢܝܐ . ܐܚܪܢܐ ܕܝܢ ܠܗܓܢܝܐ . ܘܢܩܝܦܝܢ ܠܚ̈ܕܕܐ . ܡܛܠ ܕܐܦ 19/ ܐܣܛܘ̈ܟܣܐ
ܕܡܢܗܘܢ ܡܬܝܠܕܝܢ ܢܩܝܦܝܢ ܠܚ̈ܕܕܐ . ܘܥܛܪܐ ܬܢܢܢܝܐ ܡܢ ܐܪܥܐ 20/ ܐܝܬܘܗܝ .
ܥܛܪܐ ܕܝܢ ܠܗܓܢܝܐ ܡܢ ܡ̈ܝܐ .

Text: (2a) We have demonstrated before that (1) the exhalations [*ʿeṭrē*, corr. gr. ἀναθυμιάσεις] are two, (2/3) and one of them is smoky, while the other is vapoury. They cleave to [*nqīpīn l-*] each other, because the elements from which they are generated cleave to each other. (2/3) Smoky exhalation is generated from [*īṯaw(hy) men*] earth, while vapoury exhalation is generated from water.

Šifāʾ 58.4-5 [beg. of chapter on winds]: وقد حان لنا أن نتكلم فى أمر الرياح، فنقول: كما أن المطر وما يجرى مجراه إنما يتولد عن البخار الرطب، فكذلك الريح وما يجرى مجراها يتولد عن البخار اليابس وهو الدخان.

Time has come for us to speak about the subject of winds. We say: (2) just as rain and what is analogous to it [*mā yağrī mağrāhu*] are generated from moist vapour, (3) wind and what is analogous to it are generated from dry vapour, i.e. smoke.

l. 1-2. "two exhalations", "as we have said": cf. *But*. Mete. I.i.1. and I.iii.6. above. - In I.i.1 BH, had followed Abū al-Barakāt in speaking of "vapour" (*lahgā*) and "dust" (*ḥellā*). In I.iii.6, where he was following IS, he spoke of "vapour" and "smoke" (*tennānā*). Here he follows Nic. syr. in using the term "exhalation" (*eṭrā*, corr. gr. ἀναθυμίασις), which is either "vapoury" or "smoky" (*lahgānāyā/tennānāyā*) and so we are at last given what may be seen as a definition of this key Aristotelian term, which we have already encountered several times in *But*. in the book on mineralogy (in the chapters on earthquakes, Min. II.ii.1, ii.4; II.iii.1-4; and in connection with the salinity of the sea, Min. V.ii.1, V.ii.4) and once so far in the book on meteorology (Mete. II.iv.2.1).

l. 4. ܩܪ̈ܚܐ ("tornadoes"): see comm. on III.iv.4.3 below.

III.i.1.4-9: Generation of wind

The passage is based on Fakhr al-Dīn al-Rāzī's *Mabāḥiṯ* rather than on the *Šifāʾ*. BH had used the same passage of the *Mabāḥiṯ* at *Cand*. 123.9-124.2, although the vocabulary used there differs from that used here.

Aristotle had rejected the view that wind was "moving air" (349a 16 ἀὴρ κινούμενος) and made the dry smoky exhalation the cause and

material of wind.[2] While the common-sense view that wind was moving air was not one that could be dismissed so easily,[3] it is the Aristotelian view that is followed by IS in the *Šifāʾ*. Fakhr al-Dīn was aware of the problem here (see *Mabāḥiṯ* II.190.14-16) and in his description of how winds are generated introduced the notion of wind being caused by a "surging of the air" (*tamawwuğ al-hawāʾ*), a phrase which was probably taken from the Ikhwān al-Ṣafāʾ.[4] BH, as may be seen, follows Fakhr al-Dīn al-Rāzī in using this notion and also says that the smoky exhalation *causes* wind (l. 6f., 10 *rūḥā ʿāḇdīn*), rather than that they turn into (the material of) wind. - It is not altogether clear how conscious BH was of the implications of his words here, since in the rest of the chapter he simply follows Aristotle and IS in talking of the smoky exhalation as being the material of wind (note III.i.2.1-2 below: "if the *material* of wind is smoky exhalation ...").

But. [underline: agreement with *Mabāḥiṯ*]

(1a) Wind [*rūḥā*] is generated [*metyaldā*] (1b) when bodies of smoke [lit. smokes, *tennānē*] rise [*sālqīn*] (1c) and come to the cold stratum [*ʿraqtā qarrīrtā*]. (2a) If they grow cold there, (2b) they become heavy, (2c) fall [*nāḥet*] violently downwards, (2d) cause the air to surge [*mgallʿlīn*] (2e) and [thus] cause wind [*rūḥā ʿāḇdīn*]. (3a) If they do not grow cold there, (3b) they ascend [*metʿallēn*] [further] (3c) and reach [*meṯmannʿīn*] the sphere of fire [*espēr nūrā*], which moves with the movement of the celestial sphere [*mawzaltā*]. (3d) When their ascent is obstructed [*meṯʿakkar*] by that circular and swift movement of the fire, (3e) they slide [*meštarglīn*] violently downwards (3f) and in this way cause wind.

Mabāḥiṯ II.190.19-191.5:

وتولد الرياح عن الادخنة/ ٢٠ على وجهين الاول اكثرى والثانى اقلى/ ٢١ اما الاكثرى هو انه اذا صعدت ادخنة كثيرة الى فوق فعند وصولها/ ١ الى الطبقة الباردة اما ان ينكسر حرها ببرد ذلك الهواء او لا

---

[2] See e.g. 360a 12f. ἡ δὲ ξηρὰ [sc. ἀναθυμίασις] τῶν πνευμάτων ἀρχὴ καὶ φύσις πάντων; 361a 30f.: ἐπεὶ δ' ἐστὶν ἄνεμος πληθός τι τῆς ξηρὰς ἐκ γῆς ἀναθυμιάσεως κινούμενον περὶ τὴν γῆν.

[3] Theophrastus already reverts to talking of wind as "moving air", Theoph. *De ventis* [Wimmer] 382.38f.: ἡ δὲ τούτου [sc. ἀέρος] κίνησις ἄνεμος; cf. Lettinck (1999) 160f. with the literature cited there. - See further Gilbert (1907) 533-539.

[4] Ikhwān al-Ṣafāʾ [Bustānī] II.71.3: واعلم أن الريح ليست شيئًا سوى تموّج الهواء. For the discussion of how the wind is generated from the dry exhalation, see ibid. II:71.7-14; cf. Sersen (1976) 147-149 (cf. also ibid. 133 n.385). On the relationship between Theophrastus, Kindī, Ikhwān al-Ṣafāʾ and Qazwīnī, see further Lettinck (1999) 176f.; the passage on generation of wind at Qazwīnī, *Ağāʾib* 94.26-95.4 can, in fact, be understood as a combination of the Ikhwān al-Ṣafāʾ and Fakhr al-Dīn al-Rāzī (on the relationship between Fakhr al-Dīn al-Rāzī and Qazwīnī, see comm. on Min. III.ii.3 above).

ينكسر فان/ ٢ انكسر فلا محالة يثقل وينزل (تثقل وتنزل) فيحصل من نزولها تموج الهواء فتحدث الريح/ ٣ وان لم ينكسر برودة (ببرودة) تلك الطبقة من الهواء فلا بد من ان يتصاعد الى ان يصل/ ٤ الى كرة النار المتحركة بحركة الفلك وحينئذ لا تتمكن من الصعود بسبب حركة/ ٥ النار فترجع تلك الادخنة وتصير ريحا

(1a) Generation [*tawallud*] of wind from [*ʿan*] smoke [occurs] in two ways, one being the usual way [*akṯarī*] and the other the unusual [*aqallī*]. The usual way: (1b) When much smoke [pl., *adḫina*] rises upwards, (1c) at its arrival in the cold stratum [*al-ṭabaqa al-bārida*], (1d) its heat is either broken [*yankasiru*] by the coldness of that air or it is not broken. (2a) If it is broken, (2b) (the smoke) will no doubt become heavy (2c) and descend. (2d) From its descent there arises a surge [*tamawwuǧ*] of the air (2e) and (the smoke) causes [*tuḥdiṯu*] wind [or: and wind arises (*taḥduṯu*)]. (3a) If it is not broken by the coldness [lege *bi-burūda*] of that stratum of air, (3b) (the smoke) will inevitably rise [further] (3c) and reach the sphere of fire [*kurat al-nār*], which moves with the movement of the celestial sphere [*falak*]. (3d) Then, it is unable to rise [further] because of the movement of the fire, (3e) so that that smoke [pl.] turns back (3f) and causes [*tuṣṣayyiru*] wind [or: becomes (*taṣīru*) wind].

Šifāʾ 58.6-9:

ويتولد عنه على وجهين: أحدهما أكثرى والآخر أقلى. وأما الأكثرى فإذا صعدت أدخنة كثيرة ألى فوق، ثم عرض لها أن ثقلت فهبطت لبرد أصابها، أو لأنها قد حبستها حركة الهواء العالى عن النفوذ، فرجعت تارة مطيعة لحركة ذلك الهواء فى جهة، وتارة فى جهة أخرى.

(1a) (Wind) is generated [*yatawalladu* (*tatawalladu*)] from (smoke) in two ways, one being usual and the other unusual. The usual way: (1a) when much smoke [pl.] rises upwards, (2) then it happens that it becomes heavy and descends because of a coldness which meets it; (3) or, because the movement of the upper air withholds it from penetration, so that it turns back following the movement of that air sometimes in one direction and sometimes in another.

l. 5. ܒܡܫܘܬܐ ܩܪܝܪܬܐ ("the cold stratum"): الطبقة الباردة Rāzī. - i.e. the cold stratum at the middle elevation of the sphere of air; see Mete. I.i.1.5-10 above.

l. 6. ܘܠܐܐܪ ܡܓܠܠܝܢ ("and [the bodies of smoke] cause the air to surge"): فيحصل من نزولها تموج الهواء Rāzī. - For the equivalence of syr. *gallel* with arab. *mawwaǧa*, see comm. on Min. I.ii.1.4 above (ܡܬܓܠܠܢܘܬܐ). The same words of the *Mabāḥiṯ* are translated quite differently by BH at *Cand*. 123.11: ܘܡܫܬܚܫܠܢܘܬܗ ܡܚܫܠܢܐ ܡܢܗ ܒܐܐܪ ܡܪܝܒ. Interestingly, in *Rad*. we find him combining two verbs, one derived from *maḥšōlā*, the word he uses in *Cand*., and *gallel*, which he uses here in *But*.: ܘܡܫܬܚܫܠܢܘܬܗ ܕܢܫܘܒܬܐ ܠܐܐܪ ܡܚܫܠ ܘܡܓܠܠ (*Rad*. ed. Istanbul 23.10).

l. 7. ܐܣܦܝܪ ܢܘܪܐ ("the sphere of fire"): كرة النار Rāzī. - Cf. II.iv.1. above. - Here again, the translation in *Cand*. is different (123.13): ܓܘܓܠܐ ܗܘ ܢܘܪܢܝܬܐ.

l. 8. ܒܙܘܥܐ ܗܘ ܚܪܝܦܐ ܘܚܘܕܪܢܝܐ ("circular and swift movement"): The adjective *ḥarrīpā* which has no counterpart in Rāzī is borrowed from Nic. syr.; see III.i.2.6 below.

III.i.2.1-4: Oblique movement of winds

In the passage of Nic. syr. which is copied almost verbatim here by BH, we are concerned with the question as to why winds blow not downwards but sideways. IS, on the other hand, is concerned at *Šifāʾ* 58.9ff. with the question as to why winds do not blow in the same direction as the flow of the "upper air" given that they are due to the reflection of the smoky exhalation by the "upper air" (replaced by the "sphere of fire" in Fakhr al-Dīn al-Rāzī and, following him, in *But.*, see III.i.1.4-9 above).

But. [underline: agreement with Nic. syr.; italics: agreement with *Šifāʾ*]
(1) *Some* of the ancients, *doubting this* [*kaḏ metpaššḵīn*], say: (2) "If the material [*hūlē*] of wind is the smoky exhalation, which is hot [lit. and this is hot], and what is hot rises upwards, (3) why are winds *not* reflected vertically [*ba-ṯrīṣū*] *downwards* retracing their paths [lit. in their footsteps, *ʿal ʿeqbāṯ-hēn*], but move obliquely [*zlīmāʾīṯ*]?"
Nic. syr. 28.4-6[5]:

ܡܟܝܠ ܀ ܐܠܐ ܐܢܗܘ ܕܗܘܠܐ ܕܪܘܚܐ 5/ ܐܝܬܘܗܝ ܥܛܪܐ ܬܢܢܝܐ : ܗܘ ܕܝܢ
ܫܚܝܢܐ ܐܝܬܘܗܝ : ܫܚܝܢܐ ܕܝܢ ܠܥܠ ܣܠܩ ܀ 6/ ܡܛܠܡܢܐ ܪܘܚܐ ܗܘܝܐ
ܐܠܟܣܝܐܝܬ ܡܬܬܙܝܥܝܢ ܀

Text: (2) But if the material of wind is the smoky exhalation, and this is hot, and what is hot rises upwards, (3) why do winds move obliquely?
Šifāʾ 58.9-11:

وذلك أنه ليس يلزم فى المندفع إلى فوق ما ظنه بعض المتشككين أنه إذا ضغط من فوق إلى أسفل بحركة معارضة، يكون (تكون ط) لا إلى الأسفل، بل إلى جهة، أن يلزم (تلزم ط) تلك الجهة.

That is to say, with a thing propelled upwards what some of the doubters [*mutašakkikīn*] think does not necessarily apply, [namely] that when it is pressed downwards from above by an opposing movement, which is not downwards but in a sideways direction, that it necessarily moves in that direction.

III.i.2.4-8: Oblique movement of winds (contd.): answer

But.: (1) To them the Master [Aristotle] says: (2) When the smoky exhalation rises, (3) it meets the air or the fire which is around it. (4) When it cannot penetrate them because their swift movement prevents [*qāḏem l-*] their penetration, (5) it strikes them and bounces back [*šāwar ʿāṭep̄*], (6) and makes an oblique movement, (7) just as an arrow that is shot at the ceiling of a house, if it does not stick, is reflected not vertically downwards but obliquely.

---

[5] Cf. Arist. *Mete*. 361a 22ff.; Ps.-Arist. *Prob*. 15.14; Olymp. 174.25-30, 177.31-33; Olymp. arab. 118.21-22; Takahashi (2002a) 195-198.

Nic. syr. 28.18-22[6]:

ܘܐܡܪܝܢܢ ܕܬܢܢܝܬܐ ܟܕ ܣܠܩܐ ܠܥܠ . ܦܓܥܐ ܒܐܐܪ ܕܪܕܐ ܘܟܕ ܠܐ ܡܫܟܚܐ /19
ܕܢܬܦܪܝܘܗܝ ܠܗܢܐ : ܐܘ ܡܛܠ ܕܚܪܝܦ ܙܘܥܗ ܘܩܕܡ ܙܘܥܗ ܠܬܦܪܝܗ ./20 ܐܘ ܡܛܠ
ܕܥܒܝܐ ܗܝ ܬܢܢܝܬܐ : ܘܠܐ ܡܫܟܚܐ ܕܬܣܩ ܝܬܝܪ ܡܢ /21 ܡܐ ܕܣܠܩܐ . ܐܝܟܢܐ
ܕܡܬܚܙܝܐ ܡܢ ܕܠܐ ܢܫܒܢ ܪܘܚܐ ܒܪܝܫ ܛܘܪܐ . ܢܩܫ/22 ܒܗ ܘܗܦܟ ܚܠܦ .
ܘܡܛܠܗܢܐ ܥܒܕ ܡܬܬܙܝܥܢܘܬܐ ܠܣܛܪܝܬܐ .

[18 ante ܘܐܡܪܝܢܢ: add. ܢܝܩܠܐܘܣ ܐܡܪ P || ܕܪܕܐ: ܪܕܐ P || ܘܡܛܠܗܢܐ: ܡܛܠܗܢܐ CP]

(1) We say: (2) When the smoky exhalation rises, (3) it meets the flowing air, (4) and when it cannot penetrate this (air) - either because its movement is swift and prevents its penetration, or because the exhalation is dense and for this reason is unable to rise more than it rises, as is shown by the fact that winds do not blow on tops of mountains - (5) (the exhalation) strikes (the air) and bounces back, (6) and for this reason makes an oblique movement.

l. 5. ܐܘ ܒܢܘܪܐ ܕܚܕܪܘܗܝ ("or the fire which is around it"): The addition can be explained from III.i.1.4-9 above.

l. 6. ܩܕܡ ܙܘܥܗܘܢ ܚܪܝܦܐ ܠܬܦܪܝܗܘܢ ("because their swift movement prevents their penetration"): lit. "their [air/fire's] swift movement precedes their being torn". - I translate somewhat freely with the etymology of "prevent" (praevenio) in mind. The likely source passages are: Olymp. 175.21-23: τοῦτο μὲν διὰ τὸ ὀξέως κινεῖσθαι ἐκεῖνον τὸν ἀέρα καὶ τῇ ὀξείᾳ κινήσει λανθάνοντα τὴν καπνώδη ἀναθυμίασιν καὶ μὴ διαιρούμενον δι' αὐτῆς; and Olymp. 178.2: διὰ τὸ ἐκεῖνον ὀξέως κινεῖσθαι καὶ μὴ φθάνειν διαιρεθῆναι.

l. 7-8. "just as an arrow ...": The addition was probably inspired by *Šifā'* 58.11-14: فربما أوجب هيئة صعوده وهيئة لحوق المادة به أن ينكس إلى خلاف جهة المتحرك المانع، كالسهم يصيب جسما متحركا إلى جهة فيعطفه تارة إلى جهته، ... وتارة إلى خلاف تلك الجهة، ... ("For often the condition of its ascent and the condition of the adhesion of the material to it force it to turn back in the opposite direction from that of the thing moving and detaining it, as when an arrow strikes an object moving in one direction and the object directs it sometimes in its direction ... sometimes in the opposite direction ...").

III.i.3.1-2: Reservoir theory

The material here goes back to two different passages in Arist. *Mete*., (1) 361a 25ff. where the question as to whether winds originate from above or below is discussed, and (2) 349a 17ff., where the "reservoir" theory of winds is refuted.

---

[6] Cf. Olymp. 175.19-30, 177.31-178.4; Olymp. arab. 119.11-19.

But. [underline: agreement with *Šifā*ʾ; italics: with Nic.]:

(1) Just as some thought that *the entirety* [*kullāyūṯā*] of water was confined [*ḥbīšā*] inside *the earth*, (2) others thought that *the entirety* of wind was confined inside *the earth* (3) and [that] it rose from below.

*Šifā*ʾ 66.2-4:

وكما قد اتفق أن ظن قوم أن للمياه معدنا فيه كليتها، وهو فى غور الأرض؛ كذلك قد ظن قوم أن للرياح أيضا معدنا يحصرها فى غور الأرض. فإنما تهب من هناك بقدر.

(1) Just as some came to think that there was for water [pl.] a source [*maʿdin*] which contained its entirety [*kullīya*] - and that was in the depth [*ġaur*] of the earth - (2) others thought that there was for winds too a source which confines them in the depth of the earth (3) and that they blew from there in certain amounts.

Nic. syr. 24.22-24[7]:

ܐܚܪ̈ܢܐ ܐܡܪ̈ܝܢ 23/ ܕܒܐܪܥܐ ܚܒܝܫܐ ܟܠܗ ܗܘܠܐ ܕܪܘܚܐ. ܘܠܐ ܫܪܝܪܐ.
ܐܝܟܢܐ ܕܐܦܠܐ ܟܠܗܘܢ 24/ ܡ̈ܝܐ ܕܢܗܪ̈ܘܬܐ ܚܒܝܫܝܢ ܒܐܪܥܐ. ܗܟܢܐ ܐܦܠܐ
ܟܠܗ ܐܐܪ.

(2) Others say that the whole of the material [*kullāh hūlē*] of winds is confined [*ḥbīšā*] in the earth, but this is not true; (1) just as it is not the case that the whole of the water of rivers is confined in the earth, so neither is the whole of air [confined in the earth].

l. 3. "Just as some thought ...": cf. Min. V.iv. above (on Tartarus).

III.i.3.2-4: Origin of winds

But.: (1) That this is not so, (2) but the winds begin to blow from above [lit. the origin (*šurrāyā*) of the blowing (*maššbā*) of wind is from above], (3) is proven by the fact (4) that clouds move before we become aware of the wind [lit. before the wind becomes known to us] (5) because they are affected [*ḥāšān*] by it first.

Nic. syr. 28.22-24[8]:

ܡܫܬܠܡܢ 23/ ܕܡܢ ܠܥܠ ܐܝܬܘܗܝ ܫܘܪܝܐ ܕܪܘܚܐ. ܡܬܚܙܝܐ ܗܝ ܡܢ ܗܕܐ ܕܩܕܡ ܕܬܬܝܕܥ ܠ
ܪܘܚܐ. 24/ ܡܬܬܙܝܥܝܢ ܥܢ̈ܢܐ. ܡܛܠ ܕܗܠܝܢ ܡܪܓܫܢ ܩܕܡܝܬ ܒܗ ܒܚܫܐ.

[= Paris 346, 69r.2-4: 22 ܡܫܬܠܡܢ: ܘܡܫܬܠܡܢ P]

(2) That the origin [*šurrāyā*] of wind is from above (3) is proven by the fact (4) that clouds move before the wind becomes known to us, (5) because these sense the effect [*ḥaššā*] first.

l. 3. ܫܘܪܝܐ ܕܡܫܒܐ ܕܪܘܚܐ (lit. "the origin of the blowing of wind"): ܫܘܪܝܐ ܕܪܘܚܐ ("the origin of wind") Nic.- The addition of the word *maššḇā* is due to *Šifā*ʾ 60.10 مبادئ هبوب الرياح (cf. also 66.5-6: التى منها تبتدئ الرياح فى هبوبها).

[7] Cf. Arist. 349a 25-26; Olymp. 98.5-8; 99.9-13; 168.17-19; Olymp. arab. 120.3.

III.i.3.4-9: Origin of winds (contd.)

A corresponding passage also occurs in the *Šifāʾ* (66.4-7),[8] but the passage here is based entirely on Nic. syr.

But.: (1) Nor do we perceive [*margšīnan*] the smoky exhalation as it rises (2a) because it gathers little by little [*b-qallīl qallīl metkannaš*] (2b) and not as if it were discharged [*metnappaṣ*] from some reservoir [*awṣrā*]. (3a) This is known (3b) from the fact that when winds start up they blow gently, (3c) later strongly. (4a) If all their material were really [*b-maʿbdānūtā*] confined in the earth, (4b) they ought to start up strongly, then weaken, (4c) just as when the flood-gates [*nāskē*] of the water in a cistern [*pesqīn*] are opened.

Nic. syr. 28.24-29.3[9]:

ܐܠܐ ܐܢܗܘ ܕܡܢ/25 ܬܢܢܐ ܥܛܢܢܐ ܗܘܝܢ ܪܘܚܐ ܡܛܠܡܢܐ ܠܐ ܢܦܠ ܬܚܝܬ
ܪܓܫܬܢ ܗܘ ܬܢܢܐ/26 ܥܛܢܢܐ ܗܢܐ ܕܣܠܩ. ܘܐܡܪܝܢܢ ܡܛܠ ܕܒܩܠܝܠ ܩܠܝܠ ܣܠܩ.
ܐܦܠܐ ܓܝܪ ܐܝܟ ܕܡܢ/27 ܐܘܨܪܐ ܡܕܡ ܡܬܢܦܨ. ܐܠܐ ܐܝܟ ܕܐܦ ܗܘܠܐ
ܕܢܗܪܘܬܐ ܒܩܠܝܠ ܩܠܝܠ ܡܬܟܢܫ. /28 ܐܦ ܓܝܪ ܗܘܠܐ ܕܪܘܚܐ ܠܘ ܟܠܗ
ܒܡܥܒܕܢܘܬܐ ܚܒܝܫܐ ܒܐܪܥܐ. ܘܐܝܬ ܠܟ ܕܬܚܙܐ/29 ܗܕܐ ܗܘܐ ܘܕܒܩܠܝܠ ܩܠܝܠ
ܡܬܚܠܨܐ. ܡܢ ܗܝ ܕܟܕ ܡܫܪܝܢ ܡܢ ܪܘܚܐ ܡܫܬܠܡܢܘܬܐ/30 ܢܫܒܢ. ܒܫܪܝܐ ܕܝܢ
ܒܢܝܚܘܬܐ ܢܫܒܢ. ܒܬܪ ܗܘܐ ܕܝܢ ܕܗܠܝܢ ܕܠܩܘܒܠܐ ܗܘܝܢ. ܐܠܘ/31 ܓܝܪ ܟܠܗ ܗܘܠܐ
ܕܐܠܨܝܢ ܗܘܐ ܒܡܥܒܕܢܘܬܐ ܒܐܪܥܐ. ܐܣܬܪܐ ܐܝܟܐ ܕܝܢ ܕܢܫܒܢ//1
ܒܥܘܫܢܐ ܢܫܒܢ ܘܒܬܪܟܢ ܢܬܡܚܟܢ. ܐܝܟܢܐ ܕܝܢ ܕܢܣܟܐ ܕܡܬܦܬܚܝܢ/2 ܢܩܒܐ
ܕܡܝܐ ܕܒܦܣܩܝܢ ܗܘܘ ܡܢ ܩܕܝܡ. ܒܫܘܪܝܐ ܡܢ ܢܦܩ ܒܥܘܫܢܐ ܕܝܢ. ܒܩܠܝܠ ܕܝܢ
ܩܠܝܠ ܡܬܡܚܟܝܢ.

(1) If winds arise from smoky exhalation, why does this smoky exhalation not fall under our perception [*nāḥet tḥēt rgēštan*] as it rises? (2a) We say: [This is] because (the smoky exhalation) rises little by little. (2b) For it is not discharged as if from some reservoir, (2a) but gathers little by little just like the material of rivers. For the material of the wind is not all confined in the earth in reality [*b-maʿbdānūtā*]. (3a) It is possible for you

[8] *Šifāʾ* 66.4-7: ولو كانت الأمر كذلك لكانت الرياح التي تنبعث من الأرض تبتدئ قوية ثم تضعف، كالماء المنبثق فإنه في ابتدائه يقوى ثم يضعف. وليس توجد (يوجد ط) حال الأرض التي منها تبتدئ الرياح في هبوبها هكذا، بل على عكسه، وإنما تشتد الرياح في أعلى الجو. ("If the matter were so, the winds which are discharged from the earth would be strong at the beginning and then weaken, like water discharged through a flood-gate [*al-māʾ al-munbatiq*], for (such water) is strong at the beginning and then weakens. But the condition of the earth from which winds begin to blow is not thus, but the opposite. Winds are strong in the upper part of the air").

[9] Cf. Arist. 361b 1-8; Olymp. 99.8ff., 178.31-34; Olymp. arab. 120.3-10. - The passage here: ≈ Paris syr. 346 [= P], 68r 20-21, 69r 4-18: 24-26: ܘܡܛܠܡܢܐ ܠܐ ܢܦܠ ܬܚܝܬ ܪܓܫܬܢ ܗܘ ܬܢܢܐ ܥܛܢܢܐ ܗܢܐ ܕܣܠܩ P (68r 20-21): ܬܘܒ ܕܝܢ ܐܡܪ ܢܝܩܘܠܐܘܣ ܐܢ ܗܘ ... P ܕܡܛܠܡܢܐ ܠܐ ܢܦܠ ܬܚܝܬ ܪܓܫܬܢ ܗܘ ܬܢܢܐ ܥܛܢܢܐ ܗܢܐ ܕܣܠܩ ܒܝܕ ܕܒܩܠܝܠ ܩܠܝܠ ... (69r 4-6) || 26 ܡܛܠ: ܒܝܕ P || ܩܠܝܠ: om. C || 27, 28 & 31 ܗܘܠܐ C || 29 ܕܒܩܠܝܠ P || 30 ܢܩܫܢ: ܢܫܒܢ P || 31 ܕܐܠܨܝܢ: ܕܐܠܨܝܢ P || 1 ܢܫܬܪܝܢ P.

to know this and [and to know] that (the material) gathers little by little (3b) from the fact that when winds start up they blow gently; (3c) later they strike strongly. (4a) The opposite should be the case, if all their material were really confined in the earth, (4b) I mean, that they ought to start up more strongly and then weaken. (4c) Just so, we see that when the flood-gates of water which had been gathered before are opened, (the water) flows more strongly at the beginning, and weakens little by little.

III.i.3.9-12: Origin of winds (contd.)

But.: (1) If a little wind, which is not the entirety of wind, (2) because it is confined inside the earth, (3) causes (the earth) to quake and tremble [*mnīdā lāh w-mar^c^lā*], (4) [then] if the entirety of the material of wind were confined inside the earth, (5) it would open up (the earth) with great force (6) and go out in its entirety.

Šifā' 66.7-9:

ومع ذلك فإن الريح القليلة التى ليست كلية الرياح، فقد يحدث من احتقانها فى الأرض زلزلة ورجفة. فلو كانت للرياح كلية محصورة فى الأرض، لكانت قد خسفت البقعة المنحصرة فيها، وتخلصت دفعة.

(1) Furthermore, if with a little wind which is not the entirety of winds, (2-3) earthquakes and tremors [*zalzala wa-rağfa*] occur due to its confinement in the earth, (4) then, if the entirety of winds were enclosed in the earth, (5) it would cause the place where it was enclosed to give way (6) and would be discharged all of a sudden.

l. 10-11. "causes (the earth) to shake and quake": On wind as the cause of earthquake, see Min. II. above.

III.i.4.1-4: Opposition of wind and rain

But.: (1) The material of wind is not the material of rain, (2a) because, usually, in a year abounding in rains winds occur less [lit. are fewer, *bāṣrān*] (2b) and in a year abounding in winds rains occur less [lit. are fewer], (3) but if the material for them both [lit. for them (winds) and them (rains)] were the same, (4) they would not oppose each other [*etdalqab(w) la-ḥdādē*].

Šifā' 59.17-19:

ومما يدل على أن مادة الريح غير مادة المطر، الذى هو البخار الرطب، هو أنهما فى أكثر الأمر يتمانعان. والسنة التى يكثر فيها المطر لكثرة البخار الرطب تقل الرياح، والسنة التى يكثر فيها الرياح تكون سنة جدب وقلة المطر.

(1) Among things showing that the material of wind is not the material of rain - which is moist vapour - (4) is the fact that the two usually oppose each other [*yatamāna^c^āni*]. (2a) [In] a year in which rain abounds due to the abundance of moist vapour, winds are few. (2b) A year in which winds abound is a year of drought and scarcity of rain.

III.i.4.4-5: Opposition of wind and rain (contd.)

While IS too talks of how wind and rain stop each other at *Šifā*ʾ 60.6-9, the immediate source used by BH here is Nic. syr. (which is itself the indirect source of the *Šifā*ʾ via Olymp. arab.).[10]

But.: (1) Rains dismiss winds because they extinguish [*mdaʿʿkīn*] the smoky exhalation that rises. (2) Winds stop rains because they rarefy [*mqaṭṭnān*], dissipate [*mp̄awššān*] and scatter [*mbaddrān*] vapour.

Nic. syr. 26.20-23[11]:

ܫܪܝܢ ܪ̈ܘܚܐ ܠܡܛܪ̈ܐ ܒܚܡܝܡܘܬܗܘܢ ܘܡܩܛܢܝܢ/21 ܘܡܦܘܫܝܢ ܠܠܗܓܐ ܕܝܠܗ. ܘܐܝܬ ܐܡܬܝ ܕܒܡܬܬܙܝܥܢܘܬܗܘܢ ܡܒܕܪܝܢ ܠܗ./22 ܡܛܪ̈ܐ ܬܘܒ ܫܪܝܢ ܠܪ̈ܘܚܐ. ܡܛܠ ܕܟܕ ܦܓܥܝܢ ܒܠܗܓܬܐ ܬܢܢܝܬܐ ܕܣܠܩܐ ܡܕܥܟܝܢ/23 ܠܗ.

[= Paris syr. 346, 65r 5-9: 22 ܡܛܪ̈ܐ P]

(2) Winds dismiss rain with their heat, and rarefy and dissipate its vapour; sometimes they scatter it with their movement. (1) Rains too dismiss winds, because when they meet the smoky exhalation which is rising they extinguish it.

III.i.4.5-8: Rain assisting wind

Here again, BH seems to be following Nic. syr. rather than the corresponding passage of the *Šifā*ʾ (59.19-60.2).[12]

But.: (1) Sometimes rain helps in [*lwāt*] the generation of wind (2) because it moistens and waters [*mṣabbaʿ w-margē*] the earth (3) and makes it

---

[10] Olymp. arab. 118.4-6: والمطر يحل الريح، لأن الماء المنحدر إذا لقي البخار الدخاني المتصاعد أطفأه وأجمده. وأما الرياح فتسكن المطر لأحد شيئين: إما لتلطيفها البخار الرطب بحرارتها وحلها إياه ‹أو› لأنها بحركتها تبدده ("(1) Rain dismisses wind, because when the falling water meets the rising smoky vapour, it extinguishes it and freezes it. (2) Winds cause rain to abate for one of two reasons: either through its rarefaction [*talṭīf*] and its dismissal [*ḥall*] of the moist vapour with its heat; or because it scatters it with its movement"); *Šifā*ʾ 60.6-9: وأما في أكثر الأمر فإن المطر يبل البخار الدخانى ويثقله ويجمده ويمنعه أن يصعد أو يتصل بعضه ببعض. فإذا نزل بثقله المستفاد عن الترطيب، ضعفت حركته. وكذلك الريح فى أكثر الأمر تحلل السحاب وتلطف مادته بحرارتها، أو تبدده بحركتها. ("(1) Usually, rain moistens the smoky vapour, makes it heavy, freezes it and prevents it rising or [its particles] joining together; when (smoky vapour) descends with the weight it has gained through moistening, its movement is weak. (2) Similarly, wind usually dissipates the clouds and rarefies their material with its heat, or scatters them with its movement").

[11] Cf. Olymp. 170.7-9, 13-15; Olymp. arab. 118.4-6; *Cand*. 125.1-2.

[12] *Šifā*ʾ 59.19-60.2: لكنه كثيرا ما يتفق أن يعين المطر على حدوث الريح تارة بأن يبل الأرض، فيعدها لأن يتصعد منها دخان، فإن الرطبة تعين على تحلل اليابس وتصعده، وتارة بما يبرد البخار فيعطفه ("(1) But it often happens that rain helps in [*ʿalā*] the occurrence of wind, (2) sometimes because it moistens [*balla*] the earth (3) and makes it suitable [*aʿadda*] for smoke to rise from it. For moisture helps in the dissipation [taḥallul] of dry [exhalation] and its ascent. Sometimes because it makes the vapour cold and prepares it [*ʿaṭṭafa*]").

suitable for the production of fume [*mawldā d-šemrā*], (4) just as moist pieces of wood raise more smoke.

Nic. syr. 26.28[13]:

ܡܛܪ̈ܐ ܡܘܠܕܝܢ ܪ̈ܘܚܐ܆ ܒܗܕ ܕܡܨܒܥܝܢ ܠܐܪܥܐ. ‹ܘ›ܟܕ ܡܨܛܒܥܐ ܗܝ܆ ܡܣܩܐ ܬܢܢܐ.

[= Paris syr. 346, 65r 16-17: 28 ܡܛܪܐ P]

(1) Rains produce winds (2) because they moisten [*mṣabbʿīn*] the earth (3) and when the latter is moistened, it raises smoke.

Nic. syr. 25.20-23[14]:

ܡܥܕܪ ܚܠܬܐ ܠܡܝܢܝܬܐ ܠܬܢܢܝܬܐ ܠܘܬ/21 ܗܘܝܗ̇. ܠܡܝܢܝܐ ܓܝܪ ܟܕ ܩܛܪ ܘܡܬܚܠܦ
ܠܡܝܐ. ܢܦܠ ܘܡܨܒܥ/22 ܘܡܪܘܐ ܠܗ̇ ܠܐܪܥܐ܆ ܘܥܒܕ ܠܗ̇ ܡܫܟܚܢܝܬܐ ܠܘܬ
ܡܘܠܕܐ ܕܫܡܪܐ./23 ܟܕ ܡܬܪܛܒܐ ܓܝܪ ܐܪܥܐ ܝܬܝܪ ܬܢܢܐ ܡܣܩܐ.
ܐܝܟ ܩܝܣ̈ܐ ܕܝܬܝܪ ܪܛܝܒܝܢ.

[= Paris syr. 346, 64v 21-65r 5: 22 ܕܡܪܘܝܐ: sic P, et in marg. ܗܘ ܕܡܪܘܐ; ܕܡܫܟܚܢܐ ut vid. C || 23 ܝܬܝܪܐ: ܝܬܝܪ P]

(1) The vapoury exhalation helps the smoky (exhalation) in [*lwāṯ*] [its] generation; for when vapour condenses [*qāṭar*] and is transformed into water, (2) it falls, moistens and waters [*mṣabbaʿ w-margē*] the earth, (3) and makes it suitable for production of fume [*mawldā d-šemrā*]. (4) For when earth has been moistened it raises more smoke, just like pieces of wood which are more moist.

III.i.4.8-11: Wind assisting rain

The passage of *But.* here is based mainly on Nic. syr., but it will be seen that the corresponding passage of *Šifa*ʾ helps to explain those instances where BH departs from the wording of Nic. syr.

But. [underline: agreement with Nic.; italics agreement with *Šifā*ʾ]:

(1) *Wind too* produces *rain* (2) *because* it drives *clouds* towards each other and *compacts* [*mlabbdān*] them, (3) because *the coldness of the cloud is made to withdraw* inside (the cloud) by the heat of the wind (4) and because *(wind) dissipates* *the smoky exhalation* out of the vapour, so that (the vapour) *grows cold*, thickens and falls as rain [or: and rain falls].

[13] Cf. Arist. 360b 30-32; Olymp. 170.9-13; Olymp. arab. 117.20-21. - Both the manuscript (ms. Tashkent 353b 15-16) and Badawī's text of Olymp. arab. are evidently corrupt here. With the help of Nic. syr. one might propose the following emendations: أما ‹المطر› فيولد الريح لانه اذا بل الارض ولدت بخارا كثيرا وعنه يكون تولد الريح كما قلت فيما تقدم [أما: اما cod.; إما Badawī || المطر: supplevi || فيولد: بتولد Badawī; sine punctis cod.; cf. 117.21 وأما الرياح فتولد المطر || الارض: الريح cod. et Badawī || ولدت: om. Badawī, habet cod.] ("Rain produces wind, because when it moistens the earth, (the earth) generates much vapour, and it is from (vapour) that wind is generated as I have said before").

[14] Cf. Arist. 361a 18-20; Olymp. 165.33-38; 167.27-168.12; *Cand.* 124.8-9.

Nic. syr. 26.23-27[15]:

ܪ̈ܘܚܐ ܕܝܢ ܡܥܒ̈ܕܢ ܡܛܪܐ: ܡܛܠ ܕܕܒܪܢ ܠܥܢ̈ܢܐ ܘܠܒ̈ܕܢ ܠܗܝܢ: ܐܝܟܢܐ/24
ܕܗܘܐ ܒܐܬܪ̈ܘܬܐ ܚܡܝܡ̈ܬܐ. ܐܘ ܡܛܠ ܕܥܪܩܐ ܩܪܝܪܘܬܐ ܕܠܗܓܐ ܡܢ/25
ܚܡܝܡܘܬܗܝܢ ܘܢܚܬܐ ܠܥܘܡܩܐ ܘܡܥܒܝܐ ܠܗܘܠܐ. ܐܘ ܡܛܠ ܕܡܩ̈ܠܠܢ/26
ܒܚܡܝܡܘܬܗܝܢ ܠܬܢܢܐ ܕܥܡ ܠܗܓܐ. ܟܕ ܗܢܐ ܓܝܪ ܦܪܫ ܠܠܗܓܐ/27
ܐܝܟ ܕܠܡܘܒܠܐ ܢܘܟܪܝܬܐ: ܦܐܪ ܗܘ ܠܗܓܐ ܘܡܬܥܒܐ ܘܢܦܠ ܘܗܘܐ ܡܛܪܐ.

[= Paris syr. 346, 64v 9-21: 23 ܡܛܪܐ: om. P || ܘܠܒ̈ܕܢ: ܘܕܒ̈ܪܢ P || 25 ܠܚܡܝܡܬܐ C]

(1) Winds produce rain (2) because they drive the clouds and compacts [*lābdān*] them, as happens in hot places, (3) or because the coldness of the vapour flees from their heat, sinks to the depth and thickens [*mᶜabbyā*] the material [of rain], (4) or because they rarefy with their heat the smoky exhalation which is with the vapour - for when this [smoky exhalation] leaves the vapour, as if [leaving] a strange burden [*mawblā nuḵrāytā*], the vapour grows cold, thickens and falls, and rain occurs.

*Šifā*ᵓ 60.2-6:

والريح أيضا كثيرا ما تعين على تولد المطر بأن تجمع السحاب أو بأن تقبض برودة السحاب إلى باطنٍ، للتعاقب المذكور أو تعين على تحلل ما فيه من البخار الدخانى، أو تكون متولدة عن المنفصل منه من البخار الدخانى فيبرد بانفصاله. وإن كانت باردة أعانت أيضا بالتبريد

(1) Wind too often helps in the generation of rain, (2) in that it gathers clouds, (3) or in that the coldness of the clouds withdraws towards the inside due to mutual repulsion [of heat and cold] already mentioned [*al-taᶜāqub al-maḏkūr*][16]; (4) or (wind) helps in the dissipation of the smoky vapour that is in (the clouds); or (wind) is generated from the smoky vapour that is separated from (the clouds), so that it grows cold in its separation. If it is cold, it also helps in refrigeration.

Mete. III. Section ii.: *Definition of the Directions of Winds*

The material here basically follows the *Šifā*ᵓ, although BH greatly expands the discussion of the manner in which the twelve points from which the winds blow are determined.[17] - The Greek names of the winds given in III.ii.3 are taken from Nic. syr.

	*Šifā*ᵓ	Other sources	
III.ii.1.1-3	61.12-13	Nic. 32.5-6	Twelve winds
III.ii.1.3-7	61.13-15		Definition

[15] Cf. Arist. 360b 33-361a 3; Olymp. 170.15-19; Olymp. arab. 117.21-118.4; *Cand.* 124.10-125.1.

[16] See comm. on Mete. I.iii.4.1-4 above.

[17] The manner in which IS determined these points was criticised by Fakhr al-Dīn al-Rāzī in (*Mabāḥiṯ* II.194.18-196.13; see Lettinck [1999] 186). BH seems to have taken no account of this criticism.

III.ii.2.1-12	61.16-17		(contd.)
III.ii.3.1-3	61.17-62.4		Principal winds
III.ii.3.3-8		Nic. 32 (diagram)	Greek names of winds

l. 1. ܡܛܠ ܬܚܘܡ ܦܢ̈ܝܬܐ ܕܡܫܫ̈ܒܐ ܕܪ̈ܘܚܐ: lit. "on the definition of the directions of the blowings/sources [*maššbē*] of winds": cf. *Šifā'* 61.12: المهاب المحدودة للرياح. - At III.1.3.3 above, *maššbā* corresponded to IS's *hubūb* (nomen actionis), here the phrase *penyātā d-maššbā* answers to IS's *mahabb* (nomen loci). At the corresponding place in *Cand.*, we encounter a more wordy rendition of *mahābb* as "the directions from which winds blow" (126.1: ܡܛܠ ܦܢ̈ܝܬܐ ܕܡܢܗܝܢ ܪ̈ܘܚܐ ܢܫ̈ܒܢ; cf. *Mabāḥit* II.194.10: البحث الخامس فى مهاب الرياح واساميها)

III.ii.1.1-3: Twelve winds

But. [underline: agreement with *Šifā'*; italics: with Nic. syr.]:

(1) The ancients *enumerate* [*mānēn*] twelve directions [*penyātā*] *from which winds blow*, (2) because the horizon is defined [*mettaḥḥam*] by twelve points [*tḥōmē*], (3) three easterly, three westerly, three northerly and three southerly.

Šifā' 61.12-13:

والمهاب المحدودة للرياح اثنا عشر، لأن الأفق يتحدد باثنى عشر حدا، ثلاثة مشارق، وثلاثة مغارب، وثلاث نقط شمالية، وثلاث نقط جنوبية.

(1) The source-directions [*mahabb*] defined [*maḥdūd*] for the winds are twelve, (2) because the horizon is defined [*yataḥaddadu*] by twelve points [*ḥudūd*]: (3) three easts, and three wests, and three northern points and three southern points.

Nic. syr. 32.5-6:

ܠܕܘ̈ܟܝܬܐ ܕܝܢ ܕܡܢܗܝܢ ܢܫ̈ܒܢ ܪ̈ܘܚܐ . ܘܠܡܢܝܢܐ ܕܝܠܗܝܢ ܡܢ ܪܘܫܡܐ/6 ܕܗܘ ܠܬܚܬ ܐܣܬܟܠ .

Observe the places [*dukkyātā*] from which winds blow and their number [*menyānā*] in the diagram [*rušmā*] below.[18]

III.ii.1.3-7: Definition

This definition of the points on the horizon from which the winds blow goes back to Arist. 363a 26ff. and there is a corresponding passage at Nic. syr. 32.6-11,[19] but it is mainly the *Šifā'* which is used

[18] On the diagram referred to here, see under III.ii.3. below.

[19] Nic. syr. 32.6-11: ܐܝܬ ܚܕ ܚܘܕܪܐ ܕܛܒ ܪܒ ܕܡܬܩܪܐ ܐܘܪܝܙܘܢ . ܘܠܗܢܐ/7 ܦܣܩܝܢ ܠܬܪܝܢ ܚܘܕܪ̈ܐ ܠܐܘܪܝܙܘܢ : ܘܬܚܘܡܐ ܚܘܕܪܐ ܦܐܪܐܠܠܝܠܘܣ ܕܐܪܟܛܝܩܘ ./8 ܗܘ ܕܒܗ ܕܢܦܩܝܢ ܕܝܢ ܡܢ ܟܠܗܘܢ ܐܝܟ ܕܐܪܟܛܝܩܘ ܡܬܚܙܝܢ ܠܢ . ܘܗܘ/9 ܕܢܦܩܝܢ ܕܝܢ ܡܢ ܟܠܗܘܢ ܐܝܟ ܕܐܢܛܐܪܟܛܝܩܘ ܠܐ ܡܬܚܙܝܢ ܠܢ . ܘܬܠܬܐ ܛܪܘܦܝܩܐ/10 ܕܡܬܩܪܝܢ ܠܦܪܘܫܘ ܕܒܥܠܡܐ . ܘܡܢ ܗܢܘܢ ܕܟܕ ܦܠܓܗ ܚܘܕܪܐ ܠܚܘܕܪܐ ./11 ܒܬܪܥܝܢ ܦܠܓܝܢ ܦܠܓܗ ܠܗ ܠܐܪܒܥܐ ܦܠܓܘ̈ܬܐ ܗܫܐ ܡܢ ܟܠܗ . ([7 ܚܘܕܪܐ ܠܐܘܪܝܙܘܢ cod.] ("There is a great circle [*ḥudrā d-ṭāb rab*], which is the

by BH here. - Here and in 'theory' III.ii.2 below, BH expands the corresponding passage of IS through the addition of astronomical details, as he was often found to do in Chapter Min. IV above.

But.: (1) Know that eight of them are apparent [*glēn*] and are known to all. (2) The first is the equinoctial east [*madnḥā šawyūtānāyā*] which is [at the point of ascent of] the head [*rēš*] of Aries and Libra. (3) The second is the aestival east [*madnḥā qayṭāyā*] which is the head of Cancer. (4) The third is the hibernal east [*madnḥā satwāyā*] which is the head of Capricorn. (5) The fourth is the equinoctial west which is the head of Aries and Libra. (6) The fifth is the aestival west which is the head of Cancer. (7) The sixth is the hibernal west which is the head of Capricorn. (8) The seventh is the northerly point [*tḥōmā*] which is fixed [*mettaḥḥam*] at the point [*nuqdtā*] at which the meridian [*ḥudrā ṭahrāyā*] intersects [*pāseq*] the horizon below the north pole which is visible to us [*d-methzē lan*]. (9) The eighth is the southerly point which is fixed at the point at which the meridian intersects the horizon above the south pole which is not visible to us.

Šifāʾ 61.13-15:

فالمشارق الثلاثة: مشرق الاعتدال، ومشرق الصيف، وهو مطلع نقطة السرطان، ومشرق الشتاء وهو مطلع نقطة الجدي؛ ويقابلها مغارب ثلاثة. والنقط الشمالية وجنوبية الثلاث تقاطع خط نصف النهار والأفق،

...

(2) The three easts are the equinoctial east [*mašriq al-iʿtidāl*], (3) the aestival east [*mašriq al-ṣaif*] which is the ascent-point [*maṭlaʿ*] of the point [*nuqṭa*] of Cancer (4) and the hibernal east which is the point of ascent of Capricorn. (5-7) The three wests are situated opposite [*yuqābilu*] them. (8-9) The three northern and southern points [*nuqaṭ*] are the intersection [*taqāṭuʿ*] of the meridian [*ḫaṭṭ nuṣf al-nahār*] and the horizon, ...

l. 9, 11. ܦܘܠܘܣ ܓܪܒܝܝܐ ܕܡܬܚܙܐ ܠܢ ... ܦܘܠܘܣ ܬܝܡܢܝܐ ܕܠܐ ܡܬܚܙܐ ܠܢ ("the north pole which is visible to us [*methzē lan*]", "the south pole which is not visible to us"): see comm. on III.ii.2.2 below; cf. further *De mundo* syr. 136.21-24 [< gr. 392a 1-3]: ... ܬܪ̈ܝܗܘܢ ܦܘ̈ܠܘ ... ܗܘ ܚܕ ܡܢ ܡܬܚܙܝܢܐ ܗܘ ܠܢ ... ܡܬܩܪܐ ܐܦ ܗܘ ܦܘܠܘܣ ܓܪܒܝܝܐ ("the two poles, of which one is visible to us [< ἀεὶ φανερός] and is also called the north pole").

---

horizon. This has been divided in two by the meridian [*ḥudrā ṭahrānāyā*] and the five parallel circles mentioned above [i.e. at Nic. syr. 30.11-18; cf. comm. on Min. IV.ii.1.1-7 above], namely that which is greater than all those which are always visible to us [*methawwēn lan*] [i.e. the arctic circle], that which is greater than all those which are always invisible to us [i.e. the antarctic circle], and the three turning points [*šuḥlāpē*] which are called τροπικοί in the book [i.e. the two tropics and the equator]. Because, when a circle divides [i.e. intersects] a circle, it divides it at two points, twelve points are made by these, Cf. Arist. 363a 26ff.; Olymp. 185.8-25; Olymp. arab. 125.9-126.2).

III.ii.2.1-12: Definition (contd.)

But.: The other four points out of the twelve are defined thus: When one imagines [*netrnē*] a circle that is greater than circles which are always apparent to us [*d-b-ammīnū glēn lan*] - this [circle] is one which touches the horizon from above and does not set - and another circle that is greater than circles which are always concealed from us [*d-b-ammīnū ksēn menan*] - this [circle] is one which touches the horizon from below and does not rise - then, one imagines two circles parallel to the meridian in such a way that one of them touches [*negšōp*] one side of the visible circle [*ḥudrā methazyānā*] above the horizon and the other its other side likewise above the horizon, it is apparent [*galyā d-*] that at the two points of contact [*nuqdātā da-gšāpā*] two points [*tḥōmē*] are defined [*mettaḥḥam*] on either side [lit. on the two sides] of the north. In the same way, on the southern side, one of the two circles parallel to the meridian touches one side of the invisible circle below the horizon and the other its other side likewise below the horizon, and thus at these two points of contact two other points are defined on either side of the south.

Šifā' 61.16-17:

... ونقطتا تقاطع دائرتين موازيتين لدائرة نصف النهار، مماستين للدائرتين الدائمتى الظهور والخفاء، من غير قطع.

... and the two points [*nuqṭatā*] at [lit. of] the intersection of the two circles parallel to the meridian, which touch [*mumāssataini*] the two circles which are constantly apparent and concealed [*al-dā'iratāni al-dā'imatā al-ẓuhūr wa-l-ḫafā'*], without cutting.

l. 2. ܚܘܕܪܐ ܕܪܒ ܗܘ ܡܢ ܚܘܕܪ̈ܐ ܕܒܐܡܝܢܘ ܓܠܝܢ ܠܢ "circle that is greater than circles which are always apparent to us [*d-b-ammīnū glēn lan*]": This corresponds to Aristotle's "ὁ διὰ παντὸς φανερός [sc. κύκλος]" (362b 2f.; also ὁ διὰ παντὸς φαινόμενος 363b 32). - In Nic. syr. this is rendered by "that very great circle which is always visible to us [*methzē lan*]" (ܚܘܕܪܐ ܗܘ ܕܛܒ ܪܒ ܕܒܐܡܝܢܘ ܡܬܚܙܐ ܠܢ, 30.14) and "[the circle] that is greater than all those which are always visible [*methawwēn lan*]" (ܕܪܒ ܗܘ ܡܢ ܟܠܗܘܢ ܐܝܠܝܢ ܕܒܐܡܝܢܘ ܡܬܚܘܝܢ ܠܢ, Nic. syr. 32.8). - The alteration of the verbs *methzē/methawwēn* here to *glēn* and the corresponding *lā methzē/methawwēn* to *ksēn* at line 4 below may be due to IS's *ẓuhūr* and *ḫafā'*.

III.ii.3.1-3: Principal winds

But.: (1) The best-known [*ṭbībātā*] winds are four, namely the easterly, the westerly, the northerly and the southerly. (2) These are the mothers [*emmhātā*]. (3) The other eight are generated between them (4) and are called by particular names by the Greeks.

Šifā' 61.17-62.1, 62.4:

ولهذه الرياح أسام باليونانية والعربية ليست تحضرنا الآن، والمشهورات عند العرب ريح الشمال، وريح الجنوب، والصبا وهو المشرقية، والدبّور وهو المغربية، والبواقى تسمى نكباء. ويشبه أن تكون هذه الأربع هو الغالبة، ومن الأربع الشمال والجنوب ... فتكون أمهات الرياح عنده ريحين

(4) These winds have names in Greek and Arabic, which we do not recall now. (1) The best-known [*al-mašhūrāt*] among the Arabs is the north wind, the south wind, the *ṣaban*, i.e. the east wind, and the *dabūr*, i.e. the west. (3) The rest are called side winds [*nakbāʾ*]. (2) It seems that these four winds are the predominant ones, and of the four the north and the south [in particular]. ... (2) Thereby the mothers of winds [*ummahāt al-riyāḥ*] become two. ...

III.ii.3.3-8: Greek names of winds

The Greek names for the twelve winds (or rather the eight other than those from the cardinal points) here are taken from Nic. syr. - BH himself had given a different list in *Cand.* 126.1-127.5 following *De mundo* syr. 142.30-143.19. Of such lists found elsewhere in Syriac, that in Bar Kepha, *Hex.* (ms. Paris syr. 311, 57r)[20] follows that in Jacob of Edessa, *Hex.* [Chabot] 84f., which in turn resembles the list in the Greek *De mundo* but is different from that in the Syriac *De mundo*.[21]

The orthography of the names show a great variation in the diagram and elsewhere in the text of Nic. syr. in both the Cambridge and Paris manuscripts. The fact that BH has more or less succeeded in producing the correct orthography (except for Lips, λίψ) suggests that he had a relatively good copy of Nic. syr., although we must also bear in mind the possibility that he used other sources for correcting his orthography.

But.: They call (the wind) which blows from the aestival east [i.e., roughly speaking ENE (east-north-east)] Caecias, that from the hibernal east [ESE] Eurus, that from the aestival west [WNW] Aparctias, that from the hibernal west [WSW] Lips, that between aestival east and north [NNE] Meses, that between hibernal east and south [SSE] Phoenicias, that between aestival west and north [NNW] Thrascias, and that between hibernal west and south [SSW] Libonotus.

Nic. syr. p. 32, diagram [clockwise from north][22]:
N [7]: ܙ ܓܪܒܝܐ; NNE [10]: ܝ ܡܣܝܣ; ENE [6]: ܘ ܩܐܩܝܐܣ; E [2]: ܒ ܡܕܢܚܐ; ESE [4]: ܕ ܐܘܪܘܣ; SSE [12]: ܝܒ ܦܘܢܝܩܝܐܣ; S [8]: ܚ ܬܝܡܢܐ; SSW [11]: ܝܐ ܠܝܒܘܢܘܛܘܣ; WSW [3]: ܓ ܠܝܒܣ; W [1]: ܐ ܡܥܪܒܐ; WNW [5]: ܗ ܐܪܓܣܛܝܣ; NNW [9]: ܛ ܬܪܐܣܩܝܐܣ

---

[20] The names are displaced in the diagram in Paris syr. 241, 188v; see tr. Schlimme (1977) 618f., 654.

[21] See Takahashi (2002b) 230f. - On the names of winds in Latin and Greek, see Gilbert (1907) 550f.

[22] Cf. Arist. 363a 34-b 26; Alex. Aphr. 107.28ff.; Olymp. 185.26-187.1; Olymp. arab. 127.1-9.

[C = Cantab.; p = Paris 346, fol. 61v, diagram || ܩܡܩܩܝܣ C hic et alibi: ܩܘܪܩܩܝܣ p; ܩܡܩܩܝܣ *But*. FMV (om. PLl) et p alibi || ܐܘܪܐܘܣ Cp: melius ܐܘܪܘܢܘܣ *But*. FM (om. PLl) et Cp alibi || ܢܘܛܘܣ: add. ܪܘܚܐ ܗ ܐܘܪܘܣ p (i.e. οὖρος?) || ܓ: om p || ܣܩܐܝܪܘܢ p || ܠܒܘܣ Cp: ܠܒܝܣ *But*. omnes codd.; recte ܠܒܘܣ p alibi || ܐܪܓܣܛܝܐܣ Cp: recte ܐܪܓܣܛܐܣ *But*. omnes codd.; ܐܪܓܣܛܐܣ C alibi; ܐܪܓܣܛܝܐܣ p alibi]

l. 5. ܐܦܪܩܛܝܐܣ ("Aparctias"): "Aparctias" according to Arist. 363b 14 is the wind from due north (together with "Boreas"). We see that in Nic. syr. this has replaced Aristotle's Argestes-Olympias-Sciron (363b 24).

l. 8. ܠܒܘܢܘܛܘܣ ("Libonotus"): Arist. gives no name for the SSW wind (see 363b 33). "Libonotus" is the name given for the wind from that direction in later works: *De mundo* gr. 394b 34 (syr. 143.18 ܠܒܢܘܛܘ); Alex. Aphr. 110.7; Olymp. 186.11.

Mete. III. Section iii.: *On the Variation of Winds according to Places and Seasons*

The section corresponds roughly to the latter half of IS's chapter on the winds in the *Šifā*ʾ (ed. Cairo p. 62-65) and the order of the presentation basically follows that of the *Šifā*ʾ, although we observe here again an extensive use of materials taken from Nic. syr. - The first part of the section (III.iii.1-2) is concerned mainly with the variation in the temperature of winds, while 'theory' III.iii.3 deals with the reasons as to why winds fail to occur in certain seasons of the year and 'theories' III.iii.4-6 discuss those seasonal winds which Arist. called "ἐτησίαι" (annual) and explained as being due to the movement of the sun.

	*Šifa*ʾ	Nic. syr.	
III.iii.1.1-5	62.4-9	(33.24-26)	North and south winds
III.iii.1.5-11		27.22-28.4	South wind
III.iii.2.1-2	62.17-18		East and west winds
III.iii.2.2-5	63.1-3		(contd.)
III.iii.2.5-7	(31.5-7)	33.14-24	Retention of heat in the east
III.iii.2.7-11	(31.7-9)	?	Refutation
III.iii.3.1-5		29.3-10	Absence of wind (1): summer
III.iii.3.5-10	63.13-15	29.3-4, 10-13	Absence (2): other seasons
III.iii.4.1-7	64.17-65.6	29.15-18, 28-31	Etesian winds
III.iii.4.7-9	65.8-9	30.26-29	Leuconotus
III.iii.5.1	65.16		Etesian winds (definition)
III.iii.5.1-11	65.13, 65.17-66.1	29.31-30.6, 31.10-15	Polar south winds
III.iii.6.1-7		34.18-19, 23-28	East/west etesian winds

l. 1. ܡܛܠ ܫܘܚܠܦ ܪܘܚܐ ܒܫܘܚܠܦ ܐܬܪܘܬܐ ܘܙܒܢܐ: lit. "on the variation of winds with variation of places and seasons"

III.iii.1.1-5: North and south winds

This is closer to the *Šifā*ʾ than to the corresponding passage of Nic. syr. (33.24-26).[23]

> *But*.: (1) The north wind *is cold* (2a) because it *passes* over [*ʿāḇrā ʿal*] a multitude [*suggā*] of snow-covered *mountains* [*ṭūrē mṭallḡē*] (2b) before coming to us [lit. then come to us]. (3a) If it is able to proceed further [*l-metmṯāḥū*] in a southerly direction [lit. to the region/direction (*pnīṯā*) of the south], (3b) it will no doubt [lit. it is not be surprising if, *lā dmīrā enhū d-*] grows hot through its passage over hot places. (4a) The south wind is hot (4b) because it reaches us after its passage over hot places.
>
> *Šifā*ʾ 62.4-9:
>
> فالرياح التى تأتى من ناحية الشمال، هى أبرد الرياح. وذلك لأن معنى قولنا إنها شمالية، هى أنها تكون شمالية بالقياس إلى بلادنا. وناحية الشمال منا باردة، وفيها جبال وثلوج كثيرة، فتتبرّد الرياح المارة بها إلينا. فإن جاز أن تمتد إلى ناحية الجنوب لم يبعد أن تسخن بمرورها بالبلاد الحارة. والجنوبية هو أسخن الرياح، لأنها إنما تصل إلي ديارنا وقد جاوزت بلادا محرقة حارة
>
> (1) Winds which come from the region [*nāḥiya*] of the north are the coldest of winds. (2) That is because what we mean by saying that they are northerly is that they are northerly in relation to our country [*bilād*]. The region north of us is cold and there are mountains and much snow [*ǧibāl wa-ṯulūǧ kaṯīra*] there, so that the winds passing through it [*al-mārra bihā*] towards us grow cold. (3a) If they can proceed further [*tamtaddu*] towards the region [*nāḥiya*] of the south, (3b) it is probable that [*lam yabʿud an*] they will grow hot through their passage over hot countries. (4a) The south wind is the hottest of winds (4b) because it only reaches our areas [*diyāru-nā*] after it has passed through torrid and hot countries ...

l. 4. ܒܬܪ ܡܥܒܪܢܘܬܗ ܥܠ ܐܬܪܘܬܐ ܫܚܝܢܐ ("after its passage over hot places"): "after it has passed through torrid and hot countries" IS. - Although the corresponding passage at *Mabāḥiṯ* II.196.14-16 is shorter and omits many elements which are retained in *But*., in this phrase *But*. stands closer to *Mabāḥiṯ* than to *Šifā*ʾ : لمرورها بالمواضع الحارة ("through its passage over hot places", 196.16)

III.iii.1.5-11: South wind

The passage is closely based on Nic. syr., although BH shortens the passage of Nic. syr. and rearranges the material of the parts designated (4) and (5) below to make better sense of what appears to be contradictory

---

[23] Nic. syr. 33.24-26: ܡܢܗܝܢ ܓܝܪ ܪܘܚܐ ܫܚܝܢܢ ܘܐܚܪܢܝܬܐ ܩܪܝܪܬܐ ܡܢ/25 ܥܠܬܐ ܕܐܬܪܘܬܐ ܕܥܒܪܢ ܒܗܝܢ. ܐܡܬܝ ܕܥܒܪܢ ܒܛܘܪܐ ܩܪܝܪܐ ܐܘ ܥܠ ܡܝܐ ܢܣܒܢ/26 ܡܢ ܩܪܝܪܘܬܐ ܘܩܪܝܪܬܐ ܢܫܒܢ. [26 ܡܢ: ܡܢܫܒܢ legend.?] ("[Some] winds are hot, (1) while others are cold (2) on account of the places through which they pass [*ʿāḇrān b-*]. When they pass through [*b-*] cold mountains or over [*ʿal*] water, they partake [*nāsbān*] of the coldness and blow as cold [winds]"; cf. Arist. 358a 32-b1; *Prob*. 26.15; Olymp. 161.5-7; Olymp. arab. 129.7-9).

statements in Nic. syr. whereby we are told that little exhalation rises in (5) but that much exhalation rises in (4).[24] - The same passage of Nic. syr. appears to be the source (no doubt via Olymp. arab.) of a passage occurring later on in the *Šifā*ʾ at 65.9-15 (cf. under III.iii.5. below).

But.: (1) Here the Master [Aristotle] says: (2a) If someone says (2b) "how can wind blow from that torrid region [*penyāṯā hāy yāqedtā*] which is extremely dry?", (3a) we say to him (3b) that the north wind drives vapoury exhalation there, (3c) moistens [*mṣabbʿā*] the earth (3d) and causes large quantities of [lit. much] smoke [pl.] to rise, (3e) just like moist wood. (4a) Furthermore, because the equator [lit. tropic of equality/equinox, *ṭrōpīqōs d-šawyūṯā*] is spacious and wide [*rwīḥ wa-pṯē*], (4b) large quantities of [lit. much] exhalations rise, (5a) but because of their lack of concentration [*knīšūṯā*], (5b) the winds which blow from there are weak. (6) For the north and south are like the navels of a melon [πέπων] and the isemeric tropic its middle.

Nic. syr. 27.22-28.4[25]:

ܘܐܝܟܢܐ ܢܫܒܐ ܪܘܚܐ ܡܢ ܦܢܝܬܐ ܝܩܕܬܐ ܟܕ ܝܒܝܫܬܐ/23 ܐܝܬܝܗ̇ . ܘܫܡܫܐ
ܥܒܪ ܥܠܝܗ̇ ܐܡܝܢܐܝܬ . ܘܐܡܪܝܢܢ ܕܪܘܚܐ ܓܪܒܝܝܬܐ ܟܕ/24 ܢܫܒܐ . ܕܚܝܐ ܠܬܡܢ
ܚܠܬܐ ܠܡܥܢܢܝܬܐ . ܘܗܝ ܗܢܐ ܡܪܛܒܐ ܠܐܪܥܐ ܘܡܛܠܗܢܐ/25 ܚܠܬܐ ܡܬܢܢܝܬܐ
ܡܣܩܐ ܣܓܝܐܐ . ܐܝܟ ܕܩܝܣܐ ܪܛܝܒܐ . ܡܛܠܗܢܐ ܢܫܒܐ ܒܩܝܛܐ/26 ܝܬܝܪܐܝܬ
ܢܫܒܐ ܪܘܚܐ ܗܕܐ . ܡܛܠܠܗ̇ ܕܡܫܟܚܢܝܬܐ ܗܝܕܝܟ ܪܘܚܐ ܓܪܒܝܝܬܐ/27
ܕܬܕܚܘܫܝܗ̇ ܠܗܠܝܢ ܠܠܡܥܢܐ . ܗܠܝܢ ܕܝܢ ܐܝܬ ܠ ܠܡܬܟܢܫܘ ܡܕܡܐ ܗܘ/28 ܚܠܬܐ
ܩܠܝܠܬܐ ܕܗܢ ܗܟܢ ܟܠܗ . ܘܡܬܠܠܗܝܢ ܐܦ ܬܢܢܐ ܕܗܢ ܗܟܢ ܢܦܩܢ/29 ܡܣܬܟܠ .
ܡܛܠܬܗ ܕܝܢ ܕܠܦܣܘܬܗ ܗ̄ ܟܘܢܫܐ ܕܒܘܩܬܐ . ܠܝܬܗܘܢ/30 ܐܠܝܒܢܐ ܐܠܐ
ܦܘܢܐ ܡܨܥܝܐ ܐܝܬ ܠܗ . ܟܕܡ ܡܢ ܐܪܟܐ ܡܨܥܝܬܐ ܚܠܬܐ ܡܨܥܝܐ//1
ܟܠܗ . ܕܗܢܐ ܚܕ ܚܕܡܝܢ ܡܢ ܥܘܡܩܝܢ ܠܟܘܪܝܗ̇ ܕܦܠܝܓܝܐ ./2 ܛܪܘܦܝܩܘܣ ܕܝܢ
ܐܝܣܘܡܝܪܝܢܘܣ ܠܡܨܥܬܗ .

[= Paris 346, 65v marg.: 29 ܗ̄ ܟܘܢܫܐ: om. P || 1 ܚܠܬ: om. P || ܕܦܠܝܓܝܐ: add. ܦܘܩܐ ܗ̄ ܟܘܢܫܐ in marg. C || 2 ܕܝܢ: om. C || ܐܝܣܘܡܝܪܝܢܘܣ C]

(2a) How can wind blow from the torrid region, seeing that [*kad*] it is dry and the sun constantly passes over it? (3a) We say: (3b) When the north wind blows, it drives the vapoury exhalation there (3c) and this (vapoury exhalation) moistens the earth; (3d) for this reason (the earth) raises much smoky exhalation, (3e) just like moist wood. Therefore, this wind blows much in summer, because then the north wind is able to drive the vapour

[24] The problem seems to have arisen out of an attempt to combine those passages of Arist. where we are told about the possibility of winds blowing from the equatorial region with those where we are told why they are weak and not continuous (see Arist. 362a 26-28; cf. Olymp. 182.29-183.3, 183.6-10). - A different solution from the one used by BH is adopted in Olymp. arab. where the passages corresponding to (5) and (4) here are joined by an adversative conjunction (118.19 *illā annahu*, "except that").

[25] Cf. Arist. 361a 14-21, 363a 8-20; Olymp. 170.36-171.9, 193.3-8; Olymp. arab. 118.12-20.

there. (5a) Furthermore, we might say that the smoky exhalation that rises from there is small (5b) and for this reason the winds which blow from there too are weak. (4a) Thirdly, the tropic - i.e. turning point [*šuhlāpā*] - of the equinox [*d-šawyūtā*] is not narrow but has a great width [*ptāyā saggī'ā īt leh*], (4b) so that from much earth much exhalation rises. - (6) For the north and south are like the navels of a melon [*paṭṭīḥā*], the isemeric tropic its middle.

l. 6. ܦܢܝܬܐ ܗܝ ܝܩܕܬܐ ("that torrid region"): ܦܢܝܬܐ ܝܩܕܬܐ Nic. - In Nic. syr. this answers to the κατακεκαυμένος τόπος of Arist. 363a 13 and κατακεκαυμένη ζώνη of Olymp. 193.3. The adjective is almost invariably in the form of the participle *yāqedtā* in Nic. syr. (27.17 *pnītā hāy yāqedtā*; 30.21: *zōnī yāqedtā*; 31 marg.: *ar'ā yāqedtā*, cf. Arist. 358a 15 κατακεκαυμένη γῆ; 31.9, 27: *yāqedtā*, without noun), although the participial adjective appears once at 27.13 (*ar'ā yaqqīdtā*). - The Greek word κεκαυμένη as applied to the equatorial region was also known to BH (*Rad.* [Istanbul] 15.4f. ܐܦ ܠܗ̇ ܩܐܟܐܘܡܢܐ ܐܡܪܝܢ ܡܬܘܩܕܬܐ ܡܟܢܝܢܐ ܐܝܟ ܗ̇ܘ),[26] most likely from Severus Sebokht (Sachau [1870] 128.5f.: ܗ̇ܝ ܩܐܟܐܘܡܢܐ ܡܬܩܪܝܐ ܡܬܘܩܕܬܐ).

l. 7-8. "just like moist wood": cf. III.i.4. above.

l. 10. ܫܪ̈ܝ ܦܦܘܢ ("navels of a melon [πέπων]"): ܫܪ̈ܘܗܝ ܕܦܛܝܚܐ Nic. - Both mss. of Nic. syr. have the Syriac *paṭṭīḥā* in the text, but ms. C adds ܝܘܢܐܝܬ ܗ̄ ܦܦܘܢ ("graece: πέπων") in the margin (cf. PS 3044, s.v. ܦܦܘܢ). - The comparison of the polar region to "navel" (δίκην ὀμφαλοῦ) occurs at Olymp. 171.3 (cf. also 184.4; see comm. on Min. IV.ii.1.3. [ܠܒܢܬܐ] above), but there is no mention of "melon" there.

III.iii.2.1-2: East and west winds

IS tells us that the east and west winds are temperate, but does not tell us why.[27] The explanation here may be BH's own.

But.: (1) The east and west winds are more temperate [lit. close to moderation, *qarrībān la-mmaššḥūtā*] (2) because of the equality [*šawyūtā*] of the distance and proximity [*ruḥqā w-qurbā*] of the sun to their regions.

Šifā' 62.17-18:

وأما الرياح المشرقية والمغربية فيجب أن تكون أقرب إلى الاعتدال، وأن يقع لها اختلاف كبير بسبب اختلاف البلدان الكائن بسبب البحار والجبال.

(1) The east and west winds must be closer to moderation [*aqrab ilā al-i'tidāl*] but there occur to them a large difference by reason of the difference of countries which arise by reason of the seas and mountains.

---

[26] The word is in fact written ܩܐܐܟܐܘܡܢܝ with one more *ālap* in ms. Bodl. or. 467, 7v 1, the ms. from which the Istanbul edition was copied; cf. PS 3460.

[27] Neither does Fakhr al-Dīn al-Rāzī (*Mabāḥit* II.196.20).

l. 1-2. "because of the equality ...": This is a more awkward expression than seems at first. It cannot mean much more than that the sun is not so far from the east and west as it is from the north and not so near to them as it is to the south, although BH's wording is probably influenced by his arguments in Chapter Min. IV above about moderation in temperature resulting from equal distance from heat and cold.

III.iii.2.2-5: East and west winds (contd.)

But.: (1) The east wind, however, is somewhat hotter [lit. inclines towards heat, *l-ḥammīmūṯā ṣālyā*] and the west wind somewhat colder (2) because the east wind passes over dry land [*yabšā*] which is heated by the sun before coming to us [lit. then comes to us], (3) whereas the west wind enters and passes over the sea inside the habitable world [*da-ḇ-ḡaw ʿāmartā*] and grows cold before coming to us [lit. then comes to us].

Šifāʾ 63.1-3:

والرياح المشرقية تأتينا ونحن لا على طرف البحر، مارة على اليبس متسخنة بالشمس؛ وأما المغربية فتأتينا مارة على البحر. والمشرق أسخن من المغرب لأنه أكثر يبسا وبرية، وإنما البحر فى جانبين منه فقط، وقد تتباعد العمارة عنه فيها.

(2) The east winds come to us, who are not on the edge of the sea, passing over dry land [*yabs*] and heated by the sun, (3) whereas the west winds come to us passing over the sea. (1) The east is hotter than the west because it has more dry land and desert [*barrīya*] and the sea is on two sides of it only [i.e. to the north and south], while the inhabited part [*ʿimāra*] is often far from (that sea).

l. 3. ܠܚܡܝܡܘܬܐ ܨܠܝܐ ... ܠܩܪܝܪܘܬܐ (lit. "inclines towards heat ... towards coldness"): The wording here suggests that BH may also have checked the discussion of winds in IS's *Qānūn fī al-ṭibb*: cf. *Qānūn* I.2.2.1.8 [Cairo (1987)] 122.8[28]: ولذلك كانت المغربيات اقل حراً ومن المشرقيات وأميل الى البرد ("For this reason the west winds are less hot than the east winds and incline more [*amyal*] towards coldness").

l. 3. ܕܫܚܝܢܐ ܡܢ ܫܡܫܐ ("which is heated by the sun"): متسخنة بالشمس IS. - As the text stands in the *Šifāʾ*, it is the east wind itself which is heated by the sun; in *But*. it is the land over which it passes.

l. 4. ܒܗ ܡܬܠܝܐ ܠܘܬܢ (lit. "then comes to us"): cf. III.iii.1.2 above.

l. 4-5. "the sea inside the habitable world": i.e. the Mediterranean. - Cf. V.i.1.3 above: "through it [the Pillars of Hercules] it enters into the habitable world ...". - Also *De mundo* syr. 141.1 ܝܡܐ ܗܘ ܓܘܝܐ (gr. 393b 29 ἡ ἔσω θαλάσση). - IS does not specify the sea over which the west wind travels, but the Meditteranean is the sea that lies to the west of the region where both IS and BH lived.

---

[28] = ed. Būlāq I.89.26; ed. New Delhi (1982) 153.4f.

III.iii.2.5-7: Retention of heat in the east

The Aristotelian theory reported here (< Arist. *Mete*. 364a 24-27) is one which makes no sense in the context of a "round earth" theory and has been criticised as such.[29] In his presentation of the theory BH follows Nic. syr. The theory is also mentioned by IS in the *Šifā᾽*, not in the chapter on winds but in the chapter on the habitable world, and is followed there by a refutation (*Šifā᾽* 31.5ff.).[30]

But.: (1a) The ancients say (1b) that the sun retains [*nāṭar*] the heat which it gives [*yāheḇ*] to those in the east [*madnḥāyē*] (1c) because it is [*(h)u*, copula] above the earth, (2a) but it does not retain that (heat) which which it gives [*yāheḇ*] to those in the west [*maᶜrḇāyē*] (2b) because it is [*hāwē*] below the earth.

Nic. syr. 33.14-24[31]:

ܡܛܠ ܡܢܐ ܐܬܪ̈ܘܬܐ ܕܡܥܪ̈ܒܝܐ ܩܪ̈ܝܪܢ. ܘܡܕ 15/ ܢܚ̈ܝܬܐ ܫܚ̈ܝܢܢ. ܟܕ
ܫܡܫܐ ܫܘܝܐܝܬ ܡܫܚܢ ܠܬܪ̈ܬܝܗܝܢ ܦܢܝ̈ܬܐ܀ 16/ ܐܝܟܢܐ ܕܡܟܬܪܢܐ ܫܬ ܫ̈ܥܝܢ
ܡܕܢܚܐ: ܗܟܢܐ ܘܒܡܥܪܒܐ ܫܬ ܫ̈ܥܝܢ ܡܩܘܐ. 17/ ܘܐܝܟܢܐ ܕܩܕܡ ܚܕܐ ܫܥܐ
ܕܡܕܢܚܐ ܡܫܚܢ. ܗܟܢܐ ܘܒܬܪܟܢ 18/ ܚܕܐ ܫܥܐ ܕܡܥܪܒܗ ܡܫܚܢ
ܠܡܥܪܒܐ. ܘܐܡܪܝܢܢ ܕܟܕ ܒܡܕܢܚܐ 19/ ܗܘܐ ܫܡܫܐ ܐܬܪ̈ܘܬܐ ܡܕܢ̈ܚܝܐ:
ܠܐܬܪ̈ܘܬܐ ܡܕܢ̈ܚܝܬܐ ܣܓܝ ܡܫܚܢ: ܠܐܬܪ̈ܘܬܐ ܡܥܪ̈ܒܝܬܐ 20/ ܩܠܝܠ. ܘܟܕ ܒܡܥܪܒܐ
ܗܘܐ ܗܦܟܐ ܡܢ ܗܠܝܢ. ܠܡܥܪܒܐ ܣܓܝ ܫܚܝܢܘܬܐ/ 21 ܘܠܡܕܢܚܐ
<ܝܗܒ>. ܠܡܕܢܚܐ ܩܠܝܠ ܫܚܝܢܘܬܐ. ܘܐܦ ܠܘ ܫܘܝܐܝܬ ܝܗܒ ܫܚܝܢܘܬܐ ܝܗܒ/
22 ܫܚܝܢܘܬܐ. ܐܠܐ ܠܫܚܝܢܘܬܐ ܕܝܗܒ ܠܐܬܪ̈ܘܬܐ ܕܡܕܢܚܐ ܢܛܪ. ܚܠ/
23 ܕܠܥܠ ܡܢ ܐܪܥܐ ܩܐܡ. ܠܫܚܝܢܘܬܐ ܕܝܗܒ ܠܐܬܪ̈ܘܬܐ ܡܥܪ̈ܒܝܐ ܠܐ
ܢܛܪ. /24 ܥܠ ܕܠܬܚܬ ܐܪܥܐ ܗܘܐ.

Why are regions of the west colder, and eastern (regions) hot, seeing that the sun heats the two directions equally? Just as it stays [*mḵattar*] for six hours in the east, so it remains [*mqawwē*] for six hours in the west; and just as it first heats the east for one hour during its rising [*madnḥā*], just so later it heats the west for one hour during its setting [*maᶜrḇā*]. We say: When the sun is in the east, it warms the eastern regions much and western regions little. When it is in the west, the opposite [happens], [i.e. it gives] much heat to the west and little heat to the east. Even though it gives heat equally to the two sides, (1b) it retains the heat which it gives to the regions of the east [*aṯrawwāṯā d-madnḥā*], (1c) because it remains above the earth, (2a) but it does not retain the heat which it has bestowed [*šakken*] on the western regions [*aṯrawwāṯā maᶜrḇāyē*], (2b) because it goes [*hāwē*] below the earth.

---

[29] See, e.g., Lee (1952) 193 n. c; Strohm (1970) 189f.; Lettinck (1999) 159 n. 3.

[30] IS himself accepts the different relationships to the sun as the cause of the east winds being hotter than the west at *Qānūn* I.2.2.1.8, [Būlāq] I.89.23-26; [Cairo (1987)] 122.5-8; [New Delhi (1982)] 153.1-4.

[31] Cf. Arist. 364a 24-27; Alex. Aphr. 111.2-25; Olymp. 194.14-25; Olymp. arab. 128.2-13.

Šifā᾽ 31.5-7:

والذى قيل: إن الشرقية إنما هى أسخن من الغربية، بسبب أن الغربية تكون الشمس آخذة عنها فى حركتها ومودّعة إياها، والشرقية تكون آخذة إليها فى حركتها

It has been said that the eastern [lands, sc. *bilād*] are hotter than the western because the sun begins its movement in the eastern [lands] and leaves it, and it ends its movement in the western [lands].

l. 5. ܘܩܕ̈ܡܝܐ ܐܡܪܝܢ ("The ancients say"): Although what follows at *Šifā᾽* 63.4-9 is quite different from what we have here, IS too mentions the "ancients" at *Šifā᾽* 63.4 (وكان القدماء ينسبون ...) immediately following the passage used by BH in III.iii.2.2-5 above; this may have prompted BH to include this discussion of the "ancient" theory given here.

l. 6, 7. ܠܡܕ̈ܢܚܝܐ . . . ܠܡܥܪ̈ܒܝܐ ("to those in the east", "to those in the west"): ܐܬܪ̈ܘܬܐ ܡܥܪ̈ܒܝܬܐ, ܐܬܪ̈ܘܬܐ ܡܕܢ̈ܚܝܬܐ. - Nic. syr. talks of the eastern and western "regions" (*atrawwāṯā*; so also IS *bilād*). In the text of *But*. there is no noun which can be understood, so that we have to understand "to those people in the east/west".

III.iii.2.7-11: Refutation

But.: We say that this would be true if the morning and evening of those in the east and the west were not different [*lā mšaḥlpīn (h)waw*]. This, however, is not the case, but [what is] morning for those in the east is evening for those in the west and the morning of those in the west the evening of those in the east.

Cf. *Šifā᾽* 31.7-9:

فهو كلام من لا بصر له البتة. فإن كل نقطة من الأرض تأخذ إليها الشمس، وتأخذ عنها بالسواء؛ وليس الشرق شرقا والغرب غربا، إلا بالإضافة

This is the word of one who has no discernment at all. For the sun comes to every point of the earth and departs from it equally. East is not east and west is not west except in relation [to something].

III.iii.3.1-5: Absence of wind (1): summer

But.: (1) In summer absence of winds [*glīzūṯ rūḥē*] occurs (2) because of the shortage of the exhalationary material [*ḥsīrūṯ hūlē ʿeṭrānāytā*] (3) which is dissipated [*metpawššā*] by the heat, (4) just as the flame of a lamp [*dalqā da-šrāḡā*] is extinguished by a blaze [*šalheḇīṯā*], (5) not as like is destroyed like [*law aḵman d-ḏāmyāyā men dāmyāyā meṯhabbal*], but as a lesser by a greater. (6) For the earth too, in abundance of heat [*saggīʾūṯ ḥammīmūṯā*], dries up [intr., *yāḇšā*] and does not produce smoke [*w-lā mṯannʿnā*], (7) just as a straw in a blaze is burned up before it produces smoke.

Nic. syr. 29.3-10[32]:

ܓܠܝܙܘܬ ܪܘܚܐ ܕܝܢ ܗܘܝܐ ܒܐܪܒܥܬܗܘܢ/4 ܫܘܚܠܦܐ. ܐܠܐ ܠܐ ܡܛܠ ܚܕܐ ܘܗܝ
ܗܝ ܥܠܬܐ. ܒܩܝܛܐ ܡܢ/5 ܗܘܝܐ. ܟܕ ܙܥܘܪ ܗܘ ܠܗܓܐ ܕܡܬܝܠܕ. ܘܡܬܒܕܪ
ܡܬܒܕܪܐ ܡܢ ܣܘܓܐܐ/6 ܕܫܚܝܢܘܬܐ ܘܕܥܟ: ܐܝܟ ܫܠܗܒܝܬܐ ܕܫܪܓܐ ܟܕ
ܩܪܒܐ ܠܝܩܕܐ./7 ܠܐ ܐܝܟ ܕܕܡܝܐ ܡܢ ܕܕܡܝܐ ܡܬܚܒܠ: ܐܠܐ ܐܝܟ
ܙܥܘܪܐ ܡܢ ܡܝܬܪܐ./8 ܬܘܒ ܕܝܢ ܫܡܫܐ ܙܒܢ̈ܝܢ ܣܓܝ̈ܐܢ ܡܝܒܫ ܒܣܘܓܐܐ
ܕܫܚܝܢܘܬܗ ܡܩܕܡ/9 ܠܐܪܥܐ ܩܕܡ ܕܢܬܓܡܪ ܗܘܝܗ ܕܠܗܓܐ. ܐܝܟ ܕܟܕ
ܡܬܬܪܡܐ ܬܒܢܐ ܒܫܠܗܒܝܬܐ/10 ܪܒܬܐ. ܝܒܫ ܡܢ ܩܕܡ ܕܢܬܥܛܪ ܬܢܢܐ.

[= Paris 346, 69r 18-20 (cf. also the heading at 68r 21-22): 7 P ܐܝܟ ܡܢ || ܡܢ om. P]

(1) Absence of wind [*glīzūt rūḥā*] occurs during the four seasons [*šuḥlāpē*] of the year, but not for the one and same reason. (1) In summer it occurs, (2) since the exhalation generated is little, (3) and this is dissipated by the abundance of heat (4) and is extinguished, (4) just like the flame of a lamp when it approaches a blaze, (5) not as like is destroyed like, but as a lesser by a greater. (6) Furthremore, the sun with the abundance of its heat often dries up [tr.] the earth in advance before the exhalation is formed completely, (7) just as when a straw is placed in a large blaze, it dries up before it produces smoke.

l. 1. ܒܨܝܪܘܬ ܗܘܠܐ ܠܗܓܢܝܬܐ ("the shortage of the exhalationary material"): The alteration of simple "exhalation" in Nic. (29.5: *etrā*) to "exhalatory material" and the use of the word "shortage" (*ḥsīrūtā*) which does not occur here in Nic. may be explained from *Šifā'* 63.13: والصيف تقل فيه الريح لعوز المادة ("In summer wind is rare because of the lack [*ᶜawaz*] of the material).

III.iii.3.5-10: Absence of wind (2): other seasons

But. [underline: agreement with Nic.; italics: with *Šifā'*]:

(1a) *In winter* wind is made scarce [*metgalzā*] (1b) *because of the shortage of its efficient cause* [*ḥsīrūt ᶜelltā ᶜābōdtā*], namely heat, (1c) and the exhalation is extinguished by the abundance of the chill [*saggī'ūt qurrā*]. (2a) *In spring winds are rare* [*bāṣrān*] (2b) because the exhalation is extinguished beforehand [*qādem dāᶜek*] by the chill of winter; (3a) *in autumn*, on the other hand, (3b) because the exhalation has not yet been generated *because of drought* [*yabbīšūtā*]. (4a) Absence of wind therefore occurs during [all] four seasons of the year, (4b) but *it becomes abundant* [*sāgyā*] at any time *when the causes which counteract* its *hindrances* [*ᶜellātā da-mdalqbān l-mᶜawwkānyātāh*] abound [*sāgyān*].

Nic. syr. 29.3-4, 10-13[33]:

ܓܠܝܙܘܬ ܪܘܚܐ ܕܝܢ ܗܘܝܐ ܒܐܪܒܥܬܗܘܢ/4 ܫܘܚܠܦܐ. ܐܠܐ ܠܐ ܡܛܠ ܚܕܐ ܘܗܝ
ܗܝ ܥܠܬܐ. ... ܒܣܬܘܐ ܕܝܢ ܗܘܝܐ ܓܠܝܙܘܬ ܪܘܚܐ/11 ܡܛܠܠܗ ܕܡܬܕܥܟ
ܠܗܓܐ ܡܢ ܣܘܓܐܐ ܕܩܘܪܫܐ. ܒܡܬܐܐ/12 ܕܝܢ ܗܘܝܐ ܓܠܝܙܘܬ ܪܘܚܐ ܡܛܠܠܗ
ܕܡܩܕܡ ܐܬܕܥܟ ܠܗܓܐ ܡܢ ܩܘܪܫܐ/13 ܕܣܬܘܐ. ܒܬܫܪܝܢܐ ܕܝܢ ܗܘܝܐ
ܓܠܝܙܘܬ ܪܘܚܐ. ܟܕ ܠܐ ܥܕܟܝܠ ܐܬܝܠܕ ܠܗܓܐ.

[= Paris 346, 69v 6-8]

---

[32] Cf. Arist. 361b 14-27; Olymp. 175.32-176.2, 179.25-29; Olymp. arab. 120.11-15.

[33] Cf. Arist. 361b 25-30; Olymp. 176.2-10, 180.1-10; Olymp. arab. 120.16-18.

(4a) Absence of wind occurs during the four seasons [*šuḥlāp̄ē*] of the year, but not for the one and same reason. ... (1a) In winter absence of wind occurs (1c) because the exhalation is extinguished by the abundance of the chill [*saggīʾūṯā d-qurrā*]. (2a) In spring absence of winds occurs (2b) because the exhalation has been extinguished beforehand by the winter which has preceded it. (3a) In autumn absence of winds occurs (3b) since the exhalation has not yet been generated.

Šifāʾ 63.13-15:

والشتاء تقل فيه لعوز الفاعل/ ١٤ وربما اتفق أن تكثر، إذا اتفق من الأسباب ما يضاد المانعين. وقد يتفق أيضا أن تقل/ ١٥ فى الربيع للجمود، وفى الخريف لليبس

(1a) In winter it is rare (1b) because of the lack of the efficient [cause] [*li-ʿawaz al-fāʿil*] but often it happens that (wind) (4b) becomes abundant when there arises some cause which counteract the hindrances [*min asbābin mā yuḍāddu l-mānīʿaini*]. (2a) Wind may also happen to be rare [*taqillu*] in spring (2b) because of freezing, (3a) and in autumn (3b) because of drought [*yabs*] ...

III.iii.4.1-7: Etesian winds

The passage is based mainly on the *Šifāʾ*, but BH probably also had before his eyes a passage of Nic. syr. which was itself no doubt the source of the *Šifāʾ* (via Olymp. arab.).

But. [underline: agreement with *Šifāʾ*; italics: with Nic.]:

(1) Each of the twelve winds blows when the sun inclines [*ṣālē*] in its direction, (2) but not as soon as [*law meḥdā d-*] the sun arrives in its area [*l-waʿdāh*], (3) because at the beginning of its visit [*meṯwaʿʿḏānūṯā*] it dissipates [*mp̄awšeš*] what exhalation and smoke it finds [*meškaḥ*], (4) but it can only turn the frozen moistures [lit. moistures which have been frozen] into exhalation after a while [lit. cannot (*lā meškaḥ*) exhalationise (*l-maʿṭārū*) ... except after a time]. (5) *For this reason*, wind *blows not as soon as* [*law meḥdā d-*] *(the sun)* arrives, *but after* twenty days, (6) especially [*mālōn dēn*] the south wind that blows from dry land which is dissolved with difficulty [*ʿasqāʾīṯ*]. In fact this wind can be delayed for as long as [*āp̄ w- ... ṭūḇ*] two months.

Šifāʾ 64.17-65.6:

ومن شأن الرياح الاثنتى عشرة أن تهب كل واحدة منها عند ميل الشمس ألى جهته،/ ١ ولكن ليس فى أول ما تصل إليه، وخصوصا الشمالية والجنوبية، لأن الشمالية والجنوبية/ ٢ لا تهب كما توافى الشمس ناحيتها أولا، وذلك لأن الشمس تحلل الحاصل من/ ٣ البخار والدخان لقربها، ولا يقدر على أن تحلل الجامد من الرطوبات إلى البخار/ ٤ بسرعة فى أول وصولها. وما لم تحللها وتسيلها وتبل بها الارض، لا تعد الأرض/ ٥ لأن تدخن عن الحرارة دخانا كثيرا. فأن الارضية تعين على تصعيدها مخالطة/ ٦ المائية. ولهذه العلة قد تتأخر عشرين يوما، وخصوصا الجنوبية التى لا تهب/ ٧ عند القطب، بل تهب من دون البحر من الأرض اليابسة، لأن اليابس أبطأ انحلالا./ ٨ فلذلك هذه الرياح تتأخر قريبا من شهرين

(1) It is a characteristic of the twelve winds that each of them blows when the sun inclines [*ʿinda mail al-šams*] in its direction, (2) but not as soon as it comes [*taṣilu*] to it - especially the north and the south wind - because

the north and south winds do not blow as soon as [*kamā*] the sun reaches [*tuwāfī*] their regions. (3) That is because the sun dissolves [*tuḥallilu*] what there is of vapour and smoke in its proximity, (4) but is unable to dissolve the frozen moistures into vapour with speed on its first arrival. So long as it cannot dissolve and liquefy them and moistens the earth with them, it does not prepare the earth to produce much smoke with the heat. For the mixture of water helps the earthiness in its vaporisation [*taṣᶜīd*]. (5) For this reason, they are delayed for twenty days, (6) especially the south wind - which does not blow from the [south] pole, but blows from this side of the sea, from dry land - because the dry is slower in being dissolved. For this reason, these winds are delayed for approximately two months.

Nic. syr. 29.15-18, 28-31[34]:

ܢܫܒܢ ܕܝܢ ܐܦ ܪ̈ܘܚܐ ܓܪܒܝܝܬܐ ܗ̈ܢܝܢ/16 ܕܡܬܩܪܝܢ ܫܢܬܢܝܬܐ ܢܫ̈ܒܢ . ܗ̈ܢܝܢ
ܕܢܫ̈ܒܢ ܡܢ ܒܬܪ ܥܣܪܝܢ ܝܘ̈ܡܝܢ/17 ܕܡܥܠܬܐ ܕܫܡܫܐ ܕܠܛܪܘܦܘܣ ܩܝܛܝܐ . ܡܛܠ
ܕܒܝ ܕܬܠܓܐ ܣܓܝܐܐ/18 ܐܝܬ//28 ܒܦܢܝܬܐ ܓܪܒܝܝܬܐ . ܡܫܪܐ ܒܗ ܐܬܪܐ
ܫܡܫܐ ܠܡܚܕܐ ܕܐܬܐ ܠܗ . /29 ܘܟܕ ܡܫܬܪܐ ܡܬܪܛܒܐ ܐܪܥܐ ܘܡܫܬܟܚܐ ܫܘܝܬܐ ܗܘܝܐ
ܠܡܬܝܠܕܘ ܕܐܠܘܬܐ ܬܢܢܝܬܐ . ܘܡܛܠ/30 ܗܕܐ ܠܐ ܡܚܕܐ ܗܘ ܕܥܐܠ ܫܡܫܐ
ܠܛܪܘܦܘܣ ܩܝܛܝܐ ܢܫ̈ܒܢ : ܐܠܐ ܡܢ/31 ܒܬܪ ܥܣܪܝܢ ܝܘ̈ܡܝܢ .

[29.15-18 = Paris 346, 69v 12-16; 29.28-31 = Paris 346, 70r 9-13: 28 ܒܦܢܝܬܐ ܓܪܒܝܝܬܐ: om. P || ante ܡܫܪܐ: add. ܗܟܝܠ ܐܡܪ ܢܝܩܘܠܐܘܣ P || ܒܗ ܐܬܪܐ: ܕܐܬܪܐ P || ܠܡܚܕܐ: ܠܦܢܝܬܐ ܓܪܒܝܝܬܐ P || 30 ܡܚܕܐ: ܡܢ P]

Then also these north winds called etesian [*ša(n)tānāyāṯā*] blow, which blow after twenty days after the entry [*maᶜᶜaltā*] of the sun in the summer tropic. For because there is much snow in the northern region, the sun melts it as soon as [*meḥdā d-*] it arrives there, and as it is melted, the earth is moistened and becomes suitable for generation of smoky exhalation. (5) For this reason (the etesian winds) blow not as soon as [*law meḥdā mā d-*] the sun enters the summer tropic, but after twenty days.

l. 2. ܒܡܬܘܥܕܢܘܬܗ ... ܠܡܘܥܕܗ ("in its [wind's] area ... its [sun's] visit"): إلى جهته ... تصل إليه IS. - Given the inherent sense of *waᶜdā* as "an appointed meeting-place", the two words here are probably intended to convey something of the idea of the regularity with which the sun inclines in each direction every year. Despite the reported use of *metwaᶜᶜdānūṯā* in the sense of "conjunction" in an astronomical context (PS 1067), there is no need to understand the word in a technical sense here. Cf. Mete. I.ii.1.3-4 above (ܬܘܥܕ ... ܠܡܘܥܕܐ); also BH *Asc.* [Nau] 54.13, 19 (where the verb *etwaᶜᶜad* is used simply as an equivalent of *etmaṭṭī* at 54.10, 16).

l. 3. ܠܗܘ ܡܐ ܕܡܫܟܚ ܡܢ ܐܠܘܬܐ ܘܬܢܢܐ ܡܒܕܪ ("it dissipates what exhalation and smoke it finds"): تحلل الحاصل من البخار والدخان لقربها ("dissolves what there is

---

[34] Cf. Arist. 361b 35-362a 19; Olymp. 176.27-177.13; Olymp. arab. 120.19-121.3, 122.1-16. - The whole point of the passage here should be that the sun *does not* melt the snow immediately upon its arrival. This point which is there in all the related passages (incl. Olymp. arab. 120.19-121.3) is missing in the passage of Nic. syr. as it stands. Perhaps one should assume a lacuna in the middle of the passage quoted here.

[*al-ḥaṣil*] of vapour and smoke in its proximity) IS - Given the use of *lā meškaḥ* in the sense of "cannot" in the following line, this could mean "disspates what exhalation and smoke it can", but the corresponding phrase of the *Šifā'* favours the interpretation of *meškaḥ* in the sense of "find", and of the use of *meškaḥ* in its two different senses as an intentional word-play on BH's part.

l. 6. "two months": According to Arist. the southern etesian winds begin to blow *seventy* days after the winter solstice (*Mete*. 362a 24, so also Olymp. 177.20, Nic. syr. 30.6-9).[35] This has been altered to "sixty days" in Olymp. arab. (121.12) and thence to "two months" in the *Šifā'* and *But*.

III.iii.4.7-9: White south wind (Leuconotus)

But. [underline: agreement with Nic.; italics: with *Šifā'*]:

(1) It [lit. which] *is called* "white wind" [*ḥewwārtā*] by some because it brings clear weather (2) *and* "egg wind" [*bīʿtānāytā*] by others because *it makes* [hens] pregnant without mingling [*ḥulṭānā*] of cocks [*tarnāḡlē*] with hens [*tarnāḡlāṯā*].

Šifā' 65.8-9:

وتسمى البيضاء لإحداثها الصحو، وبيضية لأن من خاصيتها أن تحبل الدجاج بيضا من غير سفاد

... (1) and are called "white winds" [*baiḍā'*] because of their causing clear weather and "egg winds" [*baiḍīya*] because they cause chicken [*duǧāǧ*] to bear eggs without copulation [*sifād*].

Nic. syr. 30.26-29 [scholion][36]:

ܐܦܐ ܗܘܠܝܐ ܠܡܠܐ/27 ܕܚܘܪܐ ܐܝܟ ܐܡܝܪܘܣ ܗ̄ ܒܝܥܬܢܝܬܐ ܚܘܪܬܐ ܡܢ ܝܠܕܐ
ܒܥܒܪܐ. ܘܐܦ/28 ܒܝܥܬܢܝܬܐ ܚܘܪܬܐ ܡܫܬܡܗܐ. ܟܕ ܗܝ ܢܥܒܕ ܠܗ̇ ܬܪܢܓܠܬܐ
ܕܒܟܐ ܡܢ/29 ܚܘܠܛܢܐ ܕܬܪܢܓܠܐ.

[= Paris 346, 70v 8-11: 26 ante ܐܦܐ: add. ܐܬܐܪ ܚܢܝܢ ܫܘܝܢ ܟܕ ܡܛܠܬܐ ܕܠܒܪ P || 27 ܐܘܦ P || 28 ܒܝܥܬܢܠܒܢܝܬܐ P]

---

[35] Nic. syr. 30.6-9: ܗܘܝܢ ܕܝܢ ܬܝܡܢܝܐ/ ܐܬܣܝܢܘܬܐ ܪܘܚܐ ܕܢܫܒܢ ܡܢ ܬܝܡܢܢ ܕܝܠܢ. ܗܢܘ ܕܝܢ ܡܢ ܛܪܘܦܝܩܘܣ/ ܐܝܣܝܡܪܝܢܘܣ. ܢܫܒܢ ܕܝܢ ܗܠܝܢ: ܡܢ ܒܬܪ ܫܒܥܝܢ ܝܘܡܝܢ ܕܡܥܠܬܗ ܕܫܡܫܐ/ ܚܣܟܐ ܕܠܛܪܘܦܝܩܘܣ ܣܬܘܝܐ. ("There occur southern etesian winds, which blow from our south - i.e. the isemeric tropic. These blow seventy days after the entry of the sun to the winter solstice").

[36] Cf. Olymp. 177.21-27; Olymp. arab. 121.16-18. - This is one of the two instances where a scholion is explicitly attributed to Ḥunain in ms. Paris syr. 346. Although the scholion as a whole appears to be based on Olymp. 177.21-27, the reference to the Homeric "ἀργεσταί νότοι" is not in Olymp. While the addition of such Homeric references was a part of the stock trade of Greek scholiasts, one should not rule out the possibility of the addition being due to Ḥunain, who was one of the few Syriac-Arabic writers known to have been capable of reciting Homer (see Strohmaier [1980]). There is at least one more identifiable Homeric quotation in the scholia to Nic. syr. at 33.5-7: ܘܢܦܠܘ ܐܟܚܕܐ ܩܘܡܝܐ ܠܘܬ ܬܝܡܢܐ ܘܡܥܪܒܝܬܐ ܚܪܝܦܬܐ ܒܕܒܪܐ ܢܫܒܬܐ: ܘܓܪܒܝܝܬܐ ܕܡܨܡܚܐ. ܕܡܥܓܠܐ ܓܠܠܐ ܪܘܪܒܐ (< *Odyssey* 5.295-296: σὺν δ' Εὖρός τε Νότος τ' ἔμπεσον Ζέφυρός τε δυσαὴς, καὶ Βορέης αἰθρηγενέτης, μέγα κῦμα κυλίνδων).

(1) The poet calls these winds "ἀργεστᾶο νότοιο",[37] i.e. the southern white winds [*taymnāyātā ḥewwārātā* < λευκόνοτοι], because they bring clear weather. (2) They are also called "hen winds" [*tarnāglṯānāyāṯā*, cf. Olymp. arab. *duğāğīya*, < ὀρνιθίαι] because when they blow hens bear eggs [*bīʿē*] without mingling with the males.

l. 9. ܡܥܒܠܢܐ ("cause to bear") sic F: ܡܥܒܠܬܐ PLl; ܡܥܒܠܢܐ M¹; ܡܥܠܢܐ VM^V - We see how the copyist of V, faced with two corrupt readings, succeeded in restoring the sense, but no quite the right word.

III.iii.5.1: Etesian winds (definition)

But.: Winds that blow with the movement of the sun [*ʿam mettzīʿānūṯ šemšā*] are called etesian [*ša(n)tānāyāṯā*].
Šifāʾ 65.16: وهذه الرياح التى تهب مع حركة الشمس تسمى الحولية
These winds which blow with the movement of the sun are called etesian [lit. annual, periodic, *ḥaulīya*]

l. 1. ܫܢܬܢܝܬܐ ("etesian"): lit. "annual". - This is the word regularly used for the gr. ἐτησίαι in Nic. syr. - There is a scholion explaining the word at Nic. syr. 29.25-26: ܫܢܬܢܝܬܐ ܡܬܩܪ̈ܝܢ / ܪ̈ܘܚܐ ܟܕ ܢܫܒܢ ܐܝܟ ܗܘ ܚܕܐ ܠܫܢܬܐ ܠܫ̈ܢܐ. ("They are called etesian winds, since they blow once [a year] and from year to year").

III.iii.5.1-11: Why the polar south winds do not reach us

According to Aristotle the south winds we know are the winds blowing not from the south pole or the winter tropic (Tropic of Capricorn) but from the "torrid zone" (κατακεκαυμένος τόπος) to the south of us (*Mete*. 362a 31-32, 363a 8-13).

But. [underline: agreement with Nic.; italics: with *Šifāʾ*]:
(1) Although the same efficient cause [*ʿāḇōḏā*] of wind, the sun, and the same *material* [*hūlē*] for it, snow, are found in both the directions [*pnīṯā*] of the north *pole* and of the south *pole*, (2) the north winds are etesian, whereas the south winds are not etesian. (3) That is to say, in summer when the sun is in the direction of the north pole, north winds blow towards us. (4) *In winter*, however, when the sun is in the direction of the south pole, south winds do not blow towards us [*lwāṯan*] (5) because we are in the north [lit. are northerners] (6) and the south winds, since they blow from far away, (7) are weakened (8) and *do not reach us*, (9) unlike the north winds which reach us with their *strength* [intact]. (10) Hence it happens that those in that other habitable world [*ʿāmartā*] opposite us, if there is a habitable world there, wonder why the south winds are etesian but the north are not.

---

[37] Cf. Homer, *Iliad* 11.306, 21.334.

Nic. syr. 29.31-30.6[38]:

31 ܐܠܐ ܡܬܕܡܪ ܐܢܫ: ܕܡܛܠܡܢܐ ܐܝܟܢܐ ܕܡܣܩܝܢ ܬܝܡܢܐ//1 ܐܬܝܣܝܬܐ
ܐܬܝܣܝܬܐ: ܠܐ ܗܘܝܢ ܐܦ ܬܝܡܢܝܬܐ ܐܬܝܣܝܬܐ. ܡܛܠ ܕܚܕ/2 ܗܘ ܚܕ ܗܘ
ܥܒܘܕܐ ܘܗܝ ܗܝ ܗܘܠܐ ܐܫܬܟܚܘ ܒܬܪܬܝܗܝܢ ܦܢܝܬܐ. ܥܒܘܕܐ ܡܢ/3
ܫܡܫܐ. ܗܘܠܐ ܕܝܢ ܬܠܓܐ. ܘܐܡܪܝܢ ܕܟܒܪ ܗܘܝܢ ܐܬܝܣܝܬܐ ܬܝܡܢܝܬܐ/4. ܐܠܐ
ܛܥܝܢ ܠܢ. ܒܗܝ ܕܡܢ ܪܚܝܩ ܢܫܒܢ ܡܬܡܚܠܢ. ܐܝܟܢܐ/5 ܕܢܬܕܡܪܘܢ ܗܢܘܢ
ܕܒܥܡܪܬܐ ܗܝ ܐܚܪܬܐ ܕܠܘܩܒܠܢ. ܕܡܛܠܡܢܐ ܬܝܡܢܐ/6 ܗܘܝܢ ܐܬܝܣܝܬܐ
ܐܬܝܣܝܬܐ ܗܘܝܢ: ܓܪܒܝܝܬܐ ܕܝܢ ܠܐ ܗܘܝܢ.

[= Paris 346, 70r 13-22: 31 ܐܠܐ ܡܬܕܡܪ ܐܢܫ: ܕܡܛܠܡܢܐ: ܘܐܡܪܝܢ ܐܚܪܢܐ ܡܛܠ ܬܝܡܢܐ ܟܠܗ ܠܡܢܐ ܕܡܛܠܡܢܐ P || ܕܡܣܩܝܢ ܬܝܡܢܐ: ܕܬܝܡܢܐ P || 5 post ܘܐܡܪܝܢ: add. ܐܝܟܢܐ ܕܐܠܗܘܢ ܪܒܝܢ P]

(2) But one might wonder why <u>the south winds</u> too <u>are not etesian</u> as <u>the north winds are etesian</u>, (1) since <u>the same efficient cause</u> [*ʿāḇōḏā*] <u>and the same material</u> <u>are found in both directions</u> - the efficient cause [being] <u>the sun</u> and the material [being] <u>snow</u>. We say: the south winds are perhaps etesian, but escape our notice [*ṭāʿyān lan*, < gr. λανθάνω], (6) because, <u>as they blow from far away</u>, (7) <u>they are weakened</u>, (10) so that <u>those in that other inhabited world opposite us</u> <u>wonder why the south winds are etesian but the north winds are not</u> etesian.

Nic. syr. 31-10-15[39]:

ܬܝܡܢܐ ܕܝܢ/11 ܡܬܝܕܥܐ ܕܠܘܬܢ ܠܘ ܡܢ ܦܢܝܬܐ ܗܝ ܕܠܘܩܒܠ ܓܪܒܝܐ ܢܫܒܐ ܒܠ/
12 ܕܡܬܡܚܠܐ ܡܛܠ ܪܘܚܩܗ ܕܐܘܪܚܐ: ܘܠܐ/13 ܡܡܛܝܐ ܕܢܬܦܫܛ ܠܘܬܢ.
ܗܟܢ ܐܦ ܬܝܡܢܐ ܓܪܒܝܬܐ. ܟܕ ܡܢ/14 ܦܢܝܬܐ ܕܓܪܒܝܐ ܢܫܒܐ.
ܡܬܡܚܠܐ ܡܛܠ ܪܘܚܩܗ ܕܐܘܪܚܐ: ܘܠܐ ܡܬܡܛܝܐ/15 ܠܟܠܗ ܥܡܪܬܐ ܕܝܠܢ.

[= Paris 346, 70v 20-26: 10 ante ܬܝܡܢܐ: add. ܟܕ ܗܟܘ ܐܡܪ ܠܘܬܗܘܢ ܐܝܟ ܕ ܠܐܬܝܣܝܐ ܬܝܡܢܐ P || 10 ܕܝܢ: ܟܕ P || 12 & 14 ܪܘܚܩܗ P]

(4) The south wind we know [*da-lwāṯan*] blows not from the region opposite the north [i.e. south pole], (7) because [the wind from there] <u>is weakened</u> (6) because of the long distance it has to travel [lit. the farness of the way] (8) and can<u>not reach us</u>. For the north wind too, although it blows from the region of the north, is weakened because of the distance it has to travel, and does not reach the whole of our inhabited world.

Šifāʾ 65.13: وأما فى الشتاء فلا يتفق أن يبلغنا ما حدثت من الرياح الجنوبية لبعد المسافة

(4) <u>In winter</u> whatever south winds there occur (8) <u>do not reach us</u> (6) because of the great distance [lit. farness of the distance].

Šifāʾ 65.17-66.1:

وكل ريح فإن قوتها فى البلاد التى تبتدئ منها، وضعفها فيما يقابلها. وأكثر الرياح هى الشمالية
والجنوبية، لوفور المواد عند كل واحد من القطبين، المواد المعدة بترطيبها الأرض لتصعيد الأدخنة،
واستحالتها رياحا.

(9) Each wind is at its strongest in [lit. its strength is in] in the country from which it starts and is at its weakest in [lit. its weakness is in] (the place) which it opposite it. (1) The most winds are northerly and southerly, because of the abundance of the materials at both <u>the poles</u>, the materials

[38] Cf. Arist. 362a 11-16; *Prob*. 26.2; Olymp. 177.13-18; Olymp. arab. 121.3-11.

[39] Cf. Arist. 362b 30-363a 9; Olymp. 184.24-33; Olymp. arab. 121.3-11.

which prepare for vaporisation of smoke and its transformation into winds by moistening the earth.

l. 10. ܐܢ ܐܝܬ ܬܡܢ ܥܡܘܪܬܐ ("if there is a habitable world there"): For the discussion of whether there is a habitable world in the southern hemisphere, see Min. IV.i.4, ii.2 above.

III.iii.6.1-7: East and west winds

But.: (1) In summer heat abounds [*sāḡyā*] in the east and dries up the exhalation in advance, (2) whereas in the west heat is moderate and causes much vapour to rise. (3) For this reason in summer east winds blow less and west winds more. (4) In winter heat is small in the west and does not cause much exhalation to rise, (5) whereas in the east heat is moderate and produces much exhalation. (6) For this reason in winter west winds blow less, and east winds more. (7) Autumn is like summer and spring is like winter.

Nic. syr. 34.18-19, 23-28[40]:

ܢܫܒܢ ܕܝܢ ܪ̈ܘܚܐ/19 ܡܢ ܡܕܢܚܝܬܐ ܒܬܪܐܐ ܘܒܣܬܘܐ.//23 ܪ̈ܘܚܐ ܕܝܢ
ܡܥܪ̈ܒܝܬܐ/24 ܒܬܫܪܝܬܐ ܘܒܩܝܛܐ. ܒܩܝܛܐ ܡܢ ܓܝܪ ܣܓܝܐܐ ܫܚܝܢܘܬܐ
ܒܡܕܢܚܐ/25 ܘܡܝܒܫܐ ܡܩܕܡܐ ܠܐܛܪܐ. ܒܡܥܪܒܐ ܕܝܢ ܡܡܨܥܝܐ ܘܡܣܩܐ
ܠܗܠܐ/26 ܣܓܝܐܐ. ܡܛܠ ܗܢܐ ܪ̈ܘܚܐ ܡܢ ܡܕܢܚܝܬܐ ܠܐ ܢܫܒܢ: ܡܥܪ̈ܒܝܬܐ
ܕܝܢ ܢܫܒܢ./27 ܒܣܬܘܐ ܕܝܢ ܡܡܨܥܝܐ ܫܚܝܢܘܬܐ ܒܡܕܢܚܐ. ܘܡܫܟܚܐ ܕܬܣܩ
ܐܛܪܐ/28 ܣܓܝܐܐ. ܒܡܥܪܒܐ ܕܝܢ ܙܥܘܪܐ ܘܠܐ ܡܫܟܚܐ ܕܬܫܒ.

[= Paris 346, 63r 5-12: 19 ܒܬܪܐܕܐ P || 24 ܣܓܝܐ P (et *But*.) || 28 ܙܥܘܪܐ: ܒܥܘܪܐ P || ܕܬܫܒ: ܕܬܫܒܐ P]

(7) East winds blow in spring and winter, and west winds in autumn and summer. (1) For in summer, heat is great [*saggīʾā* C, but *sāḡyā* P] in the east and dries up the exhalation in advance, (2) whereas in the west it is moderate and causes much vapour to rise. (3) For this reason east winds do not blow, but west winds blow. (4) In winter, (5) heat is moderate in the east and is able to produce much exhalation, (4) whereas in the west it is small [*zʿoryā*] and is unable to cause [winds] to blow.

l. 2. ܗܠܐ ("vapour"): sic here, as opposed to "exhalation" (*eṭrā*) in both Nic. syr. and *But*.

Mete. III. Section iv.: *On Cloud Wind, Whirlwind, Tornado and Eddy Wind*

In Arist. the main discussion of "cloud wind" (ἐκνεφίας), whirlwind etc. is found at *Mete*. III.i, separated from the main discussion of winds (*Mete*. II.iv-vi) by the discussions of earthquakes (II.vii-viii)

[40] Cf. Arist. 364b 23-24; Olymp. 199.4-8, 198.21-24; Olymp. arab. 131.12-20.

and thunder and lightning (II.ix),[41] and this is also the order followed in Nic. syr. - In IS's *Šifā᾽*, "cloud wind" (*rīḥ saḥābīya*) and whirlwind are treated in the middle of the chapter on winds, before the discussion the directions from which the winds blow (corr. *But.* Mete. III.ii above).[42] - The placement of the discussion here at the *end* of the chapter on winds in *But.* might be considered a compromise between the two.

Within the section, the order followed is basically that of the *Šifā᾽*, although for the actual content much use is made of Nic. syr. - The first 'theory' (III.iv.1) deals with the "cloud wind" (*rūḥā ʿnānāytā*), the remaining three (III.iv.2-4) with "whirlwinds" (τυφῶν, *zaubaʿa*, *ʿalʿālā*). - BH follows IS in giving four ways in which whirlwinds are generated. Of these the fourth type (III.iv.4.3-7, simple meeting of two winds) has no counterpart in Nic. syr. The second type (III.iv.3.1-3, collision of wind with earth and subsequent collision with another wind) corresponds most closely in meaning to that described at Nic. syr. 44.26-45.4 (where, however, Nic. syr. says "a firm object" rather than the "earth"), but that passage of Nic. syr. is used by BH in describing his first type (III.iv.2.1-5, collision of wind with a thick cloud and subsequent collision with another wind).

	Šifā᾽	Nic. syr.	
III.iv.1.1-3	60.12-15		"Cloud wind"
III.iv.1.3-7	60.15	44.18-24	(contd.)
III.iv.2.1-5	60.16-17	44.26, 44.30-45.4	Whirlwind (1): collision with cloud
III.iv.2.5-9		45.8-13	Whirlwind and other winds
III.iv.3.1-3	61.1-2		Whirlwind (2): wind striking earth
III.iv.3.3-6	61.3-4	45.13-15, 17-18	Ascent and descent
III.iv.3.6-8	61.4-5		Perseverance of whirlwinds
III.iv.4.1-3	60.17-61.1	45.4-8, 15-17	Whirlwind (3): from torn cloud
III.iv.4.3-7	61.6-8	44.23-26, (35.6-9)	Whirlwind (4): from two winds

III.iv.1.1-3: "Cloud wind" (ἐκνεφίας)

But.: (1) People today [lit. sons of our age, *bnay dāran*] call wind which generates cloud "cloud wind" [*rūḥā ʿnānāytā*]. (2) The ancients [*qadmāyē*] call not this, but wind which is ejected [*mezdanqā*] from a dense [*rṣīptā*] cloud as it bursts [*metpatqā*] "cloud wind".

---

[41] The "cloud wind" (ἐκνεφίας) is also mentioned in the section on winds (*Mete.* II.vi, 365a 1ff.).

[42] The order is retained by Fakhr al-Dīn al-Rāzī, who treats the "cloud wind" and whirlwind" in the middle of his section on winds at *Mabāḥit* II.193.14-194.10 and this order is followed by BH in his *Cand.* (123-127 on winds; 125.3-12 on whirlwinds).

Šifāʾ 60.12-15:

والرياح المولدة للسحاب تسمى رياحا سحابية، واسم الرياح السحابية يقع فى الأكثر بحسب عادتنا على هذه الرياح. وقد يقال رياح سحابية، وخصوصا فى القديم، لما كان من الرياح ينفصل عن السحاب إلى ناحية الأرض، ولأنها منضغطة مقسورة فهى قوية العصف جاعفة مغرقة.

(1) Winds which generate clouds are called "cloud winds" [*riyāḥ saḥābīya*] and the name "cloud winds" usually applies according to our usage [*bi-ḥasabi ʿādati-nā*] to these winds, (2) but "cloud winds" was said, especially in ancient times [*fī al-qadīm*], of the type of winds which was discharged [*yanfaṣilu*] from clouds in the direction of the earth. Because they are compressed and constrained [*munḍaġiṭa maqsūra*], they blow violently and are sweeping are overwhelming.

l. 1, 2-3. ܪܘܚܐ ܥܢܢܝܬܐ ("cloud wind"): رياحا سحابية IS. - This corresponds to Aristotle's ἐκνεφίας (365a 1, 3 etc.), which is regularly rendered as *rūḥā ʿnānāytā* in Nic. syr. (9 times; once as *ʿnānāytā rūḥā*; twice simply *ʿnānāytā* with ellipsis of *rūḥā*).

l. 2. ܕܡܢ ܥܢܢܐ ܪܨܝܦܬܐ ܟܕ ܡܬܦܬܚܐ ܡܬܕܚܝܐ ("which is ejected from a dense cloud as it bursts"): The two verbs used by BH here, as well as the adjective *rṣīpā*, occur together in a passage of Nic. syr. dealing with the whirlwind rather than the "cloud wind": Nic. syr. 45.4-5 (see under III.iv.4.1-3 below): ‹ܡܢ› ܠܥܠ/ ܕܝܢ ܗܘܝܢ ܟܪܟܘܬܗ . ܟܕ ܡܬܦܬܚܐ ܘܡܬܕܚܝܐ ܡܢ ܥܢܢܐ ܕܛܒ ܪܨܝܦܐ ("[The whirlwind's] spirals are from above when it bursts out and is ejected from a cloud which is very dense ...").

III.iv.1.3-7: "Cloud wind" (contd.)

But.: (1) They say that it is severer [*ʿasqā*] than other winds (2) because (other winds) [lit. those] occur with their material flowing little by little, (3) whereas (the cloud wind) [lit. this] is released suddenly, as from some reservoir. (4) (Other winds) occur when smoky exhalation strikes moving air and is reflected, (5) whereas (the cloud wind) [lit. this] occurs when it is driven out of a cloud by force (6) and for this reason blows violently.

Nic. syr. 44.18-24[43]:

ܪܘܚܐ/ 19 ܗܟܝܠ ܥܢܢܝܬܐ ܥܣܩܐ ܚܣܝܢܐ ܘܥܫܝܢܐ ܡܢ ܟܠܗܝܢ ܪܘܚܐ ܕܐܬܐܡܪ̈ܝ .
ܗܢܝܢ/ 20 ܡܢ ܓܝܪ ܟܕ ܩܠܝܠ ܩܠܝܠ ܪܕܝܐ ܗܘܠܐ ܕܗܝܢ . ܗܕܐ ܕܝܢ ܟܕ ܡܫܬܒܩܐ
ܡܬܬܣܩܪܐ/ 21 ܐܝܟ ܕܡܢ ܐܘܨܪܐ ܡܕܡ . ܘܗܢܝܢ ܗܘܝܢ ܡܢ ܟܕ ܢܩܫ ܥܛܪܐ ܒܐܐܪ ܡܬܬܙܝܥܢܐ
ܘܡܬܦܢܝܢ/ 22 ܘܚܠܦܐ ܗܘܝܢ : ܗܠܝܢ ܟܕ ܡܕܡ ܐܝܬܝܗ ܕܪܥܐ ܠܗܘܢ
ܒܩܛܝܪܐ . ܪܘܚܐ ܕܝܢ/ 23 ܗܕܐ ܥܢܢܝܬܐ . ܟܕ ܒܩܛܝܪܐ ܡܕܡ ܡܬܕܚܝܐ ‹ܡܢ›
ܥܢܢܐ ܗܘܬ . ܘܥܙܝܙܐ ܘܩܛܝܪܐ/ 24 ܐܘܚܕܢܐ ܕܬܪܥܝܬܐ ܐܢܢܝܐ ܘܡܣܬܥܪ ܐܠܦܐ .

[19 ܕܐܬܐܡܪܝ cod. || 20 ܗܘܠܐ cod. || 22 ܕܪܘܚܐ cod. || 23 ‹ܡܢ› supplevi, cf. *But.*]

(1) Cloud wind then is severer and stronger than all the winds mentioned. (2) For those occur with their material flowing little by little, (3) whereas

[43] Cf. Arist. 370b 9-10; Olymp. arab. 131.22-132.8.

the cloud wind [lit. this] [occurs] when it is released suddenly as from some reservoir. (4) Furthermore, they occur when smoky exhalation strikes moving air and is reflected, and not when something else drives them with force, (5) whereas this cloud wind occurs when it is driven out of a cloud by some force. (6) It is so violent that it loosens trees and overturns ships.

l. 4. ܐܝܟ ܕܡܢ ܐܘܨܪܐ ܡܕܡ ("as from some reservoir"): cf. III.i.3.6 above.

l. 6. ܘܡܛܠܗܕܐ ܒܚܝܠܐ ܢܫܒܐ ("and for this reason blows violently"): This addition at the end, which has no counterpart in Nic. syr., probably answers to IS's *fa-hiya qawīyat al-ʿaṣf* at the end of the passage quoted under III.iv.1.1-3 above (*Šifāʾ* 60.15).

III.iv.2.1-5: Whirlwind (1): from collision with cloud

But. [underline: agreement with Nic.; italics: with *Šifāʾ*]:

(1) If *a cloud wind*, *which is heavy and moist* and is sliding [*meštarglā*] *downwards* with force, (2) *collides with* [*nāqšā b-*] *a* thick *cloud*, (3) because of (the cloud's) density [*ʿabyūṯāh*] it does not pass through it (4) and because of its own weight does not turn back, (5) but it is carried sideways [*l-gābbā mettaytyā*]. (6) When on the side too it collides with a strong wind, (7) it *produces a downward spiral* [*krākṯā*]. (8) In this way, from a cloud wind a whirlwind [*ʿalʿālā*], which is severer [*ʿseq*] than it, is generated.

Nic. syr. 44.26, 44.30-45.4[44]:

26 ܢܫܒܐ ܕܝܢ ܗܢܐ//30 ܕܪܘܚܐ ܡܬܟܪܟܢܝܬܐ/31 ܟܕ ܡܢ ܠܥܠ ܐܝܬܝܗ ܠܬܚܬ ܒܟܪܟܬܗ
ܡܢ ܠܬܚܬ. ܐܠܐ ܡܢ ܠܬܚܬ ܗܘܝܢ ܟܘܪܟܝܗ//1 ܡܐ ܕܢܚܬܐ ܠܬܚܬ ܘܦܓܥܐ
ܒܓܘܫܡܐ ܡܕܡ ܫܪܝܪܐ. ܘܟܕ ܠܐ ܡܨܝܐ/2 ܕܬܥܒܪ ܒܗ ܡܛܠ ܥܒܝܘܬܗ.
ܐܦܠܐ ܕܬܬܗܦܟ ܡܛܠ ܕܚܝܐ ܕܪܘܚܐ/3 ܕܕܚܝܐ ܠܗ. ܡܬܬܝܬܝܐ ܠܓܒܐ. ܘܟܕ ܡܢ
ܓܒܐ ܗܘ ܬܦܓܘܥ ܒܪܘܚܐ ܚܝܠܬܢܝܬܐ/4 [ܘ]ܚܫܐ ܟܕ ܚܫ ܗܘ ܐܚܪܢܐ +ܟܕ ܗܘܐ ܕܝܢ
ܡܣܬܠܩܐ ܐܚܪܢܐܝܬ+ ܥܒܕܐ ܟܪܟܬܐ.

This spiralling wind blows with its spiral [descending] from above and [ascending] from below, but its spirals are from below, (1) when it descends downwards (2) and collides with a firm object [*gušmā meddem šarrīrā*], (3) and being unable to pass through it because of its own density [*ʿabyūṯāh*] (4) or to turn back because of the propulsion of the wind which is pushing it, (5) it is carried sideways. When on the side too it collides with a strong wind and also undergoes the same, +while this [latter] is affected in a different way [?]+ (7) it produces a spiral.

Šifāʾ 60.16-17:

والزوبعة أكثرها من الريح الساحبية الثقيلة الرطبة التى تندفع إلى فوق فتصدم سحابة فتلويها وتصرفها فتستدير نازلة؛ وهو أرداها.

[16 فوق ed. Cair.: أسفل ed. Ṭ et DSM ‖ 17 فتلويها ed. Ṭ: فتلونها ed. Cair.]

44 Cf. Arist. 370b 17-28; Olymp. 200.6-14, 200.21-201.10, 204.24-205.3, 205.6-24; Olymp. arab. 142.19-143.5.

Most whirlwinds [*zaubaʿa*] arise (1) from heavy and moist cloud wind which is driven upwards [downwards ṬDSM], (2) then collides with [*taṣdimu*] a cloud. (The cloud) twists and turns it [*tulawwī-hā wa-tuṣarrifu-hā*], and it makes a downward spiral [lit. it circulates while descending, *tastadīru nāzilatan*]. (8) This is the worst of them.

l. 5. "whirlwind, which is severer than it": This is taken from Nic. syr. 44.24. [see under III.iv.4.3-7 below]: ܚܣܝܢ ܕܝܢ ܡܢ ܗܕܐ ܥܠܥܠܐ ("Whirlwind is severer than this [sc. cloud wind]"); cf. Olymp. 201.20-22: λοιπὸν εἴπωμεν καὶ τὰς διαφορὰς ἐκνεφίου καὶ τυφῶνος. ἡ γὰρ κοινωνία κατάδηλος· οὐδὲν γὰρ ἄλλο ἐστὶν ἐκνεφίας ἢ τυφὼν ἀνήνεμος, καὶ τυφὼν οὐδὲν ἄλλο ἐστὶν ἢ ἐκνεφίας ἐπιτετάμενος. ("... whirlwind is nothing other than an extreme cloud wind").

III.iv.2.5-9: Whirlwind and other winds

But.: (1) Whirlwind does not occur with the north wind, nor cloud wind with rain. (2) Cloud wind is finer [than whirlwind], (3) so that it requires little coldness for it not to occur, (4a) whirlwind much coldness (4b) because it is denser. (5) For the heat which is in fine exhalation is cooled more easily than that in dense [exhalation].

Nic. syr. 45.8-13[45]:

ܥܡ ܓܪܒܝܐ ܕܝܢ ܒܥܢܢܝܬܐ/9 ܠܐ ܗܘܐ ܥܠܥܠܐ. ܐܦܠܐ ܥܡ ܡܛܪܐ ܓܪܒܝܐ
ܥܢܢܝܬܐ. ܓܪܒܝܐ ܓܝܪ ܥܢܢܝܬܐ ܩܛܝܢܐ ܗܝ/10 ܡܢ ܥܠܥܠܐ. ܥܠܥܠܐ ܕܝܢ ܥܒܝܐ ܗܝ.
ܒܨܝܪܐ ܓܪܒܝܐ ܓܝܪ ܥܢܢܝܬܐ ܠܐ ܩܪܝܪܘܬܐ/11 ܣܢܝܩܐ ܗܘܝܐ. ܠܥܠܥܠܐ ܕܝܢ ܣܓܝܐܐ
ܩܪܝܪܘܬܐ. ܥܠܥܠܐ ܕܝܢ ܠܐ ܩܪܝܪܘܬܐ ܡܣܬܢܩܐ./12 ܚܡܝܡܘܬܐ ܓܝܪ ܕܒܠܛܝܦܬܐ
ܡܠܦܠܐ ܗܝ ܕܠܥܠܥܠܐ ܡܬܩܪܪܐ ܡܢ ܗܝ ܕܒܥܒܝܬܐ/13 ܒܥܓܠ.

[9 ܥܢܢܝܬܐ ܓܪܒܝܐ : ܓܪܒܝܐ ܥܢܢܝܬܐ cod.]

(1) Whirlwind does not occur with the north wind, nor cloud wind with rain. For cloud wind is finer, (4b) while whirlwind is denser, (3) so that cloud wind requires little coldness in order [*lwāṯ hāy d-*] for it not to occur, (4a) whirlwind much coldness. (5) For the heat in fine exhalation is cooled more easily than that in dense exhalation.

III.iv.3.1-3: Whirlwind (2): wind striking earth

But.: (1) Whirlwind is also generated from windy material [*hūlē rūḥāytā*] (2) which is driven downwards (3) and, colliding with [*nāqšā b-*] the earth, (4) turns back [*ʿāṭpā*], (5) meets another wind of the same kind [*bat gensāh*] (6) and forms spirals [lit. undergoes convolutions, *krāḵāṯā ḥāšā*].

Šifāʾ 61.1-2:

وربما كانت الزوبعة من مادة ريحية هبطت إلى أسفل، وقرعت الأرض، ثم انثنت، فلقيتها ريح أخرى من جنسها فلوتها.

[45] Cf. Arist. 371a 3-9; Olymp. 201.30-40, 207.3-7.

(1) Often the whirlwind arises from windy material [*mādda rīḥīya*] (2) which falls downwards (3) and collides with [*qaraᶜat*] the earth. (4) It then turns back [*intanat*]. (5) Another wind of the same kind [*min ğinsi-hā*] meets it (6) and twists it.

III.iv.3.3-6: Signs of ascending and descending whirlwinds

According to Olymp. *in Mete*. 201.13-16, it is the ascending (κάτωθεν) whirlwinds whose spirals ascend and descend, while the descending (ἄνωθεν) whirlwinds have only descending spirals. As observed by Lettinck (1999) 179 n. 64, this is reversed in Olymp. arab. and *Šifāʾ*, as well as in Nic. syr. and *But*.

But. [underline: agreement with Nic.; italics: with *Šifāʾ*]:
(1) *The sign of a descending whirlwind is* (2) *that* it ascends and descends *simultaneously* with *its* spirals *like a dancer* [lit. one who is dancing]. (3) It seizes the object it meets but does not push [it] away. (4) *The sign of an ascending whirlwind is* (5) *that only* ascent *is observed in its spirals*. (6) It first pushes away the object, then [*ken*] seizes [it].

Šifāʾ 61.3-4:

وعلامة الزوبعة النازلة أن تكون لفائفها تصعد وتنزل معا، كالراقص. وعلامة الصاعدة أن لا ترى للفائفها إلا الصعود.

[4 ترى ed. Cair.: يرى ed. Ṭ]
(1) The sign of a descending whirlwind is (2) that its spirals [*lafāʾif*] ascend and descend simultaneously like a dancer. (4) The sign of an ascending [whirlwind] is (5) that you only see ascent in its spirals [or: that only ascent is seen in its spirals].

Nic. syr. 45.13-15, 17-18[46]:

ܘܐܢܗܘ ܕܡܢ ܠܥܠ ܢܣܒ ܥܠܥܠܐ ܡܬܟܪܟܢܘܬܐ. ܢܠܒܟ ܠܡܦܓܥܢܐ/14 ܕܦܓܥ ܒܗ̇. ܠܐ ܕܝܢ ܐܦ ܕܚܐ. ܐܚܪܝܢ ܐܦ ‹ܡܢ› ܠܬܚܬ ܡܩܕܡܐ ܕܚܐ ܘܡܣܬܪܝܘ ܢܠܒܟ./15 ܘܐܢ ܗܘ ܡܢ ܠܥܠ ܡܢ //17 ܢܚܬܢ/18 ܟܪܟܘܬܐ ܘܣܠܩܢ. ܐܚܪܝܢ ܡܢ ܠܬܚܬ ܣܠܩܢ ܘܠܐ ܢܚܬܢ:

[14 ܡܢ: supplevi, ܐܦ fort. delendum]
(1) If the whirlwind receives its spiralling [*metkarkānūṯā*] from above, (3) it seizes the objects it meets but does not push [them] away; (4) if from below, (6) it first pushes away and then [*hāydēk*] seizes. (1) If from above, (2) the spirals descend and ascend; (3) if from below, (5) they ascend but do not descend.

l. 5. ܡܬܚܙܝܐ ("is observed"): ترى ("you see") IS ed. Cair.: يرى ("[ascent] is seen") ed. Ṭ; cf. Olymp. arab. 143.13 رأيت. - Given the correspondence with

[46] Cf. Olymp. 201.10-20; Olymp. arab. 143.5-9, 12-14. - The *Šifāʾ* allows us to restore the two words Badawī was unable to decipher in lines 143.13 and 14 of his edition of Olymp. arab. (fol. 358b 12 in the manuscript; lege 143.13 رأيت لفائفها and 143.13-14 تلك اللفائف).

raʾaita of Olymp. arab., what IS actually wrote was probably *tarā*; BH's rendition here suggests that he read the word as *yurā*.

l. 3. ܐܝܟ ܗܘ ܕܪܩܕ ("like a dancer"): كالراقص IS. - It is tempting to posit a misreading of the verb χωρέω as χορεύω behind this phrase, although we find no support for this in Nic. syr. or Olymp. arab. 143.13-14. - Cf. Olymp. 201.13-14: ἐπὰν τὴν ἕλικα ὁρῶμεν ποτὲ μὲν κάτω, ποτὲ δὲ ἄνω χωροῦσαν.

III.iv.3.6-8: Perseverance of whirlwinds

But.: (1) The configuration of the spiral [*sukkām metkarkānūtā*] perseveres [*maggar*] in (the whirlwind) (2) because of [*mettul*] the weight of its nature [*yaqqīrūtā da-kyāneh*] and the solidity of its constitution [*qṭīrūt quyyāmeh*] due to [*mettul*] its moisture, (3) since if its constitution were fine [lit. if it were fine of constitution], its shape [*eskēmā*] would quickly leave it.

Šifāʾ 61.4-5:

وإنما يعرض لها كل ذلك التشكل، ثم يلزمها، لثقل طبعها، وثخونة جوهرها، لرطوبتها. ولو كانت لطيفة، لم يلزمها ذلك التشكل.

[4 التشكل edd. Cair. et Ṭ: الشكل SM ‖ 5 التشكل ed. Cair.: الشكل ed. Ṭ et cod. M]

(1) All this configuration [*tašakkul*] occurs to (the whirlwind) and perseveres [*yalzamu*] in it (2) because of [*li-*] the weight of its nature [*tiql ṭabʿi-hā*] and the solidity of its substance [*tuhūnat ǧauhari-hā*], because of [*li-*] its moisture. (3) If it were fine that configuration [*tašakkul*; or: shape (*šakl*)] would not persevere in it.

III.iv.4.1-3: Whirlwind (3): from torn cloud

But. [underline: agreement with Nic.; italics: with *Šifāʾ*]:

(1) Whirlwind also arises when a cloud is torn [*mettalḥā*] (2) and, because of the *crookedness* [*ʿqalqlūtā*] of those openings through which it escapes, (3) the wind comes out in a spiral [*metkarkānāʾīt*], (4) *like hair which is made curly* [*metʿasqas*] *by the tortuosity* [*ptīlūtā*] *of the pores through which it grows* [lit. pores of its growth (*mawʿītā*)].

Nic. syr. 45.4-8[47]:

‹ܡܢ› ܠܥܠ/5 ܕܝܢ ܡܩܝܡ ܒܪܒܥܘܬܗ. ܒܪ ܡܬܟܪܟܢܘܬܐ ܘܡܩܕܡܘܬܐ ܡܢ ܚܝܠܐ ܕܝܩܝܪܘܬܗ
ܥܢܢܐ ܐܦ ܗܘ/6 ܡܬܬܠܚܐ ܕܐܬܦܠܓ ܒܬܪܒܥܘܬܗ: ܥܠ ܕܝܩܝܪ ܒܟܝܢܐ. ܗܘܐ ܕܝܢ ܐܦ ܠܥܠܐ
ܕܬܬܦܠܓ ܡܬܥܣܩܐ/7 ܡܛܠ ܡܬܥܩܡܘܬܐ ܕܦܘܪܐ ܕܢܫܒܐ ܠܗ. ܐܝܟ ܕܒܣܥܪܐ ܚܠܦ ܐܝܟܐ
ܒܪܝܐ ܕܐܬܥܩܡ. ܘܡܢ/8 ܐܕܢܝ ܡܫܟܚܐ ܢܬܐܠܐ ܡܢܐ ܡܬܥܩܡܘܬܐ
ܡܬܦܬܠܢܘܬܐ.

[5 ܒܪܒܥܘܬܗ cod. ‖ 6 ܠܬܪܒܥܘܬܗ cod. ‖ 7 ܡܬܥܣܩܘܬܐ cod.corr.: ܦܬܝܠܘܬܐ cod.1]

---

[47] Cf. Arist. 370b 28-371a 1; Olymp. 200.23-201.10, 206.7-10, 206.12-15: Olymp. arab. 142.19-143.1.

Its spirals are from above (1) when it bursts out and is ejected [*metpatqā w-mezdanqā*] from a cloud which is very dense and is unable to pass through the rest of (the cloud) because it is very thick or to be reflected because of the continuity of the wind which is pushing it. As a result it is reflected in the manner mentioned, and when it undergoes this many times (3) it is shaped into a spiral [lit. spirally, *metkarkānā'īt*].

Nic. syr. 45.15-17 [scholion on 45.8 *mettappsā metkarkānā'īt*][48]:

ܗ̄ ܐܝܟ ܡܢ̈ܝܐ ܕܣܥܪܐ ܕܟܕ ܡܫܬܕܪܝܢ ܡܬܡܛܝܢ ܠܡܫܟܐ / ܘܡܛܠ ܥܒܝܘܬܗ ܠܐ ܡܨܝܢ ܕܬܪܝܨܐܝܬ ܢܥܒܪܘܢ. ܟܕ ܠܣܛܪܐ ܡܬܪܟܢ ܚܝܠܐ ܕܡܘܙܚܐ / ܕܙܘܥܗܘܢ. ܐܣܟܡܐ ܥܩܠܩܠܐ ܘܡܬܟܪܟܢܐ +ܢܩܒܥܝܢ ܠܣܥܪܐ ܘܠܗܘܢ+.

(3) i.e. like strings [*mennē*] of hair, which, because as they are sent out they reach the skin but due to (the skin's) density cannot pass straight [through it], as the force of their movement inclines to the side, they [?] impart a crooked [*ʿqalqlā*] and spiral shape to the hair and to them [?].

Šifā' 60.17-61.1:

وربما زادها تعرج المنافذ التفافا وتلولبا، كما يعرض للشعر أن ينجعد بسبب التواء منبته من المسام.

(2) Often the crookedness [*taʿarruǧ*] of the passages increase its twistedness [*iltifāf*] and spiralling [*talaulub*], (4) just as it happens to hair that it is made curly [*yanǧaʿidu*] by reason of the tortuosity [*iltiwā'*] of its place of growth [*manbat*] in the pores.

III.iv.4.3-7: Whirlwind (4): from two winds

But. [underline: agreement with *Šifā'*; italics: with Nic.]:

(1) Whirlwind also arises from the meeting of two severe winds (2) and [this] is called "qarḥā". (3) It often uproots [*ʿāqar*] *trees* from the earth (4) and *snatches* [*ḥāṭep̄*] *ships* from the sea. (5) When it fastens onto [*sābek b-*] a part [*mnāṯā*] of a cloud it appears like a dragon [*tannīnā*] flying through air. (6) Also from the meeting of two winds which are not very strong a whirlwind arises, (7) and [this] is called "zīqā".

Šifā' 61.6-8:

وقد تحدث الزوبعة أيضا من تلاقى ريحين شديدتين أو غير شديدتين. وربما كانت شديدة قوية ثابتة تقلع الأشجار وتختطف المراكب من البحر. وربما اشتملت على طائفة من السحاب أو غيره فترى كأنه تنينا يطير فى الجو.

(1) Whirlwind also sometimes arises from the meeting of two severe winds (6) or those that are not severe. Often they are severe, strong and enduring, (3) and uproot [*taqlaʿu*] trees (4) and snatch [*taḫtaṭifu*] ships from the sea. (5) Often it implicates [*ištamala*] a portion [*ṭā'ifa*] of a cloud or something else and looks as if it were a dragon [*tinnīn*] flying through the air.

Nic. syr. 44.23-26[49]:

ܘܗܘܝܐ ܡܬܝܠܕܐ/24 ܐܝܟܢܐ ܕܬܩܠܥ ܐܝ̈ܠܢܐ ܘܬܚܛܘܦ ܐܠ̈ܦܐ. ܚܣܡܐ ܕܝܢ ܡܢ ܗܢܐ ܥܠܥܠܐ. ܗܢܘ/25 ܕܝܢ ܟܒܪܐ ܕܩܘܪ̈ܝܢ ܐܦ ܐܝ̈ܠܢܐ ܡܢ ܥܩܪܝܗܘܢ ܫܠܦ. ܘܐܝ̈ܠܢܐ ܡܢ ܐܪܥܐ/26 ܕܥܩܪ. ܘܡܬܚܙܐ ܐܝܟ ܕܪܩܘܢܐ.

[48] Cf. Olymp. 201.5-9, 13.8-11.

[49] Cf. Olymp. 13.12-18, 200.12-16; Olymp. arab. 143.9-12.

(Cloud wind) is so violent that it loosens trees and overturns ships. Whirlwind is severer than this [i.e. cloudy wind]. (3-4) For (whirlwind) often snatches up [*ḥāṭep̄*] trees together with their roots and ships with those in them, and water as if in pipes.

Nic. syr. 35.6-9 [in section on winds, rather than whirlwinds][50]:

ܐܝܬ ܕܝܢ ܐܕܫܐ ܡܕܡ ܕܪܘܚܐ 7/ ܥܫܝܢܬܐ ܕܩܪܝܢ ܠܗ ܡܠܚܐ ܫܡܝܢܝܬܐ ܘܣܝܦܘܢ.
ܡܢ ܕܕܡܝܘܬ 8/ ܐܒܘܒܐ ܕܣܠܩܝܢ ܒܗܘܢ ܡܝܐ. ܘܗܟܢ ܥܫܝܢܐ. ܕܫܠܦܐ ܐܠܦܐ
ܥܡ 9/ ܓܒܪܐ ܘܐܝܠܢܐ ܘܥܡ ܥܩܪܝܗܘܢ.

[= Paris 346, 63r marg.: 7 ܠܗ: om. P || ܫܡܝܢܝܬܐ sic P: ܫܡܝܢܘܬܐ C; cf. οὐρανία Olymp. 13.16; السماوية Olymp. arab. 132.1]

There is a certain species of strong wind, which sailors call "heavenly" and "siphon" on account of [its] similarity to pipes through which water rises; it is so violent that (3-4) it snatches up a ship together with the men [on board], and trees together with [lit. and] their roots.

l. 4, 7. "qarḥā": The term *qarḥā* does not occur in Nic. syr. It does occur in *De mundo* syr., where it is coupled with *kōḵīṯā* (143.26 *qarḥā dēn w-ḵōḵīṯā* < gr. 395a 7 λαῖλαψ καὶ στρόβιλος) and is differentiated from *ᶜalᶜālā* (143.25 < gr. 395a 6 θύελλα). - According to *De mundo* syr., *ᶜalᶜālā* is an ascending whirlwind (no doubt due to misreading of gr. ἄφνω as ἄνω), while *qarḥā* and *kōḵīṯā* ascend and descend (misinterpretation of gr. κάτωθεν ἄνω). - BH has these three terms in the heading (125.3) of the section on whirlwinds in *Cand*., where he applies the terms *qarḥā* and *ḵrōḵīṯā*[51] to the type of whirlwind produced by the collision of a descending wind with a cloud (125.3-7, the process described < *Mabāḥiṯ* II.193.17-21 < *Šifāʾ* 60.16-17, corr. type (1), III.iv.2.1-5 above) and *ᶜalᶜālā* to an ascending whirlwind (125.7-8; i.e. as in *De mundo* syr.).[52] He further tells us that *kōḵīṯā* is also produced by descending wind rebounding from the earth and meeting another

---

[50] Cf. Olymp. locc. citt. supra; Olymp. arab. 132.1-2.

[51] Sic with *rēš* here edd. Bakoš (125.7) and Çiçek (95.24; < ms. Jerusalem), though with variant ܟܘܟܝܬܐ in ms. B. At *Cand.* 125.3 and 125.8, Bakoš writes ܟܘܟܝܬܐ without *rēš*, reporting no variants at 125.3, but with variant ܟܘܪܟܝܬܐ in ms. V at 125.8 (here also with *rēš* in ed. Çiçek 95.27). - *Qarḥā wa-krōḵīṯā* is also read in BH *Rad*., ed. Istanbul 23.15 (= Bodl. or. 467, 11r 15; similarly in other mss. accessible to me: Vat. syr. 169 [1330 A.D.] 14r 12; Paris syr. 213, 7r 7; Vat. Borg. syr. 145 [15/17th c.] 7r 6 has *w-ḵāruḵtā*, ܟܪܘܟܬܐ). - *De mundo* syr., though it has *qarḥā wa-kōḵīṯā* at 143.26, has the form with *rēš* at 138.9 (*w-rūḥē wa-krōḵyāṯā d-ᶜalᶜālē*; sic with *d-*; corr. gr. 392b 11 πνοαί τε ἀνέμων καὶ τυφώνων) and 146.30 (*w-mardīṯā tūḇ wa-ḵrōḵīṯā*; corr. gr. 396a 23 ῥοαί τε καὶ δῖναι). - Other than in *De mundo* syr. and BH *Cand.* and *Rad*., *krōḵīṯā/kārōḵīṯā* in the sense of "whirlwind" is known to the modern dictionaries only from older lexica (see PS 1826; Brockelmann 345).

[52] As was noted by Bakoš in his edition of *Cand*. (125f. n.1), Bar Kepha has *ᶜalᶜālā* and *qarḥā* the other way round in his *Hex*., *ᶜalᶜalā* being the descending whirlwind and *qarḥā* the ascending (ms. Paris 241, 191r b3-11; tr. Schlimme [1977] 625f., where "Wirbelwind" = *ᶜalᶜalā*; "Wirbelsturm" = *qarḥā*).

wind (< *Mabāḥiṯ* II.194.2-3 < *Šifāʾ* 61.1-2, corr. type (2), III.iv.3.1-3 above) and that the *qarḥē* uproot trees and seize ships (i.e. as here in *But.*). - The *De mundo* does not tells us anything about the damage caused by the *qarḥā* and *kōḵīṯā*. A possible source for the association of the *qarḥā* with whirlwinds that uproot trees is Jacob of Edessa, who tells us that *kōḵyāṯā w-qarḥē* (*Hex.* [Chabot] 85a 35 [ult.]) "uproot [*ʿāqrān*] firm plants from their roots" and "break large trees".[53]

l. 7. "zīqā": The ground on which BH applies the term *zīqā* to this type of whirlwind is unclear. - BH himself glosses "*b-yawmā d-zīqā*" of Ezech. 34.12 as "in the day of violent wind (*b-yawmā d-rūḥā ʿaššīntā*)" in his *Horr.* (Gugenheimer [1894] 26; tr. Dean [1930] 57). - The term *zīqā* occurs in Nic. syr. in a scholion explaining the difference between ἄνεμος and πνεῦμα, Nic. syr. 18.1-3: "Comment: '*zīqē*' [corr. ἄνεμοι] and all winds [*rūḥē*: πνεύματα]': He rightly said '*zīqē* and all winds', for winds are more general than *zīqē*; 'every *zīqā* is a wind, but not every wind is a *zīqā*'."[54] - On "zīqā" as a type of rain, see Mete. I.ii.3. above.

l. 5. ܬܢܝܢܐ ("sea monster"): تنين IS. - On the description of whirlwinds as *tinnīn* (dragon, large serpent, sea monster) in Arabic, see Sersen (1976) 116-122. - It is probably otiose to ask what exactly BH understood a *tinnīn*/*tannīnā* to be, since he is merely following IS here. - For BH's comment on the "*tannīnē rawrḇē*" of Gen. 1.21, see *Horr*. [Sprengling-Graham] 14.16-17: ܗ ܝܘܢܝܐ ܗܘ ܩܐܛܐ ܐܡܪ ܢܘܢܐ ܪܘܪܒܐ ܘܩܕܝܫܐ ܝܥܩܘܒ ܚܘܝܐ ܪܒܐ ܢܣܒ ܠܬܢܝܢܐ. ("The Greek [i.e. the Syrohexapla] says κήτη, i.e. huge fish. Holy Jacob [of Edessa] understands the *tannīnā* to be a great serpent").

---

[53] Jacob of Edessa, *Hex.* [Chabot] 85b 8-11: ܘܥܩܪܝܢ ܐܦ ܠܢܨܒܬܐ ܫܪܝܪ̈ܬܐ ܡܢ ܥܩܪ̈ܝܗܝܢ. ܘܡܬܒܪܝܢ ܠܐܝܠܢ̈ܐ ܪܘܪ̈ܒܐ ܒܥܫܝܢܘܬܐ ܠܦܘܚܐ.

[54] Cf. Alex. Aphr. 53.19-22; Olymp. 100.11-13. - The main text of Nic. syr. to which this scholion refers has *rūḥā* for ἄνεμος and *nšāḇā* for πνεῦμα (Nic. syr. 18.1: ܢܐܡܪ ܕܝܢ ܥܠ ܪܘܚܐ ܘܢܫ̈ܒܐ ܘܢܗܪ̈ܘܬܐ ܘܝܡܐ; < Arist. 349a 12-13: περὶ δὲ ἀνέμων καὶ πάντων πνευμάτων, ἔτι δὲ ποταμῶν καὶ θαλάττης λέγωμεν).

BOOK OF METEOROLOGY, CHAPTER FOUR

ON THE REMAINING PRODUCTS OF SMOKY EXHALATIONS

The phenomena discussed in this chapter, which, with the exception of thunder, were considered to be due to the combustion of the smoky exhalation, were discussed by Arist. in different parts of his *Mete*. - IS brought them together in the fifth chapter of his treatise on meteorology and it is to that chapter that this chapter of *But*. corresponds. The order of presentation within the chapter is basically that of the *Šifā*ʾ and much of the material is also borrowed from that work, although as in the previous chapter BH has made a fairly extensive use of Nic. syr., especially in the latter half of section IV.i. and throughout section IV.ii. Fakhr al-Dīn al-Rāzī's *Mabāḥiṯ* is also used at certain points in the chapter, most notably in those parts of IV.i.1 and IV.i.2 describing the mechanism by which thunder and lightning are produced.

	*Šifā*ʾ	Arist., *Mete*.	
IV.i.	67.4-70.6	II.ix.	Thunder and lightning
IV.ii.	70.6-71.3	III.i. (371a 15 ff.)	Thunderbolt and firewind
IV.iii.	71.4-74.15	I.iv.-vii.	Shooting stars, comets etc.

l. 1. ܡܛܠ ܫܪܟܐ ܕܥܒ̈ܕܐ ܚܠܓܬܐ ܬܢܢܝܬܐ : The corresponding chapter heading in the *Šifā*ʾ (67.3) is more pedestrian: "on thunder, lightning, thunderbolt ..." - The heading here is based on Nic. syr. 43.20f.: ܢܐܡܪ ܡܟܝܠ ܥܠ ܫܪܟܐ ܕܥܒ̈ܕܘܗܝ / ܕܚܠܓܬܐ ܗܕܐ . ("Let us speak from here onwards about the rest [*šarkā*] of the works [*ʿbāḏē*] of this exhalation").[1]

Mete. IV. Section i.: *On Thunder and Lightning*

The section corresponds to the first part of the corresponding chapter of the *Šifā*ʾ, where IS discusses thunder and lightning (ed. Cairo 67.4-70.6). - The order of presentation is basically that of the *Šifā*ʾ, except for the placement of the passage on thunder caused by extinction of lightning (IV.i.3.1-4) where the order of the material in Fakhr al-Dīn

[1] Cf. Arist. 370b 3f.: περὶ δὲ τῶν ὑπολοίπων εἴπωμεν ἔργων τῆς ἐκκρίσεως ταύτης, τὸν ὑφηγημένον ἤδη τρόπον λέγοντες. - In Arist. as in Nic. syr. these words follow after the discussion of thunder and lightning (*Mete*. II.ix.) and introduce the discussion of whirlwinds and thunderbolt/firewind (III.i.).

al-Rāzī's *Mabāḥit* may have had an influence. The accounts of how thunder and lightning are generated (IV.i.1.1-7, IV.i.2.1-4) are also based on the *Mabāḥit* rather than on the *Šifā'*. - In 'theory' IV.i.4 we find a typical example of the manner in which BH combines the *Šifā'* with Nic. syr., while the last part of the section dealing with the pre-Aristotelian theories on thunder and lightning (IV.i.5-6) is based largely on Nic. syr. rather than the *Šifā'* where these theories are only treated briefly.

	Šifā'	*Mabāḥit*	Nic.	
IV.i.1.1-7	(67.4ff.)	II.187.5-11	41.25-42.9	Thunder
IV.i.1.7-9	68.3-4			(contd.)
IV.i.2.1-4	(68.4-8, 10-12)	II.187.11-16		Lightning
IV.i.2.5-9	68.8-12	(II.187.14-15)		Inflammability
IV.i.3.1-4	69.9-14	(II.187.17-18)		Thund. by extinction
IV.i.3.4-9	68.12-18	(II.188.14-21)		Flames on ground
IV.i.4.1-4	68.19, 69.14-16			Lightning/thunder
IV.i.4.4-8	69.1-5		42.9-11	Vision and hearing
IV.i.4.8-11	69.6-8		42.11-14	(contd.): oars
IV.i.5.1-2	70.1-2		42.14-17	Empedocles
IV.i.5.2-4	(70.4-5)		43.2-4	Refutation
IV.i.5.4-5	70.2		42.18-19	Anaxagoras
IV.i.5.5-7	70.5-6		42.28-30	Refutation
IV.i.6.1-3	(70.2-3)		43.11-15	Cleidemus
IV.i.6.3-6			43.15-19	Refutation

IV.i.1.1-7: Thunder

As was the case with the explanation of wind (Mete. III.i.1.4-9 above) the explanation of how thunder is generated is based on Fakhr al-Dīn al-Rāzī's *Mabāḥit*, rather than on the *Šifā'*. The vocabulary used suggests that BH also consulted the corresponding passage of Nic. syr.

But. [underline: agreement with *Mabāḥit*; italics: with Nic.]:
(1a) Vapour and smoke [are] not [found] by themselves (1b) but usually the two of them rise together, mixed with each other. (2a) When this mixture reaches the cold stratum of air, (2b) it thickens, (2c) a cloud is generated from it (2d) and smoke *is confined* [*methbeš*] inside it, (3a) so that, if the smoke retains its heat [lit. persists in its heat, *b-ḥammīmūṯeh pā'eš*], (3b) it *is carried* [*mettaytē*] upwards, (3c) violently tears [*msaddeq*] the cloud (3d) and goes out. (3e) From the tearing [*sedqā*] the sound of thunder [*qāl ra'mā*] is generated. (4a) If the smoke *grows cold* [*qā'ar*], (4b) it becomes heavy (4c) and *is carried* [*mettaytē*] downwards, (4d) [also] tears the cloud (4e) and goes out, (4f) and causes the sound of thunder to be heard [*mašma'*].

Mabāḥit II.187.5-11:

قد عرفناك الفرق بين البخار والدخان وانه لا يوجد بخار خالص ولا دخان خالص بل هما فى اكثر الامر يصعدان معا فاذا ارتفع بخار مخلوط بدخان ارتفاعا يصل الى الطبقة الباردة من الهواء حتى تكاثف وينعقد سحابًا فلا محالة يحتبس ذلك الدخان فى جوف السحاب فذلك الدخان اما ان يبقى حارا او يصير باردا فان بقى حارا قصد العلو ومزق السحاب تمزيقا عنيفًا فيحصل من ذلك التمزيق الرعد وان صار باردا تثاقل وقصد السفل ومزق السحاب فحصل الرعد

(1a) We have taught you about the difference between vapour and smoke and that vapour is not found alone nor smoke alone, (1b) but usually the two rise together. When vapour mixed with smoke rises (2a) in such a way as to reach the cold stratum of the air, (2b) so that it thickens (2c) and is condensed into a cloud, (2d) then, that smoke will no doubt be confined in the interior of the cloud. (3/4) That smoke will either remain hot or become cold. (3a) If it remains hot [*baqiya ḥārran*], (3b) it will try to rise [lit. aim for the heights] (3c) and tears [*mazzaqa*] the cloud violently, (3e) and from that tearing [*tamzīq*] thunder [*ra*ᶜ*d*] arises. (4a) If it becomes cold, (4b) it will become heavy, (4c) try to descend (4d) and tears the cloud, (4f) so that thunder arises.

Nic. syr. 41.25-28, 42.1-5, 42.9[2]:

ܡܬܝܠܕ/26 ܒܪܩܐ ܒܗܢܐ ܐܘܪܚܐ. ܟܕ ܓܝܪ ܚܠܝܬܐ ܬܢܢܝܬܐ. ܡܬܚܒܫ ܠܓܘ ܡܢ
ܥܢܢܐ. ܡܬܛܝܛܝܐ/27 ܦܫ ܡܢܗ ܕܩܛܝܢ ܠܥܠܝܐ. ܘܡܛܠ ܩܛܝܢܘܬܗ ܠܥܠ ܡܬܢܦܨܐ. ܡܢܬܐ ܕܝܢ
ܕܥܢܢܐ. ܘܡܛܠ/28 ܡܬܢܦܨܘܬܗ ܕܡܫܚܢܢܘܬܐ ܠܐܩܪ ܠܗ. ܘܡܬܥܒܝܐ/[42]/1 ܕܥܢܢܐ
ܕܡܢܗ ܐܬܢܦܨܬ ܚܠܝܬܐ ܬܢܢܝܬܐ. ܟܕ ܩܐܪܐ ܡܛܠ/2 ܫܘܢܝܗ ܕܗܘ ܕܡܫܚܢ ܗܘܐ
ܠܗ ܡܬܥܒܝܐ. ܘܟܕ ܡܬܥܒܝܐ: ܡܬܥܨܪ/3 ܘܡܬܕܚܩ ܗܘ ܡܕܡ ܕܫܪܟ ܡܢܗ
ܕܚܠܝܬܐ ܗܕܐ ܬܢܢܝܬܐ. ܘܟܕ ܠܐܪܥ ܠܬܚܬ/4 ܡܬܕܚܩܐ. ܠܕܘܟܬܐ ܕܠܘܩܒܠܗ ܠܗܝ
ܕܡܢܗ ܐܬܠܚܨܬ ܡܬܛܝܛܝܐ. ܘܟܕ ܫܘܪ/5 ܗܢܐ ܚܠܝܬܐ. ܢܩܫ ܒܥܢܢܐ ܐܚܪܢܝܐ
ܘܥܒܕ ܪܥܡܐ. ܘܡܛܠ ܢܩܫܐ ܡܬܕܠܩ ://9 ܘܥܒܕ ܒܪܩܐ .

[28 ܘܡܬܥܒܝܐ : ܡܬܥܒܝܐ cod.]

Lightning is generated in this way: (2d) When smoky exhalation is confined [*methbeš*] inside [*l-gāw men*] a cloud, (3b) the finer part of it is carried [*mettaytyā*] to the upper part of the cloud, (3d) and because of its fineness is discharged [*metnappṣā*] upwards, but the part of the cloud from which the smoky exhalation is discharged, (4a) as it grows cold [*qā*ᵓ*rā*] because of the removal of that [exhalation] which was making it hot, (4b) becomes condensed. As (the cloud) becomes condensed, what there is left of this smoky exhalation is squeezed and pressed. (4c) As it is pressed downwards, (this exhalation) is carried [*mettaytē*] to the place which is opposite to that from which was squeezed. As this exhalation jumps, it strikes another cloud and causes thunder, and because of the stroke it is kindled and causes lightning.

l. 2f. ܒܪܘܥܬܐ ܩܪܝܪܬܐ ܕܐܐܪ ("the cold stratum of air"): cf. Mete. III.i.1.5 above (ܒܪܘܥܬܐ ܩܪܝܪܬܐ).

l. 2. ܟܕ ܒܚܕܕܐ ܚܠܝܛܝܢ ܣܠܩܝܢ ("they rise mixed with each other"): cf. Mete. I.i.1.5 above (ܟܕ ܡܬܚܠܛܝܢ ܒܚܕܕܐ ܡܬܣܠܩܝܢ).

[2] Cf. Arist. 369a 12-b 7; Olymp. 37.25-28; Olymp. arab. 141.14-21.

IV.i.1.7-9: Thunder (contd.)

But.: (1) If a loud [*ʿaššīnā*] sound is heard from wind tearing thin and fine air [*āʾār raqqīqā w-qaṭṭīnā*] as it blows [lit. in its blowing, *b-maššbāh*], (2) how much more [*kmā yattīrāʾīt*] should a thundering sound [*qālā rāʿmā*] be heard from that [wind] tearing a thick cloud!
Šifāʾ 68.3-4:

والريح إذا عصفت فى الهواء الرقيق اللطيف سمع لها صوت شديد، فكيف فى سحاب كثيف. فيجب أن يسمع له صوت الرعد.

(1) A strong [*šadīd*] sound is heard when wind blows violently [*ʿaṣafat*] through thin and fine air. (2) How about [when it blows through] a thick cloud? It must be that the sound of thunder [*ṣaut al-raʿd*] is heard from it.

l. 9. ܩܠܐ ܪܥܡܐ ("thundering sound") FMV: ܩܠܐ ܪܥܡܝܐ Ll. - I accept the reading of FMV as being that of the better mss. as well as being *difficilior*, taking ܪܥܡܐ as a participle (*rāʿmā*), although ܩܠܐ ܪܥܡܝܐ is read at IV.i.3.2 below.

IV.i.2.1-4: Generation of lightning

The passage is more closely based on the *Mabāḥiṯ* than on the *Šifāʾ*, although the correspondence of "collision with each other" to IS's "*iḍṭirāb*" suggests that BH also had the passage of the *Šifāʾ* before him.

But. [underline: agreement with *Mabāḥiṯ*; italics: with *Šifāʾ*]:
(1a) *Because this smoke is a fine body* (1b) in which wateriness and earthiness are *mingled* [*ḥbīkān*], (2) *the heat* and *the* mixing *movement* [*mettzīʿānūṯā mmazzḡānīṯā*] create [*ʿāḇdān*] an oily mixture [lit. a *mixture* of *oiliness*, *muzzāḡ dahhīnūṯā*] in it, (3) so that *it is inflamed* [*meštalhaḇ*] by that *intense* [*ʿaššīnā*] *movement* and the strong *collision* [*nqāšā taqqīpā*] *with each other* of the parts of clouds which are torn (4) and causes *lightning* [*barqā*].
Mabāḥiṯ II.187.11-13, 15-16[3]:

ولان هذا الدخان شىء لطيف وفيه مائية وارضية عملت فيه الحرارة والحركة والخلخلة المازجة عملا قرب مزاجه من الدهنية فهو لا محالة يشتعل بادنى سبب مشعل فكيف بالحركة الشديدة والمحاكة القوية ...
واذا كان كذلك اشتعلت تلك الادخنة من شدة محاكتها عند شدة تمزيقها للسحاب وذلك هو البرق

(1a) Because this smoke is something fine (1b) and it contains wateriness and earthiness, (2) the heat, the movement [*ḥaraka*] and the mixing rarefication [*al-ḫalḫala al-māziğa*] have worked [*ʿamilat*] in it an effect [*ʿamal*] which makes the mixture approach oiliness [*duhnīya*], (3) so that it will no doubt be inflamed by the least inflaming cause [*sabab mušʿil*]. How about with an intense [*šadīd*] movement and a strong [*qawīy*] friction? ... This being the case, those smokes are inflamed [*ištaʿalat*] by their

---

[3] For 187.14-15, see under IV.i.2.5-9 below.

intense friction as they violently tear [lit. at the intensity of their tearing of (*ʿinda šiddat tamzīqihā*)] the cloud, (4) and that is lightning [*barq*].
Šifāʾ 68.4-8, 10-12[4]:

ولأن هذا الدخان لطيف متهيء للاشتعال، فإنه يشتعل بأدنى سبب مشعل، فكيف بالحركة الشديدة والمحاكة القوية مع جسم كثيف؟ والحك أولى بالإسخان من نفس الحركة أو مثلها. وقد علم هذا فى موضع آخر، فلا عجب أن تحيله المحاكة والاضطراب والانضغاط إلى حرارة مفرطة، فيشتعل لهذه العلل نارا ويستحيل برقا. ... فكيف إذا حرك الشىء اللطيف المختلط من مائية وأرضية، عمل فيهما الحرارة ولحركة والخلخلة المازجة عملا قرب بمزاجه من الدهنية، حركة شديدة وهى مستعدة لطيفة دخانية؟

(1a) Because this smoke is fine and inflammable [lit. ready to be inflamed], (3) it is inflamed by the least inflaming cause. How about with an intense movement and a strong friction with a thick object? Friction itself is more liable to cause heat than mere movement or its like. - This was made known in another place. - (3) It is no wonder then that friction, clashing with each other [*idṭirāb*] and compression transform it into excessive heat. (The smoke) is therefore inflamed into fire by these causes (4) and becomes lightning. ... (1b) Then, what about when a fine thing mixed out of wateriness and earthiness, (2) in which heat, movement and mixing rarefication have worked in it an effect which makes the mixture approach oiliness, (3) undergoes an intense movement, when they are ready, fine and smoky?

IV.i.2.5-9: Inflammability

But. [underline: agreement with *Šifāʾ*; italics: with *Mabāḥiṯ*]:
(1) If you wish to know that fine substances [*kyānē qaṭṭīnē*] are easily lit [*psīqāʾīṯ metnabršīn*] by a little movement [*b-mettzīʿānūṯā bṣīrtā*], (2) observe what happens *when you pass your hand* [lit. what arises from the passage (*maʿbartā*) of your hand] *over a black* and smooth *objects at night*, how fine *sparks and lights* [*zalgē w-nahhīrē*] are generated. (3) *This being the case* [lit. these things being thus], what need is there to say about oily smoke which undergoes such a forceful movement [*mettzīʿānūṯā qṭīrāytā*].
Šifāʾ 68.8-12:

وإذا شئت أن تعلم أن الأشياء اللطيفة يسهل اشتعالها بأدنى حركة، فتأمل ما يحدث من إمرارك اليد على الأشياء السود فى الليل، فإنك ترى أضواء والتهابات لطيفة تحدث من تلك الحركة اللطيفة، فكيف إذا حرك الشىء اللطيف المختلط من مائية وأرضية، عمل فيهما الحرارة ولحركة والخلخلة المازجة عملا قرب بمزاجه من الدهنية، حركة شديدة وهى مستعدة لطيفة دخانية؟

(1) If you wish to know that fine things [*al-ašyāʾ al-laṭīfa*] are easily inflamed [*yashulu ištiʿālu-hā*] by the least movement [*bi-adnā ḥarakatin*], (2) observe what happens from your passing [*imrār*] the hand over black things at night. For you will see fine sparks and flames [*aḍwāʾ wa-iltihābāt*] occur from that fine movement. (3) Then, what about when a fine thing mixed out of wateriness and earthiness, in which the heat, the movement and the mixing rarefication have worked in it an effect which makes the

[4] For 68.8-10, see under IV.i.2.5-9 below.

mixture approach oiliness, undergoes [lit. moves] an intense movement [*ḥaraka šadīda*], when they are ready, fine and smoky?

Mabāḥit̲ II.187.14-15:

ويؤكد ذلك ما يحدث من الاضواء والالتهابات عند امرار اليد على الاشياء السود فى اليل مع كثافتها فكيف مع الشىء اللطيف واذا كان ذلك ...

(2) This is confirmed by the sparks and flames which arise when one passes the hand over black things at night despite their thickness. (3) Then, what about with a fine thing? This being the case ...

l. 7. ܙܠܓܐ ܘܢܗܝܪܐ ("sparks and lights"): الاضواء والالتهابات IS/Rāzī. - Given the correspondence elsewhere of *nuhrā*/*nahhīrā* to *ḍau*ʾ, *nahhīrē* here may be intended to represent *aḍwā*ʾ and *zalgē* to represent *iltihābāt*.

IV.i.3.1-4: Thunder caused by extinction of lightning

The wording of the passage corresponds with that of the *Šifā*ʾ, although the placement of the passage is probably due to the corresponding passage of the *Mabāḥit̲* (II.187.17-18), which immediately follows that corresponding to IV.i.2 above

But.: (1) Lightning sometimes becomes the cause of thunder, (2) when the inflamed smoke is extinguished in a cloud (3) and from its extinction [*duʿʿākā*] a thundery sound [*qālā raʿmāyā*] is heard, (4) just like from the extinction of coal [*gmurtā*] in water. (5) For fire in its flight [*ʿruqyā*] from water moves violently and strikes [*nāqšā*] the air with force and makes a sound.

*Šifā*ʾ 69.9-14:

وربما كان البرق أيضا سبب الرعد، فإن الريح المشتعلة تطفأ فى السحاب، فيسمع لانطفائها صوت بعده بزمان للمعنى المذكور. والسبب فى حدوث ذلك الصوت، أن السبب الأول أنه يحدث من مفاعلة ما بين النار والرطوبة حركةٌ عنيفة سريعة تكون هى سبب الصوت، كما أنا إذا أطفأنا النار فيما بين أيدينا حدث صوت دفعة، لحدوث حركة هوائية عنيفة دفعة، بقرع ذلك المتحرك سائر الهواء بحركته السريعة الصاعدة أو مائلة قرعا شديدا يحدث منه الصوت.

(1) Often lightning also becomes the cause of thunder. (2) For the inflamed wind is extinguished in the cloud, (3) so that a sound is heard due to its extinction [*inṭifā*ʾ] a while after it in the manner mentioned. [As for] the reason for the occurrence of that sound, the first reason is that from the interaction [*mufāʿala*] between fire and moisture a violent and quick movement occurs, which is the cause of the sound, (4) just as when we extinguish the fire before us a sound suddenly occurs (5) because of the sudden occurrence of the violent movement of the air, with a strong collision [*qarʿ*] of that moving (air) with the rest of the air as it moves quickly, upwards or sideways, from which (collision) a sound arises.

l. 2f. ܐܝܟܢܐ ܕܐܦ ܡܢ ܕܥܟܐ ܕܓܡܘܪܬܐ ܒܡܝܐ ("just like from the extinction of coal [*gmurtā*] in water"): The immediate source of this addition has not been identified. The comparison of the sound of thunder to the sound caused by quenching of heated iron (*parzlā maḥḥmā*) in water is found in

Theoph. *Mete*. syr., Cantab. Gg. 2.14, 351r 8-10 (ed. Daiber [1992] p. 176; corr. arab. [Ibn al-Khammār], ed. Daiber p.228, 1.10f.; and Bar Kepha, *Hex.*, Paris 241, 184v a18-20).

l. 3. ܢܘܪܐ ܒܥܪܘܩܘܬܗ ܡܢ ܡ̈ܝܐ ("fire in its flight from water ..."): cf. *Šifā'* 68.13 [between the passages used in IV.i.2 and IV.i.3.4-9]: وربما كان اشتعالها من اختناق الحر هربا من البرد ("Often the inflammation [of lightning] is due to the confinement of the heat fleeing from coldness"); also Min. III.ii.1.7 ("its wateriness being the cause of its flight from fire").

IV.i.3.4-9: Flames on ground

But.: (1) From earths which are saline or contain oily grime [lit. in whose substance [*ūsiya*] there is salt-marsh/salinity [*mālaḥtā*] or oily viscidity [*tallūšūtā dahhīntā*] fine oily exhalations rise (2) and are ignited by a little heat of the sun or of lightning [lit. solar or fulminous heat, *ḥammīmūtā bṣīrtā šemšānāytā aw barqānāytā*], (3) and bright flames are seen on the surface of the earth. (4) Because of their fineness they do not have much capacity for combustion [lit. do not cause to burn much, *law saggī mawqdān*], (5) like the fire that burns in the exhalation rising from wine [*ḥamrā*] containing salt and sal ammoniac when the flask [*šṭīptā*] is placed on coals [*gumrē*].

Šifā' 68.13-18:

وقد يعرض أن تمطر بعض البقاع التى فى جوهرها سبخة أو لزوجة دهنية، ثم تتصعد من تلك البقعة أبخرة دسمة لطيفة، فتشتعل من أدنى سبب شمسى أو برقى. ويرى على وجه الأرض شعل مضيئة غير محرقة إحراقا يعتد به للطفها، ويكون حالها كحال شعلة القطن المنفوش، بل كحال النار التى تشتعل فى بخار شراب مجعول فيه الملح والنوشادر، إذا وضعت قنينة فى جمر فبخر فقرب من بخاره سراج فاشتعل، وبقى مشتعلا مدة قيام البخار.

(1) It sometimes happens that it rains on some place in whose substance [*ǧauhar*] there is salt-marsh/salinity [*sabaḫa*] or oily viscidity [*luzūǧa duhnīya*], then from that place greasy [*dasim*] and fine vapours rise (2) and are ignited by the least solar or fulminous cause [*adnā sabab šamsī au barqī*], (3) and bright flames are seen on the surface of the earth, (4) without significant capacity for combustion [lit. without causing to burn a significant burning] because of their fineness. (5) Their condition is like the condition of a flame of carded cotton, or rather like the condition of fire which is ignited in the vapour of wine [*šarāb*], to which salt and sal ammoniac have been added [*maǧʿūl fīhi*]. When the flask [*qinnīna*] is placed on live coal [*ǧamr*], it is vaporised. [When] a lamp is brought near its vapour, (the vapour) is kindled and continues burning so long as the vapour lasts.

l. 4f. ܐܝܟ ܕܡܠܚܬܐ ܐܘ ܠܛܠܘܫܘܬܐ ܕܕܗܝܢܬܐ ܐܝܬ ܒܐܘܣܝܐ ܕܝܠܗܝܢ. (lit. "in whose substance there is salt-marsh/salinity [*mālaḥtā*] or oily viscidity"): بعض البقاع التى فى جوهرها سبخة أو لزوجة دهنية IS. - The Arabic *sabaḫa*/*sabḫa* normally means "salina", "salt marsh" ("land that exudes water and produces salt", Lane 1292) and the same applies to the Syriac *mālaḥtā* ("locus salsus, terra

salsuginosa" PS 2135). When one talks, however, of there being *sabaḫa*/*mālaḥtā* and "oily viscidity" (*luzūǧa duhnīya*/*ṭallūšūṯā dahhīntā*) in the substance (*ǧauhar*/*ūsiya*) of the earth, the word must be understood as designating an abstract quality, namely "the kind of quality found in a salina", "salinity". - I translate somewhat freely here, since "land whose substance contains the quality of a salina" will, in concrete terms, be a "salina".[5]

l. 9. ܐܡܘܢܝܩܘܢ ("sal ammoniac"): نوشادر IS. - Cf. Min. III.i.3.3 above.

IV.i.4.1-4: Lightning and thunder

But.: (1a) Lightning only occurs [lit. does not occur except] when thunder occurs with it (1b) because [it is] from wind which is constrained [*mezdarbā*] in a cloud, tears it, is ejected and escapes from it burning [that] lightning arises. (2a) Thunder can occur without lightning occurring with it (2b) when the wind is not very strong, (2c) so that it tears the cloud but does not inflame [*mnaḇršā*] the air.

*Šifā*ʾ 68.19:

ولا يكون برق إلا ومعه رعد، لأنه لا يكون إلا من ريح تضطرب فى الغمام ثم تتخلص مشتعلة.

(1a) Lightning does not occur except when thunder [occurs] with it, (1b) because it does not occur except from wind which knocks about [*taḍṭaribu*] in a cloud and then escapes burning.

*Šifā*ʾ 69.14-16:

والغالب أن مع كل برق رعدا، وإن لم يسمع. فإنه لن تنفذ فى الغيم نار متحركة إلا وهناك نشيش أو غليان أو خفق للريحية ولا يبعد أن لا يكون مع الرعد برق، فليس كلما عصفت ريح بقوة اشتعلت.

(1) As a rule [*al-ġālib an*] with every lightning there is thunder, even if it is not heard, since moving fire will not penetrate through a cloud without there being hissing, boiling or knocking due to the windiness [?]. (2a) It is not impossible [*lā yabʿudu*] that there be no lightning with thunder, (2c) since wind is not ignited [*ištaʿalat*] every time it blows with force.

l. 2. ܡܙܕܪܒܐ ("is constrained"): The word does not correspond in meaning to IS's *idṭaraba*, but is similar in sound.

l. 4. ܠܐܐܪ ܠܐ ܡܢܒܪܫܐ ("but does not inflame the air"): In the *Šifā*ʾ, it is the wind itself (whose material, we remember, is the inflammable smoky exhalation) which is set on fire. BH alters this to the inflammation of "air"; cf. IV.ii.3.1f. below, where we hear of firewind (πρηστήρ) igniting "air".

IV.i.4.4-8: Vision and hearing

But. [underline: agreement with *Šifā*ʾ; italics: with Nic.]:

(1) *Although thunder occurs* before *lightning*, (2) *we sense the lightning*

---

[5] It seems that Fakhr al-Dīn al-Rāzī too (or a subsequent copyist/editor) had difficulty in accepting *sabaḫa* as an abstract noun. The word *sabaḫa* is simply omitted at the corresponding place in the *Mabāḥiṯ* II.188.15 (بعض البقاع التى تكون فيها لزوجة دهنية).

first [*luqdām*], *then* [*īṯā*] *the thunder*, (3a) because vision requires contraposition [*lqubblāyūṯā*] and transparency, (3b) which do not occur in time, (4a) whereas hearing requires undulating movement [*mettzīʿānūṯā gallānāytā*] of the air for the sound to reach it [lit. so that the sound reaches it], (4b) and all movement [takes place] in time.

Šifāʾ 69.1-5:

لكن البرق يرى، والرعد يسمع ولا يرى، فإذا كان حدوثهما معا رؤى البرق فى الآن وتأخر سماع الرعد، لأن مدى البصر أبعد من مدى السماع، فإن البرق يحس فى الآن بلا زمان، والرعد الذى يحدث مع البرق يحس بعد زمان. لأن الإبصار لا يحتاج فيه إلا إلى موازاة وإشفاف، وهذا لا يتعلق وجوده بزمان. وأما السمع فيحتاج فيه إلى تموج الهواء، أو ما يقوم مقامه، ينتقل به الصوت إلى السمع، وكل حركة فى زمان.

But lightning is seen, whereas thunder is heard but not seen. (1) When their occurrence is simultaneous, (2) lightning is seen immediately and the hearing of the thunder is delayed (3) because the expanse [*madan*] of vision is further than the expanse [*madan*] of hearing. (2) Therefore, the lightning is sensed immediately without interval [*zamān*], whereas the thunder which occurs with the lightning is sensed after an interval, (3a) because vision does not require anything other than contraposition [*muwāzāt*] and transparency, (3b) whose being is not dependent on time. (4a) Hearing, on the other hand, requires undulation [*tamawwuǧ*] of air or a substitute, by which the sound is conveyed to hearing, (4b) and all movement is in time.

Nic. syr. 42.9-11[6]:

ܘܐܢ ܗܟܢܐ ܩܕܡ ܪܥܡܐ ܡܬܗܘܐ ܗܘܐ/10 ܘܒܬܪܗ ܗܘܐ ܒܪܩܐ. ܡܪܓܫܝܢܢ ܟܝ ܡܪܓܫܝܢܢ ܒܒܪܩܐ. ܩܕܡܐܝܬ ܒܬܪܗ ܕܝܢ/11 ܒܪܥܡܐ. ܡܛܠ ܕܚܙܬܐ ܩܠܝܠܐ ܝܬܝܪ ܡܢ ܡܫܡܥܬܐ.

(1) Although thunder thus occurs first and lightning (occurs) after it, (2) we sense the lightning first [*qaḏmāʾīṯ*] and after it [*bāṯreh*] the thunder, because vision is swifter than hearing.

IV.i.4.8-11: Vision and hearing (contd.)

But. [underline: agreement witt Nic.; italics: with *Šifāʾ*]:

(1a) *For this reason*, when an oar strikes water and goes up, and strikes and goes up again [*tūḇ*], (1b) we hear the sound of the first stroke after [seeing] the second stroke. (2) We also *see the stroke of an axe* from *a distant place before hearing the sound* [lit. hearing of the sound].

Nic. syr. 42.11-14[7]:

ܡܛܠܗܕܐ ܗܝ ܐܡܬܝ/12 ܕܡܚܝܐ ܡܠܘܚܐ. ܟܕ ܚܕ ܡܚܐ ܠܡܝܐ ܘܣܠܩ ܘܡܚܐ ܘܣܠܩ : ܡܫܬܡܥ ܕܝܢ/13 ܩܠܐ ܕܡܚܘܬܐ : ܩܕܡܝܬܐ ܒܬܪ ܬܪܬܝܢܝܬܐ ܡܚܘܬܐ. ܘܚܙܝܢ ܡܫܟܚܝܢ ܠܡܚܘܬܐ ܕܡܚܘܬܐ/14 ܡܪܚܩܬܐ.

[13 ܡܫܟܚܝܢ : ܡܫܟܚܝܢ cod.]

[6] Cf. Arist. 369b 7-9; Olymp. arab. 142.4-7; *Cand.* 118.10-119.1.

[7] Cf. Arist. 369b 9-11; Olymp. arab. 142.7-9; *Cand*. 119.1-3.

[This] is known from those who strike [water] with oars. (1a) For when the oar strikes water and goes up, and strikes and goes up again [*men-d-rēš*], (1b) it rises the second time and then we hear the sound of the first stroke.

Šifā' 69.6-8:

ولهذا ما يرى وقع الفأس، وهو إذا كان يستعمل فى موضع بعيد قبل أن يحس بالصوت بزمان محسوس القدر، وأما إذا قرب فلا يمكنك أن تفرق بين ذلك الزمان القصير وبين الآن.

(1) For this reason, (2) the fall of an axe is seen, when it is used in a far place, before the sound is sensed by an appreciable interval, but when it is near, then you cannot distinguish between that short interval and an instant [*al-ān*]

IV.i.5.1-2: Empedocles' theory

But. [undeline: agreement with Nic syr.; italics: with *Šifā'*]:

(1) Empedocles *says that* lightning is a flash [*zalgā*] of *the sun's rays* [lit. solar rays, *zallīqē šemšānāyē*] *which* are confined [*methabšīn*] in clouds and are ignited in them, (2) and [that] thunder is generated when they are extinguished [Peal].

Nic. syr. 42.14-17[8]:

14 ܐܡܦܕܘܩܠܝܣ ܕܝܢ ܐܡܪ ܕܡܢܬܐ ܡܢ ܙܠܝܩܐ ܕܫܡܫܐ/15 ܡܬܚܒܫܐ
ܒܥܢܢܐ. ܘܗܘ ܗܢܐ ܡܫܚܢܐ ܠܥܢܢܐ. ܘܟܕ ܡܫܚܢܐ ܠܗܝܢ/16 ܡܬܚܒܨܐ
ܘܡܬܥܨܝܐ ܟܕ ܡܬܥܒܝܢ ܒܪ̈ܝܬܐ ܡܢܗܝܢ܆ ܘܐܠܝܟܐ ܡܢ ܕܙܠܝܩܐ/17 ܐܡܪ
ܕܐܬܬܣܝܡ ܒܪܩܐ. ܪܥܡܐ ܕܝܢ ܐܡܪ ܕܗܘܐ ܟܕ ܡܬܕܥܟ ܗܘ ܗܢܐ ܕܐܬܚܡܡ.

[16 ܡܢ: ܡܢ cod.]

(1) Empedocles says that a part of the sun's rays [*zallīqā d-šemšā*] is confined [*methabšā*] in clouds. This makes the clouds hot and as it makes them hot, it is squeezed and pressed as the outer parts (of the clouds) become thickened. (1) He says that lightning is the flash [*zalgā*] of the rays. (2) Thunder, he says, occurs when that which has been heated is extinguished [Ethpa.].

Šifā' 70.1-2:

وقد قيل فى الرعد والبرق أقاويل، ليست بصحيحة، كمن قال: إن البرق شعاع الشمس يحتبس فى السحاب،

Incorrect things have been said concerning thunder and lightning, (1) like the one who said: lightning is the sun's rays [*šuʿʿāʿ al-šams*] which are confined [*yaḥtabisu*] in the cloud.

l. 1. ܗܡܦܕܘܩܠܝܣ ("Empedocles"): ܐܡܦܕܘܩܠܝܣ Nic. - BH write the name with initial *hē* on the two occasions where Empedocles is mentioned in *But*. Min.-Mete. (here and Min. V.ii.4.1), as well as at *But*. De plantis I.i.2 [Drossaart Lulofs-Poortman] l. 15. In his earlier works the forms with initial *ālap* appear to be more common,[9] but the form ܗܡܦܕܘܩܠܝܣ is

[8] Cf. Arist. 369b 11-19; *Cand*. 119.3-4.

[9] ܐܡܦܕܘܩܠܝܣ BH *Chron*. [Bruns-Kirsch] 19.16; [Bedjan] 18.6; *Cand*. 56.10, 136.1; ܐܡܦܕܘܩܠܝܣ *Cand*. 119.3, 151.11.

found at *Cand*. 166.9,[10] where the content of the passage corresponds to *But*. De plant. I.i.2, while at *Cand*. 151.11 (on salinity of sea, corr. *But*. Min. V.ii.4), where Bakoš adopts ܐܡ ܦ ܕܘܩܠܝܣ, ms. B has the variant ܗܡܦܕܘܩܠܝܣ. - Since in the passage quoted above and elsewhere the surviving parts of Nic. syr. invariably has the form with initial *ālap̄* and this applies also to other sources normally used by BH,[11] one has to assume another source of information which led BH to produce the form here, which is correct according to a system where Greek -ε- (also -αι-) is represented by *hē*.

IV.i.5.2-4: Refutation

But.: According to this theory, lightning and thunder ought to occur during the day and in summer when the rays are abundant.

Nic. syr. 43.2-4[12]:

ܐܝܟ ܡܠܬܗ ܕܐܡܦܐܕܘܩܠܝܣ ܒܐܝܡܡܐ /3 ܙܕܩ ܗܘܐ ܕܢܗܘܘܢ ܗܠܝܢ ܘܠܐ ܒܠܠܝܐ.
ܡܛܠ ܕܒܐܝܡܡܐ ܡܣܓܝܢ ܙܠܝܩܐ /4 ܘܒܩܝܛܐ ܝܬܝܪ ܡܢ ܕܒܣܬܘܐ. ܚܙܝܢܢ ܕܝܢ
ܕܗܘ ܕܠܩܘܒܠܐ ܗܘܝܐ ܀

According to Empedocles' theory, these things [i.e thunder and lightning] ought to occur during the day and not at night - because during the day the rays are abundant - and in summer more than in winter. We see, however, that the opposite happens.

*Šifā*ʾ 70.4-5: ولو كان البرق شعاعا استأسر في غمام، لكانت السحب الناشئة ليلا لا تبرق.

If lightning were rays imprisoned [*ista*ʾ*sara*] in a cloud, then clouds appearing at night would not cause lightning.

IV.i.5.4-5: Anaxagoras' theory

But. [undeline: agreement with Nic syr.; italics: with *Šifā*ʾ]:

Anaxagoras says that *a part of the fire of* *the αἰθήρ is confined in a cloud* and causes lightning and thunder.

Nic. syr. 42.18-19[13]:

ܐܢܟܣܐܓܘܪܣ ܕܝܢ ܐܡܪ. ܕܡܢܬܐ ܡܢ ܐܝܬܝܪ ܕܡܬܚܒܫܐ ܒܥܢܢܐ ܘܒܪܩܐ /19
ܕܐܝܬܝܪ ܚܒܝܫܐ ܒܥܢܢܐ ܘܪܥܡܐ.

[18 ܐܢܟܣܐܓܘܪܣ : ܐܢܟܣܐܓܘܪܣ cod. || ܐܡܪ : ܐܡܪ cod.]

---

[10] = Drossaart Lulofs-Poortman (1989) p. 57, l. 4. There are manuscript variants, but they also have initial *hē*.

[11] Nic. syr.: ܐܡܦܐܕܘܩܠܝܣ 42.14, 43.2, 43.8; without second *ālap̄* ܐܡܦܕܘܩܠܝܣ 20.7; adjective ܐܡܦܕܘܩܠܝܐ 39.9. - *De mundo* syr.: ܐܡܦܕܘܩܠܝܣ 154.15 (gr. 399b 15). - The forms in Michael, *Chron.* are hopelessly corrupt but they too, with one exception, have initial *ālap̄*: ܐܦܕܘܩܠܝܣ 66b 15; ܐܦܕܘܩܠܝܣ 66b 17; ܐܡ ܦ ܕܘܩܠܝܠܝܣ 69a 12; ܒܪܩܘܣܝܣ 117a 10 (*sic*, cf. tr. Chabot I.198 n.6).

[12] The point made here in Nic. syr. is not in Arist. (or Alex. Aphr.), but the idea is similar to that used in refutation of Cleidemus' theory (see IV.i.5.3-6 below).

[13] Cf. Arist. *Mete*. 369b 14-15.

Anaxagoras says that a part of the αἰθήρ is confined in clouds and causes lightning and thunder in the manner mentioned.

Šifāʾ 70.2: أو أنه قطعة من نار الأثير يختنق فيه

... or that it is a piece [*qiṭʿa*] of the fire of the aether which is confined in it.

l. 4. ܗܐܬܝܪ ("αἰθήρ"): ܐܐܬܝܪ Nic. - BH invariably writes ܗܐܬܝܪ in *But*. Min.-Mete. (invariably changed to ܐܐܬܝܪ by the copyist of V). - Nic. syr. has ܗܐܬܝܪ in its earlier parts related to the discussion of "aether" in Arist. *Mete*. I.iii. (339b 21-27; Nic. syr. 10.25, 26; 11.19; 14.7), but ܐܐܬܝܪ in the later parts related to *Mete*. II.ix. (369b 14, 20; Nic. syr. 42.18, 22, 26, 29; 43.2, 7).

IV.i.5.5-7: Refutation

Arist. in his refutation of Anaxagoras' theory talks of the natural movement of "aether" as being upwards and the contrast made there is between upward and downward movement (*Mete*. 369b 19-22; so also Alex. Aphr. 129.33-35).[14] In Nic. Syr. this is altered to a contrast between circular and straight movement.[15] In IS the contrast is between upward and downward movement as in Arist.

But. [undeline: agreement with Nic syr.; italics: with *Šifāʾ*]:

(1) This view too is not correct (2) because fire circulates [*ḥādrā*] *in its place* [*b-atrāh*] with the *circular* motion of the celestial sphere (3) and is not *driven* [*metdaḥyā*] straight [*ba-trīṣū*] towards a cloud.

Nic. syr. 42.28-30 [cf. Arist. *Mete*. 369b 19-22]:

ܠܦܘܬ ܣܘܟܠܗ ܕܐܢܟܣܓܘܪܐܣ(ܗܘܐ) 29/ ܡܫܬܟܚ ܠܓܘܫܡܐ ܕܐܐܬܝܪ : ܕܠܐ
ܚܘܕܪܢܐܝܬ ܡܫܬܪܓܓ : ܐܠܐ ܬܪܝܨܐܝܬ . ܐܢܗܘ/30 ܕܝܢ ܕܠܬܚܬ ܡܬܬܚܒܛ ܠܘܬ
ܥܢܢܐ ܬܪܝܨܐܝܬ ܡܫܬܪܓܓ : ܗܢܐ ܕܣܟܠܐ ܀

[28 (ܗܘܐ)ܕܐܢܟܣܓܘܪܐܣ : (ܗܘܐ)ܕܐܢܐܟܣܓܘܪܐܣ cod.]

According to Anaxagoras' theory, the body [*gušmā*] of the αἰθήρ is found to move (2) not in a circle [*ḥudrānāʾīt*] (3) but straight [*trīṣāʾīt*]. (3) For if it is carried downwards towards the cloud it moves straight [*trīṣāʾīt*], which is absurd.

Šifāʾ 70.5-6: وأما الجرم الأثير فلا زاج له إلى أسفل زجا بغتة، وطباعه طاف، ومحركه مدير.

[6 مدير ed. Ṭ: مدبر ed. Caír.]

And as for [the theory that it is] a bulk [*ǧirm*] of the aether, (3) there is nothing which suddenly *drives* it [*lā zāǧǧa lahu*] downwards, when its nature is to float (upwards) [*ṭāfin*] (2) and its path [*maḥrak*; or *muḥrik*: "its mover", i.e. the celestial sphere] is circular.

---

[14] Arist. 369b 21f.: τοῦ τε γὰρ κάτω φέρεσθαι τὸ πεφυκὸς ἄνω δεῖ λέγεσθαι τὴν αἰτίαν, ...

[15] The circular motion of the "aether" is also mentioned in another scholion at Nic. syr. 42.24f.: ܐܐܬܝܪ ܩܪܐ ܐܢܐܟܣܓܘܪܐܣ ܠܗܘܢ ܠܓܘܫ̈ܡܐ/25 ܕܡܬܟܪܟܝܢ ܚܘܕܪܢܐܝܬ . ("Anaxagoras calls 'aether' the bodies which go round in a circle"). - Cf. Alex. Aphr. 129.23.f.: ὅτι δὲ ᾿Αναξαγόρας τὸν αἰθέρα καὶ τὸ κυκλοφορητικὸν σῶμα πῦρ λέγει; also Alex. Aphr. 8.16-25; *De mundo* 392a 7: ἀλλὰ διὰ τὸ ἀεὶ θεῖν κυκλοφορουμένην, etc.

l. 5. ܘܐܦܠܐ ܗܢܐ ܪܢܝܐ ܫܪܝܪ ("This view too is not correct"): cf. *Šifā'* 70.1 (quoted under IV.i.5.1-2 above): أقاويل ليست بصحيحة

l. 6. ܗܘ ܙܘܥܐ ܚܘܕܪܢܝܐ ܕܥܠܘܝܬܐ ("the circular motion of the celestial sphere"): cf. Mete. III.i.1.7-9 above ("the sphere of fire, which moves with the movement of the celestial sphere. ... that circular and swift movement of the fire").

IV.i.6.1-3: Cleidemus' theory

But.: (1) Cleidemus says that lightning does not [really] exist [*law mehwā hāwē*], (2) but is merely an illusion (3) which arises from the division of a cloud, due to the reflection of the sun's rays from them [lit. due to the reflection <that occurs?> to the sun's rays from them (see comments below)], (4) just as water sprinkled from oars flashes [*mabrqīn*] at night.

Nic. syr. 43.11-15 [cf. Arist. *Mete*. 370a 10-15]:

ܘܩܠܐܘܕܝܡܘܣ ܕܝܢ ܥܠ ܪܥܡܐ ܡܢ ܐܡܪ/12 ܕܡܬܘܡ ܒܫܪܝܪܐ ܗܘܐ. ܥܠ
ܒܪܩܐ ܕܝܢ ܠܐ ܐܡܪ ܕܒܫܪܝܪܐ ܡܬܩܝܡ: ܐܠܐ ܡܚܙܝܬܐ ܒܠܚܘܕ/13 ܦܣܩ
ܕܐܬܚܙܝܘ. ܐܡܪ ܓܝܪ ܕܟܕ ܡܬܦܠܓܝܢ ܥ̈ܢܢܐ ܠܡܢܘ̈ܬܐ ܣܓܝ̈ܐܬܐ ܦܢܝܬܐ ܠܫܡܫܐ/14
ܗܘܝܐ ܠܡܢܗܘܢ. ܘܡܛܠ ܗܕܐ ܡܬܚܙܐ ܒܪܩܐ. ܐܝܟܢܐ ܕܐܦ ܟܕ ܡܬܪܣܣܝܢ/
15 ܡ̈ܝܐ ܡܢ ܡܚܘܬܐ ܕܡ̈ܩܦܐ ܒܠܠܝܐ ܡܬܚܙܝܢ ܕܡܒܪܩܝܢ.

[12 ܒܪܩܐ sub lin. ‖ 14 ܒ supra lin.]

(1) Cleidemus says concerning thunder that it truly exists [*mehwā ba-šrārā hāwē*]. (1) Concerning lightning, however, he denies that it truly exists [*metqayyam*], (2) but stated that it was merely an illusion. (3) For he says that when clouds are divided into many parts, the sun is reflected from them [lit. reflection occurs to the sun from them], [and] for this reason a lightning is seen, (4) just as when water is sprinkled by the stroke of oars at night, (the water) is seen to flash [*mabrqīn*].

l. 1. ܘܩܠܐܘܕܝܡܘܣ ("Cleidemus/Κλείδημος") scripsi: ܘܠܐܘܕܝܡܘܣ FLIM: ܘܠܝܘܕܝܡܘܣ V - Although all the manuscripts here present forms in which the *qōp̄* has been corrupted into *waw* ("and *Leodemus"), BH should have known that the name was "Cleidemus" (or, as the Syriac transliteration has it: "*Cleodemus"), since the *qōp̄* is preserved in Nic. syr. 43.11 and BH *Cand.* 119.4 (ܩܠܐܘܕܝܡܘܣ).

l. 1-2. "Cleidemus says that lightning does not [really] exist, but is merely an illusion": This goes back, ultimately, to Arist. *Mete*. 370a 10-12: εἰσὶ δέ τινες οἳ τὴν ἀστραπήν, ὥσπερ καὶ Κλείδημος, οὐκ εἶναί φασιν ἀλλὰ φαίνεσθαι.

l. 2. +ܕܡܢ ܦܘܠܓܐ ܕܥܢܢܐ ܠܦܢܝܬܐ ܗܘܝܐ ܕܠܙܠܝܩܐ ܕܫܡܫܐ ܠܡܢܗܘܢ+ - Two solutions seem possible: (1) read: ܘܡܢ ("... is merely an illusion and that as a result of [*men*] the division of a cloud reflection occurs to the sun's rays from them"); (2) read ܠܦܢܝܬܐ with mss. LIV and read ܕܠܙܠܝܩܐ ("... is merely an illusion which arises through the division of a cloud, due to reflexion, namely the reflexion of the sun's rays by them").[16] - The first option still

[16] There is a third option which, however, involves a more drastic change of word order: ܕܡܢ ܦܘܠܓܐ ܕܥܢܢܐ ܗܘܝܐ ܠܦܢܝܬܐ ܕܠܙܠܝܩܐ ܕܫܡܫܐ ܠܡܢܗܘܢ ("which arises from the division of a cloud, due to the reflexion of the sun's rays towards them").

leaves an awkward syntax, with too much read into the preposition "*men*", while the corresponding part of Nic. syr. (ܚܠܘܦܐ ܗܘܐ ܠܫܡܫܐ ܠܘܬ ܗܠܝܢ) argues against the second option. The problem here, one suspects, is due to hasty composition on BH's part. What he intended to write was probably something similar to the second option. In writing the sentence he may have copied the phrase "*ʿṭupyā* ..." from Nic., without subsequently checking to see that the syntax of the sentence as a whole made sense. - It is also not immediately clear what is reflected from what, but it seems easiest to understand the sentence to mean that "the sun's rays" (in Nic.: the sun) are reflected "from/by [*lwāt*] the cloud". Although in line 4 below, the talk is of the "vision/visual ray" being reflected "by [*men*] the cloud" "towards [*lwāt*] the sun", the point here is that Cleidemus himself did not subscribe to the "visual-ray" theory.

l. 2. ܙܠܝܩܝ ܫܡܫܐ ("the sun's rays"): ܫܡܫܐ ("the sun") Nic. - The addition of "rays" is due to *Šifāʾ* 70.2-3: وكمن قال: إنه عكس شعاعى ("and like the one who said that [lightning] is the reflection of rays").

l. 2. ܠܘܬܗܝܢ: ܠܘܬ ܗܠܝܢ Nic. - In Nic. syr., *hālēn* refers back to the *mnawwāṯā sagīʾāṯā* ("many parts") into which the cloud is divided. Due to the omission precisely of that phrase, the suffix of *lwāṯ-hēn* has no noun to refer back to in *But*.

IV.i.6.3-6: Refutation

IS gives us no refutation of Cleidemus' theory. The passage here is copied almost verbatim from Nic. syr.

But.: (1) According to this theory, lightning ought not to be seen at night, (2) because the sun is [then] below the earth (3) and vision cannot be reflected from the cloud towards it. (4) Water, however, is seen to flash at night [or: flashing water is seen ...], not during the day, (5) because the sun's light obscures its brightness during the day.

Nic. syr. 43.15-19 [cf. Arist. *Mete*. 370a 16-21]:

ܠܘܬ ܡܠܬܗ ܕܗܢܐ /16 ܗܢܐ ܗܘܐ ܕܒܪܩܐ ܠܐ ܐܝܬ ܗܘܐ ܕܢܬܚܙܐ ܒܥܕܢ ܠܠܝܐ. ܡܛܠ
ܕܬܚܝܬ ܐܪܥܐ ܗܘܐ /17 ܫܡܫܐ. ܟܕ ܕܝܢ ܬܚܝܬ ܐܪܥܐ ܗܘܐ ܠܐ ܡܫܟܚܐ ܕܡܢ
ܥܢܢܐ ܬܬܟܦܦ ܠܘܬܗ ܚܙܝܐ. /18 ܒܠܠܝܐ ܕܝܢ ܡܬܚܙܝܢ ܡܝܐ ܕܡܒܪܩܝܢ ܝܬܝܪ ܡܢ
ܕܒܐܝܡܡܐ. ܡܛܠ /19 ܕܢܘܗܪܐ ܢܘܗܪܗ ܕܫܡܫܐ ܒܐܝܡܡܐ ܠܨܡܚܘܬܗܘܢ.

[17 ܬܬܟܦܦ: ܬܬܟܦܦ cod. || ܠܘܬܗ: ܠܘܬܗ cod.]

(1) According to this theory of this man, lightning ought not to be seen at night, (2) because the sun is [then] below the earth (3) and when it is below the earth vision cannot be reflected towards it from the cloud. (4) Water, however, is seen to flash more at night than during the day, (5) because the sun's light obscures its brightness during the day.

Mete. IV. Section ii.: *On Thunderbolt and Firewind*

The section deals with the thunderbolt (κεραυνός) and "firewind" (πρηστήρ), which are discussed by Arist. in the second half *Mete*. III.1 at 371a 15-b 14. - IS discusses these briefly at ed. Cairo 70.6-71.3. - There being no obvious word in Arabic for representing Aristotle's πρηστήρ, IS has to make do with *ṣāᶜiqa* for both the κεραυνός and the πρηστήρ, and is unable to distinguish very clearly between the two.[17] BH with access to Nic. syr. can and does by using the two terms in transliteration.[18]

The order of presentation is the same in the *Šifāʾ* and Nic. syr. (IS is following Olymp. arab. here, which in turn is basically a translation of Nic. syr.). In producing a synthesis of the two, BH has made comprehensive use of Nic. syr. and has left out very little of what is in the section on thunderbolts and firewinds there (The apparent gaps, at 45.25-27, 45.33-46.5, 46.16-17, are taken up by interpolated scholia in the Cambridge manuscript).

	Šifāʾ	Nic. syr.	
IV.ii.1.1-2	70.6, (70.16-17)	45.19-21	Thunderb./firew.: Definition
IV.ii.1.2-6	70.6-9		Thunderbolt/firewind: contd.
IV.ii.2.1-9	70.12-16	45.21-25, 27-31	Types of Thunderbolts
IV.ii.3.1-3	70.16-17	45.31	Firewind
IV.ii.3.2-3	71.1-2		Firewind (contd.)
IV.ii.3.3-5		45.32, 46.5-8	Wind as material
IV.ii.3.5-9	(71.2-3)	46.8-16	Wind as material (contd.)
IV.ii.3.9-10		46.16, 17	Concluding comment

---

[17] In Olymp. arab. the two words κεραυνός and πρηστήρ are preserved in transliteration: Olymp. arab. 144.2-4: فتصير الساعقة التى تسمى قارونس من ريح سحابية تستحيل إلى طبيعة النار؛ والصاعقة المسامة فرسطير من زوبعة تستحيل إلى طبيعة النار. cf. 144.14: وأما الصاعقة التى تسمى فرسطير - Ibn al-Biṭrīq's Arabic version of Arist. *Mete*. only talks of the thunderbolt (*ṣāᶜiqa*; ed. Schoonheim [2000] p. 121f.).

[18] One is in fact rather hard put to find a genuine Syriac term for representing κεραυνός, as may be seen from the various ways in which the word is translated in *De mundo* syr.: 138.11: *barqē* (< gr. 392b 12; ibid. *paqᶜē* < γνόφοι); 141.21: *zalgē* (< 394a 18; ibid. *paqᶜē* < πρηστῆρες); 144.13, 18: *zalqā d-māḥē* (395a 22, 26); 148.30: *barqā* (397a 21); 157.8: *barqē w-raᶜmē* (401a 18; no word in syr. corresponding to κεραύνιος of gr. 401a 17). See further Hanna-Bulut (2000) German-Syriac p. 68 s.v. "Blitzschlag" (where we are given transliterated forms of κεραυνός and σκηπτός; for the latter, see *De mundo* syr. 144.27, 30 < gr. 395a 35, 38); Khaushābā-Yūkhannā (2000) 638 s.v. *al-ṣāᶜiqa* (where we are given 16 equivalents of arab. *ṣāᶜiqa*, but it is difficult to find among them a non-Greek term that refers specifically to "thunderbolts" and is represented in older Syriac literature outside the lexica).

IV.ii.1.1-2: Thunderbolt and firewind: definition

But. [underline: agreement with *Šifā*ʾ; italics: with Nic.]:
(1) *Thunderbolt* [κεραυνός] is *cloud wind* [*rūḥā ʿnānāytā*] which *is inflamed* [*d-metnabršā*]. (2) *Firewind* [πρηστήρ] is *whirlwindy* wind [*rūḥā ʿalʿālāytā*] which *is inflamed*.

*Šifā*ʾ 70.6[19]: وأما الصاعقة فإنها ريح سحابية مشتعلة
Thunderbolt [*ṣāʿiqa*] is inflamed [*muštaʿila*] cloud wind [*rīḥ saḥābīya*].

*Šifā*ʾ 70.16-17: وربما كانت سحابية زوبعية مشتعلة، وتكون من مادة كثيفة، فتكون شر الصواعق
(2) Sometimes it is an inflamed whirlwindy cloud wind [*saḥābīya zaubaʿīya*] and arise from thick material. This is the worst of thunderbolts.

Nic. syr. 45.19-21[20]:
ܩܪܘܢܘܣ ܕܝܢ ܘܦܪܝܣܛܝܪ ܗܘܝܢ ܡܢ ܢܘܒܪܫܐ ܕܗܠܝܢ./20 ܪܘܚܐ ܓܝܪ ܥܢܢܝܬܐ
ܟܕ ܡܬܢܒܪܫܐ ܥܒܕܐ ܩܪܘܢܘܣ. ܘܥܠܥܠܐ ܕܝܢ ܟܕ/21 ܡܬܢܒܪܫܐ ܥܒܕܐ
ܦܪܝܣܛܝܪ.

Thunderbolt and firewind arise from the inflammation [*nubrāšā*] of these. (1) For cloud wind, when it is inflamed [*kad metnabršā*], produces thunderbolt [κεραυνός], (2) and whirlwind [*ʿalʿālā*], when inflamed, produces firewind [πρηστήρ].

IV.ii.1.2-6: Thunderbolt and firewind: contd.

But.: (1) The material of both is denser [*ʿabyā*] than the material of lightning. (2a) For this reason, not only its light [*nuhrā*] but [also] its body [*gušmā*] reaches the earth aflame [*kad dāleq*], (2b) because of its density [*rṣīpūtā*] and the concentration [*knīšūtā*] of its earthy weight [*yuqreh arʿānāyā*], (2c) or because of the propulsion [*dḥāyā*] and compulsion [*znuqyā*] which occur to it. (3) The material of lightning, on the other hand, because of its fineness does not persist for long, but is quickly dissipated and extinguished, and its light reaches the earth without its body.

---

[19] See also *Šifā*ʾ 70.7-8, 71.1, where the definition is repeated.

[20] Cf. Arist. 371a 15-19; Olymp. 12.22-13.18; Olymp. arab. 144.2-4. - Cf. also Nic. syr. 44.6-10 [scholion]: ܥܛܪܐ ܗܟܝܠ/7 ܬܢܢܝܬܐ ܕܡܬܚܒܫܐ ܒܥܢܢܐ. ܐܘ ܡܬܒܕܪܐ: ܘܗܘܝܐ ܩܠܐ ܘܒܪܩܐ. ܐܘ/8 ܡܩܘܝܐ. ܘܗܢܐ. ܐܘ ܠܐ ܡܢܒܪܫܐ ܗܘܐ ܘܕܠܐ ܪܛܝܒܘܬܐ ܘܥܒܕܐ/9 ܪܘܚܐ ܥܢܢܝܬܐ. ܐܘ ܥܡ ܪܛܝܒܘܬܐ ܘܥܒܕܐ ܛܘܦܘܢ. ܐܘ ܡܢܒܪܫܐ ܗܘܐ [8 ܘܕܠܐ/10 ܪܛܝܒܘܬܐ ܘܥܒܕܐ ܩܪܘܢܘܣ. ܐܘ ܥܡ ܪܛܝܒܘܬܐ ܘܥܒܕܐ ܦܪܝܣܛܝܪ ܀/ ܘܕܠܐ: ܕܠܐ ܐܘ. ܗܘ cod. || 8, 9 & 10bis ܘܥܒܕ: ܘܥܒܕܐ cod. || 9 ܘܕܠܐ: ܕܠܐ ܐܘ cod.] ("Smoky exhalation, therefore, which is confined in a cloud is either dispersed [cf. Arist. 370b 5 διαχεόμενον] and becomes thunder and lightning, or is continuous. This [latter] is either not combustive [*mnabršā*] and without moisture and causes cloud wind, or with moisture and causes τυφών, or is combustive and without moisture and causes thunderbolt [κεραυνός], or with moisture and causes firewind [πρηστήρ]"; cf. Arist. 370b 4-13; Olymp. 12.22-13.18, 203.7-9, 203.12-16; Olymp. arab. 144.15-145.1).

Šifāʾ 70.6-9:

... ليست بلطيفة لطف البرق الذي لأجله لا يبقى شعاع البرق زمانا يعتد به، بل يتحلل ويطفأ، بل هي ريح سحابية مشتعلة تنتهي إلى الأرض، لا ضوؤها وحده، بل جرمها المشتعل لاستحصافه واجتماع ثقله الأرضي، أو لاضطراره إلى ذلك المأخذ والجهة، على ما نبأنا به.

(1/3) [Thunderbolts (*ṣāʿiqa*)] is not fine [*laisa bi-laṭīfa*] like the lightning, (3) whose ray for this reason does not persist for any significant time, but is dissipated and extinguished. Rather it is inflamed cloud wind (2a) which reaches the earth, not only its light, but [also] its inflamed bulk [*ǧirmu-hā al-muštaʿil*], (2b) because of its solidity [*istiḥṣāf*] and the concentration [*iǧtimāʿ*] of its earthy weight [*ṯiqlu-hu al-arḍī*], (2c) or because of its compulsion [*iḍṭirār*] towards that source and direction, in accordance with what we have said.

l. 4. ܕܘܚܩܐ ܘܙܢܘܩܝܐ ("propulsion and compulsion"): اضطرار ("compulsion") IS. - *Znuqyā* is a rare word, known to the dictionaries only from here and BH *Eth.* IV.xi.3 [Bedjan] 421.16 (see Brockelmann 201b; PS Suppl. 113). In both instances the word seems to have a sense, like IS's *iḍṭirār* (< √*ḍrr*), of "compulsive, unavoidable motion".[21]

IV.ii.2.1-9: Types of Thunderbolts (ἀργής and ψολόεις)

But. [underline: agreement with Nic. syr.; italics: with *Šifāʾ*]:

(1a) Thunderbolt which arises from finer material is called ἀργής, i.e. "white" [*ḥewwārā*], (1b) because it passes imperceptibly [*lā metraḡšānāʾīt*] through an object, (1c) so that it does not even impart colour to it. (2a) That [arising] from denser material is named [*meštammah*] ψολόεις, i.e. "touching" [*gāšōpā*], (2b) because it blackens [*mawkem*] the object it meets, (2c) but does not burn [*mawqeḏ*] it, (2d) unless perchance [*ellā kḇar*] (the object) is more compact [*aṭīm*]; (2e) a looser [*yattīr saḥḥīḥ*] object, it passes through and does not damage [*mʿīq*]. (3a) Consequently, when it meets *a shield* [*sakrā*] (3b) *it melts the silver and the copper* in it because they are solid and prevent it from passing through. (3c) The wood, however, as being something more loose, *it does not burn,* (3d) *although it may blacken it.* (4) [Similarly] *it melts the gold in a purse* [*ṣrārā*], *but it does not burn the purse.*

[21] The definition of the word as "prohibitio" given by Brockelmann for the instance in *Eth.* requires revision. At *Eth.* 421.14-17, which occurs in the chapter on confidence in God (*tuklānā*, corr. Ghazālī's *tawakkul*), BH is talking of four possible motives for human action: 1. desire for the expected good; 2. retention of good; 3. care [to avoid] the expected harm [*zhīrūṯā d-men suḡpānā mestakkyānā*]; 4. *znuqyā d-suḡpānā*. In view of the sense of *znuqyā* here in *But.*, along with BH's use of *mezdneq* at *But.* Mete. III.iv.1., IV.i.4, IV.ii.3 (see further PS Suppl. 113), I am inclined to understand the last phrase as "repulsion exerted by (certain and inevitable) harm", rather than simply "rejection [by man] of harm". Cf. the related passage at BH *Columb.* [Bedjan] 562.18-20, where BH talks of the just man confiding in God in the face of "possible harm" (*suḡpānā metmaṣṣyānā*), contrasting this with the (forced) confidence in the face of "necessary/unavoidable" [*ālṣānāyā*] harm.

Nic. syr. 45.21-25, 27-31[22]:

ܩܪܐܘܢܘܣ ܕܝܢ ܗܘ ܡܢ ܕܩܝܩܬܐ ܡܢ ܗܘܠܐ ܕܢܦܫ/22 ܡܬܝܠܕܐ . ܡܬܩܪܐ ܐܪܓܝܣ
ܐܘܟܝܬ ܚܘܪܐ . ܗܢܐ ܕܥܒܪ ܒܨܒܘܬܐ ܕܦܓܥ ܒܗ/23 ܟܕ ܠܐ ܡܬܪܓܫ
ܐܝܟܢܐ ܕܐܦܠܐ ܢܨܒܘܥܝܗ ܠܗܝ . ܗܘ ܕܝܢ ܕܡܢ ܗܘܠܐ ܕܥܒܝܬܐ/24 ܡܬܩܪܐ
ܦܣܘܠܘܐܝܣ ܐܘܟܝܬ ܡܬܩܪܒܢܐ . ܗܢܐ ܕܝܢ ܬܡܝܢ ܟܠܙܒܢ ܨܒܥ/25 ܠܨܒܘܬܐ
ܠܨܒܘܬܐ ܕܦܓܥ ܒܗܝܢ .//27 ܘܡܣܡܩ ܠܗܝܢ . ܠܐ ܕܝܢ ܡܢ ܟܠ ܙܒܢ ܡܘܩܕ :/
28 ܐܠܐ ܐܢ ܗܝ ܨܒܘܬܐ ܕܦܓܥ ܒܗ ܐܠܝܨܐ ܐܝܬܝܗ ܡܘܩܕ ܠܗ . ܐܢ ܕܝܢ
ܗܝ ܨܒܘܬܐ/29 ܪܦܝܐ ܘܕܠܝܠ ܠܐ ܡܘܩܕ ܠܗ . ܐܠܐ ܥܒܪ ܒܗ ܟܕ ܠܐ ܡܚܒܠ
ܠܗ . ܒܗܕܐ ܟܕ/30 ܦܓܥ ܒܡܓܢܐ ܠܢܚܫܐ ܡܢ ܕܒܗ ܐܝܟ ܨܒܘܬܐ ܐܠܝܨܐ
ܡܦܫܪ . ܠܩܝܣܐ ܕܝܢ/31 ܐܝܟ ܨܒܘܬܐ ܕܠܝܠܐ ܥܒܪ ܒܗ ܕܠܐ ܚܒܠܐ .

[21 & 23 ܗܘܠܐ: ܗܘ̈ܠܐ cod. (ut saepe)]

(1a) Thunderbolt which arises from finer material is called ἀργής, i.e. white. (1b) This passes through the object it meets without being perceived [*kad lā metrgeš*], (1c) so that it does not even impart colour to it. (2a) That [arising] from denser material is called [*metqrē*] ψολόεις, i.e. touching. (2b) This always gives colour to the objects it meets and blackens them. (2c) It does not, however, always burn, (2d) but if [*ellā en*] the object it meets is more compact it burns it; (2e) if it is looser [*saḥḥīḥ*] and rarer [*dallīl*] it does not burn, but passes through it while not damaging it. (3a) Consequently, when it meets a shield (3b) it melts the copper in it, as something that is more compact. (3c) The wood, however, as being something rarer, (3d) it passes [through] without damage.

*Šifā*ʾ 70.12-16:

وقوامها مع ذلك مختلف: فربما كانت ريحا سحابية سادجة، فتكون صاعقة لطيفة؛ وربما كانت لافحة فقط؛ وربما كانت سافعة اللون، وربما كانت مؤثرة فيما يقوم فى وجهها، لكنها تنفذ فى الأجسام المتخلخلة، ولا تحرقها، ولا تبقى فيها أثرا؛ وربما كانت أغلظ من ذلك فتنفذ فى المتخلخل نفوذا يبقى فيه أثر سواد، وتذيب ما تصادمه من الأجسام المتكاثفة، ولذلك ما تذيب الضباب المضببة على الترسَة ونحوها المتخذ من الفضة والنحاس، ولا تحرق الترسة، بل ربما سودتها؛ وكذلك قد تذيب الذهب فىَ الصرة ولا تحرق الصرة، إلا ما يحترق عن الذوب؛ وربما كانت شرا من ذلك؛ وربما كانت سحابية زوبعية مشتعلة، وتكون من مادة كثيفة، فتكون شر الصواعق.

Its consistency [*qawām*] neverthless varies. Often it is plain cloud wind, and it becomes a fine thunderbolt. Often it merely scorches [*lāfiḥ*]. Often it makes a black stain [*sāfiʿa*]. Often it leaves a mark [*muʾaṯṯir*] on what stands in its way. Only it penetrates loose objects, but does not burn them and does not leave a mark in them. (2) Often it is denser than that and penetrates through a loose (object) in a way which leaves a black mark in it, and melts the thick objects it strikes. (3ab) For that reason, it melts the bolts [*ḍibāb*] on shields [*tirasa*] and the like which are made from silver and copper, (3c) but does not burn [*tuḥriqu*] the shield, (3d) although it often blackens it [*sawwadat-hā*]. (4) In the same way it sometimes melts the gold in a purse [*ṣurra*], but does not burn the purse, except what is burned by melting.

---

[22] Cf. Arist. 371a 19-29; Alex. Aphr. 137.20-138.5; Olymp. 12.26-13.2, 202.1-11, 208.10-16; Olymp. arab. 144.5-13; *Cand*. 119.7-120.1.

l. 1, 3. ܐܪܓܝܘܣ ܐܘܟܝܬ ܚܘܪܐ, ܦܣܘܠܐܘܣ ܐܘܟܝܬ ܓܫܘܫܐ ("ἀργής i.e. white", "ψολόεις i.e. touching"): ܐܪܓܝܣ ܐܘܟܝܬ ܚܘܪܐ, ܦܣܘܠܐܝܣ ܐܘܟܝܬ ܓܫܘܫܐ Nic. - Nic. syr. has retained forms which are more faithful to the Greek.[23] The words occur in the same forms as in *But*. in *Cand*.,[24] so that BH must be held responsible for the corruption here. The misinterpretation, on the other hand, of ψολόεις as "touching" (as if from ψαύω, rather than from ψόλος, "soot") is there in Nic. syr. and can, in fact, be blamed on Olymp. 13.1f.: διὰ τοῦτο γὰρ καὶ ἀργῆτα κεραυνὸν εἶπε διὰ τὸ ἀργῶς καίειν, ψολόεντα δὲ διὰ τὸ ψαύοντα μελαίνειν.[25]

IV.ii.3.1-3: Firewind (πρηστήρ)

But. [underline: agreement with Nic. syr.; italics: with *Šifā*ʾ]:
Firewind, because it *arises from denser material*, burns and inflames [*mawqed wa-mšalheb*] the air.
Nic. syr. 45.31[26]: . ܦܪܣܛܝܪ ܕܝܢ ܡܘܩܕ ܘܡܫܠܗܒ ܠܐܐܪ
Firewind burns and inflames the air.
*Šifā*ʾ 70.16-17[27]: وربما كانت سحابية زوبعية مشتعلة، وتكون من مادة كثيفة، فتكون شر الصواعق
Sometimes it is an inflamed whirlwindy cloud wind and arise from thick material. This is the worst of thunderbolts.

IV.ii.3.2-3: Firewind (contd.)

In the passage of the *Šifā*ʾ quoted below IS is talking about thunderbolts in general. BH has taken some liberty in applying this to "firewinds" in particular. - For the mention of stones in thunderbolts, see Min. I.i.4 above.

---

[23] "Ψολόεις" occurs only once in the surviving text of Nic. syr.; "ἀργής" occurs twice more besides in the passage quoted above, at 45.26 where it is written ܐܪܓܝܣ, and at 46.6 where it is written ܐܪܓܝܘܣ as in *But*. and *Cand*.

[24] ܐܪܓܝܣ ed. Bakoš 119.11, but ܐܪܓܝܘܣ mss. BV and ed. Çiçek 91.7; ܦܣܘܠܐܘܣ edd. Bakoš 120.1 and Çiçek 91.15.

[25] The terms ἀργής and ψολόεις also occur in the Greek *De mundo*, but are lost in what is a problematic passage in the Syriac version: *De mundo* gr. 395a 26f. τῶν δὲ κεραυνῶν οἱ μὲν αἰθαλώδεις ψολόεντες λέγονται, οἱ δὲ ταχέως διᾴττοντες ἀργῆτες; syr. 144.18f. ܡܢܗܘܢ ܕܝܢ ܕܒܪ̈ܩܐ ܕܢܚܫܝܢ ܗܢܘܢ ܕܛܠܝܠܝܢ. ܟܒܪܝܬܢܝܐ ܡܬܩܪܝܢ. ܗܢܘܢ ܕܝܢ ܕܩܠܝܠܝܢ. ܪ̈ܗܘܛܐ ܕܠܐ ܦܣܩܝܢ ("Out of thunderbolts, those which are moist [*tallīlīn*, corr. αἰθαλώδεις?] are called 'sulphureous/fuming' [corr. ψολόεντες], those which are quick [are called] 'undividing runners' [corr. διᾴττοντες ἀργῆτες]"). - "Moist": how αἰθαλώδεις came to be translated as "moist" is unclear, but it may be noted that ἀερῶδες of gr. 395a 20 is also rendered by "moist" [*tallīl*] at syr. 144.12; cf. Ryssel (1880) I.43, PS 4437. - "Sulphureous/fuming": *De mundo* arab. F [Brafmann] 95r 9 has *qutārī*; cf. Brafmann's comm. ad loc. (p. 227). - "Undividing": presumably due to wrongly deriving ἀργής from privative ἀ- + ῥήγνυμι.

[26] Cf. Arist. *Mete*. 371a 16-17; Olymp. 202.12-14; Olymp. Arab. 144.14-15; *Cand*. 120.4-6.

[27] Cf. under IV.ii.1.1-2 above.

But.: (1) When it is extinguished, its thick material is transformed into hard bodies [*gušmē qšayyā*] in accordance with the mixture it has [*lpūt muzzāgā d-qanyā*] and falls to the earth.

Šifā' 71.1-2:

وبالجملة فالصواعق رياح سحابية مشتعلة، وربما طفئت هذه الصواعق فتستحيل أجساما أرضية بحسب المزاج الذى يكون فيها، وعلى ما اقتصصنا لك من خبرها.

In short, thunderbolts are inflamed cloud winds. (1) Often these thunderbolts are extinguished and are transformed into earthy bodies [*ağsām arḍīya*] in accordance with the mixture [*bi-ḥasabi l-mizāğ*] which is in them and in accordance with what we have told you about the matter.

IV.ii.3.3-5: Wind as material of thunderbolt and firewind

But.: (1) That these things arise from winds (2) is known from the observation [*methazyānūt*] of flames blowing through the air, (3) as was observed in the fire in the temple [*hayklā*] in Ephesus. (4) For flame is nothing other than [lit. nothing else but if] burning smoke.

Nic. syr. 45.32, 46.5-8[28]:

ܘܕܡܢ ܪ̈ܘܚܐ ܗܘܝܢ ܗܠܝܢ . ܘܠܐ ܡܢ ܐܘܣܝܐ ܐܪܥܢܝܬܐ ܐܝܟ ܕܐܣܒܪܘ ܐܢ̈ܫܝܢ/
[46] 5/ ܝܕܝܥܐ ܡܢ ܗܝ ܕܐܪܓܝܣ ܥܒܪܐ ܠܓܫܡܐ/6 ܐܟܘܬ ܕܠܐ ܡܬܪܓܫܢܘܬܐ . ܐܦ ܓܝܪ
ܫܠܗܒܝܬܐ ܥܒܪܐ ܒܐܐܪ . ܐܝܟܢܐ ܕܐܬܚܙܝܬ ܒܢܘܪܐ/7 ܕܗܝܟܠܐ
ܕܒܐܦܣܘܣ ܡܛܠ ܕܡܢ ܪܘܚܐ ܡܬܦܚܡܐ . ܐܬܚܘܝܬ ܓܝܪ ܕܠܫܠܗܒܝܬܐ/8 ܫܠܗܒܝܬܐ
ܠܐ ܐܚܪܝܢ : ܐܠܐ ܐܢ ܬܢܢܐ ܕܝܩܕ .

[32 ܗܘܝܢ : ܗܘ̈ܝܢ cod. || ܕܐܣܒܪܘ: ܕܐܣܒܪ cod. || 7 ܕܒܐܦܣܘܣ: ܕܒܐܦܣܘܣܘ cod.]

(1) That these things arise from winds and not from an earthy substance [οὐσία] as some thought (2) is known from the fact that ἀργής passes through an object without being perceived. For flame also passes through air - (3) as was seen in the fire in the temple in Ephesus - because it is formed from wind. (4) For it has been shown that flame is nothing other than burning smoke.

l. 3. ܗܠܝܢ ("these things"): i.e. thunderbolt and firewind. - Strictly speaking, in *But*., the pronouns *could* be understood as referring back to the "hard bodies" of line 2, but in Nic. syr., as well as in the corresponding passage of Arist. (371a 29 ταῦτα πάντα), the reference is to the thunderbolt and the firewind. Note also "thunderbolt and firewind" in l. 6 below.

IV.ii.3.5-9: Wind as material (contd.)

But.: (1) Before and after thunderbolt and firewind, wind blows, (2) because at the beginning and at the end wind is squeezed and discharged weakly from the cloud (3) and for this reason is not set on fire. (4) When, on the other hand, it is ejected with force [lit. "with strength of its ejection"], it is

[28] Cf. Arist. 371a 29-b 2; Alex. Aphr. 138.3-10; Olymp. arab. 144.22-145.2.

set on fire. (5) Wind in fact also occurs with thunder. (6) This often tears even the earth.

Nic. syr. 46.8-16[29]:

ܗܘ ܕܝܢ ܕܠܡܨܥܬܐ ܕܪܘܚܐ ܗܕܐ/9 ܕܡܢܗ ܗܘܝܢ ܡܫܬܪܝܘܬܐ ܘܦܪܣܛܝܪ. ܡܬܝܩܕܐ
ܒܠܚܘܕ ܡܬܝܩܕܐ. /10 ܣܘܦܝܗ̈ ܕܝܢ ܠܐ ܡܬܝܩܕܝܢ. ܘܡܛܠ ܗܢܐ ܡܢ ܩܕܡ
ܡܫܬܪܝܘܬܐ ܘܡܢ/11 ܒܬܪܗ ܗܘܝܐ ܪܘܚܐ. ܒܫܘܪܝܐ ܓܝܪ ܘܒܚܪܬܐ ܘܡܬܠܚܨܐ
ܡܫܠܝܐܝܬ ܡܬܦܪܩܐ/12 ܘܡܬܦܠܛܐ ܪܘܚܐ ܡܢ ܥܢܢܐ. ܘܡܛܠ ܗܢܐ ܠܐ
ܡܬܝܩܕܐ. ܟܕ ܕܝܢ ܕܠܘܩܒܠ:/13 ܡܛܠ ܕܚܝܠܬܢܐܝܬ ܡܬܦܠܛܐ ܡܬܝܩܕܐ.
ܟܕ ܗܟܢ ܕܥܬܝܕ ܡܫܬܪܝܘܬܐ/14 ܠܡܗܘܐ ܒܝܡܐ ܡܬܦܠܓ. ܘܐܦ ܡܢ ܒܬܪ
ܕܕܥܟ ܡܬܦܠܓ. ܠܘ ܕܝܢ ܒܠܚܘܕ/15 ܥܡ ܡܫܬܪܝܘܬܐ ܐܝܬ ܪܘܚܐ. ܐܠܐ ܘܥܡ
ܪܥܡܐ. ܗܢܐ ܕܒܙܒܢܐ ܣܓܝܐܐ ܐܦ ܠܐܪܥܐ/16 ܡܨܪܝܐ.

[9 ܗܘܝܢ : ܗܘܝܐ cod. || 14 post ܒܝܡܐ aut post ܡܬܦܠܓ: fort. ܗܘ ܝܡܐ supplend.; vide Nic. syr. 46.16]

We ought also to understand that only the middle of this wind, from which thunderbolt and firewind arise, is on fire, while its edges are not on fire. (1) For this reason wind occurs before and after thunderbolt. (2) For at the beginning and at the end wind is squeezed and dicharged weakly from the cloud (3) and for this reason it is not set on fire. (4) When, on the other hand, it is ejected with force, because it is discharged strongly [*ʿaššīnāʾīt*], it is set on fire. Consequently, when a thunderbolt is about to occur on the sea, (the sea) is split, and also after (the thunderbolt) is extinguished it is divided. (5) There is wind not only with thunderbolt but also with thunder, (6) which [wind] often even tears the earth.

Cf. *Šifāʾ* 71.2-3: وإذا أرادت صاعقة أن تصعق، تقدمتها في أكثر الأمر ريح

When a thunderbolt is about to strike, a wind in most cases precedes it.

IV.ii.3.9-10: Concluding comment

But.: So much, then [lit. these are/were] on thunderbolts and firewinds.

Nic. 46.16, 17: ܘܗܠܝܢ ‹ܗܟܝܠ› ܥܠ ܡܫܬܪܝܘܬܐ //17 ܘܦܪܣܛܝܪ. .

[16 ‹ܗܟܝܠ› addidi, cf. Nic. syr. 45.18 etc., et *But*.]

So much, then, on thunderbolt and firewind.

Mete. IV. Section iii.: *On Δοκίδες, Shooting Stars and Comets*

The section corresponds to the latter half of the chapter of the *Šifāʾ* on phenomena produced through the combustion of the smoky exhalation, where IS discusses such phenomena as shooting stars and comets (ed. Cairo 71.4-74.15). The order of presentation is that of the *Šifāʾ* and most of the material is also taken from that work. - Nic. syr. is used in

---

[29] Cf. Arist. 371b 7-14; Alex. Aphr. 138.10-20; Olymp. 208.18-26; Olymp. arab. 145.2-12, 144.16-21.

the first 'theory' of the section and the latter part of that 'theory" describing the different types of shooting stars is based mainly on Nic. syr. - There are also a number of indications that BH consulted Fakhr al-Dīn Rāzī's *Mabāḥiṯ* in composing this section. - The passage at IV.iii.4.7-10, which deals with the "column of fire", a phenomenon in which BH seems to have had a particular interest, has no counterpart in the *Šifā*ᵓ and is best explained as being based on passages taken from BH earlier works.

	*Šifā*ᵓ	Other sources	
IV.iii.1.1-5	71.4-5, 7-10	Nic. 11.19-20	Shooting stars
IV.iii.1.6-10	71.10-11	Nic. 11.21-28	Types of shooting stars
IV.iii.2.1-9	71.15-16, 72.5-17		Extinction of fire
IV.iii.3.1-4	72.18-20		contd.
IV.iii.3.4-7	73.7-14, 16-17		contd.: comets
IV.iii.3.7-10	(73.14-16)		Comet of 1264 A.D.
IV.iii.4.1-4	74.3-5		Rarity of comets etc.
IV.iii.4.4-6	74.5-6	(*Mab.* II.189.5-6)	"Terrible signs"
IV.iii.4.7	74.8-9		"Chasms" etc.
IV.iii.4.7-10		(*Mab.* II.190.6-12) Cf. *Cand.*, *Tract.*	"Column of fire"
IV.iii.4.10-11	74.14-15		Comets etc. as portents

l. 1. ܕܘܩܝܕܝܣ: < gr. δοκίδες. - The word, also used by BH at *Cand.* 120.7, is taken from Nic. syr. (see under IV.iii.1.6-10 below).[30]

l. 1. ܫܘܘܪܐ ("shooting stars"): The word is probably to be read as a nomen agentis *šāwōrā*, i.e. "jumper", "jumping star" (see PS Suppl. 330a). - The word ܫܘܪܐ (*šāwrā*, participle?) is used of a type of shooting stars at Nic. syr. 12.24,[31] while it is probable that ܫܘܪܪܐ at Nic. syr. 11.28 (see under IV.iii.1.6-10 below) should be emended to ܫܘܘܪܐ. The verb *šāwar* is used several times of the movement of shooting stars in Nic. syr. (11.23, 12.1 bis, 12.2).

IV.iii.1.1-5: Shooting stars

But. [underline: agreement with *Šifā*ᵓ; italics: with Nic.]:

(1) The material of δοκίδες, i.e. torches which are seen in *the αἰθήρ*, and the rest of these things is smoke, (2a) since vapour *sinks* [*šāḵeṯ*] *because of its weight* [*b-yaḏ yuqreh*] (2b) and does not rise [*meṯᶜallē*] there [*l-ṯammān*] (2c) but grows cold before it reaches that place [*waᶜdā*]. (3)

[30] The δοκίδες of *De mundo* gr. 392b 4, 395a 12 are rendered by *nayzḵē* in *De mundo* syr. (138.3, 145.11).

[31] Nic. syr. 12.24: ܫܘܪܐ ܕܒܙܒܙܐ ܕܢܝ̈ܙܟܐ ("the jumping type of shooting stars", corr. Arist. 344a 15 τὰς τῶν σποράδων ἀστέρων διαδρομάς).

The smoky exhalation, if fine, *is ignited* [*meštalhab*] when it reaches the αἰθήρ, (4) and because of its fineness, *the flame* [*šalhebītā*] travels through it with speed. (5) As soon as it has caught fire [*dāleq*] it is dissipated, (5) and it seems [*mestbar*] like a star [*ak kawkābā*] which jumps and falls from the sky.

Šifā' 71.4-5, 7-10:

وأما الآثار المحسوسة فى أعلى الجو فإنها متكونة من الدخان، إذ البخار لا يتصعد إلى ما هنالك لثقل حركته، ولأنه يبرد فيما دون ذلك .. ومن ذلك شهب الرجم، ومادتها أيضا البخار الدخانى اللطيف السريع التحلل. وذلك أن هذا الدخان إذا وصل إلى الجو المحرق اشتعل وسرى فيه الاشتعال كأنه يقذف، ويكون كما يشتعل يتحلل فيرى كأن كوكبا ينقذف.

(1) The phenomena observed in the upper air [*aʿlā al-ǧauw*] are formed from smoke, (2ab) since vapour does not rise to those parts [*mā hunālika*] because of the weight of its movement [*li-tiql ḥarakatihi*] (2c) and because it grows cold while in parts below it. ... (3) Hence, the material of the shooting stars [*šuhub al-ruǧum*] too is smoky vapour, which is fine and quickly dissipated. That is, this smoke, when it reaches the burning air [*al-ǧauw al-muḥriq/muḥraq*, i.e. aether], is ignited [*ištaʿala*] (4) and the flame [*ištiʿāl*] spreads through it as if it were being thrown. (5) As soon as [*kamā*] it is ignited it is dissipated, (6) so that it looks [*yurā*] as if [*ka-anna*] a star is being hurled [*yanqadifu*].

Nic. syr. 11.19-20[32]:

ܘܗܢܐ ܦܪܚ ܟܕ ܡܬܩܪܒ ܠܐܝܬܪ. ܗܘ ܕܝܢ ܪܛܝܒܐ ܡܛܠ ܝܘܩܪܗ ܫܐܟܬ ܘܩܘܝܐ ܒܐܐܪ/
ܘܦܐܫ . ܗܘ ܕܝܢ ܝܒܝܫܐ ܡܬܬܙܝܥ ܘܡܬܢܒܪܫ . ܫܠܗܒܝܬܐ ܓܝܪ ܐܝܬܝܗ̇/
ܪܬܚܐ ܕܪܘܚܐ ܝܒܝܫܬܐ .

The latter [sc. dry exhalation] floats, approaching the αἰθήρ. (2) The moist (exhalation), because of its weight [*meṭṭul yuqreh*], sinks [*šāket*] and persists in the air. (3) The dry (exhalation) moves and is kindled [*metnabraš*]. For flame [*šalhebītā*] is in fact boiling up of dry wind.

l. 1. ܕܘܩܝܕܣ ܐܘܟܝܬ ܠܡܦܐܕܐ ("δοκίδες, i.e. torches"): see under IV.iii.1.6-10 below.

l. 5. ܕܫܘܪ ܘܡܢ ܫܡܝܐ ܢܬܪ ("which jumps and falls from the sky"): ينقذف ("is hurled") IS. - For *šāwar*, see comm. on IV.iii.tit. above. - The verb *nāṭar* which occurs only here in *But*. Mete. and which does not occur in Nic. syr. is also used by BH at the corresponding place in *Cand*. 121.8: ܘܐܝܟ ܟܘܟܒܐ ܕܫܘܪ ܡܢ ܫܡܝܐ ܘܢܬܪ ܡܬܚܙܝܐ

IV.iii.1.6-10: Types of shooting stars

But. [underline: agreement with Nic.; italics: with *Šifā'*]

(1a) If the smoke is *dense*, (1b) has much ὑπέκκαυμα, i.e. *material* [*mlō'ā*], (1c) and is extensive [*mtīḥ*] in length and width, (2) a large flame, like stubble in a field [*ḥabbṭā da-b-ḥaqlā*], is seen burning and persists for days. (3) If it is extensive in length [only], things like torches [*lampīdē*]

[32] Cf. Arist. *Mete*. 341b 10-12, 18-24.

are seen. (4) If things like *sparks* [*belṣuṣyāṯā*], which are fine and sprout like hair from a single source, are seen, these are called "goats" [*ʿezzē*]. (5) If there occurs something like a stream of fire [*redyā d-nūrā*], it is named δαλός.

Nic. syr. 11.21-28[33]:

ܡܐ ܕܣܓܝܐܬܐ ܗܘܬ ܒܐܘܪܟܐ ܘܦܬܝܐ. ܡܬܚܙܝܐ/22 ܕܝܩܕܐ ܐܝܟ ܚܩܠܬܐ
ܕܚܣܟܬܐ. ܡܐ ܕܝܢ ܕܒܐܘܪܟܐ ܒܠܚܘܕ: ܕܘܩܝܕܣ/23 ܘܠܡܦܝܕܐ ܡܬܩܪܝܢ. ܡܐ
ܕܝܢ ܕܐܝܟ ܫܠܗܒܝܬܐ ܡܬܚܒܪܬܐ ܘܣܓܝܐܬܐ ܢܫܘܪ/24 ܡܢ ܚܕ ܪܝܫܐ ܥܙܐ
ܡܬܩܪܝܢ. ܗ̄ ܡܛܠ ܕܐܝܟ ܕܐܝܠܝܢ ܕܡܥܪܐ ܕܥܙܐ ܡܢ ܡܫܟܗ/25 ܕܥܙܐ ܡܚܝܕ
ܘܡܚܒܪ ܠܡܫܟܗ: ܗܟܢ ܡܚܝܕܢ ܫܠܗܒܝܬܐ ܠܢܒܪܫܬܐ܀ /26 ܡܐ ܕܝܢ ܕܐܝܟ
ܪܕܝܐ ܒܠܚܘܕ ܕܢܘܪܐ ܗܘܐ: ܕܐܠܘܣ ܡܬܩܪܐ. ܗ̄ ܠܡܦܝܕܐ. ܡܐ ܕܝܢ/27 +
ܕܒܕܘܟܝܬܐ ܣܓܝܐܬܐ: ܘܩܠܝܠ ܩܠܝܠ ܐܝܟ ܫܠܗܒܝܬܐ ܢܗܘܘܢ ܟܘܟܒܐ
ܕܢܗܠܝܢ+ :/28 ܫܘܘܪܢܐ ܡܬܩܪܝܢ. +ܗ̄ ܡܢ ܢܚܠܐ ܢܦܩܐ ܢܚܠܝܢ+ ܀

[22, 23 & 26 prim. ܡܐ ܕܝܢ: ܡܕܝܢ cod. ‖ 22 ܕܘܩܝܕܣ < δοκίδες, cf. *But*. et *Cand*. 120.7: ܕܘܩܝܕܣ cod. ‖ 26 ܫܘܘܪܢܐ cf. Nic. syr. 12.24 et *But*.: ܫܘܪܪܐ cod.]

(1) When it is great [*saggīʾṯā*] in length and width, (2) it is seen burning like stubble in a field.[34] (3) When in length only, they are called δοκίδες and torch [*lampīdā*]. (4) When they jump like many connected [*sḇīsāṯā*] sparks from a single source [*rēšā* < ἀρχή], they are called goats [< αἶγες] - i.e. because just as the hair coming out [*nāḏē*] from the skin of a goat is united and connected to its skin, so the sparks are connected to the flame [*naḇreštā*]. - (5) When it is merely like a stream of fire, it is called δαλός - i.e. torch [*lampīdā*].- When the sifted [*d-nāḥlīn*] stars become like sparks in many places and little by little, they are called jumping (stars) - i.e. +He derives "sifted" [*nāḥlīn*] from "sifting" [*nḥālā*] [?][35]+.

Šifāʾ 71.10-11:

وقد يتفق أن يبقى اشتعاله طويلا قطعة يسيرة من الزمان، وقد يكون له شرر، هذا إذا كانت المادة أكثف

Sometimes the flame [*ištiʿāl*] may remain longer, for a brief moment of time; it may have sparks [*šarar*]. (1) This happens when the material [*mādda*] is thicker.

l. 6. ܐܘܦܩܐܘܡܐ ("ὑπέκκαυμα"): cf. Arist. *Mete*. 341b 24. - The word is not found in the Cambridge manuscript of Nic. syr., but may still have been there in BH's copy of that work.

l. 8-9. ܐܝܟ ܠܡܦܝܕܐ ܡܬܚܙܝܢ ("things like torches are seen"): The verb is in agreement with *lampīdē* (m.pl.), so that we cannot translate "... appear like torches", but must understand *aḵ lampīdē* as the subject meaning "quasi/pseudo-torches". The same applies to l. 9 *aḵ belṣusyāṯā* with the

[33] Cf. Arist. *Mete*. 341b 24-35; Olymp. 37.35-38.10, 41.18-26; Olymp. arab. 95.2-9, 95.16-96.7.

[34] Arist. 341b 26f.: ὥσπερ ἐν ἀρούρᾳ καιομένης καλάμης.

[35] The phrase *d-nāḥlīn* ("which are sifted") of the main text may be connected to Arist. 341b 33 πολλαχῇ διεσπαρμένα. A later glossator may have failed to understand this and tried to derive the word from *naḥlā* in the sense of "torrent".

verbs *šāwḥān*, *metḥazyān* and *metqaryān* and l. 10 *ak redyā d-nūrā* with the verbs *nehwē* and *metkannē*.

l. 10. ܥܙܐ ("goats"): - "Goats" (*aʿnaz*) are mentioned later in the *Šifāʾ* at 73.19, in the continuation of the passage corresponding to III.iii.3.4-7 below.

IV.iii.2.1-9: Extinction of fire

The passage here may be seen as a summary mainly of the latter part of IS's discussion on the nature of fire and its extinction (*Šifāʾ* 71.15-72.17). - IS's discussion of fire was summarised by Fakhr al-Dīn al-Rāzī at the corresponding place in the *Mabāḥit* (II.189.13-190.4). BH's summary here in *But.* is different from Rāzī's and follows the text of the *Šifāʾ* itself, although BH had used Rāzī's summary in *Cand.* (131.8-132.3, in the section on the element fire, rather than that on the element air, where comets etc. are discussed).

But.: (1) Here one ought to give the cause of extinction [*duʿʿākā*] of fire. (2a) For we say that pure fire [*nūrā ḥattīttā*] glows [*dālqā*] (2b) when it is burning [*ḥābā*] with smoky material [*hūlē tennānāytā*] and not on its own, (2c) as has already been demonstrated. (3a) Therefore, its extinction occurs, (3b) either because in the region of air or water it is transformed by coldness and moisture and becomes air or something else, (3c) or because the material with which the fire burns is transformed into fieriness in its entirety, (3d) so that there remains nothing earthy at all. (3e) Thus, since the fire no longer has [lit. there remains to fire] any material with which to burn and glow, (3f) it reverts to its transparent nature (3g) and becomes invisible, (3h) and it appears as if it is already extinguished, (3i) when in truth it is not dead but is [still] alive.

Šifāʾ 71.15-16:

ويجب أن نتكلم ههنا فى علة طفوء النار، حتى يتوصل به إلى معرفة شىء مما نريد أن نقوله من هذا. فنقول: ...

[16 من هذا: من بعد ed. Ṭ]

(1) We must talk here about the cause of extinction [*ṭufūʾ*] of fire, so that one may obtain knowledge of the matter about which we wish to speak from here. (2) We say ...

Šifāʾ 72.5-17:

ولما كان الضوء، كما علمت، ليس شيئا يلزم ذات النار الصرفة، بل يعرض للنار إذا كانت متعلقة بمادة دخانية، ويكون حامل الضوء تلك المادة الدخانية، وقد ثبت هذا فيما سلف، كان طفوء النار إما بسبب فى نفس القوة الفاعلة للاشتعال والإشراق، وإما بسبب فى القوة القابلة، أعنى جوهر الدخان. فمن المعلوم أن القوة الطبيعية الفاعلة ما دامت ملاقية للمادة القابلة، فمن المستحيل أن يبطل فعلها إلا ببطلانها. فإذا بطل هذا الإشراق، فالسبب فيه لا محالة، إما من جهة الفاعل بأن تكون تلك النار قد استحالت ببرد غشيها أو رطوبة هواء أو شىء آخر، وهذا هو الطفوء الذى يكون فى حيز الهواء أو الماء بسبب البرد والرطوبة؛ وإما بسبب المادة فإنها إذا استحالت استحالة تامة إلى النارية حتى لم يبق فيها من طبيعة الأرضية شىء فبطلت الدخانية فلم يكن للنار شىء تتعلق به وتشرق فيه، بل صار الشىء كله نارا شافة، والشاف ليس يضىء بضوء نفسه. وإذا كان كذلك غابت النار عن الحس، وقيل إنها طفئت.

[7 ويكون: ويكون ed. Ṭ || 8 كان: كأن ed. Cair.]

(2a) Since the glow [*ḍau*ʾ], as you know, is not something which is inseparable from [*yalzamu*] the essence [*ḏāt*] of pure fire [*al-nār al-ṣirfa*], (2b) but occurs to fire when it is attached [*mutaᶜalliqa*] to smoky material and that smoky material becomes the bearer of the glow - (2c) this has been established before - extinction of fire is either due to a cause in the efficient force [*qūwa*] itself of the inflammation and illumination or due to a cause in the recipient force, I mean the substance of the smoke. It is known that a natural efficient force [persists] so long as it encounters recipient material and it is absurd that its action should cease except with its cessation. (3a) When, therefore, the illumination [*išrāq*] ceases, (3b) the cause must be either [i.] on the side of the efficient cause [*al-fāᶜil*] in that that fire has been transformed into air or something else by the coldness or moisture which has overwhelmed it - this being the [type of] of extinction which occurs in the region of air or water due to coldness and mosture - or [ii.] (3c) due to the material. For when it is transformed completely into fieriness (3d) so that there remains in it nothing of earthy nature and the smokiness disappears, (3e) the fire has nothing to be attached to and to glow with, (3f) but the whole thing becomes transparent fire and a transparent thing itself does not glow. (3g) This being the case, the fire is concealed from perception (3h) and is said to be extinguished.

IV.iii.3.1-4: Extinction (contd.)

But.: (1) Torches, comets etc. are not extinguished by the first cause because coldness and moisture have no power [*šulṭānā*] in that upper height [*rawmā ᶜellāyā*], (2) but by the second cause, namely, when their material is transformed completely into fieriness [lit. into complete fieriness], it becomes transparent and its glow is hidden from the eye [lit. eyes, *ḥāzyāṯā*].

Šifāʾ 72.18-20:

فهذه الشهب والكواكب وذوات الأذناب وغير ذلك يستحيل أن تطفأ وهو فى العلو بالسبب الأول، لأن البرد والرطوبة لا سلطان لها هناك، بل إنما تطفأ بالسبب الثانى وهو أن مادتها تستحيل بالكمال نارا فتشف فلا ترى ضوءا.

[18 وذوات: ذوات BDSM]

(1) It is impossible that these shooting stars [*šuhub*], stars [*kawākib*], comets etc. be extinguished by the first cause seeing that [*wa-*] they are in the upper atmosphere [*ᶜulūw*], because coldness and moisture have no power [*sulṭān*] there, (2) but they are extinguished only by the second cause, i.e. [*wa-huwa anna*] their material is transformed completely into fire, becomes transparent and is not seen as a glow.

IV.iii.3.4-7: Extinction (contd.)

But.: (1) If the material is very fine its imaginary extinction [*duᶜᶜāḵā makklānāyā*] is accelerated, (2a) but if it is thick (2b) it is retarded (2c) and its glow appears as a star with a lock of hair, a tail or a beard, and also as a horned beast and like fiery rods

*Šifā*ʾ 73.7-14, 16-17:

فإن كانت المادة لطيفة وخفيفة حتى حصل لها باللطافة أن كانت سريعة الاستحالة إلى النارية، وبالخفة أن تمكنت من الحصول فى الحيز الذى فيه النار قوية جدا، اضمحل اشتعالها دفعة وخلصت نارا، وشفت. فإن كانت المادة كثيفة وذات مدد وثقيلة، فإنها تبطئ استحالتها نارا خالصة، ولا يكون لها برد مطفئ، ولا أيضا تصعد صعودا سريعا ممعنا فى حيز النار إلى أن تبلغ المكان الشديد قوة النارية، فيعرض لذلك أن يبقى التهابها واشتعالها مدة طويلة إما على صورة ذؤابة أو ذنب، وأكثره شمالى وقد يكون جنوبيا، وإما على صورة كوكب من كواكب، كالذى ... وقد يكون على صورة لحية، أو صورة حيوان له قرون، وعلى سائر الصور

(1) If the material is fine and light so that because of its fineness it is quickly transformed [lit. it is quick of transformation] into fieriness and because of its lightness it is able to reach the region in which fire is very strong, its inflammation [*ištiʿāl*] vanishes suddenly, it becomes pure fire and it becomes transparent. (2a) If the material is thick and has reinforcement and is heavy (2b) so that its transformation into pure fire is retarded, [if] it has no coldness to extinguish it and [if] also it does not rise quickly and eagerly in the region of fire so as to reach the place which is intense in the strength of fieriness, (2c) then its combustion and inflammation will for that reason persist for a long time, either in the shape of a lock of hair [*ḏuʾāba*] or a tail - most of these being in the north, sometimes in the south - or in the shape of a star, like the one which ... (2c) Sometimes they are in the shape of a beard, or the shape of a horned animal, and other shapes.

l. 7 ܫܒܛܐ ܢܘܪܢܝܐ ("fiery rods"): This is not in IS. - The noun is presumably *šabṭā* (rod) rather than *šḇāṭā* (straight lock of hair), since "rods and fiery shoots" (or "fiery rods and shoots") are also mentioned by BH in *Rad.*, although there they are given as alternative names for "lances" (cf. comm. on Mete. II.i.1.1 above): *Rad.* [Istanbul] 24.9-11: ܗܘ ܗܟܘܬ ܢܝ̈ܙܟܐ ܘܓܐ̈ܪܐ ܘܫܒ̈ܛܐ ܘܣܘ̈ܟܐ ܢܘܪܢܝܐ ܕܡܬܚܙܝܢ. ܡܢܘ̈ܬܐ ܐܢܝܢ ܕܥܢܢ̈ܐ ܕܡܬܚܡܝܢ. ("Similarly, the lances, arrows, rods and fiery shoots that are seen are parts of clouds which are on fire"). - The four terms given here in *Rad*. are the same as the four given in *Cand*. 117.2-3, where, however, there is no question of them being "fiery". It is unclear why BH came to speak of these "rods" as being fiery when writing *Rad.* and to include them also under comets in *But*., but there may be some influence here of Abū al-Barakāt, who groups the "rod" (*ʿaṣan*) and the "mock suns" (*šumūs*) together with comets, torches etc., rather than with the rainbow and the halo, as phenomena due to ignited smoky vapour whose shapes are retained by "celestial forces" (*Muʿtabar* II.223.12-21, 226.19-22).

IV.iii.3.7-10: Comet of 1264 A.D.

Where IS had given an account of a comet he had observed, BH substitutes an account of a comet he observed himself, borrowing some of the wording from IS. - The comet BH refers to is the well-attested comet of July-October 1264, whose disappearance was associated in Europe with the death of Pope Urban IV. The same comet is mentioned by BH at *Cand*. 221.6-10, where it is associated with the death of Hulagu Khan in the following winter. The same association is made

by historians writing under Mongol rule, such as Rashīd al-Dīn, Grigor of Akner and Kirakos of Gandzak.[36]

But.: ... such as those which appeared [*eṯhawwī(w)*] in the year 1575 of Seleucus around the sign of Leo and circulated with it. They persisted for the whole summer, but having thinned out [*kaḏ etqaṭṭan(w)*] little by little were dissipated and vanished [*etpawšaš(w) w-eṯlaytī(w)*].

*Šifā*ʾ 73.14-16:

كالذى ظهر فى سنة سبع وتسعين وثلاث مائة للهجرة، فبقى قريبا من ثلاثة أشهر يلطف ويلطف حتى اضمحل، وكان فى ابتدائه إلى السواد والخضرة، ثم جعل كل وقت يرمى بالشرر ويزداد بياضا ويلطف حتى اضمحل.

[16 جعل: حصل ed. T]

... like the one which appeared [*ẓahara*] in the year 397 A.H. [1006/1007 A.D.] and persisted for nearly three months, becoming thinner and thinner [*yalṭufu wa-yalṭufu*] until it vanished [*iḍmaḥalla*]. At the beginning it was blackish and greenish, then all the time it was emitting sparks, turning whiter and becoming thinner until it vanished.

IV.iii.4.1-4: Rarity of comets etc.

But.: (1) These signs occur rarely and inconstantly (2) because it is with difficulty that [*l-maḥsen*] the smoky material of all this is formed [sufficiently] dense and compact (3) so that it is not dissipated along its path (4) and its glow persists for a long time, (5) and that it is able to be lifted up to the upper height of the αἰθήρ despite being heavy in this way. (6) For only a violent and very strong force will lift it.

*Šifā*ʾ 74.3-5:

ويقل تكون أمثال هذه الآثار، لأن يقل أن تكون مادة دخانية يتأتى لها أن تبلغ ذلك الموضع ولا تتبدد فى الطريق، وأن تكون كثافتها الكثافة التى تبقى لها مشتعلة فلن تصعّدها إلا قوة شديدة.

(1) The occurrence of the likes of these signs is rare, (2/5) because it is rare that it is easy [*yataʾattā*] for the smoky material to reach that place (3) and not be dispersed on the way, (4) for its density to be such that it remains burning, (6) and only a strong force will lift it.

IV.iii.4.4-6: "Terrible signs"

The main source for this part too is the *Šifā*ʾ, but it should be noted that BH agrees with Fakhr al-Dīn al-Rāzī in the addition of "and black".[37]

---

[36] Rashīd al-Dīn, *Ǧāmi*ᶜ *al-tawārīḫ*, ed. Quatremère (1836) I.416.5-8; Grigor of Akner, *Patmut'iwn azgin netołac'*, ed. Blake-Frye (1949) 350 [82]; Kirakos of Gandzak, *Patmut'iwn hayoc'*, tr. Dulaurier (1858) 507f. No such association is made by the Egyptian Maqrīzī (*K. al-sulūk li-ma*ᶜ*rifat duwal al-mulūk*, ed. Ziyāda-ᶜĀshūr [1934-72] I.516.13-517.1; tr. Quatremère [1838-42] I.1.241f.). See also Rada (1999/2000) 85.

[37] Bahmanyār also adds "black", *Taḥṣīl* 715.3.

But. [underline: agreement with *Šifā'*, italics: with *Mabāḥit*]:
Very *thick* and moist bodies of smoke [*tennānē*], *which are lifted up*, do not glow but become coaly [*gmurtānāyē hāwēn*] and cause *terrible red and black signs* [*āṯwāṯā dḥīlāṯā d-summāqān w-ukkāmān*] to appear in the αἰθήρ.
Šifā' 74.5-6:

وقد يعرض أن تكون أدخنة تصعد إلى الجو أكثف وأغلظ وأرطب من ذلك فلا تشتعل، بل تتجمر، فترى منها فى الجو علامات حمر هائلة.

[6 علامات حمرة فى الجو هاذلة ed. Ṭ]
Sometimes bodies of smokes [*adḫina*] rise to the air which are thicker, denser and more moist than that and do not catch fire, but turn coaly [*tataǧammaru*], and terrible red signs [*ʿalāmāt ḥumr hā'ila*] appear from them in the air.
Mabāḥiṯ II.189.5-6 [cf. *Cand*. 121.10-11]:

وقد تكون الادخنة الصاعدة غليظة فرؤيت العلامات الهائلة الحمر والسود

Sometimes the rising bodies of smoke are dense, so that there appear terrible red and black signs.

IV.iii.4.7: "Chasms" etc.

But.: They also cause illusions of chasms [*peḥtē*] in the air and dark porches [*abbūlē*] in the sky.
Šifā' 74.8-9:

وربما تفحمت وتراكمت وبقيت وخيلت أنها هوات فى الجو وأخاديد أو منافذ مظلمة فى السماء

Often they become coal-black, accumulate, persist and are imagined to be chasms [*hūwāt*] in the air and trenches [*aḫādīd*] or dark passages [*manāfiḏ*] in the sky.

l. 7. ܦܚܬܐ ("chasms"): هوات IS. - BH also mentions the "chasms" in *Cand*., although the word is written with an extra *taw* there.[38] Although these terms are missing in the Cambridge manuscript of Nic. syr., since the clause immediately preceding in *Cand*. resembles Nic. syr. 12.8-9, these terms may well have been taken from a lost passage of Nic. syr. The three phenomena themselves can be traced back to Arist. *Mete*. 342a 35-36 "χάσματά τε καὶ βόθυνοι καὶ αἱματώδη χρώματα". - The terms in the *Šifā'* are no doubt based on those in Olymp. arab., where the section on "blood-red colours" (*alwān damawīya*, < Arist. αἱματώδη χρώματα) (96.8-18) is followed by a section on the phenomenon called "chasm" (*hāwiya*, < Arist. χάσματα), where the "trenches" (*aḫādīd*, < Arist. βόθυνοι) are also mentioned (96.19-24), and then by a section on "hollows" (*taǧwīfāt*, 97.1-6).
l. 7. ܐܒܘܠܐ ܚܫܘܟܐ ("dark porches"): منافذ مظلمة ("dark passages") IS. - The lexica also allow the meaning of "column" for *abbūlā* (hence the translation of the phrase here as "dark columns in heaven" in PS Suppl. 1b, s.v.

[38] *Cand*. 121.1-2: ܗܠܝܢ ܕܡܬܩܪܝܢ ܦܚܬܬܐ ܘܥܘܡܩܐ ܘܣܓܝܐܐ ܡܚܙܝܢܐ ܕܡܐ ܘܙܗܘܪܝܬܐ ܘܐܪܓܘܢܝܬܐ ("Those which are called 'chasms', and 'depths' and many blood-red, scarlet and purple illusions").

ܐܒܘܠܐ), but the correspondence of the word to IS's *manāfiḏ* suggests that it is better understood here in the sense of "porch, vestibule". BH himself defines the word as "gate, vestibule" (*dihlīz*) at *Horr*. in Dan. 8.2 (ܐܒܘܠ ܐܘܠ), ms. Bodl. Hunt. 1, 101c 19-21: ܐܒܘܠ ܗܘ ܬܪܥܐ ܐܝܟ ܕܡܠܟܐ. ܐܘܠ ܫܡ ܢܗܪܐ.

IV.iii.4.7-10: "Column of fire"

This passage has no counterpart in the *Šifāʾ*. The phenomenon described here has similarities rather with that called *ḥarīq* ("conflagration") by Fakhr al-Dīn al-Rāzī and described by him at *Mabāḥiṯ* II.190.6-12, which occurs when "viscous and oily smoky vapour" (*buḫār duḫānī laziǧ duhnī*) rises to the "region of fire" (*ḥayyiz al-nār*) "without its connection with the earth being severed" (*min ġairi an yanqaṭiʿa ittiṣāluhu ʿani l-arḍi*) and is ignited. Since that passage occurs at the end of the part of the *Mabāḥiṯ* dealing with phenomena produced by combustion of the smoky vapour/exhalation, it helps to explain the placement of the passage of *But.* here (nearly) at the end of the chapter on such phenomena. There is, however, no mention of "sulphureous exhalation" or of "column of fire" in the *Mabāḥiṯ*.

The appearance of these two concepts in the passage of *But*. here leads us to associate it with passages in two earlier works of BH, *Cand.* 120.7-13, where we hear of the inflammation of "sulphureous exhalation" resulting in the appearance of lights (*nuhrē*) which are "like columns", and *Tract*., ms. Cantab. Add. 2003, 56v 12-16, where we are told how "columns of fire [extending] from the earth to the sky" are formed.

The passage at *Cand*. 120.7-13, in turn, corresponds in its position and parts of its contents to *Mabāḥiṯ* II.188.14-21, while the latter part of it is based on Nic. syr., but there is no mention of "columns of light" in either of these source passages. Similarly, the passage of *Tract.* corresponds in its position to Ghazālī, *Maqāṣid al-falāsifa* [Dunyā] 342.23-26, but there is no mention of "columns of fire" in Ghazālī. In other words, the notion of "columns of light/fire" is an addition by BH in both cases. The reason for this particular interest on BH's part in the "columns of light/fire" is probably to be sought in his desire to explain the "column of fire" of Exodus 13.21.[39] One suspects that it is for the same reason that the phrase "column of fire" occurs in other Oriental works dealing with Aristotelian meteorology for which

[39] As a historical event, a "likeness of the column of light" (*dmūṯ ʿammūḏā d-nuhrā*) which appeared in the sign of Virgo in Šḇāṭ 1345 [1034 A.D.] is reported at BH *Chron.* [Bedjan] 216.17.

Jewish or Christian authors were responsible, such as the Arabic version of Arist. *Mete*. by the Christian Yaḥyā b. al-Biṭrīq and the Hebrew translation of that Arabic version by Samuel b. Ṭibbon, as well as the compendium of Aristotelian meteorology attributed to Ḥunain b. Isḥāq.[40] In the Arabic version by Ibn al-Biṭrīq, it is difficult to find any points of similarity with the passages of BH's works under consideration here beyond the coincidence of the phrase "column of fire". Passages where there are at least some further points of contact with BH may be found in the *K. al-muᶜtabar* of Abū al-Barakāt, who was of Jewish origin, even if he is reported to have converted to Islam in his old age.

In *Cand*. and *Tract*. BH seems to envisage by "columns of light/fire" a phenomenon constantly in motion like shooting stars. At *Rad*. [Istanbul] 26.7-11, on the other hand, although the passage is closely based on *Cand*., as a result of the omission of various phrases which are in the corresponding part of *Cand*., BH appears to be thinking of a stationary phenomenon, and the same may be said of the passage of *But*. under examination here.

But. [underline: agreement with *Cand*.; italics: agreement with *Tract*.]:
(1a) Sometimes very thick sulphureous exhalation [*ᶜeṭrā kebrīṯānāyā*] rises from the earth (1b) and *reaches* [*metmaṭṭē*] the upper height [*rawmā ᶜellāyā*], (1c) while its root [*ᶜeqqārā*] is fixed [*qbīᶜ*] *to the earth*. (2a) When it catches fire [*dāleq*] from the αἰθήρ, (2b) there appears a *column of fire* [*ᶜammūḏā d-nūrā*] extending *from the earth to the sky*.

Cand. 120.7-13 [underline: agreement with *Mabāḥit*; italic: agreement with Nic.]:

ܡܛܠ ܕܘܩܝܕܣ ܐܘܟܝܬ ܠܡܦܐܕܐ ܘܒܠܨܘܨܝܬܐ ܕܡܬܚܙܝܢ ܒܠܠܝܐ. ܒܐܬܪ̈ܘܬܐ
ܗܠܝܢ ܕܐܝܬ ܒܗܝܢ ܟܝܢܐ ܟܒܪܝܬܢܝܐ. ܣܠܩܝܢ ܡܢܗܝܢ ܥܛܪ̈ܐ ܟܒܪܝܬܢ̈ܝܐ. ܘܢܩܦܝܢ
ܒܐܐܪ ܗܘ ܕܐܬܪܛܒ ܒܩܪܝܪܘܬܐ ܠܠܝܠܝܬܐ ܘܗܟܢܐ ܡܬܗܦܟ ܗܘ ܗܘ ܐܐܪ
ܠܟܝܢܐ ܕܗܝܢܐ ܕܩܠܝܠ ܠܡܬܕܠܩܢܘܬܐ. ܘܡܬܕܠܩ ܕܝܢ ܡܢ ܙܠܝܩ̈ܐ ܕܫܡܫܐ. ܐܝܟܢܐ
ܕܒܥܪܒܐ ܒܡܥܪܒܐ ܕܡܢ ܕܠܘܩ ܕܠܐ ܬܠܘ ܐܝܟܢܐ ܕܠܬܚܬ ܡܢܗ. ܘܡܣܟܢܐ ܠܥܠܐ
ܡܬܩܢܝܢ ܐܝܟ ܒܥܡܘܕ̈ܐ ܡܬܚܙܝܢ ܕܢܣܩܘܢ ܠܥܠܐ ܠܗܒܐ ܘܠܫܢܐ.

[120.13 ܡܬܚܙܝܢ sic PBV et Nic.: ܡܬܚܙ̈ܝܢ Bakoš et Çiçek]

On δοκίδες or torches [*lampīdē*] and sparks [*belṣuṣyāṯā*] which are seen at night. (1a) Sulphureous exhalations [*ᶜeṭrē kebrīṯānāyē*] rise from those places which have in them [*d-īṯ b-hēn*] a sulphureous substance [*kyānā kebrīṯānāyā*] and adhere to that air which has been moistened by the nocturnal coldness; and thus that air is transformed into an oily [*dahhīnā*] substance which is easily kindled, (2a) and is kindled by the rays of the

---

[40] Ibn al-Biṭrīq [Petraitis] 30.8, 11, 31.10, [Schoonheim] l. 207, 210, 222; Ibn Tibbon [Fontaine] I.321, 324, 335 (*ᶜammūd ēš*). The phrase in these instances apparently represent "φλόξ" of Arist. 341b 2, 26. - Ḥunain, *Comp*. [Daiber] 294, 303, 305, 306; Bar Kepha, *Hex*., Paris 241, 196v b25 (tr. 642.2).

stars, *as happens with a lamp that (the lamp below) catches fire* [*dālqā*] *from the light of (the lamp) above.* (2b) Thus *are seen many lights*, **like columns**, *falling* both *to the earth and to the sea.*[41]

Tract., ms. Cantab. Add. 2003, 56v 12-16:
12 ܘܬܢܢܐ ܟܕ ܡܢ ܠܐ ܡܬܟܠܝܢܘܬܐ ܡܛܐ ܠܗܘ ܚܘܕܪܐ/13 ܗܘ ܢܘܪܢܝܐ ܒܗ ܡܫܬܓܪܐ :
ܘܡܫܬܓܪܐ ܒܗ ܡܫܬܓܪܐ . ܒܕܪܓܐ ܗܘ/14 ܗܘܠܐ ܕܝܠܗ ܐܢ ܥܕܡܐ ܠܐܪܥܐ
ܗܘ . ܟܠܗܘܢ ܡܢܘ̈ܬܗ/15 ܚܕܐ ܡܢ ܚܕܐ ܒܐܡܝܢܘܬܐ ܡܫܬܓܪܢ . ܘܗܟܢܐ
ܥܡܘ̈ܕܝ/16 ܢܘܪܐ ܡܢ ܐܪܥܐ ܠܘܬ ܫܡܝܐ ܡܬܚܙܝܢ .

(1b) When smoke reaches that fiery circle [i.e. the sphere of fire] unhindered, (2a) it catches fire [*meštgar*] in it, so that, (1c) if its material [extends] to the earth, (2a) all its parts steadily catch fire from one another (2b) and in this way are seen **columns of fire** [*ʿammūḏay nūrā*] [extending] from the earth to the sky.[42]

---

[41] Cf. *Mabāḥiṯ* II.188.14-21 [< *Šifā*ʾ 68.13-18, cf. *But.* Mete. IV.i.3.4-9 above; underline: agreement with *Cand.*]: الفصل الثالث فى الانوار التى تشاهد بالليل فى بعض المواضع. اذا اصاب المطر بعض البقاع التى تكون فيها لزوجة دهنية تصعدت من تلك البقاع ابخرة دسمة لطيفة فتشتعل من ادنى سبب شمسى او برقى او من انوار الكواكب. فترى على وجه الارض شعل مضيئة غير محترقة احتراقا يعتد به للطفها ويكون حالها كحال النار التى تشتعل فى بخار شراب مجعول فيه الملح ونوشاذر اذا وضعت الفتيلة فى خمر تبخر ثم قرب من بخاره سراج فانه يشتعل ويبقى اشتعاله مدة بقاء البخار على ان الابخرة المطرية تكون الطف وارق كثيرا. ("Third Chapter: on lights [*anwār*] which are seen at night in some places. When rain falls on some places which have in them [*allatī takūna fī-hā*] an oily viscosity [*luzūğa duhnīya*], there rise from those places fine greasy vapours, which are kindled by the least solar or fulminous cause or by the fires of the stars. Then, there is seen on the surface of the earth light-emitting flames [*šuʿal*], without significant combustion [*iḥtirāq*] due to their fineness. Their condition is like that of the fire which is kindled in the vapour of wine [*šarāb*] mixed with salt and sal ammoniac, when a wick is placed in the vaporised wine [ḫamr] and a lamp is placed near its vapour so that it is kindled and the flame [*ištiʿāl*] endures as long as the vapour endures, except that the vapours generated by rain are much finer and thinner"). - Nic. syr. 12.1-4 [cf. Arist. 342a 3-5, 10-11; underline: agreement with *Cand.*]: ܙܒ̈ܢܝܢ ܣܓܝ̈ܐܢ ܗܝ/ ܢܘܪܐ ܕܡܢ ܠܬܚܬ ܡܢ ܗܝ ܕܠܥܠ ܒܕ ܫܘܪܐ : ܡܫ ܕܠܗܠ ܡܫܬܓܪܢܘܬܐ . ܐܝܟܢܐ ܕܚܙܝܢ ܕܗܘܝܐ ܒܫܪܓܐ/ ܡܫܬܓܪܢܐ . ܕܡܢ ܕܠܬܚܬ ܡܢ ܢܘܗܪܗ ܕܠܥܠ ܕܠܥܠܝܬܐ ÷ ܣܓܝ̈ܐܐ ܡܫܟܚܝܢ/ ܗܟ ܢܘܗܪ̈ܐ ܕܡܬܚܙܝܢ ܕܢܦܠܝܢ ܠܐܪܥܐ ܘܠܝܡܐ . ("Often (the fire) below is kindled by the fire above as it jumps, as we see happen with a lamp that (the lamp) below catches fire from the light of (the lamp) above. Therefore many are the lights which are seen falling to the earth and to the sea"). - *Rad.* [Istanbul] 26.7-11 [underline: agreement with *Cand.*]: ܐܠܝܟܐ . ܡܢ ܕܘ̈ܟܝܬܐ ܟܒܪܝܬܢܝܬܐ ܚܠ̈ܝܐ ܠܐܐܪ ܣܠܩܝܢ ܟܒܪܝܬܢܝܬܐ . ܘܡܢ ܐܝܠܝܢ ܒܙܠ̈ܓܐ ܕܟܘ̈ܟܒܐ ܢܗܝܪ̈ܐ ܡܫܬܓܪܝܢ . ܐܝܟ ܫܪܓܐ ܕܡܢ ܗܘ ܕܠܥܠ ܡܢܗ ܕܠܐ ܩܘܪܒܐ ܡܫܬܓܪ . ܘܗܟܢܐ ܥܡܘ̈ܕܐ ܕܢܘܗܪܐ ܒܕܘܟ ܕܘܟ ܡܬܚܙܝܢ . ("Ray: From sulphureous places sulphureous exhalations rise into the air and are kindled by the rays of the bright stars, like a lamp which catches fire from the one above it without contact. In this way **columns of light** are seen in some places").

[42] Cf. Ghazālī, *Maqāṣid al-falāsifa* [Dunyā] 342.23-26: وإن لم يضربه البرد تصاعد إلى الأثير، واشتعلت فيه النار فحصل منه نار تشاهد. وربما يستطيل بحسب طول الدخان، فيسمى كوكبًا منقضًا. ("If coldness does not strike it, (the smoke) rises to aether [*aṯīr*], fire is kindled in it and there arises from it a fire which is seen; sometimes the fire extends along the length of the smoke and is then called a shooting star [*kaukab munqaḍḍ*]"); cf. also IS, *Dāniš-nāma*, Ṭabīʿīyāt [Mishkāt] 71.1-4.

l. 8. ܠܓܬܐ ܟܒܪܝܬܢܝܬܐ ("sulphureous exhalation"): In the passage of *Cand.* quoted above, BH speaks of "sulphureous exhalations" rising from places containing "sulphureity" (*kyānā kebrītānītā*). At the corresponding place Fakhr al-Dīn al-Rāzī speaks of vapour rising from places containing "oily viscidity" (*luzūğa duhnīya*, *Mabāḥit* II.188.15), while IS, in the source passage used by Rāzī, speaks of places with "salinity and oily viscidity" (*sabaḫa wa-luzūğa duhnīya*, *Šifā'* 68.14; see comm. on IV.i.3.4-9). A possible link leading to the association of such "salinity" (*sabaḫa*) with "sulphureity" is provided by the passage of Abū al-Barakāt used as the source at Mete. I.ii.5.3-4 above: *Muʿtabar* II.216.9: "a salty, feverish [?] and sulphureous salina [*al-sabaḫa al-māliḥa al-ḥummāʾīya al-kibrītīya*]".

l. 9-10. ܥܡܘܕܐ ܕܢܘܪܐ ("column of fire"): cf. Abū al-Barakāt, *Muʿtabar* II.222.11-16:

هذا كلها تحدث فى البخار الدخانى الممتزج الصاعد الى اعالى الجو حتى تنتهى الى قرب كرة النار فتشتعل كاشتعال الدخان الصاعد بنار فوقه وعود اللهبة فيه هابطة الى المتدخن كما انك اذا اطفئت مصباحا وبقى دخانه يصعد ثم ادنيته الى مصباح آخر فوضعته تحته بحيث يصعد الدخان من المصفى الى المشتعل ترى النار تلتهب فى ذلك الدخان وتنزل الى المطفئ فى عمود دخانه فتشعله. كذلك هذه الشعل والشهب تشتعل ...

These all [i.e. shooting stars, comets, etc.] arise from mixed smoky vapour rising to the upper parts of the atmosphere until it reaches the proximity of the circle [sphere] of fire and is kindled in the same way as [when] rising smoke is kindled by fire above it and the flame in (that smoke) descends back to the source of the smoke [lit. the thing turned into smoke, *mutadakkin*] - just as, when you extinguish a lamp but its smoke continues to rise, and you then bring (the lamp) near another lamp and place (the first lamp) below (the second lamp) in such a way that smoke rises from the extinguished (lamp) to (the lamp) which is lit, you then see fire being ignited in that smoke, descending to the extinguished (lamp) in the **column** [*ʿamūd*] of its smoke and kindling it - just so ...

Muʿtabar, II.224.24-225.1[43]:

ولقد رأيت فى ليلة من الليالى المظلمة فى الحلة فى ريح عاصف انوارا كالأعمدة عظيمة جدا من الارض الى السماء ...

I once saw on a dark night in al-Ḥilla in violent wind lights [*anwār*] like very large **columns** [extending] from the earth to the sky ...

In the first of these passages the word "column" occurs in the context of a comparison of the way in which smoky exhalation is ignited to that of smoke in lamps, a comparison which we also encounter in the passage of *Cand.* quoted above. In the second, we find the phrase "from the earth to the sky", which we also find in the passages of the *Tract.* and *But.* quoted above.

[43] Cf. Lettinck (1999) 84, where there is a summary of *Muʿtabar* II.224.24-225.21.

IV.iii.4.10-11: Comets etc. as portents

But.: These signs [*āṯwāṯā*] portend [lit. indicate, *mšawdᶜān ᶜal*] winds, drought [lit. absence of rain, *glīzūṯ meṭrā*] and death-bringing [*mmīṯānē*], feverish [lit. hot] and withering [lit. dry] diseases.

Šifā' 74.14-15:

وهذه الآثار كلها تدل على الرياح وقلة الأمطار، وعلى فساد الجو ويبسه واستحراره، وعلى الأمراض الحارة اليابسة القاتلة.

All these signs [*āṯār*] portend [*tadullu ᶜalā*] winds, drought [lit. scarcity of rains, *qillat al-amṭār*], as well as corruption, desiccation and calefaction of the air and hot, dry and fatal [lit. killing, *qātila*] diseases.

BOOK OF METEOROLOGY, CHAPTER FIVE

ON GREAT EVENTS WHICH TAKE PLACE IN THE WORLD

The chapter of *But.* here corresponds to the sixth (last) chapter of the treatise on meteorology in the *Šifāʾ* (ed. Cairo. p. 75-79), a chapter in which IS begins to move beyond the traditional boundaries of Aristotelian meteorology. Of the subjects treated here by IS, periodic deluges were mentioned by Arist. in his discussion of the rivers and seas at *Mete*. I.xiv, 352a 17-b 16 and might therefore be considered to fall within the scope of Aristotelian meteorology. The possibility of certain species of animals and plants becoming extinct through the destruction wrought by such deluges then leads IS on to a discussion of spontaneous generation, a subject that belongs more properly to the scope of Aristotelian zoology (esp. *GA*), as well as to a brief discussion of how the accumulation of human knowledge begins anew after the extinction and regeneration of the human species.

The correspondences of the sections in the chapter of *But.* here to the *Šifāʾ* may be summarised as follows.

	Šifāʾ	
Section i.	75.4-76.18	Deluges
Section ii.	76.18-79.6	Spontaneous generation
Section iii.	79.6-18	Renewal of knowledge

In his reworking of this chapter of the *Šifāʾ*, BH makes a number of departures from the material he found in the *Šifāʾ*. While the first section of the chapter (V.i.) is largely a paraphrase of the corresponding part of the *Šifāʾ*, there is an insertion without a counterpart in the *Šifāʾ* at V.i.3.6-7 where BH talks of the passing away of "this world of ours" and the beginning of a "new world". Similarly much of the second section (V.ii.), although the section as a whole is considerably shorter than the corresponding part of the *Šifāʾ*, may be considered a summary of the somewhat involved discussion of spontaneous generation in the *Šifāʾ*, but here again we find an insertion at V.ii.2.10-11, where BH talks "a father and a mother" being created anew in the "ground" (*ād̠amt̠ā*) after the "universal flood". Of the two 'theories' in the third section (V.iii.), the first begins as a paraphrase of the *Šifāʾ*,

but BH moves away from the *Šifā*ʾ towards the end of the 'theory' (V.iii.1.8-9), while the renewal of speech and writing discussed in the second 'theory' (V.iii.2) is a subject that is not discussed in the corresponding part of the *Šifā*ʾ.

Out of the alterations made by BH, the insertions at V.i.3.6-7 and V.ii.2.10-11 are of particular interest in view of the vocabulary used there with its bibical connotations, and that especially in view of the fact that the whole content of the chapter is closely related to the question of the eternity of the world, a question over which Greek philosophy and the monotheistic religions were in disagreement. On first reading it may appear as if BH is supporting IS's view that the world is eternal and is using biblical vocabulary to enforce this view. A more careful reading, however, shows that this is not quite the case.

The insertions made by BH are best understood in the light of what BH had to say on the matter at *Cand.* Base II, chap. 1-2 [Bakoš] 60.4-73.14.[1] There, in chapter 1 (60.4-68.5), BH begins with the confirmation of the orthodox Christian (as well as Muslim and Jewish) view that the world is created (*hawyā*) and is not eternal *a parte ante* (*mt̲ōmāyā*, corr. arab. *azalī*). Then, in chapter 2 (68.6-73.14), BH goes on to discuss the eternity *a parte post* (*ʿālmīnāyūt̲ā*, corr. arab. *abad*)[2] and perishability (*meštaryānūt̲ā*) of the world, a question in which he sees three possibilities, namely: (a) God will preserve this world (*ʿālmā hānā*), (b) He will dissolve and annihilate (*šārē wa-mb̲aṭṭel*) it completely, or (c) He will annihilate its form (*eskēmā*) only (68.14-69.2). BH tells us furthermore that this is a question which cannot be solved through reasoning (*ḥuššāb̲ā hawnānāyā*), but only through recourse to the Scripture, and the conclusion he reaches after the examination of the scriptural and patristic evidence in *Cand.* is that the world itself is eternal *a parte post* by divine ordinance and that its form alone is destined for annihilation,[3] one of the key passage quoted in support of

---

[1] Cf. also the corresponding discussion at *Rad.* I.7.2-3 [Istanbul] 70.11-78.3.

[2] For ܥܠܡܝܢܘܬܗ of ed. Bakoš 68.6 ("constitution" tr. Bakoš), read ܥܠܡܝܢܝܘܬܗ (so ed. Çiçek 55.32).

[3] *Cand.* 73.6f.: ܐܡܪܝܢ ܕܟܠܗܝܢ ܐܟܚܕܐ ܠܒܛܠܠ ܐܕܫܢܝܘܬܗ ܡܬܚܒܠܢܝܬܐ ܕܥܠܡܐ ܗܢܐ ܚܘܝܢ ܘܡܫܘܕܥܢ. ܘܠܐ ܠܒܛܠܗ ܓܡܝܪܐ ܘܚܒܠܗ ܕܠܓܡܪ. ("We say that all these [scriptural testimonies] together show and indicate the annihilation of the perishable form [*ad̲šānāyūt̲eh met̲habblānāytā*] of this world, not its complete annihilation and its utter destruction").

this latter point being I Cor. 7.31 ("the shape [*eskēmā*] of this world passes away [*ʿāḇar*] ...").[4]

In the chapter of *But*. here,[5] we are not concerned directly with the question of the eternity of the world *a parte ante*, but only with that of eternity *a parte post*, and what BH says in the inserted passage at V.i.3.6-7 is in fact in agreement with what he had said in *Cand*., the key phrase here being *ʿāḇar eskēmeh* ("its shape will pass away") of l. 6. What BH has done through the insertion of the passage at V.i.3.6-7 is to equate this eschatological passing away of the "shape" of "this world of ours" (*ʿālman hānā*) with the possible destruction of the habitable world (*ʿāmartā*) through some future deluge or migration of the sea, or, in other words, to provide a possible philosophical-scientific expalanation for a theological-eschatological event.

As far as can be gathered from what is said in this chapter of *But*. such a destruction and renewal of the "shape" of the world could occur any number of times - and the attribution of the deluges and the migration of the sea to the configuration of the celestial bodies (V.i.3.1-3; cf. V.ii.3.8-10) implies that that is in fact the case, since astronomical configurations are recurrent. - Theology, on the other hand, envisages only one such destruction in future. What needs to be borne in mind here is the fact that *But*. is a work on philosophy and that we are concerned in this work not with theological truths but with philosophical-scientific possibilities and explanations. BH is not saying that the destruction and renewal *will* be repeated time after time, but merely that such repetition is, philosophically-scientifically speaking, a possibility.

---

[4] Quoted at *Cand*. 72.12f.; cf. *Cand*. 73.12-14: ܘܗܢܐ ܕܢܥܒܪܘܢ ܡܘܕܥ ܠܫܘܚܠܦܐ ܕܡܢ ܡܬܚܒܠܢܘܬܐ ܠܠܐ ܡܬܚܒܠܢܘܬܐ ܡܣܬܦܩ. ܘܗܢܐ ܕܐܠܗܝܐ ܦܘܠܘܣ ܣܦܩ ܠܗ ܦܘܫܩܐ ܕܡܦܫܩ. ܟܕ ܐܣܟܝܡܗ ܐܡܪ ܕܥܠܡܐ. ܘܠܐ ܥܠܡܐ. ("'They will pass away' indicates the change from perishability to inperishability, and the (word) of the divine Paul suffices to clarify this point, since he says 'the shape of the world' and not 'the world'").

[5] The situation is different with the statement at Min. V.iii.4.5-6 above ("because the world is without beginning and not created"). - For a better overview of how BH deals with the question of eternity in *But*., we still need to make a careful examination of such parts of it as the following: *But*. Ausc. phys., chap. IV (on "Time", esp. section IV.iii..: "on infinity of time", ܡܛܠ ܠܐ ܡܣܬܝܟܢܘܬܐ ܕܙܒܢܐ); *But*. De caelo, section V.vi. ("that the world is one", ܡܛܠ ܗܝ ܕܚܕ ܗܘ ܥܠܡܐ); *But*. Gen. et corr., sections I.iii.-iv. ("opinions of ancients on generation, corruption etc."; "refutation of these heretical theories").

It is in the same light that we should consider the addition at V.ii.2.10-11, which is clearly reminiscent of the birth of Adam and Eve as told in the Book of Genesis. BH is not saying that such a birth will be repeated after the "universal flood", but merely that it is a possibility. The addition is, at the same time, an attempt at a "scientific" explanation of the event told in the Book of Genesis in terms of Aristotelian-Avicennian natural philosophy.

The biblical connection is less clear for the additions BH makes at V.iii.1.8-9 and V.iii.2, but if one is correct in associating the additions here with the narrative of the Tower of Babel,[6] these additions too could be seen as an attempt to provide a scientific account of the development of human knowledge and languages that lies in the background to that narrative.

1. 1. ܡܛܠ ܥܒ̈ܕܐ ܪ̈ܘܪܒܐ ܕܗܘܝܢ ܒܥܠܡܐ: The chapter heading is based on the corresponding heading in the *Šifā*ʾ (75.3): فى الحوادث الكبار التى تحدث فى العالم

Mete. V. Section i.: *On Deluges*

With the exception of the passage at V.i.3.6-7 which has been discussed above, the order, as well as the content, of the section is basically that of the *Šifā*ʾ. - There appears to be no passages in the section that could be derived from those parts of Nic. syr. correponding to Arist. *Mete*. 352a 17-b 16, which would have had its place in the lacuna between p. 18 and 19 in the Cambridge manuscript of that work.

	*Šifā*ʾ	
V.i.1.1-2	75.4	Introduction
V.i.1.2-4	75.4-5	Definition
V.i.1.4-6	75.8-10	Causes of deluges
V.i.1.6-11	75.10-14	Types of deluges
V.i.2.1-7	75.15-76.4	Possibility of deluges
V.i.3.1-6	76.4-14	Migration of the sea
V.i.3.6-7	-	New world
V.i.3.7-11	76.15-18	Migration of the sea (contd.)

V.i.1.1-2: Introduction

But.: It accords with [*nāqpā l-*] what has been said that we should speak about great events like deluges [*ṭawpānē*] and the rest of those [phenomena]

[6] See the introductory comments on Section V.iii. below.

which are brought about [*metkayynīn*] once in a long while [*ḥdā l-nuḡrā*] in the world.

Šifā' 75.4: ومما يخلق بنا أن نتكلم فيه فى هذا الموضع أمر الطوفانات

Among what it is fitting for us [*yaḫluqu bi-nā*] to speak about in this place is the matter of deluges [*ṭūfānāt*].

V.i.1.2-4: Definition

But.: We say that a deluge is the predominance [*ʿulbānā*] of one of the four elements over the whole or part of the habitable quarter [*rubʿā metʿamrānā*] of the earth.

Šifā' 75.4-5: فنقول: إن الطوفان هو غلبة من أحد العناصر الأربعة على الربع المعمور كله أو بعضه

We say: a deluge is the predominance [*ġalaba*] of one of the four elements over the whole of part of the habitable quarter [*al-rubʿ al-maʿmūr*].

l. 4. ܪܘܒܥܐ ܡܬܥܡܪܢܐ ("habitable quarter"): الربع المعمور IS. - Cf. Min. IV.i.3 above.

V.i.1.4-6: Causes of deluges

On the influence of the stars in the transformation of elements, cf. Min. IV.i.2 above.

But.: (1a) Its efficient cause is the conjunctions [*sunehdō*] of the planets and certain fixed stars [lit. the planetary stars and some of the decans], (1b) along with [*ʿam*] the favourable disposition of the element itself [*ʿāhnūṯēh dīleh d-esṭuksā*].

Šifā' 75.8-10:

فنقول: إن السبب فى وقوع الطوفانات اجتماعات من الكواكب على هيئة من الهيئات توجب تغليب أحد العناصر فى المعمورة، قد عاونتها أسباب أرضية واستعدادات عنصرية.

We say: (1a) The cause for the occurrence of deluges is the conjunctions [*iǧtimāʿāt*] of the stars in a configuration [*hai'a*] which necessitates the predominance [*taġlīb*] of one of the elements in the habitable world, (1b) which is sometimes helped by terrestrial causes and favourable dispositions of the element [lit. elemental readinesses, *istiʿdādāt ʿunṣurīya*].

l. 5. ܣܘܢܗܕܘ ("conjunctions"): ܣܘܢܕܘ F; اجتماعات IS. - Although the form ܣܘܢܘܕܘܣ apparently prevails even in the manuscripts of BH's works (see the passages cited in PS 2674f.; Brockelmann 484a, s.v.), I accept the form with *hē* here on the strength of BH *Splend.* [Moberg] 208.26f., where BH writes ܣܘܢܗܕܘܣ and tells us, wrongly as it happens, that the *hē* indicates a *rbāṣā* (i.e. "syn-e-dos", as if the word meant "sitting together").

l. 5. ܕܩܢܘܣܐ ("fixed stars"): lit. "decans". - Cf. Mete. II.i.5.9 above.

l. 6. ܥܗܢܘܬܗ ܕܝܠܗ ܕܐܣܛܘܟܣܐ ("the favourable disposition of the element itself"): استعدادات عنصرية IS. - BH paraphrases here; at V.ii.2.2f. he has *ʿāhnūṯā esṭuksānāytā*, a phrase that mirrors the one in IS here.

V.i.1.6-12: Types of deluges

But.: (1a) A deluge of water [lit. watery deluge] occurs suddenly, (1b) through repeated [lit. many] and continuous high tides [*meleʾē saggīʾē wa-sbīsē*], (1c) through heavy and incessant rains [*meṭrē tkīḇē w-ammīnē*], (1d) or through immoderate transformation [*šuḥlāpā lā mmaššḥā*] of air into water. (2) That of fire [lit. the fiery] arises from inflammation [*nuḇrāšā*] of strong winds [*rūḥē ʿazzīzāṯā*], and this [type of deluge] is faster [to spread]. (3) That of air [lit. the airy] [arises] from severe and destructive blast [*maššḇā ʿaššīnē w-mawbdānē*] of winds. (4a) That of earth [lit. the earthy] [arises] from the diffusion [*ašīdūṯā*] of a large quantity of sand [*ḥālā*] over the habitable world [*ʿāmartā*], (4b) or severe coldness [lit. severity of coldness, *ʿušnā d-qarrīrūṯā*] which hardens animals and plants, such that their petrified likenesses are found in various places.

Šifāʾ 75.10-14:

فالمائية منها قد تقع من انتقالات البحار على صُقْع كبير دفعة، لأسباب عظيمة/١١ ريحية توجب ذلك، أو أسباب توجب شدة من المد، ومن أمطار دائمة، ولاستحالة/١٢ مفرطة تقع للهواء إلى المائية. والنارية تعرض من اشتعالات الرياح العاصفة، وهذه/١٣ أشد انتشارا. والأرضية تعرض لسيلان مفرط يقع من الرمال على برارى عامرة أو لكيفية/١٤ شديدة أرضية باردة مجمدة، مما حدثنا عنه. والهوائية تقع من حركات ريحية شديدة جدا مفسدة.

[10 البحار ed. Ṭ: البخار ed. Cair. ‖ 14 شديدة ed. Ṭ: تشتد DSM; تسيّل ed. Cair.]

(1a) The watery among them sometimes occurs [*yaqiʿu*] suddenly from transportations of seas[7] over a large area [*suqʿ*] due to great windy causes which induce that, (1b) or causes which induce intense high tide [*šidda min al-madd*], (1c) and from incessant rain [*amṭār dāʾima*], (1d) and due to excessive transformation [*istiḥāla mufriṭa*] of air into wateriness. (2) The fiery arises [*yaʿriḍu*] from inflammations [*ištiʿālāt*] of violent winds [*al-riyāḥ al-ʿāṣifa*], and this [type] is more intense in diffusion. (4a) The earthy arise due to excessive flowing [*sayalān*] of sands [*rimāl*] over inhabited plains [*barārī ʿāmira*], (4b) or due to an intense earthy cold and freezing quality [*kaifīya šadīda arḍīya bārida muǧammida*], of which we have spoken. (3) The airy arise from very strong and destructive movements of winds [*ḥarakāt rīḥīya šadīda ǧiddan mufsida*].

l. 6. ܡܠܐ̈ܐ ("high tides"): مدّ IS. - Cf. Min. V.iii.3.9 above (ܡܘܒܐ̈ ܘܡܠܐ̈ܐ, "ebb and flow").

l. 8. ܡܥܓܠ ("suddenly"): I take this adverb placed at the very end of the sentence (l. 8) with the verb *hāwē* near the beginning of the sentence (l. 6), on the assumption that it is intended to represent the *dufʿatan* of *Šifāʾ* 75.10, taken with the verb *yaqiʿu* in the same line.

l. 8-9. ܚܪܝܦܐ ܣܓܝ ("faster [to spread]"): أشد انتشارا (lit. "more intense in being diffused") IS. - The adjective *ḥarrīpā* can also mean "acute, severe", meanings which would also make good sense here, but the phrase at the corresponding place in the *Šifāʾ* suggests that the reference here is to the speed with which

[7] Reading *al-biḥār* with the Tehran edition; cf. *Šifāʾ* 76.5 (quoted under V.i.3.1-7).

the fiery deluge spreads. - Cf. also Mete. III.i.1.8 and III.i.2.6 above, where *ḥarrīpā* is used of the "swift movement" of fire.

l. 11. "..., such that their petrified likenesses are found in various places": On petrified animals and plants, cf. Min. I.i.3.1-2, Min.V.iii.2.2-3 above.

V.i.2.1-7: Possibility of deluges

But.: (1a) [With] everything which is subject to increase [*yattīrūṯā*] and decrease [*bṣīrūṯā*] - (1b) although its medium degree [*dargeh meṣ*ʿ*āyā*] and that which is close to it occur in most cases [*ak̠ da-b̠-suggā*] - (1c) the occurrence of its two extremes [*sāk̠ē*], the excessive and the deficient [*saggī yattīrā w-saggī bṣīrā*], is not impossible. (2a) Consequently, just as there are times when in some of the great habitable worlds [ʿ*āmrāṯā rawrb̠āṯā*] rain does not fall for many years, (2b) it is also possible for a profusion [*šp̈ī*ʿ*ūṯā*] of rain to occur suddenly at the other extremes, (2c) or for air to be transformed suddenly into the nature of water, (2d) and [so] for a deluge of water to occur, (2e) [and] similarly also for a deluge of fire, wind and dust.

*Šifā*ʾ 75.15-76.4:

ومما يقنع فى وجود هذه وحدوثها كثرة الأخبار المتواترة فى حديث طوفان الماء. وما يقنع فى إثبات ذلك أن الأشياء القابلة للزيادة والنقصان والقلة والكثرة، وإن كان أكثر الوجود فيها الوجودُ المتوسط بين طرفى الإفراط والتفريط وما يقرب منه، فإن طرفهما لا يخرج عن حد الإمكان. وكما قد يتفق كثيرا أن تأتى السنون على بقاع عظيمة من المعمورة فلا يكون فيها مطر البتة، وذلك فى جانب النقصان؛ فكذلك قد يمكن أن يفرط المطر دفعة واحدة، ويستحيل الهواء إلى طبيعة مائية دفعة، إذ كان ما بين الأوساط مختلفا بالزيادة والنقصان، وكذلك فى سائر الطوفانات.

Among things arguing for the existence of these [deluges] and their occurrence is the multitude of recurring reports on the legend of the deluge of water and what argues for its confirmation is the fact that (1a) [with] things which are subject to increase [*ziyāda*] and decrease [*nuqṣān*] and scarcity [*qilla*] and abundance [*kaṯra*] - (1b) even if the majority of states [*akṯar al-wuǧūd*] are the state intermediate [*al-wuǧūd al-mutawassiṭa*] between the two extremes [*tarafāni*] of excess [*ifrāṭ*] and deficiency [*tafrīṭ*] and that close to it - (1c) the extremes of these two are not beyond the limits of possibility. (2a) Just as it often happens that years pass by in large areas [*buqā*ʿ ʿ*aẓīma*] of the habitable world [*al-ma*ʿ*mūra*] and rain does not occur in them at all - that being on the side of deficiency [*nuqṣān*] -, (2b) just so, it is possible that rain become excessive all of a sudden, (2c) and that air be suddenly transformed into a watery nature, since what is in the middle varies with increase and decrease. (2e) Similarly for the rest of deluges.

V.i.3.1-7: Disappearance of habitable land due to the migration of the sea

On the migration of the sea, cf. Section Min. V.iii. above. - Most parts of the passage of *But*. here may be seen as a paraphrase of the somewhat

lengthy passage of the *Šifā᾽* quoted below. - On the last part of the passage (part (3)), see the introductory comment to Chap. Mete. V above (cf. also V.ii.1.2 below, "new creation").

But.: (1a) If the cause of the migration [*šunnāyā*] of the sea is a certain celestial configuration [*dmūṯā meddem šmayyānāytā*], (1b) like the apogee or the perigee, (1c) or the variation [*šuḥlāpā*] in the obliquity [*ṣlāyā*] of the sun['s course] whereby the zodiac coincides with [lit. will be in] the plane of the equator [lit. plane of the isemeric, *šṭīḥūṯ īsīmerīnōn*], (2a) [then] it must be that the sea will gradually conquer [*nešlaṭ ᶜal*] the habitable places [*aṯrawwāṯā metᶜamrānē*] (2b) and those [places] unsuitable for habitation [*ᶜmuryā*] will be laid bare [*netgallgūn*]. (2c) In this way the earth will be divided into sea and dry land [*yabšā*] which are not suitable for the sustenance [*quyyāmā*] of animals which breathe [*sāyqāt*] the air, (3) and this world of ours [*ᶜālman hānā*] comes to an end [*bāṭel*] and its shape [*eskēmā*] passes away [*ᶜāḇar*], until this present configuration [*hāḏē dmūṯā d-hāšā*] recurs and a new world [*ᶜālmā ḥadtā*] begins.

Šifā᾽ 76.4-14:

وإن كان ما نحدس من اتباع/ ٥ البحار لجهة من الفلك صحيحا، فيجب أن تنتقل بانتقاله حتى تعمر وقتا ما هذه النواحى التى/ ٦ لا يجوز أن تتعداها العمارة، وهو أن يحصل الموضع الناقل للبحر الأعظم بانتقاله من/ ٧ الفلك كأوج أو حضيض أو شيء آخر غيره في قرب معدل النهار، فيسيح الماء على/ ٨ المكان الذى يجب أن تكون فيه المعمورة، وينكشف قطب أو قطبان، وينتقل إليها البر/ ٩ المقابل للبحر، وهناك مانع من العمارة، فتكون الأرض مقسومة إلى بر وبحر ليس أحدهما/ ١٠ بمحتمل للعمارة بالحيوانات المتنفسة من الهواء. وكذلك إن كان حال الميل، وما نحدس من/ ١١ تغيره وزواله شيئا يثبت له حقيقة، وحتى يصح أن يكون لفلك البروج/ ١٢ انطباق أو شبه انطباق مع دائرة معدل النهار، إن جميع ذلك مما يوجب فساد العمارة، وإن لم يكن ذلك أيضا/ ١٣ بممكن: فإن ما قلناه من الإفراطات وما نصححه من إمكان انتقال البحر من ناحية قطب/ ١٤ إلى قطب غير خارجة عن الإمكان.

[5 البحار ed. Ṭ: البخار ed. Cair. || تنتقل: ينتقل edd. Cair. et Ṭ || تعمر: يعم ed. Cair.; يعمّر ed. Ṭ; يعمر BSM sec. ed. Cair. || 11 وحتى: حتى ed. Ṭ et BM]

(1a) If what we surmise about the adherence [*ittibāᶜ*] of the seas to a side of the celestial sphere [*falak*] is correct, then they will necessarily migrate [*tantaqilu*] with its movement [*intiqāl*], (2) until at some time those regions which the habitable world [*ᶜimāra*] cannot extend to become habitable - (2) that is, the place bearing the great sea will come to the vicinity of the equator [*muᶜaddal al-nahār*] (1) due to its movement [*intiqāl*] by the celestial sphere, (1b) such as the apogee and the perigee etc., (2a) so that water will flow over the place which ought to contain the habitable world [*maᶜmūra*], (2b) a pole or the two poles will be uncovered [*yankašifu*], the dry land [*barr*] which opposes the sea will move towards them, but there is a hindrance there to habitation [*ᶜimāra*]. (2c) As a result the earth will be divided into dry land [*barr*] and sea, neither of which is sustentative [*muḥtamil*] of habitation [*ᶜimāra*] by animals that breathe [*mutanaffisa min*] the air. (1c) Similarly, if the condition of the declination/obliquity [*mail*] and what we surmise about its variation [*taġayyur*] and its disappearance [*zawāl*] is something that is true, so that it is correct that the zodiac circle [*falak al-burūǧ*] will coincide [lit. will have a coincidence

(*intibāq*)] or will appear to coincide [lit. will have a semblance (*šibh*) of coincidence] with the celestial equator [*dā'irat mu'addal al-nahār*], all these are things which will bring about the destruction [*fasād*] of the habitable world; even if that too is not possible, what we have said concerning the excesses [*ifrāṭāt*] and what we have established as correct about the possibility of the migration [*intiqāl*] of the sea from the region of one pole to another are not beyond possibility.

l. 1. ܕܐܣܟܡܐ ܡܕܡ ܫܡܝܢܝܬܐ ("a certain celestial configuration"): There is no equivalent phrase in the passage of the *Šifā'* quoted above, but cf. *Šifā'* 75.8 *hai'a min al-hai'āt* (in the passage quoted under V.i.1.4-6 above).

l. 1f. ܐܝܟ ܐܦܘܓܝܘܢ . ܐܘ ܐܦܪܝܓܝܘܢ ("like the apogee or the perigee"): كأوج أو حضيض IS. - Cf. Min. IV.i.2.7 above.

l. 2. ܨܠܝܐ ܕܫܡܫܐ ("obliquity of the sun['s course]"): الميل IS. - Cf. comm. on Min. IV.ii.3.5 above; here the word *ṣlāyā* must mean not simply the distance of the sun from the celestial equator at a given time, but its maximum declination, or the obliquity of its path/the ecliptic.

l. 4. ܢܥܠܠ ("conquer"): For the metaphor of sea "conquering" land, cf. Min. IV.i.1.8 above (ܥܠܠ ܝܡܐ ܘܐܘܩܝܢܘܣ).

l. 4. ܗܢܘܢ ܠܐ ܚܫܝܚܢ ܠܡܥܡܘܪܘܬܐ ("those [places] unsuitable for habitation"): IS talks at the corresponding place (76.8) of "the pole or the two poles", which are, of course, unsuitable for habitation (note also 76.9 وهناك مانع من العمارة). Cf. Min. IV.i.3 above.

l. 5. ܚܝ̈ܘܬܐ ܡܬܢܫ̈ܡܢܝܬ ܐܐܪ ("animals that breathe the air"): الحيوانات المتنفسة من الهواء IS; cf. Min. IV.i.2.10 above (ܚܝ̈ܘܬܐ ܡܬܢܫ̈ܡܢܝܬܐ).

l. 6f. ܘܡܫܠܡ ܥܠܡܐ ܗܢܐ ܘܥܒܪ ܐܣܟܡܗ . ܥܕܡܐ ܕܗܘܐ ܕܡܘܬܐ ܕܥܠܡܐ ܡܬܚܕܬ ܘܡܠܟܘܬܐ ܚܕܬܐ ܢܫܪܐ: The balanced, chiastic structure of the sentence is worth noting: verb-subject, verb-subject, *'dammā d-* subject-verb, subject-verb.

l. 6. ܘܡܫܠܡ ܥܠܡܐ ܗܢܐ ܘܥܒܪ ܐܣܟܡܗ ("and this world of ours comes to an end and its shape passes away"): The language used is biblical (see I Cor. 7.31: ܥܒܪ ܠܗ ܓܝܪ ܐܣܟܡܗ ܕܥܠܡܐ ܗܢܐ; Mt. 24.35; I Jn. 2.17; Ps. 102.26-27, etc.); cf. BH *Cand*. 72.2-74.14.[8]

[8] Cf. further BH, *Horr*. in Mt. 24.35, ed. Carr, Syriac text 71.21-72.2: ܫܡܝܐ ܘܐܪܥܐ ܢܥܒܪܘܢ ܘܡܠܝ ܠܐ ܢܥܒܪܢ . ܗ̄ ܥܒܪܘܬܐ ܗܪܟܐ ܠܐ ܠܒܛܠܢܘܬܐ ܘܠܓܡܝܪܘܬܐ ܐܡܪ . ܐܠܐ ܠܫܘܚܠܦܐ ܐܝܟ ܗܝ ܕܥܒܪܘ ܡܢ ܒܝܫܬܐ ܘܥܒܕܘ ܛܒܬܐ ("Heaven and earth shall pass away, and my words shall not pass away. - N. He speaks of passing away here, not as a ceasing to be and complete annihilation, but as change, according to that saying, 'they passed away from evil and wrought good'", tr. Carr, p. 58); *Horr*. in Ps. 102.27, ed. de Lagarde 208.17-18: ܘܟܠܗܘܢ ܐܝܟ ܠܒܘܫܐ ܢܒܠܘܢ . ܗ̄ ܐܢ ܠܐ ܢܛܪ ܐܢܬ ܠܗܘܢ . ܘܐܝܟ ܡܪܛܘܛܐ ܢܫܬܚܠܦܘܢ . ܗ̄ ܒܬܪ ܩܝܡܬܐ ("'They [sc. heaven and earth] will be worn away like clothes', i.e., if thou dost not preserve them. 'They will be changed like a garment', i.e. after the resurrection'"; There is no comment on the first part of 102.27, ܗܢܘܢ ܢܥܒܪܘܢ ܘܐܢܬ ܩܐܡ ܐܢܬ); *Horr*. in Jes. 51.6, ed. Tullberg 29.1-2: ܡܛܠ ܕܫܡܝܐ ܐܝܟ ܬܢܢܐ ܢܥܒܪܘܢ . ܗ̄ ܐܝܟ ܬܢܢܐ ܡܒܕܪ ܐܝܟܢܐ ܐܢܐ ܠܗܘܢ.

l. 6. ܫܠܐ ("comes to an end"): cf. *buṭṭāl ʿāmartā* in l. 10 below (corr. IS *inqiṭāʿ al-ʿimāra*).
l. 7. ܘܥܠܡܐ ܚܕܬܐ ("a new world"): cf. Jes. 65.17, 66.22 ܫܡܝܐ ܚܕܬܐ ܘܐܪܥܐ ܚܕܬܐ.[9]

V.i.3.7-11: Migration of the sea (contd.)

But.: (1a) For we know for certain [lit. without doubt, *d-lā puššāḵ*] (1b) that this northern region was covered by water [*b-mayyā mḥappyā (h)wāṯ*] (1c) and for that reason [*w-ʿal-hāḏē*] mountains were formed [*eṯīled(w)*] in it, (1d) whereas now the seas are in the south [lit. are southern]. (2a) The sea is therefore liable to migration [*mšannyānā*] (2b) and it may be that one or other of its migrations will be the cause of the disappearance [*buṭṭāl*] of the habitable world. (3a) In this way after every beginning there is an end, and vice versa, (3b) though over an indeterminable number of years [lit. years which are difficult to delimit].

Šifāʾ 76.15-18:

ونحن نعلم بأقوى حدس أن ناحية الشمال كانت مغمورة بالماء حتى تولدت الجبال. والآن فإن البحار جنوبية، فالبحار منتقلة، وليس يجب أن يكون انتقالها محدودا، بل يجوز فيه وجوه كثيرة، بعضها يؤذن بانقطاع العمارة، فيشبه أن تكون فى العالم قيامات تتوالى فى سنين لا تضبط تواريخها.

(1a) We know with the highest degree of probability [*bi-aqwā ḥads*] (1b) that the region of the north was immersed with water [*maġmūra bi-l-māʾ*] (1c) until/so that [*ḥattā*] the mountains were formed [*tawalladat*], (1d) whereas now the seas are in the south [lit. are southern]. (2a) Therefore, the seas are liable to migration [*muntaqil*] (2b) and it is not necessary that their migration be limited, but many modes [*wuǧūh*] are possible and a certain mode [*baʿḍu-hā*] may herald the cessation [*inqiṭāʿ*] of the habitable world, (3) so that it is likely that there be in the world resurrections [*qiyāmāt*] (3b) which recur over years whose dates cannot be determined.

l. 8. ܘܥܠܗܕܐ ("and for that reason"): حتى IS. - It is more natural to understand IS's *ḥattā* as temporal "until" than as final "so that". BH, however, has taken the latter option. The explanation for this may be sought in Min. I.ii.1. above, where the formation of islands and mountains is attributed to the actions of the sea.
l. 11. ܒܫܢܝܐ ܕܥܣܩ ܠܡܬܚܡܘ ܡܣܟܐ (lit. "over years which are difficult to delimit"): فى سنين لا تضبط تواريخها IS. - Cf. Min. V.iii.2.5f. above ("over a long time, the number of whose years cannot be recorded in books").

ܘܡܥܒܪܢܐ ܠܗܘܢ ܠܐܡܬܝ ܕܨܒܐ. ("'For the heavens will pass away like smoke', i.e., I will cause them to pass away whenever I desire, 'but my salvation' for you 'will be for ever'", tr. Hicks, 84).

[9] Cf. further BH, *Horr*. in Jes. 65.17, ed. Tullberg 35.8-9: ܡܛܠ ܕܗܐ ܐܢܐ ܒܪܐ ܐܢܐ ܫܡܝܐ ܚܕܬܐ ܘܐܪܥܐ ܚܕܬܐ. ܗ̄ ܕܘܒܪܐ ܪܘܚܢܝܐ ܕܩܢܝܢ ܐܪܥܢܝܐ ܒܡܫܝܚܐ. ("'For behold I create a new heaven and a new earth,' i.e., the spiritual mode of life which the earthlings acquire through the Messiah", tr. Hicks, p. 91).

Mete. V. Section ii.: *On the Fact that Plants and Animals are also Generated without Reproduction*

From his discussion of deluges IS moves on to a discussion of spontaneous generation,[10] without which species of animals and plants could become extinct as a result of such deluges. Although BH makes a number of departures from the *Šifāʾ* (most notably at V.ii.2.10-11 as discussed above in the introductory comments on Mete. Chapter V), most parts of the section may be considered a summary of the corresponding discussion in the *Šifāʾ*. That BH made use of Fakhr al-Dīn al-Rāzī's summary of IS's discussion in composing his own summary may be gathered from the fact that the examples of spontaneously generated animals given at V.ii.1.3-8 agree with the examples given in the *Mabāḥiṯ* rather than with those given in the *Šifāʾ*. On V.ii.2.10-11, see the introductory comments to Chap. Mete. V above.

	Šifāʾ	*Mabāḥiṯ*	
V.ii.1.1-3	76.18-77.3	(II.218.17-20)	Spontaneous generation
V.ii.1.3-8	77.3-4	II.218.21-219.3	Examples
V.ii.2.1-3	77.4-7	-	Possibility of spont. generation
V.ii.2.3-9	77.7-78.14 (esp. 77.13-78.4)	(II.219.4-13)	contd.
V.ii.2.10-11	-	-	
V.ii.3.1-4	78.5-6	-	contd.
V.ii.3.4-8	78.14-79.3	(II.219.14-20)	Possibility of extinction
V.ii.3.8-10	79.4-6	-	contd.

l. 1. ܒܠܐ ܝܘܒܠܐ ܡܬܝܠܕܝܢ ("... are generated without reproduction [*yubbālā*]"): cf. *Šifāʾ* 77.1 تحدث بالتولد دون التوالد. ("... arise through generation [*tawallud*] without reproduction [*tawālud*]"); cf. also *Mabāḥiṯ* II.218.17f. - In discussions of spontaneous generation in Arabic the word *tawallud* (√*wld* V) was frequently used to designate "spontaneous generation" and this was contrasted with *tawālud* (√*wld* VI), used to designate the normal mode of generation through reproduction.[11] There being no easy way of rendering the two terms in Syriac, BH is forced to paraphrase these terms throughout this section.

---

[10] For surveys of materials in Arabic relating to spontaneous generation, see Ullmann (1972) 54-56, Kruk (1990).

[11] Cf. Kruk (1990) 266.

V.ii.1.1-3: Possibility of spontaneous generation

But.: (1a) There is not even one proof for the impossibility of the generation [*hwāyā*] of species [*ādšē*] after their destruction [*ḥubbālā*] in the manner of a new creation, (1b) nor for the necessity of their generation by means of seed [lit. by seminal means, *ba-znā zarʿānāyā*] and by reproduction [*yubbālā*] from a male and a female.

Šifāʾ 76.18-77.3:

وليس بمستنكر أن تفسد الحيوانات والنباتات أو أجناس منها، ثم تحدث بالتولد دون التوالد. وذلك لأنه لا برهان البتة على امتناع وجود الأشياء وحدوثها بعد انقراضها على سبيل التولد دون التوالد، فكثير من الحيوانات يحدث بتولد وتوالد.

(1a) The theory is not to be rejected [*mustankar*] that animals and plants, or [certain] kinds [*ağnās*] among them, are destroyed [*tafsidu*] and then arise through generation [*tawallud*] without reproduction [*tawālud*]. That is because there is no proof at all for the impossibility of the existence and arising [*ḥudūṯ*] of things after their extinction [*inqirāḍ*] in the manner of generation without reproduction, (1b) so that many animals arise [*yaḥduṯu*] by [spontaneous] generation [*tawallud*] and reproduction [*tawālud*].

l. 2. ܚܘܒܠܗܘܢ ("their destruction"): انقراضها IS. - Rather than use the exact equivalent of IS's *inqirāḍ*, BH opts here for *ḥubbālā* which usually corresponds to arab. *fasād* (cf. *Šifāʾ* 76.18 تفسد). The result is that the two terms used here for "generation" and "destruction" (*hwāyā*, *ḥubbālā*) are those used generally of "generation and corruption" (as in Arist. *De generatione et corruptione*).

l. 2. ܒܪܝܬܐ ܚܕܬܐ ("new creation"): cf. comm. on V.i.3.6-7 above.

V.ii.1.3-8: Examples of spontaneous generation

The examples of spontaneously generated animals given here agree exactly with those given in the *Mabāḥiṯ* rather than with those given in the *Šifāʾ*.

But. [underline: agreement with *Šifāʾ*; italics: with *Mabāḥiṯ*]:

(1) Consider how [lit. how and behold] herbs [*ʿesbē*] and trees generated without seed [*d-lā zarʿā*] are more numerous than those which are sown [*mezdarʿānē*]. (2) Among *animals*, too, (2a) *bees* are *generated from cow-dung* [*kbāyē d-tawrē*], (2b) *snakes from hair cast in water*, (2c) *scorpions from* ὤκιμον - i.e. *basil* - *and figs*, (2d) *frogs from rain* (2e) *and mice from dust*. (3a) Although all of these are [usually] generated from a father and a mother, (3b) they are also formed without them.

Šifāʾ 77.3-4:

وكذلك النبات. وقد تتحد حيات من الشعر، وعقارب من التبن والباذروج، والفأر يتولد من المدر، والضفادع تتولد من المطر؛ وجميع هذه الأشياء فلها أيضا توالد.

[3 تتحد ed. Cair.: يتخذ ed. Ṭ ‖ التبن sic edd. Cair. et Ṭ, sed التين *Mabāḥiṯ* (ed. Hyderabad) et ܬܐܢܐ *But*.]

(1) ... similarly the plants. (2b) Snakes are sometimes formed from hair, (2c) scorpions from straw [or: fig] and basil, (2e) mice are generated from mud [*madar*], (2d) and frogs are generated from rain. (3a) For all these things there is also reproduction [*tawālud*].

Mabāḥit II.218.21-219.3:

الثانى ان كثير من الحيوانات يتولد ويتوالد مثل النحل المتولد من اخثاء البقر والعقرب المتولد من التين والباذروج والحيات المتولدة من الشعر اذا القى فى الماء والفار المتولد من المدر والضفادع المتولدة من المطر فهذه الاشياء وجودها تارة بالتولد وتارة بالتوالد.

Second [sc. proof for the possibility of spontaneous generation]: (2/3) many animals are [spontaneously] generated [*yatawalladu*] and reproduced [*yatawāladu*], (2a) such as the bee generated from cow-dung, (2c) the scorpion generated from fig and basil, (2b) the snakes generated from hair when it is cast in water, (2e) the mice generated from mud, (2d) and the frogs generated from rain. (3) The existence [*wuǧūd*] of these things is (3b) sometimes by generation (3a) and sometimes by reproduction.

l. 3f. ܥܣ̈ܒܐ ܘܐܝܠܢ̈ܐ ܕܠܐ ܙܪܥܐ ܡܬܝܠܕܝܢ. ܝܬܝܪ ܡܢ ܡܙܕܪ̈ܥܢܐ ܡܣܓܐܝܢ ("herbs and trees generated without seed are more numerous than those which are sown"): Fakhr al-Dīn al-Rāzī does not mention the plants in the passage quoted above; IS does, but does not elaborate. - The expansion by BH here may be associated with *Cand.* 149.4-150.4, where, with reference to Gen. 1.12, BH talks of how the plants, like Adam and Eve, were instantly produced in their fully-developed state by divine command on the first day (before the creation of the sun, which makes them grow) and tells us furthermore that these plants included many "which are not sown" (*d-lā mezdarʿān*), such as the different types of reed (150.2f.).

l. 6. ܐܘܩܝܡܘܢ ܐܘܟܝܬ ܚܘܟܐ ("ὤκιμον - i.e. basil"): باذروج IS & Rāzī. - The same identification of gr. ὤκιμον, syr. *ḥawkā* and arab. *bādirūǧ* is found in the list of medicinal plants at BH *Cand*. 193.4 (ܐܘܩܝܡܘܢ ܚܘܟܐ ܗܘܝܘ ܒܐܕܪܘܓ),[12] as well as in the lexicon of Bar Bahlūl [Chabot] 86.14.[13]

V.ii.2.1-3: Possibility of spontaneous generation

But.: (1) Although [*kad*] [such] extraordinary and strange formation [*gbīltā nukrāyat la-ʿyādā w-aksnāytā*] is not effected constantly [*law b-ammīnū*

[12] Cf. Ḥaddād (1981/2) 511. The list of plants is missing in mss. Paris. syr. 210 and Vat. syr. 168 of *Cand*. and hence the instance here is not registered in PS, where most of the citations from *Cand*. are taken from the Parisinus.

[13] The identification of *ōqīmōn* with *ḥawkā* also occurs in Sergius of Rēšʿainā's Syriac version of Galen's *De simplicibus* (Merx [1885] 301.15: ܐܘܩܝܡܘܢ ܕܐܝܬܘܗܝ ܚܘܟܐ), as well as in Bar ʿAlī sec. PS 88, although I am unable to find the entry at the expected place in Hoffman's edition of that lexicon. - BH also mentions *ḥawkā* at *But.* De plant. IV.i.3 [Drossaart Lulofs-Poortman] l. 377 and *Cand*. [Bakoš] 170.12 (corr. Drossaart Lulofs-Poortman [1989] p. 63, l. 60) but makes no identification of it with *ōqīmōn* there.

meštammšā], (2) it is not necessarily [the case] that it does not occur at all (3) when there occur [lit. at (*ṣēḏ*) the constitution (*kuyyānā*) of] a certain configuration of the celestial sphere [*eskēmā meddem mawzaltānāyā*] (4) and a favourable disposition of the elements [*ᶜāhnūṯā esṭuksānāytā*], (5) which are realised [lit. are found] on rare occasions [*l-maḥsen*] in long periods of time.

*Šifā*ᵓ 77.4-7:

وليس إذا انقطع هذا التولد، فلم يشاهد فى سنين كثيرة، لا يوجب أن لا يكون له وجود فى الندرة، عند تشكل نادر من الفلك لا يتكرر إلى حين، واستعداد من العناصر لا يتفق إلا فى كل طرف زمان طويل

(1) Although [*iḏā*] such generation is not constant [lit. is cut, *inqaṭaᶜa*] so that it is not observed for many years, (2) it is not necessarily the case that it does not occur [lit. have an existence] on rare occasions [*fī nudra*], (3) when there occurs [lit. at, *ᶜinda*] a rare configuration of the celestial sphere [*tašakkul nādir min al-falak*], (5) which is not repeated for some time, (4) and a favourable disposition of the elements [*istiᶜdād min al-ᶜanāṣir*], (5) which only arises at every end of a long period.

l. 2f. ܥܗܢܘܬܐ ܐܣܛܘܟܣܢܝܬܐ ("favourable disposition of the elements"): استعداد من العناصر IS. - Cf. V.i.1.6 above (ܒܥܗܢܘܬܗ ܕܟܠܗ ܕܐܣܛܘܟܣܐ).

V.ii.2.3-11: Possibility of spontaneous generation (contd.)

At *Šifā*ᵓ 77.7-78.14, IS has a lengthy and rather difficult passage in which he attempts to prove the possibility of spontaneous generation of animals in terms of the Aristotelian theories concerning the formation of animals and other "homoeomerous" bodies from the mixtures of the elements (cf. Arist. *Mete*. IV; *Šifā*ᵓ, Ṭabīᶜīyāt, fann 4, maqāla 2). The passage of *But.* here, though sometimes quite different from the *Šifā*ᵓ in wording, is best understood as a summary of that passage, whereby BH seems to have taken the middle part of IS's argument (77.13-78.4) as the basis of his summary, but also takes in various elements from the passages preceding and following. - Fakhr al-Dīn al-Rāzī, too, has a summary of IS's arguments at *Mabāḥiṯ* II.219.4-13. BH's summary here is different from Rāzī's, but there are a number of phrases here which suggest that BH also consulted that passage of the *Mabāḥiṯ* when composing this passage. - On the last part of the passage (part (8)), see the introductory comments on Chap. Mete. V.

But.: (1) One must not think that the generation of an animal is only accomplished through the casting [*tarmīṯā*] of the seed in the womb [*marbᶜā*], (2) because all compounds take their origin [*qānēn rēšīṯā*] from mixture [*meṯmazzḡānūṯā*] of elements and mixture occurs through assembly [*knīšūṯā*], (3) so that, just as assembly occurs in the womb, (4) it is also possible for a drop to be assembled out of airy water in some part [*mnāṯā*]

of the earth (5) and to acquire a mixture similar to that in the belly [*karsā*], (6) and, when the conditions are favourable [lit. at the achievement of suitable condition (*ʿāhnūṯā*)], (7) for a human form [*ṣurtā nāšāytā*] to appear upon it and to grow through the agency of the giver of forms [*men yāheḇ ṣūrāṯā*]. (8) Just as in the belly, so in the ground [*adamṯā*] it is possible for a pair to be generated [*tehwē*] and for a father and a mother to be created [*netḇrōn*] at first [*šarwāṯīn*] without a father and a mother after a universal deluge [*ṭawpānā kullānāyā*].

Šifāʾ 77.13-78.4:

فإن ظن أن ذلك يمتنع، إلا فى مكان محدود وقوة محدودة كالرحم والنطفة، فإن الكلام بعد المسامحة قائم فى المزاج الذى يقع للرحم، حتى يتكون فيه ما يتكون؛ والذى يقع للنطفة، حتى يتكون منها ما يتكون. فإن الكلام فى ذلك كالكلام فى الأصل. فإن جميع هذه إنما تتكون عن امتزاج ينتهى إلى العناصر، فإن ابتداء ذلك من العناصر ثم يستحيل، والرحم مثلا ليس يفعل شيئا إلا ضبطا وجمعا وتأدية، وأما الأصل فهو الامتزاج، والامتزاج عن الاجتماع. وهذا الامتجاع كما يمكن أن يقع عن قوى جامعة فى الرحم وغيره، فلا يبعد أن يقع بأسباب أخرى، وبالاتفاق. فإنه ليس جزء من الأرض يستحيل أن يوافى جزءا من الماء، ويلتقى به على وزن معلوم؛ وليس يمتنع أن يقع ذلك الوزن ولا معاوق، فلا يحتاج إلى الصُوان. وأما القوى الفعالة فيهبها واهب القوى، إذا حصل المستعد، فيفعل بعد المزاج الأول ما يجب فى تكميل النوع من الأمزجة الثانية والثالثة، ويرفدها التدبير العالى رفدا كافيا.

(1) If it is thought that that is impossible except in a definite place and [with] a definite power [*qūwa*] such as the womb [*raḥim*] and the drop [i.e. sperm, *nuṭfa*], the statement [*kalām*] concerning the mixture which occurs in the womb stands with due allowances, so that what is generated is generated in it, and concerning (the mixture) that occurs in the drop, so that what is generated is generated from it. The statement concerning that is like the statement concerning the principle [*aṣl*]. All these are generated from mixture which terminate with the elements, and the beginning of that is from the elements and it is then transformed. Similarly the womb has no functions other than retention, assembly [*ǧamʿ*] and conveyance. (2) The principle [*aṣl*] is mixture [*imtizāǧ*] and mixture arises from assembly [*iǧtimāʿ*]. (3) Just as this assembly can be effected by the assembling forces [*quwan ǧāmiʿa*] in the womb etc., (4) it is not improbable that it should occur through other causes and through chance. There is no particle [*ǧuzʾ*] of earth which cannot meet a particle of water and combine with it in a fixed measure, (5) and it is not impossible for that measure to occur when there is no hindrance, so that there is no need for a receptacle [*ṣuwān*]. (7) As for the efficient forces [*al-quwā al-faʿʿāla*], the giver of forces [*wāhib al-quwā*] gives them (6) when a suitable [mixture] [*al-mustaʿidd*] arises, (7) then effects after the first mixing what secondary and tertiary mixings as are necessary for the achievement of the species, while the higher direction [*al-tadbīr al-ʿālī*][14] provides assistance as required.

[14] Cf. *Mabāḥiṯ* II.219.9 *al-nafs al-mudabbira*.

l. 4. ܗܘܝܐ ܚܝܘܬܐ ("generation of animal"): "Animals" are mentioned not in the passage of the *Šifā'* quoted above, but earlier on at *Šifā'* 77.11: كما أن الحيوان يتولد عن امتزاج الأخلاط بعد امتزاج العناصر.
l. 4. ܙܪܥܐ ("seed"): cf. *Šifā'* 77.12 من غير بذر أو مني.
l. 5. ܡܪ̈ܟܒܐ ("compounds"): cf. *Šifā'* 78.11: وتتركب تركبا ثانيا على أي نسبة كانت.
l. 5. ܐܣܛܘܟ̈ܣܐ ("elements"): cf. *Šifā'* 77.7, 8, 9, 11, 16 عناصر.
l. 7. ܛܘܦܬܐ ܡܢ ܡܝ̈ܐ ("drop of water"): IS has *ǧuz' min al-mā'* ("particle of water") in the corresponding clause (78.1), but *nuṭfa* ("drop") at *Šifā'* 77.13, 15, a word which in Arabic will have Koranic connotations in a context such as this (Koran 16.4; 18.37; 23.14; 36.77; 53.46; 75.37; 76.2; 80.19).
l. 8-9. ܥܕ ܫܘܡܠܝ ܚܫܝܚܘܬܐ (lit. "at the achievement of suitability"): cf. *Mabāḥit* II.219.9: على كمال استعداده.
l. 9. ܨܘܪܬܐ ܐܢܫܝܬܐ ("a human form"): IS does not mention the word "human being" (*insān*) until *Šifā'* 78.15 (see under V.ii.3.4-8 below). In the summary in the *Mabāḥit*, on the other hand, the word appears from the beginning of the passage (II.219.4).
l. 9. ܝܗܘܒ ܨܘܪ̈ܬܐ ("the giver of forms"): IS has *wāhib al-quwā* at the corresponding place (78.2), but *al-wāhib li-l-ṣuwar* at *Šifā'* 78.13; even closer to BH's phrase here is *wāhib al-ṣuwar* of *Mabāḥit* II.219.9f.
l. 10. ܐܕܡܬܐ ("ground"): The word is one that will have a biblical connotation for a Christian reader (Gen. 2.7). The same applies for the verb *netbrōn* ("to be created") of l. 11; cf. comm. on Min. III.ii.3.2 above ("nature creates").
l. 11. ܕܠܐ ܐܒܐ ܘܐܡܐ ("without a father and a mother"): cf. V.ii.1.7 above ("from a father and a mother"). - The phrase "without father and mother" will no doubt have been a stock phrase in discussions of spontaneous generation (cf. Ibn Ṭufail, *Ḥayy b. Yaqẓān* [Gauthier] 24.3f. من غير ام ولا اب), but in combination with the notion of generation in the "ground" (*ādamtā*) it is difficult not to be reminded of Adam and Eve by the phrase.
l. 11. ܒܬܪ ܛܘܦܢܐ ܟܠܢܝܐ ("after a universal deluge"): cf. V.iii.1.7 below (ܒܬܪ ܐܒܕܢܐ ܟܠܢܝܐ ܕܐܬܐ ܘܠܛܘܦܢܐ ܕܢܘܚܐ). - Once a connection has been made with the Book of Genesis through Adam and Eve, it is difficult again not to associate the phrase here with Noah's flood (Gen. 6.1-9.29), even if Noah and his family in fact survived the flood and the creation of Adam and Eve took place before the flood. It may be noted that Noah's Ark is compared to a womb at BH *Horr*. in Gen. 7.17 [Sprengling-Graham] 38.9f.: ܘܗܘܐ ܛܘܦܢܐ ܐܪܒܥܝܢ ܝܘ̈ܡܝܢ. ܗ̄ ܘܗܝܕܝܢ ܐܬܬܙܝܥܬ ܩܒܘܬܐ. ܐܝܟ ܕܐܦ ܚܒܠܐ ܒܟܪܣܐ ("'And the Deluge was forty days.' That is, and then the ark was stirred, as also the fetus in the womb", tr. Sprengling-Graham).

V.ii.3.1-4: Possibility of spontaneous generation (contd.)

But.: (1a) The mixture suitable for the reception of animal form [*ṣurtā ḥayyūtānāytā*] is better and more complete [*yattīr rēšītā wa-mšamlāytā*] (1b) if it occurs in the womb, (2a) but even if does not, (2b) it is not impossible for a mixture like it or somewhat inferior to it to be constituted through different causes.

Šifā' 78.5-6:

نعم إن كانت مثلا رحم، كان ذلك أسلس وأوفق؛ وإن لم تكن، فليس مستحيلا فى العقل أن يقع ذلك من حركات وأسباب أخرى.

(1b) Certainly, if there is, for instance, a womb, (1a) that is the easiest and the most suitable [*aslas wa-aufaq*], (2a) but even if there is not, (2b) it is not impossible to conceive [lit. in the intellect, *fī al-ʿaql*] that that arises from movements and other causes.

V.ii.3.4-8: Possibility of extinction

The passage may be seen as a summary of *Šifā'* 78.14-79.3, although the wording of the passage suggests that BH also had an eye on the corresponding passage of the *Mabāḥit̲*.

But. [underline: agreement with *Šifā'*; italics: with *Mabāḥit̲*]:

(1) *If* this were *not* the case [*ellū lā hākannā*] (2) species [*adšē*] *would be annihilated* [*etlaytī(w)*] *entirely* [*la-ğmar*] (3) and *would not* [then] be generated again and *be restored* [*pnaw*], (4) *because it is not necessary* that *from* every man *another* man *be formed*, or from every tree another tree, (5) since the casting of the drop in the womb and of the seed on the earth is *voluntary* [*ṣebyānāytā*] and possible [*metmaṣṣyānāytā*], *which usually takes place*, but *not necessary* [*ananqāytā*], (6) and there are times, *once in a long while* [*ḥdā l-nuğrā*], when something like this fails to take place [lit. does not occur].

Šifā' 78.14-79.3:

ولولا هذا لكان يجوز أن يقع للأنواع انقطاع، وذلك لأنه ليس بواجب أن يكون عن كل إنسان إنسانٌ ضرورة، ولا عن كل واحد من الناس، وكذلك عن كل شجرة؛ بل ذلك جائز أكثرى، ولا يستحيل أن يفرض وقت مّا يتفق فيه أن ينصرم كائنات من غير أن تتكون عن كل واحد منها خالف، إذ لا يوجد ولا واحد منها واجبا بالضرورة أن يتكون منه آخر. لأن الجماع الذى هو مبدأ التوالد إرادى لا ضرورى، ووقوع البذور فى البيادر طبيعى من جملة الأكثرى لا من جملة الضرورى، أو إرادى. ولا شىء من هذين يجب ضرورة، وما لا يجب ضرورة فيجوز أن يقع فى النادر بخلاف ذلك.

(1) If this were not the case [*lau lā hād̲ā*], (2) then it would be possible for extinction [*inqiṭāʿ*] to befall [certain] species [*anwāʿ*]. (4) That is because it is not necessary [*wāğib*] that there arise from every human being a human being by necessity [*ḍarūratan*], not even from any single human being, and similarly from every tree. Rather that is possible [*ğā'iz*] and usual [*akt̲arī*], and it is not impossible that a time be ordained when it will happen that the existing things [*kā'ināt*] will vanish [*yanṣarimu*] without a successor [*ḫālif*] being formed from any one of them, since there is not a single one of them from which another will necessarily be formed, (5) because the intercourse [*ğimāʿ*] which is the principle [*mabda'u*] of reproduction is voluntary [*irādī*] and not necessary [*lā ḍarūrī*], and the falling of seeds [*wuqūʿ al-bud̲ūr*] on threshing floors is natural and in the category [*ğumla*] of the usual [*akt̲arī*] and not in the category of the necessary, or is voluntary. (6) There is nothing among these two which is necessary, and with what is not necessary it is possible that its opposite happens once in a while [*fī al-nādir*].

Mabāḥit II.219.14-20:

الرابع انه لو لم يكن حدوث الانواع بالتولد ممكنا لكان يجوز ان ينقصع الانواع بحيث لا تعود البتة لانه ليس يجب ان يتولد من الشخص شخص آخر لان الجماع الذى هو مبدء التوالد ارادى لا ضرورى ووقوع البذور فى البوادر طبيعى لكنه اكثرى ولا ضرورى واذا لم يكن احد هذين ضروريا لم يكن تأدى كل شخص من النوع الى شخص آخر منه ضروريا فيجوز فى النادر حينئذ ان ينقطع فلو لم يكن حصول الانواع الا بالتوالد لكانت الانواع حينئذ تنقطع

Fourth consideration: (1) If the occurrence of species by [spontaneous] generation were not possible, (2) then it would be possible for species to become extinct [*yanqaṭiʿu*] (3) in such a way that they are not restored [*taʿūdu*] at all [*al-battata*], (4) because it is not necessary that from an individual another individual be formed, (5) since the intercourse which is the principle of reproduction is voluntary and not necessary, and the falling of seeds on threshing floors is natural but [merely] usual and not necessary. Since neither of these two is necessary, production by an individual member of a species of another individual is not necessary, (6) so that it is possible that (the species) become extinct once in a while. If the occurence of species were only by reproduction, there would be species that become extinct.

V.ii.3.8-10: Possibility of extinction (contd.)

But.: (1) If it were not for [*w-ellū lā*] the eternity of the motions of the celestial spheres [*ʿālmīnāyūt zawʿē mawzaltānāyē*] which are the causes of eternal generations [*hwāyē lʿālmīnāyē*], (2) there would be an opportunity [lit. place, *atrā*] for extinction [*metgammānūtā*] of species without restoration and generation afresh [*belʿād punnāyā wa-hwāyā men-d-rēš*].

Šifāʾ 79.4-6:

فلو لم تكن حركات ونسب عائدة من الأفلاك توجب كون أشخاص من هذه الأنواع/ ٥ مبتدأة حتى لا يكون لشىء من الأنواع انقطاع، بحيث لا يعود، لكان يجوز أن يقع/ ٦ انقطاع لا عود له، ولكان هذا الجائز قد وقع فيما لا نهاية له من قدرة الله.

[4 كون : تكون Ṭ; أن تكون DS]

(1) If the movements and recurring relationships of the celestial spheres [*ḥarakāt wa-nisab ʿāʾida min al-aflāk*] did not necessitate that individual members [*ašḫāṣ*] of these species become [*kaun*, v.l. *takawwun*] the beginnings, so that none of the species undergoes extinction [*inqiṭāʿ*] in such a way that they are not restored, (2) then, it would be possible for extinction without restoration [*inqiṭāʿ lā ʿaud lahu*] to occur, and this possibility would already have become a reality [lit. occurred] in the course of eternity [*fī-mā lā nihāya lahu*] by the omnipotence of God.

l. 10. ܡܬܓܡܡܢܘܬܐ ("extinction"): lit. being cut off. - The word, possibly a hapax legomenon (the instance recorded at PS Suppl. 75b is the one here), appears to be a calque answering to arab. *inqiṭāʿ*.

Mete. V. Section iii.: *On the Fact that Skills, Languages and Writing too are Renewed*

The development of human knowledge (ἐπιστήμη) and skills (τέχναι) is a subject discussed on a number of occasions by Arist., as, for example, at the beginning of his *Metaphysica*, while the notion of the recurrent cycles of knowledge is one that occurs several times in the Aristotelian corpus, including, in passing, at *Mete*. 339b 27-30.[15]

From his discussion of the possible extinction and renewal of species, including the human species, IS moves on to discuss briefly this matter of the development and renewal of human knowledge at the end of his treatise on meteorology (ed. Cairo 79.6-18).

In the section of *But.* here, the first 'theory' appears to be based for the most part on that passage of the *Šifāʾ*, although BH begins to move away from the *Šifāʾ* in the last part of that 'theory' (V.iii.1.8-9, "building" of knowledge). The second 'theory' (V.iii.2) then deals with the renewal of speech and writing, a subject that is not discussed by IS in his treatise on meteorology in the *Šifāʾ*, and the terms in which the subject is treated brings one more into the realm of discussions connected to the *De anima* and *Politica*.

The reason why BH makes these additions at V.iii.1.8-9 and in V.iii.2 is not altogether clear, but the allusions to Adam and Eve and to Noah's flood at V.ii.2.10-11 above suggests that a connection may have been provided by the narrative of the Tower of Babel (Gen. 11.1-9), the next major event recounted in the Book of Genesis after Noah's flood and one which has to do with "building" and with "languages" (though admittedly with their differentiation rather than with their origin).[16] It may also be remembered that BH in his *Chron.* accepts the identification of the biblical Enoch with Hermes

[15] This passage, however, is missing in the Arabic version of *Mete*. by Ibn al-Biṭrīq (ed. Schoonheim, p. 9, l. 49; similarly in the Hebrew version by Ibn Tibbon, ed. Fontaine, p. 14, I.120). For the notion elsewhere in Arist., see *De caelo* 270b 19, *Metaph*. 1074b 1-14, *Polit*. 1329b 25-29; cf. further the literature cited at Strohm (1970) 138 (comm. ad 338 b 20ff.).

[16] The latent implication could then be that the accumulation of knowledge as described at V.iii.1.8-9 is in vain if based merely on human effort (i.e. *nisi Dominus aedificaverit domum*, Ps. 127.1), although this is probably reading too much into the text of *But.* here.

Trismegistus, who is then considered among other things to be the inventor of writing (*sep̄rā wa-k̲t̲īb̲āt̲ā*) and the one who taught men the "building of cities" (*benyān md̲īnāt̲ā*).[17]

V.iii.1.1-4: Origin of knowledge

But.: (1) Every acquired skill [*ummānūt̲ā metqanyānītā*] takes its origin [*qānyā šurrāyā*] from the deliberation of the individual soul [*maḥšab̲tā d-nap̄šā qnōmāyā*] or from the divine revelation [*gelyānā allāhāytā*] which occurs to some men, (2) since the universal man [*barnāšā kullānāyā*] is something imaginary and not real [lit. is in thought (*ḥuššāb̲ā*) only and not in reality (*suᶜrānā*)] (3) and everything whose cause is created [*hwīt̲ā*, ptc. pass.] and has a beginning [*mšarryānīt̲ā*] is itself created and has a beginning.

Šifāʾ 79.6-9:

وأنت إذا تأملت الصناعات وجدتها مخترعة عن روية النفس، أو من إلهام الله؛ وأنها لا يكون مبدؤها إلا روية شخص أو إلهام شخص. فإن الكلى متوهم لا وجود له، وما مبدؤه جزئى حادث فهو حادث بعد ما لم يكن أصلا.

(1) When you examine the skills [*ṣināᶜāt*] you will find that they are invented [*muḫtaraᶜa*] by deliberation [*rawīya*] of the soul or [originate] from divine inspiration [*ilhām allāh*], and their origin [*mabdaʾu*] can only be the deliberation of an individual [*šaḫs*] or the inspiration of an individual. (2) For the universal (man) [*al-kullī*] is imaginary [*mutawahham*] and has no existence [*lā wuǧūd lahu*]. (3) Something whose origin is particular [*ǧuzʾī*] and comes into being [*ḥādiṯ*] is [itself] something that comes into being [*ḥādiṯ*] after not exiting at first.

V.iii.1.4-9: Renewal of knowledge

But.: (1) The sciences [*yad̲ᶜāt̲ā*] and the skills, therefore, have beginnings (2) and for this reason they grow gradually and acquire greater completeness [*šumlāyā yattīrā*] from one period to another. (3) Their renewal [*ḥuddāt̲ā*] testifies to the renewal of their inventor [*meškḥānā*], that is, as we have said, the first man [*barnāšā qad̲māyā*] who is created [*metb̲rē*] after the universal destruction of the species and the general deluge. (4) This man lays the foundations of the teachings and crafts [*yulpānē w-amnē*] through divine revelation or natural power of reasoning [*ḥaylā mlīlā kyānāyā*] and the "building" [*benyānā*] is perfected little by little by those who come after him.

---

[17] BH *Chron.* [Bedjan] 5.10-25. - See also *Hist. dyn.* [1958] 7.9-8.13, where BH tells us, at 7.17-10, that he had a copy of the Syriac version of Hermes' treatise addressed to Tat [Ṭāṭī], but seems otherwise to be largely dependent on Sāᶜid al-Andalusī; see the latter's *Ṭabaqāt* [Cheikho] 18.20-19.6, 39.7-16, tr. Blachère 84f., 54f. (with footnotes ad loc.); cf. EI² III.463, s.v. "Hirmis" (Plessner).

Šifā' 79.9-10:

فكل صناعة حادثة، ويدل على حدوثها تزيدها كل وقت، ويدل حدوثها على أن الناس منشأون بعد انقراض.

(1) Therefore, all skills come into being [*ḥādiṯ*]. (2) Their constant growth [lit. their growing (*tazayyud*) all the time] prove their coming-into-being [*ḥudūṯ*] (3) and their coming-into-being proves that men are brought into being [*munša'*] after extinction.

l. 6. ܚܘܕܬܐ ("renewal"): The word, if not quite the same in meaning, is etymologically cognate with IS's *ḥudūṯ* ("coming-into-being"; syr. √*ḥdt* corr. arab. √*ḥdṯ*). It is at the same time a word that is often used in connection with the Resurrection, e.g. *Cand*. X [Zigmund-Cerbü] 16.2f.: ܩܝܡܬܐ ܐܘܟܝܬ ܚܘܕܬܐ ܕܦܓܪܐ ܐܢܫܝܐ ("resurrection, i.e. the renewal of the human body"); *Rad*. [Istanbul] 74.7f.: ܡܛܠܬ ܕܥܠܡܐ ܗܢܐ ܡܬܚܕܬ ܒܠܚܘܕ ܠܫܘܠܡܐ ܘܠܘ ܠܠܐ ܡܕܡ ܓܪܝܢ ("on the fact that this world is merely renewed at the end and is not reduced to nothing").

l. 7. ܒܪܢܫܐ ܩܕܡܝܐ ("the first man"): IS constantly speaks of "men" in the plural in the corresponding passage (*nās* 79.10, 12; *al-nās al-awwalīn* 79.13), although he does say "the first man or the first men" on one occasion (*awwal al-insān au awwal al-nās* 79.14). The alteration to the singular here is probably deliberate, whereby BH will have had in mind the monogenism of mankind through Adam, as well as the notion of Christ as the first-born of the new creation. - There is a comparable instance at *But*. Eth. I.v.2; BH speaks there of the "God-Man" (*alāh barnāš*) with clear reference to Christ, where Ṭūsī in the source passage had spoken of someone who has reached the highest degree of perfection as "God's vicegerent" (*ḫalīfa-i ḫudā-i taᶜālā*)", one of God's "saints" (*auliyā-i ḫāṣṣ*) and "complete and absolute man" (*insānī-yi tāmm-i muṭlaq*) (*Aḫlāq-i nāṣirī*, ed. Mīnowī-Ḥaidarī 71.1-3; tr. Wickens 52).

l. 7. ܐܒܕܢܐ ܟܠܢܝܐ ܕܙܢܝܐ ܘܛܘܦܢܐ ܓܘܢܝܐ ("after the universal destruction of the species and the general deluge"): cf. V.ii.2.11 above (ܒܬܪ ܛܘܦܢܐ ܟܠܢܝܐ).

l. 9. ܡܫܬܟܠܠ ܒܢܝܢܐ ("the 'building' is perfected"): For a similar metaphorical use of the word *benyānā* with reference to knowledge and with the verb *šaklel* as here, see BH, *Chron*. [Bedjan] 98.14f.: ܠܘ ܥܠ ܫܬܐܣܬܐ ܐܚܪܬܐ ܣܡܘܗ. ܐܠܐ ܥܠ ܕܘܡܣܐ ܝܘܢܝܐ. ܒܢܝܢܐ ܚܟܡܬܢܝܐ ܫܟܠܠܘ ܡܗܝܡܢܘܬܐ ... ("[Arabic philosophers, mathematicians and physicians] perfected [*šaklel(w)*] the buildings of the sciences [*benyānē ḥekmṯānāyē*], placing them not on another foundation [*šeṯestā*], but on Greek foundations [*dumsē* < δόμος]")

V.iii.2.1-10: Renewal of speech and writing

It is not clear whether one should be looking for a single source that lies behind this 'theory' as a whole or should consider the 'theory' as an "original" composition by BH, composed out of materials taken from different sources. The latter possibility is made somewhat more

likely by the fact that, if we subtract from the 'theory' those considerations relating to the annihilation of the human species (which is what connects the 'theory' to what precedes), much of what remains has a close resemblance to an earlier discussion of the matter by BH at *Cand.* Base VIII (de anima), 2.8.2, where BH discusses communication (*mšawdᶜānūṯā*) by means of a) uttered words (*mellṯā nāp̄ōqtā*, ed. Bakoš [1948] 58.23-60.2), b) signs (*mšawdᶜānūṯā remzānāytā*, 60.3-12) and c) writing (*kāṯōḇūṯā*, 60.13-21).[18] The passage of *But.* here also has certain similarities with the part of IS *Šifāᵓ*, al-nafs, maqāla V, faṣl 1, where IS talks of communication (*iᶜlām*) by means of voice (*ṣaut*) and gestures (*išāra*),[19] but the agreements are not such as to allow us to propose that passage as the principal source of our 'theory'.

Comparison with BH's *Muḫtaṣar fī ᶜilm al-nafs al-insānīya* (= *Muḫt. nafs*) is also of some interest. The discussion of human language in that work (faṣl 15, ed. Sbath p. 26f.) is largely a summary of *Cand.* VIII.2.8.2. There are, however, at least two instances where we find concepts in *Muḫt. nafs* which it shares with *But.* but not with *Cand.* (see comm. on l. 6, l. 7 below). This fact is of interest in connection with the question of the authenticity of *Muḫt. nafs*, which has been thought by some to be the work of a later epitomist rather than BH himself. - It may be remembered that a similar case has been noted above in connection with BH's *Liber radiorum* (= *Rad.*), namely a passage that occurs in *But.* and *Rad.* but not in *Cand.* (see comm. on Min. II.iii.4. 4-7). - These facts put together suggest that the manner in which *Muḫt. nafs* was composed was similar to that of *Rad.* (except, naturally, for the fact that *Muḫt. nafs* is in Arabic rather than Syriac), in that both works are basically summaries of *Cand.*, but also contain some new elements, which are then sometimes taken up again by BH in *But.*

V.iii.2.1-7: Renewal of speech

But. [underline: agreement with *Cand.*]:

(1) It is obvious that writing and speech [*sep̄rē w-mellē*] disappear [*bāṭlīn*] with the disappearance of writers and speakers. (2a) Since the human individuals [*qnōmē nāšāyē*] who arise after the annihilating destruction [*ḥubbālā mawbḏānē*] have by nature [*ba-ḵyānā*] an aptitude [*ᶜāhnūṯā*] for

[18] Corr. ed. Çiçek 646-648: cf. also the brief discussion at BH *Rad.* VI.4.4 [Istanbul] 186f.

[19] Ed. Cairo [Anawati-Zayed (1972)] p. 182; corr. ed. Bakoš (1956) 199f.; ed. Rahman (1959) 203f. - Cf. also Fakhr al-Dīn al-Rāzī, *Mabāḥiṯ* II.410. - The question of human communication is not discussed at the corresponding place in *But.* at De anima V.i. (ms. F 123v-124r, cf. Furlani [1931] 43f.).

communicating to each other [their] emotions [lit. passions of the soul, *ḥaššē napšānāyē*] and thoughts [lit. movements of the mind, *zawᶜē tarᶜīṯānāyā*], (3) the likelihood is that they will at first communicate [them] with signs made with fingers [*remzē ṣeḇᶜānāyē*] and shapes made with limbs [*sukkālē haddāmāyē*]. (4a) They will then advance to a stage where [*lwāṯ hāy d- ... meṯdarrgīn*] instead of things and actions [*suᶜrānē w-sāᶜōrwāṯā*] they will use nouns and verbs [*šmāhē w-mellē*], (4b) beginning with necessary things [*ananqāyē*] and ending with less useful things [*bṣīray ḥšaḥtā*].

Cand., Base VIII [Bakoš] 59.8-14:

... ܐܠܝܨܐܝܬ ܐܣܬܢܩܘ ܒܢܝ̈ܢܫܐ ܠܥܘܕܪܢܐ ܕܚܕ ܡܢ ܚܕ̈ܕܐ . ܗܢܐ ܕܝܢ ܥܘܕܪܢܐ ܡܢ
ܡܫܘܕܥܢܘܬܐ ܗܘܐ . ܘܗܝ ܡܫܘܕܥܢܘܬܐ ܒܒܢ̈ܬ ܩܠ̈ܐ ܗܘܬ ܡܬܢܗܪܬܐ ܘܡܬܥܬܪܬܐ
ܗܘܬ . ܒܕܝܠ ܕ ܩܢܘ ܥܗܝܢܘܬܐ ܡܢ ܟܝܢܐ ܢܣܒܘ . ܐܫܟܚܘ ܘܣܡܘ ܠܟܠ ܟܝܢܐ
ܘܡܥܒܕܢܘܬܐ ܕܠܐ ܙܒܢܐ ܫܡܐ ܝܗܒܘ . ܘܠܟܠ ܡܥܒܕܢܘܬܐ ܕܥܡ ܙܒܢܐ
ܡܠܬܐ . ܘܐܪܟܒܘ ܫܡ̈ܗܐ ܘܡܠ̈ܐ ܠܐܣܪ̈ܐ ܦܐܝ̈ܐ ܘܡܛܠ ܗܕܐ

... human beings perforce require assistance from each other. This assistance arises from communication [*mšawdᶜānūṯā*] and and communication is made clearer and richer by words [*bnāṯ qālē*]. (2) Therefore, since [human beings] have received an aptitude [*ᶜāhnūṯā*] from nature, (4a) they have invented and have composed for every object [*kyānā*] and action [*maᶜbdānūṯā*] without time a noun [*šmā*], and for every action in time a verb [*melltā*], and have joined the nouns and verbs with suitable particles [*esrē*] ...

Cand., Base VIII [Bakoš] 60.3ff.:

ܡܛܠ ܡܫܘܕܥܢܘܬܐ ܪܡܙܢܝܬܐ . ܐܝܬ ܠܗ ܠܒܪܢܫܐ ܥܗܝܢܘܬܐ ܟܝܢܐܝܬ .
ܕܐܝܟܢܐ ܕܠܟܠ ܟܝܢܐ ܘܡܥܒܕܐ ܒܬ ܩܠܐ ܡܪܟܒ . ܗܟܢ ܒܐܣܟܡܐ ܡܕܡ
ܕܝܠܝܐ ܢܫܘܕܥܝܘܗܝ . ܒܥܝ̈ܢܘܗܝ ܒܣܦ̈ܘܬܗ ܘܒܨܒ̈ܥܬܗ . ܘܝܬܝܪ ܕܝܢ ܒܗ
ܒܠܚܘܕ ܗܕ̈ܡܘܗܝ ܗܢܘܢ ܕܨܒܝܢܐܝܬ ܡܬܬܙܝܥܝܢ .

On communication by signs [*mšawdᶜānūṯā remzānāytā*]: (2) Man [*barnāšā*] has the aptitude [*ᶜāhnūṯā*] by nature [*kyānāʾīṯ*] - just as he composes a word [*baṯ qālā*] for evey object and deed - (3) also to communicate it with some particular shape [*eskēmā*], with his eyes, lips, fingers, and especially with those limbs of his which can be moved at will.

l. 2f. ܟܕ ܟܝܢܐܝܬ ܩܢܝܢ ܕܠܚ̈ܫܐ ܢܦܫܢܝ̈ܐ ܘܙܘ̈ܥܐ ܬܪܥܝܬܢܝ̈ܐ ܠܡܫܘܕܥܘ ܗܘܘ ܬܕ̈ܪܐ ܢܫ̈ܐ ܒܪ̈ܢܫܐ ܒܟܝܢܐ ܩܝܡ ("... have by nature an aptitude for communicating to each other [their] emotions [*ḥaššē napšānāyē*] and thoughts [*zawᶜē tarᶜīṯānāyā*]"): cf. *Šifāʾ*, al-nafs, V.1 [Cairo] 182.1-2: ... احتاج الإنسان أن تكون له فى طبعه قدرة على أن يعلم الآخر الذى هو شريكه ما فى نفسه بعلامة وضعية ("... man needs to have in his nature the ability [*qudra*] to communicate [*yuᶜlim*] what is in his soul [*nafs*] to another, who is his companion, by means of conventional signs [*ᶜalāma waḍᶜīya*]").

l. 2. ܚ̈ܫܐ ܢܦܫܢܝ̈ܐ ("emotions"): The section (*nīšā*) immediately preceding that on "communication" in *Cand*. is concerned with "emotions" (*Cand*. VIII.2.8.1, ed. Bakoš, p. 56-58: ܡܛܠ ܚ̈ܫܐ ܢܦܫܢܝ̈ܐ ܕܕܝܠܝܢ ܠܒܪܢܫܐ; cf. *Rad*. VI.4.4. [Istanbul] 185.16).[20]

[20] ܕܕܝܠܝܢ: sic (recte) ed. Çiçek 644.6; ܕܕܝ̈ܠܐܝܢ Bakoš; ܕܕܝܠܐܝܢ P.

l. 3. ܠܚ̈ܕܕܐ ܠܡܘܕܥܘ ܘܡܢ ܚ̈ܕܕܐ ܠܡܬܝܕܥܘ ("for communicating to each other"): lit. "for making known to each other and being made known to by each other". - Cf. *Šifāʾ*, al-nafs [Cairo] 182.16: الإعلام والاستعلام.

l. 4. ܪܡܙܐ ܨܒܥܢܝܐ ("signs made with fingers", lit. "digital signs"): The phrase *remzā ṣeḇʿānāyā* is also used at BH *Splend*. [Moberg] 158.32f.: ܓܝܪ ܐܡܪܐ. ܐܝܟ ܕܡܫܘܕܥܢܘܬܐ ܠܡܠܘܫܘܬܐ ܕܚܝܘܬܐ ܕܦܘܩܕܢܐ ܘܐܝܟ ܕܒܪܡܙܐ ܨܒܥܢܝܐ ܡܫܬܡܫܝܢ. ܡܫܡܫܐ ܨܒܥܢܝܬܐ (in the explanation of the particle *gēr*; cf. PS 710, s.v. ܓܝܪ).

l. 5. ܫܡܗ̈ܐ ܘܡܠ̈ܐ : lit. "names and words", but here best taken in the technical sense of "nouns and verbs" as is made clear by the first passage of *Cand*. quoted above. These are, of course, also the senses in which the two terms are used in Syriac grammars, including those of BH himself (see, for instance, the relevant book/chapter headings at *Splend*. [Moberg] 6.1, 89.7 and *Gramm*. [Bertheau] p. 5, 42).

l. 6. ܟܕ ܡܢ ܐܢܢܩܝ̈ܐ ܡܫܪܝܢ. ܘܡܫܠܡܝܢ ܒܚܣܝܪ̈ܝ ܡܘܬܪܢܘܬܐ ("beginning with the necessary things and ending with less useful things"): This notion, which is not in *Cand*., is also found as an addition in *Muḫt. nafs*, at the end of a sentence which is otherwise a paraphrase of *Cand*. (*Muḫt. nafs*, ed. Sbath 26.12: قبالة ما تقع الحاجة إليه). - Cf. *Šifāʾ*, al-nafs [Cairo] 182.16f., and as a more remote source, Arist. *Polit*. 1329b 27-29; *Metaph*. 981b 17-20.

V.iii.2.7-10: Renewal of writing

But.: When the population [*nāšūṯā*] increases and interests [lit. wishes, *ṣeḇyānē*] become opposed, villages and cities are built. Hence vocal communication [*mšawdʿānūṯā qālānāytā*] becomes inadequate with the distance between persons and necessity [lit. cause, *ʿelltā*] calls for composition of written images [*demwāṯā meṯkaṯbānyāṯā*] instead of [spoken] nouns and verbs. Thus writing [*sep̄rē*] too is renewed and come to differ with the differences in modes of life and inhabitants [*ʿumrē w-ʿāmōrē*].

l. 7. ܡܬܚܠܦܝܢ ܨܒܝ̈ܢܐ ("interests [lit. wishes] become opposed"): also in connection with 1f. ܩܢܘ̈ܡܐ ܐܢܫ̈ܝܐ ("human individuals") above, cf. *But*. Polit. I.i.2., ms. F 217v b3-6: ܡܛܠ ܕܝܢ ܕܡܫܚܠܦܝܢ ܨܒܝ̈ܢܐ ܕܩܢܘ̈ܡܐ ܐܢܫ̈ܝܐ ("because the wishes/interests of the human individuals differ, ..."); Ṭūsī, *Aḫlāq-i nāsirī* [Mīnowī-Ḥaidarī] 250.23 اقتضای تباین همَم وآرای, 251.2 اختلاف عزایم ("diversity of purposes", "disparity of aspirations and opinions", tr. Wickens 189 fin.).

l. 7. ܩܘܪ̈ܝܐ ܘܡܕܝܢ̈ܬܐ ("villages and cities"): cf. *Muḫt. nafs* [Sbath] 26.7: اجتماع الناس الى القرى والمدن. - Here again (as in l. 6 above), the sentence in *Muḫt. nafs* may otherwise be considered a paraphrase of *Cand*. (VIII, ed. Bakoš 58.23ff.) but the words "villages and cities" are not there in *Cand*.

APPENDIX ONE

COLOPHONS OF MANUSCRIPTS

Given below are the subscriptions found in the manuscripts of *But.* accessible to me (mss. FLlMmVR and Beirut, USJ 48).

The variant readings of those parts which are repeated in the different manuscripts are discussed under no. 1, Laur. or. 83 (= F).

It will seen that these subscriptions serve to confirm, among other things, the kinship of L and l, and of M and m.

Besides the information they provide on *But.* and on the manuscripts themselves, interesting features of these colophons include the biographical notes on Barhebraeus in ms. F, the affection shown by the copyist of ms. L for his sons and the notes on the contemporary events in mss. LmR.

(a): end of *But*. Poet.; **(b)** end of *But*. De anima; **(c)** end of *But*. Theol.; **(d)** end of *But*. Polit.

1. *Florence, Laurentianus or. 83 (= ms. F)*

Of the following notes, those on fols. 191v (= c below) and 227r (= d.i-iv; *errore* "277a" Margoliouth) were reproduced earlier by Margoliouth (1887) 41f.

(a) Fol. 3r 9-11:

ܫܠܡ ܟܬܒܐ ܕܦܘܐܝܛܝܩܐ ܘܒܫܘܡܠܝܗ/10 {ܐܫܬܡܠܝ ܝܘܠܦܢܐ ܩܕܡܝܐ
ܕܠܘܓܝܩܝ{ܬܐ ܒܦܠܓܘܬܐ ܩܕ̄ ܕܟܬܒܐ ܕܚܐܘܬ ܚܟܡܬܐ/ {ܢܩ}ܦ ܠܗ ܕܝܢ
ܒܦܠܓܘܬܐ {ܬܪܝܢܝܬܐ ܟܬܒܐ ܕܫܡܥ}ܐ ܟܝܢܝܐ. ܘܠܐܠܗܐ ܫܘ̄ ܕܚܝܠ ܀

[9-10] Here ends the Book of Poetics. With its conclusion [*šumlāyā*] {comes to end the first *yulpānā*, on logic}, in the first division [*pālgūṯā*] of the book of Cream of Wisdom. [11] It is followed in the second division by {the book of} Physical Hearing [*Auscultatio physica*]. Glory to God, who has given strength.

N.B. The parts in brackets are now barely legible due to water damage. The same words as here are found at the end of ms. Laur. or. 8 (= f), except that Laur. or. 8 has ܩܕ̄ and ܫܘ̄ written *plene* and adds ܘܥܕܪ after ܕܚܝܠ at the end, followed by a note in Arabic, تم الجـز الاوّل من كتـاب زُبدَة الحكمـة في المنطق. - Lines 9-10, ܕܠܘܓܝܩܝܬܐ ... ܫܠܡ, are also in LlR; 9 ܘܒܫܘܡܠܝܗ Ffl: ܘܒܫܘܡܠܗ LR.

(b) Fol. 130r b11-14:

ܫܠܡ ܟܬܒܐ ܕܥܠ ܢܦܫܐ / ܘܒܫܘܡܠܝܗ ܐܫܬܡܠܝ ܝܘܠܦܢܐ / ܬܢܝܢܐ ܕܟܬܒܐ ܡܢ
ܟܬܒܐ ܕܚܐܘܬ / ܚܟܡܬܐ. ܘܫܘܒܚܐ ܠܐܠܗܐ ܗܘ̄ ܕܝܗܒ ܚܝܠ ܘܥܘܕܪ.

[11-14] Here ends the Book on the Soul. With its conclusion [*šumlāyā*] comes to end the second *yulpānā* of the book of Cream of Wisdom. [14] Glory to God who has given strength and help.

N.B. 11-14 ܫܠܡ ... ܚܟܡܬܐ also in LlMmV; 14 ܘܫܘܒܚܐ ... ܘܥܘܕܪ also in Mm - 12 ܘܒܫܘܡܠܝܗ Fml; ܘܒܫܘܡܠܝܗ (sic) M; ܘܒܫܘܠܡܗ LV || 14 ܚܟܡܬ̈ܐ LlmV || ܘܫܘܒܚܐ: ܫܘܒܚܐ m || ܕܝܗܒ ܚܝܠ ܘܥܘܕܪ: add. ܘܡܛܠ ܐܡܝܢ m.

(c) Fol. 191v a 5-13:

5 ܫܠܡ ܟܬܒܐ ܕܬܐܘܠܘܓܝܐ / ܘܒܫܘܠܡܗ ܐܫܬܠܡܬ ܡܢܬܐ / ܬܐܘܪܝܛܝܩܝܬܐ
ܕܦܝܠܘܣܘܦܝܐ ܡܢ / ܟܬܒܐ ܕܚܐܘܬ ܚܟܡܬܐ. ܣܛܪ ܡܢ / ܗܠܝܢ ܡܬܡܛܝܩܝܬܐ
ܕܠܡܐܡܪܐ/10 ܐܚܪܢܐ ܐܝܬܝܗܝܢ ܠܫܡ. ܟܢܘܢ ܩܕܡ / ܫܢܬ ܐܦܨܙ ܕܝܘ̈ܢܝܐ.
ܘܫܘܒܚܐ ܠܐܠܗܐ ܗܘ ܕܝܗܒ ܚܝܠ / ܘܥܘܕܪ ܒܛܝܒܘܬܗ. ܗܟܢܐ ܟܬܝܒ ܗܘܐ / ܒܐܪܟܝܬܗ
ܕܐܒܘܢ ܛܘܒܢܐ ܘܡܢܝܚܐ.

[5-8] Here ends the Book of Theology. With its completion [*šullāmā*] comes to end the theoretical part of philosophy in the book of Cream of Wisdom, [8-10] except for the mathematical [parts] which belong to another treatise. [10-12] End of Kanon I, the year 1597 of the Greeks [= Dec. 1285 A.D.]. Glory to God who has given strength and help in His goodness. - [12-13] Thus was written in the copy of our blessed and deceased father.

N.B. 5-10 ܫܠܡ ... ܠܫܡ also in LlMm - 6 ܘܒܫܘܠܡܗ FLlMm || ܐܫܬܡܠܝܬ F: ܐܫܬܠܡܬ MmLl || 8 ܚܟܡܬ̈ܐ Llm.

(d.i.) Fol. 227r b 5-10:

ܫܠܡ ܟܬܒܐ ܕܦܘܠܝܛܝܩܘܢ /6 ܘܒܫܘܡܠܝܗ ܐܫܬܡܠܝ / ܟܬܒܐ ܕܚܐܘܬ ܚܟܡܬܐ
ܒܬܡܢܝܐ / ܒܫܒܛ ܫܢܬ ܐܦܨܙ ܕܝܘ̈ܢܝܐ ܘܫܘܒܚܐ ܠܐܠܗܐ ܗܘ̄ ܕܝܗܒ ܚܝܠ ܘܥܘܕܪ/
ܒܛܝܒܘܬܗ. ܘܒܣܘܓܐ ܪ̈ܚܡܘܗܝ/10 ܕܡܣܓܝܐܝܢ ܦܝܫܐ

[5-7] Here ends the Book of Politics. With its conclusion [*šumlāyā*] comes to end the book of Cream of Wisdom. [7-8] On 8th Shebaṭ, the year 1597 of the Greeks [= Feb. 1286]. [8-10] Glory to God, who has given strength and help in His goodness and abundance of His manifold mercies.

N.B. 5-7 ܚܟܡܬܐ ... ܫܠܡ also in Ll - 5 ܕܦܘܠܝܛܝܩܘܢ Ll || 6 ܘܒܫܘܡܠܝܗ F: ܘܒܫܘܠܡܗ Ll || 7 ܚܟܡܬ̈ܐ Ll.

(d.ii.) Fol. 227r b 10-15:

ܗܟܢܐ/11 ܟܬܝܒ ܗܘܐ ܒܐܪܟܝܬܗ ܕܐܒܘܢ / ܡܦܪܝܢܐ ܩܕ̄ ܕܡܕܢܚܐ ܡܢܝܚܐ
ܡܪܝ/ ܓܪܝܓܘܪܝܘܣ ܢܬܒܣܡ ܒܒܘܣ̈ܡܐ/ ܘܒܚܕ̈ܘܬܐ ܠܐ ܡܬܦܣܩܢܐ ܘܐܡܝܢܝܐ/
15 ܥܡ ܟܐܢ̈ܐ ܒܥܠܡܐ ܕܗܘܢܐ ܡܠܐܟܝܐ/

[12 ܡܪܝ: ܡܪܝܐ Margoliouth || 13 ܒܒܘܣܡܐ Margoliouth]

Thus was written in the copy of our father, the holy Maphrian of the East, the late Mar Gregory. May he rejoice in the never-ending and eternal delights and joys together with the just in the world of the angelic intelligences.

(d.iii.) Fol. 227r, between columns, in the hand of the copyist:

ܗܢܐ ܙܒܢܐ ܕܫܘܠܡ ܣܘܢܩܣܗ ܕܟܬܒܐ ܗܢܐ ܐܝܬܘܗܝ ܒܐܝܕܐ ܕܡܣܩܒܠܢܗ ܐܒܘܢ
ܡܦܪܝܢܐ ܡܢܝܚܐ .

[ܣܘܢܩܣܗ: ܣܘܢܩܣܗ Margoliouth || ܡܢܝܚܐ: ܡܢܝܚܗ Margoliouth]

This date for the conclusion of the composition of this book is in the hand of the author, our father, the late maphrian.

(d.iv.) Fol. 227r b 16-24 (ult.):

16 ܐܢܐ ܕܝܢ ܡܣܟܢܐ ܢܓܡ ܕܡܬܩܪܐ ܩܫ̈ ܘܕܝܪ/ ܒܪ ܫܡܫ ܡܫܡܫܢܐ ܘܐܣܝܐ
ܘܢܘܡܝܩܐ/ ܒܪ ܐܣܝܐ ܡܢܝܚܐ ܐܒܘܠܦܪܓ/ ܕܡܬܝܕܥ ܒܪ ܩܣܝܣ ܕܡܢ/20
ܡܕܝܢܬܐ ܡܪܕܝܢ . ܫܠܡܬ/ ܠܟܬܒܐ ܗܢܐ ܒܝܘܡ ܬܪܝܢ ܒܫܒܐ/ ܒܬܡܢܝܐ ܒܐܝܪ
ܫܢܬ ܐܠܦܐ ܘܫܬܡܐܐ/ ܘܚܡܫܝܢ ܘܚܕܐ ܝܘܢܝܐ/ ܫܘܒܚܐ ܠܐܠܗܐ ܕܝܗܒ ܚܝܠܐ ܘܥܘܕܪܢܐ/ ܒܛܝܒܘܬܗ .

"I, wretched Najm, called priest and monk, son of Shams[1], deacon, physician and notary [νομικός],[2] son of the late physician Abu al-Faraj, known as Bar Qissīs, from the city of Mardin, completed this book on Monday, 8th Iyar, the year 1651 [= May 1340] of the Greeks. Glory be to God who has given strength and help in His goodness."

(d.v.) Fol. 227v 1-5:

1 ܐܝܬ ܗܘܐ ܠܐܒܘܢ ܡܢܝܚܐ ܡܦܪܝܢܐ ܩܕܝܫܐ ܗ̄ܘ ܒܝܬ ܝܠܕܐ ܕܢܩܝܦ ܗܘܐ ܒܗ/
ܥܠ ܐܦܩܘܬܗ ܘܫܘܢܝܗ ܠܒܝܬ ܩܕ̈ܝܫܐ . ܘܟܕ ܐܬܚܙܝ ܠܗ/ ܥܠ ܢܦܫܗ
ܗܟܢ ܐܡܪ ܀ / ܡܨܝܕܬܐ ܕܥܠܡܐ ܒܫܢܬ ܐ̄ܦ̄ܠ̄ܙ ܨܕܬܢܝ/5 ܘܒܫܢܬ ܐ̄ܦ̄ܨ̄ܙ ܣܒܪ
ܐܢܐ ܠܝ ܕܠܐ ܐܗܘܐ ܒܟܝ/

Our deceased father, the aforementioned holy maphrian had a horoscope [*bēṯ yaldā*] which was attached to him concerning his departure and transfer to the abode of the saints. Observing it, he said concerning himself: O net of the world, in the year 1537 [= 1225/6] thy snare caught me, but in the year 1597 [= 1285/6] I think I shall not be in thee.[3]

(d.vi.) Fol. 227v 6-14:

6 ܟܕ ܡܛܝܗ ܕܡܪܥܐ ܫܢܬܐ ܒܡܪܥܬܐ ܢܦܠ ܒܗܘܪܐ ܒܓܪ . ܐܬܕܒܪ ܗܘܐ . ܗܘܐ/
ܒܓܪܬ ܫܢܬܐ ܘܒܓܠܬ ܐܦܘܩܪܝܣܐ . ܘܟܕ ܗܘܐ ܡܬܢܣܒ ܗܘܐ ./ ܡ̈ܝܐ
ܐܬܕܒܪܗ . ܘܠܐܢܐ ܕܝܢ ܒܐܒܐ ܠ ܒܢܝ ܡܩܪܒ ܕܡܪܥܐ ܫܢܬܐ/ ܒܡܪܥܬܐ . ܠܐ
ܡܛܝܐ ܠܗܢܐ ܠܡܫܠܡܘ ܐܦܠܐ ܦܘܫܐ ܠܫܢܬܐ/10 ܫܒܥܝܢܐ ܕܒܝܢܬܐ ܗܢܐ ܢܝܚ
ܚܝܠܐ ܓܢܣܫ ܦܓܪܐ . ܡܢ/ ܫܘܐ . ܫܒܝܪ ܒܢ̈ܝܐ . ܡܬܝܐܒ ܘܒܢܝܢ ܡܬܥܕܠ
ܒܬܐܪܝܐ/ ܘܒܣܘܡܟܪܘܬܐ . ܕܒܢܝܗ ܐܠ ܣܒܐ ܘܐܘܫܘܬܐ ܕܠܐ ܒܪ ܚܕܪ
ܒܢܬ/ ܠܟܠܗܝ ܡܬܠܡܕ ܠܚܬܝܡܘܬܐ . ܠܚܬܝܡܘܬܐ . ܕܠܟܠ ܫܘܐܠܐ ܦܘܫܝܐ/
ܕܢܬܬ ܫܦܪ ܗܘܐ ܒܒܢ̈ܝ ܫܘܐܠܘܗ ܢܦܫ ܗܘܐ .

[6 ܒܓܪ: lege ܒܓܪܗ]

---

[1] Called "Shams al-Daula" at fol. 227v 16.

[2] On νομικοί, see Kawerau (1960) 50, with n.407.

[3] On BH's prophecy concerning his approaching death, see BH *Chron. eccl.* 465-467.

When ten months had passed of the above-mentioned year, he said: 'See, a year has passed, and astrology [lit. astronomy] has lied'. When he was thinking thus, death overtook him in the night of Tuesday, 30th Tammuz of the aforementioned year. Tongue cannot speak nor mouth narrate the attributes of this man, sweet in company and pleasant to meet, elegant in appearance, beautiful in manners, loving, and complete and perfect in theory and practice. In his time, he sowed love and concord, which was not small, between all the opponents of the Christians and the Christians. For every question he gave the answer which was the best in the eyes of those questioning.

(d.vii.) Fol. 227v 15-18:

15 [illegible]
[illegible]
[illegible]
[illegible]

[15 [illegible] supra lin. || 16 [illegible] sub lin.]

[This manuscript] was written at the request of my blood brother, Rabban Badr, the excellent priest. I [am] Najm, monk and priest in name, son of the notary, 'philosopher' and deacon, Shams al-Daula, son of the late physician Abu al-Faraj, known as Bar Burhān Bar Qissīs, from the city of Mardin. Glory be to God. Amen.

(d.viii.) Fol. 227v 19-22: In a similar but slightly different hand, no doubt, of the copyist's brother Badr

19 [illegible]
[illegible]
[illegible]
[illegible]

{As [?] at the beginning}, middle and end of the treatise, may {Christ Our Lord [?]} always be a helper and at all times a supporter of the glory of the monks, the honour {of --, and --} of the wise, Rabban Najm, our brother, who took the trouble to copy this spiritual treasure for the sake of the study and practice of philosophy.

2. *London, British Library, Or. 4079 (= ms. L)*

(a) Fol. 157r a 32-b 18:

32 [illegible]
[illegible]
[illegible]
[illegible]
[illegible]
[illegible]
[illegible]
[illegible]
[illegible]
[illegible]

[34 [illegible]: lege [illegible]]

[32-34] Here ends the Book of Poetics ... [34-37] In the blessed month of Shebaṭ, on 15th, Tuesday, in the year 2120 of the Greeks. [1-2] Of the birth of Christ Our Lord, year 1809. [3-7] This book was written by the hand of the weak [*ḥallāšā*] priest George, son of the priest Yaqo, son of the deacon Dusho, surnamed of the house of Yuḥanna, [8-14] from the village of Alqosh, the village of the prophet Nahum. May Our Lord people it with His strong right hand, and preserve its inhabitants from all harm and injury through the intercession of Our Lady Mary, Ever-Virgin and Mother of God, and of all the saints. Amen. [15-18] To Him who has helped and assisted and led me thus far be glory and honour, praise and thanksgiving for ever and ever. Amen. Amen.

(b) Fol. 238v b7-15 [7-10 [illegible] . . . [illegible] in rubric]:

[illegible] /8 [illegible] /9 [illegible]
[illegible] /10 [illegible] /11 [illegible] /12 [illegible]
[illegible] /13 [illegible] /14
[illegible] /15 [illegible]

[7-10] Here ends the Book on the Soul ... [11-15] May the grace and mercy of our God flow upon the sinner, the priest George, son of the priest Yaqo of the house of Yuḥanna, who wrote [this], now and always and for ever and ever. Amen. Amen.

(c) Fol. 285r b12-25:

12 [illegible] /. [illegible] /
[illegible] /. [illegible] /15 [illegible]
[illegible] / [illegible] /. [illegible]
[illegible] / [illegible] /20 [illegible] /.
[illegible] / [illegible] / [illegible]
[illegible] /25 [illegible]

[12-17] Here ends the Book of Theology ... [18-20] In the middle of Ab, on the eve of the feast of the Transfer [i.e. Assumption] of Our Lady Mary, Mother of God, into heaven. In the hand of the lowliest of men, priest George, the son of the priest Yaqo Yuḥanna, who wrote it in the year 2120 of the Greeks, 1809 of the Incarnation of Our Lord.

This is followed by a note (in rubric and unfortunately difficult to read on microfilm) on the death, in the above-named year, of Aḥmad Pasha of Mosul in a battle on the banks of the Zab at the hand of "[...] of the house of ᶜAbd al-Jalīl", along with a note on the birth of the copyist's son Manṣūr at Easter in that same year.

(d) Fol. 314r b 1-ult.:

1 [illegible] /. [illegible] /
[illegible] / [illegible] /5 [illegible]
[illegible] / [illegible] /. [illegible] /
[illegible] /10 [illegible]
[illegible] / [illegible] / [illegible]
[illegible] /15 [illegible]
[illegible] / [illegible] / [illegible] /

ܕܟܬܒܗ ܠܗܢܐ ܟܬܒܐ ܡܛܠܠܬܗ/ ܘܡܛܠ ܒܢܘܗܝ ܚܒܝܒܐ. ܡܫܡܫܢܐ/20 ܝܩܘ
ܘܡܫܡܫܢܐ ܡܝܟܐܝܠ/ ܘܡܫܡܫܢܐ ܝܘܣܦ. ܘܡܢܨܘܪ/ ܐܝܬܘܗܝ ܗܘܐ ܒܪ ܚܡܫܐ
ܝܪ̈ܚܝܢ ./ ܘܐܦ ܗܕܐ ܕܥ ܐܚܘܢ ܩܪܘܝܐ ./ ܕܐܪܟܝܛܘܦܐ ܕܟܬܒܬ ܡܢܗ ܗܢܐ/25
ܟܬܒܐ ܥܣܩ ܣܓܝ ܗܘܐ ܗܘܬ. ܘܠܐ ܐܢܫ ܡܫܟܚ ܗܘܐ ܠܡܩܪܐ ܒܦܫܝܩܘܬܐ/ ܕܡܛܠ ܗܘ
ܡܛܠ ܣܡܝܘܬ ܐܬܘܬܐ/ ܘܢܘܩܙܐ ܠܝܬ ܗܘܐ ܒܗ ܟܠ ܟܠܗ./ ܘܠܪܫܡܐ ܕܕܠܬ ܘܪܫ
ܒܨܚܝܚ ܡܣܝܡ/30 ܗܘܐ.

[1-3] Here ends the Book of Politics ... [3-10] In the month of Ab [Aug.], on the 29th, on the eve of the 2nd Sunday of Mar Elijah, of his antiphon: "*b-kursay dīnāḵ* ...", in the year 1809 of the Incarnation of Christ, our God. To Him be glory for ever and ever. Amen. Amen. [11-12] In the year 2120 of Alexander the Greek. [13-16] By the hand of the most wretched and lowly of men, priest in name, George, son of the priest Yaqo, son of the deacon Dusho, surnamed of the house of Yuḥanna. [17-22] The above-mentioned scribe wrote and sullied this book for himself and for his beloved sons, the deacon Yaqo, the deacon Michael, the deacon Joseph, and Manṣur [who] was five months old. [23-30] Our brother reader, know this too, that the copy from which I transcribed this book was very difficult, and nobody was able to read it with ease, because of the obscurity of the letters [*samyūt āṯwāṯā*], and [because] there were no points [*nuqzē*] at all, and the signs of *dalat* and *resh* were rarely placed.

3. *London, British Library, Or. 9380 (= ms. l)*

(a) Fol. 309r 2-5 [in rubr.]:
2 ܫܠܡ ܟܬܒܐ ܕܦܘܐܝܛܝܩܐ ܕܒܡܣܘܟܠܗ ܐܪܣܛܛܠܝܣ/ ܗ̇ܘ ܦܝܠܣܘܦܐ ܡܝܬܪܐ ܕܒܡܠܠܘܬܐ.
ܒܐܝ̈ܕܝ ܐܒܝܠܐ ܘܡܣܟܢܐ/ ܡܫܡܫܢܐ ܥܝܣܐ ܫܘܒܚܐ ܠܝܫܘܥ ܘܬܫܒܘܚܬܐ ܠܡܪܝܡ ܕܝܗܒܘ/5
ܚܝܠܐ ܘܡܥܕܪܢܘܬܐ ܠܥܒܕܗܘܢ ܐܝܢ ܘܐܡܝܢ ✤

[2-3] Here ends the Book of Poetics ... [3-5] By the hand of the lowly and poor one, the deacon ʿIsa. Glory be to Jesus and praise be to Mary, who have given strength and assistance to their servant. So be it. Amen.

(b) Fol. 453r 26-v 3:
26 ܫܠܡ/ ܟܬܒܐ ܕܥܠ ܢܦܫܐ. ܕܒܡܣܘܟܠܗ ܐܪܣܛܛܠܝܣ ܗ̇ܘ ܦܝܠܣܘܦܐ ܡܝܬܪܐ/ ܡܝܬܪܐ
ܕܬܢܝܢܬܐ ܡܢ ܟܬܒܐ ܕܐܪܝܣܛܘ ܬܕܥܬܐ ܘܥܠ ܚܛܝܐ//1 ܡܫܡܫܢܐ ܥܝܣܐ :
ܘܠܡܟܬܒܢܐ ܡܝܣܛܪ ܘܐܠܝܣ ܒܕܓ/ ܫܝܢܗ ܘܪ̈ܚܡܘܗܝ ܕܐܠܗܢ ܢܫܒܘܢ. ܗܫܐ
ܘܒܟܠܙܒܢ/ ܘܠܥܠܡ ܥܠܡܝܢ ܐܝܢ ܘܐܡܝܢ ✤

[26-28 ܬܕܥܬܐ ... ܫܠܡ et 1 ܡܝܣܛܪ ܘܐܠܝܣ ܒܕܓ in rubr. || 27-28 ܡܝܬܪܐ ܡܝܬܪܐ sic]

[26-28]. Here ends the Book on the Soul ... [28-3] May the grace and mercy of our God flow upon the sinner, the deacon ʿIsa and on the writer [*maktbānā*][4] Mr. Wallis Budge, now and always and for ever and ever. So be it. Amen.[5]

[4] Or perhaps, in this context, "the one who commissioned the manuscript", retaining the causative (Aphel) force of "*maktbānā*".

[5] ≈ ms. L fol. 238v b7-15, except for change of the copyist's name and the addition of "Mr. Wallis Budge", along with the minor alterations of ܒܡܫܠܡܢܘܬܗ to

(c) Fol. 521r 8-11:

8 ܫܠܡ ܟܬܒܐ ܕܬܐܘܠܘܓܝܐ. ܘܡܫܬܠܡܢܘܬܗ / ܐܫܬܠܡܬ ܡܢܘܬܐ ܬܐܘܪܝܛܝܩܝܬܐ
ܕܦܝܠܘܣܘܦܝܐ. ܡܢ /10 ܟܬܒܐ ܕܚܐܘܬ ܚ̈ܟܡܬܐ. ܘܟܠܗ ܡܢ ܡܠܝ ܡܫܬܡܫܢܘܬܐ
/ ܕܠܘܬ ܡܠܦܢܘܬܐ ܐܠܗܝܬܐ ܠܫܡܗ. ܨܠܘ ܥܠ ܚܛܝܐ ܕܟܬܒ.

[8-11] Here ends the Book of Theology ... [11] Pray for the sinner who wrote [this]."

(d) Fol. 562v 20-563r 24 [562v 20-27; 563r 7-11, 19-24 in rubric]

20 ܫܠܡ ܟܬܒܐ ܕܦܘܠܝܛܝܩܝ ܘܒܫܘܠܡܗ ܐܫܬܠܡ/ ܟܬܒܐ ܕܚܐܘܬ ܚ̈ܟܡܬܐ
ܡܢ ܣܘܢܗ ܕܡܪܝ ܓܪܝܓܘܪܝܘܣ/ ܝܡܐ ܕܚ̈ܟܡܬܐ. ܘܓܙܐ ܕܣܝ̈ܡܬܐ. ܘܐܘܨܪܐ
ܕܝܘܠܦܢܐ./ ܘܥܘܬܪܐ ܕܡܣ̈ܟܢܐ. ܡܦܪܝܢܐ ܡܫܒܚܐ. ܕܡܕܢܚܐ./ ܕܐܝܬܘܗܝ
ܐܒܘ ܐܠܦܪܓ̰ ܟܠܘܡܪ ܐܒܐ ܕܒܘܝܐܐ ܒܪ ܐܗܪܘܢ/25 ܐܣܝܐ ܕܡܢ ܡܠܝܛܝܢܝ
ܫܘܒܚܐ ܠܐܠܗܐ ܕܥܕܪ ܒܫܘܪܝܐ/ ܘܣܝܥ ܥܕ ܠܫܘܠܡܐ. ܚܢܢܗ ܘܪ̈ܚܡܘܗܝ
ܢܬܢܓܕܘܢ ܥܠܝܢ/ ܒܟܠܙܒܢ. ܐܝܢ ܘܐܡܝܢ: 1/ ܐܫܬܪܝ ܡܢ ܕܝܢ ܘܐܫܬܠܡ ܟܬܒܐ
ܗܢܐ ܕܚܐܘܬ ܚ̈ܟܡܬܐ/ ܒܝܪܚܐ ܒܪܝܟܐ ܟܢܘܢ ܩܕܡ ܒ̇ܝ ÷ ܒܝܘܡ ÷ ܒ̇ ܒܫܒܐ
ܕܣܘܒܪܐ/ ܕܫܢܬ ܐܦܨܒ. ܕܡܬܒܪܢܫܢܘܬܗ ܕܡܫܝܚܐ ܐܠܗܢ/ ܘܡܬܝܠܕܢܘܬܗ
ܠܡܬܒܪܢܫܢܘܬܗ ܕܡܫܝܚܐ ܐܠܗܢ ܡܢ̈ܐ/5 ܘܡܠܟܢ ܘܦܪܘܩܢ. ܕܠܗ ܫܘܒܚܐ
ܠܥܠܡܝܢ ܘܠܥ̈ܠܡܝ ܥ̈ܠܡܝܢ/ ܘܥܠܝܢ ܚܢܢܗ ܠܥܠܡ ܥܠܡܝܢ ܐܡܝܢ ÷/7
ܐܫܬܠܡ ܕܝܢ ܒܟܬܝܒܬܐ ܒܐܝܕܝ ܡܫܟܢܝܬܐ ܐܠܦܘܫ./ ܡܪܕܝܢ ܕܡܣܬܟܠ ܢܨܝܚܐ.
ܕܟܠܢܐ ܥܠ ܟܠ ܓܒ ܚܟܡܬܐ/ ܕܒܝܬ ܡܐܪܕܝܢ ܦܪܘܫܐ. ܡܢ ܚܟܡܬܐ ܕܟܘܠܗܝܢ
ܡܠܝܢ ./10 ܡܫܝܚܐ ܢܬܠ ܠܡܐܪܕܝܢ ܡܢ ܟܠ ܢܟܝܢ ܚܟܡܬܐ/ ܘܓܠܝܢܐ ܐܝܢ
ܘܐܡܝܢ ./12 ܐܫܬܠܡ ܕܝܢ ܒܩܘܫܛ ܐܝܟܐ ܕܐܫܟܚܬ. ܘܟܬܒܐ ܕܪ̈ܚܘܬܐ. ܨܠܘ
ܥܠܘܗܝ. ܡܫܡܫ ܟܬܝܒܐ. ܘܐܦܢܝܬ ܡܟܪܙܐ. ܘܡܫܡܫܢ/ ܕܟܬܒܐ ÷ ܡܠܬ ܐܠܗܐ
ܬܘܒ ܟܬܒܐ ܕܬܘܠܕܬܐ ܦܪܝܫܬܐ/15 ܕܟܬܒ ܕܡܫܝܚܐ ÷ ܡܫܝܚܐ ܡܢ ܢܦܫܗ
ܟܬܒܗ/ ܘܦܢܝܬ ܠܟܘܠܗܘܢ. ܘܠܐܒܗܝ ܘܠܐܚܘܗܝ ܠܟܘܡܪܐ/ ܕܒܝܬܐ ܬܘܠܕܬܗ.
ܘܠܝܘܚܢܢ ܥܠ ܟܠܗܘܢ/ ܒܫܘܒܚܗ ܘܒܣܘܓܐܐ ܕܫܒܝܚܐ ܘܐܝܟ ܐܝܟܢܐ ܘܐܡܝܢ/
19 ܐܫܬܠܡ ܕܝܢ ܒܝܪܚܐ ܡܒܪܟܐ ܘܣܠܝܩܐ ܡܫܡܫܢܐ ܚܒܝܒܐ/20 ܒܪ ܐܝܫܘܥ ܒܪ ܐܒܪܗܡ ܒܪ
ܡܫܡܫܢܐ ܡܩܕܫܐ ܕܡܢ ܩܪܝܬܐ/ ܐܝܬܘܬ ܕܒܐܬܪܐ ܕܦܘܠܘܣ. ܟܬܒܗ ܠܦܘܪܐ
ܢܘܪܐ/ ܕܒܥܘܢ ܠܗܘܢ ܘܠܐܡܗܘܢ ܒܝܠܕܘܬܗܘܢ ܥܡ ܟܬܒܐ/ ܕܟܬܒ ܢܦܫܗ
ܡܛܠ ܟܢܝ ܕܡܫܝܚܐ ܡܢ/ ܐܡܝܢ ÷

[20-27] Here ends the Book of Politics. With its completion comes to end the book of *Cream of Wisdom* among the writings of Mar Gregory, the sea of wisdom, the casket of treasures, the storehouse of teaching, the opulence of the poor, the glorious Maphrian of the East, who is Abu al-Faraj, i.e. the father of solace [*abā d-buyyāʾā*][6] the son of Aaron the physician, from Melitene. Glory to God, who helped at the beginning and assisted until the end. May his grace and mercy flow upon us always. So be it. Amen. [1-6] This book of *Cream of Wisdom* was begun and was finished in the blessed month of Kanon I [= December], on the 12th, on the second Monday of Advent [*subbārā*], of the year 1892 of the Incarnation of Christ, our God,

---

ܘܡܫܬܡܠܝܢ and the first ܐܡܝܢ to ܐܝܢ.

[6] Cf., bearing in mind that this ms. belonged to Budge, the discussion of the name Abu al-Faraj, at Budge (1932) xvi. The name is also given in its "Syriacised" form at Manna (1901) I.3, II.357.

our Lord, our King and our Saviour. To Him be glory, He who causes seasons to pass, while He does not pass away for ever. Amen. [7-11] It was written in the blessed and blissful [*brīḵtā wa-mḇarraḵtā*] village Alqosh, the village of the prophet Nahum, which has been built beside the monastery of Rabban Hormizd the Persian and the monastery of the Virgin Mary. May Christ guard its inhabitants from all concealed and apparent harm. So be it. Amen. [12-18] It was written in the days of the father of fathers, the chief of pastors, the one who ties crowns and binds belts, the dispenser of talents, Mar Elijah XII, Catholicos Patriarch of Babylon and of the East.[7] May Christ our Lord sustain his throne, lengthen his years and extend his days for the glory of the Catholic Church, and let him triumph over his enemies through the intercession of the apostles and the fathers. So be it. Amen. [19-24] It was written by the hand of the wretched sinner, deacon ᶜIsa, son of Isaiah [Ishaᶜya], son of the deacon Cyriacus, from the village of Eqror in the region of the Sindāyē,[8] who begs the noble readers to remember the scribe and his forebears in their prayers which are granted, so that perhaps he may obtain mercy before the judgement-seat of Christ, our Lord. Amen."

4. *Birmingam, Mingana syr. 310 (= ms. M)*

(b) Fol. 216r 7-20:

7 ܫܠܡ ܟܬܒܐ ܕܥܠ ܢܦܫܐ / ܘܫܘܟܠܠܗ ܐܫܬܟܠܠ / ܘܫܘܠܡܗ ܟܠܝܠܐ ܐܝܬܝܗ ܕܬܪܝܨܘܬܐ
ܗܟ/10 ܟܬܒܐ ܕܐܒܘܢ ܢܨܝܚܐ . ܗܟܢ / ܟܬܝܒ ܗܘܐ ܒܟܬܒܗ ܕܐܒܘܢ /
ܢܛܪܗ ܐܠܗܐ ܠܚܝܘܗܝ . ܟܒ ܠܐܒ / ܒܫܢܬ ܐܬܩܨܘ ܕܝܘܢܝܐ / ܒܡܕܝܢܬܐ ܡܘܨܠ
ܚܣܝܢܬܐ /15 ܘܫܘܒܚܐ ܠܐܠܗܐ ܥܘ ܕܝܗܒ ܚܝܠܐ ܘܥܘܕܪܢܐ / ܐܫܬܠܡ ܕܝܢ ܟܬܒܐ ܬܪܝܢܐ ܝܐ ܠܬܫܪܝܢ
ܩܕܡܝ / ܒܫܢܬ ܐܬܩܨܙ ܕܝܘܢܝܐ / ܐܫܬܠܡ ܕܝܢ ܗܢܐ ܕܝܠܢ ܬܠܝܬܝܐ ܟܙ ܠܬܡܘܙ
ܒܫܢܬ ܒܩܥܘ ܕܝܘܢܝܐ ܒܐܣܟܘܠ ܕܝܠܕܬ /20 ܐܠܗܐ ܐܡܝܢ

[7-10] Here ends the Book on the Soul ... [10-12] Thus was written in the copy [*ktāḇā*] of our illustrious father - may God preserve his life - [12-15] on 22nd Ab, the year 1596 of the Greeks [= Aug. 1285 A.D.] in the fortified city of Mosul. Glory to God who has given strength and help. [16-17] The second copy [*ktāḇā*] was finished on 11th Teshrin I, the year 1597 of the Greeks [= Oct. 1285 A.D.]. [18-20] This our own third [copy] was finished on 27th Tammuz,[9] the year 2176 of the Greeks [= July 1865] in the School of the Mother of God. Amen.

(c) Fol. 316v 3-8:

3 ܫܠܡ ܟܬܒܐ ܕܬܐܘܠܘܓܝܐ . ܘܫܘܟܠܠܗ ܐܫܬܟܠܠܬ / ܡܢܘܬܐ ܬܠܝܬܝܐ ܕܬܐܘܠܘܓܝܐ .
ܕܦܝܠܘܣܘܦܘܬܐ . ܗܟ /5 ܟܬܒܐ ܕܐܝܬܘܗܝ ܡܚܕܬܐ . ܥܒܕ ܗܟ ܡܠܟ /
ܡܬܬܓܡܝܠܬܐ ܕܠܦܝܠܘܣܘܦܘܬܐ ܐܪܝܬܐ ܠܬܡܙ . / ܒܫܢܬ ܕܐܪܒ ܝܘܢܝܐ . ܫܢܬ
ܒܩܥܘ ܕܝܘܢܝܐ / ܘܫܘܒܚܐ ܠܐܠܗܐ ܕܝܗܒ ܚܝܠܐ ܘܥܕܪܐ ܠܟܬܘܒܗ .

---

[7] Elijah XII Abū al-Yūnān: Chaldean Patriarch 1879-1894.

[8] Eqror: a village some 16 miles to the NE of Zakho (Aßfalg [1963] 48; cf. Mingana [1933] 347; Fiey [1965] 315).

[9] Mar Mellus evidently read the date as ܝܙ (17th) rather than ܟܙ (see below, Ms. Vat. 614, fol. 39v).

[3-6] Here ends the Book of Theology ... [7-8] In the middle of Ab, the year 2176 of the Greeks [= Aug. 1865]. Glory to God who has given strengthen and help in His goodness.

(d) Fol. 380r: no colophon.

5. *Birmingham, Mingana syr. 23 (= ms. m)*

(b) Fol. 75r b 10-20:

ܫܠܡ ܟܬܒܐ ܕܥܠ ܢܦܫܐ . ܘܡܫܬܡܠܝܢܘܬܗ / ܐܫܬܟܚܬ ܘܠܘܬܐ ܐܢܝܬܐ
ܕܒܪ̈ܢܫܘܬܐ . ܡܢ / ܟܬܒܐ ܕܢܣܚܢ ܬ̈ܟܬܒܬܐ . ܗܟܢ ܟܬܝܒ ܗܘܐ / ܒܟܬܒܐ
ܕܟܬܒܢ ܠܗܘܢ : ܕܒܟܬܒܐ ܕܐܒܘܢ / ܡܪܝ ܓܪܝܓܘܪܝܘܣ ܕܗܘ ܒܪ ܥܒܪܝܐ .
ܒܟܒ ܠܝܘܡ / ܐܒ . ܫܢܬ ܐܬܩܨܘ ܕܝܘ̈ܢܝܐ . ܒܡܕܝܢܬ ܡܚܣܢܬܐ ܡܘܨܠ / ܠܐܠܗܐ
ܫܘܒܚܐ . ܕܝܗܒ ܚܝܠܐ ܘܥܕܪ ܘܣܡܟ ܐܡܝܢ ./ ܐܫܬܠܡ ܕܝܢ ܗܢܐ ܟܬܒܐ ܕܒܐܝܕܝܢ ܥܠ
ܐܝ̈ܕܝܐ ܟܬܝܒܬܐ ܘܢܣܝܒܬܐ / ܫܢܬ ܒܪܗ . ܡܫܝܚܝܘܬܐ ܐܦܨܕ . ܘܝܘܡܐ ܕܚܡܫܐ / ܒܬܫܪܝܢ
ܩܕܡܝܐ ܡܢ ܐܝܕܐ ܚܛܝܬܐ ܕܡܬܝ ܡܫܡܫܢܐ / ܒܪ ܡܫܡܫܐ ܦܘܠܘܣ .
ܒܡܕܝܢܬܐ ܕܡܘܨܠ . ܒܫܠܡܐ ܐܡܝܢ .

[ad 10-14 in marg.: هذا تاريخ الكتاب القديم || 18 ܒܪܗ: ٢٢٠٥ sub lin.]

Here ends the Book on the Soul ... Thus was written in the copy [*ktāḇā*] from which we made (this) copy: 'in the copy of our father, Mar Gregory, Bar Hebraeus, 22nd Ab, the year 1596 of the Greeks, in the fortified city of Mosul. Glory to God who has given strength, help and support. Amen.' - This copy at hand was finished in the year 2205 of the Greeks, 1894 of Christ, on 5th Teshrin I, by the hand of the sinner, deacon Matthew, son of the late Paul, in the city of Mosul. In peace. Amen.

(c) Fol. 107v b 1-13

ܫܠܡ ܟܬܒܐ ܕܬܐܘܠܘܓܝܐ :/ ܘܡܫܬܡܠܝܢܘܬܗ ܐܫܬܠܡܬ ܡܢܝܢܬܐ ܬܐܘܪܝܐ ܕܬܐܘܠܘܓܝܐ :/
ܕܦܝܠܘܣܘܦܝܐ . ܡܢ ܟܬܒܐ ܕܚܐܘܬ ܕܪ̈ܒܢܝܬܐ . ܘܐܦ ܡܢ / ܥܠܝܢ ܡܬܐܘܠܘܓܝܬܐ
ܕܠܦ̈ܝܠܣܘܦܐ ܐܬܝܬܐ ܠܬܡܢ . / ܒܚܕܥܣܪ ܝܘܡܝܢ ܝܐ ܒܬܫܪܝܢ ܩܕܡ ܐܦܨܕ ܡ ܘܠܐܠܗܐ
ܫܘ ܕܝܗܒ / ܘܥܕܪ ܒܛܝܒܘܬܗ ܘܥܠܝܢ ܪ̈ܚܡܘܗܝ ܠܥܠܡܝܢ ✧ ܘܒܗܕܐ ܫܢܬܐ /
ܡܬܩܕܡܬܐ / ܕܒܥܣܪܝܢ ܘܬܡܢܝܐ . ܒܚܙܝܪܢ ܗܘܐ ܙܘܥܐ ܘܢܝܕܐ ܒܡܕܝܢܬܐ / ܕܡܠܟܐ ܕܗܝ
ܩܘܣܛܢܛܝܢܐ . ܘܢܦܠܘ ܒܬ̈ܐ ܣܓܝܐܐ ܘܦܬܩ̈ܐ ܥܠ ܐܢܫܐ ܕܒܓܘܗܘܢ ܘܡܝܬܘ
ܐܢܫܐ ܣܓܝܐܐ ܕܠܐ ܡܢܝܢ . ܘܐܬܚܪܒ / ܫܘܩܐ . ܘܗܘܐ ܙܘܥܐ ܬܡܢܝܐ ܝܘܡܝܢ .
ܘܗܘܐ ܙܘܥܐ ܐܦ ܒܝܡܐ / ܘܐܫܬܠܡܬ ܡܠܬܗ ܕܕܘܝܕ ܢܒܝܐ ܕܐܡܪ . ܕܚܙܐ ܒܐܪܥܐ
ܘܡܙܝܥ ܠܗ / ܡܛܠ ܚܛܗ̈ܐ ܕܒ̈ܢܝܢܫܐ . ܒܫܢܬܐ ܗܢܝܬܐ ܕܝܘ
٢٢٠٥/ ܒܝܘܡܝ ܡܠܟܘܬܐ ܕܥܒܕ ܐܠܡܓܝܕ ܣܘܠܛܢܐ ✧ ܫܠܡ

[1-4] Here ends the Book of Theology ... [5-6] On 11th Teshrin I, 1894 A.D. Glory to God who has given strength and help in His goodness. May His mercy be upon us for ever. [6-13] In this above-mentioned year, on 28th Ḥaziran, there was an earthquake [*zawʿā wa-nyāḏā*] in the city of the king, which is Byzantium. Many houses and buildings [*puttāqē*] fell on those who were inside them, many, countless, people died and the quarters of the city were destroyed. The tremor continued for eight days. Shaking occurred also in the sea. The word of the prophet David was fulfilled, who said: 'The Lord looks on the earth and it quakes' [Ps. 104.32] on account

of the wickedness of its inhabitants etc.[10] In the year 2205 of the Greeks, in the days of Sultan ᶜAbd al-Ḥamīd the Turk. End.

(d.i.) Fol. 127v b 29-31

ܐܝܟܢܐ ܕܚܕܐ ܡܠܚܐ . ܕܡܛܝܐ ܐܠܦܗ/ ܠܠܡܐܢܐ . ܗܟܢ ܚܕܐ ܟܬܘܒܐ .
ܒܣܘܪܛܐ/ ܐܚܪܝܐ ܕܣܪܛ .

Just as the sailor rejoices that his ship has reached harbour, so does the scribe rejoice at the last line that he has scratched.[11]

(d.ii.) Fol. 128r 1-25

ܠܐܝܩܪܐ ܘܠܫܘܒܚܐ ܕܬܠܝܬܝܘܬܐ ܩܕܝܫܬܐ . ܐܒܐ ܘܒܪܐ ܘܪܘܚܐ ܩܕܝܫܐ . ܚܕ
ܐܠܗܐ ܫܪܝܪܐ . ܕܒܬܠܝܬܝܘܬܐ ܡܬܝܕܥܐ ܘܡܬܩܪܝܐ/ ܡܫܬܒܚܐ . ܘܒܬܠܝܬܝܘܬܐ ܫܪܝܪܬܐ
ܡܫܬܒܚܐ ܘܡܫܬܓܪܐ . ܘܠܬܠܝܬܝܘܬܐ ܕܒܠܝܫܘܬܐ ܡܬܩܢܘܡܝܘܬܐ ܡܫܬܘܕܥܐ . ܐܒܐ ܝܠܘܕܐ
ܕܒܪܐ/ ܡܫܬܡܗܢܐ . ܘܒܪܐ ܝܠܝܕܐ ܡܢ ܐܒܐ ܡܬܩܕܡܐܝܬ ܘܠܐ ܙܒܢܐܝܬ . ܘܪܘܚܐ
ܩܕܝܫܐ ܠܩܢܘܡܐ ܘܠܩܢܘܡܐ ܕܝܠܗ ܡܢ ܐܒܐ ܥܠܬܢܐܝܬ/ ܘܠܗܘܢ ܡܢ ܒܪܐ ܡܬܝܕܥܢܐܝܬ .
ܬܠܝܬܝܐ ܚܕ . ܘܚܕ ܬܠܝܬܝܐ . ܚܕ ܐܠܗܐ ܫܪܝܪܐ ܕܫܠܝܐ . ܚܕ ܚܝܠܐ . ܚܕ ܫܘܠܛܢܐ . ܚܕ
ܐܘܣܝܐ . ܚܕܐ ܐܝܬܘܬܐ ./ ܚܕܐ ܪܒܘܬܐ . ܚܕܐ ܡܠܟܘܬܐ . ܚܕܐ ܐܠܗܘܬܐ .
ܚܕܐ ܡܪܘܬܐ . ܚܕܐ ܡܫܒܚܘܬܐ . ܕܬܠܝܬܝܘܬܐ ܡܫܒܚܬܐ ܢܟܬܘܒ ܘܢܫܪܪ/
ܘܢܫܠܘ ܘܢܫܬܒܚ ܠܥܠܡܥܠܡܝܢ . ܕܥܠ ܡܛܠܬܗ ܕܝܠܢ . ܘܥܠ ܬܫܒܘܚܬܗ ܕܥܠܡܝܢ
ܫܘܒܚܐ ܠܗ . ܘܩܘܠܩܠܗ/ ܡܫܒܚܝܢ . ܘܥܠ ܐܝܠܝܢ ܡܫܬܡܫܝܢ ܕܠܗ ܥܡܪܝܢ
ܘܐܝܬܝܐ ܘܡܬܩܪܐ ܒܝܬܐ ܠܥܠܡܥܠܡܝܢ ܐܡܝܢ ܀ ܫܠܡ/ ܘܐܫܬܠܡ ܗܢܐ
ܟܬܒܐ ܕܩܝܡܝܢ ܥܠ ܐܪ̈ܒܥܐ ܕܝܘܢܝܬܐ ܬܫܒܚܬܐ ܡܠܟܘܬܐ ܠܬܪܝܢܘܬܐ . ܕܟܬܒ
ܠܝܘܡܐ ܠܡܟܬܒ ܒܣܦܪܐ ./ ܘܫܘܠܡܐ ܫܡܝܐ . ܡܢ ܡܪܝ ܒܪܝܟܐ ܕܟܬܒܗ ܘܟܬܒܐ
ܘܡܛܝܒܐ ܕܡܬܚܫܒ ܒܗ ܟܬܒܐ . ܡܢܐ ܠܦܘܬ ܕܠܐ ܩܕܡ/ ܐܠܘܐ ܐܬܟܬܒ
ܟܬܝܒܬܐ ܡܢܐ ܘܐܪܝܬܐ . ܕܗܘ ܐܟܘ ܐܠܦܐ ܡܢ ܐܚܪ̈ܢܐ ܐܚܪܝܐ
ܡܬܠܛܠܝܐ : ܒܫܢܬ ܐܠܦܐ ܡܫܝܚܝܬܐ :/ ܘܒܫܢܬ ܐܠܦܐ ܘܡܐܬܐ ܘܫܬܝܢ ܘܫܒܥ
ܕܝܘܢܝܐ . ܒܝܪܚܐ ܕܐܕܪ ܒܝܘܡܐ ܘ ١٦ ܒܗ ܒܝܘܡ ܚܕ ܒܫܒܐ ܀ ܒܫܘܠܛܢܗ ܕܚܢܢ/
ܡܢ ܐܢܓܠܝܣܝܐ ܦܠܝܛܘ ܦܠܛܝܢ ܛܠܝܢܐ ܒܗ ܒܐܝܠܠ ܕܚܕܐ . ܘܒܩܘܡ
ܐܟܡ̈ܬܐ ܣܓܝܐ ܡܢ ܦܐܠܝܠܘ/ ܡܠܦܘܠܝܠܐ ܒܦܘܪܚܢܐ ܕܗܘܐ ܐܡܠܝܦܘܐ
ܕܒܦܘܪܟܐ ܕܐܢܓܠܝܣܝܐ . ܘܡܢ ܕܦܠܝܛܘܢ ܫܡܝܐ ܒܫܢܬ ܐܠܦܐ ܒܦܘܪܟܐ/
ܕܡܫܝܚ . ܘܡܢ ܦܐܠܝܠܘ ܫܡܝܐ ܐܠܦܐ ܐܠܦܐ ܒܦܘܪܟܐ ܕܡܢ ܡܫܪ ܘܦܘܪܩܢܐ .
ܒܢ ܥܒܕܐ ܕܐܟܡܬܐ . ܕܐܝܬܝܗ ܒܦܘܪܩܢܐ/ ܒܡܫܝܚܘܬܐ ܘܡܫܪܝܬܐ . ܘܡܢ
ܐܠܗܐ ܢܚܡܝܘ ܫܪܝܪܐ . ܘܢܦܪܘܩ ܥܒܕܐ ܕܠܗܘܢ . ܩܪ̈ܒܐ ܕܫܪܘܝܬܐ . ܘܫܪܝܐ
ܕܦܝܣܩܬܐ ܀/ ܕܢܚܪܘܢ ܠܥܕ̈ܬܐ ܩܪ̈ܝܬܐ . ܕܢܦܠ ܦܐܪܐ ܦ̈ܐܪܐ . ܒܫܠܡܐ
ܘܡܫܡ ܘܡܒܐܪܐ . ܐܡܝܢ ܀ ܘܒܫܘܒܚܗ ܠܡܪܐ ܫܡܝܐ ܐܡܝܢ/ ܡܡܠܠܐ ܘܡܫܒܚ ܥܠ
ܐܝܬܘܗܝ ܕܐܠܗܐ . ܘܒܠܝܐ ܡܚ̈ܝܢܐ . ܘܫܒܩ ܚܛ̈ܗܐ . ܡܪܝܡܐ ܡܫܒܚܢܐ
ܡܫܡܫܢܐ . ܒܪ/ ܦܘܠܘܤ ܡܫܝܚܐ ܒܪ ܠܚܡ̈ܐ ܐܠܠܗ . ܕܡܢ ܡܕܝܢܬܐ ܡܪܕܝܢܐ
ܕܡܪܕܝܢ . ܕܐܬܟܬܒ ܒܐܝܕܝ ܘܫܠܡܐ . ܘܟܬܒܐ ܐܝܟ ܡܢ ܟܠ/ ܐܝܟ ܘܡܫܠܡܐ
ܕܦܓܪ ܒܟܬܒܐ ܗܢܐ ܕܝܗܒ ܘܢܝܠܐ ܝܠܗܬܐ ܕܫܘܡܠܐ ܠܟܬܒܐ . ܘܐܡܪ ܡܪܢܐ
ܐܠܗܐ ܥܒܕܗ ܘܫܡܗ ܠܒܪܢܫ ܫܠܝܐ ܡܪܬܝܗܘܢ ܒܦܪܘܣܗ ܘܫܠܡܘܗܝ .
ܘܫܡܗ ܠܒܝܬܘܗܝ . ܘܐܚܝܕ ܐܠܘ ܒܡܠܟܘܬܐ ܫܡܝܢܝܬܐ . ܒܐܝܕ ܕܟܠܢܐ .
ܘܒܐܡܝܢ/ ܕܫܘܪܝܐ . ܘܒܨܠܘܬܐ ܢܠܗܘܢ ܐܠܗܐ ܡܪܚܡ ܘܟܠܗܘܢ ܩ̈ܕܝܫܐ . ܐܡܝܢ .
ܒܪܝܟ ܐܠܗܐ ܠܥܠܡ ܥܠܡܝܢ . ܘܡܫܒܚ ܫܡܗ ܠܕܪܕܪܝܢ/ ܘܐܡܝܢ ܕܟܠܗܘܢ .
ܘܐܡܪܐ ./ ܐܡܝܢ .

[4 ܚܕ ܚܝܠܐ: supra lin.]

[10] Cf. BH *But*. Min. II.iii.4.

[11] On the "scribe reaches harbour" simile, which is frequently encountered in colophons, see Brock (1995). The version here is the same as that given at Brock (1995) 200 fin., except for the verb *sraṭ*, instead of *kāṯeḇ*, at the end.

[1-7] To the glory and honour of the Holy Trinity, the Father, the Son and the Holy Spirit, the one true God, who is known in the three holy persons, is worshipped in the three venerable names and is divided into three characteristic properties. the Father, the eternal begetter of the Son, the Son begotten of the Father in eternity, not in time, and the Holy Spirit who proceeds and partakes, who truly proceeds from the Father and properly partakes of the Son, the single Three and threefold One, the one true and revered God, one nature, one power, one authority, one being, one greatness, one kingdom, one divinity, one lordship, one action, whom we glorify, extol, praise and worship as the glorious Trinity for ever and ever. With trust in Him I began, and with confidence in Him I finished these lines. By His succour we are strengthened and by His help we are sustained. To Him be glory and honour and manifold praise for ever and ever. Amen.

[7-11] This book on hand - *Cream of Wisdom*, Division Two, composed by our blessed and honoured father, the chaste monk, our lord, Mar Gregory, the Maphrian of the East, known as Barhebraeus, holy and renowned, whose like is not to be found among the teachers ancient or modern, Abū al-Faraj, son of Aaron, the Melitenite physician - was finished and completed in the year 1894 A.D., the year 2205 of the Greeks, on Sunday 16th Teshrin I.

[11-16] In the days after the decease of our lord Ignatius Patriarch Peter III, on 25th Ilul. In the days of the reverend fathers Mar Cyril Metropolitan George who has become the caretaker of the see of Antioch, of Mar Dionysius *ḥasyā* Behnam, the incumbent of the see of Mosul, and Mar Cyril *ḥasyā* Elias, the incumbent of the see of Mar Mattai and the villages, together with the rest of the fathers, the incumbents of the sees in cities and monasteries. May Lord God extend their lives and lengthen their years in times of joy and months of gladness. May they feed their flocks on pure food, so that they may bear comely fruit, thirtyfold, sixtyfold and a hundredfold. Amen.

[16-23] The one who wrote this book was a wretched man, in need of God's mercy, full of blemishes and much in debt, Matthew, deacon and *maqdšāyā*, son of the late Paul, son of Ni^{c}mat Allah, from the blessed city of Mosul, known as Athor and Nineveh. I ask every brother and teacher who comes across this book to entreat and pray for the forgiveness of the scribe, and say: Lord God, forgive and pardon your servant, the sinner Matthew, his failings and sins; and pardon his departed [forebears] and let them rest in the heavenly kingdom. With the 'so be it' of those above and the 'amen' of those below. With the remembrance of the Mother of God, Mary, and all the saints. Amen. Blessed be God for ever and ever, and glory be to His name for ages of ages. Our Father in heaven etc. Amen.

(d.iii.) Fol. 128r 24-26

ܒܫܢ̈ܝ ܢܚܘܡ / ܫܠܡ ܠܐܠܗܐ ܕܒܥܘܕܪܢܗ ܠܥܕܡܐ ܠܗܪܟܐ . ܘܫܠܡܬ ܒܫܘܒܚܐ ܒܬܫܒܘܚܬܐ
ܘܒܪܥܘܬܐ ܘܒܫܝܢܐ ܘܫܠܡܐ :/ ܘܐܡܝܢ ܐܡܝܢ ܕܠܐ ܫܘܠܡܐ ܠܟܠ ܡܢ ܕܩܪܐ .
ܘܫܘܒܚܗ ܕܫܘܒܚܢܐ ܫܒܝܚܘܬܗ ܠܐܠܗܐ ܕܝܗܒ ܚܝܠܐ :

In [the metre of] Mar Jacob: Worship to God who created the heaven in His wisdom, and decorated it with the stars, the luminaries, the sun and the moon; who made firm the earth without foundations above the river, and commanded it to bring forth its fat [Gen. 27.28] for the being which dwells upon it.

6. *Vatican, Syr. 613-615 (= ms. V)*

(-) End of *But*. Animal.: Ms. 613, fol. 236v 18-20 [following list of contents on fols. 230v-236v]:

ܫܠܡܬ ܟܬܒܬܗܘܢ ܟܘܪ̈ܣܐ ܗܠܝܢ ܒܟܒ ܐܝܪ / ܐܝܪ . ܒܟ . ܒܗ . ܒܡܕܝܢܬ ܐܬܘܪ
ܡܘܨܠ/ من كتب ايليا ميلوس وخط يده ١٨٨٧

I finished writing these quires on 22nd Iyar [May] 1887 in Mosul. *Ex libris* Elijah Mellus. By his hand. 1887.

(b) Ms. 614, fol. 39v [written at right-angles to the main text; cf. ms. M, fol. 216r]:

1 ܫܠܡ ܟܬܒܐ ܕܥܠ ܢܦܫܐ / ܘܡܫܠܡܢܘܬܗ ܐܫܬܠܡ / ܘܠܦܠܓܘܬܐ ܬܪܝܢܝܬܐ ܕܟܬܒܢܝܬܐ
/ ܡܢ ܟܬܒܐ ܕܚܐܘܬ ܕܚܟܡܬܐ 5/ ܒܟܒ ܠܐܒ ܫܢܬ ܐܬܩܨܘ / ܕܝܘ̈ܢܝܐ ܒܡܘܨܠ
ܡܕܝܢܬܐ / ܐܣܚܬܐ ܬܪܝܢܝܬܐ ܐܬܦܣܩܬ / ܕܝܢ ܟܬܒܐ ܬܪܝܢܐ ܒܝܐ ܠܬܫܪܝ / ܩܕܝܡ
ܡܢܝܢܐ ܫܢܬ ܐܬܩܨܙ 10/ ܕܝܘ̈ܢܝܐ . ܐܣܚܬܐ ܕܝܢ ܬܠܝܬܝܬܐ / ܕܩܕܡܝܢ ܐܬܟܬܒܬ ܒ
ܠܬܡܘܙ / ܒܝܙ ܫܢܬ ܒܩܥܘ ܕܝܘ̈ܢܝܐ

[1-4] Here ends the Book on the Soul ... [5-6] 22nd Ab, the year 1596 of the Greeks [= Aug. 1285 A.D.] in the city of Mosul. [7-10] The second copy [*aṣṣaḥtā*]: The second book [*ktāḇā*] was finished on 11th Teshrin I, the year 1597 of the Greeks [= Oct. 1285 A.D.]. [10-12] The third copy which is before us was written on 17th Tammuz, the year 2176 of the Greeks [= July 1865].

(c) Ms. 614, fol. 160r 13f. [following list of contents on fols. 157r-160r]:

من كتب ايليا ميلوس وخط يده نسخهُ/ فى الموصل ١٨٨٧. حُرر فى ماردين ١٨٩٢.

Ex libris Elijah Mellus. By his hand. Copied in Mosul 1887. Revised [?] in Mardin 1892.

(d.i.) Ms. 615, fol. 83r 3-14:

3 ܫܠܡ ܟܬܒܐ ܕܦܘܠܝܛܝܩܝ ܕܡܢ ܡܢܬܐ/ ܕܦܠܣܘܦܘܬܐ ܣܥܘܪܢܝܬܐ .
ܘܒܫܘܠܡܗ/5 ܫܠܡܘ ܟܠܗܘܢ ܟܬ̈ܒܐ ܕܦܠܓܘܬܐ ܬܪܝܢܝܬܐ/ ܡܢ ܟܬܒܐ ܕܚܐܘܬ
ܕܚܟܡܬܐ ܘܐܝܬܝܗܘܢ/ ܐ . ܟܬܒܐ ܕܫܡܥܐ ܟܝܢܝܐ ... 13 ܒܝܘܡ ܥܪܘܒܬܐ .
ܓ . ܒܬܡܘܙ / . ܫܢܬ ܐܦܙ ܡܫܝܚܝܬܐ . ܒܡܘܨܠ . ܫܘܒܚܐ ܠܐܠܗܐ .

[3ff.] Here ends the Book of Politics in the part on practical philosophy. With its completion comes to end all the books of the second division [*pālgūṯā*] of the book of Cream of Wisdom, which are: 1. *Auscultatio physica* ... [13-14] On Friday, 3rd Tammuz [July] 1887 A.D. in Mosul. Glory to God.

(d.ii.) Ms. 615, fol. 85r, 10-13 [following the list of contents on fols. 83v-85r]

من كتب المطران ايليا ميلوس وخط يده/ نسخه بالموصل ١٨٨٧/ ١٨٨٩/ حُرر

Ex libris Metropolitan Elijah Mellus. By his hand. Copied in Mosul 1887. Revised 1889.

7. *Manchester, John Rylands Library 56 (= ms. R)*

(-) Brief notes marking the end of each 'book' are found at the following places, on three occasions with the date of completion: Bk. I. Isag.: 12v (25th Tammuz = July, 1887); Bk. II. Categ.: 23r; Bk. III. Periherm.: 46v (10th Ab = Aug. 1887); Bk. IV. Anal.: 77v; Bk. V. Apod.: 96r (Ilul = Sept. 1887); Bk. VI. Dial.: 127r; Bk. VII. Soph.: 141v; Bk. VIII. Rhet. 176r.

(a.i) 182v b10-ult.:

10 ܫܠܡ ܟܬܒܐ ܕܦܐܘܛܝܩܐ. ܡܣܟܠܢܘܬܗ ܐܪܣܛܛܠܝܣ/ ܦܝܠܣܘܦܐ ܡܗܝܪܐ
ܘܫܠܡܘܬܐ ܆ ܒܝܪܚܐ ܕܐܝܠܘܠ/ ܒܪܝܟܐ ܆ ܟܙ ܆ ܒܗ ٢٧ ܝܘܡ ܕܬܪܝܢ ܒܫܒܐ/
ܒܗܘܐ ܕܐܙܠ ܚܣܝܐ ܕܝܠܢ ܡܪܝ ܕܝܘܢܘܣܝܘܣ ܡܛܪܦ/ ܒܗܢܡ
ܠܩܘܣܛܢܛܝܢܘܦܘܠܝܣ ܙܒܢܬܐ ܕܬܪܬܝܢ/15 ܡܛܠ ܥܕ̈ܬܐ ܕܐܬܢܣܒܝ ܡܢܢ. ܗܝ ܥܕܬܐ
ܕܝܠܕܬ ܐܠܗܐ/ ܡܪܝܡ ܒܡܚܘܪܬܐ ܕܢܓܪ̈ܐ ܒܩܠܥܐ: ܒܫܢܬ ܐܠܦܐ/ ܐܠܦ
ܘܬܡܢܡܐܐ ܘܫܒܥܐ ܘܬܡܢܝܢ ܡܫܝܚܝܐ :/ ܘܒܫܢܬ ܐܠܦܝܢ ܘܡܐܐ ܘܬܫܥܝܢ ܘܬܡܢܝܐ
ܠܝܘ̈ܢܝܐ :/ ܐܬܟܬܒ ܗܢܐ ܟܬܒܐ ܒܐܝ̈ܕܝ ܚܠܫܐ ܘܡܫܬܡܝܢܐ/20
ܡܫܡܫܢܐ ܡܬܝ ܒܪ ܡܢܝܚܐ ܦܘܠܘܣ ܒܪ/ ܢܥܡܬ ܐܠܠܗ ܕܡܢ ܡܕܝܢܬܐ
ܡܘܨܠ ܕܡܬܟܢܝܐ/ ܢܝܢܘܐ ܘܐܬܘܪ. ܡܪܝܐ ܐܠܗܐ ܢܛܪ ܠܥܡܘܪ̈ܝܗ/ ܡܢ ܟܠ
ܢܟ̈ܝܢ ܘܟܠ ܗܘ̈ܓܒܝܢ. ܒܨܠܘܬܗ ܕܡܪܬܢ ܡܪܝܡ/ ܒܬܘܠܬܐ ܕܒܟܠܙܒܢ ܝܠܕܬ ܐܠܗܐ.
ܘܕܟܠܗܘܢ ܩܕ̈ܝܫܐ ܐܡܝܢ./25 ܠܗܘ ܕܝܗܒ ܚܝܠܐ ܘܡܥܕܪܢܘܬܐ ܠܡܛܝܘܬܐ :/ ܫܘܒܚܐ
ܘܐܝܩܪܐ ܘܪܘܡܪܡܐ ܘܬܘܕܝܬܐ/ ܠܥܠܡܥܠܡܝܢ ܐܡܝܢ./27 ܐܝܢ ܘܐܡܝܢ

[15 ܕܐܬܢܣܒܝ: lege ܕܐܬܢܣ̈ܒܝ || ܗܝ ܥܕܬܐ: ܗܝ supra lin.]

[10-11] Here ends the Book of Poetics ... [11-16] In the blessed month of Ilul [= Sept.],[12] on 27th, Monday, the day on which our bishop Mar Dionysius Metropolitan Behnan went to Constantinople for the second time on account of the churches which had been taken from us, namely the church of the Mother of God, Mary, in the Quarter of the Carpenters in the Citadel.[13] [16-18] In the year 1887 of Christ, and the year 2198 of the Greeks. [19-24] This book was written by the hand of the weak and slothful deacon Matthew, son of the late Paul, son of Ni^{c}mat Allah, from the city of Mosul, known as Nineveh and as Athor. May the Lord God preserve its inhabitants from all harm and injury through the intercession of Our Lady Mary, Ever-Virgin and Mother of God, and of all the saints. Amen. [25-27] To Him who has given help and assistance and led me thus far be glory and honour, exaltation and thanksgiving for ever and ever. Amen. So be it. Amen.

---

[12] The month is given, wrongly, as Ayyar/Iyar [= May] in the following Garshuni note.

[13] Dionysius Behnam: Syr. Orth. Metr. of Mosul, 1867-1911 (Fiey [1993] 245). On the dispute over the churches in Mosul between the two West Syrian communities, see Fiey (1959) 61-63. The Church of Our Lady mentioned is the one which was later to be the Syr. Cath. cathedral and is described at Fiey (1959) 136-141 ("Ṭāhira des syriens catholiques dite «l'Église Ancienne»"). - Mar Behnam's return from his previous trip to Constantinople is noted by Matthew b. Paul in the colophon of ms. Ming. 100A, dated 3rd June 1885 (Mingana [1933] 247; cf. Macuch [1976] 191 n.59; see also Mingana [1933] 294f., 560).

(a.ii) Fol. 183r a1-b ult. [in Garshuni]:

١ قد كُمل وانجز هذا الكتاب المكنّى من واضعه ܕܚܐܘܬ/ ܚܟܡܬܐ الذى هو المنطق، كترتيب
افتكار ملّة/ السريان، من مؤالفه السيد النبيل والمعلّم الحاذق/ الفضيل، القديس العظيم، المبجّل
الكريم، الاب الطاهر/ ٥ والنجم الزاهر، مار غريغوريوس ابو الفرج، ابن/ العبرى، مفريان المشرق، قدّس
الله روحه، ونوّر/ ضريحه، وبدعاه ينفعنا، وبعلمه يرفعنا، امين./ والحمد لله رب العلمين، ولالهنا
نسجد ونسبح ونعظّم/ الان وكل اوان والى ابد الابدين امين. وكان الفراغ من/ ١٠ كاتبة هذا الكتاب،
سبعه وعشرين يوم من شهر/ اٍيار المبارك، سنه الف وثمانماية وسبعه وثمانون/ مسيحيه، الموافقه الفين
ومايه وثمانيه/ وتسعين يونانيه، فى ايام اب الابا (الآباء)، وريّس الرؤسآ (رئيس الرؤساء)/ تاج الملّة
السريانيه، وضابط الكرسى انطاكيه، مار/ ١٥ ايغناطيوس بطريرك بطرس الثالث، من اعمال/ الموصل،
ومار غريغوريوس سيدنا مطران جرجس ضابط كرسى/ القدس الشريف، ومار ديونسيوس سيدنا مطران
بهنام/ ضابط كرسي الموصل، الذى توجّه اليوم الى الاستانه/ مرة الثانيه لاجل دعوة كنيسة الطاهرة التى
اخذوها/ ٢٠ مننا ملّة البابويه، ومار قوريللوس ابينا مطران/ الياس ضابط كرسى دير مار متى والقرى،
ومار ديونسيوس/ مطران يوسف الهندى ضابط كرسى مليبار، وريس (ورئيس)/ اساقفة الهند، مع باقى
الابهات، ادامهم الرب فى/ روسنا (رؤوسنا)، وصلاواتهم (وصلواتهم) تحفظ مراعيثهم امين. وكان
المهام/ ٢٥ بكتابته وترجمته من حرف الكلدانى الى السريانى/ الصحيح، العبد الفقير من الحسنات،
الغنى بالخطات/ والزلات، الراجى العفو من مولاه؟ يوم الحكم/ والمحاسبات، الضعيف بين الشمامسه
الاسطفانوسيين// ١ الذى لا يستحق ان يكتب اسمهُ بهذا الكتاب، لاكن لاجل/ ‹صلوات› القاريين
ماتيوس ابن المرحوم بولص السريانى/ من كنش (كنيسة؟) سيدتنا مريم العذرى (العذراء) فى محلّة
القلعه/ والنجّارين فى بلد الموصل، فالمرجو من كل معلّما (معلّم)/ ٥ الذى يصادف هذه الكلمات،
لايسكب علينا اللوم عاجلاً،/بل يغتنم من الثوابات، ويقول يا ربنا يا باسط/ الارض، ورافع السمَوات،
اغفر لعبدك الخاطى متى/ كاتب الاحرف الذميمات، بمن انقبل لهم دعَا (دعاء) وطلبات،/ ولوالديه معه
بالمسامحات، امين. يا رب العالمين./ ١٠ اعلان لكلمن يقرا بهذا كتابنا فليكن معلوم، اذ/ اننا بقوّة
الله تعالى قد نقلناه من نسخة عتيقَه/ كان محرف فى حرف الكلدانى، وكما وجدنا حررنا، فانكان/ يوجد
غلطه ام نقص فهو على حاله، فنرجو من الذى/ يطالع فيه يكمّله ويصلحه، لان ليس كامل الا الله/ ١٥
الواحد. ܡܫܒܚܐ ܠܟܠ ܐܠܗܐ ܡܫܒܚ، ܘܒܟܠܗܘܢ ܩܕܝܫܐ/ ܘܩܕܝܫܬܐ ܐܡܝܢ
ܘܐܡܝܢ.

[1-9] This book has been finished and completed, which is called by its author '*d-ḥēwat ḥekmātā*', i.e. logic, according to the system of thought of the Syrian nation [*milla*]. From the works of the noble master, the learned and excellent teacher, the great saint, the venerable, the revered, the chaste father and the shining star, Mar Gregory Abu al-Faraj Ibn al-ᶜIbrī, the Maphrian of the East. May God sanctify his soul and shed light on his grave. May he help us with his prayer and raise us with his learning. Amen. Praise be to God, the Lord of the Universe. Our God we worship, glorify and magnify now and always and for ever and ever. Amen.

[9-24] The copying of this book was completed on the 27th day of the blessed month of Ayyar [May], the year 1887 of Christ, corresponding to 2198 of the Greeks, in the days of the father of fathers, the head of heads, the crown of the Syrian nation and the incumbent of the see of Antioch, Mar Ignatius Patriarch Peter III, [a native] of the province of Mosul; and of Mar Gregory *sayyidnā* Metropolitan George, the incumbent of the see of Jerusalem; and of Mar Dionysius *sayyidnā* Metropolitan Behnām, the incumbent of the see of Mosul, who has departed today for Istanbul for the second time in order to make an appeal for the Church of Our Lady [al-Ṭāhira] which the Papist nation has taken from us; and of Mar Cyril, our father, Metropolitan Elias, the incumbent of the see of Dair Mar Matta and the villages; and of Mar Dionysius Metropolitan Joseph al-Hindī, the incumbent of the see of Malabar and the head of bishops of India; along

with the rest of the fathers. May the Lord perpetuate their rule over us, and may their prayers protect their dioceses.[14] Amen.

[24-4] The one who undertook the copying and transliterating from the Chaldean into the correct Syrian script was the servant poor in merits and rich in sin and errors, one who begs forgiveness from his Lord on the day of judgement and reckoning, a weak one among the Stephanite deacons, whose name does not deserve to be written in this book, but for the sake of <the prayers of> the readers: Matthew, son of the late Paul, the Syrian, of the Church [?] of Our Lady Mary the Virgin in the Quarter of the Qal^{c}ah and the Carpenters in the city of Mosul.

[4-10] It is requested of all the learned men who come across these words that they do not hastily pour rebuke upon us, but take the opportunity for pious deeds and say: 'O Lord who levels the earth and raises the heavens, forgive your sinful servant Matthew, the writer of faulty letters - through [the intercession of] those whose prayer and requests are accepted - and his parents with him in [your] magnanimity. Amen. O Lord of the Universe.'

[10-16] A notice to all those who read this book: Let it be known that we have copied this book with the help of God from an old manuscript written in the Chaldean script, and we have written as we found it. If therefore there are errors or omissions it is as it was. We beg those who read it to complete and correct it, for noone is perfect except the one God. *Commemorata sit Deipara Maria, et omnes sancti et sanctae*. Amen. Amen.

8. *Beirut, Université St. Joseph, syr. 48*

(-) End of *But*. Anal.: Fol. 201r 11-21 (in the hand of the copyist):

ܫܠܡ ܟܬܒܐ ܕܐܢܐܠܘܛܝܩܐ/ ܩܕܡܝܐ ܘܠܐܠܗܐ ܫܘܒܚܐ ܕܚܝܠ/ ܘܥܕܪ ܒܛܝܒܘܬܗ
ܐܡܝܢ ܀ ܟܬܒܗ . ܐܢܫ ܡܚܝܠܐ/ ܚܛܝܐ ܕܘܝܐ ܡܣܠܝܐ ܕܟܠܗܘܢ ܒܢܝܢܫܐ
ܕܠܐ/ 15 ܫܘܐ ܕܢܬܟܬܒ ܫܡܗ ܐܠܐ ܡܛܠ ܨܠܘܬܐ ܕܚܘܣܝܐ/ ܕܢܠܩܘܛ ܡܢ ܦܘܡ
ܩܪܘܝܐ ܓܒܝܐ . ܕܐܝܬܘܗܝ/ ܪܒܢ ܝܘܚܢܢ ܪܘܟܘܣ . ܒܫܡܐ ܘܒܟܘܢܝܐ ܒܒܥܒܕܐ/
ܐܦܠܐ ܥܠܡܝܐ ܒܪ ܡܗܝܡܢܐ ܫܡܥܘܢ ܒܪ ܣܦܪܓܦܘ/ ܕܡܢ ܩܪܝܬܐ ܒܪܝܟܬܐ
ܘܡܒܪܟܬܐ ܡܢܓܝܫܐ ܢܝܬܒܝܗ ܠܗ/ 20 ܡܪܝܐ ܒܝܡܝܢܗ ܚܣܝܢܬܐ ܐܡܝܢ ܐܘ
ܡܪܝܐ ܥܕܪܝܢ/ ܐܝܢ ܘܐܡܝܢ

[17 ܝܘܚܢܢ ܪܒܢ: ܪܒܢ̈ ܝܘܚܢܢ̄ cod. || ܣܦܪܓܦܘ: ܦܘܪܓܦܘ ut vid.]

End of the book of Prior Analytics. Glory be to God who has given strength and help in His goodness. Amen. It was written by a certain weak one, a sinner, wretched, most despised of all, whose name does not deserve to be written but for the sake of the prayer for pardon which he might gather from the mouth of the elect readers. He is Rabban John Rokos, in name and title, but in deed not even a layman, son of the believer Simeon, son of Sprgpw [?], from the blessed and blissful village Mangeshe. May the Lord people it with His strong right hand. Amen. Lord, help us. So be it. Amen.

---

[14] مرعيث, pl. مرعيثات < ܡܪܥܝܬܐ (√$r^c y$), Graf (1954) 105. The plural *marāʿīṯ* here is ungrammatical (as if from root √$r^c t$).

Fol. 202r (in the hand of Mar Elijah Mellus):

ܦܪܢܣܐ ܠܥܘܗܕܢܐ ܚܘܒܢܝܐ ܡܢ/ ܡܪܝ ܐܠܝܐ ܝܘܚܢܢ ܡܠܠܘܣ/ ܠܘܬ ܐܒܐ
ܡܝܩܪܐ ܘܪܚܝܡܐ/ ܠܘܝܣ ܫܝܟܘ ܡܠܦܢܐ ܒܚܝܪܐ/ ܟܬܝܒ ܒܩܠܝܬܐ ܕܡܝܛܪ̈/ ܕܟܠܕܝܐ
ܒܡܪܕܝܢ ܫܢܬ/ ܐܦܨܗ ܡܫܝܚܝܬܐ

In loving remembrance from Mar Elijah John Mellus to the honoured and beloved Father Louis Cheikho, the excellent teacher. Written in the cell of the Metropolitan of the Chaldeans in Mardin. 1895 A.D.

APPENDIX TWO

GLOSSES AND MARGINAL NOTES IN MANUSCRIPTS

Of the manuscripts used for the edition of *But.* Min. & Mete., **Ll** have occasional marginal glosses giving the Arabic equivalents of the Syriac words in the text, invariably in Garshuni.

Ms. **M** has a greater number of such glosses, usually in Garshuni but occasionally also in the Arabic script.[1] Since many of these glosses coincide with those found in ms. F, and mss. M and F evidently derive independently of each other from the autograph, it may be that these glosses were already there in the autograph and, given the date of M's exemplar, they may have been added to the text by the author himself.

Ms. **F** (= Laur. or. 69 & 83) contains, besides an even greater number of such glosses (which are invariably in the Arabic script and usually interlinear rather than marginal), a large number of longer marginal notes in Arabic. There appears to be at least two different hands responsible for these notes, one of them for most of the Arabic translations of the chapter and section headings and a small number of the remaining notes (incl. the rendition of the passage at Min. I.i.2.10-12 and the note on the comet of "664 A.H." relating to Mete. IV.iii.3) and the other for the vast majority of the remaining notes. Since the note on the comet at Mete. IV.iii.3 must originate from someone who himself witnesssed the comet of 1264 A.D. nearly eighty years before F was copied, the likelihood is that the first of these hands is the same as that of the main Syriac text (Najm b. Shams of Mardin), copying together with the Syriac text the notes already found in the exemplar.[2] That

---

[1] Orthographical errors in M indicate that these glosses were written in the Arabic script in the exemplar (Min. I.i.4.7 ܢܩܛܥ = يقطع, for تقطع; Min. III.i.4.2 (ܬܠܩܛܐܪ = تلقطار, for تلقطار).

[2] It is not so easy to compare handwritings in Arabic and Syriac, but I believe the first Arabic hand has traits similar to those in the Syriac text. The size of the letters too is usually about the same as that of the Syriac text and larger than that of the second Arabic hand (except, naturally, in the interlinear glosses, which are written in smaller letters by necessity). One tell-tale difference between the two Arabic hands is the way in which the preposition فى (which appears constantly in chapter and section headings) is written. The first hand, which I wish to identify as Najm's writes this

being the case, the second hand responsible for the majority of the longer notes must be someone other than the original copyist.

On the Scholia in Ms. F.

S.E. Assemani believed the scholiast of ms. F to be "Daniel Rabbanus" (Rabban Daniel), who, as we learn from the note on ms. 69 fol. 1r, bought this manuscript in 1367, twenty-seven years after it was originally copied in 1340.[3] The note of purchase on ms. 69 fol. 1r is in Syriac, while the marginal notes in the rest of the manuscript are invariably in Arabic, so that it is difficult to confirm the identification through a comparison of the handwritings in this manuscript alone, but Assemani's identification and the identification, in turn, of our Rabban Daniel with Daniel of Mardin (1326/7-after 1382), the epitomist/translator of several of Barhebreaus' works and a man of some erudition in his own right,[4] can be confirmed through a comparison of the handwriting in the following manuscripts of Barhebraeus' works.[5]

Laur. or. 298 (olim Palat. 62), fol. 3v-80v: *Grammar*, followed (on fol. 81-82) by a short biography of Barhebraeus, with a list of his works.
Laur. or. 342 (olim Palat. 200): *Tractatus tractatuum*.
Berol. Petermann I, syr. 23 (206 Sachau): *Nomocanon*.

Of these manuscripts, the first part of Laur. 298 and Berol. Pet. I.23, finished repectively in Nisan 1671 A.Gr. (1360) and in 1685 A.Gr. (1373/4), were copied by a certain Daniel, and the hand of the Syriac text in these two manuscripts closely resembles that of the note of purchase on Laur. 69 fol. 1r, while the hand of the Arabic notes in Laur. 298, though usually neater than that of the notes in ms. F, comes

---

with a flourish to the right.

[3] Assemani (1742) 329; cf. Drossaart Lulofs (1989) 43 and 730.

[4] See Takahashi, *Bio-Biblio.*, Section I.2.7.

[5] Mss. Laur. 298 and 342 were available to me on microfilm. For checking the hand of Berol. Pet. I.23, I have used Hatch (1946) Plate CXLV. - Though not on the page reproduced by Hatch, Berol. Pet. I.23 is reported, like our ms. F and Laur. 342, to contain a large number of marginal notes, mostly in Arabic, which "könnte sehr wohl von dem Schreiber des Textes, Daniel, selbst herrühren" (catal. Sachau [1899] 683). - Among mss., incidentally, copied from copies made by Daniel are British Library, Add. 7202/1 (*Gramm*., Rosen-Forshall [1838] 94) and Paris, syr. 226 (*Nom.*, Zotenberg [1874] 174).

to resemble the latter when it becomes rougher. The scribe of Laur. 342 is not named in the manuscript as we have it today,[6] but in this manuscript, where the text of the *Tractatus* is accompanied by long marginal notes in Arabic like those found in our ms. F, the hand of the Syriac text closely resembles that of Berol. Pet. I.23 and the rougher parts of Laur. 298 (esp. fols. 81-82), while that of its Arabic notes is clearly the same as that of the Arabic notes in ms. F.[7]

In his edition of the Book of Plants in the *Butyrum*, Drossaart Lulofs was able to identify the scholia he found in his part of ms. F as quotations from Barhebraeus' source, the *Šifāʾ*.[8] This, however, is not quite the whole story, since, in the Books of Mineralogy and Meteorology at any rate, these notes include quotations not only from the *Šifāʾ* but also from other sources used by Barhebraeus, such as the works of Fakhr al-Dīn al-Rāzī (*al-Mabāḥit̠ al-mašriqīya*) and Abū al-Barakāt al-Baghdādī (*K. al-muʿtabar*).[9] The scholiast, in other words, had already undertaken the work also undertaken by the present editor and succeeded in identifying the sources used by Barhebraeus.

---

[6] Assemani (1742) 339 tells us that Laur. 342 was copied by a certain Ignatius in 1707 A.Gr. (1395/6 A.D.), referring to a note at the end of the manuscript. The hand of this note (on fol. 180 bis recto), however, is clearly different from that of the main part of the manuscript (the hand which has supplied the last part of the text of BH *Tract.*, at 175r-179v, looks similar to that of the note on fol. 180 bis r). The note, in fact, does not tell us anything about when this manuscript was copied, but tells us instead about how a number of books from the Monastery of Mar Hanania (D. al-Zaʿfarān) had to be sold to defray the cost of the reconstruction work in the wake of an attack on the monastery by Timur, the date 1707 A.Gr. mentioned there being the date of Timur's attack. Furthermore, Assemani ascribes the Arabic notes in the manuscript to a certain "Monachus Harebus". I have been unable to find this name in the microfilm of the manuscript available to me.

[7] At the end of ms. Paris syr. 244, which is a copy of BH *Asc.*, there is an autobiographical account by Daniel of Mardin recounting his imprisonment in Adar 1693 A.Gr./1382 (fol. 142v). Nau (1899) tr. ix-x (also id. [1905]) believed this note to be an autograph by Daniel. The hand of the first 41 folios of this manuscript, in fact, resembles that of Daniel of Mardin. The rest of the manuscript, however, is in a different hand and Daniel's account of his imprisonment on fol. 142v is in yet another hand, which is quite different from the hand I wish to identify as that of Daniel, while that this note is not an autograph is also suggested by its opening words ("Rabban Daniel of Mardin, monk and philosopher, recounts his suffering and says ..."; for a further copy of BH *Asc.* with Daniel's note at the end, see Mingana [1933] 584, ms. Mingana 306B/C, dated ca. 1840).

[8] Drossaart Lulofs-Poortman (1989) 43.

[9] There is also a quotation from Ibn Sīnā's *Qānūn fī al-ṭibb* at Min. IV.iv.1/4. A number of the quotations have not so far been identified.

What is more intriguing is the fact that in a significant number of cases these quotations from the *Šifā*ᵓ and other Arabic works differ from the texts of the Arabic works as we know them and in such cases they usually agree in meaning with the Syriac text of the *Butyrum*. These scholia, in other words, seem to represent a kind of "harmonisation" between the text of the *Šifā*ᵓ etc. and the *Butyrum*.

As an explanation for these facts, one might first consider the possibility that the scholia in ms. F might, in fact, have been copied from the notes Barhebraeus had taken in preparation for composing his work. In taking such notes, he may have summarised and shortened some parts of the passages he was copying, and if these notes were then used by Barhebraeus as the basis for rendering the relevant passages into Syriac, this would explain the agreement in meaning between the scholia in ms. F and the Syriac text.

A more likely explanation seems, however, to be that these scholia are indeed the result of "harmonisation" by a later scholiast. If there was anyone among the Syrian Christians in the 14th century who was in a position to identify Barhebraeus' sources, it was Daniel of Mardin, a keen student of Barhebraeus as well as of philosophy, who tells us himself that he travelled as far as Egypt for the sake of the study of philosophy. It has been found, moreover, that in his *Book of Brilliance* (*K. al-išrāq fī al-uṣūl al-dīnīya wa-l-qawāᶜid al-bīᶜīya*), a work based on the third and fourth bases of Barhebraeus' *Candelabrum*, Daniel quotes directly from the original Arabic of those parts of Fakhr al-Dīn al-Rāzī's works which were also used by Barhebraeus in his *Candelabrum*.[10] What is found, in other words, in the marginal notes of our ms. F is comparable to what Daniel has done in his *Book of Brilliance*, a conflation, in Arabic, of Barhebraeus' Syriac work with those Arabic works used as sources by Barhebraeus. Whether Daniel intended also to compose a work based on the *Butyrum* is not known, but if he did, the scholia as found in ms. F might have served as preparatory notes for such a work.

---

[10] See Sepmeijer (1994) 381, 386.

Glosses and Marginal Notes

The glosses and notes are listed below according to the places in the text of *But.* to which they relate.

While the quotations from the *Šifāʾ* and other Arabic works in the margin of ms. F are not straightforward quotations as has been explained above, these quotations nevertheless have a certain value as 14th c. witnesses for the texts in question, especially where they are in agreement with the readings of particular manuscripts. For this reason an attempt has been made below to indicate such readings in the form of an apparatus. For the sigla used for the manuscripts and editions of the *Šifāʾ*, see Introduction, Section 3.2.3a above

ad: gloss on
{-}: illegible
(؟): unclear
(-) -: outside the bracket: reading of the manuscript; inside: corrected orthography (e.g. (ثلث (ثلاث)

Min.tit.: فى المعدنيات F
Min. I.i.tit.: مباحث فى تكوّن الحجر F

Min. I.i.2.1-2: اعلم ان الارض لا تتحجر لان استيلاً اليبس لا يفيدها استمساكًا بل تفتتًا F
≈ *Šifāʾ* 3.9-10: اعلم ان F: ونقول أما فى الأكثر فإن *Šifāʾ* ‖ الارض: add. الخالصة *Šifāʾ* ‖ استمساكا F et ed. Cair.: امتساكا BDM et Holmyard (et Bahmanyār, *Taḥṣīl* 718.5) ‖

Min. I.i.2.5 [ad ܠܡܪܕܝܬܗ]: انحدار الما بسرعه F

Min. I.i.2: {-}ون {...} ميبسه ارضيه او سبب مجفف حار F
Rest illegible; cf. *Šifāʾ* 4.18-19.

Min. I.i.2.8-10: حكى الشيخ الريس انه شاهد فى الطفوليه طينًا لزجًا فى شط جيحون ثم شاهده بعد مدّة ثلث (ثلاث) وعشرون سنةً قد تحجر حجراً رخواً F. - Ibid.: وهو الطين الذى يغسل به الراس F
Cf. *Šifāʾ* 3.15-17: الطفولية F: طفولتنا *Šifāʾ* ed. Cair., pro quo طفولته expectaretur; طفوليتنا ed. Holmyard ‖ حجراً F cum D et Holmyard: تحجراً ed. Cair. ‖

Min. I.i.2.10-12: قال المولى الاجل العلامه المفريان المعظم غريغوريوس فى هذا الفصل فى زمن الصبى لما كان فى المدينه المحروسه ملطيه فخرج ذات يومٍ مع والده فوصل الى دير هناك كانوا يعمرونها فشاهد هناكَ قبال الاردخل حجر عظيم صمّ وقد جُبل به خزف وفحم F
[والده: ܐܒܗ̈ܘܗܝ (pl.) *But.* ‖ حجر عظيم صمّ: lege حجراً عظيمًا أصمّ]
The most exalted and erudite master, the revered maphrian Gregory said in this section that in (his) childhood, when he was in the city, protected [by God] of Malatya, he went out one day with his father and came to a monastery there which they were building.* He saw there, in front of the builders, a large solid stone, in which pottery and coal were mixed.
*yaʿmurūna-hā: making *dair* feminine, no doubt due to the influence of the Syriac *dairā* (common gender, but usually fem. in the sense of "monastery").

Min. I.i.2.12 [ad ܡܩܕܘܢܝܐ]: ܡܩܕܘܢܝܐ هو الفحم M; ܐܠܦܚܡ : ܐ (vocal. *sic*) L; : ܐ ܐܠܦܚܝܡ (*sic*) l
Min. I.i.3.7 [ad ܐܦܬܐ]: الخبّازين F
Min. I.i.4.1 [ad ܡܢܘܪܐܣܩܘܣ]: الصاعقه F; ܐܠܨܐܥܩܗ M; ܐܠܨܐܥܩܐ L

Min. I.i.4.3-11: [١] وحكى الشيخ ان شاهد تلميذه فى بلاد جورجانان من امر حديد نزل من الهوآ لعله يزن مايه وخميسن منًا فنفذ فى الارض ثم نبا نبوةً او {نبوتين} نبو الكره الذى يرمى بها الحايط ثم عاد فنشب فى الارض وسمع الناس لذلك صوتًا عظيمًا هايلاً [٢] وبالجهد فصلوا منه قطعةً فانفذوه الى المتولى [٣] ورام ان يطبع منه سيفًا فتعذر عليه وكأن ذلك الجوهر ملتما من اجزاء جاورسيه صغار مستديرة التصق بعضها ببعض F
Cf. *Šifā*ʾ 5.20-6.9: The beginning and the middle of the note, where it diverges from the text of the *Šifā*ʾ, look like translations back into Arabic from the Syriac text of *But*., although there too the scholiast was guided by the *Šifā*ʾ in his choice of vocabulary.
[١]: corr. *Šifā*ʾ 5.20-6.3: وحكى الشـيخ ... جـورجـانان = *But*. 3-4 || جـورجـانان F = ܓܘܪܓܢܝܐ *But*. cod. F: جوزجان ed. Cair.; جوزجانان Holmyard || نزل من الهوآ: post لعله يزن مايه وخميسن منا in *Šifā*ʾ || فنفذ F cum BṬM: فنقر ed. Cair. || الذى [post الكره]: التى *Šifā*ʾ ||
[٢]: ≈ *But*. 7-8 (corr. *Šifā*ʾ 6.3-7)
[٣]: corr. *Šifā*ʾ 6.7-9: وكأن ذلك: وحكى أن جملة ذلك *Šifā*ʾ || الجوهر: add كان *Šifā*ʾ || ملتما ut vid. F: ملتئما *Šifā*ʾ ||

Min. I.i.4.4 [ad ܟܘܪܙܣܛܢܐ (ܟܘܪܙܣܛܢܐ F)]: كرجستان F

Min. I.i.4.7 [ad ܦܘܠܝܛܘܣ]: الاله التى تقطع بها الحجاره F; ܐܠܐܠܗ ܐܠܬܝ ܢܩܛܥ ܒܗܐ ܐܠܚܓܐܪܗ M
The ungrammatical ܢܩܛܥ (يقطع), since the corruption of ܬ into ܢ is unlikely, suggests that the scribe was copying a gloss written in Arabic in his exemplar.

Min. I.i.4.11-12: ويحكى ان كثيراً من السيوف اليمانيه الجليله انما اتخذ منها من مثل هذا الحديد وشعراً العرب قد وصفوا ذلك فى شعرهم F
≈ *Šifā*ʾ 6.10-11: ويحكى: وحُدِّثت *Šifā*ʾ || الجليله F cum BDS et Holmyard: الجميله ed. Cair. || اتخذ F cum DSṬM et Holmyard: تتخذ ed. Cair. || منها من: من *Šifā*ʾ ||

Min. I.i.4.12 [ad ܩܠܩܢܬܐ]: ܐܠܩܠܩܢܬ M
Min. I.i.4.12 [ad ܩܘܩܐܢܝܣܛܘܣ]: ܒܐܣܬܪܐܣܐܝܣ M
Min. I.i.5.2 [ad ܒܘܪܘܠܐ]: ܐܠܒܠܘܪ M
Min. I.i.5.5 [ad ܩܘܪܐܠܝܘܢ/ܩܘܪܐܠܝܐ]: المرجان FM
Min. I.ii.tit.: مباحث فى تكوّن الجبال وصلوحهم et فى تكوّن الجبال ومنافعها F

Min. I.ii.1: قال الشيخ فى الشفا والاشبه ان هذا المعمور قد كان فى سالف الدهر مغموراً فى البحار فحصل هناك الطين اللزج الكثير ثم حصل الحجر بعد الانكشاف فلذلك كثرت الجبال ولهذا نجد فى كثير من الاحجار اذا كسرناها اجزا الحيوانات البحريه كالاصداف F
≈ *Mabāḥiṯ* II.208.16-20 [< *Šifā*ʾ 7.11-15]: قال الشيخ فى الشفا: deest in *Mab.* || سالف الدهر: سالف *Mab.* et *Šifā*ʾ || كان: كانت *Mab.* et *Šifā*ʾ || هذا المعمور: هذه المعمورة *Mab.* et *Šifā*ʾ cod. C; سالف الأيام Cair. || مغموراً: مغمورة *Mab.* et *Šifā*ʾ || الحجر: الزمان *Mab.* et *Šifā*ʾ cod. C; ... || ولهذا: ومما يؤكد هذا الظن انا *Mab.* || البحريه: المائيه *Mab.* et *Šifā*ʾ || التحجر *Mab.* ||

Min. I.ii.1.5 [ad ܡܬܩܦܐ]: يجمد ويعقد ut vid. F
Min. I.ii.4.1-3: ومنفعه الجبال ايضًا فى سيلان الاوديه الجاريه من العيون لان البخار اذا احتبس فى باطن الجبل فيتكاثف ببرده الحجرى ينقلب ميهًا F.
≈ *But*.

Min. I.iii.tit.: {-} وفى F
Min. I.iii.1.1-2: [ad ܟܐܦܐ ܕܐܪܥܐ] احجرة الارض F
Min. I.iii.1.1-2 [ad ܡܦܩܢܐ ܘܫܒܝܠܐ]: المنافذ والاسراب ut vid. F
Min. I.iii.2.2 [ad ܐܡܒܘܒܐ]: القنات المجرى ܐܡܒܘܒܐ F
Min. I.iii.2.2-3 [ad ܡ̈ܝܐ ܪܛܝܒܐ]: النزز F
Min. I.iii.4.3 [ܡ̈ܝܐ ܪܛܝܒܐ ܡܢ ܟܠܗܘܢ ܡܢܗ ܡܚܒܠܢ̈ܐ]: وماَ النزز ردى مضر F (ردئ مُضِرّ)
Min. II.tit.: فى الزلزله F
Min. II.i.tit.: مباحث فى ارآ القُدما على الزلزله F
Min. II.ii.tit.: مباحث فى السبب الحقيقى للزلزله F
Min. II.iii.1.12 [ad ܩܘܦܐ]: تقبض وتشجة F
Min. II.iii.4.3 [ad ܥܘܩܠܩܠܘܬܐ]: الاعوجاج والتوآ(ء) F
Min. II.iii.4.3 [ad ܫܒ̈ܘܪܐ]: منافذ F

Min. II.iii.4.4-7: ومن منافع الزلازل تفتيح مسام الارض للعيون واشعار قلوب فسقه العامه رعب الله تعالى F
= *Šifāʾ* 19.14-15

Min. III.tit.: فى الاجسام المعدنيه F; الثالث في الاجسام المتفتته ܐܠܟܐܒ (*sic*) M
Min. III.i.tit.: مباحث فى اجناس المعدنيات F
Min. III.i.1.3 [ad ܚܫܝܬܐ ܘ ܩܘܦܪܩܢܘܢ]: الزاج , الشب (*sic*) F; ܐܠܙܐܓ ܘܐܠܫܒ M
Min. III.i.3.3 [ad ܫܘܒܐ]: الشب F
Min. III.i.3.3 [ad ܐܪܡܘܢܝܩܘܢ]: نوشادر F
Min. III.i.4.1 [ad ܚܫܝܬܐ]: الزاج F
Min. III.i.4.2 [ad ܟܠܩܢܬܘܣ]: قلقند F
Min. III.i.4.2 [ad ܟܠܩܛܝܣ]: قلقطار F; ܩܠܦܛܐܪ M (i.e. misreading as قلفطار)
Min. III.i.4.5 [ad ܟܠܩܛܝܣ]: قلقند ut vid. F
Min. III.i.5.4 [ad ܫܘܒܐ]: الشب F
Min. III.i.5.4 [ad ܚܫܝܬܐ]: الزاج F
Min. III.i.5.5 [ad ܐܪܡܘܢܝܩܘܢ]: النوشادر F; ܐܠܢܘܫܐܕܪ M
Min. III.i.5.6 [ad ܐܪܡܘܢܝܩܘܢ]: المردمنج الاحمر ut vid. F; *recte* ܐܠܡܪܬܟ ܐܠܐܣܦܝܕܓ M

Min. III.ii.3: واصحاب الكيميا انما يصححون هذه الدعاوى لانهم يعقدون الزوابيق بالكباريت فيحصل لهم ظن غالب بان الاحوال الطبيعيه مقاربه للاحوال الصناعيه واعلم ان دعاويهم واحكامهم مجرّد ظنون واوهام غير متاكده بحجه اقناعيه فضلا عن حجةٍ برهانيه F

Min. III.iii.1: [١] وليس يبعد ان يحاول اصحاب الحيل حيلاً يصير (تصير) بها احوال انعقادات الزيبق بالكباريت انعقادات محسوسه بالصناعه وان لم تكن الاحوال الصناعيه على حكم الطبيعه بل تكون مشابهه [٢] واما ما يدعيه اصحاب الكميا (الكيمياء) فيجب ان تعلم انه ليس فى ايديهم ان يقلبوا الانواع قلبًا حقيقيًا بل فى ايديهم تشبيهات حسيه حتى يصبغوا الاحمر صبغًا ابيض شديد الشبه بالفضه [٣] وان يسلبوا الرصاصات أكثر ما فيها من النقص والعيوب الا ان جواهرها تكون محفوظه وانما يغلب عليها كيفيات مستفاده بحيث يغلط في امرها
Cf. *Šifāʾ* 22.11-23.3
[١]: ≈ *Šifāʾ* 22.11-13 || على حكم الطبيعه :على حكم الطبيعية et add. وعلى صحتها Cair. ||
[٢]: *Šifāʾ* 22.16-18 || بل :لكن ed. Cair. ||
[٣]: *Šifāʾ* 23.1-3 ||

Min. III.iii.2.2 [ad ܓܘ̈ܢܐ ܘܩܘܡܣܐ]: {-}اللون والرزا F
Min. III.iii.2: اللون يمكن اكتسابه والطريق اليه عسر F

Min. III.iii.2: واما سلخ هذه الاصباغ والاعراض من الروايح فهذه لا يجحده العاقل
Cf. *Šifā*ʾ 23.9-11: فهذه مما لا يجب أن يُصر : فهذه ... ‖ *Šifā*ʾ والأوزان أو كسوها add :الروايح
‖ *Šifā*ʾ على جحده

Min. III.iii.2: واما ان يكون الفصل المنوّع بسلب او بكسب فلم يتبين لى امكانه اذ لا سبيل الى حل المزاج الى المزاج الاخر فان هذه الأحوال المحسوسه يشبه ان لا تكون هى الفصول التى بها تصير هذه الاجساد انواعًا بل هى عوارض ولوازم والفصول مجهوله واذا كان الشى مجهولاً كيف يمكن ان يقصد ايجاده او افقاده
Cf. *Šifā*ʾ 23.5-9 ‖ بل بعيد عندى جوازه add :امكانه ‖ *Šifā*ʾ يسلب أو يكسى :بسلب او بكسب
‖ *Šifā*ʾ يقصد قصد :يقصد ‖ *Šifā*ʾ وفصولها :والفصول ‖ *Šifā*ʾ

Min. III.iii.3: [١] قال اصحاب الكمية فى خرافاتهم الذى (التى) لم تخرج الى الفعل [٢] لو وجدنا الكبريت الاحمر او الابيض المصفى وقدرنا على خلطه بالزيبق عملنا ذهبا او فضة وبالحقيقة ما وجدوا وما عَملُوا لما قيل من انّ القوه لا تعزم ولا توجد الا حيث توجد وعما عنه تُوجَد [٣] ومعرفه الاسباب القريبة المَتوسَطة فى هذه الاشيا متعذره علينا بل ممتنعه عما امتنع علينا وتعذّر ان نعرف السبب المزاجى والفاعلى الذى تدورت به الجوزا (الجوزة) وتطاولت به اللوزة وحلى به لحم الاتروجه (الاترجه) وحمض به شحمها [٤] بل نعرف الصوره من جهه المشاهده والافعال بالتجربه وكما لا نقدر ان نمزج من العناصر ما نتخذ منه اتروجا (اترجا) ولا رمانا كذلك لا نقدر على ان نمزج منها ذهبًا ولا فضه ومعرفه المعرفه والجهل معرفه F
Cf. Abū al-Barakāt, *Muʿtabar* II.230.2-18
[١]: = *But*.
[٢]: ≈ *Muʿt*. 230.2-4 ‖ لو وجدنا: اذا وجدوا *Muʿt*.; ܟܒܪܝܬܐ ... ܣܘܡܩܐ ܐܘ *But*. ‖ الاحمر او: الاحمر 230.1; النقى الابيض المصفى *Muʿt*. 231.2, sed الابيض المصفى: ܚܘܪܐ ܡܨܠܝܐ ܘܕܟܝܐ ܐܘ ܡܨܘܪܬܐ *But*. ‖ وقدرنا: وقدروا *Muʿt*.; ܘܐܫܟܚܢ *But*. ‖ عملنا: عملوا *Muʿt*.; ذهبا 230.2 ‖ وبالحقيقة: deest in *Muʿt*. 230.3, sed. فضة : ذهبا او فضة ‖ *But*. ܕܗܒܐ ܐܘ ܣܐܡܐ ܕ *Muʿt*.; ‖ تعزم: تعرض *Muʿt*., v.l. تعدم ‖ القوة الفعالة: القوه *Muʿt*. ‖ *But*. ܚܝܠܐ ܣܥܘܪܐ ܕ *Muʿt*.; حيث توجد: حيث يوجد *Muʿt*. ‖ عنه تُوجَد: عنه يوجد *Muʿt*. ‖
[٣]: ≈ *Muʿt*. 230.12-15 ‖ المتوسطة: والمتوسطة *Muʿt*. ‖ عما: كما *Muʿt*. ‖ تدورت به الجوزا, ‖ الجوزا: تدورت به النارنجة واحمرت *Muʿt*.; ܐܬܓܠܓܠ ܓܘܙܐ *But*. (= F marg.) ‖ وتطاولت به : وتطاولت به اللوزة وحلى به لحم الاتروجه وحمض به شحمها ‖ *sic* (cf. syr. ܓܘܙܐ) ‖ الاترجة واصفرت وحمضت به الرمانة وحلت *Muʿt*.; ܐܬܪܓܐ ܚܡܥ ܘܐܬܡܬܚ ܠܘܙܐ ܘܚܠܝ ܒܣܪܗ ܕ ܐܬܪܘܓܐ ܘܚܡܥ ܫܚܡܗ *But*. (= F marg.) ‖ الاتروجه, *sic* (cf. syr. ܐܬܪܘܓܐ) ‖
[٤]: = *Muʿt*. 230.16-18 ‖ اترجا (*sic*): اتروجا *Muʿt*. ‖

Min. III.iii.3: فهذه الفاظ تدل على اوهامٍ لا حقايق لها F
Min. III.iii.3.8 [ad ܐܬܪܘܓܐ]: ܐܬܪܘܓܐ M
Min. IV.tit.: فى احوال المسكونه F (cf. *Šifā*ʾ 24.3 فى أحوال المسكونة وأمزجة البلاد)
Min. IV.i.tit.: مباحث فى وضع (؟) الارض الطبيعى (؟) F
Min. IV.i.1.5f. [ad ܢܚܠܐ ܘܡܪܙܒܐ]: الاودية الصغار والسواقى (؟) F
Min. IV.i.1.5 [ad ܢܚܠܐ]: ܘܢܗܪܐ ܙܥܘܪܐ M
Min. IV.i.1.6 [ad ܡܪܙܒܐ]: ܐܠܟܪܘܪ ܘܐܠܡܙܐܒ M
Min. IV.i.2: اهل الطلسمات زعموا ان للفلك اقبالاً وادباراً غايه كل واحدٍ منهما ثمانيه اجزآ تتم فى ستمايه سنه واربعين سنة F
Min. IV.i.3.3 [ad ܡܘܐܙܝ]: موازى F
Min. IV.ii.tit.: مباحث فى امزجة البلدان F
Min. IV.ii.6.2 [ad ܦܘܩ]: فوق F
[Min. IV.iii.: deest titulus in marg. F]

Min. IV.iv.tit.: مباحث فى سبب الحروره والبروده فى الازمان والبلدان F

Min. IV.iv.1: س [؟] اعلم ان ليس المراد ان الشعاع شى نارى ينفصل عن الشمس ولا ان الشمس تقهر شيًا من الناريه وتنزل ولا انّ الشمس فى جوهرها حاره لوجوه كثيره فنقل منها الآن لانه لو كان احد الوجوه الثلثه لكان كلما كان البعد عن الارض اكثر حتى كان القرب اًلى الشمس اكثر فوجب ان تكون السخونة اكثر لكن التالى كاذب فان الهوا على قلل اجبال ابرد من الهوا المحيط بالارض فالمقدمات كلها كاذبه بل الشعاع شى يحدث فى المقابل للضو دفعةً اذا توسّط بينهما جسم شفاف ثم ان المستضى بلزمه ان يتسخن واذا ثبت هذا لم يكن للشعاع نزول حقيقى ولكن لما كان حدوثه عن شى عالٍ توهّم كانه ينزل واذا ثبت ذلك جهر انّ انعكاس الشعاع لا يكون امراً حقيقيًا بل وهميًا فلهذا التحقيق جَعَل الشيخ فيضان الشعاع عن الشمس وهميًا فثبت ان الشعاع ليس بجسم فهو عرض يحدث فى {--} المقابل (؟) {--} / {--} عن (؟) اجزا حارة (ناريه؟) تنزل من فوق / بل السخونه عرض (؟) تحدث فى المقابلة (؟) / {----} المضى الراويه (؟) والمتوسط بين {-----}عكاس (انعكاس؟) F
Cf. *Šifā*ʾ 27.16-28.1.

Min. IV.iv.1 or 4 (?): قال الشيخ فى القانون الصيف حار لقرب الشمس من سمت الروس وقوه الشعاع الفايض عنها الذى يتوهم انعكاسه على زوايا حاده جدًا واما ناكصًا على اعقابه فى الخطوط التى نفذ فيها فيكثف عندها الشعاع F
≈ IS, *Qānūn* I.2.2.1.3, ed. New Delhi (1982), 142.6-8 (corr. ed. Cairo [1987] 113.16-18): انعكاسه: add. فى الصيف إما editiones.

Min. IV.iv.1 or 4 (?): عر [؟] فاذا كانت حاده كان محل فعل المضى اقل فكان الفعل فيه اَقوَى واذا كانت قايمه او منفرجه كان محل الفعل اوسع فكان الفعل اضعف فانه ليس استضأة (استضاءة) البيت الصغير من (؟) السراج كاستضآءة الصحن الواسع منه F

Min. IV.iv.2.1 [ad ܚ̈ܫܘܫܘܬܐ]: الخشونات F
Min. IV.iv.2.2 [ad ܢܬ̈ܝܐ]: الناتيه F (الناتئة)
Min. IV.iv.2.2 [ad ܦܘ̈ܪܘ]: المسام F

Min. IV.iv.5: الموضع الذى يكون مسامتًا للشمس لا بد وان يكون متسخنًا ولكن ذلك الموضع له وسط ومحيط والوسط {--} امور ثلثه الاول مسامته الشمس المسخّنة الثانى كونه محاطًا بالمحيط المسخّن الثالث منع المحيط المسخّن تاثير المبرّد فالموضع الذى يقع عليه سهم الاسطوانه فى غايه السخونه فاما محيط ذلك الموضع فقد وجد فيه الوجه الاول فلم يوجد الوجهان الاخران فلذلك سخونه الاطراف ليست فىِ الافراط كسخونه الوسط بل اقلّ F
Cf. *Šifā*ʾ 28.3-10.

Min. V.tit.: فى امور البحر واحواله F
Min. V.i.tit.: مباحث فى وضع البحار F
Min. V.ii.tit.: مباحث فى ملوحة ما البحر F

Min. V.ii.1.12 [ad ܫܚܠܐ]: ترويق، تصفية F
The same definition of *šeḥlā* is found in the lexicon of Bar ʿAlī (see PS 4115).

Min. V.iii.tit.: مباحث فى تحويل البحر F
Min. V.iii.3.9 [ad ܡܕܐ ܘܡܓ̈ܠܐܐ]: الجزر والمد F
Min. V.iii.3.9-10: وسبب المد والجزر فى المشهور لطلوع القمر وغروبه اليومى والشهرى F

Mete. tit. فى الاثار العلوية F
Mete. I.i. tit.: مباحث فى كيفية تكون(؟) الغمام F
Mete. I.i.2.6 [ad ܟܘܣܘܣ]: جوبس بلده من اعمال مدينه ملطيه F

Mete. I.i.2.8 [ad ܟܦܬ]: انثنت وانحنت F
Mete. I.ii.tit.: مباحث فى المطر F
Mete. I.ii.6.3 [ad ܡܛܠܬܐ, ubi M praebet ܛܠܬܐ]: ܒܥܒܪ ܠܗܘ ܕܡܛܠ M
Mete. I.ii.6.3 [ad ܠܚܡܐ]: ܒܥܒܪܢܝܬܐ ܠܗܘ ܕܥܒܪܘܗܝ M
Mete. I.iii.tit.: مباحث فى الطل والصقيع والثلج والبرد والزوبعة والضباب F; ܥܠ ܐܘܡܢܘܬܐ ܕܥܠܝܬ ܡܛܪܐ ܕܡܢܩܘܡ M
Mete. I.iii.1.6 [ad ܡܛܪܐ]: ܐܠܝܡܣ . ܗ̄ . ܩܘܩܘ . ܛܪܐ L
Mete. I.iii.1.7 [ad ܓܠܝܕܐ]: صقيع F
Mete. I.iii.2: قطع ثلج نازله F
Mete. I.iii.2.4 [ad ܡܬܩܦܣܝܢ]: ܡܬܩܦܢܝܢ M
Mete. I.iii.2.4 [ad ܩܘܫܝܐ]: ܡܬܦܬܚܝܢ M
Mete. I.iii.2.8 [ad ܟܡܘܬܐ]: ܟܕ ܡܬܩܦܢ ܠܦܘܬ ܟܝܢܗ ܗܘ ܕܝܢ ܪܘܝܐ ܘܟܝܢܐ ܕܡܬܚܙܐ ܐܝܟ ܟܝܢܐ ܡܛܘܠ ܐܝܟ [et ܟܝܢܗ item arabice كنينه] M
Mete. I.iii.3.2 [ad ܡܬܢܦܩܝܢ]: ܡܬܢܦܩܝܢ M
Mete. I.iii.4.9 [ad ܒܡܛܪܐ ܕܢܚܬ ܒܐܪܥܐ ܐܝܠܝܢ ܕܡܬܢܥܝܢ?]: بلد {-}مه F
Mete. II.tit.: فى الامور الخياليه المشاهده فى الغمام F

Mete. II.i.tit. et II.i.1: مباحث فى تقدمه المعرفه فى السبب الفاعل للخيالات الغماميّه الهاله وقوس قزح والشمسيّات (الشميسات) والنيازك هى خيالات ومعنى الخيال هو ان يرى صوره الشى مع صورة شى مظهر له كالمراه فيظن ان الصوره حاصله فيه وليست حاصله فيه فى نفس الامر F (fol. 65r between columns to bottom left).

Mete. II.i.1: الخيال هو ان ترى صوره شىء مع صوره المراه ويظن ان تلك الصوره حاصله فى المراه مع انه لا يكون الامر كذلك F (fol. 65r, bet. columns, from middle to top).

Mete. II.i.1: [١] معنى الخيال هو ان يجد الحس شبح شى فى صوره شى اخر كما نجد صوره الانسان مع صورة المراه [٢] على معنى ان يؤدى الثانى كالمراه للاول كصوره الانسان الى البصر [٣] ولا يكون لذلك الاول انطباع حقيقى فى ماده ذلك الشى الثانى الذى يوديه ويرى معه F (fol. 65r, right margin, to bottom).
Corr. *Šifā*ʾ 40.6-8, but the middle part of the scholion agrees with *But*.
[١] = *Šifā*ʾ 40.6-7, except: فى : مع *Šifā*ʾ (ܥܡ *But.*)
[٢] not in *Šifā*ʾ; cf. *But*. II.i.1.2f.
[٣] = *Šifā*ʾ 40.7-8, except: ولا: ثم لا *Šifā*ʾ (ܘܠܐ ܟܕ *But.*) ‖ لذلك الاول : لتلك الصورة *Šifā*ʾ ‖ يوديه ويرى معه: يوديها ويرى معها *Šifā*ʾ.

Mete. II.i.1.2 [ad ܦܪܨܘܦܐ]: على وجه F
Mete. II.i.1.3 [ad ܡܫܬܠܡܢܘܬܐ]: التادى F

Mete. II.i.1.5-9: اما الحس فيدرك معًا المرآه وذلك الشى فيتخيل عنده انه مدرك صوره ذلك الشى فى المراه F
Cf. *Šifā*ʾ 40.15-16: المرآه prim.: الأملس الذى هو المرآة *Šifā*ʾ ‖ فيتخيل: فيخيل *Šifā*ʾ ‖ مدرك: يدرك *Šifā*ʾ.

Mete. II.i.2.1 [ad ܢܨܡܐ]: شبح F
Mete. II.i.2.1 [ad ܡܚܙܝܬܢܝܬܐ]: المراى F (lege المرئى)
Mete. II.i.2.1 [ad ܐܝܟ ܕܐܝܬܝܗܘܢ]: كما هو F
Mete. II.i.2.2 [ad ܡܬܛܠܠ]: illeg. F
Mete. II.i.2.2 [ad ܒܥܝܢܬܐ]: فى العين F
Mete. II.i.3.2 [ad ܡܫܦܝܐ]: مشف F
Mete. II.i.3.3 [ad ܒܥܝܢܬܐ ܡܬܢܨܡܐ]: تتشبح فى العين F

Mete. II.i.3.7 [ad ܢܬܚܙܐ]: نتخيل F
Mete. II.i.4.1 [ad ܐܬܦܢܝܘܬܐ]: الانعكاس المفرد F

Mete. II.i.6.3 [supra lin., ad ܟܕ ܢܦܠ . . . ܣܘܡܩܐ]:
F إذا وقع الضو على الغمام الاسود يرى احمر
ܟܕ ܢܦܠ ܢܘܗܪܐ ܥܠ ܥܢܢܐ ܐܘܟܡܬܐ ܡܬܚܙܝܐ ܣܘܡܩܐ
Cf. *Šifā'* 45.4-5, but: الغمام الاسود: السحابة السوداء *Šifā'* || يرى احمر: رؤيت حمراء *Šifā'* ||

Mete. II.ii.tit.: مباحث فى الهالة F
Mete. II.ii.1.1-2: الهالة وقوح قزح والشمسيّات (والشميسات) والنيازك من باب الخيالات ولا وجود لها فى نفس الامر F
Mete. II.ii.2: وسط الهالة يرى كالسواد المظلم لان الضو اللامع قوى فى الوسط فيعرض للبصر ان لا يحس بذلك الغيم الرقيق واذا لم يحس بذلك الغيم راى ذلك الموضع كالروزية (؟) النافذة الى القمر F
Mete. II.ii.3: واعلم ان المسامتة المعتبرة فى تخيل الهالة تخالف (؟) المسامتة المعتبرة فى تخيل القوس فان المسامتة المعتبرة فى تخيل الهالة مشروطة بكون الغمام متوسّطًا بين البصر وبين المبصور واما المسامتة المعتبرة فى تخيل القوس فهى مشروطة بكون البصر متوسّطًا بين المراة وبين المرأى (المرئى) F
Mete. II.ii.3.2 [ad ܡܘܠܐ]: المحور F

Mete. II.iii.tit.: مباحث فى القوس قزح F
Sic, with the article "al-".

Mete. II.iii.1.1 [ad ܡܢܘܬܐ ܡܢ ܡܝܐ]: اجزا مائية F
Mete. II.iii.1.1 [ad ܪܣܣܐ]: رشى F
Mete. II.iii.1.2 [ad ܡܢܗܪܢܐ ܘܫܦܝܐ]: شفافه وصافيه؟ F
Mete. II.iii.1.2 [ad ܡܬܓܢܒܬ]: وحصل وراء F

Mete. II.iii.1.: وكل واحد من تلك الاجزاء فى غايه الصغر فلا يودى الشكل بل يود (يؤدى) الضو ويكون ذلك مركبًا من لون المراه وضو الشمس F
= *Mabāḥiṯ* II.181.3-5, except: فى غايه الصغر: صغير *Mab.* || ويكون ذلك: ويكون ذلك اللون *Mab.* ||

Mete. II.iii.1.3 [ad . . . ܓܘܫܡܐ]: جسم كثيف اما جبل او سحاب مظلم F
Mete. II.iii.1.9 [ad ܓܠܝܐ]: ܠܗ̄ ܓܠܐ . ܠܥܠܬܐ ܣܘܡܩܐ 1
Mete. II.iii.2.3 [ad ܐܝܠܝܢ ܕܡܚܝܠܝܢ ܚܙܬܐ]: هذا كان ضعيف البصر F

Mete. II.iii.3: شكل القوس مستدير لان كل واحد من اجزا ذلك الهوا الرشى التى ينعكس عنها شعاع البصر الى الشمس وضعها هو بحيث اذا جَعَلنا الشمس مركز دائره لكان القدر الذى يقع مِن تلك الدائره فوق الارض يمرّ على تلك الاجزاء F (67r top)
From اجزا ذلك الهوا ... = *Mabāḥiṯ* II.182.10-12, except: اجزا: الاجزا *Mab.* || ذلك الهوا الرشي: deest in *Mab.*, cf. *But.* ܪܣܣܢܝܬܐ || الى الشمس: deest in *Mab.*, cf. *But.* ܠܘܬ ܫܡܫܐ || وضعها هو: وقعت *Mab.*, cf. *But.* ܣܡܢ || اذا: لو انا *Mab.* || لكان: كان *Mab.* ||

Mete. II.iii.3: السبب فى استداره هذا القوس ان الاجزا التى ينعكس عنها شعاع البصر وقعت بحيث انا لو جعلنا الشمس مركز دائره كان القدر الذى يقع من تلك الدائره فوق الارض يمر على تلك الاجزا فان كان الشمس على الافق كان الخط المار بالناظر والنير على بسيط الافق وهو المحور فيكون حينيذ بسيط الافق يقسم المنطقه بنصفين ويرى القوس نصف دائره وكلما كان الارتفاع اكثر كان القوس اصغر F (67r, between columns)
= *Mabāḥiṯ* II.182.10-15, except: السبب في: البحث الخامس عن *Mab.* || ان: هو ان *Mab.* || انا لو: لو انا *Mab.* || بسيط sec.: سطح *Mab.* || وكلما ... اصغر: deest in *Mab.*, sed cf. *Mab.* 182.15-17.

Mete. II.iii.7.7 [ad ܣܘܡܩܐ]: الاصفر F; اصفر M
Mete. II.iv.: مباحث فى بنات الشمس وحراب F (67v, between columns)

Mete. II.iv.tit.-II.iv.1.: [١] فى الشمسيات (الشميسات) والنيازك وهى خيالات فاسبابها القابله (؟) ثلثه (ثلاثة) احدها انه يحصل بقرب الشمسَ غيم كثيف صقيل يقبل فى ذاته ضو الشمس [٢] ومع الضو يودى خيال الشمس ايضا [٣] لان المراه الكبيره (المرآة الكبيرة) كما تودى اللون فهى تودى الشكل ايضا [٤] وايضا البخار اللزج اذا تصاعد وتشكّل بشكل الاستداره كما فى الاجرام الرطبه وبلغ فى صعوده الى كره النار اشتعلت النار فيه وهو مستدير فتشكل بشكل الشمس وربما كانت تلك الماده كثيفه فبقيت اياماً وليالى وتحركت بالاستداره F (67v, left to bottom)
Cf. *Mabāḥiṯ* II.184.15-185.4
[١]: corr. *Mab.* 184.15-17: والنيازك: deest in *Mab.* (cf. *But.*) ‖ وهى خيالات: deest in *Mab.* (cf. *But.* ܫܘܡܠܝ ܥܠܬܐ ܕܡܬܩܒܠܢܘܬܐ ܐܝܬ) ‖ فاسبابها القابله ثلثه: ان لها اسبابا ثلاثة *Mab.* ‖ انه: ان *Mab.* ‖ كثيف صقيل: كثيف مندمج الاجزاء صقيل *Mab.* ‖ يقبل: فيقبل *Mab.*
[٢]: ≈ *But.* II.iv.1.3-5 (cf. *Mab.* 184.17-19).
[٣]: ≈ *Mab.* 184.19-20, except: تودى bis: يؤدى bis (*sic*) *Mab.* ed. Hyderabad ‖ فهى: deest in *Mab.* ‖
[٤]: corr. *Mab.* 184.21-185.4: وايضا: وثالثها ان *Mab.* ‖ كما فى الاجرام الرطبه: على ما هو طبيعة الاجسام الرطبه فى الهواء *Mab.*; ܐܝܟ ܗܝ ܕܒܓܘܫܡ̈ܐ ܪ̈ܛܝܒܐ ܕܒܐܐܪ *But.* ‖ تلك الماده: فلا جرم يكون شكله شكل *Mab.* ‖ فتشكل بشكل: مستدير الشكل *Mab.* ‖ مستدير: بل شهورا وربما وصل الى addit: وليالى *Mab.* ‖ فبقيت: فتبقى *Mab.* (deest تلك) ‖ الماده *Mab.* ‖ وتحركت بالاستداره: وتحركت بالاستدارة فهو ايضا تحرك على الاستدارة الموضع الذى يتحرك عليه الفلك *Mab.*

Mete. III.tit.: فى الرياح F
Mete. III.i.tit.: مباحث فى تولد الرياح F
Mete. III.i.1.5 [ad ܫܘܬܐ]: طبقة F
Mete. III.i.3.9 [ad ܡ̈ܝܐ ܕܓܘܒܐ]: ما الصهريج F; ܠܐ ܐܝܬܘܗܝ ܒܥܒܪܝܐ 1
Mete. III.i.4.7 [ad ܬܢܢܐ]: الدخان F
Mete. III.ii.tit.: مباحث فى تحديد الجهات التى منها يصدُر مهبّ الرياح F
Mete. III.ii.2.5 [ad ܦܪܐܠܠܐ]: متوازيه F
Mete. III.iii.tit.: مباحث فى اختلاف الرياح باختلاف البلدان والازمان F
Mete. III.iii.4.: لان المشرق حار بالطبع لطول مسامتة الشمس اياه فى وقت ظهورها وكلما تجى فى الصعود فهى كلما لعه (طالعه؟) عليهم والمغرب بالطبع بارد لان الشمس عند وصولها اليهم تبتدى فى الانصراف والغروب فتكون الحراره ضعيفةً ففى الصيف يجتمع ان طبيعه المشرق حاره وكون الشمس فى المشرق الشماليه فتتضاعف الحراره فتحلل ماده الرياح الشرقيه فلاجل ذلك تهب الغربيه اكثر كون ان (الكون؟) الحراره المحلله لمادتها ضعيفه F
Mete. III.iv.4.3 [ad ܡܬܩܡܛ]: يتجعد F
Mete. IV.tit.: {...} عن (؟) فيما يصدر F (illeg.)
Mete. IV.i.tit.: مباحث فى الرعد والبرق F

Mete. IV.ii.tit.: مباحث فى الصاعقة وهى نوعان محرقه ولا محرقه وهى البيضا والمحرقه فانها مع احراقها تسوّد F
Cf. *But.* Mete. IV.ii.2.

Mete. IV.iii.tit.: مباحث فى الانوار المشاهده فى الليل والنازلات والمنقضه وذوات الاذناب F
الانوار المشاهده فى الليل: cf. *Mabāḥiṯ* II.188.14 الانوار التى تشاهد بالليل فى بعض المواضع.

Mete. IV.iii.3.7-10: خلف الذنب المذكور ظهر كوكب في الصيف قمريه في سنة اربع وستين وستمئه الشمس في المغرب عند العشيات وكان عظيمًا ثم بقى اياما وكان مُسامتًا لبرج الاسد وعند (؟) / كما نزلت الشمس الى الاسد اختفى وغاب كما تغيب الكواكب وبعد عده ايامٍ لمّا بعدت الشمس عن مُسامته ظهر في المشرق على هيته (هيئته) وصورته ثم لم يزل ينبسط حتى صار مثل الرماح الطوال وبعد ذلك تحلل يسير يسير الى ان فنى/ شاهدناه كما قد ذكر مصنّف هذا الكتاب يغمدهُ اللهُ برحمته F

The words "*šāhadnāhu* ..." at the end indicates that the note stems from a near contemporary of Barhebraeus who actually saw the comet in 1264 A.D. The date 664 A.H., however, is wrong, since it was in the period Ramaḍān-Ḏū al-ḥiǧǧa 662 A.H. that the comet in question was visible.
"*miṯl al-rimāḥ al-ṭiwāl*": The description of the same comet as a "long spear" also occurs in Maqrīzī (*K. al-sulūk* [Ziyāda-ᶜĀshūr] I.516.15, there in singular, *naḥw rumḥ ṭawīl*).

IV.iii.4.10-11: وهذه الاثار كلها تدل على الرياح وقله الامطار وفساد الجو ويبسه والامراض الحاده القاتله F
= *Šifā*ᵓ 74.14-15, except: وفساد: وعلى فساد *Šifā*ᵓ ‖ والامراض: وعلى الامراض *Šifā*ᵓ ‖ الحاده: الحارة, et add. اليابسة *Šifā*ᵓ (sed الحادة cod. S).

V.tit.: في الامور العظام التى تحدث في العالم F
*Šifā*ᵓ 75.3: في الحوادث الكبار التى تحدث في العالم

V.i.tit.: مباحث في الطوفانات F
V.i.1.5 [ad ܦܝ̈ܬܝܘ]: قرابات F
V.i.1.8 [ad ܝܩܕܢܐ]: اشتعال احراق F
V.i.1.9 [ad ܡܫܒܐ ܥܠܥܠܐ]: هبوب عاصف F
V.i.1.10 [ad ܐܫܝܕܘܬܐ]: انتشار, انصباب F
V.ii.tit.: مباحث في ان يتولد الشجر والحيوان من غير لقاح ولا نكاح F
V.ii.1.5 [ad ܙܒܠܐ ܕܬܘܪ̈ܬܐ]: اخثا البقر F
V.ii.1.6 [ad ܐܘܩܝܡܘܢ]: الباذروج F
V.iii.tit.: مباحث في ان الصناعات واللغات والكتابات تتجدد F

BIBLIOGRAPHY

1. *Ancient and Medieval Authors and Works*

ms.: manuscript(s) used; **ed**.: edition(s) used; **tr**.: modern translation(s) used (often with commentaries) - [A] Arabic; [E] English; [F] French; [G] German; [I] Italian; [L] Latin; [R] Russian; [S] Spanish
For further bibliographical data, see under the names of editors/translators (in case of anonymous editions, under the names of institutions responsible or places of publication) in the second part of the bibliography.
Where more than one edition/translation of the same work is named, the edition/translation underlined is the one normally used for the purpose of quotations and citations.

ABŪ AL-BARAKĀT Hibat Allāh b. ʿAlī b. Malkā al-Baghdādī, *Muʿtabar* = *K. al-muʿtabar fī al-ḥikma*: ed. Yaltkaya (1357-8 h.).

ABŪ AL-FIDĀʾ, ʿImād al-Dīn Ismāʿīl b. ʿAlī, *K. taqwīm al-buldān*: ed. Reinaud-de Slane (1840).

ALEXANDER of APHRODISIAS (Alex. Aphr.), *In Aristotelis Meteorologicorum libros commentaria*: ed. Hayduck (1899).

Pseudo-APOLLONIUS of TYANA [= Balīnūs], *K. sirr al-ḫalīqa* = *K. al-ʿilal*: ed. Weisser (1979).

ARISTOTLE (Arist.), *Mete*. = *Meteorologica*: ed. Fobes (1919); ed./tr. [E] Lee (1952); ed./tr. [F] Louis (1982); tr. [G] Strohm (1970).

-, *Meteorologica* (arab.): see Ibn al-Biṭrīq.

-, *Meteorologica* (lat.): ed. Schoonheim (1999).

-, *Meteorologica* (heb.): see Ibn Tibbon.

Pseudo-ARISTOTLE (Ps.-Arist.), *De mundo*: ed./tr. [E] Furley (1955); tr. [G] Strohm (1970).

-, *De mundo* (syr.): ed. de Lagarde (1858) 134-158; partial tr. [G] Ryssel (1880).

-, *De mundo* (arab.): ed. Brafman (Diss. 1985).

-, *De plantis* (gr.): see Nicolaus Damascenus.

-, *Kitāb al-aḥǧār*: ed./tr. [G] Ruska (1912).

[BA = Bar ʿAlī]

BAHMANYĀR b. al-Marzbān, *K. al-taḥṣīl*: ed. Muṭahharī (1970).

BAR ʿALĪ, Īšōʿ [= BA], *Lexicon*: partial ed. Hoffmann (1874).

BAR BAHLŪL, Abū al-Ḥasan [= BB], *Lexicon syriacum* [*Puššāqā d-ḥašḥāṯā suryāyāṯā w-yawnāyāṯā*]: ed. Duval (1901).

-, *K. al-dalāʾil*: ed. Ḥabbī (1987); ed. of excerpts from Theoph. *Mete*.: Daiber (1992) 201-215.

BARHEBRAEUS [= BH]

-, *Aequ*. = *De aequilitteris*/*De vocibus aequivocis* [*Mēmrā ʿal dāmyāyāṯā*]: ed. Martin (1872) II.77-126.

-, *Asc*. = *Ascensus mentis* [*Sullāqā hawnānāyā*]: ed./tr. [F] Nau (1899).

-, *But*. = *Butyrum sapientiae* [*Ḥêwaṯ ḥekmṯā*]: manuscripts: see Introduction.

-, *But*. Poetica: ed. Margoliouth (1887) 114-139.

-, *But*. De plantis: ed./tr. [E] Drossaart Lulofs-Poortman (1989) 68-112.

-, *But*. De anima: summary in Furlani (1931).

-, *But*. Oeconomica: ed./tr. [I] Zonta (1992).

-, *But*. Oeconomica, cap. III. (= Physiognomica): ed./tr. [G] Furlani (1929) 3-16.

-, *Cand.* = *Candelabrum sanctuarii* [*Mnāraṯ qudšē*]: ed. Çiçek (1997), and separate editions; tr. [A] Jijāwī (1996).
-, *Cand.* Bases I-II: ed./tr. [F] Bakoš (1930-3).
N.B. The references to this work without specification of the "Base" are to Bakoš's edition of Bases I-II.
-, *Carmina* [*K. d-mušḥāṯā*]: ed. Scebabi (1877); ed. Dolabani (1929); ms. Bodl. Hunt. 1, p. 238-253; ms. Laur. or. 298, 84r-104v.
-, *Chron.* = *Chronicon* (civile) [*Maḵtḇānūṯ zaḇnē*]: ed. Bruns-Kirsch (1789); Bedjan (1890); tr. [E] Budge (1932).
-, *Chron. eccl.* = *Chronicon ecclesiasticum*: ed./tr. [L] Abbeloos-Lamy (1872-7).
-, *Columb.* = *L. columbae* [*K. d-yawnā*]: ed. Bedjan (1898a) 519-599; ed. Cardahi (1898); tr. [E] Wensinck (1919).
-, *Denḥā* = *Letter to Catholicus Denḥā I*: ed./tr. [F] Chabot (1898).
-, *Eth.* = *Ethicon* [*K. d-īṯīqōn*]: ed. Bedjan (1898a); ed. Çiçek (1985); partial ed./tr. [E] Teule (1993) (Book I).
-, *Fabulae* = *K. d-ṯunnāyē mḡaḥḥkānē*: ed./tr. [E] Budge (1897).
-, *Ghāfiqī* = *Muntaḫab kitāb ǧāmiᶜ al-mufradāt li-Aḥmad b. Muḥammad b. Ḫalīd al-Ġāfiqī*: partial ed./tr. [E] Meyerhoff-Sobhy (1932-40) [letters *alif* to *waw*].
-, *Gramm.* = *Grammatica parva* [*K. da-ḡrammaṭīqī ba-mšuḥtā d-mār Aprēm*]: ed./tr. [L] Bertheau (1843); ed. Martin (1872) II.
-, *Hist. dyn.* = *Historia dynastiarum* [*Muḫtaṣar taʾrīḫ al-duwal*]: ed. Ṣālḥānī (1958).
-, *Horr.* = *Horreum mysteriorum* [*Awṣar rāzē*]: separate editions and ms. Bodl. Hunt. 1, p. 2-147.
-, *Ind.* = *L. indicationum et prognosticorum* [*K. d-remzē wa-mᶜīrānwāṯā*]: ms. Laur. or. 86.
-, *Muḫt. nafs* = *Muḫtaṣar fī ᶜilm al-nafs al-insānīya*: ed. Sbath (1928); Barṣaum (1938).
-, *Nom.* = *Nomocanon* [*K. d-huddāyē*]: ed. Bedjan (1898b); ed. Çiçek (1986); tr. [L] J.A. Assemani (1838).
-, *Pupill.* = *Liber pupillarum* [*K. d-ḇāḇāṯā*]: ed./tr. [G] Steyer (1908); ed./tr. [E] Janssens (1930-36).
-, *Rad.* = *Liber radiorum* [*K. d-zalgē*]: ed. Istanbul (1997); ms. Bodl. Or. 467.
-, *Serm. sap.* = *Sermo sapientiae* [*Swāḏ sōpiya*]: ed./tr. [F] Janssens (1937).
-, *Serm. sap.* (arab.): ed. Barṣaum (1940); partial ed. Platti (1988).
-, *Splend.* = *Liber splendorum* [*K. d-ṣemḥē*]: ed. Martin (1872) I; Moberg (1922); tr. [G] Moberg (1907-13).
-, *Tract.* = *Tractatus tractatuum* [*Têḡraṯ têḡrāṯā*]: mss. Cantab. Add. 2003; Laur. or. 342; British Library, Or. 4080; Berol. Sachau 211.
BAR KEPHA, Moses, *Hex.* = *Hexaemeron* [*K. da-šṯaṯ yawmē*]: ms. Paris, syr. 241 [= P]; ms. Paris, syr. 311 [= Q]; tr. [G] Schlimme (1977).
BAR SHAKKO, Severus Jacob, *Dial.* = *L. dialogorum* [*K. d-ḏiyalōḡō*]: ms. Göttingen, or. 18c; partial ed./tr. [G] Ruska (1896), id. (1897).
-, *Thes.* = *Liber thesaurorum* [*K. d-sīmāṯā*]: mss. Paris, syr. 316 [= P]; British Library, Add. 7193 [= L]. - Part IV (on Creation): summary in Nau (1896).
al-BATTĀNĪ al-Ṣābī al-Ḥarrānī, Abū ᶜAbd Allāh Muḥammad b. Jābir, *al-Zīǧ al-ṣābī*: ed./tr. [L] Nallino (1899-1907).
[BB = Bar Bahlūl]
[BH = Barhebraeus]
al-BĪRŪNĪ, Abū al-Raiḥān Muḥammad b. Aḥmad, *Āṯār* = *K. al-āṯār al-bāqiya ᶜan al-qurūn al-ḫāliya*: ed. Sachau (1923); tr. [E] Sachau (1879).

-, *Ǧamāhir* = *K. al-ǧamāhir*: ed. Krenkow (1355 h.); tr. [R] Belenickij (1963).
-, *Tafhīm* = *K. al-tafhīm li-awā'il ṣinā'at al-tanǧīm* (arab.): ed./tr. [E] Wright (1934).
-, *Tafhīm* (pers.): ed. Humā'ī (1316 h.).
-, *Taḥdīd* = *K. taḥdīd nihāyāt al-amākin li-taṣḥīḥ masāfāt al-masākin*: ed. Bulgakov (1962); tr. [E] Ali (1967); comm. Kennedy (1973).
al-BIṬRŪJĪ, Nūr al-Dīn b. Isḥāq [Alpetragius], *K. fī al-hai'a* (arab. & heb. [tr. Moshe b. Tibbon]): ed./tr. [E] Goldstein (1971).
CAUSA CAUSARUM [*K. d-ᶜellat kul ᶜellān* = *K. d-ᶜal īdaᶜtā da-šrārā*]: tr. [G] Kayser (1893); ms. Laur. or. 298 (olim palat. or. 62 Assemani).
CHRONICON miscellaneum ad annm Domini 724 pertinens: in ed. Brooks (1904) 77-155.
CHRONICON ad annum Christi 1234 pertinens, Part I: ed. Chabot (1920); tr. Chabot (1937).
al-DIMASHQĪ, Shams al-Dīn Abū ᶜAbd Allāh Muḥammad, *K. nuḫbat al-dahr*: ed. Mehren (1866); tr. [F] Mehren (1874).
DIOSCORIDES, *De materia medica*: ed. Wellmann (1914).
FAKHR AL-DĪN Muḥammad b. ᶜUmar AL-RĀZĪ, *Mabāḥiṯ* = *K. al-mabāḥiṯ al-mašriqīya fī ᶜilm al-ilāhīyāt wa-l-ṭabīᶜīyāt*, ed. Hyderabad (1343 h.).
-, *Šarḥ ᶜuyūn al-ḥikma*: ed. al-Saqā (1400 h.).
GALEN: ed. Kühn (1821-33).
-, *De simplicibus* (syr.): ed. Merx (1885).
al-GHAZĀLĪ, Abū Ḥāmid Muḥammad b. Muḥammad, *Maqāṣid* = *Maqāṣid al-falāsifa*: ed. Dunyā (1961); tr. [S] Alonso (1963).
GRIGOR of AKNER: *History of the Nation of the Archers* [*Patmut'iwn azgin netolac'*]: ed./tr. [E] Blake-Frye (1949).
HIPPOCRATES, Opera omnia: ed. Littré (1839-61).
ḤUDŪD al-ᶜĀLAM: tr. [E] Minorsky (1937).
ḤUNAIN b. ISḤĀQ al-ᶜIbādī, Abū Zaid, *Comp.* = Compendium of Arist. *Mete.* [*Ǧawāmiᶜ li-kitāb Arisṭūṭālīs fī al-āṯār al-ᶜulwīya*]: ed./tr. [G] Daiber (1975).
-, *al-ᶜAšr maqālāt fī al-ᶜain*: ed./tr. [E] Meyerhoff (1928).
-, *al-Masā'il fī al-ᶜain*: ed./tr. [F] Sbath & Meyerhoff (1938).
IBN al-BIṬRĪQ, Abū Zakarīyā' Yaḥyā, Arabic version of Arist. *Mete.* [*K. Arisṭūṭālīs fī al-āṯār al-ᶜulwīya*]: ed. Petraitis (1967); Schoonheim (1999); partial tr. [I] Baffioni (1980).
IBN al-FAQĪH al-Hamadānī, Abū Bakr Aḥmad b. Muḥammad, *K. al-buldān*: ed. de Goeje (1885).
IBN al-KHAMMĀR, Abū al-Khair al-Ḥasan b. Suwār b. Bābā, *Mutaḫayyila* = *Maqāla fī al-āṯār al-mutaḫayyila fī al-ǧaww*: ed./tr. [E] Lettinck (1999) 313-379.
-, Arabic version of Theoph. *Mete.*: ed./tr. [E] Daiber (1992) 228-245.
IBN KHURDĀDHBIH, Abū al-Qāsim ᶜUbaid Allāh b. ᶜAbd Allāh, *K. al-masālik wa-l-mamālik*: ed. de Goeje (1889).
IBN RUSTA, Abū ᶜAlī Aḥmad b. ᶜUmar, *K. al-aᶜlāq al-nafīsa*, book VII: ed. de Goeje (1891); tr. [F] Wiet (1955).
IBN SĪNĀ, Abū ᶜAlī al-Ḥusain b. ᶜAbd Allāh [= IS], *Dāniš-nāma-i ᶜalā'ī*, Ṭabīᶜīyāt: ed. Mishkāt (1951/2); tr. [F] Achena-Massé (1958).
-, *K. al-išārāt wa-l-tanbīhāt*: ed. Dunyā (1947), id. (1957-60); tr. [F] Goichon (1951).
-, *K. al-naǧāt*: ed. Dānishpazhūh (1986).
-, *Šifā'* = *K. al-šifā'*
N.B. References without further qualification are to the *fann* on mineralogy and meteorology (Ṭabīᶜīyāt, *fann* V).

-, *Šifāʾ*, Mantiq, *fann* I: ed. Madkour et al. (1952).
-, *Šifāʾ*, Ṭabīᶜīyāt, *fann* I: ed. Zayed (1983).
-, *Šifāʾ*, Ṭabīᶜīyāt, *fann* II-IV: ed. Qassem (1969).
-, *Šifāʾ*, Ṭabīᶜīyāt, *fann* V: ed. Montasir et al. (1964); partial ed./tr. [E] Holmyard-Mandeville (1927) [V/1.i.-iii., v.]; partial tr. [I] Baffioni (1980) [V/1.i.-iii., v.]; partial tr. [E] Sersen (1976) 195-212 [V/2.i.].
-, *Šifāʾ*, Ṭabīᶜīyāt, *fann* VI: ed. Anawati-Zayed (1974).
-, *Šifāʾ*, Riyāḍīyāt, *fann* IV: ed. Madwar-Ahmad (1980).
-, *Šifāʾ*, Ilāhīyāt, *fann* I: ed. Anawati-Zayed (1960).
-, *Qānūn* = *K. al-qānūn fī al-ṭibb*: ed. al-Qassh (Cairo 1987); ed. Būlāq (1294 h.) [rep. Beirut: Dār Ṣādir, sine anno]; partial ed. IHMMR (New Delhi 1982) [Bk. I.]; partial tr. [G] Hischberg-Lippert (1902) [III.3.1-3].
-, *ᶜUyūn* = *K. ᶜuyūn al-ḥikma*: ed. Ülken (1953) I.2-55; partial ed. Istanbul (1298 h.) 2-25 [Ṭabīᶜīyāt].
IBN TIBBON, Samuel, *Otot ha-Shamayim* (= Hebrew version of Arist. *Mete.*): ed./tr. [E] Fontaine (1995).
IBN ṬUFAIL, Abū Bakr Muḥammad b. ᶜAbd al-Malik, *Ḥayy b. Yaqẓān*: ed./tr. [F] Gauthier (1936).
IBN YŪNUS, Abū al-Ḥasan ᶜAlī b. Abd al-Raḥmān, *K. al-zīǧ al-kabīr al-ḥākimī*: partial ed./tr. [F] Caussin (1804).
IKHWĀN AL-ṢAFĀʾ, *Rasāʾil Iḫwān al-ṣafāʾ wa-ḫillān al-wafāʾ*: ed. Zirikli (1928); ed. Bustānī (1957); partial tr. [G] Dieterici (1876).
al-ᶜIRĀQĪ [al-Sīmāwī], Abū al-Qāsim Muḥammad b. Aḥmad, *K. al-ᶜilm al-muktasab fī zirāᶜat al-ḏahab*: ed./tr. [E] Holmyard (1923).
[IS = Ibn Sina]
ISHOᶜDAD of MERV, *Commentaries on the Gospels*: ed./tr. [E] Gibson (1911).
JĀBIR b. ḤAYYĀN: see Kraus (1942-43).
- , *K. al-īḍāḥ*: ed. Holmyard (1928) 51-58.
JACOB of EDESSA, *Hex.* = *Hexaemeron* [*K. da-šṯaṯ yawmē*]: ed. Chabot (1928); tr. [L] Vaschalde (1932); partial ed./tr. [L] Hjelt (1892).
JOB of EDESSA, *L. thesaurorum* [*K. d-sīmāṯā*]: ed./tr. [E] Mingana (1935).
JUWAINĪ, ᶜAlāʾ al-Dīn Aṭā Malik b. Muḥammad, *Tārīḫ-i ǧahāngušā*: ed. Qazwini (1912-1937); tr. [E] Boyle (1958).
al-KHARAQĪ, Abū Bakr Muḥammad b. Aḥmad, *Muntahā al-idrāk fī taqsīm al-aflāk*: excerpt in Nallino (1903-7) I.169-75.
al-KHWĀRIZMĪ, Abū ᶜAbd Allāh Muḥammad b. Aḥmad, *Mafātīḥ* = *K. mafātīḥ al-ᶜulūm*: ed. van Vloten (1895).
KIRAKOS of Ganjak, *History of the Armenians* [*Patmut'iwn hayoc'*]: tr. [E] Bedrosian (1975); partial tr. [F] Dulaurier (1858).
Ps.-MAJRĪṬĪ, *Picatrix* = *K. ġāyat al-ḥakīm wa-aḥaqq al-natīǧatain bi-l-taqdīm*: ed. Ritter (1933); tr. [G] Ritter-Plessner (1962).
al-MAQRĪZĪ, Taqī al-Dīn Aḥmad b. ᶜAlī, *K. al-sulūk li-maᶜrifat duwal al-mulūk*: ed. Ziyāda-ᶜĀshūr (1934-72); partial tr. [F] Quatremère (1838-42).
al-MASᶜŪDĪ, Abū al-Ḥasan ᶜAlī b. al-Ḥusain, *K. al-tanbīh wa-l-išrāf*: ed. de Goeje (1894); tr. Carra de Vaux (1896).
MEDICINES, THE SYRIAC BOOK OF [*K. d-sammānē*]: ed./tr. [E] Budge (1913).
MICHAEL I, Patriarch, *Chron.* = *Chronicon* [*Maḵtḇānūṯ zaḇnē*]: ed./tr. [F] Chabot (1899-1910).
NICOLAUS DAMASCENUS, *Compendium of Aristotelian Philosophy* (syr.): ms. Cantab. Gg 2.14; partial ed./tr. [E] Drosaart Lulofs (1965) [Bk. I-V]; excerpts: ms. Paris, syr. 346.

-, *De plantis* (syr., arab., heb., lat., gr.): ed. Drossaart Lulofs-Poortman (1989) [with tr. [E] of Arab. and Heb. versions].

[Nic. syr. = Syriac version of Nicolaus Damascenus, *Compendium of Aristotelian Philosophy*]

OLYMPIODORUS (Olymp.), *In Aristotelis Meteora commentaria*: ed. Stüve (1900).

Olymp. arab. = so-called Arabic version of Olympiodorus, *In Arist. Mete. comm.* [*Tafsīr Ulimfidūrūs li-kitāb Arisṭāṭālīs fī al-āṯār al-ᶜulwīya*, translated by Ḥunain b. Isḥāq and revised by Isḥāq b. Ḥunain]: ed. Badawī (1971) 83-192; ms. Tashkent 2385, fol. 347a-368a; partial tr. [I] Baffioni (1980).

PHILOPONUS, John, *In Aristotelis Meteorologicorum librum primum commentaria*: ed. Hayduck (1901).

POSIDONIUS, Fragments: ed. Edelstein-Kidd (1989); comm. Kidd (1988).

PTOLEMY, Claudius, *Alm.* = *Almagest* [*Μαθηματικὴ σύνταξις*]: ed. Heiberg (1898/1903); tr. [E] Toomer (1984).

-, *Geog.* = *Γεωγραφικὴ ὑφήγησις*: ed. Nobbe (1843-5).

al-QAZWĪNĪ, Zakarīyāᵓ b. Muḥammad b. Maḥmūd, *ᶜAǧāᵓib* = *K. ᶜaǧāᵓib al-maḫlūqāt*: ed. Wüstenfeld (1849); partial tr. [G] Ethé (1868), [G] Giese (1986).

-, *Āṯār* = *K. āṯār al-bilād*: ed. Wüstenfeld (1848).

[Ibn] al-QIFṬĪ, ᶜAlī b. Yūsuf, *Taᵓrīḫ al-hukamāᵓ* [extract by al-Zauzanī]: ed. Lippert (1903).

al-RĀZĪ, Abū Bakr Muḥammad b. Zakarīyāᵓ, *K. al-asrār*: ed. Dānishpazhūh (1964); tr. [G] Ruska (1937); partial tr. [E] Stapleton et al. (1927).

-, *K. al-madḫal al-taᶜlīmī*: tr. [E] Stapleton et al. (1927).

SĀᶜID b. Aḥmad al-ANDALUSĪ, *Ṭabaqāt* = *K. ṭabaqāt al-umam*: ed. Cheikho (1912); tr. [F] Blachère (1935).

SEVERUS SEBOKHT, *Constell.* = *De constellationibus* [*Mêmrā ᶜal demwāṯā d-mawzaltā, M. ᶜal demwāṯā hālēn d-meṯamrān d-meṯhazyān ba-šmayyā*]: partial ed. Sachau (1870) 127-134; tr. [F] and partial ed. Nau (1930-31).

STRABO: ed./tr. [E] Jones (1917-1932).

STRATO of Lampsacus: ed. Wehrli (1950).

al-TAMĪMĪ, Abū ᶜAbd Allāh Muḥammad b. Aḥmad, *K. al-muršid*: partial ed./tr. [G] Schönfeld (1976) [Chap. 14].

THEO of Alexandria, *Small Comm.* = *Small Commentary on Ptolemy's Handy Tables* [*Εἰς τοὺς προχείρους κανόνας*]: ed./tr. [F] Tihon (1978).

THEOPHRASTUS (Theoph.), Opera omnia: ed./tr. [L] Wimmer (1866).

-, Frag. = Fragments: ed./tr. [E] Fortenbaugh et al. (1993).

-, *De lap.* = *De lapidibus*: ed./tr. [L] Wimmer; ed./tr. [E] Eichholz.

-, *Metarsiologica* (*Mete.*), Fragments of Syriac version preserved in ms. Cambridge Gg 2.14: ed. Wagner-Steinmetz (1964); Daiber (1992) 176-188.

- , *Mete.* (arab.): see Ibn al-Khammār and Bar Bahlul.

al-ṬŪSĪ, Naṣīr al-Dīn Muḥammad b. Muḥammad, *Aḫlāq-i nāṣirī*: ed. Mīnowī-Ḥaidarī (1985).

-, *Risāla-i muᶜīnīya* [= *R.-i haiᵓa*]: facsimile ed. Dānishpazhūh (1335 h.š.).

-, *Taḏkira* = *al-Taḏkira fī ᶜilm al-haiᵓa* [= *al-Taḏkira al-naṣīrīya*]: ed./tr. [E] Ragep (1993).

VITRUVIUS Pollio, *De architectura*, Bk. VIII: ed./tr. [F] Callebat (1973).

al-YAᶜQŪBĪ, Aḥmad b. Abī Yaᶜqūb, *Taᵓrīḫ*: ed. Houtsma (1883); partial tr. [G] Klamroth (1888).

YĀQŪT b. ᶜAbd Allāh al-Rūmī al-Ḥamawī, *K. muᶜǧam al-buldān*: ed. Wüstenfeld (1866-1873).

2. *Modern Authors (and Institutions)*

ABBELOOS, J.B., & T. J. LAMY (1872-77, ed./tr.): *Gregorii Barhebraei Chronicon ecclesiasticum*, 2 parts (3 vols.), Louvain.

ABDOLLAHY, R. (1990): "Calendars, ii. Islamic Period", EIr IV.668-674.

ABRAMOWSKI, R. (1940): *Dionysius von Tellmahre, jakobitischer Patriarch von 818-845, zur Geschichte der Kirche under dem Islam* (AKM XXV/2), Leipzig.

ACHENA, M., & H. MASSÉ (1955-58, tr.): *Avicenne. Le livre de science*, 2 vols., Paris.

ALI, J. (1967, tr.): *The Determination of the Coordinates of Positions for the Correction of Distances between Cities, a translation from the Arabic of Al-Bīrūnī's Kitāb Taḥdīd Nihāyāt al-Amākin Litaṣḥīḥ Masāfāt al-Masākin*, Beirut.

ALONSO Alonso, M. (1963, tr.): *Algazel. Maqāṣid al-falāsifa o Intenciones de los filósofos*, Barcelona.

ANAWATI, G.C. (1996): "Arabic Alchemy", EHAS 853-885.

-, & Saᶜid ZAYED (1960, ed.): *Ibn Sīnā. Al-Shifā'. Al-Ilāhiyyāt (1) (La Métaphysique)*, Cairo.

ARMALET, Issac [Isḥāq Armala] (1937): *Al-Ṭurfa fī maḫṭūṭāt Dair al-Šarfa* [*Catalogue des manuscrits de Charfet*], Jounieh.

-, (1996) [edited by Bihnān Hindū]: *Ta'rīḫ al-kanīsa al-suryānīya* (Silsilat ta'rīḫ al-suryān 2), Jounieh.

ASSEMANI [-nus], Joseph Aloysius (1838, tr.): "Ecclesiae Antiochenae Syrorum Nomocanon a Gregorio Abulpharagio Bar-Hebraeo syriace compositus et a Iosepho Aloysio Assemano in Latinam linguam conversus", in A. Mai (ed.), *Scriptorum veterum nova collectio*, vol. X, Rome, part 2, p. 2-268.

ASSEMANI, Joseph Simonius, *BOCV = Bibliotheca Orientalis Clementino-Vaticana*, 3 vols. (4 parts), Rome, 1719-28. Rep. Hildesheim-New York 1975.

ASSEMANI, Stephanus Evodius (1742): *Bibliothecae Mediceae Laurentianae et Palatinae codicum MMS. orientalium catalogus*, Florence.

-, & Joseph Simonius ASSEMANI (1758-59): *Bibliothecae Apostolicae Vaticanae codicum manuscriptorum catalogus*, part I, vols. II-III, Rome. Rep. Paris 1926.

ASSFALG, J. (1963): *Syrische Handschriften* (VoHD 5), Wiesbaden.

AUDO, Thomas (1897): *Sīmtā d-leššānā suryāyā* [*Dictionnaire de la langue chaldéene*], Mosul. Rep. Glane/Losser 1985.

ᶜAWWĀD, Gurgis (1978): "Al-Turāṯ al-suryānī al-manqūl fī al-ᶜuṣūr al-ḥadīṯa ilā al-luġa al-ᶜarabīya", JSyrA IV.65-95.

BADAWĪ, ᶜAbdurraḥmān (1971, ed.): *Commentaires sur Aristote perdus en grec et autres épîtres*, Beirut.

BAFFIONI, C. (1980): *La tradizione araba del IV libro dei 'Meteorologica' di Aristotele* (AION.S 23), Naples.

BAKOŠ, J. (1930-3, ed./tr.): *Le Candélabre des sanctuaires de Grégoire Aboulfaradj dit Barhebraeus*, PO XXII (fasc. 4) 489-628; PO XXIV (fasc. 3) 295-439 [cited according to the continuous pagination given to the two fascicles, p. 1-285].

- (1948, ed./tr.): *Psychologie de Grégoire Aboulfaradj dit Barhebraeus d'après la huitième base de l'ouvrage le Candélabre des Sanctuaires*, Leiden.

BARNES, J. (1984, ed.): *The Complete Works of Aristotle* (The Revised Oxford Translation), Princeton.

BARṢAUM, Patr. Ignatius Ephrem I, *Lu'lu' = Al-Lu'lu' al-manṯūr fī tārīḫ al-ᶜulūm wa-l-ādāb al-suryānīya* [*Histoire des sciences et de la littérature syriaque*]. Rep. Glane/Losser 1987.

- (1938, ed.): *Risāla fī ᶜilm al-nafs al-insānīya li-l-ᶜallāma Mār Ġrīġūriyūs Ibn al-ᶜIbrī mafriyān al-mašriq*, Jerusalem.
- (1940, ed.): *Kitāb ḥadīṯ al-ḥikma li-l-ᶜallāma al-šahīr wa-ḥuǧǧat al-falāsifa al-ḫaṭīr Mār Ġrīġūriyūs Abī al-Faraǧ Ibn al-ᶜIbrī mafriyān al-mašriq al-suryānī* [*L'Entretien de la sagesse par Mar Gregorius Abulfarage Bar Hebraeus Maphrien (catholicos) syrien de l'Orient*], Homs.

BARSOM, Murad Saliba (tr.), & Athanasius Yeshue SAMUEL (ed.) (1991): *Anaphoras. The Book of the Divine Liturgies According to the Rite of the Syrian Orthodox Church of Antioch*, Lodi (N.J.).

BAUMSTARK, A., *GSL = Geschichte der syrischen Literatur mit Ausschluß der christlich-palästinensischen Texte*, Bonn 1922.
- (1894): *Lucubrationes syro-graecae* (JCPh.S 21/5), Leipzig.
- (1936): "Syrische Handschriften der Bibliothek der Erzbischöflichen Akademie in Paderborn", OrChr XXXIII (3. Ser. XI) 97-101.

BECK, E. (1955, ed.): *Des heiligen Ephraem des Syrers Hymnen de fide* (CSCO 154, syr. 73), Louvain.
- (1986): "Besrâ (sarx) und pagrâ (sōma) bei Ephräm der Syrer", OrChr LXX.1-22.

BEDJAN, P. (1890, ed.): *Gregorii Barhebraei Chronicon syriacum*, Paris.
- (1895, ed.): *Acta martyrum et sanctorum*, vol. V, Paris.
- (1898a, ed.): *Ethicon, seu Moralia Gregorii Barhebraei* and *Liber columbae seu Directorium monachorum Gregorii Barhebraei*, Paris-Leipzig.
- (1898b, ed.): *Nomocanon Gregorii Barhebraei*, Paris-Leipzig.

BEDROSIAN, R. (1975, tr.): "Kirakos Gandzakets'i's *History of the Armenians*", electronic edition accessed under www.virtualscape.com/rbedrosian.

BEHNAM, Ghrighūriyūs Būlus (1958): "Yanābīᶜ al-maᶜrifa ᶜinda Ibn Sīnā (975-1037)", MMIA XXXIII.213-237.
- (1975): "*Al-Fīziyāʾ wa-l-kīmiyāʾ fī al-muʾallafāt al-suryānīya*", JSyrA I.5-46.

BELENICKIJ, A.M., & G.G. LEMMLEJN (1963, tr.): *Abu-r-rajchan Muchammed ibn Achmed al-Biruni. Sobranie svedenij dlja poznanija dragocennostej (Mineralogija)*, Leningrad.

BERNSTEIN, G.H. (1836): *Lexicon syriacum Chrestomathiae kirschianae denuo editae accommmodatum*, Leipzig.
- (1858): "Gregorii Bar-Hebraei scholia in librum Iobi", appended to *Academia Ienensi secularia tertia celebranti gratulatur Academia Vratislaviensis*, Breslau.

BERTHEAU, E. (1843, ed./tr.): *Gregorii Bar Hebraei qui et Abulfaraǵ Grammatica linguae syriacae in metro ephraemeo*, Göttingen.

BERTHELOT, M., & P.R. DUVAL (1893): *La chimie au moyen âge*, Tome II. L'alchimie syriaque, Paris. Rep. Osnabrück-Amsterdam 1967.

BERTHELOT, M., & C.É. RUELLE (1888): *Collection des anciens alchimistes grecs*, Paris. Rep. Osnabrück 1967.

BLACHÈRE, R. (1935, tr.): *Ṣâᶜid al-Andalusî. Kitâb Ṭabaḳât al-Umam (Livre des Catégories des Nations)*, Paris.

BLAKE, R.P., & R.N. FRYE (1949, ed./tr.): "History of the Nation of the Archers (The Mongols) by Gregor of Akancᶜ, hitherto ascribed to Maγakᶜia the Monk", HJAS XII.269-399.

BOSWORTH, C.E. (1975): "The Early G͟haznavids", in *The Cambridge History of Iran*, vol. 4 (ed. R.N. Frye), Cambridge, p. 162-197.

BOYLE, J.A. (1958, tr.): *The History of the World-Conqueror by ᶜAla-ad-Din ᶜAta-Malik Juvaini. Translated from the text of Mirza Muhammad Qazvini*, Manchester.
- (1980): "Alexander and the Mongols", CAsJ XXIV.18-35.

BRAFMAN, D.A. (1985, ed.): "The Arabic 'De Mundo': an edition with translation and commentary", Diss. Duke University.

BRIGHTMAN, F.E. (1896): *Liturgies Eastern and Western*, vol. 1. Eastern Liturgies, Oxford.

BRIQUEL-CHATONNET, F. (1997): *Manuscrits syriaques de la Bibliothèque nationale de France (nos. 356-435, entrés depuis 1911), de la bibliothèque Méjanes d'Aix-en-Provence, de la biblothèque municipale de Lyon et de la Bibliothèque nationale et universitaire de Strasbourg*, Paris.

-, A. DESREUMAUX & J. THEKEPARAMPIL (1997): "Catalogue des manuscrits syriaques de la collection du Saint Ephrem Ecumenical Research Institute (Kottayam)", Muséon CX.383-446.

BROCK, S.P. (1980): "From Antagonism to Assimilation: Syriac Attitude to Greek Learning", in N. Garsoïan et al. (ed.), *East of Byzantium: Syria and Armenia in the Formative Period*, Washington D.C., p. 17-34. Rep. in Brock, *Syriac Perspectives on Late Antiquity*, London 1984, no. V.

- (1993): "The Syriac Commentary Tradition", in C. Burnett (ed.), *Glosses and Commentaries on Aristotelian Logical Texts. The Syriac, Arabic and Medieval Latin Traditions* (Warburg Institute Surveys and Texts XXIII), London, p. 3-18.

- (1995): "The Scribe Reaches Harbour", in S. Efthymiadis (ed.), *Bosporus: essays in honour of Cyril Mango* (ByF 21), Amsterdam, p.195-202. Rep. in Brock, *From Ephrem to Romanos. Interactions between Syriac and Greek in Late Antiquity*, Aldershot-Brookfield (VT) 1999., no. XVI.

BROCKELMANN, C., *GAL = Geschichte der arabischen Literatur*, 5 vols., Leiden 1937-1943.

- (1925): *Syrische Grammatik mit Paradigmen, Literatur, Chrestomathie und Glossar*, 4th ed., Berlin.

- (1928): *Lexicon syriacum*, 2nd ed., Halle.

BROOKS, E.W. (1904, ed.): *Chronica minora II*, Paris. Rep. Louvain 1955 (CSCO 3, syr. 3).

BRUNS, P.J. (1780): *De rebus gestis Richardi Angliae regis in Palaestina. Excerptum ex Gregorii Abulpharagii Chronico syriaco*, Oxford.

-, & G.W. KIRSCH (1789, ed./tr.): *Bar-Hebraei Chronicon syriacum*, 2 vols., Leipzig.

BUDGE, E.A.W. (1913, ed./tr.): *The Syriac Book of Medicines*, 2 vols., London. Rep. St. Helier-Amsterdam 1976.

- (1932, ed./tr.): *The Chronography of Gregory Abû'l Faraj, the son of Aaron, the Hebrew physician, commonly known as Bar Hebraeus*, 2 vols., Oxford-London.

BULGAKOV, P., & Ibrāhīm AḤMAD (1962, ed.): "Kitāb taḥdīd nihāyāt al-amākin li-taṣḥīḥ masāfāt al-masākin li-Abī al-Raiḥān Muḥammad b. Aḥmad al-Bīrūnī al-ḫwārazmī al-mutawaffā 440 h.", MMMA VIII.1-328.

al-BUSTĀNĪ, Buṭrus (1957, ed.): *Rasā'il iḫwān al-safā' wa-ḫillān al-wafā'*, Beirut.

CAHEN, C. (1974): "Ḳalāwdhiya", EI² IV.484.

CALLEBAT, L. (1973, ed./tr.): *Vitruve. De l'architecture. Livre VIII*, Paris.

CARDAHI, G. [Jibrā'īl al-Qardāḥī], *Lobab = Al-Lobab, seu Dictionarium syro-arabicum*, Beirut 1887-1891.

- (1898, ed.): *Abulfaragii Gregorii Bar-Hebraei Mafriani Orientis Kithâbhâ dhiyaunâ seu Liber columbae*, Rome.

CARMODY, F.J. (1952): *Al-Biṭrûjî. De motibus celorum. A critical ediction of the Latin translation of Michael Scot*, Berkeley-Los Angeles.

- (1960): *The Astronomical Works of Thabit b. Qurra*, Berkeley-Los Angeles.

CARR, W.E.W. (1925, ed./tr.): *Gregory Abu'l Faraj, commonly called Bar-Hebraeus. Commentary on the Gospels from the Horreum Mysteriorum*, London.

CARRA de VAUX, B. (1896, tr.): *Maçoudi. Le livre de l'avertissement et de la revision*, Paris.

CAUSSIN de Perceval, J.J.A. (1804): "Le livre de la Grande table hakémite, Observée par le Sheikh, l'Imam, le docte, le savant, Aboulhassan Ali ebn Abderrahman, ebn Ahmed, ebn Iounis, ebn Abdalla, ebn Mousa, ebn Maïsara, ebn Hafes, ebn Hiyan", NEMBN VII.16-240.

CHABOT, J.B. (1898, ed./tr.): "Une lettre de Bar Hébréus au Catholicos Denḥa Ier", JA IXe sér. XI.75-128.

- (1899-1910, ed./tr.): *Chronique de Michel le Syrien, Patriarche jacobite d'Antioche (1166-1199)*, 4 vols., Paris.

- (1916, ed.): *Anonymi auctoris chronicon ad annum Christi 1234 pertinens II*, Paris. Rep. Louvain 1953 (CSCO 82, syr. 37).

- (1920, ed.): *Chronicon ad annum Christi 1234 pertinens I, praemissum est Chronicon anonymum ad A.D. 819 pertinens curante Aphram Barsaum*, Paris. Rep. Louvain 1953 (CSCO 81, syr. 36).

- (1928, ed.): *Iacobi Edesseni Hexaemeron*, Paris. Rep. Louvain 1953 (CSCO 92, syr. 44).

- (1937, tr.): *Anonymi auctoris chronicon ad annum Christi 1234 pertinens I, praemissum est Chronicon anonymum ad A.D. 819 pertinens*, Paris. Rep. Louvain 1952 (CSCO 109, syr. 56).

CHEIKHO, L. (1898): *Nubḏa fī tarǧama wa-taʾālīf al-ʿallāma Ġrīġūriyūs Abī al-Faraǧ b. Ahrūn al-ṭabīb al-Malaṭī al-maʿrūf bi-Ibn al-ʿIbrī*, Beirut.

- (1912, ed.): *K. ṭabaqāt al-umam li-l-qāḍī Abī al-Qāsim Ṣāʿid b. Aḥmad b. Ṣāʿid al-Andalusī al-mutawaffā sana 462 h. (1069-1070 m.)*, Beirut.

- (1924): *Kitāb al-maḫṭūṭāt al-ʿarabīya li-katabat al-naṣrānīya* [*Catalogue des manuscrits des auteurs arabes chrétiens depuis l'Islam*], Beirut.

CHRISTENSEN, O.H. (1951, ed./tr.): "The Scholia of Bar Hebraeus on Proverbs and Job", Diss. Chicago.

ÇIÇEK, Julius Y. (1983, ed.): *Ktāḇā d-yawnā meṭṭul dubbārā d-īḥīḏāyē b-ḵaryāṯā men syāmē d-aḇūn ṭuḇtānā mār Grīgōriyōs map̄ryānā qaddīšā d-maḏnḥā d-meṯīḏaʿ Bar ʿEḇrāyā* [*Bar Hebraeus's Book of the Dove*], Glane/Losser.

- (1985, ed.), *Ktāḇā d-īṯīqōn d-ʿal myattrūṯ dubbārē men syāmē d-mār Grīgōriyōs Yōḥannān Bar ʿEḇrāyā map̄ryānā d-maḏnḥā* [*Ethicon. Christian Ethics (Morals) by the Great Syrian Philosopher and Author of Several Christian Works, Mar Gregorius Barhebraeus Catholicos of the East (1226-1286)*], Glane/Losser.

- (1986, ed.): *Huddāyē meṭṭul qānōnē ʿēdtānāyē w-nāmōsē ʿālmānāyē men syāmē d-aḇūn yadduʿtānā mār Grīgōriyōs Yōḥannān map̄ryānā mšabbḥā d-maḏnḥā d-ʿēdtā suryāytā trīṣaṯ šuḇḥā d-Anṭiyoḵ* [*Nomocanon of Bar-Hebraeus*], Glane/Losser.

- (1987, ed.): *Maḵtḇānūṯ zaḇnē d-Ḇar ʿEḇrāyā* [*The Chronography of Bar Hebraeus*], Glane/Losser.

- (1997, ed.): *Mnāraṯ quḏšē meṭṭul šeṯessē ʿēdtānāyāṯā d-yadduʿtānā Bar ʿEḇrāyā map̄ryānā d-maḏnḥā* [*Mnorath Kudshe (Lamp of the Sanctuary)*], Glane/Losser.

CLEMONS, J.T. (1966): "A Checklist of Syriac mss. in the United States and Canada", OCP XXXII.224-251, 478-522.

COAKLEY, J.F. (1993): "A Catalogue of the Syriac MSS in the John Rylands Library", BJRL LXXV.105-207.

CZEGLÉDY, K. "The Syriac Legend concerning Alexander the Great", AOH VII.231-249.

DAIBER, H. (1975, ed.): *Ein Kompendium der aristotelischen Meteorologie in der Fassung des Ḥunain ibn Isḥâq* (ASL, Prolegomena et Parerga 1), Amsterdam-Oxford.

- (1980): *Aetius Arabus. Die Vorsokratiker in arabischer Überlieferung*, Wiesbaden.

- (1991): "Nestorians of 9th Century Iraq as a Source of Greek, Syriac and Arabic. A Survey of some Unexplored Sources", ARAM (Oxford) III.45-52.
- (1992, ed./tr.): "The *Meteorology* of Theophrastus in Syriac and Arabic Translation", in W.W. Fortenbaugh & D. Gutas (ed.), *Theophrastus, His Psychological, Doxographical, and Scientific Writing* (Rutgers University Studies in Classical Humanities 5), New Brunswick-London, p. 166-293.
- (1993, ed./tr.): *Naturwissenschaft bei den Arabern im 10. Jahrhundert n. Chr. - Briefe des Abū l-Faḍl Ibn al-ᶜAmīd (gest. 360/970) an ᶜAḍudaddaula* (IPT 13), Leiden-New York-Cologne.

DĀNISHPAZHŪH [Danechpajouh], Muḥammad Taqī: (1355 h.š, ed.): *Al-Risāla al-muᶜīnīya az Ḫwāǧa-i Ṭūsī* (Intišārāt-i Dānišgāh-i Tihrān 300), Tehran.
- (1964, ed.): *Kitāb al-asrār wa-sirr al-asrār li-Abī Bakr Muḥammad b. Zakarīyā b. Yahyā al-Rāzī, bā Taǧārib-i šahriyārī* [*Al-Asrār & Sirr-al-Asrār par Muhammad Ibn Zakarīyā, avec un supplément en Persan Tajárib-e Shahriyāri*], Tehran 1343 h.š.
- (1986, ed.): *Ibn Sīnā. Al-Naǧāt min al-ġaraq fī baḥr al-ḍalālāt*, Tehran 1364 h.š.

DĀNIYĀL, Bihnām (1981/2): "Al-Šāᵓir Yaᶜqūb Sākā (1864-1931)", JIASyr VI.301-340.

DEAN, J.E. (1930, facsimile ed./tr.): "The Scholia of Bar Hebraeus on the Books of Ezekiel and Daniel", Diss. Chicago.

DEGEN, R. (1977): "A Further Manuscript of Barhebraeus' "Creme of Wisdom": Princeton, Theological Seminary, MS Nestorian 25", OrChr LXI.86-91.
- (1978): "The Oldest Known Syriac Manuscript of Ḥunayn b. Isḥāq", Symp. Syr. II.63-71.

DELAMBRE, J.B.J. (1817): *Histoire de l'astronomie ancienne*, 2 vols., Paris. Rep. New York-London 1965.

DESREUMAUX, A. [& F. BRIQUEL-CHATONNET] (1991): *Répertoire des bibliothèques et des catalogues de manuscrits syriaques*, Paris.

DIETERICI, F. (1876, tr.): *Die Philosophie der Araber im IX. und X. Jahrhundert n. Chr. aus der Theologie des Aristoteles, den Abhandlungen Alfārābīs und den Schriften der Lautern Brüder*, vol. V. Die Naturanschauung und Naturphilosophie, 2nd ed., Leipzig.
- (1886, ed.): *Die Abhandlungen der Ichwân el-Safâ in Auswahl*, Leipzig.

DIETRICH, A. (1988, ed./tr.), *Dioscurides Triumphans. Ein anonymer arabischer Kommentar (Ende 12. Jarh. n. Chr.) zur Materia medica* (AAWG.PH III/173), Göttingen.
- (1991, ed./tr.): *Dioscurides-Erklärung des Ibn al-Baiṭār. Ein Beitrag zur arabischen Pflanzensynonymik des Mittelalters* (AAWG.PH III/191), Göttingen.

DITTBERNER, H. (1971): "Zur Geschichte der Kenntnis und Ordnung der Salze", Diss. Frankfurt am Main.

DŌLABĀNĪ, Philoxenus Yōḥannān (1929, ed.): *Mušḥātā d-mār Grīgōriyōs Yōḥannān Bar ᶜEbrāyā mapryānā d-madnḥā* [*Dīwān al-ᶜallāma al-kabīr wa-l-šāᶜir al-šahīr al-failasūf al-suryānī mār Grīgūriyūs Yūḥannā Ibn al-ᶜIbrī mafriyān al-mašriq, Bar Hebraeus' Mush'hotho Book*], Jerusalem. Rep. Glane/Losser 1983.
- (1994): *Catalogue of Syriac Manuscripts in Syrian Churches and Monasteries (Dairotho w'Idotho Suryoyotho)* (Syriac Patrimony 10), Aleppo.
-, R. LAVENANT, S. BROCK & Samir Khalil SAMIR (1994): "Catalogue des manuscrits de la bibliothèque du Patriarcat Syrien Orthodoxe à Ḥomṣ (Auj. à Damas)", ParOr XIX.555-661.

DONZEL, E. van, & C. OTT (2001): "Yādjūdj wa-Mādjūdj", EI² XI.231-234.

DOZY, R.P.A. (1881): *Supplément aux dictionnaires arabes*, Leiden.

DROSSAART LULOFS, H.J. (1965, ed./tr.): *Nicolaus Damascenus on the Philosophy of Aristotle* (PhAnt 13), Leiden.
- (1985): "Aristotle, Bar Hebraeus and Nicolaus Damascenus on Animals", in A. Gotthelf (ed.), *Aristotle on Nature and Living Things. Philosophical and Historical Studies Presented to David M. Balme on his Seventieth Birthday*, Pittsburgh-Bristol, p. 345-357.
-, & E.L.J. POORTMAN (1989, ed./tr.): *Nicolaus Damascenus. De plantis. Five Translations* (ASL 4), Amsterdam-Oxford-New York.
DROWSER, E.S., & R. MACUCH (1963): *A Mandaic Dictionary*, Oxford.
DUKE-ELDER, S. (1965): *Diseases of the Eye*, Part I (System of Ophthalmology 8), London.
-, & P.A. MACFAUL (1974), *The Ocular Adnexa* (System of Ophthalmology 13), London.
DULAURIER, E. (1858, tr.): "Les mongols d'après les historiens arméniens; fragment traduits sur les textes originaux", JA 5e sér. XI.192-255, 426-473, 481-508.
DUNYĀ, Sulaimān (1947-48, ed.): *Al-Išārāt wa-l-tanbīhāt li-Abī ᶜAlī b. Sīnā*, Cairo 1366 h.
- (1957-60, ed.): *Al-Išārāt wa-l-tanbīhāt li-Abī ᶜAlī b. Sīnā maᶜ šarḥ Naṣīr al-Dīn al-Ṭūsī*, Cairo.
- (1961, ed.): *Muqaddimat Tahāfut al-falāsifa al-musammāt Maqāṣid al-falāsifa li-l-imām al-Ġazālī*, Cairo.
DUVAL, R. (1881): *Traité de grammaire syriaque*, Paris. Rep. Amsterdam 1969.
- (1901, ed.): *Lexicon syriacum auctore Hassano Bar Bahlule*, 3 vols., Paris.
EBIED, R.Y. (1974): "Some Syriac Manuscripts from the Collection of Sir E.A. Wallis Budge", Symp. Syr. I.509-539.
EDELSTEIN, L., & I.G. KIDD (1989): *Posidonius. I. Fragments*, 2nd ed., Cambridge.
EICHHOLZ, D.E. (1965, ed./tr.): *Theophrastus. De Lapidibus*, Oxford.
ETHÉ, H. (1868, tr.): *Zakarija Ben Muhammed Ben Mahmûd el-Kazwînî's Kosmographie. Die Wunder der Schöpfung*, Erster Halbband, Leipzig.
EWALD, H. (1839): "Barhebräus über die syrischen Accente", ZKM II.109-124.
FIEY, J.M. (1959): *Mossoul chrétien* (RILOB 12), Beirut.
- (1965): *Assyrie chrétienne*, vol. I-II (RILOB 22-23), Beirut.
- (1988): "Guba (Gubos)", DHGE XXII.609f.
- (1993): *Pour un Oriens Christianus novus. Répertoire des diocèses syriaques orientaux et occidentaux* (Beiruter Texte und Studien 49), Stuttgart.
FOBES, F.H. (1919, ed.): *Aristotelis Meteorologicorum libri quattuor*, Cambridge (Mass.). Rep. Hildesheim 1967.
FONTAINE, R. (1995, ed./tr.): *Otot ha-Shamayim, Samuel Ibn Tibbon's Hebrew Version of Aristoteles Meteorology* (ASL 8), Leiden-New York-Cologne.
FORTENBAUGH, W.W., P.M. HUBY, R.W. SHARPLES & D. GUTAS (1993): *Theophrastus of Eresus. Sources for his Life, Writings, Thought and Influence*, Leiden-New York-Cologne.
FRANKE, O. (1893): "Beziehungen der Inder zum Westen", ZDMG XLVII.595-609.
FREUDENTHAL, G. (1991): "(Al-)Chemical Foundations for Cosmological Ideas: Ibn Sînâ on the Geology of an Eternal World", in Sabetai Unguru (ed.), *Physics, Cosmology and Astronomy, 1300-1700: Tension and Accommodation*, Dordrecht-Boston-London, p. 43-73.
FREYTAG, G.W. (1830-37): *Lexicon arabico-latinum*, Halle.
FURLANI, G. (1917): "A Cosmological Tract by Pseudo-Dionysius in the Syriac Language", JRAS 1917.245-272.

- (1929): "Die Physiognomik des Barhebräus in syrischer Sprache, I.", ZS VII.1-16.
- (1931): "La psicologia di Barhebreo secondo il libro *La crema della Sapienza*", RSO XIII (1931-33) 24-52.
- (1932): "Barhebreo sull' anima razionale (Dal *Libro del Candelabro del Santuario*)", Or. N.S. I.1-23, 97-115.
- (1934a): "Avicenna, Barhebreo, Cartesio", RSO XIV.21-30.
- (1934b): "Di tre scritti in lingua siriaca di Barhebreo sull' anima", RSO XIV.284-308.
- (1946): "La versione siriaca del Kitāb al-Išārāt wat-Tanbīhāt di Avicenna", RSO XXI.89-101.
- (1948): "Estratti del *Libro della causa delle cause* in un manoscritto siriaco vaticano", RSO XXIII.37-45.

FURLEY, D.J. (1955, tr.): *Pseudo-Aristotle. De mundo* (in Aristotle III, LCL 400), Cambridge (Mass.)-London.

GARBERS, K. (1948): *Kitāb Kīmiyā᾿ al-ᶜIṭr wat-Taṣᶜīdāt. Buch über die Chemie des Parfüms und die Destillationen, von Yaᶜqūb B. Isḥāq al-Kindī, ein Beitrag zur Geschichte der arabischen Parfümchemie und Drogenkunde aus dem 9. Jahrh. P.C.*, Leipzig.

-, & J. WEYER (1980): *Quellengeschichtliches Lesebuch zur Chemie und Alchemie der Araber im Mittelalter*, Hamburg.

GAUTHIER, L. (1936, ed.): *Hayy ben Yaqdhân. Roman philosophique d'Ibn Thofaïl*, 2nd ed., Beirut.

GEMAYEL, N. [Nāṣir Jumayyil] (1984): *Les échanges culturels entre les Maronites et l'Europe. Du Collège Maronite de Rome (1584) au Collège de ᶜAyn-Warqa (1789)*, 2 vols., Beirut.

- (1997): *Al-Nussāḫ al-mawārina wa-mansūḫātuhum*, 2 vols., Beirut.

GIESE, A. (1986, tr.): *Al-Qazwînî. Die Wunder des Himmels und der Erde*, Stuttgart-Vienna.

GIBSON, M.D. (1911-13, ed./tr.): *The Commentaries of Isho'dad of Merv, Bishop of Ḥadatha (c. 850 A.D.)*, 5 vols., Cambridge.

GILBERT, O. (1907): *Die meteorologischen Theorien des griechischen Altertums*, Leipzig. Rep. Hildesheim 1967.

GOEJE, M.J. de (1885, ed.): *Compendium libri Kitâb al-Boldân auctore Ibn al-Fakîh al-Hamadhânî* (BGAr 5), Leiden.

- (1889, ed./tr.): *Kitâb al-masâlik wa'l-mamâlik (Liber viarum et regnorum) auctore Abu'l-Kâsim Obaidallah ibn Abdallah ibn Khordâdbeh et excerpta a Kitâb al-kharâdj auctore Kodâma ibn Djaᶜfar* (BGAr 6), Leiden.
- (1892, ed.): *Kitâb al-aflâk an-nafîsa VII auctore Abû Alî Ahmed ibn Omar Ibn Rosteh et Kitâb al-Boldân auctore Ahmed ibn abî Jakûb ibn Wâdhih al-Kâtib ak-Jakûbî*, editio secunda (BGAr 7), Leiden.
- (1894, ed.): *Kitâb at-Tanbîh wa'l-Ischrâf auctore al-Masûdî* (BGAr 8), Leiden.

GOICHON, A.-M. (1951, tr.): *Ibn Sīnā (Avicenne). Livre des directives et remarques (Kitāb al-᾿Išārāt wa l-tanbīhāt)*, Beirut-Paris.

GOLDSTEIN, B.R. (1964): "On the Theory of Trepidation according to Thābit b. Qurra and al-Zarqāllu and its Implications for Homocentric Planetary Theory", Centaurus X.232-247.

- (1971, ed./tr.): *Al-Biṭrūjī: On the Principles of Astronomy*, 2 vols., New Haven-London.

GOSHEN-GOTTSTEIN, M.H. (1979): *Syriac Manuscripts in the Harvard College Library. A Catalogue*, Missoula (MT).

GOTTHEIL, R.J.H. (1890): "Contributions to the History of Geography, II. Candelabrum sanctorum and Liber radiorium of Gregorius Bar ᶜEbhrāyā", Hebr. VII. 39-55.

GÖTTSBERGER, J. (1900): *Barhebräus und seine Scholien zur heiligen Schrift*, Freiburg.
GRAEFE, E.-[M. PLESSNER] (1966): "Haram", EI[2] III.173.
GRAF, G., *GCAL* = *Geschichte der christlichen arabischen Literatur*, 5 vols. (StT 118, 133, 146, 147, 172), Vatican City 1944-1953.
- (1954): *Verzeichnis arabischer kirchlicher Termini*, 2nd ed. (CSCO 147, subs. 8), Louvain.
GUGENHEIMER, R. (1894, ed.): *Die Scholien des Gregorius Abulfaragius Bar Hebraeus zum Buche Ezechiel*, Berlin.
ḤABBĪ, Yūsuf [Joseph] (1980): "Les Chaldéens et les Malabares au XIXe siècle", OrChr LXIV.82-107.
- (1987, ed.): *Kitāb al-dalāʾil li-l-Ḥasan b. al-Bahlūl, awāsiṭ al-qarn al-rābiᶜ al-hiğrī*, Kuwait 1408 h.
ḤABBĪ, Yūsuf, Miḫā MAQDASĪ, Bihnām DĀNIYĀL, Ḥannā JAJĪKĀ, Suhail QĀSHĀ, Buṭrus ḤADDĀD, Hurmiz ṢANĀ, Nūʾīl QAYĀBALLŪ & Īlīyā ᶜĪsā SAKMĀNĪ (1977): *Fahāris al-maḫṭūṭāt al-suryānīya fī al-ᶜIrāq* [*Catalogue of the Syriac Manuscripts in Iraq/Mḥawwyānē da-k̠t̠ib̠āt̠ā suryāyāt̠ā da-b̠gaw ᶜIrāq*], vol. 1, Baghdad.
ḤABBĪ, Yūsuf, Isḥāq SĀKĀ, Buṭrus ḤADDĀD, Bihnām DĀNIYĀL, Yūḥannā IBRĀHĪM, Yūṣuf Khīdū al-BĀZĪ & Binyāmīn ḤADDĀD (1981): *Fahāris al-maḫṭūṭāt al-suryānīya fī al-ᶜIrāq*, vol. 2, Baghdad.
ḤADDĀD, Binyāmīn (1981/2): "Faṣl fī al-nabātāt min kitāb 'Manārat al-aqdās' li-Ibn al-ᶜIbrī (1226-1286 m.)", JIASyr VI.407-515.
ḤADDĀD, Buṭrus (1985): "Min mašāhīr ḫaṭṭāṭī usrat Hūmū", JIASyr IX.167-194.
-, & Jāk ISḤĀQ (1988): *Al-Maḫṭūṭāt al-suryānīya wa-l-ᶜarabīya fī ḫizānat al-rahbānīya al-kaldānīya fī Baġdād* [*Syriac and Arabic Manuscripts in the Library of the Chaldean Monastery Baghdad/Kt̠ib̠āt̠ā suryāyāt̠ā w-ᶜarabāyāt̠ā d-b̠ēt̠ arkē d-d̠ayrāyūt̠ā d-k̠aldāyā b-B̠aġdād*] (Catalogues of the Syriac Manuscripts in Iraq 3), 2 parts, Baghdad.
HALLEUX, A. de (1987): "Les manuscrits syriaques du *CSCO*", Muséon C.35-48
HALLEUX, R., & J. SCHAMP (1985, ed./tr.): *Les lapidaires grecs (Lapidaire orphique, Kérygmes lapidaires d'Orphée, Socrate et Denys, Lapidaire nautique, Damigéron-Évax)*, Paris.
HALMA, N. (1822, ed.): *Commentaire de Théon d'Alexandrie sur le livre III de l'Almageste de Ptolémée; Tables manuelles des mouvements des astres*, Paris.
HAMBYE, E.R. (1977): "Some Syriac Libraries of Kerala (Malabar), India - Notes and Comments", in R.H. Fischer (ed.), *A Tribute to Arthur Vööbus - Studies in Early Christian Literature and Its Environment, Primarily in the Syrian East*, Chicago, p. 35-46.
HANNA, Sabo, & Aziz BULUT (2000), *Wörterbuch Deutsch-Aramäisch Aramäisch-Deutsch*, Heilbronn.
al-HASAN, A.Y., & D.R. HILL (1986): *Islamic Technology. An Illustrated History*, Cambridge-Paris.
HASCHMI, M.Y. (1966): "Die geologischen und mineralogischen Kenntnisse bei Ibn Sīnā", ZDMG CXVI.44-59.
HATCH, W.H.P. (1946): *An Album of Dated Syriac Manuscripts*, Boston.
HAYDUCK, M. (1899, ed.): *Alexandri In Aristotelis Meteorologicorum libros commentaria* (CAG III/II), Berlin.
- (1901, ed.): *Johannis Philoponi In Aristotelis Meteorologicorum librum primum commentaria* (CAG XIV/I), Berlin.

HEIBERG, J.L. (1898-1903, ed.): *Claudii Ptolemaei opera quae exstant omnia, vol. I. Syntaxis mathematica*, 2 parts, Leipzig.

HETT, W.S. (1953-57, tr.): *Aristotle. Problems*, 2 vols., Cambridge (Mass.)-London.

HICKS, J.H. (1933, ed./tr.): "The Scholia of Barhebraeus on the Book of Isaiah", Diss. Chicago.

HILL, D.R. (1996): "Engineering", EHAS 751-795.

HINZ, W. (1955): *Islamische Masse und Gewichte* (HO, Erg. 1), Leiden.

HIRSCHBERG, J., & J. LIPPERT (1902, tr.): *Die Augenheilkunde des Ibn Sina*, Leipzig.

HIRTH, F. (1890): "Zur Geschichte des Glases in China", in id., *Chinesische Studien*, Munich-Leipzig, p. 62-67.

HJELT, A. (1892, ed./tr.): "Études sur l'Hexaméron de Jacques d'Édesse, notamment sur ses notions géographiques contenues dans le 3ième traité", Diss. Helsinki [Helsingfors].

HOFFMANN, G. (1874, ed.): *Syrisch-arabische Glossen*, 1. Band: Autographie einer Gothaischen Handschrift enthaltend Bar Ali's Lexicon von Alif bis Mim, Kiel.

HOLMYARD, E.J. (1923, ed./tr.): *Kitāb al-ᶜIlm al-Muktasab fī Zirāᶜat a<u>dh</u>-<u>Dh</u>ahab. Book of Knowledge Acquired concerning the Cultivation of Gold by Abu 'l-Qāsim Muḥammad Ibn Aḥmad al-ᶜIrāqī*, Paris.

- (1928, ed.): *The Arabic Works of Jâbir Ibn Ḥayyân*, vol. 1, part 1 (Arabic texts), Paris.

HOLMYARD, E.J., & D.C. MANDEVILLE (ed./tr. 1927): *Avicennae De congelatione et conglutinatione lapidum, being Sections of the Kitâb al-Shifâ*, Paris.

HONIGMANN, E. (1929): *Die sieben Klimata und die πόλεις ἐπίσημοι. Eine Untersuchung zur Geshichte der Geographie und Astrologie im Altertum und Mittelalter*, Heidelberg. Rep. Frankfurt 1992.

HONIGMANN, E. (1954): *Le couvent de Barṣaumā et le patriarcat jacobite d'Antioche et de Syrie* (CSCO 146, subs. 7), Louvain.

HORTEN, M., & E. WIEDEMANN (1913): "Avicennas Lehre vom Regenbogen nach seinem Werke al Schifâ", Meteorologische Zeitschrift XXX, 11. 1913, 533-544 (= E. Wiedemann, *Gesammelte Schriften* II.733-744).

HOUTSMA, T. (1883, ed.): *Ibn-Wādhih qui dicitur al-Jaᶜqubī, Historiae*, Leiden.

HUMĀʾĪ, Ǧalāl (1316 h.š., ed.): *Kitāb al-tafhīm li-awāʾil ṣināᶜat al-tanǧīm taʾlīf-i ustād Abū Raihān Muḥammad b. Aḥmad Bīrūnī dar sāl 420 h.š.*, Tehran.

IHMMR = Institute of History of Medicine and Medical Research (1982, ed.): *Al-Qanun Fi' l-Tibb. Al-Shaikh Al-Raʾis Abu ᶜAli Al-Husain Bin Abdullah Bin Sina [980-1037 A.D.]*, Book I, New Delhi.

ISTANBUL (1298 h.): *Tisᶜ rasāʾil fī al-ḥikma wa-l-ṭabīᶜīyāt*, Istanbul [rep. with altered pagination, Cairo 1326 h./1908].

ISTANBUL (1997) [Yeshūᶜ b. Gabriel]: *Ktāḇā d-zalgē w-šurrārā d-šeṯessē ᶜēdtānāyāṯā men syāmē d-aḇūn qaddīšā Mār Grīḡōriyōs mapryānā d-hū Bar ᶜEḇrāyā* [*Book of Zelge by Bar-Hebreaus, Mor Gregorius Abulfaraj, the great Syrian philosopher and author of several Christian works*], Istanbul: Zafer Matbaası.

JANSSENS, H.F. (1930): "Crème de la science ou Science des sciences? - le vrai titre d'un ouvrage de Bar Hebraeus", Muséon XLIII.365-372.

- (1930-5, ed./tr.): "Bar Hebraeus' *Book of the Pupil of the Eye*" AJSL XLVII (1930/1) 26-49, 94-134; XLVIII (1932) 209-263; LII (1935) 1-21.

- (1937, ed./tr.): *L'entretien de la sagesse. Introduction aux oeuvres philosophiques de Bar Hebraeus* (BFPUL 75), Liège.

JIJĀWĪ, Dionysius Behnam (1996, tr.): *Mnāraṯ qudšē. Manārat al-aqdās li-l-ᶜallāma Mār Ġrīġūriyūs Abī al-Farağ Ibn al-ᶜIbrī mafriyān al-mašriq 1226-1286* [*Lamp of the Sanctuary by Mar Gregorios Yohanna Bar Ebroyo Maphryono d-madnho*], Aleppo.

JONES, H.L. (1917-1932, ed./tr.): *The Geography of Strabo*, Cambridge (Mass.)-London.

JOOSSE, N.P.G. (1999): "Bar Hebraeus' ܟܬܒܐ ܕܚܐܘܬ ܚܟܡܬܐ (*Butyrum sapientiae*). A Description of the Extant Manuscripts", Muséon CXII.417-458.

JOOSTEN, J. (1992): "The Negation of the Non-Verbal Clause in Syriac", JAOS CXII.584-588.

JWAIDEH, W. (1959, tr.): *The Introductory Chapters of Yāqūt's Muᶜjam al-Buldān*, Leiden.

KAHLE, P. (1936): "Bergkristall, Glas und Glasflüsse nach dem Steinbuch von el-Bērūnī", ZDMG XC.321-356.

KARIMOV, U.I. (1957): *Neizvestnoe sočinenie ar-Razi "Kniga tajny tajn"*, Tashkent.

KAUFHOLD, H. (1990): review article "Catalogues of the Syriac Manuscripts in Iraq, Baghdad ... vol. 1 und 2, ... 1977 und 1981 ...; vol. 3 ... 1988 ... Behnam Sony, Catalogue of Karakosh Manuscripts ... Bagdad 1988 ...", OrChr. LXXIV.263-266.

KAWERAU, P. (1960): *Die jakobitische Kirche im Zeitalter der syrischen Renaissance. Idee und Wirklichkeit* (BBA 3), 2nd ed., Berlin.

KAYSER, K. (1893, tr.): *Das Buch von der Erkenntniss der Wahrheit oder der Ursache aller Ursachen*, Strassburg.

KAZIMIRSKI, A. de Biberstein (1960): *Dictionnaire arabe-français*, nouvelle édition, Paris.

KENNEDY, E.S. (1973): *A Commentary upon Bīrūnī's Kitāb Taḥdīd al-Amākin. An 11th Century Treatise on Mathematical Geography*, Beirut.

- (1984): "Two Persian Astronomical Treatises by Naṣīr al-Dīn al-Ṭūsī", Centaurus XXVII.109-120.

KHALIFÉ, Ignace-Abdo, & François BAISSARI (1964): "Catalogue raisonné des manuscrits de la Bibliothèque Orientale de l'Université Saint Joseph. Manuscrits syriaques", MUSJ XL.235-285.

KHAUSHĀBĀ, Shlaimūn Īshū, & ᶜImmānū'īl baitū YŪKHANNĀ (2000): *Zahrīrā. Qāmūs ᶜarabī-suryānī/Lehksīqōn ᶜarabāyā-suryāyā*, Dohuk.

KIDD, I.G. (1988): *Posidonius. II. The Commentary*, 2 vols., Cambridge.

KILLY, H.E. (1954): "Beryll", RAC II. 157-159.

KING, D.A. (1990): "al-Mayl", EI^2 VI.914f.

KIRAZ, G.A. (1988): *ᶜIqd al-ğumān fī aḫbār al-suryān. Baḥṯ taʾrīḫī iğtimāᶜī ḥaula al-suryān fī al-diyār al-muqaddasa* [*ᶜEqqā d-bērūlē/Ikduljuman*], Glane/Losser.

KLAMROTH, M. (1878, ed.): *Gregorii Abulfaragii Bar Ebhraya in Actus apostolorum et Epistulas catholicas adnotationes* (Diss. Göttingen), Göttingen.

- (1888): "Ueber die Auszüge aus griechischen Schriftstellern bei al-Jaᶜqûbî. IV. Mathematiker und Astronomen", ZDMG XLII 1-44.

KOFFLER, H. (1932): *Die Lehre des Barhebräus von der Auferstehung der Leiber* (OrChr(R) 28/1), Rome.

KOONAMMAKKAL, T. (1994): "St. Ephrem and 'Greek Wisdom'", Symp. Syr. VI.169-176.

KRAEMER, J.L. (1989, tr.): *The History of al-Ṭabarī (Taʾrīkh al-rusūl wa'l-mulūk), Volume XXXIV. Incipient Decline*, Albany: University of New York.

KRAUS, P. (1942-43): *Jābir Ibn Ḥayyān, Contribution à l'histoire des idées scientifiques dans l'Islam.*, Cairo.

KRENKOW, F. (1355 h., ed.): *Kitāb al-ğamāhir fī maᶜrifat al-ğawāhir min taṣnīf al-ustāḏ Abī al-Raiḥān Muḥammad ibn Aḥmad al-Bīrūnī*, Hyderabad.

KROTKOFF, G. (1982): *A Neo-Aramaic Dialect of Kurdistan*, New Haven.

KRUK, R. (1990): "A Frothy Bubble: Spontaneous Generation in the Medieval Islamic Tradition", JSSt XXXV.265-282.

KUGENER, A. (1907): "Un traité astronomique et météorologique syriaque attribué à Denys l'aréopagite", in *Actes du XIVe Congrès International des Orientalistes, Alger 1905*, Paris, II.137-198.

KÜHN, C.G. (1821-33, ed.): *Claudii Galeni opera omnia*, Lepizig.

KUNITZCH, P. (1974): *Der Almagest. Die Syntaxis mathematica des Claudius Ptolomäus in arabischer-lateinischer Übersetzung*, Wiesbaden.

LAGARDE, P. de (1858, ed.): *Analecta syriaca*, Leipzig. Rep. Osnabrück 1967.

- (1879, ed.): *Praetermissorum libri duo*, Göttingen.

LANE, E.W. (1863-93): *An Arabic-English Lexicon*, London-Edinburgh. Rep. London 1984.

LANTSCHOOT, A. van (1965): *Inventaire des manuscrits syriaques des fonds vatican (490-631) Barberini oriental et Neofiti* (StT 243), Vatican City.

LEE, H.D.P. (1952, ed./tr.): *Aristotle, Meteorologica* (LCL 397), Cambridge (Mass.)-London.

LEROY, J. (1957): *Moines et monastères du Proche-Orient*, Paris.

- (1971): "La renaissance de l'Église syriaque au XIIe-XIIIe siècles", Cahiers de civilisation médiévale XIV.131-48, 239-55.

LETTINCK, P. (1999): *Aristotle's Meteorology and its Reception in the Arabic World. With an Edition and Translation of Ibn Suwār's Treatise on Meteorological Phenomena and Ibn Bājja's Commentary on the Meteorology* (ASL 10), Leiden-Boston-Cologne.

LEVI DELLA VIDA, G. (1910): "Pseudo-Beroso siriaco", RSO III.7-43, 611-612.

LEVY, J. (1867-68): *Chaldäisches Wörterbuch über die Targum und einen grossen Theil des rabbinischen Schrifttums*, 2 vols., Leipzig-London.

LIDDELL, H.G., R. SCOTT, H.S. JONES et al., *A Greek-English Lexicon, ... With a Supplement*, Oxford 1968.

LIPPERT, J. (1903, ed.): *Ibn al-Qifṭī's Taʾrīḫ al-ḥukamāʾ* [*Taʾrīḫ al-hukamāʾ wa-huwa muḫtaṣar al-Zauzanī al-musammā bi-l-Muntaḫabāt al-muntaqaṭāt min Kitāb iḫbār al-ʿulamāʾ bi-aḫbār al-hukamāʾ li-Ǧamāl al-Dīn Abī al-Ḥasan ʿAlī b. Yūsuf al-Qifṭī*], Leipzig.

LITTRÉ, É. (1839-61, ed./tr.): *Oeuvres complètes d'Hipporate*, 10 vols., Paris.

LOEHR, M. (1889, ed.): *Gregorii Abulpharagii Bar Ebhraya in Epistulas paulinas adnotationes*, Pars prior (Diss. Göttingen), Göttingen.

LOUIS, P. (1982, ed./tr.): *Aristote. Météorologiques*, 2 vols., Paris.

LÖW, I. (1881): *Aramäische Pflanzennamen*, Leipzig.

[LSJ = Liddell-Scott-Jones, *A Greek-English Lexicon*]

MACLEAN, J.A. (1901): *A Dictionary of the Dialects of Vernacular Syriac as Spoken by the Eastern Syrians of Kurdistan, North-West Persia and the Plain of Mosul*, Oxford. Rep. Amsterdam 1972.

MACOMBER, W.F. (1969): "New Finds of Syriac Manuscripts in the Middle East", in W. Voigt (ed.), *XVII. Deutscher Orientalistentag (Würzburg)*, Wiesbaden, Teil 2, p. 473-482.

MACÚCH, R. (1965): *Handbook of Classical and Modern Mandaic*, Berlin.

- (1976): *Geschichte der spät- und neusyrischen Literatur*, Berlin-New York.

MADKOUR, I., M. el-KHODEIRI, G. ANAWATI, F. el-AHWANI (1952, ed.): *Ibn Sīnā. Al-Shifāʾ. La Logique. 1. L'Isagoge (al-madkhal)*, Cairo.

MADWAR, M., & I. AHMAD (1980, ed.): *Ibn Sīnā. Al-Shifāʾ. Mathématiques. 4. Astronomie*, Cairo.

MANNA, Jacques Eugène [Yaᶜqōb Awgīn] (1900): *Haddīʾā d-leššānā ārāmāyā kaldāyā/Dalīl al-rāġibīn fī luġat al-ārāmīyīn*, Mosul. Rep. sine anno et loco.
- (1901): *Margē peġyānāyē d-mardūṯā ārāmāyē* [*Al-Murūǧ al-nuzhīya fī ādāb al-luġat al-ārāmīya/Morceaux choisis de littérature araméenne*], 2 vols., Mosul.
MARGOLIOUTH, D.S. (1887, ed.): *Analecta orientalia ad Poeticam Aristoteleam*, London. Rep. Hildesheim-New York-Zurich 2000.
MARGOLIOUTH, G. (1899): *Descriptive List of Syriac and Karshuni MSS. in the British Museum, Acquired since 1873*, London.
MARGOLIOUTH, J.P. [= J. Payne Smith] (1903): *A Compendious Syriac Dictionary Founded upon the Thesaurus Syriacus of R. Payne Smith D.D.*, Oxford.
- (1927): *Supplement to the Thesaurus Syriacus of R. Payne Smith, S.T.P.*, Oxford.
MARSH, F.S. (1927, ed./tr.): *The Book which is called the Book of Hierotheos with Extracts from the Prolegomena and Commentary of Theodosios of Antioch and the "Book of Excerpts" and other works of Gregory Bar=Hebraeus*, London-Oxford.
MARTIN, J.P.P. (1872, ed.): *Oeuvres grammaticales d'Abou'lfaradj dit Bar Hebreus*, 2 vols., Paris.
MARTÍNEZ Martín, L. (1984): "al-Madd wa 'l-Dj̲azr", EI² V.949-951.
MARZOLPH, U. (1985): "Die Quelle der Ergötzlichen Erzählungen des Bar Hebräus", OrChr LXIX.81-125.
MEHREN, A.F. (1866, ed.): *Cosmographie de Chems-ed-Din Abou Abdallah Mohammed ed-Dimichqui*, St. Petersburg.
- (1874. tr.): *Manuel de la cosmographie du moyen âge, traduit de l'arabe. "Nokhbet ed-Dahr fi ᶜadjaib-il-birr wal-bah'r" de Chems ed-Dîn Abou-ᶜAbdallah Moh'ammed de Damas*, Copenhagen.
MEISSNER, B. (1903): "Lexicographische Studien III.", ZA XVII.239-250, 271.
MERTENS, M. (1995, ed.): *Les alchimistes grecs*, Tome IV, 1ère partie, Zosime de Panopolis, *Mémoires authentiques*, Paris.
MERX, A. (1885): "Proben der syrischen Uebersetzung von Galenus' Schrift über die einfachen Heilmittel", ZDMG XXXIX.237-305.
MEYERHOFF, M. (1928, ed/tr.): *The Book of the Ten Treatises on the Eye Ascribed to Hunain Ibn Iss-hâq (809-877 A.D.) (al-ᶜAšr maqālāt fī l-ᶜain)*, Cairo.
- (1940, ed./tr.): *Šarḥ asmāʾ al-ᶜuqqār (L'explication des noms des drogues). Un glossaire de matière médicale composé par Maïmonide*, Cairo.
-, & J. SCHACHT (1931): *Galen. Über die medizinische Namen*, Berlin.
-, & G.P. SOBHY (1932-40, ed./tr.): *The Abridged Version of "The Book of Simple Drugs" of Ahmad Ibn Muhammad al-Ghâfiqî by Gregorius Abu'l-Farag (Barhebraeus)*, vol. 1, fasc. 1-4, Cairo.
MIDDELDORPF, H. (1835, ed.): *Codex Syriaco-Hexaplaris*, Berlin.
MINGANA, A. (1933): *Catalogue of Mingana Collection of Manuscripts*, vol. 1, Cambridge.
- (1935, ed./tr.): *An Encyclopaedia of Philosophical and Natural Sciences as Taught in Baghdad about A.D. 817 or Book of Treasures by Job of Edessa* (Woodbrooke Scientific Publications 1), Cambridge
MINORSKY, V. (1937, tr.): *Ḥudūd al-ᶜĀlam. 'The Regions of the World'. A Persian Geography 372 A.H.-982 A.D.* (Gibb Memorial Series, N.S. 11), Oxford-London.
MIQUEL, A. (1967-88): *La géographie humaine du monde musulman jusqu'au milieu du 11e siècle*, 4 vols., Paris.
MISHKĀT, Muḥammad (1951/2, ed.): *Ṭabīᶜīyāt-i Dāniš-nāma-i ᶜalāʾī taṣnīf-i Šaiḫ-i raʾīs Abū ᶜAlī Sīnā*, Tehran 1371 h./1331 h.š.

MOBERG, A. (1907-13, tr.): *Buch der Strahlen. Die grössere Grammatik des Barhebräus*, 2 vols., Leipzig.
- (1922, ed.): *Le livre des splendeurs. La grande grammaire de Grégoire Barhebraeus* (SHVL 4), Lund.
MONTAṢIR, ᶜAbd el-Ḥalīm, Saᶜīd ZAYED, ᶜAbdallāh ISMĀᶜĪL (ed.) & Ibrahim MADKOUR (rev. & intro.) (1964): *Ibn Sīnā. Al-Shifāʾ. La Physique V. Les Métaux et la Météorologie*, Cairo 1384 h.
MURAOKA, T. (1997): *Classical Syriac. A Basic Grammar with a Chrestomathy* (PLO N.S. 19), Wiesbaden.
MURRE-VAN DEN BERG, H.H.L. (1999): "The Patriarchs of the Church of the East from the Fifteenth to Eighteenth Centuries", Hugoye II/2.
MUṬAHHARĪ, Murtaḍā (1970, ed.): *Al-Taḥṣīl taʾlīf Bahmanyār b. al-Marzubān*, Tehran 1349 h.š.
NALLINO, C.A. (1899-1907): *Al-Battānī sive Albatenii Opus astronomicum*, 3 vols., Milan.
NAPEL, E. ten (1980), "Influence of Greek Philosophy and Science in Emmanuel bar Shahhare's Hexaemeron", Symp. Syr. III.109-118.
NAU, F. (1896): "Notice sur le Livre des Trésors de Jacques de Bartela, Évêque de Tagrit", JA 9e sér. VII.286-331.
- (1899, ed./tr.): *Le livre de l'ascension de l'esprit sur la forme du ciel et de la terre. Cours d'astronomie rédigé en 1279 par Grégoire Aboulfarag, dit Bar-Hebraeus*, 2 vols., Paris.
- (1905): "Rabban Daniel de Mardin, auteur syro-arabe de XIVe siècle", ROC X.314-318.
- (1910): "La cosmographie au VIIe siècle chez les Syriens", ROC XV.225-254.
- (1916, ed./tr.): *Documents pour servir à l'Histoire de l'Église Nestorienne*, PO XIII.111-326.
- (1930-31): "Le traité sur les «Constellations» écrit, en 661, par Sévère Sébokt évêque de Qennesrin", ROC XXVII.327-410; XXVIII.85-100.
NĀẒIM, M. (1931): *The Life and Times of Sulṭān Maḥmūd of Ghazna*, Cambridge.
NEDOSPASSOWA, M.E. (1993): "The term *ballûr* in the Arabic Bible", in R. Contini et al. (ed.), *Semitica. Serta philologica Constantini Tsereteli dicata*, Turin, p.175-181.
NEEDHAM, J., & WANG Ling (1959): *Science and Civilisation in China*, vol. 3. Mathematics and the Sciences of the Heavens and the Earth, Cambridge.
NEEDHAM, J., WANG Ling & K.G. ROBINSON (1962): *Science and Civilisation in China*, vol. 4, part 1. Physics, Cambridge.
NEEDHAM, J., & LU Gwei-Djen (1974): *Science and Civilisation in China*, vol. 5. Chemistry and Chemical Technology, part 2, Cambridge.
NEEDHAM, J., HO Ping-Yü, LU Gwei-Djen & N. SIVIN (1980): *Science and Civilisation in China*, vol. 5. Chemistry and Chemical Technology, part 4, Cambridge.
NEUGEBAUER, O. (1959): "Regula Philippi Arrhidaei", Isis L.477-478.
- (1975): *A History of Ancient Mathematical Astronomy*, 3 parts, Berlin-Heidelberg-New York.
NIESE, B. (1888-92. ed.): *Flavii Iosephi opera*, 5 vols., Berlin.
NOBBE, C.F.A. (1843-45, ed): *Claudii Ptolemaei Geographia*, 3 vols., Leipzig. Rep. 1898-1913.
NÖLDEKE, T. (1868): *Grammatik der neusyrischen Sprache am Urmia-See und in Kurdistan*, Leipzig.
- (1875): *Mandäische Grammatik*, Halle. Rep. Darmstadt 1964.

- (1898): *Kurzgefasste syrische Grammatik*, 2nd ed., Leipzig. Rep. with "Anhang" by Anton Schall, Darmstadt 1966.
- (1910): *Neue Beiträge zur semitischen Sprachwissenschaft*, Strassburg.
PANAINO, A. (1990): "Calendars, i. Pre-Islamic calendars", EIr IV.658-668.
PAYNE SMITH, J.: see J.P. Margoliouth.
PAYNE SMITH, R., *Thesaurus Syriacus*, vol. I. 1879, vol. II. 1901, Oxford.
- (1864): *Catalogi codicum manuscriptorum Bibliothecae Bodleianae pars sexta, codices syriacos, carshunicos, mendaeos complectens*, Oxford.
PELLAT, C. (1980): "al-Ḳubba, Ḳubbat al-ᶜĀlam, Ḳ. al-Arḍ, Ḳ. Arīn", EI² V.297.
PETRAITIS, C. (1967, ed.): *The Arabic Version of Aristotle's Meteorology* [*Kitāb al-āṯār al-ᶜulwīya li-Arisṭūṭālīs*], Beirut.
PINES, S. (1979): *Collected Works of Shlomo Pines*, vol. I. Studies in Abu'l-Barakāt al-Baghdādī, Physics and Metaphysics, Jerusalem-Leiden.
PLATTI, E. (1988): "«L'entretien de la sagesse» de Barhebraeus. La traduction arabe", MIDEO XVIII.153-194.
PLESSNER, M. (1965): "Ḥadjar", EI² III.29f.
- (1967): "Hirmis", EI² III.463-465.
PLOEG, J.P.M. van der (1983): *The Christians of St. Thomas in Southern India and their Syriac Manuscripts* (Placid Lecture Series 3), Bangalore.
[PS = R. Payne Smith, *Thesaurus syriacus*]
[PS Comp. = J.P. Margoliouth (1903)]
[PS Suppl. = J.P. Margoliouth (1927)]
al-QADDUMI, G.H. (2001): "Yāḳūt", EI² XI.262-263.
QASSEM, Mahmoud (1969, ed.) & Ibrahim MADKOUR (pref. & rev.): *Ibn Sīnā. Al-Shifa. Physique 2-4*, Cairo 1389 h.
al-QASSH, Idwār (1987, ed.): *Al-Qānūn fī al-ṭibb taʾlīf al-šaiḫ al-raʾīs Abū ᶜAlī al-Ḥusain b. ᶜAlī b. Sīnā*, Cairo 1408 h.
QAZWÍNÍ, Mírzá Muḥammad Ibn ᶜAbdu'l-Wahháb (1912-37, ed.): *The Ta'ríkh-i-Jahán-Gushá of ᶜAlá'u 'd-Dín ᶜAṭá Malik-i-Juwayní (composed in A.H. 658 = A.D. 1260)* (Gibb Memorial Series XVI/1-3), Leiden-London.
QUATREMÈRE, É. (1836, ed./tr.): *Histoire des mongols de la Perse écrite en persan par Raschid-Eldin*, vol. 1, Paris.
- (1838-42, ed.): *Histoire des sultans mamlouks, de l'Égypte, écrite par Taki-Eddin-Ahmed-Makrizi*, 2 vols., Paris.
RADA, W.S. (1999/2000): "Comets in Arabic Literature AD 700-1600", ZGAIW XIII.71-91.
RAGEP, F.J. (1994, ed./tr.): *Naṣīr al-Dīn Ṭūsī's Memoir on Astronomy (al-Tadhkira fī ᶜilm al-hayʾa)*, 2 vols., New York-Berlin-Heidelberg.
al-RĀWĪ, Munᶜim Mufliḥ (1984): "Usus al-ǧiyūlūǧiyā fī «al-maᶜādin wa-l-āṯār al-ᶜulwīya» li-Ibn Sīnā", MMMA XXVIII.547-564.
REINAUD, J.T., & M. de SLANE (1840, ed.): *Géographie d'Aboulféda*, Paris.
REININK, G.J. (1997): "Communal Identity and the Systematisation of Knowledge in the Syriac 'Cause of All Causes'", in P. Binkley (ed.), *Pre-Modern Encyclopaedic Texts. Proceedings of the Second COMERS Congress, Groningen, 1-4 July 1996*, Leiden-New York-Cologne, p. 275-288.
RELLER, J. (1994): *Mose bar Kepha und seine Paulinenauslegung, nebst Edition und Übersetzung des Kommentars zum Römerbrief* (GOF.S 35), Wiesbaden.
RENAN, E. (1852): *De philosophia peripatetica apud syros commentatio*, Paris.
RENAUDOT, E., *LOC* = *Liturgiarum orientalium collectio*, 2 vols., 2nd ed., Frankfurt-London 1847.

RITTER, H. (1933, ed.): *Pseudo-Mağrīṭī. Das Ziel des Weisen [K. ġāyat al-ḥakīm wa-aḥaqq al-natīğatain bi-l-taqdīm al-mansūb ilā Abī al-Qāsim Maslama b. Aḥmad al-Mağrīṭī]*, Lepzig-Berlin.
-, & M. PLESSNER (1962, tr.): *"Picatrix". Das Ziel des Weisen von Pseudo-Mağrīṭī*, London.
ROPER, G. (1992-94, ed.): *World Survey of Islamic Manuscripts*, 4 vols., London.
ROSEN, F., & J. FORSHALL (1838): *Catalogus codicum orientalium qui in Museo Britannico asservantur*. Pars prima, codices syriacos et carshunicos amplectens, London.
RUSKA, J. (1896, ed./tr.): *Das Quadrivium aus Severus Bar Šakkû's Buch der Dialoge* (Diss. Heidelberg), Leipzig.
- (1897): "Studien zu Severus bar Šakkû's 'Buch der Dialoge'", ZA XII.8-41, 145-161.
- (1912, ed./tr.): *Das Steinbuch des Aristoteles*, Heidelberg.
- (1923): *Sal ammoniacus, Nušādir und Salmiak* (SHAW.PH 1923/5), Heidelberg.
- (1924): *Arabische Alchemisten*, I. Chālid ibn Jazīd ibn Muᶜāwija, Heidelberg.
- (1926): *Tabula smaragdina. Ein Beitrag zur Geschichte der Hermetischen Literatur*, Heidelberg.
- (1934): "Die Alchemie des Avicenna", Isis XXI.14-51.
- (1934b): "Avicennas Verhältnis zur Alchemie", Fortschritte der Medizin LII.836-837.
- (1936): "al-Nūshādir" EI(D)[1] III.1045, = EI[2] VIII.148f. (1993).
- (1937, tr.): *Al-Rāzī's Buch Geheimnis der Geheimnisse* (Quellen und Studien zur Geschichte der Naturwissenschaften und der Medizin 6), Berlin.
RUSKA-[C.J. LAMM] (1960): "al-Billawr", EI[2] I.1220f.
RYSSEL, V. (1880): *Über den textkritischen Werth der syrischen Übersetzungen griechischer Klassiker*, I. Theil, Lepzig.
- (1881): "Über den textkritischen Werth der syrischen Übersetzungen griechischer Klassiker", II. Theil, in Programm des Nicolaigymnasiums in Leipzig, No. 468, Leipzig, p. 1-56.
SACHAU, E. (1870): *Inedita syriaca. Eine Sammlung syrischer Übersetzungen von Schriften griechischer Profanliteratur*, Vienna.
- (1879, tr.): *The Chronology of Ancient Nations. An English version of the Arabic text of the Athâr-ul-Bâkiya of Albîrûnî, or "Vestiges of the Past," collected and reduced to writing by the author in A.H. 390-1, A.D. 1000*, London. Rep. Frankfurt 1969.
- (1895): *Skizze des Fellichi-Dialekts von Mosul*, Berlin.
- (1899): *Verzeichnis der syrischen Handschriften der Königlichen Bibliothek zu Berlin*, Berlin.
- (1923, ed.): *Chronologie orientalischer Völker von Albêrûnî [K. al-āṯār al-bāqiya ᶜan al-qurūn al-ḫāliya taʾlīf Abī al-Raiḥān Muḥammad b. Aḥmad al-Bīrūnī al-ḫwārizmī ...]*, Neudruck, Leipzig.
SACY, A.-I. Silvestre de (1801): "Observations sur l'origine du nom donné par les Grecs et les Arabes, aux Pyramides d'Aegypte, et sur quelques autres objets relatifs aux Antiquités Aegyptiennes", Magasin encyclopédique VI.446-503. Rep. in *Silvestre de Sacy (1758-1838)* (Bibliothèque des arabisants français), vol. I, Cairo, 1905, p. 224-264.
- (1810): *Relation de l'Égypte par Abd-Allatif, médecin arabe de Bagdad [1162-1231 A.D.]*, Paris. Rep. Frankfurt 1992 (VIGAIW Geog. 9-10).
- (1826-27): *Chrestomathie arabe, ou extraits de divers écrivains arabes tant en prose qu'en vers*, 3 vols., 2nd ed., Paris.
SĀKĀ, Isḥāq (1963): *Tafsīr al-quddās bi-ḥasab ṭaqs al-Kanīsa al-Suryānīya al-Urṯūḏuksīya al-Anṭākīya*, Zahle.

- (1999, ed.): *Mēmrē mğabbyē men mušḥātā d-qaššīšā Ya^cqōḇ Sākā*, 2nd ed., Aṭšāna.
ṢALḤĀNĪ, A. (1958, ed.): *Ta'rīḫ muḫtaṣar al-duwal li-l-ᶜallāma Ġrīġūriyūs al-malaṭī al-maᶜrūf bi-Ibn al-ᶜIbrī*, 2nd ed., Beirut.
al-SAQĀ, Aḥmad Ḥijāzī (1400 h., ed.): *Šarḥ ᶜuyūn al-ḥikma ta'līf šaiḫ al-islām al-imām Faḫr al-Dīn al-Rāzī Muḥammad b. ᶜUmar b. al-Ḥusain al-mutawaffā 606 h.*, 3 vols., Cairo.
SARAU, O., & W.A. SHEDD (1898): *Catalogue of Syriac Manuscripts in the Library of the Museum Association of Oroomiah College*, Urmia.
SAUMA, A. (2003): *Gregory Bar-Hebraeus's Commentary on the Book of Kings from His Storehouse of Mysteries* (SSU 20), Uppsala.
SBATH, P. (1928, ed.): *Muḫtaṣar fī ᶜilm al-nafs al-insānīya li-Ġrīġūriyūs Abī al-Faraǧ al-maᶜrūf bi-Ibn al-ᶜIbrī [Traité sur l'âme par Bar-Hebraeus, Mort en 1286]*, Cairo.
-, & M. MEYERHOFF (1938, ed./tr.): *Le livre des questions sur l'oeil de Ḥonaïn Ibn Isḥāq (Kitāb al-masā'il fī l-ᶜain)*, Cairo.
SCEBABI, A. (1877, ed.): *Gregorii Bar-Hebraei carmina*, Rome.
SCHALL, A. (1960): *Studien über griechische Fremdwörter im Syrischen*, Darmstadt.
SCHER, Addai (1906): "Notices sur les manuscrits syriaques conservés dans la bibliothèque du couvent des chaldéens de Notre-Dame-des-Semences", JA 10e ser. VII.479-512, VIII.33-82.
- (1907): "Notice sur les manuscrits syriaques et arabes conservés à l'archevêché chaldéen de Diarbékir", JA 10e sér. X.331-362, 385-431.
- (1908): "Notices sur les manuscrits syriaques et arabes conservés dans la bibliothèque de l'évêché chaldéen de Mardin", Revue des bibliothèques (Paris) XVIII.64-95.
SCHLIMME, L. (1977, tr.): *Der Hexaemeronkommentar des Moses bar Kepha* (GOF.S 14), 2 vols., Wiesbaden.
SCHMIDT, M.G. (1999): *Die Nebenüberlieferung des 6. Buches der Geographie des Ptolemaios. Griechische, lateinische, syrische, armenische und arabische Texte*, Wiesbaden.
SCHÖNFELD, J. (1976): *Über die Steine. Das 14. Kapitel aus dem "Kitāb al-Muršid" des Muḥammad ibn Aḥmad at-Tamīmī* (Islamkundliche Untersuchungen 38), Freiburg.
SCHOONHEIM, P.L. (1999): *Aristotles' Meteorology in the Arabico-Latin Tradition* (ASL 12), Leiden-Boston-Cologne.
SEMYONOV, A.A. [et al.] (1952-87): *Sobranie vostočnych rukopisej Akademii nauk Uzbekskoj SSR*, 11 vols., Tashkent.
SERSEN, W.J. (1976): "Arab Meteorology from Pre-Islamic Times to the Thirteenth Century A.D.", Diss. London (SOAS).
SEZGIN, F., *GAS* = *Geschichte des arabischen Schrifttums*, vol. 1-9, Leiden 1967-84; Index, Frankfurt 1995; vol. 10-12, Frankfurt 2000.
SHARPLES, R.W. (1998): *Theophrastus of Eresus. Sources for his Life, Writings, Thought and Influence, Commentary volume 3.1, Sources on Physics*, Leiden-Boston-Cologne.
SHERWOOD, P. (1957): "Le fonds patriarcal de la bibliothèque manuscrite de Charfet", OrSyr II.93-107.
SIEGEL, J.L. (1928, tr.): "The Scholia of Bar Hebraeus on the Book of Psalms", Diss. Chicago.
SIGGEL, A. (1950): *Arabisch-Deutsches Wörterbuch der Stoffe*, Berlin.
SLANE, W. MacGuckin de (1883-95): *Catalogue des manuscrits arabes de la Bibliothèque Nationale*, Paris.

SONY, Behnam (1993): *Fihris maḫṭūṭāt al-Baṭriyarkīya fī Dair al-Šarfa - Lubnān* [*Le catalogue des manuscrits du Patrarcat au Couvent de Charfet - Liban*], Beirut.

SPRENGLING, M., & W.C. GRAHAM (1931, ed./tr.): *Barhebraeus' Scholia on the Old Testament*, Part I: Genesis-II Samuel, Chicago.

SPULER, B. (1961, ed.): *Wüstenfeld-Mahler'sche Vergleichungs-Tabellen zur muslimischen und iranischen Zeitrechnung mit Tafeln zur Umrechnung orient-christlicher Ären*, Wiesbaden.

- (1961b): "Die nestorianische Kirche", "Die westsyrische (monophysitische/ jakobitische) Kirche" and "Die Thomas-Christen in Süd-Indien", in *Religionsgeschichte des Orients in der Zeit der Weltreligionen* (HO I.VIII/2), p. 120-169, 170-216, 226-239.

STAPLETON, H.E., & R.F. AZO (1905): "Alchemical Equipment in the Eleventh Century A.D.", MASB I.47-70.

STAPLETON, H.E., R.F. AZO & M. HIDĀYAT ḤUSAIN (1927): "Chemistry in ᶜIrāq and Persia in the Tenth Century A.D.", MASB VIII.317-397.

STEINHART, N. (1895, ed.): *Die Scholien des Gregorius Abulfaraǵ Bar-Hebraeus zum Evangelium Lukas*, Berlin.

STEINMETZ, P. (1964): *Die Physik des Theophrastos von Eresos* (Palingenesia I), Bad Homburg-Berlin-Zürich.

STEYER, C. (1908): *Buch der Pupillen von Gregor Bar Hebräus* (Diss. Leipzig), Leipzig.

STROHM, H. (1970, tr.): *Aristoteles. Meteorologie/Über die Welt* (Aristoteles Werke in deutscher Übersetzung 12), Berlin.

STROHMAIER, G. (1980): "Homer in Bagdad", Byzantinoslavica (Prague) XLI.196-200. Rep. in id., *Von Demokrit bis Dante. Die Bewahrung antiken Erbes in der arabischen Kultur*, Hildesheim-Zürich-New York 1996, p. 222-226.

STÜVE, W. (1900, ed.): *Olympiodori In Aristotelis Meteora commentaria* (CAG XII/II.), Berlin.

TAKAHASHI, H. (2001): "Simeon of Qalᵉa Rumaita, Patriarch Philoxenus Nemrod and Bar ᶜEbroyo", Hugoye IV/1.

- (2002a): "Syriac Fragments of Theophrastean Meteorology and Mineralogy - Fragments in the Syriac version of Nicolaus Damascenus, *Compendium of Aristotelian Philosophy* and accompanying scholia", in W. Fortenbaugh & G. Wöhrle (ed.), *On the Opuscula of Theophrastus, Akten der 3. Tagung der Karl-und-Gertrud-Abel-Stiftung vom 19.-23. Juli 1999 in Trier*, Stuttgart, p. 189-225.

- (2002b): "The Greco-Syriac and Arabic Sources of Barhebraeus' Mineralogy and Meteorology in *Candelabrum of the Sanctuary*, Base II", Islamic Studies (Islamabad) XLI.215-269.

- (2002c): "Barhebraeus und seine islamischen Quellen. Têḡraṯ têḡrāṯā (Tractatus tractatuum) und Ġazālīs Maqāṣid al-falāsifa", in M. Tamcke (ed.), *Syriaca. Zur Geschichte, Theologie, Liturgie und Gegenwartslage der syrischen Kirchen. 2. Deutsches Syrologen-Symposium (Juli 2000, Wittenberg)* (Studien zur Orientalischen Kirchengeschichte 17), Münster-Hamburg-London, p. 147-175.

- (2002d): "Syriac as the 'Missing Link' in the Transmission of Knowledge - The Case of the Syriac Version of Nicolaus Damascenus' *Compendium of Aristotelian Philosophy*" (in Japanese), The Toyo Gakuho LXXXI/3.370-350 (= 023-043).

- (2003a): "Observations on Bar ᶜEbroyo's Marine Geography", Hugoye VI/1.

- (2003b): "Reception of Ibn Sīnā in Syriac. The Case of Gregory Barhebraeus", in D. Reisman (ed.), *Before and After Avicenna. Proceedings of the First Conference of the Avicenna Study Group* (IPTS 52), Leiden, p. 249-281.

-, *Bio-Biblio.* = *Barhebraeus (Bar ᶜEbroyo): a Bio-Bibliography*, to be published from Gorgias Press, Piscataway (N.J.).

TAQIZADEH, S.Ḥ. (1939-40): "Various Eras and Calendars used in the Countries of Islam", BSOAS IX.903-922, X.107-132.
- (1956): "Djalālī", EI^2 II.397-400.
TAYLOR, F.S. (1937): "The Origins of Greek Alchemy", Ambix I.30-47.
TEULE, H.G.B. (1993, ed./tr.): *Gregory Barhebraeus. Ethicon (Mēmrā I)* (CSCO 534-535, syr. 218-219), Louvain.
TIHON, A. (1978, ed./tr.): *Le "Petit commentaire" de Théon d'Alexandrie aux Tables faciles de Ptolémée* (StT 282), Vatican City.
- (1992): "Les Tables Faciles de Ptolémée dans les manuscrits en onciales (IXe-Xe siècles)", RHT XXII.47-87.
TISSERANT, E. (1941): "Syro-Malabare (Église)", DThC XIV.3089-3162.
TKATSCH, J. (1928-32): *Die arabische Übersetzung der Poetik des Aristoteles*, 2 vols., Vienna-Leipzig.
TOOMER, G.J. (1984, tr.): *Ptolemy's Almagest*, New York.
TULLBERG, O.F. (1842, ed.): *Gregorii Bar Hebraei in Jesaiam Scholia*, Upsala.
ÜLKEN, Hilmi Ziya (1953, ed.): *İbn Sina Risâleleri*, 2 vols. (İstanbul Üniversitesi Edebiyat Fakültesi Yayınları 552), Ankara.
ULLMANN, M. (1972): *Die Natur- und Geheimwissenschaften in Islam* (HO, 1. Abt., Erg. VI.2), Leiden.
- (1979): "Al-Kīmiyā", EI^2 V.110-115.
URI, J. (1787): *Bibliothecae Bodleianae codicum manuscriptorum orientalium videlicet Hebraicorum, Chaldaicorum, Syriacorum, Aethiopicorum, Arabicorum, Persicorum, Turcicorum, Copticorumque Catalogus*, Part I, Oxford [cited according to the new pagination beginning with the section on Syriac manuscripts].
VASCHALDE, A. (1932, tr.): *Iacobi Edesseni Hexaemeron seu Opus creationis libri septem* (CSCO ss. syr. ser. II. tom. LVI, Versio), Paris. Rep. Louvain 1953 (CSCO 97, syr. 48).
VLOTEN, G. van (1895, ed.): *Liber Mafâtîh al-Olûm, explicans vocabula technica scientiarum tam arabum quam peregrinorum, auctore Abû Abdallah Mohammed ibn Ahmed ibn Yûsof al-Kâtib al-Khowarezmi*, Leiden.
VÖÖBUS, A. (1978): "In Pursuit of Syriac Manuscripts", JNES XXXVII.187-193.
VOSTÉ, J.M. (1928): "Catalogue de la bibliothèque syro-chaldéenne du Couvent de Notre-Dame dès Semences près d'Alqoš, Iraq", Angelicum V.3-36, 161-194, 325-358, 481-498.
- (1929): "Manuscrits syro-chaldéens récemment acquis par la Bibliothèque Vaticane", Angelicum VI.35-46.
- (1939): "Catalogue des Manuscrits Syro-Chaldéens conservés dans la Bibliothèque Épiscopale de ᶜAqra (Iraq)", OCP V.368-406.
VULLERS, J.A. (1855-64): *Lexicon persico-latinum etymologicum*, 2 vols., Bonn.
WAGNER, E. (ed./tr.) & P. STEINMETZ (intro./comm.) (1964): *Der syrische Auszug der Meteorologie des Theophrast* (AAWLM.G 1964/1), Wiesbaden.
WALKER, H.H. (1930, ed./tr.): "The Scholia of Bar Hebraeus on the Book of Jeremiah", Diss. Chicago.
WEHR-COWAN = H. WEHR, *A Dictionary of Modern Written Arabic*, edited by J.M. COWAN, 3rd ed., New York 1976.
WEHRLI, F. (1950): *Die Schule des Aristoteles V. Straton von Lampsakos*, Basel.
WEISSER, U. (1974): "Das „Buch über das Geheimnis der Schöpfung" von Pseudo-Apollonios von Tyana - eine späthellenistische physikalische Kosmogonie", Diss. Frankfurt am Main.

- (1979, ed.): *Buch über das Geheimnis der Schöpfung und die Darstellung der Natur (Buch der Ursachen) von Pseudo-Apollonius von Tyana [Sirr al-ḫalīqa wa-ṣanʿat al-ṭabīʿa. Kitāb al-ʿilal. Balīnūs al-ḥakīm]* (Sources & Studies in the History of Arabic-Islamic Science. Natural Sciences Series 1), Aleppo.

- (1980): *Das „Buch über das Geheimnis der Schöpfung" von Pseudo-Apollonios von Tyana*, Berlin-New York.

WELLMANN, M. (1914, ed.): *Pedanii Dioscuridis Anazarbei De materia medica*, Berlin. Rep. Berlin 1958.

WENSINCK, A.J. (1919, tr.): *Bar Hebraeus' Book of the Dove, together with some chapters from his Ethikon*, Leiden.

WIEDEMANN, E., *Aufsätze* = *Aufsätze zur arabischen Wissenschaftsgeschichte* (ed. W. Fischer), Hildesheim-New York 1970.

-, *Ges. Schr.* = *Gesammelte Schriften zur arabisch-islamischen Wissenschaftsgeschichte* (ed. D. Girke & D. Bischoff) (VIGAIW B/1,1-3), Frankfurt 1984.

- (1878): "Zur Chemie der Araber", ZDMG XXXII.575-580 (= *Ges. Schr.* I.13-18).

- (1907): "Zur Alchemie bei den Arabern", Journal für praktische Chemie N.F. LXXVI.65-87, 105-123 (= *Ges. Schr.* I.177-218).

- (1909): "Über chemische Apparate bei den Arabern", in P. Diergart (ed.) *Beiträge aus der Geschichte der Chemie*, Leipzig-Vienna, p. 234-252 (= *Ges. Schr.* I.291-309).

- (1911): "Zur Chemie bei den Arabern" (Beiträge zur Geschichte der Naturwissenschaften XXIV), SPMSE XLIII.72-113 (= *Aufsätze* I.689-730).

- (1912a): "Beiträge zur Geschichte der Naturwissenschaften XXVII. 1) Geographisches von al Bêrûnî. 2) Auszüge aus Al Schîrâzî's Werk über Astronomie. 3) Ueber die Grösse der Meere nach Al Kindî. 4) Geographische Stellen aus den Mafâtîḥ", SPMSE XLIV.1-40 (= *Aufsätze* I. 776-815).

- (1912b): "Zur Mineralogie im Islam" (Beiträge XXX), SPMSE XLIV.205-256 (= *Aufsätze* I. 829-880).

- (1913): "Über optische Täuschungen nach Faḫr al Dîn al Râzî und Nasîr al Dîn al Ṭûsî" (Beiträge XXXIII), SPMSE XLV.154-167 (= *Aufsätze* II.25-38).

- (1916): "Zur Lehre der Generatio spontanea", Naturwissenschaftliche Wochenschrift (N.F.) XV.279-281 (= *Ges. Schr.* II.827-829).

WIET, G. (1955, tr.): *Ibn Rusteh. Les Atours précieux*, Cairo.

WIMMER, F. (1866, ed./tr.): *Theophrasti Eresii opera, quae supersunt, omnia*. Paris. Rep. Frankfurt 1964.

WRIGHT, R.R. (1934, ed./tr.): *The Book of Instruction in the Elements of the Art of Astrology by Abu'l-Rayḥān Muḥammad Ibn Aḥmad al-Bīrūnī*, London.

WRIGHT, W. (1870-72): *Catalogue of the Syriac Manuscripts in the British Museum Acquired since the Year 1838*, London.

-, & S.A. COOK (1901): *A Catalogue of the Syriac Manuscripts Preserved in the Library of the University of Cambridge*, 2 vols., Cambridge.

WÜSTENFELD, F. (1848, ed.): *Zakarija Ben Muhammed Ben Mahmud el-Cazwini's Kosmographie*, Zweiter Theil: *Kitāb āṯār al-bilād* (Die Denkmäler der Länder), Göttingen.

- (1849, ed.): *Zakarija Ben Muhammed Ben Mahmud el-Cazwini's Kosmographie*, Erster Theil: *Kitāb ʿaǧāʾib al-maḫlūqāt* (Die Wunder der Schöpfung), Göttingen.

- (1866-73, ed.): *Jacut's geographisches Wörterbuch*, 6 vols., Leipzig.

YALTKAYA, Şerefettin (1357-8 h., ed.): *Al-Kitāb al-muʿtabar fī al-ḥikma li-sayyid al-ḥukamāʾ auḥad al-zamān Abī al-Barakāt Hibat Allāh b. ʿAlī b. Malkā al-Baġdādī* ..., 3 vols., Hyderabad

ZAYED, Said (1983, ed.): *Ibn Sīnā. Al-Shifāʾ (Al-Ṭabīʿiyyāt) 1 - Al-Samāʿ Al-Ṭabīʿī*, Cairo.

ZIGMUND-CERBÜ, É. (1969, ed./tr.): *Le Candélabre du sanctuaire de Grégoire Aboul'faradj dit Barhebraeus. Dixième base: de la Résurrection* (PO XXXV/2), Turnhout.

al-ZIRIKLĪ, Khair al-Dīn (1928, ed.): *Rasā'il Iḫwān al-ṣafā' wa-ḫillān al-wafā'*, Cairo 1347 h.

ZIYĀDA, Muḥammad Muṣṭafā, & Sa^cīd ^cAbd al-Fattāḥ ^cĀSHŪR (1934-72, ed.): *Kitāb al-sulūk li-ma^crifat duwal al-mulūk li-Taqī al-Dīn Aḥmad b. ^cAlī al-Maqrīzī*, 4 parts, Cairo.

ZONTA, M. (1992): *Fonti greche e orientali dell' Economia di Bar-Hebraeus nell' opera "La crema della scienza"*, Naples.

- (1998): "Structure and Sources of Bar-Hebraeus' 'Practical Philosophy' in *The Cream of Science*", Symp. Syr. VII. 279-292.

ZOTENBERG, H. (1874): *Catalogues des manuscrits syriaques et sabéens (mandaïtes) de la Bibliothèque Nationale*, Paris.

3. *Manuscripts Consulted*

Given in square brackets are the principal works for which the manuscripts were consulted.

BEIRUT, Université Saint Joseph, Bibliothèque Orientale, syr. 48 [BH *But.*].

BERLIN, Staatsbibliothek zu Berlin - Preußischer Kulturbesitz, Orientabteilung, Sachau 91 (211 Sachau) [BH *Tract.*].

BIRMINGHAM, Orchard Learning Resources Centre, Mingana syr. 23 [BH *But.*].

-, Mingana syr. 310 [BH *But.*].

-, Mingana syr. 326 [BH *But.*].

-, Mingana syr. 581 [*Causa causarum*, BH *Cand.* (excerpt)].

CAMBRIDGE, University Library, Add. 2003 [BH *Tract.*].

-, Gg. 2.14, fols. 328r-385v [Nic. syr.].

FLORENCE, Biblioteca Nazionale Medicea Laurenziana, or. 69 (olim Palat. or. 186 Assemani) [BH *But.*].

-, or. 83 (olim Palat. or. 187 Assemani) [BH *But.*]

-, or. 86 (olim Palat. or. 185 Assemani) [BH *Ind.*]

-, or. 298 (olim Palat. or. 62 Assemani) [BH *Gramm.*, list of BH's works, BH *Carmina*, Anon. *Causa causarum* etc.]

-, or. 342 (olim Palat. or. 200 Assemani) [BH *Tract.*].

GÖTTINGEN, Niedersächsische Staats- und Universitätsbibliothek, orient. 18c (syr. 3) [Bar Shakko, *Dial.*].

LEEDS, University of Leeds, Brotherton Library, coll. Budge 2, fols. 114v-137v [Biography of BH by John Elijah Mellus].

LONDON, British Library, Add. 7193 [Bar Shakko, *Thes.*].

-, Or. 4080 [BH *Tract.*].

-, Or. 4179 [BH *But.*].

-, Or. 9380 [BH *But.*].

MANCHESTER, John Rylands University Library, Syr. 44 (Harris 165) [BH *But.*].

-, Syr. 56 [BH *But.*].

OXFORD, Bodleian Library, Huntingdon 1 (122 Payne Smith) [BH, *Horr.*, *Gramm.*, *But.* etc.].

-, Or. 467 (171 Payne Smith) [BH *Rad.*].

PARIS, Bibliothèque Nationale, syr. 213 [BH *Rad.*]
-, syr. 214 [BH *Rad.*]
-, syr. 241 [Bar Kepha, *Hex.*].
-, syr. 244 [BH *Asc.*].
-, syr. 311 [Bar Kepha, *Hex.*].
-, syr. 316 [Bar Shakko, *Thes.*].
-, syr. 346 [excerpts of Nic. syr.].
PRINCETON, Princeton Theological Seminary, Speer Library, Cabinet C, Nestorian 25 [BH *But.*].
TASHKENT, Uzbek Academy of Sciences, Abu Raihon Berunii nomidagi Sharq-shunoslik Instituti 2385, fols. 347r-368r [Olymp. arab.].
VATICAN, Bibliotheca Apostolica Vaticana, syr. 169 [BH *Rad.*].
-, syr. 469, fols. 132v-151v [BH *But.*].
-, syr. 603-604 [BH *But.*].
-, syr. 613-615 [BH *But.*].
-, Borg. syr. 145 [BH *Rad.*]

INDEX VERBORUM

Registered in the index are all words that occur in the text edited above with the exception of the more common prepositions, conjunctions and particles, as well as the verb *hwā* in its auxiliary usage. - The letters "N" or "DM" after an entry indicate that the word occurs in the source passage in Nic. syr. or *De mundo* syr. - The Arabic words given in brackets are those which occur at the corresponding places in the source passages of Ibn Sīnā (IS), Abū al-Barakāt (AB), Fakhr al-Dīn al-Rāzī (FR) or Naṣīr al-Dīn al-Ṭūsī (Ṭ), whereby "cf." indicates that the correspondence, either in meaning or in the position within the passage, is not so exact.

ܐ

ܐܐܪ, ܐܐܪܣ (ἀήρ) [* = هواء]: **Min. I**.ii.3.2 (IS*); iii.1.1; **II**.i.1.2 (N/IS*); i.1.3, 6 (N); i.4.4 (N); ii.1.7; ii.1.10 (N); ii.3.3 (N); ii.3.7 (IS*); iii.1.6 (N); iii.1.7 (IS جو); **IV**.ii.2.8; ii.5.6 (IS*); ii.1.2 (IS*); V.iii.2.8; **Mete. I**.i.1.6 (AB جو); i.1.8bis (sec. pl.); i.2.1, 3, 4 (IS*); i.2.5 (IS جو); ii.1.3, 4bis, 7; ii.2.6, 7 (AB*); ii.3.9; ii.5.2, 4 (AB جو); ii.6.1; ii.6.3 (AB جو); iii.2.3; iii.4.6 (IS جو); iii.5.3; iii.6.9; **II**.iii.1.1, 2 (IS*); iii.2 3 (IS جو); iii.2.5bis, 6 (N); iii.2.8 (IS*); iv.1.6 (FR*); **III**.i.1.1; i.1.6 (FR*); i.2.5 (N); iv.1.5 (N); iv.4.6 (IS جو); **IV**.i.1.3 (FR*); i.1.7 (IS*); i.3.4 (IS*); i.4.4; i.4.7 (IS*); ii.3.2, 4 (N); iii.2.3, 4 (IS*); iii.4.7 (IS جو); V.i.1.8 (IS*); i.2.6 (IS*); i.3.5 (IS*). - ܐܐܪܝܐ [* = هوائي]: **Min. IV**.iv.1.4 (AB*); **Mete. I**.iii.6.2; **V**.i.1.9 (IS*); ii.2.8. - ܐܐܪܝܘܬܐ [* = هوائية]: **Min. III**.i.3.1 (IS*); ii.1.2, 3 (AB*); ii.1.5 (IS/AB*); ii.3.7 (AB*).

ܐܒܐ, ܐܒܗܬܐ: **Min. I**.i.2.10; **III**.ii.2.7; **Mete. V**.ii.1.7; ii.2.10, 11.

√ʾbd: Pe: **Min. V**.iii.2.7 (IS هَلَكَ). - ܐܒܕܢܐ: **Mete. V**.iii.1.7. - ܡܘܒܕܢܐ: **Mete. V**.i.1.9 (IS مُفسد); iii.2.2.

ܐܒܘܒܢܐ: **Min. I**.ii.5.8.

ܐܒܘܠܐ: **Mete. IV**.iii.4.7 (IS منفذ).

ܐܒܪܐ [* = رصاص]: **Min. III**.i.5.8; ii.1.8bis, 9 (AB/IS*); ii.2.3 (IS*); ii.3.6 (IS رصاص قلعي; AB*); iii.1.9 (IS*).

ܐܓܡܐ: **Mete. I**.ii.5.1.

ܐܓܢܐ **Min. II**.ii.2.10 (N).

ܐܓܘܓܐ (ἀγωγός): **Min I**.iii.2.2.

ܐܓܘܣܛܘܣ (nom. prop. "Augustus"): **Min. IV**.i.2.6.

ܐܕܡܐܣ (ἀδάμας): **Min. III**.i.2.3 (N).

ܐܕܡܬܐ **Mete. V**.ii.2.10.

ܐܕܪܝܐܛܝܩܘܣ (nom. prop. "Adriaticus"): **Min. V**.i.2.3 (DM); i.2.4.

√ʾdš: Ethpa: **Min. III**.iii.2.2 (IS صار أنواعًا). - ܐܕܫܐ [* = نوع]: **Min. III**.i.5.8; iii.1.5 (IS*); **Mete. V**.ii.1.1 (IS جنس); ii.3.4 (IS/FR*); ii.3.10; iii.1.7. - ܐܕܫܢܝܘܬܐ: **Min. III**.iii.1.5 (IS الفصل المُنَوِّع).

ܐܘ (in sensu "quam"): **Min. V**.ii.2.7.

ܐܘܣܝܐ, ܐܘܣܝܐܣ (οὐσία) [* = جَوهَر]: **Min. I**.i.4.11; **III**.i.1.2; i.2.5bis (IS*); i.5.1, 3 (IS/AB*); ii.3.4 (IS*); **Mete. I**.ii.2.1, 2bis; iii.5.1 (IS*); **IV**.i.3.5 (IS*). - ܐܘܣܝܢܐܝܬ: **Min. I**.ii.2.1 (IS بالذات); **III**.iii.2.1 (cf. IS جوهر). - ܐܘܣܝܢܝܘܬܐ: **Min. III**.iii.2.8.

ܐܘܦܐܩܩܘܡܐ (ὑπέκκαυμα): **Mete. IV**.iii.1.6.

ܐܘܩܝܐܢܘܣ (nom. prop. "Oceanus"): **Min. V**.i.1.4 (DM).

ܐܘܩܝܡܘܢ (ὤκιμον): **Mete. V**.ii.1.6.

ܐܘܪܕܥܐ: **Mete. V**.ii.1.6 (IS/FR ضفدع).

ܐܘܪܝܙܘܢ (ὁρίζων) [* = أُفُق]: **Min. II**.iii.1.3 (N); **IV**.i.3.1; iv.2.5; **Mete. II**.iii.3.5bis (IS/FR*); **III**.ii.1.2, 9, 10 (IS*); ii.2.3, 4, 6, 7, 10, 11.

ܐܘܪܘܦܐ (nom. prop. "Europa"): **Min. V**.i.1.5.

√ʾzl: ܡܐܙܠܢܘܬܐ [* = فَلَك]: **Min. IV**.iv.2.3; **Mete. II**.iv.1.9 (FR*); **III**.i.1.8 (FR*); **IV**.i.5.6. - ܡܐܙܠܢܝܬܐ: **Mete. V**.ii.2.2 (cf. IS فلك); ii.3.9 (cf. IS فلك).

√ʾḥd: Pe: **Min. IV**.iv.4.8 (cf. AB حافظ). - Ethpe: **Min. V**.iii.2.6 (cf. IS ضَبط). - ܐܘܚܕܢܐ **Min. IV**.i.1.8.

√ʾḥr: Eshtaphal (ܐܫܬܘܚܪ): **Mete.** **III**.iii.4.7 (IS تأخّر); **IV**.iii.3.6 (cf. IS أبطأ). - ܐܚܪܢܐ: **Min.** **I**.i.3.6; **III**.iii.2.4; **V**.iii.2.12; **Mete.** **II**.iii.7.1; **V**.i.2.5. - ܐܚܪܝܐ [* = آخَر]: **Min.** **I**.i.2.5; ii.6.7 (IS*); iii.4.5; **II**.ii.1.10 (N); iii.4.5; **III**.i.5.8; iii.1.5; **IV**.i.1.5 (cf. IS غيره); i.2.8; i.4.1 (IS*); ii.1.4, 6, 7; ii.2.3; ii.2.8; ii.4.3, 6; iii.6.10; iv.3.9; iv.4.2 (AB*); **V**.ii.1.1 (IS*); ii.2.3; ii.3.1 (IS*); iii.1.2, 6, 7; iii.2.8, 9; iv.3.3; **Mete.** **I**.i.3.5; ii.2.1 (AB غير); ii.4.5bis, 6 (AB*); **II**.i.1.2 (IS*); i.5.7, 8 (IS*); i.6.3 (IS*); ii.3.2; ii.4.1, 4; iii.2.9 (IS*); iii.7.9 (IS*); **III**.i.1.3; i.3.2; ii.2.1, 3, 6, 7, 10, 11, 12; ii.3.2; iii.4.8; iii.5.10 (N); iv.3.2 (IS*); **IV**.ii.3.5 (N); iii.1.2; iii.2.5 (IS*); **V**.ii.3.3 (IS*); ii.3.5, 6 (FR*). - ܐܚܪܝܬܐ: **Min.** **I**.i.5.10; **II**.i.4.9 (N). - ܐܚܪ̈ܢܝܢ, ܐܚܪ̈ܢܝܬܐ: **Min.** **I**.iii.4.4. - ܡܫܬܘܚܪ (Shaphel ptc. pass.): **Mete.** **I**.iii.1.1 (IS مُتباطئ).

ܐܛܠܢܛܝܩܘܣ (nom. prop. "Atlanticus"): **Min.** **V**.i.1.1 (DM); i.3.1.

√ʾṭm: ܐܛܡܘܣ: **Min.** **V**.iii.2.10; **Mete.** **IV**.ii.2.5 (N).

ܐܬܪܘܓܐ: **Min.** **III**.iii.3.8 (AB أُترُجّة); iii.3.10 (AB أُترُجّ).

ܐܝܒܪ̈ܝܐ (nom. prop. "Iberus"): **Min.** **V**.i.3.9. - ܐܝܒܪܝܐ (nom. prop. "Iberia/Iberi"): **Min.** **V**.i.2.7.

ܐܓܘܦܛܘܣ (nom. prop. "Aegyptus"): **Min.** **V**.i.3.4.

ܐܝܛܐ (εἶτα): **Min.** **I**.i.1.3; i.4.6 (IS ثم); **II**.iii.3.10 (N ܡܢ ܒܬܪ); **III**.i.3.3 (IS ثم); **V**.i.3.5; **Mete.** **III**.iii.2.5; **IV**.i.4.5 (N ܒܬܪܗ ܕܝܢ); **V**.iii.2.5.

ܐܝܛܠܝܐ (nom. prop. "Italia"): **Min.** **V**.i.3.10.

ܐܝܟܢܝܘܬܐ: **Mete.** **I**.i.tit. (IS كيفية); **II**.i.1.4.

ܐܝܠܢܐ [* = شجرة/شجر]: **Min.** **I**.i.5.6; **Mete.** **II**.i.4.3bis, 6 (IS*); i.5.5; **III**.iv.4.4 (N/IS*); **V**.ii.1.3; ii.3.6bis (IS*).

ܐܝܡܡܐ [* = نهار]: **Min.** **II**.iii.1.10 (N); **IV**.ii.4.1 (IS*); .ii.4.2, 4; iii.2.3; iv.3.1, 2 (AB*); iv.3.7, 8bis (AB*); iv.4.1; iv.4.3, 4, 5bis (AB*); iv.4.10; **Mete.** **II**.ii.2.6; iv.1.8 (FR يوم); **IV**.i.5.3 (N); i.6.6bis (N).

ܐܝܢܐ ܕܗܘ ("quivis"): **Mete.** **III**.iii.3.9.

ܐܝܟܢܝܘܬܐ: **Min.** **I**.i.1.4 (كيفية); i.2.8.

ܐܝܣܡܪܝܢܘܣ (ἰσημερινός) [* = معدل النهار]: **Min.** **IV**.i.3.2; ii.1.2 (IS*); iii.5.2; **Mete.** **V**.i.3.3 (IS*). - ܐܝܣܡܪܝܢܝܐ: **Mete.** **III**.iii.1.10 (N).

ܐܟܪ̈: **Min.** **IV**.ii.6.9.

ܐܝܬ: headings passim; **Min.** **I**.i.3.1, 2; i.5.2; ii.6.6; **II**.i.4.2; ii.1.6; ii.2.4; **III**.ii.3.7; **IV**.i.2.2; i.4.10; **V**.i.1.6; ii.3.5 (N); **Mete.** **I**.ii.5.4; **III**.iii.5.10; **IV**.i.2.8; i.3.5, 8; iii.3.2, 6; **V**.ii.3.9. - ܐܝܬ ܐܡܬܝ ܕ: **Min.** **I**.i.1.9; **II**.ii.3.5; **Mete.** **II**.iii.7.9; **III**.i.4.5; **IV**.i.3.1 (IS ربّما); iii.4.7; **V**.i.2.4; ii.3.8. - ܐܝܬܘܗܝ, ܐܝܬܝܗ, ܐܝܬܝܗܘܢ etc.: **Min**.tit.; **I**.i.4.1; **II**.ii.1.1; ii.3.4 (N); ii.4.1; **III**.i.2.6; i.3.3 (supplevi); i.3.5; i.5.1; **IV**.i.3.7; **V**.i.1.5; i.2.6; ii.4.1 (N); ii.4.5, 6, 8; iii.2.4; iii.3.6; iv.1.2, 5bis, 7; iv.3.1, 4, 5; **Mete**. tit.; **I**.ii.2.2, 5; ii.5.5, 7; iii.2.5; iii.3.5, 9; iii.5.1; **II**.i.1.2; i.2.1; i.3.4; ii.1.1 (N); ii.1.2; ii.3.2; ii.3.3 (IS كان); iv.3.2; **III**.i.1.1, 3; i.2.2 (N); i.3.3 (N); i.3.9 (neg. corr. IS ليس); i.4.1; ii.2.2, 4; iii.5.3; iv.3.3, 5; **IV**.i.5.1 (N); ii.1.1; iii.1.2; **V**.i.1.3, 4; ii.3.9; iii.1.3.

ܐܟܘܬ (+ suff.): **Min.** **II**.iii.2.2; **IV**.i.2.4; **V**.iv.3.6; **Mete.** **V**.ii.3.3.

√ʾkm: Aph: **Mete.** **IV**.ii.2.4, 8 (N/IS سَوَّد). - ܐܘܟܡܐ [* = أسود]: **Mete.** **II**.i.4.8, 10 (cf. IS سَواد); i.6.4, 5 (IS*); ii.2.1, 4 (IS*); iii.5.4bis (N); iii.7.7 (IS*); **IV**.i.2.7 (IS/FR*); iii.4.6 (FR*). - ܐܘܟܡܘܬܐ: **Mete.** **II**.ii.2.4, 5; iii.5.6 (IS سَواد); iii.7.5 (IS ظُلمة).

ܐܟܣܩܘܣܛܐ (ἑξηκοστή): **Min.** **IV**.ii.3.8.

ܐܟܣܢܝܐ: **Mete.** **V**.ii.2.1.

ܐܟܣܐܪܡܐ (ἔξαρμα): **Mete.** **II**.iii.3.7.

√ʾlh: ܐܠܗܐ: **Min.** **II**.iii.4.6 (IS الله). - ܐܠܗܝܐ: **Min.** **IV**.i.2.9 (IS إلهي); **Mete.** **V**.iii.1.2 (cf. IS الله); iii.1.8.

ܐܠܟܣܢܕܪܝܐ (nom. prop. "Alexandrinus"): **Min.** **IV**.i.2.5.

ܐܠܢܝܐ (nom. prop. "Alanus"): **Min.** **V**.i.3.9.

√ʾlp: ܐܠܦܐ ("navis"): **Min.** **V**.ii.3.3 (N); iii.2.5; **Mete.** **III**.iv.4.5 (N/IS مَركَب). - ܐܠܦܐ ("mille"): **Mete.** **IV**.iii.3.8.

√ʾlṣ: Pe: **Min.** **IV**.iii.1.6; iii.5.5; iii.6.7; **V**.iii.1.1 (cf. IS واجب); **Mete.** **II**.i.7.2 (IS

(وَجَبَ); **V**.i.3.3; ii.2.2 (IS أوجبَ); ii.3.5 (FR وَجَبَ; cf. IS واجب). - ܐܠܝܨ: **Min. V**.i.1.2 (DM); iii.3.3 (cf. IS ضَاق); **Mete. I**.ii.1.2 (cf. AB مُضيق); ii.1.6 (AB ضيق). - ܐܠܝܨܘܬܐ: **Min. IV**.i.4.2 (IS وجوب); **Mete. V**.ii.1.2.

√ʾmm: ܐܡܐ, ܐܡܗܬܐ: **Mete. III**.ii.3.2 (IS أمّ); **V**.ii.1.7; ii.2.11bis.

√ʾmn: Paiel (ܗܝܡܢ): **Min. III**.iii.1.3 (cf. IS تصديق). - ܡܗܝܡܢ [* = دائم]: **Min. II**.iii.4.2 (N); **IV**.ii.6.6 (IS*); **V**.i.1.7 (IS*). - ܡܗܝܡܢܐܝܬ: **Min. IV**.iv.5.3 (AB أبدًا); **Mete. I**.ii.3.7 (cf. DM ܡܗܝܡܢܐ). - ܒܐܡܝܢܘ: **Min. I**.ii.1.4; **IV**.iii.6.8 (AB في كل زمان); **Mete. II**.iii.8.7; **III**.ii.2.2, 4 (cf. IS دائم); **IV**.iii.4.1; **V**.ii.2.1. - ܐܡܝܢ: **Mete. V**.iii.1.9. - ܐܘܡܢܐ: **Min. I**.i.4.9; I.iii.1.8. - ܐܘܡܢ̈ܝܬܐ: **Min. III**.iii.1.4. - ܐܘܡܢܘܬܐ [* = صناعة]: **Min. III**.iii.1.1, 2 (IS*); iii.2.4; **Mete. V**.iii.tit.; iii.1.1, 4 (IS*).

√ʾmr: Pe [* = قَال]: **Min. I**.i.3.7; i.4.1; i.5.1; ii.5.4, 5; **II**.i.1.3; i.2.1, 5 (N); i.3.1; i.4.1, 6, 7; ii.1.1 (N); iii.4.1 (N); iii.4.2; **III**.iii.3.1; **IV**.i.4.7; ii.2.1; ii.3.1; ii.6.6, 8; iii.1.3 (Ṭ*); iii.6.2; iv.2.4; **V**.ii.2.3; ii.4.1 (N/IS*); ii.4.2; iii.1.8; iii.2.1; iv.1.5; iv.2.1; **Mete. I**.i.1.3; ii.3.2 (AB*); ii.6.1; iii.6.1; **II**.i.1.5; i.3.1, 7; ii.2.4; ii.4.3 (IS ذكر); ii.4.4 (IS حَكَى); ii.4.8; iii.2.1; iii.7.1, 10; iv.2.3 (IS*); **III**.i.1.2; i.2.1; i.2.4 (N); iii.1.5bis, 6; iii.2.6, 7; iv.1.3; **IV**.i.2.8; i.5.1 (N/IS*); i.5.4 (N); i.6.1 (N); iii.2.1 (IS*); **V**.i.1.2 (IS*); iii.1.6. - Ethpe: **Min. IV**.iii.3.1; iii.4.1; iii.5.1; **Mete. V**.i.1.1. - ܐܡܝܪ (ptc. pass.): **Min. I**.i.4.11; **Mete. II**.iii.6.1.

ܐܡܪܐ ("Aries"): **Min. IV**.ii.3.4; ii.6.3; iii.1.4; **Mete. III**.ii.1.4, 6.

ܐܢܐ ("ego"): **Min. I**.i.2.10; **IV**.ii.6.5; iii.6.2; **Mete. I**.i.2.6, 8; iii.4.9; **II**.iii.7.10, 11.

ܐܢܟܣܓܘܪܣ (nom. prop. "Anaxagoras"): **Min. II**.i.1.2 (N); **Mete. IV**.i.5.4 (N).

ܐܢܟܣܝܡܢܣ (nom. prop. "Anaximenes"): **Min. II**.i.3.1 (N/IS اراكيماس).

ܐܢܠܘܓܝܐ (ἀναλογία): **Min. V**.ii.4.4 (N/IS تأويل).

ܐܢܩܠܣܝܣ (ἀνάκλασις): **Mete. II**.i.4.1 (N); i.4.7.

ܐܢܒܝܩ: **Min. I**.ii.5.5, 8 (IS انبيق).

ܐܢܛܝܘܟܝܐ (nom. prop. "Antiochia"): **Min. IV**.iii.6.11; **Mete. I**.iii.4.9.

ܐܢܛܝܦܪܘܢ (nom. prop. "Antipheron"): **Mete. II**.iii.2.3 (N).

ܐܢܟ: **Min. III**.i.5.8; ii.3.8 (IS أنّك).

ܐܢܢܩܐ (< ἀνάγκη): **Mete. V**.ii.3.7 (IS/FR ضَرُوري); iii.2.6.

√ʾnp: ܐܦܐ [* = وجه]: **Min. I**.ii.4.5; iii.1.6; iii.1.8 (IS*); iii.2.2; **II**.ii.1.4 (FR*); ii.2.3; ii.4.7 (N); iii.1.9, 11 (IS*); iii.2.8 (IS*); **IV**.iv.2.6; iv.4.7; **V**.iii.3.2 (IS*); **Mete. I**.iii.2.7 (AB*); **IV**.i.3.7 (IS*). - ܒܐܦܝ: **Mete. I**.i.3.3 (IS قُدّام). - ܠܐܦܝ: **Min. II**.i.1.6 (N); iii.1.6; **V**.i.2.4, 5; **Mete. I**.ii.1.2, 5 (AB إلى); **III**.i.1.6, 9; i.2.2, 3 (N); i.2.8; **IV**.i.1.4, 6.

√ʾnš: ܐܢܫܐ ("homo"): **Min. I**.iii.2.2; **IV**.i.4.10, 11. - ܐܢܫ, ܐܢܫܝܢ ("quidam"): **Min. I**.ii.1.7; **II**.iii.4.1 (N); **III**.iii.2.3; **IV**.ii.2.6; ii.6.8; **V**.ii.1.9; iv.3.8; **Mete. I**.ii.2.3 (AB انسان); ii.3.3 (AB ناس); **II**.ii.4.3; iii.1.8 (IS انسان); iii.1.9; iv.2.3 (IS بعضهم); **III**.i.2.1 (IS بعض); i.3.1 (IS قوم); iii.1.5; iii.4.7; **V**.iii.1.2. - ܐܢܫ̈ܐ: **Min. IV**.i.4.9; ii.2.4; **V**.iii.1.3; iv.3.7; **Mete. V**.ii.2.9; iii.2.2. - ܐܢܫܘܬܐ: **Min. V**.i.1.7; **Mete. V**.iii.2.7. - ܒܪܢܫܐ, ܒ̈ܢܝܢܫܐ [* = انسان]: Mete. I.iii.1.5; II.i.1.3 (IS*); II.iii.2.4 (N); V.ii.3.5bis (IS*); V.iii.1.2bis, 7.

ܐܫܝܬܐ, ܐܫܝܐ [* = حائط]: **Min. I**.i.4.5, 6 (IS*); **IV**.iv.5.2 (AB جدار); **Mete. II**.i.4.5 (IS*); iii.2.8, 9 (IS*).

√ʾsy: ܐܣܝܘܬܐ [* = طبّ]: **Min. I**.ii.6.8 (IS*); **II**.iii.3.5 (IS صناعة الطبّ).

ܐܣܛܕܝܘܢ, ܐܣܛܕ̈ܝܐ (στάδιον): **Mete. II**.ii.4.6 (IS اسطاذيا); iv.2.7 (IS فرسخ).

ܐܣܛܘܟܣܐ [* = عُنصر]: **Min. I**.i.1.7; **III**.ii.tit.; ii.2.1 (IS*); iii.3.10 (AB*); **IV**.i.1.4, 7 (IS*); i.2.2; **V**.iv.3.4; **Mete. I**.i.1.6; **V**.i.1.3 (IS*); i.1.6 (cf. IS عُنصري); ii.2.5. - ܐܣܛܘܟܣܢܝܐ: **Mete. V**.ii.2.3 (cf. IS عُنصر).

ܐܣܝܐ (nom. prop. "Asia"): **Min. V**.i.1.4.

ܐܣܟܡܐ (σχῆμα) [* = شَكل]: **Min. II**.ii.3.3 (N); **V**.ii.4.2 (N); iii.2.10; **Mete. I**.iii.4.10; **II**.i.5.2 (IS*); i.9.1, 2bis (N/IS*); ii.1.8 (FR*); ii.1.9; iii.3.1;

iv.1.3, 5 (IS/FR*); iv.1.6 (FR*); **III**.iv.3.8 (IS تشكُّل, v.l. شَكل); **V**.i.3.6; ii.2.2 (IS تشكُّل).

ܐܣܦܘܓܐ (σπόγγος): **Min. V**.ii.1.7.

ܐܣܦܢܝܐ (nom. prop. "Hispania"): **Min. V**.i.3.11.

ܐܣܦܝܪܐ (σφαῖρα) [* = كُرة]: **Min I**.i.4.5 (IS*); **IV**.i.3.1, 4; ii.1.1 (IS*); iv.5.3 (AB*); **V**.ii.1.9, 10 (IS*); iv.1.7; **Mete. III**.i.1.7 (FR*). - ܐܣܦܝܪܝܐ: **Min. II**.i.4.5 (N); **IV**.i.1.7 (cf. IS كُرية); **Mete. I**.iii.4.7 (N; cf. IS استدارً); **II**.i.5.4 (IS مُستدير). - ܐܣܦܝܪܝܘܬܐ: **Min. IV**.i.1.5 (IS كُرية, تدوير).

√ʾsr: Ethpe: **Min. V**.ii.3.7 (N/IS مكتوف).

√ʾpy: ܐܦܝܐ: **Min. I**.i.3.7.

ܐܦܪܩܛܝܐܣ (ἀπαρκτίας): **Mete. III**.ii.3.5 (N).

ܐܦܘܕܘܣ (ἔφοδος): **Min. IV**.i.2.6.

ܐܦܘܕܝܩܣܝܣ, ܐܦܘܕܝܩܣܝܐ (ἀπόδειξις): **Mete. II**.i.9.5; **V**.ii.1.1 (IS برهان).

ܐܦܘܛܠܣܡܛܝܩܘ (ἀποτελεσματικοί): **Min. IV**.i.2.3.

ܐܦܘܓܝܘܢ (ἀπόγειον) [* = أوج]: **Min. IV**.i.2.7 (IS*); **Mete. V**.i.3.2 (IS*).

ܐܦܘܟܐ, ܐܦܘܟܝܣ (ἐποχή): **Min. IV**.i.2.7 (IS مكان).

ܐܦܣܘܣ (nom. prop. "Ephesus"): **Mete. IV**.ii.3.4 (N ܐܦܣܣܘܣ).

ܐܦܪܝܓܝܘܢ (περίγειον) [* = حضيض]: **Min. IV**.i.2.7 (IS*); i.4.4; **Mete. V**.i.3.2 (IS*).

√ʾsr: ܐܣܝܪܐ: **Mete. III**.i.3.6 (N); iv.1.4 (N).

ܐܪܐ (ἄρα): **Min. III**.ii.1.2; ii.2.7; **IV**.ii.4.6; ii.5.8; iii.4.3.

ܐܪܐܒܝܐ (nom. prop. "Arabia"): **Min. V**.i.3.3.

ܐܪܓܘܢܐ: **Mete. II**.iii.4.4 (IS أرجوانية); iii.5.7 (IS أرجواني); iii.7.4bis; iii.7.6 bis (IS أرجواني); iii.7.9 (IS أرجوانية). - ܐܪܓܘܢܝܬܐ: **Mete. II**.iii.6.4; iii.7.5 (supplevi); iii.7.6 (cf. IS أرجوانية).

ܐܪܓܝܣ (ἀργής): **Mete. IV**.ii.2.1 (N ܐܪܓܝܣ).

ܐܪܕܠܐ: **Min. I**.i.2.11.

√ʾrḥ: ܐܘܪܚܐ: **Min. II**.ii.1.3; **Mete. IV**.iii.4.3.

ܐܪܝܐ ("Leo"): **Min. IV**.ii.5.2, 4 (IS الأسد); **Mete. IV**.iii.3.9.

√ʾrk: ܐܪܝܟ: **Min. II**.iii.1.3 (N/IS مستطيل); **V**.iii.1.2; **Mete. I**.ii.1.8 (cf. AB أطول); **V**.ii.2.3 (IS طويل). - ܐܪܝܟܘܬܐ: **Min. IV**.ii.4.3, 4; iv.3.9; iv.4.1. - ܐܘܪܟܐ: **Min. IV**.i.3.6; i.3.7 (N/IS طول); **Mete. I**.iii.3.4 (cf. AB طولي); **IV**.iii.1.7, 8 (N).

ܐܪܡܢܝܩܘܢ (ἀρμενιακόν, recte ἀμμωνιακόν) [* = نوشادر]: **Min. III**.i.3.3 (IS*); i.3.4 (supplevi, IS*); i.5.5 (IS*); i.5.7; **Mete. IV**.i.3.8 (IS*).

ܐܪܣܢܘܝܛܝܣ (nom. prop. "Arsinoitis"): **Min. V**.iii.2.2.

ܐܪܣܢܝܩܘܢ (ἀρσενικόν): **Min. III**.i.5.6 (IS زرنيخ); i.5.7.

√ʾrʿ: ܐܪܥܐ [* = أرض]: **Min. I**.i.1.8; i.2.1 (IS*); i.2.6 (IS*); i.3.5 (IS*); i.4.5, 6 (IS*); ii.1.3, 5 (AB*); ii.1.6; ii.2.2 (IS*); ii.2.3 (IS*); ii.4.5; ii.5.1, 7 (IS*); ii.6.4 (IS*); iii.1.1; iii.1.3 (IS*); iii.1.5, 6; iii.1.7 (IS*); iii.2.1, 2; iii.2.4bis (IS*); iii.2.6 (AB*); iii.3.2, 3, 4, 5, 7 (AB*); **II**.tit.; i.tit.; i.1.1 (IS*); i.1.3bis, 4, 8 (N); i.2.1; i.2.2, 4bis, 5 (N); i.3.1; i.3.3bis (N); i.4.3, 4, 5, 7, 8 (N); ii.1.1 (N); ii.1.4 (FR*); ii.1.6 (N/IS*); ii.1.7; ii.1.9 (IS); ii.2.3; ii.2.6; ii.2.8 (IS*); ii.3.3 (N); ii.3.6, 8; ii.4.1; ii.4.7 (N); iii.1.2 (N); iii.1.4 (N/IS*); iii.1.5, 8; iii.1.11 (IS*); iii.2.4; iii.2.8 (IS*); iii.3.2 (N); iii.4.3 (N); iii.4.5 (IS*); iii.4.7; **IV**.i.tit.; i.1.1, 2, 4, 7, 8 (IS*); i.3.1, 4, 10, 11; i.4.3bis (IS*); ii.1.1, 3, 5 (IS*); iii.3.4bis, 5; iv.2.4, 6; iv.4.7bis (AB*); iv.5.3 (AB*); **V**.i.1.1; i.3.2, 5, 7, 8bis; ii.3.8; ii.4.1 (N/IS*); ii.4.7 (IS*); iv.1.2, 4, 7; iv.2.2, 3; iv.3.3; **Mete. I**.i.1.1; i.1.4bis, 8 (AB*); ii.1.1; ii.4.2, 3 (AB*); ii.5.3 (AB*); ii.6.4 (IS*); iii.2.5, 6, 7 (AB*); iii.4.5 (N); iii.4.7; iii.6.2 (cf. IS أرضي); iii.6.3; **II**.iii.3.4, 6 (FR*); **III**.i.1.3 (N); i.3.1, 2 (N/IS*); i.3.8 (N); i.3.10 (IS*); i.4.7 (N); iii.1.7 (N); iii.2.6, 7 (N); iii.3.4 (N); iii.4.5 (IS*); iv.3.2 (IS*); iv.4.4; **IV**.i.3.4; i.3.7 (IS*); i.6.4 (N); ii.1.4, 6 (IS*); ii.3.3; ii.3.9 (N); iii.4.8, 9, 10; **V**.i.1.4; i.3.5 (IS*); ii.2.7; ii.3.6. - ܐܪܥܢܝܐ: **Mete. IV**.ii.1.4 (IS أرضي). - ܐܪܥܢܝܐ [* = أرضي]: **Min. I**.ii.1.9 (AB*); **III**.i.2.5 (IS*); ii.3.5 (IS*); **IV**.iv.1.3

(AB*); **Mete. I**.iii.6.4; **IV**.iii.2.6 (IS*); **V**.i.1.10 (IS*). - ܐܪܥܢܝܬܐ: **Min. I**.i.2.7 (IS أرضية). - ܐܪܥܢܝܬܐ [* = أرضية]: **Min. I**.i.3.2; ii.6.2 (IS*); iii.4.2, 4 (IS*); **III**.i.1.5 (IS*); i.3.1 (IS*); i.3.4 (IS*); ii.1.1, 5 (IS*); **V**.ii.1.3 (IS*); ii.3.1 (IS*); **Mete. IV**.i.2.2 (IS/FR*).
√ʾšd: ܐܫܝܕܘܬܐ **Min. III**.ii.1.3 (AB ميعان); **Mete. V**.i.1.10 (IS سَيَلان).
√ʾty: Pe: **Min. IV**.iii.6.4 (AB وَرَدَ); **Mete. II**.i.3.8 (N); **V**.iii.1.9. - Ettaph: **Min. V**.iii.3.7; **Mete. III**.iv.2.3 (N); **IV**.i.1.5, 6 (N).
ܐܬܐ, ܐܬܘܬܐ [* = عَلامة]: **Mete. I**.iii.6.8 (IS*); **III**.iv.3.3, 4 (IS*); **IV**.iii.4.1 (IS أثر); iii.4.6 (IS*); iii.4.10 (IS أثر).
ܐܬܐܝܠ: **Min. I**.ii.5.7 (IS قرع).
√ʾtr: ܐܬܪܐ: **Min. I**.i.4.4 (IS بلاد); i.4.7; ii.4.3; iii.3.4; **II**.i.4.10 (N); ii.4.2 (IS بلاد); iii.2.3; **IV**.i.2.10 (IS مكان); i.4.6, 10; ii.tit. (IS بلاد); ii.3.2 (IS موضع); ii.4.6; ii.6.7, 8 (IS بلاد); iii.3.2; iii.4.4; iii.6.8; iv.3.1 (AB موضع); iv.3.7, 8; iv.4.5; iv.5.4 (AB صُقع); iv.5.7, 8; **V**.iii.2.8; **Mete. I**.i.2.5 (IS موضع); i.2.10; ii.1.4; ii.4.4 (AB بلاد); ii.5.2 (AB بلاد); ii.5.5 (AB أرض); **II**.iii.3.8 (IS بلاد); **III**.iii.tit.; iii.1.3, 4 (IS بلاد); **IV**.i.5.6; iii.2.3 (IS حَيِّز); **V**.i.3.4; ii.3.10. - ܐܬܪܢܝܬܐ: **Min. IV**.iv.tit.

ܒ

√bʾr: ܒܐܪܐ, ܒܝܪܐ ("puteus") [* = بئر]: **Min. I**.iii.1.6 (IS*); iii.2.1bis; iii.2.5 (AB*); iii.2.9; iii.3.6 (AB*); iii.4.5 (IS*); **II**.ii.4.2 (IS*).
√bʾš: ܒܐܫ: **Min. III**.ii.3.4 (IS غير نقى); ii.3.5, 6, 8, 9 (IS ردى).
ܒܒܠܝܐ (nom. prop. "Babylonii"): **Min. I**.ii.1.7.
√bgn: Pa: **Min. IV**.ii.6.9 (IS شَكَّا).
ܒܓܠܠ [* = أنّ]: **Min. I**.ii.2.4; **II**.iii.2.4; **III**.i.4.4 (IS*); iii.1.3 (IS*); iii.2.3; iii.3.9; **IV**.i.1.4; i.4.4; iii.5.2, 6; iv.3.3; iv.4.6; **Mete. I**.ii.3.2 (AB*); **II**.i.6.2; i.9.3 (IS*); ii.1.4; iii.3.4 (IS/FR*); iii.4.2 (IS فلذلك); iii.5.3, 6 (IS*); **III**.iv.2.7 (N); **IV**.i.1.4; i.2.3 (FR*); i.4.4; ii.2.6 (N); iii.2.3 (IS*); **V**.i.2.3; ii.2.6.
√bdq: Pa: **Min. II**.iii.2.8 (IS دلّ); **Mete. I**.iii.5.4, 5 (IS أنذر). - Ethpa: **Mete. II**.iii.2.2 (IS ظهر); iii.8.4.
√bdr: Pa: **Mete. III**.i.4.5 (N).
√bhq: Aph: **Mete. II**.ii.2.3ter.
√bhr: ܒܗܝܪ: **Mete. II**.iii.8.4, 7 (IS ضعيف اللون).
√bht: Pe: **Min. IV**.iii.6.2.
√bwb: ܒܘܒܬܐ [* = عين]: **Mete. II**.i.2.2, 3; i.3.3 (IS*); i.3.3; i.3.4, 6 (IS*); i.4.2, 5.
ܒܘܙܢܛܝܐ (nom. prop. "Byzantium"): **Min. V**.i.2.5; iii.3.4.
ܒܘܟܪܐ (nom. prop.): **Min. IV**.ii.6.8 (IS بخارا). - ܒܘܟܪܝܐ: **Min. IV**.ii.6.9.
ܒܘܠܓܪܐ (nom. prop. "Bulgarus"): **Min. V**.i.3.10.
√bwn: ܒܘܢܐ: **Min. IV**.i.4.11.
√bḥn: ܒܘܚܢܐ: **Mete. II**.i.3.1.
√bḥr: ܒܚܝܪ: **Min. III**.iii.2.4.
√byʿ: ܒܝܥܬܢܝܬܐ: **Mete. III**.iii.4.8 (IS بيضية).
√byt: ܒܝܬܘܬܐ: **Mete. III**.i.2.7. - ܒܝܬ ܡܫܘܬܐ: **Mete. III**.iv.2.4.
√bṭl: Pe: **Min. II**.i.4.9 (N); **III**.iii.2.8; **Mete. I**.iii.2.7 (AB انقطع); **II**.iii.3.8 (IS/FR لا كان); **V**.i.3.6; iii.2.1. - Pa: **Mete. III**.i.4.5. - ܒܘܛܠܐ: **Mete. V**.i.3.10 (IS انقطاع); iii.2.1. - ܡܒܛܠܢܐ: **Mete. I**.i.3.5.
ܒܛܢܬܐ: **Min. I**.ii.5.9.
ܒܠܢܐ (βαλανεῖον) [* = حَمّام]: **Mete. II**.iii.1.9 (IS*); iii.2.8 (IS*).
√bṭn: Pa: **Mete. III**.iii.4.9.
√blʿ: Ethpe: **Min. V**.iv.3.2.
ܒܠܚܘܕ: **Min. I**.ii.6.3; **Mete. II**.i.9.2; ii.1.8; **V**.ii.tit. (IS دُون); ii.1.8; ii.3.10.
ܒܠܘܝܬܐ: **Min. II**.ii.2.9 (N); **Mete. IV**.iii.1.9 (N/IS شرّ).
√bny: Ethpe: **Min. I**.i.2.11; **Mete. V**.iii.2.8. - ܒܢܝܢܐ: **Min. I**.ii.1.8, 10 (AB بُنيان); **Mete. V**.iii.1.9.
√bsr: ܒܣܪܐ: **Min. III**.iii.3.8; **IV**.iii.6.11.
ܒܣܬܪܐ: **Min. V**.iii.3.8. - ܒܣܬܪ (praepos.) [* = وراء]: **Mete. II**.i.8.4 (IS*); iii.1.2, 5 (IS*).
√bʿy: Pe: **Min. I**.ii.1.8 (AB أراد); **IV**.i.1.2; **Mete. II**.iii.7.10.
√bʿr: ܒܥܝܪܐ: **Min. IV**.i.4.11.
√bṣr: Pe [* = قَلَّ]: **Min. I**.iii.4.7; **II**.i.4.9 (N ܒܨܝܪ ܗܘܐ); ii.4.3 (IS*); **V**.ii.2.8; iii.2.4; **Mete. III**.i.4.2, 3 (IS*); iii.3.7

(IS*). - ܒܨܝܪܐ (adj.) [* = قليل]: **Min. III**.ii.1.2, 4 (AB*); **IV**.i.3.9; iii.3.5; iv.4.3, 5; **V**.iv.2.3, 5; **Mete. I**.ii.5.2; iii.1.2 (IS*); iii.3.10 (cf. IS قلّ); **III**.i.3.9 (IS*); **IV**.i.2.5 (IS أدنى); i.3.6 (IS أدنى); **V**.i.2.3 (cf. IS تفريط); iii.2.6. - ܒܨܝܪܐ (adv.): **Min. II**.iii.2.4, 7 (N; cf. IS قلّما); **IV**.ii.5.2, 4bis; iii.3.7; iv.1.6; **Mete. I**.ii.5.5 (cf. AB أقلّ); **II**.i.4.8, 10 (cf. IS أقلّ); iii.3.9 (cf. IS قلّة); iv.2.3 (cf. IS قلّ); iv.3.8 (IS قلّما); **III**.iii.6.3, 6. - ܒܨܝܪܐܝܬ: **Mete. II**.iii.8.2, 10 (IS على سبيل الندرة ,في الندرة). - ܒܨܝܪܘܬܐ: **Min. II**.iii.2.6 (IS قلّة); **III**.ii.1.3; **Mete. V**.i.2.1 (IS قلّة/نقصان). - ܒܘܨܪܐ: **Min. IV**.i.2.8; ii.3.7; iv.3.4.

√bqy: Ethpa: **Mete. IV**.i.2.6 (IS تأمّل).

√brr: ܒܪ̈ܝܪܐ: **Min. III**.iii.3.8; **Mete. II**.iii.6.2bis, 3, 4, 5; iii.7.2. - ܒܪ̈ܝܪܐܝܬ: **Min. III**.iii.1.3. - ܠܒܪ: **Min. V**.i.3.1; ii.1.12 (N); iv.1.7.

√bry: Pe: **Min. III**.ii.3.2. - Ethpe: **Mete. V**.ii.2.11; iii.1.7. - ܒܪܝܬܐ: **Mete. V**.ii.1.2.

ܒܪܐ ("filius"): ܒܢܝ̈ ܐܢܫܐ: **Min. I**.i.4.7. - ܒܢܝ̈ ܙܘܓ: **Mete. III**.iv.1.1. - ܒܪܬ ܫܡܫܐ [* = شُمَيسات ,شُمَيس]: **Mete. I**.iii.6.7 (IS*); **II**.i.1.1 (N/IS*); ii.1.1; iv.tit.; iv.1.9; iv.2.1 (IS*). - ܒܪܬ ܓܢܣܗ: **Mete. III**.iv.3.2 (IS من جنسه). - ܒܪܬܐ ("ovum"): **Min. V**.ii.3.2 (N/IS بيض). - ܒܪܢܫܐ: vide sub ܐܢܫܐ.

ܒܪܘܠܐ [* = بلّور]: **Min. III**.i.5.2 (AB*); **Mete. II**.i.8.4 (IS*); iii.1.4 (IS*).

ܒܪܒܪܐ (nom. prop. "Barbarus"): **Min. V**.i.3.3.

√brd: ܒܪܕܐ [* = بَرَد]: **Mete. I**.iii.tit.; iii.3.2 (IS/FR*); iii.3.3, 10; iii.4.2; iii.4.3 (IS*); iii.4.5 (N/IS*); iii.4.8 (IS بَرَدة); iii.4.10; iii.6.6. - ܒܪ̈ܕܢܐ: **Mete. I**.iii.4.7.

ܒܪܘܐܐ (nom. prop. "Beroea/Aleppo"): **Mete. I**.iii.4.9.

√brḥ: ܡܒܪܚ ,ܡܒܪܚܘ [* = مشف]: **Min. IV**.iv.1.4 (AB شفاف); iv.2.3; **Mete. II**.i.3.2 (IS*); i.8.1, 3 (IS*); iii.1.2 (IS*); **IV**.iii.2.7 (IS شافّ); iii.3.4 (cf. IS شفّ). - ܡܒܪܚܢܐܝܬ: **Mete. II**.i.8.1 (IS مشفًا). - ܡܒܪܚܢܘܬܐ: **Mete. IV**.i.4.6 (IS إشفاف).

ܒܪܛܢܝܐ (nom. prop. "Britannia"): **Min. V**.i.3.9.

ܒܪܡ [* = لكن]: **Min. I**.i.2.10; **II**.iii.1.5 (N); **III**.ii.2.6; ii.3.4 (IS*); **IV**.i.4.10; ii.2.4; iii.1.5 (FR*); iv.1.2; **V**.iii.1.2; iii.2.5; iv.3.3; **Mete. I**.iii.5.1; **II**.ii.3.4 (IS*); iii.8.1; iv.3.2 (IS إلا أن); **III**.iii.2.2; iii.4.2 (IS*); **V**.i.2.2; i.3.11.

√brq: Aph: Mete. IV.i.6.3, 5 (N). - ܒܪ̈ܩܐ [* = برق]: **Min. II**.ii.3.9bis (N); **Mete. IV**.i.tit.; i.2.4 (IS/FR*); i.3.1 (IS*); i.4.1, 3bis (IS*); i.4.5bis (N); i.5.1 (N/IS*); i.5.4, 5 (N); i.6.1, 4 (N); ii.1.3 (IS*); ii.1.5. - ܒܪ̈ܩܢܝܬܐ: **Mete. IV**.i.3.6 (IS بَرقي).

√bšl: Pa: **Min. V**.ii.1.7. - Ethpa: **Min. V**.ii.1.4, 5 (cf. IS طَبَخ); ii.3.10 (N).

ܓ

ܓܐܘܓܪܐܦܝܐ (γεωγραφία): **Min. V**.i.3.12.

ܓܐܘܡܛܪܝܐ (γεωμετρία): **Mete. II**.i.9.5 (IS صناعة الهندسة).

ܓܠܛܝܩܘܣ (nom. prop. "Galaticus"): **Min. V**.i.2.3 (DM).

ܓܐܪܐ: **Mete. III**.i.2.7 (IS سَهم).

√gby: ܓܒܝܐ: **Min. III**.iii.2.6.

√gbl: Ethpe: **Min. I**.i.5.6 (N); **Mete. I**.iii.2.2 (IS تخلّق); **V**.ii.1.8. - ܓܒܝܠ (ptc. pass.): **Min. I**.i.2.12. - ܓܒܝܠܘܬܐ: **Mete. V**.ii.2.1 (cf. IS تولّد). - ܓܒܘܠܐ: **Min. III**.iii.3.7 (AB فاعلي). - ܡܬܓܒܠܢܘܬܐ: **Mete. I**.iii.3.1 (FR انخلاق).

√gbr: ܓܒܪܐ: **Min. IV**.ii.6.8.

√gdy: Pe: **Min. II**.ii.2.7; **Mete. III**.i.2.2 (N). - ܓܕܝܐ ("Capricornus") [* = جدى]: **Min. IV**.i.4.4; i.4.7 (IS*); ii.6.3, 4; **V**.iii.1.9; **Mete. III**.ii.1.6, 7 (IS*).

√gdš: Pe [* = عَرَض]: **Min. II**.i.1.1 (IS*); ii.2.1 (IS*); iii.2.5, 7 (N/IS*); iii.3.4bis (IS*); **Mete. II**.iii.2.3 (N); **III**.iii.5.9; **IV**.ii.1.5. - ܓܕܫܐ: **Min. V**.iii.3.1 (cf. IS عَرَض). - ܓܕܫ̈ܢܝܬܐ: **Min. I**.ii.2.3 (IS بالعَرَض); **III**.iii.2.2.

√gww: ܓܘܝܐ: **Min. II**.ii.2.6; iii.1.4; **IV**.iii.1.1 ("communitas"); **V**.ii.1.10 (IS باطن). - ܓܘܗ: **Min. I**.ii.4.2, 4; iii.1.2; iii.2.1; iii.3.5; **II**.ii.1.2; ii.1.8; iii.1.8; iii.3.3; **III**.i.2.4 (N); **IV**.i.1.1 (IS في ضمن); iii.6.5; **Mete. I**.i.1.1; **II**.i.1.8 (IS في); **III**.i.3.1, 2 (IS في غـور); i.3.10, 11

(IS في); i.4.10 (IS إلى باطن); iii.2.5; **IV**.i.1.3 (FR في جوف). - ܓܘܫܐ: **Min.** V.i.3.6; **Mete.** **II**.iii.6.2bis, 3bis, 5; iii.7.4.

√gwy: ܓܘܝܢܝܐ: **Min.** **I**.i.1.1 (IS عامّ); V.ii.4.7; **Mete.** V.iii.1.7. - ܓܘܝܐܝܬ: **Min.** **II**.i.4.3 (N).

ܓܘܒܘܣ (nom. prop. "Gubos"): **Mete.** I.i.2.7, 8.

ܓܘܓ (nom. prop. "Gog"): **Min.** V.i.3.7.

ܓܘܪܓܢܐ (nom. prop.): **Min.** **I**.i.4.4 (IS جورجان).

√gwz: Pe: **Min.** V.iii.1.5.

ܓܘܙܐ: **Min.** I.i.5.3 (N) [lectio incerta]; **III**.iii.3.7.

√gwḥ: Pe: **Min.** V.i.3.2; iii.1.6; iv.2.2. - Aph: **Min.** **II**.ii.2.4 (cf. IS تَفَجَّرَ); iii.4.5.

√gwn: ܓܘܢܐ [* = لون]: **Min.** **I**.i.3.8 (IS*); **III**.iii.2.2, 7 (IS صبغ); **Mete.** **II**.i.5.6 (IS*); i.6.2, 3 (IS*); i.7.4 (IS*); i.8.5; i.9.1, 2 (N/IS*); ii.4.8 (IS*); iii.1.7 (IS*); iii.2.1 (N/IS*); iii.2.9 (IS*); iii.4.1bis (IS*); iii.4.3, 5; iii.5.1 (IS*); iii.5.4, 7bis; iii.5.8 (IS*); iii.7.5 (supplevi); iii.7.8, 9; iv.1.4 (IS/FR*): iv.3.2 (IS*); **IV**.ii.2.2. - ܡܓܘܢ (Pa ptc. pass.): **Mete.** **II**.i.8.4 (IS ذو لون). - ܡܓܘܓܢ (Pauel ptc. pass.): **Mete.** **II**.ii.4.9; iii.1.3 (IS مُلَوَّن).

ܓܘܢܝܐ, ܓܘܢܝܬܐ (γωνία) [* = زاوية]: **Mete.** I.iii.3.5; iii.4.6 (IS*); **II**.i.5.3 (IS*).

ܓܝܚܘܢ (nom. prop.): **Min.** **I**.i.2.9 (IS جيحون).

√gzr: ܓܙܪܬܐ: **Min.** **I**.ii.1.6; V.i.1.1; i.1.6; i.2.1; i.3.9.

√gyd: ܓܝܕܐ: **Min.** **I**.i.5.2 (N).

√gll ("undus"): Pa: **Mete.** **III**.i.1.6 (cf. FR تَمَوَّجَ). - ܓܠܠܐ: **Mete.** **IV**.i.4.7 (cf. IS تَمَوَّجَ). - ܡܓܠܠܐ: **Min.** V.iii.3.3 (IS مُمَوَّج). - ܡܓܠܠܢܘܬܐ: **Min.** **I**.ii.1.4 (cf. AB تمويج).

√gll ("rotundus"): Ethpalpal: **Min.** **III**.iii.3.7 (AB تَدَوَّرَ); **Mete.** **I**.iii.3.5; iii.4.6. - ܓܠܝܠ: **Min.** **I**.i.4.10 (IS مستدير). - ܓܠܝܠܘܬܐ: **Mete.** **I**.iii.3.3 (cf. AB دارَ). - ܓܝܓܠܐ: **Mete.** **II**.iii.1.6 (IS آلة).

√gly: ܓܠܝܐ: **Min.** **II**.ii.1.9; **IV**.iii.3.4; iii.5.4; iv.5.6; **Mete.** **II**.iii.8.7; **III**.ii.1.4; ii.2.2 (cf. IS ظهور); ii.2.7; V.iii.2.1. - ܓܠܝܐܝܬ: **Mete.** **II**.iii.8.4. - ܓܠܝܢܐ: **Min.** **IV**.iv.3.9; **Mete.** V.iii.1.2 (cf. IS إلهام); iii.1.8.

√glg: Ethpa: **Mete.** V.i.3.5 (IS انكَشَفَ).

√gld: Aph (intr.): **Mete.** **I**.iii.2.3 (IS جَمَدَ); iii.4.1, 2, 4bis (IS جَمَدَ); **III**.iii.4.3 (cf. IS جامد). - Aph (trans.): **Mete.** **I**.iii.3.2 (cf. IS جَمَدَ intr.); iii.3.8 (IS أجمَدَ). - ܡܓܠܕܢܐ: **Min.** **IV**.i.3.11; V.iii.2.10. - ܡܓܠܕܢܘܬܐ: **Mete.** **I**.iii.1.6 (cf. IS جَمَدَ).

√glz: Ethpe: **Mete.** **III**.iii.3.5. - ܓܠܝܙ: **Min.** **III**.ii.1.2. - ܓܠܝܙܘܬܐ: **Min.** **II**.i.3.4 (N); iii.1.1 (N/IS فقدان); iii.3.2 (IS فقدان); **III**.ii.1.3 (AB عَدَم); **Mete.** **I**.i.1.9 (AB غيبة); **III**.iii.3.1, 9 (N); **IV**.iii.4.11 (IS قلّة).

√gmm: ܡܓܡܓܡܢܘܬܐ: **Mete.** V.ii.3.10 (IS انقطاع).

√gmr: ܓܡܝܪܐܝܬ: **Mete.** **IV**.iii.3.3 (cf. IS بالكمال). - ܠܓܡܪ: **Mete.** V.ii.3.4 (FR البَتَّة).

ܓܡܘܪܬܐ, ܓܡܘܪܐ: **Mete.** **IV**.i.3.3; i.3.9 (IS جمر). - ܡܓܡܪܢܝܬܐ: **Mete.** **IV**.iii.4.6 (cf. IS تجمّر).

√gnb: ܓܢܒܐ: **Min.** **I**.i.3.9 (IS وجه); **IV**.i.1.1 (IS جانب); iii.5.6; V.iii.1.5bis, 7bis (cf. IS جهة); iii.3.4, 9; **Mete.** **I**.ii.4.6; I.ii.5.5 (AB جهة); **II**.iv.3.7; **III**.ii.2.6, 7, 8, 9, 10bis, 12; iv.1.3bis (N). - ܠܘܬ ܓܢܒ: **Min.** **I**.i.2.11; **II**.iii.1.3; **Mete.** **II**.ii.2.3 (IS في جنب); iv.3.2 (IS في جنبة).

√gnḥ: ܓܢܘܚܐ: **Min.** **II**.ii.4.9.

[√gns] (< γένος): ܓܢܣܐ [* = جنس]: **Min.** **III**.i.tit.; i.1.1 (IS قسم); **Mete.** **II**.iii.4.3 (IS*); **III**.iv.3.2 (IS*).

√gᶜy: ܓܥܝܬܐ: **Min.** **II**.ii.3.2 (N).

√gpn: ܓܦܬܐ: **Min.** **IV**.iii.6.11.

√grby: ܓܪܒܝܐ [* = شمال]: **Min.** **IV**.i.2.4 (IS*); i.3.7, 10; i.4.10; iii.3.3; V.i.2.4, 6, 7bis, 10; **Mete.** **I**.iii.4.8; **II**.iv.2.3, 6 (cf. IS شمالي); **III**.ii.2.8; ii.3.6, 7; iii.1.10 (N). - ܓܪܒܝܝܐ [* = شمالي]: **Min.** **IV**.i.3.3, 5, 11; ii.1.3, 7; ii.2.4; iii.5.6; V.i.1.5; i.2.2 (DM); **Mete.** **I**.ii.4.5 (AB*); **II**.iii.3.8 (cf. IS شمال); iv.2.8 (IS*); **III**.ii.1.2, 8 (IS*); ii.1.9; ii.3.1 (cf. IS شمال); iii.1.1 (cf. IS شمال); iii.1.6 (N); iii.5.3; iii.5.4 (N); iii.5.5bis, 7, 8; iii.5.11 (N); iv.2.5 (N); V.i.3.8 (cf. IS شمال).

√grm: ܓܪܡܬܐ: **Min.** **V**.iii.2.2 (cf. IS رَميم). - ܓܪܡܘܕܐ: **Min.** **I**.ii.1.8 (AB نوًى).
√grs: Ethpa: **Min.** **V**.iii.2.7.
√gšš: ܓܫܫܘܬܐ: **Min.** **II**.ii.3.8 (N).
√gšm: ܓܘܫܡܐ [* = جسم]: **Min.** **I**.i.1.2; I.i.1.6; i.3.6; ii.6.1 (IS*); **II**.ii.1.6 (N); ii.3.4 (N); **III**.tit.; i.1.1 (IS*); i.1.1; i.2.8; i.4.2, 3 (IS جَسَد); i.5.1 (IS*); iii.tit.; iii.1.4; iii.2.1 (IS جَسَد); iii.2.7; iii.3.5; **IV**.iv.1.3 (AB*); iv.2.1, 2; **Mete.** **II**.i.3.2; i.6.1 (IS*); i.7.1bis; i.8.1, 4 (IS*); iii.1.3; iii.2.7; iii.8.3 (IS شيء); iv.1.6 (FR جـــرم); **III**.iv.3.4, 6 (N); **IV**.i.2.1 (FR شيء); i.2.6 (IS/FR شيء); ii.1.3, 6 (IS جرم); ii.2.2, 4, 5 (N); ii.3.2 (IS*).
√gšp: Pe: **Min.** **I**.ii.5.7; **Mete.** **III**.ii.2.3, 4; ii.2.6, 10 (cf. IS مُمَاس). - ܓܫܦܐ: **Mete.** **III**.ii.2.8, 12. - ܓܫܦܬܐ: **Mete.** **IV**.ii.2.3 (N).

ܕ

ܕܐܠܘܣ (δαλός): **Mete.** **IV**.iii.1.10 (N).
√dbq: Pe: **Min.** **I**.ii.1.6; ii.1.9 (AB التَبَس); **IV**.i.1.8 (cf. IS مُلحق); **Mete.** **I**.i.3.5 (cf. IS إلصاق). - ܕܒܝܩ (ptc. pass.): **Min.** I.i.4.10 (cf. IS التَصَق).
√dbr: ܕܒܘܪܝܬܐ: **Mete.** **V**.ii.1.4 (FR نحل).
√dgl: ܕܓܠܐ: **Mete.** **II**.i.2.3.
√dhb: ܕܗܒܐ [* = ذهب]: **Min.** **III**.i.5.2 (AB*); i.5.8; ii.3.4 (IS*); iii.1.8 (IS*); iii.3.4, 11 (AB*); **Mete.** **IV**.ii.2.8.
√dhn: ܕܗܢܐ [* = دُهني]: **Min.** **I**.i.4.2 (N); **III**.i.5.5 (IS*); **Mete.** **IV**.i.2.8 (cf. IS دُهنية); i.3.5 prim. (IS*); i.3.5 secund. (IS دَسَم). - ܕܗܝܢܘܬܐ [* = دُهنية]: **Min.** **III**.i.2.7 (IS*); i.3.2 (cf. IS دُهني); **Mete.** **IV**.i.2.3 (IS/FR*).
√dwb: Pe: **Min.** **I**.iii.3.6 (AB ذاب).
ܕܘܕܘܢ (nom. prop. "Dodona"): **Min.** **V**.ii.3.8 (N).
√dwl: Aph: **Min.** **II**.ii.1.6 (IS حرّك).
ܕܘܩܝܣ (δοκίς, δοκίδες): **Mete.** **IV**.iii.tit.; iii.1.1.
√dwr: ܕܘܪܐ: **Min.** **I**.i.2.10; **IV**.ii.2.7. - ܕܘܪܝܐ: **Min.** **IV**.ii.2.4, 7. - ܕܝܪܐ: **Min.** **V**.iii.1.2; **Mete.** **III**.iv.1.1.
√dḥy: Pe: **Min.** **II**.ii.1.10 (N); ii.2.3 (IS سَاق); iii.1.4 (N); **V**.iii.3.3 (cf. IS اندفاع); iii.3.8; **Mete.** **I**.ii.4.3; ii.6.5 (cf. IS اندفاع); iii.3.4; **III**.i.4.9 (N); iii.1.7 (N); iv.3.4, 5 (N). - Ethpe: **Min.** **I**.i.5.10; iii.1.3; **II**.ii.1.2; **V**.iii.3.7; **Mete.** **I**.ii.6.2; **III**.iv.1.6 (N); iv.3.1; **IV**.i.5.7 (cf. IS زَجّ). - ܕܚܝܐ: **Mete.** **IV**.ii.1.4.
√dḥl: Pa: **Min.** **II**.iii.4.6. - ܕܚܝܠ [* = هائل]: **Min.** **I**.i.4.6 (IS*); **II**.ii.3.1 (IS*); **Mete.** **I**.iii.6.8 (IS*); **IV**.iii.4.6 (IS*).
√dḥn: ܕܘܚܢܐ: **Min.** **I**.i.4.10 (cf. IS جاورسي).
ܕܝܐܩܠܐܣܝܣ (διάκλασις): **Mete.** **II**.i.4.4 (N); i.4.8. - ܕܝܐܩܠܐܣܛܝܩܝܬ: **Mete.** **II**.iii.2.9.
ܕܝܠܢܝܐ: **Min.** **V**.iii.1.1 (cf. IS اختصاص); **Mete.** **II**.i.2.4; **III**.ii.3.3. - ܕܝܠܢܝܐܝܬ: **Min.** **II**.i.4.3 (N?).
ܕܝܡܘܩܪܝܛܘܣ (nom. prop. "Democritus"): **Min.** **II**.i.2.1 (N).
√dkk: ܕܘܟܬܐ [* = موضع]: **Min.** **I**.ii.2.5, 6; ii.6.2 (IS*); iii.2.1; iii.3.1; iii.4.5; **II**.i.4.2; iii.2.1 (N/IS بلاد); **IV**.i.1.8; **V**.ii.2.1 (IS*); iii.1.1, 2 (IS*); iii.3.3; iv.3.3; **Mete.** **I**.ii.2.8bis (AB*); ii.6.2 (AB*); **II**.i.2.4 (IS*); iii.1.5 (IS مكان); **IV**.i.4.10 (IS*). - ܕܘܟ ܕܘܟ: **Min.** I.i.3.1. - ܒܕܘܟ ܕܘܟ: **Mete.** **V**.i.1.12. - ܒܕܘܟ ܕܘܟ ... ܒܕܘܟ ܕܘܟ: **Min.** **I**.iii.2.8, 9. - ܠܕܘܟ: **Min.** **II**.i.4.3; **IV**.iv.5.2.
√dky: Pa: **Min.** **III**.iii.1.8.
ܕܟܝܐ: **Min.** **III**.ii.3.1 (IS نَقِي); ii.3.1 (IS لا درن); ii.3.3; iii.3.2 (AB نَقِي). - ܕܟܝܘܬܐ: **Min.** **III**.ii.3.2 (IS نقاء).
√dkr: Aph: **Min.** **I**.i.4.12. - ܕܟܪܐ: **Mete.** **V**.ii.1.3.
√dll: ܕܠܝܠ ("paucus"): **Min.** **IV**.i.4.10 (IS لا يعتدّ بها). - ܕܠܝܠܐܝܬ ("facile"): **Min.** **II**.ii.4.7 (N); **V**.ii.3.5 (N); **Mete.** **I**.iii.4.4; **III**.iv.2.9 (N).
√dly: Pe: **Min.** **I**.iii.1.8; **V**.iii.1.4 (IS استَقَى). - Ethpa: **Min.** **I**.iii.4.5 (IS نُزِح).
ܕܠܐ ("sine"): **Min.** **II**.ii.3.5.
√dlḥ: Pe: Mete. I.ii.2.6 (cf. AB كَدِر). - Pa: Min. II.iii.4.6. - ܕܠܝܚ: Min. V.ii.2.4; Mete. II.iii.1.5 (IS كدر). - ܕܠܝܚܘܬܐ: Mete. I.ii.2.7bis (AB كدر).
√dlq: Pe [* = اشتَعَل]: **Mete.** **IV**.i.3.8 (IS*); ii.1.3 (IS*); iii.1.5 (IS*); iii.2.2 (cf. IS ضوء); iii.2.7 (IS أشرق); iii.4.5 (IS*); iii.4.9. - ܕܠܩܐ: **Mete.** **III**.iii.3.2 (N); **IV**.iii.3.4 (IS ضوء); iii.3.7; iii.4.3 (cf. IS مُشتعل).

[√dlqb]: Pauel: **Mete. III**.iii.3.10 (IS ضَادَّ). - Ethpaual (ܐܬܕܠܩܒ): **Mete. III**.i.4.3 (IS تَمَانَعَ); **V**.iii.2.7.

√dmy: Pe: **Min. III**.iii.1.7, 8 (cf. IS شبه); **IV**.iii.6.9 (cf. AB لَ); **Mete. I**.ii.2.7 (cf. AB لَ); **II**.iv.3.2; **III**.i.1.3, 4 (cf. IS جَرَى مجراه); iii.1.10 (N); **V**.ii.2.8; iii.2.4. - Ethpe/Ethpa: **Min. III**.iii.1.2 (cf. IS مُشابه). - ܕܡܝܢܐ: **Min. II**.ii.3.4 (N); **Mete. II**.iii.7.8 (IS مُتشابه); **III**.iii.3.3bis (N). - ܕܘܡܝܐ: **Min. I**.i.3.1; **III**.iii.2.4; **V**.ii.4.3 (N); **Mete. II**.iv.1.2 (cf. FR كَمَا). - ܕܡܘܬܐ: **Min. III**.iii.3.8 (cf. AB صورة); **Mete. II**.i.3.2 (IS صورة); iii.2.6, 7; iii.8.5; **IV**.iii.3.7; **V**.i.1.11; i.3.1, 7; iii.2.9. - ܕܡܘܬ (praepos.) [* = لَ]: **Min. II**.ii.1.5 (IS*). - ܡܕܡܝܢܘܬܐ: **Min. III**.iii.1.6 (IS تشبيه). - ܡܬܕܡܝܢܘܬܐ: **Min. III**.iii.tit.; iii.2.1.

√dmr: ܕܡܝܪ: **Mete. II**.ii.4.4 (IS يعيد); **III**.iii.1.3 (IS بَعُدَ). - ܕܘܡܪܐ: **Min. IV**.iii.6.1.

√dnb: ܕܘܢܒܐ: **Mete. IV**.iii.3.6 (IS ذَنَب). - ܕܘܢܒܬܐ: **Min. I**.ii.5.8 (IS ذَنَب).

√dnḥ: Pe: **Min. IV**.iv.1.3 (AB صَدَرَ); iv.2.3; **Mete. I**.i.1.3 (AB أشرَقَ); **III**.ii.2.5; **V**.ii.2.9. - ܕܢܚܐ: **Min. IV**.iv.3.2 (AB طلوع); **V**.iii.3.10. - ܕܢܚܘܬܐ: **Mete. II**.i.3.3 (IS وقوع). - ܡܕܢܚܐ [* = مَشرق]: **Min. IV**.i.3.2, 6; **V**.i.2.6; i.3.2, 6bis, 7; **Mete. III**.ii.1.4, 5bis (IS*); ii.3.3, 4, 6bis; iii.6.1, 4 (N). - ܡܕܢܚܝܐ [* = مشرقي]: **Min. V**.i.1.4; **Mete. I**.ii.4.5 (AB شرقي); **III**.ii.1.2 (cf. IS مَشرق); ii.3.1 (IS*); iii.2.1, 2, 3 (IS*); iii.2.6, 8, 9, 10; iii.6.3, 6 (N).

√dᶜk: Pe [* = طَفِئَ]: **Min. I**.i.4.3 (IS*); **Mete. III**.iii.3.2, 6, 7 (N); **IV**.i.3.2 (IS*); i.5.2 (cf. N Ethpa); ii.1.6 (IS*); ii.3.2 (IS*); iii.2.8 (IS*); iii.3.2 (IS*). - Pa: **Mete. III**.i.4.4 (N). - ܕܘܥܟܐ: **Mete. IV**.i.3.2bis (IS انطفاء); iii.2.1, 3 (IS طفوء); iii.3.5.

√dᶜt: ܕܘܥܬܐ [* = عَرَق]: **Min. V**.ii.4.1 (N/IS*); ii.4.4 (IS*).

√dpp: ܕܦܦܐ: **Min. IV**.ii.1.5, 6bis (IS دفّي); ii.2.2, 3, 4, 6.

√dqn: ܕܩܢܐ: **Mete. IV**.iii.3.6 (IS لحية).

ܕܝܩܢܐ (δεκανός): **Mete. II**.i.5.9 (IS ثابتة); **V**.i.1.5.

√drg: Ethpa: **Mete. V**.iii.2.6. - ܕܪܓܐ: **Min. IV**.iii.6.2; **Mete. V**.i.2.1. - ܕܘܪܓܐ: **Min. II**.iii.3.4 (IS تدريج).

√drk: Ethpe: **Mete. II**.i.5.3, 5 (IS حُسَّ). - Aph: **Min. III**.iii.2.7 (cf. IS محسوس); iii.3.9; **Mete. II**.i.1.8 (IS أدرَكَ); iv.3.4. - ܡܕܪܟܢܘܬܐ: **Mete. II**.i.1.2.

ܗ

ܗܐ ("ecce!"): **Min. V**.iii.2.9 (IS هو ذا); **V**.iv.3.2; **Mete. II**.i.2.5; II.ii.2.5; **III**.iii.2.9; **V**.ii.1.3.

√hgg: Pa: **Mete. II**.i.6.2 (IS خَيَّلَ); iii.2.10; **IV**.iii.4.7 (cf. IS خُيِّلَ). - ܗܓܝܓܐ, ܗܓܓܬܐ [* = خيالة/خيال]: **Mete. II**.i.tit.; i.1.1, 2 (IS*); i.8.2, 3 (IS*); ii.1.1 (N); ii.1.5 (FR*); ii.1.6 (IS/FR شَبَح); ii.1.7 (FR*); ii.4.2; iii.1.1; iii.1.10 (IS*); iii.8.3, 7 (IS*); iv.1.3 (FR*); iv.3.1 (IS*); **IV**.i.6.1 (N). - id., pl. ܗܓܓܘܬܐ: **Mete. II**.iv.1.1, 5 (IS خيالات). - ܗܓܓܝܐ: **Mete. II**.tit. - ܗܓܓܝܐ [* = خيالي]: **Mete. II**.i.6.4 (IS*); iii.4.3 (IS*). - ܗܓܓܝܐܝܬ: **Mete. I**.iii.6.7; **II**.i.4.3, 6. - ܡܗܓܓܢܘܬܐ: **Mete. II**.i.4.1.

ܗܓܡܘܢ (ἡγεμών): **Min. I**.i.4.8 (cf. IS سلطان).

ܗܕܝܘܛܐ (ἰδιώτης): **Mete. I**.ii.2.2 (AB الدَهماء).

ܗܕܡܐ: **Mete. V**.iii.2.4.

ܗܘ: passim. - ܗܘ ܒܗ ܗܘ ("idem"): **Min. I**.i.3.6; **Mete. I**.ii.2.2; **III**.iii.5.1, 2 (N). - ܗܘܝܘ: **Min. II**.ii.1.3; **III**.iii.1.4; **V**.ii.1.3; iv.3.5; **Mete. II**.iii.5.7 (IS هو); **V**.iii.1.6.

√hwy: Pe: **Min. I**.i.1.7; i.2.1; i.2.2 (IS تَكَوَّنَ); i.3.2, 4, 8; i.3.9 (IS تكوَّنَ); i.4.2; i.5.1 (N); i.5.7, 9; ii.1 (cf. IS تكوَّن); ii.1.2, 7; ii.1.10 (AB صَارَ); ii.2.4, 5, 6; ii.4.3 (IS استَحَالَ); ii.6.2; iii.1.1, 2; iii.1.5 (cf. IS حَدَثَ); iii.2.3 (IS تَوَلَّدَ); iii.3.1; **II**.i.2.5bis (N); i.3.4 (N); i.4.2, 8bis, 9 bis (N); i.4.11; ii.2.5; ii.2.7 (IS كَانَ); ii.2.8, 10 (N); ii.3.2, 3, 9 (N); ii.4.2, 3; ii.4.5 (N); ii.4.8; iii.tit.; iii.1.1, 7 (N); iii.1.7 (IS كَانَ); iii.1.10 (N); iii.2.1 (N/IS كَانَ); iii.2.4 (N); iii.2.7 (N/IS عَرَضَ); iii.2.8 (N); iii.3.1 (IS كَانَ); iii.3.4bis; iii.4.1, 2, 4 (N); **III**.i.4.3 (cf. IS كَونُ); i.5.8; ii.3.5, 6 (IS كَانَ); iii.2.5, 7; **IV**.i.1.1

(IS كَانَ); i.1.5 (IS اِستحَالَ); i.1.6 (IS وَقَعَ); i.1.7 (IS اِستحَالَ); i.2.8, 9, 10 (IS كَانَ); i.4.8; IV.ii.3.7; ii.4.5; ii.5.7; iii.1.6; iii.3.2, 4; iii.5.3, 5, 7; iii.6.6; iv.1.7 (AB كَانَ); iv.3.5; iv.5.3, 5, 6; V.ii.1.2, 5; ii.2.2 (IS صَارَ); ii.3.10 (N); iii.1.1, 7, 8, 9, 10; iii.2.1, 2, 4, 5; iv.2.4; iv.3.9bis; **Mete.** I.i.1.7; i.2.2 (cf. IS تكوُّن); i.2.4, 6; i.3.2; ii.1.4, 9; ii.2.3 (AB كَانَ); ii.5.1 (AB كَانَ); ii.6.6; iii.1.7; iii.2.1; iii.3.6bis (IS كَانَ); iii.3.8; iii.4.3 (IS كَانَ); II.i.1.5; i.3.2bis, 4; i.4.2; i.6.4; i.8.1 (IS كَانَ); i.8.4bis, 5; i.9.5; ii.4.1 (IS تَوَلَّدَ); ii.4.2, 6bis (IS كَانَ); iii.1.2, 3; iii.3.5 (IS/FR كَانَ); iii.3.8, 9 (IS حَدَثَ); iii.7.4, 7; iii.7.8 (IS كَانَ); iii.8.1, 8, 9; iii.8.10 (IS تَوَلَّدَ); iv.2.3, 4, 5; iv.2.8, 9 (IS صَارَ); iv.3.3 (IS كَانَ); iv.3.7; iv.3.8 (IS كَانَ); III.iii.2.7 (N); iii.3.1, 9 (N); iii.5.4bis (N); iii.5.5, 6; iii.5.11bis (N); iv.1.4, 6bis (N); iv.2.5; iv.2.6, 7 (N); iv.3.1 (IS كَانَ); iv.4.1; iv.4.3, 7 (IS حَدَثَ); IV.i.1.3; i.1.6 (FR حَصَلَ); i.2.6 (IS حَدَثَ); i.3.1 (IS كَانَ); i.4.1bis (IS كَانَ); i.4.3ter; i.4.5 (N); i.4.6; i.5.2, 3 (N); i.6.1bis, 2, 4 (N); ii.2.1, 5 (N); ii.3.1 (IS كَانَ); ii.3.3 (N); ii.3.9; iii.1.10; iii.2.4, 5, 8; iii.3.4; iii.4.1 (cf. IS تكوُّن); iii.4.6; V.tit. (IS حَدَثَ); i.1.6 (IS وَقَعَ); i.1.8 (IS عَرَضَ); i.2.2, 5, 6; i.3.3, 10; ii.2.2 (cf. IS كان له وجود); ii.2.6, 7, 10; ii.3.2bis, 4; ii.3.5 (IS كَانَ; FR تَوَلَّدَ); ii.3.7, 8; iii.1.2; iii.2.2. - ܗܘܐ (ptc. pass.): **Min.** V.iv.3.5, 6; **Mete.** V.iii.1.3, 4 (cf. IS حادث). - ܗܘܝܐ: **Min.** I.i.tit.; i.1.3 (IS كَـونٌ); ii.tit.; II.ii.4.8; III.ii.2.6 (IS تكوُّن); iii.1.4bis (IS كَـونٌ); **Mete.** I.i.tit. (IS تولّد); II.iii.8.2, 5 (IS تكوُّن); III.i.4.6 (N); V.i.2.2; ii.1.1 (IS حُدُوث); ii.1.2; ii.2.4; ii.3.9, 10.

ܗܘܠܐ, ܗܘܠܐ (ὕλη) [* = مادّة]: **Min.** I.ii.3.3; II.ii.2.2 (IS*); III.i.2.4 (IS*); i.3.5; **Mete.** I.iii.3.10; III.i.2.2 (N); i.3.7 (N); i.3.11; i.4.1bis (IS*); i.4.3; iii.3.1; iii.5.2 (N); iv.1.4 (N); iv.3.1 (IS*); IV.ii.1.2bis, 5; ii.2.1, 3 (N); ii.3.1 (IS*); ii.3.2; iii.1.1; iii.2.2, 5, 7 (IS*); iii.3.3, 5 (IS*); iii.4.2 (IS*).

√hwn: ܗܘܢܐ: **Min.** III.iii.1.3; iii.2.3.

ܐܣܘܦܘܣ (nom. prop. "Aesopus"): **Min.** V.ii.2.3.

ܐܘܪܘܣ (εὖρος): **Mete.** III.ii.3.4 (N).

ܐܘܪܝܦܘܣ, ܐܘܪܝܦܘ (εὔριπος): **Min.** V.iii.3.4, 6.

ܗܘܪܩܢܝܐ (nom. prop. "Hyrcania"): **Min.** V.i.2.6.

ܗܛܢܐ (nom. prop. "Aetna"): **Min.** I.i.5.8 (N?); II.ii.2.10 (N).

ܗܝܕܝܢ: **Min.** II.iii.1.8; IV.iii.5.4; iv.5.2.

ܗܝܕܝܟ: **Min.** IV.iv.1.7; **Mete.** I.ii.3.5; iii.1.5; iii.2.8; iii.4.2.

ܗܝܟܠܐ: **Mete.** IV.ii.3.4 (N).

ܗܝܡܢܘܬ: **Min.** II.iii.2.7; **Mete.** III.i.1.9; ii.2.7, 11.

ܗܟܢ: **Min.** II.i.4.8 (N ܗܟܢܐ); IV.iv.3.9 (AB كذلك); **Mete.** II.iii.8.3.

ܗܟܢܐ [* = كـذلك]: **Min.** II.ii.2.6; ii.2.10 (N); IV.i.1.8; i.2.9 (IS*); i.2.10; i.3.4; ii.2.1, 2; ii.3.5; iii.6.1; iv.4.4; iv.5.8; V.iii.1.6 (cf. IS*); iii.3.9; iv.3.2, 10; **Mete.** I.ii.6.4; II.i.2.5; i.4.3, 6; ii.2.6; iii.1.7 (IS*); iii.3.2 (FR بحَـيث); iv.2.8; III.i.3.3; ii.2.1, 8, 11; iii.2.9; iv.2.4; IV.i.2.8; iii.2.6; iii.4.3; V.i.3.5, 10; ii.2.10; ii.3.4, 10.

√hlk: ܡܬܗܠܟܢܘܬܐ: **Min.** V.i.3.8 (Biruni مسلوك).

ܗܡܕܐܢ (nom. prop. هَمَـــذان): **Mete.** II.ii.4.8.

ܗܡܦܕܘܩܠܝܣ (nom. prop. "Empedocles"): **Min.** V.ii.4.1 (N ܐܡܦܕܘܩܠܝܣ, IS انبادقليس); **Mete.** IV.i.5.1 (N ܐܡܦܕܘܩܠܝܣ).

ܗܢ (= ܗܢܐ): **Min.** IV.iii.5.5; **Mete.** II.ii.3.3 (IS هذا); IV.i.1.2.

ܗܢܕܘ (nom. prop. "India"): **Min.** V.i.3.6bis; ii.3.3 (N); iii.3.10.

√hpk: Pe: **Min.** III.ii.2.2; V.iv.2.2: **Mete.** II.i.6.1 (IS انعَكَسَ); III.i.2.8; V.i.3.7. - ܗܦܟܐ: **Min.** IV.iii.6.7; iv.3.10 (AB ضِـــدّ); **Mete.** III.iii.5.8; V.i.3.11. - ܗܦܝܟܘܬܐ: **Min.** V.iii.3.8; iv.1.9.

ܐܟܠܝܦܣܝܣ (ἔκλειψις): **Min.** II.iii.2.1 (N ܐܟܠܝܦܣܝܣ, IS كسوف).

√hrr: Aph: **Min.** IV.iii.6.4 (cf. AB ضارَّ).

ܗܪܟܐ ܬܘܒ: **Min.** IV.i.4.11; ii.3.8; **Mete.** I.i.1.7; ii.6.6.

ܗܪܟܐ: **Min.** I.ii.5.3; IV.ii.3.8; iii.4.1; **Mete.** I.iii.1.7; II.i.3.7; i.9.4; iii.7.1; III.iii.1.5; iii.5.9; IV.iii.2.1 (IS ههنا); V.iii.2.8.

ܗܪܡܢ (= arab. هَرمَان): **Min.** V.iii.2.12.

ܗܪܩܠܝܣ (nom. prop. "Hercules"): **Min.** **V**.i.1.2 (DM).
ܗܪܣܝܣ, ܗܐܪܣܝܣ (αἵρεσις) [* = مذهب]: **Mete.** **II**.i.1.5 (IS*); i.2.1 (IS*); i.3.1 (IS*); i.3.6.
ܗܫܐ: **Mete.** **V**.i.3.7; i.3.9 (IS الآن).
ܐܬܪ (αἰθήρ): **Min.** **I**.i.4.4 (IS هواء); **Mete.** **II**.iv.1.5; **IV**.i.5.4 (N ܐܐܪ, IS أثير); iii.1.1, 4 (N/IS جو); iii.4.4; iii.4.6 (IS جو); iii.4.9.

ܘ

√wly: Pe: **Min.** **IV**.i.4.8; **Mete.** **II**.i.4.8, 9; iii.1.3; iii.7.4, 6 (cf. IS أولى); **IV**.i.1.9; iii.2.1 (cf. IS يجب أن); **V**.ii.2.3.
√wᶜd: Ethpa: **Mete.** **I**.ii.1.4. - ܡܘܥܕܐ: **Mete.** **I**.ii.1.3; **III**.iii.4.2; **IV**.iii.1.3. - ܡܬܘܥܕܢܘܬܐ: **Mete.** **III**.iii.4.2.

ܙ

√zbn: ܙܒܢܐ ("temps") [* = زمان]: **Min.** **I**.ii.1.6 (cf. AB*); ii.1.9; iii.4.2, 4 (IS مُدّة); **II**.i.3.4bis (N); iii.tit.; iii.1.2; iii.2.9 (N); **IV**.iii.3.5; iii.5.5; iii.6.11; iv.3.1bis (AB*); iv.3.5bis; **V**.iii.2.5, 12; **Mete.** **I**.ii.3.8; **III**.iii.tit.; iii.3.9, 10; iii.4.4; **IV**.i.4.6, 8 (IS*); **V**.iii.1.5bis. - ܙܒܢܬܐ, ܙܒܢܝܢ ("fois"): **Min.** **I**.i.4.5; **V**.ii.1.5; iii.1.4; **Mete.** **I**.i.2.3. - ܒܙܒܢ: **Min.** **II**.ii.2.8 (N). - ܒܙܒܢ . . . ܒܙܒܢ: **Min.** **II**.ii.2.1 (IS ربما ... ربّما); ii.3.1 (N/IS ربّما); ii.4.4; **IV**.ii.4.6; **V**.iii.3.5; **Mete.** **II**.i.5.1, 6 (IS تارةً ... تارةً). - ܙܒܢܝܢ ܣܓܝܐܐ: **Min.** **II**.iii.4.2 (N); **V**.iii.3.6 (IS ربّما); **Mete.** **III**.iv.4.4; **IV**.ii.3.9 (N). - ܟܠܙܒܢ: **Mete.** **I**.i.2.5 (IS كل مرّة). - ܙܒܢܢܝܐ: **Min.** **IV**.iv.tit. (AB زماني).
√zgg: ܙܓܓܝܬܐ: **Min.** **III**.ii.3.8 (cf. IS/AB صَرّ). - ܙܓܓܝܬܐ [* = زُجاج]: **Min.** **I**.i.5.5 (N); ii.5.6 (IS*); **III**.i.5.3 (AB*); **Mete.** **II**.iii.2.8.
√zdq: Pe: **Min.** **II**.i.4.8, 10 (N); **IV**.i.2.10; **Mete.** **I**.i.1.3; **III**.i.3.8 (N); **IV**.i.5.3 (N); i.6.3 (N). - ܙܕܩܐ: **Min.** **I**.i.1.1.
ܙܘܕܝܩܘܣ (ζῳδιακός): **Mete.** **V**.i.3.3 (IS فلك البروج).
ܙܘܢܐ (ζώνη): **Mete.** **II**.ii.3.2, 5 (IS منطقة).
√zwᶜ: Pe: **Min.** **II**.iii.4.7. - Ethpe [* = تَحَرَّكَ]: **Min.** **I**.ii.1.5 (AB*); **II**.i.1.2 (IS*); **V**.iii.3.1, 3 (cf. IS حَرَكَة); iii.3.9; iv.1.6, 8, 9; **Mete.** **I**.ii.1.2, 6; **II**.i.4.7; **III**.i.1.8 (FR*); i.2.4 (N); i.3.4 (N); **IV**.i.3.4 (cf. IS حَرَكَة). - Aph [* = حَرَّكَ]: **Min.** **I**.ii.1.4 (AB*); **II**.i.1.2 (IS*); i.1.8 (N); i.2.7 (N); i.3.3 (cf. N ܙܘܥܐ ܕܐܪܥܐ); i.4.4, 5 (N); ii.1.4 (N/FR*); ii.3.6 (IS زَلْزَلَ); iii.1.9 (N). - Ethpalpal (ܐܙܕܥܙܥ): **Mete.** **II**.i.4.9. - ܙܘܥܐ [* = حَرَكَة]: **Min.** **II**.i.1.1 (IS*); i.3.1; i.3.4 (N); i.4.8 (N); i.4.10 (IS زلزلة); ii.3.7bis (cf. IS تموُّج); ii.4.2 (IS زلزلة); iii.tit.; iii.1.10; iii.2.1 (N/IS زلزلة); iii.2.5; iii.2.7 (IS زلزلة); iii.2.9 (N); **IV**.i.2.1 (IS*); **Mete.** **I**.i.3.6 (IS*); ii.3.5bis (cf. AB تنقل); **II**.i.4.7, 9; **III**.i.1.7, 8 (FR*); i.2.6 (N); **IV**.i.2.3 (IS/FR*); i.5.6; **V**.ii.3.9 (IS*); iii.2.3. - ܙܘܥܐ ܕܐܪܥܐ [* = زلزلة]: **Min.** **I**.i.1.8; **II**.i.tit.; i.1.3 (N); i.2.1; i.2.4 (N ܙܘܥܐ ܙܘܥܐ ܕܐܪܥܐ); ii.2.1 (IS*); ii.4.1 (IS*); ii.4.4 (N); iii.1.1 (N/IS*); iii.3.1 (IS*); iii.4.1 (N); iii.4.4 (IS*). - ܙܥ ܐܪܥܐ: **Min.** **II**.iii.1.8. - ܡܙܕܥܙܥܢܘܬܐ: **Mete.** **I**.ii.2.7 (AB حركة). - ܡܬܬܙܝܥܢܐ: **Mete.** **III**.iv.1.5 (N). - ܡܬܬܙܝܥܢܘܬܐ [* = حركة]: **Min.** **I**.i.1.1 (IS*); iii.4.1 (IS*); **II**.ii.2.5 (IS*); iii.2.3 (IS*); **Mete.** **I**.ii.3.1bis (AB*); iii.3.3 (AB*); **III**.i.2.7 (N); iii.5.1 (IS*); **IV**.i.2.2 (IS/FR*); i.2.5, 8 (IS*); i.4.7, 8 (IS*).
√zwp: Pa: **Min.** **II**.i.4.1; **V**.iv.1.4.
ܙܘܦܬ [* = زئبق]: **Min.** **III**.i.5.7; ii.tit.; ii.1.1 (IS/AB*); ii.1.9bis (AB*); ii.2.1, 3, 4 (IS*); ii.2.7; ii.3.1, 3, 4, 5 (IS*); ii.3.6 (IS/AB*); ii.3.8 (IS*); iii.1.1 (IS*); iii.3.3 (AB*).
ܙܝܒܐ: **Mete.** **I**.ii.3.7 (DM); **Mete.** **III**.iv.tit.; iv.4.7.
√zky: Pe: **Mete.** **I**.ii.3.5.
√zlg: ܙܠܓܐ: **Mete.** **IV**.i.2.7; i.5.1 (N).
√zlḥ: Ethpe: **Mete.** **II**.iii.2.1 (N/IS انتشر); **IV**.i.6.3 (N).
√zlm: ܙܠܡ: **Min.** **V**.i.2.3 (cf. DM ܙܠܝܡܘܬܐ); **Mete.** **III**.i.2.7 (N). - ܙܠܡܬܐ: **Mete.** **III**.i.2.4, 8 (N).
√zlq: ܙܠܩܐ [* = شُعَاع]: **Min.** **II**.iii.3.2 (IS*); **IV**.iv.1.2, 3, 5, 7 (AB*); iv.2.3, 4; iv.3.2, 3bis (AB*); iv.3.5, 9; iv.5.2, 3, 5, 6 (AB*); **Mete.** **I**.i.1.3 (AB*); i.1.7 (cf. AB شعاعي); ii.1.1; **II**.i.1.5, 6 (IS*); i.4.2, 4; ii.2.2 (IS*); iii.2.7 (IS*);

iii.3.2 (FR*); iii.8.2; **IV**.i.5.1 (N/IS*); i.5.3 (N); i.6.2.
√zmm: ܙܡܝܡܐ: **Min. II**.ii.3.1 (IS دَوِيَ).
ܙܡܝܡܘܬܐ: **Mete. I**.iii.1.7.
ܙܢܐ: **Min. I**.i.2.3 (IS وجه); i.2.3 (IS سبيل); i.2.5; **II**.ii.1.8 (IS حُكم); **IV**.i.2.7; **V**.ii.4.4; **Mete. II**.i.1.3; i.4.1,4 (N); i.6.4; **III**.ii.2.5; **V**.i.2.6; ii.1.2 (IS سبيل); ii.1.2.
√znq: Ethpe: **Mete. III**.iv.1.2 (N); **IV**.i.4.2; ii.3.7 (N). - ܐܙܕܢܩܘܬܐ: **Mete. IV**.ii.1.4 (cf. IS اضطرار).
√zʿr: Pe: Mete. I.ii.1.3 (cf. AB صَغُرَ); I.iii.4.5 (IS صَغُرَ). - ܙܥܘܪܐ [* = صغير]: **Min. I**.i.4.8; i.4.10 (IS*); **V**.i.1.7; i.2.2 (DM); iv.3.9; **Mete. I**.ii.3.7 (DM); ii.3.8bis; iii.6.2; **II**.i.4.7 (IS*); i.5.2 (IS أصغر); i.9.1 (N/IS*); ii.1.7 (FR*); iii.3.7 (IS نَقَصَ); iv.2.8; iv.3.4 (IS*); **III**.iii.3.3 (N); iii.6.4 (N); iv.2.7 (N). - ܙܥܘܪܘܬܐ: **Min. I**.i.2.10; **Mete. I**.iii.1.3 (cf. IS صغير).
√zrb: Pe: **Min. II**.i.2.4 (cf. N ܙܪܝܒܐ); **Mete. I**.ii.6.5 (cf. IS انضغاط). - Ethpe: **Min. I**.ii.4.5; **V**.iii.3.4 (cf. FR انضَغَطَ); **Mete. IV**.i.4.2 (cf. IS اضطرب).
√zrʿ: ܙܪܥܐ: **Mete. V**.ii.1.4; ii.2.4; ii.3.6 (IS بَذر). - ܙܪ̈ܥܘܢܐ: **Mete. V**.ii.1.3. - ܙܪܥܘܢܝܬܐ: **Mete. V**.ii.1.4.
√zrq: ܙܪܝܩ: **Min. I**.iii.2.4 (cf. IS مُنتشر); **Mete. II**.iii.1.2 (IS مُنتشر).

ܚ

ܚܐܝܠܐ: **Min. I**.ii.4.5; iii.1.3; iii.2.3 (IS بقوة); **II**.ii.1.2; ii.2.3; ii.4.8; **V**.iii.3.3 (IS بقوة); **Mete. I**.ii.3.6 (DM); **III**.i.1.6, 9; i.3.11; iv.1.6; **IV**.i.1.5 (cf. FR عَنيف); i.3.4 (cf. IS شديد). - ܚܐܝܠܬܢܐ: **Min. II**.iii.2.5.
ܚܐܝܠܘܬܐ: **Min**.tit.; **Mete**. tit.
√ḥbb: Pe: **Mete. II**.iv.1.7 (FR اشتَعَلَ); **IV**.iii.2.2, 5, 7. - ܚܒܝܒܐ: **Min. IV**.iv.1.7. - ܚܒܝܒܘܬܐ: **Mete. IV**.iii.1.7 (N).
√ḥbk: ܚܒܝܟ (ptc. pass.) [* = خَالَطَ]: **Min. III**.i.2.5 (IS*); i.3.6 (IS*); ii.1.1 (IS*); **Mete. IV**.i.2.2 (cf. IS مُختلط).
√ḥbl: Ethpa: **Min. I**.i.1.3; **Mete. III**.iii.3.3 (N). - ܚܘܒܠܐ: **Min. I**.i.1.3 (IS فَسَاد); **III**.iii.2.8; **V**.ii.4.7; iii.2.7; **Mete. V**.ii.1.2 (cf. IS فَسَدَ); iii.2.2. - ܡܚܒܠܢܘܬܐ: **Min. I**.i.1.2.
√ḥbṣ: Ethpe: **Min. II**.i.2.4 (N).
√ḥbr: Ethpa: **Mete. II**.iii.7.10. - ܚܒܪܐ: **Mete. I**.i.2.6; **II**.iii.7.2 (IS صَاحِب).
√ḥbš: Pe: **Min. I**.ii.5.6 (IS حَقَنَ); ii.6.4; **II**.iii.1.1 (N; cf. IS احتباس); **IV**.ii.1.4, 5 (IS أحَاطَ); **Mete. I**.i.3.4; iii.2.6 (AB حَبَسَ). - Ethpe: **Min. I**.ii.3.1 (cf. IS احتقان); ii.4.2; ii.4.4 (cf. IS محتقن); ii.5.2; **II**.iii.1.5 (cf. IS حَبَسَ); iii.2.4; iii.3.3 (N; cf. IS حَبَسَ); **Mete. IV**.i.1.4 (N/FR احتَبَسَ); i.5.2 (N/IS احتَبَسَ); i.5.5 (N/IS اختنق). - ܚܒܝܫ (ptc. pass.): **Min. I**.iii.3.2; **II**.i.4.6 (N); ii.4.1 (IS محتقن); iii.1.8; **V**.ii.2.1; **Mete. III**.i.3.1, 2 (N; cf. IS حَصَرَ); i.3.7 (N); i.3.10 (cf. IS احتقان); i.3.11 (IS محصور). - ܚܒܘܫܝܐ: **Min. IV**.i.1.8. - ܚܒܘܫܝܘܬܐ: **Min. IV**.i.4.3.
√ḥd: ܚܕ: **Min. I**.i.2.3, 12; i.3.7, 9; ii.4.3; **II**.iii.4.4; **III**.i.4.3 (IS أحد); **IV**.i.2.5; i.4.2; ii.3.7, 8; ii.4.3; iii.6.10, 11bis; **V**.i.1.1; ii.2.7; iii.1.4, 7; **Mete. I**.ii.6.1; iii.4.7, 8; **II**.i.1.5; i.7.1 (IS واحد); i.7.3; ii.3.1; iii.6.1bis; iii.7.7 (IS واحد); **III**.i.1.1; i.4.3; ii.2.5, 6, 9bis; **IV**.iii.1.9 (N); **V**.i.1.2; i.1.3 (IS أحد); ii.1.1; ii.3.8. - ܚܕ ܒܚܕ: **Min. II**.ii.4.8; **V**.iv.3.9; **Mete. II**.ii.4.6 (IS قَلَّمَا); iii.8.1 (N/IS في الأحيان); iii.8.9 (cf. IS قَلَّ); **IV**.iii.4.1 (cf. IS قَلَّ). - ܚܕܚܕ [* = كل واحد]: **Min. IV**.i.1.2; i.3.3; ii.1.3, 5 (IS*); **Mete. II**.ii.1.7bis (FR*); **III**.iii.4.1 (IS*); **IV**.i.1.1. - ܐܟܚܕܐ [* = معًا]: **Min. III**.ii.1.6; **Mete. II**.i.1.8 (IS*); i.2.3; i.3.5; i.6.5 (IS*); i.7.4; ii.1.8; ii.4.4 (IS*); **III**.iv.3.4 (IS*); **IV**.i.1.2 (FR*). - ܚܕܝܐܝܬ: **Mete. III**.iii.4.2 (IS في أول); iii.4.4 (N); **IV**.iii.1.5 (IS كَمَا). - ܚܕܢܝܘܬܐ: **Mete. II**.i.4.1. - ܚܕ̈ܢܐ [* = بعض ... بعض]: **Min. I**.i.4.10 (IS*); **II**.i.2.4; **III**.ii.1.6; iii.tit.; **IV**.i.1.4 (IS*); i.2.2; **Mete. I**.ii.1.8; ii.4.1 (AB*); iii.2.1, 4bis; iii.3.5; **II**.ii.4.4; iii.6.3; iv.1.8; iv.2.6; **III**.i.4.4, 9; **IV**.i.1.2; i.2.4; **V**.iii.2.3bis. - ܠܚܘܕ: **Min. III**.iii.3.8. - ܒܠܚܘܕ: **Min. II**.ii.1.8; **III**.i.1.4 (IS وَحدَ); i.2.2 (cf. IS إنَّمَا); i.5.5 (IS وَحدَ); iii.1.3, 6; iii.2.5; iii.3.5 (AB لا ... ألا); **IV**.ii.2.7; ii.4.5; ii.6.4; **Mete. I**.ii.2.6; ii.5.1 (AB لا ... ألا); **II**.iii.8.7 (cf. IS إنَّمَا); **III**.iv.3.5 (IS لا ... ألا); **IV**.i.6.2 (N); ii.1.3 (IS وَحدَ); ii.2.8;

V.iii.1.3. - ܡܚܕܪ (Pa ptc. pass.): **Min.** V.ii.3.3 (N).

√ḥdr: Pe: **Min.** IV.i.1.2 (IS مُحيط); V.i.3.2, 6; **Mete.** II.iii.1.6; IV.i.5.6; iii.3.9. - ܚܕܪ: **Min.** II.i.4.7; V.i.1.1. - ܚܘܕܪܐ [* = دائرة]: **Min.** IV.i.3.1, 2; i.3.7, 9 (IS دَوْر); ii.1.2bis (IS*); ii.1.4; ii.6.4, 5; iv.5.4, 6 (AB*); **Mete.** I.ii.1.2, 5 (cf. AB مُحيط); II.ii.1.2 (IS*); ii.2.3; ii.3.3, 5bis, 6 (IS*); iii.3.3bis (FR*); iii.5.6 (FR منطقة); iii.6.1, 2, 3bis, 4; iii.7.2, 3; iv.3.4 (IS*); III.ii.2.2bis, 3, 4, 5, 6, 9, 10 (IS*). - ܚܘܕܪܐ ܡܨܥܬ ܝܘܡܐ: **Mete.** III.ii.1.8, 10 (IS خط نصف النهار); ii.2.5, 9 (IS دائرة نصف النهار). - ܡܚܘܕܪܢܐ: **Mete.** II.ii.1.9; II.iii.3.1 (cf. FR استدارة); iv.1.6 (cf. FR استدارة); iv.3.3; III.i.1.8; IV.i.5.6 (IS مُدير). - ܡܚܘܕܪܢܐܝܬ: **Mete.** II.iv.1.9 (cf. FR على الاستدارة). - ܚܕܪ, ܚܕܪܝ (praepos.) [* = حول]: **Min.** IV.i.4.7; **Mete.** II.ii.1.3 (IS*); ii.1.9; ii.4.6, 8 (IS*); iii.1.8 (IS حَوالى); III.i.2.5; IV.iii.3.8.

√ḥdt: Ethpa: **Mete.** V.iii.tit.; iii.2.10. - ܚܕܬ: **Min.** I.i.2.10; **Mete.** V.i.3.7; ii.1.2. - ܚܘܕܬܐ: **Mete.** V.iii.1.6bis (cf. IS حُدوث).

√ḥwy: Pa: **Min.** I.i.4.8; II.i.4.7 (cf. N ܐܬܚܘܝ); **Mete.** I.iii.6.9; II.i.8.5; i.9.1; iv.1.5; iv.3.6; IV.iii.4.6 (cf. IS رُؤي). - Ethpa [* = رُؤي]: **Min.** II.i.4.6; iii.2.3; III.ii.2.7; IV.iv.1.4 (AB ظَهَر); iv.2.1; iv.3.10; V.i.3.12; **Mete.** I.iii.6.6 (IS تَراءَى); II.i.4.8, 9; i.6.4 (IS*); ii.1.6 (IS/FR*); ii.2.1; ii.2.2, 4 (IS*); iv.1.7; iv.2.7; III.iv.3.5 (IS*); iv.4.6 (IS*); IV.iii.2.3; iii.3.8 (IS ظَهَر); iii.4.10. - ܡܚܘܝܢܘܬܐ: **Min.** IV.i.4.2 (IS برهان). - ܡܬܚܘܝܢܘܬܐ: **Min.** IV.iv.3.5. - ܬܚܘܝܬܐ: **Min.** I.ii.5.5; **Mete.** II.iii.1.4.

ܚܘܝܐ, ܚܘܝܬܐ: **Mete.** V.ii.1.5 (IS/FR حَيّة).

√ḥwb: Pe: **Min.** III.iii.2.6; IV.iv.4.11; **Mete.** II.iv.2.2.

√ḥwg: ܚܘܓܬܐ [* = هالة]: **Min.** IV.ii.1.4 (IS مُحيط); **Mete.** I.iii.6.7 (IS*); II.i.1.1 (N/IS*); ii.tit.; ii.1.1; ii.1.2 (IS*); ii.2.1; ii.3.1bis, 2, 5 (IS*); ii.4.1, 2, 4, 5, 6, 8 (IS*); iii.4.1 (IS*); iii.4.4 (N); iv.1.7 (FR كُرة).

ܚܘܟܐ: **Mete.** V.ii.1.6 (IS/FR باذروج).

√ḥwl: ܚܠܐ: **Mete.** V.i.1.10 (IS رَمْل). - ܚܠܢܝܐ: **Min.** I.iii.2.6 (AB رملي).

√ḥws: ܚܣܝܢܐ: **Mete.** I.ii.3.7 (DM). - ܚܣܝܢܐܝܬ: **Min.** I.i.4.10; i.5.7; III.i.2.5 (cf. IS شديد); i.3.1 (cf. IS شديد); ii.3.7 (cf. IS شديد).

√ḥwr ("vidit"): Pa: **Min.** II.iii.4.6. - ܚܘܪܐ **Mete.** II.iii.2.4 (N).

√ḥwr ("albus fuit"): ܚܘܪ [* = أبيض]: **Min.** III.ii.3.1 (IS*); iii.1.7 (IS*); iii.3.2 (AB*); **Mete.** II.i.4.8, 10 (cf. IS بياض); iii.4.1 (IS*); iii.4.5 (N); iii.4.7; III.iii.4.7 (N/IS*); IV.ii.2.1 (N). - ܚܘܪܘܬܐ: **Min.** III.ii.1.5 (IS/AB بَياض); **Mete.** I.iii.2.3 (DM).

√ḥzy: Pe [* = رأى]: **Min.** I.i.2.4; i.2.8 (IS شاهد); i.2.11; i.3.7 (IS*); i.4.3; IV.ii.6.8 (IS شاهد); iii.6.11; iv.1.5 (AB*); V.ii.2.6; **Mete.** I.i.2.4 (IS شاهد); i.2.7 (cf. IS شاهد); ii.2.4bis (AB*); iii.4.9; II.i.1.9 (IS أدرك); i.5.2 (IS*); i.5.7 (cf. IS رُؤي); i.8.2 (IS*); ii.2.4bis, 5, 6; ii.4.4bis, 8 (IS*); iii.1.10; iii.2.4 (N); iii.2.6, 7; iv.3.5 (cf. IS رُؤي); IV.i.4.11 (cf. IS رُؤي). - Ethpe [* = رُؤي]: **Min.** I.i.4.1 (N); ii.1.7; II.iii.1.3 (N/IS*); V.iv.3.3; **Mete.** I.ii.3.9; I.iii.6.9; II.tit.; i.2.2, 3 (IS*); i.4.3, 6; i.5.4 (IS*); i.5.6; i.7.2bis; i.8.1 (IS رُؤي?); i.8.3 (IS*); i.9.3 (IS*); ii.1.2 (IS*); ii.1.5 (FR*); iii.1.7; iii.4.7; iii.5.3 (IS*); iii.5.4, 5 (N); iii.5.6 (IS*); iii.5.8; iii.6.1, 2; iii.7.11; iv.1.2; iv.3.3 (IS*); III.ii.1.9, 11; IV.i.3.7 (IS*); i.6.4, 5 (N); ii.3.5 (N); iii.1.1; iii.1.8 (N); iii.1.9, 10; iii.3.6. - ܚܙܝܐ [* = بَصَر]: **Mete.** II.i.1.4; i.1.4, 5, 6 (IS*); i.3.2 (IS*); i.3.7 (cf. N ܚܙܝܬܐ); i.5.1 (IS*); i.8.2 (IS*); i.9.5 (IS*); ii.1.7; ii.3.2 (IS*); iii.3.2 (FR*); iii.5.3, 5 (IS*). - ܚܙܝܬܐ [* = بَصَر]: **Min.** II.ii.3.9 (N/IS*); **Mete.** II.iii.2.2, 5 (N/IS*); iii.8.6; iv.1.4; IV.i.4.6 (IS إبصار); i.6.5 (N). - ܚܙܬܐ: **Mete.** IV.iii.3.4. - ܚܙܘܐ [* = راء]: **Mete.** II.i.2.5; i.4.7, 9; i.5.8 (IS*); ii.1.3 (FR*); ii.3.3, 4 (IS*); iv.3.6 (IS ناظر). - ܡܚܙܝܬܐ [* = مرآة]: **Min.** IV.iv.1.5, 7 (AB*); **Mete.** II.i.1.3, 7, 8bis, 9 (IS*); i.2.4 (IS*); i.3.4, 5; i.4.7; i.6.1 (IS*); i.7.2, 5, 6; i.8.5 (IS*); i.9.1; iii.1.2; iii.8.4; iv.1.4 (FR*). -

ܡܚܣܢܬܢܐ̈: **Mete. II**.i.8.2. - ܡܬܚܣܢܢܐ [* = مَـرني]: **Mete. II**.i.1.5, 7; i.2.1 (IS*); i.2.2 (IS*); i.2.4; i.3.1, 2 (IS*); i.3.8; i.4.7, 9; i.5.1; ii.1.5; ii.3.3 (IS*); iii.2.6; **III**.ii.2.6, 10; **IV**.iii.2.8. - ܡܬܚܣܢܢܘܬܐ: **Mete. IV**.ii.3.4.
√ḥzb: ܚܙܒܐ: **Min. II**.ii.1.5 (IS خابثة).
√ḥzq: ܚܙܩܐ̈: **Min. I**.i.2.2 (IS استمساك).
√ḥḥr: ܚܘܪܬܐ: **Min. I**.i.3.7bis (IS رغيف).
√ḥṭṭ: ܚܛܘܛܐ: **Min. I**.i.3.9 (cf. IS تخطيط).
√ḥṭp: Pe: **Mete. III**.iv.3.4, 6 (N); iv.4.5 (N/IS اختَطف).
√ḥyy: Pe: **Mete. IV**.iii.2.9. - ܚܝ: **Min. III**.i.2.7 (IS حَيّ). - ܚܝܐ̈: **Min. V**.iii.1.2 (cf. IS عُمر). - ܚܝܘܬܐ [* = حيوان]: **Min. I**.i.3.1 (IS*); i.3.8 (IS سَـبُع); **IV**.i.2.10 (IS*); i.4.11 (IS*); iii.6.3 (AB*); **V**.ii.3.6 (N/IS*); ii.4.8; iii.2.2 (IS*); **Mete. IV**.iii.3.7 (IS*); **V**.i.1.11; i.3.5 (IS*); ii.tit.; ii.1.4 (FR*); ii.2.4. - ܚܝܘܬܢܐ̈: Mete. V.ii.3.1.
√ḥyl: ܚܝܠܐ [* = قـوة]: **Min. I**.i.2.6 (IS*); i.3.4 (IS*); i.5.10 (N); iii.1.5; iii.1.7 (cf. IS*); iii.2.4 (IS*); iii.4.6; **II**.ii.1.6 (cf. IS أقـوَى); ii.4.7; **III**.i.4.1, 3, 4, 5 (IS*); ii.3.2 (IS*); iii.3.4 (AB*); iii.3.7; **IV**.iii.2.2; **V**.iv.3.7; Mete. **IV**.iii.4.4 (IS*); **V**.iii.1.8. - ܚܝܠܬܢܐ̈: **Min. III**.i.5.1 (IS/AB قوِيّ); **Mete. I**.ii.3.1 (cf. AB قوة); iii.2.2 (cf. FR شديد). - ܚܝܠܬܢܐܝܬ: **Mete. II**.iii.5.3 (IS أقوَى).
[√ḥyn]: ܡܚܝܢ (Pa ptc. pass.): **Min. IV**.i.4.11; **Mete. II**.iii.7.7 (IS مُناسب).
√ḥkk: ܚܟܘܟܐ: **Mete. I**.iii.4.6 (IS احتكاك).
√ḥkm: ܚܟܝܡ: **Min. II**.iii.4.4. - ܚܟܡܬܐ: **Min**.tit.; **IV**.i.2.9 (IS حكمة); **Mete**. tit.
√ḥll: ܚܠܠܐ: **Min. I**.iii.1.2. - ܚܠܝܠ: **Min. IV**.iv.1.6 (AB مُـقـعَّـر); **V**.ii.1.9; **Mete. II**.iv.3.7. - ܚܠܝܠܘܬܐ: **Min. V**.iv.1.4; **Mete. II**.i.5.3. - ܚܠܐ: **Mete. I**.i.1.4 (AB غُبار).
√ḥly: Ethpa: **Min. III**.iii.3.8 (AB حَلا). - ܚܠܐ [* = عـذب]: **Min. V**.ii.1.2 (IS*); ii.1.8; ii.1.11bis (N/IS*); ii.2.1, 2 (IS*); ii.3.2, 9 (N).
√ḥlṭ: Pa: **Min. III**.iii.3.3 (cf. AB خلطَ). - Ethpa: **Min. I**.ii.6.1 (cf. IS خلطَ); **II**.i.2.3; **Mete. I**.i.1.5 (cf. IS مُختلط). - ܚܠܝܛ (ptc. pass.): **Mete. IV**.i.1.2 (FR مـخلوط). - ܚܠܛܐ: **Mete. IV**.i.1.2. - ܚܘܠܛܢܐ̈: **Min. I**.ii.6.2; iii.4.2, 4 (IS اختلاط); **III**.iii.3.3; **V**.ii.1.1 (IS مُخالطة); ii.1.2, 3 (cf. IS خَالطَ); ii.4.5bis (cf. IS خَـالطَ); **Mete. II**.iii.4.3 (cf. IS اختَلطَ); **III**.iii.4.8 (N/IS سفاد).
√ḥlp: Ethpa: **Min. IV**.i.2.2 (IS تَبَـدَّلَ); **V**.iii.2.9. - Shaphel: **Min. III**.i.1.5 (IS أحَالَ); iii.1.5 (cf. IS أسلب); **IV**.i.2.7 (cf. IS مُتغيِّر); **Mete. I**.iii.1.3. - Eshtaphal [* = اسـتـحَـالَ]: **Min. I**.i.5.4; **III**.ii.2.6 (IS اختلف); **V**.iii.2.8 (IS تَغيَّرَ); **Mete. IV**.ii.3.3 (IS*); iii.2.4, 6 (IS*); iii.3.4 (IS*); **V**.i.2.6 (IS*); iii.2.10. - ܡܫܚܠܦ (Shaphel ptc. pass.): **Min. I**.ii.2.4 (IS مـخـتـلـف); **II**.ii.3.4bis (N); **Mete. III**.iii.2.8. - ܫܘܚܠܦܐ: **Min. II**.ii.3.3 (N); **III**.ii.2.6 (IS اخـتـلاف); **IV**.i.1.4 (cf. IS استحَالَ); i.2.2 (IS إحالة); **Mete. II**.iii.5.2 (IS اختلاف); iv.2.8; **III**.iii.tit. bis; **V**.i.1.7 (IS استحالة); i.3.2 (IS تَغيُّر); iii.2.10.
√ḥmm: Pe: **Min. II**.ii.1.1 (N?); **IV**.iii.3.6; iv.1.2; **Mete. I**.ii.5.4; **III**.iii.1.3 (IS سَخُنَ); iii.2.4 (cf. IS مُتسخن). - Aph: **Min. IV**.ii.3.5, 7; ii.5.2bis, 4, 6; iii.3.7; iii.5.6; iv.5.1 (AB أوجب الحرَّ). - ܚܡܝܡ [* = حارّ]: **Min. III**.i.3.5 (IS*); **IV**.iii.3.6; iii.4.3; iv.1.1 (AB*); iv.5.6; iv.5.7 (AB*); iv.5.8, 9; **Mete. I**.ii.4.1 (AB*); ii.5.2 (AB*); ii.6.4 (cf. IS حـرارة); **III**.i.2.2bis (N); iii.1 3 (IS*); iii.1.4 (IS أسخَنُ); iii.1.4 (IS*); **IV**.iii.4.11 (IS*). - ܚܡܝܡܘܬܐ [* = حرارة]: **Min. I**.i.2.7 (IS*); i.3.5; i.5.4 (N); ii.1.1 (IS حـرّ); ii.3.2 (IS*); ii.4.4 (IS*); ii.5.8; **II**.ii.1.2 (N); iii.1.11 (IS حـرّ); iii.3.2 (IS*); **III**.i.3.3 (IS*); ii.1.7 (AB حـرّ); ii.2.2, 3 (cf. IS حُمَّى); **IV**.ii.3.2, 9; ii.4.2, 4, 7; ii.5.7 (cf. IS تسخُن); iii.2.2, 4; iii.3.1; iii.6.3, 4 (AB حرّ); iii.6.5 (AB*); iii.6.6 (AB حرّ); iv.tit. (AB حرّ); iv.1.2, 6 (AB*); iv.3.3 (AB*); iv.3.3, 6 (AB حرّ); iv.4.1; iv.4.8 (AB*); iv.4.9 (AB حـرّ); iv.4.11; **Mete. I**.i.1.4, 6 (AB*); ii.1.1 (AB*); ii.5.3 (AB*); ii.6.3 (AB*); iii.6.1 (IS*); **II**.iv.2.2; iv.3.9; **III**.i.4.10 (N); iii.2.3 (cf. IS أسخَنُ); iii.2.6 (N); iii.3.2, 4 (N); iii.3.6; iii.6.1, 2, 4, 5 (N); iv.2.8 (N); **IV**.i.1.4 (cf. FR حـارّ); i.2.2 (IS/FR*); i.3.6. - ܚܘܡܐ [* = حرّ]: **Min. IV**.i.3.8, 9; i.4.4 (cf. IS تسخين); i.4.7 (IS*); ii.2.2; ii.5.8, 9; ii.6.2, 6, 8, 9 (IS*); iii.1.7 (FR*);

iii.4.2, 4, 5; iii.5.6, 7bis (FR*); iv.3.1 (AB*); iv.4.2, 4 (AB*); iv.4.5, 11. - ܚܡܝܪܘܬܐ: **Min. II**.iii.4.6. - ܚܡܝܪܬܐ: **Min. IV**.ii.5.5; iv.5.2; **Mete. I**.i.1.9.
√ḥmᶜ: Pe: **Min. III**.i.3.1 (IS تَخَمَّرَ). - ܚܡܝܥܘܬܐ: **Min. III**.i.3.2 (IS تخمير).
√ḥmṣ: Aph: **Min. III**.iii.3.8 (AB حَمَّضَ).
√ḥmr: ܚܡܪܐ: **Mete. IV**.i.3.8 (IS شَراب).
√ḥmš: ܚܡܫܐ, ܚܡܫ [* = خمسة]: headings passim; **Min. IV**.ii.1.1 (IS*); **V**.iii.1.8; **Mete. II**.ii.4.5 (IS*); **IV**.iii.3.8. - ܚܡܝܫܝܐ: **Mete.** tit.; **III**.ii.1.6. - ܚܡܫܡܐܐ: **Mete. IV**.iii.3.8.
ܚܢ ("nos"): **Mete. III**.iii.5.7.
√ḥsn: ܠܚܣܝܢ: **Min. I**.i.4.7 (cf. IS ما ... إلا بالجهد); **Mete. IV**.iii.4.1; **V**.ii.2.3 (cf. IS في نُدرة).
√ḥsr: Pe: **Min. V**.ii.1.6. - ܚܣܝܪ: **Mete. V**.ii.3.3. - ܚܣܝܪܐܝܬ: **Mete. I**.ii.2.6. - ܚܣܝܪܘܬܐ: **Mete. III**.iii.3.1; iii.3.6 (IS عوَز). - ܚܘܣܪܢܐ: **Min. II**.ii.2.1, 4 (cf. IS ضارّ).
√ḥpp: ܚܦܦܐ: **Min. I**.i.2.9 (cf. IS غُسل).
√ḥpy: Pa: **Min. III**.iii.1.9; **Mete. II**.i.7.5; ii.2.2 (IS أخفى); **IV**.i.6.6 (N). - Ethpa: **Mete. IV**.iii.3.4. - ܡܚܦܝ (Pa ptc. pass.): **Min. IV**.i.4.9; **Mete. V**.i.3.8 (IS مغمور). - ܡܚܦܝܢܢܐ: **Min. IV**.i.1.9. - ܡܚܦܝܢܘܬܐ: **Min. IV**.iv.3.5. - ܚܦܝܐ: **Min. IV**.iv.3.10.
√ḥpr: Pe: **Min. V**.iv.3.8. - Ethpe: **Min. I**.ii.2.5 (IS/FR انحفر); iii.3.3, 4 (cf. AB حَفَر); **II**.ii.4.3 (IS حُفَر). - ܚܦܝܪ (ptc. pass.): **Min. I**.ii.2.6, 7. - ܚܦܘܪܐ: **Min. I**.iii.2.8 (AB محتفر). - ܡܬܚܦܪܢܐ [* = معدني]: **Min. I**.i.1.6; ii.6.1 (IS*); **III**.tit.; i.1.1 (IS*); ii.2.1.
√ḥṣp: ܚܨܝܦܐ: **Min. I**.i.2.12; ii.5.6 (IS خزف).
√ḥql: ܚܩܠܐ: **Mete. IV**.iii.1.7 (N).
√ḥrr: Ethpa: **Min. IV**.i.1.9. - ܚܐܪܐ: **Min. II**.i.1.7bis (N); ii.3.3 (N); ii.3.5; iii.3.2 (N); iii.4.3 (N); **Mete. III**.iv.4.2.
√ḥry: Ethpe: **Min. III**.iii.2.3. - ܚܪܝܢܐ: **Min. IV**.iii.4.1.
√ḥrb: Ethpa: **Min. V**.iii.2.7.
√ḥrk: ܡܚܪܟܢܐ: **Min. IV**.i.4.4.
√ḥrs: ܚܪܘܫܘܬܐ: **Min. IV**.iv.2.1 (cf. AB خَشن); **Mete. II**.i.5.5 (IS خُشُونة).
√ḥrᶜ: ܚܘܪܥܐ [* = أصفر]: **Min. III**.iii.1.8 (IS*); **Mete. II**.iii.7.7 (IS*); iii.7.9.
√ḥrp: ܚܪܝܦ: **Min. II**.ii.3.8 (N); **Mete. III**.i.1.8; i.2.6 (N); **V**.i.1.9.
√ḥrt: Pe: Min. I.ii.2.5 (cf. IS حفّار).
√ḥšš: Pe: **Min. I**.i.3.5; i.5.9 (N); **III**.i.2.4; **V**.iii.3.8; iv.1.3, 9; **Mete. III**.i.3.4 (cf. N ܚܫܝܫ); iv.3.4; **IV**.i.2.9. - ܚܫܐ: **Min. I**.i.5.2 (N); **II**.i.4.2 (N); **Mete. V**.iii.2.2. - ܚܫܝܫܘܬܐ: **Min. I**.i.1.4 (IS انفعال).
√ḥšb: ܚܘܫܒܢܐ: **Mete. V**.iii.1.3 (cf. IS مُسْتَوفٍ). - ܚܘܫܒܢܐ: **Min. IV**.i.2.6. - ܡܚܫܒܬܐ **Mete. V**.iii.1.1 (IS رَوِيّة).
√ḥšḥ: Pe: **Min. I**.ii.1.10; ii.3.1 (cf. IS مَنفَعة); ii.4.1; **IV**.ii.2.4; **Mete. V**.i.3.6. - Ethpa: **Min. III**.iii.3.2; **V**.ii.4.3 (N); **Mete. II**.i.3.7 (IS استَعمَل). - ܚܫܚܬܐ: **Min. I**.ii.tit.; ii.6.7; **Mete. V**.iii.2.6.
√ḥšk: Pe: **Mete. II**.i.6.2 (IS كان مُظلم). - ܚܫܘܟܐ (subst.) [* = ظُلمة]: **Mete. II**.i.6.3 (IS*); iii.4.3 (IS*). - ܚܫܘܟܐ (adj.) [* = مُظلم]: **Mete. II**.iii.1.4 (IS*); **IV**.iii.4.7 (IS*). - ܚܫܘܟܘܬܐ: **Mete. II**.iii.4.5.
√ḥšl [mḥšl]: Ethpalpal (ܐܬܡܚܫܠ): **Min. V**.iv.1.3, 8.
√ḥtt: ܚܬܝܬ: **Min. I**.i.2.1 (IS خالص); ii.6.3; **II**.ii.tit.; **Mete. II**.i.3.1; **IV**.iii.2.2 (IS صرف). - ܚܬܝܬܐܝܬ: **Min. I**.i.1.3; **Mete. II**.i.1.4 (cf. IS حقيقي). - ܚܬܝܬܘܬܐ: **Mete. II**.i.3.6.

ܛ

ܛܪܛܪܘܣ (Τάρταρος): **Min. V**.iv.tit.; iv.1.2, 6; iv.2.2; iv.3.1, 4. - ܛܪܛܪܐ: **Min. V**.iv.1.8.
ܛܪܢܛܝܢܐ (nom. prop. "Tarentinus"): **Mete. II**.iii.2.3 (N).
√ṭbb: ܛܒܐ: **Min. I**.ii.5.3 (IS معروف); **IV**.ii.2.5; **V**.i.1.4; Mete. **III**.ii.3.1 (IS مشهور). - ܛܒܒܘܬܐ: **Mete. II**.i.3.7 (cf. IS أشهر وأعرف).
√ṭbl: ܛܒܠܝܬܐ: Min. IV.ii.1.3, 6, 7 (IS طبلي); IV.ii.2.1.
√ṭbᶜ: Pe ("mersus est"): **Min. V**.ii.3.7 (N/IS رَسَب). - Pe ("impressit"): **Min. I**.i.3.8 (cf. IS مرقوم). - Ethpe: **Min. V**.ii.3.5 (N Pe); **Mete. II**.i.2.2bis, 3 (cf. IS مُنطبع). - ܛܒܝܥ (ptc. pass.): **Mete. II**.i.1.4 (cf. IS انطباع); i.2.4 (IS مُنطبع).
ܛܠܠܝܢܘܣ (τελλίνη): **Min. V**.iii.2.3.
√ṭhr: ܛܗܪܐ [* = نصف النهار]: **Min. II**.iii.1.10 (N); **IV**.ii.5.4, 5; iv.5.4, 6,

8; **Mete. II**.iii.3.8 (IS*); iv.3.7, 8 (IS*). - ܛܡܣܝܪܐ: **Min. IV**.iii.2.4. - ܣܘܪܝܐ ܛܘܪܣܝܐ: vide sub ܣܘܪܝܐ.
√ṭwb: Ethpa: **Mete. I**.i.2.7. - ܛܒ (adj., "bonus"): **Min. III**.ii.3.4, 6 (IS جَيِّد). - ܛܒ (adv., "valde"): **Min. III**.ii.1.1; **V**.ii.3.6; **Mete. III**.iii.1.6; **IV**.iii.4.5.
√ṭwḥ: ܛܘܚܐ: **Mete. I**.ii.1.8 (AB مسانة).
√ṭws: Pe: **Mete. III**.iv.4.6 (IS طار).
√ṭwp: Pe: **Min. V**.ii.3.2bis, 7 (N). - ܛܘܦܬܐ, ܛܘܦܐ: **Min. I**.ii.5.9; **Mete. I**.ii.3.5 (AB قطرة); **V**.ii.2.7 (cf. IS نُطفة). - ܛܘܦܢܐ [* = طوفان]: **Min. V**.iii.1.9; iii.2.7 (IS*); **Mete. V**.tit.; i.1.1, 3, 6 (IS*); i.2.6, 7 (IS*); ii.2.11; iii.1.7.
√ṭwr: ܛܘܪܐ [* = جَبَل]: **Min. I**.tit.; i.1.7; ii.tit.; ii.1 (IS*); ii.2.1, 6; ii.3.1, 3 (IS*); ii.4.1bis, 2, 4; ii.5.1 (IS*); ii.5.4; ii.5.5, 7 (IS*); ii.6.3, 7 (IS*); **II**.i.3.2 (N/IS*); i.4.9, 11 (N); iii.4.7; **IV**.iv.1.2; iv.5.2 (AB*); **V**.i.3.8; iii.1.9, 11; iii.2.9 (IS*); **Mete. I**.i.2.3 (IS*); i.2.6, 7 (cf. IS*); i.3.3 (IS*); ii.2.3 (AB*); ii.5.7 (AB*); ii.6.1; ii.6.5 (IS*); iii.4.8 (IS*); iii.4.9; **II**.i.5.5; iii.1.3 (IS*); **III**.iii.1.1 (IS*); **V**.i.3.9 (IS*). - ܛܘܪܐ ܕܩܘܩܐ (nom. prop.): **Min. V**.i.3.2. - ܛܘܪܝܐ: **Min. I**.ii.3.2; ii.5.3 (IS جبلي).
ܛܘܪܣܢܝܐ (nom. prop. "Tyrrhenia"?): **Min. I**.i.5.6 (N).
ܛܘܪܢܘܣ (τόρνος): **Min. I**.i.4.7 (IS مقطع).
ܛܘܪܩܝܣ (nom. prop. "Turci"): **Min. V**.i.3.8.
√ṭyn: ܛܝܢܐ [* = طين]: **Min. I**.i.2.3, 4, 9 (IS*); ii.1.2 (IS*); **II**.i.1.5. - ܛܝܢܢܐ: **Min. I**.iii.2.6bis (AB طيني). - ܛܝܢܢܘܬܐ [* = طينية]: **Min. I**.iii.2.7 (AB*); **III**.iii.3.8 (IS*).
√ṭkn: Ethpa: Min. I.iii.1.8.
√ṭks: ܡܛܟܣܢܘܬܐ: Min. I.i.1.5; IV.i.1.3 (IS نظام).
√ṭll: ܛܠܠܐ [* = ظِلّ]: **Mete. I**.iii.tit.; iii.1.1, 5, 6 (IS*); iii.1.5, 6; iii.1.7 (N/IS*); iii.6.6 (IS*). - ܛܠܠܐ: **Min. IV**.iv.5.2 (AB ظِلّ).
√ṭly: ܛܠܝܐ (adj.): **Min. I**.i.2.8 (cf. IS طفولة). - ܛܠܝܐ (nom.): **Min. I**.i.5.3 (N).
√ṭlq: Ethpe: **Min. V**.ii.2.7; **Mete. I**.iii.5.5.
√ṭlš: ܛܠܘܫܐ [* = لزج]: **Min. I**.i.2.3 (IS*); ii.1.2 (IS*); **Mete. II**.iv.1.5 (FR*). - ܛܠܘܫܘܬܐ: **Mete. IV**.i.3.5 (IS لُزوجة).
√ṭmm: ܛܡܝܡܐ: **Mete. IV**.ii.2.7.
√ṭmʾ: ܛܡܐܐ: **Min. III**.ii.3.5 (IS دَنِس); ii.3.6 (IS نَجِس).
√ṭss: Ethpa [* = انطَرَقَ]: **Min. III**.i.1.4 (AB*); i.2.3 (N?); i.2.7 (IS*). - ܠܐ ܡܬܛܣܣܢܘܬܐ [* = مُنطرق]: **Min. III**.i.2.1 (IS*); i.2.2 (IS ما ينطرق); i.2.5 (IS*); i.5.1 (IS ما ينطرق; AB*); ii.tit.
√ṭʿy: Pe: **Mete. II**.iii.1.4 (IS غَلِطَ). - ܛܥܝܐ ("planeta"): **Min. V**.iii.1.8; **Mete. V**.i.1.5.
√ṭʿn: ܛܥܝܢܐ: **Min. II**.i.4.4 (N); **Mete. I**.ii.2.1 (AB حَمَلَ). - Ethpe: **Min. II**.i.1.3 (N).
[√ṭps]: Pa: **Min. V**.iv.3.10.
√ṭry: Ethpe: **Min. I**.i.4.6.
ܛܪܘܦܝܩܘܣ (τροπικός): **Mete. III**.iii.1.8, 10 (N).
√ṭrp: ܛܪܦܐ: **Min. II**.i.1.4 (N); **Mete. II**.i.4.6.

ܝ

√ybl: Pa [* = أدّى]: **Mete. II**.i.1.6; iii.8.6 (IS*); iv.1.4 (FR*). - Ethpa: **Mete. II**.i.1.7; i.8.2; iii.2.2, 9. - Aph: **Min. I**.i.4.8 (IS أنفذ); **Mete. II**.ii.1.7, 8 (FR أدّى); ii.4.3 (cf. IS تأدية). - ܡܘܒܠܐ [* = تَوَالُد]: **Mete. V**.ii.tit. (IS*); ii.1.3 (IS*). - ܡܝܒܠܢܘܬܐ **Mete. II**.i.1.3 (cf. IS أدّى).
√ybš: Pe: **Min. V**.iii.1.5 (cf. IS جَفَّ); **Mete. I**.ii.5.4; **III**.iii.3.4. - Pa: **Min. II**.iii.1.12 (N; cf. IS تجفيف); **Mete. III**.iii.6.1 (N). - Ethpa: **Min. V**.iii.1.11; iii.2.1. - ܝܒܝܫܐ [* = يابس]: **Min. I**.i.4.3 (IS*); **II**.i.2.5 (N); ii.2.4 (IS*); **Mete. I**.iii.6.4; **III**.i.1.3 (IS*); iii.1.6 (N); iii.4.6 (IS*); **IV**.iii.4.11 (IS*). - ܝܒܝܫܘܬܐ [* = يَبَس]: **Min. I**.i.2.2 (IS*); **II**.i.2.5; i.3.3 (N); iii.2.8 (IS*); **III**.i.1.5 (IS*); i.3.6 (IS*); iii.3.8 (IS*). - ܝܒܝܫܬܐ [* = بَرّ]: **Min. IV**.i.2.9 (IS*); **V**.iii.1.7, 8 (IS*); iii.2.1; **Mete. III**.iii.2.3 (IS يَبَس); **V**.i.3.5 (IS*).
√yd: ܐܝܕܐ: **Min. I**.iii.1.6; **III**.iii.1.6 (IS يد); V.iv.3.7, 10; **Mete. IV**.i.2.6 (IS يد). - ܒܐܝܕܐ ܒܐܝܕܐ: **Min. I**.ii.1.2 (IS قليلا قليلا); ii.1.9; **Mete. IV**.iii.3.9; V.i.3.3; iii.1.5.
√ydy: Aph (ܐܘܕܝ): **Mete. II**.i.2.1. - ܬܘܕܝܬܐ **Mete. II**.i.2.3.

√ydᶜ: Pe: **Min. II**.iii.4.4; **III**.iii.2.9; iii.3.6 (AB عَـرَفَ); **IV**.ii.2.8; **V**.iii.3.10; **Mete. II**.iii.7.10 (IS عَـرَفَ); **III**.ii.1.3; **IV**.i.2.5 (IS عَلَمَ); **V**.i.3.7 (IS عَلَمَ). - Ethpe: **Mete. II**.i.5.8; **III**.i.3.4 (N). - Shaphel [* = دَلَّ]: **Min. II**.iii.2.6 (N/IS*); **Mete. II**.iv.2.1, 3 (IS*); iv.2.3 (cf. IS دلالة); **IV**.iii.4.11 (IS*); **V**.iii.1.6 (IS*); iii.2.3, 4. - Eshtaphal: **Mete. V**.iii.2.3, 5. - ܝܕܝܥ (ptc. pass): **Min. I**.i.2.8 (neg. corr. IS مجهول); iii.3.2; **II**.ii.4.1 (cf. IS دليل); **III**.ii.2.1, 5; iii.3.6; **IV**.ii.2.4, 6; iv.1.1; **V**.iv.2.3; **Mete. I**.iii.4.4; **II**.i.2.3; **III**.i.3.6; ii.1.4; **IV**.ii.3.3 (N). - ܝܕܥܬܐ: **Min. I**.i.1.1 (cf. IS تعـــريف); i.1.5; **III**.iii.3.6 (AB معرفة); **Mete. V**.iii.1.4. - ܡܫܘܕܥܢܘܬܐ: **Mete. V**.iii.2.8.

√yhb: Pe: **Min. I**.i.2.2; ii.6.6; **II**.iii.2.3; **III**.iii.1.8; **V**.ii.2.4; **Mete. III**.iii.2.6, 7 (N); **IV**.iii.2.1. - ܝܗܝܒ (ptc. pass.): **Mete. II**.iii.5.2. - ܡܘܗܒܬܐ (nom.): **Min. I**.ii.6.6; **Mete. V**.ii.2.9 (IS واهب).

√ywm: ܝܘܡܐ [* = يَوم]: **Min. IV**.ii.3.6, 8; iii.6.7 (AB*); **Mete. I**.ii.2.4 (AB*); iii.2.6 (AB*); **III**.iii.4.5 (N/IS*); **IV**.iii.1.8. - ܝܘܡܢܝܐ: **Min. V**.iii.3.10; **Mete. I**.iii.1.1 (IS يومي).

ܝܘܢܝܐ (nom. prop. "Graecus"): **Mete. III**.ii.3.2 (cf. IS يونانية).

√yld: Pe: **Mete. I**.ii.2.1. - Ethpe [* = تَوَلَّدَ]: **Min. I**.i.1.4; ii.3.4 (IS*); ii.4.2; ii.6.4 (IS*); iii.4.2 (cf. IS متولد); **II**.ii.1.2 (FR*); ii.1.5 (IS*); ii.1.8 (cf. IS حَدَثَ); iii.2.9 (N); **IV**.ii.6.6; **V**.ii.3.7, 9 (N); **Mete. I**.i.1.1, 4; **II**.iii.4.4 (IS*); iii.5.8 (IS*); iii.7.8 (IS*); **III**.i.1.2, 4 (IS*); i.1.4 (cf. FR تولُّد); ii.3.2; iii.3.8 (N); **V**.i.3.8 (IS*); ii.tit. (cf. IS تولُّد); ii.1.4; ii.1.5 (cf. FR متولد); ii.1.7. - Aph: **Min. II**.iii.4.3; **Mete. III**.i.4.8 (N); iii.6.5 (N); iv.1.1 (IS مُوَلِّد). - ܝܠܕܐ: Mete. IV.tit. - ܡܘܠܕܐ: Mete. III.i.tit. (FR تولُّد); III.i.4.7 (N). - ܡܘܠܕܢܐ: **Min. I**.iii.4.4.

√ylp: Pa: **Min. I**.i.1.8. - Ethpa: **Min. II**.iii.3.5. - ܝܘܠܦܢܐ [* = علم]: **Min. I**.ii.6.7 (IS*); **Mete. II**.i.9.5 (IS*); **V**.iii.1.8. - ܡܠܦܢܐ: **Min. IV**.ii.3.1. - ܡܠܦܢܘܬܐ: **Min. I**.i.1.5; **Mete. II**.i.tit. (IS تعليم); iii.5.1.

√ymm: ܝܡܐ [* = بحر]: **Min. I**.i.5.7 (N); ii.5.9 (IS*); **IV**.i.2.9; ii.2.9; **V**.tit. (IS*); i.tit.; i.1.1 (DM); i.1.3; i.2.1, 5; ii.tit.; ii.1.4; ii.1.6bis; ii.1.10 (N); ii.2.1 (IS*); ii.2.3, 4, 8; ii.3.1 (N/IS*); ii.3.3, 4bis (N); ii.4.1 (N/IS*); ii.4.5, 6 (IS*); iii.tit.; iii.1.1, 3 (IS*); iii.4.7, 8; iii.2.1, 4; iii.3.1 (IS*); iii.3.6, 7, 8; iv.3.4, 6; **Mete. I**.ii.4.1, 3bis, 6 (AB*); ii.5.1, 3, 5, 6 (AB*); ii.6.2, 4 (AB*); ii.6.5; **II**.iii.2.1 (N); **III**.iii.2.4 (IS*); iv.4.5 (IS*); **V**.i.3.1, 4, 5, 9bis (IS*). - ܝܡܐ ܐܘܩܝܢܘܣ (nom. prop.): **Min. V**.i.3.3; ii.3.5 (N). - ܝܡܐ ܕܓܠܝܢܐ (nom. prop.): **Min. V**.i.3.9. - ܝܡܐ ܕܦܢܛܘܣ (nom. prop.): **Min. V**.i.3.5; ii.3.3 (N); iii.3.9. - ܝܡܐ ܕܣܘܦ (nom. prop.): **Min. V**.i.3.4. - ܝܡܐ ܕܓܠܛܝܐ (nom. prop.): **Min. V**.i.3.5; iii.3.9. - ܝܡܝܐ: **Min. V**.iii.2.2 (cf. IS بحر); **Mete. I**.ii.5.6 (AB بحري). - ܝܡܬܐ: **Min. V**.i.2.6; ii.3.3, 4bis (N); ii.3.5 (IS بُحيرة); iii.2.3, 4.

√ymn: ܝܡܝܢܐ: **Mete. II**.iv.3.2 (IS يمنة). - ܬܝܡܢܐ ("regio meridionalis") [* = جَنُوب]: **Min. IV**.i.2.5 (IS*); i.3.6, 9; i.4.3 (IS*); i.4.7; iii.5.6; **V**.i.2.1; i.3.1, 6; **Mete. II**.iv.2.4, 7 (cf. IS جـنـوبـي); **III**.ii.2.12; ii.3.6, 8; iii.1.2 (IS*); iii.1.10 (N). - ܬܝܡܢܐ (nom. prop.): **Min. I**.i.4.11 (cf. IS يمني). - ܬܝܡܢܝܐ [* = جنوبي]: **Min. IV**.i.3.3; ii.1.3, 7; ii.2.6; **V**.i.1.5; **Mete. I**.ii.4.4 (AB*); **II**.iv.2.8 (IS*); **III**.ii.1.3, 10 (IS*); ii.1.11; ii.2.9; ii.3.2 (cf. IS جنوب); iii.1.4 (IS*); iii.4.5 (IS*); iii.5.3; iii.5.4 (N); iii.5.6, 7bis; iii.5.11 (N); **V**.i.3.9 (IS*).

√ysp: Aph: **Min. I**.i.5.1; I.ii.2.7. - Ettaph: **Min. I**.ii.1.6; **IV**.iv.3.6; **V**.ii.2.4; **Mete. V**.iii.1.5 (cf. IS تَزَيُّد). - ܬܘܣܦܬܐ: **Min. IV**.i.2.8; ii.3.5; iv.3.3 (AB زيادة).

√yᶜy: ܡܥܝܢܘܬܐ: **Mete. III**.iv.4.3 (cf. IS مَنبَتٌ).

√ypy: Aph: **Min. V**.iii.1.7 (IS انحَسَمَ).

√yqd: Pe: **Min. I**.i.5.8 (N); **III**.i.1.3 (AB احـــتَـرَقَ); **IV**.iii.3.6; **Mete. III**.iii.3.5; **IV**.i.4.2 (IS اشتعل); ii.3.5 (N); iii.1.8 (N). - Aph: **Min. IV**.iv.1.6 (cf. AB إحـراق); iv.1.8 (AB أحـرق); i.3.8 (cf. IS مُـحـرق); ii.2.4, 8, 9 (N/IS أحرق); ii.3.1 (N). - ܝܩܝܕ (ptc.): **Mete. III**.iii.1.6 (N). - ܝܩܕܐ: **Min.**

III.ii.3.1, 3 (IS مُحرَق); ii.3.5 (IS فيه قوة احتراقية); **V**.ii.1.3 (IS مُحترق); ii.4.5 (FR مُحترق). - ܝܩܕܢܐ: **Min. V**.iii.1.10; **Mete. IV**.ii.3.4 (N). - ܡܬܝܩܕܢܐ: **Min. II**.ii.2.6 (IS مُحرَق); **IV**.i.3.9; ii.2.2.

ܝܩܘܢܬܐ: **Min. III**.i.5.2 (AB ياقوت).

[√yqn]: Ethpa: **Mete. II**.i.3.3 (IS تَشَبَّع). - ܝܘܩܢܐ (εἰκών) [* = صورة]: **Mete. II**.i.1.2 (IS شَبَح); i.1.2, 3 (IS*); i.1.6; i.2.1 (IS شَبَح); i.2.4 (IS*); i.3.3, 4bis (IS*); i.7.2, 4, 5 (IS شَبَح); i.8.5.

√yqr: Pe [* = ثَقُلَ]: **Mete. I**.ii.1.6, 7; ii.3.5; ii.6.7 (IS*); **III**.i.1.6 (FR*); **IV**.i.1.6 (FR تَاقَلَ). - ܝܩܝܪ [* = ثقيل]: **Min. II**.ii.1.7 (N); ii.3.8; **V**.ii.3.2 (IS*); ii.3.3 (N); ii.3.6; **Mete. I**.iii.3.9 (IS*); **III**.iv.2.1 (IS*); **IV**.iii.4.3. - ܝܩܝܪܘܬܐ: **Mete. III**.iv.2.3; iv.3.6 (IS ثقل). - ܝܘܩܪܐ [* = ثقل]: **Min. I**.iii.1.7 (IS*); **III**.ii.1.2 (AB*); **V**.iii.3.5; **Mete. I**.i.3.1; **IV**.ii.1.4 (IS*); iii.1.2 (N/IS). - ܡܝܩܪ̈ܢܐ: **Mete. I**.i.3.6.

√yrḥ: ܝܪܚܐ [* = شهر]: **Min. IV**.ii.6.9 (IS persicè ماه); iii.6.8 (AB*); **Mete. III**.iii.4.6 (IS*). - ܝܪܚܢܐ: **Min. V**.iii.3.10. - ܝܪ̈ܚܢܝܐ: **Min. IV**.iii.6.9 (AB شهوري).

√yrk: Pe: **Min. III**.iii.3.7 (AB تَطاوَلَ); **IV**.iv.3.1 (AB طَالَ).

√yrq: Aph: **Min. III**.i.4.6 (IS اخضَرَّ). - ܝܘܪܩܘܬܐ: **Mete. II**.i.4.6 (IS خُضرة).

√yrt: ܡܝܪܬܘܬܐ: **Min. II**.ii.1.5 (IS عصير).

ܝܬ: **Min. II**.i.3.1; **V**.ii.1.1 (IS نفس); iv.3.8, 10; **Mete. II**.ii.1.2 (N); ii.1.5 (IS/FR نفس); iv.1.3 (IS نفس; FR ذات).

√ytb: Ethpa: **Min. IV**.i.4.6 (cf. IS عمارة).

√ytr: ܝܬܝܪ (adj.): **Min. I**.i.5.1; ii.6.2; iii.1.4, 5; **II**.iii.2.5 (cf. IS شدّة); **III**.ii.1.2, 5; **IV**.i.4.6; iii.1.2; iii.2.3; iv.4.5; **Mete. I**.i.3.2; ii.2.4; **III**.i.4.8 (N); **V**.i.2.3 (cf. IS إفراط); iii.1.5. - ܝܬܝܪܐܝܬ (adv.) [* = elative]: **Min. I**.i.1.6; iii.2.7; iii.2.8; iii.4.1, 3; **II**.ii.4.7; iii.tit.; iii.2.8; iii.2.9 (N); **III**.i.3.4 (IS*); ii.2.4; ii.3.2; **IV**.ii.3.2 (IS*); ii.3.9; ii.5.1, 3, 6, 7; iii.6.3 (AB*); iv.1.1, 5bis; iv.4.4; iv.4.6bis (AB*); iv.4.8; iv.5.1; iv.5.7 (AB*); iv.5.8bis; **V**.ii.1.2 (IS*); ii.1.8; ii.3.1 (IS*); **Mete. I**.ii.1.8bis (AB*); ii.5.6; ii.6.1, 4; iii.4.3 (IS أكثر); iii.4.8; **II**.i.4.8, 10 (cf. IS أشدّ); i.7.5 (cf. IS أقوى); iii.3.10 (cf. IS كثرة); iii.4.4, 6, 7; iii.8.8 (cf. IS شديد); iv.2.4 (cf. IS اشتدّ); **III**.ii.2.2, 3; iii.6.3, 6; iv.1.3 (N); iv.2.6, 8, 9 (N); **IV**.ii.2.1, 3, 5bis, 7 (N); ii.3.1; **V**.i.1.9; ii.1.4; ii.3.2 (IS*). - ܝܬܝܪܐܝܬ: **Min. II**.ii.4.2 (cf. IS كَثُرَ); iii.1.11; iii.2.2 (N/IS خصوصًا); **IV**.iv.4.1; **Mete. I**.ii.2.6; **II**.i.7.4 (cf. IS أقوى); iv.3.8 (IS لا سيّما); **IV**.i.1.8. - ܝܬܝܪܘܬܐ: **Min. II**.iii.2.6, 7; **IV**.iii.3.1, 3 (cf. FR شديد); iii.4.6; iv.1.5, 6; **Mete. V**.i.2.1 (IS كثرة/زيادة). - ܝܘܬܪܢܐ: **Min. II**.ii.2.1, 2 (cf. IS نافع); iii.4.4 (IS مَنفَعة). - ܡܝܬܪܬܐ [* = أفضل]: **Min. I**.iii.4.2 (IS*); **III**.ii.3.2 (IS*); **IV**.iii.1.1. - ܡܝܬܪܬܐ ܕܐܢܫܐ (i.e. Avicenna): **Min. I**.i.3.6; **Mete. II**.iii.7.1.

ܟ

√kʾb: ܟܐܒܐ: ܟܐܒ ܦܪ̈ܩܬܐ (i.e. ἀρθρῖτις [νόσος]): **Min. I**.i.5.2 (N).

ܟܐܘܢܝܐ (nom. prop. "Chaonia"): **Min. V**.ii.3.8 (N).

ܟܐܠܣܝܠܝܘܣ: vide ܟܠܣܝܠܝܘܣ.

ܟܐܠܝܬܘܣ: vide ܟܠܝܬܘܣ

ܟܐܡܬ: **Min. II**.ii.4.5; iii.4.7; **IV**.iii.6.10; **V**.ii.4.4; **Mete. II**.iii.2.5; iv.3.6; **III**.iii.5.4; **IV**.iii.3.3 (IS وهو أن).

ܟܐܦܐ [* = حَجَر]: **Min. I**.i.tit.; i.2.1; i.2.3 (IS*); i.2.4bis (IS*); i.2.10, 12; i.3.1, 8; i.5.1, 4 (N); i.5.6; i.5.8bis (N); ii.1.1 (IS*); ii.1.8 (AB*); ii.1.10 (AB صَخر); **III**.i.1.2 (IS*); i.1.2 (AB*); i.4.1 (IS*). - ܟܐ̈ܦܐ: Min. I.ii.5.5. - ܟܐܦ̈ܢܝܐ: **Min. I**.i.4.2 (N); ii.2.5 (IS/FR حجرى); ii.4.3; **III**.i.5.3; **Mete. V**.i.1.11. - ܟܐܦܢܝܘܬܐ: **Min. I**.i.3.2 (IS حجرية); i.3.3.

√kʾr: Pe: **Min. II**.iii.4.7.

ܟܐܪܨܝܢܝ (= خارصينى): **Min. III**.i.5.9.

√kby: ܟܒ̈ܝܐ: **Mete. V**.ii.1.5 (FR خشي).

ܟܒܪ: **Min. I**.i.5.10 (N); **III**.i.1.6; **V**.ii.4.3; **Mete. IV**.ii.2.5, 7.

ܟܒܪܝܬܐ [* = كبريت]: **Min. III**.i.1.2 (IS*); i.3.1 (IS*); i.5.6 (IS*); i.5.7; ii.1.8 (IS/AB*); ii.2.6, 7; ii.3.1, 2, 4, 5, 6, 9 (IS*); iii.1.1 (IS*); iii.3.2 (AB*). - ܟܒܪܝܬܢܝܐ: **Mete. I**.ii.5.4 (AB كبريتي); **IV**.iii.4.8. - ܟܒܪܝܬܢܝܘܬܐ [* = كِبريتية]: **Min. III**.i.4.1, 3 (IS*).

√kbš: Ethpe: **Mete. I**.ii.4.2 (AB كُبس).
ܟܒܘܢ: ܟܒܘܢܘ: **Min. II**.i.4.7; **Mete. IV**.iii.2.8.
√khn: ܟܘܡܪܐ: **Min. V**.i.1.7.
√kww: ܟܘܬܐ: **Mete. II**.iii.2.8 (IS كُوَّة).
√kwkb: ܟܘܟܒܐ [* = كوكب]: **Min. IV**.i.2.1 (IS*); **V**.iii.1.8; **Mete. I**.iii.6.8; **II**.ii.4.5 (IS*); **IV**.iii.tit.; iii.1.5 (IS*); iii.3.6; **V**.i.1.5 (IS*). - ܟܘܟܒܐ ܕܩܘܡܬܐ [* = ذات الأذناب]: **Mete. I**.iii.6.8 (IS*); **IV**.iii.3.1 (IS*).
ܟܘܟܒܘܬܐ: **Mete. I**.iii.2.8 (DM).
√kwn: Ethpa (ܐܬܟܝܢ): **Mete. IV**.iii.4.2; **V**.i.1.2; ii.3.4. - ܡܟܝܢ (Aph ptc. pass.): **Min. III**.iii.3.5; **V**.iii.1.1; **Mete. II**.i.7.1 (cf. IS من شأنه أن); ii.4.7 (cf. IS بَلَغ). - ܟܝܢܐ [* = طبيعة]: **Min. I**.i.1.2; i.2.6; i.4.2 (N); **III**.ii.3.2; iii.1.2; iii.1.4 (IS*); iii.3.2; **IV**.i.1.2 (IS*); iv.4.1, 2; iv.4.8 (AB طبع); **V**.iii.1.1 (IS طبع); iii.3.1 (IS طبع); iv.1.6; iv.3.10; **Mete. I**.i.1.8; ii.6.6; **III**.iv.3.7 (IS طبع); **IV**.i.2.5 (IS شيء); iii.2.7; **V**.i.2.6 (IS*); iii.2.3. - ܟܝܢܝܐ [* = طبيعى]: **Min. I**.ii.6.7 (IS*); **I**.iii.3.5; **III**.i.1.6 (IS*); **IV**.i.tit.; i.1.2, 3 (IS*); i.2.10 (IS*); **V**.iii.1.8. - ܟܝܢܝ̈ܐ: **Mete. II**.i.3.1 (IS طبيعيون). - ܟܝܢܝܬܐ: **Min**.tit.; Min. **I**.i.1.1 (IS طبيعيات). - ܟܝܢܐܝܬ: **Min. V**.iv.1.8. - ܟܝܢܘܬܐ: **Mete. II**.iii.8.8; **V**.ii.2.2. - ܬܟܝܢܘܬܐ: **Min. IV**.i.1.4 (cf. IS شأن).
ܟܘܘܬܐ: **Mete. II**.iv.3.7. - ܟܘܘܬܘܬܐ: **Min. V**.iv.1.4; **Mete. II**.i.5.3 (IS تقبيب).
ܟܘܫܝܬܐ (nom. prop. "Chusita, Aethiops") [* = الحَبَشَة]: **Min. IV**.ii.6.8 (IS*); iii.4.5; **V**.i.3.3; **Mete. I**.ii.6.4 (IS*).
ܟܝܘܣ (nom. prop. "Chios"): **Min. V**.i.1.6.
√kll: Eshtaphal: **Mete. V**.iii.1.9. - ܟܠ [* = كُلّ]: **Min. I**.iii.4.1; **II**.i.4.3, 4 (N); **III**.i.2.1 (IS جميع); ii.2.1 (IS جميع); ii.2.2; ii.2.5 (IS*); **IV**.i.1.1 (IS جميع); i.3.5bis; i.3.7 (N); ii.3.2; ii.4.1, 2, 6; iii.1.1, 5, 6; iii.4.4; iii.6.8 (AB*); iii.6.9; iv.3.1 (AB*); iv.3.7; iv.4.4; iv.5.4 (AB*); iv.5.8; **V**.i.1.1, 7; i.3.10; ii.4.7; iv.3.1; **Mete. II**.i.7.4; ii.1.8; iii.7.1; **III**.i.3.7 (N); ii.1.4; **IV**.i.4.8 (IS*); iii.3.9; iii.4.2; **V**.i.1.3 (IS*); i.3.10; ii.1.7; ii.2.5; ii.3.5bis (IS*); iii.1.1, 3. - ܟܠܗ: **Min. I**.i.1.9. - ܟܠܗܘܢ: **Min. V**.iv.1.7. - ܗܕܐ ܟܠ ("universum"): **Min. IV**.i.1.3 (IS الكلّ); i.4.6. - ܗܕܐ ܟܠܗ ("tantum, adeo ut"): **Min. IV**.i.4.6; **V**.ii.3.6. - ܟܠܢܫ: vide ܐܢܫܘܬܐ. - ܟܠܡܕܡ: vide ܡܕܡ. - ܟܠܚܕ: vide ܚܕ. - ܟܠܝܬܐ [* = كُلّية]: **Min. III**.i.3.4 (IS*); **Mete. III**.i.3.1, 2, 10, 11, 12 (IS*); **IV**.iii.2.6 (cf. IS تامّة). - ܟܠܢܝܐ: **Min. IV**.iii.5.2 (cf. FR كلّه); **Mete. V**.ii.2.11; iii.1.2 (IS كلّي); iii.1.7. - ܟܠܢܝܘܬܐ: **Min. V**.iv.1.2, 5; iv.3.1, 4. - ܟܠܝܠܐ: **Min. IV**.iii.6.7 (AB جُملة).
√kly: Pe: **Min. IV**.ii.2.8; **Mete. I**.i.3.3 (cf. IS مانع); ii.3.2 (AB قاوم); **IV**.ii.2.7. - Ethpe: **Min. I**.iii.1.4; **Mete. I**.i.3.1. - ܟܠܝܢܐ: **Min. I**.ii.3.1.
ܟܠܩܝܛܝܣ, ܟܠܩܝܛܝܣ (χαλκῖτις): **Min. III**.i.4.2, 5 (IS قلقطار); i.5.5.
ܟܠܩܢܬܘܣ, ܟܠܩܢܬܘܣ (χάλκανθος): **Min. III**.i.4.2 (IS قلقند); i.4.6.
ܟܡܝܪܘܬܐ: **Min. V**.ii.2.7; iv.2.1, 6; **Mete. II**.i.5.8; iii.3.7.
ܟܡܝܠܝܐ: **Min. III**.i.5.6; iii.1.5 (IS كيمياء); iii.3.1.
ܟܢ: **Min. I**.i.2.5; i.4.5; **II**.ii.3.7; **V**.i.3.4, 9; ii.1.8; **Mete. I**.i.2.6; ii.3.3 (AB ف); **III**.i.3.8; ii.2.5; iii.1.2; iii.2.4; iv.3.6. - ܟܢܬܐ: **Mete. III**.i.3.7 (N).
√kny: Ethpa: **Min. I**.i.1.8, 9; **Mete. I**.ii.3.9; **III**.iv.4.7; **IV**.iii.1.10.
√knp: ܟܢܦܐ: **Mete. II**.iii.1.6 (IS جَناح).
√knš: Ethpa: **Min. I**.ii.4.3; **V**.ii.2.7; iii.1.9, 10; **Mete. I**.ii.1.5; iii.1.5 (cf. IS اجتماع); iii.2.1 (IS اجتمع); **III**.i.3.5 (N); **V**.ii.2.8. - ܟܢܝܫܐܝܬ: **Min. II**.i.4.6 (N). - ܟܢܘܫܝܘܬܐ [* = اجتماع]: **Mete. I**.iii.6.1; **III**.iii.1.9; **IV**.ii.1.4 (IS*); **V**.ii.2.5 (IS*). - ܟܢܘܫܝܐ: **Mete. I**.iii.3.3. - ܟܘܢܫܐ: **Mete. I**.ii.1.6.
√ksy: Pa: **Min. IV**.i.4.2 (cf. IS مغمور). - ܡܟܣܝܐ (ptc. pass.): **Min. III**.iii.2.9 (cf. IS مجهول); **Mete. III**.ii.2.4 (cf. IS خفاء).
√ksn: ܟܣܢܐ: Min. I.i.5.5 (N).
ܟܐ ("ubi"): **Min. III**.iii.3.5; **IV**.ii.1.6; ii.3.6.
√kpp: ܟܦܝܦ: **Mete. II**.iv.3.6 (IS مُتحدّب); iv.3.6. - ܟܦܝܦܘܬܐ: **Mete. II**.iv.3.5. - ܟܦܟܦܐ: **Mete. II**.iii.3.3 (FR قَدر); iv.3.3 (IS تطعة). - ܟܦܬܐ, ܟܦܐ: **Min. V**.iii.3.5.
√kpy: Pe: **Mete. I**.i.2.8.
√kpn: ܟܦܢܐ: **Min. V**.iii.2.7.

√kpr: Pe: **Mete. II**.iii.1.10 (cf. IS مُسِحَ).
√kry: Ethpe: **Min. IV**.iv.3.2 (cf. AB قَصُرَ). - ܟܪܝܐ (ptc. pass.): **Min. I**.iii.4.3 (IS قَصُرَ); **IV**.iv.4.10.
√krh: ܟܘܪܗܢܐ: **Mete. IV**.iii.4.11 (IS مَرَض). - ܡܟܪܗܢܐ: **Min. V**.ii.4.7.
√krk: Pe: **Mete. II**.iv.1.9. - ܟܪܟܬܐ: **Mete. III**.iv.2.4 (N); iv.3.2 (cf. IS لَوَّى); iv.3.3, 5 (IS لفـانـة). - ܡܬܟܪܟܢܐܝܬ: **Mete. III**.iv.4.2 (N). - ܡܬܟܪܟܢܘܬܐ: **Mete. III**.iv.3.6.
√krkm: Ethpalpal: **Min. III**.i.4.5 (IS اصفَرَّ).
√krs: ܟܪܣܐ, ܟܪܣܬܐ: **Min. IV**.iii.6.5 (cf. AB بواطن الأبدان); **Mete. V**.ii.2.8, 10.
√krt: ܟܪܬܐ: **Mete. II**.iii.4.4; iii.7.7 (IS لون كَـراثي); iii.7.9. - ܟܪܬܢܝܐ: **Mete. II**.iii.5.8 (IS كَرَاثي); iii.6.5.
√kšṭ: Ethpe: **Mete. III**.i.2.8.
√ktb: ܟܬܝܒ (ptc. pass.): **Min. V**.iv.1.1. - ܟܬܒܐ: **Min**.tit.; **I**.i.1.7; **V**.i.3.12; iii.2.6; Min. fin.; **Mete**. tit. - ܟܬܝܒܬܐ: **Min. V**.iii.2.9, 12 (IS كِـتـابة). - ܡܟܬܒܢܘܬܐ: **Mete. V**.iii.2.9.
√ktr: Pa: **Min. I**.ii.6.2 (cf. IS إقـامـة). - ܟܘܬܪܐ: **Min. II**.iii.4.3 (N).

ܠ

ܠܐ: passim. - ܠܐ ܗܘܐ (= ܠܘ): **Min. I**.i.2.2; **II**.i.4.1.
√lbb: ܠܒܐ, ܠܒܝܒܘܬܐ: **Min. II**.iii.4.5 (IS قلب).
√lbd: Pa: **Mete. III**.i.4.9 (cf. N Pe). - Ethpe/Ethpa: **Min. I**.ii.3.3 (cf. IS عاقد); iii.1.2. - ܠܒܝܕ (ptc. pass.) [* = كَثِيف]: **Min. II**.i.1.6; **Mete. I**.ii.2.5; **III**.iv.2.2; **IV**.i.1.8 (IS*); ii.3.2; iii.4.5 (IS*); iii.4.8. - ܠܒܝܕܘܬܐ: **Min. II**.ii.1.4 (FR كثافة); **Mete. I**.ii.4.2 (AB تراكُم); ii.5.2 (cf. AB مُنعقد).
√lbn: ܠܒܢܬܐ, ܠܒܢܐ: **Min. I**.i.5.6 (N?).
√lbš: ܠܒܝܫ (ptc.): **Min. IV**.ii.6.10. - ܠܒܘܫܐ: **Min. IV**.ii.6.10.
√lhb: Shaphel: **Mete. IV**.ii.3.2 (N). - Eshtaphal [* = اشتَعَلَ]: **Mete. IV**.i.2.4 (IS/FR*); i.3.1, 6 (IS*); iii.1.4 (IS*). - ܡܫܬܠܗܒܢܘܬܐ: **Mete. III**.iii.3.2, 5 (N); **IV**.i.3.7 (IS شُعلة); ii.3.4, 5 (N); iii.1.4 (N/IS اشتعال); iii.1.7.
√lhg: Ethpa: **Min. I**.iii.1.2; iii.2.5 (IS تَبَخَّرَ); **V**.ii.2.2 (IS تَحَلَّلَ بخاراً). - ܠܗܓܐ [* = بخار]: **Min. I**.ii.3.1 (IS*); ii.4.2 (IS*); ii.5.2; ii.5.6 (IS*); ii.6.1 (IS*); iii.1.4 (IS*); iii.2.3 (IS*); iii.3.1; iii.4.6 (IS*); **V**.ii.1.8; ii.2.6, 8, 9; iv.2.4; **Mete. I**.i.1.5 (AB*); i.2.1 (IS جوهر بخاري); i.2.5 (IS*); i.3.1 (IS*); i.3.4; ii.1.1, 5, 7 (AB*); ii.2.5; ii.4.1 (AB*); ii.5.1 (AB*); ii.6.3 (AB*); ii.6.5 (IS*); iii.1.1 (IS*); iii.2.5, 8 (AB*); iii.3.9 (IS*); iii.4.3; iii.6.3, 5 (IS*); **II**.iv.1.5 (FR*); iv.1.8; **III**.i.4.5; i.4.10 (N); iii.6.2 (N); **IV**.i.1.1 (FR*); iii.1.2 (IS*). - ܠܗܓܢܝܐ [* = بخارى]: **Min. II**.ii.2.2 (IS*); **Mete. III**.i.1.2 (N); iii.1.7 (N).
ܠܘܝܐ: **Min. III**.iii.3.7.
√lzz: Pa: **Min. II**.iii.2.8 (cf. IS كَثُفَ). - ܠܙܝܙ: **Mete. II**.iv.1.8 (cf. FR كثيف).
√lḥm: Pe: **Min. I**.ii.6.8; **Mete. II**.iii.1.2.
ܠܝܒܘܐ (nom. prop. "Libya"): **Min. V**.i.1.6.
ܠܝܒܘܢܘܛܘܣ (λιβόνοτος): **Mete. III**.ii.3.8 (N).
ܠܝܛ ܠܝܛ: **Mete. III**.iv.3.7.
√lyl: ܠܠܝܐ, ܠܠܝܘܬܐ [* = ليلة/ليل]: **Min. II**.iii.1.8 (N); **IV**.ii.4.1 (IS*); ii.4.2, 4; iii.2.1 (FR*); iv.2.3, 4; iv.3.7, 8bis; iv.4.6 (AB*); iv.4.10; **Mete. I**.iii.1.2 (IS*); iii.1.6; **II**.ii.2.5; iii.8.1, 9 (IS*); iv.1.9 (FR*); **IV**.i.2.7 (IS/FR*); i.6.3, 4, 5 (N). - ܠܠܝܢܐ: **Mete. II**.iii.4.6.
ܠܝܦܪܐܘ (nom. prop. "Liparaei"): **Min. II**.ii.2.10 (N).
ܠܝܒܣ (λίψ): **Mete. III**.ii.3.5 (N).
ܠܒܩܐ: **Mete. II**.iii.2.1 (N/IS مِجداف); **IV**.i.4.8 (N); i.6.3 (N).
ܠܝܬ: **Min. I**.ii.6.3; iii.1.5; **II**.i.4.10 (N); **IV**.ii.2.8; iv.2.6; **V**.iv.3.7; **Mete. V**.ii.1.1. - Ethpa (ܐܬܠܝܬ): **Min. IV**.iv.5.2 (cf. AB عَدِمَ); **Mete. IV**.iii.3.10 (IS اضمَحَلَّ); **V**.ii.3.4 (FR انقَطَعَ; cf. IS انقطاع).
ܠܝܬܘܕܢܕܪܘܢ (λιθόδενδρον): **Min. I**.i.5.5 (N?).
ܠܟ: **Min. I**.i.3.7; i.4.3; **II**.iii.4.6; **V**.iv.1.2; **Mete. II**.ii.2.4.
ܠܡܝܢܐ (λιμήν): **Min. V**.i.1.3 (DM).
√lmd: ܡܠܡܕ (Pa. ptc. pass.): **Min. I**.i.4.10 (IS ملتئم); **Mete. I**.iii.2.4. - ܡܬܠܡܕܢܘܬܐ: **Min. I**.i.4.4; **Mete. I**.i.2.9.

ܠܡܦܐܕܐ (λαμπάς): **Mete. IV**.iii.1.1; iii.1.8 (N); iii.3.1.
√lʿz: ܠܥܙܐ: **Min. V**.iii.2.8 (IS لغة).
√lšn: ܠܫܢܐ: **Min. V**.i.2.4, 5; i.3.4; **Mete. V**.iii.tit.

ܡ

ܡܐܐ ("centum"): **Min. I**.i.4.4.
ܡܐܢܐ: **Min. I**.i.5.3 (N); **Mete. I**.i.2.9; **II**.iii.4.6bis.
ܡܓܘܓ (nom. prop. "Magog"): **Min. V**.i.3.7.
ܡܓܢܝܣܝܐ (μαγνησία): **Min. III**.i.1.3.
ܡܕܝܢ: **Min. IV**.ii.3.9; ii.4.2; iii.1.5; iii.3.6; iii.4.4; iv.5.4; **V**.ii.1.2; iv.1.8; iv.2.6; iv.3.4; **Mete. I**.ii.2.5; **III**.iii.3.9; **V**.i.3.9 (IS ف); iii.1.4 (IS ف).
ܡܕܝܢܝܐ (nom. prop. "Midianita"): **Min. IV**.ii.6.9 (IS بدوي).
ܡܕܡ [* = شيء]: **Min. I**.i.3.4 (IS بعض); i.5.9; iii.4.7; **II**.ii.1.6 (IS*); **III**.i.4.2 (IS بعض); iii.1.5; **IV**.i.1.5; i.2.8; i.3.4; iv.3.8; **V**.i.1.3 (DM); iii.1.1 (IS*); ii.2.9; iv.3.2; **Mete. I**.i.2.1; i.3.4; ii.2.4; ii.4.4; iii.3.8 (IS*); **II**.i.1.2 (IS*); i.6.5 (IS*); i.8.5; ii.2.5; iii.4.3 (IS*); iii.7.8 (cf. IS واحد); **III**.i.3.6; iv.1.4 (N); **IV**.ii.3.5 (N); iii.2.4, 6 (IS*); **V**.i.1.4 (IS بعض); i.1.5; i.2.4; i.3.1; i.3.9 (IS بعض); ii.2.2, 7; ii.3.3.
√mdn: ܡܕܝܢܬܐ: **Min. II**.ii.2.9 (N); **IV**.ii.6.8; **Mete. V**.iii.2.7.
ܡܐܘܛܝܣ (nom. prop. "Maeotis"): **Min. V**.iii.2.3.
ܡܝܛܐܠܠܘܢ, ܡܝܛܐܠܠ (μέταλλον): **Min. III**.i.4.4. - ܡܝܛܐܠܠܝܩܘ (μεταλλικοί): **Min**. tit.; I.i.1.5; **III**.i.1.1; ii.2.7; Min. fin. - ܡܝܛܐܠܠܩܢܐ [* = معدني]: **Min. I**.i.2.6 (IS*); i.3.4 (IS*); **III**.i.tit.; i.5.1 (IS*).
ܡܝܛܐܦܘܪܐ (μεταφορά): **Min. V**.ii.4.2 (N).
ܡܝܛܐܘܪܘܠܘܓܝܐ (μετεωρολογία): **Mete. I**.i.1.2.
ܡܣܘܣ (μέσης): **Mete. III**.ii.3.6 (N).
ܡܐܘܛܝܣ (nom. prop. "Maeotis"): **Min. V**.i.2.6.
ܡܘܙܐ: **Min. III**.iii.1.9.
ܡܘܪܐ, ܡܘܪܣ (μοῖρα): **Min. IV**.i.2.4, 5; iii.5.2.
√mwt (myt): Pe: **Mete. IV**.iii.2.9. - ܡܡܝܬܢܝܬܐ: **Min. IV**.i.3.8; **V**.ii.4.8; **Mete. IV**.iii.4.11 (IS قاتل).
√mzg: Pa: **Min. III**.iii.3.11 (AB مَـزَجَ). - Ethpe/Ethpa: **Min. I**.ii.1.5 (AB امتَـزَجَ); **IV**.ii.4.2; iii.6.7; iv.3.7; **Mete. II**.iv.1.8. - ܡܙܝܓ (ptc. pass.): **Min. III**.ii.3.7 (IS مُخالط; cf. AB امتزاج/مزاج); **Mete. I**.iii.6.3, 4 (cf. IS مُختلط). - ܡܡܙܓ (Pa ptc. pass.): **Min. IV**.i.4.8; ii.3.2 (cf. IS اعـتـدال); iii.4.5. - ܡܡܙܓܘܬܐ: **Min. IV**.ii.2.8; ii.4.5; iii.4.4. - ܡܡܙܓܢܐ: **Mete. IV**.i.2.2 (IS/FR مَازج). - ܡܘܙܓܐ [* = مزاج]: **Min. I**.i.1.4; **III**.i.5.3 (IS/AB*); ii.1.5 (AB امتزاج); iii.3.7 (cf. AB مزاجي); iii.3.10; **IV**.ii.tit. (IS*); **Mete. IV**.i.1.3 (IS/FR*); ii.3.2 (IS*). - ܡܘܙܓܢܐ: **Mete. I**.ii.5.4 (AB مزاجي). - ܡܬܡܙܓܢܘܬܐ [* = امـتـزاج]: **Min. I**.ii.6.5 (IS*); **III**.ii.1.4 (IS/AB*); **Mete. V**.ii.2.5, 6, 8 (IS*); ii.3.1, 3.
√mḥy: Pe: **Mete. I**.iii.1.2 (IS أصـاب); **IV**.i.4.8, 9 (N). - Ethpe: **Min. III**.i.2.4 (N). - ܡܚܘܬܐ: **Mete. IV**.i.4.9bis, 10 (N).
√mḥl: Ethpa: Mete. III.i.3.8 (N); III.iii.5.8. - ܡܚܝܠ [* = ضَـعيف]: **Min. II**.ii.3.1; iii.1.4bis (N); **III**.i.5.2 (IS سـخـيف/ضـعـيف); ii.2.3; ii.3.9 (IS*); **IV**.iii.2.2; **Mete. I**.ii.3.2 (cf. AB ضُعف); iii.3.8 (IS*); **II**.i.7.4 (cf. IS ضُعف); iii.2.2 (N; cf. IS ضَـعُـفَ); **III**.iii.1.9 (N). - ܡܚܝܠܐܝܬ: **Min. II**.iii.1.5; **Mete. II**.iii.5.5; **III**.i.3.7 (N); **IV**.ii.3.7 (N). - ܡܚܝܠܘܬܐ: **Mete. II**.iii.2.5 (N).
√mṭy: Pe: **Min. III**.iii.1.2; iii.1.5; **IV**.i.3.10; iii.6.6; iii.6.10 (AB أدرك); iv.5.2; **Mete. I**.i.1.6 (cf. AB وُصول); i.2.9; iii.2.2 (FR وَصَلَ); iii.3.2 (FR وَصَلَ); **III**.iii.1.2; iii.2.4, 5 (IS أتى); iii.4.2, 4 (IS وَصَلَ); iii.5.8, 9 (N Ethpa/IS بَلَغَ); **IV**.ii.1.3, 6 (IS انتَهى); iii.2.3, 4 (IS وَصَلَ). - Ethpa: **Min. II**.i.2.3; **Mete. I**.ii.1.3; **II**.iv.1.7 (FR بَلَغَ); **III**.i.1.5 (cf. FR وُصول); iii.1.5 (IS وَصَلَ); **IV**.i.1.3 (FR وَصَلَ); i.4.7; iii.4.9.
√mṭr: Aph: **Min. V**.iii.1.6; **Mete. I**.ii.4.5 (AB أمطَرَ). - ܡܛܪܐ [* = مَطَر]: **Min. I**.iii.3.6 (AB*); **II**.i.2.2 (N); i.3.2, 4 (N); **V**.iii.1.4; iv.1.6; **Mete. I**.i.2.5, 9; i.tit.;

ii.1.6 (AB*); ii.2.1, 2, 4 (AB*); ii.2.6; ii.3.1, 2; ii.3.3, 4, 5, 6 (AB*); ii.3.6 (DM*); ii.4.2, 4; ii.5.1 (AB*); ii.5.2; ii.5.6 (AB*); ii.6.3; ii.6.4 (AB*); ii.6.5 (IS*); iii.1.7 (N/IS*); iii.5.4, 5 (IS*); iii.6.5 (IS*); **II**.iv.2.1, 4 (IS*); **III**.i.1.2 (IS*); i.4.1, 2, 3 (IS*); i.4.4, 5, 6 (N); i.4.8, 11 (N/IS*); iv.2.5 (N); **IV**.iii.4.11; **V**.i.1.7; i.2.5bis (IS*); ii.1.6 (IS/FR*). - ܡܝܠܢܝܐ: Min. II.i.3.4 (N).

√my: ܡ̈ܝܐ [* = ماء]: **Min. I**.i.2.5 (IS*); i.2.6; i.3.3 (IS*); ii.1.3, 4bis, 5bis, 8 (AB); ii.2.5 (IS*); ii.2.6; ii.4.3bis (IS*); ii.5.4; iii.1.1 (IS*); iii.1.2, 3; iii.1.4, 6 (IS*); iii.2.2; iii.2.2 (IS*); iii.3.1, 2; iii.3.3, 5, 6 (AB*); iii.4.1, 3; iii.4.5bis (IS*); iii.4.6; **II**.i.1.4 (N); i.2.1; i.2.2bis, 3 (N); i.2.6bis; i.4.6 (N); ii.1.7; iii.2.2 (N/IS*); **III**.i.3.5 (IS*); **IV**.i.1.1bis, 3, 5, 6, 8 (IS*); i.2.8 (IS*); i.4.1, 3 (IS*); i.4.9; **V**.ii.tit.; ii.1.1, 4, 5 (IS*); ii.1.6bis, 8bis; ii.1.10; ii.2.1 (IS*); ii.2.5, 7; ii.3.1bis (N/IS*); ii.3.6, 8, 9 (N); ii.4.5, 6 (IS*); iii.1.4 (IS*); iii.1.4, 7; iii.2.4; iii.3.4bis, 7, 8; iv.1.2, 3, 5, 6, 8; iv.2.1, 3, 5, 6; iv.3.1; **Mete. I**.i.1.4, 5, 8 (AB*); i.2.1, 2 (IS*); i.2.9; ii.1.1; ii.2.7 (AB*); iii.1.3bis (IS*); iii.4.4 (IS*); iii.6.2; **II**.i.4.2, 3, 5bis (IS*); iii.1.1, 6, 8 (IS*); iii.2.1 (N/IS*); iii.3.6, 7; **III**.i.1.1 (N); i.3.1 (N/IS*); i.3.9 (N); **IV**.i.3.3bis; i.4.8 (N); i.6.3, 5 (N); iii.2.4 (IS*); **V**.i.1.8 (cf. IS مائية); i.2.6bis (IS*); i.3:8 (IS*); ii.1.5 (FR*); ii.2.7 (IS*). - ܡܝ̈ܢܝܐ [* = مائى]: **Min. III**.i.2.5 (IS*); V.iv.3.4; **Mete. I**.ii.6.6; iii.1.5; iii.2.4; iii.6.2; V.i.1.6. - ܡܝܢܝܘܬܐ [* = مائية]: **Min. III**.i.1.4, 5 (IS*); i.2.6 (IS جوهر مائى); i.3.1 (IS*); ii.1.1 (AB*); **Mete. II**.iii.4.2, 4. - ܡܝܢܝܘܬܐ [* = مائية]: **Min. III**.ii.1.4, 5, 7 (IS/AB*); **Mete. IV**.i.2.1 (FR/IS*).

√mkk: Pe: **Min. I**.iii.4.3. - ܡܟܝܟ: **Min. I**.iii.2.2.

√mll: Pa: **Mete. V**.i.1.2 (IS تكلّم). - ܡܡܠܠ: **Mete.** V.iii.1.8. - ܡܠܬܐ: **Mete. V**.iii.2.1. - ܡܠܠܘܬܐ: **Min. I**.i.1.3; ii.5.4; **IV**.iii.6.1; V.ii.4.2 (cf. IS كلام); iv.1.1; iv.2.1; **Mete. II**.iii.7.1; **IV**.i.5.3 (N); i.6.3 (N); **V**.iii.2.1, 6, 9. - ܡܡܠܠܘܬܐ **Mete. I**.i.1.2.

√mly: Pe: **Min. II**.ii.2.9 (N). - Eshtaphal: **Min. I**.ii.6.5; **Mete. V**.ii.2.4. - ܡܠܝܐ: **Min. II**.i.2.2 (N). - ܡܠܝܘܬܐ: **Mete. II**.iii.8.7, 8 (IS تَبَدُّر; cf. N ܡܠܝܐ). - ܡܠܘܐܐ [* = مادّة]: **Min. I**.ii.1.6; iii.1.3, 5; iii.2.3 (IS*); iii.4.6 (IS*); **II**.ii.1.3; V.iii.1.3 (cf. IS استمدّ); **Mete. I**.i.3.2 (IS*); iii.1.2 (IS*); iii.1.4; iii.6.5, 7 (IS*); **IV**.iii.1.6 (IS*). - ܡܠܝ̈ܐܐ: **Min.** V.iii.3.9; **Mete. V**.i.1.6 (IS مَدّ). - ܡܫܡܠܝܢܐ: **Mete. V**.ii.2.9; iii.1.5. - ܡܫܡܠܝ̈ܢܐ: **Min. V**.ii.4.6 (FR غاني; cf. IS غاية). - ܡܫܬܡܠܝܢܐ: **Mete. V**.ii.3.2. - ܡܫܬܡܠܝܢܐܝܬ **Mete. I**.ii.2.5; iii.4.2.

ܡܠܘܢ (μᾶλλον) [* = خصوصًا]: **Min. I**.i.3.3; ii.5.2; **II**.ii.2.7; **IV**.i.2.2 (IS*); iv.1.6 (AB*); **Mete. I**.iii..5.3 (IS*); iii.6.1; **III**.iii.4.5 (IS*).

ܡܠܘܥܐ: **Min. IV**.ii.6.3, 4bis, 5; **Mete. IV**.iii.3.8.

√mlḥ: Ethpa: **Min. V**.ii.4.5, 6 (IS تَمَلَّحَ). - ܡܠܚܐ [* = ملح]: **Min. I**.i.2.7 (IS*); **III**.i.1.2 (IS*); i.3.4 (supplevi, IS*); **V**.ii.1.5 (IS*); ii.3.10 (N); **Mete. IV**.i.3.9 (IS*). - ܡܠܚ̈ܢܐ [* = ملحى]: **Min. III**.i.5.4 (IS*). - ܡܠܝܚܘܬܐ [* = ملحيَة]: **Min. III**.i.4.1, 3 (IS*). - ܡܠܝܚ [* = مَالِح]: **Min. V**.ii.1.9; ii.2.2 (IS*); ii.3.6, 8 (N); ii.4.2 (N; cf. IS ملوحة); **Mete. I**.ii.5.4 (AB*). - ܡܠܝܚܘܬܐ [* = ملوحة]: **Min. V**.ii.tit.; ii.1.4 (cf. IS ملح); ii.3.1 (IS*); ii.4.6 (IS*). - ܡܠܚܐ ("nauta"): **Min. V**.ii.1.6. - ܡܠܚܬܐ: **Mete. IV**.i.3.4 (سَبَخَة).

ܡܠܝܠܝܣ (nom. prop.): **Min. I**.i.2.11.

ܡܠܝܣܛܐ (μάλιστα): **Min. V**.iii.3.3 (IS خصوصًا).

√mlk: ܡܠܟܘܬܐ: **Min. IV**.i.2.6.

ܡܢܗ ܘܠܗ: **Mete. IV**.i.1.1 (cf. FR خَالِص); IV.iii.2.2.

√mny: Pe: **Mete. III**.ii.1.1 (cf. N ܡܢܝܢܐ). - ܡܢܝܢܐ: **Min. I**.ii.1.3; V.iii.2.6. - ܡܢܬܐ [* = جُزء]: **Min. I**.i.1.9; i.2.2; i.4.7 (IS شيء); ii.1.4, 9 (AB*); ii.2.2 (IS طائفة); ii.2.3 (IS/FR*); **II**.i.1.1 (IS*); i.1.4 (IS جَنَبَة); i.1.5, 6 (N); i.4.4 (N); ii.1.7; **III**.i.2.7; ii.1.6; **IV**.i.1.4, 6 (IS*); i.3.4; ii.1.3 (IS طائفة); **V**.ii.2.5; iii.2.11; iv.3.5, 6; **Mete. I**.ii.1.3, 8 (AB*); ii.4.2; iii.1.4, 5 (IS*); iii.2.1 (FR*); iii.2.3; iii.3.1 (FR*); iii.4.2; iii.6.2bis, 4bis; **II**.i.5.4

(IS*); i.7.3; ii.1.4; ii.1.6 (FR*); ii.4.3 (IS*); iii.1.1 (IS*); iii.3.1, 4 (FR*); iv.1.8; **III**.iv.4.5 (IS طائفة); **IV**.i.2.4; i.5.4 (N/IS قطعة); **V**.ii.2.7. - ܡܢܬܢܝܘܬܐ [* = جُزئي]: **Min. I**.ii.6.7bis (IS*); **II**.iii.3.5 (IS*).
ܡܢܝܐ (μνᾶ): **Min. I**.i.4.5 (IS مِن). - ܡܢܝܢܐ: **Mete. I**.iii.4.8 (IS مِن).
ܡܢܝܢܘܬܐ: **Min. III**.iii.2.2.
√mnᶜ: Ethpa: **Mete. I**.i.1.7 (AB انتَهى); **III**.i.1.8 (FR وَصَلَ).
√msy: ܡܬܡܣܝܢܘܬܐ: **Min. I**.iii.4.3, 4 (IS عفونة).
√mṣy: Ethpe: **Min. I**.i.4.9; **III**.iii.2.9 (IS أمكَنَ); iii.3.3 (AB قَدَرَ); **V**.iv.3.8; **Mete. III**.iii.1.2 (IS جازَ); **IV**.iii.4.4. - ܡܬܡܨܝܐ (ptc. pass.): **Min. II**.i.4.4 (N); ii.1.9; **III**.iii.1.1; iii.3.6 (neg. corr. AB امتنع); iii.3.10, 11 (AB قَدَرَ); **IV**.iv.2.4; **Mete. I**.i.2.2; **II**.i.7.2; ii.4.1 (IS أمكَنَ); **IV**.i.4.3; i.6.5 (N); **V**.i.2.5 (IS أمكَنَ); i.3.9 (cf. IS جَــازَ); ii.2.7, 10; ii.3.3 (neg. corr. IS مستحيل). - ܡܬܡܨܝܐ ܣܓܝ: **Min. II**.ii.1.6 (cf. IS أقوى على). - ܠܐ ܡܬܡܨܝܐ ܣܓܝ: **Min. I**.iii.1.7 (cf. IS ناقـصـة القـوة عن). - ܠܐ ܡܬܡܨܝܢܘܬܐ: **Min. III**.iii.3.6 (neg. corr. AB مـمـتـنـع); **V**.iv.1.5; **Mete. V**.ii.3.7. - ܡܬܡܨܝܢܘܬܐ: **Min. III**.iii.2.3; **IV**.ii.2.7.
√mṣᶜ: Ethpa: **Mete. II**.ii.1.3 (FR تَوَسَّطَ). - ܡܨܥܝܐ: **Min. I**.i.2.4; **IV**.ii.1.6; ii.2.2; **Mete. I**.i.2.1 (IS مُـتَـوَسِّـط); **II**.iii.6.4; iii.7.3, 5; **V**.i.2.1 (IS مُـتَـوَسِّـط). - ܡܨܥܝܘܬܐ: **Mete. II**.iii.5.4 (N). - ܡܨܥܬܐ [* = وَسَطٌ]: **Min. I**.i.3.7 (IS*); iii.3.1; **IV**.i.2.9; iv.1.7 (AB*); iv.5.4, 6 (AB*); **Mete. I**.i.3.4; ii.2.4; **II**.ii.2.1bis (cf. IS داخل); ii.4.3 (IS*); iii.5.8 (cf. IS بين); **III**.iii.1.11 (N). - ܡܨܥܬ (praepos. "inter"): **Min. I**.ii.1.4; Mete. I.iii.2.3.
ܡܨܪܝܢ (nom. prop. "Aegyptus"): **Min. II**.i.4.11 (N); **V**.iii.2.1, 10 (IS مصر).
√mrr: ܡܪܝܪ: **Min. V**.ii.1.2; ii.1.3 (IS مُرّ). - ܡܪܝܪܘܬܐ: **Min. V**.ii.4.5 (FR مرّة).
√mry: ܡܪܝܐ: **Min. III**.iii.1.1 (IS صاحب); iii.2.3.
√mrg: ܡܪܓܢܝܬܐ: **Mete. I**.ii.5.7 (AB مَرجي?).
√mrḥ: ܡܡܪܚ (ptc.): **Min. IV**.iii.6.1.
√mšḥ: Pe: **Min. V**.iv.3.8. - Ethpe: **Mete. II**.ii.4.5 (IS قُدِّرَ). - ܡܬܡܫܚܢܘܬܐ: **Min. IV**.ii.5.6; **Mete. II**.i.5.1 (IS مـقـدار). - ܡܫܚܬܐ: **Mete. III**.iii.6.2, 5 (N). - ܠܐ ܡܬܡܫܚܢܐ: **Mete. V**.i.1.8 (IS مُـفـرط). - ܡܫܘܚܬܢܘܬܐ [* = اعتدال]: **Min. IV**.iii.6.7, 8 (AB*); **Mete. III**.iii.2.1 (IS*). - ܠܐ ܡܫܘܚܬܢܘܬܐ: **Min. IV**.i.3.8 (N); ii.2.4 (N).
ܡܬܡܫܚܢܘܬܐ: **Min. V**.iv.3.5, 6.
√mtḥ: Pe: **Min. IV**.i.3.6. - Ethpe [= امتدَّ]: **Min. II**.iii.1.6 (N); **V**.i.3.4; **Mete. II**.i.1.7 (IS*); i.4.2, 5; **III**.iii.1.2 (IS*). - ܡܬܝܚ (ptc. pass.): **Mete. IV**.iii.1.7, 8; iii.4.10. - ܡܬܚܐ: **Min. I**.ii.1.2 (cf. IS مدّة); **Mete. II**.i.5.7; **V**.ii.2.3 (cf. IS زمان).
√mtl: Aph: **Min. V**.ii.2.3.

ܢ

√nbh: Pe: **Min. IV**.iv.1.2 (AB صَدَرَ).
√nbᶜ: Pe: **Min. I**.ii.4.5; iii.1.3; **V**.iii.1.6. - ܡܒܘܥܐ: **Min. I**.i.5.7 (N); iii.tit.; **V**.iii.1.11.
√nbrš: Pa (ܢܒܪܫ): **Mete. IV**.i.4.4. - Ethpa (ܐܬܢܒܪܫ): **Mete. IV**.i.2.6 (cf. IS اشتعال); ii.1.1, 2 (N/IS اشـتـعـال): ii.3.8bis (N). - ܢܒܪܫܬܐ: **Mete. V**.i.1.8 (IS اشتعال).
√ngd: Pe: **Min. I**.i.1.6.
√ngr: Pe: **Min. I**.iii.4.4 (IS طال). - Aph: **Min. IV**.iii.1.5; iii.6.3 (cf. AB دَوَام); iv.4.7 (cf. AB بقاء); **Mete. III**.iv.3.6 (IS لَزِمَ). - ܢܓܝܪܐ: **Min. V**.iii.2.5; **Mete. I**.iii.4.6 (cf. N ܢܓܝܪܘܬܐ). - ܢܓܝܪܘܬܐ: **Min. IV**.ii.5.1 (cf. IS مـداومـة); ii.5.9; iii.3.2, 6; iv.3.3 (AB دَوَام); iv.3.4. - ܢܓܝܪܐܝܬ: **Min. I**.ii.6.2; **Mete. IV**.iii.4.3. - ܠܐ ܣܓܝ ܢܓܝܪܐܝܬ: **Mete. V**.i.1.2; ii.3.8 (IS/FR في النادر).
√ndy: ܢܕܝܬܐ [* = نُتوء]: **Min. IV**.i.1.7 (IS*); iv.2.2 (AB*).
√ndr: Pe: **Min. I**.i.2.5.
√nhr ("luxit"): Pe: Min. IV.iv.1.5; IV.iv.2.3. - Aph: Min. IV.iv.2.3. - ܢܘܗܪܐ [* = ضَـوء]: **Min. IV**.iv.2.1 (AB نور); **Mete. II**.i.3.4 (IS*); i.6.1, 2, 3, 4 (IS*); ii.1.8 (FR*); ii.1.9; ii.2.5; iii.8.2, 7 (cf. IS نَـيِّـر); iii.8.8 (cf. IS إضـاءة); iv.1.3bis (IS/FR*); **IV**.ii.1.3, 6 (IS*). - ܢܘܗܪܢܐ: **Mete. II**.i.6.2 (IS نَيِّر); ii.4.3. - ܢܗܝܪ [* = مُضيء]: **Mete. I**.iii.6.8; **II**.i.3.8 (N); i.6.1 (IS*); i.7.5 (cf. IS ضوء); iii.4.1

(IS مُنير); iii.4.3 (cf. IS ضوء); iii.4.5; iii.5.4 (N); iii.7.3; **IV**.i.1.7; i.3.7 (IS*). - ܢܗܝܪܐܝܬ: **Min. V**.i.3.12. - ܢܗܝܪܘܬܐ: **Mete. IV**.i.6.6 (N).

√nhr ("flumen"): ܢܗܪܐ [* = نهر]: **Min. I**.ii.4.6; ii.5.3, 9 (IS واد); **V**.ii.2.1, 4, 9; ii.3.3; ii.3.9 (N); iii.1.3, 4, 5, 6 (IS*); iii.1.11; iii.3.2 (IS واد); iv.3.1, 2; **Mete. I**.ii.5.1. - ܢܗܪܘܢܐ: **Min. V**.i.2.5.

√nwd: Aph: **Min. II**.i.1.8 (N); iii.1.5; iii.2.4; **Mete. III**.i.3.10 8 (cf. IS زلزلة). - ܡܢܕܐ [* = زلزلة]: **Min. I**.i.3.5 (IS*); ii.2.1 (IS*); **II**.tit. (IS*); i.1.1 (IS*); i.2.4 (N); i.2.5; i.4.2, 11; ii.tit.; ii.3.1; ii.3.5; ii.3.7 (IS*); ii.4.3 (IS*); iii.1.7 (IS*); iii.2.4 (IS*).

√nwḥ: Pe: **Min. II**.iii.4.1 (N); **Mete. I**.ii.6.4 (neg. corr. AB دام). - ܡܬܢܝܚܐܝܬ: **Mete. I**.ii.3.7 (DM); iii.1.4.

ܢܘܢܐ: **Min. V**.ii.3.7, 9 (N).

√nwr: Ethpa: **Min. II**.ii.2.5bis (IS اشتعَل); **Mete. IV**.i.5.2. - ܢܘܪܐ [* = نار]: **Min. I**.i.5.9 (N); ii.5.8; **II**.ii.1.8bis (IS*); ii.2.6 (IS*); ii.2.9, 10 (N); **III**.ii.1.7; ii.2.4; **Mete. I**.i.1.6 (AB*); **II**.iv.1.7bis (FR*); **III**.i.1.7, 9 (FR*); i.2.5; **IV**.i.3.3; i.3.8 (IS*); i.5.4 (IS*); i.5.6; i.6.6 (N); iii.1.10 (N); iii.2.1bis, 5, 7 (IS*); iii.4.10; **V**.i.2.7. - ܢܘܪܢܝܐ [* = ناري]: **Min. II**.ii.1.9; **III**.i.3.5 (cf. IS نارية); ii.2.5 (IS*); ii.3.3 (IS*); **V**.iii.2.10; **Mete. I**.iii.6.4; **IV**.iii.3.7; **V**.i.1.8 (IS*). - ܢܘܪܢܝܘܬܐ [* = نارية]: **Min. I**.i.4.2 (IS*); **III**.i.3.3 (supplevi, IS*); ii.1.2, 3 (AB*); **Mete. IV**.iii.2.5 (IS*); iii.3.3 (IS نار).

√nḥl: ܢܚܠܐ: **Min. I**.ii.2.3, 5; **IV**.i.1.6.

√nḥš: ܢܚܫܐ [* = نُحاس]: **Min. III**.i.4.5 (IS*); i.5.8; ii.3.5 (IS*); iii.1.7; **Mete. IV**.ii.2.6 (N/IS*).

√nḥt: Pe [* = نزَل]: **Min. I**.ii.5.3; iii.3.7 (AB*); **II**.i.2.2 (N); **Mete. I**.tit. (IS*); i.2.5, 9; ii.1.5 (AB هَبَط); ii.1.7; ii.2.3, 6; ii.3.2; ii.3.4 (AB*); ii.3.6; ii.3.7, 8 (DM); ii.4.2; ii.6.3; iii.1.4 (IS*); iii.1.6; iii.2.3 (IS*); iii.2.8 (DM); iii.3.2 (FR*); iii.3.4 (AB هَبَط); iii.4.3; iii.4.5, 7 (IS*); iii.5.3 (IS انحدَر); **III**.i.1.6 (FR*); i.4.11 (N); iv.3.3, 4 (IS*); **V**.i.2.5. - Ethpa: **Min. I**.iii.4.7. - Aph: **Mete. I**.ii.4.4. - ܡܚܬܬܐ: **Mete. I**.ii.1.5; ii.3.1 (AB نزول).

√nṭp: Pe: **Min. I**.i.2.5bis; ii.5.9. - Aph: **Mete. I**.ii.1.7. - ܢܛܘܦܬܐ: **Min. I**.ii.5.10; **Mete. I**.ii.1.9 (AB قَطرة); ii.3.6, 7, 9 (DM); iii.2.2 (IS/FR حَبّ); iii.3.1 (IS/FR حَبّ); **V**.ii.3.6.

√nṭr: Pe: **Min. II**.iii.4.4 (N); **V**.ii.2.4; **Mete. I**.ii.5.7 (AB حافظ); **III**.iii.2.6, 7 (N). - ܢܛܝܪ (ptc. pass.): **Min. III**.iii.2.2 (IS محفوظ).

ܢܝܙܟܐ [* = نيزك]: **Mete. II**.i.1.1 (IS*); ii.1.1; iv.tit.; iv.3.1 (IS*); iv.3.7.

ܢܝܠܘܣ (nom. prop. "Nilus"): **Min. V**.i.3.2.

√nkl: ܢܟܠܬܐ: **Mete. IV**.iii.3.5.

ܢܡܘܣܐ (νόμος): **Min. IV**.iv.4.2.

√nkr: ܢܘܟܪܝܐ: **Mete. V**.ii.2.1.

√nsy: ܢܣܝܘܢܐ [* = تَجربة]: **Min. I**.ii.1.7; **II**.iii.3.5 (IS*); **III**.iii.2.5; iii.3.9 (AB*); **V**.ii.2.6. - ܢܣܝܢܘܬܐ: **Min. V**.iii.3.5. - ܡܣܐܬܐ (constellatio "Libra"): **Min. IV**.ii.3.4; ii.6.4; iii.1.4; **Mete. III**.ii.1.4, 6.

√nsb: Pe: **Min. I**.i.5.4 (N); **II**.i.4.3 (N); **V**.iii.1.3. - Ethpe: **Min. V**.ii.4.4.

√nsk: ܢܣܟܐ: **Mete. III**.i.3.9 (N).

√npl: Pe [* = وَقَع]: **Min. I**.i.2.6; i.4.4 (IS نزَل); i.4.6; **II**.i.1.6 (N); i.2.7 (N); i.3.2 (N/IS سَقَطَ); i.4.10 (N); **IV**.iv.2.5; **V**.iii.1.11; **Mete. I**.iii.4.10; **II**.i.6.3 (IS*); ii.4.1 (IS*); iii.2.8 (IS*); iii.3.3 (FR*); **IV**.ii.3.3. - ܡܦܘܠܬܐ: **Min. I**.i.4.1 (N); **II**.i.3.3 (IS سقوط).

√npḥ: Pe: **Min. V**.ii.4.7; **Mete. II**.iii.1.8 (IS نَفَخ). - ܡܦܘܚܬܐ: **Min. II**.ii.2.6 (IS نَفخ). - ܡܦܚܬܐ: **Min. I**.i.5.3 (N).

√npṣ: Ethpe/Ethpa: **Min. I**.i.5.8 (N?); **Mete. I**.ii.6.7; **III**.i.3.6 (N). - ܢܦܘܨܐ: **Mete. IV**.ii.3.8 (N).

√npq: Pe: **Min. I**.i.2.10; **II**.ii.1.4 (cf. FR خروج); ii.1.6; ii.3.6; iii.1.12; **III**.ii.3.7 (cf. AB أخرَج); iii.3.1; **V**.i.2.4, 5; iv.2.1, 4, 5; **Mete. II**.i.1.5 (IS خرَج); i.3.7 (N); i.4.2, 4; **III**.i.3.12 (IS تخلّص); iv.4.2; **IV**.i.1.5, 7. - Aph: **Min. I**.iii.2.2; **II**.ii.2.3; **III**.ii.3.8; iii.3.9 (AB أخرج propono). - ܢܦܩܬܐ: **Min. III**.iii.2.5. - ܡܦܩܐ: **Min. III**.ii.3.8. - ܡܦܩܢܐ: **Min. II**.ii.2.7 (IS خروج); ii.4.3 (IS مَخلص).

√npš: ܢܦܫܐ: **Min. V**.iv.3.9; **Mete. V**.iii.1.1 (IS نفس). - ܢܦܫܬܐ: **Mete. V**.iii.2.2.

√nṣb: ܢܨܝܒ: **Min. IV**.iii.6.5 (AB غَرِيزِي). - ܢܨܒܬܐ [* = نَبَات]: **Min. I**.i.3.1 (IS*); **IV**.iii.6.3 (AB*); **V**.ii.4.8; **Mete. V**.i.1.11; ii.tit.
√nṣp: ܢܨܝܦܐ [* = نَاصِع]: **Min. III**.ii.3.3 (cf. IS*); **Mete. II**.iii.5.3 (IS*); iii.7.3 (IS*).
√nqb: ܢܩܒܬܐ: **Mete. V**.ii.1.3.
√nqd: ܢܘܩܕܬܐ: **Min. IV**.iv.2.6; **Mete. III**.ii.1.8, 10 (IS نُقطة); ii.2.7, 11.
√nqp: Pe: **Min. IV**.iv.1.6; iv.3.2, 3; iv.4.1; **Mete. I**.i.1.2; I.i.3.5 (cf. IS إلحاق); V.i.1.1. - Aph: **Min. I**.i.1.2. - ܢܩܝܦ: **Mete. II**.iii.7.3. - ܢܩܝܦܘܬܐ: **Mete. II**.iii.6.3.
√nqš: Pe: **Min. II**.ii.3.4 (N); **Mete. III**.i.2.6 (N); iv.1.5 (N); iv.2.2 (N/IS صَدَم); iv.2.4 (N); iv.3.2 (IS قَرَعَ); **IV**.i.3.4 (cf. IS قرع). - ܢܩܫܐ: **Mete. IV**.i.2.3 (cf. IS اضطراب). - ܢܩܫܢܝܬ: **Min. II**.ii.4.5 (N).
ܢܬܚܐ: **Mete. IV**.i.4.10 (IS نَأس).
√nšš: ܢܫܝܫ: **Min. III**.ii.3.9.
√nšb: Pe [* = هَبَّ]: **Min. II**.iii.1.4 (N); **V**.iii.3.2 (IS عَصَفَ); V.iii.3.6; **Mete. I**.i.3.2; ii.5.5 (cf. AB هبوب); ii.5.6 (AB*); **III**.i.3.7 (N); ii.1.1 (N; cf. IS مَهَبّ); ii.3.3; iii.1.5, 10 (N); iii.4.1 (IS*); iii.4.5 (N); iii.4.6 (IS*); iii.5.1 (IS*); iii.5.6, 7; iii.5.8 (N); iii.6.3, 6 (N); iv.1.7 (N); **IV**.ii.3.4, 6. - ܡܢܫܒܢܐ: **Min. I**.ii.2.4 (cf. IS نَسَّافة/FR هبوب); **Mete. I**.ii.3.6 (AB مَهَبّ); ii.4.6 (AB هبوب); **III**.i.3.3; ii.tit. (cf. IS مَهَبّ); **IV**.i.1.7 (cf. IS عَصَفَ); **V**.i.1.9.
√ntl: vide ܢܬܠ.
√nṭʿ: Pe: **Min. V**.iii.3.5.
√ntp: Pe: **Min. II**.i.2.6 (N); iii.1.11 (cf. IS جَذَبَ). - Ethpe: **Min. I**.iii.4.5 (IS جُلِبَ); **II**.i.2.6 (N).
√ntr: Pe: **Mete. IV**.iii.1.5.

ܣ

ܣܐܡܐ (nom. prop.): **Min. V**.i.3.5.
ܣܐܘܢܐ [* = نَضَّة]: **Min. III**.i.5.8; ii.3.2 (IS*); iii.1.7, 8 (IS*); iii.3.3, 11 (AB*); **Mete. IV**.ii.2.6 (IS*). - ܣܐܘܢܝܐ: **Min. V**.i.3.2.
ܣܐܡܘܣ (nom. prop. "Samos"): **Min. V**.i.1.6.
√sbk: Pe: **Min. III**.ii.2.5 (IS عَلِقَ); **Mete. III**.iv.4.5 (cf. IS اشتَمَلَ).
√sbs: ܣܒܣܘܣ: **Min. II**.iii.4.2 (N); **Mete. V**.i.1.7.
√sbr: Pe: **Min. I**.iii.3.2; **Mete. I**.ii.2.2; **II**.i.1.9; iii.2.4 (N Aph); **III**.i.3.1bis (IS ظَنَّ). - Ethpe: **Min. III**.iii.1.3; **Mete. IV**.iii.1.5; iii.2.8. - Aph: **Mete. II**.ii.2.5; V.ii.2.3 (cf. IS ظَنَّ). - ܡܣܬܒܪܢܘܬܐ: **Min. IV**.i.4.2 (cf. IS أغلب الظن). - ܡܣܬܒܪܢܐ: **Min. IV**.iii.5.8.
√sgy: Pe: **Min. I**.ii.4.4; **Mete. I**.ii.3.4; ii.4.2 (AB اشتدَّ); ii.6.5 (IS كَثُرَ); iii.2.7; **II**.iii.3.6; **III**.iii.3.10bis (IS كَثُرَ); iii.6.1 (N?); V.iii.2.7. - ܣܓܝܐܐ [* = كثير]: **Min. I**.ii.1.2 (IS*); ii.2.3; ii.5.6 (IS*); iii.2.3 (IS*); **II**.i.3.3; ii.2.6; ii.4.3 (IS*); ii.4.4 (cf. IS كَثُرَ); iii.1.8 (N); iii.2.2 (IS*); iii.2.3 (N); **III**.i.3.2; i.3.4 (IS*); ii.1.4; iii.1.3; **IV**.i.4.11; ii.3.6, 7; ii.6.2 (IS شديد); iv.4.3; **V**.i.1.4, 7; ii.2.5bis, 6, 8; iii.1.4; iii.2.9 (cf. IS كثير من); iii.3.10; iv.2.4, 5; **Mete. I**.i.2.4 (cf. IS كثيراً); i.3.1; ii.5.5 (AB*); ii.6.3; iii.1.4; **II**.i.7.1 (IS*); iii.8.6 (IS شديد); **III**.i.4.2bis (cf. IS كَثُرَ); iii.1.7, 9 (N); iii.3.3 (N); iii.6.2, 4, 5 (N); iv.2.8 (N); **IV**.i.5.3 (N); iii.1.6; **V**.i.1.6, 10; ii.1.4. - ܣܓܝ [* جداً]: **Min. I**.i.4.10; iii.3.4; **II**.ii.2.4; ii.2.7 (IS*); ii.4.8; **III**.i.3.6 (IS*); iii.1.7, 8 (cf. IS شديد); **IV**.ii.3.5; ii.6.1; iii.1.5; iii.2.2; iii.3.4, 5bis; iii.4.3bis; **Mete. I**.i.1.6; i.2.5 (cf. IS شديد); ii.1.7; ii.3.8; ii.5.2 (AB elative); ii.6.3; iii.3.10; iii.4.5; **II**.i.5.7 (IS*); iii.8.1, 5, 9; iv.1.8; iv.2.5; **III**.iv.4.6; **IV**.i.3.8 (cf. IS يعتد به); i.4.4; ii.1.5 (cf. IS يعتد به); iii.3.5; iii.4.5, 8; **V**.i.2.3bis. - ܣܓܝܐܘܬ: **Min. I**.ii.6.1. - ܣܓܝܐܘܬܐ: **Min. II**.iii.4.2 (N); **III**.iii.2.4; **IV**.iii.4.5; **V**.iv.2.4, 6; **Mete. II**.iv.2.1 (IS وُفور); **III**.iii.3.4, 6 (N). - ܣܘܓܐܐ [* = أكثر]: **Min. I**.i.2.2; ii.3.3 (IS*); ii.5.1 (IS*); ii.6.4 (IS*); iii.2.5 (IS*); **III**.i.1.6 (IS*); i.2.2 (IS*); iii.1.9 (IS*); **Mete. III**.iii.1.1; V.i.2.4. - ܒܝܬ ܣܘܓܐܐ: **Min. I**.i.2.1 (IS في الأكثر); iii.3.2; **II**.iii.1.2 (cf. IS كثيراً); **Mete. II**.ii.4.7 (IS في الأكثر); **III**.i.4.1 (IS في أكثر الأمر); **IV**.i.1.1 (FR في أكثر الأمر); **V**.i.2.2 (cf. IS أكثر); ii.3.7 (cf. IS أكثري).

√sgp: ܡܣܓܦܢܘܬܐ: **Min. IV**.ii.6.3 (cf. IS أَبْتَلَى).
√sdn: ܣܕܢܐ: **Min. III**.i.2.4 (N).
√sdq: Pa: **Mete. IV**.i.1.5, 6 (FR مَزَّقَ); i.1.9; i.4.2. - Ethpa: **Mete. IV**.i.1.4. - ܣܘܕܩܐ: **Mete. IV**.i.1.5 (FR تمزيق).
√sdr: Pe: **Mete. II**.iii.7.2 (IS أَتَى).
√shd: Pe: **Mete. III**.i.3.3 (N).
√shr: ܣܗܪܐ [* = قَمَر]: **Min. II**.iii.2.1 (N); **V**.iii.3.10; **Mete. II**.i.5.9 (IS*); ii.1.3 (IS*); ii.1.4ter, 5, 6, 7 (FR*); ii.1.9bis; ii.3.2; iii.4.5; iii.8.1, 8, 9bis (N/IS*); iv.1.2 (FR*).
√swk: Pa: **Min. V**.iii.1.2. - ܣܘܟܐ: **Min. III**.iii.2.4; **Mete. II**.i.9.2, 3; ii.3.1 (cf. IS انتَهَى); ii.4.5 (IS قُطر); iii.6.4, 5; **V**.i.2.3 (IS طَرَف); V.i.2.5. - ܣܘܦ . . . ܠܐ: **Min. IV**.ii.2.6; **Mete. I**.ii.6.4; **IV**.iii.2.6; **V**.ii.2.2. - ܡܣܘܟܐ: **Mete. V**.i.3.11.
√swm: Pe: **Min. II**.i.1.2 (cf. IS نَسَبَ); i.2.1; i.3.1; **IV**.i.2.6; **V**.ii.1.7; **Mete. V**.iii.1.9; iii.2.6. - Ethpe: **Min. V**.ii.1.5; **Mete. IV**.i.3.9 (IS وُضِعَ). - ܣܝܡ (ptc. pass.): **Min. I**.ii.1.4; ii.5.4; **IV**.i.3.5; IV.ii.2.5; **V**.iv.1.7; **Mete. II**.i.1.8; i.7.3bis; ii.1.6; iii.3.2. - ܣܝܡ ܕܣܝܡ: **Min. III**.iii.2.8. - ܣܝܡܬܐ: **Min. I**.i.1.8; **IV**.i.tit.; ii.5.5; **V**.i.tit.; **Mete. I**.ii.4.6; **II**.ii.1.5; iii.5.2 (وَضع); iv.2.8.
ܣܘܢܘܕܘܣ (σύνοδος): **Mete. V**.i.1.5 (IS اجتماع).
√swp: Pe: **Min. V**.iii.2.8. - ܣܘܦܐ [* = طَرَف]: **Min. IV**.iv.1.7 (AB*); iv.5.6; iv.5.7 (AB*). - ܣܘܦܢܝܘܬܐ: **Mete. II**.iii.5.7.
ܣܘܣ: vide ܣܘܣ ܢܗܪܐ.
√swq: ܣܝܩܐ (ptc.): **Mete. V**.i.3.5 (IS مُتَنَفِّس). - ܣܘܩܐ: **Min. IV**.i.2.10 (cf. IS نسيم).
ܣܘܪܛܝܣ (nom. prop. "Syrtis"): **Min. V**.i.2.2 (DM).
√shh: ܣܚܝܚ: **Min. IV**.iv.1.4 (AB لطيف); **Mete. I**.iii.2.4; **IV**.ii.2.5, 7 (N).
ܣܛܝܠܣ (nom. prop. "στήλαι"): **Min. V**.i.1.2, 5 (DM); i.3.1, 11.
√str: ܣܛܪܐ: **Min. I**.ii.2.2. - ܣܛܪ: **Min. I**.ii.6.8 (IS وغيره); **IV**.ii.2.9; **Mete. IV**.iii.3.1 (IS وغير ذلك). - ܣܛܪ ܡܢ: **Min. III**.i.2.3 (N); iii.2.8; **IV**.i.4.2 (IS إلا); ii.5.8; **Mete. III**.iii.4.8 (N/IS من غير).
√syb: ܣܝܒܐ (in sensu "Avicenna"): **Min. I**.i.4.3; **IV**.iii.1.1. - ܣܝܒܐ ܪܝܫܢܐ (= الشـــيخ الرئيس): **Min. I**.i.2.8; i.3.6; **IV**.ii.3.1; **Mete. II**.ii.4.8; iii.7.1. - ܣܝܒܢܝܐ (i.e. "Avicennianus"): **Min. III**.tit.
√syp: ܣܝܦܐ: **Min. I**.i.4.8, 11 (IS سيف).
ܣܝܩܠܝܐ (nom. prop. "Sicilia"): **Min. V**.i.1.6. - ܣܝܩܠܝܘܣ: **Min. V**.i.2.4 (DM).
√skl: Ethpe: **Min. V**.iii.2.13.
[√skm]: Ethpa: **Mete. I**.iii.3.2; **II**.iv.1.6 (FR تَشَكَّلَ). - ܡܣܟܡ (Pa ptc. pass.): **Mete. II**.i.9.4 (IS مُشَكَّل). - ܡܣܟܡܢܘܬܐ: **Mete. III**.iv.3.6 (IS تشكُّل); **V**.iii.2.4.
√skr: ܣܟܪܐ: **Mete. IV**.ii.2.6 (N/IS تُرس).
ܣܠܘܩܘܣ (nom. prop. "Seleucus"): **Mete. IV**.iii.3.8.
√slq: Pe [* = صَـعِـدَ]: **Min. II**.i.1.6 (N); ii.4.7 (N); **III**.ii.1.6; **V**.ii.2.8, 9 (N); **Mete. I**.i.3.1; ii.1.1 (AB*); ii.4.1 (AB*); iii.2.5 (AB*); iii.5.4 (cf. IS مُتصعِّد); **III**.i.1.5 (FR*); i.2.4 (N); i.3.2; i.3.5 (N); i.4.4, 9 (N); iv.3.3, 5 (IS*); **IV**.i.1.2 (FR*); i.3.5 (IS تصـعّـد); i.3.8; i.4.9bis (N); iii.4.8. - Ethpa [* = تَصَعَّدَ]: **Min. I**.ii.3.2 (IS*); **III**.i.3.5 (IS*); **Mete. I**.i.1.5; **II**.iv.1.5 (FR تصـاعـد); **III**.i.1.1; **IV**.iii.4.4 (IS صَـعِـدَ). - Aph: **Min. II**.ii.2.9 (N); **V**.ii.2.6; **Mete. I**.i.1.5 (cf. AB صَعِدَ); iii.6.3 (IS أصعَدَ); **III**.i.4.8 (N); iii.1.7 (N); iii.6.2, 4 (N); **IV**.iii.4.4 (IS صَعَّدَ). - ܡܣܩܬܐ [* = صُعُود]: **Mete. I**.i.3.6 (cf. IS مُصَعَّد); iii.1.2 (IS*); **III**.i.1.9 (FR*); iv.3.5 (IS*). - ܡܣܩܢܘܬܐ: **Mete. I**.i.3.3 (IS صـعـود); ii.1.2 (cf. AB صعود). - ܡܣܩܢܘܬܐ ܕܠܥܠ: **Min. III**.ii.1.7 (cf. AB صعود).
√smdr: ܣܡܕܪܐ: **Min. IV**.iii.6.11.
√sml: ܣܡܠܐ: **Mete. II**.iv.3.3 (IS يَسرة).
√smq: Aph: **Min. III**.i.4.5 (IS احمرَّ); ii.2.3 (IS يُرى محمرا); ii.2.5. - ܣܘܡܩܐ [* = أحمر]: **Min. III**.iii.1.7 (IS*); iii.3.2 (AB*); **V**.i.3.3; ii.3.6; **Mete. II**.i.6.4 (IS*); iii.4.4 (IS*); iii.5.3 (IS حُمرة); iii.5.5 (N); iii.5.6 (IS حُمرة); iii.6.2; iii.7.3 (IS حُمرة); iii.7.6 (IS*); iii.7.6 secund. et tert. (cf. IS نُصـوع); iii.7.8 (IS حُـمـرة); **IV**.iii.4.6 (IS*). - ܣܘܡܩܘܬܐ: **Min. III**.ii.2.4 (IS حُمرة); **Mete. II**.iii.7.4; iii.7.5 (supplevi).

√snq: ܡܣܢܩ (ptc. pass.) [* = اِحتَاج]: **Min.** I.ii.6.1 (IS مُحتَاج); **Mete.** II.iii.8.2, 5 (IS*); III.iv.2.7 (N); IV.i.4.6, 7 (IS*).
√sʿy: ܡܣܥܝܢܘܬܐ: **Min.** IV.ii.3.3bis; ii.4.1; ii.5.1; ii.6.1; iii.tit.; iii.1.2, 3 (T حُكم); iii.2.1; iii.3.1; iii.4.1.
√sʿr: Pe: **Min.** II.ii.4.9. - ܡܣܥܪܢܐ: **Mete.** II.ii.1.2 (N); V.iii.1.3 (cf. IS وجـود); iii.2.5. - ܡܣܥܪܢܘܬܐ: **Mete.** V.iii.2.5.
ܣܥܪܐ ("crines") [* = شَعر]: **Mete.** III.iv.4.2 (N/IS*); IV.iii.1.9 (N); V.ii.1.5 (IS/FR*).
√spq: Pe: **Min.** V.iii.1.3; **Mete.** II.iii.8.6. - ܣܦܝܩܘܬܐ: **Min.** V.iv.1.4.
√spr: ܣܦܪܐ: **Mete.** V.iii.2.1. - ܣܦܪܐ: **Mete.** V.iii.tit.; iii.2.1, 10. - ܣܦܪܐ: **Min.** I.i.2.8.
ܣܩܘܬܝܐ (nom. prop. "Scythes"): **Min.** IV.ii.6.7 (IS تُرك); iii.4.6; V.i.3.10.
√sql: Ethpe: **Min.** IV.iv.1.5 (cf. AB صقال). - ܡܣܩܠ (ptc. pass.) [* = صَقيل]: **Min.** IV.iv.1.4 (AB*); iv.2.1 (AB*); **Mete.** II.i.1.6 (IS*); i.8.1 (IS*); iii.8.3, 6; iv.1.2 (FR*).
√sry: Pe: **Min.** V.ii.4.7 (IS أجَنّ). - ܣܪܝܐ: **Min.** III.ii.3.9 (IS مُنتَن); V.ii.3.5.
√srd: ܣܪܕܘܢܐ: **Min.** II.iii.4.5.
ܣܪܕܘܢܝܩܘܣ (nom. prop. "Sardonicus"): **Min.** V.i.1.3 (DM).
√srṭ: ܣܘܪܛܐ [* = خَطّ]: **Mete.** II.i.3.8 (N); ii.3.3, 4 (IS*); iv.3.5. - ܡܣܪܛܢܘܬܐ: vide … - ܣܪܛܢܐ ("Cancer") [* = سَرَطان]: **Min.** IV.ii.3.6; ii.5.2, 3 (IS*); ii.6.3, 5; iii.5.3; V.iii.1.10; **Mete.** III.ii.1.5, 7 (IS*).
√srn: ܣܪܢܐ: **Mete.** II.ii.3.2bis, 4 (IS محور).
√stw: ܣܬܘܐ [* = شِتَاء]: **Min.** II.iii.2.4. 6 (N/IS*); IV.ii.6.3, 5; iii.3.4; iii.5.3, 6 (FR*); iii.6.4, 6 (AB*); iv.3.4; iv.4.7 (AB*); iv.4.9, 11; V.iii.1.9; V.iv.2.5; **Mete.** I.iii.3.6, 7 (IS*); iii.3.7 (cf. IS شـتـوي); II.iii.3.9 (IS*); III.iii.3.5 (N/IS*); iii.3.7 (N); iii.5.6; iii.6.3, 5, 7 (N). - ܣܬܘܝܐ: **Min.** IV.ii.4.4; ii.6.10; iii.3.3; i.4.8, 10; **Mete.** I.i.1.9 (cf. AB شتاء); II.iii.3.8 (cf. IS شتاء); III.ii.1.5, 7 (cf. IS شتاء); ii.3.4, 5, 6, 8.
√str: Pa [* = سَتَّرَ/سَتَرَ]: **Min.** IV.iv.2.4; **Mete.** II.ii.1.4 (FR*); ii.4.7 (IS*).
[√stt]: ܣܬܬܘܬܐ: **Min.** V.iv.1.3.

ܥ

√ʿbb: ܥܒܒܐ: **Min.** II.i.2.2, 3, 7 (N); V.i.1.2 (DM); i.2.2, 3 (DM); i.3.3.
√ʿby: Ethpa: **Mete.** I.i.2.3 (IS تَقَبَّض); III.i.4.11 (N). - ܥܒܝܐ [* = كثيف]: **Min.** V.ii.1.8; ii.1.12 (N); ii.2.2 (IS*); ii.3.2, 6 (N); **Mete.** I.iii.5.2, 3; II.iv.2.2; III.iv.2.8, 9 (N); IV.ii.1.2; ii.2.3 (N); ii.3.1 (IS*); iii.1.6 (IS أكثَف); iii.3.5 (IS*); iii.4.2. - ܥܒܝܘܬܐ: **Mete.** III.iv.2.2 (N).
√ʿbd: Pe: **Min.** I.i.5.8 (N); ii.2.3; II.i.2.4 (N); ii.1.5; ii.3.4 (N); III.ii.1.8 (AB جَعَلَ); ii.3.4; iii.2.9 (cf. IS إيجاد); iii.3.4bis (AB عَمِلَ); iii.3.10; IV.iv.3.4 (AB أوجب); iv.5.1; V.i.1.4; i.3.3, 4, 5, 8; ii.1.9 (N); ii.4.8; **Mete.** I.iii.3.8, 9 (IS فَعَلَ); II.ii.1.9; iii.3.3 (FR جَعَلَ); iii.7.2; III.i.1.7 (FR أحدَثَ); i.1.10 (FR صَيَّرَ); i.2.7 (N); i.4.7 (N); iii.4.8 (N; cf. IS إحداث); iv.2.4 (N); IV.i.1.3 (IS/FR عَمِلَ); i.2.4; i.3.4; i.5.5 (N). - Ethpe: **Min.** I.i.4.8. - ܥܒܕܐ: **Min.** III.iii.1.2bis. - ܥܒܘܕܐ [* = فَاعِل]: **Min.** I.ii.2.2 (IS*); III.iii.3.4 (AB فَـعَّـال); **Mete.** II.i.tit. (IS*); III.iii.3.6 (IS*); iii.5.2 (N); V.i.1.5. - ܥܒܘܕܘܬܐ: **Min.** I.i.1.4 (IS فعل). - ܥܒܕܢܘܬܐ: **Min.** III.iii.3.1; iii.3.9 (AB فـعـل); IV.i.2.2 (IS تأثير); iv.3.9; **Mete.** II.i.8.2, 3 (IS فعل); III.i.3.8 (N).
√ʿbṭ: ܥܒܛܐ: **Mete.** II.i.5.5.
√ʿbr: Pe [* = مَرَّ]: **Min.** I.ii.1.3; II.i.1.7bis (N); IV.i.4.8; ii.3.4, 6; iii.1.4; iv.5.4, 5, 7; V.i.2.4; i.3.1, 4, 6bis, 7, 9, 10; ii.1.12; iii.2.5; iv.3.7; **Mete.** II.iii.3.4 (FR*); III.iii.1.2 (IS*); iii.2.4, 5 (IS*); iv.2.2 (N); IV.ii.2.2, 5 (N); V.i.3.6. - ܡܥܒܪܢܐ: **Min.** V.iii.3.6, 7. - ܡܥܒܪܢܘܬܐ: **Mete.** III.iii.1.3, 4 (IS مرور); IV.i.2.6 (IS إمرار); ii.2.7.
√ʿgl: ܥܓܠ: **Min.** V.ii.2.7; **Mete.** II.iv.3.9. - ܒܥܓܠ: **Mete.** I.iii.4.1; IV.ii.1.6.
ܥܕܡܐ: **Min.** III.i.2.7.
√ʿdn: ܥܕܢܐ: **Min.** IV.ii.5.6, 7 (FR وقت); **Mete.** I.i.2.9; II.iii.8.8; iv.3.7.
√ʿdr: Pa: **Mete.** III.i.4.6 (N).

√ʿhn: Pe: **Min. I**.iii.3.1; **IV**.i.4.8; ii.5.6 (cf. IS استعداد); **Mete. V**.i.3.4; ii.3.1. - ܡܥܬܕܢܐ: **Mete. II**.iii.8.8 (cf. IS استعداد); **III**.i.4.7 (N). - ܡܥܬܕܢܘܬܐ [* = استعداد]: **Min. I**.ii.6.5 (IS*); **II**.i.4.2 (N); **IV**.ii.2.7; **Mete. II**.iii.8.6 (IS*); **V**.i.1.5 (IS*); ii.2.2 (IS*); ii.2.9 (cf. IS مُستعدّ); iii.2.2.

√ʿwd: ܥܘܕܢܐ: **Min. IV**.ii.2.5; **Mete. V**.ii.2.1.

√ʿwk: ܡܥܘܟܢܐ: **Mete. III**.iii.3.10 (IS مَانِع).

√ʿwl: ܥܘܠܐ: **Min. II**.iii.4.7. - ܥܘܠ: **Min. II**.iii.4.6 (IS فَاسِق).

√ʿwq: Aph: **Mete. IV**.ii.2.5 (N).

√ʿwr: Ethpe (ܐܬܬܥܝܪ): **Mete. II**.iii.1.9 (cf. IS مُنتبِه).

√ʿzz: Pe: **Min. IV**.i.4.4; ii.5.8 (cf. FR أقوى); iii.2.3; **Mete. IV**.i.4.4. - ܥܫܝܢ: **Min. I**.ii.2.2; **IV**.i.4.7 (cf. IS اشتدّ); ii.5.9; **Mete. III**.iv.4.7: **V**.i.1.9 (IS عاصف). - ܥܫܝܢܐ: **Min. I**.i.3.5; **Mete. III**.iii.5.9 (cf. IS قُوّة); **IV**.ii.3.8 (N).

ܥܙܐ: **Mete. IV**.iii.1.10 (N).

√ʿṭl: ܥܛܠܢܘܬܐ: Min. III.i.2.3 (IS بعُسر).

√ʿṭp: Pe [* = انعكس]: **Min. IV**.iii.6.5 (cf. AB انعكاس); iv.1.7 (AB*); **V**.iii.3.8; **Mete. II**.i.1.7 (IS*); i.4.3, 5; iii.2.5, 6; iii.2.9 (IS*); iii.3.2 (FR*); iii.5.3, 5 (cf. IS انعكاس); iii.8.3, 4 (IS*); **III**.i.2.3; i.2.6 (N); iv.1.5 (N); iv.2.3; iv.3.2 (IS انثنى); **IV**.i.6.5 (N). - Aph: **Min. III**.iii.1.6 (cf. IS أكسي). - ܡܥܛܦܢܘܬܐ: **Min. V**.iv.1.3. - ܥܛܦܐ: **Mete. II**.i.4.2, 4; **IV**.i.6.2 (N).

√ʿṭr: Aph: Mete. III.iii.4.4. - ܥܛܪܐ [* = بُخار]: **Min. II**.ii.1.2 (FR*); ii.4.7 (N); iii.1.2, 5 (N); iii.1.11 (IS*); iii.2.5 (N/IS*); iii.2.7; iii.3.3 (N); iii.4.3 (N); **V**.ii.1.3; ii.1.12 (N); ii.4.5; **Mete. II**.iv.2.1 (IS*); **III**.i.1.1 (N); i.2.1, 4 (N); i.3.5 (N); i.4.4 (N); i.4.10 (N/IS*); iii.1.6, 9 (N); iii.3.7, 8bis (N); iii.4.3 (IS*); iii.6.1, 4, 5 (N); iv.1.5 (N); iv.2.8 (N); **IV**.tit. (N); i.3.5, 8 (IS*); iii.1.3 (IS*); iii.4.8. - ܥܛܪܢܝܬܐ: **Mete. III**.iii.3.1.

ܥܝܠܡ (nom. prop.): **Min. V**.i.3.5; iii.3.9.

√ʿyn: ܥܝܢܐ ("oculus") [* = عين]: **Min. I**.i.5.3 (N); **Mete. II**.ii.2.6; iii.1.9 (IS*). - ܥܝܢܐ, ܡܥܝܢܐ ("fons") [* = عين]: **Min. I**.ii.4.2; ii.5.1, 3, 8 (IS*); iii.1.1, 4 (IS*); **II**.ii.2.3 (IS*); iii.4.5 (IS*); **V**.iii.1.4, 5, 6 (IS*). - ܡܥܝܢܐ: **Min. I**.tit.; i.1.8; iii.4.6 (IS عين); **V**.ii.3.8, 9 (N); iv.2.2.

√ʿkr: Ethpa: **Mete. III**.i.1.9.

√ʿll: Pe: **Min. I**.iii.1.2; **III**.iii.2.2; **V**.i.1.3; ii.1.11 (N); ii.2.9; iv.2.2, 4, 5; **Mete. I**.ii.2.3 (AB دَخَل); **III**.iii.2.5. - Ettaph: **Min. I**.i.5.6. - ܥܠܠܐ: **Mete. I**.iii.2.3. - ܥܠܬܐ [* = سبب]: **Min. I**.ii.2.1 (IS*); **II**.i.1.1 (IS*); i.1.3 (N/IS علة); i.2.1; i.3.1; i.4.3 (N); ii.tit.; ii.1.9; ii.2.7 (IS*); ii.4.1 (IS*); iii.3.1 (IS*); iii.4.3 (N); **III**.ii.1.2, 5, 6, 7; iii.3.5 (AB*); **IV**.i.2.7 (IS*); ii.2.8; iv.tit. (AB*); iv.4.1 (AB*); **V**.ii.1.3; ii.4.6 (FR*); iii.3.9; **Mete. I**.iii.2.3; iii.3.3; **II**.i.tit. (IS*); i.3.3 (IS علـة); iii.5.1 (IS*); iii.7.10 (IS*); **III**.iii.3.6; iii.3.10 (IS*); **IV**.i.3.1 (IS*); iii.2.1 (IS علة); iii.3.1, 3 (IS*); **V**.i.1.4 (IS*); i.3.1, 10; ii.3.3 (IS*); ii.3.9; iii.1.4; iii.2.9. - ܥܠܠܬܐ, ܥܠ̈ܠܬܐ (?): **Min. V**.i.1.7.

√ʿly: Pa: **Min. I**.iii.4.8. - Ethpa: **Mete. I**.i.2.6, 8; **III**.i.1.7 (FR تَصَاعَدَ); **IV**.iii.1.2 (IS تَصَعَّدَ). - ܥܠܝܐ: **Min. II**.i.1.5 (N); **IV**.i.3.1, 3, 5; i.4.1; **Mete. I**.i.1.9; ii.1.7 (cf. AB عَـــالٍ); **II**.iii.5.2 (IS أعلى); **IV**.iii.3.2 (cf. IS عُلُوّ); iii.4.4, 8. - ܠܥܠ: **Min. II**.i.1.7bis (N); iii.1.6bis (N); **IV**.iv.2.5; **V**.ii.1.7; iv.1.6, 7, 8; **Mete. I**.i.2.8; i.3.2; ii.6.1; **II**.iii.3.4, 6; **III**.i.2.2 (N); i.3.3 (N); ii.1.10; ii.2.3, 6, 7; iii.2.6 (N); **IV**.i.1.4 (cf. FR عُلُوّ). - ܠܥܠ ܡܢ: vide ܡܢܝܢܐ. - ܥܠܘܗܝ ܡܢܝܢܐ: vide ܡܢܝܢܐ.

√ʿlb: Pe: **Min. I**.i.2.7; **IV**.ii.4.3, 4. - ܡܥܠܒܢܘܬܐ: **Mete. V**.i.1.3 (IS غَلَبَة).

√ʿlm: ܥܠܡܐ [* = عالم]: **Min. I**.i.1.2; ii.5.3 (IS*); **V**.iv.3.5, 6; **Mete. V**.tit. (IS*); i.1.2; i.3.6, 7. - ܡܢ ܥܠܡ: **Min. II**.i.4.10. - ܥܠܡܢܝܐ: **Mete. V**.ii.3.9. - ܥܠܡܢܝܘܬܐ: **Mete. V**.ii.3.9.

√ʿlʿl: ܥܠܥܠܐ [* = زوبعة]: **Mete. III**.i.1.4; iv.tit.; iv.2.5bis, 7 (N); iv.3.1, 3, 5 (IS*); iv.4.1; iv.4.3, 7 (IS*). - ܥܠܥܠܢܐ: **Mete. IV**.ii.1.2 (IS زَوبعي).

√ʿmm: ܥܡܡܐ: **Min. II**.iii.4.5 (IS عامّة); **V**.iii.2.6 (IS أمّة).

√ʿmd: Pe: **Min.** **I**.i.4.5 (IS تَقِدُ); **III**.i.2.4 (N); **Mete.** **III**.i.2.8. - ܡܥܡܘܕܝܬܐ: **Mete.** **IV**.iii.4.9.
√ʿmṭ: ܡܥܡܛܐ: **Mete.** **II**.i.7.5; iii.1.5.
√ʿmṣ: Pa: **Mete.** **II**.ii.2.6.
√ʿmq: Ethpa: **Min.** **I**.iii.2.7. - ܥܡܝܩ: **Min.** **I**.ii.2.3; **I**.iii.2.8; **V**.iii.2.4; **Mete.** **II**.i.5.6 (cf. IS أشدّ). - ܥܘܡܩܐ: **Min.** **I**.ii.2.7; ii.5.7 (IS تعر); iii.3.5, 7; **II**.ii.2.2; ii.4.6, 7 (N); iii.2.1 (IS غور); **V**.iii.2.1; iii.3.2 (IS تعر); iv.3.7. - ܥܡܝܩܘܬܐ: **Min.** **II**.i.2.2 (N).
√ʿmr: Pe: **Min.** **II**.i.1.5 (N). - Ethpe: **Min.** **IV**.i.4.5; ii.2.2. - ܥܘܡܪܐ: **Min.** **IV**.i.4.9 (IS عمارة); ii.2.8; **V**.iii.2.10. - ܥܡܘܪܐ (nom. "incola"): **Min.** **IV**.iii.4.2. - ܥܡܘܪܐ: **Min.** **II**.iii.4.7; **IV**.ii.2.5bis, 6; ii.6.2; iv.5.5, 6; **V**.iii.2.10. - ܥܡܘܪܬܐ: **Mete.** **V**.i.3.4 (cf. IS عمارة). - ܥܡܪܬܐ: **Min.** **I**.i.1.8; **IV**.tit. (IS مسكونة); i.3.7; **V**.i.1.3; **Mete.** **III**.iii.2.5; iii.5.9, 10 (N); **V**.i.1.10 (IS براري عامرة); i.2.4 (IS معمورة); i.3.10 (IS عمارة). - ܐܬܪܐ ܕܥܡܝܪ: **Min.** **IV**.i.3.7 (N Ethpe); **Mete.** **V**.i.1.4 (IS معمور); i.3.4 (cf. IS معمورة).
√ʿnn: Pa: **Min.** **V**.iii.1.5. - ܥܢܢܐ [* = سحابة/سحاب]: **Min.** **I**.ii.3.3bis (IS*); **II**.iii.1.2 (IS*); iii.1.7 (N); **V**.ii.2.2 (IS*); **Mete.** **I**.tit. (IS*); i.tit. (IS*); i.1.3; i.1.7 (AB*); i.2.2, 4 (IS*); i.2.6, 7; ii.2.1, 3 (AB*); ii.2.5; ii.2.7, 8 (AB*); ii.3.4; ii.4.3; ii.6.1; ii.6.6 (IS غمام); iii.1.1 (IS*); iii.2.2; iii.3.1, 8; iii.4.1, 5, 7 (IS*); iii.5.1 (IS غمام); iii.5.2 (IS*); iii.5.2; iii.6.5 (IS*); iii.6.7; **II**.tit.; i.6.3 (IS*); ii.1.3 (IS*; FR غيم); ii.2.2 (IS*); ii.4.1bis (IS*); ii.4.7 (IS*); iii.1.3, 5 (IS*); iii.5.2 (IS*); iii.8.3, 5, 8 (N); iv.1.1 (FR غيم); iv.1.2; iv.2.5 (IS*); iv.3.9 (IS*); **III**.i.1.2; i.3.4 (N); i.4.9bis (N/IS*); iv.1.1, 2 (IS*); iv.1.6 (N); iv.2.2 (N/IS*); iv.4.1 (N); **IV**.i.1.3, 5, 6 (FR*); i.1.8 (IS*); i.3.2 (IS*); i.4.2 (IS غمام); i.4.4; i.5.2, 5 (N/IS*); i.5.7 (N); i.6.2, 5 (N); ii.3.7 (N). - ܥܢܢܢܐ: **Mete.** **II**.i.tit.; **III**.iv.4.5 (cf. IS سحاب); **IV**.i.2.4. - ܥܢܢܬܐ ܕܩܫܬܐ: vide sub ܩܫܬܐ.
ܥܢܬܐ: **Mete.** **I**.iii.4.10.
√ʿnb: ܥܢܒܬܐ, ܥܢܒܐ: **Min.** **IV**.iii.6.11.
√ʿsb: ܥܣܒܐ: **Mete.** **V**.ii.1.3.
√ʿsq: ܥܣܝܩ: **Min.** **I**.i.3.2; **II**.ii.4.6 (N); **Mete.** **III**.iv.1.3 (N); iv.2.5 (cf. N ܥܣܝܩ); **V**.i.3.11. - ܥܣܩܘܬܐ: **Mete.** **III**.iii.4.6.
√ʿsr: ܥܣܪܝܢ: **Min.** **I**.i.2.9; i.4.4; **IV**.ii.3.8; **Mete.** **III**.iii.4.5 (N/IS عشرون).
√ʿpy: ܥܦܝܦܐ: **Min.** **IV**.iii.5.2 (FR ضعف); iv.4.3.
√ʿpr: ܥܦܪܐ [* = تُراب]: **Min.** **I**.iii.1.7 (IS*); **Mete.** **I**.ii.2.7 (AB*); ii.5.3 (AB*); **V**.i.2.7; ii.1.7 (IS/FR مَدَر).
√ʿṣr: Pe: **Min.** **V**.ii.1.8; **Mete.** **I**.i.2.9. - Ethpe: **Mete.** **IV**.ii.3.7 (N). - ܥܨܪܐ: **Min.** **III**.ii.3.7 (AB عصر).
√ʿqb: Pa: **Mete.** **I**.i.1.3. - ܥܩܒܬܐ: **Mete.** **III**.i.2.3.
√ʿqbr: ܥܩܒܪܐ: **Mete.** **V**.ii.1.7 (IS/FR فأر).
√ʿql: ܥܩܠܩܠܬܐ: **Min.** **II**.iii.4.3 (N); **Mete.** **III**.iv.4.1 (IS تَعرُّج; cf. N ܥܩܠܩܠܐ).
√ʿqs: Ethpalal (ܐܬܥܩܣܩܣ): Mete. **III**.iv.4.3 (IS انجَعَدَ).
√ʿqr: Pe: **Mete.** **III**.iv.4.5 (IS قَلَعَ). - ܥܩܪܐ: **Mete.** **IV**.iii.4.9.
√ʿqrb: ܥܩܪܒܐ: **Mete.** **V**.ii.1.6 (IS/FR عَقرَب).
√ʿry: Pe: **Min.** **II**.i.2.3 (N).
√ʿrr: ܥܪܘܪܐ: **Mete.** **I**.iii.tit.; iii.5.2bis; iii.6.6.
√ʿrb: Pe: **Mete.** **III**.ii.2.3. - ܥܪܒܐ: **Min.** **V**.iii.3.10. - ܡܥܪܒܐ [* = مَغرب]: **Min.** **IV**.i.3.2, 6; **V**.i.1.2 (DM); i.3.1; i.3.10, 11; **Mete.** **III**.ii.1.6, 7bis (IS*); ii.3.4, 5, 7, 8; iii.6.2, 4 (N). - ܡܥܪܒܝܐ [* = مغربي]: **Min.** **V**.i.1.5; **Mete.** **I**.ii.4.5 (AB غربي); **III**.ii.1.3 (cf. IS مَغرب); ii.3.1 (IS*); iii.2.1 (IS*); iii.2.3, 4 (IS*); iii.2.7 (N); iii.2.8, 9, 10; iii.6.3, 6 (N).
√ʿrp: ܥܪܘܦܝܐ: **Min.** **III**.iii.2.6. - ܡܥܪܦܝܐ: **Min.** **III**.iii.2.6.
√ʿrpl: ܥܪܦܠܐ [* = ضَباب]: **Mete.** **I**.ii.2.4 (AB*); iii.tit.; iii.5.1 (IS*); iii.5.3bis; iii.6.6; **III**.i.1.2. - ܥܪܦܠܝܐ: **Min.** **II**.iii.1.7 (IS ضبابي).
√ʿrq: ܥܪܘܩܝܐ: **Min.** **III**.ii.1.7; **Mete.** **IV**.i.3.3. - ܥܪܘܩܝܬܐ [* = طَبَقة]: **Mete.** **III**.i.1.5 (FR*); **IV**.i.1.2 (FR*).
√ʿšn: Pe: **Min.** **IV**.ii.4.7; iii.1.7; iv.3.1, 2 (AB اشتَدَّ); iv.4.9 (AB غَلَبَ); iv.4.11; **Mete.** **I**.iii.2.8. - ܥܫܝܢ [* = شديد]: **Min.**

I.i.3.5; II.ii.2.5 (IS قَوِيَ); ii.4.8; iii.1.4 (N); iii.1.6; iii.2.1´ (N); IV.ii.3.10; **Mete.** I.iii.3.7 (IS*); II.ii.2.2 (cf. IS قُوَّة); IV.i.1.8 (IS*); i.2.3 (IS/FR*); V.i.1.9 (IS*). - ܥܫܝܢܘܬܐ: **Min.** II.ii.4.3; **Mete.** III.i.3.7, 8 (N). - ܥܫܝܢܐ [* = شدّة]: **Min.** I.i.3.4 (IS*); II.iii.1.11 (IS*); **Mete.** III.iv.2.1; IV.i.3.3 (cf. IS عنيف); V.i.1.10 (cf. IS شديد).

√ʿtd: ܡܥܬܕ (Pa ptc. pass.): **Min.** I.iii.1.6; V.iv.3.6, 8, 10. - ܥܬܝܕܬܐ: **Mete.** II.i.tit. (cf. IS مُقدّمات).

√ʿtq: ܥܬܝܩܐ: **Min.** I.ii.5.4; II.i.tit.; IV.i.2.3; V.iii.2.11; **Mete.** I.i.1.2; III.i.2.1.

√ʿtr: ܥܬܝܪ: **Mete.** II.iii.8.2.

ܦ

√pʾy: ܦܐܝܐ: **Min.** I.i.1.5; V.ii.4.3.

ܦܠܣܛܝܢܝ (nom. prop. "Palaestine"): **Min.** V.ii.3.5 (N/IS فلسطين).

√pgʿ: Pe [* = لَقِيَ]: **Min.** II.ii.2.2; III.ii.3.3; IV.ii.2.7; V.ii.2.6; **Mete.** II.i.1.6; i.1.7 (IS*); III.i.2.5 (N); iv.3.2 (IS*); iv.3.4 (N); IV.ii.2.4, 6 (N). - ܦܓܥܬܐ: **Mete.** III.iv.4.3, 6 (IS تلاق).

√pgr: ܦܓܪܐ [* = بَدَن]: **Min.** II.ii.1.10; iii.3.5 (IS*); III.i.5.7; IV.ii.6.6 (IS*); iii.6.4 (AB*).

√pdd, pwd: Pe: **Mete.** II.i.5.1 (cf. IS غَلَطَ); V.iii.2.8.

√phy: ܦܗܝܬܐ: **Mete.** tit. (IS الآثار العلوية); I.i.1.2; Mete. fin.

ܦܝܕܘܢ (nom. prop. "Phaedo"): **Min.** V.iv.1.1; iv.2.1.

ܦܪܝܦܛܛܝܩܘ (περιπατητικοί): **Min.** IV.ii.1.1; ii.3.1; **Mete.** II.iii.7.2 (IS مَشّاؤون).

ܦܘܐܝܛܝܐ (ποιητής): **Min.** I.i.4.12 (IS شَاعِر); V.ii.4.3 (N; cf. IS شعري).

ܦܘܐܝܡܐ, ܦܘܐܝܡܛܐ (ποίημα): **Min.** I.i.4.12 (IS شعر).

ܦܘܕܓܪܐ (ποδάγρα): **Min.** I.i.5.2 (N).

√pwg: Ethpa: **Min.** IV.iii.6.5.

ܦܘܢܛܘܣ (nom. prop. "Pontus"): **Min.** V.i.2.5.

ܦܘܢܝܩܝܣ (φοινικίας): **Mete.** III.ii.3.7 (N).

ܦܘܪܡܝܕܝܣ (πυραμίδες): **Min.** V.iii.2.11.

ܦܘܪܘܣ, ܦܘܪܘ (πόρος) [* = مَسَام]: **Min.** I.iii.1.1; iii.3.6; II.iii.4.5 (IS*); IV.iv.2.2 (AB*); V.ii.1.10, 11 (N); iv.2.2; **Mete.** III.iv.4.3 (IS*).

ܦܘܪܝܛܝܣ (πυρίτης): **Min.** III.i.1.3.

√pwš: Pe [* = بَقِيَ]: **Min.** I.ii.2.6 (IS*); V.ii.2.3 (IS*);´ **Mete.** II.i.8.3; iv.1.9 (FR*); IV.i.1.4 (FR*); iii.2.6, 7 (IS*).

√pḥḥ: ܦܚܝܚ [* = متخلخل]: **Min.** I.i.2.10 (IS رَخو); ii.5.6 (IS*); ii.6.4; II.i.1.4 (N); iii.2.1 (N/IS*).

√pḥm: Ethpa: **Mete.** II.iii.4.7. - ܦܚܡܐ: **Min.** IV.iv.2.5; **Mete.** I.iii.1.7 (IS نسبة); II.iv.2.6; V.iii.2.5, 9. - ܦܘܚܡܐ: **Min.** IV.iv.3.7, 8 (cf. AB قياس).

√pḥr: ܦܘܚܪܐ: **Min.** I.i.2.3´(IS تفخير).

√pḥt: Pa: **Mete.** III.i.3.12 (IS خَسَفَ). - ܦܚܬܐ: **Mete.** IV.iii.4.7 (IS هُوَّة).

ܦܝܠܘܣܘܦܐ (φιλόσοφος): **Min.** V.ii.4.3 (N; cf. IS فلسفي). - ܦܝܠܘܣܘܦܝܐ (φιλοσοφία): **Mete.** I.i.1.1.

ܦܝܣܐ (πεῖσις/πεῖσαι): **Mete.** II.iii.7.2 (+ ܚܕܐ: IS أَقنَعَ).

ܦܦܘܢ (πέπων): **Mete.** III.iii.1.10 (N?).

ܦܠܛܘܢ (nom. prop. "Plato"): **Min.** V.iv.1.1.

√plg: Pa: **Min.** IV.ii.1.1; ii.2.1. - Ethpa: **Min.** IV.i.3.4; Mete. V.i.3.6 (IS كان منقسمًا). - ܦܠܓܐ [* = نُصف]: **Min.** IV.i.3.1, 3bis; iv.2.5; iv.5.3 (AB*); **Mete.** I.i.2.7; II.iii.3.5, 6bis (FR*). - ܦܠܓܘܬܐ [* = نُصف]: **Min.** IV.i.3.7 (IS*); ii.6.5; **Mete.** II.ii.3.6 (IS*). - ܦܘܠܓܐ: **Min.** II.iii.1.5; IV.ii.1 (IS قسم); ii.1.2, 4 (IS قطعة); ii.2.1.

√plhd: Ethpa: **Min.** IV.iv.2.2 (cf. AB تفرّق). - ܦܘܠܗܕܐ: **Min.** I.ii.6.3.

ܦܘܠܘܣ (πόλος): **Min.** IV.i.3.11; **Mete.** III.ii.1.9, 11; iii.5.3bis (IS قُطب); iii.5.5.

√plḥ: ܦܠܚܐ: **Min.** III.i.5.6; iii.1.5 (IS صاحب); iii.3.1.

√plṭ: Pe: **Min.** II.ii.3.5; **Mete.** III.iv.4.2; IV.i.4.2 (IS تَخَلَّص).

√pm: ܦܘܡܐ: Min. V.i.1.2 (DM); Mete. II.iii.1.8 (IS فم).

√pny: Pe: **Mete.** IV.iii.2.7; V.ii.3.5 (FR عَادَ). - ܦܢܝܬܐ: **Min.** I.i.3.4 (IS بُقعة); IV.i.2.8 (IS جهة); iv.3.1; **Mete.** I.ii.3.6 (AB جهة); ii.4.6; II.iv.2.6; III.ii.tit.; ii.1.1; iii.1.2 (IS ناحية); iii.1.6 (N); iii.2.2;

iii.4.1 (IS جهة): iii.5.3 (N); iii.5.5, 6; V.i.3.8 (IS ناحية). - ܦܢܝܬܐ: **Mete.** V.ii.3.10 (IS عود).

ܦܢܛܣܝܐ (φαντασία): **Mete. II**.ii.1.8.

ܦܣܝܠܐܘܣ (ψολόεις): **Mete. IV**.ii.2.3 (N ܦܣܘܠܐܘܣ).

√psl: ܦܣܝܠܘܬܐ (sc. ܟܐܦܐ): **Min. I**.i.2.11.

√psq: Pe: **Min. I**.i.4.7 (IS فَصَلَ); **IV**.i.3.2, 4; **Mete. I**.ii.3.3 (AB قطَعَ); **II**.iii.3.5 (FR قَسَمَ); **III**.ii.1.8, 10 (cf. IS تقاطع). - Ethpe: **Min. IV**.iv.2.2 (cf. AB انقطاع); V.iii.2.11; **Mete. I**.ii.3.2; ii.3.3 (AB انقَطَعَ). - ܡܦܣܩܐ (Pa ptc. pass.): **Mete. I**.iii.2.4 (cf. DM ܡܦܣܩܬܐ). - ܦܘܣܩܐ: chapter and sections headings passim. - ܦܣܘܩܐ: **Mete. IV**.i.6.2.

ܦܝܣܝܢ ("piscina"): **Mete. III**.i.3.9.

√p^{c}r: ܦܥܪܐ: **Min. I**.iii.1.1; iii.3.6. - ܦܥܪܬܐ: **Min. II**.iii.2.1 (N).

√pqd: Pe: **Min. I**.i.4.8 (IS رام).

√pq^{c}: ܦܩܥܬܐ: **Min. I**.ii.3.4; **IV**.i.1.6; iv.1.1.

√pry: ܦܐܪܐ: **Min. IV**.iii.6.10 (AB ثمر).

√prd: ܦܪܕܬܐ, ܦܪܕܐ: **Min. I**.i.4.9; **Mete. I**.iii.3.8; iii.4.2, 6; iii.4.8 (IS قطعة); iii.4.9.

ܦܪܘܛܐܣܝܣ (πρότασις): **Mete. II**.i.9.4 (cf. IS مقدمات وتوطئات).

ܦܪܙܠܐ [* = حَديد]: **Min. I**.i.4.4 (IS*); i.4.9; ii.5.5 (IS*); **III**.i.2.4 (N); i.4.4 (IS*); i.5.2 (AB*); i.5.8; ii.3.6 (IS*).

ܦܪܝܣܛܝܪ, ܦܪܝܣܛܝܪܣ (πρηστήρ): **Mete. I**.iii.6.8; **IV**.ii.tit.; ii.1.1 (N); ii.3.1 (N); ii.3.6, 10 (N).

√prk: Ethpalpal: **Min. II**.i.3.2 (N); **III**.i.1.3 (AB تَفَتَّتَ). - ܦܘܪܟܟܐ: **Min. I**.i.2.2 (IS تفتُّت). - ܡܬܦܪܟܟܢܐ: **Min. III**.i.5.2 (AB ما ينكسر).

ܦܪܠܠܝܠܐ (παράλληλος) [* = مُوازٍ]: **Min. IV**.i.3.3; ii.1.2 (IS*); **Mete. III**.ii.2.5, 9 (IS*).

ܦܪܢܓܝܐ (nom. prop. "terra Francorum"): **Min. V**.i.3.11.

√prs ("expandit"): ܦܪܝܣ (ptc. pass.): **Min. II**.i.1.4 (N).

√prs (< πόρος): Ethpa: **Min. I**.iii.1.8. - ܦܘܪܣܐ [* = حيلة]: **Min. III**.i.1.6 (IS*); i.2.1 (IS*); iii.1.1 (IS*).

√prs^{c}: Ethpalpal (ܐܬܦܪܣܥ): **Mete. I**.iii.4.3 (cf. IS تَخَلخَلَ). - ܡܦܪܣܥ (ptc. pass.) [* = مُتَخَلخِل]: **Min. II**.i.1.5 (IS*); **III**.i.5.3 (AB*); **Mete. I**.iii.2.4.

√pr^{c}: Pe: **Min. I**.i.1.5.

ܦܪܥܘܬܐ: **Mete. I**.iii.3.3; iii.4.5.

ܦܪܨܘܦܐ (πρόσωπον): **Mete. V**.iii.2.8.

√prq: Pe: **Min. III**.i.2.6 (IS بَرِئَ). - Ethpa: **Min. III**.ii.1.6; **Mete. I**.ii.1.2 (AB تَفَرَّقَ). - ܦܘܪܩܢܐ: **Min. I**.ii.6.3; **Mete. I**.ii.1.3.

√prš: Pe: **Min. IV**.ii.1.3, 5 (IS فَصَلَ); **Mete. II**.iii.1.5 (IS مَيَّزَ). - Ethpe: **Mete. II**.iv.2.6 (IS تَمَيَّزَ). - Pa: **Mete. II**.i.9.3 (cf. IS انفسم). - ܦܪܝܫ (ptc. pass.): **Mete. II**.ii.3.1 (cf. IS خَالَفَ). - ܦܘܪܫܢܐ: **Min. I**.ii.6.8 (IS تفصيل); **Mete. II**.i.9.2 (cf. IS قَسَّمَ); iii.7.8 (IS انفصال). - ܦܘܪܫܢܘܬܐ: **Min. III**.iii.2.1, 6, 8 (IS فَصل).

√pšš: Pauel [* = حَلَّلَ]: **Min. II**.iii.1.10 (N); **Mete. II**.ii.4.7 (IS*); **III**.i.4.5 (N); i.4.10 (cf. IS تحلُّل); iii.4.3 (IS*). - Ethpaual [* = تَحَلَّلَ]: **Min. I**.ii.4.4; I.ii.5.2; iii.2.5 (IS*); **II**.ii.1.5; **Mete. I**.i.2.2 (IS*); ii.1.4 (AB تَشَتَّتَ); **II**.iv.3.9 (cf. IS حَلَّلَ); **III**.iii.1.2 (N); iii.4.6 (cf. IS انحلال); **IV**.ii.1.6 (IS*); iii.1.5 (IS*); iii.3.10; iii.4.3 (IS تَبَدَّدَ). - ܦܘܫܫܐ: **Min. II**.iii.2.7 (IS تحليل); **Mete. I**.iii.5.5 (cf. IS تَحَلَّلَ); **II**.iv.2.2.

√pšṭ: Ethpe: **Min. I**.ii.3.4 (IS تَوَجَّهَ); **V**.ii.2.7. - ܦܫܝܛ: **Min. I**.i.3.3. - ܦܫܝܛܐ ("stultus"): **Min. I**.iii.3.2; **III**.iii.2.5.

√pšk: Ethpa: **Mete. III**.i.1.1 (cf. IS مُتشكّك); iii.5.9. - ܦܘܫܟܐ: **Mete. V**.i.3.7.

√pšq: ܦܫܝܩ: **Min. I**.i.3.3. - ܦܫܝܩܐܝܬ: **Min. III**.i.5.4, 6 (IS بِسُهُولَة); **Mete. IV**.i.1.5 (cf. IS سَهُلَ).

√pšr: Pe: **Mete. I**.iii.4.5 (IS ذابَ). - Pa: **Mete. IV**.ii.2.6, 8 (N/IS أذابَ). - Ethpe/Ethpa [* = ذابَ]: **Min. I**.i.5.8 (N); i.5.10 (N?); **III**.i.1.3 (AB*); i.1.6 (IS*); i.2.1, 2 (IS*); ii.2.1 (cf. IS ذائب); ii.2.3 (IS كان ذوبه); ii.2.4 (IS*). - ܡܬܦܫܪܢܐ [* = ذائب]: **Min. III**.i.1.2 (IS*); i.4.2 (IS*); i.4.4; ii.2.7. - ܡܬܦܫܪܢܘܬܐ: **Min. III**.ii.2.2 (IS ذَوب).

√pty: ܦܬܝܐ (ptc. pass.): **Min. II**.i.1.3 (N); **Mete. III**.iii.1.8. - ܦܬܝܘܬܐ: **Min. V**.ii.2.5. - ܦܬܝܐ [* = عَرض]: **Min. II**.i.4.5 (N); ii.4.5, 6 (N); **IV**.i.3.6; i.3.9 (IS*); iii.5.1 (FR*); **Mete. I**.iii.3.4 (cf. AB عَرضي); **II**.iii.7.8 (IS*); **IV**.iii.1.7 (N).

√pth: Ethpe: **Mete. III**.i.3.9 (N). - ܦܬܝܚ (ptc. pass.): **Min. V**.i.1.3 (DM); **Mete. II**.i.5.7 (cf. IS أُثِلَ). - ܦܘܬܚܐ: **Min. II**.iii.4.5 (IS تفتيح).
√ptk: ܦܬܝܟ: **Mete. II**.iii.4.3. - ܦܬܝܟܘܬܐ: **Mete. II**.iii.5.1.
√ptl: ܦܬܝܠܘܬܐ: **Mete. III**.iv.4.3 (IS اِلتِواء).
√ptq: Ethpe: **Mete. III**.iv.1.2 (N).

ܨ

√ṣʾy: ܨܐܐ: **Mete. II**.iii.4.6bis. - ܨܐܘܬܐ: **Min. III**.iii.1.9 (cf. IS عيب).
√ṣby: Pe: **Mete. IV**.i.2.5 (IS شَاءَ). - ܨܒܘܬܐ: **Min. V**.iv.3.9. - ܨܒܝܢܐ: **Mete. V**.iii.2.7. - ܨܒܝܢܝܐ: **Mete. V**.ii.3.7 (IS/FR إرادي).
√ṣbᶜ ("mersit"): Pe: **Min. III**.iii.1.8 (IS صَبَغَ). - Pa: **Mete. III**.i.4.6 (N); iii.1.7 (N). - ܨܘܒܥܐ [* = صبغ]: **Min. III**.iii.1.7, 8 (IS*); **Mete. II**.i.5.6 (IS*).
√ṣbᶜ ("digitus"): ܨܒܥܬܐ: **Mete. V**.iii.2.4.
√ṣdd: Aph: **Mete. II**.iii.2.3 (N).
√ṣwb: ܨܘܒܐ: **Mete. I**.ii.6.5.
ܨܘܝܒܘܬܐ: **Mete. IV**.iii.3.6 (IS ذُؤابة). - ܨܘܝܒܢܝܐ: **Mete. IV**.iii.tit. - ܟܘܟܒܐ ܨܘܝܒܢܝܐ: vide sub ܟܘܟܒܐ.
√ṣwr: ܨܘܪܬܐ: **Min. I**.i.3.8; ii.6.6 (IS صورة); ii.6.6; **Mete. II**.iii.2.4 (N); iii.8.4; **V**.ii.2.9; ii.3.1.
ܨܝܢ (nom. prop. "Sina"): **Min. III**.i.5.8. - ܨܝܢܝܐ: **Min. V**.i.3.7.
√ṣḥy: ܨܚܘܐ [* = صَحو]: **Mete. I**.i.2.4, 8 (IS*); iii.5.4 (IS*); **III**.iii.4.7 (N/IS*).
√ṣll: ܨܠܠ: Mete. II.iii.1.2 (IS صاف).
√ṣly: Pe: **Mete. I**.ii.3.6; **III**.iii.4.1 (cf. IS مَيَلَ). - ܨܠܐ (ptc. pass.): **Mete. II**.iii.5.6; iii.7.4bis; iii.7.5 (supplevi); **III**.iii.2.3. - ܨܠܝܢܐ [* = مَيل]: **Min. IV**.ii.3.5, 7, 8 (IS*); iii.5.2 (FR*); **Mete. V**.i.3.2 (IS*).
√ṣmd: Ethpa: **Min. IV**.iv.2.2 (AB اتَّصَلَ); **Mete. I**.ii.1.8 (cf. AB اتّصال); iii.2.1; iii.4.2. - ܡܨܛܡܕܢܘܬܐ: **Mete. I**.iii.3.1 (FR اجتماع).
√ṣᶜr (√ṣmᶜr): Ethpamal (ܐܨܛܡܥܪ): **Min. IV**.iii.1.3.
√ṣpr: ܨܦܪܐ: **Min. IV**.iv.5.8 (AB غداة); **Mete. II**.iv.3.8 (IS طلوع); **III**.iii.2.8, 9bis. - ܨܦܪܝܬܐ: **Mete. II**.iii.1.9 (IS غداة).
√ṣrr: ܨܪܪܐ: **Mete. IV**.ii.2.8bis (IS صُرَّة).
√ṣry: Pe: **Min. I**.ii.2.4; ii.4.5; iii.1.3; iii.1.7 (IS شَقَّ); iii.2.3 (IS خَرَقَ); **II**.ii.1.5 (IS شَقَّ); ii.2.8; **Mete. III**.i.2.5 (N); **IV**.i.1.8; **IV**.i.4.4. - ܨܪܝܐ: **Min. I**.iii.1.5; **II**.ii.1.4; ii.3.6; iii.2.3 (IS خَرق); **Mete. III**.i.2.6 (N).
√ṣrp: ܨܪܦܐ [* = شَبّ]: **Min. III**.i.3.4 (IS*); i.5.4 (IS*).

ܩ

ܩܐܪܘܬܐ (κηρός): **Min. V**.ii.1.9 (N/IS شمع).
√qby: Pe: **Mete. II**.iii.1.7. - ܩܒܝܘܬܐ: **Min. I**.ii.5.4.
√qbl: Pa: **Min. III**.iii.2.3; **V**.ii.1.7; **Mete. II**.iv.1.3 (IS/FR قَبِلَ). - ܡܩܒܠܐ: **Min. I**.ii.6.6; **II**.i.4.2 (N); **IV**.i.4.5; ii.5.7; **Mete. II**.iv.2.4; **V**.ii.3.1. - ܡܩܒܠܢܐ: **Min. I**.ii.5.9; **Mete. V**.i.2.1 (IS قابل). - ܠܘܩܒܠ ܠܘܩܒܠ ܠܘܩܒܠܐ: **Min. II**.i.4.7; **IV**.iii.4.1; **Mete. II**.i.1.8 (cf. IS قَابَلَ); i.3.2 (cf. IS قَابَلَ); i.3.4, 5 (cf. IS محاذاة); ii.1.4, 6 (cf. FR قَابَلَ); ii.4.5 (cf. IS حَاذَى); iii.1.7 (cf. IS وَازَى); iii.1.8 (IS حِذاءَ); iii.2.8; iv.3.7; **III**.iii.5.10 (N). - ܕܠܘܩܒܠܐ: **Min. IV**.ii.3.1; iii.6.2. - ܠܘܩܒܠܝܘܬܐ: **Mete. IV**.i.4.6 (IS مُوازاة). - ܕܠܘܩܒܠܢܐ: **Mete. I**.i.3.4 (IS مُتَقابل). - ܩܘܒܠܢܐ: **Min. II**.iii.1.4 (N).
√qbᶜ: ܩܒܝܥ: **Min. III**.iii.3.3; **Mete. II**.i.2.5; **IV**.iii.4.9. - ܩܒܝܥܐ: **Min. IV**.i.2.3 (IS ثابتة).
√qdm: Pe: **Min. II**.i.4.10 (N); ii.3.1; ii.3.8 (cf. IS أَسبَقَ); **Mete. I**.i.3.4 (cf. IS مُتقدِّم); **III**.i.2.6 (N); iii.3.7 (N); iii.6.1 (N); **IV**.iii.2.3 (cf. IS فيما سلف). - Ethpa: **Min. IV**.iii.3.3 (cf. FR سَابِق). - ܩܕܡܝܐ [* = أوّل]: chapter, section and theory headings passim; **Min. I**.i.1.2 (IS أولي); i.1.4 (IS عنصري); **II**.i.4.1; **IV**.ii.1.1; ii.3.3; ii.5.6, 8 (FR*); iii.1.3 (Ṭ*); iv.4.4; **Mete. II**.i.1.3, 4; i.3.6; iii.5.1; **III**.ii.1.1, 4; iii.2.5; iv.1.1 (cf. IS في القديم); **IV**.i.4.10 (N); iii.3.1 (IS*); **V**.iii.1.7. - ܩܕܡܝܬ: **Min. I**.i.1.6; **Mete. III**.i.3.4 (N ܩܕܡܝܐܝܬ); iv.3.5 (N ܩܕܡܐܝܬ). - ܒܩܕܡܝܐ: **Mete. V**.iii.2.4. - ܠܘܩܕܡ: **Min. I**.i.2.4 (IS أولاً); **II**.ii.3.7 (cf. N ܩܕܡܐܝܬ); ii.3.9 (N); **Mete.**

I.i.1.3; IV.i.4.5 (cf. N ܩܢܛܪܘܢܝܬܐ). - ܡܩܢܛܪ: Min. I.i.5.10 (N); V.iii.2.1.
ܩܕܣܐ (κάδος): Min. V.ii.1.6, 7.
ܩܢܛܪܘܢ, ܩܢܛܪܘܢܐ (κέντρον) [* = مركز]: Min. IV.i.4.6 (IS*); V.iv.1.2, 6, 7; Mete. I.ii.1.2, 6 (AB*); II.ii.3.3, 4 (IS*); iii.3.3 (FR*).
ܩܐܩܝܐܣ (καικίας): Mete. III.ii.3.4 (N).
ܩܪܐܘܢܘܣ (κεραυνός) [* = صاعقة]: Min. I.i.4.1 (N); Mete. IV.ii.tit.; ii.1.1 (N/IS*); ii.2.1 (N); ii.3.6, 10 (N).
√qwy: Pa [* = قوّي]: Min. V.ii.1.12 (N); iv.3.3; iv.3.9; Mete. IV.ii.1.5 (IS*); iii.1.8; iii.3.9 (IS*); iii.4.3 (IS*).
√qwl: ܩܠܐ [* = صوت]: Min. I.i.4.6 (IS*); II.ii.3.1 (N/IS*); ii.3.2, 4 (N); ii.3.7 (N/IS*); III.ii.3.8; Mete. IV.i.1.5, 7; i.1.8, 9 (IS*); i.3.2, 4 (IS*); i.4.8, 11 (IS*). - ܩ̈ܠܐ: Mete. V.iii.2.8.
√qwm: Pe: Min. I.iii.1.6; iii.4.5 (cf. IS راكد); II.ii.2.9 (N); IV.i.4.2 (IS قَامَ); iv.2.6; Mete. I.i.3.3 (cf. IS وقوف); i.3.4 (cf. IS واقف). - Pa: Mete. II.ii.1.2 (N). - ܩܝܡܬܐ: Mete. II.iv.3.5 (IS مَقَام). - ܩܘܝܡܐ: Min. I.iii.1.4 (IS راكد/N?). - ܩܘܝܡܘܬܐ: Min. I.ii.2.4; III.iii.2.2, 7; Mete. I.iii.5.1 (IS قِوَام); III.iv.3.7bis (IS جوهر); V.i.3.5. - ܡܩܝܡܢܘܬܐ: Min. III.iii.2.8. - ܡܩܝܡܬܐ: Min. V.i.1.2.
ܩܘܓܟܘܠܝܐ (κογχύλια): Min. V.iii.2.3.
ܩܘܦܪܘܣ (nom. prop. "Cyprus"): Min. V.i.1.7.
ܩܘܪܐܠܝܘܢ (κοράλλιον): Min. I.i.5.4 (N).
√qtl: ܩܛܝܠ (ptc. pass.): Min. III.ii.3.9 (IS قتل propono).
√qtm: ܩܛܡܐ: Min. II.ii.2.8; V.ii.1.4 (IS رماد).
√qtn: Pa: Mete. III.i.4.5 (N). - Ethpa: Min. I.iii.4.1 (cf. IS لطف); Mete. I.i.2.2; IV.iii.3.9 (cf. IS لَطُفَ). - ܩܛܝܢ [* = لطيف]: Min. III.i.3.5 (IS*); ii.1.1 (IS*); ii.3.3 (IS*); V.i.2.4; i.3.4; ii.1.2 (IS رقيق); ii.1.11 (N); ii.2.2 (IS ألطف); Mete. I.iii.5.2, 3; II.iii.1.6; III.iv.2.6, 8 (N); iv.3.7 (IS*); IV.i.1.7 (IS*); i.2.1 (IS/FR*); i.2.5 (IS*); i.2.7 (IS*); i.3.5 (IS*); ii.2.1 (N); iii.1.3 (IS*); iii.1.9; iii.3.5 (IS*). - ܩܛܝܢܘܬܐ: Min. III.iii.2.4; Mete. I.iii.1.3; IV.i.3.7 (IS لُطف); ii.1.5 (cf. IS لُطف); iii.1.4.
√qtr: Pe (intrans.) [* = انعقد]: Min. I.i.2.7 (IS*); ii.1.5 (cf. AB انعقاد); II.iii.1.7 (N); III.i.1.4 (cf. IS جمود); i.2.6, 7 (IS جمد); i.3.3 (IS*); i.4.4; V.ii.1.5 (IS*); ii.3.10 (N); Mete. I.ii.6.1; ii.6.6 (IS*); IV.i.1.3 (FR كاثف). - Pe (trans.) [* = عَقَدَ]: Min. III.i.3.6 (cf. IS انعقد); ii.1.8 (AB*); ii.2.6 (IS*); III.iii.1.1 (cf. IS انعقادات). - Aph: Mete. I.iii.1.2 (IS عَقَّدَ). - ܩܛܝܪ (ptc. pass.): Min. III.ii.1.9 (AB منعقد). - ܩܛܝܪܐܝܬ: Min. V.iv.1.9. - ܩܛܝܪܘܬܐ: Mete. III.iv.3.7 (cf. IS ثُخونة). - ܩܛܝܪܐ ("necessitas, violentia"): Mete. III.iv.1.6 (N). - ܩܛܝܪܬܐ [* = شديد]: Mete. IV.i.2.9 (IS*); iii.4.4 (IS*). - ܩܛܝܪܐ: Min. I.i.2.5 (IS جمود); Min. I.i.5.4 (N); III.ii.3.10 (IS انعقاد).
√qyt: ܩܝܛܐ [* = صيف]: Min. II.iii.2.6 (N/IS*); IV.ii.6.4, 5; iii.5.4, 7bis (FR*); iii.6.6bis (AB*); iv.3.4; iv.4.9, 11; V.iii.1.10; iv.2.3; Mete. I.iii.3.6, 9 (IS*); iii.4.3 (IS*); II.iii.3.9bis (IS*); III.iii.3.1 (N); iii.5.5; iii.6.1, 3, 6 (N); IV.i.5.3 (N); iii.3.9. - ܩܝܛܝܐ [* = صيفي]: Min. IV.ii.4.4; iii.3.1; iv.4.6 (cf. AB صيف); iv.4.8, 10; Mete. I.ii.5.1 (AB*); III.ii.1.5, 7 (cf. IS صيف); ii.3.3, 5, 6, 7.
ܩܝܢܕܘܢܘܣ (κίνδυνος): Min. IV.iii.6.6 (cf. AB لا سَلم).
√qys: ܩܝܣܐ: Min. I.ii.5.6 (IS خشب); Mete. III.i.4.7 (N); iii.1.8 (N); IV.ii.2.7 (N).
√qll: ܩܠܝܠ: (adj.): Min. II.ii.3,7; V.ii.3.3 (N). - ܩܠܝܠ (adv.): Min. I.iii.2.9; iii.3.3; IV.iv.1.6; Mete. II.iii.2.3 (N). - ܒܩܠܝܠ ܩܠܝܠ: Mete. III.i.3.5 (N); iv.1.4 (N); V.iii.1.9.
ܩܠܝܕܝܡܘܣ (nom. prop. "Cleidemus"): Mete. IV.i.6.1 (N).
ܩܠܘܕܝܐ (nom. prop. "Claudia"): Mete. I.i.2.7.
ܩܠܝܡܐ (κλίμα): Min. IV.i.3.5, 6; i.4.10; ii.2.5; ii.3.5, 9; ii.4.3; iii.2.4; iii.3.7; iii.5.1, 3, 4; iv.4.2, 4, 5 (cf. AB اقليم).
√qlᶜ: ܩܘܠܥܐ: Min. I.i.5.7 (N).
√qny: Pe: Min. I.i.5.10; iii.1.4; III.i.3.3; i.4.4bis (IS استفاد); Mete. I.ii.6.6; V.ii.2.5, 8; V.iii.1.2, 6. - Aph: Mete. IV.ii.2.3 (N). - ܡܩܢܐ (ptc. pass.): Min. I.i.5.10; iii.1.5; iii.2.3; iii.4.7; II.iii.2.1; III.i.4.2 (IS فيها); ii.3.3 (IS فيه); IV.i.1.4;

iii.2.2; **Mete.** I.iii.1.8; iii.5.2 (cf. IS له); II.i.7.4; iii.8.6; IV.ii.3.3 (IS نيها); iii.1.6; V.iii.2.3. - ܕܝܬܩܢ̈ܐ: **Mete.** V.iii.1.1.

ܩܢܘܡܐ: **Mete.** I.iii.4.9; V.iii.2.1. - ܩܢܘ̈ܡܝܐ: **Mete.** V.iii.1.1 (cf. IS شخص).

ܩܢܘܦ (nom. prop., titulus libri): **Min.** IV.i.2.5.

ܩܢܛܘܢ: vide ܩܣܛܘܢ.

ܩܢܝܕܘܣ (nom. prop. "Cnidus"): **Min.** I.i.5.7 (N).

ܩܣܦܝܐ (nom. prop. "Caspia"): **Min.** V.i.2.6.

√qᶜy: ܩܥܝܐ: **Min.** II.ii.3.6 (cf. IS مُصَوِّت). - ܩܥܬܐ: **Min.** II.ii.3.5.

√qpd: Pe: **Min.** II.iii.1.12 (N).

ܩܦܠܐܘܢ (κεφάλαιον): chapter headings passim.

√qps: Ethpe/Ethpa: **Mete.** III.i.4.9 (IS قَبَضَ).

√qrr: Pe [* = بَرَدَ]: **Min.** II.iii.1.7 (N); IV.iii.3.5bis: **Mete.** I.i.1.7 (AB*); i.2.4 (IS*); ii.1.4, 7 (AB*); ii.2.5 (AB*); ii.3.4; ii.4.1 (AB*); ii.6.6 (IS*); III.i.1.6, 7; i.4.11 (N/IS*); iii.2.5; IV.i.1.6 (N/FR صار باردًا); iii.1.3 (IS*). - Aph: **Min.** II.ii.1.9 (N); IV.iii.3.4. - Ethpaual: **Min.** V.ii.3.10 (N); **Mete.** III.iv.2.9 (N). - ܩܪܝܪ [* = بَارِد]: **Min.** I.i.4.3 (IS*); IV.iii.3.6; iv.4.8 (cf. AB برودة); **Mete.** I.i.1.8; I.i.2.3, 5 (IS*); ii.1.3, 4; ii.5.2 (cf. AB بَرد); ii.6.1; iii.4.4 (IS*); III.i.1.5 (FR*); iii.1.1 (IS أَبرَدُ); IV.i.1.3 (FR*). - ܩܪܝܪܘܬܐ [* = بَرد]: **Min.** I.i.2.7; ii.3.2 (cf. IS بارد); ii.4.2; iii.1.2; II.iii.2.6 (IS*); iii.3.2 (N/IS*); iii.3.3, 4 (IS*); III.i.1.4 (IS*); i.2.6 (IS*); i.3.3 (IS*); IV.ii.3.2; ii.4.2, 5, 7; iii.2.1 (FR برودة); iii.2.4; iii.3.3 (FR*); iii.6.4bis, 6 (AB*); iv.tit. (AB*); iv.3.6 (AB*); iv.4.7 (AB*); iv.4.8 (AB برودة); iv.4.9 (AB برود); iv.4.10, 11; **Mete.** I.i.1.8 (AB*); i.3.5 (IS*); ii.1.5 (AB*); iii.1.2 (IS*); iii.1.6; iii.2.2 (FR*); iii.3.2 (FR*); iii.3.7 (IS*); III.i.4.9 (N/IS برودة); iii.2.3; iv.2.7, 8 (N); IV.iii.2.4 (IS*); iii.3.2 (IS*); V.i.1.11 (cf. IS بارد). - ܩܘܪܐ [* = بَرد]: **Min.** II.iii.1.9 (IS*); iii.2.5 (IS*); IV.i.3.8, 10; ii.2.1; ii.6.2, 7, 9 (IS*); iii.4.4, 6; iv.3.2 (AB*); iv.4.9 (AB*); **Mete.** I.i.1.9; iii.2.6, 8; III.iii.4.6, 7 (N).

√qry: Pe: **Min.** III.i.5.6; **Mete.** III.ii.3.4; iv.1.1, 3 (cf. IS سُمِّيَ); V.iii.2.9. - Ethpe [* = سُمِّيَ]: **Min.** I.i.5.6; V.i.1.4; i.2.2; i.2.3 (DM); iii.2.9 ("lectus est"); iii.2.11; iii.2.13 ("lectus est"); **Mete.** I.i.1.2; ii.3.7; iii.2.9; iii.6.3, 5; II.iv.3.1; III.ii.3.3; iii.4.7 (IS*); iii.5.1 (IS*); IV.ii.2.2 (N); iii.1.10 (N).

√qrb: Pe: **Min.** IV.i.3.11; **Mete.** II.i.2.5. - Pa: **Min.** II.ii.1.10 (N). - ܩܪܝܒ [* = قريب]: **Min.** III.iii.3.5 (AB*); IV.i.3.10; ii.5.3, 5; iv.1.2 (AB*); V.ii.3.5, 8 (N); **Mete.** I.ii.4.2 (AB*); ii.5.6 (AB مُقارب); iii.4.7; II.i.6.1 (IS*); ii.4.2 (IS أقربُ); iii.4.4; iii.5.2 (IS أقربُ); iii.7.3 (cf. IS أقربُ); iv.2.6 (IS*); III.iii.2.1 (IS أقربُ); V.i.2.2 (cf. IS قَرُبَ). - ܩܪܝܒܐܝܬ: **Min.** I.i.1.6. - ܩܪܝܒܘܬܐ: **Min.** IV.i.4.3 (cf. IS أقرب). - ܩܪܒܐ: **Min.** V.iii.2.7. - ܩܘܪܒܐ [* = قُرب]: **Min.** IV.i.4.9; ii.2.3; ii.3.7; ii.5.1, 2, 8, 9; iii.1.6 (FR ما قريب); iii.2.3; iii.3.2, 6, 7; **Mete.** I.iii.2.6 (AB*); II.ii.4.6 (cf. IS قريب); iv.1.2 (FR*); III.iii.2.2. - ܩܘܪܒܐ: **Mete.** II.i.2.5.

ܩܪܒܘܢܝܐ (καρβώνια): **Min.** I.i.2.12.

√qrḥ: ܩܪܚܐ: **Mete.** III.i.1.4; iv.tit.; iv.4.4.

ܩܝܪܘܢܐ (κηρίον): **Mete.** II.iii.1.8 (IS شَمعة).

ܩܪܝܬܐ ܩܘܪܐ **Mete.** V.iii.2.7.

√qrn: ܩܪܢܐ, ܩܪܢܬܐ: **Mete.** I.iii.4.10. - ܩܪ̈ܢܢܐ **Mete.** IV.iii.3.7 (cf. IS له قرون).

ܩܪܝܢܐ [* = صَنيع]: **Mete.** I.iii.tit.; iii.1.6, 7 (IS*); iii.6.6 (IS*).

√qšy: Ethpa: **Min.** I.i.2.5; i.3.5; II.i.1.6. - ܩܫܝܐ [* = صُلب]: **Min.** I.i.2.12; ii.2.2 (IS قَوِيَ); ii.5.1, 5 (IS*); II.ii.2.6; III.i.1.3 (AB*); ii.2.3; IV.ii.2.2; **Mete.** I.iii.1.6; iv.2.4 (N); iv.4.4 (IS شديد); IV.ii.3.2. - ܩܫܝܘܬܐ: **Min.** I.i.4.9. - ܡܩܫܝ̈ܢܐ: **Mete.** V.i.1.11 (IS مُجَمَّد).

ܩܬܛܝܣܐ (καθετίσαι?): **Min.** IV.iv.5.1. - ܩܬܛܝܣܝܘܬܐ (< καθετικός): **Min.** IV.i.2.1 (IS مُسَامَتَة).

√qšt: ܩܫܬܐ [* = قوس قُزَحَ]: **Mete.** I.iii.6.7 (IS*); II.i.1.1 (IS*); ii.1.1; ii.3.1 (IS*); ii.3.4, 5 (IS قوس); ii.4.8 (IS*); iii.tit.; iii.1.5bis; iii.1.7 (IS قوس); iii.3.1 (FR

قوس); iii.3.7, 9 (IS قوس); iii.4.1, 2; iii.5.1; iii.6.1, 2bis, 3, 4, 5; iii.8.1 (N/IS قوس); iii.8.2, 5; iii.8.9 (N/IS قوس); iv.3.2 (IS*). - ܡܩܛܪܢܐ: **Mete. II**.iii.1.1; iii.1.10 (IS قوسي); iii.2.2, 10.

√qtr: ܩܛܪܐ: **Mete. II**.i.5.5.

ܪ

√rbb: ܪܒ (adj.) [* = كبير]: **Min. I**.ii.1.1 (IS*); **II**.i.3.3 (cf. IS قوي); ii.2.3; ii.4.9; iii.1.8 (N); **IV**.i.2.2; i.2.8 (IS عظيم); iv.4.3, 4, 6bis (AB طويل); **V**.i.1.6; i.2.2 (DM); ii.1.4; iii.1.9, 10; iii.2.5; **Mete. I**.ii.1.8 (AB*); ii.3.6 (N/DM); iii.3.2 (IS/FR*); iii.4.7 (N/IS*); iii.4.9; **II**.i.4.9 (IS عظيم); i.5.1 (IS أعظم); i.7.1 (IS عظيم); i.9.1; ii.4.2 (IS أعظم); iv.1.4 (FR*); iv.3.4 (IS*); **III**.i.3.12; ii.2.2, 3; **IV**.iii.1.7; **V**.tit. (IS*); i.1.1; i.2.4 (IS عظيم). - ܪܒܐ (= "Aristoteles"): **Min. I**.i.4.1; i.5.1; **II**.i.4.1, 8; ii.1.1; ii.2.8; **IV**.ii.4.2; **V**.iii.1.8; iii.2.1; iv.1.4; **Mete. II**.i.3.6 (IS المعلم الأول); ii.2.4; iii.2.1; **III**.i.2.4; iii.1.5. - ܪܒܘܬܐ: **Min. IV**.iii.6.2; **IV**.iv.2.6. - ܪܒܘ, ܪܒܘܬܐ ("decem milia"): **Min. I**.ii.6.4, 5. - ܪܒܬ: **Min. IV**.iii.3.4; **Mete. I**.ii.3.4.

√rby: Pe: **Min. I**.ii.1.9 (AB عَظُمَ); **Mete. I**.ii.1.6 (AB كَبُرَ). - Ethpa: **Min. IV**.ii.6.6 (IS نَشَأ); **Mete. V**.ii.2.9.

√rbᶜ ("quattuor"): ܐܪܒܥܐ, ܐܪܒܥ [* = أربعة]: headings passim; **Min. III**.i.1.1 (IS*); i.5.6; **IV**.i.3.4; ii.1.2; ii.3.3, 9; **Mete. III**.i.2.1; ii.3.1; iii.3.9 (N); **V**.1.1.3 (IS*). - ܪܒܝܥܝܐ: headings passim; **Min. IV**.ii.6.1; iii.4.1; **Mete. III**.ii.1.6. - ܐܪܒܥܝܢ: **Min. IV**.i.2.4; iii.5.1; **Mete. II**.ii.4.5 (IS أربعين). - ܪܘܒܥܐ [* = رُبع]: **Min. IV**.i.3.5 (IS*); i.3.9 (IS*); i.4.1 (IS*); ii.6.4; **Mete. V**.i.1.4 (IS*).

√rbᶜ ("procubuit"): ܡܪܒܥܐ [* = رَبَض]: **Mete. V**.ii.2.4, 7 (IS*); ii.3.2, 6 (IS*).

√rgg: Ethpalpal: **Min. II**.i.2.6 (cf. N ܪܓܝܓܘܬܐ).

√rgy: Aph: **Mete. III**.i.4.6 (N).

√rgz: ܪܘܓܙܐ: **Min. II**.iii.4.6.

√rgl: Eshtaphal (ܐܫܬܪܓܠ): **Mete. III**.i.1.9; iv.2.1. - ܪܓܠܐ: **Min. I**.i.5.2 (N). - ܪܓܠܬܐ: **Min. I**.ii.2.6; ii.4.1, 6. - ܪܓܠܐ: **Min. IV**.i.1.6. - ܡܪܓܠܢܐܝܬ: **Min. II**.ii.4.6.

√rgš: Aph: **Min. II**.ii.3.7, 10 (N); **IV**.ii.6.3; ii.6.7 (IS أحسّ); ii.6.7; iii.4.5; **Mete. I**.iii.1.5 (IS أحسّ); **III**.i.3.5; **IV**.i.4.5 (N). - ܪܓܫܐ [* = حِسّ]: **Mete. II**.i.1.2 (IS*); i.9.3 (IS*); iv.3.4. - ܪܓܫܬܐ: **Min. III**.iii.2.7; iii.3.9; **Mete. II**.iii.1.4. - ܡܪܓܫܢܐ: **Min. IV**.iii.4.4. - ܡܪܓܫܢܘܬܐ: **Min. IV**.iii.4.2bis, 3. - ܡܬܪܓܫܢܐܝܬ: **Mete. I**.iii.1.4 (cf. IS أحسّ); **IV**.ii.2.2.

√rdy: Pe: **Min. I**.ii.4.5; ii.5.1; iii.1.1 (cf. IS سَيَال); iii.1.6; iii.2.4; iii.4.1 (cf. IS سَيَال); **II**.iii.2.2 (N/IS جَرَى); **V**.i.3.1, 2; ii.3.4bis (N); iii.3.4; iv.2.6bis; **Mete. III**.iv.1.4 (N); **IV**.iii.1.4 (IS سَرَى). - ܪܕܝܐ: **Min. I**.ii.4.1, 6bis; iii.2.5; iii.4.4; **V**.iii.3.8; iv.1.3, 9; **Mete. IV**.iii.1.10 (N). - ܪܕܘܝܐ: **Min. I**.ii.1.8 (AB جارٍ). - ܡܪܕܝܬܐ: **Min. I**.iii.1.4.

√rhb: ܡܫܪܗܒ (Saphel ptc. pass.): **Mete. IV**.iii.3.5 (cf. IS سريع). - ܪܗܝܒ: **Min. II**.iii.2.3. - ܪܗܝܒܘܬܐ: **Min. IV**.ii.3.4 (IS سُرعة); iii.1.4; **Mete. IV**.iii.1.4.

√rhṭ: Pe: **Min. II**.i.1.7 (N); **Mete. I**.iii.6.8.

ܪܘܕܘܣ (nom. prop. "Rhodus"): **Min. V**.i.1.6.

√rwḥ: ܪܘܚܐ [* = ريح]: **Min. I**.ii.1.4 (AB*); ii.2.2 (IS*); ii.2.4 (IS/FR*); **II**.ii.1.3; ii.1.5 (IS*); ii.1.7 (N/IS*); ii.1.9 (N); ii.2.4 (IS مادّة ريحية); ii.2.8 (N); ii.3.5 (IS*); ii.4.1, 4 (IS*); iii.1.1 (N/IS*); iii.1.3 (N/IS*); iii.1.8; iii.1.10 (N); iii.2.2 (N); iii.2.4; iii.2.9 (N); iii.4.1, 3 (N); **V**.iii.3.1 (IS*); iii.3.7; **Mete. I**.i.3.2, 3 (IS*); ii.2.8 (AB*); ii.3.1, 2, 3bis, 5bis (AB*); ii.4.1, 3, 4, 6 (AB*); ii.5.5, 6 (AB*); ii.6.2; ii.6.5 (IS*); iii.3.4 (AB*); iii.6.8 (IS*); **III**.tit. (IS*); i.tit. (FR*); i.1.3 (IS*); i.1.5, 6, 10 (FR*); i.2.2, 3 (N); i.3.2 (N/IS*); i.3.3, 4, 6 (N); i.3.9, 10, 11 (IS*); i.4.1, 2, 3 (IS*); i.4.4, 5, 6 (N); i.4.8, 10 (N/IS*); ii.tit.; ii.1.1 (N/IS*); ii.3.1 (IS*); iii.tit.; iii.1.1 (IS*); iii.1.3; iii.1.5, 6, 9 (N); iii.2.1 (IS*); iii.3.1 (N); iii.3.5, 7; iii.3.9 (N); iii.4.1 (IS*); iii.4.5, 7; iii.5.1 (IS*); iii.5.2, 3 (N); iii.5.5, 6, 7, 8; iii.5.10 (N); iii.6.3, 6 (N); iv.1.1, 2 (IS*); iv.1.3,

4 (N); iv.2.5 (N); iv.3.2 (IS*); iv.4.2; iv.4.4, 6 (IS*); **IV**.i.1.7 (IS*); i.4.2 (IS*); i.4.3; ii.1.2; ii.3.3, 6, 7, 9 (N); iii.4.10 (IS*); **V**.i.1.8, 10 (IS*); i.2.7. - ܪܘܚܐ ܥܢܢܝܬܐ [* = ريح سحابية]: **Mete. III**.iv.tit.; .iv.1.1, 2 (IS*); .iv.2.1 (IS*); .iv.2.4; .iv.2.6bis (N); **IV**.ii.1.1 (N/IS*). - ܪܘܚܢܐ, ܪܘܚܢܘܬܐ: **Min. III**.i.5.7. - ܪܘܝܚܐ: **Mete. III**.iv.3.1 (IS ريحي). - ܪܝܚܐ: **Min. III**.ii.1.7 (IS رائحة); **V**.ii.4.7. - ܪܘܚ: **Min. II**.ii.3.6 (IS واسع); **Mete. I**.ii.1.2, 5 (cf. AB سَعَة); **III**.iii.1.8.

√rwm: Pe: **Mete. II**.iii.3.9 (cf. IS ارتفاع). - Aph: **Min. I**.iii.1.7; **Mete. I**.iii.6.4. - ܪܡ: **Min. I**.iii.2.1; iii.3.3 (AB عـال). - ܪܘܡܐ: **Min. I**.ii.2.1 (IS ارتفاع); ii.2.6 (cf. IS مرتفع); ii.2.7; iii.4.7; **II**.ii.4.6; **Mete. I**.i.3.1; ii.1.7; iii.5.3 (IS عُلُوّ); **II**.i.5.8bis, 9; iii.3.7 (cf. IS/FR ارتفاع); **IV**.iii.3.2 (cf. IS عُلُوّ); iii.4.3, 8. - ܡܪܝܡܘܬܐ: **Min. I**.iii.4.6.

√rzp: ܪܙܦܐ: **Mete. I**.ii.3.8.

√rḥy: ܪܚܝܐ, ܪܚܝܘܬܐ: **Mete. II**.iii.1.6 (IS رَحَى propono).

√rḥm: ܪܚܡ: **Min. IV**.iii.1.1. - ܪܚܡܬܐ: **Min. IV**.iii.1.2.

√rḥq: Aph: **Min. IV**.ii.6.1; iii.1.5; **Mete. II**.i.2.6; i.6.2 (IS بَعُدَ). - ܪܚܝܩ [* = بعيد]: **Min. IV**.i.3.8 (N); ii.2.3; ii.6.4, 5; iv.1.1; **Mete. I**.ii.4.3 (AB*); ii.5.3 (AB*); ii.6.2 (AB*); iii.4.5 (N/IS*); **II**.i.5.2, 4, 8 (IS*); i.6.5 (IS*); i.7.3 (cf. IS بُعد); ii.4.3 (IS أبعَدَ); iii.4.2 (IS*); iii.5.5 (IS أبعَـدُ); iv.2.5 (cf. IS بُعـد); **IV**.i.4.10 (IS*). - ܪܚܝܩܘܬܐ: **Mete. I**.i.1.9; **V**.iii.2.8. - ܪܘܚܩܐ: **Min. IV**.ii.2.2; iii.5.4, 5 (FR بُعـد); **Mete. II**.i.5.5; **III**.iii.2.2; iii.5.8 (N). - ܪܘܚܩܐ: **Mete. II**.i.2.5.

√rṭb: Ethpe/Ethpa: **Min. II**.ii.1.1 (N). - ܪܛܝܒ [* = رَطب]: **Min. I**.ii.3.1 (IS*); **Mete. I**.iii.3.9 (IS*); **II**.ii.1.3 (FR*); iii.1.1 (IS*); iv.1.6 (FR*); iv.2.2 (IS*); **III**.i.1.2 (IS*); i.4.8 (N); iii.1.8 (N); iv.2.1 (IS*); **IV**.iii.4.5 (IS*). - ܪܛܝܒܘܬܐ [* = رطوبة]: **Min. I**.i.5.4 (N); **II**.i.2.6 (N); iii.2.6 (IS*); **III**.i.5.4, 5 (IS*); **V**.ii.4.4 (IS*); **Mete. II**.iii.1.9 (IS*); **III**.iii.4.3 (IS*); iv.3.7 (N); **IV**.iii.2.4 (IS*); iii.3.2 (IS*).

√ryš: ܪܝܫܐ: **Min. I**.i.2.9 (IS رأس); **II**.i.3.2 (N/IS قُلّة); i.4.9 (N); **IV**.i.4.4; ii.3.6; ii.5.2bis (IS نقطة); ii.5.3; ii.6.2; iii.5.3; iv.1.2; **Mete. I**.i.2.3 (IS أعلى); i.2.6; ii.2.3; iii.4.10; **III**.ii.1.4, 5bis, 6, 7bis (IS نقطة); **IV**.iii.1.9 (N). - ܠܥܠ ܡܢ ܪܝܫܐ [* = مُسامتة]: **Min. IV**.i.4.8; ii.3.4, 6; ii.5.1 (IS*); iii.1.4, 6 (FR*); iii.2.3; iii.3.2; iv.5.1 (cf. AB*); iv.5.4, 5 (AB مُسامت لرأس); iv.5.7; **Mete. II**.iii.3.7. - ܪܝܫܐ ܚܠܝܡܐ [* = مُسامتة]: **Min. IV**.ii.6.2 (IS*); iii.1.5 (FR/Ṭ*). - ܡܪܝܫܢܘܬܐ: **Mete. V**.ii.3.10. - ܪܝܫܢܝܐ: **Min. I**.i.4.11 (IS جَليل); **Mete. V**.ii.3.2. - ܪܝܫܝܬܐ: vide sub ܥܠܬܐ. - ܪܝܫܝܘܬܐ: **Mete. V**.ii.2.5 (cf. IS أصل). - ܡܢ ܪܝܫܝܬܐ: **Min. I**.iii.3.1 (AB أبداً).

√rkk: Ethpe/Ethpa: **Min. III**.i.2.2 (IS لانَ); i.2.3 (N?). - ܪܟܝܟܐ: **Min. I**.ii.2.4 (IS رخو/FR لَيِّن).

√rkb: Ethpa: **Mete. II**.i.6.3 (IS تَرَكَّبَ). - ܡܪܟܒ (Pa ptc. pass.): **Min. III**.i.1.5; i.3.2; i.4.1 (IS مُركَّب); **Mete. II**.iii.6.6. ܡܪܟܒܐ: **Min. I**.i.3.2; **Mete. I**.iii.6.3, 5; **V**.ii.2.5. - ܪܘܟܒܐ [* = تركيب]: **Min. III**.i.5.3 (IS/AB*); **Mete. V**.iii.2.9. ܪܟܒܐ: **Mete. I**.iii.4.10.

√rmy: Pe: **Min. I**.ii.1.8 (AB ألقى); **V**.ii.1.7. - Aph: **Min. V**.ii.1.10 (N). - Ettaph: **Mete. V**.ii.1.5 (FR أُلقيَ). - ܡܬܪܡܝܢܘܬܐ: **Mete. V**.ii.2.4; ii.3.6.

√rmz: ܪܡܙܐ: **Mete. V**.iii.2.4.

√rmn: ܪܘܡܢܐ: **Min. III**.iii.3.10 (AB رُمّان).

√rmš: ܪܡܫܐ: **Min. IV**.iv.5.9 (AB عَشية); **Mete. II**.iv.3.8bis (IS غروب); **III**.iii.2.8, 9, 10.

√rny: Pe: **Mete. II**.i.3.7 (N). - Ethpe: **Mete. III**.ii.2.2, 5. - Ethpa: **Min. IV**.iii.6.3.

√rss: ܪܣܝܣܐ: **Mete. I**.ii.3.8 (N); **II**.iii.1.6 (IS رَشّ). - ܪܣܝܣܘܬܐ: **Mete. II**.iii.1.1, 2 (cf. IS رَشّ); iii.2.8 (IS رشّي); iii.3.1; iii.4.2, 4.

√rsm: Ethpe: **Mete. II**.iii.1.7 (IS انطَبَعَ). - ܪܘܫܡܐ: **Mete. I**.ii.3.9.

√rᶜᶜ: ܪܥܝܥ: **Min. III**.iii.1.3; **V**.iv.2.3.

√rᶜy: Ethpa: **Min. IV**.ii.3.1; iii.6.1. - ܪܥܝܢܐ: **Min. IV**.i.2.3; **V**.iv.2.3; **Mete. IV**.i.5.5. - ܡܬܪܥܝܢܘܬܐ: **Min. II**.i.tit.; i.4.1; **IV**.ii.3.3. - ܡܬܪܥܝܢܝܬܐ: **Mete. V**.iii.2.3.

√rʿl: Aph: **Min. IV**.ii.6.9 (IS ارتَعَدَ); **Mete. III**.i.3.11 (cf. IS رَجفة). - ܪܥܠܢܐܝܬ: **Min. II**.ii.4.4 (N).
√rʿm: ܪܥܡܐ [* = رَعد]: **Min. II**.iii.3.9, 10 (N); **V**.iii.1.9; **Mete. IV**.i.tit.; i.1.5, 7 (FR*); i.3.1 (IS*); i.4.1, 3 (IS*); i.4.4, 5 (N); i.5.2 (N); i.5.4, 5 (N); ii.3.8 (N). - ܪ̈ܥܡܐ: **Mete. IV**.i.3.2. - ܪܥܡ (ptc.?): **Mete. IV**.i.1.9 (cf. IS رَعد).
√rʿt: ܪܥܬܐ: **Mete. I**.iii.2.5 (cf. DM ܪܥܬܬܘܬܐ).
√rpy: ܪܦܝܐ [* = رَخو]: **Min. I**.i.2.4 (IS*); ii.5.2; ii.6.4; iii.2.4 (IS*); **III**.ii.3.9.
√rṣn: Pe: **Min. V**.ii.1.10 (N). - ܪ̈ܨܝܢܐ: ܚܕ ܪܨܝܢܐ [* = نَزّ]: **Min. I**.iii.2.3 (IS*); iii.4.3 (IS*).
√rṣp: Ethpe: **Min. I**.ii.4.3 (IS تَكاثَفَ); **II**.iii.3.3 (N); **Mete. I**.i.1.7; i.2.3 (IS اجتمع); i.2.4 (IS انقبض); iii.4.2. - Aph: **Mete. I**.iii.1.2 (IS كَثَّفَ). - ܪܨܝܦܐ [* = كَثيف]: **Min. II**.i.1.6 (N); **IV**.iv.1.3 (AB*); **Mete. I**.ii.2.5; **II**.iii.8.5; iv.1.1 (FR*); **III**.iv.1.2 (N); **IV**.iii.4.2. - ܪܨܝܦܘܬܐ: **Mete. I**.ii.3.4; **IV**.ii.1.4 (IS استحصاف). - ܡܪܨܦܢܘܬܐ: **Min. II**.iii.1.9 (IS تخصيف).
√rqq: Pa: **Min. I**.i.3.7 (cf. IS مُرَقَّق). - ܪܩܝܩ [* = رَقيق]: **Min. II**.iii.1.3 (N); **Mete. II**.ii.1.3 (FR*); ii.2.2 (IS*); ii.4.7 (IS*); iv.3.9 (IS*); **IV**.i.1.7 (IS*).
√rqd: Aph: **Mete. III**.iv.3.3 (cf. IS راقص).
√ršm: Ethpe: **Mete. II**.i.3.5 (cf. IS ارتَسَمَ). - ܡܬܪܫܡܢܘܬ ܨܠܡܘܬܐ: **Mete. II**.i.3.3 (cf. IS ألقى الشبح).
√rtt: ܪܬܝܬܐܝܬ: **Min. II**.ii.4.5 (N).

ܫ

√šʾl: ܫܐܝܠܐ (ptc. pass.): **Min. III**.iii.2.6.
√šbb: ܫܒܒ: **Mete. I**.ii.5.7 (cf. AB بقُرب).
ܫܒܐ (nom. prop.): **Min. V**.i.3.5.
√šbḥ: ܡܫܒܚ: **Min. IV**.iii.1.1; iii.6.1.
√šbṭ: Pa: **Min. I**.i.4.9. - ܫܒܛܐ: **Mete. II**.iv.3.1 (DM); **IV**.iii.3.7.
√šbl: ܫܒܝܠܐ: **Min. I**.iii.2.1.
ܫܒܠܬܐ ("Virgo"): **Min. IV**.ii.5.4 (IS السنبلة).
√šbʿ: ܫܒܥܐ, ܫܒܥ: headings; **Min. III**.i.5.7; **IV**.i.3.5; ii.2.5; ii.4.3; iii.5.1, 4; **Mete. II**.ii.4.3 (IS سبعة). - ܫܒܝܥܝܐ: **Min. IV**.i.4.10; iii.5.1, 4; iv.4.2; **Mete. III**.ii.1.8. - ܫܒܥܝܢ: **Mete. IV**.iii.3.8.
√šbq: Pe: **Min. II**.iii.1.12 (N); **IV**.iv.2.4; **Mete. III**.iv.3.7. - ܫܒܝܩܐ: **Mete. II**.iv.3.1 (N).
√šgr: Ethpe: **Min. I**.ii.1.2 (IS غـــانص); **III**.ii.2.4.
√šdy: Pe: **Min. I**.i.5.7 (N); **V**.i.2.5; ii.2.1, 5; ii.3.9 (N); iv.3.2.
√šhy: ܫܚܝܐ: **Min. V**.i.3.8 (Biruni خَرب). - ܫܚܝܢܘܬܐ: **Min. IV**.iii.2.2.
√šwy: Pe: **Min. IV**.i.4.5; IV.iii.6.1; **Mete. II**.iv.2.4. - ܫܘܐ (ptc. pass.): **Min. I**.iii.3.5; **II**.i.4.8 (N); **IV**.ii.4.2 (IS مُساو); iii.5.5, 7; iv.3.5; **V**.iv.2.6; **Mete. II**.iii.7.9. - ܫܘܝܐܝܬ: **Mete. II**.i.7.2. - ܫܘܝܘܬܐ: **Min. IV**.ii.5.5; iii.5.6; **Mete. III**.iii.1.8 (N); iii.2.1. - ܫܘܐ ܒܫܘܝܘܬܐ ܕܫܘܝܘܬܐ [* = خط الاســـتـــوا]: **Min. IV**.i.3.10; i.4.9 (IS*); ii.1.6; ii.3.2 (IS*); ii.3.9; ii.4.1; ii.4.5, 8; ii.6.1; iii.1.3 (Ṭ*); iii.2.1 (FR*); iii.3.3, 7; iii.4.2; iii.5.3, 5, 7 (FR*). - ܫܘܝܢܘܬܐ: **Mete. III**.ii.1.4, 6 (cf. IS اعتدال).
√šwḥ: Pe: **Mete. IV**.iii.1.9.
√šwʿ: ܫܘܥܐ: **Min. I**.i.5.7. - ܫܘ̈ܥܐ: **Min. I**.iii.2.6 (AB صخرى). - ܫܥܬܐ: **Min. IV**.iv.4.3, 4.
√šwp: Ethpe: **Mete. I**.iii.3.5.
√šwr: Pe: **Min. I**.i.4.5bis (IS نَبا); **Mete. III**.i.2.6 (N); **IV**.iii.1.5. - ܫܘ̈ܪܐ: **Mete. IV**.iii.tit.
√šḥl: ܫܚܠܐ: **Min. I**.i.5.3 (N); **V**.ii.1.12 (N/IS رشح).
√šḥn: Ethpa: **Min. II**.ii.1.10 (N). - ܫܚܝܢ: **Mete. I**.iii.4.4 (IS حارّ). - ܡܫܚܢܢܘܬܐ: **Min. IV**.iii.3.4, 5 (FR تسخُن).
√šḥr: ܫܚܘܪܐ [* = زاج]: **Min. III**.i.1.3; i.4.1 (IS*); i.4.3 (IS*); i.5.4 (IS*).
√šṭḥ: ܫܛܝܚ: **Min. II**.i.4.5 (N); **Mete. II**.i.5.3 (IS مُسَطَّح). - ܫܛܝܚܘܬܐ [* = سَطح]: **Min. I**.ii.1.6; **IV**.i.1.7; ii.1.4, 5 (IS*); iv.1.3 (AB*); **Mete. II**.i.5.4; i.7.3; iii.3.5 (FR*); iv.3.6; **V**.i.3.3.
√šṭp: ܫܛܝܦܘܬܐ: **Mete. IV**.i.3.9 (IS تَنّينة).
√škḥ: ܐܫܟܚ ("invenit") [* = وَجَدَ]: **Min. I**.i.2.9; ii.1.8; iii.2.8 (AB*); **II**.ii.1.3; ii.3.6 (IS*); ii.4.4; **III**.iii.3.3, 4 (AB*); **V**.ii.1.8; **Mete. III**.iii.4.3. - ܐܫܟܚ ("potuit"): **Min. I**.i.4.7; ii.6.6; iii.1.6; iii.4.8 (cf. IS قَدَرَ); **II**.i.1.8 (N); iii.1.10 (N); **III**.iii.1.7; **V**.iii.3.8: **Mete. III**.i.2.5

(N); iii.4.4. - Ethpe [* = وُجِـدَ]: **Min.** **I**.i.3.6; I.iii.2.6, 7 (AB*); iii.3.3, 4, 5 (cf. AB وَجَدَ); **III**.iii.3.5 (AB*); **IV**.i.1.3 (cf. IS وجود); iii.6.8 (cf. AB وَجَدَ); iv.3.9 (cf. AB وَجَدَ); **V**.ii.1.9; ii.1.11 (N); iii.2.3, 10 (IS*); iv.1.6; **Mete.** **I**.ii.4.4; **V**.i.1.12; ii.2.3. - ܡܫܟܚ (ptc.): **Min.** **IV**.ii.5.9; iii.6.8, 9 (cf. AB وَجَـدَ). - ܡܫܟܚܢܐ ("inventor"): **Mete.** **V**.iii.1.6. - ܡܫܟܚܢܐ ("possibilis"): **Mete.** **V**.i.2.3 (cf. IS إمكان). - ܠܐ ܡܫܟܚܢܘܬܐ: **Mete.** **V**.ii.1.1 (IS امتناع).

√škr: ܫܟܝܪ: **Min.** **V**.iv.1.9.

√škt: Pe: **Mete.** **IV**.iii.1.2 (N).

√šly: ܫܠܐ (ptc. pass.): **Min.** **V**.iii.3.1 (IS سَاكِن). - ܫܠܝܐ: ܡܫܠܝܐ: **Min.** **I**.ii.2.2 (IS دفعة). - ܡܫܠܝ [* = دفعةً]: **Min.** **I**.i.3.5 (IS*); i.4.3; ii.1 (IS*); **II**.iii.3.2 (IS*); iii.3.3 (IS بغتةً); iii.3.4 (IS*); iii.4.1; **Mete.** **I**.i.2.4 (IS*); iii.2.8; **III**.iv.1.4 (N); **V**.i.1.8 (IS*); i.2.5 (IS دفعةً واحدةً); i.2.6 (IS*).

√šlb: ܫܠܒܐ: **Min.** **IV**.i.1.5.

√šlṭ: Pe: **Mete.** **V**.i.3.4. - ܫܠܝܛܐ: **Min.** **IV**.i.1.8; **Mete.** **IV**.iii.3.2 (IS سلطان).

ܫܠܘܦܚܬܐ: vide √nph.

√šlm: Pe: **Min**. fin.; **Mete**. fin. - Pa: **Min.** **V**.i.3.11; **Mete.** **V**.iii.2.7. - ܫܠܝܡ (adj.): **Mete.** **II**.ii.3.5 (cf. IS تَمَّ). - ܫܘܠܡܐ: **Mete.** **IV**.ii.3.7 (N); **V**.i.3.10. - ܡܫܠܡܢܘܬܐ: **Mete.** **II**.i.9.4.

√šm (√šmh): Ethpa (ܐܫܬܡܗ): **Min.** **III**.i.5.9; **V**.i.1.1 (DM); iii.2.12; **Mete.** **I**.ii.3.8 (N); **III**.iv.4.4; **IV**.ii.2.4. - ܫܡܐ, ܫܡܗܐ: **Min.** **I**.i.1.8, 9; **Mete.** **III**.ii.3.2 (IS اسم); **V**.iii.2.5, 9.

√šmy: ܫܡܝܐ [* = سماء]: **Min.** **IV**.iv.2.5; **V**.iii.1.5, 6 (IS*); **Mete.** **III**.i.2.7 ("tectum"); **IV**.iii.1.5; iii.4.7 (IS*); iii.4.10. - ܫܡܝܢܝܐ [* = سمائي]: **Min.** **IV**.iv.1.4 (AB*); **Mete.** **I**.iii.6.1 (IS*); **V**.i.3.1.

√šmᶜ: Pe: **Min.** **V**.iii.2.12; **Mete.** **I**.iii.4.8; **IV**.i.10 (N). - Ethpe [* = سُـمِعَ]: **Min.** **I**.i.4.7 (cf. IS سَـمِعَ); **Mete.** **IV**.i.1.8, 9 (IS*); i.3.2 (IS*). - Aph: **Mete.** **IV**.i.1.7. - ܫܡܥܐ: **Mete.** **IV**.i.4.7 (IS سَمع); i.4.10. - ܡܫܡܥܢܘܬܐ: **Min.** **II**.ii.3.8, 9 (N/IS سَمع).

√šmr: Pa: **Mete.** **II**.i.7.4 (IS أرسَلَ). - Ethpa: **Min.** **V**.iii.3.2 (IS انبَعَثَ); **Mete.** **III**.iv.1.5 (N). - ܫܡܪܐ: **Mete.** **III**.i.4.7 (N).

√šmš: Ethpa: **Mete.** **V**.ii.2.1. - ܫܡܫܐ [* = شمس]: **Min.** **I**.i.2.3 (IS*); **II**.i.1.5; ii.1.1 (N); iii.1.9 (N); iii.3.1; **IV**.i.4.3, 5 (IS*); i.4.8; ii.2.2, 3 (N); ii.3.3, 6; ii.5.1, 3, 8; ii.6.1 (IS*); iii.1.4 (Ṭ*); iii.2.3; iii.3.2, 6; iii.5.2; iii.5.3, 4, 6 (FR*); iv.1.1, 3 (AB*); iv.2.4, 6; iv.3.2 (AB*); iv.5.1, 3, 4, 5, 7 (AB*); **V**.ii.1.5 (IS*); ii.2.6; iv.2.5; **Mete.** **I**.i.1.3; **II**.ii.2.5; ii.3.4 (IS*); ii.4.6, 8 (IS*); iii.1.7, 8 (IS*); iii.2.7; iii.3.2; iii.3.3, 4, 7 (FR*); iii.3.9 (IS*); iii.4.2; iii.5.2 3, 5, 6 (IS*); iv.1.1 (IS*); iv.1.3, 4, 7 (FR*); iv.2.2; iv.2.3, 6, 7bis (IS*); iv.3.2; iv.3.9 (IS*); **III**.iii.2.2; iii.2.4 (IS*); iii.2.6 (N); iii.4.1, 2 (IS*); iii.5.1 (IS*); iii.5.2 (N); iii.5.5, 6; **IV**.i.6.2, 4, 6 (N); **V**.i.3.2. - ܫܡܫܢܝܐ: **Mete.** **I**.iii.6.1; **IV**.i.3.6 (IS شمسي); i.5.1 (cf. IS شمس).

√šn: ܫܢܬܐ [* = سنة]: **Min.** **I**.i.2.9 (IS*); ii.1.3; I.ii.6.5; **IV**.i.2.4, 5, 6; ii.4.1, 3, 6; iii.6.8, 9 (AB*); **V**.iii.2.6; **Mete.** **III**.i.4.2bis (IS*); iii.3.9 (N); **IV**.iii.3.8 (IS*); **V**.i.2.4 (IS*); i.3.11 (IS*). - ܫܢܬܢܝܐ: **Min.** **IV**.iii.6.9 (AB سنوي); **Mete.** **III**.iii.5.1 (IS حَولي); iii.5.4bis, 11 (N).

√šnn: ܫܢܝܢܬܐ: **Min.** **IV**.i.1.7.

ܫܢܬܐ ("somnus"): **Mete.** **II**.iii.1.9 (IS نوم).

√šny (šgny): Pa [* = انتَقَلَ]: **Min.** **IV**.i.2.4; ii.3.4, 8; ii.6.2 (IS*); iii.1.4; **V**.ii.3.4 (N); iii.1.2, 3 (IS*); iii.2.8 (IS*); **Mete.** **I**.ii.2.8 (AB انجَرَّ); **II**.iv.2.7. - Ethpahli (ܐܫܬܢܝ): **Min.** **I**.i.3.8; **III**.iii.2.3; **V**.ii.1.1 (IS تَغَيَّرَ). - ܫܘܢܝܐ: **Min.** **V**.ii.4.3 (N); iii.tit.; **Mete.** **V**.i.3.1. (cf. IS انتَقَلَ); i.3.10 (IS انتقال). - ܫܘܚܠܦܐ: **Min.** **I**.i.3.2, 3 (IS استحالة). - ܡܫܬܢܝܢܐ: **Mete.** **V**.i.3.9 (IS مُنتَقِل).

√šnz: ܫܢܙܐ: **Min.** **IV**.i.4.5 (IS خروج).

√šᶜᶜ: ܫܥܥ: **Mete.** **II**.i.5.6; **IV**.i.2.7.

√šᶜy: ܫܥܝܘܬܐ: **Min.** **III**.iii.3.1.

ܫܥܬܐ: **Min.** **V**.ii.1.10, 11 (N).

√špp: Aph: **Min.** **IV**.ii.6.5.

√špy: Ethpa: **Min.** **V**.ii.1.4 (cf. IS صَفَّى). - ܫܦܝܐ: **Min.** **III**.ii.1.5 (cf. IS صفاء); **V**.ii.2.3.

√špl: ܡܫܦܠ (Pa ptc. pass.): **Min. I**.iii.3.4. - ܫܦܠܐ: **Min. I**.ii.5.1.
√špʿ: Pe: **Min. I**.iii.3.6; **V**.iii.1.4, 7. - ܫܦܝܥܘܬܐ: **Mete. V**.i.2.5.
√špr: ܫܦܪܐ: **Min. II**.iii.1.9 (N/IS غداة). - ܫܦܪܐ: **Min. IV**.iii.2.4. - ܫܦܝܪܘܬܐ: **Min. III**.ii.1.4 (cf. AB جُودة).
√šrr: Pe: **Mete. III**.iii.2.8. - Pa: **Min. I**.i.1.3; **IV**.ii.3.3; **Mete. II**.iii.1.4. - Ethpa: **Min. IV**.iii.4.4. - ܫܪܝܪ: **Min. II**.i.4.1; **III**.iii.2.6; **Mete. IV**.i.5.5. - ܫܪܝܪܐ: **Min. III**.iii.1.4; iii.3.4; **IV**.iii.1.2; iii.5.8; **Mete. I**.ii.3.3; **IV**.iii.2.8. - ܫܘܪܪܐ: **Min. I**.i.5.1; **III**.ii.1.5. - ܫܪܝܪܐܝܬ: **Mete. II**.iv.1.8. - ܫܪܐ ("umbilicus"): **Mete. III**.iii.1.10 (N).
√šry: Pe: **Min. II**.i.4.6; **III**.i.5.4 (IS حلّ); **IV**.iii.1.2, 3 (T ردّ); iii.5.3; iv.4.2; **Mete. III**.i.4.4 (N). - Ethpe: **Min. II**.i.3.2 (N); **III**.i.5.5 (IS انحلّ); **IV**.iii.2.1. - Pa: **Min. IV**.iii.6.10 (AB بدأ); **V**.i.3.11; **Mete. I**.iii.5.4 (cf. IS مُبتدئ); **III**.i.3.6, 8 (N); **V**.i.3.7; iii.2.6. - ܫܪܐ (ptc. pass.): **Min. III**.ii.1.8 (AB ذائب). - ܫܪܝܐ: **Min. III**.i.4.2 (IS حلالة); **IV**.iii.tit.; iii.3.1; **Mete. II**.iv.2.2. - ܫܪܝܘܬܐ: **Min. I**.i.5.2, 3 (N). - ܫܪܝܬܗܘܢ: **Mete. V**.ii.2.11. - ܫܘܪܝܐ: **Min. IV**.i.2.6; iii.5.1; **Mete. III**.i.3.3 (N); iii.4.2; **IV**.ii.3.6 (N); **V**.i.3.10; iii.1.2 (IS مَبدأ). - ܡܫܪܝܢܘܬܐ: **Mete. V**.iii.1.3, 4, 5 (cf. IS حادث). - ܫܪܝܛ: **Min. IV**.ii.4.5. - ܫܪܝܛܝܢ: **Mete. I**.iii.4.3. - ܫܪܝܛܘܬܐ [* = خريف]: **Min. II**.iii.2.8 (N); **IV**.iv.3.6; **Mete. I**.iii.3.6 (IS*); iii.4.1 (IS*); **III**.iii.3.8 (N/IS*); iii.6.6 (N).
√šrb: ܫܪܒܐ [* = أمر]: **Min. I**.i.1.1 (IS*); i.1.2; i.1.5; **II**.i.4.8; **IV**.tit. (IS حال); i.4.3 (IS*); **V**.tit. (IS حال); **Mete. I**.i.1.1; II.tit.; **V**.tit.; i.1.1; i.2.1 (IS شيء).
√šrg: ܫܪܓܐ: **Mete. II**.iii.1.8 (IS سراج); **III**.iii.3.2 (N).
√šrk: Pe: **Min. I**.iii.2.7; **III**.i.2.6; **V**.ii.1.8. - ܫܪܟ: **Min. V**.ii.1.2; iv.3.10. - ܫܪܟܐ: **Min. II**.iii.2.9 (N); **IV**.i.1.7; **Mete. I**.i.1.8; **II**.iv.1.6; **III**.iv.1.3; **IV**.tit. (N); iii.1.1; **V**.i.1.2.
√šrq: ܫܪܩܘܩܝܬܐ: **Min. II**.ii.3.2 (N).
√štt: ܫܬܐ, ܫܬ: headings passim. - ܫܬܝܬܐ: **Mete. III**.ii.1.7. - ܫܬܐܣܬܐ: **Min. IV**.i.2.3. - ܫܬܐܣܬ: **Min. IV**.iv.4.3.
ܫܬܐܣܘܬܐ: **Min. I**.ii.5.4; **Mete. V**.iii.1.8.
√šty: ܡܫܬܝܬܐ: **Min. V**.ii.1.6.

ܬ

ܬܐܘܢ (nom. prop. "Theo"): **Min. IV**.i.2.5.
ܬܐܘܪܝܐ (θεωρία): headings passim.
√tʾm: ܬܐܡܐ ("Gemini"): **Min. IV**.ii.5.3 (IS الجوزاء). - ܬܐܡܘܬܐ: **Mete. V**.ii.2.10.
ܬܐܢܬܐ, ܬܐܢܐ: **Mete. V**.ii.1.6 (FR تين).
√tbn: ܬܒܢܐ: **Mete. III**.iii.3.5 (N).
√tbr: Ethpe: **Min. V**.iii.1.11; **Mete. I**.iii.2.6; iii.3.5; iii.4.6 (cf. N Pa); **II**.iii.2.2, 4 (N).
√tgr: Ethpa: **Min. III**.i.4.5 (IS استفاد).
ܬܪܬܐ [* = رَبيع]: **Min. I**.iii.3.6; **II**.iii.2.8 (N); **IV**.ii.4.5; iii.6.9 (AB*); iv.3.6; **Mete. I**.iii.3.6 (IS*); iii.4.1 (IS*); **III**.iii.3.7 (N/IS*); iii.6.7 (N). - ܬܪܬܝܢܐ: **Min. IV**.ii.4.6.
√thy: Pa: **Min. II**.iii.2.5.
√twb: ܬܝܒܘܬܐ: **Min. V**.iii.3.9.
ܬܘܠܐ (nom. prop. "Thule"): **Min. V**.i.3.9.
√twr: ܬܘܪܐ: **Min. II**.ii.3.2 (N); **Mete. I**.iii.4.10; **V**.ii.1.4 (FR بَتَر). - ܬܘܪܐ (nomen constellationis): **Min. IV**.ii.5.3 (IS الثور).
√thb: Pe: **Min. I**.iii.2.4; **IV**.iv.4.9.
√thm: Pa: **Mete. II**.i.9.3. - Ethpa: **Mete. II**.i.9.2 (cf. IS تَحَدَّدَ); **III**.ii.1.2 (IS تَحَدَّدَ); ii.1.9, 11; ii.2.1, 8, 12. - ܡܬܚܡ (Pa ptc. pass.): **Min. I**.iii.4.7 (IS محدود). - ܬܚܘܡܐ [* = حدّ]: **Min. I**.iii.4.7 (IS*); **Mete. III**.ii.1.2 (IS*); ii.1.8, 9; ii.2.1, 8, 12. - ܬܚܘܡܝܐ: **Mete. III**.ii.tit. (cf. IS محدود).
√tht: ܡܬܚܬ (Pali ptc. pass.): **Mete. I**.i.1.6; ii.5.3 (AB أخفض). - ܬܚܬܝܐ: Min. II.i.1.4 (IS سافل); IV.i.3.2, 3; IV.i.4.1; Mete. I.i.1.9; II.ii.4.2 (IS تحتاني); II.iii.5.5 (IS أسفل). - ܠܬܚܬ [* = تحتَ]: **Min. II**.i.1.7 (N); iii.1.6 (N); **V**.iv.1.9; **Mete. I**.ii.2.3 (AB*); iii.5.4 (IS أسفل); **II**.iii.3.6; **III**.i.1.6, 9; i.2.3 (IS أسفل); i.2.8; i.3.2; ii.2.4, 10, 11; iv.2.1 (N/IS v.l. إلى أسفل); iv.3.1 (IS إلى أسفل); **IV**.i.1.6 (cf. FR سُفل). - ܠܬܚܬ ܡܢ: **Mete. III**.iv.2.4.

√ttr: ܬܪܬܪܐ: **Min.** **V**.ii.2.3.
√tkb: ܬܟܒ: **Mete.** **V**.i.1.7. - ܬܟܒܐܝܬ: **Mete.** **I**.iii.2.8 (DM). - ܬܟܒܘܬܐ: **Min.** **III**.iii.2.5.
√tkl: ܬܟܠ: **Mete.** **II**.iii.7.11 (cf. IS قَنَعَ).
√tll: ܬܠܠ: **Mete.** **I**.iii.6.2 (cf. IS بَلَّة). - ܬܠܠܘܬܐ: **Mete.** **I**.ii.5.7 (AB نَدَى).
√tly: Pe: **Min.** **I**.ii.2.2 (IS رَفَعَ).
√tlg: ܬܠܓܐ [* = ثَلْجٌ]: **Min.** **I**.iii.3.6 (AB*); **IV**.iv.4.7 (AB*); **Mete.** **I**.i.2.5 (cf. IS ثَلَجَ); ii.2.6; iii.tit.; iii.1.7 (N/IS*); iii.2.1 (IS*); iii.2.3 (DM); iii.2.5, 6, 7 (AB*); iii.2.8 (DM); iii.3.8 (IS*); iii.6.5 (IS*); **III**.iii.5.2 (N). - ܬܠܓܢܝܐ: **Mete.** **I**.ii.5.7 (AB حامل للثلوج). - ܡܬܬܠܓ (Pa ptc. pass.): **Mete.** **III**.iii.1.1 (cf. IS ثَلَجَ).
√tlḥ: Pe: **Min.** **II**.ii.1.8; **Mete.** **IV**.ii.3.9 (N). - Ethpe/Ethpa: **Min.** **V**.iii.1.10; **Mete.** **III**.iv.4.1.
√tlt: ܬܠܬܐ [* = ثلاثة]: headings passim; **Min.** **I**.tit.; i.2.9; **IV**.ii.1.4 (IS*); **V**.i.2.2; **Mete.** **II**.i.1.5 (IS*); **II**.i.3.1; **III**.ii.1.2bis, 3bis (IS*). - ܬܠܝܬܝܐ: headings passim; **Min.** **II**.i.4.7; **IV**.ii.5.1; iii.3.1; **Mete.** **II**.iii.5.1 (cf. IS ثلاثة); **III**.ii.1.5. - ܬܠܬܡܐܐ: **Min.** **IV**.iv.4.4.
√tmh: ܬܡܗ: **Min.** **I**.i.5.9 (N); **III**.iii.1.6.
√tmn: ܬܡܢܝܐ, ܬܡܢܝܬܐ: headings; **Min.** **IV**.i.2.4bis; **Mete.** **III**.ii.1.3; ii.3.2. - ܬܡܢܝܢܐ: **Mete.** **III**.ii.1.9. - ܬܡܢܝܢ: **Min.** **IV**.i.2.5.
ܬܡܢ: **Min.** **IV**.i.4.4; ii.2.7; ii.3.4; ii.4.2; ii.6.6 (IS هنالك); iii.1.7; **Mete.** **I**.i.1.7; **I**.ii.6.3; **II**.ii.2.2: **III**.i.1.5, 7; iii.1.7, 9 (N); iii.5.10; **IV**.iii.1.3.
√tmr: ܬܡܪܬܐ, ܬܡܪܐ: **Min.** **I**.ii.1.8 (AB تمر).
√tnn: Pe: **Min.** **II**.iii.4.7. - Pa: **Mete.** **III**.iii.3.4, 5. - ܬܢܢܐ [* = دخان]: **Min.** **II**.ii.1.3 (N); ii.1.9; ii.2.5 (IS*); **III**.i.3.5 (IS*); **Mete.** **I**.iii.6.5, 7 (IS*); **III**.i.1.5 (FR*); i.4.8 (N); iii.1.7 (N ܥܛܪܐ ܕܒܣܡܐ); iii.4.3 (IS*); **IV**.i.1.1, 3, 4, 6 (FR*); i.2.1 (IS/FR*); i.2.8 (cf. IS دخانى); i.3.1; ii.3.5 (N); iii.1.2, 6 (IS*); iii.4.5 (IS*). - ܬܢܢܢܝܐ [* = دخانى]: **Min.** **II**.ii.1.2 (FR*); ii.1.8 (cf. IS دخان); iii.1.2, 5 (N); iii.2.5 (N/IS*); iii.3.3 (N); **V**.ii.1.3; ii.1.12 (N); ii.4.6; **Mete.** **III**.i.1.3 (N/IS*); i.2.1, 4 (N); i.3.5 (N); i.4.4 (N); i.4.10 (N/IS*); iv.1.5 (N); **IV**.tit.; iii.1.3 (IS*); iii.2.2 (IS*); iii.4.2 (IS*). - ܬܢܝܢܐ: **Mete.** **III**.iv.4.5 (IS تنين).
√tny: Pa: **Min.** **I**.i.2.8; i.4.3; **II**.ii.2.8. - ܬܪܝܢ, ܬܪܬܝܢ [* = dual]: headings passim; **Min.** **I**.i.2.3; i.4.5; **II**.iii.4.4; **IV**.i.3.1, 3; i.3.8 (N); i.4.1; ii.1.2bis (IS*); ii.1.4; ii.1.5 (IS*); ii.2.1, 3bis; ii.5.5; iii.2.1; iii.5.5; **V**.i.2.1 (DM); i.2.1; iii.2.10 (IS*); iii.3.4, 6, 9; **Mete.** **I**.i.2.2 (IS*); iii.3.3 (AB*); **II**.i.2.1; i.3.5; iii.3.5 (FR*); iii.5.2 (IS*); iii.5.7; iii.6.1, 4, 5; **III**.i.1.1 (N); ii.2.5 (IS*); ii.2.7, 8bis, 9, 11, 12bis; iii.4.6 (IS*); iii.5.3 (N); iv.4.4, 6 (IS*); **IV**.i.1.2 (FR*); ii.1.2; **V**.i.2.2. - ܬܪܝܢܐ [* = ثاني]: headings passim; **Min.** **II**.i.4.6; **IV**.ii.3.5, 9; ii.4.1; ii.5.7bis; iii.2.4; **Mete.** **II**.i.1.3, 4 (IS*); **III**.ii.1.5; **IV**.i.4.9; iii.3.3 (IS*). - ܬܪܝܢܘܬܐ: **Mete.** **II**.i.4.4. - ܬܪܥܣܪ, ܬܪܬܥܣܪܐ [* = اثنا عشر]: **Mete.** **III**.ii.1.1, 2 (IS*); ii.2.1; iii.4.1 (IS*).
√tnr: ܬܢܘܪܐ: **Min.** **I**.i.3.9 (IS تنور).
√tql: Pe [* = وَزَنَ]: **Min.** **I**.i.4.4 (IS*); **Mete.** **I**.iii.4.8 (IS*).
√tqn: Ethpa: **Min.** **I**.i.4.12.
√tqp: Pe: **Min.** **IV**.iv.4.6 (cf. AB شدّة); iv.4.10; **Mete.** **I**.iii.2.7 (AB اشتدّ). - ܬܩܝܦ [* = قَوِيَ]: **Min.** **II**.ii.2.7 (IS*); ii.4.8; **Mete.** **IV**.i.2.4 (IS/FR*); iii.4.5. - ܬܘܩܦܐ: **Mete.** **I**.i.3.5 (IS شدّة).
ܬܪܐܣܩܝܐܣ (θρασκίας): **Mete.** **III**.ii.3.7 (N).
√trb: ܬܪܒܐ: **Min.** **III**.iii.3.8.
ܬܪܢܓܠܐ: **Mete.** **III**.iii.4.8 (IS دجاج). - ܬܪܢܓܠܬܐ: **Mete.** **III**.iii.4.9 (N).
√trᶜ: Pe: **Min.** **I**.iii.2.1. - Ethpe: **Min.** **IV**.i.1.5.
√trṣ: ܬܪܝܨ [* = مستقيم]: **Min.** **II**.iii.1.7; **IV**.ii.1.4, 5 (IS*); **Mete.** **II**.i.3.7 8 (N); ii.1.5 (cf. IS/FR استقامة); iv.3.3, 5, 6 (IS*). - ܬܪܝܨܐܝܬ: **Mete.** **II**.i.7.3; **II**.ii.1.6. - ܬܪܝܨܘܬܐ: ܒܬܪܝܨܐ: **Min.** **II**.iii.1.6 (N ܬܪܝܨܐܝܬ); **III**.i.2.3, 8; **Mete.** **IV**.i.5.6 (N ܬܪܝܨܐܝܬ).
√tšᶜ: ܬܫܥܐ: **Mete.** **II**.i.tit.

Reverse Index (Arabic-Syriac)

سَبَبٌ: ܥܠܬܐ. - (سبخ) سَبَخَة: ܡܠܚܘܬܐ. - سبُع: ܚܝܘܬܐ؛ سَبعَة: ܫܒܥܐ. - (سبق) أسبَقُ: [ܩܕܡ]؛ سَابِق: [ܐܬܩܕܡ]. - (سبل) سَبِيل: ܐܘܪܚܐ. - سَتَرَ: ܚܦܐ. - (سحب) سَحَابٌ: ܥܢܢܐ؛ ريح سحابية: ܪܘܚܐ ܥܢܢܝܬܐ. - (سخف) سَخِيف: ܡܚܝܠ. - سَخُنَ: ܫܚܢ؛ أسخَنُ: ܫܚܝܢ؛ تسخين: [ܫܘܚܢܐ]؛ تسخُن: ܡܫܚܢܢܘܬܐ، [ܫܚܝܢܘܬܐ]. - (سرج) سراج: ܫܪܓܐ. - (سرط) سَرَطان: ܣܪܛܢܐ. - (سرع) سَرِيع: [ܡܣܪܗܒ]؛ سُرعَة: ܪܗܝܒܘܬܐ. - سَرَى: ܪܕܐ. - سَطحٌ: ܫܛܝܚܘܬܐ؛ مُسَطَّح: ܫܛܝܚ. - (سفد) سفَاد: ܙܘܘܓܐ. - (سفل) سَافِل: ܬܚܬܝܐ. - سَقَطَ: ܢܦܠ؛ سقوط: ܡܦܘܠܬܐ. - (سقي) استَقَى: ܕܠܐ. - (سكن) سَاكن: ܫܪܝܐ؛ مسكونة: ܥܡܪܬܐ. - (سلط) سُلطان: ܫܘܠܛܢ، ܫܘܠܛܢܐ. - سَلَفَ: [ܩܕܡ]. - (سلك) مسلوك: ܡܬܗܠܟܢܐ. - (سلم) لا سلِمَ: [ܣܘܠܘܓܝܣܡܘܣ]. - (سم) اسمٌ: ܫܡܐ. - (سمّ) مَسَامٌ: ܦܘܪܘܣ. - (سمت) مُسَامَتَة: ܡܬܟܘܢܢܘܬܐ، ܠܥܠ ܡܢ ܪܝܫܐ، ܥܠܘܗܝ ܪܝܫܐ. - سُمِعَ: ܐܫܬܡܥ؛ سَمعٌ: ܫܡܥܐ، ܡܫܡܥܢܘܬܐ. - (سمو) سَمَاء: ܫܡܝܐ؛ سماني: ܫܡܝܢܐ. - سَمَّى: ܐܬܩܪܝ، [ܩܪܐ]. - سَنَة: ܫܢܬܐ؛ سنوي: ܫܢܬܢܝܐ. - (سنبل) السُنبُلَة: ܫܒܠܬܐ. - سَهُلَ: ... سُهُولة: [ܦܫܝܩܘܬܐ] - سَهمٌ: ܓܐܪܐ. - سَوِدَ: ܐܘܟܡ؛ أسود: ܐܘܟܡܐ؛ سَوَاد: ܐܘܟܡܘܬܐ. - (سوف) مَسَافَة: ܛܘܪܚܐ. - (سوق) سَاقَ: ܫܩܐ. - (سوى) مُساوٍ: ܫܘܐ؛ استواء: ܫܘܝܘܬܐ؛ لا سيّما: ܝܬܝܪܐܝܬ. - سَيفٌ: ܣܝܦܐ. - (سيل) سَيَّال: [ܪܕܝܐ]؛ سَيَلان: ܐܫܕܘܬܐ.

شَأنٌ: [ܨܒܘ، ܡܨܒܝܢܘܬܐ]. - شَبٌّ: ܫܒܩܐ. - (شبع) تَشَبَّعَ: ܐܬܣܒܥ؛ شَبَعٌ: ܣܒܥܐ، ܣܒܥܢܐ، [ܡܣܒܥܢܘܬܐ]. - شَبَهٌ: [ܕܡܘܬܐ]؛ تشبيه: ܡܕܡܝܢܘܬܐ؛ مُشابه: [ܐܬܕܡܝ]؛ مُتشابه: ܕܡܝܐ. - (شتّ) تَشَتَّتَ: ܐܬܒܕܪ. - (شتو) شتاء: ܣܬܘܐ. - شَجَرٌ: ܐܝܠܢܐ. - شَخَصَ: [ܚܘܪܐ]. - (شدّ) اشتَدَّ: ܥܫܢ، ܚܣܢ، ܬܩܦ [ܚܙܩ]؛ شَديد: ܥܫܝܢܐ، ܚܣܝܢ، ܡܬܩܦܢܐ، ܩܫܝܐ، [ܣܓܝܐ، ܣܓܝܐܝܬ، ܢܨܝܚܐ، ܥܙܝܙ، ܚܣܝܢ]؛ شدّة: ܥܫܝܢܘܬܐ، ܬܘܩܦܐ، [ܚܣܝܢ، ܬܩܝܦ]. - شَرَرٌ: ܙܠܝܩܘܬܐ. - (شرب) شَراب: ܫܬܝܐ. - (شرق) أشرَقَ: ܕܢܚ، ܙܠܓ؛ مَشرق: ܡܕܢܚܐ؛ شَرقي، مَشرقي: ܡܕܢܚܝܐ. - (شعّ) شُعَاع: ܙܠܝܩܐ. - شَعرٌ: ܣܥܪܐ؛ شعر: ܦܘܐܝܣܝܣ؛ شاعر: ܦܘܐܝܛܐ. - (شعل) اشتَعَلَ: ܕܠܩ، ܝܩܕ، ܐܫܬܠܗܒ، ܐܬܝܩܕ، ܐܬܚܡܝ؛ شُعلة: ܫܠܗܒܝܬܐ؛ اشتعال: ܫܠܗܒܝܬܐ، ܝܩܕܢܐ؛ مُشتعل: [ܕܠܝܩܐ]. - (شفّ) شافّ، شَفّاف، مشف: ܡܚܘܝܢ؛ مشفًا: ܡܚܘܝܢܐܝܬ؛ إشفاف: ܡܚܘܝܢܘܬܐ. - (شقّ) شَقٌّ: ܨܪܝܐ. - (شكّ) مُتشكّك: [ܐܬܦܠܓ]. - (شكل) تَشَكَّلَ: ܐܣܬܟܡ؛ شَكلٌ: ܐܣܟܝܡܐ؛ تشكُّل: ܐܣܬܟܡܐ، ܡܣܬܟܡܢܘܬܐ؛ مُشَكَّل: ܡܣܟܡ. - (شكو) شَكَا: ܐܬܟܪܝ. - شَمسٌ: ܫܡܫܐ؛ شمسي: ܫܡܫܢܝܐ؛ شُمَيس: ܫܡܝܫ. - شَمعٌ: ܩܐܪܘܬܐ؛ شَمعَة: ܩܢܕܝܠܐ. - (شمل) اشتَمَلَ: [ܚܒܫ]. - (شمل) شمَال: ܓܪܒܝܐ؛ شمالي: ܓܪܒܝܝܐ. - (شهد) شَاهَدَ: ܚܙܐ. - شَهرٌ: ܝܪܚܐ؛ شهوري: ܝܪܚܢܝܐ؛ مشهور: ܝܕܝܥ. - (شيأ) شَاءَ: ܨܒܐ؛ شيء: ܣܘܥܪܢܐ، ܟܝܢܐ، ܡܕܡ، ܨܒܘܬܐ.

صَبَغَ: ܨܒܥ؛ صبغٌ: ܓܘܢܐ، ܨܘܒܥܐ. - (صحب) صَاحِب: ܚܒܪܐ، ܡܪܐ، [ܩܠܝܢܐ]. - صَحوٌ: ܨܚܘܐ. - صَخرٌ: ܟܐܦܐ؛ صَخرِى: ܟܐܦܢܝܐ. - صَدَرَ: ܕܢܚ، ܢܦܩ. - (صدق) تصديق: [ܫܪܪ]. - صَدَمَ: ܢܩܫ. - صرٌّ: [ܠܚܚܢܝܐ]؛ صُرَّة: ܨܪܪܐ. - صرف: ܫܚܠܦ. - صَعِدَ: ܣܠܩ؛ صَعَّدَ، أصعَدَ: ܐܣܩ؛ تَصَعَّدَ: ܐܣܬܠܩ، ܐܬܥܠܝ؛ صُعُود: ܣܘܠܩܐ، ܡܣܩܬܐ، [ܡܣܬܠܩܢܘܬܐ]. - (صعق) صَاعِقَة: ܩܪܐܘܢܘܣ. - صَفَرَ: ܫܪܩ؛ صفير: ܫܪܩܐ. - (صفر) اصفَرَّ: ܐܬܝܪܩ؛ أصفر: ܝܘܪܩܐ. - (صفو) صَفَّى: [ܐܫܬܦܝ]؛ صفاء: [ܫܦܝܐ]؛ صَاف: ܫܦܝܐ. - صُقعٌ: ܐܬܪܐ؛ صَقيع: ܓܠܝܕܐ. - (صقل) صقَّال: [ܐܬܡܪܩ]؛ صَقيل: ܡܪܝܩ. - صُلب: ܩܫܝܐ. - (صنع) صناعة: ܐܘܡܢܘܬܐ. - (صوب) أصَابَ: ܡܛܐ. - صَوتٌ: ܩܠܐ؛ مُصَوِّت: ܩܠܢܝܐ. - (صور) صُورَة: ܕܡܘܬܐ، ܨܘܪܬܐ، ܨܘܪܬܐ. - (صير) صَارَ: ܗܘܐ؛ صَيَّرَ: ܥܒܕ. - صَيفٌ: ܩܝܛܐ؛ صيفي: ܩܝܛܝܐ.

(ضبّ) ضَبَاب: ܥܪܦܠܐ؛ ضبابي: ܥܪܦܠܝܐ. - ضَبطٌ: [ܐܬܐܚܕ]. - (ضدّ) ضَادٌّ: ܕܠܩܒܠ؛ ضِدٌّ: ܣܩܘܒܠܐ. - (ضرّ) ضارٌّ: [ܒܝܫ، ܣܘܪܚܢܐ]؛ ضَروري: ܐܢܢܩܝܐ؛ اضطرار: [ܐܘܠܨܢܐ]. - (ضرب) اضطَرَبَ: [ܐܙܕܘܥ]؛ اضطراب: [ܢܘܥܐ]. - (ضعف) ضَعيف: ܡܚܝܠ؛ ضعيف الون: ܦܚܘܪ؛ ضعفٌ: ܐܥܦܐ. - (ضغط) انضَغَطَ: ܐܬܕܚܩ. - ضِفدِعٌ: ܐܩܘܪܕܥܐ. - (ضمحل) اضمَحَلَّ: ܐܬܛܠܩ. - ضمن: ܓܘܐ. - ضَوءٌ: ܙܠܝܩܐ، ܢܘܗܪܐ؛ مُضيء: ܢܗܝܪ. - ضِيق: ܐܠܝܨ

طبٌّ: ܐܣܝܘܬܐ. - طَبَخَ: [ܐܬܒܫܠ]. - طبعٌ، طَبيعَة: ܟܝܢܐ؛ طبيعى: ܟܝܢܝܐ؛ مُنطبع: ܛܒܝܥ [ܐܬܛܒܥ]. - (طبق) طَبَقَة: ܪܘܟܒܬܐ. - (طبل) طبلي: ܛܒܠܢܝܐ. - طَرَفٌ: ܣܘܦܐ، ܦܘܩܐ. - (طرق) انطَرَقَ: ܐܬܛܪܩ؛ مُنطرق: ܡܬܛܪܩܢܐ. - طَفِيَ: ܕܥܟ؛ طفوء، انطفاء: ܕܘܥܟܐ. - (طفل) طُفُولَة: [ܛܠܝܘܬܐ]. - طلٌّ: ܛܠܐ. - (طلع) طلوع: ܕܢܚܐ، [ܡܕܢܚܐ]. - (طوف) طُوفَان: ܛܘܦܢܐ؛ طائفَة: ܡܢܬܐ. - (طول) طالَ: ܢܓܪ؛ طالَ، تطاوَلَ: ܐܬܢܓܪ؛ طويل، مستطيل: ܐܪܝܟܐ، [ܢܓܝܪ]؛ طُولٌ:

INDEX LOCORUM

Registered below are the ancient and medieval works cited in the introduction, commentary and appendices.

Passages cited in the introduction and the main part of the appendices are referenced to the pages where the citations occur. Passages cited in the commentary and the list of glosses and marginal notes in Appendix II are referenced to the passages of *But*. to which the commentary or the marginal note relates, whereby "A" stands for the Book of Mineralogy, "B" for the Book of Meteorology and "M" for marginal note.

Bold type indicates passages quoted at length (in original and/or translation) and compared with the text of *But*. in the commentary.

Sacra Scriptura

Gen. 1.12	BV.ii.1.3f.
Gen. 2.7	BV.ii.2.10
Gen. 6.1-9.29	BV.ii.2.11
Gen. 8.8	p. 6n
Gen. 11.1-9	BV.iii.
Exod. 13.21	BIV.iii.4.7-10
II Reg. 18.17 (Hex.)	AI.iii.2.2
II Reg. 20.20 (Hex.)	AI.iii.2.2
Job. 14.19	p. 47
Ps. 18.8	AII.iii.4.4-7
Ps. 36.10	p. 5
Ps. 51.7	p. 5n
Ps. 60.4	AII.iii.4.4-7
Ps. 72.10	AV.i.3.5
Ps. 102.26-27	BV.i.3.6
Ps. 104.32	p. 44, 56
Ps. 104.32	AII.iii.4.4-7
Jes. 6.2-3	p. 5n
Jes. 6.6	p. 5
Jes. 7.15, 22	p. 6
Jes. 65.17, 66.22	BV.i.3.7
Mt. 24.35	BV.i.3.6
Lc. 2.40	p. 6
I Cor. 7.31	BV.
I Cor. 7.31	BV.i.3.6
Col. 3.11	AIV.ii.6.7, 8
I Jn. 2.17	BV.i.3.6

Abū al-Barakāt al-Baghdādī
K. al-muᶜtabar fī al-ḥikma
Ed. Yaltkaya (1357-58 h.), vol. II

189-191	AIV.iv.1.1-8n
202.6	AIV.iv.tit.
202.7-16	AIV.iv.1.1-8
202.16-17	AIV.iv.2.1-2
202.22-203.6	AIV.iv.3.1-7
203.6-8, 10-11	AIV.iv.3.7-10
203.8-10, 11-15	AIV.iv.4.1-5
203.15-24	AIV.iv.4.5-11
204.2-11	AIV.iv.5.1-2
204.11-205.5	AIV.iv.5.3-9
206.7-16	AIV.iii.6.3-11
208.18-22	AI.ii.1.3-7
208.18-209.2	AI.ii.1.7-10
210.8-10	AI.iii.1.1-4n
211.11-15	AI.iii.3.6-7
211.13-17	AI.iii.2.5-9
211.17-21	AI.iii.3.1-5
211.21-22	AI.iii.2.5-9
213.7f.	BI.i.1.3-5
213.8-16	BI.i.1.5-7
213.8-11	BI.ii.1.1-9
213.16-19	BI.i.1.7-10
214.6-7	BI.i.1.5-7
214.6-13	BI.ii.1.1-9
214.13-17	BI.iii.2.5-8
214.19-22	BI.iii.3.3-5
215.12-21	BI.ii.2.1-8
215.21-24	BI.ii.3.1-6
215.24-216.6	BI.ii.4.1-6
216.6f.	BI.ii.5.1f.
216.8-13	BI.ii.5.2-7
216.9	BIV.iii.4.8
216.12	BI.ii.5.1
216.14-21	BI.ii.6.1-4
222.11-16	BIV.iii.4.9f.
223.12-21	BIV.iii.3.7
224.24-225.1	BIV.iii.4.9f.
226.19-22	BIV.iii.3.7
230.1-3	AIII.i.5.1-6
230.2-18	MAIII.iii.3
230.4	AIII.ii.1.7
230.5	AIII.ii.1.4

230.6 AIII.ii.3.8n
230.8 AIII.ii.2.7f.
230.17-19 AIII.ii.1.1-7
230.19-21 AIII.ii.1.7-9
230.21-231.3 AIII.iii.3.1-4
231.3f. AIII.iii.3.4f.
231.8-10 AIII.ii.3.6-9
231.10f. AIII.ii.3.8-10n
231.12-18 AIII.iii.3.5-11
231.20-232.23 AIII.iii.
232.24ff. AIII.iii.

Abū al-Fidā°, *Taqwīm al-buldān*
Ed. Reinaud-de Slane (1840)

35.9-12 AV.i.3.9n

Abū al-Qāsim al-°Irāqi (al-Sīmāwī)
K. al-°ilm al-muktasab fi zirā°at al-ḏahab
Ed. Holmyard (1923)

7.7 AIII.i.5.6-9n

Acta martyrum et sanctorum
Ed. Bedjan (1895)

V.73.10 AIII.iii.2.6n

Alchemical texts (Syriac) edited by
Berthelot-Duval (1893)

passim AV.ii.5.5, 7, 9f.
3.10; 4.9 AIII.i.1.3n
6.14f. AIII.i.1.2-4n
36.4 AIII.i.5.7
Tr. 75 n.2, 123f. AIII.i.4.2n

Alexander Aphrodisiensis
Comm. in Arist. Mete.

8.16-25 BIV.i.5.5-7n
53.19-22 BIII.iv.4.7n
78.18-22 AV.ii.2.3f.
86.20-24 AV.ii.1.6-9n
101.12ff. AIV.ii.1.1-7
107.28ff. BIII.ii.3.3-8n
110.7 BIII.ii.3.8
111.2-25 BIII.iii.2.5-7n
114.17 AII.i.1.4-8n
114.21-23 AII.i.1.3
116.10f. AII.i.4.6f.n
116.10-117.2 AII.ii.1.1-6n
116.21-34 AII.ii.1.6-8n
116.32f. AII.i.4.6f.n
116.34-117.2 AII.iii.1.1f.n
117.2-9 AII.iii.1.2-7n
117.9-22 AII.iii.1.8-12n
117.23-118.14 AII.iii.2.1-4
118.15-18 AII.iii.2.8f.n
118.18ff. AII.iii.2.4-8n
119.23-28 AII.iii.1.7f.n
120.13-19 AII.iii.1.2-7n
120.20-121.28 AII.iii.3.1-5n
121.29-122.14 AII.iii.4.1-4n
122.14-123.1 AII.ii.3.1-4n
122.23-28 AII.ii.3.7-10n
125.17-35 AII.ii.4.4-8
129.23.f. BIV.i.5.5-7n
129.33-35 BIV.i.5.5-7
137.20-138.5 BIV.ii.2.1-9n
138.3-10 BIV.ii.3.3-5n
138.10-20 BIV.ii.3.5-9n
147.32 BII.iii.2.3

Ps.-Apollonius of Tyana (Balīnūs)
K. sirr al-ḫalīqa
Ed. Weisser (1979)

III.2.3 (229.11-13) AIII.ii.3.6-9n
III.4-5 (p. 246ff.) AIII.ii.
III.4.4 (252.4) AIII.ii.3.6-9n

Aristotle, *Auscultatio physica*

195b 2, 197a 24 AIII.iii.3.5

Aristotle, *De caelo*

II.iv, IV.v AII.i.4.6f.
270b 19 BV.iii.n
293b 34 AIV.ii.1.3
294b 13ff. AII.i.1.3
297b 17ff. AII.i.4.5f.
304a 12 AV.iii.2.10n
304a 14ff. AV.iii.2.11n
306b 32f. AV.iii.2.10n

Aristotle, *Meteorologica*

- 36, 37
338a 20-26 AI.i.1.1-7
338a 26 BI.i.1.1f.
338a 26-339a 5 AI.i.1.7-9

Aristotle, *De anima*

Aristotle, *De memoria*

Aristotle, *Historia animalium*

519b 19 AI.i.5.3n
552b 10 AIII.i.4.2

Aristotle, *De generatione animalium*

735b 21 BI.iii.2.3-5
784b 7 AIV.iii.6.5

Aristotle, *Metaphysica*

981b 17-20 BV.iii.2.6
1014a 5, 1044b 1 AIII.iii.3.5
1074b 1-14 BV.iii.n

Aristotle, *Ethica Nicomachea*

1142a 28 BII.i.9.1-4n
1161a 15 AV.ii.4.3

Aristotle, *Politica*

1329b 25-29 BV.iii.n
1329b 27-29 BV.iii.2.6

Aristotle, *Poetica*

1457b 6ff. AV.ii.4.3

Ps.-Aristotle, *De mundo* (Greek)

392a 7 BIV.i.5.5-7n
393b 5 AV.i.2.6f.

Ps.-Aristotle, *De mundo*, Syriac version
Ed. de Lagarde (1858)

136.21-24 (392a 1-3) BIII.ii.1.9, 11
138.3 (392b 4) BII.i.1.1
138.3 (392b 4) BIV.iii.tit.n
138.9 BI.iii.1.6f.
138.9 (392b 11) BIII.iv.4.4, 7n
138.11 (392b 12) BIV.ii.n
138.18f. (392b 21f.) AV.i.1.1n
139.16-21 (393a 16-21) AV.i.1-4.
139.23-140.1 (393a 23-28) AV.i.2.1-4
139.29 (393a 27) AV.i.3.9
140.8f. (393b 4) AV.i.3.3n
140.15, 20 (393b 9, 17) AV.i.3.9
140.27 AV.i.2.4
141.1 (393b 29) BIII.iii.2.4f.
141.20, 28 BI.iii.1.6f.
141.21 (394a 18) BIV.ii.n
141.29f. (394a 26f.) BI.ii.2.1
142.2-5 (394a 30-32) BI.ii.3.6-9
142.5-7 (394a 32-35) BI.iii.2.3-5
142.9f. (394a 36-b 1) BI.iii.2.8
142.30-143.19 (394b 19-35) BIII.ii.3.3-8
143.25f. (395a 6f.) BIII.iv.4.4, 7
143.26 (395a 7) BI.iii.2.8
143.25f. (395a 6f.) BIII.iv.4.4, 7
144.12 (395a 20) BIV.ii.2.1, 3n
144.13, 18 (395a 22, 26) BIV.ii.n
144.18f. (395a 26f.) BIV.ii.2.1, 3n
144.22 (395a 28) BI.iii.6.1
144.22-25 (395a 28-32) BII.ii.1.1f.n
144.24, 28 (395a 30, 35) BII.i.1.1
144.27, 30 (395a 35, 38) BIV.ii.n
144.28f. (395a 35f.) BII.iv.3.2
145.11 (395b 12) BII.i.1.1
145.11 (395a 12) BIV.iii.tit.n
145.17-20 (395b 18-21) AII.ii.2.6
145.21 (395b 21) AII.ii.2.8-10n
146.5f. AII.ii.1.3f.
146.10 (396a 3f.) AII.ii.4.6
146.11-12 (396a 4-6) AII.ii.4.8f.n
146.13-14 (396a 6-7) AII.ii.2.3f.
146.18-22 (396a 11-15) AII.ii.3.5f.
146.30 (396a 23) BIII.iv.4.4, 7n
154.15 (399b 15) BIV.i.5.1n
155.24 (400a 33) AII.ii.2.8-10nf
157.8 (401a 18) BIV.ii.n

Ps.-Aristotle, *De mundo*
(Arabic versions)

- (393b 4) AV.i.3.3n
- (394a 30-32) BI.ii.3.6-9n
- (395a 26f.) BIV.ii.2.1, 3n
- (396a 3f.) AII.ii.4.6

Ps.-Aristotle, *De plantis*:
see Nicolaus, *De plantis*

Ps.-Aristotle. *Problemata*

10.43 AI.i.5.3n
15.14 BIII.i.2.1-4n
26.2 BIII.iii.5.1-11n
26.15 BIII.iii.1.1-5n

Ps.-Aristotle, *K. al-aḥǧār*
Ed. Ruska (1912)

99f. AIII.i.5.3n
112.10 AIII.ii.3.3n
123.1f. AIII.ii.3.6-9n

Athenaeus

42C (= Theoph. Frag. 214A) AI.iii.n

Bahmanyār b. al-Marzbān: *K. al-taḥṣīl*
Ed. Muṭahharī (1970)

715.3 BIV.iii.4.4-6n
716.5 AI.iii.1.1-4n
720.5f. AIII.ii.1.7-9

Bar ᶜAlī, *Lexicon*
Ed. Hoffman (1874)

123 (no. 3421) AIII.ii.3.8n

Bar Bahlūl, *Lexicon*
Ed. Duval (1901)

19.22 etc. AIII.i.5.6
86.14 BV.ii.1.6
257.27 AIV.i.2.3n
297.1 AIII.i.3.3
468.12f. AII.ii.4.8f.n
675.3 AIII.ii.3.8n
883.3 AIII.i.1.3n
859.5f. AIII.i.4.2
898.26ff. AIII.i.4.2n
899.2f. AIII.i.4.2n
960.16 AII.iii.2.8
970.3 AIII.i.1.3n
970.15f. AIII.i.1.3n
1268.10-14 AIII.ii.3.3n
1423.17f. AIII.iii.2.6n
1797.8ff. AIII.i.4.2n

Barhebraeus, *De aequilitteris*
Ed. Martin (1872)

line 1049 AII.ii.4.8f.n

Barhebraeus, *Ascensus mentis*
Ed. Nau (1899)

\- AIV.i.2.5
\- AIV.i.3.1
2.8-10 p. 6n
7.13-19 AV.iii.2.10
10.15f. AIV.iv.2.6
14.21 BII.i.1.1
p. 12f. AIV.i.2.2-8n
15.16-18 AIV.ii.3.8
17.20-18.13 AIV.ii.3.5n
19.11-20.3 AIV.i.3.1
21.3-5 BII.iii.3.7
21.9 etc. BII.iii.3.7
p. 22-27 AIV.i.4.3-4
54.10, 13, 16, 19 BIII.iii.4.2
77.14 AIV.i.3.5
103.1 AIV.i.2.2-8n
103.22-104.5 AIV.i.2.2-8
106.22f. p. 46
128.1-14 AIV.i.3.1-4
128.20-129.18 AIV.i.3.4-11
134.4-16 AV.i.3.1-11
134.6 AV.i.3.9-10
134.7 AV.i.2.6f.n
134.8 AV.i.3.8
134.19-20 AV.i.1.1n
135.4 AV.i.1.4
135.3-6 AV.i.1.4-6
135.16-136.11 AV.i.3.1-11
135.21 AV.i.3.5
136.21-137.5 AV.i.2.4f.
137.3 AV.i.1.4
137.3 AV.i.2.3-4n
137.5-9 AV.i.2.5-7
138.17 AV.i.2.6f.n
140.6, 19 AV.i.2.6f.n
140.13-15 AV.i.1.4
140.13-15 AV.i.2.3-4n
p. 141 AIV.ii.3.5
p. 141f. AIV.i.3.5
142.16 AIV.iii.5.1
144.20-145.4 AIV.ii.3.3-10
145.3-7 AIV.ii.5.1-5
145.7-9 AIV.ii.4.1-7
145.13-22 AIV.iii.5.1-8
146.20-147.2 AIV.i.4.11n
147.6-10 AIV.iii.6.3-11
190.20 AV.i.2.6f.n
194.12 AV.i.2.6f.n
199.17-19 AIV.i.2.2-8n
200.15-201.2 AIV.ii.6.9n
201.ult. BII.iv.2.7n

Barhebraeus, *Butyrum sapientiae*

Proem	p. 4f.
Isag. II.ix.	AIV.iii.3.4
Ausc. phys. I.v.1	AIII.iii.2.2
Ausc. phys., chap. IV	BV.n
De caelo II.i.4	AIV.i.2.2-8n
De caelo V.iv.6	AII.i.4.6f.
De caelo V.v.1-2	AII.i.4.5f.
De caelo V.vi.	BV.n
Gen. et corr. I.ii.4-III.iii.3	p. 16
Gen. et corr. I.iii.-iv.	BV.n
Gen. et corr. II.v.	AIII.iii.
Gen. et corr. II.v.3	AIII.iii.3.1
Gen. et corr. III.i.5	AIII.i.3.3
Gen. et corr. III.i.5	AIII.i.5.3
Gen. et corr. III.i.8	AIII.i.1.3
Gen. et corr. III.i.8	AIII.i.3.3
Gen. et corr. IV.iii.	AIII.iii.
Gen. et corr. III.iii.1	AIII.i.5.2n
Gen. et corr. III.iii.3	AIII.i.5.2n
Gen. et corr. III.iii.6	AIII.i.5.2n
Gen. et corr. III.iv	AII.iii.2.8
Gen. et corr. IV.iii.1	AIII.iii.1.5
De plantis I.i.2	BIV.i.5.1
De plant. II.ii.3	AI.ii.5.8n
De plant. III.i.1	AIV.iii.6.9
De plant. III.ii.3	AIV.iii.6.5
De plant. IV.i.3	BV.ii.1.6n
De plant. IV.ii.3	AIV.iii.6.5
De animalibus II.iii.2	AI.ii.5.8n
De animalibus III.i.6	AIII.i.4.2
De anima I.ii.1	AV.iii.2.11
De anima III.i.2	AI.ii.1.4n
De anima III.iii-iv	BII.i.n
De anima V.i.	BV.iii.2.1-10n
Metaph. VI.iii.1	AI.ii.1.4n
Econ. II.3.3	AV.i.3.9-10n
Eth. I.i.2	p. 61n
Eth. I.iii.4	AIV.ii.6.7, 8
Eth. I.iii.4	AIV.i.4.11
Eth. I.v.2	BV.iii.1.7
Eth. II.iii.2	AIV.ii.6.7, 8
Econ. II.3.3	AV.i.2.6f.
Polit. I.i.2	BV.iii.2.7
Polit. I.iii.6	AV.i.3.7n
Polit. II.iii.3	AIV.ii.6.7, 8
Colophons	p. 9, 10n, 585-600

Barhebraeus, *Candelabrum sanctuarii*
Proem, ed. Bakos (1930)

27.11f.	p. 6n

Barhebraeus, *Candelabrum sanctuarii*
Base II ed. Bakos (1930-33)

56.10	BIV.i.5.1n
60.4-73.14	BV.
72.2-74.14	BV.i.3.6
82-104	p. 43n
84.1-4	AI.ii.2.3-7n
84.6	AIII.i.3.3
84.6-85.3	AIII.i.1.2-4n
86.1-87.3	AIII.i.5.1-6
86.3	AIII.i.5.9n
86.8-87.1	AIII.i.2.3f.n
87.3	AIII.i.3.3
87.3	AIII.i.4.2n
87.4-88.1	AIII.ii.2.7f.
88.1f.	AIII.ii.2.1-5n
88.3	AIII.ii.1.7-9
88.3-11	AIII.ii.3.1-10
88.7	AIII.i.5.9n
89.5	AIII.i.4.2n
89.5-10	AV.i.5
89.6	AI.i.5.6
90.8, 9, 11	BII.iv.2.7n
91.11-92.11	AIV.i.3.1-4
94-102	AIV.i.3.5
95.2-7	AIV.i.4.11
99.2f.	AV.i.3.7
97.1	AV.i.3.5
98.10-11	AIV.iii.6.3-11
100.7	AIV.i.4.10
101.6-102.1	AIV.i.4.9-11
109-130	p. 43n
111.9f.	BI.i.1.1f.
111.11	BI.iii.tit.
111.11-113.10	p. 44n
112.7f.	BI.iii.5.2f.
112.8-10	BI.iii.2.1-3
112.10-113.2	BI.iii.3.1f.n
113.2f.	BI.iii.4.5-7
113.6	Mete I.iii.tit.
113.7f.	BI.iii.1.6f.
113.8-10	BI.ii.3.6-9
113.9	BI.ii.3.7
113.10	BI.iii.2.8
114.1f.	BII.iii.8.3-6

114.4	BI.iii.6.7
114.5-8	BII.iii.8.1-10n
114.11-115.4	BII.iii.1.1-5
115.4f.	BII.iii.5.2-5n
115.9-116.3	BII.iii.8.1-10
117.2	BII.i.1.1n
117.2f.	BIV.iii.3.7
117.9-11	BII.ii.1.1f.n
117.11 etc.	BII.iii.2.2
118.10-119.1	BIV.i.4.4-8n
119.1-3	BIV.i.4.8-11n
119.3f.	BIV.i.5.1f.n
119.4	BIV.i.6.1
119.7-120.1	BIV.ii.2.1-9
120.4-6	BIV.ii.3.1-3n
120.7	BIV.iii.tit.
120.7-13	BIV.iii.4.7-10
121.1f.	BIV.iii.4.7
121.8	BIV.iii.1.5
121.10f.	BIV.iii.4.4-6
122.10-12	AIV.iv.2.1-6n
123-127	BIII.iv.n
123.6-124.8	p. 44n
123.9-124.2	BIII.i.1.4-9
124.2-5	BIII.i.n
124.8-9	BIII.i.4.5-8n
124.10-125.1	BIII.i.4.8-11n
125.1f.	BIII.i.4.4-5n
125.3-12	BIII.iv.n
125.3-8	BIII.iv.4.4, 7
125.3, 8	BI.iii.2.8
126.1	BIII.ii.tit.
126.1-127.5	BIII.ii.3.3-8
127.6-10	AII.ii.1.1-6n
128.1-3	AII.iii.4.1-4n
128.3-5	AII.ii.3.1-4n
128.3-5	AII.ii.3.5f.
128.5-7	AII.ii.2.8-10n
128.8f.	AII.ii.2.3f.
128.9-129.1	AII.ii.4.4-8
129.1	AII.ii.4.8f.
129.1-3	AII.i.1.3f.n
129.3	AII.i.1.4
129.3-5	AII.i.1.4-8n
129.5-7	AII.i.2.1-7n
129.7-9	AII.i.3.1-4n
129.9f.	AII.ii.1.7f.
129.10-130.2	AII.iii.1.8-12n
130.2f.	AII.iii.2.4-8
130.2f.	AII.iii.2.8f.n
131.8-132.3	BIV.iii.2.1-9
134.5-9	AV.i.3.9-10
136.1	BIV.i.5.1n
149.4-150.4	BV.ii.1.3f.
151.3-5	AV.iv.1.1-5
151.6-8	AV.ii.2.1-3
151.8-11	AV.ii.1.6-9
151.11	BIV.i.5.1
152.1-3	AV.ii.4.6-8
152.3-4	AV.ii.3.1f.
152.5-7	AV.ii.3.3-5n
152.8	AV.iii.tit.
152.8-11	AV.iii.1.3-8n
152.8-153.1	AV.iii.
152.11-152.2	AV.iii.2.1-3
153.3f.	AV.iii.3.1f.
153.3-154.2	AV.iii.
153.4	AV.iii.3.4
153.5-8	AV.iii.3.4-9
153.8-11	AV.iii.3.9f.
154.2-3	AV.i.1.1
154.3	AV.i.tit.
154.3-9	AV.i.1.1-4
154.9	AV.i.1.4
154.9	AV.i.2.3-4n
154.9-155.1	AV.i.2.4f.
155.1	AV.i.1.4
155.1	AV.i.2.3-4n
155.1-5	AV.i.1.4-6
155.3-4	AV.i.1.6f.
155.5-157.11	AV.i.3.1-11
156.1f.	AV.i.3.5
156.4	AV.i.3.5
156.12	AV.i.2.6f.n
157.5	AV.i.3.8
157.11-158.2	AV.i.1.1
158.3-12	AV.i.2.5-7
160-162	AIV.i.3.5
160.12	AV.i.3.5
161.10	AV.i.2.3-4n
162.6	AV.i.2.6f.n
170.12	BV.ii.1.6n
193.4	BV.ii.1.6
198.9f.	AIV.i.2.2-8n
203.1	AII.ii.3.5, 6n
221.6-10	BIV.iii.3.7-10
247.11	AIII.i.4.2n

Barhebraeus, *Candelabrum sanctuarii*
Base IV, partial ed. Nau (1916)

154.7	AV.i.2.6f.n

Barhebraeus, *Candelabrum sanctuarii*
Base VIII, ed. Bakos (1948)

56-58	BV.iii.2.2f.
58.23-60.2	BV.iii.2.1-10
58.23ff.	BV.iii.2.7
59.8-14	BV.iii.2.1-7
60.3-21	BV.iii.2.1-10
60.3ff.	BV.iii.2.1-7

Barhebraeus, *Candelabrum sanctuarii*
Base X, ed. Zigmund-Cerbü (1969)

16.2f.	BV.iii.1.6
44.22	AIV.ii.6.7, 8

Barhebraeus, *Carmina*
Ed. Dolabani (1929)

46.7	p. 5n
95.22, 97.10, 165.17	AI.i.2.2n

Barhebraeus, *Chronicon*
Ed. Bedjan (1890)

5.10-25	BV.iii.n
8.3f., 5	AV.i.3.6n
18.6	BIV.i.5.1n
24.10	AI.iii.2.2
30.9	AV.i.3.7n
54.17-25	AIV.i.2.2-8
66.3, 6	AV.i.2.6f.
98.14f.	BV.iii.1.9
155.17	AV.i.3.5n
216.17	BIV.iii.4.7-10n
217.15	AV.i.3.7n
219.15, 220.23	p. 48n
306.18	AII.ii.4.8f.n
378.8	AII.ii.4.8f.n
402.27	AII.ii.4.8f.n
414.7f.	AV.i.1.7n
419.26	AV.i.2.6f.
432.4	AII.ii.4.8f.n
504 24	AV.i.1.7n
555.7	AV.i.3.7n
560.20	AV.i.3.7n
579.23	AV.i.3.7n
585.2	AV.i.3.7n
586.9	AV.i.3.7n
598.8	AII.ii.4.8f.n

Barhebraeus, *Chronicon ecclesiasticum*
Ed. Abbeloos-Lamy (1872-77)

I.379.10f.	AV.iii.2.10
I.379.14-17	AV.iii.2.10n
II.411.13-15	p. 56n
II.465.20-467.8	p. 59n
II.477.13-16	p. 5n, 7

Barhebraeus, *Liber columbae*
Ed. Bedjan (1898)

522.8, 19f.	p. 6n
562.18-20	BIV.ii.1.4n
578.5-7	p. 7n
580.20-581.3	p. 46

Barhebraeus, *Letter to Catholicus Denḥā*
Ed. Chabot (1898)

line 965	AI.i.2.2n

Barhebraeus, *Ethicon*
Ed. Bedjan (1898)

2.6	p. 6n
318.13-16	AIII.iii.n
421.14-17	BIV.ii.1.4
450.15-451.9	p. 46
453.18-454.10	p. 46
452.18-453.3	AV.ii.4.6-8
453.6	AV.i.3.5
453.3-10	AV.i.3.1-11
453.12	AV.i.1.4
453.13f.	AV.i.1.4-6
453.17f.	AV.i.1.7
453.20	AV.i.3.5

Barhebraeus
Muntaḫab kitāb ǧāmi[c] *al-mufradāt li-Aḥmad b. Muḥammad al-Ġāfiqī*
Partial ed. Meyerhoff-Sobhy (1932-40)

p. 92, no. 201	AI.i.4.10n
p. 112, no. 238	AI.i.4.10n

Barhebraeus, *Liber grammaticae*
Ed. Bertheau (1843)

p. 5, p. 42	BV.iii.2.5

Barhebraeus, *Liber Hierothei*
Excerpt apud Marsh (1927)

165*.18f.	p. 6n

Barhebraeus, *Historia dynastiarum*
(*Muḫtaṣar ta'rīḫ al-duwal*)
Ed. Ṣālḥānī (1858)

7.9-8.13	BV.iii.n
77.12, 187.1f., 189.25	p. 48n
73.10	AIV.i.2.2-8n
82.15f.	p. 38n

Barhebraeus, *Horreum mysteriorum*

in Gen. 1.9	p. 47
in Gen. 1.9	AI.n
in Gen. 1.9	AV.i.3.3n
in Gen. 1.9	AV.i.1.4
in Gen. 1.9	AV.i.2.3-4n
in Gen. 7.17	BV.ii.2.11
in Gen. 9.13	p. 47
in Job. 14.19	p. 47
in Ps. 72.10	AV.i.3.5n
in Ps. 102.27	BV.i.3.6n
in Jes. 6.3	p. 5n
in Jes. 6.6	p. 5n
in Jes. 7.15	p. 6n
in Jes. 51.6	BV.i.3.6n
in Jes. 65.17	BV.i.3.7n
in Ezech. 34.12	BIII.iv.4.7
in Ezech. 38.2	AV.i.3.7n
in Dan. 8.2	BIV.iii.4.7
in Mt. 24.35	BV.i.3.6n
in Lc. 2.40	p. 6n
in Act. 2.10	AV.i.1.4
in Act. 2.10	AV.i.2.3-4n

Barhebraeus
Liber indicationum et prognosticorum
Ms. Laur. or. 86

64v-65r	p. 42n
132r b12f.	p. 8n

Barhebraeus
Muḫtaṣar fī ʿilm al-nafs al-insānīya
Ed. Sbath (1928)

26.7	BV.iii.2.7
26.12	BV.iii.2.6

Barhebraeus
Nomocanon (*Liber directionum*)
Ed. Bedjan (1898)

2.2	p. 6n

Barhebraeus, *Liber radiorum*
Ed. Istanbul (1997)

1.8f.	p. 5n
2.1	p. 6n
11-18, 19-27	p. 44n
12.1	AIII.ii.2.7f.
12.8	AIII.i.5.9n
15.4f.	BIII.iii.1.6
16.9	AIV.i.2.2-8n
17.5	AV.i.1.4
17.5	AV.i.2.3-4n
17.15	BII.iv.2.7n
21.7f.	p. 44n
22.15	BI.iii.1.6f.
22.16	BI.ii.3.7
22.16-23.17	p. 44n
22.17	BI.iii.2.8
23.10	BIII.i.1.6
23.15	BIII.iv.4.4, 7n
24.9	BII.i.1.1
24.9-11	BIV.iii.3.7
26.7-11	BIV.iii.4.7-10
27.3-15	AII.iii.4.4-7n
27.16-19	AII.iii.4.4-7
27.16-19	AV.iii.3.9f.
35.17f.	AV.i.1.1n
35.18-37.6	AV.i.3.1-11
36.10	AV.i.3.5
37.2-3	AV.i.3.8
37.3	AV.i.3.9-10
37.11-17	AV.i.1.4-6
37.12	AV.i.1.4
37.12	AV.i.2.3-4n
38.6-10	AV.i.2.4f.
38.10-16	AV.i.2.5-7
38.10	AV.i.2.3-4n
70.11-78.3	BV.n
74.7f.	BV.iii.1.6
185.16	BV.iii.2.2f.
186f.	BV.iii.2.1-10n

Barhebraeus, *Sermo sapientiae*
Ed. Janssens (1937)

46.8f.	p. 4n
78.4	AIII.i.1.6
78.5-79.3	p. 41
79.2	AII.tit.n

Barhebraeus, *Liber splendorum*
Ed. Moberg (1922)

2.9, 21f. p. 6n
6.1, 89.7 BV.iii.2.5
28.14f. AI.iii.2.2
158.32f. BV.iii.2.4
174.31 AI.i.2.2n
208.26f. BV.i.1.5

Barhebraeus, *Tractatus tractatuum*
Ms. Cantab. Add. 2003

3r 4, 5 p. 6n
50r-57v p. 42n
55r 26-55v 15 p. 44n
55v 5f. BI.ii.2.1
56v 12-16 BIV.iii.4.7-10
57v 4 AIII.ii.2.7f.

Bar Kepha, *Hexaemeron*
Ms. Paris syr. 241

1v b5f, b17, b21 AIII.iii.2.6n
162r 5 AI.i.2.3n
164r a30 AIII.i.5.9n
170r a2-6 AIV.iii.6.3-11
172r 29-v3 AV.i.1.1n
172v a14 AV.i.2.1-2n
182r b19-26 BI.iii.2.3-5
184v a18-20 BIV.i.3.2f.
188v BIII.ii.3.3-8n
191r b3-11 BIII.iv.4.4, 7n
191r b12f. etc. AII.tit.n
191r b20-22 AII.i.1.5f.n
191v a11-13 AII.ii.1.5
191v a12-13 AII.i.1.4-5, 6n
191v a15-18 AII.i.1.4-5, 6n
191v b21 AII.iii.2.3
192r a3-7 AII.ii.1.5
192r a5-7 AII.i.1.4-5, 6n
192r b 5-192v a 12 AII.iii.4.4-7
192v b 12-16 AII.ii.3.5f.
193r a1-11 AII.iii.3.1
196v b4ff. BI.ii.3.7n
196v b25 BIV.iii.4.7-10

Bar Kepha, *Hexaemeron*
Ms. Paris syr. 311

37v a5 AI.i.2.3n
38b b 24 AIII.i.5.9n

42r b32 AV.i.2.3-4n
42v a1, 6, 24 AV.i.2.3-4n
46r a 3 AV.i.2.3-4n
46r a5-6, b6-33 AV.i.3.3
46r a 6-10 AV.i.1.1
46r a19 AV.i.2.1-2n
57r BIII.ii.3.3-8
59b a 10-13 AII.ii.3.5f.

Bar Shakko, *Liber dialogorum*
Ms. Göttingen, Or. 18c

300r-304r p. 41n
303v a6-b9 AIII.i.5.1-6
303v a14f. AIII.i.5.7n
303v b4 AIII.i.4.1
303v b9-304a 11 AIII.ii.3.1-10
304r a3 AIII.i.5.9n

Bar Shakko, *Liber thesaurorum*
Ms. British Library, Add. 7193

70r b20f. AV.i.1.1n
70v a27 AV.i.2.1-2n

Bar Shakko, *Liber thesaurorum*
Ms. Paris syr. 316

170r 3 AIII.i.5.9n
172v 15-18 AV.i.1.1n
173r 7 AV.i.2.1-2n
173v 11-174r 11 AV.i.3.3

Battānī, *K. al-zīǧ al-ṣābī*
Ed. Nallino (1899-1907)

25.16 AV.i.3.9n
27.9f. AV.i.2.5n
190.11, 15 AIV.i.2.3n

Ps.-Berosus
Ed. Levi della Vida (1910)

15.8f. AV.i.1.1n

Bīrūnī
K. al-āṯār al-bāqiya ʿan al-qurūn al-ḫāliya
Ed. Sachau (1923)

28.5-9 AIV.i.2.2-8n
326.1ff. AIV.i.2.2-8n

Bīrūnī, *K. al-ǧamāhir*:
Ed. Krenkow (1355 h.)

229.17-18 AIII.ii.2.7f.
251.12-15 AI.i.4.3-11

Bīrūnī
K. al-tafhīm li-awā'il ṣinā'at al-tanǧīm
Ed. Wright (1934)

101.4f. AIV.i.2.3n
101.5f. AIV.i.2.2-8n
121.5f. AV.i.tit.
121.11-12 AV.i.3.8
122.15-123.2 AV.i.2.4f.
123.2-4 AV.i.1.4-6
123.5 AV.i.1.1-4n
125.5-7 AIV.i.4.11
143.10-15 AV.i.3.5
144.14f. AV.i.3.7
145.11 AIV.i.4.9-11n
211.9, 10 AV.i.3.9-10

Bīrūnī
K. al-tafhīm, Persian version
Ed. Humā'ī (1316 h.)

132.11f. AIV.i.2.3n
168.10-13 AV.i.2.4f.
168.13-15 AV.i.1.4-6
200.11f. AIV.i.4.9-11n

Biṭrūjī *K. fī al-hai'a*
Ed. Goldstein (1971)

173.13ff. AIV.i.2.2-8n
174.1, 179.2 AIV.i.2.2-8n
174.ult. AIV.i.2.3n

Causa causarum
Ms. Laur. or. 298

132r b 23 AV.i.3.3n

Causa causarum
Tr. Kayser (1893)

348-351 AIII.i.5.9n

Chronicon ad 724 pertinens
Ed. Brooks (1904)

351.7 AV.i.2.3-4n

Chronicon ad 1234 pertinens
Ed. Chabot (1920-26)

II.284.18 AIII.iii.2.6n

Dimashqī, Shams al-Dīn Muḥammad
Nuḫbat al-dahr
Ed. Mehren (1866)

48.17f. AIII.i.5.6-9n
56.7 AIII.ii.2.7f.

Dionysius of Tellmaḥrē
Chronicle, apud Michael

- AV.iii.2.10

Dioscorides, *De materia medica*
Ed. Wellmann (1914)

- AIII.i.1.2-4
5.121 AI.i.5.5

Ephrem, *Hymnus de fide*
Ed. Beck (1955)

II.24 p. 7n

Fakhr al-Dīn al-Rāzī
K. al-mabāḥiṯ al-mašriqīya
Ed. Hyderabad (1343 h.), vol. II.

142.8f. AV.ii.4.5
142.11 AV.ii.4.6
142.16-143.4 AV.iii.
142.16 AV.iii.tit.
142.18-143.4 AV.iii.1.3-8n
143.5f. AV.iii.3.1f.
143.5-9 AV.iii.
143.6 AV.iii.3.4
143.8f. AV.iii.3.9f.
173.5-8 BI.iii.2.1f.
177.13-14 BII.i.2.3-6n
177.19 BII.i.8.1-5n
178.17-179.4 BII.ii.1.3-9
179.5ff. BII.ii.2.1-4n
180.19-181.5 BII.iii.1.1-5
182.9 BII.iii.7.7
181.3-5 MBII.iii.1
182.10-12 BII.iii.3.1-4
182.10-12 MBII.iii.3
182.10-15 MBII.iii.3
182.12-17 BII.iii.3.4-8

182.18-21	BII.iii.3.8-10
184.10-14	BII.iii.8.1-10n
184.15-185.4	BII.iv.1.1-5
184.15-185.4	MBII.iv.tit.-1
186.16	BII.i.1.1
187.5-11	BIV.i.1.1-7
187.11-13, 15f.	BIV.i.2.1-4
187.14f.	BIV.i.2.5-9
187.17-18	BIV.i.3.1-4
188.14-21	BIV.iii.4.7-10
188.14	MBIV.iii.tit.
188.15	BIV.i.3.4f.n
189.5f.	BIV.iii.4.4-6
189.13-190.4	BIV.iii.2.1-9
190.6-12	BIV.iii.4.7-10
190.13	BIII.i.tit.
190.14-16	BIII.i.1.4-9
190.19-191.5	BIII.i.1.4-9
193.14-194.10	BIII.iv.n
193.17-21	BIII.iv.4.4, 7
194.2f.	BIII.iv.4.4, 7
194.10	BIII.ii.tit.
196.20	BIII.iii.2.1-2n
199.2	AIV.ii.tit.
199.3	AIV.ii.1.1
199.16-200.9	AIV.ii.5.1-5n
200.10-17	AIV.ii.5.5-9
200.20-201.9	AIV.ii.3.3-10n
201.2	AIV.ii.3.6
201.3-5	AIV.ii.3.3-10n
201.11-16	AIV.ii.6.1-5n
201.18-202.18	AIV.iii.5.1-8
201.21-202.17	AIV.iii.3.1-7
202.19-21	AIV.iii.1.3-7
203.1-8	AIV.iii.2.1-4
203.18-204.14	AIV.iii.4.1-6
205.19-21	AII.ii.1.1-6
207.13-16	AI.i.3.1-6n
207.17f.	AI.i.3.6-9n
207.19-21	AI.i.4.1-3n
208.1	AI.i.4.4f.n
208.9-12	AI.ii.2.1-3n
208.13-16	AI.ii.2.3-7
208.16-20	MAI.ii.1
210.5-211.5	AIII.i.5.1-6
210.8-14	AI.ii.3.1-3n
211.4	AIII.i.4.2n
211.8f.	AIII.i.5.6-9n
211.8	AIII.i.5.9n
213.1f.	AIII.ii.1.5n
213.4f.	AIII.ii.2.1-5n
213.6f.	AIII.ii.1.7-9
213.11-20	AIII.ii.3.1-10
213.15	AIII.i.5.9n
214.13	AIII.i.4.2-6
214.13	AIII.i.4.2n
214.16	AIII.i.4.7
215.4-216.4	AIII.iii.3.1-11
215.18f.	AIII.iii.3.2
218.17f.	BV.ii.1.1
218.21-219.3	BV.ii.1.3-8
219.4-13	BV.ii.2.3-11
219.14-20	BV.ii.3.4-8
410	BV.iii.2.1-10n

Galen
Ed. Kühn (1821-33)

14.771	AI.i.5.3n
15.658	AV.ii.4.5

Galen, *De simplicibus*
Syriac version (Sergius of Rēšᶜainā)
Ed. Merx (1885)

301.15	BV.ii.1.6n

Ghazālī
Maqāṣid al-falāsifa
Ed. Dunyā (1961)

340.25	p. 44n
342.23-26	BIV.iii.4.7-10

Grigor of Akner
Patmut'iwn azgin netołac'
Ed. Blake-Frye (1949)

350 [82]	BIV.iii.3.7-10

Heraclitus (Diels-Kranz)

Frag. 61 (I.164.6-8)	AV.ii.2.3f.

Herodotus

I.179	AI.ii.1.7-10

Hippocrates

Aër. 9	AI.i.5.3
Aff. 30-31	AI.i.5.2f.
Epid. 6.1.5	AV.ii.4.5
Nat. Hom. 12	AI.i.5.3
Nat. Hom. 14	AI.i.5.2f.

Homer

Il. 2.243 etc. AV.ii.4.3
Il. 11.306, 21.334 BIII.iii.4.7-9n
Od. 5.295-296 BIII.iii.4.7-9n

Ḥudūd al-ᶜālam
Tr. Minorsky (1937)

§2.2 (p. 52) AV.i.3.10n
§4.25 (p. 59) AV.i.3.9n

Ḥunain b. Isḥāq
al-ᶜAšr maqālāt fī al-ᶜain
Ed. Meyerhoff (1928)

131.11, 132.7 AI.i.5.3n

Ḥunain b. Isḥāq
Ǧawāmiᶜ li-kitāb Arisṭūṭālīs fī al-āṯār al-ᶜulwīya
Ed. Daiber (1975)

224-229 AII.iii.3.1
237 AII.ii.4.6n
294, 303, 305, 306 BIV.iii.4.7-10

Ḥunain b. Isḥāq
al-Masāʾil fī al-ᶜain
Sbath-Meyerhoff (1938)

§121, §137, §140 AI.i.5.3n

Ibn al-Biṭrīq
Arabic version of Arist., *Mete.*
Ed. Petraitis (1967)

- BI.ii.3.6-9
62.6 AV.ii.1.9
80.9 AII.ii.4.6n

Ibn al-Biṭrīq
Arabic version of Arist., *Mete.*
Ed. Schoonheim (1999)

line 49 BV.iii.n
lines 207, 210, 222 BIV.iii.4.7-10
line 768 AII.iii.2.1-4
line 802 AII.iii.3.1
p. 121f. BIV.ii.n

Ibn al-Faqīh, *K. al-buldān*
Ed. de Goeje (1885)

83.18 AV.i.3.9-10n
205.9-11 AIII.iii.3.8, 10

Ibn al-Khammār
Maqāla fī al-āṯār al-mutaḫayyila fī al-ǧaww
Ed. Lettinck (1999)

356.3-13 BII.iii.5.2-5n
357.21-358.2 BII.i.5.5f.
359.15-17 BII.ii.2.4-6

Ibn Khurdādhbih
K. al-masālik wa-l-mamālik
Ed. de Goeje (1889).

70.1-6 AV.iii.3.9f.n
92.4 AV.i.3.9-10n

Ibn Rusta
K. al-aᶜlāq al-nafīsa, Book VII
Ed. de Goeje (1891)

98.17f. AV.i.3.9-10

Ibn Sīnā
Dāniš-nāma-i ᶜalāʾī, Ṭabīᶜīyāt
Ed. Mishkāt (1951/2)

71.1-4 BIV.iii.4.7-10n
73-78 AIII.i.1.2-4n
73.9f. AIII.n
74.3 AIII.i.5.3
76.5f. AIII.i.2.6f.n
76.6-77.4 AIII.i.1.6
77.3-78.3 AIII.i.1.4-6n

Ibn Sīnā, *K. al-išārāt wa-l-tanbīhāt*
Ed. Dunyā (1957-60)

286ff. p. 42n
903.4f. p. 8n

Ibn Sīnā, *K. al-naǧāt*
Ed. Dānišpažūh (1986)

314.7ff., 317.6ff. AIII.n

Ibn Sīnā, *K. al-qānūn fī al-ṭibb*
Ed. Qašš (Cairo 1987)

113.16-18 MAIV.iv.1/4
113.18-25 AIV.iv.n
120.14-121.1 AIV.ii.n
120.16-21 AIV.ii.5.1-5
122.5-8 BIII.iii.2.5-7n

122.8 BIII.iii.2.3
p. 494 AIII.i.4.2
559.23 AIII.ii.3.1n
709.2 AIII.i.4.2n
988.17f. AI.i.5.3n
990.18f. AI.i.5.3n

Ibn Sīnā, *K. al-qānūn fī al-ṭibb*
Partial ed. New Delhi (1982)

142.6-8 MAIV.iv.1/4
142.9-27 AIV.iv.n
151.4-10 AIV.ii.5.1-5
151.11-16 AIV.ii.3.3-10
153.1-4 BIII.iii.2.5-7n
153.4f. BIII.iii.2.3

Ibn Sīnā
K. al-šifāʾ, al-Madḫal
Ed. Madkour et al. (Cairo 1952)

11.12f. p. 12n

Ibn Sīnā
Šifāʾ, al-Afᶜāl wa-l-infiᶜālāt
Ed. Qassem (Cairo 1969)

201.4-10 AI.i.1.1-7
205.10-12 AV.ii.1.1-4
205.14-206.1 AV.ii.1.4-5
206.15-17 AV.ii.1.9-12
207.1-4 AV.ii.2.1-3
207.2-3 AV.ii.2.4-7
207.11f. AV.ii.3.1-2
207.12f. AV.ii.3.5-8
207.16f. AV.ii.4.1-4
207.17-208.1 AV.ii.4.4-6
208.2-4 AV.ii.4.6-8
208.11-13 AV.iii.1.1-3
208.13-209.11 AV.iii.1.3-8
209.13f. AV.iii.2.1-3
209.15-210.4 AV.iii.3.5-9
210.4-6 AV.iii.2.9-13
210.7-8 AV.iii.3.1-2
230.17 AIII.i.5.3
232.17 AIII.i.1.3
245.3ff. AIII.i.5.2n

Ibn Sīnā, *Šifāʾ*, al-Maᶜādin
Ed. Monṭasir et al. (Cairo 1964)

3.9f. AI.i.2.1f.
3.9f. MAI.i.2.1f.
3.10f. AI.i.2.5-8
3.10-15 AI.i.2.2-5
3.15-17 AI.i.2.8-10
3.15-15 MAI.i.2.8-10
4.1-13 AI.i.2.5-8
4.18f. AI.i.2.2-5
4.18f. MAI.i.2
5.1-7 AI.i.3.1-6
5.10-13 AI.i.3.6-9
5.14-16 AI.i.4.1-3
5.20-6.9 AI.i.4.3-11
5.20-6.9 MAI.i.4.3-11
6.10f. AI.i.4.11f.
6.10f. MAI.i.4.11f.
6.15-16 AI.ii.1.1-3
6.16-1 AI.ii.2.1-3
6.18-7.8 AI.ii.2.3-7
7.9-13 AI.ii.1.1-3
7.11-15 MAI.ii.1
10.4-7 AI.ii.3.1-3
10.8-11 AI.ii.4.1-6
10.13-11.2 AI.ii.5.1-4
10.18-11.2 AI.ii.4.1-6
11.3-8 AI.ii.5.4-10
11.8-12 AI.ii.5.1-4
11.13-17 AI.ii.3.1-3
12.7f. AI.ii.3.3f.
12.8-11 AI.ii.6.1-6
12.12f. AI.ii.6.6-8
13.6f. AI.iii.1.1-4
13.8-10 AI.iii.1.4-6
13.11-15 AI.iii.1.6-8
13.15f. AI.iii.2.1f.
13.16-14.1 AI.iii.4.1-4
14.1-3 AI.iii.2.2-5
14.3f. AI.iii.4.1-4
14.5-10 AI.iii.4.5-8
15.4f. AII.i.1.1f.
15.5f. AII.ii.1.2
15.6f. AII.ii.1.5f.
15.8f. AII.ii.1.8f.
16.1-4 AII.ii.1.6-8
17.1 AII.i.1.2f.
17.1 AII.i.4.3n
17.2-4 AII.i.1.4-8
17.3f. AII.ii.1.3f.
17.8-10 AII.ii.2.1-4
17.10-14 AII.ii.2.4-8
17.14f. AII.ii.3.1-4
17.15f. AII.ii.3.5f.

17.17-19	AII.ii.4.1-4
17.19f.	AII.iii.1.1f.
17.20-18-2	AII.iii.1.2-7
18.3-4	AII.iii.1.7-8
18.6-8	AII.iii.1.8-12
18.8	BI.iii.4.1-4n
18.9-11	AII.iii.2.1-4
18.14-18	AII.iii.2.4-8
18.18	AII.iii.2.8f.
18.19-19.2	AII.iii.3.1-5
19.4-9	AII.ii.4.4-8
19.11-14	AII.ii.3.7-10
19.14f.	AII.iii.4.4-7
19.14f.	MAII.iii.4.4-7
20.4f.	AIII.i.1.1f.
20.5-9	AIII.i.5.1-6
20.9-14	AIII.i.2
20.15-18	AIII.i.1.4-6
20.16	AIII.i.2.7
21.1	AIII.i.1.1
21.1f.	AIII.i.3.3-6
21.3f.	AIII.i.3.1-3
21.5-9	AIII.i.4
21.10-14	AIII.ii.1.1-7
22.11-23.3	MAIII.iii.1
21.14-17	AIII.ii.1.7-9
21.15-18	AIII.ii.2.1-5
21.18f.	AIII.ii.2.5
21.19-22.2	AIII.ii.2.6f.
22.2-7	AIII.ii.3.1-6
22.8f.	AIII.ii.3.6-9
22.9-11	AIII.ii.3.8-10
22.11-15	AIII.iii.1.1-4
22.16-23.2	AIII.iii.1.5-9
23.2-5, 7-11	AIII.iii.2.1-9
23.5-7	AIII.iii.1.5-9
23.5-9	MAIII.iii.2
23.9-11	MAIII.iii.2
24.3	AIV.tit.
24.3	MAIV.tit.
24.7-10	AIV.i.1.1-3
24.10-18	AIV.i.1.3-9
24.18f.	AIV.i.2.1f.
24.19-25.1	AIV.i.2.2-6
25.1	AIV.i.2.7f.
25.9f.	AIV.i.2.8-10
25.13-17	AIV.i.3.4-11
25.17-18	AIV.i.4.1-3
25.20-26.6	AIV.i.4.3-9
26.6-13	AIV.i.4.9-11
26.17-27.8	AIV.ii.1.1-7
27.9	AIV.ii.1.1
27.9-13	AIV.ii.3.1-3
27.16-28.1	AIV.iv.
27.16-28.1	MAIV.iv.1
28.3-10	AIV.iv.
28.3-10	MAIV.iv.5
28.13-29.3	AIV.ii.5.1-5
28.20-29.2	AIV.ii.5.5-9
29.4-15	AIV.ii.3.3-10
29.7-8, 13	AIV.ii.4.1-7
29.16-30.1	AIV.ii.6.1-5
30.1-11	AIV.ii.6.5-9
30.13f.	AIV.ii.4.6
31.5-7	BIII.iii.2.5-7
31.7-9	BIII.iii.2.7-11

Ibn Sīnā, *Šifāʾ*, al-Āṯār al-ᶜulwīya
Ed. Montaṣir et al. (Cairo 1964)

35.3	BI.tit.
35.6-9	BI.i.2.1-5
35.11-12	BI.i.2.5-6
35.12-15	BI.i.2.6-10
35.15-20	BI.i.3.1-5
36.1-5	BI.ii.6.4-7
36.5	BI.ii.1.6
36.5	BI.ii.3.6-9
36.7-9	BI.iii.1.1-6
36.9	BI.iii.1.6f.
36.10f.	BI.iii.2.1-3
36.11f.	BI.iii.1.7f.
36.12f.	BI.iii.3.1f.
36.13-15	BI.iii.3.5-10
36.16-37.5	BI.iii.4.1-4
37.12-14	BI.iii.4.5-7
38.1f.	BI.iii.4.8-10
38.2f.	BI.iii.3.5-10
38.7-9	BI.iii.5.3-5
38.9	BI.iii.1.7f.
39.1-6	BI.iii.6.1-5
39.6	BI.i.1.5
39.10-13	BI.iii.6.5-9
40.3f.	BII.i.tit.
40.5f.	BII.i.1.1
40.6-9	BII.i.1.2-4
40.8-10	BII.i.2.3-6
40.11	BII.i.1.4f.
40.12-16	BII.i.1.5-9
40.15f.	MBII.i.1

66.7-9	BIII.i.3.9-12
67.3	BIV.tit.
67.4ff.	BIV.i.1.1-7
68.3f.	BIV.i.1.7-9
68.4-8, 10-12	BIV.i.2.1-4
68.8-12	BIV.i.2.5-9
68.13	BIV.i.3.3
68.13-18	BIV.i.3.4-9
68.13-18	BIV.iii.4.7-10n
68.14	BIV.iii.4.8
68.19	BIV.i.4.1-4
69.1-5	BIV.i.4.4-8
69.6-8	BIV.i.4.8-11
69.9-14	BIV.i.3.1-4
69.14-16	BIV.i.4.1-4
70.1f.	BIV.i.5.1f.
70.1	BIV.i.5.5
70.2	BIV.i.5.4f.
70.2f.	BIV.i.6.2
70.4f.	BIV.i.5.2-4
70.5f.	BIV.i.5.5-7
70.6	BIV.ii.1.1f.
70.6-9	BIV.ii.1.2-6
70.7f.	BIV.ii.1.1f.n
70.12-16	BIV.ii.2.1-9
70.16-17	BIV.ii.1.1f.
70.16-17	BIV.ii.3.1-3
71.1f.	BIV.ii.3.2f.
71.1	BIV.ii.1.1f.n
71.2f.	BIV.ii.3.5-9
71.4-5, 7-10	BIV.iii.1.1-5
71.10f.	BIV.iii.1.6-10
71.15f.	BIV.iii.2.1-9
72.5-17	BIV.iii.2.1-9
72.18-20	BIV.iii.3.1-4
73.7-14, 16-17	BIV.iii.3.4-7
73.14-16	BIV.iii.3.7-10
73.19	BIV.iii.1.10
74.3-5	BIV.iii.4.1-4
74.5f.	BIV.iii.4.4-6
74.8f.	BIV.iii.4.7
74.14-15	BIV.iii.4.10f.
74.14f.	MBIV.iii.4.10f.
75.3	BV.tit.
75.3	MBV.tit.
75.4	BV.i.1.1f.
75.4f.	BV.i.1.2-4
75.8-10	BV.i.1.4-6
75.8	BV.i.3.1
75.10-14	BV.i.1.6-11
75.15-76.4	BV.i.2.1-7
76.4-14	BV.i.3.1-6
76.15-18	BV.i.3.7-11
76.18-77.3	BV.ii.1.1-3
77.3f.	BV.ii.1.3-8
77.4-7	BV.ii.2.1-3
77.7-78.14	BV.ii.2.3-9
77.13-78.4	BV.ii.2.3-9
78.5f.	BV.ii.3.1-4
78.14-79.3	BV.ii.3.4-8
79.4-6	BV.ii.3.8-10
79.6-9	BV.iii.1.1-4
79.9f.	BV.iii.1.4-9
79.13, 14	BV.iii.1.7

Ibn Sīnā, *Šifā᾿*, al-Nafs
Ed. Anawati-Zayed (Cairo 1974)

p. 182	BV.iii.2.1-10
182.1f.	BV.iii.2.2f.
182.16	BV.iii.2.3
182.16f.	BV.iii.2.6

Ibn Sīnā, *Šifā᾿*, ᶜIlm al-hai᾿a
ed. Madwar-Ahmad (Cairo 1980)

83.4-5	AIV.i.3.1-4n
83.5-12	AIV.i.3.4-11n
96.3-8	AIV.ii.n

Ibn Sīnā, *Šifā᾿*, Ilāhīyāt
Ed. Anawati-Zayed (Cairo 1960)

23.5-6	AI.i.1.7-9

Ibn Sīnā, *ᶜUyūn al-ḥikma*
Ed. Ülken (1953)

29.19-30.4	AIII.n
30.1-3	AIII.i.1.3

Ibn Sīnā (Avicenna)
De congelatione et conglutinatione lapidum
Ed. Holmyard-Mandeville (1927)

53.2	AIII.ii.3.3n

Ibn Tibbon, Samuel, *Otot ha-Shamayim*
Ed. Fontaine (1995)

I.120	BV.iii.n
I.321, 324, 335	BIV.iii.4.7-10

II.173 AV.ii.1.9
II.489-492 AII.iii.1.6, 7n
II.511 AII.ii.4.6n

Ibn Ṭufail, *Ḥayy b. Yaqẓān*
Ed. Gauthier (1936)

24.3f. BV.ii.2.11

Ibn Yūnus, Abū al-Ḥasan ᶜAlī
K. al-zīǧ al-kabīr al-ḥākimī
Apud Caussin (1804)

117.13 AIV.i.2.3n

Ikhwān al-Ṣafāᵓ, *Rasāᵓil*
Ed. Bustānī (1957)

II.59.2f. AIII.iii.1.5f.n
II.71.3 BIII.i.1.4-9n
II.106.14ff. AIII.ii.3.1-10n
II.108.9f. AIII.i.1.3n
II.109.1 AIII.i.1.3n
II.116.5-9 AIII.ii.3.1n
II.116.13ff. AIII.ii.3.1-10n
II.119.17-21 AIII.ii.3.6-9n
II.121.3 AIII.ii.2.7f.n
II.121.7 AIII.i.1.3n
II.121.15ff. AIII.i.1.2-4n

Isaac of Antioch
De signo quod apparuit in caelo

- AII.ii.4.8f.n

Ishoᶜdad, *in Evang.*
Ed. Gibson (1911)

II.96.11-15 (Mt. 13.8) AII.i.4.3

Jābir b. Ḥayyān, corpus attributed to
Excerpts in Kraus (1942/3)

II.19 (*K. al-ḫawāṣṣ al-kabīr*) AIII.i.5.6-9n
II.19 n.1 AIII.i.5.7n
II.19, 22 (*K. al-sabᶜīn*) AIII.i.5.6-9n
II.22 n.7 (*K. al-ḫamsīn*) AIII.i.1.2-4n

Jābir, *K. al-īḍāḥ*
Ed. Holmyard (1928)

54.11-14 AIII.ii.3.1n

Jacob of Edessa, *Hexaemeron*
Ed. Chabot (1928)

53a 34-b 1 AIII.i.5.6-9n
53b 1 AIII.i.5.9n
53b 17f. AIII.i.4.2n
53b 31. AI.iii.2.2n
84f. BIII.ii.3.3-8
85a 35, b8-11 BIII.iv.4.4, 7
99b 3, 35 AV.i.2.3-4n
100a 14 AV.i.2.5
100a 18f. AV.i.2.1-2n
100b 2, 7 AV.i.2.3-4n
100b 18-28 AV.i.2.5
100b 25 AV.i.2.5
100b 32f. AV.i.2.6f.
101b 2-102b 4 AV.i.3.3
113b 36f. AV.i.3.2

Job of Edessa, *Book of Treasures*
Ed. Mingana (1935)

405a 5f. AI.i.2.3n

Josephus, *Antiquitates judaicae*
Ed. Niese (1888-92)

I.6.1 AV.i.3.7n

Juwainī, Alāᵓ al-Dīn Aṭā Malik
Tārīḫ-i ǧahān-gušā
Ed. Qazwini (1912-1937)

I.60.1f. AV.i.1.7n

Kharaqī, Abū Bakr Muḥammad
Muntahā al-idrāk fī taqsīm al-aflāk:
Apud Nallino (1903-7)

I.175.13f. AV.i.1.1n

Khwārizmī, Abū ᶜAbd Allāh
Muḥammad
Mafātīḥ al-ᶜulūm
Ed. van Vloten (1895)

257.6 AI.ii.5.8n
258.6f., 12f. AIII.i.5.6-9
260.3, 260.7, 261.1 AIII.i.1.2-4

Kifāyat al-taᶜlīm
Apud Sezgin, *GAS* VI.102

- AIV.i.2.3n
- AIV.i.2.5

Kirakos of Gandzak, *Patmut'iwn hayoc'*
Tr. Dulaurier (1858)

507f. BIV.iii.3.7-10

Koran

16.4 etc. BV.ii.2.7

Ps.-Majrīṭī
Picatrix (*K. ġāyat al-ḥakīm*)
Ed. Ritter (1933)

78.12 AIV.i.2.3n

Maqrīzī
K. al-sulūk li-maʿrifat duwal al-mulūk
Ed. Ziyāda-ʿĀshūr (1934-72)

I.516.13-517.1 BIV.iii.3.7-10n
I.516.15 MBIV.iii.3.7-10

Medicines, The Syriac Book of
Ed. Budge (1913)

31.13 AIII.ii.3.8
63.20f., 64.11 AIII.i.4.2n
68.9 etc. AI.i.5.3n
87.7 AIII.i.4.2n
90.20 AIII.i.4.2n
201.ult. AIII.ii.3.3n

Michael Syrus, *Chronicon*
Ed. Chabot (1899-1910)

8a 34, 9b 3 AV.i.3.6n
51b 24 AI.iii.2.2
64a 30f. AV.i.3.7n
66b 15 etc. BIV.i.5.1n
149b 29 AV.i.2.3-4n
411b 6, 9 AV.i.2.3-4n
511b 1 AIII.iii.2.6n
515c 3 AV.i.2.3-4n
526b 35f. AV.iii.2.10
526b ult.-527a 4 AV.iii.2.10n
532a 7 AIII.iii.2.6n
566a 31 AV.i.3.7n
Tr. I.258 AV.i.3.6n

Nicolaus Damascenus
Compendium of Aristotelian Philosophy
(Syriac)

3.1 BII.ii.2.5
5.5 AII.i.1.4-5, 6
8.17ff. AII.i.4.5f.
9.11-15 AV.iii.2.10
10.11 etc. Mete.tit.
10.16 BI.i.1.1f.
10.17-20 AI.i.1.7-9
10.25f. etc. BIV.i.5.4
11.19f. BIV.iii.1.1-5
11.21-28 BIV.iii.1.6-10
11.23, 28 BIV.iii.tit.
12.1-4 BIV.iii.4.7-10n
12.1, 2 BIV.iii.tit.
12.8f. BIV.iii.4.7
12.24 BIV.iii.tit.
13.20-24 AIV.iv.2.1-6n
15.6f. AV.i.1.1n
15.9-11 BI.ii.3.6-9
15.14 etc. BI.iii.1.6f.
16.16-19 BI.iii.1.7f.
17.9-12 BI.iii.4.5-7
18.1 BIII.iv.4.7n
18.1-3 BIII.iv.4.7
18.13ff. AII.i.4.6f.
18.14f., 17f. AII.i.4.6
19.1-24 AV.iv.n
19.1.ff. AII.i.4.6f.n
19.24-28 AV.ii.2.3f.n
19.28-20.2 AV.ii.2.8f.
20.6-21 AV.ii.4.1-4
20.7 BIV.i.5.1n
21.6f. AV.ii.3.1f.
21.7f. AV.ii.3.3-5n
23.6-11 AV.ii.1.9-12
23.13-18 AV.ii.3.3-5
23.18-20 AV.ii.3.1f.
23.22-26 AV.ii.3.5-8
24.1-4 AV.ii.3.8-10
24.4-6 AV.ii.1.4f.
24.20f. BIII.i.n
24.22-24 BIII.i.3.1-2
25.10-15 AII.ii.1.9f.n
25.20-23 BIII.i.4.5-8
26.20-23 BIII.i.4.4-5
26.23-27 BIII.i.4.8-11
26.28 BIII.i.4.5-8
27.13, 17 BIII.iii.1.6
27.22-28.4 BIII.iii.1.5-11
28.4-6 BIII.i.2.1-4
28.18-22 BIII.i.2.4-8
28.22-24 BIII.i.3.2-4
28.24-29.3 BIII.i.3.4-9

29.3-4	BIII.iii.3.5-10
29.3-10	BIII.iii.3.1-5
29.10-13	BIII.iii.3.5-10
29.15-18, 28-31	BIII.iii.4.1-5
29.25f.	BIII.iii.5.1
29.31-30.6	BIII.iii.5.1-11
30.6-9	BIII.iii.4.6
30.11-18	AIV.ii.1.1-7
30.14	BIII.ii.2.2
30.18-25	AIV.i.3.4-11
30.18-25	AIV.ii.2.1-4
30.21	BIII.iii.1.6
30.26-29	BIII.iii.4.7-9
31.8-10	AIV.i.3.4-11
31.9, 27	BIII.iii.1.6
31.10-15	BIII.iii.5.1-11
31.13	BII.ii.2.5
32.5f.	BIII.ii.1.1-3
32.6-11	BIII.ii.1.3-7
32.8	BIII.ii.2.2
32 (diagram)	BIII.ii.3.3-8
33.14-24	BIII.iii.2.5-7
33.24-26	BIII.iii.1.1-5
34.18f., 23-28	BIII.iii.6.1-7
35.6-9	BIII.iv.4.3-7
35.12f.	AII.i.1.2-8
35.20f.	AII.i.1.2-8
35. 26-31	AII.i.1.2-8
35.30	AII.tit.n
35.31-36.7	AII.i.4.1-6
36.11-13	AII.i.4.1-6
36.13-19	AII.i.2
36.15	AII.tit.
36.20f.	AII.i.4.6f.
36.21-23, 30	AII.i.3
36.24ff.	AII.i.4.6f.n
36.28	AV.ii.2.3
36.28f.	AV.iii.2.11n
36.30-37.4	AII.i.4.7-11
37.4-10	AII.ii.1.1-6
37.10	AII.tit.n
37.11-19	AII.ii.1.6-8
37.14-16	AII.ii.1.5f.n
37.19-20	AII.iii.1.1-2
37.20-24	AII.iii.1.2-7
37.25-30	AII.iii.1.8-12
37.30-38.1	AII.iii.2.1-4
38.1-6	AII.iii.2.8-9
38.2f.	AII.iii.2.1
38.3f.	AII.iii.2.1-4n
38.6-10	AII.iii.2.4-8
38.22f.	AII.ii.2.8-10
38.24-25	AII.ii.3.5f.
38.25-39.2	AII.iii.1.7f.
39.6-16	AII.ii.1.9f.
39.9	BIV.i.5.1n
39.16-27	AII.iii.1.2-7
39.27-40.3	AII.iii.3.1-5
40.3-7	AII.iii.4.1-4
40.7-12	AII.ii.3.1-4
40.13-17	AII.ii.3.7-10
40.18-23	AII.ii.2.2-4
41.8-12	AII.ii.4.4-8
41.14-17	AII.ii.4.4-8
41.25-42.9	BIV.i.1.1-7
42.9-11	BIV.i.4.4-8
42.11-14	BIV.i.4.8-11
42.14 etc.	BIV.i.5.1n
42.14-17	BIV.i.5.1f.
42.18-19	BIV.i.5.4f.
42.18 etc.	BIV.i.5.4
42.24f.	BIV.i.5.5-7n
42.28-30	BIV.i.5.5-7
43.2-4	BIV.i.5.2-4
43.11-15	BIV.i.6.1-3
43.15-19	BIV.i.6.3-6
43.20f.	BIV.tit.
44.6-10	BIV.ii.1.1f.n
44.18-24	BIII.iv.1.3-7
44.23-26	BIII.iv.4.3-7
44.24	BIII.iv.2.5
44.26-45.4	BIII.iv.2.1-5
45.4-8	BIII.iv.4.1-3
45.4f.	BIII.iv.1.2
45.8-13	BIII.iv.2.5-9
45.13-15	BIII.iv.3.3-6
45.15-17	BIII.iv.4.1-3
45.17-18	BIII.iv.3.3-6
45.19-21	BIV.ii.1.1f.
45.21-25, 27-31	BIV.ii.2.1-9
45.26	BIV.ii.2.1, 3n
45.31	BIV.ii.3.1-3
45.32, 46.5-8	BIV.ii.3.3-5
46.6	BIV.ii.2.1, 3n
46.8-16	BIV.ii.3.5-9
46.16, 17	BIV.ii.3.9f.
46.18	BII.i.1.1
47.3-5	BII.i.4.1-6n
47.3-5	BII.ii.1.1f.
47.5-9	BII.i.4.1-6

47.9-16 BII.iii.8.1-10
48.11-12, 23 BII.iii.8.3-6
48.23-26 BII.i.9.1-4
48.28 BII.ii.1.8
48.30-49.2 BII.i.9.1-4n
49.4 BII.ii.1.8
49.6-8 BII.i.7.5
49.17-20 BII.i.3.6-8
50.25 etc. BII.ii.1.2
51.17 BII.iii.2.2
51.24, 26 AI.iii.1.8
52.4-9 BII.iv.2.2
52.13f., 22f. BII.iii.8.3-6n
52.25-26 BII.iii.8.3-6n
53.6-14 BII.iii.2.2-7
53.14-16 BII.iii.2.1-2
53.16f., 30f. BII.iii.8.3-6n
53.32-54.3 BII.iii.4.1-5
54.3-7 BII.iii.5.2-5
54.9, 21, 31 BII.iii.4.4
54.31 BII.ii.2.5
56.28 AIII.i.5.6n
57.17-19 AIII.i.2.4
58.13-16 AI.i.5.6-8
58.25f. AI.i.4.1-3
58.27-29 AI.i.5.1-4
58.29f. AI.i.5.4-6
59.3-13 AI.i.5.8-10
59.14f. AIII.i.2.4
59.16 AIII.i.5.6n
59.17, 19 AIII.i.2.3
59.19-23 AIII.i.5.2
60.2, 5-7 AIII.i.2.3f.
60.7, 24 AIII.i.5.6n
60.12 AIII.i.2.3

Nicolaus Damascenus (Syriac)
Excerpts in ms. Paris syr. 346

61v (diagram) BIII.ii.3.3-8
65r 16-17 BIII.i.4.5-8
65v marg. BIII.iii.1.5-11
68r 20-21 BIII.i.3.4-9n
68r 21-22 BIII.iii.3.1-5
69r 4-18 BIII.i.3.4-9n
69r 18-20 BIII.iii.3.1-5
69v 6-8 BIII.iii.3.5-10
69v 12-16 BIII.iii.4.1-5
70r 9-13 BIII.iii.4.1-5
70r 13-22 BIII.iii.5.1-11
70v 8-11 BIII.iii.4.7-9

Nicolaus Damascenus: lost passages?

AI.ii.4.1-6; I.iii.1.1-4; I.iii.1.4-6; I.iii.1.6-8V.ii.1.6-9; II.ii.4.8-9; IV.ii.2.4-9; IV.iv.2.1-6; V.ii.2.1-3; V.ii.2.4-7; V.ii.2.8-9; V.iii.1.1-3; V.iii.1.8-11; V.iii.2.1-3; V.iii.2.3-6; V.iii.3.4-9; BI.ii.6.4-7; II.i.5.1-9; II.ii.2.4-6; II.iii.4.5-7; II.iii.5.7; II.iii.6.1-6; IV.iii.4.7

Nicolaus Damascenus
De plantis, Arabic version
Ed. Drossaart Lulofs-Poortman (1989)
[in brackets: corresponding passage of the Greek Ps.-Aristotle, *De plantis*]

§150 (822b 31-35) AII.ii.2.3f.
§154 (823a 17-27) AII.i.1.4
§159 (823b 11ff.) AI.ii.1.3-7
§178 (825a 5-8) BI.iii.2.3-5

Olympiodorus, *In Arist. Mete. comm.*
Ed. Stüve (1900)

3.30-33 AI.i.1.7-9n
12.11-13 AII.ii.3.1-4n
12.22-13.18 BIV.ii.1.1f.n
12.26-13.2 BIV.ii.2.1-9n
13.1f. BIV.ii.2.1, 3
13.12-18 BIII.iv.4.3-7n
37.25-28 BIV.i.1.1-7n
37.35-38.10 BIV.iii.1.6-10n
41.18-26 BIV.iii.1.6-10n
68.13-28 AIV.iv.2.1-6
81.20-32 AV.ii.3.3-5n
86.10-12 BI.ii.3.6-9
98.5-8 BIII.i.3.1-2n
98.19-22 AII.ii.1.9f.n
99.8ff. BIII.i.3.4-9n
99.9-13 BIII.i.3.1-2n
100.11-13 BIII.iv.4.7n
103.25-29 AI.ii.4.1-6
111.30-11 AV.iii.1.8-11
113.25-33 AV.iii.2.5-9
116.10-15 AV.iii.2.1-3
118.30-34 AV.ii.2.3f.
123.2-5 AV.iii.2.3-6
127 24 AI.iii.1.6
127.37-128.8 AV.iv.3.6-10
128.35-129.4 AV.iii.3.4-9

240.10-18 BII.iii.5.2-5n
243.30-32 BII.iii.4.5-7
307.13, 18-20 BI.iii.2.3

Olympiodorus
In Arist. Mete. comm., Arabic version
Ed. Badawī (1971)

95.2-9 BIV.iii.1.6-10n
95.16-96.7 BIV.iii.1.6-10n
96.8-18 BIV.iii.4.7
96.19-24 BIV.iii.4.7
97.1-6 BIV.iii.4.7
100.1 BI.ii.6.6
103.22 AV.iii.2.1-3n
104.5 AV.iii.2.1-3n
104.7f. AV.iii.2.1-3
104.13 AV.iii.1.2
104.13-21 AV.iii.2.5-9
105.4-6 AV.iv.5f.
106.8-11 AV.iv.1.1-5
106.11-13 AV.iv.1.5-9
106.13-15 AV.iv.2.1-6
106.17-19 AV.iv.3.1-4
106.24-107.3 AV.ii.2.1-3n
107.9f. AV.ii.2.1-3n
107.19 AI.iii.1.4
108.2-15 AV.iv.3.6-10
108.16ff. AV.iii.3.2-4n
108.24-109.2 AV.iii.3.2-4
109.2-7 AV.iii.3.4-9
109.19-20 AV.ii.2.4-7
109.20-110.1 AV.ii.2.8f.
111.22-112.2 AV.ii.3.3-5n
112.8f. AV.ii.4.1-4n
113.19-114.2 AV.ii.4.6-8
114.4 AV.ii.4.6-8n
114.17f. AV.ii.1.9-12n
114.17f. AV.ii.3.1f.n
114.18-21 AV.ii.2.1-3
115.2-6 AV.ii.3.5-8n
115.16-21 AV.ii.3.8-10n
116.1-4 AV.ii.1.4f.n
117.20-21 BIII.i.4.5-8n
117.21-118.4 BIII.i.4.8-11n
118.4-6 BIII.i.4.4-5n
118.12-20 BIII.iii.1.5-11n
118.19 BIII.iii.1.5-11n
118.21-22 BIII.i.2.1-4n
119.11-19 BIII.i.2.4-8n
120.3 BIII.i.3.1-2n
120.3-10 BIII.i.3.4-9n
120.11-15 BIII.iii.3.1-5n
120.16-18 BIII.iii.3.5-10n
120.19-121.3 BIII.iii.4.1-5n
121.3-11 BIII.iii.5.1-11n
121.12 BIII.iii.4.6
121.16-18 BIII.iii.4.7-9n
122.1-16 BIII.iii.4.1-5n
122.17-21 AIV.ii.1.1-7
122.21-123.12 AIV.ii.2.1-4n
123.14-124.2 AIV.ii.2.4-9
124.2ff. AIV.i.3.4-11n
125.9-126.2 BIII.ii.1.3-7n
127.1-9 BIII.ii.3.3-8n
128.2-13 BIII.iii.2.5-7n
129.7-9 BIII.iii.1.1-5n
131.12-20 BIII.iii.6.1-7n
131.22-132.8 BIII.iv.1.3-7n
132.1f. BIII.iv.4.3-7n
133.12-13 AII.i.1.3f.n
133.13-17 AII.i.1.4-8n
133.7-11 AII.i.3.1-4n
133.20-134.4 AII.i.2.1-7n
134.5-11 AII.ii.1.1-6n
134.12-20 AII.i.4.7-11n
135.1-19 AII.i.4.1-6
135.20f. AII.i.4.6f.n
135.22-136.3 AII.ii.1.6-8n
136.3-6. AII.iii.1.1f.n
136.6-8 AII.iii.1.2-7n
136.8-15 AII.iii.1.8-12n
136.15-22 AII.iii.2.1-4n
136.23f. AII.iii.2.8f.n
136.23-137.3 AII.iii.2.4-8n
137.9f. AII.ii.2.8-10n
137.17-19 AII.iii.1.7f.n
137.21-24 AII.ii.1.9f.n
138.6-13 AII.iii.1.2-7n
138.13-17 AII.iii.3.1-5n
138.18-139.2 AII.iii.4.1-4n
139.3-9 AII.ii.3.1-4n
139.10-17 AII.ii.3.7-10n
139.18-140.6 AII.ii.2.2-4
140.17-141.2 AII.ii.4.4-8
140.22 AII.ii.4.8f.
141.14-21 BIV.i.1.1-7n
142.4-7 BIV.i.4.4-8n
142.7-9 BIV.i.4.8-11n
142.19-143.5 BIII.iv.2.1-5n
142.19-143.1 BIII.iv.4.1-3n

143.5-9, 12-14	BIII.iv.3.3-6
143.9-12	BIII.iv.4.3-7n
144.2-4	BIV.ii.n
144.2-4	BIV.ii.1.1f.n
144.5-13	BIV.ii.2.1-9n
144.14	BIV.ii.n
144.14f.	BIV.ii.3.1-3n
144.15-145.1	BIV.ii.1.1f.n
144.16-21	BIV.ii.3.5-9n
144.22-145.2	BIV.ii.3.3-5n
145.2-12	BIV.ii.3.5-9n
145.4f.	BII.iii.8.1-10n
145.15-22	BII.i.4.1-6n
145.22-146.10	BII.iii.2.1f.n
145.22-146.5	BII.iii.2.2-7n
146.11-15	BII.iii.8.3-6
146.15-19	BII.i.9.1-4n
149.17	BII.ii.4.6
153.10-19	BII.iii.8.1-10n
155.4-7	BII.iii.4.1-5n
155.14-18	BII.iii.4.5-7
157.4, 7-13	BII.iii.6.1-6
157.13-158.9	BII.iii.6.1-6n
159.17-19	BII.i.5.5f.
160.18-20	BII.ii.2.4-6

Olympiodorus
In Arist. Mete. comm., Arabic version
Ms. Tashkent 2385

353b 15f.	BIII.i.4.5-8n

Olympiodorus
Comm. on the Organon (Syriac)
Apud Wright (1870-2) 776a 17f.

-	Mete.tit.

Pliny, *Naturalis historia*

11.208	AI.i.5.3n
20.9	AI.i.5.2f.
28.212	AI.i.5.3n
35.166	AI.i.5.6
35.167	AI.i.5.6n
35.167	AI.i.5.7
35.182	AI.ii.1.7-10n

Posidonius
Ed. Edelstein-Kidd (1989)

Frag. 237 (= Strabo 12.1.67)	AI.i.5.6n

Ptolemy
Syntaxis mathematica (*Almagest*)
Ed. Heiberg (1898-1907)

78.22-88.4 (II.1)	AIV.i.3.1-4n
88.5-19 (II.1)	AIV.i.3.4-11
101-117 (II.6)	AIV.i.3.5
107-108 (II.6)	AIV.ii.3.5
103.5-10 (II.6)	AIV.ii.3.3-10
110.15 (II.6)	AIV.iii.5.1
111.6 (II.6)	AIV.i.4.10

Ptolemy, *Geographia*
Ed. Nobbe (1843-45)

II.3.3 etc.	AV.i.3.9
IV.8.3, 6	AV.i.3.2
VII.3.2 etc.	AV.i.3.9
VII.5.9	AV.i.2.6f.

Qazwīnī, Zakarīyā᾽ b. Muḥammad
Aǧā᾽ib al-maḫlūqāt
Ed. Wüstenfeld (1849)

94.26-95.4	BIII.i.1.4-9n
149.4-8, 12	AII.ii.1.1-6n
203.29-204.4	AIII.i.5.1-6
204.1, 205.4	AIII.i.5.9n
204.29-205.8	AIII.ii.3.1-10
205.7	AIII.ii.3.1
203 ult. f.	AIII.i.5.6-9n
207.20-22	AIII.ii.3.6-9n
209.19-23	AI.i.3.6-9n
209.ult.	AI.i.4.4f.n
243.8	AIII.ii.1.5
243.penult.	AIII.ii.3.3n

Qazwīnī, *Āṯār al-bilād*
Ed. Wüstenfeld (1848)

164.20-26	AIII.iii.3.8, 10n
410.11f.	AIV.i.4.9-11n

[Ibn al-] Qifṭī
Ta᾽rīḫ al-hukamā᾽
Ed. Lippert (1903)

108.9	AIV.i.2.2-8n

Quṭb al-Dīn al-Shīrāzī, *Nihāyat al-idrāk*
Partial tr. Wiedemann (1912)

31	AV.i.1.1n

Rashīd al-Dīn
Ǧāmiᶜ al-tawārīḫ
Partial ed. Quatremère (1836)

I.416.5-8	BIV.iii.3.7-10

Rāzī, Abū Bakr Muḥammad
K. al-asrār
Ed. Dānishpazhūh (1964)

2.11-13	AIII.i.5.6-9
3.13	AIII.ii.3.3n
2.14-17	AIII.i.1.2-4
4.24-5.1	AIII.i.4.2n
5.3f.	AIII.i.4.4
5.10f.	AIII.i.4.2-6n
8.17, 9.1, 13f., 16, 18	AI.ii.5.8n
15.15, 16.5, 18.21	AIII.ii.3.8n
16.19	AIII.ii.3.8

Rāzī, *al-Madḫal al-taᶜlīmīl*
Tr. Stapleton et al. (1927)

p. 345f.	AIII.i.5.6-9
p. 348f.	AIII.i.1.2-4
p. 349	AIII.i.4.2n
p. 363	AIII.i.5.6-9n

Sāᶜid al-Andalusī, *Ṭabaqāt al-umam*
Ed. Cheikho (1912)

18.20-19.6, 39.7-16	BV.iii.n
40.13	AIV.i.2.5
40.14	AIV.i.2.2-8n
40.14	AIV.i.2.3n

Seneca, *Quaestiones naturales*

3.20.3	AI.i.5.6

Sergius of Rēshᶜainā
Apud Sachau (1870)

225.17	AIV.i.2.2-8n

Severus Sebokht, *De constellationibus*
Excerpt apud Sachau (1870)

127.7	AIV.ii.1.1
127.7-128.1	AIV.ii.2.1-4
128.5f.	BIII.iii.1.6
128.10-21	AIV.ii.n

Severus Sebokht, *De constellationibus*
Apud Nau (1910)

232	BII.iii.3.7
237, 240	p. 45n
240	AIV.i.2.2-8n

Strabo

12.1.67 (= Posid., Frag. 237)	AI.i.5.6n
16.1.5, 14	AI.ii.1.7-10n
16.1.15	AI.ii.1.7-10
16.2.43	AI.i.5.3
17.1.38-39	AV.iii.2.2

Strato
Ed. Wehrli (1950)

Frag. 56	AIII.i.2.4

Ṭabarī, *Taʾrīḫ al-rusūl wa-l-mulūk*
Ed. de Goeje (1879-98)

III.1395	AV.i.3.5n

Tamīmī, Abū ᶜAbd Allāh Muḥammad
K. al-muršid, Chapter XIV
Ed. Schönfeld (1976)

39	AIII.i.5.3n

(Ps.)-Thābit b. Qurra
De motu octavae spherae

-	AIV.i.2.2-8n

Theo Alexandrinus
Small Comm on Ptolemy's Handy Tables
Ed. Tihon (1978)

200.2f.	AIV.i.2.2-8n
236.4-237.2	AIV.i.2.2-8

Theophrastus, *De lapidibus*
Ed. Wimmer (1866)

§3 (341.2-6)	AI.i.5.3n
§9-22 (342.7-343.47)	AIII.i.1.3
§9 (342.11f.)	AIII.i.1.6n
§41 (346.16f)	AIII.i.2.4

Theophrastus, *De ventis*
Ed. Wimmer (1866)

382.38f.	BIII.i.1.4-9n

Theophrastus, *Metarsiologica*
Syriac version
Ms. Cantab. Gg 2.14/ed. Daiber (1992)

351r 8-10 BIV.i.3.2f.
353r 24-26 BI.iii.2.3-5n

Theophrastus, *Metarsiologica*
Arabic version by Ibn al-Khammār
Ed. Daiber (1992)

1.10f. BIV.i.3.2f.
9.8-11 BI.iii.2.3-5n
15.2-7 AII.i.n
15.4f. AII.i.1.5f.n
15.26-35 AII.ii.4.4-8n

Theophrastus, *Metarsiologica*
Arabic version by Bar Bahlūl
Ed. Daiber (1992)

15.16-21 AII.ii.4.4-8n

Theophrastus, Fragments
Ed. Fortenbaugh et al. (1993)

214A AI.iii.n

Ṭūsī, Naṣīr al-Dīn Muḥammad
Aḫlāq-i nāṣirī
Ed. Mīnowī-Ḥaidarī (1985)

39.12 p. 61n
71.1-3 BV.iii.1.7
250.23, 251.2 BV.iii.2.7

Ṭūsī, *Risāla-i muᶜīnīya*
Ed. Dānishpazhūh (1335 h.š.)

60.22 AV.i.3.12
64.11-65.10 AIV.ii.n

Ṭūsī, *al-Taḏkira fī ᶜilm al-haiʾa*
Ed. Ragep (1993)

125.1 AIV.i.2.3n
245.19-247.1 AIV.i.3.1-4n
249.1 AV.i.3.3
249.6 AV.i.3.12
249.7-15 AIV.i.4.3-9
251-253 AIV.i.3.5
251.1-5 AIV.i.3.4-11
251.14 AIV.i.3.1-4n
253.8 AIV.iii.5.1
257.6-11, 15-17 AIV.ii.3.3-10
257.10-13 AIV.ii.5.1-5
257.13-15 AIV.ii.4.1-7
257.18-20 AIV.iii.1.3-7
257.21-24 AIV.iii.5.1-8
257.24 AIV.iii.6.3-11
259.3-5 AIV.i.4.11n

Vitruvius

I.5.8, VIII.3.8 AI.ii.1.7-10n

Yaᶜqūbī, *Taʾrīḫ*
Ed. Houtsma (1883)

159.11f. AIV.i.2.3n

Yāqūt, *Muᶜǧam al-buldān*
Ed. Wüstenfeld (1866-73)

I.21.13-16 AV.i.2.4f.
I.34.21, 35.9f. AIV.i.4.9-11n
IV.264.19-265.5 AIII.iii.3.8, 10n

INDEX RERUM

The index covers proper names and matters of interest in the text edited above. The references are to the page numbers in the translation.

ἀδάμας 101
Adriatic Gulf 127
Aesop 129
αἰθήρ 81, 165, 185, 189, 191
ἀνάκλασις 153
Alans 129
alchemists 103, 107, 109
alembic 85
Aleppo (Beroea) 147
alum 101, 103
Anaxagoras 91, 93, 185
Anaximenes 91
Antioch 121, 147
Antipheron of Tarentum 161
Aparctias 173
apogee, solar 111, 193
Arabia 127
ἀργής 187
Aries 115, 117, 119, 171
Aristotle 81, 93, 95, 131, 133, 135, 157, 161, 169, 175
ἀρσενικόν 103
Arsinoitis 133
Asia 127
astrologers 111
Atlantic 127
Augustus 111
Avicennian 119

Babylonians 83
Berbers 127
Bukhara 117
Bulgars 129
Byzantium 127, 135

Caecias 173
Cancer 115, 117, 121, 133, 171
Capricorn 113, 117, 133, 171
Caspia, Lake of 127
cataclysm 133
χάλκανθος 103
χαλκῖτις 103
Chaonia 131
chasms etc. in the sky 191
China 103
Chinese 129
Chios 127
Claudia 139
Cleidemus 185
clime, the first 125
clime, the second 115, 119
clime, the seventh 113, 121, 125
climes, the seven 113, 115, 117, 119
cloud wind 179, 181, 187
clouds, formation of 139
Cnidus 83
column of fire 191
Comet of 1575 A.Gr. (1264 A.D.) 191
comets 149, 189, 191
conflagration 133
cucurbit 85
Cushites 117, 119, 127
Cyprus 127

δαλός 189
declination of the sun 115, 121
deluges 193
Democritus 91
dew 145, 149
διάκλασις 153, 161
disciples of Barhebraeus 139
divisions of the earth 115
Dodona 131
δοκίδες 189

earth, natural position of 111
earthquake, definition of 91
earthquake, true causes 93
earthquakes, ancient theories on 91
earthquakes, benefits and harms 93
earthquakes, benefits of 99
earthquakes, types of 95
earthquakes, when they occur 97
east wind 175, 179
eclipse 99
egg wind 177
Egypt 93, 127, 133
elements, transformation of 111, 193
Empedocles 131, 185
equator, climate of 115, 117, 119, 121
eternity 197
Ethiopia 145
Etna 83, 95

εὔριπος 135
Europe 127
Eurus 173
ever-invisible (antarctic) circle 173
ever-visible (arctic) circle 173
ἔξαρμα 161

famine 133
fire, extinction of 189
firewind 149, 187, 189
fixed stars, trepidation of 111
fog 147, 149
Franks, Land of 129
frost 145, 149

Galatian Gulf 127
Gemini 117
geography 129
geometry 155
Gihon 79
glass 81, 85, 103, 161
glutinous clay 79, 83
God 99
Gog and Magog 129
great summer 133
great winter 133
Greeks 173
Gubos 139
Gūzgānāyē 81

habitable world 79, 113, 127, 175, 193, 195
hail, generation of 147
hailstones, size of 147
halo, formation of 157
haloes, two or more 159
Hamadan 159
hearing and vision 95, 185
Hercules, Pillars of 127
horizon 97, 113, 123, 161, 171, 173
Hyrcania, Lake of 127

Iberia, Lake of 127
Iberians 129
Ibn Sīnā 79, 81, 115, 117, 119, 159, 163
imitations of gold and silver 107, 109
implanted heat 121
India, Inner 129
isemeric circle 113, 115, 121
isemeric tropic 175
Italy 129
Īyār (month) 117

κεραυνός 187
khārṣīnī 103
κογχύλια 133
κοράλλιον 81

lance 151, 157, 165, 167
Leo 117, 191
Libonotus 173
Libra 115, 117, 119, 171
Libya 127
lightning, generation of 183
Liparians 95
Lips 173
λιθόδενδρον 81

Maeotis 127, 133
μαγνησία 101
medicine 85, 99
Mediterranean 175
Melitene 79
mercury 103, 105, 107, 109
meridian circle 171, 173
Meses 173
μεταλλικοί 79, 101, 105
metals (fusible bodies) 101, 103, 105
metals (malleable bodies) 101, 103, 105
metals, parents of 105
μετεωρολογία 139
Midianite 117
minerals, classification of 101
mist 147, 149
mock sun 149, 151, 157, 165
Mountain of Silver 127
mountains, formation of 83
mountains, uses of 85

Nile 127
noise (with earthquake) 95
north wind 143, 175, 177, 179, 181

Oceanus 127
optics 155

Palestine 131
perigee, solar 111, 193
Peripatetics 115, 163

Phoenicias 173
Plato 135
Pontus 127
πρηστήρ 187
ψολόεις 187
πυραμίδες 133
πυρίτης 101

qarṣānā (frost) 145
qarḥā 181
quarters of the earth 113, 193

rain and wind, relationship of 143; see also "wind and rain"
rain, generation of 141
rain, types of 143
rainbow, colours of 161, 163
rainbow, formation of 159
rainbow, lunar 165
rainbow, shape of 161
Red Sea 127, 131
Rhodes 127
rock crystal 103, 155, 159
rsīsā (type of rain) 143
rsāmā (type of rain) 143
rzāpā (type of rain) 143

sal ammoniac 101, 103, 185
salts 101
Samos 127
Sardinian Gulf 127
Scythian 117, 119, 129
Sea of Britain 129
Sea of Elam 127, 135
Sea of India 129, 131, 135
Sea of Reed 127
sea, migration of 133, 193
seawater, purpose of its salinity 131
seawater, salinity of 129
seepage water 87, 89
Sheba and Saba 127
shooting stars 149, 189
Sicilian Gulf 127
Sicily 127
skills, origin of 197
snow, generation of 145
snowstorm (*kōḵīṯā*) 145
south wind 143, 175, 177, 179
Spain 129
speech, origin of 199
spontaneous generation 195
spring (water) 79, 83, 85, 87, 89, 95, 99, 131, 133, 135

στῆλαι 127, 129
stones, artificial 83
stones, as material of mountains 83
stones, formation of 79
stones, formed in bodies 81
stones, from animals and plants (fossils) 81
stones, from Etna (lava) 83
stones, with thunderbolts (meteorite) 81
sulphurs 101, 107
sun, not hot 123
Syrtes 127

Tartarus 135, 137
Taurus 117
τελλίναι 133
Theo Alexandrinus 111
Thrascias 173
Thule 129
thunder, generation of 183
thunderbolt 81, 187, 189
tides 135
torches (in air) 189, 191
tornadoes (*qarḥā*) 169
Turks 129
Tyrrhenia 83

ὑπέκκαυμα 189

Virgo 117
vision, errors of 153
vision, theories of 151
vitriol 101, 103

wars 133
well water 87, 89
west wind 143, 175, 179
whirlwind 169, 179, 181
white wind 177
wind and rain, relationship of 171, see also "rain and wind"
wind, generation of 169
winds, absence of 177
winds, etesian 177, 179
winds, the twelve 171, 177
world, eternity of 137
writing, origin of 199

Yemen 81

zīqā (type of rain) 143
zīqā (whirlwind) 181
zmāytā (frost) 145

INDEX NOMINUM

The index covers those names mentioned in the introduction, commentary and appendices other than those registered in the Index Locorum and Index Rerum.
Bold type indicates instances where the item in question is discussed in detail.

Abbeloos, J.B. 29n, 405n
ᶜAbd Allāh Fāᶜūr 31
ᶜAbd al-Ḥamīd, sultan 594
Abharī, Athīr al-Dīn 8
Abramowski, R. 405n
Abū al-Barakāt al-Baghdādī **53f.**, 67, 69, 203, 603
Abyssinian 62, 351, 374, 379
Acheron 412
Adam 581
Adam and Eve 564, 573, 576, 579
Aden 382
Adler, J.G.C. 211n
Afram, G. 24, 28
Agrippa, astronomer 327n
Aḥmad Pasha of Mosul 589
ᶜAkk 382n
Albion 383
alchemists 48, 57
Aleppo 3, 19, 58
Alexander of Aphrodisias 204
Alexander Özmen 24
Alexander the Great 376
Alexandria 373
Alfred of Sarashel 13n
Alqosh 16, 18, 29, 31, 34, 592
Amid: see Diyarbakır
Amū Daryā 213, 216
Anawati, G.C. 311n, 312n
Anaxagoras 48n, 362n
Andalus 374, 380
Antioch 3, 58, 374
Antipheron of Oreus 470
Antoninus Pius 326n
Antonius Sionita 21
Arab 62
Aristyllus 327n
Aristotle 12-14, 36, 48, 51, 54, 66, 67, 204
Ps.-Aristotle
De mundo 39n
De mundo (syr.) **37**, 41, 42, 43, **55**, 57, 60, 67
De mundo (arab.) 37n
De plantis 13n
Armalet, I. 21n, 24, 32
Armenia 377
Armenian (numerals) 20n
Assemani, J.S. 7, 31n, 41n, 268n
Assemani, S.E. 15, 21, 32, 602n, 603n
Aßfalg, J. 31n, 592n
Audo, T. 5n, 476
Avars 384
ᶜAwwād, G. 18n

Baars, W. 21n
Babel, Tower of 564, 579
Babylon 225
Badawī, ᶜA. 39n, 407n, 498n, 522n
Badr b. Shams 588
Baffioni, C. 292n, 311n
Baḥrain 382n
Baissari, F. 22
Bakoš, J. 43n, 211n, 307n, 390n, 398n, 525n, 537, 562n, 583n
Balkhī school, geographers 57, 371
Baltic Sea 383
Bar Bahlūl 37n
Barhebraeus
De aequilitteris 19
Ascensus mentis 6n, 44, **45f.**, 56, 57, 60, 370
Candelabrum sanctuarii 6n, 13, **43f.**, 45, 47n, 48, 51-53, 55, 56, 62, 63n, 64, 204, 370
Carmina 5n, 19
Carmen de divina sapientia 19, 22n
Carmen de perfectione (*Mêmrā zawgānāyā*) 18n, 19, 22n
Chronicon 19, 31, 57
Chronicon ecclesiasticum 6n, 7, 8n, 19
L. columbae 6n, 19, **46f.**
Ethicon 6n, 27n, 29, **46**, 370
L. grammaticae 19, 49n, 602
L. Hierothei 6n
Historia dynastiarum 51

Horreum mystriorum 5n, 6n, **47**, 48
L. indicationum 8n, **42f.**
"*Laughable Stories*" 62n
Nomocanon 6n, 19f., 27n, 602
L. pupillarum 6
L. radiorum 5n, 6n, 27n, **44f.**, 46, 56, 370
Sermo sapientiae 4, 6, 8, 30n, **42**, 65
L. splendorum 6n, 15n, 17
Tractatus tractatuum 4, 6, 7n, 8, 21, **42**, 44, 64, 602
Bar Kepha, *Hexaemeron* 37, 38, **40**, 41, 44, **55f.**, 57, 370
Bar Maᶜdanī 19
Bar Ṣalībī 40
Barṣaum, I.E. 15n, 19n, 20, 24, 25, 26, 27
Bar Shahhārē, Emmanuel 40
Bar Shakko
Liber dialogorum **41**, 43, 56
Liber thesaurorum **41**, 56, 370
Barsom, M.S. 5n
Baselios Augen 24n
Baṣra 379
Battānī 371
Baumstark, A. 18n, 19n, 37n, 40n, 41n
Bedjan, P. 46n
Bedouin 62
Behnām (Behnān) Āṯōrāyā 17
Behnām, Būlus 311n
Belenickij, A.M. 215
Bengal, Sea of 410
Berbers 374
Berbers, Sea of 380
Bernstein, G.H. 19, 47n
Berthelot, M. 231n, 232n, 285n
Ps.-Berosus 39
Bīrūnī, *K. al-tafhīm* 43f., 57, 371
Biscay, Bay of 383
Black Sea 376
Borsun Kalesi 424
Bosporus 407, 408n
Bosworth, C.E. 216n
Boyle, J.A. 382n
Brafman, D.A. 37n, 545n
Brahe, Tycho 324n
Briquel-Chatonnet, F. 18n, 24n, 29n, 30n, 32
Brightman, F.E. 5n
Brock, S.P. 7n, 68n, 594n
Brockelmann, C. 8n, 210n, 316n, 543n, et passim
Bruns, P.J. 19, 326n, 536n
Budge, E.A.W. 16, 18, 19, 27n, 326n, 590, 591n
Bulut, A. 541n
Byzantium 374, 593

Cahen, C. 424n
Cardahi, G. 268n, 316n, 476
Carmody, F.J. 324n
Carr, W.E.W. 19
Causa causarum 32, 40
Chabot, J.-B. 31, 39n, 317n, 537n
Chalcedon (Calchedon) 407n
Chalcedonian (Calchedonian) 408n
Chalcis 408n
Charybdis 390n
Cheikho, L. 21n, 22, 600
Cheops and Chephren 405
Chinese 380
Çiçek, Y.Y. 43n, 46n, 562n, 583n
Claudia, town 58
Clazomenae 245
Clemons, J.T. 29
Cnidus 218, 221
Coakley, J.F. 23, 28
Cocytus 412
Constantinople 374, 376, 597
Coptic (numerals) 20n
Cushite 62, 335
Cyril Elias, Metr. of Mar Mattai 595, 598
Czeglédy, K. 383n

Daiber, H. 37n, 40n, 41n, 244n, 319, 436n
Damascus 3
Daniel of Mardin 16, 48n, 602-604
Dāniyāl, B. 23
David, psalmist 593
David b. Safar of Mangish 22n
Dāwud b. Abī al-Munā al-Qillithī 25
Degen, R. 29f., 31n
Delambre, J.B.J. 326n
Democritus 48n, 362n, 392, 406
Denḥā (Ṣaifī al-Ṣalaḥī?) 19f.
Desreumaux, A. 21n, 24n, 29, 30n
Dhofar 382

Dieterici, F. 285n, 305n, 306n, 307n, 310n, 314n
Dietrich, A. 217, 299n
Dionysius (Abraham of Mardin?) 20
Dionysius ᶜAbd al-Nūr Aṣlān 26, 28
Dionysius Behnām, Metr. of Mosul 595, 597, 598
Dionysius Joseph, Metr. of Malabar 598
Dionysius of Tellmaḥrē 405
Ps.-Dionysius Areopagita 39
Dittberner, H. 283n, 286n, 290n, 292n, 298n
Diyarbakır 21, 26
Dolabani, Philoxenus John 24, 25, 26, 27, 28
Donzel, E. van 382n
Dozy, R.P.A. 292n, 296n, 299n
Drossaart Lulofs, H.J. 12f., 15, 16, 18n, 26, 27, 28, 33n, 51, 68n, 405, 436, 536, 602n, 603
Drowser, E.S. 268n
Duke-Elder, S. 219n
Duval, R. 210n, 231n, 232n, 237n, 285n, 299n

Ebied, R.Y. 18n, 27n
Egypt 374
Elijah XII, patriarch 592
Elijah Homo 31n
Empedocles 48n, 390n
Enoch 579
Ephrem, copyist 21
Ephrem Gurǧo 24
Eqror 592
Erythrian Sea 381
Ethé, H. 295n, 307n
Ethiopia 382, 432
Euphrates 381, 424
Euripus 408
Eusebius b. Yonan of Mangish 22n
Ewald, H. 19n

Fakhr al-Dīn al-Rāzī 41, 42, 43, **55**, 56n, 66, 67, 69, 203, 603, 604
Fārs, Sea of 379
Fathi-Chelhod, J. 24n
Fiey, J.M. 17n, 18n, 19n, 23n, 26n, 27n, 29n, 31n, 424n, 592n, 597n
Fontaine, R. 245, 271n
Forshall, J. 602n
Franke, O. 296n
Franks 3, 380
Franks, Land of 374
Freudenthal, G. 211n, 329
Freytag, G.W. 308n
Furlani, G. 32n, 39n, 406, 444n, 582n

Gabriel Behnān Māwṣlāyā 17
Gabriel b. Ḥadbšabbā 29
Gabriel b. Khaushaba b. Joseph 29
Gabriel Sionita 21n
Galatian Sea/Gulf 381, 383
Garbers, K. 232n, 284n, 285n, 295n, 297n, 298n, 300, 302n, 307n, 309n, 310n, 314n
Gate of Iron [Darband] 377
Gemayel, N. 21n
George b. Yaqo Yuḥanna 16, 589f.
Georgia 383
Ghazālī 3
Maqāṣid al-falāsifa 42, 44n
Gibson, M.D. 251n
Giese, A. 295n, 307n
Gilbert, O. 234n, 452n, 460n, 490n, 503n
Giza 405
Goldstein, B.R. 324n
Goshen-Gottstein, M. 30n
Göttsberger, J. 19, 47, 48n, 68n
Graf, G. 21n, 27n, 599n
Graham, W.C. 19n
Gregory George, Metr. of Jerusalem 598
Gubos 3, 58
Guzzāyē (Ghuzz, Oghuz) 383n

Ḥabbī, Y. 16n, 22n, 27n, 40n
Ḥaddād, Binyāmīn 573n
Ḥaddād, Buṭrus 16n, 17f., 18n, 22n, 30, 31n
Ḥaḍramaut 382
Halleux, A. de 30
Halma, N. 325
Ham 31, 382n
Hamadan 465
Hambye, E.R. 23
Hanna, S. 541n
Hasan, A.Y. 231n, 237n
Hatch, W.H.P. 30, 602n
Hellespont 376
Heraclitus 390
Herat 216
Hermes Trismegistus 579

Hezekiah 238
Hill, D.R. 231n, 237n
Ḥilla 559
Himyarites 380
Hinz, W. 217n, 424n, 485
Hipparchus 327n
Hippocrates 488n
Hirth, F. 296n
Hoffmann, G. 573n
Holmyard, E.J. 214n, 217, 286n, 288, 292n, 298n, 302n, 303n, 306n, 307n, 308n, 309n, 310n, 314n
Holophernes 382n
Homer 514n
Honigmann, E. 329n, 332n
Honigmann, E. 424
Horten, M. 452n, 462n, 464n, 479n, 486n
Hulagu 553
Hun 335, 351, 380, 382, 383n
Ḥunain b. Isḥāq 21
K. al-aġḏiya 31n
Ǧawāmiᶜ li-k. Arisṭūṭālīs fī al-āṯār al-ᶜulwīya 40
Huntingdon, R. 19

Iberia 383
Ibn al-Biṭrīq 37
Ibn Ḥauqal 371
Ibn al-Khammār 37n, 38, 52, 204
Ibn Rusta 371
Ibn Sīnā 4, 48, 53, 61
Dāniš-nāma-i ᶜAlā'ī 42
K. al-išārāt wa-l-tanbīhāt 8, 42f.
K. al-šifā' 11-14, 36, 43, **48-50**, 51f., 54, 57, 58, 66, 67, 69, 203
Ibrāhīm, Yūḥannā 18
Ierne 383
Ignatius Peter III, pariarch 595, 598
Ikhwān al-Ṣafā' 41, 490n
Īl-Khāns 3, 383n
Indians 380
Indian Ocean 381, 410
ᶜIsa b. Isaiah b. Cyriacus 18, 590, 592
Isagoge 68
Isḥāq, J. 16n, 17f., 22n, 30
Isḥāq b. Ḥunain 13
Isles of the Blessed 374
Israel, sons of 379
Israel Audo 28
Iṣṭakhrī 371
Istanbul 598
Īsū 335
İzolu 424

Jacob of Edessa **39**, 44, 370
Janssens, H.F. 7n, 8n, 21, 65
Japheth, son(s) of 31, 382, 383n
Jerusalem 238
Job of Edessa **40**, 41
John b. Joseph of Zāz 28
John VIII Hormizd 29n
John Rokos of Mangish 22, 599
Joosse, P. 15n, 21n
Joosten, J. 210n
Joseph Audo 17
Joseph b. Cyriacus 16
Joseph Īberāyā (al-Kurjī) 19f.
Joseph II Sliwa 22, 68n
Jūzjān 215
Jūzjānī, Abū ᶜUbaid Allāh 216
Jwaideh, W. 335n

Kale 424
Karot/Karawat family 24n
Kaufhold, H. 17n
Kawerau, P. 587n
Kayser, K. 31n
Kazimirski, A. de B. 308n
Khalifé, I.-A. 22
Kharaqī 371
Khaushābā, S.I. 541n
Khazar 377
Khurāsān 350
Khwarizmī, Abū ᶜAbd Allāh Muḥammad 57
Kindī 490n
King, D. 345n
Kiraz, G.A. 19n
Kirsch, G.W. 326n, 536n
Klamroth, M. 329n
Koffler, H. 41n, 47n
Koonammakkal, T. 7n
Kottayam 24n
Kraus, P. 284n, 298n, 299n, 300n, 307n, 308n
Kraemer, J.L. 382n
Krotkoff, G. 31n
Kruk, R. 571n
Kūfa 401
Kugener, A. 39n

Kunitzch, P. 328

Lagarde, P. de 37n, 296n, 297n
Lamy, T. 29n, 405n
Lane, E.W. 217n
Lantschoot, A. van 22, 27f., 29
Laqabin 3
Lavenant, R. 25, 26, 27
Lee, H.D.P. 335n, 439n, 509n
Lemmlejn, G.G. 215
Leroy, J. 17n, 21n
Lettinck, P. 37, 38n, 39n, 41n, 265n, 335n, 347n, 389n, 397n, 439n, 440n, 453n, 454n, 464, 490n, 499n, 509n, 522, 559n
Levi della Vida, G. 39n
Levy, J. 268n
Löw, I. 217n

MacFaul, P.A. 219n
MacLean, J.A. 16n, 29n
Macomber, W.F. 17n, 21, 22, 26, 28n, 30n
Macúch, R. 18n, 19n, 20n, 22n, 27n, 268n, 597n
Maghrib 373
Maḥmūd of Ghazna 216
Manna, J.E. 31, 40n, 60n, 591n
Manuel Holobolus 13n
Manuscripts
Aleppo (?): 24
ᶜAmadiya: 18n
ᶜAqra 26: 29n
ᶜAqra 35, 89: 18n
Batnaya 7: 18n
Beirut, USJ, syr. 48: 4, 7n, **22**, 34, 599f.
Berlin, or. fol. 3122: 29n
Berlin, Pet. syr. I.23: 602f.
Berlin, Sachau 226: 31n
Berlin, alii: 18n
BL Add. 7202: 602n
BL Or. 4079 (= L): 4, 7n, 11, **16**, 18, 23, 31-34, 73-76, 588f., 601, 606, 610
BL Or 9380 (= l): 7n, **18**, 32-34, 73-76, 590-592, 601, 606, 612
Bodl. Hunt. 1: 4, 8, **19f.**, 22, 33
Bodl. Or. 467: 507n, 525n
Cantab. Add. 2003: 6n
Cantab. Add. 2811: 18n
Cantab. Gg. 2.14: 12f., 52, 53n, 204
Charfeh, Raḥmani 540, 541, 563: **31**
CSCO, syr. 4: 18n
CSCO, syr. 6, 21: 31n
CSCO, syr. 22: **30f.**
Diyarbakır (?): 32
Diyarbakır, Chald. 32: **21**, 26, 34
Diyarbakır, Chald. 33: **22**, 26, 34
Diyarbakır, Chald. 33 (?): **25**, 33
Diyarbakır, Chald. 34: **26**, 34
Dohuk 2, 6: 29n
Dohuk 14, 34: 18n
John Rylands, syr. 44: **28f.**
John Rylands, syr. 56: 4, 7n, 16, **23**, 33f., 597-599
Kandanad: 24
Konat 229: **23**, 24, 26n, 34
Laur. or. 37, 10, 6, 8: **21**, 33, 585
Laur. or. 69 & 83 (= F): 4, 8, 9, 10n, 11, **15f.**, 21, 32f., 49, 73-76, 585, 601-613
Laur. or. 86: 8n
Laur. or. 298: 32, 49n, 602f.
Laur. or. 342: 602f.
Leeds, syr. 3: 27n
Mar Antonius 22, 713: 29n
Mar Antonius 169-172: 68n
Mar Antonius 173: **30**
Mar Antonius 177: **17**
Mar Antonius 178: **17f.**
Mar Antonius 495: 16n
Mar Antonius, alii: 18n
Mardin, Chald. 47, 52: 27n
Mardin, Chald. 56-57 = Vat. syr. 603-604
Mardin, Chald. 58-60 = Vat. syr. 613-615
Midyat, Mar Barṣawma: 28
Mingana, syr. 23: 7n, 23n, 25n, 26n, 28, 33
Mingana, syr. 44A: **20f.**, 24, 33n
Mingana, syr. 109-111: 16n
Mingana, syr. 306: 603n
Mingana, syr. 309: 26n
Mingana, syr. 310 (= M): 8, 9 16, 25, **26f.**, 28, 32f., 73-75, 601, 606-608, 610, 612
Mingana, syr. 326: 4, **22**, 33
Mingana, syr. 433: 68n
Mingana, syr. 460: **31**

Mingana, syr. 558: 24n
Mingana, syr. 594: 31n
Mingana, alii: 18n, 29n, 31n
Mosul, Syr. Orth. 1.63: **18**
N.D. des Semences 62: **18**
Paderborn, syr. 3: 18n
Paris, syr. 210: 673n
Paris, syr. 213: 525n
Paris, syr. 226: 602n
Paris, syr. 244: 603n
Paris, syr. 346: 45n, 53n, 514n
Paris, syr. 384: **30**
Paris, syr. 423: 30n
Paris, syr. 425: 29n
Paris, alii: 31n
Princeton, Nest. 25 (= P): 7n, 17, **29f.**, 32-34, 73-76
Princeton, Nest. 34a: 30n
Sākā 28: **23**, 33n
Strasbourg 4132-33: 18n
Syr. Orth. Patr. 6/2: **25**, 26, 33, 74n
Syr. Orth. Patr. 6/3: **26**, 33n
Syr. Orth. Patr. 6/5: 23n, 26n, **27**, 33
Tashkent 2385: 39n, 498n
Telkeph: 16n
Urmia 64-68: 7n, **17**, 30
Urmia 177: 30n
Vat. syr. 168: 573n
Vat. syr. 169: 525n
Vat. syr. 175: 31n
Vat. syr. 469 (= W): **29**, 33f., 73-76
Vat. syr. 573: 29n
Vat. syr. 598-599: 18n
Vat. syr. 603-604: 4, 7n, 16, **22**, 23, 34
Vat. syr. 612: 27n
Vat. syr. 613-615 (= V): 7n, 16, **27f.**, 32, 34, 73-75, 596
Vat. Borg. syr. 145: 525n
Manzāyē 380
Marāgha 3
Mardin 21, 24, 27n
Margoliouth, D.S. 8, 11, 15n, 21, 585, 586, 587
Margoliouth, G. 16
Margoliouth, J.P. 18, 48n, 292n, 316, 351, 367
Maronite College, Rome 21n
Martin, J.P.P. 19
Martínez, I. 410n
Marzolph, U. 62
Matthew b. Paul 23, 26n, 27, 28, 33f., 593, 595, 597, 599
Mattias Nayis 24n
Maximus Planudes 13n
Mediterranean Sea 370, 372, 373, 374, 375, 376, 383, 384, 410
Meissner, B. 296n
Melitene 3, 58, 424
Mellus, John Elijah 16, 22, 26n, 27f., 34, 74, 358n, 429, 592n, 596, 600
Menelaus 327n
Mertens, M. 231n, 232n
Meyerhoff, M. 219n, 292n
Midianite 62
Mingana, A. 7n, 17n, 20, 22, 23n, 24n, 26, 27n, 28, 31, 592n, 597n, 603n
Miquel, A. 383n, 410n
Moberg, A. 19n
Monasteries
Dair al-Za^c^farān 20, 21, 603n
Mar Abel and Abraham (Midyat) 32n
Mar Barṣawma 424
Mar Giwargis (near Mosul) 17n
Mar Mattai 3, 56n
Notre Dame des Semences 17, 27n, 34
Rabban Hormizd 15, 27n, 30, 34, 592
Sergius and Bacchus (Sargīsiya) 424
Mongols 3, 380, 382
Moon, Mountains of 379, 380
Mosul 3, 8, 9, 21, 23, 26, 27, 28, 56n, 68, 593, 596, 599
Muqaddasī 371
Muraoka, T. 210n
Murghāb, river 216
Murre-van den Berg, H. 22n, 29n
Mutawakkil, caliph 382n

Nahum, prophet 592
Najaf 401
Najm b. Shams of Mardin 15, 587, 588, 601
Nallino, C.A. 326n
Napel, E. ten 40n
Nau, F. 32, 38n, 45n, 46n, 326n, 327, 330n, 336n, 373, 375n, 429n, 472, 603n
Nāẓim, M. 216n
Nedospassowa, M.E. 297n
Needham, J. 296n, 299
Neugebauer, O. 324n, 325n, 326n, 329n

Nicolaus Damascenus 12-14
Compendium 12-14, 36, **37-39**, 43, 47, **51-53**, 54, 60, 66, 67, 69, 203f.
De plantis 12
Noah 6n, 576, 579
Nöldeke, T. 29n, 210n, 237n, 251n, 268n, 378n
North Sea 383
Nubians 374, 379

Oghuz: see "Guzzāyē"
Olympiodorus
In Arist. Mete. 38, 52, 203f.
In Arist. Mete. (arab.) 38f., 52, 204
Ott, C. 382n
Oxus 213

Palestine 374
Pampakuda 24n
Payne Smith, R. 19f., 398n, et passim
Pellat, C. 330n
Persian (Arabian) Gulf 410
Peter b. Ḥasan al-Ḥadathī 32n
Pharan 380
Pharaoh 379
Philip Arrhidaeus 324n
Phlegethon: see "Pyriphlegethon"
Pines, S. 53n
Plessner, M. 580n
Ploeg, J.P.M. van der 23, 24n, 27n
Pognon, H. 30n
Proba 24, 68n
Ptolemy 330-333, 377
Almagest 44
Geography 39n, 45
Handy Tables 45, 56f.
Puteoli 221
Pyriphlegethon 412

Qazwīnī, Zakarīyā' b. Muḥammad 41, 490n
Qipchaqs 377, 380
Qulzum, Sea of 381
Qumair 380
Quṭrabbul 26

Rada, W.S. 554n
Ragep, F.J. 324n, 332n, 336n, 344
Rāzī, Abū Bakr 57
Reller, J. 40n, 41n
Reinink, G. 40n
Renan, E. 21, 68
Renaudot, E. 4n
Rome 374
Rosen, F. 602n
Roper, G. 24n
Ruelle, E. 232n
Rūm 374
Ruska, J. 41n, 232n, 233n, 285n, 290, 290n, 291n, 292n, 293n, 295, 295n, 299n, 300n, 305, 312n
Ryssel, V. 32n, 37n, 545n

Sabra, S. 24n
Sachau, E. 336n, 602n
Sacy, A.-I. S. de 295n, 296n, 299n, 405n
Sākā, Isḥāq 5n, 23
Sākā, Yaᶜqūb 23
Ṣanᶜa 382
Sarandīb 380, 381
Sarau, O. 7n, 17, 30
Sauma, A. 19
as-Sawwas, Y.M. 24n
Sbath, P. 219n
Scher, Addai 17, 18n, 21, 22, 26, 27
Scher, Elijah 18
Schlimme, L. 38n, 41n, 300n, 372n, 503n, 525n
Schmidt, M.G. 39n
Schönfeld, J. 219n, 285n
Scripture
Psalms 5
Isaiah 5, 6
Luke 6
Scythian 62, 383n
Seleucia-Ctesiphon 380
Seljuks 383n
Sepmeijer, F. 604n
Seraphim 5
Sersen, W.J. 41n, 490n, 526
Severus Sebokht 24, 45, **56**, 60
Sezgin, F. 38n, 312n, 326n, 327n, 328, 329n
Sharples, R. 265n
Shedd, W.A. 7n, 17, 30
Shem 31, 382n
Shirwān 377
Sidon 374
Siggel, A. 285n, 296n, 299n
Simeon, Metr. of Diyarbakır 21
Širo Suyu 424

Slav 351, 384
Sony, B. 31
Spain 383
Sprengling, M. 19n
Spuler, B. 27n, 351n
Steinmetz, P. 37n, 220, 224n, 285n
Stoics 329, 488n
Strohm, H. 335n, 509n, 579n
Strohmaier, G. 514n
Sulaimān, copyist 20
Syria 374
Syria, Sea of 373

Ṭabaristān 377
Tabrīz 3
Takahashi, H. 4n, 8n, 12n, 38n, 39n, 41n, 42n, 44n, 45n, 46n, 62n, 63n, 67n, 69n, 207, 218n, 261n, 266n, 285n, 287n, 296n, 332n, 351n, 371n, 373n, 374n, 375n, 376n, 377n, 378, 378n, 379n, 382n, 383n, 428n, 492n, 503n, 602n
Tanais (Don) 378
Taprobane 45, 381
Taqizadeh, S.H. 351n
Tartarus 253n
Tat 580n
Thābit b. Qurra 13
Thekeparampil, J. 24n
Theo Alexandrinus 48, **56f.**, 61
Theophrastus 66, 490n
Mete. 37, 40
Mete. (arab.) 37n
Mete. (syr.) 37, 40n
Thomas Rokos Khanjarkhān 22n
Tibetan 380
Tigris 381
Tihon, A. 325n, 328, 329n
Timocharis 327n
Timothy Eugene: see Baselios Augen
Timur 603n
Ṭirani (Taprobane) 381
Tiruvalla 24n
Tisserant, E. 27n
Tkatsch, J. 8n
Trebizond 376
Trichur 27n
Tripoli 3
Tughuzghuz 384
Tullberg, O.F. 5n
Turk 62, 351, 382, 383, 384n
Ṭūsī, Naṣīr al-Dīn 4, 8n, 55, 66
Aḫlāq-i Nāṣirī 12, 14
Taḏkira fī ʿilm al-haiʾa 43, 45f.
Tyre 374
Tyrrhenia 218, 221

Ullmann, M. 232n, 282n, 293n, 295n, 300, 312n, 571n
Umbria 386
Urban IV, pope 553
Uri, J. 19
Varangian (Baltic) Sea 383
Vaschalde, A. 39n
Vincent, monk 18
Volga 377
Vööbus, A. 21n
Vosté, J.M. 16n, 17, 18, 27n, 29n, 30
Vullers, J.A. 285n

Wagner, E. 37n
Warang 335
Wehr, H. 299n
Weisser, U. 299n, 300n
Weyer, J. 284n, 285n, 295n, 298n, 300, 302n, 307n, 309n, 310n, 312n, 314n
Wiedemann, E. 231n, 232n, 233n, 281n, 284n, 285n, 296n, 300, 320n, 373n, 452n, 462n, 464n, 479n, 486n
Wright, W. 420

Yamāma 382n
Yāqūt 371
Yazdegerdian (calendar) 351
Yemen 382
Yūkhannā, E. 541n
Yūra 335

Zāb, river 589
Zakho 592n
Zonta, M. 12, 31n
Zotenberg, H. 602n

L.D.S. et B.V.M.

ܫܘܒܚܐ ܠܐܠܗܐ ܘܫܝܢܐ ܘܡܪܚܡܬܐ